PHYSICS AND CHEMISTRY OF THE EARTH

VOLUME 11

ORIGIN AND DISTRIBUTION OF THE ELEMENTS

ORIGIN AND DISTRIBUTION OF THE ELEMENTS

Editor

L. H. AHRENS

University of Cape Town, South Africa

Editorial Assistant

Mme G. PROTAS

C.R.P.G. Nancy-Vandoeuvre, France

PROCEEDINGS OF THE SECOND SYMPOSIUM

PARIS, MAY 1977

PUBLISHED WITH FINANCIAL HELP FROM UNESCO, PARIS

PERGAMON PRESS

OXFORD . NEW YORK . TORONTO . SYDNEY . PARIS . FRANKFURT

U.K.	Pergamon Press Ltd., Headington Hill Hall, Oxford OX3 0BW, England
U.S.A.	Pergamon Press Inc., Maxwell House, Fairview Park, Elmsford, New York 10523, U.S.A.
CANADA	Pergamon of Canada, Suite 104, 150 Consumers Road, Willowdale, Ontario M2 J1P9, Canada
AUSTRALIA	Pergamon Press (Aust.) Pty. Ltd., P.O. Box 544 Potts Point, N.S.W. 2011, Australia
FRANCE	Pergamon Press SARL, 24 rue des Ecoles, 75240 Paris, Cedex 05, France
FEDERAL REPUBLIC OF GERMANY	Pergamon Press GmbH, 6242 Kronberg-Taunus, Federal Republic of Germany

First edition 1979

British Library Cataloguing in Publication Data

Origin and distribution of the elements.
—(International series in earth sciences; Vol. 34).
1. Geochemistry—Congresses
I. Ahrens, Louis Herman II. Series
551.9 QE515 78–40583

ISBN 0–08–022947–6

Published as Volume 11 of the journal *Physics and Chemistry of the Earth* and supplied to subscribers as part of their subscription.

Printed and bound at William Clowes & Sons Limited Beccles and London

INTERNATIONAL SERIES IN EARTH SCIENCES
EDITOR: DEAN EARL INGERSON
Department of Geology, University of Texas, Austin, Texas, USA.

Volume 34
ORIGIN AND DISTRIBUTION OF THE ELEMENTS

Vol. 1	BENIOFF et al	Contributions in Geophysics in honor of Beno Gutenberg
Vol. 2	SWINEFORD	Clays and Clay Minerals
Vol. 3	GINZBURG	Principles of Geochemical Prospecting
Vol. 4	WAIT	Overvoltage Research and Geophysical Applications
Vol. 5	SWINEFORD	Seventh National Conference on Clays and Clay Minerals
Vol. 6	TYUTYUNOV	Introduction to the Theory of the Formation of Frozen Rocks
Vol. 7	KRINOV	Principles of Meteorites
Vol. 8	NALIVKIN	The Geology of the USSR
Vol. 9	SWINEFORD	Eighth National Conference on Clays and Clay Minerals
Vol. 10	POKORNY	Principles of Zoological Micropalaeontology Vol. 1
Vol. 11	SWINEFORD	Ninth National Conference on Clays and Clay Minerals
Vol. 12	SWINEFORD	Tenth National Conference on Clays and Clay Minerals
Vol. 13	BRADLEY	Eleventh National Conference on Clays and Clay Minerals
Vol. 14	ROSENQVIST & GRAFF-PETERSON	International Clay Conference Vol. 1
Vol. 15	COLOMBO & HOBSON	Advances in Organic Geochemistry 1962
Vol. 16	BREGER	Organic Geochemistry
Vol. 17	HELGESON	Complexing and Hydrothermal Ore Deposition
Vol. 18	BATTEY & TOMKEIEFF	Aspects of Theoretical Mineralogy in the USSR
Vol. 19	BRADLEY	Twelfth National Conference on Clays and Clay Minerals
Vol. 20	POKORNY	Principles of Zoological Micropalaeontology Vol. 2
Vol. 21	ROSENQVIST & GRAFF-PETERSON	International Clay Conference Vol. 2
Vol. 22	YERMAKOV et al	Research on the Nature of Mineral-forming Solutions
Vol. 23	BRIDGE & BRIDGE	Clays and Clay Minerals
Vol. 24	HOBSON & LOUIS	Advances in Organic Geochemistry 1964
Vol. 25	BRADLEY	Thirteenth National Conference on Clays and Clay Minerals
Vol. 26	BAILEY	Fourteenth National Conference on Clays and Clay Minerals
Vol. 27	BAILEY	Fifteenth National Conference on Clays and Clay Minerals
Vol. 28	ANDREEV et al	Transformation of Petroleum in Nature
Vol. 29	RAMDOHR	The Ore Minerals and their Intergrowth
Vol. 30	AHRENS	The Origin and Distribution of the Elements
Vol. 31	SCHENK & HAVENAAR	Advances in Organic Geochemistry 1968, 4th International Meeting
Vol. 32	HOBSON	Advances in Organic Geochemistry, 3rd International Meeting
Vol. 33	VON GAERTNER & WEHNER	Advances in Organic Geochemistry 1971, 5th International Meeting

Contents

SECTION IV . DISTRIBUTION AND BEHAVIOUR OF TRACE ELEMENTS

SECTION V . GEOCHEMISTRY OF THE SURFACE

SECTION VI . GEOCHEMISTRY OF THE OCEANIC AND CONTINENTAL CRUST

SECTION IX . GEOCHEMISTRY AND HEALTH

Preface

SHORTLY after the formation of the International Association of Geochemistry and Cosmochemistry some 12 years ago it co-sponsored with the assistance of UNESCO an international symposium Origin and Distribution of the Elements. This was held at UNESCO Headquarters in May 1967. Since then, IAGC has sponsored or co-sponsored several international symposia and conferences and obtained financial support for some of them. At times critics may feel that the organization—geochemistry and cosmochemistry—is too widely dispersed. Indeed, this may be true in part, yet many of the symposia since 1977 have been directed to specific themes and a few on more general ones and it is in this coupled sense that IAGC has strength and appeal. Our first conference involved a wide range of scientific fields and at the time the past President of the International Union of Geological Sciences, Professor T. W. Barth, stated: "Here at this symposium in Paris, a whole spectrum of sciences has been united." The contents of the Proceedings Volume of the first Paris Conference reflects both this statement and many of the thoughts expressed above.

Ten years after the first conference a second one on the same theme was held in May 1977, also at UNESCO Headquarters. Scientists from many countries and working in a wide variety of fields gathered and as a result much cross-fertilization took place. Aside from the fruitful co-operation between IAGC and UNESCO on both occasions, I wish to add another comment, namely an association with CRPG at Nancy-Vandoeuvre. On the previous occasion the late Professor M. Roubault, Director of the above institute, played a very active role in leading to the success of the symposium and likewise on the second occasion, the current Director, Doctor H. de la Roche, assisted most efficiently with the help of Doctor J. Goni from the BRGM, ensured that the second symposium was successful. I wish in particular to refer to the assistance given by Mme G. Protas who for both symposia acted as a secretarial assistant and in addition as an editorial assistant for both the Proceedings Volumes. Many thanks are due also to each of the nine section chairmen who acted as co-editors.

The First Proceedings Volume was published by Pergamon Press and on the second occasion also the same publishing house is producing the volume. As before, I would like to express my appreciation for the ready co-operation of this publishing house.

L. H. AHRENS

Department of Geochemistry,
University of Cape Town

SECTION I

COSMOCHEMISTRY

Study of Some Processes of Nucleosynthesis in the Evolving Galaxy

A. K. LAVRUKHINA and R. I. KUZNETSOVA

V. I. Vernadsky Institute of Geochemistry and Analytical Chemistry, U.S.S.R. Academy of Sciences, Moscow, U.S.S.R.

Abstracts

Using semiempirical dependences of $\mathbf{s}_i$ and A_{product} from A_{target} and E_p obtained from the accelerator experiments, the calculations of the rate of the isotope formations of ^{9}Be, ^{40}K, ^{50}V and "passing" nuclides arising from the solar system matter irradiated by cosmic rays were made. The calculations were carried out at the various irradiation conditions corresponding to the astrophysical processes arising in all stages of evolution of the solar system matter.

The agreement between observed abundances of the nuclides mentioned above and their production in the fast particle reactions was received for the first time in the frames of the model of exploded supernova (or galactic nucleus). The integral flux value was determined as $I_p(E_p \geq 25\ MeV) = 5 \times 10^{21}$proton/cm^2. The irradiation time was $\geq$ 10 days. The fast particle reactions have been preceded the processes of the quick neutron capture (*r*-process). They have run in two acts strongly different in time. The first explosion of supernova (galactic nucleus) where about 99% of "passing" nuclides were synthesized took place about 10.6 milliard years ago, the second one about 4.6 milliard years ago.

In nucleosynthesis, chemical elements are formed in the stars. This process occurs by definite physico-chemical conditions which are thermonuclear reactions (reactions of hydrogen-, helium-, oxygen burning, *s*-, *r*-processes) and these reactions lead to the synthesis of all chemical elements (Burbidge, Burbidge *et al*., 1957; Trimble, 1975). However, one major problem in the theory of nucleosynthesis lies in explaining the formation of a group of stable elements with proton-rich isotopes. These isotopes are neither formed during thermonuclear reactions of hydrogen-, helium-, oxygen-, carbon-, and silicon burning in the interior parts of the stars, nor during processes of radiation capture of neutrons. Owing to that, they are named as "by-passed isotopes".

Various nuclear reactions leading to the synthesis of "by-passed isotopes" have been extensively investigated (Franck-Kamenetskii, 1961; Truran and Cameron, 1972; Audouze and Truran, 1975). However, till now we have not found natural conditions which are suitable for their synthesis. The possibility of their synthesis by (p, γ)-, (p, n)- and (γ, n)-reactions (in *p*-process) occurring in the hydrogen-rich supernova envelope has been discussed for a long time (Burbidge *et al*., 1957; Truran and Cameron, 1972). Recently, Audouze and Truran (1975) quantitatively reexamined this model and showed that the synthesis of most "by-passed isotopes" in amounts corresponding to their solar abundances in the *p*-process a temperature of not less than 2×10^9 °K and a density of 10^4 g cm^{-3} are necessary. However, the existing models of supernova cannot predict the appearance of such conditions in the evolving envelope. A more promising model appears to be that of Colgate (1975), in which the "by-passed isotopes" are formed in *p*-process under conditions preceding the origin of the shock wave in supernova. In this model, there is a difficulty, however, concerning the rise of the necessary temperature for *p*-process, which could have preceded the shock wave (Weaver and Chapline,

1974). Investigation of the possibility of a synthesis of the "by-passed isotopes" has shown that in processes of weak interactions, the smooth curve of abundances of "by-passed isotopes" cannot be reproduced. [Two such cases involve:

(1) "by-passed isotopes" in strongly preevolved but dynamically stable stars in processes of weak interaction (Reeves and Stewart, 1965), and
(2) models involving processes of weak interaction together with reactions of the capture neutron-, proton-, and α-particles, as well as photo-disintegrations, occurring at the stage of oxygen burning (Arnould, 1976).] This mechanism may be of importance only in some specific cases—in particular for ^{164}Er synthesis.

Nuclear reactions in the zone of oxygen burning lead to substantial amounts of passing isotopes. However, one succeeds in bringing them into agreement with the observed abundances only when suitable physical-chemical conditions and concentrations of target-nuclei are selected for various mass regions ($98 \geq Ap$; $102 \leq Ap \leq 144$; $144 \leq Ap \leq 196$). The possibility of producing "by-passed isotopes" in spallation reactions with cosmic rays in the atmosphere of young stars, in the interstellar medium and the matter ejected during the explosion of nova and supernova was discussed by Franck-Kamenetskii (1961). Audouze (1970) examined this model quantitatively. The estimation has shown that the fluxes responsible for the formation of light elements can contribute appreciably only to the synthesis of the heavy elements ^{184}Os, ^{190}Pt and ^{196}Hg.

According to recent ideas in catastrophic events (explosion of supernova, of galaxy nuclei and others) where the main synthesis proceeds, shock waves arise, the spreading of which leads to collective acceleration of particles (Gurevich and Rumyantsev, 1973; Colgate, 1975).

The collective acceleration of particles results simultaneously in the generation of cosmic rays and spallation reactions of atomic nuclei of the supernova matter with accelerated particles. The resulting isotopes may serve as indicators of astrophysical conditions and processes occurring in these catastrophic events. Further evolution of the matter of the solar system in the interstellar space is connected with its constant irradiation by cosmic rays of various origin. Effects of such irradiation are displayed in the form of products of nuclear reactions and in the form of tracks of radiation—chemical disturbances of the cosmic matter.

In the present paper, on the basis of the most recent data on the abundances of chemical elements (Cameron, 1973) and spallation cross-section, the attempt is made to estimate the contribution of spallation process to the abundances of elements of solar system matter at all possible stages of its irradiation by cosmic rays (Table 1).

Quantitative estimations of the effects of spallation reactions caused by cosmic rays in various cosmic objects required experimental investigations on their modelling (Lavrukhina, 1972). Targets of different chemical composition were irradiated by protons with energies ranging from 0.1 to 20 GeV and spallation cross sections were estimated by radio-chemical method. A survey of data on spallation cross-sections allowed us to establish general regularities on the process of spallation. This is described by the following relations (Lavrukhina and Kuznetsova, 1967 and 1969):

$$\sigma_{Ai} = f(A_{\text{target}}, A_{\text{product}}, E_p).$$

These relations were used for quantitative estimations of the contribution of spallation reactions in nucleosynthesis of "by-passed isotopes".

The excitation functions of the "by-passed isotopes" and the radioactive isobars being converted to "by-passed isotopes" were calculated from the target-nuclei, i.e.

TABLE 1. STAGES OF IRRADIATION OF SOLAR SYSTEM MATTER BY FAST PARTICLES

Kind of matter	Source of particles	Observed effects
Shock waves during explosion of galaxy nuclei or the supernova. $\rho \gg 10^{12}$ atoms cm^{-3}	Collective acceleration of particles $I_1 = 5\times10^{21}$ nucleon cm^{-2} $I_2 = 5\times10^{19}$ nucleon cm^{-2}	Synthesis of isotopes Be, B, D, of by-passed isotopes, ^{26}Al, ^{129}J. $T \lesssim 10$ days
Dilating envelope of supernova $\rho \geq 1\times10^{12}$ atoms cm^{-3}	Cold acceleration of particles $I_p = 6\times10^{15}$ particles cm^{-2}	Excess of ^{22}Ne in silicate C1 carbonaceous chondrites $T \lesssim 20$ years
Gas-dust protosolar nebula $\rho \ll 1\times10^{12}$ atoms cm^{-2}	Galactic cosmic rays $I = 5$ particles star (Charahchjan and Charahchjan, 1973)	Indetectable small $2\times10^8 \lesssim T \ll 5\times10^9$ years
Gas-dust protoplanetic cloud $\rho > 1\times10^{12}$ atoms cm^{-2}	Solar cosmic rays of high intensity	Chemical evolution of the cloud. Dissipations of hydrogen and volatile components. Irradiation of dust $T \lesssim 10^8$ years
Planetesimals. $R > 1$ m, $\rho \geq 2$g·cm^{-3}	The same	Not detected. $T \lesssim 10^8$ years
Meteorites and surface layers of the Moon and the planets $\rho \geq 1$ g cm^{-3}	Solar and galactic cosmic rays $I_{solar} = 33$ protons cm^{-2} s^{-1}. $I_{galac.} = 5$ particles cm^{-2} s^{-1}.	Change of the isotopic composition of insert gases of rare elements. Formation of radioisotopes. Radiation disturbances, erosion $T < 5\times10^9$ years

$A_{prod.} \leq A_{target} \leq A_{prod.} + 10$. In the $E_p < 50$ MeV region, a substantial contribution to nuclear interactions is made by simple nuclear reactions of the type (p, xn), (p, pxn) and (p, 2pxn). The cross-sections of these reactions were determined from experimental values which were obtained (Lavrukhina, 1972) by the interpolation method, and were calculated according to ratios found by Silberberg and Tsao (1970). The values for threshold energies of spallation reactions were estimated by the method described by Mattauch *et al.* (1965). The shape of the energetic spectrum of primary protons in the $E > 300$ MeV region was considered to be analogous to the energy spectrum of non-modulated galactic cosmic rays (Lavrukhina and Ibraev, 1969). The $E < 300$ MeV spectrum is also analogous to the solar proton spectrum of the 19th solar cycle (Ebeouglu and Wainio, 1966). The cross-section of each reaction was averaged according to the energetic spectrum of cosmic rays, as well as averaged according to the isotopic composition of the irradiated matter.

Spallation Reactions in Meteoritic Bodies

The contribution of spallation reactions by galactic cosmic rays has been estimated and their flux has been determined from ^{22}Na contents in freshly fallen meteorites:

$$I_p\ (E_p > 200\ \text{MeV/nucleon}) = 0.41\ \text{particles cm}^{-2}\ \text{sec sterad}$$

(Lavrukhina and Ibraev, 1969). Our calculations indicate that the fraction of cosmogenic isotopes produced is equivalent to 3×10^{-5} of their recent abundances. Thus, the nuclear reactions occurring in meteorites during their irradiation by cosmic rays could not lead to the change of the primary isotopic composition of non-volatile elements.

Spallation Reactions in the Protosolar Nebula

According to modern ideas, the most probable model of the primary solar system is a considerably flattened nebula with mass of (2 to 4) × 10^{33} g or (1.01 to 2) solar masses. Its density during compression increased from 10^{-24} g cm^{-3} to 10^{-12} g cm^{-3}. The matter of the solar system at this stage was irradiated by cosmic rays. We have carried out quantitative estimates of these effects. Owing to the low density ($\rho \ll 10^{12}$ atoms cm^{-3}), the contributions from secondary neutrons and ionization losses were neglected. The integral flux of protons was calculated from the observed ^{9}Be abundance; $I_p(E_p > 5\,\text{MeV}) = 6.9 \times 10^{19}$ protons cm^{-2}. The results obtained show that the observed abundances of "by-passed isotopes" cannot be explained by spallation reactions in the protosolar nebula.

Spallation Reactions in the Protoplanetary Cloud

For an explanation of the observed deuterium in meteorites, Audouze (1970) has suggested a model of the protoplanetary cloud irradiation after the hydrogen and helium loss. Integral intensity of solar cosmic rays is equal to 1.5 × 10^{20} protons cm^{-2}. We have used this model and calculated the amounts of "by-passed isotopes" being generated in the protoplanetary cloud (Lavrukhina *et al.*, 1974). Calculations have shown that during irradiation of the protoplanetary cloud, it is possible to obtain ^{9}Be, ^{50}V and ^{40}K in amounts comparable with their cosmic abundances. The contribution of the cosmogenic component to the ^{184}Os and ^{190}Pt contents is 15% and 30%, respectively, and the cosmogenic contribution for the other "by-passed isotopes" is 1.5 to 3.0% of their cosmic abundances. The calculated amounts of ^{124}Xe and ^{126}Xe are on the average higher by an order of magnitude than their contents in the C1 carbonaceous chondrites. This suggests strongly that the protoplanetary cloud, during the period of its matter differentiation, could not have been irradiated by such a flux of protons. It is three orders of magnitude higher than the present flux (Lavrukhina and Ustinova, 1971), even at an irradiation duration of 10^{8} years.

Spallation Reactions in the Supernova

Recent data of Gurevich and Rumyantsev (1973) and Colgate (1975) indicate that the supernova and the explosions of galaxy nuclei and pulsars were accompanied by a collective acceleration of particles up to relativistic energies. We have made calculations for the model, when spallation reactions occur in the supernova matter. This model has allowed us to obtain for the first time satisfactory agreement with the observed abundances of "by-passed isotopes" (Table 2, col. IV, Fig. 1), under the following irradiation conditions:

1. *Intensity of the flux of protons.* To explain the cosmic abundances of eighteen "by-passed isotopes" in the region of $A \geq 113$, it is necessary to have an integral proton flux equivalent to $I_p(E_p > 25\,\text{MeV}) = 5 \times 10^{21}$ protons cm^{-2}. A flux of such high intensity may only be expected in catastrophic events (supernova or explosion of the galaxy nuclei), where an irradiation of matter could occur prior to the formation of the bodies in the solar system.

TABLE 2. THE OBSERVED AND CALCULATED ABUNDANCES OF "BY-PASSED ISOTOPES"

Isotope	$N_{calcul.}/N_{observ.}$ I	II	III
^{9}Be	1.0	1.0	162
^{9}Be$^{(b)}$	—	—	1.0
^{10}Be$^{(a)}$	(0.01)	(0.1)	—
^{26}Al$^{(a)}$	(0.15)	(10)	(10)
^{40}K$^{(c)}$	3×10^{-3}	0,58	1.0
^{50}V	0.12	0.58	30.7
^{50}V$^{(d)}$	—	—	0.9
^{129}J$^{(a)}$	2.4×10^{-5}	1.7×10^{-5}	(1.4×10^{-5})
^{74}Se	—	—	0.1
^{78}Kr	—	—	0.3
^{84}Sr	—	—	0.4
^{92}Mo	—	—	0.03
^{94}Mo	—	—	0.01
^{96}Ru	—	—	0.14
^{98}Ru	—	—	0.02
^{102}Pd	—	—	0.3
^{106}Cd	—	—	0.07
^{108}Cd	—	—	0.24
^{112}Sn	—	—	0.09
^{114}Sn	—	—	0.25
^{115}Sn	—	—	0.51
^{113}In	—	—	0.6
^{120}Te	—	—	0.7
^{124}Xe	3×10^{-3}	11.3	0.7
^{126}Xe	8×10^{-3}	15.2	1.0
^{130}Ba	2.6×10^{-3}	1.5×10^{-2}	0.6
^{132}Ba	5×10^{-3}	2.4×10^{-2}	0.8
^{138}La	4.5×10^{-2}	2.3×10^{-2}	0.8
^{136}Ce	4.6×10^{-3}	3.0×10^{-2}	0.6
^{138}Ce	7.5×10^{-3}	3.4×10^{-2}	1.1
^{152}Gd	4×10^{-3}	2×10^{-2}	0.7
^{156}Dy	6×10^{-2}	3×10^{-2}	1.2
^{158}Dy	6.5×10^{-3}	3×10^{-2}	1.1
^{162}Er	5.5×10^{-3}	3×10^{-2}	1.1
^{168}Yb	6×10^{-3}	2×10^{-2}	1.0
^{174}Hf	7×10^{-3}	2×10^{-2}	1.3
^{180}W	1×10^{-2}	3.5×10^{-2}	1.5
^{184}Os	3×10^{-2}	0.14	6.3
^{184}Os$^{(g)}$	—	—	1.0
^{190}Pt	7×10^{-2}	0.33	12.1
^{100}Pt$^{(g)}$	—	—	1.1
^{196}Hg	7×10^{-3}	2.6×10^{-2}	0.9

Notes to Table 2

I. Spallations reactions in protosolar nebula $I_p(E_p > 5\ \text{MeV}) = 7\times10^{19}$ protons cm^{-2}. Composition (Cameron, 1973).

II. Spallation reactions in protoplanete cloud, $I_p(E_p > 25\ \text{MeV}) = 1.5\times10^{20}$ protons cm^{-2}. Composition of non-volatile ones (Cameron, 1973); abundances of C, O, N and noble gases are analogous with their contents to C1 carbonaceous chondrites.

III. Spallation reactions in the supernova.

(a) Quantities of radioactive nuclei in atoms/10^6 atoms Si to the end of nuclear processes; duration of irradiation was: I—10^8 years; II—10^7 years.

(b) ^{9}Be burned out in star of second generation.

(c) ^{40}K decays with $T_{1/2} = 1.25\times10^9$ years.

(d) ^{50}V was formed by the ^{56}Ni spallation.
^{56}Ni$(p, 6pn)^{50}$V; $\sigma(E_p = 660\ \text{MeV}) = 0.01$ mbarn.

(g) Only isotopes formed in *s*-process have taken part in spallation reactions of formation of ^{184}Os and ^{190}Pt.

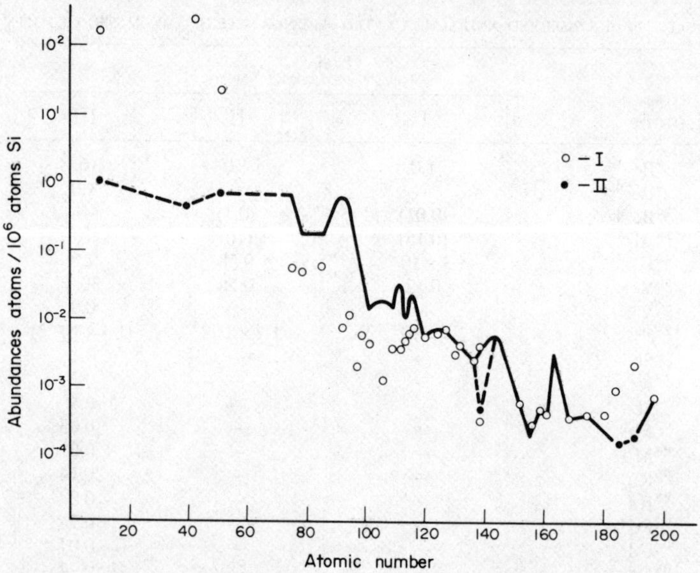

FIG. 1. Relative abundances of "by-passed isotopes" in the primary cosmic matter (in atoms for 10^6 atoms of Si) continuous curve and their contents formed in spallation reactions. I. Spallation reactions in supernova. $I_p(E_p > 25\ \text{MeV}) = 5 \times 10^{21}$ protons cm^{-2}. II. Data obtained under the following conditions: (1) ^{9}Be burns out in a star of second generation. (2) With account of the radioactive decay of ^{40}K with $T_{1/2} = 1.25 \times 10^9$ years. (3) ^{50}V was formed by the ^{56}Ni spallation. (4) Only isotopes formed in *s*-process have taken part in spallation reactions of formation of ^{184}Os and ^{190}Pt.

2. *Duration of the irradiation process.* This was determined in accordance with the cosmic abundance of ^{50}V. The ^{50}V cosmic abundance indicates that the spallation reactions accompanied a process of silicon burning (Arnett, 1969), where ^{56}Ni was produced. The calculated value for ^{50}V agrees with its observed abundance only when this isotope has been formed by the ^{56}Ni spallation. In the case of ^{50}V production from the process of ^{56}Fe spallation, we obtained a value of $N_{\text{calcul.}}/N_{\text{observ.}} = 30.7$. In accordance with the half-life values of ^{56}Ni and ^{56}Co, the duration of the spallation reactions is estimated as less than 10 days ($T_{\text{spall.}} < 10$ days).
3. *Sequence of processes of nucleosynthesis.* It results from the observed abundance of the heaviest isotopes of ^{184}Os and ^{190}Pt. This is only possible to coordinate with the calculated values under the condition that isotopes formed in the *r*-process did not participate in spallation reactions. Apparently, the *r*-process followed the spallation process or the *r*-process took place in pulsars.
4. *Time and number of nucleosynthesis acts.* The ^{40}K amount obtained in the spallation reactions agrees only with its recent abundance under the assumption that its main part had been formed 10.6 b.y. ago(^{40}K with $T_{1/2} = 1.25$ b.y.) and the last supernova (4.7 b.y. ago) added less than 1% of the cosmic abundance of ^{40}K. The ^{9}Be anomaly may be explained only under the condition that the matter synthesized 10.6 b.y. ago and has passed through an evolution stage in a star of second generation, where isotopes of light elements were burned out. The observed amounts of light elements were formed during the last supernova which added less than 1% of the "by-passed isotopes" and the observed abundance of ^{9}Be.

References

ARNETT, W. D. (1969) Explosive nucleosynthesis in stars. *Astrophys. J.* **157**, 1369–1380.

ARNOULD, M. (1976) Possibility of synthesis of proton-rich Nuclei in highly evolved stars II. *Astron. Astrophys.* **46**, 117–125.

AUDOUZE, J. (1970) Some nucleosynthesis effects of energetic proton fluxes on steller surfaces and on the solar system. *Astron. Astrophys.* **8**, 436–457.

AUDOUZE, J. (1970) Formation of deuterium in early solar system. *Mém. Soc. r. Sci., Liège* **19**, 255–259.

AUDOUZE, J. and TRURAN, J. W. (1975) The *p*-process nucleosynthesis in post shock supernova envelope environment. *Astrophys. J.* **202**, 204–213.

BURBIDGE, E. M., BURBIDGE, G. R., FOWLER, W. A. and HOYLE, F. (1957) Synthesis of the elements in stars. *Rev. Mod. Phys.* **29**, 547–650.

CAMERON, A. G. W. (1973) Abundance of the elements in the solar system. *Space Sci. Rev.* **15**, 121–146.

CHARAHCHJAN, A. N. and CHARAHCHJAN, T. N. (1973) Investigation of the cosmic rays. *Trans. Phys. Ind. Acad. Sci.* **64**, 3–48.

COLGATE, S. A. (1975) Supernova and the origin cosmic rays. (I), (II). "Origin cosmic rays". *Proc. NATO Adv. Study Inst., Durham, 1974,* Dordrecht–Boston, 425–444, 447–466.

COLGATE, S. A. (1975) Shock-wave thermalization. *Astrophys. J.* **195**, 493–498.

EBEOGLU, D. B. and WAINIO, K. M. (1966) Solar proton activation of the lunar surface. *J. Geophys. Res.* **71**, 5863–5872.

FRANCK-KAMENETSKII, D. A. (1961) Origin of chemical elements. *Astron. J.* **38**, 91–98.

GUREVICH, L. E. and RUMYANTSEV, A. A. (1973) The cosmic rays of superhigh energies. *Astrophys. Space Sci.* **22**, 79–85.

LAVRUKHINA, A. K. and KUZNETSOVA, R. I. (1967) Analysis cross-section of spallation products of different elements by protons with $E^p = 0$, 1–24 Gev. *Materials of the Xth Conf. on Nuclear Spectroscopy and the Theory of Deformed Nuclei*, Dubna JINR, 127–132.

LAVRUKHINA, A. K. and KUZNETSOVA, R. I. (1969) Calculation spallation products with mass-number near to the mass-number of target-nuclear, Coll. papers *Nuclear Spectroscopy*, Dubna JINR, 125–128.

LAVRUKHINA, A. K. and IBRAEV, T. A. (1969) On the intensity and energetic spectrum galactic cosmic rays in the solar system. *Trans. of the All-Union Conf. on Physics of Cosmic Rays* (Tashkent, October 7–16, 1968), Part II. Moscow, Phys. Inst. Acad. Sci.

LAVRUKHINA, A. K. and USTINOVA, G. K. (1971) Solar proton medium flux constancy over a million years. *Nature* **232**, 462–463.

LAVRUKHINA, A. K. (1972) *Nuclear Reactions in Cosmic Bodies*, M., "Nauka".

LAVRUKHINA, A. K., KUZNETSOVA, R. I. and RUPP, E. (1974) Nuclear reactions in Solar system matter. *Meteoritic* **33**, 14–22.

MATTAUCH, J. H. E., THIELL, W. and WAPSTRE, A. H. (1965) *Nucl. Phys.* **67**, 1–33.

REEVES, H. and STEVART, P. (1965) Positron-capture processes as a possible source of the *p*-elements, *Astrophys. J.* **141**, 1432–1438.

SILBERBERG, R. and TSAO, C. H. (1970) Partial cross-section in high-energy nuclear reactions for targets heavier than nickel. Preprint 20390, Lab. Cosmic Rays Phys., Naval Res. Laboratory, Washington.

TRIMBLE, V. (1975) The origin and abundances of the chemical elements. *Rev. Mod. Phys.* **47**, 877–976.

TRURAN, J. W. and CAMERON, A. G. W. (1972) The *p*-process in explosive nucleosynthesis. *Astrophys. J.* **171**, 89–92.

WEAVER, T. A. and CHAPLINE, G. F. (1974) Dissipation in supernova shock waves. *Astrophys. J.* **192**, L57–L60.

Some Long-lived and Stable Nuclides Produced by Nuclear Reactions

REISUKE GENSHO, OSAMU NITOH, TOMOKO MAKINO and MASATAKE HONDA

Institute for Solid State Physics, University of Tokyo, Roppongi, Minato-ku, Tokyo, Japan

Abstract

Productions of Be-10, Al-26, stable Mg isotopes, Mn-53, Cr-53, Ni-59, and Nb-92 have been studied in variable targets bombarded by high-energy protons and neutrons. The half-lives of Be-10, Ni-59 and Nb-92 have been determined with chemically separated products directly by a combination of data of mass spectrometry and radioactivity counting.

The production cross-sections have also been measured for the productions of Al-26, Mg isotopes, Mn-53 and Ni-59 in aluminium and iron targets by protons and alpha particles up to 50 MeV.

The productions of these nuclides and their decay products in extraterrestrial materials are discussed.

Long-lived radioactivities and their decay products in nature are useful in cosmochemical studies. The half-lives and production rates of many such isotopes have been investigated by mass spectroscopies, radioactivity measurements and other methods. By highly sensitive mass spectroscopies, isotopic compositions of elements and/or number of atoms can be determined in extremely small size samples. Hence the production cross-sections could be obtained directly, and combining this with the activity measurements will give half-lives of radioactivities. To this date many nuclides are still obscure in respect to their half-lives. In this paper we present some recent results obtained in our laboratory and elsewhere and discuss briefly some future problems on: (1) Mn-53 and Cr-53, (2) Al-26 and Mg-26, and (3) Nb-92 and Zr-92.

(1) *Mn-53 and Cr-53*

Production cross-sections of Mn-54 ($t_{1/2} = 312$ d) and Mn-53 ($t_{1/2} = 3.7 \times 10^6$ y; Honda and Imamura, 1971) induced in a cyclotron-bombarded iron target have been measured in the region of 18–52 MeV. The target arrangement is illustrated in Figure 1. About 10^{17} protons of 52 MeV were available at the Institute for Nuclear Studies, University of Tokyo. The target sheets were prepared from a commercially available pure iron metal, "Puron", supplied by Johnson Matthey Co. The manganese fraction containing typically 10^7 dpm of Mn-54 and 1 dpm of Mn-53 was extracted from bombarded samples in the carrier-free form and it was purified by a wet chemical method (Honda and Imamura, 1971). A silica gel method using a rhenium single filament was applied for the solid source surface ionization, and an improved pulse-counting method was used to measure 10^{10}–10^{11} atoms of Mn isotopes in the presence of about 10^{14} atoms of natural Mn-55 (Gensho *et al.*, 1972) (Fig. 2). The procedure was checked

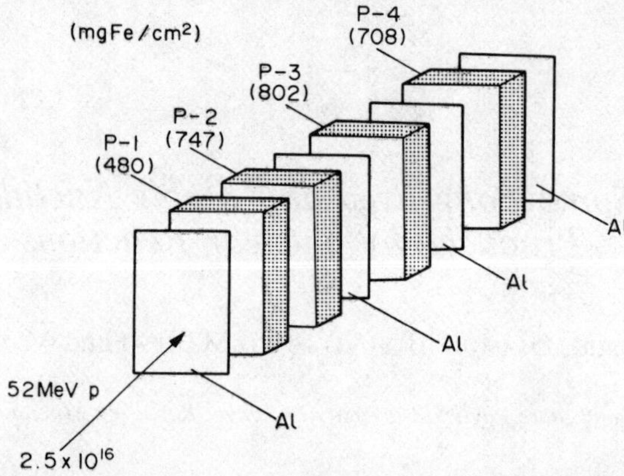

FIG. 1 Stack arrangements of iron targets and aluminium foil monitors.

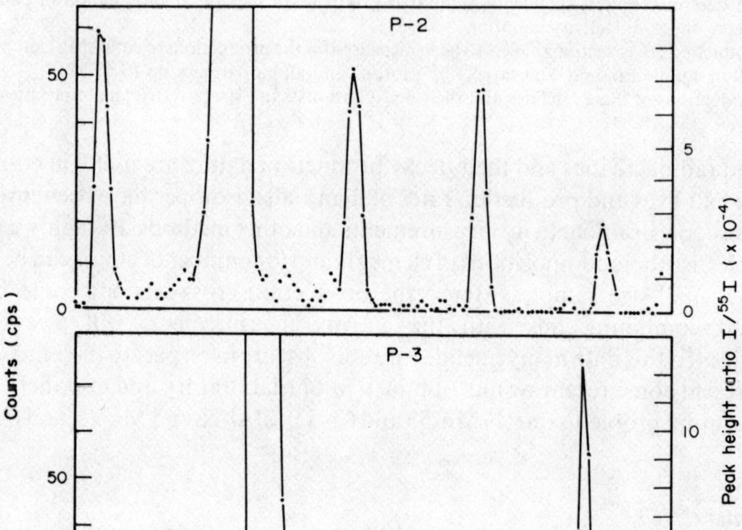

FIG. 2 Examples of mass spectra of Mn extracted from proton bombarded iron targets: P2:40–47 MeV, P3:31–39 MeV. Some corrections were necessary to reduce contributions from Fe-54, Cr-53 and Cr-54.

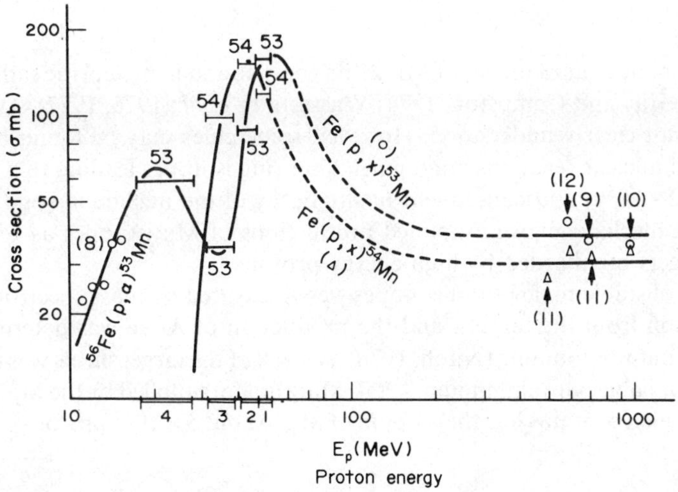

FIG. 3 Excitation functions of Fe(p, x)Mn-53 + Mn-54. 115 ± 12, 143 ± 15, 95 ± 10 and 3 mb were obtained for production cross-sections of Mn-54; 150 ± 20, 86 ± 10, 33 + 4 and 65 ± 7 mb were for Mn-53 in 51–47, 47–40, 39–31 and 28–18 MeV respectively. Literature values are also plotted in other regions. (8) (Furukawa and Shizuri, 1972) by neutron activation for Mn-53 determination; (9) (Honda and Imamura, 1971; Matsuda *et al*, 1971); (10) Honda unpublished, thick target data; σ(Mn-53)/σ(Mn-54) is a mean of 2-18 cm depth data; (11) (Honda and Lal, 1960, 1964); (12) (Perron, 1975, 1976). This figure has been reproduced from Gensho *et al.* (1972) after some minor modifications.

with a known amount of artificial carrier-free Mn-54 source and natural Mn-53 samples. Absolute amounts of each isotope were obtained from the disintegration rate of Mn-54, the neutron activation analysis for Mn-55, and the isotopic ratio of Mn-54 to Mn-53. The excitation functions of Fe(*p*,*x*)Mn-53 + Mn-54 obtained are shown in Figure 3.

The results were used to calculate the production rates of Mn-53 as well as Mn-54 induced by solar and galactic cosmic rays in targets containing iron (Imamura *et al.*, 1974). Older estimates (Reedy and Arnold, 1972; Tanaka *et al.*, 1972) have been revised by the recent calculations using this experiment.

Currently we are extending our work to measure natural Cr-53 as a decay product of Mn-53 in primitive samples (Table 1). So far no detectable enrichment in the Cr-53 exceeding 0.1% has been found in chromium extracted from some HCl and HF leached solutions and chromite fractions obtained from Allende meteorite. The normalizations for Cr-53/Cr-52 were made using Cr-54 and also Cr-50.

TABLE 1. Cr-53 IN METEORITIC SAMPLES

	Mn/Cr	Cr-53/Cr-52* $\times 10^{-2}$
Reagent chromium		11.337 ± 5
Allende fraction		
1. HCl soluble	12.4	11.324 ± 5
2. HF soluble	0.18	11.342 ± 7
3. residue	0.02	11.334 ± 6

* Normalized to Cr-54/Cr-52 = 2.822 × 10^{-2}; error: $2 \times [\Sigma\Delta^2/n(n-1)]^{1/2}$.

(2) *Al-26 and Mg-26*

Recently anomalous enrichment of Mg-26 has been found in meteoritic samples with high Al/Mg ratios (Gray and Compston, 1974; Wasserburg *et al.*, 1976, 1977). At this time, the mechanism is not clearly understood. However, some clues may be found by considering proton-induced nuclear reactions among neighbouring isotopes leading to a production of Al-26($t_{1/2} = 7.2 \times 10^5$y) and consequent *in situ* decay of the nuclide in solid materials. To understand the mechanism, we examined productions of Mg isotopes as well as Al-26 in aluminium targets bombarded by high-energy protons.

Productions of stable magnesium isotopes were measured by mass spectrometry in a Mg extracted fraction from the targets, and the production of Al-26 was determined by non-destructive gamma-ray counting (Nitoh, 1976). A stack of the target sheets was prepared from commercially available pure aluminium metal (Marz grade) supplied by the Material Research Corp. U.S.A., which contains less than 1 ppm of Mg. About 5×10^{17} protons of 15 MeV were

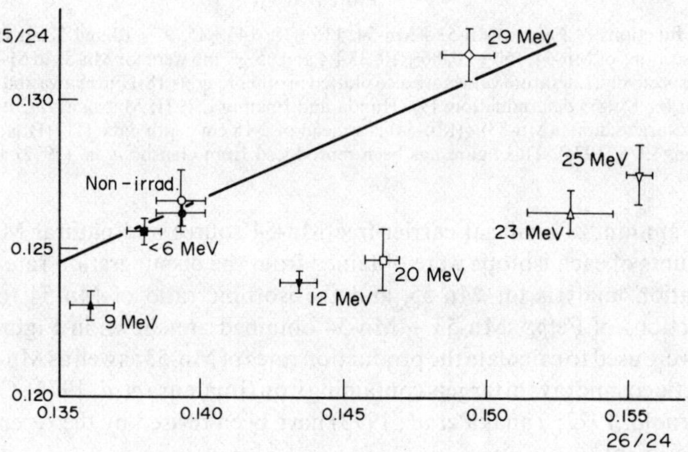

FIG. 4 Isotopic compositions of Mg separated from Al targets.

available at the Research Institute of Physics and Chemistry, Wako, Saitama, Japan; 2×10^{17} protons of 52 MeV were also used at the Institute for Nuclear Studies. Magnesium containing a few nanogram (10^{-9} g) of induced products was separated quantitatively and purified for the mass spectrometry. A few tenths of ppm of laboratory contamination of Mg were recovered along with the induced products, and the amounts were determined by atomic absorption. The silica gel technique was applied in rhenium single-filament method. Isotopic compositions found in the samples are as shown partly in Fig. 4. From these data excitation functions have been obtained as illustrated in Fig. 5. Although the production of Mg-24 via Al(p, α) (Q: + 1.6 MeV) is high in the lower energy region of 10–15 MeV, the Al-26 and Al-26m via (p, pn)(Q: − 13.1 and − 13.3 MeV respectively) and direct Mg-26 production via ($p, 2p$)(Q: − 8.3 MeV) are predominant in the region of 12–30 MeV.

Heymann and Dziczkaniec (1976) have already suggested that Al-26 production via Mg-26 (p, n)(Q: −4.8 MeV) in 5 MeV region could be responsible and the reaction could occur extensively in a gaseous phase of the early solar system just before the condensation. We are investigating the possibility of treating aluminium as the target in primitive solid particles of

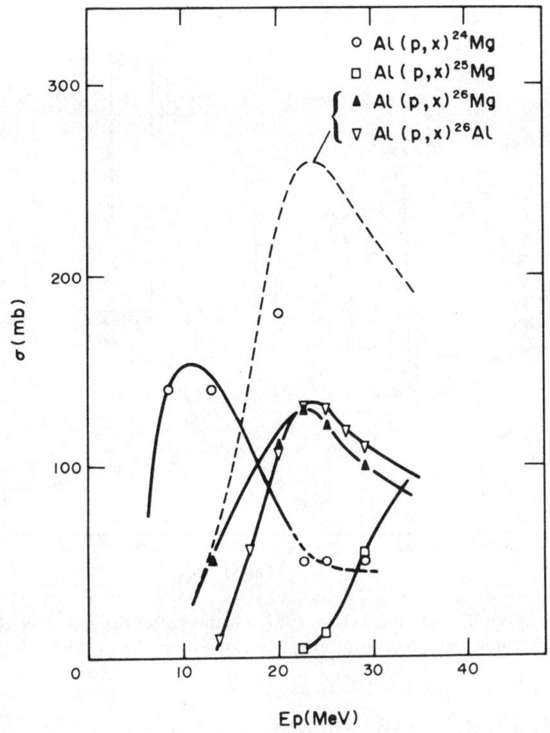

FIG. 5 Excitation functions of Al(p, x)Mg isotopes +Al-26 (Nitoh, 1976).

refractory aluminium compounds. The dose of protons may be much lower than the case of gaseous target, or may be 10^{20} p/cm^2 of higher than 10 MeV.

(3) *Nb*-92 *and Zr*-92

Long-lived nuclides of half-life longer than 10^8 years are expected to be extant activity, or still living nuclide from the time of nucleosynthesis. The half-life of Nb-92 was estimated by Knight (1972) to be 1.8×10^8 years with an uncertainty of a factor of less than 5. Apt *et al.* (1974) measured the activities due to Nb-92 and Nb-94 ($t_{1/2} = 2.03 \pm 0.16 \times 10^4$ y; Schuman and Goris, 1959) in natural niobium and calculated the half-life of Nb-92 to be 1.6×10^8 y, provided the former activity could be assumed to be primordial. If the half-life is really longer than 10^8 years, then an isotopic enrichment of Zr-92, the daughter of Nb-92, in zirconium would be expected to occur in a wide range of natural samples. Minster and Allègre (1976) found, however, no detectable enrichment in Zr-92 in some terrestrial and extra-terrestrial samples. We have measured the half-life of Nb-92 to resolve this question.

The long-lived isotopes of niobium were produced from molybdenum by fast neutron irradiation in a pile (Knight, 1972). Their radioactivities were determined by gamma-ray spectroscopy as a function of time and the atomic abundances were determined by mass spectrometry. During 1974–75, 6 g of molybdenum metal was irradiated in the JMTR, Japan Material Testing Reactor, Oh-arai Ibaraki pref., and subjected to a total thermal neutron

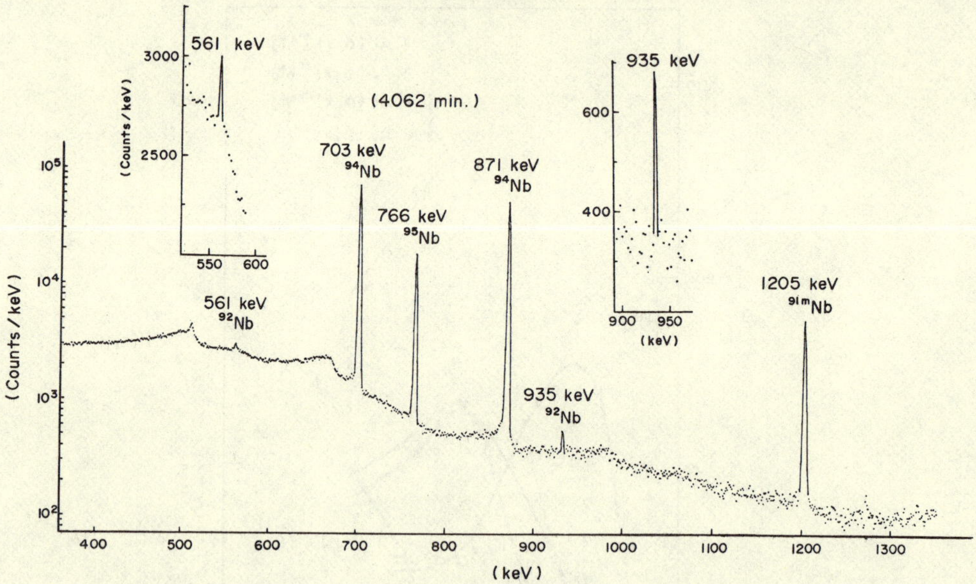

FIG. 6 Gamma-ray spectra of Nb-92 and others (Feb. 1977, 730 days after the end of irradiation). Relatively higher background at 561 keV γ is mainly due to compton spectra of 871 keV γ of Nb-94.

fluence of 7×10^{20} n/cm^2. After cooling for about 1 year, the target was dissolved in nitric acid and carrier-free niobium was separated from molybdenum and zirconium by anion exchange and extraction. Aliquots of about 30×10^{-9} g and 3×10^{-9} g of Nb-92 were used for the gamma-ray counting and mass spectrometry respectively. Gamma-rays of Nb-92, 935 keV and 561 keV (both 100%) were measured by a Ge(Li) detector after 2 years from the end of the irradiation. In the gamma-ray spectra, other peaks due to Nb-91m (1205 keV), Nb-94 (703 and 871 keV), both 100% (Schuman and Goris, 1959), and Nb-95 (766 keV), were also observed. Respective regions of the spectra obtained are shown in Fig. 6. The ratio of two long-lived activities, Nb-92/Nb-94, was found to be 0.0040 ± 0.0003. Mass spectrometric measurements were carried out by surface ionization method using a single rhenium filament (Fig. 7). Contributions of the Mo^+ beam due to impurities in the rhenium filament and of the Zr^+ beam from the sample impurity were corrected using empirical data from filament blanks and zirconium fraction separated from the target respectively. Both isotopic compositions were natural, and for corrections to $m/e = 92$ and 94 peaks Mo-95 and Zr-90 peaks were used as the references. The abundance ratio of Nb-92/Nb-94 was 6.6 ± 0.5.

As shown in Figs. 6 and 7, intercomparison of Nb-92 and Nb-94 activities and abundances yielded a value $(3.3 \pm 0.5) \times 10^7$ years (Makino and Honda, 1977) for the half-life of Nb-92. Besides, absolute counting data of Nb-92, using a calibrated Mn-54 standard, and dilution analysis for Nb-92 spiked with natural Nb-93 yielded the same result.

Recently we learned that Nethaway *et al.* (1977) have also obtained 3.2×10^7 years with an uncertainty of 30–50% for the same determination, by independent measurements of cross-section for Nb-93(n, 2n)Nb-92 + Nb-92m and the radioactivities. In our experiment, the abundance ratio of Nb-91/Nb-94 was found to be 1.6 ± 0.2, and the X-ray spectra, Zr Kα, of the sample were also observed. The latter could be attributed to the radioactivity of Nb-91. Based

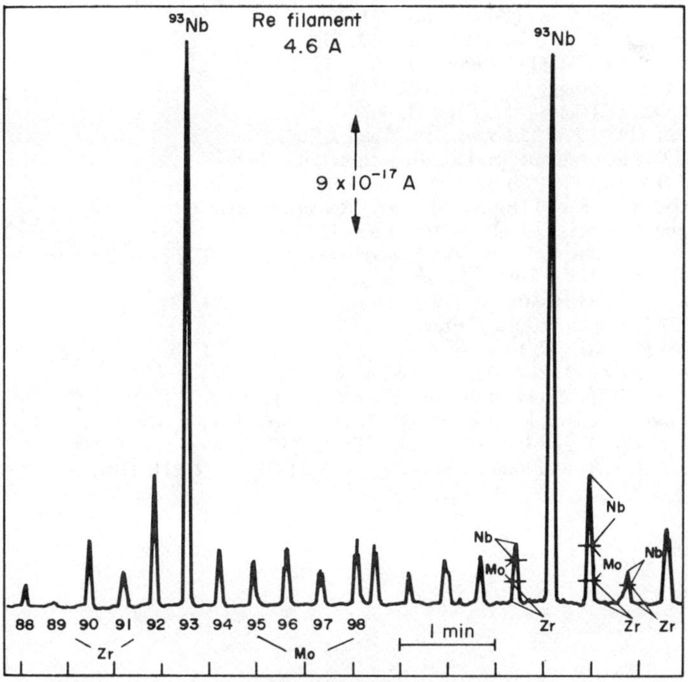

FIG. 7 Mass spectra of Nb isotopes. In this spectra, peaks at $M/e+93$, more than half of 92 and of 91, and some fractions of 94 are indicating the contributions of respective Nb isotopes. In this sample, the abundances were found to be 1.6 ± 0.2:6.6 ± 0.5:51:1 for the ratios of Nb-91, -92, -93 and -94 respectively. With higher filament temperature, interferences due to Zr decrease and filament blank of Mo increases.

on the two sets of data 0.8×10^3 years was calculated for the half-life of Nb-91, the other member of long-lived Nb isotopes.

As the half-life of Nb-92 has been determined to be much shorter than 10^8 years, it is impossible to expect any primordial extant Nb-92 in nature. It is, however, still possible to detect an excess of Zr-92 as a daughter of extinct activity of Nb-92 in primitive meteorite minerals where the content of Nb relative to Zr is high. If we assume the production rate of Nb-92 to be equal to 0.1 or lower relative to Nb-93, then in the formation interval of $\leq10^8$ y, and if we apply a simple continuous model for nucleosynthesis, the initial ratio of Nb-92/Nb-93 at the condensation of primitive solar nebula turns out to be $\leq10^{-4}$. This conclusion can also be reached using a direct comparison of a well-known extinct radioactivity, I-129 ($t_{1/2}=1.6\times10^7$ y), avoiding a strong model dependency. On the other hand, cosmic abundance ratio of Zr to Nb is about 30. Therefore, in bulk sample of ordinary chondrite the enrichment in Zr-92 may be $\leq2\times10^{-5}$ which is within a limit of experimental uncertainty of mass spectrometry for Zr (Minster and Allègre, 1976). The situation may become more promising when we can separate a high Nb/Zr phase from some primitive meteorites.

References

APT, K. E., KNIGHT, J. D., CAMP, D. C. and PERKINS, R. W. (1974) *Geochim. Cosmochim. Acta* **38**, 1485.

FURUKAWA, M. and SHIZURI, K. (1972) Abstract of 16th Symposium on Radiochemistry, Niigata, Japan.

GENSHO, R., TAKEUCHI, S. and HONDA, M. (1972) *Mass Spectroscopy* **20**, 211.

GRAY, C. M. and COMPSTON, W. (1974) *Nature* **251**, 495.
HEYMANN, D. and DZICZKANIEC, M. (1976) *Sci.* **191**, 79.
HONDA, M. and IMAMURA, M. (1971) *Phys. Rev.* C, **4**, 1182.
HONDA, M. and LAL, D. (1960) *Phys. Rev.* **118**, 1618.
HONDA, M. and LAL, D. (1964) *Nucl. Phys.* **51**, 363.
IMAMURA, M. *et al.* (1974) *Proc. 5th Lunar Sci. Conf.* **2**, 2093.
KNIGHT, J. D. (1972) Los Alamos Sci. Lab. Report, LADC–72–209.
MAKINO, T. and HONDA, M. (1977) *Geochim. Cosmochim. Acta*, in press.
MATSUDA, H., UMEMOTO, S. and HONDA, M. (1971) *Radiochim. Acta* **15**, 51.
MINSTER, J. F. and ALLÈGRE, C. J. (1976) *Meteoritics* **11**, 336.
NETHAWAY, D. R., PRINDLE, A. L. and VAN KONYENBURG, R. A. (1977) private communication, Mar. 1977.
NITOH, O. (1976) Thesis, Univ. Tokyo, Japan.
NITOH, O., GENSHO, R. and HONDA, M. (1976) *Mass Spectroscopy* **24**, 237.
PERRON, C. (1975) Thesis, Univ. Orsay, France.
PERRON, C. (1976) *Phys. Rev.* C, **14**, 1108.
REEDY, R. C. and ARNOLD, J. R. (1972) *J. Geophys. Res.* **77**, 537.
SCHUMAN, R. P. and GORIS, P. (1959) *J. Inorg. Nucl. Chem.* **12**, 1.
TANAKA, S., SAKAMOTO, K. and KOMURA, K. (1972) *J. Geophys. Res.* **77**, 4281.
WASSERBURG, G. J., LEE, T. and PAPPANASTASSIOU, D. A. (1976) *Meteoritics* **11**, 384.
WASSERBURG, G. J., LEE, T. and PAPPANASTASSIOU, D. A. (1977) *Ap. J.* **211**, 1107.

Origin of Differentiated Meteorites

G. W. WETHERILL

Department of Terrestrial Magnetism, Carnegie Institution of Washington, 5241 Broad Branch Road, N.W., Washington, D.C. 20015, U.S.A.

and J. G. WILLIAMS

Jet Propulsion Laboratory, 4800 Oak Grove Drive, Pasadena, Ca. 91103, U.S.A.

Abstract

The larger asteroids with low inclination in the inner portion of the asteroid belt are evaluated as sources of differentiated meteorites (irons, stony irons and achondrites). It is found that the calculated earth impact rate is in agreement with the observed flux of these bodies. The frequent occurrence of elapsed times before impact of 0–1000 m.y. is in agreement with the observed typical exposure ages of iron meteorites, and the observed occurrence of some very young (< 100 m.y.) exposure ages. The absence of many > 1000 m.y. iron meteorite exposure ages and > 50 m.y. achondrite exposure ages must be attributed to collisional destruction as the calculations predict a significant number of these older ages. The calculated geocentric velocities are almost always low enough to permit penetration of the atmosphere. The distribution of elongation of the radiants shows an excess in the direction of earth's anti-apex, but not sufficient to produce a significant afternoon excess. Although contributions from other sources are possible, these objects appear to be the most plausible sources of differentiated meteorites. Chemical and isotopic data on these bodies suggest a relationship to the earth and moon, and it may be that they represent residual material from earth's accretional zone.

1. Introduction

Meteorites may be divided into two principal classes: undifferentiated and differentiated. The undifferentiated meteorites, also known as chondrites, have preserved to a remarkable degree the relative proportions of the non-volatile elements in the early solar system, but differ from one another in the extent to which the volatile element abundances have been preserved (e.g. Anders, 1971). The differentiated meteorites, which include the irons, stony-irons and achondrites, have undergone further chemical fractionation, usually as a consequence of crystal–liquid fractionation, and in several ways are analogous to igneous rocks from the earth and moon. Oxygen isotope studies (Clayton *et al.*, 1976) show that these two classes of meteorites are unlikely to be genetically related, and their principal common property may simply be that they had orbits which intersected that of the earth, even though the complex dynamic processes which led to their being in these orbits were quite different.

Most of our understanding of processes occurring in the primordial solar system has been obtained by studies of these two classes of meteorites. The chondrites have preserved relics of those processes which would be destroyed by magmatic differentiation, whereas the differentiated meteorites record the existence and nature of these processes in the early solar system, and are relevant to our understanding of the planetesimals which accumulated to form the earth and moon.

It is interesting that so much has been learned from the study of these objects even though our knowledge of their source in the present solar system is very incomplete. Chondrites could be derived from comets (e.g. Wetherill, 1971, 1974a), asteroids (Chapman, 1976; Anders, 1975; Zimmerman and Wetherill, 1973), or from the earth-crossing or near earth-crossing Apollo and Amor objects (Anders, 1964; Simonenko, 1976; Wetherill, 1976a). It is also unclear to what extent the Apollo and Amor objects are of asteroidal origin (Levin *et al.*, 1976) or represent the non-volatile residue of cometary nuclei (Öpik, 1963; Wetherill, 1976a). On the other hand, with the exception of Öpik (1966) it has generally been agreed that the differentiated meteorites are of asteroidal origin. The principal basis for this, other than preconceived views regarding the chemical history of comets, has been the long ($\sim 10^8$–10^9 y) exposure ages of iron meteorites which are very difficult to reconcile with the expected dynamic lifetime ($\sim 10^6$–10^7 y) of cometary material in the inner solar system (Anders and Arnold, 1965; Wetherill, 1971, 1972). More recently, strong evidence for an asteroidal origin of a major class of differentiated meteorites, the polymict-brecciated basaltic achondrites, or howardites, has been obtained from their similarity to lunar regolith breccias (Rajan, 1974).

This paper will discuss a dynamic theory for the derivation of differentiated meteorites from the inner edge of the asteroid belt. It is fundamentally that of Arnold (1965) in which relatively low velocity collision ejecta (i.e. <1 km/s) from asteroids with perihelion near Mars achieve the large acceleration (~ 5 km/s) necessary to become earth-crossing by the random walk accumulation of Mars perturbations. The present work extends that of Arnold by using cratering studies of Gault *et al.*, (1963) and O'Keefe and Ahrens (1976) to estimate the yield of meteorites from observed asteroids. It also provides a way to circumvent a difficulty in obtaining achondrites by the Mars perturbation mechanism: the long time previously required to reorient a meteorite's orbit from perihelion near Mars to aphelion near Mars. This resulted in a minimum time of $\sim 5 \times 10^8$ y for reaching earth-crossing (Wetherill and Williams, 1968), by which time collisional destruction of meteorite-size silicate bodies is nearly inevitable (Wetherill, 1967; Dohnanyi, 1969; Gault, 1969). This minimum delay is found to disappear when the effects of the secular resonance ν_6 (Williams, 1969, 1971) are included. The importance of this resonance in the production of asteroidal meteorites was emphasized by Williams (1973). The present theory differs from that of Williams in that earth-crossing primarily results from the non-linear combination of these resonant Jupiter and Saturn secular perturbations with those resulting from close encounters to Mars. In Williams' theory resonant perturbations led directly to earth-crossing, which can occur, but less frequently. The present treatment is quite different from that of Anders and Mellick (1969) who also used secular perturbations to facilitate earth-crossing. Their use of the secular free oscillations, implicit in their fundamental equations (1) through (5), is mistaken however. These free oscillations possess a symmetry which synchronizes the variation of the argument of perihelion with the oscillations in eccentricity and inclination (Kozai, 1962; Williams, 1969). This symmetry is not merely a curious result of numerical integration over a finite time span, but an invariant in the same theory on which these equations are based, and its disregard introduces an internal inconsistency into the calculations. Therefore their conclusion that large asteroids such as 2 Pallas are deep Mars-crossers in the sense that close encounters with Mars result from the large free oscillations is eccentricity of Pallas is incorrect (see discussions in Wetherill, 1974 a, b).

In contrast, the forced oscillation terms in the secular perturbation theory do not possess this symmetry property. However, for most orbits they are of small amplitude and cannot have much effect on the subsequent evolution of asteroidal collision debris. However, the forced oscillations can reach high amplitudes in the vicinity of surfaces in *a, e, i* space for which the rates of precession of the nodes and arguments of perihelion are equal to eigenfrequencies occurring in the general theory of planetary secular perturbations (Williams, 1969). One of these resonant surfaces "ν_6" plays a major role in the present treatment (Fig. 1).

The mechanism discussed here has been presented at the Lyon I.A.U. Colloquium (Wetherill,

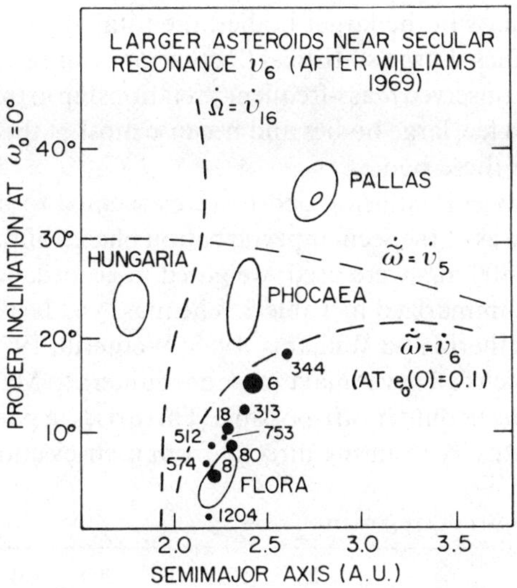

FIG. 1. Observed distribution of large asteroids in the inner portion of the asteroid belt and in the vicinity of the ν_6 resonance. The large open circle approximately define the limits of the Hungaria and Flora regions of the asteroid belt, and the Phocaea and Pallas families.

1976b). The methods of calculation are discussed in the published proceedings of that conference. A number of relevant points are also discussed in this earlier publication which will not be repeated here. The principal new contribution presented in the present paper is the replacement of preliminary results of the previous paper with more extensive calculations. The plausibility of the earlier tentative conclusions is confirmed by this further work.

2. Ejection of Collision Debris into Mars-crossing Orbits

As reported earlier, estimates of cratering yields have been calculated for large asteroids in the vicinity of Mars and the ν_6 resonance by use of the asteroidal collision theory (Wetherill, 1967) and experimental and theoretical cratering data (Gault *et al.*, 1963; O'Keefe and Ahrens, 1976). The calculated yield from the more important sources, as well as from several less productive objects in deeper Mars-crossing orbits, are given in Table 1. Because of the

TABLE 1. OBSERVED ASTEROIDS MOST PRODUCTIVE OF EJECTA IN MARS-CROSSING ORBITS (together with representative smaller bodies)

Object	Spectral class	Diameter (km)	Present perihelion (A.U.)	Δ_{Mars} (A.U.)	$\Delta_{\text{resonance}}$ (A.U.)	Calculated yield (g/yr)
6 Hebe	S	197	1.94	.141	.13	4.60×10^9
8 Flora	S	150	1.86	.024	.16	1.96×10^9
18 Melpomene	S	141	1.80	.023	.16	2.96×10^9
80 Sappho	S?	95	1.84	.095	.14	6.91×10^8
313 Chaldaea	C?	40?	1.95	−.039	.19	1.28×10^9
344 Desiderata	?	58	1.79	.015	.17	2.75×10^9
512 Taurinensis	?	16	1.64	−.145	.10	7.93×10^7
574 Reginhild	?	7	1.71	−.118	.19	1.13×10^7
1204 Renzia	?	8	1.60	−.127	.23	1.39×10^7

concentration of asteroidal mass in the larger bodies, the total yield will be similar to that obtained from the largest bodies. Large statistical fluctuations can be expected in the yields from these objects because the observed mass-frequency relationship in the asteroid belt results in dominance of the yield by a few large bodies and because most of the ejected mass derives from rare large collisions on these bodies.

In the calculation of these yields collision debris were assumed to be ejected in random directions and at random phases of the secular perturbation phases of their sources. Ejection velocities from 25 m/s up to 800 m/s were used, weighted in accordance with the cratering data referred to above, and summarized in Table 2. The mass yields were calculated by the methods used previously (Wetherill and Williams, 1968; Wetherill, 1976a). Although many of the sources are in orbits which can never make close encounters to Mars, some of the ejecta will be in orbits for which such encounters are possible. This arises in part because of changes in semi-major axis and eccentricity resulting directly from their ejection velocity following

TABLE 2. DATA USED TO CALCULATE YIELDS

Ejection velocity (nıs)	25	50	100	200	300	500	800
Fraction of ejecta at that velocity	.25	.37	.19	.13	.007	.0036	.0014

Yield $=3\times10^4 R^{2.5} V^2$ g/yr (see Wetherill, 1976),
where R = radius (km), V = velocity relative to circular orbit at random point of ejection.

collision. More important, if ejection occurs near the minimum in the secular oscillation in eccentricity of the source object, and the semi-major axis is changed to be closer to the resonant value, then the increased amplitude of the oscillations in eccentricity will cause the maximum eccentricity of the fragment to be larger. The combined effects of the decreased semi-major axis and the increased eccentricity can thereby decrease the perihelion of the fragment sufficiently to produce Mars-crossing. It is sometimes supposed that mere approach to the resonant value of the semi-major axis is sufficient to achieve this reduction in perihelion. This is not the case; ejection near the maximum phase in eccentricity will cause the increased amplitude to produce a smaller minimum eccentricity, leaving the maximum eccentricity unchanged. In any case, with random ejecta, a calculable fraction will be ejected at phases favorable for the production of small perihelia, and the remainder will simply contribute to the ordinary collisional steady-state population of small bodies in the asteroid belt.

The results of these ejecta calculations are given in Fig. 2. Material having negative values of the "minimum possible distance from Mars" can make close encounters to Mars and undergo further orbital evolution as a consequence of Mars-crossing. This is described in the next section. The total annual yield of these Mars-approaching fragments in the meteorite-size range (100 g to 10^6 g) is calculated to be $\sim 10^{10}$ g/y. The total flux of differentiated meteorites on the earth is $\sim 10^7$ to 10^8 g/y, so if this material is transferred to earth-crossing orbits with reasonable efficiency and with little collisional destruction during the time needed to complete this transfer, then this material will be adequate in quantity to provide the terrestrial flux of differentiated meteorites. Of course it is not known if these asteroids actually correspond in composition to these meteorites. Spectrophotometric data, however, show that several of the principal sources listed in Table 1 correspond to S-type asteroids which by comparison with terrestrial materials of geochemically plausible composition are believed to represent a mixture of metallic iron with silicates of composition similar to those of the achondrites (Chapman *et al.*, 1975).

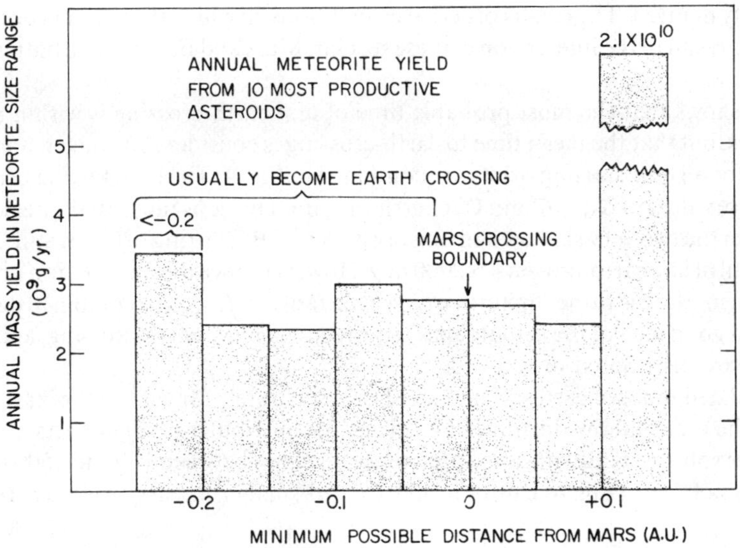

FIG. 2. Calculated annual yield of meteoritic fragments from the asteroids shown in Fig. 1, as a function of the minimum possible encounter distance from Mars (Δ_{Mars}).

3. Subsequent Orbital Evolution of the Fragmented Material

It was shown in the previous section that a significant quantity of debris from the inner edge of the asteroid belt will be placed in orbits which at least some of the time will have perihelia within the orbit of Mars, and the nodes of which will occasionally intersect the orbit of Mars. Close approaches to Mars will then occur at a calculable rate, and the further orbital evolution can be followed by the Monte Carlo method described previously (Wetherill, 1976b) which includes the effect of the ν_6 resonance.

These calculations have now been carried out for four starting orbits with different values of the initial minimum possible distance from Mars (Δ_{Mars}) and the initial distance of the semi-major axis from the resonant value ($\Delta_{resonance}$). The calculated subsequent evolution of the orbits of these bodies illustrate well the "synergistic mechanism" qualitatively proposed earlier (Wetherill, 1974a). Close approaches to Mars are initially rare (i.e. once in $\sim 10^7$ years), requiring proper orientation of the orbits of Mars and the asteroid as well as favourable values of the forced oscillation phases. Perturbations following these rare close approaches will often random walk the semi-major axis closer to the resonance, causing a slight increase in maximum eccentricity. The increase in maximum eccentricity will only be small at first because the perturbation occurs near the phase at which the eccentricity is at a maximum. However, this small increase in eccentricity decreases the minimum perihelion distance sufficiently to decrease the time between close approaches to $\sim 10^6$ years, and to permit close approaches at phases further from maximum eccentricity. Random walk toward the resonance can then occur on a more rapid time scale, the eccentricity can build up sufficiently to produce earth-crossing at the maximum phase in eccentricity. Earth perturbations will then often reduce the semi-major axis well below the resonant value, and earth-crossing will be "frozen in" with the eccentricity at least temporarily fixed near the maximum value. Of course, earth perturbations can also random walk the semi-major axis back into resonance, and there is then a chance to lose the fragment temporarily to pure Mars-crossing. This chain of events is shown schematically in Fig. 3. The results of actual calculations, in which these events occurred as the spontaneous result of the interaction of these secular, Mars and Earth perturbations, are given in Figs. 4 to 8.

Figure 4 shows that the most probable time of first earth-crossing is within the first 100 million years, but that the mean time to earth-crossing is considerably longer, about 600 m.y. The results for all four starting orbits are combined in Fig. 4. The two starting orbits with the more negative values of Δ_{Mars} (C and D) contribute about twice as much to the peak at less than 100 m.y. than the two very shallow starting orbits A and B. Starting orbits A and B contribute 80% of the tail of late earth-crossers > 2000 m.y. However, since the distribution of Δ_{Mars} for the ejecta shown in Fig. 2 is rather uniform over a wide range in Δ_{Mars}, the combined data shown in Fig. 4 gives a good idea of the actual times required to achieve earth-crossing, and shows that these times are distributed over a wide range of values.

The calculated times of earth-impact are shown separately for each of the starting orbits in Figs. 5–8. The very shallow starting orbit A (Fig. 5) contributes earth impacts over the entire history of the solar system, with some preference for times between 100 and 2500 m.y. Slightly deeper values of Δ_{Mars} (Fig. 6) causes a shift toward smaller impact times, whereas the even

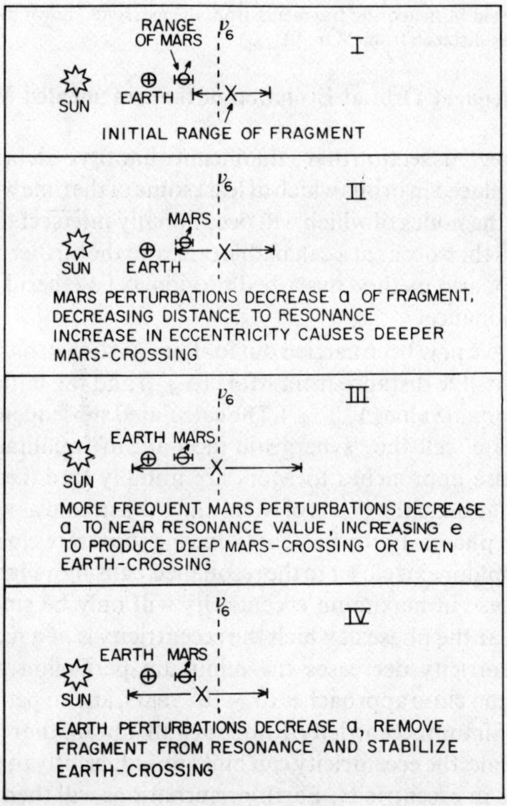

FIG. 3. Schematic description of the combined effect of secular, Mars and Earth perturbations in transferring a fragment from: (I) An occasional very shallow Mars-crosser to (II) an orbit with perihelion frequently well inside Mars orbit to (III) an orbit with semi-major axis near the resonant value and semi-major axis large enough to produce earth-crossing to (IV) an relatively stable earth-crossing orbit with semi-major axis inside the resonant value.

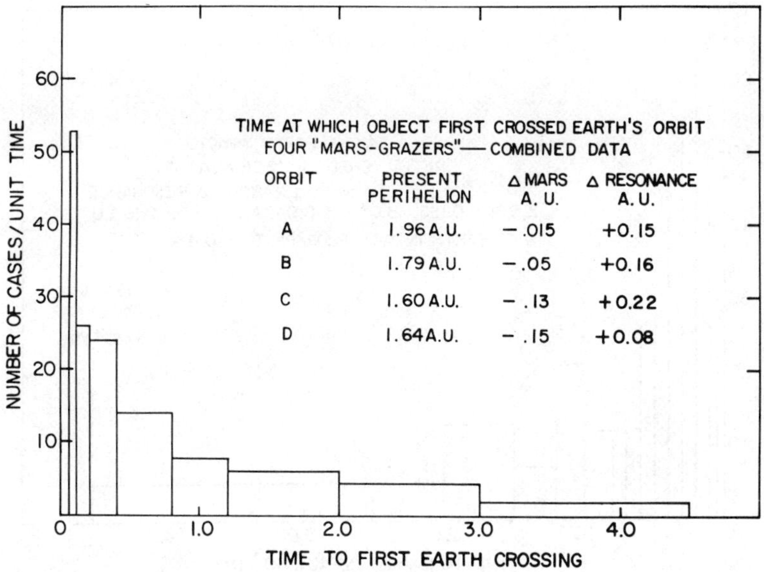

FIG. 4. Calculated distribution of times at which fragments in the designated initial orbits first become earth-crossers.

deeper initial orbits C and D (Figs. 7 and 8) will primarily generate impacts within 1000 m.y., and some impacts within the first 100 m.y. are to be expected.

These calculated impact times may be compared with the measured cosmic-ray exposure ages for iron meteorites shown in Fig. 9 (Voshage, 1967). The typical exposure ages of $\lesssim 10^9$ y

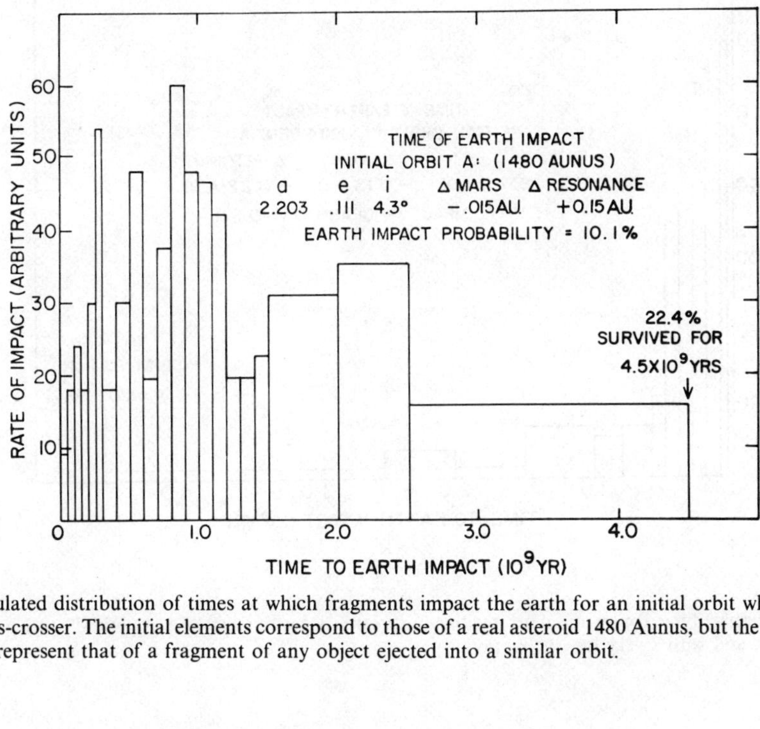

FIG. 5. Calculated distribution of times at which fragments impact the earth for an initial orbit which is a very shallow Mars-crosser. The initial elements correspond to those of a real asteroid 1480 Aunus, but the result should be taken to represent that of a fragment of any object ejected into a similar orbit.

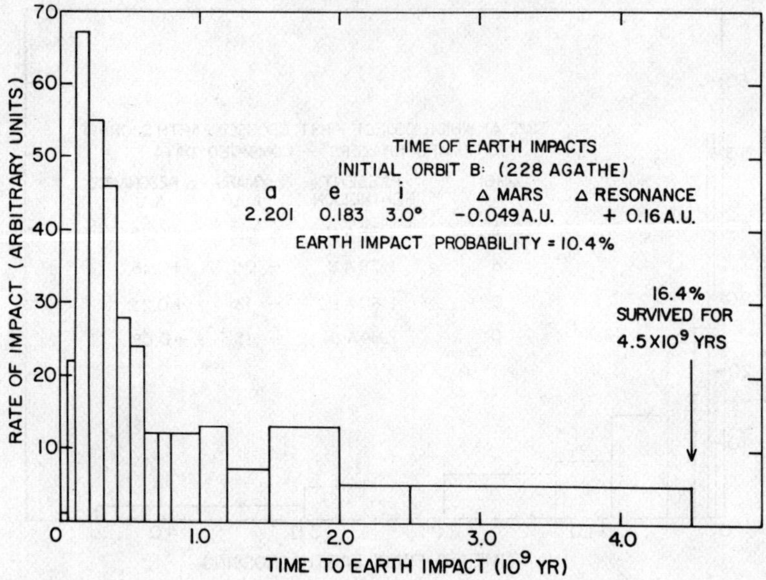

FIG. 6. Calculated distribution of times of earth impact for a slightly more negative initial value of Δ_{Mars}. A peak at ~300 m.y. now appears in the distribution, but longer times are still probable.

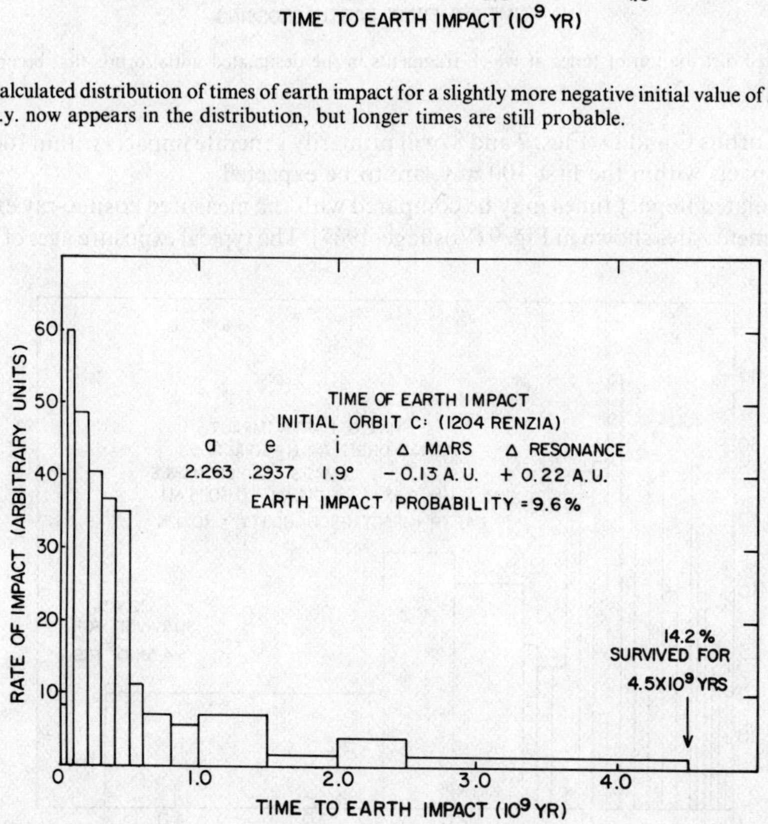

FIG. 7. Earth impact times for an initial orbit with a fairly large negative value of Δ_{Mars} initially rather distant from the resonance, and with a very low inclination.

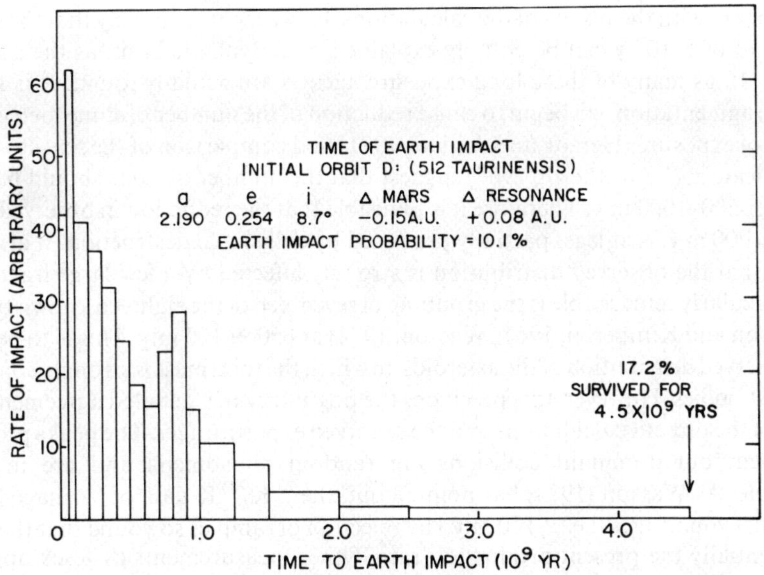

FIG. 8. Earth impact times for an initial orbit with a fairly large negative value of Δ_{Mars} and initially rather close to the resonance. Most of the impacts occur within 1000 m.y. The larger number of ~800 m.y. impact times, compared to Fig. 7, is a result of the higher inclination. The larger value of impact times < 50 m.y. is the result of the smaller $\Delta_{resonance}$.

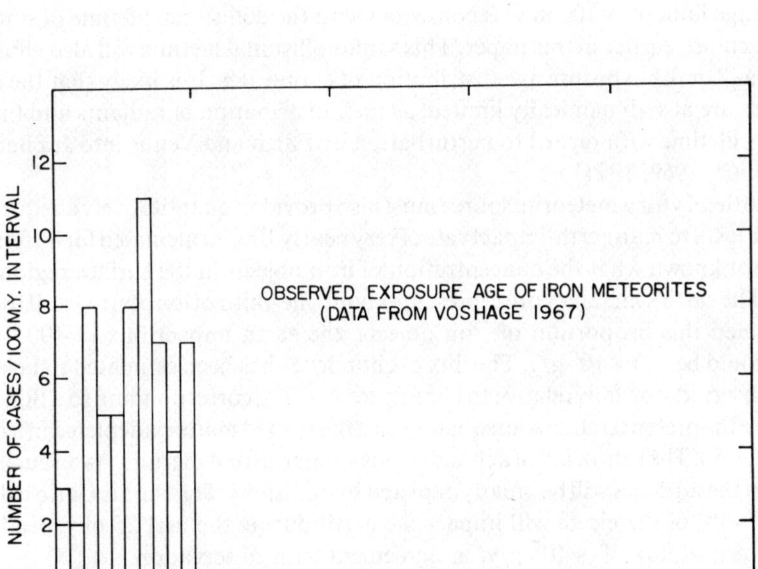

FIG. 9. Observed exposure ages of iron meteorites.

are in agreement with the results of the calculations. However, it is unlikely that the scarcity of exposure ages of $>10^9$ y can be entirely explained on a dynamic basis, as the calculations predict 5 to 10 as many of these long exposure ages as are actually found. It is likely that collisional fragmentation will begin to cause reduction of the number of stone meteorites after $\sim 2 \times 10^7$ y of exposure (Herzog and Anders, 1971) and comparison of the tensile strength of iron and silicate rock (Wetherill, 1967) suggest that the number of irons should be similarly reduced after 500–1000 m.y. Therefore it is plausible that the reduction in observed exposure ages above 1000 m.y. is at least partially produced by collisional destruction. It also must be recognized that the observed distribution is strongly affected by a few large fragmentation events. Particularly remarkable is the grouping of seventeen of the eighteen known type IIIAB irons (Wasson and Kimberlin, 1967; Wasson, 1974) at 650 $\pm$ 100 m.y. This is to be expected from the observed distribution of the asteroids, in which the total mass is strongly concentrated in the largest bodies. However, this precludes the possibility of a detailed agreement between the results of the present calculations and the observed exposure ages. The peaks produced by the infrequent but dominant collisions are random phenomena and are in principle unpredictable. As Wasson (1974) has pointed out, the $^{40}K/^{41}K$ data of Voshage (1967) are biased against young ages (i.e. $\lesssim 100$ m.y.) by rejection of samples so young that the $^{40}K/^{41}K$ ratio is essentially the present production rate. Other measurements (Vilcsek and Wänke, 1963; Cobb, 1967; Begemann, 1970) show that a larger number of iron objects with these short exposure ages exist, and furthermore their existence is predicted by the present theory.

In contrast, the achondrites have short exposure ages, mostly less than 20 m.y., but ranging up to 100 m.y. for the large enstatite achondrite Norton County. If these are derived from the asteroids under consideration here, then some mechanism must exist to eliminate the older ages. Collisional destruction of the fragments is the most plausible mechanism to accomplish this, and the age limit of ~ 100 m.y. is consistent with the collisional lifetime of small silicate bodies, as discussed earlier in this paper. This same collisional lifetime will also eliminate any long-lived "tail" in the exposure age distribution of chondrites. It is likely that the chondrite exposure ages are also dynamically limited, as their distribution of radiants and times of fall imply a short lifetime with regard to perturbation by Earth and Venus into Jupiter-crossing (Wetherill, 1968, 1969, 1971).

A plausible theory for a meteorite source must also provide a quantitatively adequate source. As shown in Figs. 4 to 8, an earth-impact rate of very nearly 10% is calculated for all four starting orbits. It is not known what the concentration of iron objects in the surface regions of these asteroids is. The earth's metallic core represents about one-third of its total mass. If the asteroid ejecta contained this proportion of iron objects, the earth impact flux of 100–10^6 g iron meteorites would be $\sim 3 \times 10^8$ g/y. The flux of chondrites has been estimated to be $\sim 10^8$ g/y. The rate of observed iron falls relative to chondrites is $\sim 7\%$, corresponding to a flux of 7×10^6 g/y. Therefore the present source is adequate even if the ejected matter is depleted in iron objects by a factor of ~ 50. The rate of fall of achondrites is similar to that of irons. As discussed earlier, the number of these bodies will be greatly depleted by collisions. The Monte Carlo calculations predict that $\sim 1\%$ of the ejecta will impact the earth during the first 25 m.y., leading to an expected terrestrial flux of $\sim 10^7$ g/y, in agreement with observation.

In order for a stone meteorite to survive passage through the atmosphere, it is necessary that the fragments enter the atmosphere at less than about 22 km/s. The calculated geocentric velocity distribution for starting orbit D is shown in Fig. 10. A geocentric velocity of 19 km/s corresponds to an entry velocity of 22 km/s when the earth's gravitational acceleration is

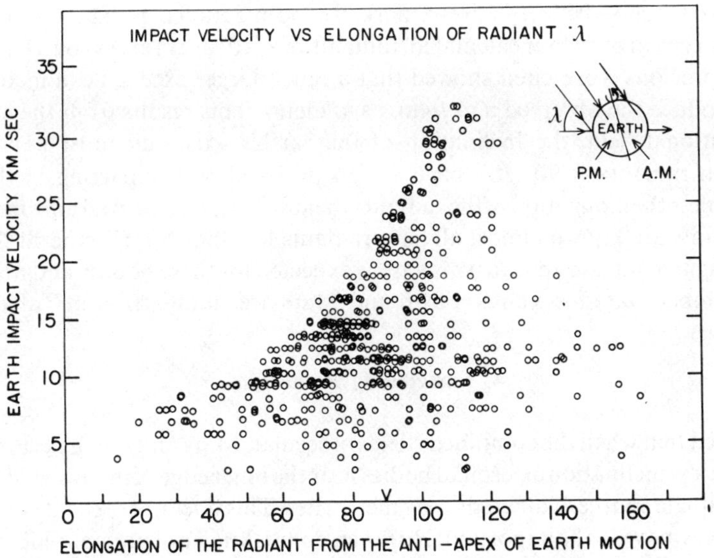

FIG 10. Distribution of geocentric velocities vs. elongation of the radiant from the anti-apex of earth's motion. Although the radiants are predominantly $<90°$, the excess of small radiants is not sufficient to produce a pronounced afternoon excess. The initial orbit is that of orbit D (Fig. 8) but the distribution is insensitive to the choice of initial orbit.

included. The great majority of the calculated impacts are of lower velocity, and atmospheric destruction would not be expected to decrease the impact flux significantly.

Figure 10 also gives the calculated distribution of the elongation of the true geocentric radiant (λ). As discussed elsewhere (Wetherill, 1969, 1972) this distribution determines the distribution of the expected time of day of the meteorite falls. A significant excess of afternoon

TABLE 3. FRACTION IMPACTING THE TERRESTRIAL PLANETS

Starting orbit	Percent		
	Venus	Earth	Mars
A	9.8	10.1	12.1
B	9.4	10.4	12.2
C	9.9	9.6	14.5
D	10.5	10.1	9.5

TABLE 4. R.M.S. VELOCITY OF IMPACTS (KM/S)

Starting orbit	Venus	Earth	Mars
A	19.2	15.1	7.3
B	22.2	16.8	7.7
C	22.0	16.3	•7.35
D	20.2	15.9	8.60

falls, as observed for chondrites, requires a marked concentration of radiants with $\lambda < 90°$. For the case shown, typical of all four calculated, radiants of $\leq 90°$ are in excess by a factor of ~ 1.6. However, the previous work cited showed that a much larger excess, i.e. a factor of ~ 5, is required to produce the observed a.m./p.m. asymmetry. This results from the effects of the earth's gravitational field, the inclination of the earth's axis, and most important, that meteorites with radiants $< 90°$ still have a high probability of inpacting in the morning hemisphere, unless the elongation of the radiant is near 0°. In fact, observed a.m. meteorite falls for which radiants are known almost all have radiants less than 90° (Wetherill, 1969, 1972). Therefore no significant a.m./p.m. asymmetry is expected for these bodies, and none is found. Velocity and impact rate for Venus, Earth and Mars are summarized in Tables 3 and 4.

4. Concluding Remarks

It is concluded that when the combined effects of secular, Mars and Earth perturbations are considered, the low inclination asteroidal bodies near the inner edge of the asteroid belt seem to provide an adequate source of differentiated meteorites. This does not preclude the possibility that other sources may also be important. However, none is known at present which will provide so great a yield and be consistent with the observed iron meteorite exposure ages. A more complete theoretical treatment of this problem would also include the effects of the ν_{16} resonance (Fig. 1). This effect is currently being studied. The principal effect of this resonance is to change the amplitude of oscillations in inclination rather than those of eccentricity, and therefore does not directly change the perihelion distance of fragments. Although conclusions must be tentative at this stage, it does not appear likely that this resonance will have major consequence for the starting orbits considered here.

On rather minimal theoretical grounds, it has been speculated (Wetherill, 1976c, 1977 a, b) that some of the asteroids identified here as sources of differentiated material may be residual planetesimals remaining from the formation of the earth and moon. If so, this is consistent with theories that require early pre-accretion igneous differentiation to explain the difference between the composition of the earth and moon (e.g. Kaula, 1977).

References

ANDERS, E. (1964) Origin, age, and composition of meteorites. *Space Sci. Rev.* **3**, 583–714.

ANDERS, E. (1971) How well do we know "cosmic" abundances? *Geochim. Cosmochim. Acta* **35**, 516–522.

ANDERS, E. (1975) Do stony meteorites come from comets? *Icarus* **24**, 363–371.

ANDERS, E. and ARNOLD, J. R. (1965) Age of craters on Mars. *Science* **149**, 1494–1496.

ANDERS, E. and MELLICK, P. J. (1969) Orbital clues to the nature of meteorite parent bodies. In *Meteorite Research*, P. M. MILLMAN, ed., Dordrecht, Reidel, 559–572.

ARNOLD, J. R. (1965) The origin of meteorites as small bodies, 2, the model; 3, general considerations. *Astrophys. J.* **141**, 1536–1547.

BEGEMANN, F., VILCSEK, E., NYQUIST, L. E. and SIGNER, P. (1970) Exposure history of the Pitts meteorite. *Earth Planet. Sci. Lett.* **9**, 317–321.

CHAPMAN, C. R. (1976) Asteroids as meteorite parent-bodies: the astronomical perspective. *Geochim. Cosmochim. Acta* **40**, 701–719.

CHAPMAN, C. R., MORRISON, D. and ZELLNER, B. (1975) Surface properties of asteroids: a synthesis of polarimetry, radiometry, and spectrophotometry. *Icarus* **25**, 104–130.

CLAYTON, R. N., ONUMA, N. and MAYEDA, T. K. (1976) A classification of meteorites based on oxygen isotopes. *Earth Planet. Sci. Lett.* **30**, 10–18.

COBB, J. C. (1967) A trace element study of iron meteorites. *J. Geophys. Res.* **72**, 1329–1341.

DOHNANYI, J. W. (1969) Collisional model of asteroids and their debris. *J. Geophys. Res.* **74**, 2531–2554.

GAULT, D. E. (1969) On cosmic ray exposure ages of stone meteorites. *Meteoritics* **4**, 177.

GAULT, D. E., SHOEMAKER, E. M. and MOORE, H. J. (1963) Spray ejected from the lunar surface by meteoroid impact. *NASA TN* 1767, 39 pp.

HERZOG, G. F. and ANDERS, E. (1971) Radiation age of the Norton County meteorite. *Geochim. Cosmochim. Acta* **35**, 239–244.

KAULA, W. M. (1977) On the origin of the moon, with emphasis on bulk composition. *Proc. 8th Lunar Sci. Conf.* 321–331.

KOZAI, Y. (1962) Secular perturbations of asteroids with high inclination and eccentricity. *Astron. J.* **67**, 591–598.

LEVIN, B. J., SIMONENKO, A. N. and ANDERS, E. (1976) Farmington meteorite: A fragment of an Apollo asteroid? *Icarus* **28**, 307–324.

O'KEEFE, J. D. and AHRENS, T. J. (1976) Impact ejecta on the moon. *Proc. 7th Lunar Sci. Conf.* 3007–3025.

ÖPIK, E. J. (1963) Survival of comet nuclei and asteroids. *Advan. Astron. Astrophys.* **2**, 219–262.

ÖPIK, E. J. (1966) The stray bodies in the solar system 2. The cometary origin of meteorites. *Adv. Astron. Astrophys.* **4**, 302–336.

RAJAN, R. S. (1974) On the irradiation history and origin of gas-rich meteorites. *Geochim. Cosmochim. Acta*, **38**, 777–788.

SIMONENKO, A. N. (1976) The meteorite orbit: Amor asteroids are a principal source of meteorites. Unpublished ms. presented at: I.A.U. Colloquium 39, Lyon.

VILCSEK, E. and WÄNKE, H. (1963) Cosmic-ray exposure ages and terrestrial ages of stone and iron meteorites derived from Cl^{36} and Ar^{36} measurements. In *Radioactive Dating*, pp. 381–393, I.A.E.A., Vienna.

VOSHAGE, H. (1967) Bestrahlungsalter und Herkunft der Eisenmeteorite. *Z. Naturforsch* **22a**, 477–506.

WASSON, J. T. (1974) Chapter XI, *Meteorites*, Springer, New York.

WASSON, J. T. and KIMBERLIN, J. (1967) The chemical classification of iron meteorites. II. Irons and pallasites with germanium concentrations between 8 and 100 ppm. *Geochim. Cosmochim. Acta* **31**, 2065–2093.

WETHERILL, G. W. (1967) Collisions in the asteroid belt. *J. Geophys. Res.* **72**, 2429–2444.

WETHERILL, G. W. (1968) Time of fall and origin of stone meteorites. *Science*, **159**, 79–82.

WETHERILL, G. W. (1969) Comments on paper by D. E. Fisher and M. F. Swanson, Frequency distribution of meteorite-earth collisions. *J. Geophys. Res.* **74**, 4402–4405.

WETHERILL, G. W. (1971) Cometary versus asteroidal origin of chondritic meteorites. In *Physical Studies of the Minor Planets*, T. GEHRELS, ed., 447–460, NASA Sp–267–687 p.

WETHERILL, G. W. (1972) Origin and age of chondritic meteorites (in Russian). In *Recent Contributions to Geochemistry and Analytical Chemistry*, A. I. TUGARINOV, ed., Nauka, Moscow, 22–34, Also in English translation, pp. 21–37, John Wiley & Sons, New York, 1975.

WETHERILL, G. W. (1974a) Solar system sources of meteorites and large meteoroids. *Ann. Rev. Earth Planet. Sci.* **2**, 303–331.

WETHERILL, G. W. (1974b) Problems associated with estimating the relative impact rates on Mars and the moon. *The Moon* **9**, 227–231.

WETHERILL, G. W. (1976a) Where do the meteorites come from? A reevaluation of the earth-crossing Apollo objects as sources of stone meteorites. *Geochim. Cosmochim. Acta* **40**, 1297–1317.

WETHERILL, G. W. (1976b) Fragmentation of asteroids and delivery of fragments to earth. In *Comets, Asteroids, Meteorites*. A. H. DELSEMME, ed., pp. 283–291. Univ. of Toledo (Ohio) Press, 587 p.

WETHERILL, G. W. (1976c) The role of large bodies in the formation of the earth and moon. *Proc. 7th Lunar Sci. Conf.* 3245–3257.

WETHERILL, G. W. (1977a) Evolution of the earth's planetesimal swarm subsequent to the formation of the earth and moon. In *Lunar Science*, VIII, 1005–1007, Lunar Science Institute, Houston.

WETHERILL, G. W. (1977b) Evolution of the earth's planetesimal swarm subsequent to the formation of the earth and moon. *Proc. 8th Lunar Sci. Conf.*, 1–17.

WETHERILL, G. W. and WILLIAMS, J. G. (1968) Evaluation of the Apollo asteroids as sources of stone meteorites. *J. Geophys. Res.* **73**, 635–648.

WILLIAMS, J. G. (1969) Secular perturbations in the solar system. Ph.D. dissertion, U.C.L.A., 273 pp.

WILLIAMS, J. (1971) Proper elements, families, and belt boundaries. In *Physical Studies of Minor Planets*, T. GEHRELS, ed., 177–181, NASA SP–267.

WILLIAMS, J. G. (1973) Meteorites from the asteroid belt? *EOS* **54**, 233.

ZIMMERMAN, P. D. and WETHERILL, G. W. (1973) Asteroidal source of meteorites. *Science* **182**, 51–53.

Halogens in Meteorites and their Primordial Abundances

GERLIND DREIBUS, B. SPETTEL and H. WÄNKE

Max-Planck-Institut für Chemie (Otto-Hahn-Institut), 6500 Mainz, B.R.D.

Abstract

We have developed a procedure which allows to measure all the halogen elements F, Cl, Br and I on the same sub-sample. With this technique we have determined the halogen elements in both undifferentiated and differentiated meteorites. Normalized to Si = 10^6 atoms we deduce the following values for the primordial abundance of the halogen elements: F = 780, Cl = 4740, Br = 9.2 and I = 1.27, respectively. In the case of fluorine our results are generally considerably lower than those of previous workers; only for carbonaceous chondrites similar low F concentrations were reported by Goldberg *et al.*, (1974).

The literature data on the halogen concentrations in meteorites and, in turn, the values for the primordial abundances of these elements show great scatter. Even the values for the ratios of halogen elements are less certain as it would be desirable. Aside from experimental difficulties, the consensus in the past has been that sample inhomogeneity is the major problem especially in the case of chlorine (Reed, 1971). The ratio of the halogen elements could be of importance in respect of several geochemical questions.

We have developed a procedure to measure all the halogen elements on the same sub-sample. The employed analytical technique was neutron activation analysis for Cl, Br and I in conjunction with a specific ion electrode for the determination of fluorine (Dreibus *et al.*, 1976).

After irradiation of the samples (50–100 mg) with thermal neutrons, pyrohydrolysis was applied for the quantitative extraction of all the halogens in a quartz apparatus shown in Fig. 1.

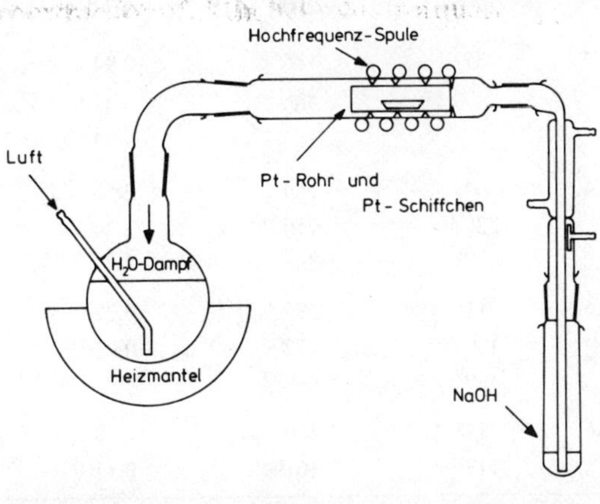

FIG. 1. Quartz apparatus for pyrohydrolysis of rock samples for the extraction of halogens. The pulverized samples were mixed with V_2O_5 and heated by high-frequency induction in a platinum crucible.

The extraction time used was 12 min. The volatile halogen compounds were absorbed in a dilute sodium hydroxide solution and the γ-ray spectra of ^{18}F, ^{38}Cl, ^{82}Br and ^{128}I were measured with a Ge(Li)-detector. After the radioactivity measurements, the native fluorine was determined with a specific fluoride electrode.

To prove the applicability of our method, the halogens were determined in some U.S.G.S. standard rocks. Table 1 shows the comparison of our values with those of the compilation by Flanagan (1973). The agreement in most cases is excellent.

TABLE 1. HALOGEN CONCENTRATION IN GEOLOGICAL STANDARD ROCKS AS DETERMINED IN THIS WORK

Standard rocks	F (ppm)	Cl (ppm)	Br (ppm)	I (ppm)
W–1	205	203	0.21	0.07
lit*	250	200	0.40	0.03
AGV–1	430	118	0.345	—
lit*	435	110	0.5	—
BCR–1	473	50.5	0.077	—
lit*	470	50	0.15	<1
BM	258	74	0.70	—
lit*	250	100	—	—

*Flanagan F. J. (1973).

TABLE 2. COMPARISON OF THE RESULTS OF THIS WORK WITH DATA FROM THE LITERATURE

Meteorite	Type	F (ppm)	Cl (ppm)	Br (ppm)	I (ppm)	
Ivuna	C1	63	698	3.91	1.73	this work
		70(1)	750(2)	5.1(2)	1.13(3)	
			280(3)	11 (3)		
				2.35(5)		
Orgueil	C1	54	678	2.53	0.65	this work
		74(1)	780(2)	3.6(2)	0.4 (2)	
		206(4)	210(3)	2.0(3)	0.23(3)	
		190(6)	290(6)	1.32(5)		
Murray	C2	37	186	0.94	0.66	this work
			200(2)	0.45(2)	0.30(2)	
				0.53(5)		
Abee	E4	64	560	2.56	0.45	this work
		228(4)	750(2)	3.34(2)	0.18(3)	
		280(6)	432(3)	3.5 (3)		
Mocs	L6	11	95	0.27	0.08	this work
		119(4)	72(2)	0.03(2)	0.01(2)	
		160(6)	220(6)			
Bruderheim	L6	11	100	0.20	0.04	this work
		117(4)	101(2)	0.17(2)	0.01(2)	

(1) Goldberg *et al.* (1974); (2) Goles *et al.* (1967; (3) Reed *et al.* (1966); (4) Reed (1964); (5) Krähenbühl *et al.* (1973); (6) Greenland *et al.* (1965).

Table 2 gives a comparison of our halogen results in several types of meteorites with those reported by other workers. For all samples (carbonaceous, enstatite and ordinary chondrites) our F values are much lower than those from previous investigations. Our fluorine values are about a factor of 4 to 10 lower (Dreibus *et al.*, 1973, 1976) as compared with the data of Reed (1964). The method employed by Reed for fluorine determination was activation analysis using the reaction $^{19}F(\gamma, n)$ ^{18}F. Greenland and Lovering (1965) determined fluorine spectrographically by means of the calcium fluoride band emission, and their values are similar to those of Reed. Only Goldberg *et al.*, (1974), using the 19 F(p, $\alpha\gamma$) ^{16}O resonant nuclear reaction, reported fluorine concentrations in carbonaceous chondrites, which lie in the range of our values. We have performed numerous experiments to test the reliability of our fluorine analysis and we are confident on the accuracy of our results.

Our data for the other halogens fall within the wide range of previous determinations. For chlorine we find a reasonable agreement with the data of Goles *et al.*, (1967). We have listed in Table 3 the results of our studies on chondrites. The precision is 5% for Cl and Br and 10% for F and I. Among each group of carbonaceous chondrites (C1, C2 and C3) the F, Cl and Br data of the individual samples differ only slightly from each other. At least in carbonaceous chondrites the halogens seem to be distributed uniformly both within the same meteorite and within the groups. Measurements on four samples of Orgueil from three different sources gave similar results (Table 3). Krähenbühl *et al.* (1973) have pointed out previously that C1 chondrites are more uniform and homogeneous than some authors have contended.

As it has been observed for other volatile elements by Anders (1964) and Larimer and Anders (1967), our halogen contents like other volatile elements in the enstatite chondrite Abee of the

TABLE 3. HALOGEN CONCENTRATIONS IN CHONDRITES

Meteorite	Type	F (ppm)	Cl (ppm)	Br (ppm)	I (ppm)	Cl/F	Cl/Br
Ivuna	C1	63	698	3.91	1.73	11	180
Orgueil (1)	C1	52	849	2.75	0.55	16	310
Orgueil (2)	C1	52	580	2.61	0.42	11	220
Orgueil (3)	C1	59	601	2.51	0.65	10	240
Orgueil (4)	C1	54	680	2.23	0.63	13	310
Mean Orgueil		54	678	2.53	0.56	13	270
Mean C1		59	688	3.22	1.15	12	230
Murray	C2	37	186	0.94	0.66	5	200
Murchison	C2	38	180	0.61	0.19	5	300
Mean C2		38	183	0.78	0.43	5	250
Allende	C3	25	237	0.79	0.16	9	300
Vigarano	C3	22	149	0.85	0.30	7	180
Warrenton	C3	30	226	0.82	0.11	8	280
Mean C3		26	204	0.82	0.19	7	250
Karoonda	C4	22	215	0.54	0.52	10	400
Abee	E4	64	560	2.56	0.45	9	220
Kesen	H4	8	90	0.36	0.06	11	250
Charwallas	H6	17	180	0.54	0.09	11	330
Bjurböle	L4	8	68	0.59	0.07	9	115
Mocs	L6	11	95	0.27	0.08	9	350
Bruderheim	L6	11	100	0.20	0.04	9	500
Leedey	L6	18	112	0.19	0.09	6	590
Alfianello	L6	9	132	0.18	0.11	15	730

TABLE 4. ABUNDANCE OF HALOGENS (Si = 10^6 atoms)

Literature		F	Cl	Br	I
This work		780	4740	9.2	1.27
Reed *et al.*	(1966)	2120	1570	6.63	0.48
Goles *et al.*	(1967)	—	5835	11.94	0.84
Goldberg *et al.*	(1974)	1024	—	—	—
Suess–Urey	(1956)	1600	8850	13.4	0.80
Suess–Zeh	(1973)	—	—	18.0	1.20
Cameron	(1968)	3630	1970	20.6	1.41
Cameron	(1973)	2450	5700	13.5	1.09
Solar photosphere					
Engvold–Hauge	(1974)	1000	8000	—	—
Pagel	(1977)	1000	6300		

petrologic type 4 differ only slightly from the C1 values; whereas ordinary chondrites have lower halogen contents. We do not see any correlation of the halogens with the petrologic type.

The Cl/F and the Cl/Br ratios of the carbonaceous chondrites, the enstatite chondrites and the ordinary chondrites show only small variations with the exception that the Cl/Br ratios in the L6 chondrites are higher than the average ratio. Our Cl/F ratios of about 10 are in accordance with the Cl/F ratio of 15 in the apatite composition measured by Van Schmus and Ribbe (1969) in H6, L6 and LL6 chondrites.

C1 chondrites are generally used to infer the abundances of the solar system, as they represent the least fractionated and most homogeneous group of meteorites. Elemental abundances are conventionally normalized to Si. In Table 4 we have summarized our values for the primordial abundances as deduced from our analytical data of Orgueil and Ivuna (relative to Si = 10^6 atoms). For comparison the data of other authors are also included in Table 4. In the case of fluorine, our values agree well with those of Goldberg *et al.* (1974). For the other halogens we find a good agreement with the data of Goles *et al.* (1967). The close match of our data with those of the solar photosphere may be fortuitous as the solar abundance of fluorine is only known within an uncertainty of a factor of two. The comparison is even worse in the case of chlorine. However, the almost identical abundance ratio (Cl/F = 6.1) deduced from our measurements and that in the solar photosphere (Cl/F = 6.3; Pagel, 1977) seems more significant.

The depletion factor f has been defined as the ratio of the abundance (concentration normalized to Si) in C2, C3, etc., divided by the C1 abundance. For the carbonaceous chondrites, Anders (1964) had advocated constant depletion factors for all volatile elements. We have calculated the halogen depletion factors for C2 and C3 chondrites using the mean values of Table 3 and compared them with those from the literature (Table 5).

TABLE 5. DEPLETION FACTORS FOR C2 AND C3 CHONDRITES

	C2				C3
	this work	Goles	Krähenbühl	Goldberg	this work
F	0.49			0.67	0.26
Cl	0.20	0.19			0.18
Br	0.18	0.07	0.19		0.16
I	0.28	0.56			0.13

For C2 chondrites the lowest depletion (depletion factor $f = 0.49$) is found for fluorine. For the other halogen elements Cl, Br and I a higher depletion is indicated from their lower depletion factors of 0.20, 0.18 and 0.28, respectively. Larimer and Anders (1967) and Krähenbühl *et al.* (1973) found a mean depletion factor of 0.5 for all volatiles (excluding Br). Goldberg *et al.* (1974) found a lower fluorine depletion similar to ours. The chlorine depletion factor of Goles *et al.* (1967) is identical with ours, whereas their bromine depletion factor of 0.07 and iodine depletion factor of 0.56 differ significantly from our values. For bromine we find a good agreement with Krähenbühl *et al.* (1973).

Recently, we investigated the chemistry of the achondrites. The results obtained so far on the halogen contents are from four eucrites, two howardites and one diogenite. These are listed in Table 6 together with halogen concentrations in lunar and terrestrial basalts. The differentiated meteorites relative to C1 chondrites have much lower concentrations for Cl, Br and I. Whereas the F content is that of C1 chondrites. The three halogen elements Cl, Br and I are highly depleted, but their ratios are identical to those in C1. This suggests strongly that they did not fractionate from each other. Magmatic differentiated rocks, in particular achondrites, lunar mare basalts from the Apollo 11, 15 and 17 missions and terrestrial basalts have Cl/Br ratios around the C1 ratio.

The Cl/F ratios are low by about a factor of 5 for diogenites and about 200 for Apollo 17 mare basalts relative to the C1 ratio. The howardites are thought to be mechanical mixtures of eucrites and diogenites as previously suggested by Jérome and Goles (1971) and McCarthy *et al.* (1972). This has been strongly supported by mixing computation model published recently by Dreibus *et al.* (1977). Our halogen data for howardites fall in between those of the eucrites and the diogenites, thus their origin as mechanical mixtures of eucrites and diogenites is also reflected from these data.

TABLE 6. HALOGEN CONCENTRATIONS IN ACHONDRITES

	F (ppm)	Cl (ppm)	Br (ppm)	I (ppm)	Cl/F	Cl/Br
Eucrites						
Jonzac	15	5.1	0.021	0.01	0.34	240
Juvinas	20	16	0.11	0.07	0.80	150
Béréba	53	16	0.102	0.13	0.30	160
Pasamonte	52	26	0.149	0.06	0.50	170
Mean Eucrites	35	16	0.096	0.07	0.50	170
Howardites						
Malvern	15	16	0.172	—	1.1	90
Yamoto 1	6	7.5	0.074	—	1.3	100
Mean Howardites	10.5	12	0.12	—	1.2	100
Diogenites						
Johnstown	2	5.5	0.051	0.02	2.8	100
Lunar Mare Basalts (mean values)						
Apollo 11	75	19	0.077	0.086	0.25	250
Apollo 15	31	5.3	0.026	0.013	0.17	200
Apollo 17	66	4.3	0.017	—	0.07	250
Terrestrial Basalts (mean values)	342	111	0.33	0.07	0.1–1.0	10^2–10^3

Acknowledgements

The samples were activated in the TRIGA-research reactor of the Institut für Anorganische Chemie und Kernchemie der Universität Mainz. We wish to thank the staff of the TRIGA- research reactor and the members of the staff of our Institut, in particular Mr. P. Deibele and Mrs. A. Delnef. For supplying the samples of carbonaceous chondrites Type 1 and achondrites we are grateful to the following donors: Ivuna and Orgueil (4)—Dr. K. Fredriksson, U.S. National Museum, Smithsonian Institution, Washington; Orgueil (1) and (2) and Béréba—Prof. J. Fabries, Museum National d'Histoire Naturelle, Paris; Jonzac—Dr. G. Kurat, Naturhistorisches Museum, Vienna; Malvern—Prof. L. H. Ahrens, University of Cape Town; Yamato 1—Prof. M. Shima, Tokyo.

References

ANDERS, E. (1964) Origin, age, and composition of meteorites. *Space Sci. Rev.* **3**, 583–714.

CAMERON, A. G. W. (1968) A new table of abundances of the elements in the solar system. 125–143. In Origin and Distribution of the Elements, L. H. AHRENS, ed., Pergamon Press, 1178 p.

CAMERON, A. G. W. (1973) Abundances of the elements in the solar system. *Space Sci. Rev.* **15**, 121–146.

DREIBUS, G., PALME, H. and WÄNKE, H. (1973) Fluorine in meteorites. *Meteoritics*, **8**, 29.

DREIBUS, G., SPETTEL, B. and WÄNKE, H. (1976) Determination of lithium and halogens and the significance of lithium to the understanding of cosmochemical processes. *Proc. 1976 Int. Conf. Modern Trends in Activation Analysis*, GDCH, München, 1022–1031.

DREIBUS, G., KRUSE, H., SPETTEL, B. and WÄNKE, H. (1977) The bulk composition of the moon and the eucrite parent body. *Proc. Lunar Sci. Conf. 8th*, in press 211–227.

ENGVOLD, O. and HAUGE, O. (1974) Report 39. Inst. Theor. Astrophys., Oslo.

FLANAGAN, F. J. (1973) 1972 values for international geochemical reference samples. *Geochim. Cosmochim. Acta* **37**, 1189–1200.

GOLDBERG, R. H., BURNETT, D. S., FURST, M. J. and TOMBRELLO, T. A. (1974) Fluorine concentrations in carbonaceous chondrites. *Meteoritics* **9**, 347–348.

GOLES, G. G., GREENLAND, L. P. and JÉROME, D. J. (1967) Abundances of chlorine, bromine and iodine in meteorites. *Geochim. Cosmochim. Acta* **31**, 1771–1787.

GREENLAND, L. P. and LOVERING, J. F. (1965) Minor and trace element abundances in chondritic meteorites. *Geochim. Cosmochim. Acta* **29**, 821.

JÉROME, D. J. and GOLES, G. G. (1971) A re-examination of relationships among pyroxene-plagioclase achondrites. *Activation Analys. in Geochem. and Cosmochem*, Universitetsforlarget, Oslo, 261–266.

KRÄHENBÜHL, U., MORGAN, J. W., GANAPATHY, R. and ANDERS, E. (1973) Abundance of 17 trace elements in carbonaceous chondrites. *Geochim. Cosmochim. Acta* **37**, 1353–1370.

LARIMER, J. W. and ANDERS, E. (1967) Chemical fractionations in meteorites. II Abundance patterns and their interpretation. *Geochim. Cosmochim. Acta* **31**, 1239–1270.

MCCARTHY, T. S., AHRENS, L. H. and ERLANK, A. J. (1972) Further evidence for the mixing model for howardite origin. *Earth Planet. Sci. Lett.* **15**, 86–93.

PAGEL, B. E. J. (1977) Solar abundances. A new table (October 1976). This volume, pp. 79–80. *Origin and Distribution of the Elements*, Int. Ass. Geochem. Cosmochem. and UNESCO, 14–15.

REED, G. W., Jr. (1964) Fluorine in stone meteorites. *Geochim. Cosmochim. Acta* **28**, 1729–1743.

REED, G. W. and ALLEN, R. O. (1966) Halogens in chondrites. *Geochim. Cosmochim. Acta* **30**, 779–800.

REED, G. W. (1971) Chlorine (17). In *Handbook of Elemental Abundances in Meteorites*, B. MASON, ed., Gordon & Breach Science Publ., 143.

SUESS, H. E. and UREY, H. C. (1956) Abundances of the elements. *Rev. Mod. Phys.* **28**, 53–74.

SUESS, H. E. and ZEH, H. D. (1973) The abundances of the heavy elements. *Astrophys. Space Sci.* **23**, 173.

VAN SCHMUS, W. R. and RIBBE, P. H. (1969) Composition of phosphate minerals in ordinary chondrites. *Geochim. Cosmochim. Acta* **33**, 637–670.

Noble Gases in Chondritic Polymict Breccias: Clues to their Origin

LUDOLF SCHULTZ

Max-Planck-Institut für Chemie, D-6500 Mainz, B.R.D.

Abstract

Studies of individual samples from solar gas-rich meteorites yield clues as to the exposure to solar particle irradiation and the exposure to galactic cosmic rays during the time when the clast was a constituent of a regolith and part of the meteoroid in the interplanetary space. Also time limits can be deduced for the formation of polymict chondritic breccias which are not directly measurable by radiometric methods.

Noble gas analyses of samples taken from the matrix and from different clast of the St. Mesmin and Weston chondrites gave the following results: (1) The cosmic ray exposure of individual parts of one meteorite is sometimes complex; at least two-step irradiations are possible. (2) Young and old xenoliths can occur in the same meteorite. Specifically, the following evolutionary stages can be anticipated: (a) short exposure of the fine-grained matrix material to solar irradiation; (b) exposure of some of the xenoliths to galactic cosmic rays (duration of the order of 10^6 years); (c) compaction of the different constituents to a polymict breccia; (d) cosmic-ray exposure of the meteoroid after ejection from its parent body.

All observations can be understood if these meteorites originate from an asteroidal regolith. The time of formation is in some cases recent, i.e. the formation took place less than 1.4 billion years ago.

Introduction

About thirty ordinary chondrites are known to have suffered before compaction a surface exposure, whereby part of their constituents were exposed directly to the solar wind. These meteorites present a brecciated structure (Wahl, 1952) which is emphasized by differences in colour, the so-called dark–light structure. The fine-grained dark matrix is the host of the trapped solar gases whereas the light clasts embedded into the matrix contain only the radiogenic and spallogenic gas components. A review about the discovery and explanation of noble gases in polymict meteoritic breccias has been given by Wilkening (1971). Trapped solar gases are found in different abundances among the groups of ordinary chondrites: H ≈ 12%, L ≈ 1.7%, and LL ≈ 8% (Schultz *et al.*, 1971b).

Meteoritic gas-rich breccias in a number of ways resemble low-grade lunar breccias. This similarity is taken as an indication that the meteoritic breccias originated in surface layers of planetary bodies. However, there are significant differences among them in the concentrations of trapped solar gases and the length of the cosmic-ray exposure ages or the degree of other irradiation features. Both similarities and differences can be expected among planetary regoliths because the evolution of the surface layer of a planet depends on many parameters which differ among the bodies in question, e.g. physical data of the planet, composition and conditions of radiation and particle interactions with the surface.

On the other hand, a comparison of irradiation effects in meteorites and lunar breccias may also reveal the processes acting in surface layers of the chondritic parent bodies. We have therefore studied the noble gases in different clasts and matrix samples of several polymict chondritic breccias (Schultz *et al.*, 1972; Schultz, 1973; Schultz and Signer, 1977). These measurements were carried out in cooperation with the track-group at the Museum National d'Histoire Naturelle in Paris (Drs. J. C. Lorin, P. Pellas and G. Poupeau) who measured particle tracks in the same samples.

This paper focuses on two aspects: (1) the formation time of the chondritic breccias, and (2) the galactic cosmic-ray exposure ages of matrix and different clasts of one individual chondrite. This discussion is mainly based on measurements on samples from the St. Mesmin and Weston chondrites.

Gas-retention Ages in St. Mesmin

St. Mesmin is classified as an LL-chondrite with extremely complex textural and chemical relationships. Of special interest is an intensely shocked and black clast of H-group material (Pellas, 1973), because this xenolith may date the admixture of this particular material to the rest of St. Mesmin (Schultz and Signer, 1973).

The brecciation event which consolidated the chondritic gas-rich meteorites from a planetary (asteroidal) regolith does not generally outgas the meteoritic material. This is supported by the high concentrations of solar trapped ^{4}He which is less retentative than radiogenic ^{40}Ar (Huneke *et al.*, 1969; Schultz *et al.*, 1971a). A direct dating of the lithification event by radioactive methods is therefore not possible. Also, the unique fission track dating method to determine the compaction time of soil grains into a breccia (Price *et al.*, 1975) has not been applied to solar gas-rich chondritic breccias.

An alternative possibility to determine the compaction time is the dating of different xenoliths in one meteorite. If their ages agree among one another a long-time coexistence is emphasized. On the other hand, if a young fragment is found in old material, then the age of the young xenolith gives an upper limit for the time of its admixture.

Table 1 summarizes the radiogenic gas concentrations, K-contents and the calculated gas-retention ages of eleven LL-clasts of St. Mesmin, and two samplings of the H-group xenolith.

The K-Ar and the U, Th–He ages of this H-group xenolith are much shorter than the gas-retention ages of all other clasts. It is unlikely that *in situ* diffusive loss of radiogenic gases affected the gas-retention ages without affecting the surrounding material. Such a loss is expected to also cause a significant discordance between the U, Th–He age and the K—Ar age,

TABLE 1. GAS-RETENTION AGES OF CLASTS FROM THE ST. MESMIN LL-CHONDRITE

	Radiogenic nuclides ^{4}He (10^{-8} ccSTP/g)	Radiogenic nuclides ^{40}Ar (10^{-8} ccSTP/g)	K (ppm)	Retention age U, Th–^{4}He† (10^{9} years)	Retention age K–Ar (10^{9} years)
Mean of eleven LL-group clasts	1480 ±350	6250 ±810	930‡ ±25	3.75 ±.66	4.33 ±.26
H-group xenolith a	386 ±50	724 ±10	913 ±25	1.29 ±.26	1.35 ±.06
b	337 ±30	750 ±24	913 ±25	1.14 ±.21	1.37 ±0.8

†Calculated with: U=(12±1) ppb; Th/U=3.6.
‡Mean of four clasts (Dodd and Jarosewich, 1976).

and a considerable loss of solar trapped gases still present in the vicinity of the H-group xenolith. Thus, the occurrence of a xenolith with low retention age within material of distinctly older ages has to be taken as strong indication that the H-group xenolith was incorporated into the St. Mesmin breccia not more than 1.4 Gy ago.

Exposure Ages of Individual Parts of Brecciated Chondrites

The meteoroids are meter-sized bodies which are exposed to galactic cosmic rays during their flight in space. The energetic particles of this irradiation produce spallogenic (cosmogenic), noble gas and other nuclides within the meteorite. Their isotopic composition is different to other gas components. The spallogenic ^{3}He, ^{21}Ne and ^{38}Ar especially have been used to calculate cosmic-ray exposure ages (see, for example, Anders, 1964; Wänke, 1966).

It has been shown that two chondritic polymict breccias—Weston and St. Mesmin—each contain one clast with an irradiation history of at least two steps: a long irradiation interval during the flight of the meteoroid and a shorter irradiation of the individual clast before compaction (Schultz *et al.*, 1972; Schultz and Signer, 1977).

The detection of pre-irradiations by galactic cosmic rays in various constituents of a chondrite is only possible if the depth dependence of spallogenic gases within the meteorite is considered (Wright *et al.*, 1973; Schultz and Signer, 1976). This is achieved using the variation with depth of the spallogenic ratios ^{3}He/^{21}Ne or ^{22}Ne/^{21}Ne. Figure 1 shows the variation of ^{3}He/^{21}Ne with the total concentration of ^{21}Ne in St. Mesmin samples. The depth effect of spallogenic gases is valid for those data points which fall along a straight line. An additional exposure time in space would move the line to the right in Fig. 1, but the slope would not change. Different exposure times of individual xenoliths within one breccia can be recognized in this diagram: samples with preirradiations will fall to the right of the line. This diagram is also relatively insensitive to chemical variations among xenoliths which could produce different concentrations of spallogenic nuclides. Due to the relatively small target dependency of the spallogenic ^{3}He production as compared to that of ^{21}Ne, the chemical differences among clasts of one chondritic breccia influence only the location of the data point on the curve.

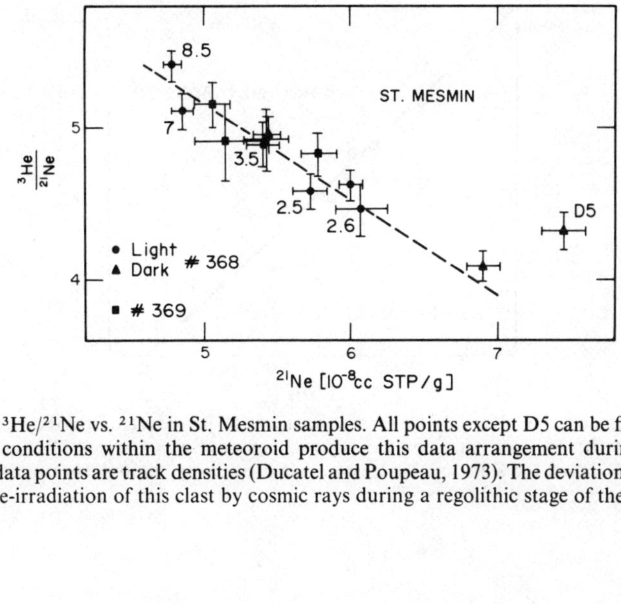

FIG. 1. Spallogenic ^{3}He/^{21}Ne vs. ^{21}Ne in St. Mesmin samples. All points except D5 can be fitted to a straight line. Variable shielding conditions within the meteoroid produce this data arrangement during its flight in space. Numbers on some data points are track densities (Ducatel and Poupeau, 1973). The deviation of D5 from this line is interpreted by a pre-irradiation of this clast by cosmic rays during a regolithic stage of the St. Mesmin material.

TABLE 2. COSMIC-RAY EXPOSURE AGES OF INDIVIDUAL CLASTS AND THE MATRIX OF ST. MESMIN AND WESTON
The matrix value of Weston has been calculated from the spallogenic ^{3}He using a production rate of 2.39×10^{-8} ^{3}HeSTP/g My

Meteorite	Exposure age (My)
St. Mesmin	
Mean of twelve clasts	10.3 ±.3
clast D 5	11.8 ±.5
Weston	
Mean of five clasts	23.3 ±1.3
clast L 3	27.3 ±2.0
matrix	28.2 ±1.3

The use of isotope correlations plots (Eberhardt *et al.*, 1970) also enables one to calculate an exposure age of matrix even when large amounts of solar trapped gases are present. Such a correlation diagram is shown in Fig. 2 for grain size sample fractions from Weston meteorite (Schultz *et al.*, 1975). The data fit a straight line (goodness of fit: $r^2 = 0.94$). From the intercept, a spallogenic ^{3}He concentration is calculated as $(67.6 \pm 3.1) \times 10^{-8}$ ccSTP/g.

A summary of measured exposure ages of the different parts of St. Mesmin and Weston is given in Table 2.

Most measured clasts of each meteorite have the same exposure ages. However, one clast of St. Mesmin and one clast and the matrix of Weston have higher exposure ages relative to the other clasts. This observation can be only explained by invoking an additional exposure to galactic cosmic rays prior to the compaction of the material and prior to its ejection from the meteorite parent body.

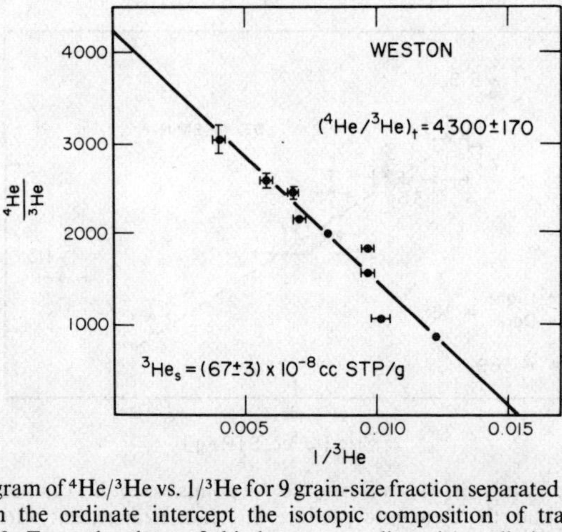

FIG. 2. Correlation diagram of ^{4}He/^{3}He vs. 1/^{3}He for 9 grain-size fraction separated from the matrix of the polymict breccia Weston. From the ordinate intercept the isotopic composition of trapped helium is calculated as $(^4He/^3He)t = 4300 \pm 170$. From the slope of this least-square line, the spallation ^{3}He content in the grain-size fractions is evaluated to be $(67 \pm 3) \times 10^{-8}$ ccSTP/g.

Discussion

Only three out of about thirty polymict brecciated chondrites with trapped solar noble gases have been investigated in detail with regard to their rare gases in a larger number of clasts. All three meteorites—St. Mesmin, Weston and Djermaia (Pellas, 1973)—show at least in one clast evidence for a complex irradiation history. The different stages of irradiation are:

1. Parts of the fine-grained matrix were exposed to the solar corpuscular irradiation before compaction. The duration of this exposure was short compared to the galactic cosmic-ray exposure. It can be limited to less than a few million years. Poupeau *et al.* (1974) concluded that the solar wind exposure of aubritic material was very short (< 100 years), based on the present-day solar wind fluxes.

   An important question refers to the time when the solar particle irradiation has occurred. Several considerations tend to place this exposure between 3.8 and 4.4 G.y. ago (see, for example, Poupeau *et al.*, 1974). Such an exposure to solar wind ions early in the history calls for a storage of this irradiated material in a shielded position within the parent body until the rock became a meteoroid. However, the occurrence of a young xenolith in the St. Mesmin chondrite indicates that this polymict breccia became a distinct assemblage later than 1.4 G.y. ago. Thus, the exposure to solar wind ions could have been occurred during a relatively late period in the history of chondritic matter.
2. Some of the xenoliths now forming clasts in polymict brecciated meteorites were exposed to galactic cosmic rays. Other xenoliths were shielded during this period. The duration of this exposure is estimated of the order of a few million years for the three meteorites. This calculation is based on the assumption that the irradiation parameters (flux, shielding) are the same as for the pre-irradiation and the irradiation during the flight in space as a meteoroid.
3. Cosmic-ray exposure of the meteoroid, during its flight in space, is calculated for stony meteorites. The exposures yield values generally less than 50 m.y. The exposure ages of H-group chondrites cluster around 4 m.y. The lifetime of meteoritic matter as a meteoroid is less than 1% of the total age of this material. This cosmic-ray exposure age dates the ejection of the meteoroid from its parent body.

A comparison of lunar irradiation features with the irradiation stages of brecciated chondrites yield some information about the differences between the lunar regolith and the surface layers of the meteoritic parent bodies. The concentration of solar noble gases in lunar fines is generally larger than in meteoritic polymict breccias. Only some meteorites such as Fayetteville and Weston show solar gas concentrations which are comparable to those in lunar soils or soil breccias. From track studies, however, the percentage of irradiated crystals and thus the intensity of irradiation is larger for lunar regolith materials (e.g. Dran *et al.*, 1972). In addition, the degree of metamorphism is also different.

The galactic cosmic-ray exposure ages of regolith soils from various missions is typically 400 m.y. However, individual grains of a given soil have different exposure ages (e.g. Kirsten *et al.*, 1972; Arvidson *et al.*, 1975). The exposure ages of lunar rocks vary from 1 to 700 m.y. In many cases these exposure ages exceed those of meteoritic breccias.

Noble gases in different clasts of one lunar breccia have been studied only in two cases: 15015 and 60016 (The European Consortium, 1974). Different clasts of breccia 15015 yielded exposure ages between 250 and 1290 m.y. These data indicate a long pre-irradiation of some parts of the breccia prior to compaction, and the pre-irradiations are much longer than those found in polymict chondritic breccias.

The above discussed observations can be explained if it is assumed that the polymict brecciated chondrites originated from a relatively young asteroidal regolith. The pre-exposure of some clasts in these meteorites have a somewhat longer life-time as being small particles in this asteroidal regolith. The solar gases are the result of a comparatively recent bombardment by the solar wind. The differences in exposure ages and noble gas concentrations between chondrites and lunar breccias are primarily caused by differences of the regoliths from which these rocks originated. In this model the differences in the gravity fields on the respective bodies many account for different turnover rates and rates of ejection of the relative fragile breccias. The galactic cosmic-ray exposure age of meteorites is determined by their life-time in the interplanetary space. However, at least for polymict brecciated chondrites the exposure age may be the sum of the regolith life-time and the duration of the meteoroid's journey in space.

Acknowledgements

The author gratefully acknowledges the close cooperation and continual discussions with Professor Peter Signer, Zürich, throughout the major portion of this investigation. The author also thanks Dr. P. Pellas, Paris, and his colleagues for valuable discussions, Dr. J. C. Laul, Richland, for reading the manuscript, and Mrs. S. Link for the preparation of the manuscript.

References

ANDERS, E. (1964) Origin, age and composition of meteorites. *Space Sci. Rev.* **3**, 583–714.

ARVIDSON, R., CROZAZ, G., DROZD, R. J., HOHENBERG, C. M. and MORGAN, C. J. (1975) Cosmic ray exposure ages of features and events at the Apollo landing sites. *The Moon* **13**, 259–276.

DODD, R. T. and JAROSEWICH, E. (1976) Olivine microporphyry in the St. Mesmin chondrite. *Meteoritics* **11**, 1–20.

DRAN, J. C., DURAUD, J. P., MAURETTE, M., DURRIEU, L., JOURET, C. and LEGRESSUS, C. (1972) Track metamorphism in extraterrestrial breccias. *Proc. 3rd Lunar Sci. Conf.*, MIT Press, 2883–2903.

DUCATEL, A. and POUPEAU, G. (1973) The cosmic-ray track record of the Saint -Mesmin LL-chondrite. *Meteoritics* **9**, 336.

EBERHARDT, P., GEISS, J., GRAF, H., GRÖGLER, N., KRÄHENBÜHL, U., SCHWALLER, H., SCHWARZMÜLLER, J. and STETTLER, A. (1970) Trapped solar wind noble gases, exposure age and K/Ar-age in Apollo 11 lunar fine material. *Proc. Apollo 11 Lunar Sci. Conf.* 1037–1070.

THE EUROPEAN CONSORTIUM AND FRIENDS (1974) The history of Lunar Breccia 15015. In: *Lunar Science* V, The Lunar Science Institute, Houston, 217–219.

HUNEKE, J. C., NYQUIST, L. E., FUNK, H., KÖPPEL, V. and SIGNER, P. (1969) The thermal release of rare gases from separated minerals of the Mócs meteorite. In: *Meteorite Research*, P. M. MILLMAN, ed., Reidel, 901–921.

KIRSTEN, T., DEUBNER, J., HORN, P., KANEOKA, I., KIKO, J., SCHAEFFER, O. A. and THIO, S. K. (1972) The rare gas record of Apollo 14 and 15 samples. *Proc. 3rd Lunar Sci. Conf.*, MIT Press, 1865–1889.

PELLAS, P. (1973) Irradiation history of grain aggregates in ordinary chondrites. Possible clues to the advanced stages of accretion. In: *From Plasma to Planet*, A. ALVIUS, ed., Wiley, 65–92.

POUPEA, G., KIRSTEN, T., STEINBRUNN, F. and STORZER, D. (1974) The records of solar wind and solar flares in Aubrites. *Earth Planet. Sci. Lett.* **24**, 229–241.

PRICE, P. B., HUTCHEON, I. D., BRADDY, D. and MACDOUGALL, D. (1975) Track studies bearing on solar-system regoliths. *Proc. Lunar Sci. Conf. 6th*, Pergamon, 3449–3469.

SCHULTZ, L. (1973) Die Bestrahlungsgeschichte xenolithischer Chondrite. *Habilitationsarbeit ETH Zürich*, pp. 108.

SCHULTZ, L. and SIGNER, P. (1973) Compaction time of dark–light structured chondrites. *Fortschr. Miner.* **50**, (Beiheft 2), 127–128.

SCHULTZ, L. and SIGNER, P. (1976) Depth dependence of spallogenic helium, neon, and argon in the St. Severin chondrite. *Earth Planet. Sci. Lett.* **30**, 191–199.

SCHULTZ, L., FRICK, U. and SIGNER, P. (1971a) Nachweis des Diffusionsverlustes von Sonnenwind—^{4}He im Mondstaub. *Helv. Phys. Acta* **44**, 614–616.

SCHULTZ, L., FRICK, U. and SIGNER, P. (1975) Noble gases in grain size fractions of Weston: A comparison with lunar soil data. *Meteoritics* **10**, 483.

SCHULTZ, L., SIGNER, P., LORIN, J. C. and PELLAS, P. (1972) Complex irradiation history of the Western chondrite. *Earth Planet Sci. Lett.* **15**, 403–410.

SCHULTZ, L., SIGNER, P., PELLAS, P. and POUPEAU, G. (1971) Assam: A gas rich hypersthene chondrite. *Earth Planet. Sci. Lett.* **12**, 119–123.

SCHULTZ, L. and SIGNER, P. (1977) Noble gases in the St. Mesmin chondrite: Implications to the irradiation history of a brecciated meteorite. *Earth Planet. Sci. Lett.* **36**, 363–371.

WAHL, W. (1952) The brecciated stony meteorites and meteorites containing foreign fragments, *Geochim. Cosmochim. Acta* **2,** 91–117

WÄNKE, H. (1966) Meteoritenalter und verwandte Probleme der Kosmochemie. *Fortschr. Chem. Forsch.* **7**, 322–408.

WILKENING, L. L. (1971) Particle track studies and the origin of gas-rich meteorites. Publication No. 11, Center of Meteorite Studies, Arizona State University, Tempe, pp. 36.

WRIGHT, R. J., SIMMS, L. A., REYNOLDS, M. A. and BOGARD, D. D. (1973) Depth variation of cosmogenic noble gases in the ~150 kg Keyes chondrite. *J. Geophys. Res.* **78**, 1308–1318.

Solar Wind Nitrogen and Indigenous Nitrogen in Lunar Material

O. MÜLLER

Max-Planck-Institut für Kernphysik, P. O. B. 103980, 69 Heidelberg, Federal Republic of Germany

Abstract

There is ample evidence that the main source of the abundant light elements in lunar, fine-grained regolith samples is the solar wind. The extra-lunar contribution is easy to detect, because lunar igneous rocks are, in general, strongly depleted in light volatile elements, such as H, He, C, N and Ne. We report on chemically bound nitrogen contents in various lunar samples from the Apollo 11 through Apollo 17 landing sites by using the Kjeldahl method. This technique detects only nitrogen in a bound state and discriminates against molecular nitrogen N_2 of atmospheric origin. We have found by the analysis of grain-size fractions, by leach-experiments, by comparison with trapped noble gases and with agglutinate contents that the main amount of the nitrogen in the fines is derived from the solar wind by ion implantation and is present primarily in a bound state. Leach-experiments on Apollo 15 and Apollo 17 fines indicate two chemical forms of nitrogen: ammonium- and nitride-nitrogen. Nitrogen and carbon chemistry in lunar fines appear to be analogous in respect to the hydrogen compounds ammonia and methane, and the presence of nitrides and carbides. Hot atom chemistry probably causes the synthesis of those compounds on the surface area of fine grains. Implanted nitrogen is better retained in fine-grained material than carbon and the light noble gases and is, therefore, a useful reference in terms of solar elemental abundances. Carbon to nitrogen ratios of lunar fines are interpreted as lower limits for the abundance ratio of these elements in the sun.

Bound nitrogen contents have also been determined in several lunar igneous rocks and U.S.G.S. standard rocks using an optimized, program-controlled Kjeldahl method. This technique is superior to the previously applied conventional Kjeldahl method with respect to sensitivity and reproducibility for nitrogen analysis. Most of the lunar igneous rocks analyzed in this work are quite uniform and low in bound nitrogen contents; the range is only about 3 to 8 ppm N for samples of various types. Lunar fines and soil breccias contain, however, a factor of 10 to 20 more nitrogen than igneous rocks. The small nitrogen contents of the rocks imply that indigenous nitrogen and/or ammonia were at very low concentrations at the time of mineral and rock formation in the early history of the moon. This indicates either severe *a priori* nitrogen depletion of the moon's parent material or rigorous loss of nitrogen and other volatile elements in the early history of the moon.

Introduction

The moon's surface is a unique target in earth-near space to study a series of physical and chemical processes which have taken place during geological history (Fig. 1). Main events are the bombardment by meteoritic and cometary matter of a wide size range. Micro-meteoritic impacts are mainly responsible for the erosion on the lunar surface. This process also causes agglutinate formation; their concentration is a measure of soil maturity. High-energy particles of the solar and galactic cosmic radiation produce spallation nuclei and fossil tracks, from which the cosmic-ray-exposure history can be deduced. The decay of natural-radioactive isotopes in lunar material provides the basis for radiometric formation ages of rocks and minerals. The solar wind implants the abundant elements of the sun, mainly H, C, N, O and noble gases, in the surface areas of fine-grained regolith material, the lunar soil or

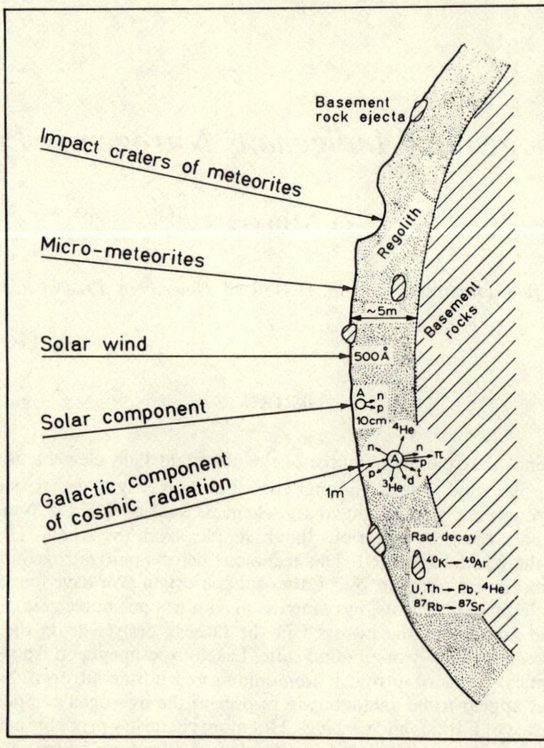

FIG. 1. Physical processes at the lunar surface. Larger impacting bodies generate deeper craters in the basement rocks (schematic graph).

fines. The depth of penetration of this low energetic solar radiation only amounts to a few hundred angstroms. The physical and chemical interactions of the solar wind plasma with the moon's surface material are of central interest in this paper.

Before the Apollo and Luna missions and the return of lunar samples to earth, meteorites have been the only extra-terrestrial material available for thorough investigation in the laboratory. Meteorite research has provided profound knowledge of the early history of our solar system in various aspects, among which solar abundances of the elements, their origin and distribution are of great importance. Carbonaceous-chondritic meteorites, dominantly type C1, represent the most primitive, least fractionated meteorite bodies of the planetary system. They serve as basis for the abundances of the heavy-volatile chemical elements in the solar system. The light, volatile elements hydrogen, helium, carbon, nitrogen, oxygen and neon are, however, strongly depleted in meteorites, on earth and moon relative to abundances of the sun. The solar H/Si atomic ratio is about 3×10^4 (Cameron, 1973), whereas this ratio amounts to only 0.14 in the earth's crust and about 2 in carbonaceous chondrites (Mason, 1971).

Abundances of the light elements relative to 10^6 silicon atoms in the solar system and in C1 chondrites are compiled in Table 1, and demonstrate the severe depletion in primitive meteoric matter due to volatility and chemically inert character of those elements.

The returned lunar fine-grained soil samples, and the successful solar-wind composition experiments by the group of Berne University, in which an aluminum foil was exposed on the

TABLE 1. ABUNDANCES OF THE LIGHT, ABUNDANT ELEMENTS IN THE SOLAR SYSTEM (Cameron, 1973) AND IN CARBONACEOUS CHONDRITES OF TYPE C1 (Mason, 1971); $Si = 10^6$

Element	Solar system	Carbonaceous chondrites, type C1
H	3.18×10^{10}	2×10^6
He	2.21×10^9	$\sim(0–10)$
C	1.18×10^7	7.1×10^5
N	3.74×10^6	4.9×10^4
O	2.15×10^7	7.5×10^6
Ne	3.44×10^6	$\sim(0–4) \times 10^{-2}$

lunar surface to trap solar wind particles, provided new and more reliable data on solar abundances of He, Ne and Ar and their isotopic composition.

The present work deals with the light element geochemistry in lunar samples with emphasis on solar wind-implanted nitrogen. Furthermore, the significance of indigenous nitrogen in lunar igneous rocks is discussed. The article summarizes results and interpretations of work which we have performed in the past years in our laboratory.

Aims of Nitrogen Investigation in Lunar Samples

Three main sources for nitrogen on the moon's surface had to be presumed:

1. Primordial, indigenous nitrogen which was chemically bound or physically dissolved in magmatic rocks at the time of their crystallization.
2. Nitrogen of impacting meteoritic matter.
3. Nitrogen by implantation of solar wind ions. The penetration of those particles with energies of about 1 Kev/nucleon amounts to about 500 Å. Solar wind particles should, therefore, be correlated with the surface area of lunar fines. The studies have revealed that this source is mainly responsible for the occurrence and enrichment of light elements on the lunar surface. Nitrogen-bearing minerals were not detected in any lunar sample so far.

The following aims have been of interest:

1. More accurate knowledge of solar elemental abundances by studying solar wind-implanted elements in lunar fines.
2. Determination of chemically bound nitrogen contents in various lunar samples of the Apollo 11 through 17 missions by applying the conventional Kjeldahl–Nessler method and an optimized, program-controlled, more sensitive Kjeldahl method.
3. Comparison of bound nitrogen contents with total nitrogen contents in individual lunar samples to find out the portion of bound nitrogen.
4. Study of the grain size dependence of bound nitrogen contents to prove the solar wind origin of nitrogen.
5. Are trapped noble gases, e.g. ^{36}Ar, correlated with implanted nitrogen? If yes, this would provide further evidence for the solar wind origin of nitrogen.
6. Investigation of the retention behavior of implanted nitrogen by comparison of $^{4}He/N$, $^{20}Ne/N$ and $^{36}Ar/N$ ratios in lunar fines with the abundance ratios of the sun.

7. Performance of leach experiments to study formation reactions of nitrogen compounds in the surface area of mineral grains of fines, and to compare these compounds with those of carbon.
8. Comparison of C/N ratios in lunar fines with the spectroscopically measured ratio of the solar photosphere.

Analytical Procedure

The suitability of the Kjeldahl method for the determination of chemically bound nitrogen in terrestrial and lunar samples has been established in earlier works: Wlotzka (1961), Hintenberger *et al.* (1970), Müller (1972, 1973).

Sensitivity and reproducibility for nitrogen analyses have been improved by the development of an optimized Kjeldahl method (Werner and Tölg, 1975). It has successfully been applied to lunar igneous rocks of low N-contents in the ppm-range by Müller *et al.* (1976). The Kjeldahl technique detects only nitrogen in a bound state and discriminates against molecular N_2 of atmospheric or indigenous origin. The principle of the Kjeldahl method and sample preparation are as follows. Bound nitrogen is converted to NH_4^+-cation by acid hydrolysis with highly pure HF and H_2SO_4 in platinum crucible in a clean, ammonia-free environment. The sample break-up is, therefore, performed in a lucite box that is ventilated with clean air. After a Kjeldahl distillation the ammonia is determined spectrophotometrically with Nessler's reagent (K_2HgI_4). Blank analyses are run parallel to the lunar and standard samples. Sample weights of about 100 mg are used for a single analysis of lunar bulk fines (< 1mm) and the grain size fractions. Some of the lunar soil samples have been analyzed in duplicates which agreed to about ±5 ppm N. The fines < 1 mm have been sieved in the presence of pure ethanol into several grain-size fractions using calibrated stainless-steel sieves.

Both the conventional and optimized Kjeldahl techniques have been applied to lunar igneous rocks. However, due to systematic errors and relatively high and scattered blanks associated with the conventional Kjeldahl method and due to the limited amounts of sample available, it formerly was not possible to determine nitrogen contents of < 10 ppm in lunar rocks. The optimized, program-controlled Kjeldahl technique is performed in a closed, micro-circulating distillation system, and the ammonium is titrated in a specially designed absorption vessel with coulometrically generated hypobromite OBr^- using biamperometrical endpoint indication. The sensitivity for nitrogen detection is improved at least by a factor of 5 to 10 compared to the conventional Kjeldahl method.

Total nitrogen contents of lunar samples have been analyzed by various techniques. Goel *et al.* (1975, and previous papers) have applied the sensitive ^{14}N (n, p) ^{14}C neutron activation reaction to the analysis of a larger number of lunar, meteoritic and terrestrial samples. Gibson and Moore (1970) have destructed the rock samples by heating them to about 2400° C in a graphite crucible containing powdered graphite. These authors use a LECO Nitrox-6 gas chromatographic analyzer for total nitrogen determination. C. B. Moore and co-workers have applied this method to the nitrogen analysis of a large number of lunar samples. Kaplan *et al.* (1970) and Chang *et al.* (1971) have used stepwise pyrolysis, subsequent acid hydrolysis and complete combustion in oxygen to determine C, N and S in lunar samples. Becker and Clayton (1975) have analyzed total nitrogen contents (and isotopic compositions) by pyrolysis of the samples at about 1350° C under vacuum. Their N results on lunar igneous rocks agree best with ours.

Results and Discussion

Bound Nitrogen Contents of Lunar Fines

The results of chemically bound nitrogen concentrations of Apollo 11 through Apollo 17 lunar fines are shown in Fig. 2. The total range is from 23 ppm to 124 ppm N. Data on individual fines and on grain-size fractions were presented and discussed in previous papers by Müller (1972, 1973, 1974, 1976). Some aspects of bound nitrogen contents in Apollo 17 fines from various sampling areas shall briefly be discussed. The Apollo 17 fines contain, in the mean, the lowest amounts of bound nitrogen (23 to 73 ppm N) among all Apollo fines analyzed by us so far. The varying nitrogen contents are functions of the exposure age and the maturity of the soil samples. Nitrogen contents of individual Apollo 17 soils nicely correlate with their state of maturity. McKay *et al.* (1974) divided the Apollo fines samples into mature, submature and immature groups by plotting graphic mean grain size versus graphic standard deviation. The average agglutinate content increases from immature to submature and mature soils being 16.3, 36.5, 46.7%, respectively. The mature Apollo 17 soils 72 501, 72 701, 76 501 and 78 501 have nitrogen contents of 70, 73, 63 and 73 ppm N, respectively, the submature soil 71 501 51 ppm N, and the immature soils 74 241 and 75 061 23 and 42 ppm N, respectively, showing strong relation of nitrogen content of soils with the state of maturity. Similarly, nitrogen contents of Apollo 16 fines are positively correlated with agglutinate contents as shown by Goel *et al.* (1974).

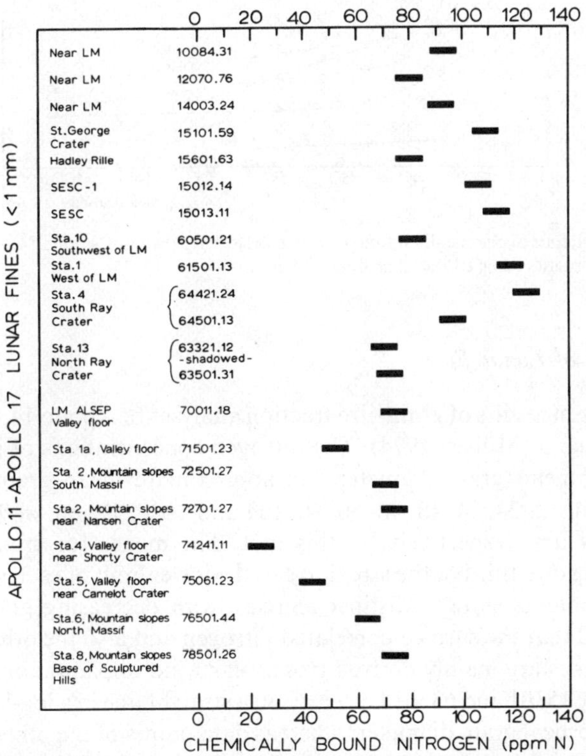

FIG. 2. Chemically bound nitrogen contents (in ppm N) in Apollo 11 through Apollo 17 lunar fines < 1 mm with error bars.

Comparison of bound nitrogen data for Apollo 17 fines with those of Petrowski *et al.* (1974) who used vacuum pyrolysis technique for the detection of total nitrogen shows satisfactory agreement, yet our numbers are systematically lower. Five soils can be compared; the numbers in parentheses are own values: 70011, 77 ppm N (73 ppm N); 71501, 60 (51); 72701, 91 (73); 75061, 49 (42); 76501, 73 (63). This indicates that the Kjeldahl method detects the main amount of nitrogen present in lunar fines, which proves that the impinging solar wind-nitrogen is capable to form chemical compounds in the surface area of fine-grained material. For further discussion see Müller (1974).

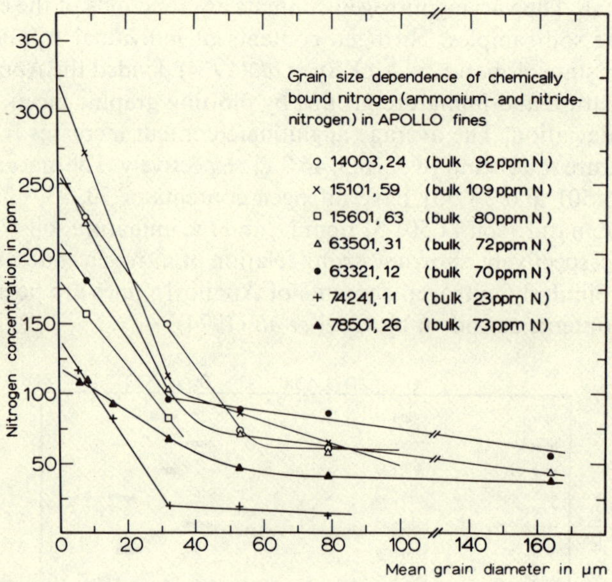

FIG. 3. Grain size dependence of chemically bound nitrogen in sieve fractions of Apollo 14, 15, 16 and 17 lunar fines. The "bulk" nitrogen contents refer to those of fines <1 mm.

Grain-size Analysis of Lunar Fines

The bound nitrogen results of grain-size fraction analyses of Apollo 14, 15, 16 and 17 fines are illustrated in Fig. 3 (Müller, 1974). The nitrogen concentrations of sieve fractions are plotted versus the mean grain diameter. As approximate mean grain diameter of the fractions 3–12, <24, 12–24, 24–48, 48–60, 60–109 and 109–272 μm, we have used 6, 8, 16, 32, 53, 79 and 164 μm, respectively, in this plot. The means for grain sizes have been calculated by adding one-third of the size range to the lower limit of each grain-size fraction. Bound nitrogen contents show a distinct increase with decreasing grain size. It is now commonly accepted that the surface-correlated nitrogen and also the other light, abundant elements in lunar fines are mainly derived from solar wind implantation. Among the fines shown in Fig. 3 only 15101 and 63501 lie nearly on a straight line in a log-log plot of nitrogen content versus reciprocal grain diameter, whereas data points of the other fines show larger scatter. The gradients of the lines are rather similar, except soil 74241, for which the gradient is much steeper.

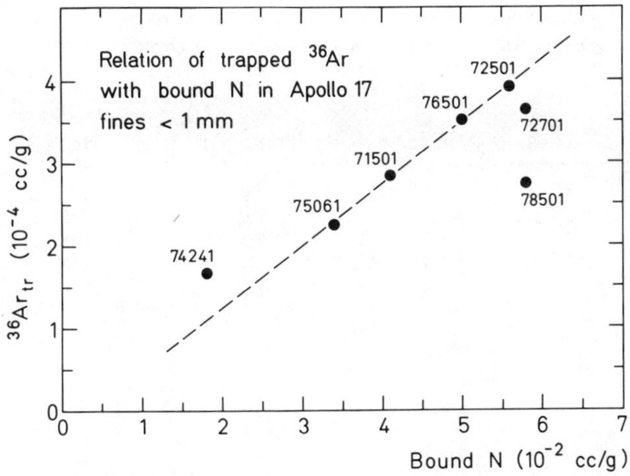

FIG. 4. Correlation of trapped ^{36}Ar with chemically bound nitrogen in Apollo 17 fines < 1 mm. Contents are given in cm^3 STP/g.

Nitrogen and Trapped Noble Gases in Lunar Fines

Besides grain-size analysis, correlation of nitrogen with trapped noble gases can provide further evidence for solar wind origin of nitrogen in fines. In Fig. 4 we plotted trapped ^{36}Ar versus bound nitrogen for seven Apollo 17 fines. Most of the fines are well correlated; fines 78501 strongly deviate from the dashed line due to a smaller $^{36}Ar/N$ ratio compared to other fines. Trapped ^{4}He and ^{20}Ne are not well related with bound nitrogen, mainly because of the large diffusion loss of the light noble gases. Noble gas data were measured parallel to nitrogen on the same samples by Hübner *et al.* (1975).

Nitrogen as Reference Element for Solar Abundances

In previous reports we have shown by means of the $^{4}He/N$ and $^{20}Ne/N$ ratios in lunar fines that nitrogen is more tightly retained because of its chemically active nature than are the noble gases (Müller, 1972, 1973, 1976). Nitrogen is, therefore, useful as a reference element for solar abundances; it is better retained than carbon. In Table 2 the abundance of chemically bound nitrogen in the <24 μm grain-size fraction of Apollo 14 and Apollo 16 fines is compared with the contents of trapped solar wind noble gases. The calculated $^{4}He/N$ ratios of 0.38 to 0.85 are lower by a factor of 1555 to 695 than the solar ratio of Cameron (1973). The $^{20}Ne/N$ ratios of $(0.82 \text{ to } 1.6) \times 10^{-2}$ are lower by a factor of 100 to 50 than the solar ratio. Assuming complete retention of implanted nitrogen, we can estimate that only about 0.1% of solar ^{4}He and 1 to 2% of solar ^{20}Ne are retained in lunar fines. The loss of the lighter noble gases from fine-grained material by diffusion and erosion processes is also evident when comparing $^{4}He/^{20}Ne$ and $^{20}Ne/^{36}Ar$ ratios of lunar fines with the ratios found in the Solar Wind Composition (SWC) experiments by Geiss *et al.* (1971, 1972). Provided that the solar $^{20}Ne/N$ ratio of about 0.8 is true, the depletion factor of ^{4}He in lunar fines relative to N can also be estimated by referring to the $^{4}He/^{20}Ne$ ratio of the SWC experiment. Noble gas data of fines in Table 2 are from Kirsten *et al.* (1972, 1973).

TABLE 2. THE SIGNIFICANCE OF SOLAR WIND-DERIVED NITROGEN AS REFERENCE ELEMENT IN TERMS OF SOLAR ABUNDANCES DEDUCED FROM ITS RELATION WITH SOLAR WIND NOBLE GASES

Lunar fines grain size fraction $<24\ \mu m$	$^{4}He/^{20}Ne$	$^{20}Ne/^{36}Ar$	$^{4}He/N$	$^{20}Ne/N$ $\times 10^{-2}$
14003,24	54	3.6	0.85	1.6
64421,24	49	2.0	0.62	1.3
67601,16	46	2.5	0.38	0.82
63321,12	43	2.7	0.48	1.1
63501,31	46	2.9	0.50	1.1
Solar wind composition experiment—Apollo 14, Geiss *et al.* (1971)	550	37	—	—
—Apollo 16, Geiss *et al.* (1972)	570	29	—	—
Solar abundances, Cameron (1973)	722	31	591	82

Leach Experiments on Lunar Fines

Leach experiments on fines were performed to study the chemical nature of bound nitrogen (Müller, 1974). The results compiled in Table 3 show that a substantial amount of bound nitrogen is easily leached in the form of ammonia by treating soils with 2 N H_2SO_4 at 20°C. This indicates that NH_3 molecules and/or NH_4^+ cations are liberated during break-up of grain surfaces, being probably originally present in minerals and glasses of fines. These

TABLE 3. LEACH EXPERIMENTS ON LUNAR FINES AND SOME TEST MATERIALS
Agent: 2N H_2SO_4, 20°C, 1 hour; grain size $<24\ \mu m$: 15 min. The numbers for leach and residue portion sum up to approximately 100%.

Sample N content	Bound Nitrogen Leach (rel. percent)	Bound Nitrogen Residue (rel. percent)	Elements determined Na, Mg, Al, Ca, Fe: Leach (rel. percent for each element)	Elements determined Na, Mg, Al, Ca, Fe: Residue (rel. percent for each element)
15101 < 1 mm 109 ppm N	61	46	~37	~66
15101 < 24 μm 272 ppm N	41	66	~33	~57
78501 < 1 mm 73 ppm N	37	50	~36	~71
78501 < 24 μm 109 ppm N	27	74	~30	~73
72501 < 1 mm 70 ppm N	86	14	~31	~68
Na–Ca–silica–Glass 2.1 wt.% N	1.1	99	Na: 4 Ca: 0	96 100
Si_3N_4 40 wt.% N	0	100	—	—
TiN 23 wt.% N	0	100	—	—

species may have been formed by the interaction of solar wind nitrogen with hydrogen by hot atom chemistry. It cannot be excluded that ammonia is derived from easily hydrolyzable nitrides of lunar fines. Bound nitrogen measured in the residue after leaching lunar fines is interpreted as nitride-nitrogen, which is hydrolyzable only with difficulty.

Test experiments were performed on powdered samples of a sodium–calcium–silica glass containing 2.1 wt.% bound nitrogen, on silicon nitride and titanium nitride. In the glass sample, only 1.1% of the total bound nitrogen was detected in the leach fraction; the nitrides showed no nitrogen in the leach. Nitrogen in the glass is present in the form of nitride–, NH_2– and NH–groups. A NH–vibration band at about 3 μm was detected by infrared spectrometry (Mulfinger and Franz, 1965). The low amount of leachable nitrogen in the glass indicates that NH– and NH_2–nitrogen is scarcely attacked in our leaching procedure.

In aliquots of both the leach solution and residue solution five elements were determined, which were mainly indicative for the olivine (Mg, Fe) and plagioclase (Na, Al, Ca) contents of fines. The results in Table 3 show that about 30 to 37% of each element were dissolved with 2 N H_2SO_4 at 20°C, meaning that all mineral components are broken up about equally. In summary, the leach experiments indicate that nitrogen and carbon chemistry in lunar fines appear to be analogous in respect to the hydrogen compounds ammonia and methane, and the presence of nitrides and carbides. Indigenous methane and carbides were detected in Apollo 11 and Apollo 12 fines by Abell *et al.* (1970) and Chang *et al.* (1971).

Solar Wind Carbon and Nitrogen

Because of its chemically active character, solar wind-implanted carbon has also a high retention on grain surfaces, which is comparable to that of nitrogen. DesMarais *et al.* (1974) calculated the overall retention for carbon to be 0.65 relative to 1.0 for nitrogen. The C/N atomic ratios of Apollo 17 fines listed in Table 4 range from 1.68 to 2.80, and are in the average distinctly higher than those previously calculated for Apollo 14, 15 and 16 fines, which vary between 1.35 and 1.75 (Müller, 1973). When comparing C/N ratios of lunar fines with the ratio of 4.2 of the solar photosphere found by Lambert (1968) from a reliable analysis of atomic and molecular spectra, it turns out that C/N ratios of fines are consistently lower than that of the sun. Volatile carbon compounds such as methane appear, therefore, to be preferentially lost from fine-grained lunar material by diffusion and possibly during reworking of the regolith. If ammonia were actually present in fines, a comparable loss of nitrogen should be expected. This is apparently not the case on the basis of C/N ratios. We interpret, therefore, the substantial amounts of leachable nitrogen in fines (Table 3) as being

TABLE 4. TOTAL CARBON AND CHEMICALLY BOUND NITROGEN DATA, AND C/N ATOMIC RATIOS IN APOLLO 17 LUNAR FINES AND IN THE SOLAR PHOTOSPHERE
Carbon data: Moore *et al.* (1974). Solar photosphere data: *Lambert* (1968).

FINES <1 mm	70011	71501	72501	72701	74240† 74241	76501	78501	Solar photosphere (C = 100)
C (ppm)	120	75	125	140	55	120	170	
C (10^{18}at./g)	6.0	3.7	6.3	7.0	2.8	6.0	8.5	100
N (ppm)	73	51	70	73	23	63	73	
N (10^{18}at./g)	3.1	2.2	3.0	3.1	1.0	2.7	3.1	24
C/N (at./at.)	1.94	1.68	2.10	2.26	2.80	2.22	2.74	4.2

† Unsieved fines.

present as NH_4^+ cations and/or easily hydrolyzable nitrides, the diffusion rate of which is much smaller than that of ammonia.

Solar wind implantation, trapping efficiencies, loss mechanism, molecule formation and isotopic alteration of light elements on the lunar surface are, at present, not fully understood. Simulated solar wind-implantation experiments with the labeled ions ^2H, ^{13}C, ^{15}N have provided more insight into the complex processes on the very surface of the moon (Pillinger *et al.*, 1972; Chang *et al.*, 1973; Wszolek *et al.*, 1974). Moreover, model calculations for the accumulation of carbon in the lunar regolith as a function of time and the corresponding evolution of its ^{13}C content (Kerridge *et al.*, 1974) and proposed mechanisms for isotopic evolution of solar wind nitrogen (Kerridge *et al.*, 1977) are a promising approach for better understanding the light element chemistry on the moon's surface.

Nitrogen in Lunar Igneous Rocks

Lunar igneous rocks contain much less bound nitrogen contents than fines. The difference amounts to a factor of 10 to 20. For most of the rock samples analyzed with the conventional Kjeldahl method so far in our laboratory, only an upper limit of about 10 ppm N could be given, because of lack of material and limited sensitivity of this procedure. However, the optimized Kjeldahl technique is capable to analyze nitrogen contents in the ppm range by use of about 150 mg sample size (Müller *et al.*, 1976). The analyses of U.S.G.S. standard rocks diabase W-1 and granite G-2 yielded mean values of about 14 ppm and 8 ppm N with a relatively small standard deviation. Wlotzka (1961) found likewise 14 ppm N in diabase W-1.

Table 5 and Fig. 5 show the results for lunar materials of various sample type and for two standard rocks. Most of the lunar igneous rock samples, analyzed with the optimized

TABLE 5. CHEMICALLY BOUND NITROGEN CONTENTS IN VARIOUS APOLLO LUNAR SAMPLES (Apollo 11, 12, 14, etc.) AND IN TWO TERRESTRIAL STANDARD ROCKS (W–1, G–2) USING THE OPTIMIZED KJELDAHL METHOD

Sample no.	Sample type	Chem. bound N (ppm) by the optimized Kjeldahl method
10057,80	Vesicular basalt, Apollo 11	6.3; 4.2 (duplicates)
12053,96	Porphyritic basalt, Apollo 12	3.1; 5.0
12063,112	Microgabbro, Apollo 12	< 10 (conventional method)
70215,21	Fine-grained basalt, Apollo 17	3.0; 3.0
70215,53	Powdered split of Moore and Lewis (1976)	5.7; 8.2
77017,32	Crushed anorthositic gabbro, brecciated and invaded by glass	5.0; 5.1
77017,47	Powdered split of Moore and Lewis (1976)	7.0; 6.5
79155,24	Homogeneous gabbro, partially coated with glass	4.5; 4.5
14303,13	Clastic rock, Apollo 14	7.9; 8.8
15012,14	Fines, SESC–1, Apollo 15	110
15013,11	Fines, SESC, Apollo 15	93; 94
W–1	U.S.G.S. diabase W–1 (mean of 4 analyses)	13.8
G–2	U.S.G.S. granite G–2 (mean of 3 analyses)	8

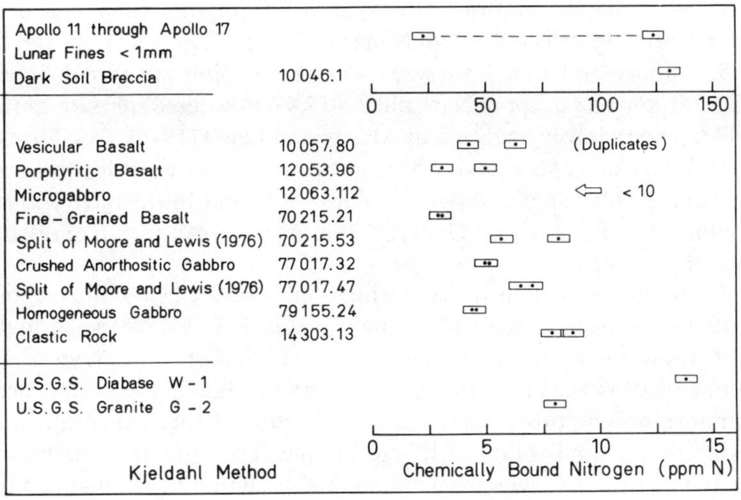

FIG. 5. Chemically bound nitrogen contents (in ppm N) in various lunar samples with error bars. The lunar rocks and the terrestrial standard rocks diabase W-1 and granite G-2 were dominantly analyzed in replicates by using the optimized Kjeldahl method (Müller *et al.*, 1976). For comparison, the results on lunar fines < 1 mm and on dark soil breccia 10046,1 are shown. Note the two different scales.

method, are quite uniform in chemically bound nitrogen contents. The total range is only about 3 to 8 ppm for samples of various types and from different landing sites. The variation of duplicate analyses is rather small, as is shown in Table 5. The small nitrogen concentrations of igneous rocks imply that indigenous nitrogen and/or ammonia were very low in abundance at the time of mineral and rock formation billions of years ago. If ammonia and/or nitrogen in reducing environment had been present, chemical reactions leading, for example, to nitride formation would have taken place. Till now, no nitrogen-bearing minerals have been observed in any lunar sample. The low nitrogen contents indicate either severe *a priori* nitrogen depletion of the moon's parent material or rigorous loss of nitrogen and other volatile elements in the early history of the moon. Obviously, magmatic gases, if present at all, were low in nitrogen and nitrogen compounds at the time of formation of lunar rocks. In a previous work, active and inert gases were released from several lunar samples, among them vesicular and dense rocks, by crushing at room temperature in a stainless-steel apparatus. Independent crushing and gas analyses using on-line gas chromatography and mass spectrometry were performed (Funkhouser *et al.*, 1971). The composition of gases released from those rocks is variable and consists of H_2, N_2, CH_4 and the noble gases, all in parts per billion quantities. These gases are probably representative of the magmatic gas phase, and the very small amounts of molecular nitrogen detected by crushing conform with the low bound nitrogen contents of lunar igneous rocks.

Some of the rock samples analyzed in this work can be compared with nitrogen data of other authors to illustrate agreement and discrepancy in some of the results obtained so far. Basalt 10057 yields 6.3 and 4.2 ppm bound nitrogen in duplicate analyses with the optimized Kjeldahl method. Hintenberger *et al.* (1970) found ≤ 5 ppm bound nitrogen in this rock by using a stepwise chemical dissolution technique. For microgabbro 12063 we obtained < 10 ppm N with the conventional Kjeldahl method; Becker and Clayton (1975) reported 0.6 and 1.0 ppm N for two splits of this rock. For two of the three analyzed Apollo 17 rocks nitrogen data also exist from other laboratories. Goel *et al.* (1975) found 21 ppm N, Moore and Lewis (1976) 88 ppm N, and we detected 3 ppm N in split 70215,21, and 5.7 ppm and 8.2 ppm N in split 70215,53. Moore and Lewis analyzed 45 ppm N while we found 5ppm N in split 77017,32, and 7.0 ppm and 6.5 ppm N in split 77017,47. Powdered splits of samples 70215,53 and 77017,47 were previously analyzed by Moore and Lewis (1976). By courtesy of NASA, we recently had the chance to analyze the same sample splits with the optimized Kjeldahl method (see Table 5). Our data of Apollo 15 fines 15012 and 15013 agree within about 15% with those obtained by Kothari and Goel (1972) using (n, p)-activation of nitrogen (98 ppm and 110 ppm N, respectively).

According to results on igneous rocks, the high-temperature extraction technique (Moore and Lewis, 1976) and neutron activation analysis (Goel et al. 1975) yield higher nitrogen contents than those methods using vacuum fusion (Becker and Clayton, 1975) or the Kjeldahl technique (this work). The Kjeldahl method including the sample decomposition applied is suitable for quantitive determination of bound nitrogen in silicate materials, and we are convinced of the reliability of this technique. The optimized method is especially qualified for the analysis of low nitrogen amounts. It should be emphasized that nitrogen data for individual lunar soils, which were analyzed by different methods, are generally in reasonable agreement, but the severe discrepancy between results for rock samples is not evident in our opinion. We propose, therefore, interlaboratory comparison of nitrogen contents in various standard materials and also in lunar rock samples. Proper evaluation of the light element abundance introduced into lunar fines from extra-lunar sources, mainly the solar wind, strongly depends on reliable knowledge of the levels of those elements in rock samples.

Acknowledgements

We are thankful to the National Aeronautics and Space Administration for providing the lunar samples. The collaboration with Professor G. Tölg, Dr. E. Grallath, Mr. W. Werner and Miss G. Wolf, Max-Planck-Institut für Metallforschung in Schwäbisch Gmünd, is highly appreciated. The skillful assistance of Mr. D. Kaether, Max-Planck-Institut für Kernphysik in Heidelberg, is gratefully acknowledged. I would like to thank Professor W. Gentner for his continued interest in this work.

References

ABELL, P. I., EGLINTON, G., MAXWELL, J. R., PILLINGER, C. T. and HAYES, J. M. (1970) Indigenous lunar methane and ethane. *Nature* **226**, 251–252.

BECKER, R. H. and CLAYTON, R. N. (1975) Nitrogen abundances and isotopic compositions in lunar samples. *Proc. Lunar Sci. Conf. 6th*, Vol. 2, 2131–2149, Pergamon Press.

CAMERON, A. G. W. (1973) Abundances of the elements in the solar system. *Space Sci. Rev.* **15**, 121–146.

CHANG, S., KVENVOLDEN, K., LAWLESS, J., PONNAMPERUMA, C. and KAPLAN, I. R. (1971) Carbon, carbides, and methane in an Apollo 12 sample. *Science* **171**, 474–477.

CHANG, S., MACK, R., GIBSON, E. K., Jr. and MOORE, G. W. (1973) Simulated solar wind implantation of carbon and nitrogen ions into terrestrial basalt and lunar fines. *Proc. Lunar Sci. Conf. 4th*, Vol. 2, 1509–1522, Pergamon Press.

DESMARAIS, D. J., HAYES, J. M. and MEINSCHEIN, W. G. (1974) Retention of solar wind-implanted elements in lunar soils (abstract). In: *Lunar Science*, V, 168–170, The Lunar Science Institute, Houston.

FUNKHOUSER, J., JESSBERGER, E., MÜLLER, O. and ZÄHRINGER, J. (1971) Active and inert gases in Apollo 12 and Apollo 11 samples released by crushing at room temperature and by heating at low temperatures. *Proc. Lunar Sci. Conf. 2nd*, Vol. 2, 1381–1396, MIT Press.

GEISS, J., BUEHLER, F., CERUTTI, H., EBERHARDT, P. and MEISTER, J. (1971) The solar-wind composition experiment. *Apollo 14 Preliminary Science Report*, NASA SP-272, 221–226.

GEISS, J., BUEHLER, F., CERUTTI, H., EBERHARDT, P. and FILLEUX, Ch. (1972) Solar wind composition experiment. *Apollo 16 Preliminary Science Report*, NASA SP-315, 14, 1–14, 10.

GIBSON, E. K., Jr. and MOORE, C. B. (1970) Inert carrier-gas fusion determination of total nitrogen in rocks and meteorites. *Anal. Chem.* **42**, 461–464.

GOEL, P. S., SHUKLA, P. N., KOTHARI, B. K. and GARG, A. N. (1974) Solar wind as source of nitrogen in lunar fines (abstract). In: *Lunar Science*, V, 270–272, The Lunar Science Institute, Houston.

GOEL, P. S., SHUKLA, P. N., KOTHARI, B. K. and GARG, A. N. (1975) Total nitrogen in lunar soils, breccias and rocks. *Geochim. Cosmochim. Acta* **39**, 1347–1352.

HINTENBERGER, H., WEBER, H. W., VOSHAGE, H., WÄNKE, H., BEGEMANN, F. and WLOTZKA, F. (1970) Concentrations and isotopic abundances of the rare gases, hydrogen and nitrogen in Apollo 11 lunar matter. *Proc. Apollo 11 Lunar Sci. Conf.*, Vol. 2, 1269–1282, Pergamon Press.

HÜBNER, W., KIRSTEN, T. and KIKO, J. (1975) Rare gases in Apollo 17 soils with emphasis on analysis of size and mineral fractions of soil 74241. *Proc. Lunar Sci. Conf. 6th*, Vol. 2, 2009–2026, Pergamon Press.

KAPLAN, I. R., SMITH, J. W. and RUTH, E. (1970) Carbon and sulfur concentration and isotopic composition in Apollo 11 lunar samples. *Proc. Apollo 11 Lunar Sci. Conf.*, Vol. 2, 1317–1329, Pergamon Press.

KERRIDGE, J. F., KAPLAN, I. R. and LESLEY, F. D. (1974) Accumulation and isotopic evolution of carbon on the lunar surface. *Proc. Lunar Sci. Conf. 5th*, Vol. 2, 1855–1868, Pergamon Press.

KERRIDGE, J. F., KAPLAN, I. R., LINGENFELTER, R. E. and BOYNTON, W. V. (1977) Solar wind nitrogen: mechanisms for isotopic evolution (abstract). In: *Lunar Science* VIII, 537–539, The Lunar Science Institute, Houston.

KIRSTEN, T., DEUBNER, J., HORN, P., KANEOKA, I., KIKO, J., SCHAEFFER, O. A. and THIO, S. K. (1972) The rare gas record of Apollo 14 and 15 samples. *Proc. Lunar Sci. Conf. 3rd*, Vol. 2, 1865–1889, MIT Press.

KIRSTEN, T., HORN, P. and KIKO, J. (1973) ^{39}Ar–^{40}Ar dating and rare gas analysis of Apollo 16 rocks and soils. *Proc. Lunar Sci. Conf. 4th*, Vol. 2, 1757–1784, Pergamon Press.

KOTHARI, B. K. and GOEL, P. S. (1972) Total nitrogen abundances in five Apollo 15 samples (Hadley-Apennine region) by neutron activation analysis. In *The Apollo 15 Lunar Samples*, 282–283. The Lunar Science Institute, Houston.

LAMBERT, D. L. (1968) The abundances of the elements in the solar photosphere. I. Carbon, nitrogen and oxygen. *Monthly Notices R. Astron. Soc.* **138**, 143–179.

MASON, B. (1971) *Handbook of Elemental Abundances in Meteorites*, B. MASON, ed., Gordon & Breach. See *ibid.*: "Hydrogen" by I. R. KAPLAN, "The Inert Gases" by D. HEYMANN. "Carbon" by G. P. VDOVYKIN and C. B. MOORE, "Nitrogen" by C. B. MOORE, "Oxygen" by W. D. EHMANN.

MCKAY, D. S., FRULAND, R. M. and HEIKEN, G. H. (1974) Grain size and the evolution of lunar soils. *Proc. Lunar Sci. Conf. 5th*, Vol. 1, 887–906, Pergamon Press.

MOORE, C. B. and LEWIS, C. F. (1976) Total nitrogen contents of Apollo 15, 16 and 17 lunar rocks and breccias (abstract). In *Lunar Science*, VII, 571–573. The Lunar Science Institute, Houston.

MOORE, C. B., LEWIS, C. F. and CRIPE, J. D. (1974) Total carbon and sulfur contents of Apollo 17 lunar samples. *Proc. Lunar Sci. Conf. 5th*, Vol. 2, 1897–1906, Pergamon Press.

MÜLLER, O. (1972) Chemically bound nitrogen abundances in lunar samples, and active gases released by heating at lower temperatures (250 to 500°C). *Proc. Lunar Sci. Conf. 3rd*, Vol. 2, 2059–2068, MIT Press.

MÜLLER, O. (1973) Chemically bound nitrogen contents of Apollo 16 and Apollo 15 lunar fines. *Proc. Lunar Sci. Conf. 4th*, Vol. 2, 1625–1634, Pergamon Press.

MÜLLER, O. (1974) Solar wind nitrogen and indigenous nitrogen in Apollo 17 lunar samples. *Proc. Lunar Sci. Conf. 5th*, Vol. 2, 1907–1918, Pergamon Press.

MÜLLER, O. (1976) Zur Geochemie der leichten Elemente, insbesondere des Stickstoffs, in Mondgesteinen. *Habilitationsschrift*, Universität Heidelberg.

MÜLLER, O., GRALLATH, E. and TÖLG, G. (1976) Nitrogen in lunar igneous rocks. *Proc. Lunar Sci. Conf. 7th*, Vol. 2, 1615–1622, Pergamon Press.

MULFINGER, H. O. and FRANZ, H. (2965) Über den Einbau von chemisch gelöstem Stickstoff in oxydischen Glasschmelzen. *Glastechn. Berichte* **38** 235–242.

PETROWSKI, C., KERRIDGE, J. F. and KAPLAN, I. R. (1974) Light element geochemistry of the Apollo 17 site. *Proc. Lunar Sci. Conf. 5th*, Vol. 2, 1939–1948, Pergamon Press.

PILLINGER, C. T., CADOGAN, P. H., EGLINTON, G., MAXWELL, J. R., MAYS, B. T., GRANT, W. A. and NOBES, M. J. (1972) Simulation study of lunar carbon chemistry. *Nature, Phys. Sci.* **235**, 108–109.

WERNER, W. and TÖLG, G. (1975) Optimierung der Kjeldahl-Methode zur Bestimmung von Stickstoff in Reinstmetallen. *Z. Anal. Chem.* **276**, 103–110.

WLOTZKA, F. (1961) Untersuchungen zur Geochemie des Stickstoffs. *Geochim. Cosmochim. Acta* **24**, 106–154.

WSZOLEK, P. C., WALLS, F. C. and BURLINGAME, A. L. (1974) Simulation of lunar carbon chemistry in plagioclase (abstract). In: *Lunar Science*, V. 854–856. The Lunar Science Institute, Houston. See also *Proc. Lunar Sci. Conf. 5th*, Vol. 2, 1747–1762 and 1763–1784 (1974).

Solar Wind Carbon Chemistry as revealed by Lunar Sample Analysis

C. T. PILLINGER

Planetary Sciences Unit, Department of Mineralogy and Petrology, University of Cambridge, Cambridge, England

Abstract

Knowledge of solar wind carbon chemistry is restricted to what has been learned from the analysis of lunar samples. The study of total carbon abundances, the search for the existence of distinct carbon compounds, and the investigation of carbon isotope ratios have demonstrated the behaviour of a reactive element implanted from the solar wind into the surface of an atmosphereless planet, and provided information about the evolution of the lunar regolith.

Introduction

As yet there has been no direct measurement of carbon in the solar wind although its occurrence was possibly recognized by the Vela satellites (Bame *et al.*, 1970). Its existence in the flux, however, may be inferred on the basis of solar abundances. The Apollo and Luna samples of lunar material have provided the first opportunity to study solar wind derived carbon. Because atmospheric and magnetic shielding on the moon is minimal, exposed surfaces of lunar soil grains must accumulate by ion implantation, concentrations of carbon which are easily detectable in the absence of large organic molecules of biogenic or abiogenic origin. A number of papers presented at the Apollo 11 Lunar Science Conference recognized the opportunities provided by an abiogenic moon and studies of solar wind carbon have blossomed since 1969 (Eglinton *et al.*, 1974). In the first instance, it was necessary to demonstrate that the few hundred parts per million of carbon associated with lunar soils were, in fact, predominantly of solar wind origin. Unlike the rare gases which are the most extensively studied solar wind elements in lunar samples and meteorites, carbon is reactive. Therefore, solar wind carbon may undergo reactions with other solar wind atoms or elements present in the target mineral grains. Considerable effort has been made to identify and measure the concentration of various compounds in lunar samples which derive from implanted carbon in addition to estimating total elemental abundance and isotopic composition. Two other active elements, hydrogen (DesMarais *et al.*, 1974) and nitrogen (Müller, 1979 of solar wind origin, are found in lunar soil grains and their behaviour should be considered in analogy to carbon.

Experimental Methods

Total carbon abundances in lunar samples are measured by combustion in a flowing oxygen atmosphere at *ca.* 1600°C followed by gas chromatographic detection of CO_2 (Moore *et al.*,

1970) or by pyrolysis directly or in steps to *ca.* 1220–1300°C. The carbon species released by pyrolysis, mainly CH_4, CO and CO_2, are detected and quantitated by low resolution mass spectrometry (LRMS) (Gibson and Johnson, 1971; Chang *et al.*, 1974a; Kerridge *et al.*, 1975), high resolution mass spectrometry (HRMS) (Holland *et al.*, 1972) or by helium ionization gas chromatography (DesMarais *et al.*, 1973c). Chemically combined carbon has been studied extensively by characterizing gaseous species released by acid dissolution of samples, preferably with deuterium-labelled reagents (Abell *et al.*, 1970a). Reaction products of acid treatments are investigated using LRMS (Chang *et al.*, 1974 a, b; Kerridge *et al.*, 1975), HRMS (Holland *et al.*, 1972) or by gas solid chromatography (Cadogan *et al.*, 1973a). Carbon isotope ratios are measured (Chang *et al.*, 1974a, b; Kerridge *et al.*, 1975; Epstein and Taylor, 1975; DesMarais *et al.*, 1975) by conversion of carbon containing entities to Co_2. This gas is introduced into a low resolution mass spectrometer from a constant pressure inlet via a capillary leak and compared to CO_2 from a standard carbonate (PDB). Isotope ratios are quoted as enrichments ($\delta^{13}C\%$) relative to the standard, where

$$\delta^{13}C\permil = \left(\frac{(^{13}C/^{12}C)\text{ sample} - (^{13}C/^{12}C)\text{ standard}}{(^{13}C/^{12}C)\text{ standard}}\right) \times 1000$$

Very recently a static mass spectrometric method has been developed which allows the detection and quantitation of carbon compounds in amounts of the order 10^{-12} g (Gardiner *et al.*, in preparation). It is applicable to acid dissolution and pyrolysis studies and, to a limited extent, to isotopic ratio determinations; however, as yet, it is little exploited (Gardiner *et al.*, 1977).

The analysis of tiny amounts of carbon compounds in an environment dominated by biologically derived organic materials has necessitated the adoption of special procedures to minimize contamination (Abell *et al.*, 1970b). The understanding of carbon chemistry has required subdivision of lunar soils into many fractions according to size and particle type. In order to handle the small amounts of sample available without appreciable contamination or degradation, new techniques such as sieving in liquid argon (DesMarais *et al.*, 1972) have been developed, and established separation methods based on magnetic and density properties modified (Cadogan, 1972).

Total Carbon

Total carbon abundances for many lunar samples are now available. Particularly productive has been the group of Moore whose data (Moore, 1975) are summarized in Fig. 1. The excess abundance of carbon in fines and breccias over igneous rocks is attributed mainly to solar wind bombardment of fine material of high surface area. Although other extralunar sources of carbon, e.g. carbonaceous chondrites impacting on the lunar surface, could contribute to the excess, the absolute identification of material from this source has proved very elusive. It is difficult to invoke mechanisms for the retention of carbon in the vapour phase as a result of impact. A possibility might include carburization of metallic iron (Jull *et al.*, 1975), however, the observed analytical data do not appear to favour such a mechanism (Gardiner *et al.*, 1977). The contribution of meteoritic carbon to the lunar regolith may be of the order of 10% of the total carbon (Hayes, 1972; Bibring *et al.*, 1974a) or less.

Most data concerning total solar wind carbon has come from the analysis of lunar soils by pyrolysis. It is possible to distinguish the lunar indigenous carbon of igneous origin from that

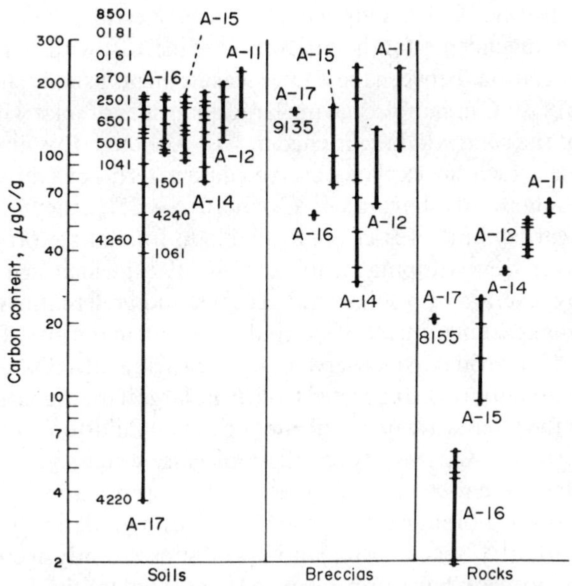

FIG. 1. A comparison of the total carbon abundances of the Apollo 17 lunar samples with those of other missions. The four-digit numbers in the figure refer to the lunar sample inventory number (from the Apollo 17 preliminary science report NASA SP-330).

added by solar wind bombardment on the basis of the temperature release profile, which suggests that igneous carbon is not liberated until very high temperature, *ca.* 1300°C (Gibson and Moore, 1973). This distinction is seldom made because the solar wind carbon is usually far in excess of contributions from other sources.

The analysis of sieved fractions of lunar soils clearly shows that carbon increases in abundance in the finest particle sizes (Kaplan *et al.*, 1970) in keeping with a surface location expected to result from implantation. The approach has been adopted by many groups but most effectively utilized by DesMarais *et al.* (1973 a, b; 1975; Basu *et al.*, 1975) who interpreted the results in terms of both surface and volume-related components of solar wind derived material by plotting carbon concentration for various size fractions against the reciprocal of average particle radius. Volume-related components arise because of recycling of very fine grains with exposed surfaces into complex aggregated particles known as glassy agglutinates or microbreccias (Cadogan *et al.*, 1972). The limited number of samples studied seem to have approximately the same surface carbon concentration (DesMarais *et al.*, 1975) which may be indicative of a saturation effect. From the grain-size dependence plots it is possible to calculate approximate surface carbon concentrations of *ca.* 8×10^{15} atoms/cm^2. This value is higher than an estimate of 7×10^{14} atoms/cm^2 made by depth profiling using a nuclear resonance reaction (Filleux *et al.*, 1977). Calculations using a solar wind flux of 4×10^4 atoms/cm^2/s suggest that particle surfaces reach saturation equilibrium as a result of sputtering effects at between 10^{14} (Bibring *et al.*, 1975) and 3×10^{15} atoms/cm^2 (Jull and Pillinger, 1977) depending on the magnitude of the sputtering rate employed (see below).

Several groups (Kaplan *et al.*, 1970; Epstein and Taylor, 1970; Oro *et al.*, 1970) noted that bulk lunar soils and breccias from the Apollo 11 site were enriched in ^{13}C relative to the PDB standard and relative to the lunar igneous rocks. The isotopic analysis of many samples has confirmed these initial impressions (Epstein and Taylor, 1971, 1972, 1973; Kaplan and

Petrowski, 1971; Smith *et al.*, 1973; Chang *et al.*, 1974 a, b,; Kerridge *et al.*, 1975; Kaplan *et al.*, 1976). In general there is a tendency for the enrichment to increase with increasing total carbon content, although correlations between these two measurements usually show only a relatively poor fit. The increase of $\delta^{13}C$ in samples having large quantities of solar wind carbon does not necessarily imply that the solar wind is enriched in ^{13}C. Although Epstein and Taylor (1972) have presented a model which can explain observed data in terms of a mixture of two or three components with the solar wind C having a $\delta^{13}C$ of $+10$ to $+25‰$, they and other authors are in agreement that a variety of processes are available to fractionate original solar wind C. Thermal diffusion, hydrogen stripping (Kaplan *et al.*, 1970) which involves giving carbon enhanced volatility by reaction with solar wind protons, and preferential sputtering (Epstein and Taylor, 1972) a momentum transfer effect, could all result in the loss of the light relative to the heavy isotope and provide the observed $\delta^{13}C$ enrichments. Of the three processes considered, preferential sputtering is expected to give the largest fractionation factor (Pillinger *et al.*, 1976a). Both of the alternative mechanisms depend on diffusion and enrichment will be related to $\sqrt{(M_1/M_2)}$ where M_1 and M_2 are the molecular weights of the diffusing species.

A complicating factor in understanding bulk carbon isotope ratios is the observation by DesMarais *et al.* (1975) that although total carbon increases systematically proportional to $1/r$ in soils 14240 and 72501, $\delta^{13}C$ reaches a maximum and subsequently decreases for very small grains. A number of hypotheses have been advanced to account for the data (DesMarais *et al.*, 1975). They include (i) some kind of preferential contamination, (ii) loss of an isotopically heavy phase which was surface located and susceptible to hydrolysis and (iii) incomplete fractionation of carbon present in selectively retentive iron-rich phases. A further alternative is that freshly implanted and relatively unfractionated (isotopically light) solar wind atoms at the surface of grains could substantially outweigh the older fractionated material of the volume-related component in fine soil material (Pillinger *et al.*, 1976a).

A number of attempts have been made to study various fractions of carbon compounds released by stepwise pyrolysis (Kaplan and Petrowski, 1971; Chang *et al.*, 1974a). Clear evidence of a variety of phases having different isotopic compositions may be discerned, however, the data are difficult to interpret because the process of gas release could be contributing to the isotopic fractionation.

At the present time the carbon isotopic ratio of pristine solar wind carbon remains an unanswered question.

Chemical Characterization of Solar Wind derived Compounds

Deuterated acid dissolution studies of lunar soils have revealed that at least 25% of the carbon in lunar soils is in some chemically combined state. This method was originally devised (Abell *et al.*, 1970a) to resolve between the origins proposed for the series of low molecular weight hydrocarbons released by protonated acids from lunar materials which, it had been suggested, could be products of a reaction between solar wind carbon and hydrogen (Abell *et al.*, 1970b) or decomposition of meteoritic carbide (Chang *et al.*, 1970). On treatment with deuterated acids, bulk lunar soils release both hydrocarbons and deuterocarbons, the latter being the more abundant by a factor of 3 or 4. Clear evidence may be presented to suggest that the hydrocarbon fractions (of which only CH_4, C_2H_4 and C_2H_6 are present in sufficient quantities to allow systematic study) originate from an interaction of two solar wind elements by a mechanism as yet undefined. Numerous investigators (see *inter alia* Cadogan *et al.*, 1972;

Holland *et al.*, 1972; DesMarais *et al.*, 1973a; Chang *et al.*, 1974a) have shown that (i) CH_4 and C_2H_6 are surface-correlated gases, (ii) abundances of both species are directly related to accepted measures of lunar surface exposure including rare gas concentration, solar flare cosmic-ray track density and radiation damage. Isotopically labelled methane can be formed from ^{13}C and D ions implanted into lunar materials and synthetic analogues during laboratory simulation studies of solar wind bombardment (Pillinger *et al.*, 1972; Bibring *et al.*, 1974b). The formation of trapped hydrocarbons may be an important pathway in the fractionation of carbon isotopes as outlined above. For the Apollo 11 and 12 samples there was an indication of some relationship between $\delta^{13}C$ for total carbon and methane content (Cadogan *et al.*, 1971). The isotopic ratio of CH_4 released by pyrolysis has been measured in a limited number of cases (Chang *et al.*, 1974a) and seems to be enriched in ^{13}C by *ca.* 30–40‰. The substantial enrichment is in keeping with ideas that the CH_4 currently in the samples is the residuum of much larger quantities of gas which had diffused away or outgassed during agglutination.

Establishing the origin of the deuterocarbons released by acid has proven to be an intriguing problem, the solution of which has implications not only for the understanding of solar wind chemistry, but also in terms of the magnetic and optical properties of lunar soils. The latter both depend on the occurrence of finely disseminated iron droplets in glassy agglutinates. The majority of deuterocarbons released by acid (C_2D_2 being the exception) may be explained as reaction products of carbon in solid solution in α-iron (Jull *et al.*, 1976). The idea that some form of iron carbide (either cohenite or cementite) could be responsible was a misconception. The use of the term carbide to describe the deuterocarbon precursor was common in the literature before 1975; its continued usage is proving difficult to discourage. A much preferable notation is hydrolysable carbon, or C_{hyd} as a shortened form.

Since hydrocarbons and deuterocarbons are measured simultaneously, it is possible to perform correlation studies similar to those carried out for CH_4 and C_2H_6 using CD_4, by far the most abundant deuterocarbon, as representative of C_{hyd}. Thus, it can be shown that C_{hyd} in bulk soils increases with decreasing particle size (Cadogan *et al.*, 1972; Holland *et al.*, 1972; Chang *et al.*, 1974a); however, the relationship between this species and other measures of exposure history is indirect (Pillinger *et al.*, 1973, 1974). In fact, the relationship between CD_4 and exposure is only revealed when variations in the chemical composition, particularly the FeO content, of lunar soils are taken into consideration (Figs. 2 and 3). The concentration of C_{hyd} and

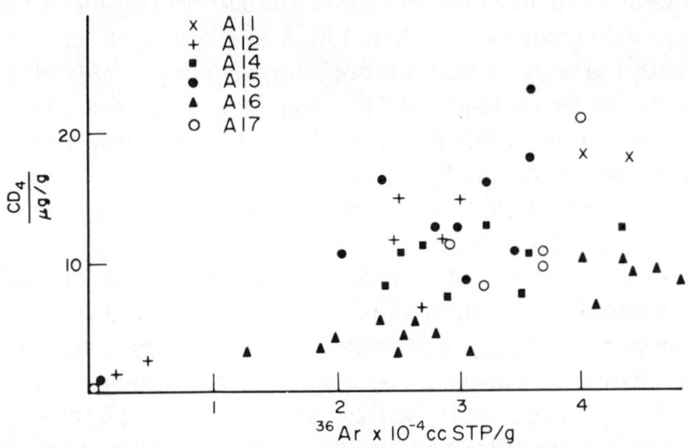

FIG. 2. Carbon as CD_4 versus ^{36}Ar concentration in Apollo 11–17 fines (from Pillinger *et al*, 1974).

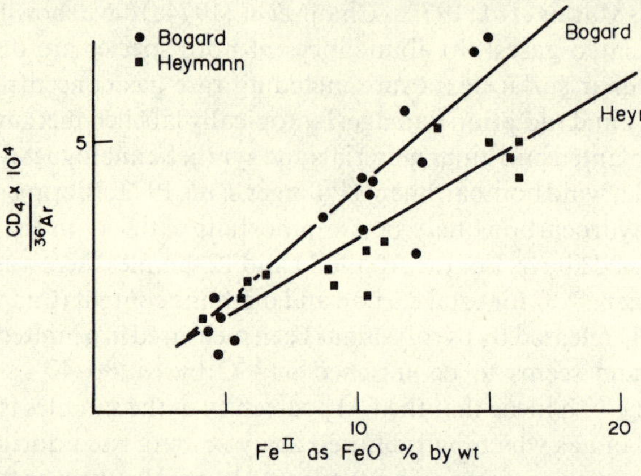

FIG. 3. Carbon as CD_4 normalized to constant ^{36}Ar concentration in Apollo 11–17 fines versus total iron content as FeO (from Pillinger *et al.*, 1974).

the abundance of finely divided (Fe°) or superparamagnetic iron as measured by ferromagnetic resonance spectroscopy (Pillinger *et al.*, 1974) or magnetic susceptibility (Pillinger *et al.*, 1977; Gardiner *et al.*, 1977) are, however, directly related. By analogy Fe° content must also be related to both ^{36}Ar and bulk ferrous iron content; this is readily demonstrable (Pillinger *et al.*, 1974). On the basis of these interrelationships, it seems apparent that finely divided iron is formed by some exposure induced reduction of lunar silicates as suggested by Housley *et al.* (1972, 1973) and hence meteoritic metal as a source of hydrolysable carbon is precluded.

The actual nature of the exposure mechanism producing iron metal from silicate has been the subject of considerable debate. Possibilities which have been recently reviewed by several authors (Pillinger *et al.*, 1976b; Yin *et al.*, 1976; Housley and Grant, 1976; Hapke, 1977) include chemical reductions by solar wind hydrogen, preferential sputtering (non-stoichiometric removal of atoms during erosion of surface by solar wind bombardment), preferential deposition of sputtered atoms, reduction in gas clouds produced by impact or shock-induced transformations. Finely divided iron was first observed by transmission electron microscopy during investigation of Apollo 11 samples (Agrell *et al.*, 1970); however, even after 8 years of study only a limited amount of information about the nature of the metal is available, viz. (i) its size distribution, predominantly less than 130 Å in diameter (Stephenson, 1970), (ii) its structure, apparently the body centred cubic or α-form (Pillinger *et al.*, 1975; Tsay *et al.*, 1974) and (iii) its carbon content, *ca.* $0.12 \pm 0.04\%$ by weight (Gardiner *et al.*, 1977). To decide among mechanisms for production of metal, all this limited information needs to be taken into consideration. The concentration of carbon as C_{hyd} promises to be a particularly valuable indicator. It seems to be possible to reproducibly obtain a non-equilibrium composition of carbon in iron; $0.12 \pm 0.04\%$ by weight is greater than the solid solubility limit for α-iron. Proposed mechanisms must account for such an apparent paradox; only solar wind chemical reduction or preferential sputtering may be able to do so (Gardiner *et al.*, 1977).

The carbon isotope ratio of C_{hyd} ought to be an informative parameter for determining the origin of the phase. Estimates which have been made appear to indicate little or no enrichment relative to bulk carbon (Chang *et al.*, 1974a; Kerridge *et al.*, 1975). The procedures used to make the measurements could not resolve between CH_4 and CD_4 and hence a mixture of trapped gas and reaction product has been studied. More recently an attempt has been made to investigate

CD_4 directly using static mass spectrometry (Gardiner *et al.*, 1977). The preliminary results suggest that C_{hyd} could be more enriched in ^{13}C than was previously thought.

Attempts to produce C_{hyd} during laboratory simulation of lunar processes have so far been unrewarded (Pillinger *et al.*, 1972; Davis *et al.*, 1975; Bibring *et al.*, 1974b). However, it is thermodynamically predictable that preferential spluttering should reduce only iron of the major rock-forming elements (Pillinger *et al.*, 1976b). Yin and his co-workers (Yin *et al.*, 1972, 1975) have demonstrated using ESCA, that bombardment of iron oxides and Fe^{II} containing silicate by chemically unreactive rare gases causes the formation of Fe° to a depth of at least 20 Å. The difficulties which have been experienced in generating C_{hyd} during similar solar wind simulation experiments may due to surface reoxidation which occurs when irradiated samples are taken from the vacuum system for analysis by methods designed to detect hydrolysable carbon (Davis *et al.*, 1975).

Both trapped hydrocarbons and C_{hyd} have been studied in highly magnetic agglutinate fractions separated from bulk soil (Cadogan *et al.*, 1973a, b, 1973; Chang *et al.*, 1974a; Bibring *et al.*, 1974a; Gardiner *et al.*, 1977). A mixing model based on data obtained from separation studies of lunar soils 10086 and 60501 demonstrates that these grains together with microbreccias are the major location for both forms of carbon in lunar soil (Cadogan *et al.*, 1973a). The same particles also concentrate the bulk of solar wind carbon (DesMarais *et al.*, 1973 a, b, 1975; Basu *et al.*, 1975). From a comparison of agglutinate and bulk soil data it is apparent that during formation of agglutinates, carbon species are lost in the order $CH_4 >$ non-chemically combined solar wind carbon $> C_{hyd}$ (Billetop *et al.*, 1976) as predicted by Gibson and Moore (1973). In comparison to rare gases C_{hyd} is particularly well retained during agglutination and thus may be a good indicator of the size distribution of material which must be aggregated together to give agglutinates (Gardiner *et al.*, 1977).

A number of minor components, DCN, C_2D_2 and CS_2, are released in 1–2 ppm amounts by acid dissolution. Unfortunately they are relatively little investigated, although all are thought to derive from implanted solar wind carbon (Wszolek *et al.*, 1973). It has been shown that ^{13}C atoms implanted into plagioclase react to give $^{13}C_2H_2$ on acid dissolution (Bibring *et al.*, 1974b), and some form of ionic carbide (possibly CaC_2) may be the precursor in both the laboratory experiment and in the case of acid treated lunar samples which release acetylene (Wszolek *et al.*, 1973). The cyanide phase may be an important precursor of the few ppb of amino acids which have been recognized in acid hydrolysates (Fox and Harada, 1973).

Acid dissolution with H_2SO_4 (Chang *et al.*, 1974 a, b; Kerridge *et al.*, 1975) also appears to release quite large quantities of CO_2, up to about 30 ppm. This product is little studied, probably because of the difficulties in obtaining acid free from absorbed atmospheric carbon dioxide. The contamination problem may also be the reason why the acid released CO_2 does not correlate with other forms of lunar carbon. It seems fairly certain that at least a portion of the CO_2 is indigenous to the lunar soil, because carbon and oxygen isotopic measurements performed on the gas give ratios distinguishable from terrestrial contamination (Kerridge *et al.*, 1975). It has been suggested that the precursor of the CO_2 is a carbonate (Chang *et al.*, 1974 a, b) although a solar wind reaction product should not be ruled out (Chang *et al.*, 1974b), since some of the ^{13}C implanted into minerals is released by acid as $^{13}CO_2$ (Bibring *et al.*, 1974b).

Very little is known about the chemical state of the major portion of solar wind carbon. For example, it is not apparent whether carbon, which has come to rest in the radiation-damaged layer of target materials after ion implantation, exists as isolated atoms, amorphous carbon or graphite, or whether it is bonded to oxygen, silicon or metal ions. Answers to these questions may not be easily obtained.

Models for Solar Wind Carbon Accumulation at the Lunar Surface

By necessity, empirical models describing the accumulation and distribution of solar wind carbon in lunar soils are still at a primitive stage. In general, first-stage processes (initial implantation of carbon, the chemical reactions involved, etc.) are less well understood than second-stage events such as the cycling of exposed particles into agglutinates. Experiments may be designed to test hypotheses, but most investigators (Pillinger *et al.*, 1972; Bibring *et al.*, 1974b; Chang and Lennon, 1975; Davis *et al.*, 1975) recognize that often the conditions employed bear only a cursory resemblance to the actual conditions which prevail in the lunar environment. Two of the major difficulties (although not the only ones) which will need to be overcome are bombardment of targets by ion fluxes containing more than, one element and duplication of the agglutinate formation process. Because ion implantation equipment is expensive, heavily utilized and in the hands of investigators not having a lunar commitment, even relatively simple fundamental data such as projected ranges for carbon in various minerals and glasses are not yet available. However, using well-established theoretical treatments (Lindhard *et al.*, 1963) such information can be calculated with a reasonable degree of certainty (Zinner and Walker, 1975; Jull and Pillinger, 1977). Projected range depends on average atomic number and density of target material, therefore carbon of solar wind energy should penetrate to a variety of depths in different lunar materials. For example, calculations suggest that it may be implanted up to 50% further into glass of plagioclase composition than into ilmenite. Thus, different lunar particle types may reach different equilibrium saturation limits if exposed to the solar wind long enough for the loss of previously implanted solar wind atoms, due to sputtering, to become approximately equal to the rate of addition of fresh material (Jull and Pillinger, 1977). Variations of this sort may be able to account for the fluctuations in total carbon content observed between mature samples from different Apollo landing sites. So far, the systematic survey of the carbon content of size differentiated pure mineral fractions has been prevented by contamination problems and a lack of equipment sufficiently sensitive to study carbon in sub-milligram samples. However, by assuming some value for the integrated solar wind sputtering rate, and calculated solar wind carbon projected ranges, it is possible to estimate equilibrium saturation concentrations for any target material including an arbitrary substance of average lunar soil composition (Jull and Pillinger, 1977). Estimates of sputtering rate have been the cause of considerable debate (Zinner *et al.*, 1976) but low values *ca.* 0.02 to 0.04 Å/y (Zinner and Walker, 1975; McDonnell, 1977) seem to be becoming most generally accepted. Using such low values, it can be shown that the total abundance of carbon in a typical mature lunar soil could not be accumulated entirely from solar wind implantation, unless it had a surface area equivalent to spherical particles *ca.* 3.5 μm average radius or *ca.* 2 μm in radius if losses, which must occur during agglutination, are considered (Jull and Pillinger, 1977). Since the mean particle size of a mature soil is *ca.* 50–70 μm, the simple model calculations readily demonstrate the importance of the very fine grains with highly exposed surfaces as the constituent material of agglutinates and recycling processes including reagglutination (Basu, 1977).

The incorporation of surficial carbon into agglutinates is the basis of all models so far suggested for the redistribution of carbon and its compounds in lunar soil. DesMarais *et al.* (1973a, b) have interpreted their data on the distribution of carbon within different size fractions to indicate that the relocation of carbon by agglutination and comminution occurs more rapidly than the initial implantation. Thus, the *y*-axis intercept of plots of total carbon versus reciprocal grain size, when corrected for igneous C, give a measure of exposure age of the sample. Kerridge *et al.* (1974) have developed mathematical expressions for the accumulation and isotopic evolution of carbon on the lunar surface and used data from the Apollo 16 samples to impose constraints on various equilibrium constants derived. Although the model provides a reasonable fit with the data, one of the most serious drawbacks, which the authors themselves point out, is the long time which samples require to reach values for total C and $\delta^{13}C$ typical of mature soils.

Models which are pertinent to the compounds of carbon are even less well defined than those for total C. A major unresolved question concerns the mechanism of formation of the finely divided iron which is location of C_{hyd}. Pillinger *et al.* (1976b) have calculated that preferential sputtering could lead to the generation of sufficient metal provided accumulation took place at the surface of very fine grains. Such a model is in keeping with surface area limitations imposed by total C content considerations (Jull and Pillinger, 1977). If iron is to be formed by a preferential sputtering mechanism, then the agglutination of fine well-exposed particles is necessary, not only from the point of view of generating sufficient metal, but also to ensure that the metal formed at the very surface of grains on the lunar surface survives in the earth's atmosphere (Davis *et al.*, 1975; Pillinger *et al.*, 1976b). Other circumstantial evidence in favour of a preferential sputtering origin for metallic iron, is that the carbon concentration in volume% (rather than wt.%) for the metal is the same order of magnitude as the volume concentrations calculated for carbon atoms implanted into a silicate grain, which has reached equilibrium saturation (Gardiner *et al.*, 1977). A factor in the further development of the preferential sputtering model may be the precise measurement of the isotopic composition of C_{hyd} which is currently under investigation (see above). The model suggests that carbon isotopes fractionated by sputtering at the very surface of grains, in the region where iron might be being formed, could be expected to be substantially enriched in ^{13}C (Pillinger *et al.*, 1976b).

Conclusions

There now appears to be little doubt that the majority of carbon in lunar soil samples has been accumulated from the solar wind. The chemical nature of a small fraction of the carbon has been established with some certainty and reactions with other solar wind atoms and matrix elements are indicated. A large portion of the carbon has not been characterized. This is not a trivial problem and some innovation is likely to be necessary before further advances can be made. Because fractionation mechanisms are active, the original isotopic ratio of the solar wind has been obscured. Models relevant to the data which have been obtained are being developed but they suffer from a deficiency of fundamental information. Some input from solid state physicists could alleviate this problem whilst improved sensitivity and better control of contamination would allow relevant data to be gleaned from sample analysis. The retentivity of carbon via chemical compounds suggests that its study could provide important information concerning the exposure of grains to the space environment and of secular variations in solar wind parameters. As yet, there have been few attempts (Begemann and Heinzinger, 1968; Abell *et al.*, 1970b) to study solar wind carbon in gas-rich meteorites; however, it is hoped that there will be some activity in this area in the future.

Acknowledgements

The support of the Science Research Council (SG/D/00772) is gratefully acknowledged.

Thanks are due to my colleagues A. J. T. Jull, L. R. Gardiner and M. R. Woodcock for the use of their data and many stimulating discussions.

References

ABELL, P. I., EGLINTON, G., MAXWELL, J. R., PILLINGER, C. T. and HAYES, J. M. (1970a) Indigenous lunar methane and ethane. *Nature* **226**, 251–252.

ABELL, P. I., DRAFFAN, G. H., EGLINTON, G., HAYES, J. M., MAXWELL, J. R. and PILLINGER, C. T. (1970b) Organic analysis of the returned Apollo 11 lunar sample. *Proc. Apollo 11 Lunar Sci. Conf.* 1757–1773.

AGRELL, S. O., SCOON, J. H., MUIR, I. D., LONG, J. V. P., MCCONNELL, J. D. C. and PECKETT, A. (1970) Observations on the chemistry, mineralogy and petrology of some Apollo 11 lunar samples. *Proc. Apollo 11 Lunar Sci. Conf.* 93–128.

BAME, S. J., ASBRIDGE, J. R., HUNDHAUSEN, A. J. and MONTGOMERY, M. D. (1970) Solar wind ions: $^{56}Fe^{+8}$ to $^{56}Fe^{+12}$, $^{28}Si^{+7}$, $^{28}Si^{+8}$, $^{28}Si^{+9}$, and $^{16}O^{+6}$. *J. Geophys. Res.* **75**, 6360–6365.

BASU, A. (1977) Steady state, exposure age and an agglutinate growth model for lunar soils. *Proc. Lunar Sci. Conf. 8th.* 3617–3632.

BASU, A., DESMARAIS, D. J. HAYES, J. M. and MEINSCHEIN, W. G. (1975) Integrated investigation of the mixed origin of lunar sample 72161, II. *The Moon* **14**, 129–138.

BEGEMANN, F. and HEINZINGER, K. (1968) Content and isotopic composition of carbon in the light and dark portions of gas-rich chondrites. In *Meteorite Research* P. M. MILLMAN, ed. D. REIDEL, 87–82.

BIBRING, J. P., BURLINGAME, A. L., LANGEVIN, Y., MAURETTE, M. and WSZOLEK, P. C. (1974a) Simulation of lunar carbon chemistry: II lunar winds contribution. *Proc. Lunar Sci. Conf. 5th.* 1763–1784.

BIBRING, J. P., BURLINGAME, A. L., CHAUMONT, J., LANGEVIN, Y., MAURETTE, M. and WSZOLEK, P. C. (1974b) Simulation of lunar carbon chemistry: I solar wind contribution *Proc. Lunar Sci. Conf. 5th* 1747–1762.

BIBRING, J. P., BURLINGAME, A. L., LANGEVIN, Y., MAURETTE, M. and VASSENT, B. (1975) Solar wind and solar flare maturation of the lunar regolith. *Proc. Lunar Sci. Conf. 6th* 3471–3493.

BILLETOP, M. C. J., EGLINTON, G., GARDINER, L. R., JULL, A. J. T. and PILLINGER, C. T. (1976) Pyrolysis-acid dissolution studies of lunar carbon chemistry. *In Lunar Science* VII, 61–63 (The Lunar Science Institute, Houston).

CADOGAN, P. H. (1972) Studies in Lunar Carbon Chemistry. Ph.D. Thesis. University of Bristol.

CADOGAN, P. H., EGLINTON, G., FIRTH, J. N. M., MAXWELL, J. R., MAYS, B. J. and PILLINGER, C. T. (1972) Survey of lunar carbon compounds. The carbon chemistry of Apollo 11, 12, 14 and 15 samples. *Proc. Lunar Sci. Conf. 3rd* 2069–2090.

CADOGAN, P. H., EGLINTON, G., GOWAR, A. P., JULL, A. J. T., MAXWELL, J. R. and PILLINGER, C. T. (1973a) Location of methane and carbide in Apollo 11 and 16 fines. *Proc. Lunar Sci. Conf. 4th* 1493–1508.

CADOGAN, P. H., EGLINTON, G., MAXWELL, J. R. and PILLINGER, C. T. (1973b) Carbon chemistry of the lunar surface. *Nature* **231**, 29–31.

CADOGAN, P. H., EGLINTON, G., MAXWELL, J. R. and PILLINGER, C. T. (1973c) Distribution of methane and carbide in Apollo 11 fines. *Nature Phys. Sci.* **241**, 81–82.

CHANG, S., LAWLESS, J., ROMIEZ, R., KAPLAN, I. R., PETROWSKI, C., SAKAI, H. and SMITH, J. W. (1974a) Carbon, nitrogen and sulphur in lunar fines 15012 and 15013: Abundances, distribution and isotopic compositions. *Geochim. Cosmochim. Acta* **38**, 853–872.

CHANG, S., LENNON, K., GIBSON, E. K. (1974b) Abundances of C, N, H, He and S in Apollo 17 soils from Stations 3 and 4: Implications for solar wind exposure ages and regolith evolution. *Proc. Lunar Sci. Conf. 5th* 1785–1800.

CHANG, S. and LENNON, K. (1975) Implantation of carbon and nitrogen ions into lunar fines: trapping efficiencies and saturation concentrations. *Proc. Lunar Sci. Conf. 6th* 2179–2188.

CHANG, S., SMITH, J. W., KAPLAN, I., LAWLESS, J., KVENVOLDEN, K. A. and PONNAMPERUMA, C. (1970) Carbon compounds in lunar fines from Mare Tranquillitatis-IV. Evidence for oxides and carbides. *Proc. Apollo 11 Lunar Sci. Conf.* 1857–1869.

DAVIS, P. R., EGLINTON, G., PILLINGER, C. T., ERENTS, S. K., MCCRACKEN, G. M., NAZIRI, H. STEPHENSON, A. HOUSLEY, R. M. (1975) Studies of the formation of acid-hydrolysable carbon species and finely divided iron by simulated solar wind implantation. *Proc. Lunar Sci. Conf. 6th.* 2163–2178.

DESMARAIS, D. J., BASU, A., HAYES, J. M. and MEINSCHEIN, W. G. (1975) Evolution of carbon isotopes, agglutinates and the lunar regolith. *Proc. Lunar Sci. Conf. 6th* 2353–2373.

DESMARAIS, D. J., HAYES, J. M. and MEINSCHEIN, W. G. (1972) Pyrolysis study of carbon in lunar fines and rocks. *The Apollo 15 Lunar Samples*, pp. 294–297 (The Lunar Science Institute, Houston).

DESMARAIS, D. J., HAYES, J. M. and MEINSCHEIN, W. G. (1973a) The distribution in lunar soil of carbon released by pyrolysis. *Proc. Lunar Sci. Conf. 4th* 1543–1558.

DESMARAIS, D. J., HAYES, J. M. and MEINSCHEIN, W. G. (1973b) Accumulation of carbon in lunar soils. *Nature Phys. Sci.* **246**, 65–68.

DesMarais, D. J., Hayes, J. M. and Meinschein, W. G. (1973c) Techniques for the analysis of gases sequentially released from lunar samples. *Analytical Methods Developed for Application to Lunar Sample Analysis* ASTM STP 539. American Society for Testing and Materials.

DesMarais, D. J., Hayes, J. M. and Meinschein, W. G. (1974) The distribution in lunar soil of hydrogen released by pyrolyses. *Proc. Lunar Sci. Conf. 5th* 1811–1822.

Eglinton, G., Maxwell, J. R. and Pillinger, C. T. (1974) Carbon chemistry Carbon chemistry of the Apollo lunar samples. *Topics in Current Chemistry* **44**, 83–113, Springer-Verlag.

Epstein, S. and Taylor, H. P. (1970) The concentration and isotopic composition of hydrogen, carbon and silicon in Apollo 11 lunar rocks and minerals. *Proc. Apollo 11 Lunar Sci. Conf.* 1085–1096.

Epstein, S. and Taylor, H. P. (1971) O^{18}/O^{16}, Si^{30}/Si^{28}, D/H and C^{13}/C^{12} ratios in lunar samples. *Proc. Lunar Sci. Conf. 2nd* 1421–1441.

Epstein, S. and Taylor, H. P. (1972) O^{18}/O^{16}, Si^{30}/Si^{28}, C^{13}/C^{12} and D/H studies of Apollo 14 and 15 samples. *Proc. Lunar Sci. Conf. 3rd* 1429–1454.

Epstein, S. and Taylor, H. P. (1973) The isotopic composition and concentration of water, hydrogen and carbon in some Apollo 15 and 16 soils and in the Apollo 17 orange soil. *Proc. Lunar Sci. Conf. 4th* 1559–1575.

Epstein, S. and Taylor, H. P. (1975) Investigation of the carbon, hydrogen, oxygen and silicon isotope and concentration relationships on the grain surfaces of a variety of lunar soils and in some Apollo 15 and 16 core samples. *Proc. Lunar Sci. Conf. 6th* 1771–1798.

Filleux, C., Tombrello, T. A. and Burnett, D. S. (1977) Direct measurement of surface carbon concentrations. *Lunar Science* VIII, 296–298.

Fox, S. W. and Harada, K. (1973) Accumulated analyses of amino acid percusors in returned lunar samples. *Proc. Lunar Sci. Conf. 4th* 2241–2248.

Gardiner, L. R., Woodcock, M. R., Pillinger, C. T. and Stephenson, A. (1977) Carbon chemistry and magnetic properties of bulk and agglutinate size fractions from soil 15601. *Proc. Lunar Sci. Conf. 8th* 2817–2839.

Gibson, E. K., and Johnson, S. M. (1971) Thermal analysis-inorganic gas release studies of lunar samples. *Proc. Second Lunar Sci. Conf. 2nd* 1351–1366.

Gibson, E. K. and Moore, G. W. (1973) Carbon and sulphur distributions and abundances in lunar fines. *Proc. Lunar Sci. Conf. 4th* 1577–1586.

Hapke, B. (1977) The albedo of lunar soils. In *Lunar Science* VIII, 398–400 (The Lunar Science Institute, Houston).

Hayes, J. M. Extralunar sources for carbon on the Moon. *Space Life Sci.* **3**, 474–483.

Holland, P. T., Simoneit, B. R., Wszolek, P. C. and Burlingame, A. L. (1972) Study of carbon compounds in Apollo 12 and 14 lunar samples. *Space Life Sci.* **3**, 551–561.

Housley, R. M., Grant, R. W. and Abdel-Gawad, M. (1972) Study of excess Fe metal in the lunar fines by magnetic separation, Mössbauer spectroscopy, and microscopic examination. *Proc. Lunar Sci. Conf. 3rd* 1065–1076.

Housley, R. M., Grant, R. W. and Paton, N. E. (1973) Origin and characteristics of excess Fe metal in lunar glass welded aggregates. *Proc. Lunar Sci. Conf. 4th* 2737–2749.

Housley, R. M. and Grant, R. W. (1976) ESCA studies of the surface chemistry of lunar fines. *Proc. Lunar Sci. Conf. 7th* 881–889.

Jull, A. J. T., Davis, P. R., Eglinton, G., Gardiner, L. R., Maxwell, J. R., Pillinger, C. T., Biggar, G. M., Humphries, D. and Batts, B. D. (1975) Simulation of lunar processes: II Redistribution of carbon in the lunar regolith during meteorite impact. In *Lunar Science* VI, 457–459 (The Lunar Science Institute, Houston).

Jull, A. J. T., Eglinton, G., Pillinger, C. T., Batts, B. D. and Biggar, G. M. (1976) Identity of lunar hydrolysable carbon. *Nature* **262**, 566–567.

Jull, A. J. T. and Pillinger, C. T. (1977) The effects of sputtering on solar wind atom accumulation. *Proc. Lunar Sci. Conf. 8th* 3817–3833.

Kaplan, I. R., Smith, J. W. and Ruth, E. (1970) Carbon and sulphur concentration and isotopic composition in Apollo 11 lunar samples. *Proc. Apollo 11 Lunar Sci. Conf.* 1317–1329.

Kaplan, I. R. and Petrowski, C. (1971) Carbon and sulphur isotope studies on Apollo 12 lunar samples. *Proc. Lunar Sci. Conf. 2nd* 1397–1406.

Kaplan, I. R., Kerridge, J. F. and Petrowski, C. (1976) Light element geo-chemistry of the Apollo 15 site. *Proc. Lunar Sci. Conf. 7th* 481–492.

Kerridge, J. F., Kaplan, I. R. and Lesley, F. D. (1974) Accumulation and isotopic evolution of carbon on the lunar surface. *Proc. Lunar Sci. Conf. 5th* 1855–1868.

Kerridge, J. F., Kaplan, I. R., Petrowski, C. and Chang, S. (1975) Light element geochemistry of the Apollo 16 site. *Geochim. Cosmochim. Acta* **39**, 137–162.

Linhard, J., Scharff, M. and Schiott, H. E. (1963) Range concepts and heavy ion ranges (notes on atomic collisions). *Kgl. Danske Vid. Seisk., Math.-Fys. Medd.* **33** (14), 1–42.

McDonnell, J. A. M. (1977) Accretionary particles: production and equilibrium on 12054. *Lunar Science* VIII, 643–645 (The Lunar Science Institute, Houston).

Moore, C. B. (1975) A check list of total carbon and sulphur analyses made by the Arizona State University lunar sample analysis group between October 1969 and May 1974. Center for Meteorite Studies, Arizona State University, no. 94.

MOORE, C. B., GIBSON, E. K., LARIMER, J. W., LEWIS, C. F. and NICHIPORUK, W. (1970) Total carbon and nitrogen abundances in Apollo 11 lunar samples and selected achondrites and basalts. *Proc. Apollo 11 Lunar Sci. Conf.* 1375–1382.

MÜLLER, O. (1979) Solar wind nitrogen and indigenous nitrogen in lunar material. This volume, pp. 47–59.

ORO, J., UPDEGROVE, W. S. and GIBERT, J. (1970) Organogenic elements and compounds in type C and D lunar samples from Apollo 11. *Proc. Apollo 11 Lunar Sci. Conf.* 1901–1920.

PILLINGER, C. T., CADOGAN, P. H., EGLINTON, G., MAXWELL, J. R., MAYS, B. J., GRANT, W. A. and NOBES, M. J. (1972) Simulation study of lunar carbon chemistry. *Nature Phys. Sci.* **235**, 108–109.

PILLINGER, C. T., BATTS, B. D., EGLINTON, G., GOWAR, A. P., JULL, A. J. T. and MAXWELL, J. R. (1973) Formation of lunar carbide from lunar iron silicates. *Nature Phys. Sci.* **245**, 3–5.

PILLINGER, C. T., DAVIS, P. R., EGLINTON, G., GOWAR, A. P., JULL, A. J. T. and MAXWELL, J. R. (1974) The association between carbide and finely divided metallic iron in lunar fines. *Proc. Lunar Sci. Conf. 5th* 1949–1961.

PILLINGER, C. T., DAVIS, P. R., GARDINER, L. R., NAZIRI, H. and CHAMPNESS, P. E. (1975) High voltage electron microscope observations of finely divided iron droplets in glassy agglutinates. *Proc. Lunar Sci. Conf. 6th* 2343–2352.

PILLINGER, C. T. JULL, A. J. T. EGLINTON, G. and EPSTEIN, S. (1976a) A model for the isotopic evolution of light elements in lunar soil. *Lunar Science* VII, 703–705 (the Lunar Science Institute, Houston).

PILLINGER, C. T., GARDINER, L. R. and JULL, A. J. T. (1976b) Preferential sputtering as a method of producing metallic iron inducing major element fractionation and trace element enrichment. *Earth Planet. Sci. Lett.* **33**, 289–299.

PILLINGER, C. T., EGLINTON, G., GOWAR, A. P. and JULL, A. J. T. (1977) Carbon chemistry of the Luna 16 and 20 samples. *Phil. Trans. Roy. Soc. Lond.* A, **284**, 145–150.

SMITH, J. W., KAPLAN, I. R. and PETROWSKI, C. (1973) Carbon, nitrogen, sulphur, helium, hydrogen and metallic iron in Apollo 15 drill stem fines. *Proc. Lunar Sci. Conf. 4th* 1651–1656.

STEPHENSON, A. (1970) Single domain grain distributions. I. A method for the determination of single domain grain distributions. *Phys. Earth Planet. Interiors* **4**, 353–360.

TSAY, F. D. and LIVE, D. H. (1974) Ferromagnetic resonance studies of thermal effects on lunar metallic Fe phases. *Proc. Lunar Sci. Conf. 5th* 2737–2746.

WSZOLEK, P. C., SIMONEIT, B. R. and BURLINGAME, A. L. (1973) Studies of magnetic fines and volatile-rich soils: Possible meteoritic and volcanic contributions to lunar carbon and light element chemistry. *Proc. Lunar Sci. Conf. 4th* 1693–1706.

YIN, L. I., GHOSE, S. and ADLER, I. (1972) Investigation of a possible solar wind darkening of the lunar surface by photoelectron spectroscopy. *J. Geophys. Res.* **77**, 1360–1367.

YIN, L. I., TSANG, T. and ADLER, I. (1975) ESCA studies on solar wind reduction mechanisms. *Proc. Lunar Sci. Conf. 6th* 3277–3284.

YIN, L. I., TSANG, T. and ADLER, I. (1976) On the ion bombardment reduction mechanism. *Proc. Lunar Sci. Conf. 7th* 891–900.

ZINNER, E. and WALKER, R. M. (1975) Ion probe studies of artificially implanted ions in lunar samples. *Proc. Lunar Sci. Conf. 6th* 3601–3617.

ZINNER, E., WALKER, R. M., CHAUMONT, J. and DRAN, J. C. (1976) Ion probe analysis of artifically implanted ions in terrestrial samples and surface enhanced ions in lunar sample 76215, 77. *Proc. Lunar Sci. Conf. 7th* 953–984.

Search for ^{59}Ni of Extraterrestrial Origin in Deep Sea Sediments

KAZUO YAMAKOSHI and SHOHEI YANAGITA

Cosmic Ray Laboratory, University of Tokyo, Tanashi, Tokyo 188, Japan

Abstract

Ni-59 was measured with an extremely low background X-ray spectrometer in 0.5 g of nickel which was isolated chemically from deep sea sediments. The obtained specific activity of Ni-59 in the sediments was (0.26 $\pm$0.14) dpm/g Ni. However, the Ni-59 activity induced by proton and alpha particle irradiation in outer space with a few assumptions can be estimated as 13.0 dpm/g Ni in cosmic dusts. The maximum contribution of extraterrestrial nickel to the nickel component in deep sea sediments is $\sim$2%.

1. Introduction

Many investigators have studied in various ways the accretion of extraterrestrial matter on the earth. One plausible approach is to find the changes of chemical compositions of terrestrial samples by accretion of cosmic matter. Another approach is to detect cosmic-ray-produced nuclides borne in accreted cosmic matter on the earth.

There are two promising sources for relatively large quantities of accreted cosmic matter; the polar ice sheets of Antarctica and Greenland and deep sea sediments. They are isolated from sources of air- and water-borne terrestrial contaminants.

McCorkell *et al.* (1967) filtered a large volume of Greenland glacier ice and they calculated the accretion rate of extraterrestrial matter from Al-26. R. Bibron *et al.* (1974) collected snow samples at East Antarctica and determined Mn-53 activity using the neutron-activation technique.

In 1950 Pettersson and Rotschi noticed an anomalously large nickel content in pelagic deep sea sediments and suggested an extraterrestrial origin for this. However, their suggestion received several objections (Smales and Wiseman, 1955; Smales *et al.*, 1957), and some estimates on the contribution of extraterrestrial nickel have been published (Yamakoshi and Tazawa, 1971). However, no decisive interpretation has been completed until now.

Barker and Anders (1968), based on iridium and osmium contents in deep sea sediments, estimated the accretion rate of extraterrestrial matter. Merrihue (1964) found He-3 and Ar isotope anomalies in the magnetic fraction of deep sea sediments.

Detections of the cosmic-ray-produced radionuclide Al-26 in deep sea sediments have been carried out by Amin *et al.* (1966), Wasson *et al.* (1967), Tanaka *et al.* (1968) and Reyss *et al.* (1976). The Al-26 data of Reyss *et al.* can be interpreted as mostly produced in the atmosphere.

Mn-53 and Ni-59 in deep sea sediments, produced mainly from iron contained in cosmic dusts by solar proton and alpha particle irradiation, have been studied only preliminarily (Hasegawa *et al.*, 1968). Our purpose is to investigate the origin of excess nickel in sediments and to recognize the accretion of extraterrestrial matter via this Ni-59 study.

In this paper we also discuss the experimental data and the expected values calculated and based on the newly obtained alpha particle-induced cross-section; Fe (α, X) Ni-59 (Yanagita, Yamakoshi and Gensho, 1978.

2. Sample Description

The deep sea sediments were obtained during the cruise KH-67-5 of the *Hakuho-Maru*† at Station 17 (0°01.3′S, 150°04.2′E) from a depth of 5245 metres. The depth of the dredged red clay below the surface layer of the sediment is unknown, but it could rarely have exceeded 15 cm because of the nature of the sampling dredger. Unfortunately there are no data on the sedimentation rate in this region. The nickel content of this dredged red clay was observed as 230 ppm.‡ Prior to any chemical procedures, the red clay sample was heated to 110°C in a constant-temperature oven for a day and then crushed into fine grains using a pottery mortar.

3. Chemical Procedures

The red clay sample of 3.14 kg (dry weight) was dissolved in HF and HNO_3 and evaporated some degree of dryness. The residue was fumed to dryness with conc. H_2SO_4 and then leached with conc. HCl.

The hydroxide aggregate was precipitated from the solution with NaOH. Ferric iron was extracted with isopropyl ether. Titanium and manganese were removed by hydroxide precipitation with excess NH_4OH, while nickel remained in the solution. The nickel was then purified with anion-exchange resin and dimethylglyoxime. The extracted nickel was electroplated onto an area of 312 cm^2 in a pair of semicylindrical sample holders, which were made of OFHC copper (oxygen free, high conductive copper). The amount of nickel electroplated onto the sample holders was estimated at (455 ± 21) mg, using the atomic absorption method.

4. Counting

The detector assembly used in this work was reported by Takahashi *et al.* (1975). A proportional counter of 500 mm in length and 100 mm in diameter was also made of OFHC. The resolution was approximately 18% at FWHM of Ni-59 peak. The sample holders were semicylindrical formed and also made of OFHC. The two semicylindrical holders were combined to form a normal cylinder which slide into the body of the counter keeping contact with its inner surface, onto which the nickel had been electroplated. The signal pulses were fed into a 1024 channel pulse height analyzer. The main proportional counter was surrounded with a multianode GM guard counter system. The gain of the system was fairly stable, a drift of more than 4∼5 channels being rarely observed.

For the background counting ∼450 mg nickel (reagent grade) was used as a blank sample, which was electroplated onto the sample holders. The extracted nickel was electroplated onto

†*Hakuho-Maru:* a researching vessel of the Ocean Research Institute, University of Tokyo.

‡The following elements were found in the sediments as here: [Fe] = 6.01%, [Mn] = 0.124%, [Co] = 411 ppm and [Si] = 23.02%.

the sample holders and the nickel thickness was 1.45 (mg/cm^2). The self absorption loss was estimated as approximately 16%.

5. Estimation of Production Rate of Ni-59 in Cosmic Dusts

In general the following processes have been considered to contribute to the production of Ni-59 in cosmic matter:

(a) proton reactions, Ni-60 (p, pn), Ni-60 (p, 2n) Cu-59→Ni-59 and Co-59 (p, n);
(b) alpha particle reactions, Fe-56 (α, n), Fe-57 (α, 2n) and Fe-58 (α, 3n);
(c) spallation of Cu-63, etc.;
(d) radiative capture, Ni-58 (n, γ).

For larger bodies such as meteorites, the main contribution comes from the processes (c) and (d). Especially the radiative capture process is dominant in stony meteorites. But for smaller dust particles, the situation is very different. The contribution from the processes (a) and (b) is dominant and these are of comparable magnitude.

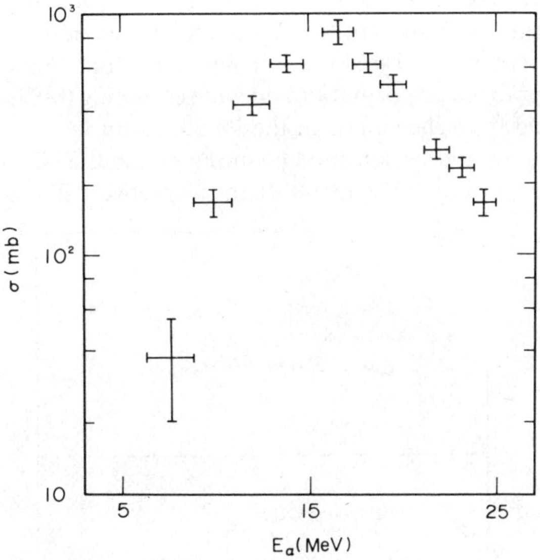

FIG. 1. Production cross-section of ^{59}Ni from alpha particle bombardment of natural iron (Yanagita *et al.*, 1978).

The recent excitation function (Yanagita *et al.*, 1978) of Ni-59 production from natural iron bombarded by alpha particles with energies ranging from 7 ~ 25 MeV is shown in Fig. 1. The solar proton flux was estimated from the studies of Mn-53 and Al-26 activity profiles in lunar samples (Nishiizumi *et al.*, 1976). At a value of $R = 100$ MV for the characteristic rigidity, the solar proton flux is

$$J(>10 \text{ MeV}, 4\pi) = 70 \text{ (protons/cm}^2\text{, s)}.$$

The ratio of proton to alpha particle flux used in this work is 20 ± 4 (van Hollenbeke, 1975). According to Lal and Venkatavaradan (1967), proton-induced Ni-59 activity can be recalculated using the above mentioned solar proton flux, and is 3.05 dpm/g Ni. The alpha particle induced Ni-59 activity is estimated at 9.90 dpm/g Ni using newly obtained cross-section (Yanagita *et al.*, 1978). The total value of Ni-59 activity induced by proton and alpha particles in cosmic dusts is approximately 13.0 dpm/g Ni or 0.175 dpm/g dust.

In the above calculation, it is assumed that the Ni-59 activity in cosmic dusts had been in secular equilibrium for a long time at 1 A.U. from the sun. We have also assumed 25% Fe, 1.35% Ni and 520 ppm Co as chemical abundances of cosmic dusts.

6. Results and Interim Conclusions

The observed counts in the gross, background and net spectra were smoothed in every five channels and are plotted in Fig. 2. The counting times were 45,390 min for the gross countings and 11,185 min for the background counting. When the background counts were subtracted from the gross counts, some broad peaks remained, which can probably be attributed to unknown radioimpurities in the counting sample. Also the Ni-59 peak was not clearly identifiable around 6.9 keV. Had the net spectrum in the wide energy range been obtained, then the interpolated curve attributed to unknown radioimpurities could have been drawn for the Ni-59 peak and thereby the net amount of Ni-59 could have been obtained. However, in this study only the spectrum between 5 ~ 9 keV was obtained. Therefore, "the mean level" counts, which were calculated from the counts of the both sides but excluding the Ni-59 region in the net spectrum, were subtracted from the counts in the Ni-59 region.

From the results, the gross and background counting rate in the Ni-59 region (6.9 ± 0.62 keV) were (0.208 ± 0.002) cpm and (0.128 ± 0.002) cpm, respectively. The mean level counts in

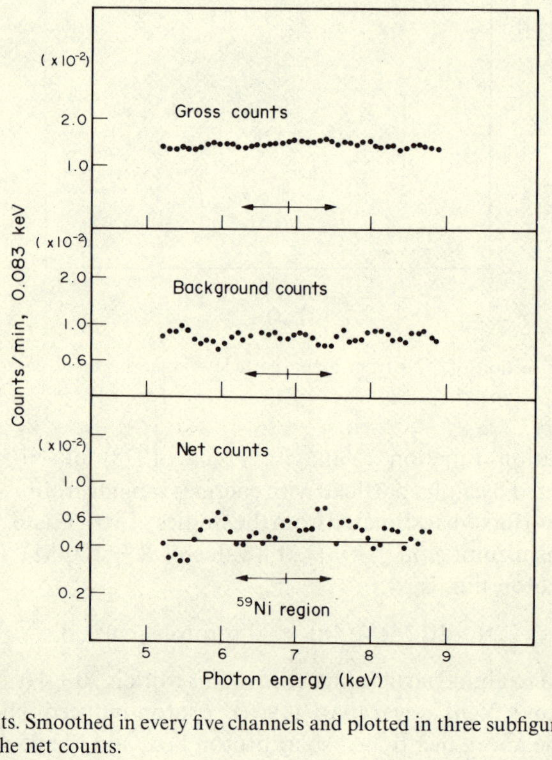

FIG. 2. The obtained results. Smoothed in every five channels and plotted in three subfigures; the gross counts, the background counts and the net counts.

the Ni-59 region were obtained as (0.072 ± 0.014) cpm. The error includes one standard deviation of counting statistics and the scattering from the mean value.

The Ni-59 counting rate was obtained as $(8.9 \pm 4.6) \times 10^{-3}$ cpm in the nickel sample. This corresponds to the specific activity of Ni-59 as (0.26 ± 0.14) dpm/g Ni, using a counting efficiency of 7.4% including 2π geometry, fluorescent yield and self-absorption loss. Comparing the obtained results with the expected value calculated in Section 5, we note that the contribution of accreted nickel of extraterrestrial origin to the nickel component in deep sea sediments is $\sim 2\%$ at the maximum.

The following problems remain: (a) to reduce the background counting rate of the detector system using the pulse shape discrimination technique (Lewyn, 1970); (b) to eliminate unknown impurities from the counting sample (recycling process); and (c) to obtain information to correct Ni-59 decayed fractions in the dredged sediments. For problem (c), the determination of Be-10 and Mn-53 in the same sediments is in progress.

7. Acknowledgements

The authors wish to express their thanks to Drs. S. Tanaka, M. Imamura, T. Inoue, T. Shimamura and K. Nogami for their helpful information and discussions. The authors are also indebted to Professor H. Hasegawa and Dr. Y. Tazawa of Kyoto University for their continuous encouragements. Many thanks are due to Drs. H. Takahashi and H. Mabuchi for kindly permitting us to use their low-level proportional counter system at Mt. Nokogiriyama. The authors are also grateful to Professor N. Nasu, Director of the deep sea expedition with the *R.V. Hakuho-Maru*, for the samples used in this work.

References

AMIN, B. S., KHARKAR, D. P. and LAL, D. (1966) Cosmogenic ^{10}Be and ^{26}Al in marine sediments. *Deep Sea Res.* **13**, 805.

BARKER, J. L., Jr. and ANDERS, E. (1968) Accretion rate of cosmic matter from iridium and osmium contents of deep sea sediments. *Geochim. Cosmochim. Acta* **32**, 627.

BIBRON, R., CHESSELET, R., CROZAZ, G., LEGER, G., MENNESSIER, J. P. and PICCIOTTO, E. (1974) Extraterrestrial ^{53}Mn in Antarctic ice. *Earth Planet. Sci. Lett.* **21**, 109.

HASEGAWA, H., YAMAKOSHI, K., NOMA, M. and MAIHARA, T. (1968) Cosmic ray produced ^{59}Ni and its astrophysical and geophysical implications. *Can. J. Phys.* **461**, S 930.

HOLLENBEKE, M. A. I. van (1975) Relative abundance of proton to helium nuclei in solar cosmic ray events. *Proc. Int. Conf. Cosmic Rays* **5**, 1563.

LAEVASTU, T. and MELLIS, O. (1955) Extraterrestrial material in deep sea deposits. *Trans. Amer. Geophys. Union* **36**, 385.

LAL, D. and VENKATAVARADAN, S. (1967) Activation of cosmic dust by cosmic ray particles. *Earth Planet. Sci. Lett.* **3**, 299.

LEWYN, L. L. (1970) Pulse shape discriminator for X-ray proportional counter background reduction. *Nucl. Instr. Meth.* **82**, 138.

MCCORKELL, R., FIREMAN, E. L. and LANGWAY, C. C. Jr. (1967) Aluminium-26 and beryllium-10 in Greenland ice. *Science* **158**, 1690.

MERRIHUE, C. (1964) Rare gas evidence for cosmic dust in modern Pacific red clay. *Ann. New York Acad. Sci.* **119**, 351.

NISHIIZUMI, K., IMAMURA, M., HONDA, M., RUSS, G. P. III, KOHL, C. P. and ARNOLD, J. R. (1976) ^{53}Mn in the Apollo 15 and 16 drill stems: Evidence for surface mixing. *Proc. Lunar Sci. Conf. 7th* 41.

PETTERSSON, H. and ROTSCHI, H. (1950) Nickel contents of deep sea deposits. *Nature* **166**, 308.

REYSS, J. L., YOKOYAMA, Y. and TANAKA, S. (1976) Aluminium-26 in deep sea sediments. *Science* **193**, 1119.

SMALES, A. A. and WISEMAN, I. D. H. (1955) Origin of nickel in deep sea deposits. *Nature* **175**, 464.

SMALES, A. A., MAPPER, D. and WOOD, A. J. (1957) The determination, by radioactivation, of small quantities of nickel, cobalt and copper in rocks, marine sediments and meteorites. *Analyst* **82**, 75.

TAKAHASHI, H., MABUCHI, H., HAMAGUCHI, H. and SHIMOKOSHI, F. (1975) An extremely low level X-ray spectrometer. *Nucl. Instrum. Meth.* **125**, 293.

TANAKA, S., SAKAMOTO, K., TAKAGI, J. and TSUCHIMOTO, M. (1968) Al-26 and Be-10 in marine sediments. *Science* **160**, 1348.

WASSON, J. T., ALDER, B. and OESCHGER, H. (1967) Aluminium-26 in Pacific sediments; implications. *Science* **155**, 446.

YAMAKOSHI, K. and TAZAWA, Y. (1971) Deposition of extraterrestrial nickel in marine sediments. *Nature* **233**, 542.

YANAGITA, S., YAMAKOSHI, K. and GENSHO, R. (1978) Production of ^{59}Ni from alpha bombardment of iron. *Nucl. Phys.* A **303**, 254.

Solar Abundances. A New Table (October 1976)

B. E. J. PAGEL

Royal Greenwich Observatory, Sussex, England

with a preamble by L. H. AHRENS

Preamble

Professor B. E. J. Pagel has kindly prepared a table of solar abundances. He states as follows:

"Most are derived from observations of photospheric absorption lines using a model of the solar atmosphere to predict the profiles and total intensities of lines as a function of the abundance of the element relative to hydrogen, since hydrogen forms the continuous background. The precision varies widely owing to varying degree of availability of suitable (atomic or molecular) solar lines and laboratory or theoretical data, mainly oscillator strengths.

"Some elements are undetectable in the photospheric spectrum, but can be seen in emission from prominences and from the ultra-violet and visible emission spectra of the corona or detected in the form of energetic nuclei from solar flares. Above 15 MeV per nucleon, the relative abundances of He, C, N, O, Ne, Mg, Si and Ar are fairly constant and could be representative of photospheric abundances; these data have been used for He, Ne and Ar together with prominence data for He and Coronal soft X-ray data for Ne and Ar. The numbers for F and Cl are based on observations of absorption bands of HF and HCL in sunspot spectra. Data from the solar wind still seem somewhat difficult to interpret and have not been used.

"A more detailed description, with a full set of references, can be found in the article by Ross and Aller,* but I have updated or otherwise modified some of their numbers in the light of a discussion held by Commission XII of the IAU in September 1976."

Introduction to the Table

The accompanying list gives logarithm of abundance relative to H = 12 or Si = 6.

Where two decimal places are given, the estimated error is ± 0.1 dex,

Where one decimal place is given, the estimated error is ± 0.2 dex,

One colon indicates that the estimated error is ± 0.3 dex,

Two colons indicate that the estimated error is $\geq \pm$ 0.3 dex.

A full list of references can be found in J. E. Ross and L. H. Aller, *Science*, **191** 1223 (1976), but some of the numbers and estimated errors are different from those in that list.

*See reference on Table.

	SOURCE			
	Element (1)	Photospheric absorption lines (2)	Other (3)	Abundance Normalised to Si=6 (4)
1.	H	(12.00)		10.40
2.	He		10.8	
3.	Li	1.0		−0.6
4.	Be	1.2		−0.4
5.	B	2.4:		0.8:
6.	C	8.66		7.06
7.	N	7.93		6.33
8.	O	8.88		7.28
9.	F		4.6:	3.0:
10.	Ne		7.8	6.2
11.	Na	6.29		4.69
12.	Mg	7.6		6.0
13.	Al	6.4		4.8
14.	Si	7.60		(6.00)
15.	P	5.5		3.9
16.	S	7.2		5.6
17.	Cl		5.4::	3.8::
18.	Ar		6.4:	4.8:
19.	K	5.15		3.55
20.	Ca	6.30		4.70
21.	Sc	3.0		1.4
22.	Ti	4.8		3.2
23.	V	4.0		2.4
24.	Cr	5.7		4.1
25.	Mn	5.4		3.8
26.	Fe	7.5		5.9
27.	Co	5.0		3.4
28.	Ni	6.3		4.7
29.	Cu	4.1		2.5
30.	Zn	4.4		2.8
31.	Ga	2.8		1.2
32.	Ge	3.4		1.8
37.	Rb	2.6		1.0
38.	Sr	2.9		1.3
39.	Y	2.1::		0.5::
40.	Zr	2.8::		1.2::
41.	Nb	2.0::		0.4::
42.	Mo	2.2::		0.6::
44.	Ru	1.8::		0.2::
45.	Rh	1.5::		−0.1::
46.	Pd	1.5::		−0.1::
47.	Ag	0.9		−0.7:
48.	Cd	2.0		0.4:
49.	In	1.7	1.5	0.0:
50.	Sn	2.0::		0.4::
51.	Sb	1.0::		−0.6::
55.	Cs		<1.9	<0.3
56.	Ba	2.1		0.5
57.	La	1.1		−0.5:
58.	Ce	1.6		0.0
59.	Pr	0.8		−0.8
60.	Nd	1.2:		−0.4:
62.	Sm	0.7:		−0.9:
63.	Eu	0.7::		−0.9::
64.	Gd	1.1::		−0.5::
66.	Dy	1.1::		−0.5::
68.	Er	0.8::		−0.8::
69.	Tm	0.3		−1.3
70.	Yb	0.8::		−0.8::
71.	Lu	0.8:		−0.8:
72.	Hf	0.9:		−0.7:
74.	W	0.8:		−0.8:
75.	Re	<−0.3:		<−1.9
76.	Os	0.7::		−0.9::
77.	Ir	2.2:		0.6:
78.	Pt	1.8		0.2
79.	Au	0.1:		−1.5:
80.	Hg	<2.1		<0.5
81.	Tl	0.9:		−0.7:
82.	Pb	1.9		0.3
83.	Bi	<1.9		<0.3
90.	Th	0.2		−1.4
92.	U	<0.6		<−1.0

SECTION II

PLANETOLOGY

The Chemical Evolution of the Galaxy and Isotopic Ratios in the Solar System

JEAN AUDOUZE

Département de Radio Astronomie, Observatoire de Meudon
and
Laboratoire René Bernas du C.S.N.S.M., Orsay, France

Abstract

The relationship between the present knowledge of the isotopic ratios in the solar system and that of the chemical evolution of the galaxies is tentatively described here: without the wealth of information regarding the chemical composition of the solar system one would have been unable to develop the current nucleosynthetic theories and the age of the Galaxy would not be determined by cosmochemistry techniques. The models of chemical evolution are necessary for an understanding of how the composition of the solar system material has been achieved.

The isotopic ratios of the solar system (earth, moon, meteorites, planets) have been extensively studied by geochemists and more generally by cosmochemists. From such analyses one has been able to determine, for instance, the age of the solar system (cosmochronology) or to disentangle the complicated chemical or physical processes occurring during the condensation and the formation of the solar system.

The purpose of this paper is to relate the present knowledge of these isotopic ratios in the solar system to the more general questions of the understanding of (1) the nucleosynthetic processes explaining the formation of the various chemical elements and (2) of the chemical (or nuclear) evolution of the whole Galaxy. In other words for a given element or isotope one tries to determine where it is formed (nucleosynthesis) and how its abundance evolves with time in the different regions of the Galaxy such as the solar neighborhood and the galactic center.

It is important to note at this point that the solar system is indeed the only sample of material for which the isotopic ratios could have been determined for all the ninety observed elements. It would not have been possible to set up the so-called tables of "cosmic abundances" without the careful isotopic determinations in carbonaceous chondrites. Although there is an increasing number (especially now) of isotopic ratios which have been determined in the interstellar medium (millimetric or UV techniques), the galactic cosmic rays, or in some cold stars (IR techniques) (Table 1), there are a large number of isotopes which have only been observed in the terrestrial, lunar and meteoritical samples, and for which special nucleosynthetic processes have been proposed. This is the case, for example, for the *p*-process elements which are the isotopes of heavy elements (from Se to Hg) with a nucleus relatively rich in protons. Unlike *s*- or *r*-process elements, the *p*-process elements cannot be formed by neutron absorption. This clearly shows the importance of the

TABLE 1. PRESENT DETERMINATIONS OF THE ISOTOPIC RATIOS OUTSIDE THE SOLAR SYSTEM

	Stars	Interstellar medium	Galactic cosmic rays
D/H	+	++	++
$^3He/^4He$	+	+	++
$^7Li/^6Li$	+		++
$^{11}B/^{10}B$			++
$^{12}C/^{13}C$	++	++	++
$^{14}N/^{15}N$	+	++	++
$^{16}O/^{17}O/^{18}O$	+	++	
$^{32}S/^{33}S/^{34}S$		++	

+ = Upper (or lower) limit. ++ = Observed value.

determinations of isotopic ratios in the solar system for the knowledge of the nucleosynthesis and the chemical evolution of the Galaxy.

Before the presentation of what is meant in the galactic evolution studies, it may be worth recalling what are the main nucleosynthetic processes which have been proposed to account for the observed element abundances. The current nucleosynthetic theories can be summarized as follows:

1. For the lightest elements D, ^3He, ^4He and possibly ^7Li, the most likely formation process is now considered to be nucleosynthesis at high temperature ($T > 10^{10}$ K) and high density ($\rho > 10^6$–10^{10} g cm^{-3}) occurring during the early phases of an expanding universe (see the calculations of big bang nucleosynthesis performed, for example, by Wagoner, 1973).
2. The light elements Li, Be and B (with the possible exception of ^7Li) have a very low abundance in comparison with that of their neighbors (Li Be B/CNO $\sim 10^{-5}$) and can be synthetized by spallation reactions induced by galactic cosmic rays bombarding the interstellar atoms (see, for example, Meneguzzi *et al.*, 1971).
3. The elements of atomic mass ($A > 12$) are produced in the nucleosynthetic processes occurring inside the stars. Such processes also explain physical properties such as their luminosity or the time scale of their evolution: the fusion thermonuclear reactions which are responsible for the formation of the $12 < A < 56$ elements† are also the main sources of the stellar energy and luminosity. One can distinguish between the nucleosynthetic processes occurring during the normal phases of stellar evolution and the explosive nucleosynthesis occurring in the late phases of stellar evolution and generally leading to the end or the death of the considered objects (see, for example, the supernova phenomenon—notice that the nova phenomenon is an explosive process but which is generally recurrent and does not constitute the end of the evolution of the considered object). Examples of quiet processes are, the CNO-cycles which explain the formation of ^{14}N and ^{13}C from ^{12}C and ^{16}O, the quiet hydrogen, helium (carbon, oxygen) burning which accounts for the formation of a part of ^4He, ^{12}C, ^{16}O, ^{20}Ne . . . up to the iron peak and the *s*-process nucleosynthesis (slow neutron absorption). The explosive nucleosynthesis is important for the formation of the rare isotopes of the $A < 56$ elements, the processes leading to the iron peak, and the *p*- and *r*-process nucleosynthesis (which are respectively the fusion or photodisintegration processes

† He is indeed the main product of the stellar nucleosynthesis but the stellar production is only about one fourth of the primordial production.

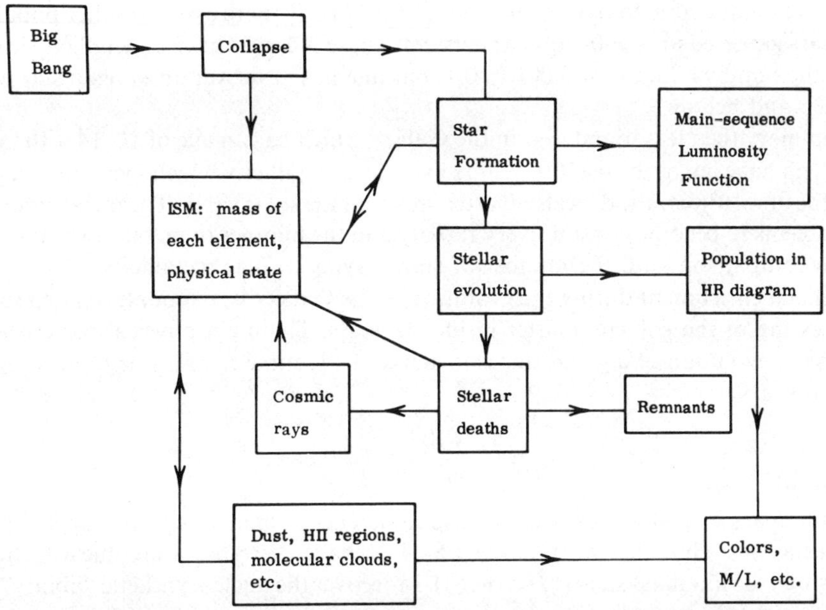

FIG. 1. Flowchart, as a guide to the relations between some processes and parameters that affect galactic evolution or that may be observed as constraints on evolutionary models. The arrows indicate the following processes: starting from protogalactic gas, stars are born, evolve, and die; they can be observed individually if they are nearby or, otherwise, in integrated light. Stellar deaths release again gas with changed composition to the interstellar medium (ISM) and perhaps produce cosmic rays, which give rise to interstellar nucleosynthesis. The mass and physical state of the interstellar medium affect the stellar birth-rate as well as the observed colors and luminosities (by gaseous emission and extinction by dust). Gas flows and stellar motions to and from the system under study are not indicated but may be very important (from Audouze and Tinsley, 1976).

responsible for these rare and by-passed† isotopes and the rapid processes of neutron absorption).

A useful concept for the following discussion is that of the primary and secondary isotopes. The primary isotopes are those which can be directly produced in a (still hypothetical) metal-free object. Such is the case, for instance, for ^{4}He, ^{12}C, ^{16}O, ^{20}Ne . . . ^{56}Fe. The secondary isotopes can only be produced in second generation objects where primary isotopes exist as a seed. The best examples are ^{13}C, ^{14}N, ^{15}N, ^{17}O, ^{18}O and more generally the rare isotopes up to Fe and all the isotopes of the nuclei $A > 56$. It is interesting to note that many elements have both primary and secondary isotopes. Such is the case, for instance, for C, O, Ne, Mg.

The various models of the chemical evolution of the Galaxy are attempts to integrate the nucleosynthetic processes with the different mechanisms by which the galactic matter evolves. Figure 1 displays the flow chart for the processes that affect the galactic evolution: star formation and deaths, cosmic rays produced by supernovae.

As is well known, our Galaxy is represented by a disc in which the gas and the stars are mainly distributed in spiral arms which start from the galactic center or galactic nucleus. Those stars which have an age comparable to or shorter than that of the Sun (which is not itself in a spiral arm now) comprise the population I stars which have a metal $(Z > 2)$

† By-passed by the neutron absorption mechanisms.

abundance comparable to that of the Sun. In the Galaxy there exists another population of older stars gathered in ~200 globular clusters. These old stars have a metallicity lower than that of the Sun by a factor of 0.001 to 0.1, but one has never yet observed a star with only hydrogen and helium.

In summary then it is found that in the Galaxy which has an age of 10–14 × 10^9 years the stars which have an age of $t \sim 10^{10}$ years have a low metallicity which increases very rapidly within the first 10^9 years and levels off after, at about the solar value. The metal content in the Galaxy seems to become constant very rapidly and the number of metal-poor stars is rather small in comparison with the number of stars having a solar metallicity.

The metal enrichment during the evolution of the Galaxy is, or course, related to the rate of processing of the galactic matter inside the stars. The main physical parameter which governs the evolution of a given star is its mass: the lifetime t_m of a star of mass m in solar mass units is:

$$t_m \approx \frac{12 \times 10^9}{m^3} + 10^6 \text{ years.}$$

When the mass of a star increases the available energy increases as m but its luminosity increases as m^4. The massive stars ($M > 5$ M⊙) have therefore a lifetime which is much shorter than the low mass stars ($M < 5$ M⊙). Moreover the nucleosynthetic "ability" of a star increases with its mass: when the mass increases the inner temperature increases also. Thus,

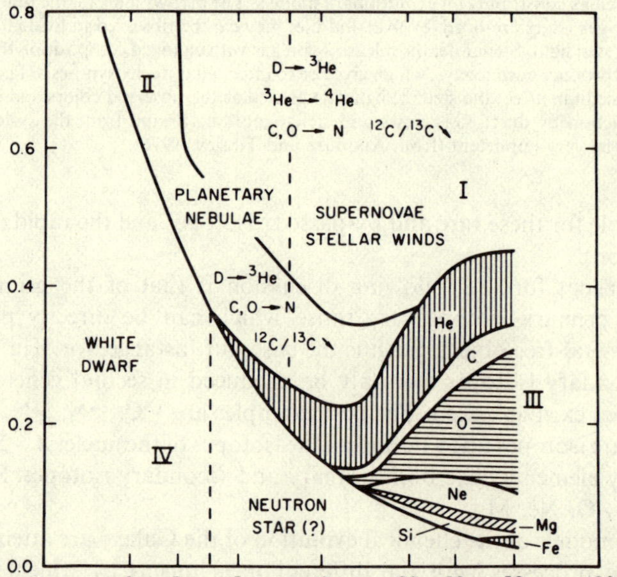

FIG. 2. Illustrative, and highly uncertain, stellar mass fractions containing different elements at the time where the mass is returned, via planetary nebulae, stellar winds or supernovae, to the interstellar medium. Zones I and II retain most of the original stellar composition, except for some destruction of light elements (including complete conversion of D to ^{3}He in both zones, and ^{3}He to ^{4}He in zone I), conversion of ^{12}C and ^{16}O to ^{14}N, and enrichment in ^{13}C. In zone III, some He produced during H burning is ejected without further processing, as are heavier elements from later stages of nucleosynthesis. "Si" refers to elements from Si to Ca, and "Fe" refers to the iron peak elements. The material in zone IV is locked into remnants. This figure has been derived, with modifications based on other sources, from Talbot and Arnett (1973, 1974) (from Audouze and Tinsley, 1976).

because the fusion reactions are temperature dependent (because of the Coulomb barrier which exists between charged nuclei) they occur between heavier elements when the temperature increases. The low mass stars end up their lives quietly as planetary nebulae and leave a white dwarf as a remnant (Fig. 2). The main nucleosynthetic contribution of the low-mass stars are the transformation of ^{12}C and ^{16}O into ^{13}C and ^{14}N through the CNO cycle reactions which, as we will see later, can affect the isotopic ratio of C and N. In contrast the massive stars explode at the end of their life as supernovae and leave a neutron star (pulsar) or possibly a black hole for the heaviest. From Figure 2 one notes that the relative amount of interstellar gas decreases with time since when a star is formed its evolution terminates in a remnant which ceases to participate in the transfer processes between the gaseous and the stellar phases. (The gas density is only $<10^{-2}$ in the galactic center and ~ 0.1 in the solar neighborhood.)

With the help of a diagram such as that displayed in Fig. 2 one is able to construct models of chemical evolution to follow the evolution of the metallicity and of the gaseous and/or the stellar phases.

As a first approximation one can construct the so-called "simple" model which can be treated analytically. In this model, well described by Pagel and Patchett (1975), one makes the following assumptions:

1. At the beginning of the galactic evolution ($t = 0$) one assumes that all the galactic material is in the gaseous form (the gas density $\mu(0) = 1$, the stellar mass density $S(0) = 0$) and that the metallicity $Z(0) = 0$.
2. The region of the Galaxy for which the model is proposed (here it is the solar neighborhood) is assumed to have an homogeneous (well mixed) composition at all times and to be closed (no matter is exchanged with the surroundings).
3. The rate of star formation $\mathrm{d}S/\mathrm{d}t$ is assumed to be proportional to the gas density:

$$\frac{\mathrm{d}S}{\mathrm{d}t} = \nu\mu^{\alpha} \quad \text{with } 1 < \alpha < 2.$$

4. The initial mass distribution (IMF) of the stars is assumed to follow the "Salpeter law":

$$\phi(m)\,\mathrm{d}m \propto m^{-\beta}\,\mathrm{d}m \quad \text{with } \beta \sim 2.35$$

   and to be independent of time. The Salpeter law is deduced from the magnitude distribution (called also luminosity function) of the nearby stars.
5. The galactic matter which does not go into a white dwarf or a neutron star is considered to be instantaneously recycled by the stars which are assumed here to have a life-time very short in comparison with the life-time of the Galaxy. This assumption is valid for the stars of mass $M > 3M_{\odot}$ ($t \lesssim 10^8$ years) which are those which contribute most to the metal enrichment.

With these assumptions the variations of the metallicity and the fraction of stars are respectively:

$$Z = p \ln\left(\frac{1}{\mu}\right)$$

where p is a constant parameter called the *yield.* It represents the fraction of metal released in the gas per unit of mass *locked into stars*. This parameter is only meaningful in the context of the instant recycling approximation,

and $$\frac{S}{S_1} = \frac{1-\mu_1^{z/z_1}}{1-\mu_1}$$

$Z_1 S_1$ and μ_1 are respectively the metal, the stellar and the gaseous mass fraction observed now:

$Z_1 = 0.02$, $S_1 = 0.90$ and $\mu_1 = 0.10$. It is to be remembered that if the exponent $\alpha = 1$, $\mu = .\exp—(t/\tau)$.

From these simple equations one can show that the metallicity grows too rapidly as a function of time. The model assumes also that $Z = 0$ for $t = 0$ (which is not observed). Moreover, the predicted number of metal-poor stars is too large in comparison with the number observed.

To resolve the discrepancy between the results of such a simple model and the observations several models have been developed in which some of the above constraints have been relaxed:

For instance, the problem of the relative small number of low-metal stars can be solved if one assumes that:

— the initial mass function varies and was relatively richer in massive stars at the beginning of the galactic evolution (VIMF);
— there was a first generation of massive stars which rapidly enriched the gas in metals with a mass abundance of 0.001 to 0.01. This is known as the prompt initial enrichment (PIE) assumption;
— the stars form preferentially in regions where the metallicity is higher (which increases the opacity and therefore the cooling mechanisms and the star-formation process). This is known as the metal-enhanced star formation (MESF) mechanism.

In order to avoid an excessive variation of the metal abundance one can assume that the region is open and can accrete some metal-free external matter (infall of extragalactic material with primordial abundances) or that the metals which are formed in massive stars can be consumed either by some black holes or comets or any solid bodies.

If we consider now the evolution of some of the isotopic ratios between elements such as C, N and O (Table 2), which have been determined both in the solar system and in the interstellar gas, we are obliged to relax the instant recycling approximation since low-mass stars ($M < 3M_\odot$) can enrich the interstellar gas into ^{13}C and ^{14}N which are secondary isotopes relative to ^{12}C and ^{16}O which are primary. Models in which one specifically takes into account the lifetime of the stars have been developed by our group (Vigroux *et al.*, 1976, for

TABLE 2. MEASURED ISOTOPIC RATIOS

Isotopic ratios	$^{12}C/^{13}C$	$^{12}C^{18}O/^{13}C^{16}O$	$^{12}C^{17}O/^{12}C^{18}O$	$^{12}C^{15}N/^{13}C^{14}N$
Solar system (age 4.6×10^9 years)	89	0.178	0.185	0.32
Solar neighborhood (age ~0)	45	0.074	0.22	0.19
Galactic center (age ~0)	25	0.04	0.31	<.026

the evolution of the $^{12}C/^{13}C$ ratio, Audouze *et al.*, 1977, for that of the $^{14}N/^{15}N$ ratio). We refer to Vigroux *et al.* (1976), where the specific but still simple formalism can be found.

As an illustration, the evolution of the ^{12}C, ^{13}C and ^{14}N abundances in the solar neighborhood are shown in Fig. 3. One can see that when the evolution time t is $>2\times10^9$ years the abundances of the secondary isotopes ^{14}N† and ^{13}C increase significantly more rapidly than that of ^{12}C which is primary.

One can expect, therefore, that since the formation of the solar system there has been an enrichment in secondary isotopes relative to that of primary isotopes. This is because the low-mass stars which are mainly responsible for the formation of these isotopes *now* release more mass into the interstellar gas than the massive stars.

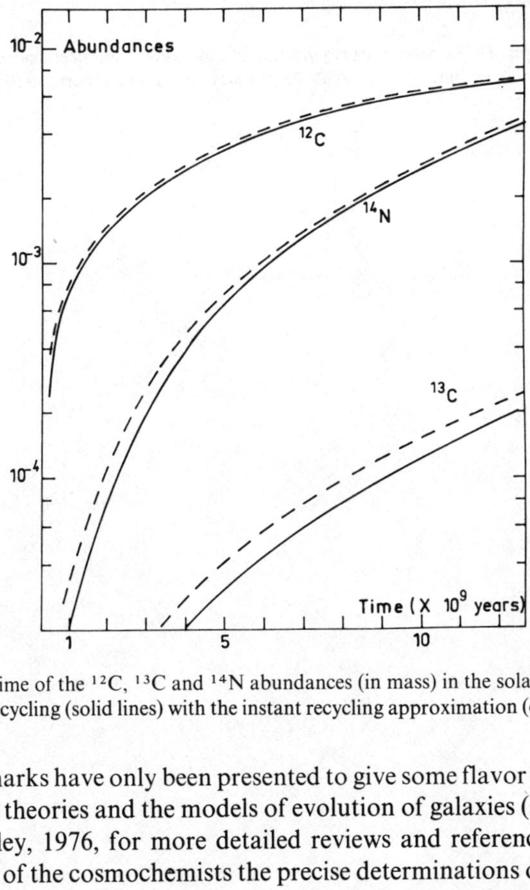

FIG. 3. Evolution with time of the ^{12}C, ^{13}C and ^{14}N abundances (in mass) in the solar neighbourhood ($\nu = 0.25$, $\delta = 0$) without instant recycling (solid lines) with the instant recycling approximation (dashed lines) (from Vigroux *et al.*, 1976).

All the above remarks have only been presented to give some flavor of the present status of the nucleosynthetic theories and the models of evolution of galaxies (see Trimble, 1975, and Audouze and Tinsley, 1976, for more detailed reviews and references). As stated before, thanks to the effort of the cosmochemists the precise determinations of the isotopic ratios in the solar system are the main input parameters of any current nucleosynthesis theory and the cosmochronology studies are one of the basic tools (with the age of the globular clusters and the study of the expansion of the universe) to estimate the age of the universe and of the Galaxy. On the other hand, the current theories of the chemical evolution of the Galaxy aid in understanding such things as the processes responsible for the observed isotopic ratios and in estimating a possible age or formation location for material showing anomalous isotopic composition such as those presented in the next paper (R. N. Clayton, this conference).

† See note added in proofs.

References

ALLOIN, D., COLLIN-SOUFFRIN, S., JOLY, M. and VIGNOUX, L. (1978) Submitted to *Astron, Astrophs.*

AUDOUZE, J., LEQUEUX, J., ROCCA-VOLMERANGE, B. and VIGROUX, L. (1977) In *CNO Isotopes in Astrophysics*, J. AUDOUZE, ed., Reidel Pub. Co. 155.

AUDOUZE, J. and TINSLEY, B. M. (1976) *Ann. Rev. Astron. Astrophs.* **14**, 43.

MENEGUZZI, M., AUDOUZE, J. and REEVES, H. (1971) *Astron. Astrophs.* **15**, 337.

PAGEL, B. E. J. and PATCHETT, B. E. (1975) *M.N.R.A.S.* **172**, 13.

PIEMBERT, M. (1978) To be published in Liège Conference Proceedings.

TALBOT, R. J. and ARNETT, W. D. (1973) *Ap. J.* **186**, 51.

TALBOT, R. J. and ARNETT, W. D. (1974) *Ap. J.* **190**, 605.

TRIMBLE, V. (1975) *Rev. Mod. Phys.*, **47**, 877.

VIGROUX, L., AUDOUZE, J. ànd LEQUEUX, J. (1976) *Astron. Astrophs.* **52**, 1.

WAGONER, R. V. (1973) *Ap. J.* **179**, 343.

† *Note added in proofs (August 1978)*: Since the presentation of this paper, it has been realized from observations of the N/O ratio in nearby galaxies that ^{14}N is partly of primary origin (Peimbert, 1978; Alloin *et al.*, 1978).

Lunar Highland Chronology

T. KIRSTEN

Max-Planck-Institut für Kernphysik, Heidelberg, Germany

Abstract

The distribution of lunar highland rock ages is of twofold significance. At first, the highest ages provide a lower limit of ~4.4 b.y. for the time of the initial differentiation of the lunar crust. Secondly, the ages reflect the collisional history of the moon in the impact-dominated period between 4.4 and 3.8 b.y. ago in which the large multiring-basins were formed. The way in which rock ages respond to impacts is rather complex and it is therefore not possible to date a basin or another lunar feature from a single rock age. Instead, rock age distributions must be considered. More than 150 highland rock ages have been dated, mainly by the K–Ar (Ar^{39}–Ar^{40}) method. The generally broad distribution of ages for breccias from a common landing site as well as the coexistence of various ages within the same breccia favor a "crater-dominated chronology" in which partial age resetting is achieved by numerous local craters in a statistical manner. This is in accordance with petrological observations which indicate that the highland breccias contain large proportions of local material reworked by secondary impacts. A statistical analysis of all available K–Ar data reveals distribution maxima at 3.9 b.y. for A 14 Fra Mauro highland basalts which are considered to be Imbrium ejecta; at 4.0 b.y. for Apollo 16 rocks from the Descartes region in the Central Highlands and, indistinguishable therefrom, also at 4.0 b.y. for the A 17 Taurus Littrow (Serenitatis) rocks. However, the age distribution is bimodal and a secondary maximum occurs for A 17 rocks at ~4.25 b.y. Evidently, the time between 3.8 and 4.0 b.y. has been a very active period in shaping the lunar crust and the major basin-forming collisions occurred within that interval. However, the frequent occurrence of rock ages between 4.0 and 4.3 b.y. raises serious objections against a purely basin-dominated chronology. Instead, a rapidly decreasing cratering rate seems to have enabled a statistically increasing number of rocks to survive the bombardment as time proceeded. The obviously limited resetting power of basin excavations prevents establishing a detailed absolute basin chronology for the sequence Serenitatis–Nectaris—Humorum–Crisium–Imbrium–Orientale as inferred from remote crater counts. However, all basins must be older than 3.8 b.y. since this is the age of Taurus Littrow mare basalt fillings which are not covered by Orientale ejecta. Most likely all circular basins formed within less than 150 m.y. approximately from 3.8 to 3.92 b.y.

Introduction

One major goal of lunar exploration is the reconstruction of the earliest periods of the moon's planetary evolution in time. Since the highlands are the dominant and also the most original lunar landform, radiometric dating of highland rocks is particularly rewarding.

Highland materials are derived from the differentiated lunar crust but they are not preserved as bedrocks but form layered pyroclastic sediments from various ejecta blankets. Most highland samples are mechanically deformed or shocked polymict breccias which were produced by the intense meteorite bombardment of the early lunar crust. In particular, bodies of asteroidal size have formed the large circular basins and deposited extended ejecta blankets of fragmented crustal materials all over the moon. Consequently, highland rock ages reflect primarily the collisional history of the moon during the impact-dominated period which followed the initial crustal differentiation. In addition, the highest highland rock ages provide also a lower limit for the time of the moonwide differentiation itself.

Rocks and Formations

The major highland rock types are anorthosites, anorthositic gabbros and Al-rich basalts, including the alkali-rich KREEP-basalts. Geochemical considerations suggest that KREEP-rocks (enriched in *K*, *R*are *E*arth *E*lements, *P*, also in U, Th) must have concentrated *below* the anorthositic crust in the process of crustal differentiation, and hence their present coexistence with anorthosites is a result of late mechanical mixing. All metamorphic grades are observed as a result of shock and thermal metamorphism.

The way in which rock ages respond to impacts is rather complex and it is therefore not possible to date a basin or another lunar feature from a single rock age. Instead, the significance of age distributions with respect to possible resetting mechanisms has to be considered (Horn and Kirsten, 1977). An additional problem arises from the difficulty to assign a particular rock to the ejecta blanket of a particular basin. Such assignments are usually based on photogeological plausibility but they may fail occasionally. For the lunar highland regions visited so far, the relations between selenological formations and the presumably related basins are as follows: Fra Mauro Formation (A14)-Imbrium basin; Apennine Mountains (A15)-Imbrium basin; Descartes Mountains, Cayley formation (A16)-Nectaris?, Imbrium?, Orientale basin??; Massifs and Sculptured Hills (A17)-Serenitatis; Apollonius Mountains (L20)-Fecunditatis basin?, Crisium basin?

Judged simply from proximity of Apollo stations to basins it is without doubt that almost all material from the highland near the Apollo 15 site has been somehow involved in the Imbrium event, whereas the Apollo 17 materials derive almost exclusively from the Serenitatis basin. For the other stations a more composite derivation is likely (McGetchin *et al.*, 1973) and evidence is growing that the contribution of local material reworked by secondary projectiles from the basins at a landing site is large (Dence *et al.*, 1974; Stöffler *et al.*, 1974).

Dating

The two major dating techniques applied to highland rocks are Rb-Sr and K-Ar. From a technical point of view, it is difficult to date anorthosites by the Rb-Sr method because of their extremely low Rb/Sr ratios (10^{-2}–10^{-3}) whereas the alkali-rich KREEP-basalts are more amenable to Rb-Sr dating.

Both rock types can be dated by the ^{39}Ar-^{40}Ar method, and anorthosites are exclusively dated in this way. The ^{39}Ar-^{40}Ar method is a modified version of K-Ar dating in which it is possible to account for Ar-diffusion losses by an incremental heating technique after labelling the K-bearing phases which are responsible for the gas release at different temperatures by conversion of a known proportion of ^{39}K into ^{39}Ar by fast neutron activation (Turner, 1976). With this method it is possible to date very small (mm-sized) rock fragments ("coarse fines") as well as large rocks. Apart from making possible better age statistics, this is also important since the cumulative grinding of rocks in the lunar regolith may preferentially remove the oldest large rocks.

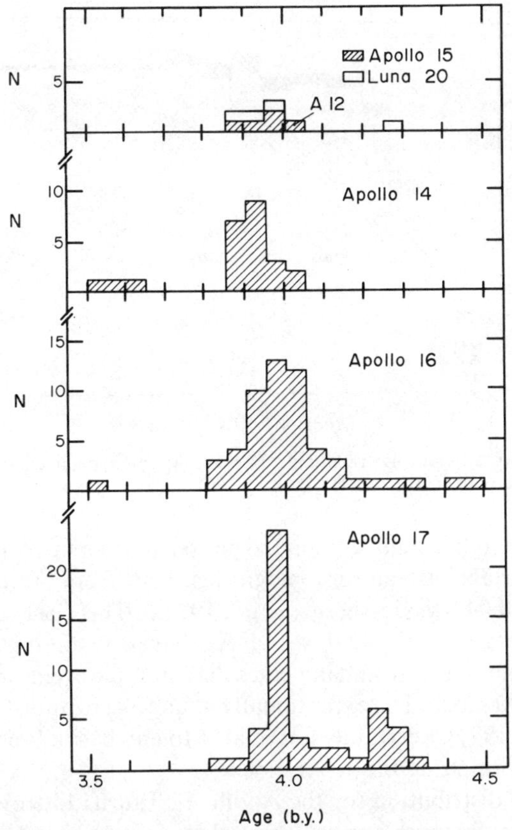

FIG. 1. Histogram of ^{39}Ar–^{40}Ar highland rock ages for the various landing sites. Literature survey after the 8th Lunar Science Conference, 1977. Note the deficient resolution of the distribution maxima for the different landing sites and the frequent occurrence of rocks with ages in excess of 4 b.y.

Age Date

So far, reliable radiometric ages have been obtained for approximately 140 highland rocks of which the majority are ^{39}Ar-^{40}Ar ages. In those cases for which intercomparisons with Rb-Sr are possible the agreement is generally very satisfying. Figure 1 is a histogram of 138 ^{39}Ar-^{40}Ar highland rock ages from a recent literature survey (see Kirsten (1978) for original references). They are separately displayed for each sampling site. The Apollo 14 Fra Mauro highland basalt ages cluster around 3.93 b.y. and range from 3.87 to 4.04 b.y. A subdivision into KREEP- and non-KREEP-basalts reveals that KREEP-basalts are systematically ~ 50 m.y. younger than non-KREEP-basalts. This distinction alone indicates that the Imbrium event did not completely reset all excavated rocks. The three exceptionally young samples are petrologically unrelated to the Fra Mauro basalts and must have been thrown into the sampling site from nearby mare regions.

Apollo 15 ANT-rocks, also associated to Imbrium, have ages similar to Apollo 14 breccias. The age distribution for Apollo 16 breccias from the Descartes region in the Central Highlands, centered around 3.98 b.y., is very broad and ranges from 3.8 to 4.5 b.y. Ages in excess of 4.1 b.y. are frequent and demonstrate that some solidified rocks could statistically

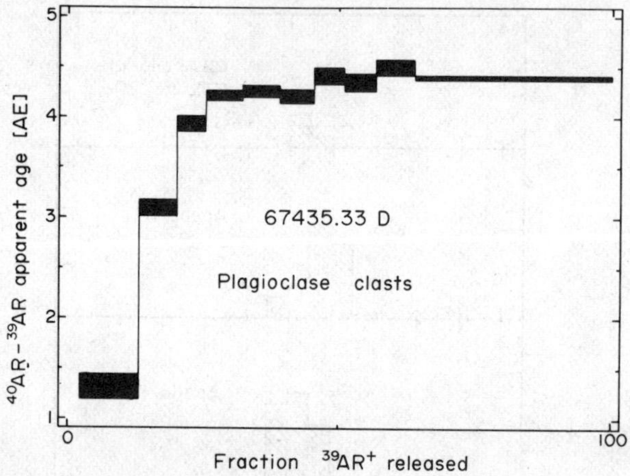

FIG. 2. Apparent ^{40}Ar–^{39}Ar ages versus fractional ^{39}Ar release for a plagioclase clast from breccia 67435 (Jessberger *et al.*, 1977). The plateau age is 4.41 ± 0.04 b.y.

survive throughout the collisionally dominated period, at least up to 4.4 b.y. As an example, Fig. 2 shows the Ar-Ar age plateau for a plagioclase clast from breccia 67435, with a well-defined age of 4.41 ± 0.04 b.y. (Jessberger *et al.*, 1977a). This clast coexisted in the breccia with other clasts which are only ~4.0 b.y. old. Age diversity among different clasts of one breccia is a rather common feature and indicates that multiple events must be responsible for the resetting of the K-Ar clock. The exceptionally young age of anorthosite 60015 (3.50 b.y., Schaeffer and Husain, 1974) is certainly not related to any basin. It demonstrates that local cratering is capable of resetting the K-Ar clock.

The peak of the age distribution for the Apollo 17 Taurus Littrow rocks (associated to Serenitatis basin) is more pronounced than that for Apollo 16, but its position is

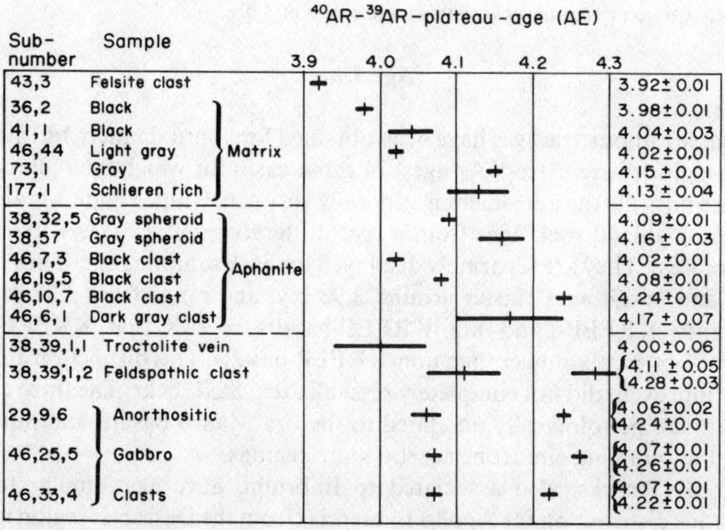

FIG. 3. Compilation of ^{39}Ar–^{40}Ar ages for seventeen subsamples of A 17 brecciȧ 73215 (Staudacher *et al.*, 1977; Jessberger *et al.*, 1977a). The total age spread is comparable to that of all Apollo 17 rocks together (see Fig. 1).

indistinguishable from the Apollo 16 age distribution maximum and the total age range is just as broad and populated as that for Apollo 16 samples.

The histogram does not include seventeen ages for individual constituents of breccia 73215 (Fig. 3). Subsamples of this single rock cover the whole rock age range defined by all Apollo 17 samples. The significance of this observation will be discussed below.

Luna 20 highland rocks from the Apollonius mountains, which may be Crisium ejecta, yield ages between 3.85 b.y. and 4.3 b.y.

Basins or Local Cratering?

From the breccious nature of the samples and the observation of various degrees of shock effects it is clear that the age distributions outlined above must be related to impact processes in one way or the other. However, several resetting mechanisms could be responsible.

1. Ages are dominated by large basin-forming events, which provide enough energy to reset essentially all excavated samples. Ejecta related to one event should have essentially identical ages. A long subsequent cooling time in thick hot ejecta blankets is excluded from the lack of thermal equilibration in many highland breccias. Consequently, the age of a basin should be given by a δ-age distribution. A variant of this model is that the basin-forming event caused a sudden uplift and cooling of deeply seated rocks which had previously been stored at temperatures exceeding the blocking temperature of the radiometric clock. Samples from shallower depths could then have higher ages and the age of the basin would be given by a distribution maximum immediately before a sharp cut-off towards low ages.
2. Basin excavations have *not* dominated the equilibration of the isotopic systems in competition with the effects of numerous local craters. In this case, age distributions should be broad, more symmetric, and indistinguishable for different landing sites. The distribution maximum is not related to any particular basin but is statistically interpreted as the result of the decreasing cratering rate in the early lunar history. As time proceeded, an increasing number of rocks could survive the bombardment.
3. A synthesis of models 1 and 2 in which the crater-dominated statistical age distribution is overprinted by a fair number of rocks which were reset during basin formation. The pattern could be bimodal or similar to case 2, in which case the interpretation of the distribution maximum would be different.

The key question in judging about these alternatives is the resetting power of basin-forming events. This shall be discussed by means of a case study: Apollo 17 light gray breccia 73215. For typical regolith breccias, the coexistence of clasts with different ages within one sample has been explained by mild secondary local compaction of rocks from different basins, all of which occur at the same site and each of them having its distinct basin age. However, such argument cannot be made for breccia 73215. This rock is exceptional in that it is not a regolith breccia but it was formed in one single event, presumably Serenitatis excavation (James *et al.*, 1975; Jessberger *et al.*, 1976). Its clasts were imbedded into a rapidly cooling groundmass and subsequent metamorphism is excluded from circumstantial evidence (James, 1976). The sample contains a felsic inclusion, which is partially a glass and which was certainly above the argon-retention temperature when the breccia formed. Under these circumstances its age, 3.92 ± 0.04 b.y. by ^{39}Ar-^{40}Ar, and 3.90 ± 0.05 b.y. by Rb-Sr

(Compston *et al.*, 1977) yields—in a straightforward manner which is exceptional rather than typical—the age of Serenitatis basin, ~3.91 b.y. (provided the sample assignment is correct).

If felsite was reset, what about the other breccia constituents, such as fragment-laden groundmass or igneous ANT-clasts, all of them definitely involved in the basin excavation and not affected afterwards? As many as sixteen of such subsamples have been dated, and the age diversity is overwhelming (Fig. 3). *None* has an age as low as the breccia-forming event. Instead, ages cover the total age range covered by all other Apollo 17 samples taken together (Fig. 1). Quite clearly, the resetting power of the basin-forming event was indeed very limited. Crystalline fragments as small as 1 mm and less were excavated, incorporated in a condensing impact vapor or impact melt and still retained at least partially their pre-basin age. What is proven for the subsamples of this well-studied rock must also hold for the rock age distributions of (possibly) unrelated regolith breccias as depicted in Fig. 1 and a purely basin-dominated chronology should be abandoned the sooner the better, also in accordance with the notion that the age distributions for individual landing sites correspond to the expected patterns outlined for model 2 (or 3) rather than 1 (Fig. 1). Further support is provided by petrological observations which indicate that highland breccias contain large proportions of local materials reworked by secondary impacts (Dence *et al.*, 1974; Stöffler *et al.*, 1974).

Highland Chronology

Having noticed that local cratering rather than individual basins have dominated the age distributions, and taking into account that local cratering is an unselective moonwide process, we may plot ages from all landing sites together for a common interpretation (Fig. 4). The broad age distribution ranging all the way up to 4.5 b.y. is in general the result of statistical survival under the influence of a rapidly decreasing cratering rate within the first 700 m.y. of the moon's history. The whole process is best considered as the final stage of the accretion of the moon itself.

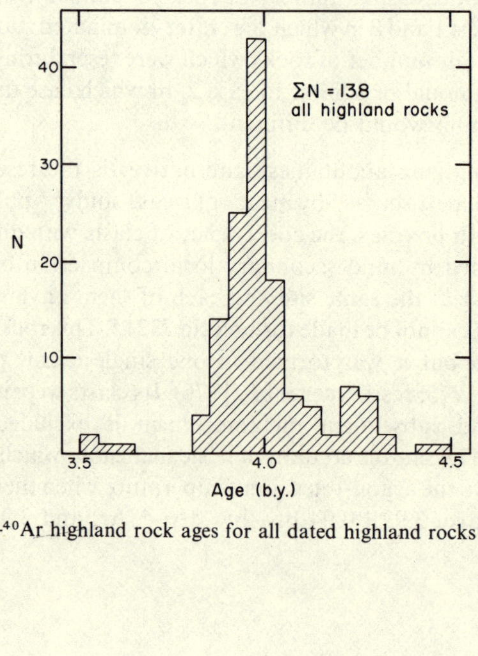

FIG. 4. Histogram of ^{39}Ar–^{40}Ar highland rock ages for all dated highland rocks combined (see also Fig. 1).

The sad consequence of the foregoing discussion is that there is no straightforward way for a detailed absolute basin chronology. No matter how well rock ages are determined, they do—in general—not relate to the time of basin formation. A few conclusions can nevertheless be drawn. A lower age limit for the period of basin-forming impacts follows from the well-determined age of the subfloor basalt in the Taurus Littrow valley, 3.8 b.y. (Kirsten and Horn, 1974). This old mare surface postdates the youngest basin (Orientale) since no ejecta debris is found on it. Furthermore, from the 73215 data discussed above, it must be concluded that Serenitatis was excavated 3.91 ± 0.04 b.y. ago. A necessary consequence of this date is that only very few of the common Apollo 17 rocks were resetted during Serenitatis excavation since their age distribution clusters around 4.0 rather than 3.9 b.y. (Fig. 1). Since Serenitatis is stratigraphically the oldest circular basin, it follows that twelve others have formed within a period of 100–150 m.y. only (Hartmann and Wood, 1971).

One more time mark may be extracted from the Apollo 14 and Apollo 15 data. This age distribution has sharper cut-offs at both ends than Apollo 16 and Apollo 17 (Fig. 1). This seems to indicate that model 3 is applicable in the case of the largest basin on the moon, Imbrium. Its age would then be near to the lower cut-off of Apollo 14 and Apollo 15 ages, ~ 3.85 b.y.

The evidence for the formation of all the major circular basins within 100–150 m.y. is supported by the Pb-isotope systematics of lunar highland rocks. A large volatilization of lead is inferred from the lower intercept of the whole rock U-Pb concordia diagram at 3.9 ± 0.1 b.y. (Tera and Wasserburg, 1974). From this observation, a "lunar cataclysm" ~ 3.9 b.y. ago has been proposed by these authors in which the impact rate was appreciably higher than between 4.0 and 4.3 b.y. ago. However, the frequent occurrence of rock ages between 4.0 and 4.4 b.y. indicates that such an astrophysically problematic supposition, as obvious as it may seem since "the nearest event looks like the only event while the past is obscured", is in fact not required. Records of the preceding period are largely erased but punctually documented by the very old (>4.1 b.y.) rocks. Notwithstanding, the uniform lead concordia intercept near 3.9 b.y. is supported by the narrow K-Ar basin time interval.

Early Crustal Differentiation

Rb-Sr studies have revealed that the early differentiation of the moon into an anorthositic crust, a KREEP layer and a mafic cumulate lasted from about 4.6 to 4.4 b.y. The evidence consists in internal isochron ages of

4.55 ± 0.1 b.y. for dunite 72417 (Papanastassiou and Wasserburg, 1975),
4.60 ± 0.1 b.y. for troctolite 76535 (Papanastassiou and Wasserburg, 1976),
4.49 b.y. for plagioclase from 76055 (Tera *et al.*, 1974) and
4.43 ± 0.05 b.y. for noritic fragments from 77215 (Nunes *et al.*, 1976).

In addition, model ages for lunar soils are 4.4 ± 0.2 b.y. and for ANT-rocks 4.6 ± 0.16 b.y. (Schonfeld, 1976). Furthermore, the U-Pb evolution of most terra rocks yields an upper intercept with concordia at ~ 4.42 b.y., indicating that a major fractionation between U and Pb occurred on a moonwide scale at this time, and some rocks (68415, 67559) have even concordant U-Pb ages at 4.42 b.y. (Tera and Wasserburg, 1974). The Ar-Ar age distribution shown in Fig. 1 fits nicely into this pattern. The oldest surviving rocks are in fact 4.47 b.y.

(fragments from 65015, Jessberger *et al.*, 1974) and 4.42 b.y. for plagioclase clasts from breccia 67435 (Jessberger *et al.*, 1977) (Fig. 2). Thus, K-Ar dating confirms the existence of a differentiated crust as early as ~4.45 b.y. ago. The subsequent 650 m.y. were dominated by impacts. After 3.8 b.y. ago the cratering rate dropped below the critical level which could do any appreciable resetting. At about the same time, mare volcanism took over as the dominant geological process on the moon.

References

COMPSTON, W., FOSTER, J. and GRAY C. (1977) Rb-Sr systematics in clasts and aphanites from Consortium breccia 73215. *Lunar Science*, VIII, 199–201.

DENCE, M., GRIEVE, R. and PLANT, A. (1974) The Imbrium basin and its ejecta (Abstract). *Lunar Science*, V. 165–167.

HARTMANN, W. K. and WOOD, C. A. (1971) Moon: Origin and evolution of multi-ring basins. *The Moon*, **3**, 3–78.

HORN, P. and KIRSTEN, T. (1977) Lunar highland stratigraphy and radiometric dating. *The Moon—A New Appraisal from Space Missions and Laboratory Analyses, Philos. Trans. r. Soc. London* **A285**, 391–395.

JAMES, O. and Consortium 73215 (1975) Consortium studies of matrix of light gray breccia 73215. *Proc. Lunar Sci. Conf. 6th* **1**, 547–577.

JAMES, O. (1976) Petrology and aphanitic lithologies in Consortium breccia 73215. *Proc. Lunar Sci. Conf. 6th* 2145–2178.

JESSBERGER, E. K., HUNEKE, J., PODOSEK, F. and WASSERBURG, G. (1974) High resolution argon analysis of neutron irradiated Apollo 16 rocks and separated minerals. *Proc. Lunar Sci. Conf. 5th* **2**, 1419–1449.

JESSBERGER, E. K., KIRSTEN, T. and STAUDACHER, Th. (1976) Argon-argon ages of Consortium breccia 73215. *Proc. Lunar Sci. Conf. 7th* 2201–2215.

JESSBERGER, E. K., DOMINIK, B., KIRSTEN, T. and STAUDACHER, Th. (1977a) New ^{40}Ar–^{39}Ar ages of Apollo 16 breccias and 4.42 AE old anorthosites. *Lunar Science*, VIII, 511–519.

JESSBERGER, E. K., KIRSTEN, T. and STAUDACHER, Th. (1977) One rock and many ages—further K-Ar data on Consortium breccia 73215. *Proc. Lunar Sci. Conf. 8th.* **2**, 2567–2580.

KIRSTEN, T. (1978) Time and the solar system. In *The Origin of the Solar System* S. F. DERMOTT, ed., J. Wiley & Sons, Publ. pp. 267–346.

KIRSTEN, T. and HORN, P. (1974) Chronology of the Taurus Littrow region III: Ages of mare basalts and highland breccias and some remarks about the interpretation of lunar highland rock ages. *Proc. Lunar Sci. Conf. 5th* **2**, 1451–1475.

MCGETCHIN, T. R., SETTLE, M. and HEAD, W. (1973) Radial thickness variation in impact crater ejecta: Implication for lunar basin deposits. *Earth Planet. Sci. Lett.* **20**, 226–236.

NUNES, P., NAKAMURA, N. and TATSUMOTO, M. (1976) 4.4 b.y. old clast in boulder 7, Apollo 17. *Lunar Science* VII, 631–632.

PAPANASTASSIOU, D. and WASSERBURG, G. (1975) Rb-Sr study of a lunar dunite and evidence for early lunar differentiates. *Proc. Lunar Sci. Conf. 6th* **2**, 1467–1489.

PAPANASTASSIOU, D. and WASSERBURG, G. (1976) Early lunar differentiates and lunar initial ^{87}Sr/^{86}Sr. *Lunar Science* VII, 665–667.

SCHAEFFER, O. A. and HUSAIN, L. (1974) Chronology of lunar basin formation. *Proc. Lunar Sci. Conf. 5th* **2**, 1541–1555.

SCHONFELD, E. (1976) Rb-Sr evolution of the lunar crust. *Lunar Science* VII, 773–775.

STAUDACHER, Th., JESSBERGER, E. and KIRSTEN, T. (1977) ^{40}Ar–^{39}Ar age systematics of Consortium breccia 73215 II. *Lunar Science* VIII, 896–898.

STÖFFLER, D., DENCE, M., ABADIAN, M. and GRAUP, G. (1974) Ejecta formations and pre-impact stratigraphy of lunar and terrestrial craters: Possible implications for the ancient lunar crust (Abstract). *Lunar Science* V, 746–748.

TERA, F. and WASSERBURG, G. (1974) U–Th–Pb systematics of lunar rocks and inferences about lunar evolution and the age of the moon. *Proc. Lunar Sci. Conf. 5th* **2**, 1571–1599.

TERA, F., PAPANASTASSIOU, D. and WASSERBURG, G. (1974) Isotopic evidence for a terminal lunar cataclysm. *Earth Planet. Sci. Lett.* **22**, 1–21.

TURNER, G. (1976) Potassium–argon chronology of the moon. *Physics and Chemistry of the Earth.* In press.

The Earth–Moon System. Chemistry and Origin

HEINRICH WÄNKE and GERLIND DREIBUS

Max-Planck-Institut für Chemie (Otto-Hahn-Institut), 6500 Mainz, B.R.D.

Abstract

In all lunar highland samples a Mg-rich component is observed. As these samples consist of only three major components and the components vary in their abundances, the composition of the Mg-rich component can be calculated with certain assumptions. Wänke *et al.*, (1977) have recently shown that the Mg-rich component probably consists itself of two portions of different origin, i.e. true-primary matter which was added to the moon when a solid feldspathic crust had already formed and material derived from residual liquids after the separation of olivine and plagioclase. The composition of the primary matter shows a striking similarity to that of the earth's mantle, except that the refractory elements seem to be somewhat enriched and all elements more volatile than Si are considerably depleted.

The element ratios of the two strongly correlated elements W and La as well as the P/Li ratio indicate that both W and P are highly depleted on the moon. The depletion for W relative to La (both elements belong to the group of refractory elements) is in fact identical to the one observed for terrestrial samples. The only feasible process for the depletion of W and P is removal with metallic iron. From measurements of the partition coefficients of W between silicate phase and metallic NiFe we know that a metal phase of 26% would be required to explain the observed depletion. This by far exceeds the amount tolerable from the lunar density. We conclude that the metal responsible for the removal of W from the lunar silicates is in the core of the earth. Hence, models in which the moon is formed from material evaporated from the earth's mantle are favoured.

A number of authors have tried to estimate the bulk composition of the moon, using various indirect methods. In almost all lunar highland samples, a Mg-rich component is observed, which consists as we will show mainly of primary matter. According to our interpretation (Wänke *et al.*, 1974a, 1975, 1976) this primary component never underwent large-scale magmatic fractionations and was added to the moon after a solid and differentiated feldspathic crust had been formed already (Fig. 1). We have considerable evidence for the assumption that the composition of the primary matter found in the lunar highlands to be representative for the whole moon. Hence, it seems we have a more direct access to the lunar bulk composition (Dreibus *et al.*, 1977).

The lunar highlands are old. The highest ^{39}Ar-^{40}Ar ages of about 4.2 G.y. for lunar highland rocks (Schaeffer *et al.*, 1976) can be taken as lower limit and it may be that the lunar crust formed close to 4.6 G.y. (Papanastassiou and Wasserburg, 1976). Hence it is, in fact, not surprising that the accumulation process was still continuing at a time when a solid crust had already formed.

Most of the lunar highland samples can be explained as mechanical mixtures of three components:

1. a trace element-enriched component (KREEP) with Fe/Mg$>$1;
2. a trace element-poor anorthositic component;
3. a Mg-rich component with Fe/Mg$<$1.

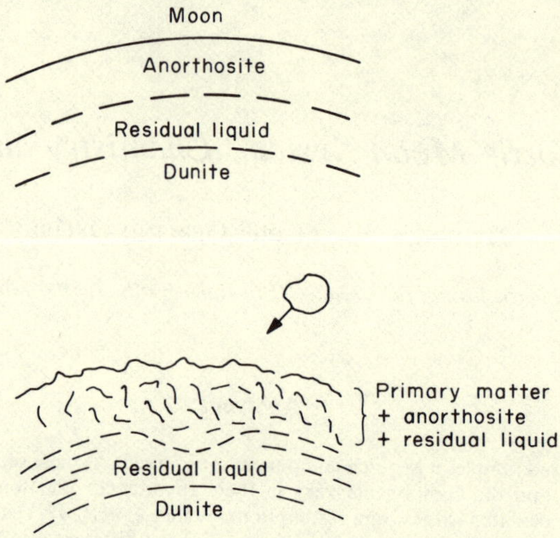

FIG. 1. Evolutionary model of the outer region of the moon. (Simplified version of the model proposed by Wood *et al.*, 1970, and Ryder and Wood, 1977.)

Exceptions are the lunar dunites and troctolites; their contribution to average highland soils and breccias can only be small and by no means can these rocks be the source of the Mg-rich component. All lunar highland samples show a nearly identical Fe/Sc ratio of about 4200, while dunite 72417 has a Fe/Sc ratio of about 5 times higher. A similar argument can be made from the Mg/Cr ratio. The origin of most of the lunar highland rocks as mechanical mixtures is clearly demonstrated by the fact that these rocks contain metallic nickel-iron up to more than 1% per weight, which would immediately segregate in any magmatic processes.

Wänke *et al.*, (1976) showed that the composition of the Mg-rich component can be computed from a ternary mixing diagram with KREEP- and lunar anorthosites as the other end-members. In this computation we used the KREEP-rich breccia 14321 as the KREEP-end-member.

We will use here another approach. Palme and Wänke (1975) have underlined the fact that all highland samples have an identical pattern of incompatible elements, which is identical to that of the KREEP-rich sample 14321. Hence, using La as one of the incompatible elements as a measure of the KREEP-component, we can subtract this component and have reduced the number of components to two. Figure 2 shows the two-component mixing diagram obtained in this way. We restrict the computation to KREEP-poor samples (i.e. to samples with La $<$ 25 ppm) and took only samples for which we have data from our laboratory. The mixing diagram was computed from the composition of the mixtures without making assumptions as to the possible end-members.

There is little doubt about the nature of the left end-member, the lunar anorthosites. As we suspected a primary origin of the Mg-rich component we took the two strictly refractory elements Al and Sc to define this component. There is overwhelming evidence that the refractory elements (see Table 1) did not fractionate from each other, during condensation and accretion but stayed together as a group of thirty-seven elements. Hence, the position of the primary component was calculated for the location where the Al/Sc ratio becomes identical to the C1 ratio of 1370. The resulting composition of the Mg-rich component is

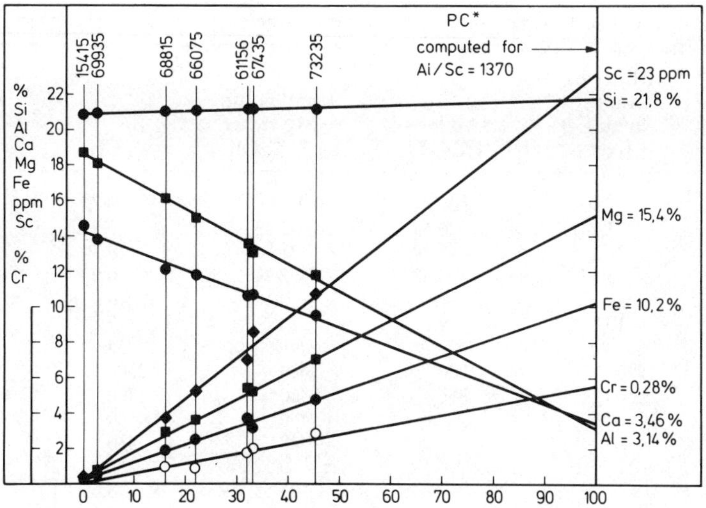

FIG. 2. After subtraction of KREEP (see Wänke *et al.*, 1977), the binary mixing diagram was computed using only those samples which were also used for the ternary diagram of Wänke *et al.* (1976), i.e. samples with La <25 ppm and samples which are clear differentiates were excluded. The mixing diagram was computed from the composition of the mixtures, the highland breccias without making assumptions as to the possible end-members. All data from the Mainz laboratory (Wänke *et al.*, 1974a, 1975, 1976).

TABLE 1. GROUP OF REFRACTORY ELEMENTS CONDENSING WITHOUT FRACTIONATION
Larimer and Anders (1970); Grossman (1972, 1973); Wänke *et al.* (1974a)

Elements:	Al, Ca, Ti, (Mg, Si) Be, Sc, V, Sr, Y, Zr, Nb, Mo, Ru, Rh,
	Ba, REE, Hf, Ta, W, Re, Os, Ir, Pt, Th and U.

listed in Table 2. The ratio Al/Ca, two other refractory elements, corresponds exactly to the C1 ratio. The obtained Mg/Si and Mg/Cr ratios are also close to the C1 values, the deviation of the Mg/Si ratio seems nevertheless significant.

Figure 1 is a sketch of an evolutionary model for the outer regions of the moon to a depth of perhaps 60 km, which are supposed to have been totally molten in the early days of the lunar history. On cooling in a crystallization process after separation of plagioclase and olivine the residual liquid will show for many elements abundance ratios similar to those of the original primary matter. The mafic elements Ti, Fe and Sc would be enriched relative to Mg as compared to the primary matter and all incompatible elements will be even more enriched, of course. In fact, the ferroan anorthosites found by Dowty *et al.*, (1974) may be related to such residual liquids.

In our mixing model we have treated the anorthosites as an independent component, hence, material originating from the residual liquid can be described as primary matter minus olivine.

TABLE 2. CALCULATION OF THE COMPOSITION OF THE PRIMARY COMPONENT IN LUNAR HIGHLAND SAMPLES

	Mg-rich comp. metal free	Olivine† (72417)	Prim. comp =Mg-rich C +31% Oli.	Cl	Prim. comp. normalized to Si and Cl
Mg %	15.9	27.4	19.5	9.25	1.00
Al	3.24	0.64	2.43	0.82	1.42
Si	22.5	18.70	21.30	10.20	1.00
Ca	3.57	0.82	2.71	0.90	1.44
Ti	—	—	0.12‡	0.041	1.42
Fe (FeO)	10.5	9.1	10.1	17.8	—
V ppm			86§	55.4	0.74
Cr	2900	2200	2680	2300	0.56
Mn	1600	850	1360	1700	0.38
Na			450*	4800	0.045
K			48*	480	0.048
Rb			0.12*	2.2	0.026
Sr			25.5‡	8.6	1.42
Ba			7.1‡	2.4	1.42
Sc	23.6	4.3	17.5	5.9	1.42
La			0.68‡	0.23	1.42
Eu			0.15‡	0.05	1.42
Zr			12.2‡	4.1	1.42
Th			0.102‡	0.035	1.42
U			0.027‡	0.009	1.42
Mg/Si	0.71	1.47	0.91	0.91	1.00
Mg/Cr	55	126	73	40	1.79
Ca/Al	1.10	1.28	1.11	1.10	1.01

Mg-rich component (PC*). Composition derived from the binary mixing diagram (Fig. 2).
† Lunar dunite 72417. Composition derived from the analysis of Laul and Schmitt (1975) and LSPET (1973).
‡ Calculated relative to Al assuming Cl abundance ratios.
§ Calculated relative to Al assuming abundance ratios equal to those in Allende inclusions (Wänke *et al.*, 1974b).
* From element correlations observed in lunar samples: Na/Eu=3000 (Wänke *et al.*, 1974a), K/La=70 and K/Rb=400 (Wänke *et al.*, 1973).

The lunar dunites like sample 72417 may very likely represent material from the olivine cumulates. The Al/Sc ratio of 1490 measured by Laul and Schmitt (1975) for sample 72417 is almost identical with the C1 ratio of 1370, and furthermore, the absolute concentration of these two elements is small. Thus, the Mg-rich component as derived from our mixing computations is probably again a mixture of true primary matter and material originating from residual liquids as it is indicated by the low Mg/Si ratio of 0.71 found for the Mg-rich component compared to the Mg/Si ratio of 0.91 in C1 chondrites.

Compositionally the residual liquid can be regarded as primary matter minus olivine and anorthosite. Hence, we have to add olivine to our Mg-rich component to correct for the contribution of material from the residual liquid and will in this way raise the Mg/Si ratio to the C1 value.

Taking the lunar dunite 72417 to be representative for the olivine cumulates and assuming the C1 value of the Mg/Si ratio to be valid for the primary matter we calculate an amount of 31% olivine, which we have to add to raise the Mg/Si ratio from 0.71 to 0.91.

The composition of the primary component is also given in Table 2. The resulting normative mineral composition is listed in Table 3.

TABLE 3. NORMATIVE MINERAL COMPOSITION OF THE PRIMARY COMPONENT OF THE LUNAR HIGHLANDS IN wt% (METAL FREE) CALCULATED FROM CHEMICAL COMPOSITION GIVEN IN TABLE 2

	%		%
Ab	0.46	Fs	8.43
An	13.2	Fo	34.2
Wo	2.98	Fa	10.8
En	29.4	Cm	0.60

The weakest point in our approach is the assumption: the Mg/Si ratio in the primary component to be identical to the C1 value. Even the C1 ratio is not known with the precision one would like. We have chosen the C1 value of Von Michaelis *et al.*, (1969) who have analysed eight carbonaceous chondrites. They found almost identical mean values of 0.91 for C1, C2 and C3 chondrites. Teschke and Wänke (1974) found a lower value for the C1 chondrite Orgueil (Mg/Si = 0.86), and an increasing trend of this ratio for C2 and C3 chondrites (0.88, 0.93 respectively). The most comprehensive analyses of Wiik (1963) yielded for Orgueil 0.90, for eight C2 chondrites a lower mean value of 0.88 (neglecting the exceptional value of 1.04 for Santa Cruz) and a higher mean value for the C3 chondrites of 0.92.

However, ordinary chondrites have considerably lower Mg/Si ratios of about 0.82 as do the enstatite chondrites (Mg/Si = 0.71). Hence, there seems to be a parallelism to the depletion of refractory elements observed in ordinary and enstatite chondrites. Magnesium seems to be more refractory than Si as it is to be expected from the condensation temperatures.

We have concluded that assuming a Mg/Si ratio of 0.91 for the primary matter is only justified if this matter shows C1 abundances for all elements less volatile than Si. In fact, relative to C1 chondrites we deduce from Table 2 a small enrichment of refractories, which could mean a Mg/Si ratio above the C1 value and in turn slightly higher amounts of olivine in the primary matter.

Can we assume that the primary matter of the outer regions of the moon is identical with the composition of the bulk moon? The mare basalts are generally thought to originate from the deep interior of the moon (Ringwood, 1977); nevertheless, we observe identical ratios for pairs of correlated elements for highland samples and mare basalts. For example, the K/La ratio is 70 for both mare basalts and highland samples. The value of 80 is found for the Fe/Mn ratio in all lunar samples and the Fe/Sc ratio of the less fractionated low-Ti mare basalts is again identical to that of highland samples. Thus, we have considerable evidence for the assumption that the primary matter of the lunar highlands is compositionally identical with the material from which the mare basalts were derived. The further extrapolation to the bulk moon seems to be well justified.

In Table 4 we compared the results about the bulk composition of the moon with recent estimates of other authors. (For detailed discussion of our approach see Dreibus *et al.*, 1977, and Wänke *et al.*, 1977.) The agreement between our results and the data of Ringwood (1977) and Taylor (1976) is striking, considering the completely different ways of approach. For all major elements the differences are without any doubt within the uncertainties of each of the three calculations. Agreement exists not only for the concentrations of refractory

TABLE 4. BULK COMPOSITION OF THE MOON.
Comparison of our data with those of other authors and with the composition of the Eucrite Parent Body (EPB) the earth's mantle and chondrites.

	Moon (metal free)			EPB (met. free)	Earth (met. free)					
	This work	Taylor (1976)	Ringwood (1977)	Dreibus *et al.* (1977)	Ringwood (1977)	C1	C3	H	L	E6
Si %	21.3	21.4	20.9	21.5	21.1	10.2	15.4	17.0	18.9	19.8
Mg	19.5	19.5	20.1	19.6	23.0	9.25	14.3	13.6	14.9	14.0
Al	2.43	3.2	1.96	1.58	2.06	0.82	1.37	1.07	1.19	0.85
Ca	2.71	3.5	2.43	1.89	2.22	0.90	1.53	1.19	1.28	0.90
Fe (FeO)	10.1	8.2	10.8	11.21	6.14	17.8	19.8	8.0	9.7	—
Cr	0.27	0.13	0.27	0.46	0.21	0.23	0.34	0.34	0.38	0.33
Mn	0.136	0.10	—	0.25	—	0.17	0.14	0.23	0.25	0.18
Na ppm	450	—	370	540	3000	4800	1590	5700	6500	5200
K	48	100	52	54	—	480	210	830	860	700
Sc	17	19	—	11.4	—	5.9	10.1	7.6	8.2	7
U ppb	27	40	26	17	26	9	20	13	14	8
Ca/Al	1.11	1.09	1.24	1.20	1.08	1.10	1.12	1.11	1.08	1.06
Mg/Si	0.91	0.91	0.96	0.91	1.09	0.91	0.93	0.80	0.79	0.71
$\frac{100\,MgO}{MgO+FeO}$ (mol%)	82	85	81	80	89	54	62	80	78	—

TABLE 5. ELEMENTAL ABUNDANCES NORMALIZED TO Si AND Cl

	Moon (met. free)			EPB (met. free)	Earth (met. free)				
	This work	Taylor (1976)	Ringwood (1977)	Dreibus *et al.* (1977)	Ringwood (1977)	C3	H	L	E6
Mg	1.00	1.00	1.06	1.00	1.20	1.02	0.88	0.87	0.78
Cr	0.56	0.27	0.57	0.95	0.44	0.98	0.89	0.89	0.74
Al	1.42	1.86	1.17	0.92	1.21	1.11	0.78	0.78	0.53
La	1.42			0.89		1.50	0.81	0.84	0.56
U	1.42	2.12	1.41	0.92	1.40	1.47	0.83	0.81	0.46
Mn	0.38	0.28		0.71		0.55	0.81	0.79	0.55
Na	0.045		0.038	0.053	0.30	0.22	0.71	0.73	0.56
K	0.048	0.099		0.053		0.29	1.04	0.97	0.75
$\frac{100\,MgO}{MgO+FeO}$ (mol%)	82	85	81	80	89	62	80	78	

elements, but also for the mole ratio MgO/MgO + FeO. The value for this ratio of 0.82 is identical to the value observed in ordinary chondrites. Except for volatile and moderately volatile elements the bulk composition of the moon seems to be relatively close to that proposed for the earth's mantle.

During the last years a moon enriched in refractory elements was popular among many scientists (Anderson, 1973; Ganapathy and Anders, 1974; Wänke *et al.*, 1974a, 1975; Dreibus *et al.*, 1976). Aside from volatile and moderately volatile elements as well as metals and sulfides, it seems that we are coming back to a nearly chondritic moon (Table 5).

We have already stressed the point that all refractory elements always appear in strictly chondritic abundance ratios. A direct confirmation of this statement is given by the fact that pairs of correlated refractory elements are found in C1 ratios in all lunar samples (Wänke *et al.*, 1973). This holds also for the achondrites and, to a lesser extent, because of larger magmatic fractionations, also for the earth. Only W, both a highly refractory and incompatible element, shows a different behaviour. Tungsten shows an excellent correlation with La, however, it is depleted by a factor of 19 relative to the C1 abundance (Fig. 3). Wänke *et al.*, (1973 and 1975) suggested that the reason for this depletion is the siderophile tendency of W, which led to a considerable removal by metallic iron. It was also shown that terrestrial samples as well as eucrites and howardites plot along the same correlation line as the lunar samples, in spite of the fact that the earth contains about 30% of metallic iron, while the upper limit of a metallic lunar core of 400 km radius (Wiskerchen and Sonett, 1977) corresponds to a metal content of the moon of 2.8%. Rammensee and Wänke (1977) have

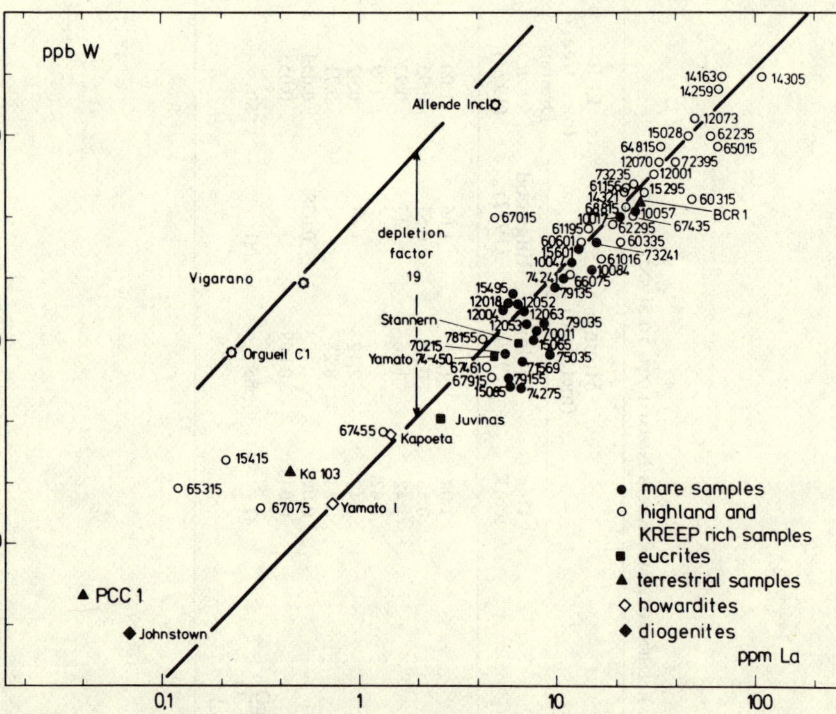

FIG. 3. W–La correlation for different lunar, achondritic and terrestrial samples, as already published by Wänke *et al.* (1973). All data from the Mainz laboratory, Wänke *et al.* (1977) and references therein.

recently determined the partition coefficient of W between metal and silicates for different temperatures, composition and oxygen fugacities. For basalts with 13% FeO and a Ni concentration in the metal of 6%, they obtained at 1300°C a value of 53 for the partition coefficient of V_W between metal and silicate. With this value a required metal content of 26% is calculated to account for the observed depletion of tungsten. As W condenses far above Fe and Ni together with the other refractory elements, besides its removal by metallic iron in a planetary body, there seems no other feasible process to explain its depletion (Rammensee and Wänke, 1977).

Considering only the less fractionated low-Ti mare basalts and KREEP-poor highland samples we find for P a similar behaviour as for W. The constant and identical P/Li ratio for lunar samples and for eucrites far off the C1 value also indicates that in the parent material of both objects large amounts of P were removed by metallic iron (Fig. 4). As was the case for W, no problem exists in the case of the eucritic parent body, because this object may contain the required amount of metallic iron contrary to the restricted amount of metallic iron in the moon.

Dreibus *et al.*, (1976, 1977) proposed that the metal/silicate fractionation took place within the proto-earth and that the metal responsible for the removal of W and P from the lunar silicates is in the earth's core. They also emphasized the close similarity of the composition of the primary matter of the lunar highlands to that of the earth's mantle and the fact that the large difference in the total Fe content of moon and earth arises solely from the difference in the metal content. The total Fe content of the earth slightly exceeds the chondritic level, while the moon is far below the chondritic value.

Dreibus *et al.*, (1976, 1977) concluded further that models put forward by Ringwood (1960, 1970), Wise (1963, 1969) and O'Keefe (1969) in which the moon accreted from material condensing from a "planetary nebula" around the earth formed by evaporation of a considerable fraction of the proto-earth or by a more direct fission process would be in accordance with the observations discussed above as well as with the important observation of Clayton and Mayeda (1975) according to which, earth and moon lie on the same oxygen isotope fractionation line.

Ringwood and Kesson (1977) have shown that in addition to W and P the abundances of Co, Ni, Ir, Os, S and Se are similar in ocean-floor tholeiites and low-Ti mare basalts and that

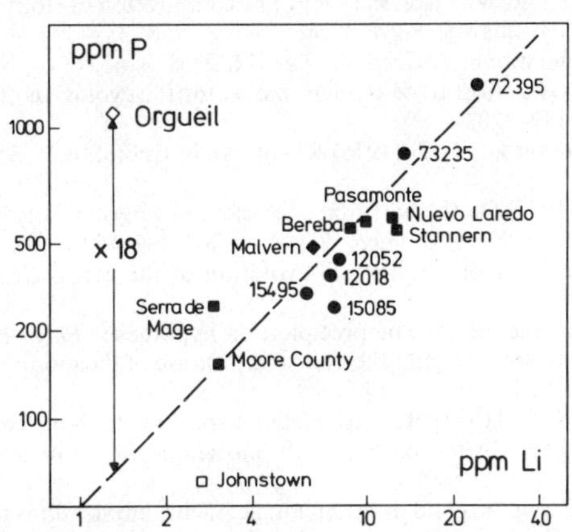

FIG. 4. P as Li in low Ti-mare basalts, KREEP-poor highland rocks, and in achondrites. The figure is taken from Dreibus *et al.*, (1977) for references of the data used see this paper.

they believe that this similarity extends to their respective source regions in the earth's mantle and lunar interior. These authors suggest that the abundances of the above siderophile elements in the earth's mantle have been determined by the interaction of several complex processes unique to the earth and they conclude that "the moon was derived from the earth's mantle after the earth's core had formed". The higher depletion of the moon for volatiles compared to the earth is explained by the recondensation process at some distance from the earth under unique conditions which permitted further loss of volatile components.

Acknowledgements

We wish to thank the staff of our Institute, in particular Mr. P. Deibele, Mrs. A. Delnef and Miss H. Prager. We are grateful to the Deutsche Forschungsgemeinschaft for financial support.

References

ANDERSON, D. L. (1973) The composition and origin of the Moon. *Earth Planet. Sci. Lett.* **18**, 301–316.

CLAYTON, R. N. and MAYEDA, T. K. (1975) Genetic relations between the moon and meteorites. *Proc. Lunar Sci. Conf. 6th* 1761–1769.

DREIBUS, G., SPETTEL, B. and WÄNKE, H. (1976) Lithium as a correlated element, its condensation behaviour, and its use to estimate the bulk composition of the moon and the eucrite parent body. *Proc. Lunar Sci. Conf. 7th* 3383–3396.

DREIBUS, G., KRUSE, H., SPETTEL, B. and WÄNKE, H. (1977) The bulk composition of the moon and the eucrite parent body. *Proc. Lunar Sci. Conf. 8th* 211–227.

DOWTY, E., PRINZ, M. and KEIL, K. (1974) Ferroan anorthosite: A widespread and distinctive lunar rock type. *Earth Planet. Sci. Lett.* **24**, 15–25.

GANAPATHY, R. and ANDERS, E. (1974) Bulk compositions of the moon and earth, estimated from meteorites. *Proc. Lunar Sci. Conf. 5th, Geochim. Cosmochim. Acta*, **2**, 1181–1206.

GROSSMAN, L. (1972) Condensation in the primitive solar nebula. *Geochim. Cosmochim. Acta* **36**, 567–619.

GROSSMAN, L. (1973) Refractory trace elements in Ca–Al-rich inclusions in the Allende meteorite. *Geochim. Cosmochim. Acta* **37**, 1119–1140.

LARIMER, J. W. and ANDERS, E. (1970) Chemical fractionations in meteorites. III. Major element fractionations in chondrites. *Geochim. Cosmochim. Acta* **34**, 367–387.

LAUL, J. C. and SCHMITT, R. A. (1975) Dunite 72417: A chemical study and interpretation. *Proc. Lunar Sci. Conf. 6th* 1231–1254.

LSPET (1973) Apollo 17 samples. Chemical and petrographic description. *Science* **182**, 659–672.

VON MICHAELIS, H., AHRENS, L. H. and WILLIS, J. P. (1969) The composition of stony meteorites II. The analytical data and an assessment of their quality. *Earth Planet. Sci. Lett.* **5**, 387–394.

O'KEEFE, J. A. (1969) Origin of the moon. *J. Geophys. Res.* **74**, 2758–2767.

PALME, H. and WÄNKE, H. (1975) A unified trace-element model for the evolution of the lunar crust and mantle. *Proc. Lunar Sci. Conf. 6th* 1179–1202.

PAPANASTASSIOU, D. A. and WASSERBURG, G. J. (1976) Rb–Sr age of troctolite 76535. *Proc. Lunar Sci. Conf. 7th* 2035–2054.

RAMMENSEE, W. and WÄNKE, H. (1977) On the partition coefficient of tungsten between metal and silicate and its bearing on the origin of the moon. *Proc. Lunar Sci. Conf. 8th* 399–409.

RINGWOOD, A. E. (1960) Some aspects of the thermal evolution of the earth. *Geochim. Cosmochim. Acta* **20**, 241–259.

RINGWOOD, A. E. (1970) Origin of the moon: The precipitation hypothesis. *Earth Planet. Sci. Lett.* **8**, 131–140.

RINGWOOD, A. E. (1977) Basaltic magmatism and the bulk composition of the moon. I. Major and heat producing elements. The Moon **16**, 389–423.

RINGWOOD, A. E. and KESSON, S. E. (1977) Basaltic magmatism and the bulk composition of the moon. II. Siderophile and volatile elements in the moon, earth and chondrites: Implications for lunar origin. The Moon **16**, 425–464.

RYDER, G. and WOOD, A. (1977) Serenitatis and imbrium impact melts: implications for lunar crustal composition and stratigraphy (abstract). The Lunar Science Institute, Houston. In *Lunar Science* VIII, 826–828.

SCHAEFFER, O. A., HUSAIN, L. and SCHAEFFER, G. A. (1976) Ages of highland rocks: The chronology of lunar basin formation revisited. *Proc. Lunar Sci. Conf. 7th* 2067–2092.

TAYLOR, S. R. (1976) Geochemical constraints on the composition of the moon. *Proc. Lunar Sci. Conf. 7th* 3461–3477.

TESCHKE, F. and WÄNKE, H. (1974) Major element analysis of extraterrestrial rock samples with 14 MeV neutrons. *Radiochem. Radioanal. Lett.*, **18**, 341–348.

WÄNKE, H., BADDENHAUSEN, H., DREIBUS, G., JAGOUTZ, E., KRUSE, H., PALME, H., SPETTEL, B. and TESCHKE, F. (1973) Multielement analyses of Apollo 15, 16 and 17 samples and the bulk composition of the moon. *Proc. Lunar Sci. Conf 4th* 1461–1481.

WÄNKE, H., PALME, H., BADDENHAUSEN, H., DREIBUS, G., JAGOUTZ, E., KRUSE, H., SPETTEL, B., TESCHKE, F. and THACKER, R. (1974a) Chemistry of Apollo 16 and 17 samples: Bulk composition, late stage accumulation and early differentiation of the moon. *Proc. Lunar Sci. Conf. 5th, Geochim. Cosmochim. Acta* 1307–1335.

WÄNKE, H., BADDENHAUSEN, H., PALME, H. and SPETTEL, B. (1974b) On the chemistry of the Allende inclusions and their origin as high temperature condensates. *Earth Planet. Sci. Lett.* **23**, 1–7.

WÄNKE, H., PALME, H., BADDENHAUSEN, H., DREIBUS, G., JAGOUTZ, E., KRUSE, H., PALME, C., SPETTEL, B., TESCHKE, F. and THACKER, R. (1975) New data on the chemistry of lunar samples: Primary matter in the lunar highlands and the bulk composition of the moon. *Proc. Lunar Sci. Conf. 6th* 1313–1340.

WÄNKE, H., PALME, H., KRUSE, H., BADDENHAUSEN, H., CENDALES, M., DREIBUS, G., HOFMEISTER, H., JAGOUTZ, E., PALME, C., SPETTEL, B. and THACKER, R. (1976) Chemistry of lunar highland rocks: A refined evaluation of the composition of the primary matter. *Proc. Lunar Sci. Conf. 7th* 3479–3499.

WÄNKE, H., BADDENHAUSEN, H., BLUM, K., CENDALES, M., DREIBUS, G., HOFMEISTER, H., KRUSE, H., JAGOUTZ, E., PALME, C., SPETTEL, B., THACKER, R. and VILCSEK, E. (1977) On the chemistry of lunar samples and achondrites. Primary matter in the lunar highlands: A re-evaluation. *Proc. Lunar Sci. Conf. 8th* 2191–2213.

WIIK, H. B. (1963) in B. MASON: The carbonaceous chondrites. *Space Sci. Rev.* **1**, 621–646.

WISE, D. U. (1963) An origin of the moon by rotational fission during formation of the earth core. *J. Geophys. Res.* **68**, 1547–1554.

WISE, D. U. (1969) Origin of the moon from the earth: Some new mechanisms and comparisons. *J. Geophys. Res.* **74**, 6034–6045.

WISKERCHEN, M. J. and SONETT, C. P. (1977) A lunar metal core? (abstract). The Lunar Science Institute, Houston. In *Lunar Science* VIII, 1020–1022.

WOOD, J. A., DICKEY, J. S., MARVIN, U. B. and POWELL, B. N. (1970) Lunar anorthosites and a geophysical model of the moon. *Proc. Apollo 11 Lunar Sci. Conf.* 965–988.

Sondage à Distance de l'Atmosphère des Planetés Géantes par Spectroscopie de Fourier

M. Combes

Groupe "Planètes", Observatoire de Meudon, 92190 Meudon, France

Abstract

The determination of abundance and isotopic ratios in the Jovian planets is of particular interest since they represent more than 98% of the non-solar mass in the solar system and are expected to be a good image of the primordial nebula. In the past few years numerous new results have been obtained mainly by means of Fourier Transform spectroscopy in the infrared spectral range. A brief review is given of the definitely or tentatively identified molecular species. The significance of the derived abundance and isotopic ratios is discussed, taking into account the different mechanisms of lines formation involved in remote sensing methods. Possible future studies of the Jovian planets are examined.

L'objet de cette communication est de faire le point de nos connaissances sur la composition chimique et les rapports élémentaires et isotopiques dans l'atmosphère des planètes géantes. Il ne sera question que de Jupiter et de Saturne, nos connaissances sur Uranus et Neptune sont en effet très limitées.

La simple identification des principaux constituants atmosphériques fut une opération difficile dont les progrès furent très lents. En 1863, l'existence d'une atmosphère sur Jupiter fut mise en évidence. Mais c'est seulement *70 ans* plus tard que fut obtenue l'identification du méthane et de l'ammoniac et *un siècle* plus tard l'identification du constituant *le plus abondant*, l'hydrogène moléculaire. Il y a seulement *5 ans* en 1972, il n'avait toujours pas été identifié de constituants autres que H_2, CH_4 et NH_3.

I. Nouvelles Identifications Moléculaires

Par contre dans les cinq dernières années les progrès ont été véritablement spectaculaires Sur Jupiter les molécules *certainement* ou seulement *probablement* identifiées, sont aujourd'hui: H_2, CH_4, NH_3, l'hélium, C_2H_2, C_2H_6, la phosphine, la vapeur d'eau, CO, CO_2, HCN, le germane GeH_4; et les isotopes HD, $^{13}CH_4$, CH_3D, $^{15}NH_3$. Les tableaux 1 et 2 montrent que l'activité dans ce domaine est très vive et que de nombreuses équipes ont participé à ce travail de découverte. Je me permettrai d'indiquer brièvement qu'elle a été la contribution du "Groupe Planètes" de l'Observatoire de Meudon à cet effort.

Grâce à l'interféromètre de Pierre Connes et Jean Pierre Maillard, nous avons pu confirmer la présence de $^{13}CH_4$ en identifiant 5 multiplets de la bande $3\nu_3$ (de Bergh *et al.*, 1976). A 10 microns à l'aide d'un interféromètre que nous avons construit nous avons observé l'acétylène C_2H_2 et l'éthane C_2H_6 (Combes *et al.*, 1974).

TABLEAU 1. **Jupiter**

He	584 Å	1974	Judge and Carlson
HD	7460 Å (4–0) P(1)	1973	Trauger *et al.*
$^{13}CH_4$	1.1 μ	1972	Fox *et al.*
	$3\nu_3$-band	1974	De Bergh *et al.*
CH_3D	5μ	1972	Beer *et al.*
H_2O	5μ	1975	Larson *et al.*
CO	5μ	1975	Beer *et al.*
PH_3	5μ	1977	Larson *et al.*
	10μ	1976	Ridgway *et al.*
	10μ	1976	Combes *et al.*
C_2H_2	10μ	1974	Ridgway *et al.*
		1974	Combes *et al.*
		1976	Aumann and Orton
C_2H_6	10μ	1974	Ridgway
		1974	Combes *et al.*
		1976	Tokunaga *et al.*
Tentative identification:			
CO_2	5μ	1975	Larson *et al.*
HCN	5μ	1976	Larson *et al.*
GeH_4	5μ	1976	Larson *et al.*
$^{15}NH_3$	10μ	1976	Lacy *et al.*
		1977	Encrenaz *et al.*

TABLEAU 2. **Saturn**

$^{13}CH_4$	1.1μ $3\nu_3$-band	1975	Combes *et al.*
C_2H_6	10μ	1975	Tokunaga *et al.*
Tentative identification:			
C_2H_4	10μ	1975	Encrenaz *et al.*
		1974	Gillett *et al.*
PH_3	10μ	1975	Encrenaz *et al.*
		1975	Bregman *et al.*

Nous avons également observé la phosphine PH_3 et l'isotope de l'ammoniac $^{15}NH_3$ (Encrenaz *et al.*, 1977b).

Sur Saturne, nous avons découvert $^{13}CH_4$ (Combes *et al.*, 1975) et, mais cela reste à confirmer, l'éthylène C_2H_4 et la phosphine PH_3 (Encrenaz *et al.*, 1975).

Il faut remarquer que les identifications les plus nombreuses, obtenues par les différentes équipes, ont été faites dans l'infrarouge entre 1 et 25 microns et par interférométrie de Fourier, ce qui montre la richesse de ce domaine spectral et la puissance de cette technique.

II. Difficultés de la Spectroscopie à Distance

La détermination des rapports d'abondance revient en définitive à la comparaison de raies spectrales d'éléments différents. La question fondamentale est donc de savoir si ces raies sont ou non comparables, si les processus de formation sont similaires et surtout si les raies sont formées dans la même région de l'atmosphère. S'il n'en est pas ainsi les rapports obtenus ne sont pas significatifs.

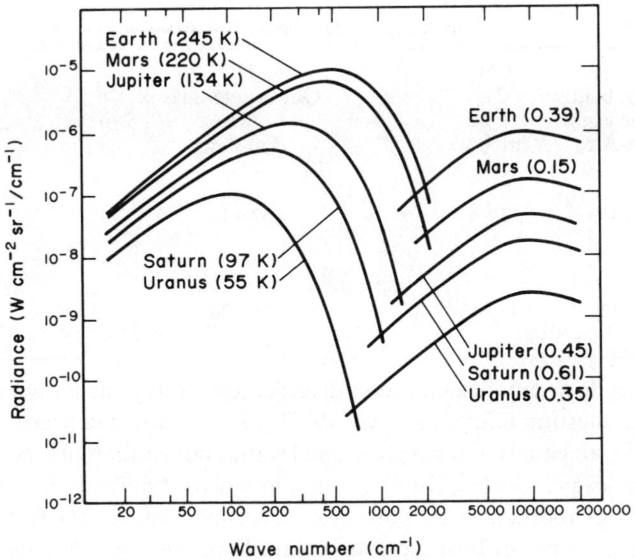

FIG. 1. Aspect général du spectre des planètes.

La figure 1 montre l'aspect général du spectre des planètes géantes. Dans le visible et le proche IR on observe la lumière solaire réfléchie par la planète. Les raies sont formées par diffusion. Leur formation dépendra essentiellement de l'intensité du coefficient d'absorption et de la distribution des aérosols et des nuages dans l'atmosphère. Dans l'IR moyen et lointain, on observe le rayonnement thermique propre à l'atmosphère. Les raies observées dépendront surtout du profil vertical de température de l'atmosphère.

III. Le Rapport H/He

Le rapport hydrogène/hélium est d'un intérêt cosmologique capital. Mais sa mesure est très difficile y compris dans le Soleil où elle reste très imprécise. La meilleure détermination de ce rapport devrait être obtenue sur les planètes géantes qui ont, semble-t-il, conservé la valeur du rapport H/He de la nébuleuse primitive.

Diverses méthodes, telles la détection directe à 584 Å, les occultations stellaires ou l'estimation de l'abondance d'hélium par l'élargissement des raies spectrales qu'il provoque, n'ont pas fourni de résultats suffisamment précis.

C'est par l'analyse du spectre infrarouge lointain, dû aux collisions hydrogène-hydrogène et hydrogène-hélium que l'on obtient les meilleurs résultats. Selon les travaux de Houck *et al.* (1975), de Orton (1975) et de D. Gautier *et al.* (1977) il apparait que le rapport $H_2/(H_2+He)$ sur Jupiter est de 0.88 ± 0.06. La valeur solaire est à peu près 0.85 ± 0.05. Pour Saturne, il n'y a pas encore de résultat.

IV. Le Rapport C/H

En ce qui concerne le rapport C/H la situation est moins favorable. La difficulté est de déterminer quelles sont les raies de CH_4 et de H_2 qu'il est raisonnable de comparer. Nous

TABLEAU 3. CH_4 AND H_2 ABUNDANCES

	CH_4 weak bands in the visible (m-Am)	CH_4 $3\nu_3$ 1.1μ (m-Am)	H_2 Quadruple lines (km-Am) 4–0	3–0	2–0	H_2 Dipole bands (km-Am) 2–0	1–0
Jupiter	~150	52 ± 6	67 ± 17	67 ± 17	$17\pm^{6}_{3}$	$56\pm^{12}_{6}$	~25
Saturn	~150	$59\pm^{15}_{7}$	78 ± 20	76 ± 20	$25\pm^{9}_{6}$	$63\pm^{13}_{8}$	$25\pm^{10}_{9}$

disposons en effet de nombreuses mesures divergentes. Nous nous sommes livrés à une étude critique de cette question (Combes *et al.*, 1977b). Les résultats concernant les abondances sont résumés ici (tableau 3). On constate que les plus courtes longueurs d'onde conduisent aux abondances les plus fortes. De plus à une longueur d'onde donnée les raies les moins intenses donnent les abondances les plus fortes. Ceci résulte des effets de la diffusion: les raies intenses sont formées par diffusion primaire dans la partie supérieure de l'atmosphère. Les raies faibles sont formées par diffusion multiple dans une couche atmosphérique plus épaisse et plus profonde. De plus la diffusion est plus intense aux courtes longueurs d'onde.

Notre conclusion est la suivante: le meilleur choix est de comparer l'abondance de CH_4 déduite de la bande $3\nu_3$ et celle de H_2 déduite de la bande dipolaire (2–0) ou de la bande quadrupolaire (4–0).

Pour Jupiter et pour Saturne, on obtient un rapport C/H de $4.2\pm1.6\times10^{-4}$.

V. Le Rapport N/H

La détermination du rapport N/H est très difficile car l'ammoniac condense à la basse température de l'atmosphère de Jupiter et Saturne. Le rapport de mélange NH_3/H_2 ne dépend alors que de la température et non du rapport N/H dans l'atmosphère profonde. Cependant, nous avons montré (Owen *et al.*, 1976; Encrenaz *et al.*, 1977a) que la bande NH_3 à 1.56 microns permettait sur Jupiter de sonder la zone convective au-dessous des nuages d'ammoniac dans une zone de l'atmosphère où le rapport NH_3/H_2 est constant. Nous en avons déduit un rapport N/H égal à 1/3 du rapport solaire. Cette valeur n'est qu'une limite inférieure. En effet, une fraction de l'azote est fixée dans les nuages de l'atmosphère profonde sous forme de composés du soufre, de l'ammoniac et de l'eau.

Cette méthode n'est pas utilisable pour Saturne où la zone de condensation de l'ammoniac est beaucoup plus étendue. La seule manière de sonder l'atmosphère au-dessous des nuages de NH_3 est d'utiliser les mesures aux longueurs d'ondes centimétriques. Orhing et Lacser (1976) trouvent que leurs mesures ne sont pas incompatibles avec le rapport N/H solaire.

VI. Le Rapport D/H

L'identification de CH_3D sur Jupiter a été obtenue par Beer *et al.* (1972). C'était la première découverte du deutérium dans un objet astronomique. Beer et Taylor (1973a) en ont déduit un rapport D/H de $5.1\pm2.2\times10^{-5}$. Récemment, Beer et Taylor (1977) ont indiqué que des valeurs plus élevées devaient être considérées comme plus probables.

Par ailleurs, Trauger *et al.* (1973) ont identifié la raie dipolaire $P_4(1)$ de HD. Par comparaison avec la raie quadrupolaire $S_4(1)$ de H_2, ils en ont déduit le rapport D/H. Cependant McKellar *et al.* (1976) ont montré que l'intensité de la raie de HD utilisée par Trauger *et al.* (1973) était erronée. Leur mesure conduit à un rapport D/H de $5.6 \pm 1.4 \times 10^{-5}$.

Ces valeurs élevées impliquent l'une et l'autre un enrichissement de l'atmosphère jovienne en deutérium par rapport à la nébuleuse primitive. Cependant nous avons montré (Combes *et al.*, 1977b) que la raie extrêmement faible de HD ne peut vraisemblablement pas être comparée, de façon significative, à la raie quadrupolaire $S_4(1)$ de H_2 (du fait du poids très différent de la diffusion dans les deux cas) et qu'une valeur plus significative du rapport D/C peut être déduite de la comparaison de HD et de bandes faibles de CH_4 observées dans le spectre visible par Lutz *et al.* (1976). Cette valeur, associée à la valeur du rapport C/H donnée plus haut (§ IV) fournit une valeur de D/H: $1.6 \pm 0.7 \times 10^{-5}$. Cette valeur est en bon accord avec la valeur prévue pour la nébuleuse primitive (Geiss *et al.*, 1972). Elle n'implique pas de phénomènes d'enrichissement de l'atmosphère jovienne en deutérium. La valeur déduite de l'observation de CH_3D peut s'expliquer par une liaison privilégiée du deutérium au méthane favorisée par catalyse dans les nuages de Jupiter (Beer et Taylor, 1973b).

VII. Le Rapport $^{12}C/^{13}C$

Audouze (1977) a montré dans une communication précédente que le rapport $^{12}C/^{13}C$ est un puissant indicateur de l'évolution chimique de la galaxie, et l'intérêt qu'il y a à déterminer sa valeur dans la nébuleuse primitive. La question de base est de décider si le rapport terrestre est ou non représentatif de la nébuleuse primitive.

La figure 2 montre le résultat de notre analyse du rapport $^{12}C/^{13}C$. La comparaison directe de la bande $3\nu_3$ très intense de $^{12}CH_4$ et de la bande très faible de $^{13}CH_4$ ne conduit pas à un résultat fiable à cause des phénomènes de diffusion. Mais la méthode que nous avons utilisée est indépendante de la diffusion et de la structure de l'atmosphère. Elle fournit directement le rapport $^{12}C/^{13}C$ sans passer par l'abondance absolue des constituants. Nous obtenons une valeur comparable à la valeur solaire avec une précision analogue (Combes *et*

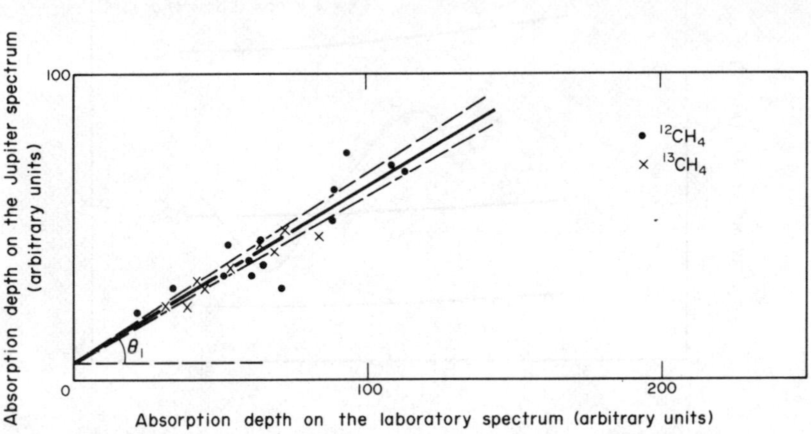

FIG. 2. Détermination du rapport $^{12}C/^{13}C$ dans l'atmosphère de Jupiter. Cette figure montre que le rapport $^{12}C/^{13}C$ dans l'atmosphère de Jupiter ne diffère pas significativement du rapport tellurique.

al., 1977a): soit $89 \pm ^{12}_{10}$ pour Jupiter et $89 \pm ^{25}_{18}$ pour Saturne. Des valeurs similaires sont donc obtenues en divers points du système solaire, dans des conditions physiques très différentes, à partir de molécules différentes. Le fait que l'on obtienne cette valeur aussi pour les planètes géantes est le meilleur argument en faveur de notre conclusion: le rapport $^{12}C/^{13}C$ obtenu est représentatif de la nébuleuse primitive.

VIII. Le Rapport $^{14}N/^{15}N$

Le rapport $^{14}N/^{15}N$ est aussi un indicateur de l'évolution chimique de la galaxie. La meilleure détermination de sa valeur pourrait être fournie par les planètes géantes. On voit sur la figure 3 une raie faible dans le flanc d'une raie forte de l'ammoniac au voisinage de 11 microns. Lacy *et al.* (1976) l'ont identifié à la raie sP(3) de la bande ν_2 de $^{15}NH_3$. Selon ces auteurs, cette observation est compatible avec une valeur tellurique du rapport $^{14}N/^{15}N$. Mais comme il a été dit plus haut, les déterminations d'abondance dans l'infrarouge moyen dépendent essentiellement du profil vertical de température de l'atmosphère. En utilisant le profil thermique actuellement le plus plausible, nous avons montré (Encrenaz *et al.*, 1977b) que le meilleur accord avec les données de Lacy *et al.*, implique un excès de ^{15}N d'un facteur 5 par rapport à la valeur terrestre. Il nous semble cependant que cet excès apparent de ^{15}N pourrait être dû à la présence à cette fréquence de las sous-bande P(9) de la bande ν_2 de PH_3.

La figure 4 montre un spectre de Jupiter au voisinage de 10 microns que nous venons d'obtenir. Le dépouillement n'en est pas achevé, en particulier il n'est pas corrigé des absorptions telluriques. Il montre l'émission de C_2H_2 et de C_2H_6 et la bande ν_2 de l'ammoniac en absorption. Il montre aussi plusieurs absorptions dûes à PH_3 et en particulier la branche Q de la bande ν_2. La présence de la phosphine à 10 microns sur Jupiter semble donc définitivement établie. Elle peut donc être partiellement responsable de l'absorption

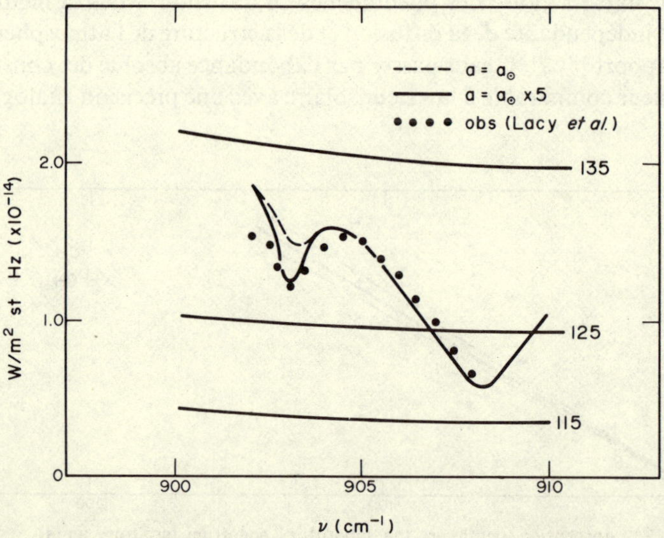

FIG. 3. Evaluation du rapport $^{14}N/^{15}N$ dans l'atmosphère de Jupiter, à partir des observations de Lacy *et al.* (1976) et du profil vertical de température de Orton (1975).

dans la région étudiée par Lacy *et al.* (1976). Cependant ce spectre semble établir également la présence de $^{15}HN_3$ dont on voit à 959 cm^{-1} la branche Q de la bande ν_2. L'étude détaillée de ce spectre donnera, on peut l'espérer, une meilleure estimation du rapport $^{14}N/^{15}N$ et du rapport P/H. Rappelons, en ce qui concerne P/H que Ridgway *et al.* (1976) à 10 microns et Larson *et al.* (1977) à 5 microns ont estimé que leurs observations sont compatibles avec un rapport P/H solaire.

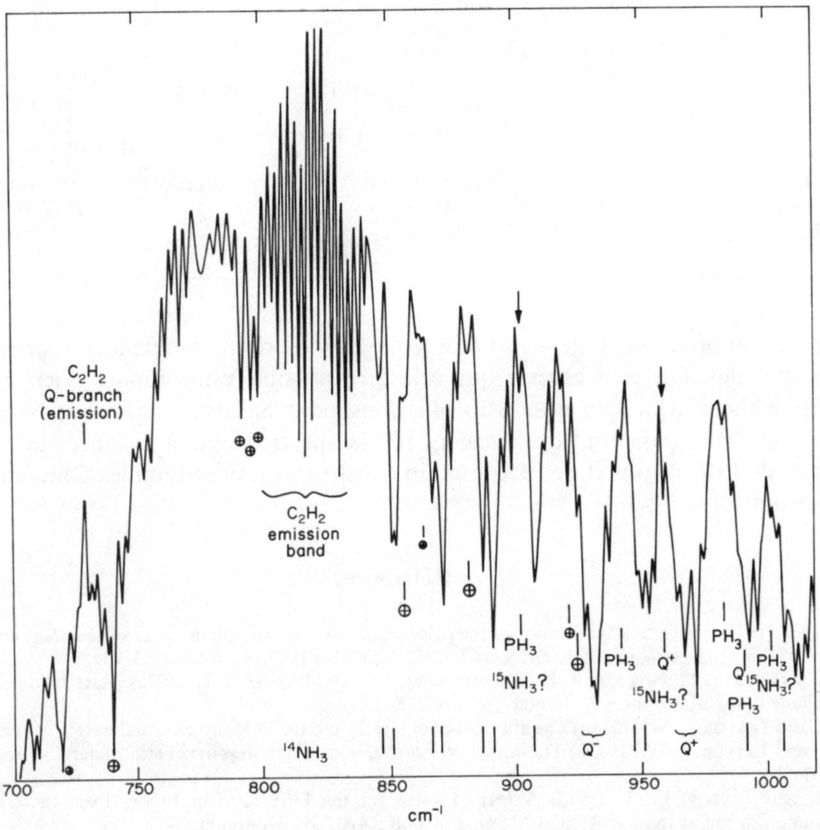

FIG. 4. Spectre de Jupiter de 720 cm^{-1} à 1000 cm^{-1} obtenu par le Groupe "Planètes" (Observatoire de Meudon). Ce spectre n'est pas corrigé des absorptions telluriques (⊕). Outre les bandes moléculaires intenses de $^{14}NH_3$ et C_2H_6, il semble que des bandes de C_2H_2, PH_3 et $^{15}NH_3$ sont présentes.

Conclusion

Le tableau 4 donne le bilan de cette étude. Seules les déterminations des rapports H/He sur Jupiter et $^{12}C/^{13}C$ sur les deux planètes sont raisonnablement précises. Elles sont en accord avec les rapports solaires. Le rapport C/H bien qu'imprécis est comparable aux récentes déterminations du rapport solaire (Pagel, 1977). Les rapports N/H et P/H ne sont encore qu'indicatifs. Le rapport D/H est selon nous voisin de la valeur prévue pour la nébuleuse primitive (Geiss *et al.*, 1972).

TABLEAU 4.

	Firm results	Tentative results	Should be obtained in the near future	
$\frac{H_2}{H_2+He}$	Jupiter: 0.88 ± 0.06		Saturn	(Inversion of IR data)
$\frac{^{12}C}{^{13}C}$	Jupiter: 89^{+12}_{-10} Saturn: 89^{+25}_{-18}			
C/H		Jupiter: 4.2×10^{-4} Saturn: 4.2×10^{-4} $(\sim(C/H)_\odot)$		
N/H		Jupiter: $> \frac{1}{3}(N/H)_\odot$	Saturn	(Inversion of mm and cm data)
P/H		Jupiter: $\sim(P/H)_\odot$	Saturn	10μ–ν_2 band (Far IR rotational spect.)
D/H		Jupiter: $(1.6 \pm 0.7) \times 10^{-5}$	Saturn	(HD visible line)
$\frac{^{14}N}{^{15}N}$			Jupiter	(10μ–ν_2 bands of $^{14}NH_3$ and $^{15}NH_3$)

Bien sûr, un progrès décisif, serait d'obtenir des mesures d'abondances *in situ* à l'aide de sondes spatiales. On peut espérer que cela sera possible pour Jupiter d'ici une dizaine d'années mais le délai sera beaucoup plus long pour Saturne, Uranus et Neptune. Les progrès nouveaux passeront donc encore par la spectroscopie à distance. Les domaines spectraux de l'IR moyen et de l'IR lointain constituent aujourd'hui les domaines les plus prometteurs.

Références

AUDOUZE, J. (1977) The chemical evolution of the galaxy and isotopic ratios in the solar system. *Second Symposium on the Origin and Distribution of the Elements*, IAGC, Paris, May 1977, this volume, 83–90.

BEER, R., FARMER, C. B., NORTON, R. H., MARTONCHIK, J. V. and BARNES, T. G. (1972) Observation of deuterated methane in the atmosphere of Jupiter. *Science* **175**, 1360.

BEER, R. and TAYLOR, F. W. (1973a) The abundance of CH_3D and the D/H ratio in Jupiter. *Astrophys. J.* **179**, 309.

BEER, R. and TAYLOR, F. W. (1973b) The equilibration of Deuterium in the Jovian atmosphere. *Astrophys. J.* **182**, L131.

BEER, R. and TAYLOR, F. W. (1977) A revised value for the D/H ratio in Jupiter from the CH_3D phase. Communication at the 8th Annual Meeting of D.P.S.–A.A.S., Honolulu.

BERGH, C. DE, MAILLARD, J. P., LECACHEUX, J. and COMBES, M. (1976) A study of the $3\nu_3$ CH_4 region in a high resolution spectrum of Jupiter recorded by FTS. *Icarus*, **29**, 307.

COMBES, M., ENCRENAZ, T., VAPILLON, L., ZEAU, Y. and LESQUEREN, C. (1974) Confirmation of the identification of C_2H_2 and C_2H_6 in the Jovian atmosphere. *Astron. Astrophys.* **34**, 33.

COMBES, M., DE BERGH, C., LECACHEUX, J. and MAILLARD, J. P. (1975) Identification of $^{13}CH_4$ in the atmosphere of Saturn. *Astron. Astrophys.* **40**, 81.

COMBES, M., MAILLARD, J. P. and DE BERGH, C. (1977a) Evidence for a Telluric value of the $^{12}C/^{13}C$ ratio in the atmospheres of Jupiter and Saturn. *Astron. Astrophys.* **61**, 531.

COMBES, M., ENCRENAZ, T. and OWEN, T. (1977b) On the D/H ratio in the atmosphere of Jupiter. To be published in *Ap. J.* (1 April 1978).

ENCRENAZ, T., COMBES, M., ZEAU, Y., VAPILLON, L. and BEREZNE, J. (1975) Tentative identification of C_2H_4 in the spectrum of Saturn. *Astron. Astrophys.* **42**, 355.

ENCRENAZ, T. and COMBES, M. (1977a) The N/H and $^{14}N/^{15}N$ Jovian ratios: a tentative determination from infrared observations. Communication at the 8th Annual Meeting of D.P.S.–A.A.S., Honolulu.

ENCRENAZ, T., COMBES, M. and ZEAU, Y. (1977b) New Observations of Jupiter in the 800–1000 cm^{-1} range. *Infrared Physics*, **17**, 551.

GAUTIER, D., LACOMBE, A. and REVAH, I. (1977) The thermal structure of Jupiter from IR spectral measurements by means of a filtered iterative method. *J. Atm. Sci.* **34**, 7, 1130.

Geiss, J. and Reeves, H. (1972) Cosmic and solar system abundances of deuterium and helium 3. *Astron. Astrophys.* **18**, 126.

Houck, J. R., Pollack, J. B., Schaack, D., Reed, R. A. and Summers, A. (1975) Jupiter: its infrared spectrum from 16 to 40 microns. *Science* **189,** 720.

Larson, H. P., Treffers, R. R. and Fink, U. (1977) Phosphine in Jupiter's atmosphere: the evidence from high-altitude observations at 5 micrometers. *Astrophys. J.* **211**, 972.

Lutz, B. L., Owen, T. and Cess, R. D. (1976) Laboratory band strengths of methane and their application to the atmospheres of Jupiter, Saturn, Uranus, Neptune and Titan. *Astrophys. J.* **203**, 541.

McKellar, A. R. W., Goetz, W. and Ramsay, D. A. (1976) The rotation–vibration spectrum of HD. *Astrophys. J.* **207**, 663.

Ohring, G. and Lacser, A. (1976) The ammonia profile in the atmosphere of Saturn from inversion of its microwave emission spectrum. *Astrophys. J.* **206**, 622.

Orton, G. S. (1975) The thermal structure of Jupiter. *Icarus*, **26**, 125.

Owen, T., McKellar, A. R. W., Encrenaz, T., Lecacheux, J., de Bergh, C. and Maillard, J. P. (1976) A study of the 1.56μ NH_3 band on Jupiter and Saturn. *Astron. Astrophys.* **54**, 291.

Pagel, B. E. J. (1977) Solar abundances. *Second Symposium on the Origin and Distribution of the elements*, IAGC, Paris, May 1977, this volume, 80–81.

Ridgway, S. T., Wallace, L. and Smith, G. R. (1976) The 800–1200 cm^{-1} absorption spectrum of Jupiter. *Astrophys. J.* **207**, 1002.

Trauger, J. T., Roesler, F. C., Carleton, N. P. and Traub, W. A. (1973) Observation of HD on Jupiter and the D/H ratio. *Astrophys. J.* **184**, L137.

Isotopic Anomalies in the Early Solar System

ROBERT N. CLAYTON

The Enrico Fermi Institute, University of Chicago, U.S.A.

Abstract

The primitive meteorites known as "carbonaceous chondrites" provide the best available information concerning the nuclear composition of the matter in the solar system at the time of formation of the sun and planets. Several light elements show "isotopic anomalies" which appear to result from major nucleosynthetic events occurring just before collapse of the solar nebula. For example, the earliest-condensed mineral grains contain excesses of ^{16}O up to 5%, while maintaining "normal" $^{17}O/^{18}O$ ratios. Magnesium contains ^{26}Mg excesses up to 10%, which are attributable to the decay of 7×10^5 year ^{26}Al. Neon has "anomalous" excesses of ^{22}Ne. Although the evidence is less clear-cut, there may also be nucleosynthetic excesses of ^{13}C and ^{15}N. Nuclear processes occurring *within* the solar system are inadequate to account for most of these observations. The isotopic anomalies have been interpreted as evidence for a supernova "trigger" for formation of the solar system.

The ^{16}O "anomaly" is not restricted to the primitive meteorites, but is found to varying degrees in all meteorites, reflecting the inhomogeneous distribution of the ^{16}O-rich component in the early solar system. This provides the basis for a genetic classification of meteorites and planets.

Variations in isotopic abundances in meteorites may result from several processes, most of which have been understood for many years. These include both nuclear processes and mass-dependent fractionation processes. The nuclear processes include radioactive decay of both long-lived and short-lived parents, and cosmic-ray reactions involving either protons or neutrons. We must consider also the implantation of ions derived from the sun, where isotopic abundances may have been modified by nuclear reactions. The mass-dependent processes are especially important for volatile light elements, which are not quantitatively condensed in the meteorites.

I shall consider as "isotopic anomalies" only those variations which are not understood in terms of known solar system processes. These then result either from processes outside the solar system (or prior to the existence of the solar system) or from processes which operated within the early solar system but are no longer active. Isotopic anomalies of this sort have been observed for a few elements, beginning with the discovery of so-called neon-E by David Black (1972). Observation of substantial ^{16}O excesses in primitive meteorites by Clayton *et al.* (1973) established the fact that the solar nebula was not isotopically homogenized prior to formation of the meteorites, and made plausible the suggestion of a separate nucleosynthetic event as the source of this anomaly. Then the discovery of ^{26}Mg-excesses by Gray and Compston (1974) and by Lee and Papanastassiou (1974), followed by the demonstration by Lee *et al.* (1976) of the origin of this anomaly from decay of ^{26}Al within the solar system, established a short time-scale for the interval between the last major nucleosynthesis and solar system condensation.

Besides these examples, there may also be anomalies in carbon (Clayton, 1963) and nitrogen (Kung, 1977), but the availability of only two stable isotopes makes an

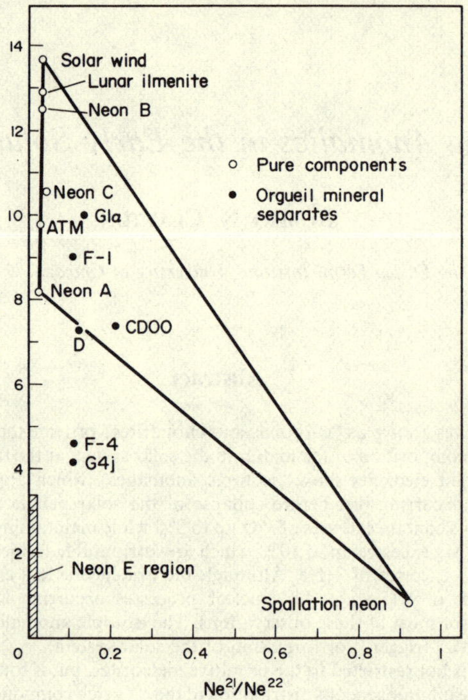

FIG. 1. Isotopic compositions of neon from separated fractions of Orgueil (Eberhardt, 1974) in comparison to major neon components observed in the earth, moon and other meteorites (Black, 1972). Subsequent measurements by Eberhardt have set a limit on $^{20}Ne/^{22}Ne$ of neon-E to ≤ 1.5.

unambiguous identification difficult. Recent measurements of silicon isotope ratios by Epstein and Yeh (1977) are inconclusive, in that sizable variations have been observed in one primitive meteorite, but they may be due to mass-fractionation rather than nuclear processes. Searches for isotopic anomalies in heavy elements have usually given negative results. Krypton and xenon have very complicated isotopic behavior, which is not completely understood, and which may contain nucleosynthetic information.

One of the main lessons learned in the last few years is that the interesting effects are rarely observed by analyzing meteorites as whole rocks, and are generally revealed only through careful sampling of specific phases, particularly phases from primitive meteorites which have undergone a minimum of chemical processing since their initial condensation.

The anomalous neon component is observed in the type C1 and C2 carbonaceous chondrites. The most extensively studied is Orgueil, as seen in the data of Eberhardt (1974). The three-isotope plot (Fig. 1) separates the various components located in different sites within the meteorite, primarily a so-called "solar" component, B, a "planetary" component, A, a cosmic-ray spallation-produced component, and the ^{22}Ne-rich neon-E. Further analytical refinement has produced an even purer neon-E fraction (by step-wise heating) supporting the conjecture that the anomaly is the result of the addition of pure ^{22}Ne. Most current ideas on the origin of the ^{22}Ne, and its strong enrichment relative to the other isotopes, involve a radioactive ^{22}Na precursor, which condensed into solid particles before decay to ^{22}Ne. This ^{22}Na may have been produced either within the solar system by a proton irradation or outside the solar system in a supernova explosion. In either case some

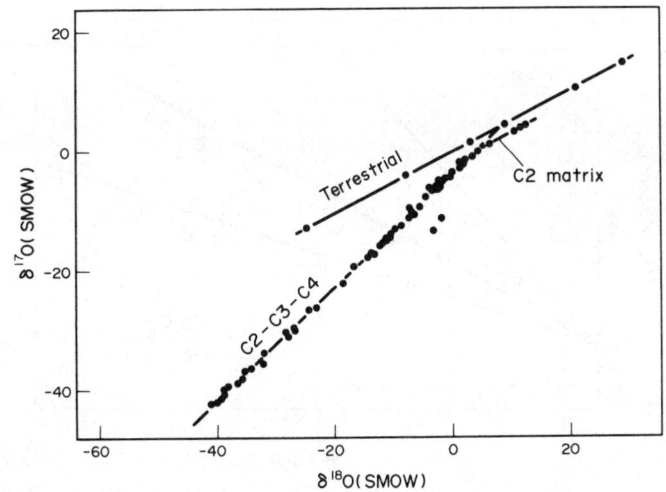

FIG. 2. Isotopic compositions of oxygen from high-temperature components of carbonaceous chondrites. The line C2–C3–C4 is a mixing line between the principal solar system reservoir at the upper right and a ^{16}O-rich exotic component at the lower left. The maximum observed ^{16}O excess is 5%.

remarkably rapid processes must be involved since the half-life of ^{22}Na is only 2.6 years! It is perhaps easier to accommodate this short time-scale into a supernova model than into a solar-system model.

The oxygen isotope anomaly is also best demonstrated on a three-isotope graph (Fig. 2). Here the well-known mass-fractionation effects appear as a single line with the predicted slope of 1/2, corresponding to a one mass-unit difference on the ordinate and a two mass-unit difference on the abscissa. The anomalous oxygen is found in the high-temperature condensate phases of C2 and C3 carbonaceous chondrites. The observed slope of the line indicates addition of a component highly enriched in ^{16}O, which may indeed be pure ^{16}O. This is a very plausible suggestion, since ^{16}O is synthesized in separate environments and by different processes than ^{17}O and ^{18}O, so that any single nucleosynthetic event could easily produce ^{16}O alone. The amount of the ^{16}O excess is found to be up to 5%. Since ^{16}O in normal oxygen is already the most abundant nuclide in the sample, this addition represents a very large number of atoms. The number is so great that no nuclear process acting within the solar system has been found which could account for the observed oxygen effects without having enormous effects on the isotopic compositions of the other light elements (D. D. Clayton *et al.*, 1977). We are, therefore, led to conclude that the synthesis of ^{16}O occurred outside the solar system in another star. The ^{16}O may have been brought into the solar system in the form of solid interstellar grains, in which case they might be very old, or it may have been introduced in gaseous form from a nearby supernova explosion just prior to solar system formation. We previously preferred the former interpretation (Clayton *et al.*, 1973) but more recent evidence favors the latter (Clayton and Mayeda, 1977).

The ^{16}O anomaly is so large that it can be observed in whole-meteorite samples, and even in whole-moon or whole-earth samples. It is found that the parent-bodies of the different types of meteorite have acquired different proportions of the ^{16}O-rich component. Thus, the oxygen isotopic composition serves to label the different bodies and to allow identification of objects formed from a common source material and to distinguish them from objects formed elsewhere in the solar system (Clayton *et al.*, 1976). It can hardly be coincidental that the

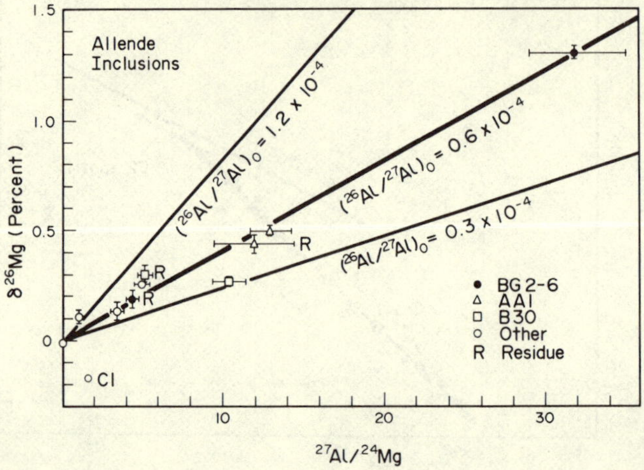

FIG. 3. ^{26}Al–^{26}Mg isochron diagram for several samples from Allende (Lee *et al.*, 1976). Correlation of excess ^{26}Mg with ^{26}Al/^{26}Mg implies *in situ* decay of ^{26}Al after solidification of the minerals.

earth and moon have identical oxygen isotopic compositions, which differ from all the major meteorite types. They must have had a common origin (Clayton and Mayeda, 1975).

Magnesium isotopes give more information on the time-scale of nucleosynthesis. Measurements on the same high-temperature condensate minerals which show ^{16}O-excesses also show ^{26}Mg excesses. When minerals separated from single centimeter-sized inclusions from the Allende meteorite are analyzed, a linear correlation is found between the excess ^{26}Mg and the ^{27}Al content of the minerals (Lee *et al.*, 1976). This isochron indicates that the ^{26}Mg formed by decay *in situ* of ^{26}Al. Since these inclusions are clearly formed within the solar system, the ^{26}Al must have been present at the time of their condensation. Since the half-life of ^{26}Al is only 720,000 years, the nuclear event which produced it could not have preceded condensation in the solar system by more than a few million years. Although there are plausible proton irradiation processes by which ^{26}Al might have been produced within the early solar system (Heymann and Dziczkaniec, 1976), the anticipated concomitant effects in other elements have not been observed (D. D. Clayton *et al.*, 1977). It is clear that further observations in other elements are needed to distinguish between a solar system or extra-solar system origin for the ^{26}Al.

If production of ^{26}Al within the solar system can be ruled out, we can then conclude that it was formed in a supernova explosion just prior to collapse of the solar nebula. This is the "supernova trigger" hypothesis which has been discussed by Cameron and Truran (1977) and by Sabu and Manuel (1976), and which has some new astronomical support from the observations of Herbst and Assousa (1977) of star formation in the shock-front surrounding a half-million-year-old supernova remnant.

It is clear that the observational evidence is not yet all in. Further measurements are necessary on elements with known anomalies to exploit the details of their distributions to understand both the nucleosynthetic processes which produced the anomalies, and the early solar system chemical processes which produced the meteorites and planets.

Searches for effects in other light elements are under way in several laboratories. It will be especially important to seek correlations between effects in different elements.

Studies are also needed on the dynamical aspects of the detonation of a supernova on a cloud of gas and dust. If indeed such an explosion did lead to collapse of the solar nebula, its effects on the chemical and physical properties of the solar system must have been profound.

Acknowledgement

This work was supported in part by National Science Foundation grant EAR-74-19038.

References

BLACK, D. C. (1972) On the origins of trapped helium, neon and argon isotopic variations in meteorites. II. Carbonaceous meteorites. *Geochim. Cosmochim. Acta* **36**, 377–394.

CAMERON, A. G. W. and TRURAN, J. W. (1977) The supernova trigger for formation of the solar system. *Icarus* **3**, 447–461.

CLAYTON, D. D., DWEK, E. and WOOSLEY, S. E. (1977) Isotopic anomalies and proton irradiation in the early solar system. *Ap. J.* **214**, 300–315.

CLAYTON, R. N. (1963) Carbon isotope abundance in meteoritic carbonates. *Science* **140**, 192–193.

CLAYTON, R. N., GROSSMAN, L. and MAYEDA, T. K. (1973) A component of primitive nuclear composition in carbonaceous meteorites. *Science* **182**, 485–488.

CLAYTON, R. N. and MAYEDA, T. K. (1975) Genetic relations between the moon and meteorites. Proc. Lunar Sci. Conf. 6th, *Geochim. Cosmochim. Acta*, Suppl. 6, Vol. 2, 1761–1769.

CLAYTON, R. N., ONUMA, N. and MAYEDA, T. K. (1976) A classification of meteorites based on oxygen isotopes. *Earth Planet. Sci. Lett.* **34**, 209–224.

CLAYTON, R. N. and MAYEDA, T. K. (1977) Correlated oxygen and magnesium isotope anomalies in Allende inclusions. I. Oxygen. *Geophys. Res. Let.* **4**, 295–298.

EBERHARDT, P. (1974) A neon-E rich phase in the Orgueil carbonaceous chondrite. *Earth Planet. Sci. Lett.* **24**, 182–187.

EPSTEIN, S. and YEH, H. W. (1977) The δ^{18}O, δ^{17}O, δ^{30}Si and δ^{29}Si of oxygen and silicon in stony meteorites and Allende inclusions (abstract). *Lunar Science* VIII, 287–289. The Lunar Science Institute, Houston.

GRAY, C. M. and COMPSTON, W. (1974) Excess ^{26}Mg in the Allende meteorite. *Nature* **251**, 495–497.

HERBST, W. and ASSOUSA, G. E. (1977) Observational evidence for supernova-induced star formation (abstract). *Lunar Science* VIII, 427–429. The Lunar Science Institute, Houston.

HEYMAN, D. and DZICZKANIEC, M. (1976) Early irradiation of matter in the solar system: magnesium (proton, neutron) scheme. *Science* **191**, 79–81.

KUNG, C. C. (1977) Nitrogen contents and isotopic compositions in stony meteorites. Ph.D. thesis, University of Chicago.

LEE, T. and PAPANASTASSIOU, D. A. (1974) Mg isotope anomalies in the Allende meteorite and correlation with O and Sr effects. *Geophys. Res. Lett.* **1**, 225–228.

LEE, T., PAPANASTASSIOU, D. A. and WASSERBURG, G. J. (1976) Demonstration of ^{26}Mg excess in Allende and evidence for ^{26}Al. *Geophys. Res. Lett.* **3**, 109–112.

SABU, D. D. and MANUEL, O. K. (1976) Xenon record of the early solar system. *Nature* **262**, 28–32.

The Main Features of the Geochemistry of Lunar Rocks

V. L. BARSUKOV and L. V. DMITRIEV

V. I. Vernadsky Institute of Geochemistry and Analytical Chemistry, U.S.S.R. Academy of Sciences, Moscow, U.S.S.R.

Abstract

Statistical cluster analysis of 635 individual bulk chemical compositions of lunar rocks distinguishes three major chemical groups which correspond respectively to highland rocks, high titanium mare basalts and mare basalts with moderate titanium contents.

Petrochemical variations are plotted on "A–S" diagrams and on "any major oxide–S" diagrams, where $A = Al_2O_3 + CaO + Na_2O + K_2O$ and $S = SiO_2$—$(FeO + MgO + MnO)$.

Considering firstly the K_2O—S variations and the SiO_2—MGO–and Al_2O_3—S variations, eight chemical subgroups are distinguished in the lunar rocks. Each subgroup is defined by variation curves on chemical diagrams and on an Ol—An—Q ternary normative diagram.

The petrogenetic meaning of these characteristic trends are discussed. Also, the similarities with or differences from terrestrial basic magmatism are demonstrated.

The evidence for two different (continental and mare) early stages of crustal formation on the moon and terrestrial planets is emphasized and it is considered to be impossible to construct, at the present time, a single universal model for the formation of the earth's crust.

Seven years ago scientists obtained the first opportunity to study directly rocks of another planetary body, when space probes delivered to the earth the first samples of rocks from the surface of the moon.

Because of the use of modern means of investigation a huge volume of facts about lunar rocks was accumulated during this relatively short period.

We have carried out a statistical cluster analysis of the analytical data on lunar igneous rocks (635 individual bulk chemical analyses) (refs 1–13 and others) and revealed three major groupings which are statistically significantly different from one another with respect to the following selected coefficients: Al_2O_3/CaO; $FeO/(FeO + MgO)$ and $TiO_2/(TiO_2 + FeO)$ (Figs. 1 and 2).

One grouping corresponds to highland rocks, the second to high-titanium mare basalts with 10–15% TiO_2, and the third to mare basalts with moderate titanium contents (2–6% TiO_2). A special index on a three-dimensional diagram designates a few compositions which plot outside of the three groupings. These compositions correspond to the ultramafic and acid rocks of the moon.

In order to ascertain the petrochemical variations within each of the three groups, we have used the A–S diagram, where $A = Al_2O_3 + CaO + Na_2O + K_2O$ and $S = SiO_2$—$(FeO + MgO + TiO_2 + MnO)$.[14,15] This diagram is preferred because all major chemical elements can be plotted on one plane and the position of any figurative point on such diagram is invariant with respect to other elements. Using this diagram the material balance within the rocks series can be calculated for hypothetical forms of genetic interrelation. It is seen on the A–S diagram that each of the groups form a rather narrow, elongated field with a

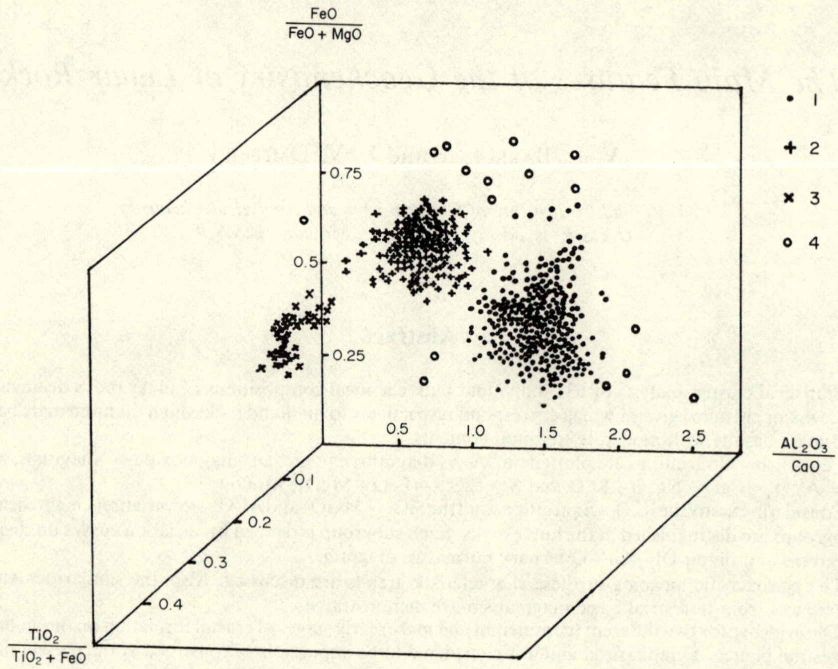

FIG. 1. Volumetric image of compositions of the main groups of lunar magmatic rocks on the diagram $FeO/(FeO+MgO)$—Al_2O_3/CaO—$TiO_2/(TiO_2+FeO)$ (wt. %). Legend: (1) highland rocks; (2) mare basalts with moderate titanium contents; (3) high-titanium mare basalts; (4) acid and ultramafic rocks.

definite slope. The graphs of paired correlation in "S—any main chemical element" coordinates show that within any of the groups distinguished, the variation of basicity (parameter S) leads to a regular change of any element content. The graph of the distribution of alkaline elements is of special interest. It was found that in each of the three groups there are some samples which are markedly enriched in alkalies (Fig. 4). These rocks were separated into subgroups and further statistical operations were carried out on them as individual groups. Furthermore, within the mare basalts with moderate titanium and alkali contents it was possible to separate three subgroups differing from one another in terms of silica, magnesium and alumina contents.

Thus, based on their petrochemical variations, the lunar igneous rocks were divided into eight groups:

(1) Gabbroic anorthosites.
(2) Alkali-enriched gabbroic and basaltic rocks corresponding to KREEP basalts in composition.
(3) Mare basalts with moderate titanium and alkali contents and high alumina contents.
(4) Mare basalts with moderate titanium and alkali contents and high silica contents.
(5) Mare basalts with moderate titanium and alkali contents and high magnesia contents.
(6) Mare basalts with moderate titanium and high alkali contents. These rocks according to their composition correspond to the lunar "high-alumina basalts".

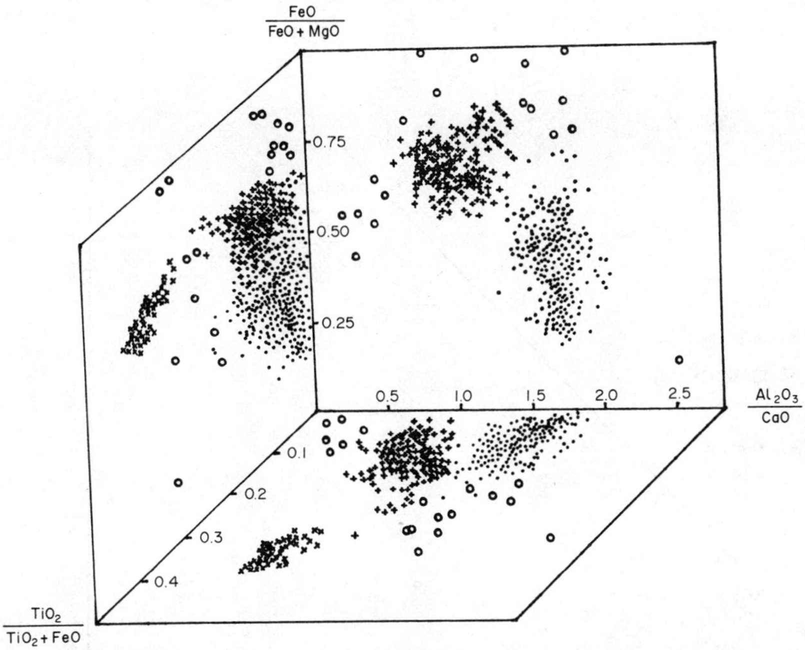

FIG. 2. Space image of the main groups of lunar magmatic rocks projected onto the three planes of Fig. 1. For legend see Fig. 1.

(7) High-titanium mare basalts with moderate alkali contents.

(8) High-titanium mare basalts with high alkali contents.

The statistical analysis enables one to turn for the first time from the qualitative estimation of rock compositional variations (orientation of composition fields) to a quantitative one based on the plotting of variation curves of the paired correlation of major chemical elements contents and compositional change due to the rock basicity (Figs. 5–9). This procedure has shown the individual character of petrochemical variations within each of the groups distinguished.

The position of the variation curves of lunar rocks in relation to the average compositions of possible terrestrial analogues indicated the similarity of the KREEP basalts and terrestrial oceanic tholeiites.[16] The lunar "mare basalts" in terms of "terrestrial" nomenclature would be better named picrites than basalts. But even terrestrial picrites are different from lunar "mare basalts" because of the high titanium content and sharply increased iron content of the latter rocks.

Lunar anorthosites are significantly enriched in Al_2O_3 and CaO in comparison with terrestrial ones. The similarity and differences of compositions of basic magmatic rocks of the earth and the moon are worthy of discussion in connection with the following consideration.

Melting experiments of basalts in vacuum (up to 10^{-7}) show the regular change of the melt composition towards significant enrichment in TiO_2, Al_2O_3, CaO, MgO and depletion in K_2O, Na_2O, SiO_2 that characterize the difference between the lunar and terrestrial basalts.[17] At the present state of knowledge one can only speak about the tendency of change of the basalt melt composition in a vacuum. However, if one takes into account that the lunar vacuum is 10^{-13} rather than 10^{-7}, its influence should probably be considered. At the same time these experiments indicate that conclusions concerning the compositional differences of

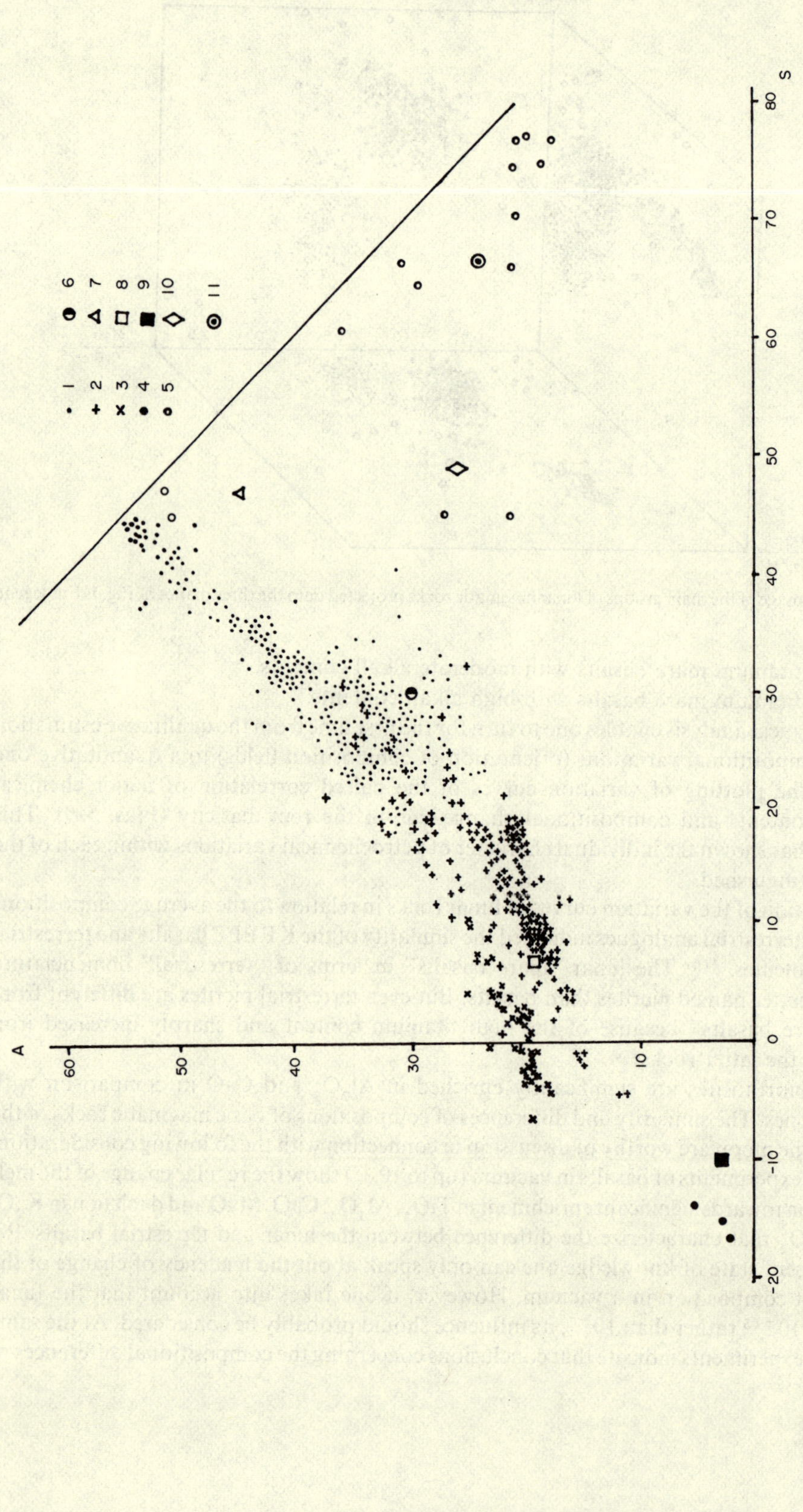

FIG. 3. Position of the main rock groups and types in A–S coordinates (see text). Lunar rocks: (1) highland rocks; (2) mare basalts with moderate TiO_2 content; (3) high-titanium mare basalts; (4) ultramafic rocks; (5) acid rocks. Terrestrial rocks; (6) oceanic tholeiite;[16] (7) anorthosite; (8) picrite; (9) harzburgite; (10) andesite; (11) granite; refs (7–11) according to R. A. Daly, 1933.

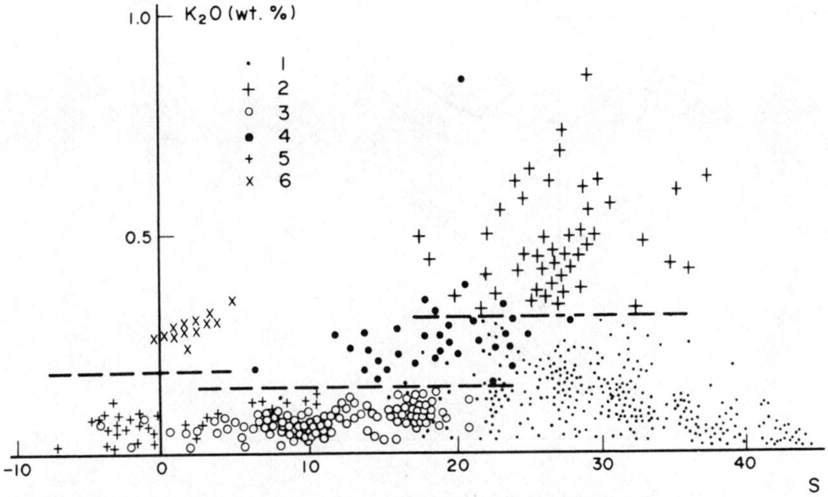

FIG. 4. Correlation of K_2O–S for lunar magmatic rocks. (1) Gabbroic anorthosites; (2) KREEP basalts; (3) mare basalts with moderate titanium and alkalies contents; (4) mare basalts with moderate titanium content and high-alkali content; (5) mare basalts with high-titanium and moderate alkali contents; (6) mare basalts with high-titanium and high-alkali contents.

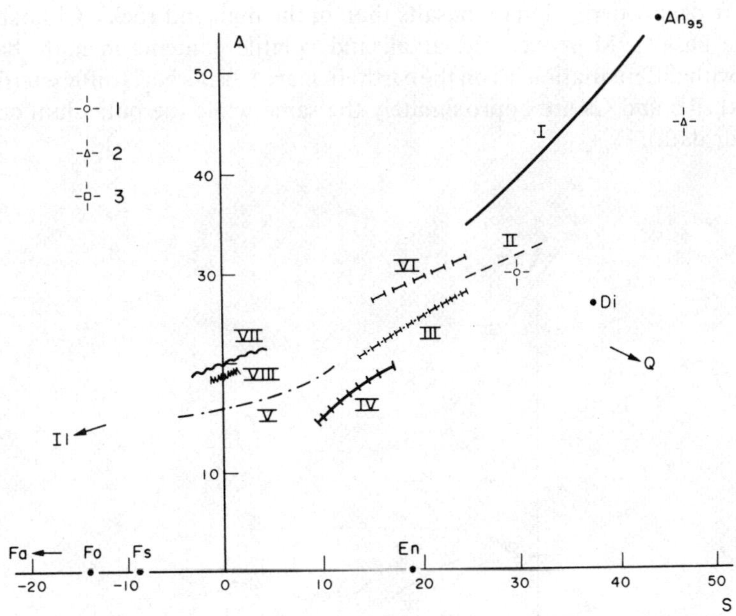

FIG. 5. Variation lines of lunar magmatic rocks in A–S coordinates (wt. %). Roman numerals correspond to ordinal numbers of the groups (see text). (1) Average composition of oceanic tholeiite.[16] (2) Anorthosites according to Daly. (3) Picrites according to Daly.

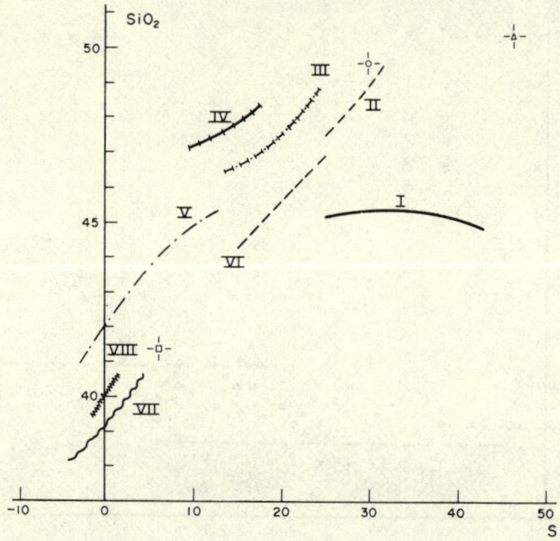

FIG. 6. Variation lines in SiO–S coordinates (wt. %). For legend see Fig. 5.

the source material of terrestrial and lunar basalts must be viewed with caution. This difference (initial depletion of the moon in volatiles) could be false and the deficiency of volcanic products in volatiles could be mostly secondary due to the difference in magmatic environment but not to the difference in the initial composition.

It should not be forgotten that in all highland rocks the vesicules are abundant. It should also be noted that the contents of such volatiles as S, Se, Te, F, Cl, Br are significantly higher in the more deeply derived mare basalts than in the highland rocks. Cl-apatite is found in such mare basalts. Moreover, the alkali and volatile contents in mare basalts increase regularly with differentiation, as on the earth. In mare basalts Na is different (from terrestrial analogues), Rb and Cs are approximately the same while the potassium content is often higher (our data).

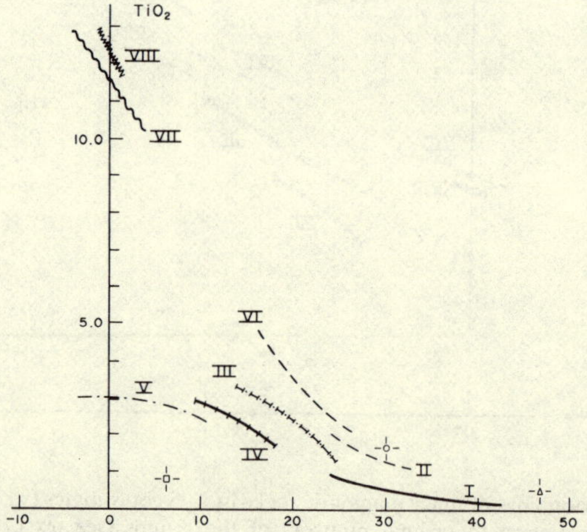

FIG. 7. Variation lines in TiO_2–S coordinates (wt. %). For legend see Fig. 5.

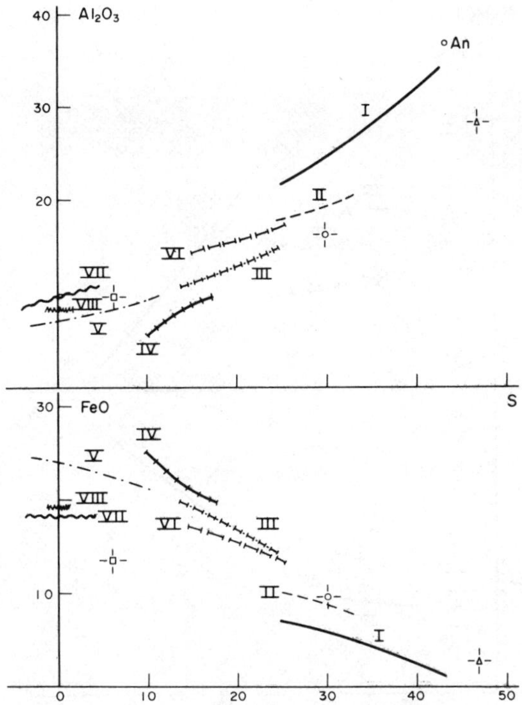

FIG. 8. Variation lines in Al_2O_3–S and FeO–S coordinates (wt. %). For legend see Fig. 5.

The statistical analysis of the compositional variations in lunar magmatic rocks also enables one to consider the results from a petrological point of view.

For this purpose the normative mineral calculations (CIPW method) of the eight groups of lunar magmatic rocks distinguished were carried out and these were plotted on the ternary Ol–An–Q diagram used by many authors for petrological considerations in relation to lunar rocks [18] (Fig. 10).

Based on the position of variation lines in this diagram some conclusions can be drawn.

As noted earlier, the variation line of highland gabbroic anorthosites and anorthosites is located in the high-temperature region. The generation of rocks in this series is due to fractionation and accumulation of anorthite, as is confirmed by the presence of an Eu-maximum in these rocks. A minor part of the highland rocks belonging to this series were possibly generated by spinel fractionation (spinel troctolites).

KREEP basalts also lie in the high-temperature region, but their compositions are closer to the Ol–An–Sp peritectic. The compositional variations are also due to the fractionation of plagioclase.

The compositional change in mare basalts with moderate titanium contents must be related to the existence of two different principal primary melts. One is located in the olivine field and the second in the plagioclase field. Correspondingly the compositional variations of the first are due to fractional crystallization of olivine and the variations of the latter to plagioclase crystallization along with the evolution of the differentiates towards the Ol–An–Sp cotectic.

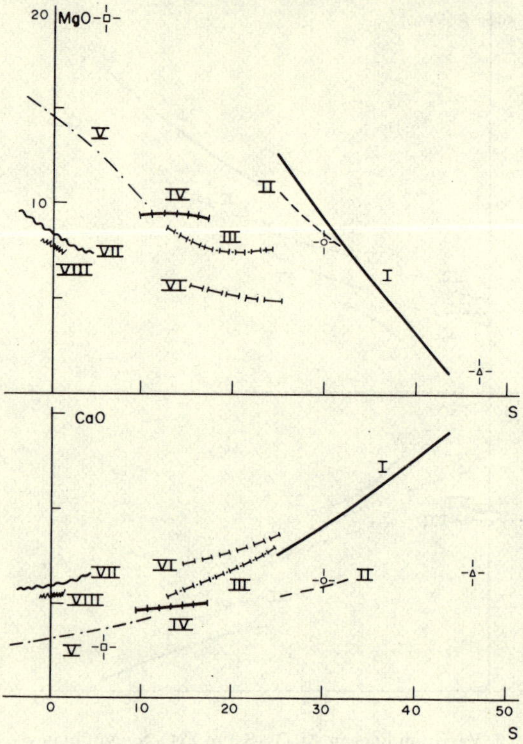

FIG. 9. Variation lines in MgO–S and CaO–S coordinates (wt. %). For legend see Fig. 5.

Basalts with high-titanium contents, and especially "alkaline" high-titanium basalts, probably represent the lowest-temperature differentiation series of lunar magmatic rocks since the variation lines of these rocks are significantly displaced towards the Ol–Opx–An eutectic.

The high-alumina basalts probably represent the most high-temperature series of lunar mare basalts. This series terminates far away from the Opx–An cotectic line, but its succession represents the beginning of the high-titanium series. On all graphs (Figs. 5–9) high-titanium basalts continue the variation lines of high-alumina basalts so it is doubtful that the above-mentioned peculiarity is a random event. There are probably several possible interpretations of the genetic interrelations between these rock groups. One interpretation would be that both series were derived from the same melt. High-titanium basalts may represent lower-temperature products of fractionation of this melt and high-alumina basalts represent higher-temperature ones. According to this scheme the initial melt should be unstable and the compositional gap between its derivatives may be due to liquid immiscibility in such a melt.

The compositions not involved within the main groups were distinguished using the statistical analysis of lunar magmatic rocks. These compositions correspond to lunar ultramafic rocks and granites. Both of these rock types are rarely found so that there are few data for a statistical estimation of their compositions and especially for the study of their

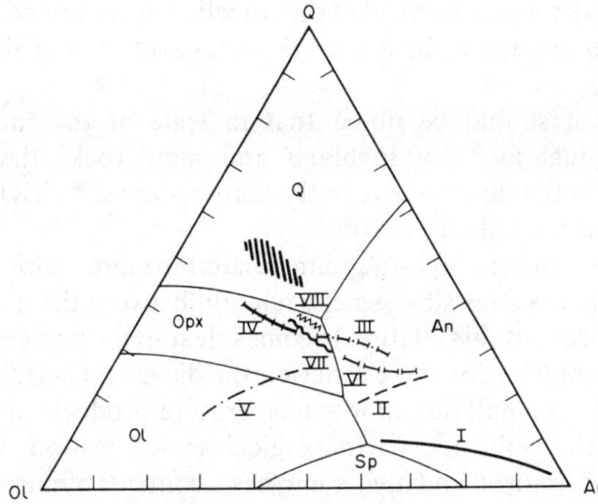

FIG. 10. Variation lines of lunar magmatic rocks on the ternary diagram olivine (Ol)-quartz (Q)-anorthite (An). For legend see Fig. 5. The field of the compositions of interstitial liquid is crossed (see text).

genetic relations with other magmatic rocks. However, in spite of the fragmentary character of the data on acid rocks compositions the material available has a particular interest. Most of the acid rocks are described in the literature as a component of mesostasis of basalts and highland gabbroic rocks whence their genetic role becomes rather clear.[19–23] Let us dwell a little on this problem.

It was found that the mesostasis of mare basalts and highland gabbroic rocks is represented by a two-phase glass which was product of the solidification of two immiscible melts; the most important phase (by volume) is near to a pyroxenite composition and the second phase to an alaskite granite composition.

The possibility of immiscibility of the melt which is near in composition to the mesostasis was confirmed by experiments. It was determined that the "granitic" liquid comprised 5 to 10% of total volume of the immiscible melts which agrees with the estimation of the actual modal phase composition of the mesostasis.

Taking into account the results of these experiments and based on the modal composition of coexisting phases in the mesostasis we have calculated the most probable average composition of the primary immiscible melts. Crystallization of these melts represents the final magmatic stage of the evolution of lunar mare and highland rocks.

Both compositions are rather close to each other and their position on the diagram Ol–An–Q is placed near the Ol–Opx–An eutectic (Fig. 10).

The results obtained are interesting first of all as evidence for the possibility and reality of two processes of generation of the so-called "basalt-derived" granitic rocks by crystallization differentiation and by liquid immiscibility. The role of the latter in the generation of terrestrial granitic rocks has evidently been underestimated.

The results obtained are also interesting because, on taking them into account, one can understand the process of crystallization of lunar rocks up to its final stage. Apparently the differentiation in the series from anorthosites to gabbroic rocks was not finished at the Ol–Sp–An peritectic but continued along the Ol–An cotectic to produce residual melts with normative compositions characteristic of the Ol–Opx–An eutectic. The differentiation of the mare basalts apparently was not finished at the cotectic but continued to the eutectic. This is probably the common feature of the magmatic series evolution of the lunar highland and mare rocks.

At the same time it should be noted that in spite of the similarity of normative compositions of residual melts of highland and mare rocks the highland rocks are significantly enriched in titanium relative to the mare rocks (8.7% TiO_2 in highland residual melts and 2.60% in mare residual melts).

At first sight it seems casual as high-titanium basalts are mare rocks. On the other hand, we have previously mentioned a possible genetic relation between the high-titanium and high-alumina basalts. The sense of this relation becomes clear in the process of the analysis of the position of the variation lines for these series on the diagram Ol–An–Q. If such relation is real then the titanium accumulation in low-temperature products of differentiation of the proposed primary melt could be due to plagioclase segregation into high-temperature products as well as to lower iron and higher magnesia contents of minerals in highland rocks and high-alumina basalts. The latter circumstance should prevent titanium dispersion and promote its concentration in the process of crystallization differentiation.

And this is just the very situation. The mineral composition of the highland and mare rocks is significantly different. First of all, the minerals of the highland rocks are relatively stable in composition whereas the minerals of mare rocks are more variable in their composition. For example, the range of the plagioclase compositions is An 95 ± 2 in highland rocks and An 73 to An 93 (averaging An 83) in mare rocks.

Olivines of highland rocks are distinguishable from mare rocks by a significantly lower content of the fayalite component (Fa < 30% and 30–100%).

Pyroxenes of highland rocks are mainly orthorhombic (bronzite at most) and clinopyroxenes are rare. In mare rocks clinopyroxene is the main rock-forming mineral and is commonly represented by augite and pigeonite. Diopside as well as orthopyroxene are absent in mare rocks. Like olivine, augite and pigeonite of highland rocks differ from mare rocks in their high magnesia content.

Spinel is represented by two types: chrom-aluminoferous and titaniferous. The titaniferous spinel of highland rocks is titaniferous chromite whereas mare rocks contain both titaniferous chromite and ulvöspinel.

The ilmenites of highland rocks are different from ilmenites of mare rocks. The first contain high magnesia (> 20% $MgTiO_3$) and low iron contents and the latter display the opposite situation.

In addition to the above-mentioned rock-forming minerals the lunar rocks also contain troilite (FeS) which is more abundant in mare rocks whereas accessory Fe metal is more common in highland rocks. Highland rocks often contain armalcolite $(FeMg)Ti_2O_5$ in which the contents of chromium and zirconium sometimes reach 4–10%. Phosphates are also found in highland (whitlockite) and mare rocks (Cl-apatite). Rare grains of tridymite or cristobalite have also been noted in mare basalts.

Highland and mare rocks of the moon are also different in rare-elements contents. Highland rocks are relatively rich in Ni, Ba, Th, Ce and relatively deficient in Cr, V, Sc, Cu, Y. Within the mare basalt series ranked from ancient to rather young (3.8–3.1 b.y.) a regular decrease of TiO_2 contents (from 13 to 1.5%) is observed. Simultaneously the contents of SiO_2 (39–47%) and siderophile elements (Cr, Ni, Co, Y) increase. The comparison of high-titanium and low-titanium basalts reveals the higher abundance of alkaline (K, Na, Rb, Li), lithophile (Ba, Sc, Zr, Hf, Ta, Th, Y, Ce) and chalcophile (Cu, Zn, Pb) elements in the first ones, although the general level of alkaline, rare earth and ore elements contents in selected groups of lunar rocks is near to the appropriate level in terrestrial analogues. For the present only a group of siderophile elements represent an unexplained exception: in lunar rocks the interrelations between members of this group are sharply different from terrestrial ones. In lunar rocks the contents of Ni and V are strongly lowered and consequently the V/Cr ratio is significantly lower (0.1 vs. 1.0–1.4) and the Co/Ni ratio is higher (1.2–1.7 vs. 0.4) compared to terrestrial basaltic rocks. The causes of such a distribution of siderophile elements is still unclear.

Thus, it is now possible to say that the main aim of the first period of the study of lunar rocks (to determine the main types, mineral and elemental composition and to reveal the degree of similarity and difference between lunar and terrestrial rocks) is much nearer to being met, although the described rocks probably do not exhaust the lunar rocks population.

On the other hand, the discovery of some stages in the history of the lunar crust formation has forced us to consider a number of new problems concerning the early stages in the history of the earth and in the history of the formation of the terrestrial crust.

The data in the comparative planetology field accumulated up to the present moment indicate that the main factors of evolution during the early stages of the terrestrial planets "life" were almost the same and that the early history of the moon and other planets with an abortive type of evolution can be used as an analogue to develop models for the early evolution of the earth.

This approach leads to a statement on the importance of the accretion energy in the total balance of planetary energy, which is commonly not taken into account in models of the earth's evolution. Once again the problem of the possibly anorthositic character of the primary terrestrial crust arises and of its important role in the most early, large-scale and planetwide process of the degassing of the upper mantle and the formation of the atmosphere and hydrosphere. This approach also leads to the idea of the primary character of ancient "continental" crust and the secondary character of "oceanic" crust and suggests the possibility that the generation of their difference began in the earliest stages of the earth's "life".

At the present moment it appears that the development of a single universal model for the formation of the terrestrial crust is not possible. Rather, based on the presence of two early stages of crust formation ("continental" and "mare") on the moon and the terrestrial planets, the problem appears to be, to develop at least four models of terrestrial crust formation:

1. Crust with inherited primary "continental" structure.
2. Crust with inherited primary "mare" structure.
3. "Continental" crust transformed into a "mare" crust.
4. "Mare" crust transformed into a continental crust.

The closely related problem is the division of the terrestrial crust into regions having any of the four above-mentioned types of crustal evolution.

It is also clear that there is a need for more extensive studies not only of the vertical but also of the horizontal inhomogeneity of the upper mantle composition under the regions differing in the types of crustal evolution, as well as of the characterization of the differences in ore-productivity of the crust and the mantle of these regions which could be important in prospecting. Once again the role of primary ring and linear structures in the distribution of magmatism and ore deposits and the problem of defining and prospecting for these structures has arisen.

Finally, the radical problem has arisen concerning the possible differences in the composition of the upper and lower mantle, and the possibility of significant differences in the composition of the upper mantle and the primary matter from which our earth originated.

On the whole we must agree with the words of H. H. Schmitt, the first geologist walking on lunar surface, who was perfectly right in saying: "The moon is now a pitted and dusty window into the earth's own origin and evolution."

References

1. *Proceedings Apollo 11 Lunar Science Conf. Geochim. Cosmochim. Acta,* Suppl. 1, vol. 1, 2, 1970.
2. *Proceedings Second Lunar Science Conf. Geochim. Cosmochim. Acta,* Suppl. 2, vol. 1. 2. 1971.
3. *Proceedings Third Lunar Science Conf. Geochim. Cosmochim. Acta,* Suppl. 3, vol. 1, 2, 1972.
4. *Proceedings Fourth Lunar Science Conf. Geochim. Cosmochim. Acta,* Suppl. 4, vol. 1, 2, 1973.
5. *Proceedings Fifth Lunar Science Conf. Geochim. Cosmochim. Acta,* Suppl. 5, vol. 1, 2, 1974.
6. *Proceedings Sixth Lunar Science Conf. Geochim. Cosmochim. Acta,* Suppl. 6, vol. 1, 2, 1975.
7. *Proceedings Seventh Lunar Science Conf. Geochim. Cosmochim. Acta,* Suppl.7, vol. 1, 2, 1976.
8. Apollo 15 lunar samples. *Science,* **175** (1972).
9. Apollo 16 lunar samples. *Science,* **179** (1973).
10. Apollo 17, Results, *EOS,* **56** no. 6, 595 (1973).
11. VINOGRADOV, A. P. The preliminary data on lunar soil transported by "Luna 16". *Geochimia,* no. 3 (1971).
12. VINOGRADOV, A. P. The preliminary data on lunar soil transported by "Luna 20". *Geochimia,* no. 7 (1972).
13. *Lunar soil from Mare Fecunditatis.* Nauka, Moskva, 1974.
14. DMITRIEV, L. V. To the question about origination of the ultramafic rocks from Mid-Indian ocean ridge. *Geochimia,* no. 10 (1969).
15. DMITRIEV, L. V. and SHARASKIN, A. YA. The petrography and petrochemistry of the igneous rock from Arabia–India oceanic ridge. In *The Investigation on the Problem of the Rift Zones of World Ocean,* vol. 2, Nauka, Moskva, 1972.
16. DMITRIEV, L. V., SHARASKIN, A. YA. and GARANIN, A. V. The main features of the ocean floor magmatism. In *Problems of Petrology,* Nauka, Moskva, 1976.
17. YAKOVLEV, O. I., KOSOLAPOV, A. I., KUSNETSOV, A. V. and NUSINOV, M. D. The results of the studying of the fraction evaporization of the basaltic melt in vacuum. *Dokl. Akad. Nauk S.S.S.R.,* **206,** no. 4 (1972).
18. WALKER, D., LONGHI, J. and HAYS, J. F. Experimental petrology and origin of the Fra Mauro rocks and soils. *Proc. Third Lunar Sci. Conf. Geochim. Cosmochim. Acta,* Suppl. vol. 3, MIT Press, 1972.
19. ROEDDER, E. and WEIBLEN, P. W. Lunar petrology of silicate melt inclusions, Apollo 11 rocks. *Proc. of Apollo 11 Lunar Sci. Conf.,* vol. 1, 1970.
20. MELSON, W. G., MASON, B., NELEN, J. and JACOBSON, S. Apollo 14 basaltic rocks. Abstr. paper submitted to the Third Lunar Sci. Conf., 1972.
21. DUNGAN, M. A., POWELL, B. N. and WEIBLEN, P. W. Petrology of some Apollo 16 feldspatic basalts: KREEP melt rocks? Abstr. paper submitted to the Sixth Lunar Sci. Conf., 1975.
22. ROEDDER, E. and WEIBLEN, P. W. Silicate liquid immiscibility in lunar magmas, evidences the melt inclusions in lunar rocks. *Science,* **167,** no. 3919 (1970).
23. MASSION, P. J. and KOSTER VAN CROSS, A. F. Liquid immiscibility in silicate. *Nature Phys. Sci.* **245,** no. 143 (1973).

SECTION III

HYDROSPHERE, ATMOSPHERE

The Archean Atmosphere of the Earth (Aldan Time)

E. A. KULISH

The Far-Eastern Institute of Mineral Resources, Khabarovsk, 680038, U.S.S.R.

Abstract

The Archean atmosphere has left certain traces in the sediments during the accumulation of primary sedimentary and volcanic–sedimentary rocks of the Aldan shield. The analysis of these traces makes it possible to establish the composition of the atmosphere of that time. Wide variation in Fe_2O_3: FeO from ∞ to O, presence of Mn^{3+} and SO_4^{-2}, and considerable development of carbonates and organic carbon in the rocks suggest the presence of free oxygen in the atmosphere. It was generated by photolytic organisms and photo-dissociation of water in the upper layers of the atmosphere. The presence of oxygen in the atmosphere excludes the presence of carbon dioxide, ammonia, hydrocarbon, free hydrogen and hydrogen sulphide. The nature of the carbonaceous rocks, Mg:Ca in the sediments, and the wide development of organisms in the Aldean Archean indicate an important role of CO_2 in the atmosphere. Precipitation of carbonates excludes the presence of "acid" gases (hydrogen chloride, sulphur trioxide, etc.) in the Archean atmosphere. As compared with the present, the Early Archean atmosphere contained considerably more carbon dioxide and water vapour, and very little oxygen. Nitrogen was the major component. The atmosphere of the earth has undergone irreversible changes, repeatedly renewing its composition as the result of the exchange of its gases with the lithosphere, hydrosphere and space, and the influx of gases from the mantle and deeper zones.

The complex of rocks of the Early Archean of the Aldan shield represents the deeply metamorphosed strata of primary sedimentary, sedimentary–volcanic, and volcanic rocks that formed in different lithological conditions of the mobile geosynclinal region. Lithological analysis of the strata provides data concerning the composition of the Archean atmosphere of that time, approximately 3.5 billion years.

The sediments that accumulated in the process of sedimentation came into direct contact with the Archean atmosphere or with its gases dissolved in surface waters.

Indications of the nature of the atmosphere are the oxidation-reduction processes acidic-alkaline conditions, the nature and rate of weathering and the sedimentation of carbonates and other compounds which are sensitive to the Eh and pH of the environment.

The atmosphere, hydrosphere and climate are closely interrelated. At the specific stage of the development of the earth they considerably influenced the character of sedimentation. It is quite natural to expect that the influence of the atmosphere has left some traces in the Archean sediments. It can be brought to light by lithological analysis of the Precambrian metamorphic rocks, by determination of the original pre-metamorphic nature of the rocks and terrains and by reconstruction of the conditions and nature of sedimentation at that period of time.

Taking into consideration all these phenomena and using as an example the Early Archean formations of the Aldan shield the Early Archean atmosphere has been reconstructed as far as possible and its role in the process of sedimentation elucidated. The problem of sources of gases and of the origin of the pre-Archean atmosphere has not been dealt with. The latter was considered in detail by A. P. Vinogradov (1964), V. V. Rubi (1957) and H. C. Urey (1952).

One of the Precambrian problems is the determination of the oxygen content of its atmosphere (Vinogradov, 1967; Sidorenko, 1967; Rubi, 1957; Kulish, 1970, 1973).

The most reliable information about the role of oxygen in the Early Archean atmosphere can be derived from the reconstruction of oxidation-reduction processes which took place in the Early Archean sediments. Very effective, in this respect, is the study of the molecular ratios Fe_2O_3: FeO (coefficient "b"), taking into account that the oxidized sediments preserve their oxygen during metamorphism up to ultrametamorphism (Rankama and Sahama, 1950; Dobretsov *et al.*, 1970). G. P. Eugster (1961) and other investigators drew attention to the fact that the rate of Fe oxidation does not depend upon the P–T conditions or the metamorphic facies. The oxygen was highly inert during regional metamorphism, because in the Aldan rocks graphite, pyrite, galena, chalcopyrite and other sulphides are situated closely to, or are even in contact with hematite, magnetite and silicates containing ferric oxide. Oxygen is thus an inactive element during metamorphism, with FeO and Fe_2O_3 appearing as independent components. Even if the oxidation-reduction reactions took place during metamorphism, they were of local character and the distinctions in the grade of oxidation of components do not diminish from bed to bed. The inert behaviour of oxygen and sulphur was noted, in particular, in the formations of the Tajodzni iron deposit occurring in the Iengr series, in which is observed the paragenesis magnetite + pyrrhotite and, sometimes, magnetite + pyrite.

Terrigenous–sedimentary rock masses of the Iengr series were derived from rocks of granitic and dioritic composition, near-shore sediments and volcanogenous–sedimentary material (Kulish, 1973). According to A. A. Beus (1972) and others the main magmatic rocks of intrusive series have the following Fe_2O_3: FeO ratios: in granites – 0.28, in diorites – 0.20, in gabbro – 0.19, in tonalites – 0.23, and in peridotite – 0.12. But it is known that the formation of acidic and intermediate intrusive rocks takes place under the direct influence of the crystalline (metamorphic) base of geological structures, and the latter had already been influenced by the atmosphere. This can be concluded from wide variations of Fe_2O_3:FeO both in metamorphic rocks of crystalline base in any region, and in granites, diorites and other intrusive rocks of acidic and intermediate composition, occurring in the lower and upper structural horizons. That is why the values of Fe_2O_3:FeO in gabbros and peridotites—the mean value 0.14 for variations 0.11 to 0.18—can be taken with reasonable reliability as the indicator of the degree of iron oxidation in the lower horizons of the earth.

If the atmosphere did not contain oxygen, then the conditions at the surface of the earth were reducing or neutral, and Fe_2O_3 in sediments will be not more than 0.18 with negligible variations. In the case of an oxygen-rich atmosphere, oxidizing conditions, like at present, should prevail at the surface of the earth, except for local reductive conditions where decomposition of sulphides and organic matter takes place. Fe_2O_3:FeO in sediments will be more than 0.14 (in average), and the variations in both directions will be wider. It should be taken into account that the original sediments in the Early Archean probably contained higher contents of Fe_2O_3, a part of which was later reduced to FeO during diagenesis and regional metamorphism by carbon and sulphide-bearing matter of the rocks and the traces of which are present everywhere in the studied formations.

The high degree of iron oxidation brought about the Iengr series of hematite quartzites, gneisses and slates, which form conformable beds and intercalations among other pararocks. Hematite rocks and ores often form thick (up to 15 m) and rather extended deposits and horizons (up to 12 km). In the basin of the Aldan river the author has found conglomerates with hematite-bearing cement and pebbles, and in the basin of the Iengr river—metamorphosed bauxites with hematite. Sometimes the hematite rocks are closely interbedded with magnetite bearing rocks, in which the iron is oxidized to a lesser degree. In places there is fine alternation of hematite, magnetite, cordierite and biotite beds having the thickness up to 3 m, i.e. there exists fine intercalations of rocks, in which iron is oxidized to different degrees. The example of intercalation of beds of variously oxidized rocks is the biotite–hypersthenic slate with graphite, hematite and sulphides, in which hematite, graphite, graphite–sulphide, biotite, hypersthene and hypersthene–biotite beds alternate.

In highly aluminiferous rocks of the Aldan complex (288 analyses; Kulish, 1973) with FeO (in per cent) $\bar{X} = 3.97$, ($\sigma_x = 3.53$, min $= 0.02$, max $= 17.04$) and with Fe_2O_3 $\bar{X} = 2.24$ ($\sigma_x = 2.63$, min $= 0.01$, max $= 20.15$) the value of Fe_2O_3: FeO is: $\bar{X} = 1.66$ ($\sigma_x = 7.66$, min $= 0.0007$, max $= 70.0$). These values show large differences in the degree of iron oxidation in the Early Archean metapelites compared with that of the Phanerozoic clays. The mean value "b" of metapelites greatly surpasses that of the series of magmatic rocks, which is the consequence of the considerable oxidation of iron during the formation of these sediments. Even greater variations can be seen if we consider other types of rocks of the Aldan complex. In quartzites with the average value of "b" = 0.68 the variations are 0.02–32.70; hematite quartzites and rocks have "b" $\rightarrow \infty$, and in garnetiferous, sulphide and cordierite rocks "b" $\rightarrow 0$. Average analyses of the main typès of the rocks of the Aldan shield also show differences in the degree of iron oxidation: "b" $= 0.09 - 5.5$.

The differences in the degree of iron oxidation are characteristic not only for lumps and intercalations, where they are most evident, but also for concrete types of rocks, as well as for complexes of rocks from various parts of the shield, various beds, horizons, suites and series. The contrast in "b" value diminishes in the direction of bigger stratigraphic and regional subdivisions and changes from 0.19 (primary pelagic sediments with considerable addition of organogenous matter) to 0.43 (primary terrigenous–pelagic strata). The "b" of the Aldan complex (= 0.32) is higher than that of a number of magmatic rocks. The level of oxidation of iron in the Phanerozoic geosynclinal strata, which can be compared with that of the Aldan complex, is 0.42. This suggests that the Archean metasediments were oxidized, but to a lesser degree than in the Phanerozoic because the content of oxygen in the Archean atmosphere was less than in the more recent atmosphere.

Some information in this respect can be obtained from the comparison of Mn:Fe in metapelites and clays. With time it steadily decreases (Ronov and Magdisov, 1970). This is explained by the fact that iron, which oxidizes more readily than manganese with the increase of free oxygen in atmosphere, precipitated more intensively than manganese. In particular MnO:(Fe_2O_3+FeO) = 0.028 in highly aluminiferous metapelites, and in the clays of the Phanerozoic it varies from 0.007 to 0.013.

In the Okurdan and Kabactin mountains (lower course of the Iengr river) hematite quartzites are genetically related to manganiferous sediments containing Mn^{+3} and Mn^{+2}. Manganese oxide is contained in viridine, manganous hematite and rare piedmontite.

Baritiferous ($BaSO_4$) formations of primary sedimentary origin are in the Hematite deposit among hematite quartzites. In the section of the Fedorov suit of the Iengr series and in the suites of the Dzeltulin series the lodes of various scapolitic slates, calcifiers and marbles, containing up to 90% of scapolites. are developed among carbonaceous, magnesian–limestone strata. The content of SO_4 in scapolites sometimes reaches 3.5%, which permits considering them to be sulphatic scapolites. In the pre-metamorphic period these rocks were sulphate-bearing (gypsum, mirabilite, glauberite, etc.) limestones, marls, clays, etc. The presence in primary sediments of sulphur with the highest degree of oxidation known in nature can only be explained by the presence of free oxygen. Study of the isotopic composition of sulphur in sulphides from the Precambrian primary sedimentary rocks shows that sulphate was one of the important components of sea water of that period (Holland, 1971; Vinogradov, 1971). Mass precipitation of carbonaceous sediments in the strata of the Fedorov, Kurena, Sutama and other suites of the complex is an objective criterion of the presence of free oxygen and appreciable amounts of carbon dioxide in the atmosphere of the Archean of the Aldan.

Local deposits of graphite- and sulphide-bearing rocks occur through the whole section of the Aldan complex. In the Dzeltulin series these rocks form rather thick and extended horizons. Sulphides are represented by pyrite, chalcopyrite, secondary covellite, and rare galena and sphalerite. The rocks are characterized by the very low oxidation of iron; cordierite, garnet and hypersthene are developed in them. Magnetite is very rare and hematite is unknown. The primary composition of the graphite- and sulphide-bearing rocks is diverse: siliceous, silty, argillaceous, marly and carbonaceous formations. Study of graphite-bearing rocks of the Aldan complex shows that the graphite and other carboniferous compounds are of organic nature, and are the remnants of living organisms, photolytic included (Kulish, 1973; S. A. and A. V. Sidorenko, 1975). The presence of organic matter in the fine-grained pelagic sediments suggests that plankton in the Archean were of great importance. The amount of the Archean rocks containing organic graphite is comparable, in general, with similar Phanerozoic formations. Lithological analysis of graphite- and sulphide-bearing rocks shows that the main condition of their sedimentation is that they formed in the conditions of hydrosulphuric poisoning, in which reducing conditions were maintained by the presence of great amounts of organic matter, sulphides and S^{-2} ions. Very possibly, considerable amounts of sulphides formed because of diagenetic reduction of sea water sulphates and during decomposition of organic matter. These conditions were created locally, at various stratigraphic levels, predominantly in abyssal and stagnant parts of the sea basin. The peculiarities of sedimentation in the parts of the basin with hydrosulphuric poisoning and reducing conditions show that one of the conditions of their preservation was the lack of access to free oxygen.

A high degree of iron oxidation is observed in the primary sedimentary rocks of the Aldan Archean as compared with the hypogene series of intrusive rocks: a wide variation of Fe_2O_3:FeO from completely oxidized (hematite rocks) to completely reduced iron (cordierite, sulphide, garnet, hypersthene, graphitic rocks) as the consequence of considerable differentiation of oxidation-reduction processes during sedimentation; occurrence of rocks with Mn^{+3}, SO_4^{-2} (viridin, barite, scapolite, etc.). This and the development of photolytic organisms suggest that a certain amount of free oxygen was present in the Aldan Archean atmosphere (Kulish, 1970). Otherwise these phenomena would be entirely excluded. Oxygen took part in the processes of sedimentation and along with surface and ground waters in the formation of the crusts of weathering (hematite in corundites and conglomerates) of source areas. Taking into consideration the lowering of Fe_2O_3:FeO in marine sediments with the movement to the depth of the basin it can be said that oxygen generally influenced near-surface waters and inshore areas, steadily reducing its influence on abyssal sediments.

Wide variations in primary sedimentary rocks of the Aldan Archean and in Fe_2O_3:FeO from 0 to infinity are the consequence of considerable differentiation of the processes of iron oxidation in sediments that came into contact with the atmosphere during weathering, transportation and sedimentation, as well as with atmospheric gases, dissolved in hydrosphere.

That atmospheric oxygen was possibly present in the atmosphere before the Aldan Archean (i.e. during formation of the rocks of the source area which later participated in formation of sedimentary rock mass) is also shown by the presence of hematite quartzites in the pebbles of the Aldan conglomerates. It should be noted that hematite in pebbles might also be secondary (martite) and have formed during oxidation of magnetite in the process of weathering of original rocks and transporting their fragments.

Free oxygen is generated by photolytic organisms that were developed in abundance and everywhere during the accumulation of the Aldan complex sediments. They were extremely developed in near-shore and intramarine areas of basins. The organisms were short-lived, and the accumulation of organic matter was seasonal, because only in this case do the period of their maximum development and maximum death coincide. Periodical, seasonal accumulation of organic matter in relatively abyssal sediments can be ensured only by the development of photolytic organisms, generating oxygen. It is probable that these organisms were the simplest algae, developed in near-shore areas of the basin, because seasonal variations are not displayed with depth (Kulish, 1973).

Taking into consideration the wide development of organic matter in the terrigenous formations (graphite of biogenic origin in conglomerates, metaarkoses and metabauxites) of the Aldan complex, it is reasonable to suppose that already in the Aldan time (Lower Archean) life was also developed at the surface of the earth. Probably these were, first of all, lichens, fungi, and possibly primitive psilophytaleans and moss. If we assume their rapid development in the then widely developed moistened zones, the biogenic generation of oxygen will be significant. Wide development of organic life in the Precambrian, the Archean included, was established by S. A. and A. V. Sidorenko (1975).

Photo-dissociation of H_2O and CO_2 in the upper zones of the atmosphere under the influence of ultraviolet radiation of the sun was an essential generator of oxygen, with the subsequent escape of that into the outer space.

Similar conclusions about the presence of free oxygen in the Precambrian atmosphere were drawn by Sidorenko (1967), Rubi (1957), Vinogradov (1967), Holland (1971), Chamberlin (1949) and others.

The presence in the atmosphere of free oxygen excluded the presence of carbon dioxide, ammonia, hydrocarbon, free hydrogen and hydrogen sulphide that probably were present in the early history of the development of the earth. At the same time it is known that the above-mentioned gases escape as products of volcanic activity and through other channels (fractures, etc.) in relatively small quantities. Appearing at the surface of the earth the gases were oxidized by atmospheric oxygen to carbonic-acid gases (carbon dioxide, hydrocarbons), nitrogen (ammonia), sulphur trioxide (hydrogen sulphide), water (hydrogen), etc. Carbonic-acid gas was partly being bound by the precipitating carbonates, and sulphur trioxide, taking into consideration its chemical properties, entered completely into reaction giving sulphates of alkalies, alkaline–earth and other elements.

The important role of CO_2 in the atmosphere is shown by the slightly increased acidity of the hydrosphere, by the increased role of the soda component in sea waters, which ensured considerable sedimentation of calcite and dolomite, and by higher Mg:Ca in the Aldan strata, as compared with the Phanerozoic sediments, under which conditions the $MgCO_3$ saturation is attained more rapidly than for $CaCO_3$. It is also shown by the intensive accumulation of carbonaceous and carbonaceous–silicate (calcareous and calcareo-magnesian) sedimentary rocks.

Ronov (1959) has shown the direct relation between the sedimentation of carbonaceous rocks and the rocks containing organic matter and the accumulation of volcanic sediments. The factor of connection is the content of carbon dioxide in atmosphere, which increased with the increase of intensity of volcanism.

The author has determined (1976) that 3.24% of essentially carbonaceous rocks, 2.64% of carbonaceous–silicate rocks and not less than 14.93% of the rocks containing in various degree (up to 10%) carbonates were present in the Aldan complex in the pre-metamorphic period. Graphitiferous gneisses and slates representing metamorphosed rocks, which earlier contained organic matter, total in the complex approximately 1%. Considerable variations in the content of these rocks across the stratigraphic section are noted; this indicates the variations in the amount of carbon dioxide in the Archean atmosphere, just as occurred in the Phanerozoic.

The ratio Ca:Mg in carbonaceous rocks also serves as an indicator of the carbon dioxide content in atmosphere (Ronov, 1959).

For the Aldan complex as a whole CaO = 5.86%, MgO = 3.64% and Ca:Mg = 1.61. In marbles Ca:Mg = 4.9 and in calcifiers – 3.5 giving a mean of 4.5; Ca:Mg in separate samples and horizons varies from 0.09 to 174.0 and in major units (series, suites) from 1.65 to 13.10.

Carbonaceous rocks of the Phanerozoic also have a variable Ca:Mg ratio which in general, increases from 3.0 in the Cambrian to 30.0–88.0 in the Mesozoic and Cenozoic (Ronov, 1959). The data indicate a higher content of carbon dioxide in the Archean as compared with the Phanerozoic, and also confirm the variation of carbon dioxide in the Archean atmosphere.

Proceeding from the supposition that precambrian sea water was possibly saturated with amorphous silica containing SiO_2 on the order of 100 mg 1, H. D. Holland suggests that in this case CO_2 pressure should have been higher than the present by one or two orders of magnitude, in order to ensure the stability of dolomite with magnesium silicate (Holland, 1971).

As is known, the increase in solubility of CO_2 in water causes the decrease in the concentration of CO_3^{-2} ions and thus the dissolution of earlier deposited calcite. Thus the carbon dioxide content in the Archean atmosphere, though considerable, was not overwhelming.

The content of other "acid" gases (hydrogen chloride, sulphur trioxide), which are of an aggressive nature, was insignificant since the resulting increased acidity of the hydrosphere would make it impossible for carbonates to precipitate. This is not recorded in the Archean strata of the Aldan shield and other regions of the Globe. The "acid" gases escaping in the process of volcanic activity dissolved in water (some of them such as hydrogen sulphide were subjected to an intermediate stage of oxidation and were neutralized by carbonates, aluminosilicates, etc.). Chlorous, sulphite, fluoric and the like compounds of alkalies, alkali earths and other metals formed during this process. The pH was regulated and maintained at a definite level by carbonate–bicarbonate and aluminosilicate buffers.

It is well known that a vast hydrosphere (rivers, lakes, seas, oceans, etc.) was present on the Globe in the Archean, and that it took an active part in the processes of sedimentation, in the processes of the formation of the sedimentary coating of the earth. On the basis of the climatic differentiation of the surface of the earth (humid and arid, tropical, moderate and polar zones, etc.) one can conclude that the earth in the Archean did not have the continuous layer of clouds, i.e. that the volume of water vapours in the then atmosphere was almost the same as at present.

On the basis of a number of indirect indications present in the Archean sediments it is reasonable to suggest that the temperature range of the Archean can be compared, without going into details, with the present, though there were, to be sure, differences in the average annual temperatures (in the Archean they were somewhat higher).

Nitrogen was, in all probability, the major component of the Archean atmosphere, and produced the corresponding atmospheric pressure which made it possible to maintain under the above-mentioned temperature limits of the surface of the earth the proportions of water, water vapours and the amount of carbon dioxide dissolved in waters, such that there was no "hothouse effect" caused by clouds, and such that there was no increase of acidity in water, blocking or slowing down the carbonate precipitation. Other data about the role of nitrogen and inert gases (neon, argon, helium, etc.) in the processes of sedimentation and their presence in the Archean atmosphere are lacking.

The foregoing data show that the Early Archean atmosphere took an active part in the processes of sedimentation, often determining a number of conditions and peculiarities of sedimentation. Certain quantities of the atmospheric elements (CO_2, S, F, O, B, etc.) became the components of sedimentary rocks, being preserved or buried in them for the definite time. The latter is specifically indicated by the fact that the content of carbon, chlorine, boron, sulphur and oxygen in the Precambrian strata and the Phanerozoic formations is several times as much as in the intrusive rocks of mantle origin.

The phenomena reflecting the influence of atmosphere on the Aldan sedimentary rocks during their formation (the Early Archean) can also be seen in other Early Archean complexes of the Globe, which supports their global character and, specifically, the fact of intermixture and relative monotony of the atmosphere of that time.

The study of the atmospheric influence upon the sedimentary rock masses of the crust makes it possible to draw the conclusion that beginning from the Early Archean the atmosphere of the earth had acquired the characteristics of the modern atmosphere, and had a nitrogen–carbon dioxide–oxygenous composition. In the process of the evolution of the atmosphere the content of carbon dioxide decreased because of being bound in the formation and accumulation in the crust of carbonaceous, carbonate-bearing sediments, coals, oils, organic gases and sediments containing organic matter. The content of oxygen increased, predominantly because of photolytic activity of the plant life. The Archean atmosphere, as compared with the present, contained considerably more carbon dioxide and water vapour, and very little oxygen. Nitrogen was the major component. In the process of its evolution the atmosphere of the earth has undergone irreversible changes, repeatedly renewing its composition as the result of the exchange of gases with the lithosphere, hydrosphere and space, and the influx of gases from the mantle and deeper zones (Vinogradov, 1964, and others).

References

BEUS, A. A. (1972) *Geochemistry of the Lithosphere*, Nedra, Moscow, 296 pp.

CHAMBERLIN, R. T. (1949) Geologic evidence on the evolution of the Earth's atmosphere. In *The Atmosphere of the Earth and Planets*, University of Chicago Press, **3**, 248–257.

DOBRETSOV, N. L. *et al.*(1970) *Metamorphic Facies*, Nedra, Moscow, 432 pp.

EUGSTER, G. P. (1961) Reduction and oxidation reaction during metamorphism of rocks. *Geokhimitcheskie issledovania, Moscow*, IL, 495–527.

HOLLAND, H. D. (1971) On some limitations of the composition of atmosphere and oceans during Precambrian. *Megd. Geok. Kongress, Moscow*, **2**, 952–953.

KULISH, E. A. (1970) Behaviour of oxygen in the archean aldonian atmosphere. *Dokl. Akad. Nauk. S.S.S.R.* **191**, no. 4, 921–923.

KULISH, E. A. (1973) *Lithology of Alumina-rich Metamorphic Rocks of the Lower Precambrian of the Aldonian Shield*, Khabarovsk, 370 pp.

KULISH, E. A. (1976) Regional geochemistry and different rocks types of the lower archean of the aldonian complex. *Lithology and mineralogy of metamorphic complexes*, Khabarovsk, 96–151.

RANKAMA, K. and SAHAMA, T. G. (1950) *Geochemistry*, University of Chicago Press, 912 pp.

RONOV, A. B. (1959) On the post-Precambrian geochemical evolution of the atmosphere and lithosphere. *Geokhimiya*, no. 5, 397–409.

RONOV, A. B and MIGDISOV, A. A. (1970) Evolution of the chemical composition of the rocks in the shields and sediment cover of the Russian and North-American platforms. *Geokhimiya*, no. 4, 403–438.

RUBI, V. V. (1957) Evolution of hydrosphere and atmosphere as shown by a special study of the probable composition of ancient atmosphere. *Zemnaia Kora*, IL, 650–671.

SIDORENKO, A. V. (1967) Some questions related with the study of Precambrian. In *Probl. orad. Geol. dokem.*, Nedra, Moscow, **2**, 5–13.

SIDORENKO, Sv. A and SIDORENKO, A. V. (1975) *The Organic Substances of Precambrian Sedimentary and Metamorphic Rocks*, Nauka, Moscow, 140 pp.

UREY, H. C. (1952) On the early chemical history of the earth and the origin of life. *Nat. Acad. Sci. Proc.* **38**, 351–362.

VINOGRADOV, A. P. (1964) The gazeous regime of earth. In *Minia zennoi Korou*, Nauka, Moscow, **2**, 5–21.

VINOGRADOV, A. P. (1967) *Introduction to the Geochemistry of Oceans*, Nauka, Moscow, 215 pp.

VINOGRADOV, V. I. (1971) The isotopic composition of sulfur as indicator of its constancy in time. *Megd. Geok. Kongress*. Moscow, **2**, 956–957.

Chemical Model for Origin and Distribution of Elements in Salts and Brines during Evaporation of Waters. Application to some Saline Lakes of Tibesti, Chad

J. Y. GAC,† A. AL-DROUBI,‡ H. PAQUET,‡ B. FRITZ,‡ and Y. TARDY‡

†*O.R.S.T.O.M. 24, rue Bayard, 75008 Paris, France*
‡*Institut de Géologie, 1, rue Blessig, 67000 Strasbourg, France*

Abstract

During evaporation of natural waters, many possibilities of successive salt parageneses and of variations in chemical facies of residual solutions are offered. Chemical model based on equilibrium reactions can be very useful for explaining the complexity of natural geochemical paths and distribution of elements in salts and brines. This model calculates the distribution of complex aqueous species, ionic strength, activity coefficients, saturation of solutions with respect to salts and amounts of minerals precipitated along progressive evaporation of water.

The model was applied to saline lakes of Ounianga Kebir in Tibesti (northern Chad), which occupy several depressions within the Nubian sandstone formation. In the initial diluted waters springing from sandstone all the major elements are present in the same order of magnitude. After concentration by evaporation, brines become strongly depleted in calcium and magnesium, initially enriched in sodium and bicarbonate and finally in potassium and chloride. Salt parageneses control chemical composition of waters, the calculated chemical model being in excellent accordance with natural observations. Silica and magnesium concentrations are controlled by an early precipitation of Mg-montmorillonite; magnesium molality is then regulated by the formation of huntite and dolomite; calcium is controlled by calcite, huntite, dolomite, pirssonite and gaylussite; sodium is first trapped in trona or natron, then by thenardite; potassium is finally picked up by sylvite; pH rises until reaching a maximum at about 10.

Behavior and distribution of elements in salts and brines can be discussed considering the relative proportions of these elements in the initial solutions and the order in which salts precipitate.

Introduction

In natural cases, when waters evaporate, many possibilities of successive salt parageneses and variations in chemical facies of residual solutions are offered. Distribution and behavior of the elements are chiefly determined by the relative amount of these elements in the initial solutions and by the nature and the order of salts precipitating (Van'Thoff *et al.*, 1912; d'Ans, 1933; Lotze, 1957; Borcher and Muir, 1964; Jones, 1965, 1966; Jones *et al.*, 1966, 1969; Hardie, 1968; Bradley and Eugster, 1969; Truesdell and Jones, 1969; Hardie and Eugster, 1970; Lerman, 1967, 1970; Eugster, 1970; Braitsh, 1971; Lerman and Jones, 1973; Phillips and Van Denburgh, 1971; Jones *et al.*, 1976).

Predictions of such behaviors can be approached using equilibrium reactions as proposed by Garrels and Mackenzie (1967), Hardie and Eugster (1970), Wood (1975), Fritz (1975), Al-Droubi (1976), Al-Droubi *et al.* (1976a, b, 1977), Gac *et al.* (1977) have shown that chemical models can be very useful to understand natural processes and to explain observed salt parageneses and chemical compositions of brines.

As an example of application, the geochemistry of Ounianga lakes was studied because of the large variety of salts encountered and of the presence in the initial solution of all the major anions, cations and silica in almost similar proportions.

Saline Lakes of Tibesti

1. Localization

In the northern part of Chad Republic, near the Lybian frontier and between Tibesti and Ennedi massives, saline lakes of Ounianga Kebir and Serir occupy several depressions within the Nubian sandstone formation. The most extended one, Lake Yoa, is surrounded by sandstone cliffs from which diluted waters spring in many points. Under very arid conditions, waters are evaporated and concentrated in the lake and salts are deposited on the edges (Fig. 1).

2. Salt Paragenesis

Forty samples collected on beaches were analyzed by means of X-ray diffraction. The different minerals found in the saline paragenesis are the following:

chlorides: halite (Nacl) and sylvite (KCl);
sulfates: thenardite (Na_2SO_4);
sulfates–carbonates: burkeite ($Na_6CO_3(SO_4)_2$);
carbonates: calcite ($CaCO_3$), aragonite ($CaCO_3$), dolomite ($CaMg(CO_3)_2$), huntite ($CaMg_3(CO_3)_4$), pirssonite (Na_2CO_3, $CaCO_3$, $2H_2O$), trona (Na_2CO_3, $NaHCO_3$, $2H_2O$), thermonatrite (Na_2CO_3, H_2O), nahcolite ($NaHCO_3$) and small amounts of gaylussite (Na_2CO_3, $CaCO_3$, $5H_2O$), magnesite ($MgCO_3$) and siderite ($FeCO_3$);
silicates: montmorillonite ($Mg_3Si_4O_{10}(OH)_2$ supposed)

On the surface, from the lake to the cliff, the following mineral sequence is observed:

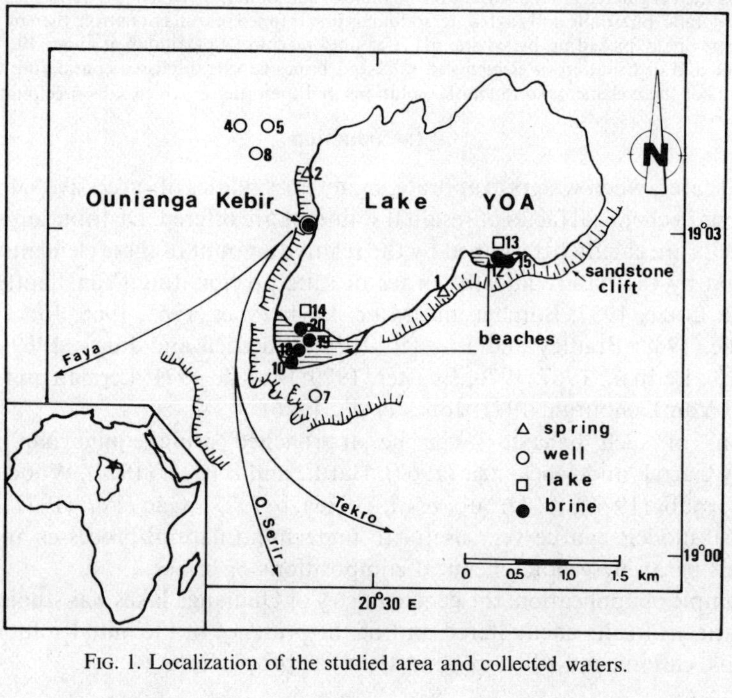

FIG. 1. Localization of the studied area and collected waters.

montmorillonite, calcium carbonates, calcium–magnesium carbonates, sodium carbonates, sodium sulfates, sodium and potassium chlorides.

In depth (60 cm) from the lake to the cliff, an inverse sequence is found: sodium chlorides and sulfates, sodium carbonates, sodium–calcium carbonates, magnesium–calcium carbonates, calcium carbonates and montmorillonite.

The sequences of salts as well as chemical compositions of waters indicate that in depth, near the water table, the gradient of concentration increases from the cliff to the lake. In contrast, at the surface of the beaches, the gradient of concentration increases from the lake to the cliff (Table 1).

TABLE 1. DISTRIBUTION OF SALTS IN BEACHES OF LAKE YOA

	Lake		Cliff
Surface	Aragonite–calcite Mg–Montmorillonite	huntite natron burkeite nahcolite	trona thenardite sylvite halite
Surface	Dolomite Pirssonite	dolomite pirssonite gaylussite thenardite halite	dolomite pirssonite trona thenardite halite
Depth	Pirssonite Gaylussite Thenardite Halite Sylvite	pirssonite halite calcite	aragonite–calcite Mg–Montmorillonite

3. Behavior of Silica and Magnesium

Chemical analyses were performed on the soluble fraction of salts samples. Magnesium appears as expected, in samples in which dolomite and huntite are present, in the median part of the sequence. Moreover, magnesium is still abundant in calcium carbonate facies associated with montmorillonite and located in surface, near the lake and in depth near the cliff. These samples do not contain magnesium carbonates but yield the highest amount of soluble silica. It must be emphasized that magnesium and silica may be associated in a Mg-bearing montmorillonite, as already shown in the Lake Chad area (Tardy *et al.*, 1974; Al-Droubi, 1976; Gac *et al.*, 1977). This magnesium–silica association is confirmed by the chemical evolution of waters.

4. Chemical Evolution of Natural Solutions

Twenty samples were collected in the springs, wells, lake and brines. Chemical analyses are given in Table 2 and are plotted on Fig. 2, molality of sodium being chosen as a linear function of the concentration factor. It is shown that potassium, chloride and sulfate molalities increase almost like the sodium one until a concentration factor of about 3500. Magnesium and calcium molalities increase and then decrease, after having reached a concentration factor of about 10, while silica and alkalinity go on to increase but more

TABLE 2. CHEMICAL COMPOSITION OF WATERS (mmoles/kg H_2O)

Location of water samples		No.	Conc. factor	pH	Alkalinity (meq.)	Cl	SO_4	Na	K	Ca	Mg	SiO_2	log f_{CO_2}	Ionic strength
Springs														
	KS_1	1	1.0	7.0	1.50	0.16	0.29	1.19	0.18	0.18	0.26	0.31	−2.04	0.0029
	KS_2	2	1.6	6.8	1.45	1.10	0.63	1.98	0.22	0.42	0.27	0.27	−1.86	0.0046
	KS_3	9	25.7	9.7	12.80	1.00	10.82	30.72	2.42	0.20	0.25	0.39	−4.20	0.0445
	KS_4	3	1.7	7.1	1.90	0.46	0.98	2.03	0.59	0.60	0.18	0.62	−2.05	0.0058
	KS_5	6	5.4	7.9	3.60	1.56	1.51	6.45	0.61	0.35	0.19	0.73	−2.59	0.0098
Wells														
	KP_6	7	5.7	7.1	5.00	1.38	3.58	6.75	0.88	0.70	2.25	1.08	−1.66	0.0177
	KP_7	5	3.9	7.1	2.80	1.88	2.43	4.62	0.26	0.80	1.45	0.40	−1.90	0.0127
	KP_8	8	6.9	7.1	3.90	4.93	3.44	8.25	0.66	1.65	1.25	0.32	−1.77	0.0197
	KP_9	4	2.0	6.9	2.64	0.93	1.15	2.42	0.31	0.90	0.68	0.45	−1.72	0.0080
Lakes														
	KL_{10}	14	945.6	9.9	440.00	292.10	241.75	1130.00	81.29	0.12	0.12	5.58	−3.69	1.0838
	KL_{11}	13	596.0	9.7	300.00	204.90	134.75	712.00	58.80	0.12	0.12	2.99	−3.37	0.7257
	K_{18}	11	96.3	9.3	45.00	23.33	27.22	115.00	7.43	0.50	0.11	1.06	−3.20	0.1362
	K_{19}	17	2711.0	9.7	1460.00	1115.70	436.00	3240.00	191.50	—	—	9.15	−3.08	2.7267
	K_{20}	16	2502.0	9.7	1360.00	1065.00	378.30	2990.00	176.04	—	—	9.52	−3.10	2.5445
Brines														
	KPL_{12}	20	5023.0	9.8	2035.50	2048.90	1203.36	6002.61	555.55	0.18	0.06	17.48	−3.29	4.8730
	KPL_{13}	19	3715.0	9.7	1225.50	639.00	914.50	4440.00	198.50	0.30	0.10	19.56	−2.97	3.3286
	KPL_{14}	18	3263.0	10.0	2130.00	618.71	691.50	3900.00	222.40	0.15	0.08	21.64	−3.53	3.0092
	KPL_{15}	10	60.7	10.3	65.00	12.17	1.25	72.50	6.98	0.17	0.52	2.04	−4.63	0.0903
	KPL_{16}	15	2276.0	9.8	1860.00	1002.10	206.50	2720.00	127.60	0.07	0.06	5.49	−3.16	2.5324
	KPL_{17}	12	102.5	7.9	75.00	62.74	13.02	122.50	13.40	0.52	0.16	3.18	−1.37	0.1633

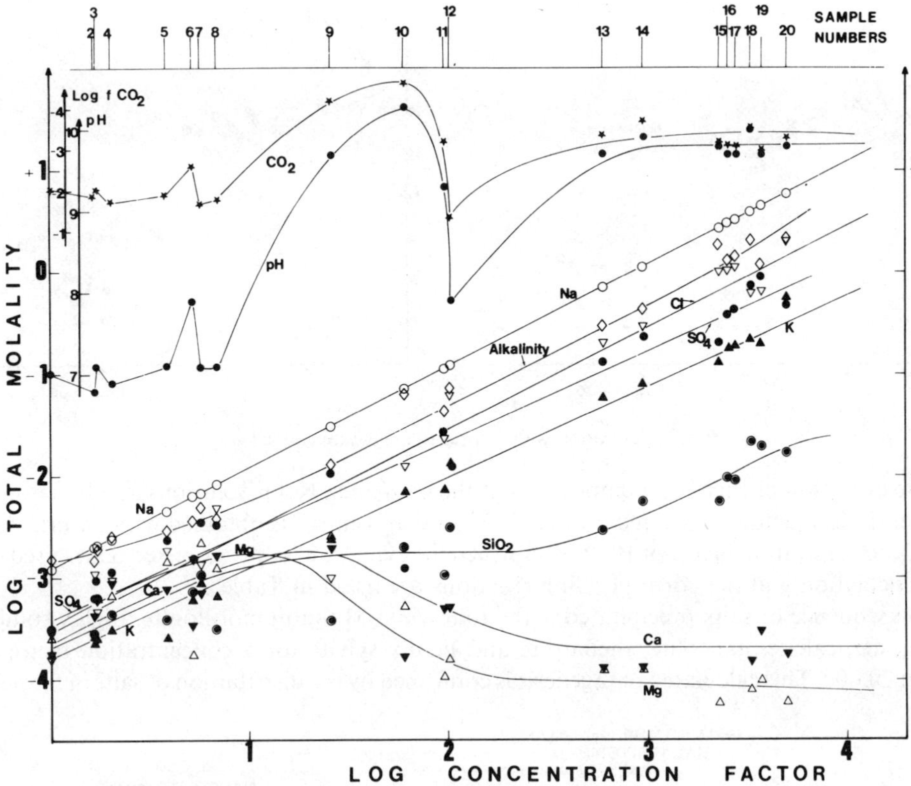

FIG. 2. Evolution of the concentration of elements as a function of concentration factor.

weakly. It must be noted that magnesium and silica together are affected before calcium depletion, which is interpreted as a formation of Mg-bearing silicate preceding calcite precipitation. The equivalent partial pressure of CO_2 is equal to $10^{-1.8}$ atm in the spring and well waters and tends progressively to a value of $10^{-3.5}$ atm, when waters evaporate at the contact with atmosphere. In a general way, pH increases and then stabilizes for a concentration factor of about 3500. At this stage (pH = 9.6), total molality of silica increases, which corresponds to the formation of $H_3SiO_4^-$ in solution. Calcium and magnesium molalities are very low in waters, according to saturation with respect to Mg-or Ca-carbonates (Fig. 3).

Simulation Model of Evaporation

A computer program called EVAPOR (Fritz, 1975; Al-Droubi, 1976) derived from the PATH CALC of Helgeson *et al.* (1970) was used to calculate the evolution of chemical composition of solutions which concentrate by evaporation. Starting from an initial composition (in this case, the spring water no. 1), the program computes, for incremented concentration factors, the ionic strength of the solution, the distribution of the aqueous complex species, the activities of the ions and the amount of minerals precipitated when saturation points are reached.

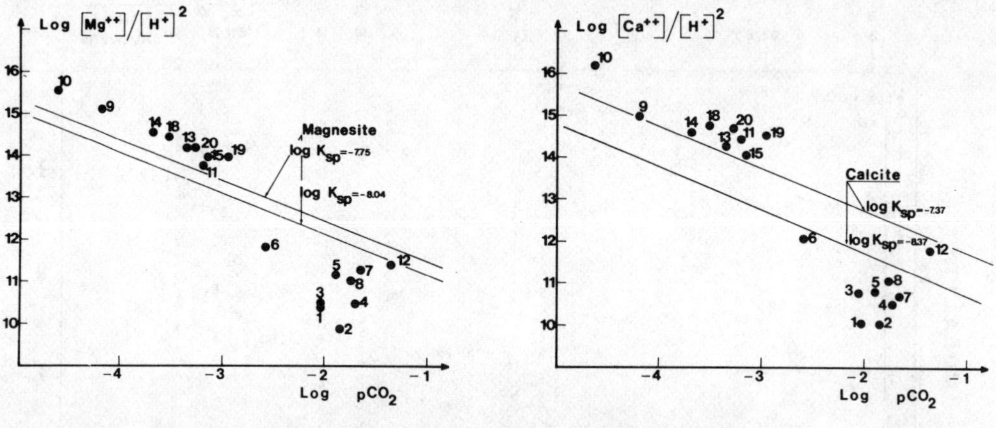

FIG. 3. Saturation with respect to magnesite and calcite.

The evolution of chemical composition of the Ounianga Kebir solutions as a function of concentration factor is reported on Fig. 4. A good agreement is obtained with natural data until a concentration factor of 1000, after which the evolution is only expected. Data used for the calculation and the corresponding reactions are listed in Table 3.

The sequence of salts precipitated is the following: Mg-montmorillonite, highly soluble magnesite, calcite, nahcolite, thenardite and finally sylvite for a concentration factor of about 20,000. This calculated paragenesis is confirmed by the distribution of salts in the field.

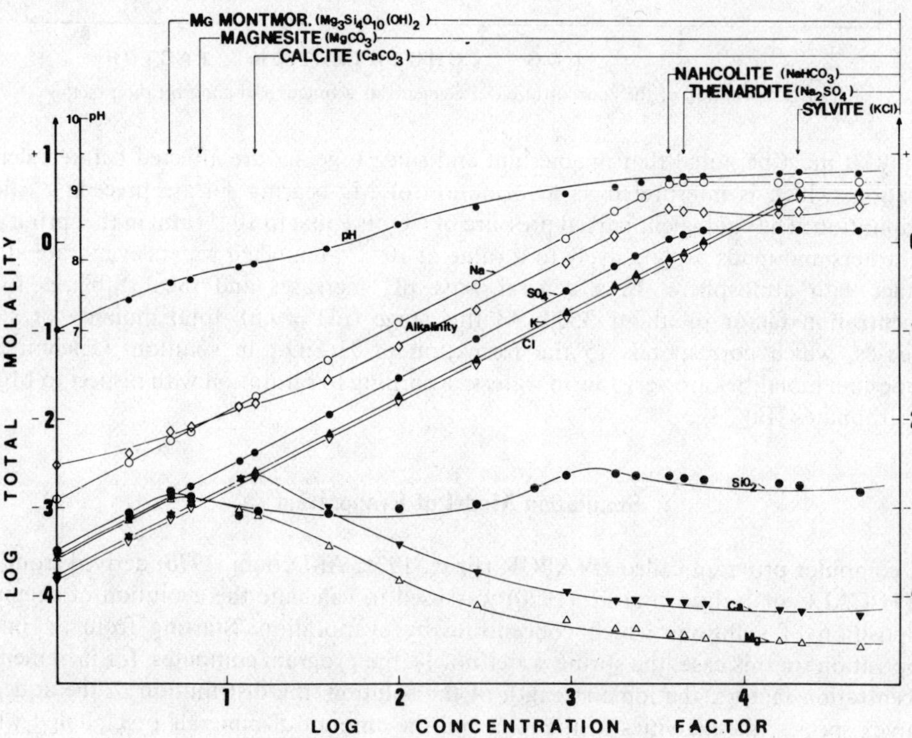

FIG. 4. Chemical evolution during simulated evaporation of the spring water no. 1.

TABLE 3. EQUILIBRIUM CONSTANTS FOR MINERALS AT 25°C

Mineral		Expression		Value
Mg–montmorillonite:		$3 \log [Mg^{2+}]/[H^+]^2 + 4 \log [H_4SiO_4]$	=	+25.0
Amorphous silica	:	$\log [H_4SiO_4]$	=	−2.70
Calcite	:	$\log [Ca^{2+}]/[H^+]^2 + \log f_{CO_2}$	=	+10.76
Magnesite	:	$\log [Mg^{2+}]/[H^+]^2 + \log f_{CO_2}$	=	+10.38
Gypsum	:	$\log [Ca^{2+}] + \log [SO_4^=] + 2 \log [H_2O]$	=	− 4.85
Thenardite	:	$2 \log [Na^+] + \log [SO_4^=]$	=	−0.87
Trona	:	$3 \log [Na^+]/[H^+] + 2 \log f_{CO_2} + 4 \log [H_2O]$	=	+24.89
Sylvite	:	$\log [K^+] + \log [Cl^-]$	=	+0.78
Halite	:	$\log [Na^+] + \log [Cl^-]$	=	+1.58
Hydrophyllite	:	$\log [Ca^{2+}] + 2 \log [Cl^-]$	=	+11.49

The problem of an early precipitation of Mg-montmorillonite (in the form of a highly soluble talc $Mg_3Si_4O_{10}(OH)_2$) may be discussed. In fact, in the initial solution the magnesium molality is higher than the calcium one, but after evaporation it becomes lower. Furthermore, magnesium depletion occurs before the calcium one, but together with the beginning of silica consumption. This is interpreted as a precipitation of Mg-bearing silicate, as already shown (Gac *et al.*, 1977), and calculated as a highly soluble talc. The different possibilities offered are presented on Fig. 5 and it seems obvious that only the hypothesis no. 4 fits with natural observations: Mg-montmorillonite is the first mineral precipitated, followed by calcite and afterwards by magnesite (or Mg-rich carbonates).

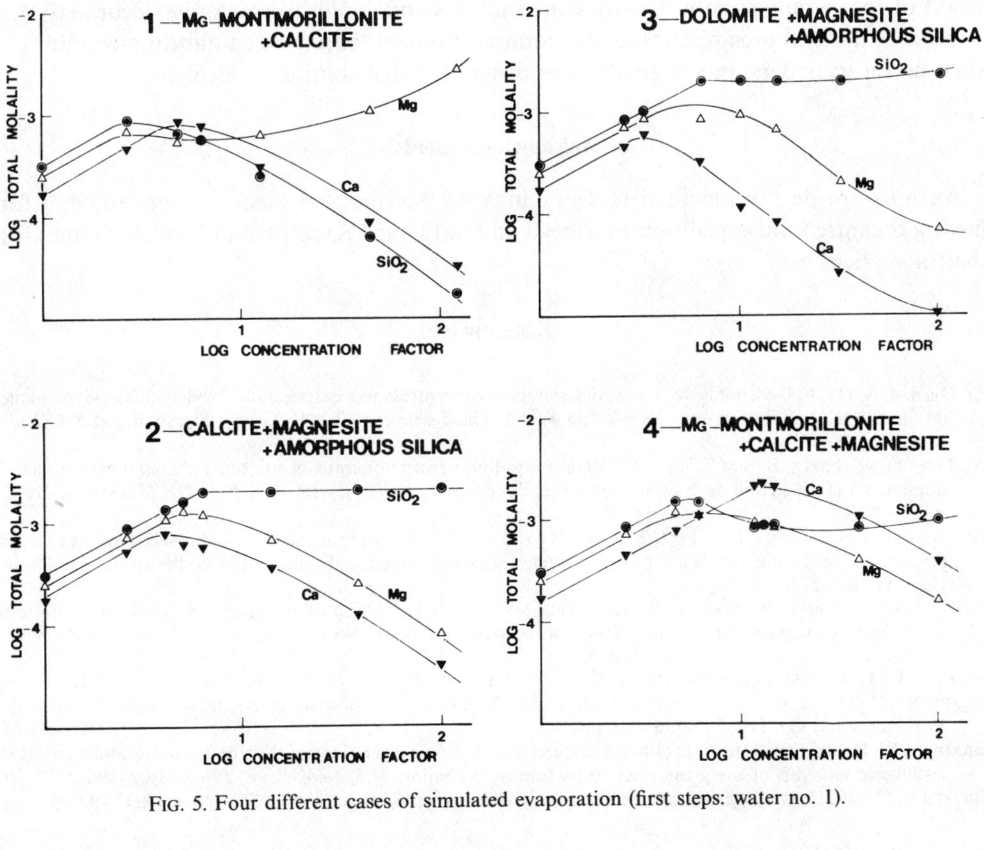

FIG. 5. Four different cases of simulated evaporation (first steps: water no. 1).

Discussion and Concluding Remarks

Along evaporation steps of natural waters, concentration of soluble species can fluctuate: generally all the molalities begin to increase during the initial stages, but afterwards, some of them may go on to increase, some others may decrease and later on increase again. Each case yields a different pattern depending on the nature, on the order of salt precipitating and on the relative amount of these species in the initial solutions. Attempts at predicting such behaviors were proposed for many years such as those of Eaton (1950), Garrels and Mackenzie (1967), Hardie and Eugster (1970), Van Beek and Van Breemen (1973). In the situations in which many elements are involved, we think that computer program calculations are needed for solving correctly the problem (Al-Droubi *et al.*, 1977).

The good agreement with natural observations such as these described here or these already presented for the Lake Chad system (Al-Droubi, 1976; Al-Droubi *et al.*, 1976b; Gac *et al.*, 1977) and for the Euphrate River (Al-Droubi *et al.*, 1977) permits the extrapolation of the calculations to a great number of natural environments.

In nature, salts precipitated are generally complex ones: magnesian calcite, dolomite, huntite, pirssonite, gaylussite, glauberite, burkeite, lecontite and above all montmorillonites, etc. The example presented shows that a fairly good pattern is obtained by using simple salts for which solubility products are empirically corrected (with respect to pure minerals), taking into account oversaturations commonly found in nature (Table 3).

In many cases, geochemical evolution of solutions and brines may be interpreted as the result of near equilibrium water–rock interactions and could be used quite accurately as a witness of mineral precipitations. The computer model, based on equilibrium reactions is then useful to understand or predict the origin and distribution of elements.

Acknowledgements

Authors are deeply indebted to Captain Yves Merliet and General Negue-Djogo for having organized the expedition to Tibesti and to Gérard Krempp and Daniel Million, for their analytical help.

References

Al-Droubi, A. (1976) Géochimie des sels et des solutions concentrées par évaporation. Modèle thermodynamique de simulation. Application aux sols salés du Tchad. Thèse Université Louis Pasteur, Strasbourg et *Sci. Géol.*, Mém. no. 46, 177 pp.

Al-Droubi, A., Fritz, B. and Tardy, Y. (1976a) Equilibres entre minéraux et solutions. Programmes de calcul appliqués à la prédiction de la salure des sols et des doses optimales d'irrigation. *Cah. ORSTOM, sér. Pédol.* XIV, 13–38.

Al-Droubi, A., Cheverry, C., Fritz, B. and Tardy, Y. (1976b) Géochimie des eaux et des sels dans les sols des polders du lac Tchad: Application d'un modèle thermo-dynamique de simulation de l'evaporation. *Chem. Geol.* **17**, 165–177.

Al-Droubi, A., Fritz, B., Gac, J. Y. and Tardy, Y. (1977) Prediction of the chemical evolution of natural waters during evaporation. *2nd Int. Symp. on Water–Rock Inter.*, Strasbourg, section II: Salts and brines, 13–22.

Beek, C. G. E. M. Van and Bremmen, N. Van (1973) The alkalinity of alcali soils. *J. Soil Sci.* **24**, 129–136.

Borchert, H. and Muir, R. O. (1964) *Salt Deposits. The Origin, Metamorphism and Deformation of Evaporites*, Van Norstrand Co. Ed., London, 338 pp.

Bradley, W. H. and Eugster, H. P. (1969) Geochemistry and paleolimnology of the trona deposits and associated authigenic minerals of the green river formation of Wyoming. *U.S. Geol. Surv., Prof. Paper*, 496 B, 71 pp.

Braitsch, O. (1971) *Salt Deposits, their Origin and Composition*, Springer-Verlag Ed., New York, 297 pp.

D'Ans, J. (1933) *Die Lösungsgleichgewichte des Systems der ozeanischen Salzablagerrungen*, Berlin: Kaliforschungsanstalt, Text 254 pp., Atlas 88 pp.

Eaton, F. M. (1950) Significance of carbonates in irrigation waters. *Soil Sci.* **69**, 123–133.

Eugster, H. P. (1970) Chemistry and origin of the brines of Lake Magadi, Kenya. *Mineral. Soc. Amer.*, Spec. paper, **3**, 213–235.

Fritz, B. (1975) Etude thermodynamique et simulation des réactions entre minéraux et solutions. Application à la géochimie des altérations et des eaux continentales. *Sci. Géol.*, Mém. no. 41, 152 pp.

Gac, J. Y., Al-Droubi, A., Fritz, B. and Tardy, Y. (1977) Geochemical behaviour of silica and magnesium during the evaporation of waters in Chad. *Chem. Geol.* **19**, 215–228.

Garrels, R. M. and Mackenzie, F. T. (1967) Origin of the chemical compositions of some springs and lakes. In *Equilibrium Concept in Natural Water Systems*, Gould, ed., Amer. Chem. Soc., Ser. 67, 222–242.

Hardie, L. A. (1968) The origin of the recent non-marine evaporite deposit of saline water, Inyo County, California. *Geochim. Cosmochim. Acta* **32**, 1279–1301.

Hardie, L. A. and Eugster, H. P. (1970) The evolution of closed-basin brines. *Mineral. Soc. Amer.* Spec. Paper **3**, 273–290.

Helgeson, H. C., Brown, T. H., Nigrini, A. and Jones, T. A. (1970) Calculation of mass transfer in geochemical processes involving aqueous solutions. *Geochim. Cosmochim. Acta* **34**, 569–592.

Jones, B. F. (1965) The hydrology and mineralogy of deep springs lake, Inyo County, California. *U.S. Geol. Surv., Prof. Paper*, 502 A, 56 pp.

Jones, B. F. (1966) Geochemical evolution of closed basin water in the western great basin. *2nd Symp. on Salt, Northern Ohio, Geol. Soc. Rau*, Ed. 1, 181–200.

Jones, B. F. and Van Denburgh, A. S. (1966) Geochemical influences on the chemical character of closed lakes. *Int. Ass. Sci. Hydrol., Symp. of Garda* **70**, 435–446.

Jones, B. F., Van Denburgh, A. S., Truesdell, A. H. and Rettig, S. L. (1969) Interstitial brines in playa sediments. *Chem. Geol.* **4**, 253–262.

Jones, B. F., Eugster, H. P. and Rettig, S. L. (1976) Hydrochemistry of the Lakē Magadi basin, Kenya (in press).

Lerman, A. (1967) Model of a chemical evolution of a chloride lake. The Dead Sea. *Geochim. Cosmochim. Acta* **31**, 2309–2330.

Lerman, A. (1970) Chemical equilibria and evolution of chloride brines. *Mineral. Soc. Amer.*, Spec. Paper **3**, 291–306.

Lerman, A. and Jones, B. F. (1973) Transient and steady state salt transport between sediments and brines in closed lakes. *Limnology and Oceanography* **18**, 72–85.

Lotze, F. (1957) *Steinsalz und Kalisalz.* I. Teil: *Gebrüder Bonträger*, Berlin, Nikolassee, 465 pp. (2nd ed., 1960).

Phillips, K. N. and Van Denburgh, A. S. (1971) Hydrology and geochemistry of Abert, Summer and Goose lakes and other closed basin lakes in South Central Oregon. *U.S. Geol. Surv., Prof. Paper*, 502 B, 86 pp.

Tardy, Y., Cheverry, C. and Fritz, B. (1974) Néoformation d'une argile magnésienne dans les dépressions interdunaires du lac Tchad. Application aux domaines de stabilité des phyllosilicates alumineux, magnésiens et ferrifères. *C. R. Acad. Sci., Paris* **278**, série D, 1999–2002.

Truesdell, A. H. and Jones, B. F. (1969) Ion association in natural brines. *Chem. Geol.* **4**, 51–62.

Vant'Hoff, J. H. *et al.* (1962) *Untersuchungen über die Bildungsverhältnisse der ozeanischen Salzablagerungen, insbesondere des stabfurter Salzlagers*, Leipzig, Akademische Verlagsgesellschaft, 374 pp.

Wood, J. R. (1972) Prediction of mineral solubilities in concentrated brines: a thermodynamic approach. Ph.D., Baltimore, 121 pp.

Discussion

R. A. Berner:

Why didn't you consider dolomite in your model?

B. Kubler:

Same remark as for Berner's paper: the crystallographic control could change drastically the behavior of the elements (Mg-calcite with different amount of substitution for the pair Mg and Ca) aragonite, H_2S and so on for Iron monosulfide.

J. Y. Gac:

We did not calculate the solubility of different Mg-calcite because we could not have a field control of the Mg variation in calcites (mixed with many minerals) and because we are not yet able to simulate solid solution precipitation. Furthermore, we simply simulated a highly soluble calcite (log $Kps =$ 7.37) and magnesite for taking into account of paragenesis calcite–dolomite–magnesite.

E. N. PENTCHEVA:

(a) Le modèle est-il appliqué pour l'étude des saumures marines où les stades concentration-cristallisation diffèrent de votre exposé?

(b) Le comportement hydrochimique simultané des éléments tels que Ba, Sr, Rb, Cs, Li a-t-il été observé par les auteurs?

J. Y. GAC:

Oui, le modèle est tout à fait applicable (et déjà appliqué) aux saumures d'origine marine. Non les éléments comme le baryum, le strontium, le lithium n'ont pas été pris en compte.

SECTION IV

DISTRIBUTION AND BEHAVIOUR OF TRACE ELEMENTS

Trace Element Distribution in Rock-Forming Silicates. The Alkali and Alkaline Earths

J. T. IIYAMA

Geological Institute, Faculty of Science, University of Tokyo, Hongo, Bunkyo-ku, Tokyo, 113 Japan

Abstract

Trace element partition between silicates and hydrothermal solution is controlled by the thermodynamic properties of the silicate solid solution formed by the substitution of its major constituent M by a trace ion A. Experimental studies using feldspars, micas, leucite and nepheline are summarized in this paper.

The difference $\Delta r = r(\mathrm{M})-r(\mathrm{A})$ of the ionic radii of M and A compared to the dimension d of the sites for M in the silicate is the most determining factor for the variation of the function of distribution $D(\mathrm{AR/MR})$ with the composition x of the solid solution.

(1) The substitution with a cation having small Δr realizes a random distribution of the A atom among the M sites in the silicate structure. The variation of $D(\mathrm{AR/MR})$ with an increase in A in the mineral is caused by the enthalpy of mixing of the silicate solid solution.

(2) The presence of two non-equivalent sites for A in the silicate strongly modifies $D(\mathrm{AR/MR})$ with an increase of A in the mineral.

(3) Substitution with $|\Delta r|/d$ greater than 0.07 causes a local lattice deformation around the sites substitute d by A. This local deformation rapidly changes the $D(\mathrm{AR/MR})$.

Variation of the partition relation in cases (2) and (3) is caused principally by the non-ideal entropy of mixing.

(4) In some cases, especially with micas, the initial pattern of trace element distribution with local lattice deformation is degenerated by further substitution of M by A.

(5) For a given mineral species, a smaller Δr gives a stronger tendency to concentrate A into the silicate.

(6) The difference in ionic charge of M and A (M^{+} and A^{++}) amplifies the relation stated in (5).

(7) Slight variation in the constitution of the silicate radical can strongly modify the partition relation. This fact should be taken into consideration when the trace element partitioning of complex silicates such as biotite is discussed.

Introduction

All rock-forming materials, such as minerals, magmas, hydrothermal solutions and amorphous colloidal matter, are considered to be solutions. Even minerals with fairly constant compositions, such as quartz or corundum for example, should be treated as solutions when minor impurities in their structures are discussed.

The study of the distribution of trace elements in the earth should therefore begin with a thorough understanding of these different kinds of solutions.

Chemical thermodynamics permits us to reasonably describe properties of solutions in relation to the distribution. Crystallochemistry helps us to understand their thermodynamic parameters. A combination of these two factors permits us to predict the distribution tendency of a given trace element among rock-forming phases of known structures.

Natural minerals are too complicated to attain this goal through studies on them alone. It is necessary to first use artificial minerals whose composition can be controlled at the experimentor's discretion. This study is concerned with minerals of simple composition and later treats those of a more complicated composition, as a closer approach to natural systems.

The purpose of this paper is to summarize the behaviour, as observed from experiment, of trace alkali and alkaline earth elements in the principal tecto and phyllo-silicates.

Thermodynamics of Trace Element Distribution

Since the first observation by Wyart and Sabatier (1956) and confirmed by Orville (1962, 1963) of the Na–K ion exchange in the system alkali feldspar–hydrothermal (Na, K) Cl solution, many authors have shown that the reaction is a common feature in other principal tecto- and phyllosilicates [plagioclases: Wyart and Sabatier (1962, 1965, 1974), Iiyama *et al.* (1963), Iiyama (1966), Orville (1963); mica: Iiyama (1964), and feldspathoids: Debron *et al.*, (1961), Debron (1962, 1965), Wellman (1970), for example].

As will be seen in the works reviewed in this paper and the study by Beswick (1973), the principle of ion exchange equilibrium can be used astutely to describe the partitioning of a trace element A between a mineral MR and hydrothermal solution.

The distribution can be expressed by an equation

$$MR + A^+ = AR + M^+ \tag{1}$$

for M and A of same valency.

The cases with M and A of different valency can be treated in similar but slightly different way. Theory for these cases will not be repeated here for simplicity.

At equilibrium (temperature T°K; pressure P kbars) we have the relation

$$\frac{X(\mathrm{AR})}{X(\mathrm{MR})}\cdot\frac{\gamma(\mathrm{AR})}{\gamma(\mathrm{MR})}\cdot\frac{m(\mathrm{M})}{m(\mathrm{A})}\cdot\frac{\gamma'(\mathrm{M})}{\gamma'(\mathrm{A})} = k \tag{2}$$

where $X(i)$ and $\gamma(i)$ designate respectively the molar fraction and activity coefficient of the constituent i of the mineral. $m(j)$ and $\gamma'(j)$ are the molarity and the activity coefficient of the ion j in solution. The constant k is related to the free energy difference ΔG° of the pure constituents of reaction (1).

$$\mathbf{R}T\cdot\ln k = -\Delta G^\circ \tag{3}$$

$$\text{where } \Delta G^\circ = [\Delta G^\circ(\mathrm{AR}) - \Delta G^\circ(\mathrm{A})] - [\Delta G^\circ(\mathrm{MR}) - \Delta G^\circ(\mathrm{M})].$$

As has been stated by many authors (Debron, 1965; Thompson and Waldbaum, 1968; Perchuk and Andrianova, 1968; Delbove 1971), the ratio $\gamma'(\mathrm{M})/\gamma'(\mathrm{A})$ can be considered as a constant or even equal to unity under ordinary conditions of hydrothermal experiments.

The ratio $\Gamma(\mathrm{AR/MR}) = \gamma(\mathrm{AR})/\gamma(\mathrm{MR})$ is a function of the composition of the crystalline solution. It is assimilated to a constant $\Gamma^\circ(\mathrm{AR})$ when $X(\mathrm{AR})$ tends to be 0. The form of the function $\Gamma(\mathrm{AR/MR})$ depends on the nature of the solid solution.

Taking these properties of the two phases into account and designating the composition of the solid by $x_{xal} = X(\mathrm{AR})/X(\mathrm{MR})$ and that of the solution by $v_{\mathrm{aq}} = m(\mathrm{A})/m(\mathrm{M})$, relation (2) is transformed simply as

$$v_{\mathrm{aq}} = \mathbf{K}^{\circ} \cdot x_{xal} \cdot \Gamma(\mathrm{AR/MR}) \tag{4}$$

where $\mathbf{K}^{\circ} = 1/k$, namely, $RT \cdot \ln \mathbf{K}^{\circ} = \Delta G^{\circ}$.

For a very low concentration of A, the crystal may behave as an ideal dilute solution, and we have a linear relation.

$$v_{\mathrm{aq}} = \mathbf{K} \cdot x_{xal} \tag{5}$$

with $\mathbf{K} = \mathbf{K}^{\circ} \cdot \Gamma^{\circ}(\mathrm{AR})$

Observation of the variation of v_{aq} as a function of x_{xal} permits us to obtain fundamental information on the distribution of element A between the mineral MR and aqueous solution.

The exchange isotherm in the very low concentration range of A in the system indicates the value of **K** from the linear relation (5). Deviation of the isotherm from this relation at higher concentrations determines the function $\Gamma(x_{xal})$.

The establishment of a relation similar to (4) for another mineral MR′ permits us to estimate the distribution of element A among two minerals MR and MR′.

It should be mentioned here that the function of the distribution $D(\mathrm{AR/MR}) = \mathbf{R}T \cdot \ln (x_{xal}/v_{\mathrm{aq}})$ is sometimes used to study the mixing properties of solid solution (Thompson and Waldbaum, 1968; Delbove, 1971; Lagache and Sabatier, 1975).

The function $D(\mathrm{AR/MR})$ has an advantage when the exchange isotherm is established from $X(\mathrm{AR}) = 0$ to $X(\mathrm{AR}) = 1$. In effect, the integration $\int_0^1 D(\mathrm{AR/MR})\, \mathrm{d}X(\mathrm{AR})$ gives the ΔG° of the relation (3) and permits an estimate of the magnitude of $\mathbf{R}T \cdot \ln \Gamma^{\circ}(\mathrm{AR})$ from the observed values of **K**.

$$\int_0^1 D(\mathrm{AR/MR}) \cdot \mathrm{d}X(\mathrm{AR}) = -\Delta G^{\circ} = -[(\Delta G^{\circ}(\mathrm{AR}) - \Delta G^{\circ}(\mathrm{A})) - (\Delta G^{\circ} - \Delta G^{\circ}(B))] \tag{6}$$

and

$$\mathbf{R}T \cdot \ln \mathbf{K} = \Delta G^{\circ} + \mathbf{R}T \cdot \ln \Gamma^{\circ}(\mathrm{AR}). \tag{7}$$

Experiments based on these principles were carried out at the Centre de Recherches sur la Synthèse et Chimie des Minéraux, C.N.R.S., Orléans, France, by G. Sabatier, M. Lagache (actually in Ecole Normale Supérieure de Paris), M. Volfinger, J. Roux, F. Delbove and the present author.

The experimental procedure used in these studies has been described before (Iiyama, 1972, 1974; Lagache and Sabatier, 1973, for example) and is not repeated here. It should, however, be mentioned that the trace element A used in the experiments was generally marked by its radioisotopes. This procedure permitted us to simplify the determination of a very small amount of A present in the solid and aqueous phases.

Besides obtaining information on the exchange of many trace element–mineral pairs, this study aimed at clarifying the contribution of crystallochemical factors to the excess free energy of mixing of the crystalline solutions. Aspects of the excess free energy of mixing at infinite dilution were explored principally by G. Sabatier, M. Lagache and F. Delbove. Those of the excess entropy of mixing were studied by R. Roux, M. Volfinger and the present author.

As the numerical data of the experiments have already been published on several occasions, the present paper synthesizes the different aspects of the distribution by using, within the limits of possibilities, graphical representation.

Exchange Isotherm and Nature of Solid Solution

Figures 1, 2 and 3 represent the three typical cases of the exchange isotherms observed in the experiments. Because of a wide range of variations in x_{xal} and v_{aq} studied, the isotherm is constructed by plotting $\log_{10} x_{xal}$ against $\log_{10} v_{aq}$. The following three types of isotherms were found.

Type (A): General Case (Fig. 1)

The analysis of an isotherm similar to the Fig. 1 is fundamental.

A straight line *a–b*, having a slope of 45° against the horizontal axis, appears in the range of low x_{xal}, and verifies the relation $v_{aq} = \mathbf{K} \cdot x_{xal}$. The intersection of the line with the vertical axis of the figure determines the constant **K**.

From point *b* to higher x_{xal}, the isotherm deviates below the linear relation. The trace element A in this range shows a tendency to be concentrated more easily in the crystal than in region *a–b*.

At point *c*, another solid phase containing M and A appears in the solid. The isotherm indicates, as verified by the phase rule, constant v_{aq} for variable x_{aq}.

Analysis of the deviation of the isotherm from the extension of the line *a–b* permits us to calculate the excess free energy of the solid solution.

Adopting the model of regular solution, the exchange isotherm showing a deviation below the linear relation gives a positive interaction parameter *W*. The excess molar free energy of the solid solution is approximated as

$$\Delta G^{ex} = W \cdot X\,(\mathrm{AR}) \cdot X\,(\mathrm{MR}). \tag{8}$$

From this relation, the Γ° (AR) of the relation (5) is determined as exp (W/RT).

The limit in x_{xal} of the linear relation of the exchange isotherm depends on *W*. A smaller *W* gives a higher limit x_{xal} for ideality.

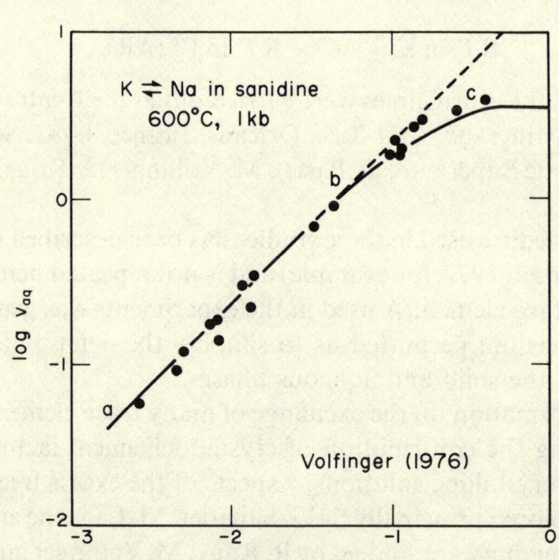

FIG. 1. Isotherm of the ion exchange K^+–Na^+ in sanidine at 600°C, 1 kbar.

The order of magnitude of W of the observed isotherms is $0 \sim 5$ kcal/mole and the limit of ideality (linear relationship in the isotherm) is always greater than 10^{-2} in x_{xal}.

The distribution of A atom in the structure of the crystal can be supposed, in this case, as random among M atoms.

Type (B): Isotherms with Noticeable Excess Entropy of Mixing (Fig. 2)

Isotherms analogous to Fig. 2 are characterized by an upward deviation from the linear relation observed in the range a–b of low x_{xal}. The crystal in the non-ideal part behaves as if it is saturated with the trace element A.

An isotherm of this type is characteristic of the exchanges Rb and K in nepheline (Roux, 1971, 1974), Rb and Cs in albite (Roux, 1971), Sr in albite and sanidine (Iiyama, 1974), Ba in sanidine (Iiyama, 1974) and Li in muscovite (Volfinger, 1969).

The deviation of the isotherm from the linear relation between x_{xal} and v_{aq} is often observed at a very low concentration of A in the crystal. In some cases, it was observed at $x_{xal} < 10^{-3}$.

Rb and K in nepheline, especially that of Rb, can be explained by the preferred fixation of Rb in the larger sites of Na in the structure of the mineral.

As the distribution of A atoms in the structure having two kinds of sites for M is no longer at random, the entropy of mixing of the solid solution cannot be assimilated to that of the ideal solution which has random distribution.

The excess entropy of mixing in this case depends on the partition of A among two kinds of sites (M_1 and M_2) for M. If the A atom had no preference for being fixed in M_1 or M_2, the resulting excess entropy of mixing will be 0. The exchange isotherm in this case will not differ so much from that of type A.

If the A atom is fixed only into the M_1 sites and not in the M_2 sites, the resulting molar entropy of mixing will have the form

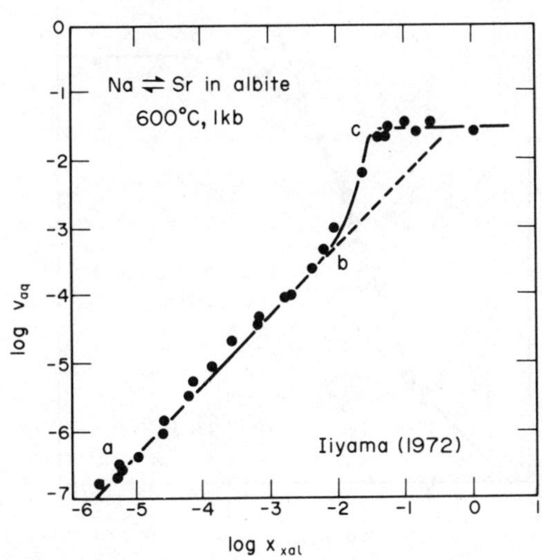

FIG. 2. Isotherm of the ion exchange Na^{+}–Sr^{+2} in albite at 600°C, 1 kbar.

$$\Delta S^{\text{mix}} = -\mathbf{R}\cdot\left[X(\text{AR})\cdot\ln\frac{X(\text{AR})}{m}-(m-X(\text{AR}))\cdot\ln\left(1-\frac{X(\text{AR})}{m}\right)\right] \tag{9}$$

where $m = M_1/(M_1+M_2)$, proportion of the number of the M_1 sites to the whole number of M sites in the crystal of MR.

A difference of the entropy of mixing for this type of solid solution from that ΔS^{id} of the ideal one is evident as

$$\Delta S^{\text{id}} = -\mathbf{R}\cdot[X(\text{AR})\cdot\ln X(\text{AR})-(1-X(\text{AR}))\cdot\ln(1-X(\text{AR}))] \tag{10}$$

An analysis performed in a different way, but producing the same results, was performed for the Rb–Na exchange in nepheline (Roux, 1971, 1974).

This kind of exchange isotherm is also observed in other silicates, where the two M sites in the structure cannot be distinguished. The model for substitutions accompanying local lattice deformation was proposed to explain the situation (Iiyama, 1974).

In this model, the M sites around a site M substituted by A are deformed and other A atoms cannot enter the deformed zone containing $(r+1)$ sites.

The resulting entropy of mixing is expressed by

$$\Delta S^{\text{mix}} = -\mathbf{R}\cdot\left[X(\text{AR})\cdot\ln X(\text{AR})+\frac{(1-(r+1)X(\text{AR}))}{r+1}\cdot\ln(1-(r+1)X(\text{AR}))\right] \tag{11}$$

The crystal will be saturated with A when $X(\text{AR})$ attains $1/(r+1)$.

An analysis of this type of isotherm was performed and the size of the locally deformed zone was estimated. The results showed $r \doteqdot 15$ for Li in muscovite, 181 for Sr in sanidine, 10 for Ba in sanidine, 33 for Sr in albite and 342 for Rb in albite.

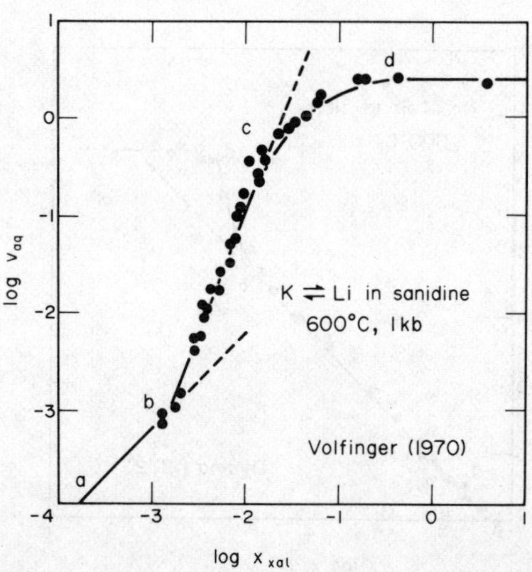

FIG. 3. Isotherm of the ion exchange K^+–Li^+ in sanidine at 600°C, 1 kbar.

Type (C): Substitution with Degenerating Local Deformation (Fig. 3)

An isotherm of this type was often observed in micas and was also noticed in the case of Li in sanidine (Iiyama and Volfinger, 1976). It is characterized by an *S*-shaped isotherm.

The equilibrium of the distribution proceeds in the same manner as in the preceding case up to a certain value of x_{xal}, (*a–c* in Fig. 3), but the slope of the curve drops again at higher concentrations instead of an increase in the case of Type B.

This case can be explained by the fact that the locally deformed M sites have some tolerance for receiving the trace element. The degeneracy of the "forbidden zone" of substitution was tentatively measured by introducing a factor *F* which is defined by the relation

$$r = r_0 \cdot [1 - F \cdot r \cdot X(\mathrm{AR})/(1-(r+1)\cdot X(\mathrm{AR}))] \tag{12}$$

where r_0 designates the size of the "forbidden zone" at $x_{xal} \to 0$.

The analysis showed $r_0 = 8385$ with $F = 0.69$ for Cs and $r_0 = 244$ with $F = 1$ for Na in muscovite at 600°C 1 kbar. The analysis was performed also for the other cases. Fixation of Li in phlogopite and muscovite enters into this category of substitution. It is difficult, however, in these cases, to judge if Li is fixed solely in the interlayer K sites or some of it is fixed in the octahedral sites. The latter case should be treated as a special case of substitution in multi-sited minerals.

Constant K and the Nature of Minerals

Relations (5) and (7) indicate that the constant **K** is composed of two terms related to ΔG° of the reaction (1) and the excess chemical potential $\mu^{ex}(\mathrm{AR})_0$ of AR at infinite dilution $= \mathbf{R}T \cdot \ln \Gamma^\circ(\mathrm{AR})$.

The study to distinguish ΔG° and $\mu^{ex}(\mathrm{AR})_0$ from the constant **K** of the distribution of Na, K, Rb and Cs in alkali feldspars (sanidine, albite and Rb-feldspar) by Lagache and Sabatier (1973) shows that ΔG° varies from 0.23 kcal/mole for the Rb/sanidine pair (−0.23 kcal/mole for K/(Rb-feldspar)) to 1.93 kcal/mole for K/albite. Therefore, ΔG° does not vary greatly from one trace element/mineral pair to the other, provided that the comparison is restricted to the same group of minerals.

On the contrary, the term $\mathbf{R}T \cdot \ln \mathbf{K}$ varies from 0.66 kcal/mole for the K/(Rb-feldspar) pair to 9.72 kcal/mole for Cs/albite.

This relation shows that the contribution of $\Gamma^\circ(\mathrm{AR})$ to the constant **K** is significant. $\mu^{ex}(\mathrm{AR})_0$ in the feldspar solid solution varies regularly with the variation of the difference in ionic radii of the element A and M (Fig. 4 of the paper by Lagache and Sabatier, 1973).

Although a study to separate ΔG° from $\mathbf{R}T \cdot \ln \mathbf{K}$ has not been carried out, the same tendency of **K** as for $\mu^{ex}(\mathrm{AR})_0$ is verified for diverse micas and feldspathoids. Figures 4 (feldspars), 5 (micas) and 6 (leucite and nepheline) show the effect of the parameter ρ on $\mathbf{R}T \cdot \ln \mathbf{K}$ for the affinity of the element in the structure of mineral MR. The parameter was proposed (Roux *et al.*, 1971) to explain the variation of the limit of $X(\mathrm{AR})$ at which the linear relation (5) of the isotherm ceases. It was defined by the ratio of the difference of ionic radii $\Delta r = r(\mathrm{M}) - R(\mathrm{A})$ to the size $d_{\mathrm{M-O}}$ of the sites for the M atom in the structure of the mineral.

$$\rho = (r(\mathrm{M}) - r(\mathrm{A}))/d_{\mathrm{M-O}} \tag{13}$$

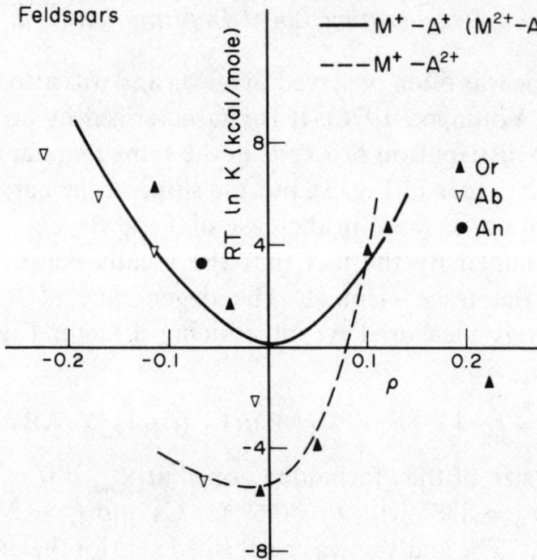

FIG. 4. Variation of $\mathbf{R}T{\cdot}\ln \mathbf{K}$ of exchange reactions as a function of the parameter ρ in feldspars at 600 C, 1 kbar.

In most cases the average interatomic distance M–O in the structure of the mineral was used to calculate the parameter $d_{\text{M-O}}$.

$d_{\text{Na-O}}$ of the larger sites was taken, however, in the case of nepheline, in which K and Rb atoms are principally fixed into these sites.

The curves $\mathbf{R}T{\cdot}\ln \mathbf{K}$ vs. ρ pass a minimum at $\rho_{\text{opt}} \doteqdot 0.00 \sim -0.07$. The constant $\mathbf{K}$ increases rapidly with a small increase in ρ and the trace elements increase their tendency to remain in the hydrothermal phase.

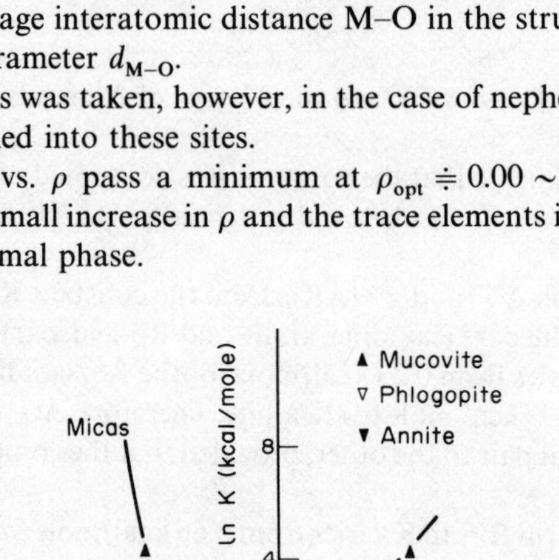

FIG. 5. Variation of $\mathbf{R}T{\cdot}\ln \mathbf{K}$ of exchange reactions as a function of the parameter ρ in micas at 600°C, 1 kbar.

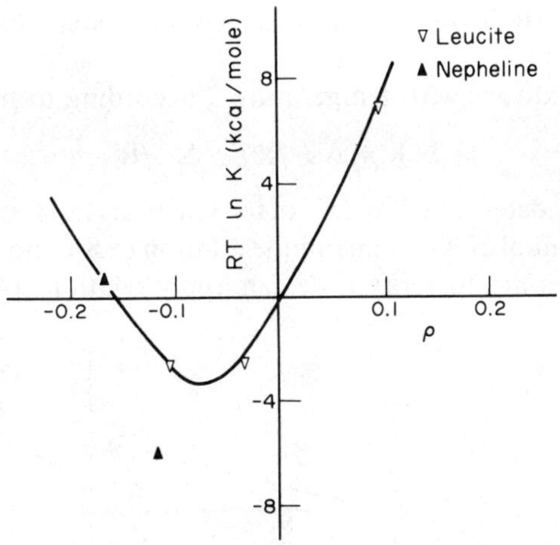

FIG. 6. Variation of $\mathbf{R}T{\cdot}\ln \mathbf{K}$ of exchange reactions as a function of parameter ρ in feldspathoids at 600°C.

The minimum values of **K** for monovalent A–monovalent M exchange pairs were found with Rb–K in feldspar (**K** = 2.9), Rb–K in trioctahedral mica (**K** = 0.44 for annite), K–Na in nepheline (**K** = 0.033) and Cs–K in leucite (**K** = 0.25). The particular tendency of Rb to be fixed more preferentially in K-silicates rather than other elements is important.

The A–M exchanges with Li were excluded from this comparison. The constant **K** of this element is entirely outside that of the general tendency. Because of its small ionic radius, the fixation of Li in the silicate structure induces local lattice deformation around the substituted sites. In addition, some of the Li will be fixed in the octahedral sites of the mica structure. The behaviour of this element in silicate structures cannot be treated in the same manner as other elements.

The variation of $\mathbf{R}T\cdot \ln \mathbf{K}$ as a function of ρ is asymmetric. Though data concerning the other elements $r_A < r_M$ are not abundant, the curve has a steeper slope in the region for $\rho < \rho_{opt}$. This aspect suggests that it is easier to substitute a large cation M by a small A than to replace a small M by a large A.

Micas show the most pronounced asymmetry while the curve is nearly symmetric in the case of feldspars.

The above-mentioned tendency (dependence of $\mathbf{R}T\cdot \ln \mathbf{K}$ on ρ) is also verified in the case of substitutions of monovalent M by bivalent A^{++} for feldspars.† The effect of ρ on $\mathbf{R}T\cdot \ln \mathbf{K}$ is more marked in this case than the case of the monovalent A–monovalent M exchange.

Except in the case of the fixation of Ca in sanidine, the alkaline earth trace elements are preferentially fixed in the crystalline phase.

The contribution of the charge of the electron of cation A on the interaction between M and A will give a negative sign of $\mu^{ex}(\mathrm{AR})_0$ and lowers the free energy, $\mathbf{R}T\cdot \ln \mathbf{K}\cdot \Gamma(\mathrm{AR/MR})$.

† In this case, the ion exchange isotherm is defined by

$$\frac{X(\mathrm{AR})}{(X(\mathrm{MR}))^2}\cdot\frac{\gamma(\mathrm{AR})}{(\gamma(\mathrm{MR}))^2}\cdot\frac{(m(\mathrm{M}))^2}{m(\mathrm{A})} = k.$$

Variation of the Constant K with Temperature

The constant **K** should vary with temperature T according to the relation:

$$\ln \mathbf{K} = \Delta H^\circ/\mathbf{R}T - \Delta S^\circ/\mathbf{R} \tag{14}$$

if the constant **K** is solely determined by ΔG° of the reaction (1). However, the contribution of the excess chemical potential of AR at an infinite dilution to **K** is more important than that of ΔG°. It is therefore necessary to verify if we can apply relation (14).

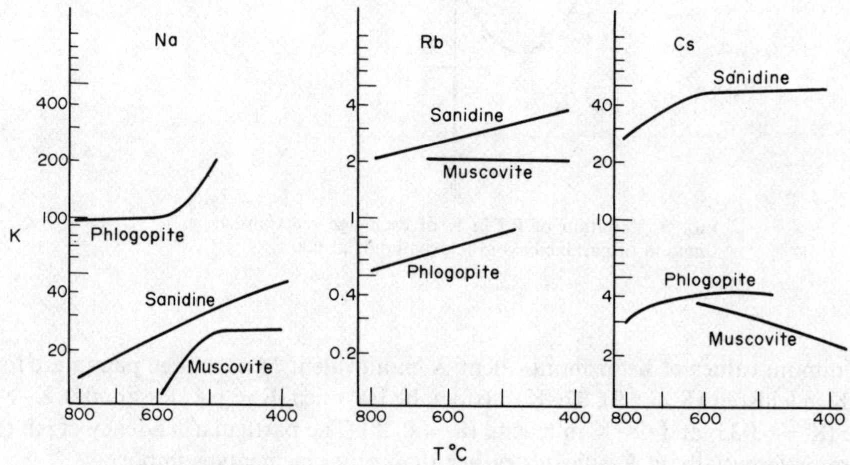

FIG. 7. Variation of the constant **K** of ion exchange reactions as a function of temperature. Cases of Rb, Na and Cs in muscovite, phlogopite and sanidine. (Volfinger, 1976, partially modified, Iiyama and Volfinger, 1976.)

This verification was performed in the case of mica and feldspars by Volfinger (1976). This analysis (Fig. 7) has shown that the fixation of Rb into muscovite, phlogopite and sanidine follows this relation well. The estimated ΔH° and ΔS° of reaction (1) between pure end members is estimated as -1.2 kcal/mole and 0.6 cal/deg mole for sanidine, 0 kcal/mole and 1.4 cal/deg mole for muscovite and -2.4 kcal/mole and -3.5 cal/deg mole for phlogopite. On the other hand, fixation of Cs and Na in these minerals does not follow relation (14) at all.

The situation is comprehensible if we consider that the mode of fixation of the trace element in the crystal structure is often accompanied by local lattice deformation. The degree of local lattice deformation will depend strongly on the amplitude of the thermal agitation of atoms in the structure. The variation of the size r of the locally deformed zone will not be a simple function of temperature. The resulting standard entropy ΔS° and enthalpy ΔH° of the exchange reaction is not constant in many cases.

A small excess chemical potential μ^{ex} $(\mathrm{R}b\mathrm{R})_0$ of Rb in many silicates explains why the variation of constant **K** for this element satisfies relation (14).

The Influence of the Composition of the Silicate Radical on the Constant K

The strong dependency of the constant **K** on the excess chemical potential of mixing suggests that the influence of the silicate radical R on this parameter should be carefully examined.

This examination was furnished by trioctahedral and dioctahedral micas (Volfinger, 1974, 1975).

A study performed at 600°C, 1 kbar with Cs and Rb showed that this influence is very strong. The constant **K** for Rb–K exchange in pholologopite with the theoretical composition $K[Mg_3\,Al\,Si_3\,O_{10}\,(OH)_2]$ is 0.56, and it changes to 1.9 when the composition of the silicate radical changes to $[Mg_{1\cdot95}Al_{0\cdot825}O_{0\cdot23}]\,[Al_{1\cdot375}Si_{2\cdot625}]\,O_{10}(OH)_2$. The constant **K** for Cs in these two micas varies from 4.2 to 17. Variation of the constant **K** for Cs in phlogopite (**K** = 4.2) and for annite (**K** = 1.8) is also worthy of note.

The substitution of a part of Mg and Si in phlogopite by Al renders the constant **K** close to that of the dioctahedral micas.

Comparison of the partition relation of trace elements in different kinds of rocks should therefore be undertaken with great care, especially in the cases where complex silicates such as micas and amphiboles are concerned.

It should be noted that variation of **K** with the composition of the silicate radical is much greater than the variation of **K** as a function of temperature for a given mineral.

Constant K and the Composition of the Host Mineral

It is rare that feldspars, feldspathoids and micas are found in nature in the state of their pure Na or K end-members. They are solid solutions between these two end-members.

If this solid solution ((A, Na, K)R) is ideal with respect to the three end-members, the constant **K** defined by

$$\frac{m(\mathrm{A})}{m(\mathrm{Na})+m(\mathrm{K})} = \mathbf{K}\cdot\frac{X\,(\mathrm{AR})}{X\,(\mathrm{NaR})+X\,(\mathrm{KR})}\ \text{for}\ X\,(\mathrm{AR})\rightarrow 0$$

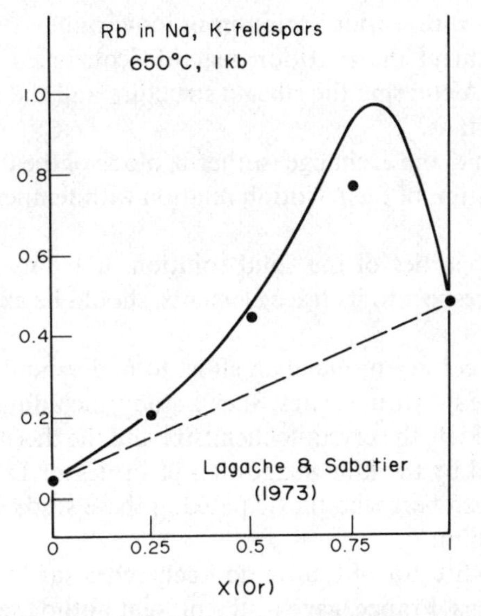

FIG. 8. Effect of the composition of binary solid solution albite–sanidine on the constant $1/\mathbf{K}$ as defined in the text.

will vary linearly between the constant **K** for the K end-member to that for the Na end-members.

However, solid solution is not ideal and there are interactions between the three constitutent "molecules" so that the excess chemical potentials of these molecules also contribute to the constant **K** for the trace element A.

Experimental work on this problem was initially performed by Lagache (1971). She found that the free energy of fixation of Rb in the sanidine–albite solid solution passes a minimum at Or_{75} Ab_{25}. The constant 1/**K** is 0.48 for pure sanidine and 0.04 for pure albite, whereas it is 0.77 for a feldspar of intermediate composition.

The thermodynamic analysis performed by Sabatier (1971) adopting the ternary regular symmetric solution model and that by Lagache and Sabatier (1973) adopting the method of Kohler (1960) explained the observed relation for Rb and for Cs (Fig. 8).

The variation of 1/**K** having a maximum at the intermediate composition ($Lc_{0\cdot35}$ $Pol_{0\cdot65}$) was also found in the case of the fixation of Rb in the leucite–pollucite solid solution series (Lagache, 1974 a, b).

It is therefore concluded that the partition of a trace element should be studied with great care; for example, the great affinity of Rb for K-feldspar is more accentuated by the feldspar Or_{75} Ab_{25} than by pure K-feldspar.

Concluding Remarks

Experimental studies of the partitioning of trace elements in silicate minerals have been performed since 1968 to clarify the fundamental factors controlling their distribution.

To avoid consideration of the complicated properties elements, such as electronegativity and the electronic configuration of outer shells in the structure of atoms, studies with alkalis and alkaline earths in tecto- and phyllosilicates were first examined. These elements have the advantage of furnishing us with a wide variation in ionic radii.

The observed equilibrium of the partition reaction convinced us that the interaction between the trace element A entering the silicate structure and the major element M in the host mineral is very important.

The constant **K**, the form of the exchange isotherm, mode of the distribution of A atom in the structure, and the variation of the partition relation with temperature are controlled by this interaction.

The thermodynamic properties of the solid solution, not only in respect to its major constituents but also with respect to its trace elements, should be explored in more detail in future for other elements.

At the same time it is necessary to make an effort to understand the modification of the silicate structure around the substituted sites. Such a study, including the elastic properties of silicates, would help to elucidate the crystallochemistry and the thermodynamics of minerals.

This review was prepared by the kind suggestion of Professor D. M. Shaw of McMaster University (Canada). All members who participated in these studies presented here wish to express their gratitude to him.

Dr. G. Sabatier, former director of Centre de Recherches sur la Synthese et Chimie des Mineraux, C.N.R.S., Orléans, France, gave to the present author very kind permission and encouragement to present this review in spite of the writer being in Japan, and he here expresses his hearty thanks to him.

The writer would also like to express his gratefulness to his colleagues, Drs. F. Delbove, M. Lagache, M. Volfinger and J. Roux, for their frank and friendly discussions during participation with the studies.

References

BESWICK, A. E. (1973) An experimental study of alkali metal distribution in feldspars and micas. *Geochim. Cosmochim. Acta* **37**, 183–208.

DEBRON, G. (1965) Contribution à l'etude des réactions d'échange des ions alcalins et alcalinoterreux dans les feldspathoïdes. *Bull. Soc. Franç. Min. Crist.* **88**, 69–96.

DEBRON, G., DONNAY, G., WYART, J. and SABATIER, G. (1961) Réaction d'échange des ions sodium par les ions potassium dans la néphéline. Application à l'étude du système néphéline–kalsilite. *C.R Acad. Sci. Paris* **252**, 1255–1257.

DELBOVE, F. (1971) Equilibre d'échange d'ions entre feldspaths alcalins et halogénures sodi-potassiques fondus. Application au calcul des propriétés thermodynamiques de la série des feldspaths alcalins. *Bull. Soc. Franç. Min. Crist.* **94**, 456–466.

GHELIS, M. and LAGACHE, M. (1972) Etude de l'equilibre entre les solutions hydrothermales et les feldspaths de la Série K Al Si_3O_8–Rb Al Si_3O_8 à 600° sous une pression de 1000 bars. *Bull. Soc. Franç.* Min. Crist. **95**, 157–158.

IIYAMA, J. T. (1964) Etude de réaction d'échange d'ions Na–K dans la série muscovite–paragonite. *Bull. Soc. Franç. Min. Crist* **87**, 532–541.

IIYAMA, J. T. (1966) Contribution à l'étude des équilibres subsolidus du système ternaire orthose–albite–anorthite à l'aide des réactions d'échange d'ions Na–K au contact d'une solution hydrothermale. *Bull. Soc. Franç. Min. Crist.* **89**, 442–454.

IIYAMA, J. T. (1972) Fixation des éléments alcalinoterreux Ba, Sr et Ca dans les feldspaths, étude expérimentale. *Proc. 24th Int. Geol. Congress*, Sect. **10**, 122–130.

IIYAMA, J. T. (1974a) Behaviour of trace elements in feldspars under hydrothermal conditions. *The Feldspars*, W. S. MACKENZIE and J. ZUSSMAN, (eds.,). Manchester Univ. Press, 552–573.

IIYAMA, J. T. (1974b) Substitution, déformation locale de la maille et équilibre de distribution des éléments en traces entre silicates et solution hydrothermale. *Bull. Soc. Franç. Min. Crist.* **97**, 143–151.

IIYAMA, J. T. and VOLFINGER, M. (1976) A model for trace element distribution in silicate structures *Min. Mag.* **40**, 555–564.

IIYAMA, J. T., WYART, J. and SABATIER, G. (1963) Equilibre des feldspaths alcalins et des plagioclases à 500, 600, 700 et 800°C sous une pression d'eau de 1000 bars. *C.R. Acad. Sci., Paris*, **256**, 5016–5020.

KÖHLER, F. (1960) Zur Berechnung der thermodynamischen Daten lines ternaren Systems aus den Zugehörigen binaren System. *Monat. Chem.* **91**, 738–740.

LAGACHE, M. (1968) Etude expérimentale de la répartition des éléments traces entre la leucite, l'orthose et des solutions hydrothermales. Le rubidium à 600°C. *C.R. Acad. Sci. Paris* **267**, 141–144.

LAGACHE, M. (1969) Etude expérimentale de la répartition des éléments traces sodium et cesium entre la leucite, orthose et dans des solutions hydrothermales à 600°C. *C.R. Acad. Sci., Paris* **268**, 1241–1243.

LAGACHE, (1971) Etude expérimentale de la répartition du cesium entre les feldspaths sodi-potassiques et des solutions hydrothermales à 700°C, 1 kbar. *C.R. Acad. Sci., Paris* **272**, 1328–1330.

LAGACHE, M. (1974a) Calcul des propriétés thermodynamiques des solutions ternaires de leucite (K, Rb, Cs) par le modèle de Köhler. Application expérimentale dans le cas où l'un des composants est à l'étate de traces. *C.R. Acad. Sci., Paris* **278**, 2497–2500.

LAGACHE, M. (1947b). Utilisation du modèle de F. Köhler pour le calcul de la distribution d'un élément en traces entre une solution solide binaire et une solution hydrothermale. Application à quelques tecto-silicates. *Bull. Soc. Franç. Min. Crist.* **97**, 117–121.

LAGACHE, M. and SABATIER, G. (1973) Distribution des éléments Na, K, Rb et Cs à l'état de trace entre feldspaths alcalins et solutions hydrothermales à 650°C 1 kbar: données expérimentales et interprétation thermodynamique. *Geochim. Cosmochim. Acta* **37**, 2617–2640.

ORVILLE, P. M. (1962) Alkali metasomatism and feldspars. *Norsk. Geol. Tidsskr.* **42**, 283–316.

ORVILLE, P. M. (1963) Alkali ion exchange between vapor and feldspar phases. *Amer. J. Sci.* **261**, 201–237.

ORVILLE, P. M. (1972) Plagioclase cation exchange equilibria with aqueous chloride solution: results at 700°C and 2000 bars in the presence of quartz. *Amer. J. Sci.* **272**, 234–272.

PERCHUK, L. L. and ANDRIANOVA, Z. S. (1968) Experimental and theoretical investigations of mineral equilibria. Nauka, publisher, 37–72.

ROUX, J. (1971) Fixation du rubidium et du césium dans la néphéline et dans l'albite à 600° C dans les conditions hydrothermales. *C.R. Acad. Sci., Paris* **272**, 1469–1472.

Roux, J. (1974) Etude des solutions solides des néphélines (Na, K) Al SiO_4 et (Na, Rb) Al SiO_4. *Geochim. Cosmochim. Acta*, **38**, 1213–1244.

Roux, J., Volfinger, M., Lagache, M., Iiyama, J. T., Sabatier, G. and Wyart, J. (1971) Ideal solid solution formation by substitution of alkali and alkali earth elements in silicates. *Miner. Soc. Japan, Special Papers* **1**, 214–221.

Sabatier, G. (1971) Théorie de la distribution d'un élément entre solution hydrothermale et solution solide. Application à la distribution du césium dans la série des feldspaths alkalins. *Bull. Soc. Franç. Min. Crist.* **94**, 451–455.

Thompson, J. B. and Waldbaum, D. R. (1968) Mixing properties of sanidine crystalline solutions (1) Calculation based on ion exchange data. *Amer. Miner.* **53**, 1965–1999.

Volfinger, M. (1969) Partage de Rb et Cs entre sanidine, muscovite, et solution hydrothermale à 600°C et 1000 bars. *C.R.Acad. Sci., Paris* **269**, 1–3.

Volfinger, M. (1970) Partage de Na et Li entre sanidine, muscovite et solution hydrothermale à 600°C et 1000 bars. *C.R. Acad. Sci., Paris* **271**, 1345–1347.

Volfinger, M. (1974) Effet de la composition des micas trioctaédriques sur les distributions de Rb et Cs à l'état de traces. *Earth Planet. Sci. Lett.* **24**, 299–304.

Volfinger, M. (1975a) Effet de la composition des micas, dioctaédriques sur les distributions des éléments alcalins en traces. *Bull. Soc. Franç. Min. Crist.* **98**, 49–52.

Volfinger, M. (1975b) Effet de la température sur les distributions de Na, Rb et Cs entre la sanidine, la muscovite, la phlogopite et une solution hydrothermale sous une pression de l kbar. *Geochim. Cosmochim. Acta*, **40**, 267–282.

Wellman, T. R. (1970) The stability of sodalite in a synthetic syenite and aqueous chloride fluid system. *J. Petrology* **11**, 49–71.

Wyart, J. and Sabatier, G. (1956a) Mobilité des ions alcalins et alcalinoterreux dans les feldspaths. *Bull. Soc. Franç. Min. Crist.* **80**, 574–581.

Wyart, J. and Sabatier, G. (1956b) Transformation mutuelle des feldspaths alcalins. Reproduction du microlin et de l'albite. *Bull. Soc. Franç. Min. Crist.* **80**, 574–581.

Wyart, J. and Sabatier, G. (1962) L'equilibre des feldspaths et des felspathoides en présence de solutions sodi-potassiques. *Norsk. Geol. Tidsskr.* **42**, 317–329.

On Rare-Earth Element Behavior in Igneous Rocks

LARRY A. HASKIN

Departments of Earth and Planetary Sciences and Chemistry and McDonnell Center for the Space Sciences, Washington University, St Louis, Mo. 63130, U.S.A.

Abstract

Trends of the past decade in geochemical investigations of trace elements are reflected in studies of rare-earth elements (REE). Substantial improvements have been made in accuracy, sensitivity and convenience of REE analysis, mainly through improved techniques and equipment for neutron-activation analysis and mass spectrometric isotope dilution. The most detailed studies have been on mafic rock systems, with a shift away from broad surveys to detailed examinations of single regions or formations. Ocean ridge tholeiitic basalts have been shown to derive from mantle sources partly depleted by earlier melting. Associated basalts deviating from typical ocean ridge tholeiite have been attributed to invading plumes of more primitive mantle matter. Lunar mare basalts and highland KREEP basalts have been characterized as having REE distributions roughly analogous to terrestrial ocean and continental distributions. Models for genesis of mafic liquids as products of partial melting or fractional crystallization have been developed and applied. These models are dependent on values for distribution coefficients. Such values are not well known or the factors affecting them well characterized. Experimental techniques for estimating distribution coefficients are still being developed and tested. Present models cannot yet account quantitatively for the REE concentrations and distributions observed in even the best understood natural systems. The semiquantitative results on a variety of mafic systems nevertheless constitute important gains in understanding of both trace element behavior and igneous rock petrogenesis.

Introduction

The purpose of this paper is to convey an impression of the development of our understanding of trace-element behavior during igneous rock formation since the 1967 conference on the Origin and Distribution of the Elements. It is impossible to review the progress broadly, so this discussion is restricted to rare-earth elements, mainly in mafic igneous rocks (REE). A more extensive review of REE geochemistry has been prepared by Haskin and Paster (1978). The REE are especially valuable for geochemical study because of their mutually similar behavior and, most recently, because of data on Sm–Nd isotopic systematics. REE abundance distributions have become an important tool, as part of multielement analysis on single samples, to be used along with major element composition, field relationships, petrographic characterization, experimental petrology and isotopes, to name some of the key pieces of information needed to provide correct solutions to geochemical and petrological riddles.

REE studies in igneous rocks have benefitted strongly from improved analytical techniques, especially mass spectrometric isotope dilution and neutron activation analysis. It is now possible to obtain the same values for REE, within stated analytical uncertainties, from several laboratories, as indicated by the results for two aliquants of lunar basalt 14053,

TABLE 1. ppm REE IN BASALT 14053

	A	B		A	B
La	12.8±.2	13.0	Tb	1.62±.08	—
Ce	36.4±.8	34.5	Dy	11.1±.4	10.5
Nd	22±1	21.9	Ho	2.1±.2	—
Sm	6.50±.06	6.56	Er	—	6.51
Eu	1.23±.03	1.21	Yb	6.1±.3	6.00
Gd	8.5±.2	8.59	Lu	0.89±.01	—

A. P. Helmke, neutron activation analysis.
N. J. Hubbard, Mass spectrometric isotope dilution.

one done by NAA, the other by MSID (Table 1). The improvements in analysis are a legacy of the Apollo lunar sample program. Along with accuracy, sensitivity and ease have also improved, as indicated, for example, by analyses of 155 1–2-mm fragments of Apollo 17 soils (Blanchard *et al.*, 1975).

To discuss REE geochemistry, the now conventional comparison diagram is used. REE abundances in a sample are divided, element by element, by the abundances in some standard material, usually chondrites. As seen in Fig. 1, the relative REE abundances in chondritic meteorites are within error of measurement the same as those in the sun

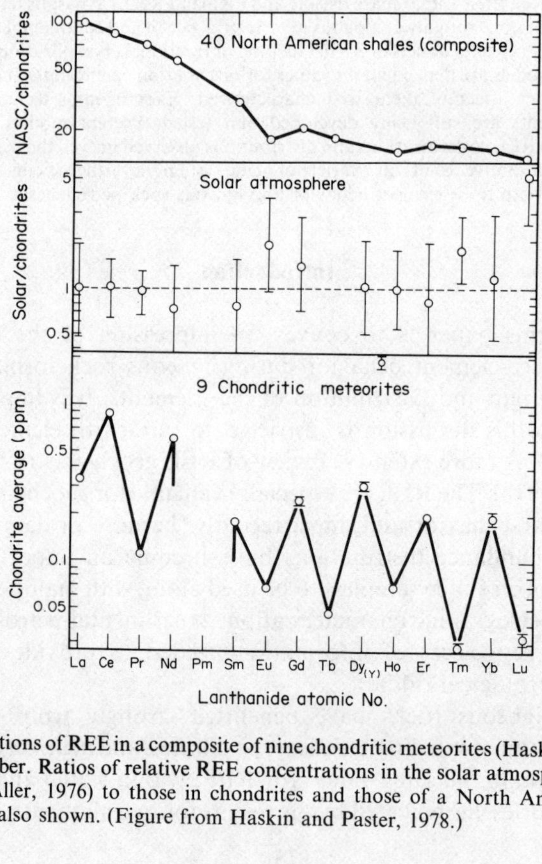

FIG. 1. Average concentrations of REE in a composite of nine chondritic meteorites (Haskin *et al.*, 1968) are plotted against REE atomic number. Ratios of relative REE concentrations in the solar atmosphere (adapted from data compiled by Ross and Aller, 1976) to those in chondrites and those of a North American shales composite (Haskin *et al.*, 1968) are also shown. (Figure from Haskin and Paster, 1978.)

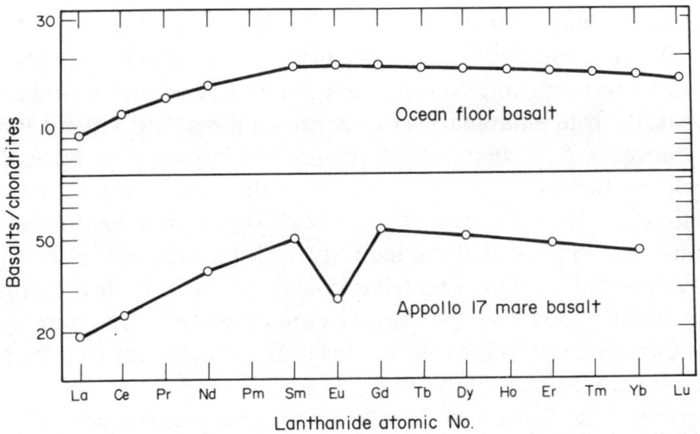

FIG. 2. Comparison diagram for a typical mid-ocean ridge basalt (idealized average, data of Frey *et al.*, 1974) and for a lunar mare basalt (Shih *et al.*, 1975). (Figure from Haskin and Paster, 1978.)

(normalized here to La = 1) and perhaps therefore represent those of the parent material for planets. Most materials at Earth's surface are enriched in all REE compared with chondrites, but more so in light REE (LREE), e.g. the North American Shales Composite (Fig. 1). Eu behaves anomalously when it is partially reduced from the typical lanthanide 3+ oxidation state to the 2+ state under common natural conditions.

Mafic Igneous Rocks

Much of the gain in knowledge of REE behavior has involved mafic rocks. The light-lanthanide depleted rocks from the ocean ridges (Fig. 2) were interpreted by Gast (1968) and others to have been derived by partial melting of mantle sources that had previously lost a

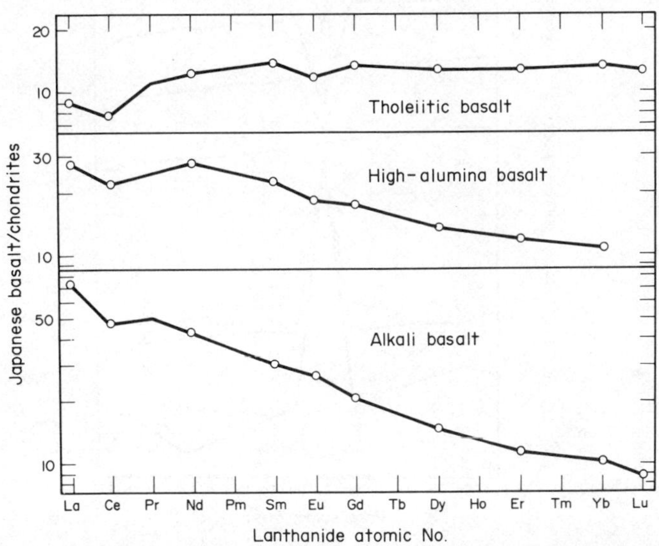

FIG. 3. Comparison diagram for Japanese basalts (Masuda, 1968). (Figure from Haskin and Paster, 1977.)

substantial fraction of their incompatible elements. Present evidence (e.g. Kay *et al.*, 1970) indicates that such sources predominate for ocean floor igneous rocks, and that these sources are separate from those producing continental volcanics or oceanic islands. It is interesting to observe that basalts from lunar maria (Fig. 2) are similarly depleted in LREE and the best current interpretation is that their source regions are products of an earlier episode of melting that depleted them substantially in incompatible elements (e.g. Nava and Philpotts, 1973). The analogy between the two sets of rocks should not be pushed very far. It is nevertheless amusing to suggest that the naming of Moon's dark areas by the ancients was very clever after all, even if the lunar seas never contained water. Note the strong depletion in Eu, a consequence of Moon's very reducing condition; this depletion appears to be a characteristic of the source region for the lunar basalts with the missing Eu probably in the more feldspathic highlands (e.g. Shih *et al.*, 1975).

The transition from LREE-depleted basalts of the ocean floor igneous suite to the LREE-enriched volcanics of continents is nicely illustrated by the basalts of Japan (e.g. Masuda, 1968), with the LREE-depleted basalts (Fig. 3) coming from the Pacific Ocean side, those most enriched in LREE from the Asian continent side. Schilling and coworkers (e.g. Schilling, 1975) have made a case for mantle plumes, bringing magma from sources not previously depleted in compatible elements, to account for lavas of islands well isolated from continents. They postulate mixing of such magmas with those of depleted sources to form the LREE-enriched basalts sometimes found along ocean ridges or to account for transitions in basalt type such as those found in Iceland. Hermann and coworkers (1976) have recently

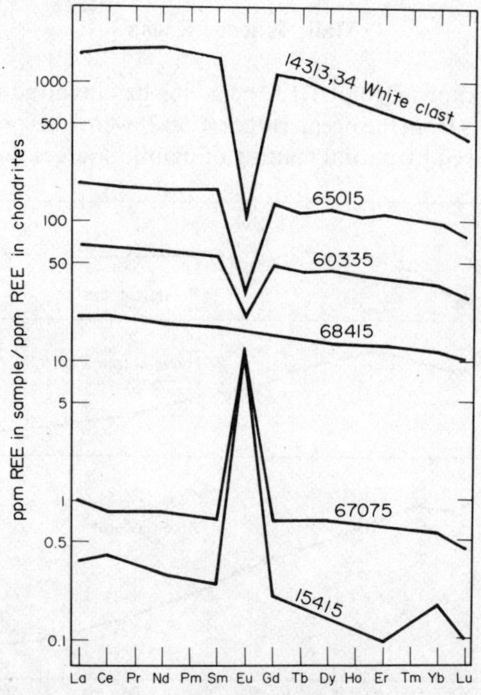

FIG. 4. Comparison diagram showing extremes of REE concentrations found in samples from the lunar highlands (adapted from Haskin *et al.*, 1973).

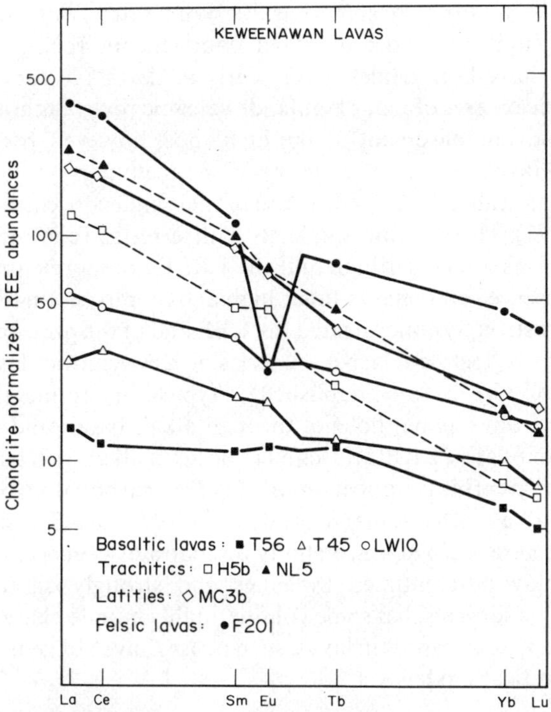

FIG. 5. Comparison diagram for Keweenawan volcanics, North Shore group, northeastern Minnesota (Haskin, Brannon and Green, unpublished data).

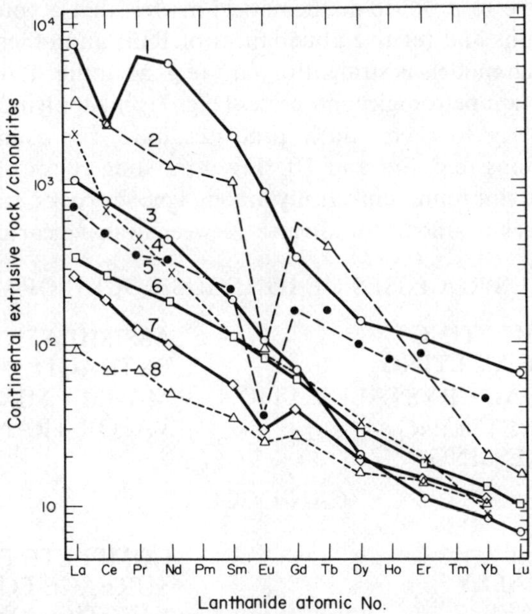

FIG. 6. Comparison diagram for strongly alkaline extrusive rocks: 1, carbonatite; 2, kimberlite; 3–8, alkali-rich volcanics. (Figure from Haskin and Paster, 1978.)

shown that REE distributions in some basaltic komatiites from the ancient Barberton greenstone belt are similar to those of ocean floor igneous rocks, suggesting that some mantle sources may have been depleted very early in Earth's history.

Petrographically clear cases of lunar highlands volcanic rocks cannot be found because of the intense meteoroid bombardment of the highlands; however, rocks such as poikilitic breccia 65015 (e.g. Haskin *et al.*, 1973) may well be metamorphosed highland volcanics. (Such compositions are called KREEP because of their high concentrations of K, REE and P; Hubbard *et al.*, 1971). There is some similarity with terrestrial continental volcanics in that the REE concentrations are rather high and the LREE are more enriched than the heavy ones. The present observed extremes in lunar highlands composition (Fig. 4) show evidence of processes that very strongly concentrated the REE and of the presence of considerable Eu in massive cumulate plagioclase feldspar. A series of Keweenawan lavas now under study (Haskin, Brannon and Green, to be published) is typical of continental volcanics (Fig. 5); these samples range from olivine tholeiite through alkali basalt and andesite to rhyolite. More extreme cases of relative LREE enrichment occur in alkali-rich basalts and kimberlites (Fig. 6); these data are from Loubet *et al.* (1972) (carbonatites), Fesq *et al.* (1975) (kimberlite), and for alkali-rich basalts (Kay and Gast, 1973, and Gerasimovskii *et al.*, 1972). The important point here is the absence of any discontinuity between lightly differentiated tholeiites and strongly differentiated tholeiites and strongly differentiated alkali and volatile-rich lavas. This suggests that some volatile, highly mobile phase in the mantle greatly influences relative REE abundances in lavas; such phases have not generally been considered in quantitative theoretical models yet.

Mathematical Models for Trace-element Partitioning

It should be possible to produce mathematical models that account quantitatively for observed concentrations and relative abundances of REE and other trace elements. The chemical basis for such models is straightforward (e.g. McIntire, 1963); the difficulty is in recognizing all important petrological processes (Fig. 7) and applying the chemical aspects of the models properly to them. Such processes have, for example, yielded strong geochemical correlations (e.g. Sm and Hf, Fig. 8) among rocks of diverse types. This particular correlation, not found universally among igneous rocks, supports field evidence for a genetic relationship among the diverse Keweenawan volcanics mentioned earlier.

SOME PROCESSES OF IGNEOUS ROCK FORMATION

PARTIAL MELTING	ASSIMILATION
COMPLETE MELTING	EXTRACTION
FRACTIONAL CRYSTALLIZATION	MAGMA MIXING
CRYSTAL SETTLING	VAPOR TRANSPORT
FILTER PRESSING	

CAN OCCUR

SINGLY	COMPLETE EQUILIBRIUM
SEQUENTIALLY	SURFACE EQUILIBRIUM
IN PARALLEL	KINETIC CONTROL

FIG. 7. Summary of petrological processes to be considered in modelling of trace-element partitioning.

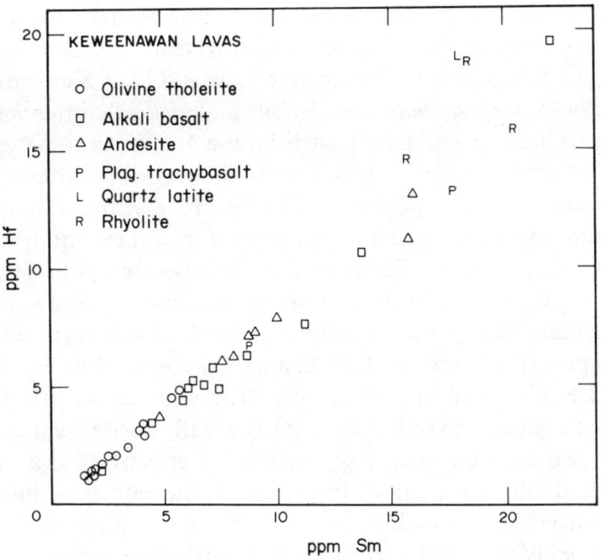

FIG. 8. Variation diagram, Sm and Hf, Keweenawan volcanics, North Shore, Lake Superior; ○, olivine tholeiite; □, alkali basalt; △, andesite; R, rhyolite; P, plagioclase trachybasalt (Haskin, Brannon and Green, unpublished data).

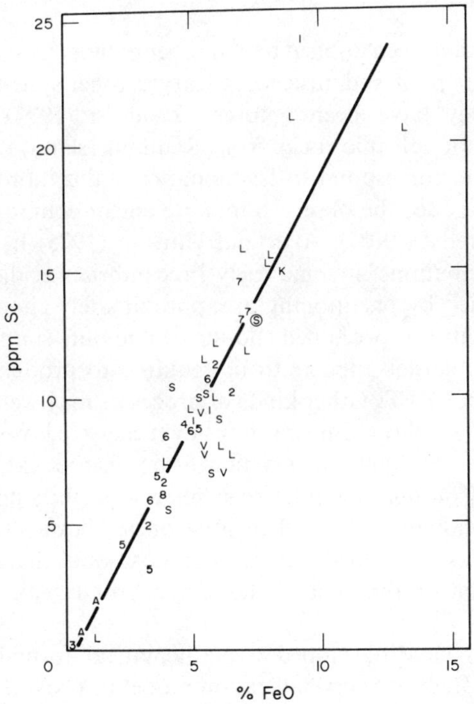

FIG. 9. Variation diagram, Sc and FeO, lunar highland rocks and soils (Haskin *et al.*, 1973). (S, soils; A, anorthosites; V, very high alumina "basalts"; K, KREEP; members, 1–2 mm soil fragments, all from Apollo 16; L, 2 mm fragments from Luna–16.)

Strong correlations between major and trace elements have been discovered (e.g. Sc and FeO in lunar highland rocks, Fig. 9) and the absence of universal correlation between any pair of major and trace elements has been reconfirmed (e.g. Sc and FeO, Keweenawan basalts, Fig. 10). *Proper models must describe both correlations and lack of correlations.*

Present quantitative models emphasize partitioning according to a Nernst distribution, which states simply that the ratio of concentrations between two equilibrium phases is constant (for constant T, P, composition). The Nernst distribution is used directly for complete equilibrium partitioning or incrementally for surface-equilibrium partitioning, when the interior of a crystallizing solid becomes quickly isolated from the cooling melt, or in some other manner depending on the process being modelled (e.g. Shaw, 1970). REE tend to prefer liquid rather than solid phases; different minerals accept them selectively (Fig. 11). Garnet eagerly accepts HREE, but not LREE, apatite accepts all REE; feldspars accept Eu better than other REE; Ca-rich pyroxene accepts REE more readily than Ca-poor pyroxene or olivines, etc. For fractional crystallization, REE usually become enriched in the residual liquid. To separate two trace elements from each other effectively in a simple partitioning process, a substantial difference in distribution coefficients is required; in fractional crystallization all trace elements with $D \lesssim 0.1$ effectively remain in the liquid phase; those with $D \gtrsim 10$ are all quickly extracted into the crystallizing solid. (In this case, the value of D is the weighted average of D values for all crystallizing minerals).

Applications of Trace-element Partitioning Models

Partitioning models have been applied to a number of igneous systems with useful results; they have been applied in poor and misleading ways in other systems, mainly when obvious constraints of petrology have been ignored. Zielinski (1975) has shown that REE distributions in a suite of volcanic rocks from Reunion Island, ranging from olivine-rich alkali basalt to trachyte, correspond to fractional crystallization of common minerals in amounts required to produce the observed major element compositions. Severe fractional crystallization is required ($\sim$90%). Arth and Hanson (1975) have argued that strongly fractionated REE distributions in some early Precambrian acidic volcanic and intrusive rocks might be explained by partitioning an approximately chondritic REE distribution against garnet. If the mantle is presumed chondritic and only simple silicate liquid–mineral partitioning is allowed, garnet appears to be required to produce any REE distribution substantially enriched in LREE. Other kinds of processes may well be important, however. Helmke and Haskin (1973) showed how groups of successive lavas from Steens Mountain, Oregon, might be related by fractional crystallization. Rather extensive crystallization (up to 60%) was required to produce cumulative solids with slightly positive Eu anomalies and corresponding residual liquids with small negative ones. They also showed how mixtures of crystallized solid and parent liquid yielded the compositional characteristics often found in gabbros, with REE relative abundances showing a broad peak with a maximum in the vicinity of Gd.

Allegre *et al.* (1977) have developed a mathematical formalism for optimizing the parameters in a simple fractional crystallization model to provide the best fit to data on a rock system. They discuss the information that can be obtained from a variety of trace elements with differing geochemical behaviors. Nesbitt and Sun (1976) have emphasized the importance of trace-element ratios in understanding igneous rock petrogenesis.

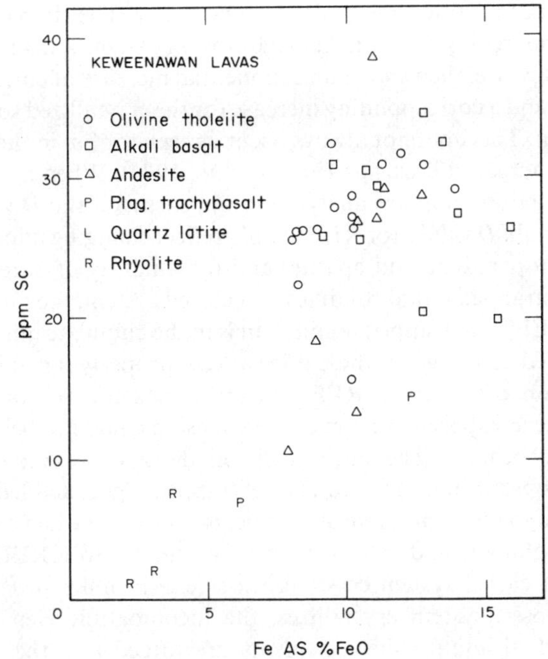

FIG. 10. Variation diagram, Sc and FeO, Keweenawan volcanics, North Shore, Lake Superior; ○, olivine tholeiite; □, alkali basalt; △, andesite; R, rhyolite; P, plagioclase trachybasalt (Haskin, Brannon and Green, unpublished data).

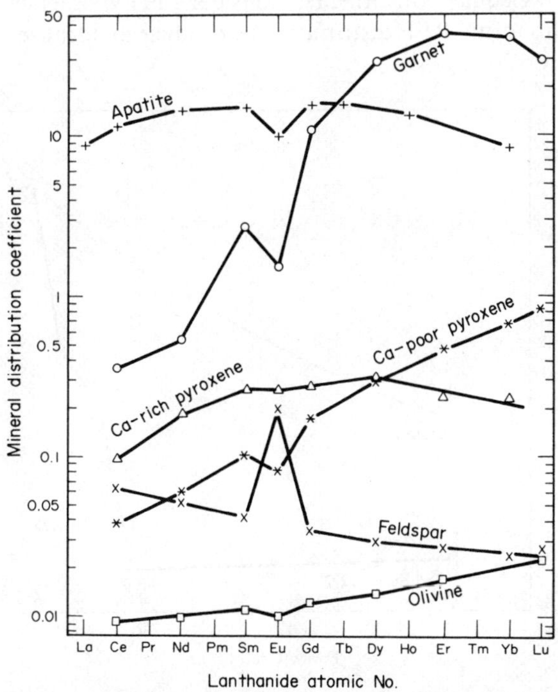

FIG. 11. Typical D values for REE (Schnetzler and Philpotts, 1970, except apatite, Paster *et al.*, 1974). (Figure from Haskin and Paster, 1978.)

Proper application of partitioning models can be tricky. The simplest application of the model to a system undergoing fractional crystallization assigns a distribution coefficient to the crystallizing solid phase, then shows an exponential increase of concentration for a REE in the residual liquid and a corresponding increase for the crystallized solid, but beginning at a lower concentration. This does not always occur as seen for Sm in the Kiglapait intrusion (Fig. 12) or in the Skaergaard intrusion (Paster *et al.*, 1974). When minerals from different zones, or levels of fractionation, are analyzed, it appears as if the D values were changing strongly. In fact, the bulk D values for REE mainly reflect changing mineralogy (e.g. onset of crystallization of clinopyroxene and apatite) and the quantity of parent liquid trapped in each zone. When the trapped liquid solidifies as a closed system, the final growth of crystals incorporates more REE from trapped liquid than is in the cumulate minerals. The model for the Skaergaard layered series, when these effects were properly taken into account, gave a reasonable description of observed REE concentrations for constant values of D and demonstrated that the unexposed portion of the intrusion could not be nearly as large as the 70% to 80% previously believed. The much smaller hidden zone was independently confirmed by gravity studies. Proper attention to presence of trapped liquid enabled Haskin *et al.* (1974) to show that a lunar highlands cumulate of plagioclase and olivine had a parent liquid similar to that of average highlands matter rather than one similar to KREEP.

The importance of closed system crystallization (e.g. Helmke *et al.*, 1972) is too often overlooked. As a closed system crystallizes, the incompatible elements become highly concentrated in residual liquid. These elements are forced into the major crystals least reluctant to accept them. This occurs until the concentrations of the formerly incompatible elements become high enough to precipitate exotic mineral phases in which they are essential, or at least welcome, constituents. Consider the hypothetical crystallization of a mafic liquid with a chondritic REE distribution to produce an idealized ecologite consisting

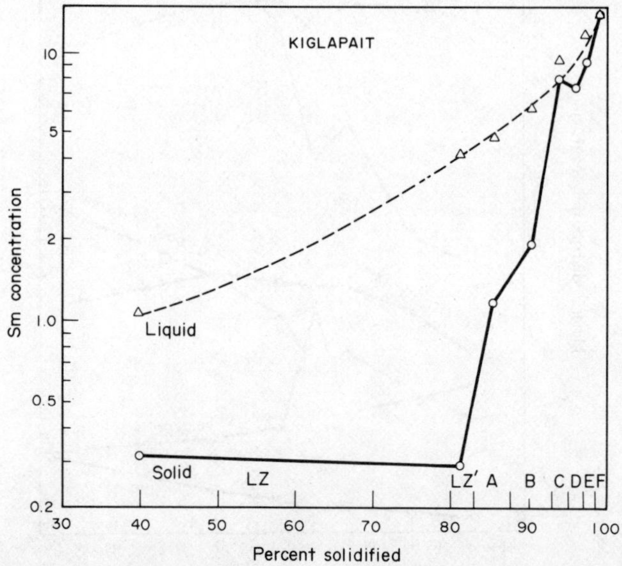

FIG. 12. Sm concentrations in the layered series of the Kiglapait intrusion as a function of fraction solidified; upper curve, residual liquid; lower curve, precipitated solid (Schauwecker, Jacobs and Haskin, unpublished data).

only of garnet and clinopyroxene (no mesostasis) (Fig. 13). The HREE are in garnet, the LREE in pyroxene. A comparison of the pyroxene concentrations with those of the whole rock might suggest that pyroxene preferred the LREE, but as the distribution coefficients indicate, neither mineral prefers LREE, they are concentrated in the clinopyroxene because that mineral can accept them more easily than garnet can.

It is important to test models, however sound they may seem, on systems whose petrological nature is well understood. Haskin and Korotev (1977) have made such a test of the closed-system model on a single sample of lunar basalt. They analyzed separated minerals and mesostasis and showed by a mixing model that they could match the composition of the whole rock very closely in terms of the minerals and mesostasis. They used the closed-system description in conjunction with D values from the literature on terrestrial rocks to test whether the mineral compositions could be described by the model. Because the agreement was reasonable, they then reversed the procedure, assumed validity of the model, and calculated average distribution coefficients for the minerals during solidification of the rock.

Current models for trace-element partitioning as applied so far have worked well in some cases but by no means all cases; indicating that all important processes have not yet been properly described. Frey and Green (1974) have studied peridotite nodules as possible sources for basalts. One class of nodules studied had low REE concentrations and a distribution that might serve as a parent to ocean floor basalts. The other type had lower concentrations of heavy REE, but higher concentrations of light REE as if parental to alkali basalts. Surprisingly, the more refractory, alkali and calcium-poor, residue-like nodules, were the most enriched in the light REE, just the opposite of what would be required to make the two basalt types. In this and other works, Frey and coworkers have demonstrated that major and trace elements are not closely coupled in the basalt source regions.

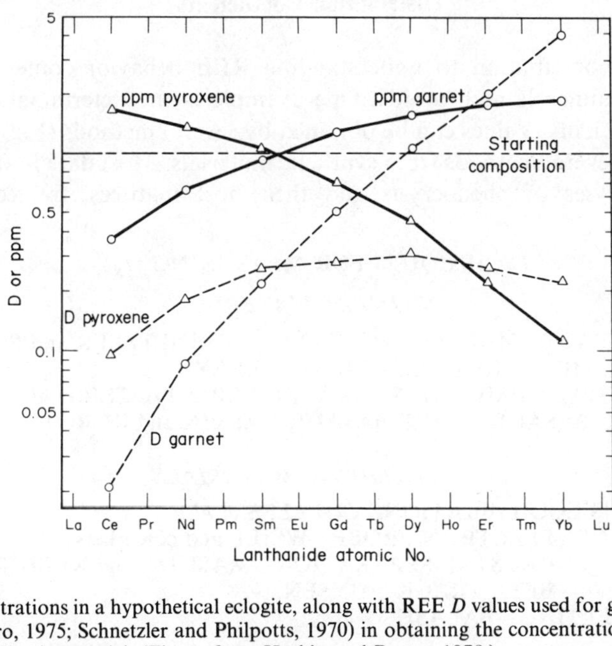

FIG. 13. REE concentrations in a hypothetical eclogite, along with REE D values used for garnet and clinopyrene (Shimizu and Kushiro, 1975; Schnetzler and Philpotts, 1970) in obtaining the concentrations through use of the closed-system crystallization model. (Figure from Haskin and Paster, 1978.)

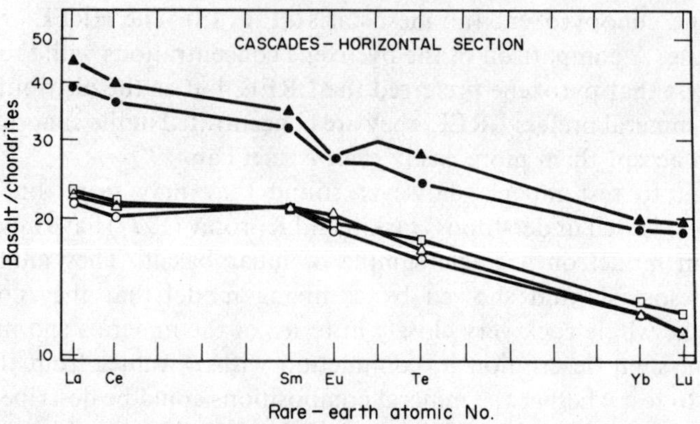

FIG. 14. Comparison diagram for REE in five samples of a single Cascades mesa basalt flow (lateral sampling) (Brannon, Haskin and McBirney, unpublished data).

Furthermore, observed variations in trace element concentrations in terrestrial lavas themselves are not entirely compatible with fractionation models as presently applied. A horizontal segment of a single flow of Cascades basalt (Fig. 14) (Brannon, McBirney and Haskin, to be published) shows substantial differences in REE distribution. The change in ratio of La to Sm would correspond to $\sim 60\%$ fractional crystallization of pure clinopyroxene, clearly impossible, especially since major element concentrations and mineralogy hardly vary. In some igneous systems, alteration can be a complicating factor (e.g. Hallman and Henderson, 1977).

Distribution Coefficients

An important contribution to understanding REE behavior comes from studies of synthetic silicate minerals and liquids. One example is the determination of values for distribution coefficients. Values can be obtained by several methods (Fig. 15) from natural materials and by several methods from synthetic materials. Most data in the literature were obtained by analyses of phenocrysts and their host matrices, treated as equilibrium

TECHNIQUES FOR MEASURING D_{REE}

NATURAL MATERIALS

PHENOCRYST—MATRIX—VOLCANICS, SCHNETZLER and PHILPOTTS
PARTIAL MELT—RESIDUE—LIZARD, FREY
SEQUENTIAL CUMULATES—SKAERGAARD, PASTER *et al.*
APHYRIC BASALT—MARE BASALT, HASKIN and KOROTEV

SYNTHETIC MATERIALS

AQUEOUS EQUILIBRATION—CULLERS *et al.*
DRY MELT—ELECTRON PROBE—WEILL and coworkers
—CRYSTAL SEPARATION—MASUDA and KUSHIRO
—BETA TRACK—MYSEN
ZONE MELTING—BALASHOV *et al.*

FIG. 15. Summary of methods used for estimating or determining *D* values.

solid–liquid pairs. Results from this method were first reported at the last Paris conference by Schnetzler and Philpotts (1967). Frey (1969) obtained values for distribution coefficients by treating the Lizard high-temperature peridotite as a residue and nearby gabbro as the liquid from partial melting. Paster *et al.* (1974) obtained values from Skaergaard minerals and estimates of residual liquid concentrations. Haskin and Korotev (1977) obtained them from an aphyric lunar basalt as shown earlier.

Results from natural materials show consistently the preferences of rock-forming minerals for certain of the REE but offer little information for explaining the spread of values. In addition to equilibrium effects of T, P, and composition, there may be kinetic effects in natural systems as shown, for example, by Albarede and Bottinga (1972). Laboratory experiments offer the chance to study the effects of these parameters. An equilibration of REE tracers between an aqueous fluid and silicate crystals or silicate liquid was tried by Cullers *et al.* (1973). Values obtained fell within the range found for natural systems, but Zielinski and Frey (1974) were unable to match the results and there are other limitations to the method, which has not yet been adequately tested. The most trustworthy experimental values obtained so far have come from water-free systems. Weill, Drake and coworkers (e.g. Drake and Weill, 1975) used the electron probe to determine concentrations in crystals and glass in REE-doped silicates; this method requires substantially larger concentrations of REE than are present in natural systems, inviting deviations from dilute solution behavior. Some workers have tried lower REE concentrations and analysis of crystals and glass by mass spectrometric isotope dilution; failure to remove last traces of glass from crystals can seriously affect results (e.g. Masuda and Kushiro, 1970; Shimizu and Kushiro, 1975). The most recent technique, developed by Mysen and Seitz (e.g. Mysen, 1976) uses beta tracks from radioactive tracers for determining concentrations, thus allowing very low concentrations of dopant. Early results by this technique indicated that levels of REE ($\sim$0.1 wt.%) in synthetic silicates made for microprobe and EPR measurements do indeed violate dilute solution conditions. However, the results of Cullers *et al.* (1970) indicated that mafic minerals equilibrated with aqueous fluid accept REE to concentrations of parts-per-thousand without observable change in behavior, so more studies are needed to clarify this problem.

EPR Spectroscopy

Electron paramagnetic resonance (EPR) spectroscopy was used by Morris (1975) to indicate the nature of the sites occupied in synthetic silicate crystals by Gd^{3+} and Eu^{2+} ions. Except in feldspar, Gd^{3+} occupied normal lattice sites whenever Ca was a major constituent of the mineral. Charge compensation apparently occurred remote from the site of the ion, as it had no effect on the spectrum. In feldspar the spectrum for Gd^{3+} was the same as that found for Gd in silicate glass, apparently because the 3+ ion was very small for the site or because the high charge disrupted the crystal structure at that site. Eu^{2+} occupied Ca sites in all minerals containing them. Both Gd^{3+} and Eu^{2+} tended to cluster in pure Mg ortho- or metasilicates. Morris and coworkers (1974a, b) also studied effects of oxygen fugacity and bulk composition on the relative amounts of Eu^{2+} and Eu^{3+} in silicate glasses. They found a strong effect of bulk composition on the mole fraction of Eu^{2+} in the melt at constant oxygen fugacity. Fraser (1975) has suggested that this results from amphotoeric behavior of REE^{3+} oxides.

It is a general characteristic of the past 10 years' progress in understanding the behavior of REE and other trace elements in igneous systems that new experimental and analytical chemical techniques were important. These techniques are dependent upon expensive, sophisticated, complex electronic instrumentation. Advances in knowledge during the next decade will surely also depend on the further use and development of such instruments.

Acknowledgement

The author thanks the New Holland Publishing Company for permission to reprint several of the figures from the *Handbook of Rare Earths,* volume 2 (Haskin and Paster, 1978).

References

ALBAREDE, F. and BOTTINGA, Y. (1972) Kinetic disequilibrium in trace element partitioning between phenocrysts and host lava. *Geochim. Cosmochim. Acta* **36**, 141–156.

ALLEGRE, C. J., TREUIL, M., MINSTER, J. F., MINSTER, B. and ALBAREDE, F. (1977) Systematic use of trace element in igneous process. Part 1: Fractional crystallization processes in volcanic suites. *Contr. Mineral Petrol.* **60**, 57–75.

ARTH, J. G. and HANSON, G. N. (1975) Geochemistry and origin of the early Precambrian crust of Northeastern Minnesota. *Geochim. Cosmochim. Acta,* **39**, 325–362.

BLANCHARD, D. P., KOROTEV, R. L., BRANNON, J. C., JACOBS, J. W., HASKIN, L. A., REID, A. M., DONALDSON, C. H. and BROWN, R. W. (1975) A geochemical and petrographic study of 1–2 mm fines from Apollo 17. *Proc. Lunar Sci. Conf. 6th,* 2321–2341.

CULLERS, R. L., MEDARIS, L. G. and HASKIN, L. A. (1970) Gadolinium distribution between aqueous and silicate phases. *Science* **169**, 580–583.

CULLERS, R. L., MEDARIS, L. G. and HASKIN, L. A. (1973) Experimental studies of the distribution of rare earths as trace elements among silicate minerals and liquids and water. *Geochim. Cosmochim. Acta* **37**, 1499–1512.

DRAKE, M. J. and WEILL, D. F. (1975) Partition of Sr, Ba, Ca, Y, Eu^{2+}, Eu^{3+}, and other REE between plagioclase feldspar and magmatic liquid: an experimental study. *Geochim. Cosmochim. Acta* **39**, 689–712.

FESQ, H. W., KABLE, E. J. D. and GURNEY, J. J. (1975) Aspects of the geochemistry of kimberlites from the Premier mine, and other selected South African occurrences with particular reference to the rare earth elements. *Phys. Chem. Earth,* **9**, 687–707.

FREY, F. A. (1969) Rare earth abundances in a high-temperature peridotite intrusion. *Geochim. Cosmochim. Acta* **33**, 1429–1447.

FREY, F. A., BRYAN, W. B. and THOMPSON, G. (1974) Atlantic Ocean floor: geochemistry and petrology of basalts from legs 2 and 3 of the Deep-Sea Drilling Project. *J. Geophys. Res.* **79**, 5507–5527.

FREY, F. A. and GREEN, D. H. (1974) The mineralogy, geochemistry and origin of herzolite inclusions in Victorian basanites. *Geochim. Cosmochim. Acta* **38**, 1023–1059.

GAST, P. W. (1968) Trace element fractionation and the origin of tholeiites and alkaline magma types. *Geochim. Cosmochim. Acta* **32**, 1057–1086.

GERASIMOVSKII, V. I., BALASHOV, YU A. and KARPUSHINA, V. A. (1972) *Geokhimiya* no. 5, 515–530.

HASKIN, L. A., HASKIN, M. A., FREY, F. A. and WILDEMAN, T. R. (1968) Relative and absolute terrestrial abundances of the rare earths. In *Origin and Distribution of the Elements*, L. AHRENS, ed., Pergamon Press, Oxford, *Int. Ser. Mon. Earth Sci.* **30**, 899–912.

HASKIN, L. A., HELMKE, P. A., BLANCHARD, D. P., JACOBS, J. W. and TELANDER, K. (1973) Major and trace element abundances in samples from the lunar highlands. *Proc. Lunar Sci. Conf. 4th,* 1275–1296.

HASKIN, L. A. and KOROTEV, R. L. (1977) Test of a model for trace element partition during closed-system solidification of a silicate liquid. *Geochim. Cosmochim. Acta* **41**, 921–939.

HASKIN, L. A. and PASTER, T. P. (1978) Geochemistry and mineralogy of the rare earths. In *Handbook of Rare Earths,* vol. 2, L. EYRING and K. GSCHNEIDER, eds., North Holland (in press).

HASKIN, L. A., SHIH, C. Y., BANSAL, B. M., RHODES, J. M., WIESMANN, H. and NYQUIST, L. E. (1974) Chemical evidence for the origin of 76535 as a cumulate. *Proc. Lunar Sci. Conf. 5th* 1213–1225.

HELLMAN, P. L. and HENDERSON, P. (1977) Are rare-earth elements mobile during spilitization? *Nature* **267**, 38–40.

HELMKE, P. A. and HASKIN, L. A. (1973) Rare earths, Co, Sc, and Hf in Steens Mountain basalts. *Geochim. Cosmochim. Acta* **37**, 1513–1529.

HELMKE, P. A., HASKIN, L. A., KOROTEV, R. L. and ZIEGE, K. E. (1972) Rare earths and other trace elements in Apollo 14 samples. *Proc. Lunar Sci. Conf. 3rd* 1275–1292.

HERMANN, A. G., BLANCHARD, D. P., HASKIN, L. A., JACOBS, J. W., KNAKE, D., KOROTEV, R. L. and BRANNON, J. C. (1976) Major, minor, and trace element compositions of peridotitic and basaltic Komatiites from the Precambrian crust of South Africa. *Contr. Mineral. Petrol.* **59**, 1–12.

HUBBARD, N. H., MEYER, C., JR. GAST, P. W. and WIESMANN, H. (1971) The composition and derivation of Apollo 12 soils. *Earth Planet. Sci. Lett.* **10**, 341–350.

KAY, R. W. and GAST, P. W. (1973) The rare earth content and origin of alkali-rich basalts. *J. Geol.* **81**, 653–682.

KAY, R. W., HUBBARD, N. L. and GAST, P. W. (1970) Chemical characteristics and origin of Ocean Ridge Volcanic rocks. *J. Geophys. Res.* **75**, 1585–1613.

LOUBET, M., BERNAT, M., JAVOY, M. and ALLEGRE, C. (1972) Rare earth contents in carbonatites. *Earth Planet. Sci. Lett.* **14**, 226–232.

MASUDA, A. (1968) Geochemistry of lanthanides in basalts of central Japan. *Earth Planet. Sci. Lett.* **4**, 284–292.

MASUDA, A. and KUSHIRO, I. (1970) Experimental determination of partition coefficients of ten rare-earth elements and barium between clinopyroxene and liquid in the synthetic silicate system at 20 kilobar pressure. *Contr. Mineral. Petrol.* **26**, 42–49.

MCINTIRE, W. L. (1963) Trace element partition coefficients—a review of theory and applications to geology. *Geochim. Cosmochim. Acta* **27**, 1209–1264.

MORRIS, R. V. (1975) Electron paramagnetic resonance study of the site preferences of Gd 3+ and Eu 2+ in polycrystalline silicate and aluminate minerals. *Geochim. Cosmochim. Acta* **39**, 621–634.

MORRIS, R. V. and HASKIN, L. A. (1974) EPR measurement of the effect of glass composition on the oxidation states of europium. *Geochim. Cosmochim. Acta* **38**, 1435–1445.

MORRIS, R. V., HASKIN, L. A., BIGGAR, G. M. and O'HARA, M. J. (1974) Measurement of the effects of temperature and partial pressure of oxygen on the oxidation states of europium in silicate glasses. *Geochim. Cosmochim. Acta* **38**, 1447–1459.

MYSEN, B. (1976) Partitioning of samarium and nickel between olivine, orthopyroxene, and liquid; preliminary data at 20 kbar and 1025°C. *Earth Planet Sci. Lett.* **31**, 1–7.

NAVA, D. F. and PHILPOTTS, J. A. (1973) A lunar differentiation model in light of new chemical data on Luna 20 and Apollo 16 soils. *Geochim. Cosmochim. Acta* **37**, 963–973.

NESBITT, R. W. and SUN, S. S. (1976) Geochemistry of Archean spinifex-textured peridotites and magnesian and low-magnesian tholeiites. *Earth Planet Sci. Lett.* **31**, 433–453.

ROSS, J. E. and ALLER, L. H. (1976) The chemical composition of the sun. *Science* **191**, 1223–1229.

SCHILLING, J. G. (1975) Azores mantle blob: rare-earth evidence. *Earth Planet. Sci. Lett.* **25**, 103–115.

SCHNETZLER, C. C. and PHILPOTTS, J. A. (1968) Partition coefficients of rare earth elements and barium between igneous matrix material and rock-forming mineral phenocrysts. I. In *Origin and Distribution of the Elements*, L. H. AHRENS ed., 929–938.

SCHNETZLER, C. C. and PHILPOTTS, J. A. (1970) Partition coefficients of rare earths between igneous matrix material and rock-forming mineral phenocrysts. II. *Geochim. Cosmochim. Acta* **34**, 331–340.

SHAW, D. M. (1970) Trace element fractionation during anatexis. *Geochim. Cosmochim. Acta* **34**, 237–243.

SHIH, C.-Y., HASKIN, L A., WIESMANN, H., BANSAL, B. M. and BRANNON, J. C. (1975) On the origin of high-Ti mare basalts. *Proc. Lunar Sci. Conf. 6th.* 1255–1285.

SHIMIZU, N. and KUSHIRO, I. (1975) The partitioning of rare earth elements between garnet and liquid at high pressures: preliminary experiments. *Geophys. Res. Lett.* **2**, 413–416.

ZIELINSKI, R. A. (1975) Trace element evaluation of a suite of rocks from Reunion Island, Indian Ocean. *Geochim. Cosmochim. Acta* **39**, 713–734.

ZIELINSKI, R. A. and FREY, F. A. (1974) An experimental study of the partitioning of a rare earth element (Gd) in the system diopside-aqueous vapor. *Geochim. Cosmochim. Acta* **38**, 545–565.

Variation in Trace Element Partition (Crystal/Magma) as a Function of Crystal Growth Rate

PAUL HENDERSON

Department of Mineralogy, British Museum (Natural History), London SW7 5BD, U.K.

and

C. T. WILLIAMS

Department of Geology, Trinity College, Dublin 2, Ireland

Abstract

Fast crystal-growth rates can lead to the disequilibrium partition of trace and major elements in igneous rocks and to modification of the compositional path taken by an intrusive, fractionating magma. Kinetic variations in partition coefficients may be used to determine crystal-growth rates in selected cases, by the Burton, Prim and Slichter theoretical model (1953). Fission track analysis has been used to study the variation in uranium concentration in olivines of different morphologies from the Rhum layered intrusion, Scotland, and the BPS model used to assess the average crystal growth rate for each morphological type. The growth rate for the harrisitic morphology was approximately five times faster than that for the equant morphology. Reliable data, especially on diffusion coefficients of appropriate elements, are needed before absolute growth rates may be calculated.

The thermodynamic equilibrium distribution of trace elements between growing crystals and the parent melt will only prevail when the growth rate is slow in relation to the diffusion rate of a trace element (Fig. 1a). If the growth rate is relatively fast then the concentration of the element in a narrow liquid layer around the crystal will be significantly higher than in the bulk liquid when the partition coefficient ($k_o = C_s/C_1$) is <1 (Fig. 1b) and vice versa when $k_o > 1$. The existence of the concentration gradient is reflected by an increased (for $k_o < 1$) or decreased (for $k_o > 1$) concentration of the element in the crystal (e.g. $C_1 > C_s' > C_s$ in Figs. 1a and 1b). The measured, or effective, partition coefficient ($k_{ef} = C_s'/C_1$) tends to unity with increasing growth rate, and reaches a steady state at this limit (Fig. 1, b and c).

Any effect on element distribution through variation in crystal growth rate is most likely to be apparent where the growth rates are considered to have been fast as, for example, in olivine growth in many komatiites. Furthermore, the important work of Lofgren *et al.* (1975), Donaldson *et al.* (1975) and others has shown that crystal morphology is dependent upon the cooling rate of the parent melt. Thus, the differences in morphology and trace element composition of a mineral will be expected to be a function of the crystal growth rate. Variations in morphology of crystals that have grown from the same parent magma do occur in some extrusive and intrusive igneous rocks. A good example of the latter is the layered intrusion of Rhum, Inner Hebrides, Scotland, where olivine (and sometimes plagioclase) can

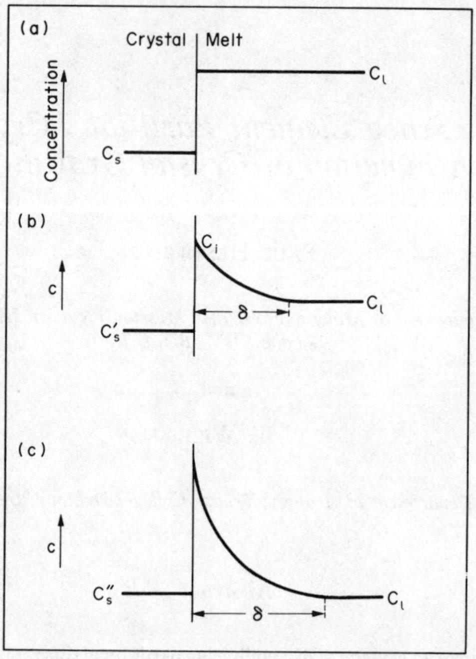

FIG. 1. Concentration profiles of trace element at crystal/liquid interface. (a) Equilibrium distribution at infinitely slow crystal growth rate; $k_o = C_s/C_1$. (b) Typical distribution during finite crystal growth; $C'_s/C_1 = K_{ef}$, $k_o \approx C'_s/C_i$, (c) Steady-state distribution at very fast crystal growth rate; $k_{ef} = C''_s/C_1 = 1$. Note: $C''_s > C'_s > S_s$.

exhibit a wide morphological range even in the same hand specimen. This type of material is suitable for the study of possible variations in composition due to crystal growth rate.

There is a paucity of data on the diffusion rates of elements in magmas, but published results (e.g. Medford, 1973; Seitz, 1974) suggest that the diffusion rates of highly charged, incompatible ions (e.g. of Th, or REE) are relatively slow and so their distribution is the most likely to be affected by variations in mineral growth rate. Elements which have very large, or very small, equilibrium partition coefficients would be the most suitable for initial study as changes in element partition will be more readily discernible. However, some of these ions (e.g. Ni^{2+} which partitions strongly into olivine) may prove to have relatively fast diffusion rates, so their distribution will not be affected except by the fastest growth rates.

In so far as crystal growth rate can affect trace element distribution it is a parameter that should be considered in geothermometry, geobarometry and in the theoretical modelling of fractional crystallisation processes, especially where the incompatible elements are involved. There is also the likelihood that major element concentrations may be affected (e.g. aluminium incorporation into pyroxenes). It is important, therefore, to assess quantitatively the effect that mineral growth variations have on element distribution. Furthermore, as more data become available on diffusion coefficients, it will be possible to use the observed changes in element partition to calculate crystal growth rates. The variation of uranium concentration with the morphology of olivine crystals of the Rhum layered intrusion is reported here and a preliminary attempt made to determine the growth rate for each morphology from the variation in partition of uranium.

Experimental Method

The determination of compositional variations of incompatible elements raises some analytical problems because their concentration in the mineral will be very low (a few ng/g). In experimental systems the use of radioactive tracers and beta-track mapping (e.g. for samarium distribution determination) is a suitable method, but it cannot be applied directly to natural systems. The use of particle track analysis for the determination of the distribution of uranium and/or boron, however, is applicable to natural systems and, in the case of uranium, offers a very sensitive method (Fleischer *et al.*, 1975; Seitz and Hart, 1973). This technique was adopted for the work reported here.

An external solid state nuclear track detector (SSNTD) was used to record fission fragments of ^{235}U in the samples and in a N.B.S. standard glass (SRM-614). Polished thin sections of the samples and standards in contact with the SSNTD (Makrofol KG) were irradiated with a total thermal neutron dose of 2×10^{17} n cm^{-2}. After irradiation the SSNTD was etched by 6.25 M NaOH, which had previously been saturated with etch products (Peterson, 1970) at 70°C for 10 min, and the revealed tracks counted using a transmitted light microscope at $\times 400$ magnification. The U concentration (U_{min}) is calculated from the formula (after Fleischer *et al.*, 1975):

$$U_{min} = U_s \cdot \frac{d_s}{d_{min}} \frac{R_s}{R_{min}} \frac{I_s}{I_{min}} \tag{1}$$

where d is track density,

R represents the effective range of fission fragments (mg cm^{-2}); and I is the isotopic abundance ratio: $^{235}U/^{238}U$ (where $I_s = 1.173 \times 10^{-3}$, Carpenter, 1972, and $I_{min} = 7.26 \times 10^{-3}$).

Subscripts s and min refer to the standard and mineral respectively. Values for R_s and R_{min} were obtained from Mory *et al.* (1970).

In order to relate the tracks to their source minerals, a 4-mm^2 grid was inscribed on the detector prior to irradiation and a photographic contact print obtained of the rock/detector couple (Williams, 1976).

The accuracy of the U determinations was estimated by analysing glasses, standardized against international rock standards by use of the delayed neutron analytical technique, to be better than 5%. Precision was of the order of 1.5% at the 2 sigma level.

Results

The range and mean of the uranium concentrations in the olivine crystals of different morphologies are given in Table 1. Low uranium concentrations and, therefore, low track densities necessitated counting of tracks from a number of crystals of the same morphology in each rock to make the determinations statistically significant. Because of the low concentrations it was not possible to detect any zoning with respect to uranium in the olivine crystals. A total of fifty-one determinations were made on a variety of morphological types (see Table 1). The majority of the crystals studied here were in the compositional range Fo_{84-86}. In some cases olivines with differing morphologies occurred in the same hand specimen, but the major element compositions of the ones studied were virtually identical.

TABLE 1. URANIUM CONCENTRATION IN RHUM OLIVINES

Morphology	No. of analyses	U ng g^{-1} Range	Mean	Standard deviation
Poikilitic	6	4.30–4.52	4.42	0.09
Equant	13	4.88–5.33	5.00	0.43
Rounded	9	5.17–6.40	5.92	0.43
Irregular	8	5.13–6.71	5.99	0.42
Tabular	9	6.35–8.03	7.16	1.08
Harrisitic	6	8.21–8.43	8.34	0.06

Some descriptions of the olivine morphologies may be found in Wager and Brown (1968) and Donaldson (1974).

The results show that some morphological types have distinct uranium concentrations. Harrisitic (or "crescumulate") olivines contain approximately twice as much U as poikilitic ones. Similar concentrations, however, are found in the rounded and irregular morphologies.

On the basis of microprobe analyses, the nickel contents of the olivines show no significant variation with morphology.

The variations in uranium concentrations could be a result of the different growth rates of the morphological types, the harrisitic olivines having developed faster than the others. Alternatively, some change in uranium concentration could come about by post-depositional reaction of the olivine either with intercumulus liquid and/or with other cumulus phases. Textural evidence suggests that the harrisitic olivines have not re-equilibrated or reacted after formation (Wager and Brown, 1968). Furthermore, different morphological types occur in the same hand specimen yet still retain distinct uranium concentrations. Therefore, it seems reasonable to assume that post-depositional reaction has not been widespread, although it cannot be excluded in all the cases studied here. If the uranium concentration was set simply by the crystal growth rate and the amount of uranium in the parent magma, then the variation in uranium may be used to establish the actual growth rates for each morphological type. In order to do this we have used the theoretical work of Burton *et al.* (1953).

Theory of Burton, Prim and Slichter (B.P.S.)

Burton *et al.* (1953) solved a one-dimensional steady-state diffusion equation expressing conservation of the solute across the crystal-melt interface, coupled to a boundary condition that in a liquid layer immediately around the crystal, molecular diffusion is the sole mass transfer process while beyond this layer the element concentration in the bulk liquid is maintained at a uniform level by mixing. The derived equation is:

$$k_{ef} = \left[1+\left(\frac{1}{k_o}-1\right)\exp\left(-Y\frac{\delta}{D}\right)\right]^{-1} \tag{2}$$

where k_{ef} if the effective or measured partition coefficient (C_s'/C_1, Fig. 1b),

k_o is the equilibrium partition coefficient (C_s/C_1, Fig. 1a),
Y is the linear growth velocity of the crystal melt interface (cm s^{-1}),
D is the diffusion coefficient of the trace element in the melt (cm^2 s^{-1}),
δ is the effective thickness of the diffusion boundary layer.

The value of δ depends upon the D value of the trace element, the kinematic viscosity of the melt and the degree of convection, or mixing in the melt, and possibly to a small extent on the crystal growth rate.

Although the B.P.S. model is not strictly applicable to systems other than those with well-behaved growth (e.g. see discussion in Carruthers, 1975 and references quoted therein) it may be applicable to a number of igneous systems where $\delta/D \gg 1$ for a particular trace element. Where the B.P.S. model is used to determine crystal growth rates from changes in element partition, it is readily applicable only to crystals of equant morphology. In other cases an "average" growth rate of the various faces of the crystal will be obtained, but when growth is highly anisotropic, as in some spinifex-textured olivines, errors in the use of the B.P.S. model could be large. However, if the dimensions of the crystals are obtained, it will be possible to identify this situation and in some cases to correct for the anisotropy.

At the present time the problems in the application of the B.P.S. model (and most other crystal growth models) are compounded further by lack of data on the required parameters, particularly of diffusion coefficients. However, with much attention being paid now to obtaining these data it is to be hoped that this problem will be reduced significantly in the near future. In the present work values of δ and D are required:

- δ. Limited data on thorium (Seitz, 1974) and selected major elements (Donaldson, 1975; Bottinga *et al.*, 1966; Fidzuk and Henderson, unpublished data) suggest that δ is frequently in the range 4 to 25 μm. A common value of 10 μm is assumed here.
- D. There are few published data on diffusion rates in magmas (e.g. Medford, 1973). Seitz (1974) has calculated, from diffusion profiles, that D_{Th} in a diopside–albite–anorthite (50:25:25 wt%) melt is 2×10^{-8} cm^2 s^{-1} at about 1350°C, and 20 kbar. No data appear to be available for D_U in basaltic melts so the value is assumed to be similar to that for Th, i.e. $\sim 10^{-8}$.

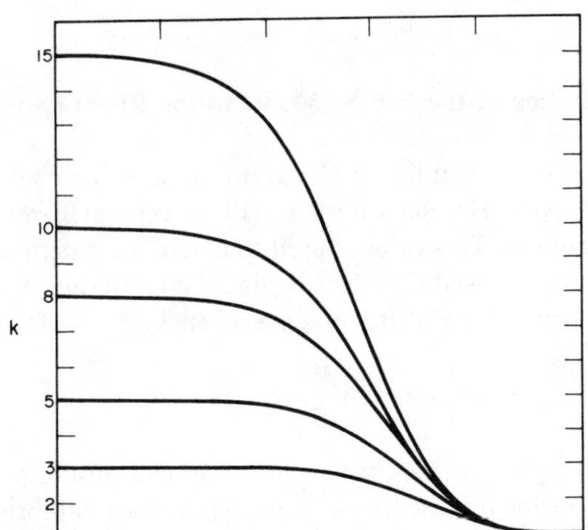

FIG. 2. Variation in partition coefficient, k (with $k \geq 1$), as a function of crystal growth rate as calculated from the B.P.S. model (see text) for δ/D value of 10^5 cm^{-1} s.

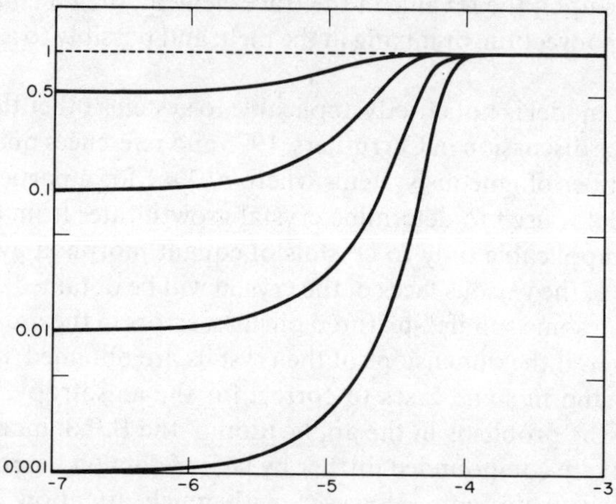

FIG. 3. As for Fig. 2, but for k values ≤ 1.

Thus, δ/D for uranium in the system studied here is taken to be 10^5 cm^{-1} s at the temperature of crystallization. This figure may need to be modified in the light of new data on diffusion coefficients. However, it will be noted from the crystal-growth equation that determination of *relative* growth rates does not require a knowledge of the value of δ/D provided that it remains constant for the different velocities. A δ/D value of 10^5 cm^{-1} s has been used to construct k_{ef} *versus* growth-rate curves for different equilibrium partition coefficients, as determined from the B.P.S. model (Figs. 2 and 3). It is seen that growth rates of between 10^{-7} and 10^{-4} cm s^{-1} are distinguishable from variations in appropriate partition coefficients. Elements that diffuse faster than uranium may be used to distinguish growth rates faster than 10^{-5} cm s^{-1}.

Application of the B.P.S. Model to the Rhum Olivines

In order to apply the B.P.S. formulation, the uranium partition coefficient, k_{ef}, is required for each morphological type. Henderson *et al.* (1971) estimated that the Rhum magma contained 0.4 μg g^{-1} uranium. This value is used to derive the partition coefficients given in Table 2, although it should be noted that the calculated growth rate, Y, is insensitive to errors in the trace element concentration of the magma when $k_o \ll 1$, as from equation (2):

$$\ln \frac{C_s}{C_s'} \approx -Y\frac{\delta}{D}, \text{ when } k_o \ll 1. \tag{3}$$

The calculated, relative growth rates are given in Table 2 on the basis that the uranium content of the poikilitic olivine represents the thermodynamic equilibrium concentration for uranium. (If $k_o < 0.011$ then the differences in the growth rates of the olivine morphologies are decreased.) The relative rates are calculated from $\ln k_{ef}(1-k_o)/k_o(1-k_{ef})$. Lack of data on the appropriate values of D_U and δ_U prevent the calculation of accurate, absolute growth rates, although the rates given in Table 2 for $\delta/D = 10^5$ cm^{-1} s are considered to be of the right order of magnitude.

TABLE 2. RELATIVE AND ABSOLUTE GROWTH RATE OF RHUM OLIVINES

Morphology	Mean U ($ng\ g^{-1}$)	K_{ef}	Relative growth rate	Absolute growth rate, ($cm\ sec^{-1}$) ‡
Poikilitic	4.42	0.011†	<1	$<1.2\times10^{-6}$
Equant	5.00	0.012(5)	1	1.2×10^{-6}
Rounded	5.92	0.015	2.5	3.1×10^{-6}
Irregular	5.99	0.015	2.5	3.1×10^{-6}
Tabular	7.16	0.018	4.0	4.8×10^{-6}
Harrisitic	8.34	0.021	5.3	6.4×10^{-6}

$\dagger = k_o$. ‡for $\delta/D = 10^5\ cm^{-1}\ s$.

The preliminary evidence on the constancy of Ni contents of the olivine crystals of differing morphologies suggests that δ/D for Ni is $2\times10^3\ cm^{-1}\ s$ and so if $\delta_{Ni} = 10\ \mu m$ then $D_{Ni} \approx 5 \times 10^{-7}\ cm^2\ s^{-1}$ for this system.

The estimated growth rates presented here support the findings of Donaldson (1974) and the related suggestions of Drever and Johnston (1967) especially in that the harrisitic and tabular olivines have grown at fast rates compared with the other olivine morphologies. The work also shows the important role that crystal growth rate can play in determining trace-element distribution in igneous systems.

References

BOTTINGA, Y., KUDO, A. and WEILL, D. (1966) Some observations on oscillatory zoning and crystallization of magmatic plagioclase. *Amer. Miner.* **51**, 792–806.

BURTÔN, J. A., PRIM, R. C. and SLICHTER, W. P. (1953) The distribution of solute from crystals grown from the melt. Part 1: Theoretical. *J. Chem. Phys.* **21**, 1987–1999.

CARPENTER, B. S. (1972) Determination of trace concentrations of boron and uranium in glass by the nuclear track technique. *Anal. Chem.* **44**, 600–602.

CARRUTHERS, J. R. (1975) Crystal growth from the melt. In *Treatise on Solid State Chemistry*, N. B. HANNAY, ed., vol. 5, *Changes of State*. Plenum Press.

DONALDSON, C. H. (1974) Olivine crystal types in harrisitic rocks of the Rhum pluton and in Archean spinifex rocks. *Geol. Soc. Amer. Bull.* **85**, 1721–1726.

DONALDSON, C. H. (1975) Calculated diffusion coefficients and the growth rate of olivine in a basalt magma. *Lithos* **8**, 163–174.

DONALDSON, C. H., USSELMAN, T. M., WILLIAMS, R. J. and LOFGREN, G. E. (1975) Experimental modeling of the cooling history of Apollo 12 olivine basalts. *Proc. Lunar Sci. Conf. 6th*, 843–869.

DREVER, H. I. and JOHNSTON, R. (1967) Picritic minor intrusions. In *Ultramafic and Related Rocks*, P. J. WYLLIE, ed., Chap. 3, pp. 71–86, John Wiley.

FLEISCHER, R. L., PRICE, P. B. and WALKER, R. M. (1975) *Nuclear Tracks in Solids, Principles and Applications*. Univ. of California Press.

HENDERSON, P., MACKINNON, A. and GALE, N. H. (1971) The distribution of uranium in some basic igneous cumulates and its petrological significance. *Geochim. Cosmochim. Acta* **35**, 917–925.

LOFGREN, G. E., DONALDSON, C. H. and USSELMAN, T. M. (1975) Geology, petrology and crystallisation of Apollo 15 quartz-normative basalts. *Proc. Lunar Sci. Conf. 6th*, 79–99.

MEDFORD, G. A. (1973) Calcium diffusion in a mugearite. *Can. J. Earth Sci.* **10**, 394–402.

MORY, J., DEGUILLEBON, D. and DELSARTE, G. (1970) Mesure du parcours moyen des fragments de fission avec le mica comme detecteur—influence de la texture cristalline. *Rad. Effects* **5**, 37–40.

PETERSON, D. D. (1970) Improvement in particle track etching in Lexan polycarbonate film. *Rev. Sci. Instr.* **41**, 1214–1255.

SEITZ, M. G. (1974) Promotion of kinetic disequilibrium of trace thorium. *Carn. Inst. Washington Yr. Bk.* **73**, 551–553.

SEITZ, M. G. and HART, S. R. (1973) Uranium and boron distributions in some oceanic ultramafic rocks. *Earth Planet. Sci. Lett.* **21**, 97–107.

WAGER, L. R. and BROWN, G. M. (1968) *Layered Igneous Rocks*, Oliver & Boyd (Edinburgh).

WILLIAMS, C. T. (1976) Geochemical studies on cumulus phases of some layered igneous rocks. Unpublished Ph.D. thesis, Univ. of London.

Rare-earth Elements in Apatite from Layered Norites and Iron-Titanium Oxide Ore-bodies Related to Anorthosites (Rogaland, S.W. Norway)

IWAN ROELANDTS and JEAN-CLAIR DUCHESNE

Géologie-Pétrologie-Géochimie, Université de Liège, Sart Tilman, Belgium

Abstract

Rare-earth elements (REE) were investigated by neutron-activation analysis in nine apatites and seven coexisting plagioclase cumulate minerals at three different levels of the Bjerkrem–Sogndal layered lopolith (Rogaland, S.W. Norway), a massif which displays the complete andesine anorthosite–charnockite series of rocks.

The REE (except Eu) behave as hygromagmatophile elements (as defined by Treuil) in the course of the fractional crystallization process which has given rise to the massif. The concentration is indeed inversely proportional to F, the fraction of residual liquid. This characteristic permits placing an upper limit to the value of the partition coefficients of REE between apatite and magma. They must approximately be equal or less than 5. The apatite–plagioclase distribution coefficients show little variation during differentiation. This can be explained by an open system crystallization of the intercumulus liquid and permits calculating the plagioclase-magma partition coefficients. Values similar to those of Paster *et al.* (1974) are obtained.

A range of possible D_{Eu}^{apat} is calculated, together with the corresponding D_{Eu}^{plag} at three different levels of the intrusion. Drake's (1975) geobarometer based on Eu in plagioclase is used to estimate the fugacity of oxygen. There is good agreement with the values obtained with the Buddington and Lindsley geobarometer. A continuous decrease of f_{O_2} throughout the whole sequence cannot be invalidated.

REE were also determined in apatite from five Fe–Ti oxide ore-bodies occurring in the same magmatic province. The consistency between the trace element behaviour in the oxides and the REE in the apatite permits identifying fractional crystallization as controlling the generation of the ore-bodies.

Introduction

Quantitative modelling of trace element behaviour in magmatic processes constitutes one of the most powerful methodological tools geochemistry has supplied to petrology. However, the method, to be fully efficient, requires knowledge of partition coefficients of trace elements between minerals and magma.

Direct measurement of partition coefficients of traces between phenocrysts and matrix in experimental systems or in natural rocks is the generally recommended method of determination. However, it has been suggested (Duchesne, 1971; Paster *et al.*, 1974) to treat this question by studying minerals in magmatic cumulate series of layered intrusions. Indeed, the very slow cooling of these intrusions and the amount of monomineralic cumulates suggest that ideal conditions of equilibrium and complete separation of the liquid from the solid are approached.

The present paper describes an application of this method to apatite, a mineral commonly accessory in rocks but rich in trace elements such as the REE. The data obtained by

Nagasawa (1970) for REE partition coefficients for apatite in dacites show very high values (as much as 100). Several authors have pointed out the possible—and so far neglected—role of this mineral in balancing the effect of crystallization of silicate minerals (e.g. Zielinski and Frey, 1970). Others have questioned the high values: Paster *et al.* (1974) have calculated that values of 8 to 16 lead, in the Skaergaard intrusion, to the model most consistent with the petrological data; Sun and Hanson (1976) inferred values of about 5 or less on the basis of apatite–clinopyroxene partition coefficients.

Our purpose is to show that low values (5–10) for REE partition coefficients in apatite can be deduced from the study of a layered intrusion, namely the Bjerkrem–Sogndal lopolith. These results will also be applied to some apatite-rich, Fe–Ti oxide ore-bodies.

The first author (I.R.) is mainly responsible for the analytical work, the second (J. C. D.) has collected the samples and led the petrological and geochemical study.

Experimental

Mineral fractions were purified by means of heavy liquids and a magnetic separator. Extreme caution was exercised in the separation of plagioclase to avoid contamination by apatite. Heavy liquid density was adjusted as close as possible to that of the plagioclase to be separated and the fractions were leached with hot 4 N HCl.

The neutron-activation procedures were performed at the Mineralogical Geological Museum, University of Oslo, Norway.

For apatite, REE were determined non-destructively by Ge(Li) γ-spectrometry with thermal (La, Ce, Nd, Eu, Tb, Dy, Yb and Lu) and epithermal (Sm, Gd, Ho, Er) reactor neutrons on two independent fractions of the mineral; a known amount of each REE was used as monitor samples (Brunfelt and Roelandts, 1974). The precision ($t_{0.95}s$) estimated from replicate measurements was better than 5% for La, Sm, Eu, Dy, 10–15% for Ce, Tb, Yb and about 20% for Nd, Gd, Ho, Er and Lu, within the range of concentrations studied here.

As for plagioclase, due to the extremely low concentrations of REE (ppb to ppm range), a radiochemical group separation was carried out after a 7-day thermal neutron activation. An oxidant alkaline fusion of the irradiated sample was followed by leaching with water. The insoluble hydroxide fraction was dissolved in HCl and analysed for eight REE by γ-spectrometry (Roelandts, 1975). The USGS reference sample BCR-1 was used as the standard. The working conditions were chosen to give approximately the same precision as that obtained for apatite, i.e. 15–20% for Nd, Tb, Yb and Lu, and about 5–10% for the other measured REE. Single instrumental NAA was also used for analysing two plagioclase samples (7969 and 6675: see Table 1) with less precision and higher detection limit.

The Bjerkrem–Sogndal Lopolith

The Bjerkrem–Sogndal (Bk–Sg) massif situated in the anorthositic province of Egersund (Rogaland, S.W. Norway) is a synkinematic layered lopolith (Michot, 1960, 1965). It displays the complete anorthosite–charnockite series of rocks and results from a gravity differentiation process. Micro- and macrotextures typical of igneous layered cumulates (Wager and Brown, 1968) characterize the rocks (lamination, small-scale rhythmic layering, banding, etc.). The lower part of the massif is essentially formed by anorthosites, leuconorites

TABLE 1. REE CONTENTS (ppm) IN APATITE AND PLAGIOCLASE

	Bjerkrem–Sogndal lopolith														
	Rhythm III								Rhythm IV						
	JCD6674		7969		7970	JCD6675	Average		JCD6444		JCD6695		JCD66212	Average	
	Apat	Plag	Apat	Plag	Apat	Plag	Apat	Plag	Apat	Plag	Apat	Plag	Apat	Apat	Plag
La	144	2.13	149	1.93	162	1.84	152	1.97	216	2.98	248	3.19	234	233	3.09
Ce	429	3.25	414	3.27	451	3.66	431	3.39	606	4.94	686	5.30	637	643	5.12
Nd	532	1.3	494		485		504	1.3	479	2.1	703	2.5	660	614	2.3
Sm	96.5	.19	95.6	.16	98.7	.16	96.9	.17	130	.33	146	.29	127	134	.31
Eu	20.2	1.19	19.4	1.29	21.1	1.48	20.2	1.32	25.9	1.66	26.4	2.05	26.7	26.3	1.86
Gd	86.0		85.3		84.5		85.3		125		124		162	137	
Tb	13.5	.015	11.5	<.02	12.8	.016	12.6	.016	16.1	.034	19.2	.020	16.3	17.2	.027
Dy	51.4		53.1		51.9		52.1		65.3		78.4		67.9	70.5	
Ho	12.2		9.2		9.3		10.2		10.3		14.7		15.2	13.4	
Er	22.2		19.8		21.5		21.2		26.0		33.8		36.4	32.1	
Yb	10.0	.020	9.9	<.1	13.5	<.1	11.1	.02	17.8	.043	17.4	.036	20.7	18.6	.039
Lu	1.0	<.004	1.0	<.006	1.2	<.006	1.1		1.9	.007	2.1	.006	1.8	1.9	.006
La/Yb							14	99						13	79
Eu/Eu*							0.68	28						0.61	22
An %		44		42		45		44		44		41			43

	Bjerkrem–Sogndal lopolith							Fe–Ti oxide ore bodies					
	Rhythm V												
	JCD6460		JCD66103	JCD6458	JCD66102N	Average		Kydl. 66181	Rödem. II 6615	Kaknud. 6611	Eigeröy 66150	Hesnes 6633	
	Apat	Plag	Apat	Apat	Plag	Apat	Plag	Apat	Apat	Apat	Apat	Apat	Chondrite
La	344	4.03	337	381	3.68	354	3.86	204	269	286	820	951	0.33
Ce	947	6.24	918	972	6.07	946	6.16	519	729	779	2192	2649	0.88
Nd	652	2.6	658	816	2.3	709	2.45	(732)	609	626	1563	2108	0.60
Sm	180	.33	175	190	.31	182	.32	136	158	165	421	466	0.181
Eu	31.4	3.51	32.6	34.6	3.59	32.9	3.55	30.8	30.5	32.3	43.7	44.5	0.069
Gd	144		173	183		167		120	168	175	400	423	0.249
Tb	26.7	.015	23.7	25.1	.023	25.2	.019	17.7	22.8	24.3	64.5	67.9	0.047
Dy	99.8		94.3	102		98.7		87.0	92.8	99.3	273	300	0.29
Ho	14.8		21.9	19.4		18.7		18.3	18.1	24.9	55.2	55.1	0.070
Er	49.3		43.2	36.1		42.9		32.3	33.8	43.5	131	142	0.200
Yb	25.0	.041	23.7	23.7	.034	24.1	.038	19.5	24.7	24.1	73.8	83.9	0.20
Lu	2.4	.004	2.1	2.3		2.3	.004	2.7	2.1	2.1	9.0	10.4	0.034
La/Yb						15	102	10.5	10.9	11.9	11.1	11.3	
Eu/Eu*						.59	43.	0.72	0.57	0.58	0.32	0.30	
An %		38		36	39		37						

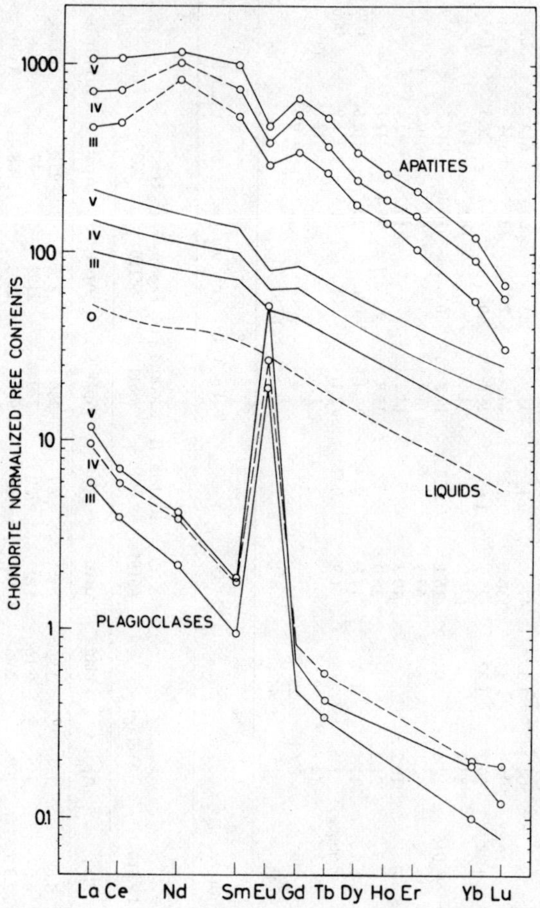

FIG. 1. Chondrite-normalized REE patterns in average apatites and plagioclases coexisting at three different levels of the Bjerkrem–Sogndal massif. Parental liquid (O) and successive residual magmatic liquids (III, IV, V) calculated. by quantitive modelling using REE* partition coeficients of Table 3 and $D_{Eu}^{apat} = 5.8$.

and norites. It has been divided into five macroscale rhythmic units (I to V) (Michot, 1960) which are anorthositic or leuconoritic at the base and grade upward into more mafic terms. They have been interpreted by Duchesne (1972a) as the result of repeated influxes of fresh magma into the magmatic chamber during the crystallization process.

The evolution of the mineralogy (Michot, 1965; Duchesne, 1972a, b) is as follows. Plagioclase (An_{50-45}, unzoned), orthopyroxene and hemo-ilmenite, in the average weight proportion of 74%, 16% and 10%, respectively (Duchesne, 1978), characterize the anorthosites and leuconorites of rhythms I, II, bases of III and of IV. Magnetite and apatite together with clinopyroxene first appear in the middle and upper part respectively of III. The same association occurs in the norites of IV and V and grades into jotunitic cumulates, which constitute the rapid transition towards acidic rocks. Apatite-bearing norites show an average weight proportion of 45% plagioclase + 22% orthopyroxene + 9% clinopyroxene + 4% apatite + 20% oxides (Duchesne, 1978). The An content of plagioclase decreases regularly from III to V. Though plagioclase is always granulated and this phenomenon often blurs the typical texture of cumulates, the plagioclase has originally accumulated as primocrysts. In rhythms III and IV, pyroxenes form euhedral primocrysts, but in V, orthopyroxene (inverted pigeonite) shows a poikilitic structure. The oxides always fill the interstices between the silicate minerals. Coarse apatite crystals lie parallel to the general lamination. Smaller ones may also be included in the oxides or in the poikilitic pyroxenes.

REE in Apatite and Plagioclase

The results obtained are listed in Table 1. It should be observed that different apatites and different plagioclases from any one rhythm are remarkably similar. We can therefore neglect the internal variations in each rhythm and consider only the average values per rhythm. The chondrite-normalized values are plotted in Fig. 1. For apatite and plagioclase, points are situated on smooth curves, except for Eu which shows a distinct negative anomaly in apatite and a high positive anomaly in plagioclase.

The following remarks can be made:

1. For apatite, parallel curves are obtained (leading to very constant La/Yb ratios) except for Eu whose anomaly increases with the concentration and for Nd which deviates the more from a mean curve the lower its content. A systematic analytical error is suspected for this last element.† This does not allow the rigorous definition of a normal distribution pattern of the light REE (LREE) in the Nd region and makes impossible the assessment of a possible Ce negative anomaly, as suggested by Puchelt and Emmerman (1976). In any case, the factors which are responsible for the Ce depletion according to these authors cannot have operated here. Ce-rich minerals, such as monazite and allanite, are absent and the evolution of the rocks is characterized by low and decreasing oxygen fugacity (see below) with no indication of a previous stage of evolution in oxidizing conditions. In the following, we shall not take into account the Nd content in apatite.
2. REE*‡ contents in apatite show a progressive enrichment from rhythm III to rhythm V, concomitant with the differentiation process. Between the two rhythms, the REE* average enrichment factor is 2.0 ± 0.2. The Eu anomaly defined as the Eu/Eu* ratio (Eu* being obtained by linear interpolation between Sm and Gd) slightly decreases.
3. For plagioclase, the parallelism of the curves is also striking for the LREE. The La–Sm average enrichment factor between rhythms III and V is 1.9 ± 0.1, i.e. close to that observed for apatite. The heavy REE (HREE) pattern for plagioclase is not as well defined as it is for apatite; three elements only (Tb, Yb and Lu) have been measured, the contents are close to the detection limit and the precision of the determination is rather poor. The fact that the plagioclases from rhythm IV have Tb, Yb and Lu contents which are higher than those of plagioclase from rhythm V may be attributed to these factors. Consequently we shall not draw any geochemical conclusion from the behaviour of

† Counting rates for the ^{147}Nd photopeak at 531.0 keV are low; as for the more sensitive 91.1 keV line of ^{147}Nd, it shows an important background, whose subtraction from the peak area can introduce a systematic error, the more important as the Nd contents are lower.

‡ REE* is used here to designate the REE, except Eu.

these three elements in plagioclase. It can, however, be noted that from rhythm III upwards the contents in these elements increase and the La/Yb ratio, i.e. the slope of the distribution pattern, remains approximately constant. We thus feel allowed to extrapolate the conclusions drawn from the LREE to the whole REE* in plagioclase. As for Eu in plagioclase, the content and the amplitude of the positive anomaly increase continuously between rhythms III and V.

TABLE 2. APATITE-PLAGIOCLASE DISTRIBUTION COEFFICIENTS

	Rhythms				
$D^{apat-plag}$	III	IV	V	s	Average
La	77	75	92	7.3	81
Ce	127	126	154	9.3	136
Sm	570	430	570	56	523
Yb	560	480	630	82	557

s = pooled standard deviation.

4. The distribution coefficients of La, Ce, Sm and Yb between apatite and plagioclase ($D^{apat-plag}$) were calculated for each rhythm and are listed in Table 2. For a given element, the different values vary somewhat between the three rhythms but the differences are not statistically significant as can be shown by using a *t*-test of comparison. The $D^{apat-plag}$ can thus be considered constant and averaged. This constancy which expresses the simultaneous increase of the REE* in both minerals could be assigned, not to differentiation, but to a proportional increase of the two sets of mineral–magma partition coefficients D^{apat}_{REE*} and D^{plag}_{REE*}. This explanation can be rejected. Indeed, the strikingly differing abundances and distribution patterns in the two minerals reflect major differences in cristallochemical control of the REE contents by those minerals. Variation in a mineral would very unlikely lead to a proportional variation in the other. We infer from this that the constancy of the apatite–plagioclase partition coefficient $D^{apat-plag}$ is due to the constancy of the mineral-magma partition coefficients D^{apat} and D^{plag}. It should be noted that the major chemical compositions of the minerals vary little (from An_{45} to An_{37}) or not at all (apatite). We can conclude that the REE enrichment between rhythms III and V both in apatite and plagioclase is due to the increase of these elements in the *magma* during the crystallization of the massif. The enrichment factor of the REE* in the magma between III and V is thus equal to those which were calculated for apatite and plagioclase here above and approaches 2. These considerations also imply that the two minerals remained in equilibrium with the magma during crystallization of the cumulate. In other words, the rock minerals have formed through a process of *adcumulus growth* (Wager and Brown, 1968) in open system with the overlying magma.

On the other hand, it has been demonstrated (Duchesne, 1971, 1978) that the differentiation process in the massif follows the Rayleigh Law (Neumann *et al.*, 1954)

$$c_{i,\rho} = c_{i,o} F_{\rho}^{\bar{D}_i - 1} \tag{1}$$

where $c_{i,o}$ and $c_{i,\rho}$ are the concentrations of a given trace element i in the parental magma and in the residual one at stage ρ respectively, $F\rho$ is the weight fraction of magma remaining at stage ρ and $\bar{D}_i$ is the bulk (or effective) mineral-magma distribution coefficient

$$\bar{D}_i = \sum_j \mu_j D_i^j \tag{2}$$

where μ_j is the weight fraction of a given mineral j and D_i^j the distribution coefficient of the element i between mineral j and the magma ($D_i^j = c_i^j/c_i$).

As shown by Anderson and Greenland (1969) if $\bar{D}_i$ is negligible relatively to 1, equation (1) becomes

$$c_{i,\rho} = \frac{c_{i,o}}{F_\rho} \tag{3}$$

and $c_{i,\rho}$ is inversely proportional to F_ρ.

The elements showing this property have been called *hygromagmatophile* by Treuil (Treuil and Varet, 1973; Treuil and Joron, 1975) who has demonstrated that a hygromagmatophile behaviour still holds within a few percent for $\bar{D}$ values up to 0.2.

Modelling of the Sr–Ca and K–Rb behaviours in the Bk–Sg differentiation (Duchesne, 1978) shows that the beginning of crystallization of apatite as a cumulate mineral (i.e. in the upper part of rhythm III) corresponds to $F_{\text{III}} = 0.47$ and that the end of rhythm V is characterized by value of $F_{\text{V}} = 0.21$. The ratio of concentrations of a hygromagmatophile element in the liquid (or in the minerals in equilibrium with the liquid) at stages III and V, i.e. $c_{i,\text{V}}/c_{i,\text{III}}$, is equal to $F_{\text{III}}/F_{\text{V}}$, that is 2.2.

Since there is a close agreement between this value and the enrichment factors based on the apatite–plagioclase evolution the REE* behaviour between III and V is hygromagmatophile and consequently the $\bar{D}_{\text{REE*}}$ are ≤ 0.2.

The average normative composition of the apatite-bearing rocks indicates an apatite content of 4% (μ_{apat}) (Duchesne, 1978) and application of equation (2), in which the most important contribution to the $\bar{D}_i$ is $\mu_{\text{apat}}\, D_{\text{REE*}}^{\text{apat}}$ gives a maximum value of $D_{\text{REE*}}^{\text{apat}}$ around 5. A lower limit to the $D_{\text{REE*}}^{\text{apat}}$ variation is given by the $D_{\text{REE*}}^{\text{plag}}$ values which are related to the former through $D_{\text{REE*}}^{\text{apat-plag}}$ values. $D_{\text{REE*}}^{\text{apat}}$ equal to half the values calculated by Paster *et al.* (1974) for the Skaergaard have been adopted here (Table 3); they are distributed around 5 and lead to acceptable $D_{\text{REE*}}^{\text{plag}}$ values though somewhat lower than those reported by Paster *et al.* (1974).

The values obtained permit us to calculate the REE* contents in the residual magmas corresponding to rhythms III, IV and V, and application of equation (3) leads to the REE* contents in the parental magma (Fig. 1). It should be noted that equation (3) can be strictly applied between the base of the massif and the top of III, because, since apatite is absent from the association of the cumulate minerals, the $\bar{D}_{\text{REE*}}$ are very low and the REE* behaviour is strongly hygromagmatophile.

TABLE 3. REE* DISTRIBUTION COEFFICIENTS FOR APATITE AND PLAGIOCLASE

	La	Ce	Nd	Sm	Eu*	Gd	Tb	Ho	Yb	Lu
D^{apat}	4.3	5.5	6.9	7.3	(7.5)	7.9	7.7	6.6	4.0	(3.25)
$D^{\text{apat-plag}}$	81	136		523					557	
D^{plag}	.053	.040	(.022)	.014	(.013)	(.011)	(.010)	(.009)	.007	(.0065)

Eu* is obtained by linear interpolation between Sm and Gd.
Values in brackets are obtained by interpolation after smoothing the distribution coefficient-atomic number curves.

Europium cannot be considered hygromagmatophile. The rocks are rich in plagioclase and the high Eu anomaly in plagioclase gives high D_{Eu}^{plag} and $\bar{D}_{Eu}$ values. Furthermore, contrary to the REE*, D_{Eu}^{plag} varies with the An content of the plagioclase, with the oxygen fugacity of the magma (Philpotts and Schnetzler, 1972; Philpotts, 1970; Weill and Drake, 1973; Drake, 1975) and also with the composition of the magma (Morris *et al.*, 1974). The first two factors must be taken into account in the Bk–Sg massif: the An content of plagioclase decreases from 50 to 37 from the beginning of the evolution to the upper part of V (Duchesne, 1978) and the appropriate use of the Buddington and Lindsley geothermo-barometer (Duchesne, 1972a) has shown that the f_{O_2}–T conditions were $10^{-11\pm1}$ atm and 900–975°C as soon as magnetite appears in the sequence of rocks (in the middle of rhythm III) and that f_{O_2} decreases with T further on. The effects of composition of the magma are more difficult to assess. Morris *et al.* (1974) could detect them experimentally with synthetic liquids of extreme compositions but with natural magmas no dependance could be observed. It is therefore assumed here that, within the range of composition of the successive liquids, the influence of this factor is negligible in regard to that of the first two factors.

The available data for Bk–Sg allow us to calculate the D_{Eu}^{plag} variations. The following equations give the evolution of the Eu anomaly in the magmatic liquids between the initial stage (0) and rhythm III (eq. 4) and between rhythms III and V (eq. 5).

$$(\mathrm{Eu/Eu^*})_{l,\mathrm{III}} = (\mathrm{Eu/Eu^*})_{l,0} F_{\mathrm{III}}^{(\bar{D}_{\mathrm{Eu},0} - \bar{D}_{\mathrm{Eu^*},0})}, \tag{4}$$

$$(\mathrm{Eu/Eu^*})_{l,\mathrm{V}} = (\mathrm{Eu/Eu^*})_{l,\mathrm{III}} (F_{\mathrm{V}}/F_{\mathrm{III}})^{(\bar{D}_{\mathrm{Eu,III}} - \bar{D}_{\mathrm{Eu^*,III}})}. \tag{5}$$

The equilibrium between plagioclase, apatite and magmatic liquids in stages III and V is expressed by equations (6) and (7):

$$D_{\mathrm{Eu,III}}^{\mathrm{plag}} = \frac{\mathrm{Eu_{plag,III}}}{\mathrm{Eu_{apat,III}}} D_{\mathrm{Eu,III}}^{\mathrm{apat}}, \tag{6}$$

$$D_{\mathrm{Eu,V}}^{\mathrm{plag}} = \frac{\mathrm{Eu_{plag,V}}}{\mathrm{Eu_{apat,V}}} D_{\mathrm{Eu,V}}^{\mathrm{apat}}. \tag{7}$$

Since $D_{Eu} \simeq D_{Eu^*}$ in minerals other than plagioclase and $D_{Eu^*}^{plag}$ is negligible compared to D_{Eu}^{plag}, $\bar{D}_{Eu} - \bar{D}_{Eu^*} \simeq \mu_{plag} D_{Eu}^{plag}$. In the present case, as indicated above, μ_{plag} is equal to 0.74 and 0.45 for rhythms III and V, respectively. F_{III} and F_{V} are equal to 0.47 and 0.21, respectively.

In order to solve equations (4) to (7) the following hypotheses can be introduced:

1. $(\mathrm{Eu/Eu^*})_0 = 1$, i.e. the initial magma has no Eu anomaly, as is the case in the Hidra massif (Duchesne *et al.*, 1974).

2. $D_{\mathrm{Eu,III}}^{\mathrm{apat}} = D_{\mathrm{Eu,V}}^{\mathrm{apat}}$ = constant. It has already been shown that $D_{\mathrm{REE^*}}^{\mathrm{apat}}$ is constant. Since Eu^{3+} is mainly incorporated by apatite (more than 95% of the total Eu using the method of calculation of Philpotts, 1970), the variations of the Eu^{2+}/Eu^{3+} ratio in the magma are not likely to be recorded by apatite.

Resolution of the system gives the values of the four unknowns:

$$D_{\mathrm{Eu},0}^{\mathrm{plag}} = 0.57, D_{\mathrm{Eu,III}}^{\mathrm{plag}} = 0.45, D_{\mathrm{Eu,V}}^{\mathrm{plag}} = 0.75 \text{ and } D_{\mathrm{Eu}}^{\mathrm{apat}} = 6.95.$$

Evaluation of the oxygen fugacity can be obtained by applying the experimental relation of Drake (1975):

$$\log f_{O_2} = -4.60 \log \left(\frac{Eu^{2+}}{Eu^{3+}}\right)_{plag} - 3.86$$

in which Eu^{3+} is calculated by

$$Eu^{3+}_{plag} = \frac{Eu_l - Eu_{plag}/D^{plag}_{Eu^{2+}}}{1/D^{plag}_{Eu^{3+}} - 1/D^{plag}_{Eu^{2+}}}$$

where, according to Philpotts (1970), $D^{plag}_{Eu^{2+}} = D^{plag}_{Sr}$. The D^{plag}_{Sr} values obtained by quantitative modelling of the Sr–Ca evolution (Duchesne, 1978), i.e. 2.0, 2.3 and 2.7 for stages 0, III and V respectively, were introduced in the equations to calculate the f_{O_2} values at the different stages (see Table 4, line 1).

The results obtained are semi-quantitative. Drake's geobarometer leads in the present case to f_{O_2} values differing by two orders of magnitude if the Eu^{2+}/Eu^{3+} ratio of the liquid is used instead of that of the plagioclase. It should, however, be noted that where the oxide mineral assemblage permits calculating f_{O_2} (i.e. in rhythm III), the value obtained (10^{-11}atm) is similar to that based on Eu in plagioclase. There is also a good agreement in the fact that the f_{O_2} calculated by the two methods decrease from rhythm III to rhythm V, a normal trend for fractional crystallization in a closed system (Osborn, 1959).

The calculation, however, gives unexpected results for the initial stage (stage 0): $D^{plag}_{Eu,0}$ is higher than $D^{plag}_{Eu,III}$ and f_{O_2} is lower than at stage III, which gives an evolution reversed to the trends found in the rest of the sequence. The f_{O_2} at stage 0 cannot be calculated by means of the association of the oxides because magnetite is absent up to rhythm III. It is thus not possible to confirm the trend between 0 and III by an independent method. In fact, this anomaly could be due to a lack of precise evaluation of the factor F_V introduced in equation (5) and could have no geochemical and petrological meanings. The method used to determine factor F_V (Duchesne, 1978) indeed permits a small interval of variation about the value 0.21. If a small decrease of F_V—from, say, 0.21 to 0.20—is admitted, a new set of values can be calculated (Table 4, line 2) for the D^{plag}_{Eu} and D^{apat}_{Eu} partition coefficients as well as for the f_{O_2}. It is observed that the major effect of the decrease is to modify the values at stage 0 without much change at stages III and V. By decreasing sufficiently F_V, it is thus possible to increase f_{O_2} values at stage 0 in order to obtain a set of continuously decreasing f_{O_2} values, i.e. a normal f_{O_2} evolution throughout the whole sequence (Table 4, lines 3 and 4). The set of values with $F_V = 0.186$ and $D^{apat}_{Eu} = 5.80$ is eventually preferred because it gives f_{O_2} conditions at the base of the massif close to those of the Ni–Ni0 buffer, assuming an intrusion temperature of 1000–1100°C.

The Iron–Titanium Oxide Ore-bodies

The Fe–Ti oxide ore-bodies studied here stake out the northern limit of the Håland massif (S. Egersund). They constitute small discordant lenticular masses or veins either within the Håland massif (Hesnes, Eigeröy, Rödemyr II) or at the limit with a complex unit called by Michot (1955) the norito-granitic zone (Kydlandsvatn, Kaknuden). A magmatic origin is unquestionable for the first ones but a metasomatic origin has been put forward for the last ones by Michot (1956). Field relationships are, however, ambiguous and available geochemical data (Duchesne, 1973) have already shown the great similarities between these

TABLE 4. APATITE AND PLAGIOCLASE Eu PARTITION COEFFICIENTS AND FUGACITIES OF OXYGEN

	Initial Stage O					Stage III				Stage V			
F_V	D_{Eu}^{apat}	D_{Eu}^{plag}	$\left(\frac{Eu^{2+}}{Eu^{3+}}\right)_{plag}$	$\left(\frac{Eu^{3+}}{Eu}\right)_l$	$-\log f_{O_2}$	D_{Eu}^{plag}	$\left(\frac{Eu^{2+}}{Eu^{3+}}\right)_{plag}$	$\left(\frac{Eu^{3+}}{Eu}\right)_l$	$-\log f_{O_2}$	D_{Eu}^{plag}	$\left(\frac{Eu^{2+}}{Eu^{3+}}\right)_{plag}$	$\left(\frac{Eu^{3+}}{Eu}\right)_l$	$-\log f_{O_2}$
0.21	6.95	0.57	60	0.72	12.1	0.45	42	0.80	11.3	0.75	79	0.72	12.6
0.20^{1}	6.34	0.41	39	0.79	11.2	0.41	38	0.82	11.1	0.69	69	0.75	12.3
0.19^{2}	6.00	0.32	28	0.84	10.5	0.39	35	0.83	11.0	0.65	65	0.76	12.2
0.18^{6}	5.80	0.27	23	0.87	10.1	0.38	34	0.83	10.9	0.63	62	0.77	12.1

ore-bodies and magmatic ones. The following data will also plead in favour of a similar origin.

The situation and description of the ore-bodies have been given by Hubaux (1960) and Duchesne (1973).* In Kydlandsvatn, the association of the oxides is hemo-ilmenite + homogeneous Ti-poor magnetite (Type II of Duchesne, 1973). In the other bodies a homogeneous ilmenite coexists with Ti-magnetite (type III). In the former, apatite-bearing ores are an exception and the amount of apatite is approximately 20 vol.%. In the latter, apatite is always present (nelsonite) and in Hesnes and Eigeröy its content approaches 30–35 vol.%, a possible eutectical composition according to Philpotts (1967). A deuteric readjustment of the chemical composition of the oxides has taken place in all ore-bodies down to T–f_{O_2} conditions indicated by the Buddington and Lindsley geothermo-barometer (Table 4).

Trace elements in magnetite and ilmenite (Duchesne, 1973 and Table 5) show large variations. Ore-bodies with Ti-rich magnetite (i.e. Hesnes and Eigeröy) have V, Cr, Ni, Co contents lower and a Zn content higher than those with Ti-poor magnetite. Similar trends have been observed in the Bk–Sg oxides from the bottom to the top of the massif (Duchesne, 1972a), which justifies the use of trace elements in the oxides as a semi-quantitative differentiation index.

REE in Apatites

The results for apatites in the ore-bodies are included in Table 1 and plotted on Fig. 2. If Nd is excepted for the same reasons as those already developed, the REE* show very regular patterns with nearly constant La/Yb ratios. The similarities between Hesnes and Eigeröy as well as between Kaknuden and Rödemyr II are striking. In the latter case it considerably strengthens the hypothesis of a common origin of the two occurrences. The Eu anomaly increases with the REE content. This evolution is similar to what can be anticipated by means of the differentiation index based on the oxides (Table 5) and also to the Bk–Sg trend. The

TABLE 5. TRACE-ELEMENT CONTENTS (ppm) IN Fe–Ti OXIDE MINERALS†

	Kydlandsvatn		Rödemyr II	Kaknuden		Eigeröy		Hesnes	
	Ilm	Mt	Mt	Ilm	Mt	Ilm	Mt	Ilm	Mt
V	2500	6800	6800	400	7500	100	2600	130	2200
Cr	280	4400	700	—	500	—	2000	—	70
Zn	—	320	780	—	1400	—	3000	—	1400
Ni	270	550	200	—	250	—	90	—	60
Co	150	150	130	80	140	40	90	40	60
Hem or Usp‡	21.0	3.2		3.4	13.1	2.3	41.5	5.1	22.2
T°C§	500–550°			600°		670°		680°	
f_{O_2} (atm)§	?			10^{-21}		10^{-20}		10^{-18}	

† Data from Duchesne (1973) except Rödemyr II.

‡ Hem and Usp (mol. %) correspond to the hematite content of the ilmenite and ulvöspinel content of unoxidized magnetite.

§ T° and f_{O_2}, calculated according to the Buddington and Lindsley geothermo-barometer, are the conditions of deuteric equilibrium.

— Not detected, i.e. less than 80 ppm Zn, 70 ppm Cr, 30 ppm Ni.

*Except for Rödemyr II which is situated within the Håland massif and thus distinct by its geological situation and mineral paragenesis from the Rödemyr Mine mentioned by these authors.

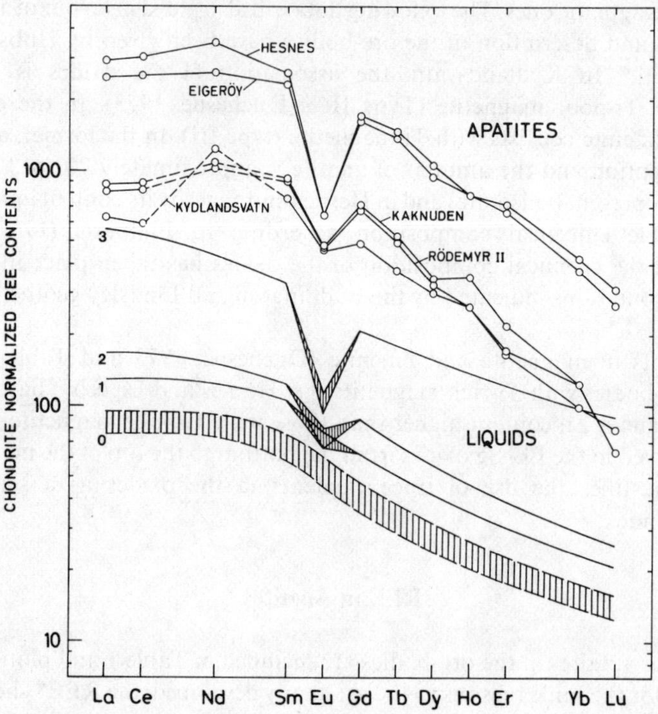

FIG. 2. Chondrite-normalized REE patterns in apatites from Fe–Ti oxide ore-bodies and in magmatic liquids 1, 2, 3, corresponding to Kydlansvatn, Kaknuden–Rödemyr II and Hesnes–Eigeröy respectively. Two extreme values are indicated for Eu in the liquids assuming D_{Eu}^{apat} equal to 6.95 or 5.80 (see Table 4). The same D_{Eu}^{apat} and corresponding D_{Eu}^{plag} values (Table 4) lead to a range of possible parental liquids (0) (vertically hatched).

consistency between trace element behaviour in the oxides and the REE in apatites thus suggests that fractional crystallization, in which plagioclase plays an important role, is a possible mechanism for the generation of the ore-bodies. Segregation of Fe, Ti-rich liquids (filter-press) has taken place at successive stages of the magmatic differentiation of the parental magma of the anorthosite series. This is consistent with the hypothesis already put forward by Duchesne (1973).

The REE content of the liquids in equilibrium with apatites of Kydlandsvatn, Kaknuden–Rödemyr II and Eigeröy–Hesnes (Fig. 2) has been calculated using D_{REE^*} values of Table 3. Small negative Eu anomalies appear in the Kydlandsvatn liquid if D_{Eu}^{apat} values within the range 6.95 and 5.80, taken from Table 4, are used. They can be explained using a similar equation to (4) and the proper D_{Eu}^{plag} values, by crystallization from the initial stage to F values between 0.58 and 0.74 of a hypothetical plagioclase-rich cumulate ($\mu_{plag} = 0.80$) now represented by the anorthositic massifs. The extreme values for the REE in the initial liquid are plotted on Fig. 2. F values for the Hesnes–Eigeröy liquid vary between 0.18 and 0.24 and for Kaknuden–Rödemyr II, between 0.49 and 0.62.

During this first interval of crystallization, a hemo-ilmenite-rich liquid could be segregated and formed ore-bodies of Type I (hemo-ilmenite without magnetite) (Duchesne, 1973). It is to be recalled that occurrence of hemo-ilmenite as the only oxide is also characteristic of the beginning stage of the Bk–Sg crystallization.

Acknowledgements

This paper has greatly benefited from an exhaustive review by D. M. Shaw. I. R. wishes to thank K. S. Heier, J. Dons and A. O. Brunfelt for providing laboratory facilities and helpful guidance during a period of work at the Mineralogical Geological Museum, University of Oslo. This research has been partly supported by the Belgian F.N.R.S.

References

ANDERSON, A. T. and GREENLAND, L. P. (1969) Phosphorous fractionation diagram as a quantitative indicator of crystallization differentiation of basaltic liquid. *Geochim. Cosmochim. Acta,* **33,** 493–505.

BRUNFELT, A. O. and ROELANDTS, I. (1974) Determination of rare earths and thorium in apatites by thermal and epithermal neutron-activation analysis. *Talanta,* **21,** 513–521.

DRAKE, M. J. (1975) The oxidation state of europium as an indicator of oxygen fugacity. *Geochim. Cosmochim. Acta,* **39,** 55–64.

DUCHESNE, J. C. (1971) Le rapport Sr/Ca dans les plagioclases du massif de Bjerkrem–Sogndal (Norvège méridionale) et son évolution dans la cristallisation fractionnée du magma plagioclasique. *Chem. Geol.* **8,** 123–130.

DUCHESNE, J. C. (1972a) Iron–titanium oxide minerals in the Bjerkrem–Sogndal massif, Southwestern Norway. *J. Petrology,* **13,** 57–81.

DUCHESNE, J. C. (1972b) Pyroxènes et olivines dans le massif de Bjerkrem–Sogndal (Norvège méridionale). Contribution à l'étude de la série anorthosite–mangérite. *24th Int. Geol. Congress, Montreal,* sect 2, 320–328.

DUCHESNE, J. C. (1973) Les gisements d'oxydes de fer et titane dans les roches anorthositiques du Rogaland (Norvège méridionale). In: *Les Roches plutoniques dans leurs rapports avec les gîtes minéraux*, Coll. E. Raguin. Masson, Paris, 403 pp.

DUCHESNE, J. C. (1977) Quantitative modeling of Sr, Ca, Rb and K in the Bjerkrem–Sogndal layered lopolith (S.W. Norway). *Contr. Miner. Petrol.* **66,** 175–184.

DUCHESNE, J. C., ROELANDTS, I., DEMAIFFE, D., HERTOGEN, J., GIJBELS, R. amd DE WINTER, J. (1974) Rare-earth data on monzonoritic rocks related to anorthosites and their bearing on the nature of the parental magma of the anorthositic suite. *Earth Planet. Sci. Lett.* **24,** 325–335.

HUBAUX, A. (1960) Les gisements de fer titané de la région d'Egersund, Norvège. *N. Jb. Miner., Festband Ramdohr,* **94,** 926–992.

MICHOT, P. (1955) Anorthosites et anorthosites. *Acad. Roy. Belgique, Bull. Cl. Sci.*, série 5, **41,** 275–294.

MICHOT, P. (1956) Les gisements de minerais noirs de la région d'Egersund *Ann. Soc. Géol. Belgique,* **79,** 183–202.

MICHOT, P. (1960) La géologie de la catazone: Le problème des anorthosites, la palingenèse basique et la tectonique catazonale dans le Rogaland méridional (Norvège méridionale). *Norges Geol. Unders.* **212g,** 1–54.

MICHOT, P. (1965) Le magma plagioclasique. *Geol. Runds.* **54,** 956–976.

MORRIS, R. V., HASKIN, L. A., BIGGAR, G. M. and O'HARA, M. J. (1974) Measurement of the effects of temperature and partial pressure of oxygen on the oxidation states of europium in silicate glasses. *Geochim. Cosmochim. Acta,* **38,** 1447–1459.

NAGASAWA, H. (1970) Rare-earth concentrations in zircons and apatites and their host dacites and granites. *Earth Planet. Sci. Lett.* **9,** 359–364.

NEUMANN, H., MEAD, J. and VITALIANO, C. J. (1954) Trace element variation during fractional crystallization as calculated from the distribution law. *Geochim. Cosmochim. Acta,* **6,** 90–99.

OSBORN, E. F. (1959) Role of oxygen pressure in the crystallization and differentiation of basaltic magma. *Am. J. Sci.* **257,** 609–647.

PASTER, T. P., SCHAUWECKER, D. S. and HASKIN, L. A. (1974) The behaviour of some trace elements during solidification of the Skaergaard layered series. *Geochim. Cosmochim. Acta,* **38,** 1549–1577.

PHILPOTTS, A. R. (1967) Origin of certain iron-titanium oxide and apatite rocks. *Econ. Geol.* **62,** 303–315.

PHILPOTTS, J. A. (1970) Redox estimation from a calculation of Eu^{2+} and Eu^{3+} concentrations in natural phases. *Earth Planet. Sci. Lett.* **9,** 257–268.

PHILPOTTS, J. A. and SCHNETZLER, C. C. (1970) Phenocryst–matrix partition coefficients for K, Rb, Sr and Ba, with applications to anorthosite and basalt genesis. *Geochim. Cosmochim. Acta,* **34,** 307–322.

PHILPOTTS, J. A. and SCHNETZLER, C. C. (1972) Large trace cation partitioning in igneous processes. *24th Int. Geol. Congress, Montreal,* sect. 10, 51–59.

PUCHELT, H. and EMMERMAN, R. (1976) Bearing of rare earth patterns of apatites from igneous and metamorphic rocks. *Earth Planet. Sci. Lett.* **31,** 279–286.

ROELANDTS, I. (1975) Contribution au dosage par activation neutronique des terres rares et d'autres éléments en trace dans les roches magmatiques. Doctorat thesis, University of Liège (unpublished).

SUN, SHEN-SU and HANSON, G. N. (1976) Rare earth element evidence for differentiation of McMurdo Volcanics, Ross Island, Antarctica. *Contr. Miner. Petrol.* **54,** 139–155.

TREUIL, M. and JORON, J. L. (1975) Utilisation des éléments hygro magmatophiles pour la simplification de la modélisation quantitative des processus magmatiques. Exemples de l'Afar et de la dorsale médioatlantique. *Soc. Ital. Mineral. e Petrol., Milano,* **31,** (1), 125–174.

TREUIL, M. and VARET, J. (1973) Critères volcanologiques, pétrologiques et géochimiques de la genèse et de la différenciation des magmas basaltiques: exemple de l'Afar. *Bull. Soc. Géol. France,* 7e s., **15**(5–6), 506–540.

WAGER, L. R. and BROWN, G. M. (1968) *Layered Igneous Rocks.* Oliver and Boyd Ltd., London, 588 pp.

WEILL, D. F. and DRAKE, M. J. (1973) Europium anomaly in plagioclase feldspar: experimental results and semi quantitative model. *Science,* **180,** 1059–1060.

ZIELINSKI, R. and FREY, F. (1970) Gough Island: Evaluation of a fractional crystallization model. *Contr. Miner. Petrol.* **29,** 242–254.

Progress in the Knowledge of Indicator Elements

E. SCHROLL

Geotechnical Institute, BVFA Arsenal, Vienna, Austria

Abstract

The development of the search for indicator elements is based on the statistical intercorrelation of chemical data, depending on a physicochemical base. The correlation between chemical and isotopic data in form of "indicator parameters" may be successful in the case of genetically complex geological bodies.

1. Introduction

Since V. M. Goldschmidt (1942) created the idea of "guide elements" (Leitelemente), which should help to find out the genetic relations between different rocks and the origin of minerals, rocks and mineral deposits, there was a constant discussion about the use of indicator elements. In the beginning the idea of guide elements appeared to be the "philosopher's stone" of the geochemist followed by critical and sceptical valuations (Hegemann, 1943; Hegemann and Albrecht, 1954; Haberlandt, 1952, Haberlandt and Schroll, 1954; Jedwab, 1953; Shaw, 1964). However, the application was not always successful.

"General statements" can not be derived from the indicator elements alone. They may differ both in quality and quantity in certain geological units, petrogenetic provinces or ore deposits (Rösler and Lange, 1972, 1975). Mercer (1976) does not mention indicator elements of metal deposits of sedimentary rocks, except that the Co/Ni ratio in pyrites may be indicative of a hydrothermal environment (Hegemann, 1943). The identification of indicator elements, which should give significant geological information has failed in many cases, because the generalizations that have taken place were based on insufficient local knowledge.

At the same time the term "indicator element" (often called "pathfinder element" too) is used in exploration geochemistry in another sense for such elements which constitute halos around mineral deposits. As the practical results of geochemical prospecting shows that use of such "indicator elements" is valid. However, the ore is found mostly by indications caused by secondary processes of weathering. The problem of indicator elements related to primary processes is connected much more with the original idea of "guide elements".

The modern knowledge about indicator elements is based on more accurate chemical analysis, statistical evaluation methods, new possibilities of the crystallochemical localization of trace concentrations and the better known physicochemical background of element distribution in mineral associations.

2. Methods in "Indicator Element" Research

The research for indicator elements proposes the application of analytical methods, which comply with the following requirements for the analytical method (Hahn-Weinheimer and Johanning 1963):

(a) sensitive enough for the determination of an element in all samples taken in consideration,
(b) good reproducibility which is smaller than the differences of concentration between the investigated samples, and
(c) sufficiently accurate to permit comparison of the results.

Initially the use of one element as a guide was considered (*one-element method*). Theoretically such a method may be effective only for minerals, and must be questioned in the case of a two-or-multiphase system like rock samples, especially if the amount of the host mineral shows significant dispersion. Therefore the *two-element method* is preferred. Moreover, the character of the indicator elements can be assessed by statistical methods, such as regression and correlation analysis. Three elements can be selected in a similar practice like the application of the "two-element method", using the ratio of an element pair and the concentration of a third element (such as K/Rb–Ti or Ca/Sr–Mg) or two ratios formed by three components (such as Tb/La–Tb/Ca, Möller *et al.*, 1975) or a triangular projection (such as Ti–Zr–Y or Ti–Zr–Sr, Pearce and Cann, 1973). The graphical representation for four components are analogous. More elements can be placed in form of line diagrams. Special displays can be made for lanthanides, such as the group of coherent elements. The *multi-element methods* need the statistical evaluation of n-dimensional systems, such as multi-variate techniques, discriminatory analysis, cluster analysis, main component analysis or factor analysis.

The indicator element can be selected mostly according to the crystalchemical relationship, such as K/Rb, Ca/Sr, Zr/Hf, Cl/Br ("crystallochemical method"). The experience has shown that generally such element ratios are useful and reliable parameters (see Schroll, 1965). Main constituents of host minerals may be used also as a kind of internal standard, such as the potassium of the clay component of carbonate rocks, showing no diadochy (K/B, K/Zn, etc.).

The multi-element analysis enables the statistical selection of indicator elements, which are typical for mineral association ("Typochemical method"†). The more chemical components can be determined the better. The multi-element analysis can be emphasized as the best method for the general geochemical characterization of geological materials.

One of the most remarkable results is the classification of diamonds of alluvial deposits with respect to their source using analytical values of thirty-nine elements (Watterson *et al.*, 1976).

However, the empirical search for indicator elements can be concluded successfully only if the geochemical distribution of these elements is led back to a physicochemical base.

The application of partition coefficients (distribution coefficients) for the investigation of the distribution of elements using the model of two phases, such as melt-crystal, solution-crystal or between coexisting minerals brought great progress (McIntyre, 1963, see Hawskins and other publications, *ibid.*). Although the geological processes are mostly complex and

† "Typochemical" in the sense of Jedwab (1953).

their physicochemical conditions are not ideal, it is possible to get indications about equilibrium states, concentrations in initial melts or ore solutions, temperatures, pH–Eh conditions, etc.

The simplest geochemical system is the relation between a solution being nearly constant in its composition like seawater, which forms salt deposits during evaporation or reacts with sediments over a limited range of temperatures. Therefore it is clear that the use of indicator elements can be successful for the indication of paleosalinity or the chemofacies of marine sediments (reviews, Ernst, 1970 and Schroll, 1976). The use of boron as an indicator element can be explained experimentally and theoretically by the sorption processes of clay minerals.

The application of petrological systems is in development. The modern theories about the composition of the upper mantle and about the origin of the magmatic rocks would be impossible without the knowledge of partition coefficients.

However, the genesis of batholiths of hydrothermal ore deposits is often very difficult to investigate. The diversity and the complexity of the processes cause interfacies and render the selection of indicator elements very difficult. The solution can be found only by comprehensive investigations of all geochemical parameters of a geological body, including the isotopic geochemistry (such as of H, C, O, S, Sr and Pb).

3. Indicator Parameter

The use of "petrochemical parameters" is well known. Extensive geochemical data received for the characterization of geological bodies need a condensation in form of significant "*indicator parameters*". The use of both chemical concentrations and isotopic ratios—such as the relation between sulphur isotope ratio and trace element concentration in sulphide minerals—justifies also the new term.

Complex geological processes must be broken down into suitable factors which can be characterized by indicator parameters, such as for the reconstruction of the primary source material (source rock), for the transport medium, for the supply from the wall rocks, in which the geological body is formed, for the conditions of the geochemical barriers (physicochemical and biochemical parameters), which have caused the primary mineralization, and for changes by secondary processes†.

Conclusion

As a conclusion it can be pointed out that the use of indicator elements and indicator parameters, following the original idea of guide elements, gives us a real qualitative and quantitative understanding of geochemical processes forming rocks and mineral deposits. There is no doubt that this is one of the keys for improving the knowledge of the mineral resources of our world. As a first step geochemical data should be improved for such global comparisons.

†As example a study with the UNESCO-IUGS-Correlation Project No. 7316 (Diagnostic features of metallization occurrences in limestones and dolomites) about the correlation between chemical and isotopic composition of lead–zinc metallizations in the Eastern Alps.

References

DEGENS, E. T. (1968) *Geochemistry of Sediments, a Brief Survey,* Prentice Hall, Englewood Cliffs, 282 p.

ERNST, W. (1970) *Geochemical Facies Analysis, Methods in Geochemistry and Geophysics,* Elsevier, Amsterdam–London–New York, 152 p.

GOLDSCHMIDT, V. M. (1924) Geochemische Verteilungsgesetze der Elemente II. Videnskap, Skrifter, 1. math. naturv. Kl. No. 4, 36.

HABERLANDT, H. (1952) Die systematischen Grundlagen und die Fortschritte in der Erforschung der Spurenelemente als geochemische Leitelemente in Mineralien und Erzen. *N. Jb. Min. Abh.* **31** 76–78.

HABERLANDT, H. and SCHROLL, E. (1954) Uber den Wert oder Unwert der Spurenelement-analyses für die Lagerstättenforschung, Minerogenese und Petrogenese. *Tschermarks. Miner. Petrogr. Mitt.* **5**, 110–125.

HAHN-WEINHEIMER, P. and JOHANNING, H. (1963) Geochemische Untersuchunge an magmatisch differenzierten Gesteinskörpern des Südschwarzwaldes l. *Contr. Mineral. Petr.* **9**, 175–190.

HEGEMANN, F. (1943) Die geochemische Bedeutung von Kobalt und Nickel im Pyrit. *Z. angew.* Min. **4**, 122–239.

HEGEMANN, F. and ALBRECHT, F. (1954) Zur Geochemie oxydischer Eisenerze. *Chemie d. Erde* **17** 81–103.

JEDWAB, J. (1953) Sur la définition des éléments typochimiques. *Bull. Soc. belge Géol. Paléont. Hydrol.* **62**, 173–179.

MCINTYRE, W. L. (1963) Trace element partition coefficients—review of theory and applications to geology. *Geochim. Cosmochim. Acta* **27**, 1209–1264.

MERCER, W. (1976) Minor elements in metal deposits in sedimentary rocks—a review of recent literature. In *Handbook of Strata-Bound and Stratiform Ore Deposits,* K. H. WOLF, ed., Vol. 2. Geochemical Studies, Elsevier, Amsterdam, 1–28.

PEARCE, J. A. and CANN, J. R. (1973) Tectonic setting of basic volcanic rocks determined using trace analysis. *Earth Planet. Sci. Lett.* **19**, 290–300.

RÖSLER, H. J. and LANGE, H. (1972) *Geochemical Tables,* Elsevier, Amsterdam–London–New York, 486 pp.

SCHROLL, E. (1968) Abundances of the chemical elements in the main rock types of the lithosphere in relation to a system of correlations. In *Origin and Distribution of Elements,* L. H. AHRENS, ed., Pergamon Press, Oxford, 599–618.

SCHROLL, E. (1975, 1976) *Analytische Geochimie.* Bd. 1. *Methodik,* 294 p., Bd. II *Grundlagen und Anwendungen,* Enke Stuttgart, 374 p.

SHAW, D. M. (1964) *Interprétation géochimique des éléments en traces dans les roches cristallines.* Masson, Paris, 237 p.

WATTERSON, J. I. W., RASMUSSEN, S. E., ERASMUS, C. S., SELLSCHOP, J. P. D. and HAWKINS, D. M. (1976) Multielement neutron activation analysis and the recognition of geochemical patterns for mineral exploration. *Extend. Abstracts Int. IUPAC-Symp. on Analytical Chemistry in Exploration, Mining and Processing of Materials.* Johannesburg, 214–224.

Microcomponents as Indicators of the Differentiation of Alkaline Magmatic Series

L. N. KOGARKO

V. I. Vernadsky Institute of Geochemistry and Analytical Chemistry, U.S.S.R. Academy of Sciences, Moscow, U.S.S.R.

Abstract

The distinction of two types of differentiation of alkaline magmatic series (1 related to the evolution of alkaline basaltic magma and 2 strongly undersaturated with silica, ultrabasic alkaline) is confirmed by the distribution of trace elements. Final products of both series are characterized by residual melts oversaturated in alkalis (agpaitic).

The principal difference in the course of differentiation of alkaline basaltic and ultrabasic alkaline series lies in the presence of plagioclase in the alkaline basaltic series, while there is melilite instead of plagioclase in the ultrabasic series.

The utilization of distribution and partition coefficients of Ca, Sr, Ba, REE shows that the ultimate products of the differentiation of alkaline basalts should be characterized by high Ca/Sr, Ca/Ba, La/Yb and Sm/Eu ratios.

Calculations based on the mass balance equation and on the distribution coefficients of trace elements permit the estimation of the composition of residual liquids, which appear in the course of differentiation of alkaline basaltic series. These are oversaturated in alkalis. The distribution of trace elements in the series fits well to the model of fractional crystallization differentiation.

Geological, geochemical and petrological studies (Kogarko, 1977) demonstrate that agpaitic nepheline syenites are formed at the last stages of differentiation of strongly silica-undersaturated magmas of mantle origin. Their composition may vary from nepheline-normative olivine alkaline basalts to melilite olivine nephelinites.

The agpaitic character of alkaline basaltic magmas arises as a result of the simultaneous action of plagioclase and pyroxene. On the contrary, the evolution of strongly silica-undersaturated feldspar-free series is characterized by the operation of a pyroxene effect alone. The differences between the residual liquids of these two series should be strongly marked by elements incorporated into plagioclase and pyroxene: Sr and Ba belong among such components.

The distribution coefficients of strontium and calcium ($D_{Sr/Ca}$)† for monoclinic pyroxene-magma and feldspar-magma are quite different:

$$D^{Pl}_{Sr/Ca} = 1.6 \quad D^{FS}_{Sr/Ca} = 5 \quad \text{(for anorthoclase)},$$

$$D^{Cpx}_{Sr/Ca} = 0.04$$

† $D_{Sr/Ca} = \frac{C^X_{Sr} C^L_{Ca}}{C^L_{Sr} C^X_{Ca}}$ where C^X is concentration of the element in crystals and C^L is concentration in silicate melt.

according to the averaged data by Berlin and Henderson (1968). Strontium will be more strongly enriched in the residual liquid of the series in which monoclinic pyroxene alone is crystallizing (melilite basalt→phonolite series) than in the case of simultaneous crystallization of clinopyroxene and plagioclase (alkali olivine basalt→phonolite series).

In fact, Sr/Ca ratios in rocks of the Ilimaussag massif (Sr/Ca = 0.006) which are related to the differentiation of alkaline basalt magmas are one order of magnitude lower than in rocks from the Lovozero (Sr/Ca = 0.07) and Khibina (Sr/Ca = 0.08) massifs, which are related to differentiation of melilite basalt melt. We may consider these relationships in a more quantitative aspect.

The change in Ca/Sr ratio in the melt during the fractional crystallization may be calculated by the following equation (Ryabchikov, 1965):

$$(\mathrm{Sr/Ca})^L = \mathrm{Sr}^\circ/\mathrm{Ca}^\circ\left(\frac{m^L_{\mathrm{Ca}}}{m^T_{\mathrm{Ca}}}\right) D^{\mathrm{Comb}}_{\mathrm{Sr/Ca}} - 1 \tag{1}$$

where Ca° and Sr° are the concentrations in the initial magma; m^L_{Ca} is the mass of calcium in the residual liquid at a given moment; m^T_{Ca} is the total mass of calcium in the system under consideration:

$$D^{\mathrm{Comb}}_{\mathrm{Sr/Ca}} = \Gamma^{\mathrm{Pl}}_{\mathrm{Ca}} D^{\mathrm{Pl}}_{\mathrm{Sr/Ca}} + \Gamma^{\mathrm{Cpx}}_{\mathrm{Ca}} D^{\mathrm{Cpx}}_{\mathrm{Sr/Ca}} + \Gamma^{\mathrm{Fs}}_{\mathrm{Ca}} D^{\mathrm{Fs}}_{\mathrm{Sr/Ca}}$$

is the bulk distribution coefficient, where $\Gamma^{\mathrm{Pl}}_{\mathrm{Ca}}$, $\Gamma^{\mathrm{Cpx}}_{\mathrm{Ca}}$ and $\Gamma^{\mathrm{Fs}}_{\mathrm{Ca}}$ are the proportions in which calcium is taken up during crystallization by plagioclase, clinopyroxene and anorthoclase respectively

$$(\Gamma^{\mathrm{Pl}}_{\mathrm{Ca}} + \Gamma^{\mathrm{Cpx}}_{\mathrm{Ca}} + \Gamma^{\mathrm{Fs}}_{\mathrm{Ca}} = 1).$$

TABLE 1. THE COMPOSITIONS OF ROCKS USED FOR THE CALCULATIONS OF VARIATIONS IN Ca/Sr RATIOS DURING THE FORMATION OF AGPAITIC NEPHELINE SYENITES

Oxides	Alkali olivine basalt, Hualalai, Hawaii	Melilite–olivine nephelinite, Honolulu, Hawaii	Mean composition of the Lovozero alkaline massif
SiO_2	46.53	35.83	53.84
Al_2O_3	14.31	10.86	17.05
Fe_2O_3	3.16	6.43	6.42
FeO	9.81	10.17	1.54
MnO	0.18	0.23	0.30
MgO	9.54	11.09	1.00
CaO	10.32	12.21	1.57
Na_2O	2.85	5.45	9.63
K_2O	0.84	1.82	5.16
H_2O^-	—	0.27	—
H_2O^+	0.08	0.72	—
TiO_2	2.28	2.64	0.91
CO_2	—	—	—
P_2O_5	0.28	1.38	—
Total	100.24	99.78	100
Reference	Yoder and Tilley, 1962	Tilley *et al.*, 1965	Gerasimovsky *et al.*, 1966

Analyses of alkaline olivine basalt from Hualalai, Hawaii (Yoder and Tilley, 1962), and melilite olivine nepheline from Honolulu, Hawaii (Tilley, Yoder and Schairer, 1965), were used for the composition of primary magmas. The average composition of the Lovozero massif (similar to melanocratic foyaite) was used for the composition of the agpaitic phonolite eutectic (Gerasimovsky *et al.*, 1966). The analyses of these rocks are given in Table 1.

The value m^L_{Ca}/m^T_{Ca} was calculated by the formula†

$$\frac{m^L_{Ca}}{m^T_{Ca}} = \frac{M^L}{M^T}\frac{C^{-L}_{Ca}}{C^{\circ}_{Ca}}$$

For this calculation, it is necessary to estimate the value m^L/m^T for the liquid having the composition corresponding to the nepheline syenite minimum. We can take for this estimation the ratio of concentrations of some microcomponent which is not incorporated by any crystallizing phases. For consideration of the processes of partial melting of mantle, and for early stages of the crystallization differentiation of basaltic magma, the contents of K, Ti and P are commonly used for computations of this kind (Anderson and Greenland, 1969). It is also possible to use information on the distribution of Zr in rocks. According to Brooks (1969) this element cannot be accommodated by the crystalline lattices of rock-forming minerals. In our case Ti and P are not suitable elements indicators, because the petrological investigation of layered basic intrusions show that these components are captured by cumulus titanomagnetite, ilmenite and apatite long before the melt composition reaches the nepheline syenite melting minimum. The ratio of K concentrations may not be used for the series alkaline basalt–trachyte–phonolite because this element is rather markedly fixed by solid phases at the stage of the formation of syenitic rocks. On the other hand, the data on K may yield realistic estimates for the series melilite basalt–nephelinite–phonolite, because substantial transfer of this element into solid phases commences only at the crystallization stage of the nepheline syenitic eutectic. In fact the ratio of K concentrations in alkaline olivine basalts to its mean content in the rocks of the Ilimaussag alkaline massif is close to 0.11, while the ratio of Zr concentrations in the same rocks approaches 0.05.

Taking into account the fact that a substantial part of the K was taken up by solid phases during the formation of syenitic rocks, we accepted for an estimate of the fraction of the remaining liquid at the beginning of the nepheline syenites crystallization stage the last value (0.05).

The ratios of K and Zr concentrations in melilite basalts and the contents of these elements in the rocks of the Khibina pluton are rather similar, 0.39 for K and 0.32 for Zr.‡ We accepted for the following calculations 0.05 for the series alkaline olivine basalt–phonolite and 0.35 for the series melilite basalt–phonolite. By using the compositions given in Table 1 we may find values m^L_{Ca}/m^T_{Ca} equal to 0.008 and 0.02, respectively.

During calculation of the proportions of separation of calcium-containing minerals from the melt the value

$$\frac{\Gamma^{Pl}_{Ca}+\Gamma^{Fs}_{Ca}}{\Gamma^{Cpx}_{Ca}}$$

was assumed to be equal to the ratio of the Ca contents in normative anorthite and diopside

† M^L is the total mass of liquid phase remaining at a given moment, M^T is the total mass of the system.

‡ For the Lovozero massif these ratios are 0.35 for potassium and 0.05 for zirconium, which is probably due to the substantial degree of differentiation of the magma in the feeding chamber already at the stage of the crystallization of nepheline–syenitic eutectics.

in the composition of primary magma, while the Γ_{Ca}^{Fs} value was estimated from the following formula derived from the consideration of mass balance in the system:

$$\Gamma_{Ca}^{Fs} = \frac{m_{Ca}^{Fs}}{m_{Ca}^{Fs} + m_{Ca}^{Pl} + m_{Ca}^{Cpx}} = \frac{C_{Ca}^{Fs}(C_K^0 - \frac{M^L}{M^T} \cdot C_K^L)}{C_K^{Fs}(C_{Ca}^{\circ} - \frac{M^L}{M^T} \cdot C_{Ca}^L)},$$

C_{Ca}^{Fs}/C_K^{Fs} being accepted to be equal to 1 as it follows from the data on anorthoclases from kenyites (Berlin, Henderson, 1969).

These calculations yielded for the series alkali olivine basalt–agpaitic phonolite the following values:

$\Gamma_{Ca}^{Cpx} = 0.50$; $\Gamma_{Ca}^{Pl} = 0.44$; $\Gamma_{Ca}^{Fs} = 0.06$ whence at

$D_{Ca}^{Cpx-L} = 0.04$; $D^{Pl-L} = 1.6$ and $D_{Sr/Ca}^{Fs-L} = 5$ (see above)

we obtain $D_{Sr/Ca}^{Comb} = 1.06$. For the series melilite basalt–agpaitic phonolite $\Gamma_{Ca}^{Cpx} = 1$ and therefore the bulk distribution coefficient is equal to $D_{Sr/Ca}$ for clinopyroxene.

$$D_{Sr/Ca}^{Comb} = D_{Sr/Ca}^{Cpx-L} = 0.04.$$

The value $C_{Sr}^{\circ}/C_{Ca}^{\circ} = 0.006$ which we also need for calculation with the above equation was inferred from the data on the contents of chemical elements in basic rocks (Vinogradov, 1956). It may be noted that the most melanocratic effusives from the East African rift which satisfy a number of criteria for primary magmas are characterized by Sr/Ca ratios close to this value (Gerasimovsky and Polyakov, 1974).

For the series alkali olivine basalt→agpaitic phonolite calculations carried out by the same method with $D_{Sr/Ca}^{Comb} = 0.96$; $m_{Ca}^L/m_{Ca}^T = 0.008$; $C_{Sr}^{\circ}/C_{Ca}^{\circ} = 0.0065$ yielded $C_{Sr}^L/C_{Ca}^L = 0.048$: for the series melilite olivine nephelinite→agpaitic phonolite the value is equal to $C_{Sr}^L/C_{Ca}^L = 0.33$ (with $D_{Sr/Ca}^{Comb} = 25$; $m_{Ca}^L/m_{Ca}^T = 0.02$; $C_{Sr}^{\circ}/C_{Ca}^{\circ} = 0.0065$). In the real rocks of these series the Sr/Ca ratio should decrease to some extent, in comparison with the calculated values, due to the crystallization of alkali feldspar at the stage of crystallization of nepheline syenitic melt.

Therefore, the calculated ratio of the final product of the olivine alkali basalt series is 10 to 100 times greater than that of the melilite olivine nephelinite series. This relation is corroborated by the difference in Sr/Ca ratios in agpaitic nepheline syenites of Kola peninsula and South Greenland. This is a rather convincing argument in favour of a relationship between the nepheline syenites in the Gardar alkaline province and olivine alkali basaltic magma, and between the rocks of the Lovozero and Khibina massifs and nephelinitic melts.

There are representatives both of the nephelinitic series and differentiates of the less silica-undersaturated olivine alkali basaltic magma among the effusives of the East African rift zone. Accordingly we find phonolites with both high Sr/Ca ratios (usually associated with the earlier nephelinites, e.g. phonolites from Ol-Donyo-Lengai—Sr/Ca = 0.062; Gerasimovsky and Polyakov, 1974) and with rather low Sr/Ca values (in these cases there are substantial masses of trachytes among the effusives, i.e. there are indications of a connection with less undersaturated primary magmas), e.g. in phonolites from Mount Suswa the Sr/Ca ratio is equal to 0.005 (Gerasimovsky and Polyakov, 1974). Alkaline acid effusives from the East African rift zone whose origin is related to the differentiation of alkali basaltic magma are invariably characterized by low Sr/Ca ratios, e.g. comendite near Lake Naishwa has Sr/Ca = 0.004, and pantellerite from the same locality has Sr/Ca = 0.006 (Gerasimovsky and

Polyakov, 1974). Therefore, Sr is an extremely sensitive indicator permitting one to distinguish the final products of the fractionation of primary magmas undersaturated in silica to different extents.

Distribution coefficients of Ba and Sr are close in both pyroxene and plagioclase (Philpotts and Schnezler, 1970; Gast, 1968; Berlin and Henderson, 1969) which should lead to a closeness of the ratio of these elements in the final products of differentiation of the alkali basalt–phonolitic and the nephelenite–phonolitic series. In fact, Sr/Ba ratios do not vary profoundly from one nepheline syenitic massif to another and also among various members of effusive series (Ilimaussag—Sr/Ba = 0.52; Lovozero–Sr/Ba = 0.87; Khibina—Sr/Ba = 0.90; melanophelinites of the East African rift—Sr/Ca = 1.30; phonolites—Sr/Ba = 0.96; pantellerites—Sr/Ba = 0.75—Gerasimovsky and Polyakov, 1974). On the other hand, the Ca/Ba ratio as well as the Sr/Ca ratio may be used as an indicator of the association of nepheline syenites with comagmatic series undersaturated in silica to a varying degree (e.g. Ilimaussag—Ca/Ba = 372; Lovozero—Ca/Ba = 12; Khibina—Ca/Ba = 11; Gerasimovsky, 1974).

The behaviour of rare-earth elements also differs during the crystallization of clinopyroxenes and plagioclases: distribution coefficients of elements of the Ce and Yt groups are less than unity for clinopyroxenes and more than unity for plagioclase (Gast, 1969; Schnetzler and Philpotts, 1970; Dudas, 1971). Because of this, the light REE should accumulate more strongly in residual products than the heavy REE, during the fractionation of a plagioclase-free series. The relation is less evident in the case of an alkali basalt magma from which significant amounts of plagioclases are crystallizing. In this connection it is interesting to note that the Lovozero rocks are more strongly enriched in light REE (La/Yb = 48, Gerasimovsky *et al.*, 1966) in comparison with Ilimaussag rocks (La/Yb = 25; Gerasimovsky, 1974).

Discussing the behaviour of the REE during the fractionation of alkaline magma it is worth while to consider in more detail the variation of Eu concentrations. This element, in contrast to the other REE, exists in magmatic processes in both the trivalent and the divalent state. Eu^{2+}, being very similar in size to Sr, is rather intensely transferred into the solid phase during the crystallization of plagioclase. Because of this it may be supposed that the final products of the differentiation of alkali basaltic magmas (crystallizing plagioclase) will be characterized by a stronger Eu anomaly than the residual melts arising from the cooling of melilite nephelinitic magmas (which separated no plagioclase). This conclusion is consistent with the assumption of the association of Ilimaussag rocks with the fractionation of alkali basaltic magma (Eu/Sm = 0.13—Gerasimovsky, 1974) and the affiliation of Lovozero agpaitic rocks with the differentiation of more strongly silica undersaturated alkaline-ultrabasic melts; Eu/Sm = 0.25 for differentiated complex of the Lovozero pluton (Gerasimovsky *et al.*, 1966).

The distribution of the REE in the effusives of the East African rift zone reveals similar trends: in phonolites from Ol-Donyo-Lengai (most probably related to the differentiation of melanophelinitic magmas) La/Yb = 55; Eu/Yb = 0.18, while for pantellerite near Lake Naiwasha (a correlation with the differentiation of alkali basaltic magma is assumed) La/Yb = 26; Eu/Sm = 0.06.

Therefore, the geochemistry of Sr, Ba and REE in alkaline rocks confirm the conclusion that agpaitic magmas may arise at the last stage of differentiation of an extensive spectrum of primary magmas, ranging from nepheline-normative alkali basalts to melilite nephelinites. Agpaitic nepheline syenites from Greenland belong, according to the observed trace-element contents to the first of the above groups, while agpaitic rocks from the Kola Peninsula belong to the second group.

References

ANDERSON, A. T. and GREENLAND, L. P. (1969) Phosphorus fractionation diagram as a quantitative indicator of crystallization differentiation of basaltic liquids. *Geochim. Cosmochim. Acta* **33**, 493–505.

BERLIN, R. and HENDERSON, C. (1968) A reinterpretation of Sr and Ca fractionation trends in plagioclases from basic rocks. *Earth. Planet. Sci. Lett.* **4**, 79–83.

BERLIN, R. and HENDERSON, C. (1969) The distribution of Sr and Ba between the alkali feldspar, plagioclase and groundmass phases of porphyritic trachytes and phonolites. *Geochim. Cosmochim. Acta* **33**, 247–255.

BROOKS, C. K. (1969) On the distribution of zirconium and hafnium in the skaergaard intrusion, East Greenland. *Geochim. Cosmochim. Acta* **33**, 357–374.

GAST, P. (1968) Trace element fractionation and the origin of tholeiitic and alkaline magma types. *Geochim. Cosmochim. Acta* **32**, 1057–1086.

GERASIMOVSKY, V. I., VOLKOV, V. P., KOGARKO, L. N., POLYAKOV, A. I., SAPRYKINA, T. V. and BALASHOV, YU. A. (1966) *Geochemistry of the Lovozero Alkaline Massif*, Nauka, Moscow.

GERASIMOVSKY, V. I. (1974) In *Alkaline Rocks*, Sorensen, ed., 402–411.

GERASIMOVSKY, V. I. and POLYAKOV, A. I. (1974) *Geochemistry of Volcanic Rocks from Rift Zones of East Africa. East African Rift System*, Nauka, Moscow, 193 pp.

KOGARKO, L. N. (1977) *Problems of the Genesis of Agpaitic Magmas*, Nauka, Moscow, 294 pp.

PHILPOTTS, J. A. and SCHNETZLER, C. C. (1970) Phenocryst-matrix partition coefficients for K, Rb, Sr and Ba with applications to anorthosite and basalt genesis. *Geochim. Cosmochim. Acta* **34**, 307–322.

TILLEY, C. E., YODER, H. S. and SCHAIRER, J. F. (1965) Melting relations of volcanic tholeiite and alkali rock series. *Carn. Inst. Wash.* no. 64, 69–82.

VINOGRADOV, A. P. (1956) Average contents of chemical elements in the major types of igneous rocks of earth's crust. *Geokhimiya* no 1.

YODER, H. S. and TILLEY, C. E. (1962) Origin of basalt magmas: an experimental study of natural and synthetic rock systems. *J. Petrol.* **3**, 342–532.

Strontium Isotope Geochemistry in Syenite–Alkaline Granite Complexes

PHILIPPE VIDAL,† LAURE DOSSO,† PETER BOWDEN‡ and JEAN LAMEYRE§

† Centre Armoricain d'Etude Structurale des Socles (C.N.R.S.), Université de Rennes, France.
‡ Department of Geology, University of St. Andrews, Scotland.
§ Laboratoire de Pétrologie, E.R.A. 381 (C.N.R.S.), Université de Paris, VI, France.

Abstract

Nigerian younger granites and Kerguelen Islands (Indian Ocean) Miocene granite–syenite ring complexes are compared. Both show a negative correlation between K and Rb. This trend, which follows Shaw's hydrothermal–pegmatitic trend, indicates the importance of late magmatic and post-magmatic processes, particularly in a continental setting. Initial $^{87}Sr/^{86}Sr$ ratios which vary widely in Nigeria (up to 0.752) remain constant in Kerguelen (0.7057) are consistent with the idea that syenite–granite magmas may originate from the mantle. Although it is tempting to attribute the isotopic variations in Nigeria to some form of interaction of the mantle magma with the continental crust, an increase of the Sr isotopic ratio in the magma itself is proposed here to account for the continental $^{87}Sr/^{86}Sr$ ratio discrepancies.

1. Introduction

Initial $^{87}Sr/^{86}Sr$ ratios have been commonly used to constrain petrogenetic models. They allow mantle-derived rocks, with very low ratios, to be distinguished from rocks which are the product of a crustal reworking, with high ratios (Faure and Hurley, 1963; Faure and Powell, 1972).

Alkaline and peralkaline magmas are sometimes thought to originate from the crust but more often from the upper mantle (Powell and Bell, 1974; Bowden, 1974). However, the role of continental crustal contamination in the magma genesis is still in debate. Contamination cannot be observed when $^{87}Sr/^{86}Sr$ initial ratios in these rocks are lower than about 0.705 and therefore consistent with a mantle origin. In other cases, high ratios (>0.710) are considered as due to a crustal origin, or to a large sialic assimilation by a mantle-derived magma. A special case is that of peralkaline plutons subjected to a post-emplacement thermal history: their high initial $^{87}Sr/^{86}Sr$ ratio may be related to internal strontium isotopic rehomogenizations (Delhal *et al.*, 1971).

We believe that the comparison of data obtained from similar complexes located in totally different geologic settings, such as those in Nigeria and in the Kerguelen Islands, Indian Ocean, may give some interesting insights on their genesis and a better understanding of syenite–alkaline granite complexes Sr isotope geochemistry. Both Nigerian and Kerguelen syenite–granite ring complexes have similar petrographic and chemical characters. In particular, they have the same K vs. Rb pattern and very similar trace elements contents, like high Rb content and extremely low content in Ba, Sr (in some cases, Sr content lower than 1 ppm). But initial $^{87}Sr/^{86}Sr$ ranges are quite different.

II. Geological Settings

(a) Nigeria

Over forty individual Mesozoic ring complexes, emplaced into a late Precambrian igneous and metamorphic "basement", have been recognized in Nigeria (Bowden and Turner, 1974). They constitute the southern part of a 200-km-wide zone of ring-complex development located adjacent to the 9th meridian extending from northern Niger to the margin of the Benue trough in Nigeria. According to Bowden *et al.* (1976) the occurrence of these ring-structures in this part of West Africa may be related to successive periods of mid-plate magmatism and mineralization from Ordovician times which progressively migrated southwards until the initial opening of the south Atlantic in the vicinity of the Gulf of Guinea.

Although the disposition of the ring-complexes may be controlled to some extent by the Pan-African structural trends in the "basement", the subvolcanic centres vary greatly in size and complexity. Typically in Nigeria they are circular in outline, 10–25 km in diameter with concentrically arranged ring-dyke intrusions of fayalite–hedenbergite granites and peralkaline granites containing acmitic pyroxene and riebeckitic amphibole. Hastingsite granites and biotite granites form a more abundant associated series of concentric intrusions, but some biotite granites may exhibit shallower dipping contacts associated with late intrusive stocks or cupolas. It is primarily within these biotite granites that ore deposits of columbite, cassiterite, sphalerite, etc., are located both in disseminated form and as veins associated with substantial wall-rock alteration.

The proportion of oversaturated syenite is curiously low although its volcanic equivalent has been recognized in a number of northern complexes. In fact there is ample evidence to indicate that the majority of the Nigerian ring-complexes represent the roots of volcanoes which erupted voluminous quantities of rhyolite as lava and ignimbrite, associated with minor but important occurrences of trachyte and basalt. Unfortunately the actual magmatic composition of the basaltic rocks preserved through caldera collapse is difficult to establish because of the degree of hydrothermal modification from the sub volcanic granite intrusions. It appears, therefore, that during Triassic to Jurassic times there was continued sequential magmatic activity in Nigeria as chains of volcanoes were fed by repeated pulses of dominantly syenite-related granitic magma. The actual life time of one volcanic centre is difficult to determine but with overlapping centres some renewal and extension of activity may have continued for several million years. Certainly the hydrothermal mineralization testifies to the continued process of evolution to lower temperatures.

(b) Kerguelen

Located in the southern Indian ocean, between longitudes 68°40′ and 70°30′E and latitudes 48°30′ and 49°45′S, the Kerguelen archipelago represents the emerged parts of the Kerguelen–Gaussberg aseismic plateau.

Some plutons are intruded into the Miocene volcanic and volcano-sedimentary formations which constitute a major part of the archipelago. The most important one is situated in the south-west, in the Rallier du Baty peninsula.

Recent work on the Rallier du Baty peninsula (Marot and Zimine, 1976) has revealed the existence of five ring-complexes. In each case, the arrangement of chilled margins and the nature of enclaves indicate that the sequence of instrusions is centripetal. Furthermore, the contact relationships between the centres show that the magmatic activity has migrated northwards with time. The ring-complex diameter decreases from south (oldest centre) towards north (youngest centre). Finally, the regular rhythmic alternation of basic cone sheets and acid ring dykes has been pointed out (Marot and Zimine, 1976).

Basalts of tholeiitic affinity have been recognized in the lowermost volcanic series (Watkins *et al.*, 1974) and alkaline affinities in the more recent ones. In the Rallier du Baty ring-complexes, one finds a series going from olivine gabbros (the oldest ones) to granitic rocks. This sequence is practically the same as in the alkali volcanic series of oceanic islands. However, syenitic rocks largely predominate and, furthermore, one can observe in the youngest rocks a clear trend towards biotite bearing aluminous compositions lacking pyroxenes and sodic amphiboles.

The southern centre is composed of syenites in which the first phases are represented by fayalite (Fa 88–90), augite and ilmenite. Perthites, which make up nearly the whole of these rocks, have crystallized at the same time as aegyrinic clinopyroxenes and titanomagnetites. Zircon, aenigmatite, magnesio-richterite evolving towards ferro-richterite, quartz, riebeckite-arfvedsonite and Fe-biotite occur towards the end of crystallization.

In granitic types, the early crystallization of perthite is accompanied by Fe–Ti oxides whereas zircon, aenigmatite, chevkinite, aegyrine, ferrorichterite, riebeckite-arfvedsonite constitute the last stages of magmatic crystallization. Abundant fluorite fills microlitic cavities and flakes of molybdenite occur in fissures. There is a trend towards transsolvus compositions and there are important manifestations of hydrothermal activity.

The absence of any nodules of continental character leads one to reject any hypothesis in which the Kerguelen–Gaussberg plateau could be a continental fragment (Zhivago, 1965; Wilson, 1965; Schlich, 1974; Watkins *et al.*, 1974). This latter hypothesis presented by Wilson (1963), Dietz (1966), Francis and Rait (1967) was supported for a while by Schlich and others (1971) and taken up by Nougier and Lameyre (1973), who emphasized the analogy existing between the Rallier du Baty pluton and continental anorogenic complexes.

III. K/Rb and Sr^{87}/Sr^{86} Data

(a) Nigeria

Van Breemen *et al.* (1975) and Bowden *et al.* (1976) have shown that the radiometric ages of Nigeria vary from late Triassic in the north to late Jurassic in the south. Over a time period of approximately 50 m.y. the Nigerian anorogenic granites have evolved as roots of chains of volcanoes younging progressively southwards towards the Benue. This systematic variation in age is probably related to a thermal anomaly in the mantle over which the African plate was quasi-statically positioned during Jurassic times.

It has been known for a long time that rubidium is greatly enriched in many Nigerian anorogenic granites. In contrast the amount of potash does not vary widely, so that the K/Rb ratio or a plot of K vs. Rb chiefly reflects the degree of rubidium enrichment in these rocks. However, reporting the data of Butler *et al.* (1962) and Bowden and Turner (1974) a negative correlation between K and Rb appears (Fig. 1). For all the Nigerian sub volcanic centers, the $^{87}Sr/^{86}Sr$ initial ratios vary widely from 0.705 to 0.752 (Fig. 2).

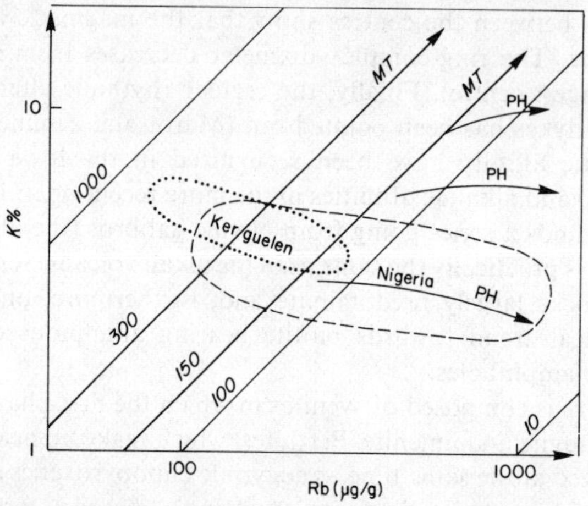

FIG. 1. K vs. Rb diagram for Nigerian and Kerguelen granite-syenite complexes. MT: Shaw's Main Trend. PH: Shaw's pegmatitic hydrothermal trend.

(b) Kerguelen

Three $^{87}Sr/^{86}Sr$ ratios on ring A (Fig. 4) of the Rallier du Baty Peninsula southern complex have been determined by C. Hedge (Watkins *et al.*, 1974). We recently engaged in a geochemical and isotopic study of the Rallier du Baty complexes (Dosso, 1977; Marot and Zimine, 1976; Lameyre *et al.*, 1976; Vidal *et al.*, in preparation).

The same abnormal trend as in Nigeria is found between K and Rb. This trend exists at the whole Peninsula scale (Fig. 1). It is found also inside a particular complex, and even in one single ring (Fig. 3). This leads to a strong increase of K/Rb as a function of K, from about 100 to 700.

Within one single ring-complex (southern), initial $^{87}Sr/^{86}Sr$ ratios derived from the whole-rock isochrons of each individual ring are almost identical (Fig. 4). This ratio is 0.7057.

IV. Discussion

Earlier workers such as Butler *et al.* (1962) attributed such Rb concentrations to extreme differentiation of a parent fayalite granite magma although the major element compositions do not vary widely amongst the granite types. More recently Bowden and van Breemen (1970) and Bowden and Turner (1974) have suggested that Nigerian granites containing high Rb (up to 1500 ppm) and low K/Rb ratios (down to 35) have undergone substantial recrystallization in the subsolidus during albitization. The trend of K/Rb ratios to low values in Nigerian granites is believed to represent a degree of post-magmatic readjustment in response to (auto) metasomatic fluids. It is perhaps significant that the Nigerian anorogenic granites but also Kerguelen ones, plot within the field of Shaw's pegmatitic–hydrothermal group (Fig. 1), thus emphasizing the importance of certain geochemical readjustments as a granite cools and consolidates. It is interesting to notice that this trend exists in other anorogenic complexes, e.g. in Corsica and Massachusetts (Vidal *et al.*, in press).

Isotopic ratios can be explained as follows:

(1) An origin in the continental crust could readily account for very high initial isotopic

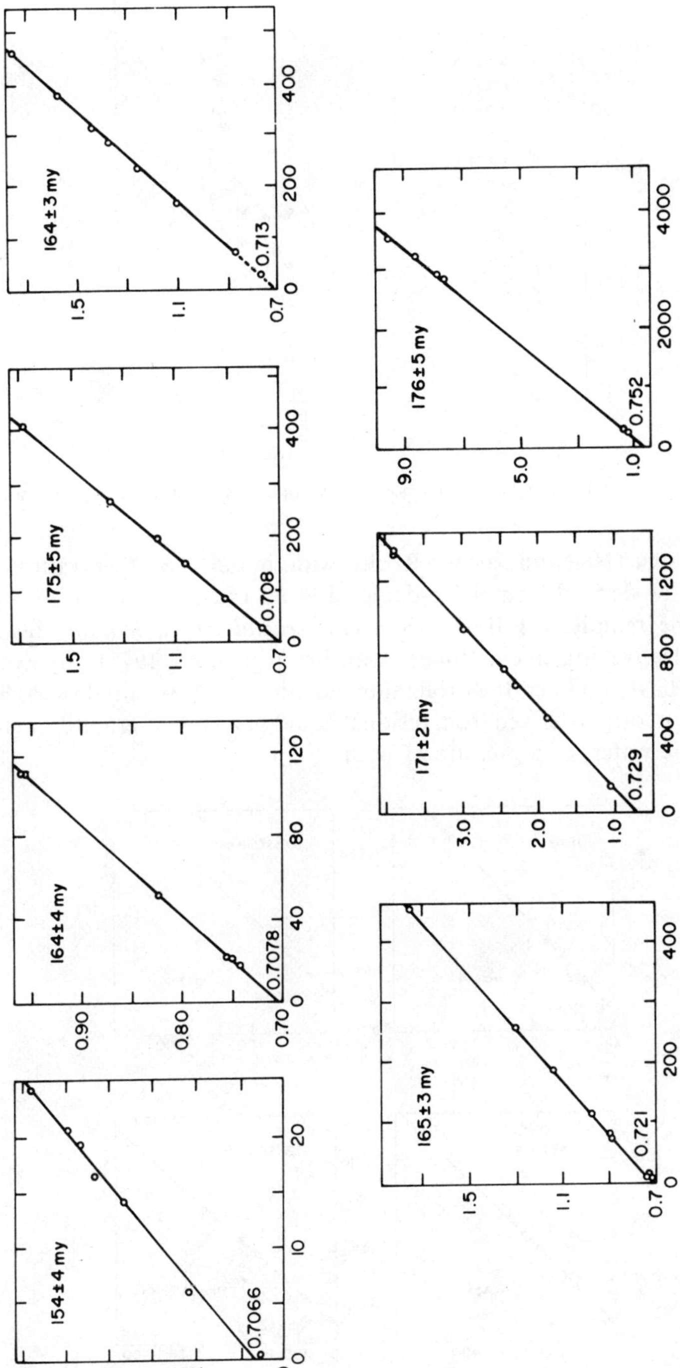

FIG. 2. $^{87}Sr/^{86}Sr$ vs. $^{87}Rb/^{86}Sr$ whole-rock diagrams for Nigerian Mesozoic granites (after van Breemen *et al.*, 1975).

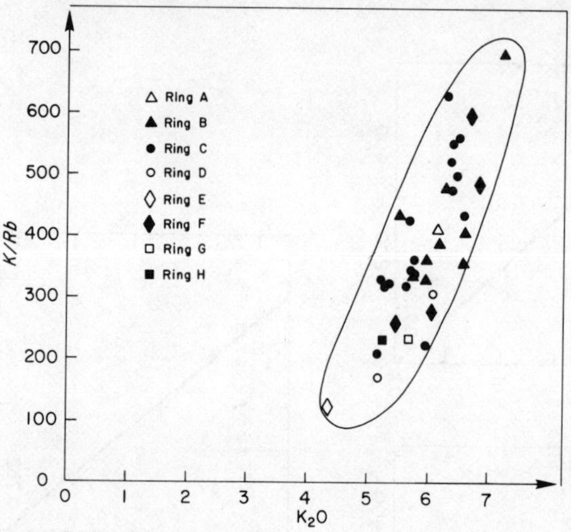

FIG. 3. K/Rb vs. K_2O diagram for Rallier du Baty Southern complex. A: outermost ring. H: innermost ring.

ratios found in some Nigerian rocks. Rocks with initial $^{87}Sr/^{86}Sr$ ratios as low as 0.705–0.706 could be derived from a Rb-depleted lower crust.

(2) Origin in the mantle. (a) *With continental crustal contamination*—by radiogenic strontium, with selective crustal enrichment (van Breemen *et al.*, 1975)—by hydrothermal fluids, which may have modified the original magmatic $^{87}Sr/^{86}Sr$ initial ratio. Such fluids could be magmatic in origin derived from residual liquid of a syenite-granite magma or from circulating meteoric water, or a mixture of both.

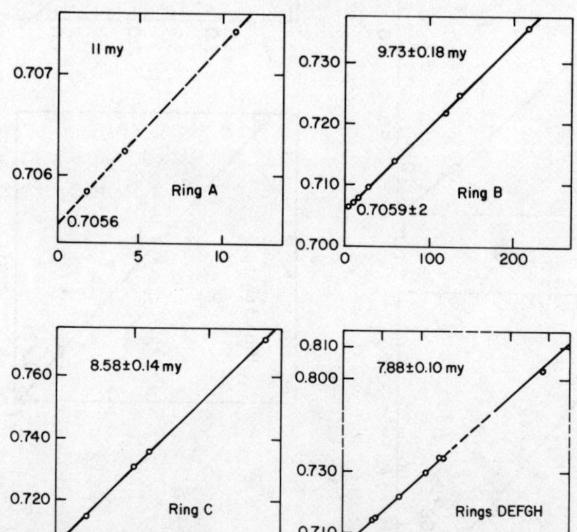

FIG. 4. $^{87}Sr/^{86}Sr$ vs. $87Rb/^{86}Sr$ whole-rock diagrams for Rallier du Baty Peninsula (Kerguelen Islands) Southern complex. Ring A: outermost ring. Ring H: innermost ring.

It is conceivable that these fluids could contain substantial ^{87}Rb as well as ^{87}Sr which may account for the rather curious correlation between the initial ^{87}Sr/^{86}Sr ratio and the ^{87}Rb/^{86}Sr which rise concomitantly in the Nigerian granites (Fig. 2). The highest initial ^{87}Sr/^{86}Sr ratio is found in the granites with the lowest K/Rb ratio together with an enriched assemblage of the "incompatible" trace elements. Significant mineralization shows that hydrothermal fluid–rock interaction has occurred in the Nigerian younger granite province.

In Nigeria, age relationships do not preclude the possibility that the already cooled rings were not submitted to metasomatic processes with Rb (and ^{87}Sr) introduction. But in Kerguelen, very strong constraints can be given for the aquisition of the chemical characters of the different rings, due to the existence of an individual isochron on each ring. The Rb introduction in hydrothermal conditions could modify the chemistry of only the cooling liquid, without modifying the former rings already solidified; otherwise, a strong scattering between two reference isochrons at 11 and 7.9 m.y. would be expected.

(b) *Without continental crust contamination.* In Kerguelen islands, nearly all of the basic volcanic rocks from the whole archipelago have the same initial ratios as the granitic and syenitic rocks (Dosso, 1977). This value of 0.7057 belongs to the most Sr radiogenic islands group, (Sun, Hanson, 1975) and is consistent with a petrogenesis involving mantle sources not depleted in trace elements.

The lowest values from Nigeria complexes may also be explained in the same way. For the high isotope ratios, one has to consider that the increase of ^{87}Sr/^{86}Sr ratios could take place in the magma itself before solidification: if the magma remains liquid for a rather long time, the very high Rb/Sr ratio (up to 1200 in Nigeria) can explain some very high initial ^{87}Sr/^{86}Sr ratios. For example, the ^{87}Sr/^{86}Sr of a liquid with a Rb/Sr ratio of 500 grows from 0.705 to 0.750 in about only 2 m.y.; we suggest that this process, which has been underestimated in Sr isotope geochemistry, could better explain the apparent correlation between the initial ^{87}Sr/^{86}Sr ratio and the mean Rb/Sr ratio observed in Nigerian complexes (Fig. 2) rather than a crustal origin or a strong crustal contamination.

In Kerguelen, the rather low and identical initial ratios can be explained by a short lifetime in the liquid state for the rocks corresponding to each ring. Constraints can be given to this time span for each liquid if its Rb/Sr ratio is known, in order to have an increase of the initial strontium isotopic ratio of less than 3×10^{-4}, because all isotopic initial ratios are different by less than this amount.

We do not have enough information on these Rb/Sr mean ratios, but with the available data it is possible to suggest that lifetimes in the liquid state fall between 200,000 years and less than one million years. Furthermore, age differences between the four rings imply that all the rings did not come from one single magma chamber having such a high Rb/Sr ratio.

In Nigeria the liquid state lifetime could be greater (some million years) in order to explain some very high ratios. In the case of Nigeria, we cannot exclude the possibility of crustal interaction in some form. This is shown by the common-lead data, which give model ages in some cases similar to the late Precambrian surrounding rocks (Tougarinov *et al.*, 1968).

In Kerguelen granites, syenites and basic plutonic and volcanic rocks have the same range of lead isotopic compositions: data for ^{207}Pb/^{204}Pb vs. ^{206}Pb/^{204}Pb show a linear correlation which give a 2.2 b.y. age. This trend is very similar to those found for oceanic island basalts; so contamination by an old sialic component does not seem to be involved (Dosso, 1977).

In conclusion, it appears that the peculiar major and trace element distribution in Nigerian and Kerguelen alkaline ring-complexes are isotopically consistent with an original mantle origin, but that some processes of modification have taken place to a larger extent in the continental setting. The exact nature of this process remains to be clearly identified.

Acknowledgements

We thank B. Bonin and B. M. Jahn for helpful discussions and comments.

References

BOWDEN, P. (1974) Oversaturated alkaline rocks: granites, pantellerites and comendites. In *The Alkaline Rocks*, H. SORENSEN, ed., Wiley Interscience, 109–123.

BOWDEN, P. and VAN BREEMEN, O. (1970) Isotopic and chemical studies on younger granites from Northern Nigeria. *Proc. Conf. African Geology*, 105–120.

BOWDEN, P., VAN BREEMEN, O., HUTCHINSON, J. and TURNER, D. C. (1976) Paleozoic and Mesozoic age trends for some ring complexes in Niger and Nigeria. *Nature* **259**, 297–299.

BOWDEN, P. and TURNER, D. C. (1974) Peralkaline and associated ring complexes in the Nigeria–Niger Province, West Africa. In *The Alkaline Rocks*, H. SORENSEN ed., Wiley-Interscience, 330–351.

BUTLER, J. R., BOWDEN, P. and SMITH A. Z. (1962) K/Rb ratios in the evolution of the younger granites of northern Nigeria. *Geochim. Cosmochim. Acta* **26**, 89–100.

DELHAL, J., LEDENT, D., PASTEELS, P. and VENIER, J. (1971) Etude du comportement isotopique des systèmes Rb/Sr et U/Pb dans le granite hyperalcalin de Noqui (République Démocratique du Congo et Angola). *Ann. Soc. Géol. Belge* **94**, 223–236.

DIETZ, R. (1966) Passive continents, spreading sea floors and collapsing continental rises. *Amer. J. Sci.* **264**, 3, 177–193

DOSSO, L. (1977) Données isotopiques (Sr, Pb) sur le volcanisme et le plutonisme des îles Kerguelen. Le problème de la contamination crustale. Thèse 3e cycle, Rennes.

FAURE, G. and HURLEY, P. M. (1963) The isotopic composition of strontium in oceanic and continental basalts: application to the origin of igneous rocks. *J. Petrology* **4**, 31–50.

FAURE, G. and POWELL, J. L. (1972) *Strontium Isotope Geology*, Springer Verlag, Berlin–Heidelberg–New York.

FRANCIS, T. J. G. and RAIT, R. W. (1967) Seismic refraction measurements in Southern Indian Ocean. *J. Geophys. Res.* **72**, no.12, 3015–3041.

LAMEYRE, J., MAROT, A., ZIMINE, S., CANTAGREL, J. M., DOSSO L. and VIDAL, Ph. (1976) Chronological evolution of the Kerguelen Islands syenite–granite ring complex. *Nature* **263**, 306–307.

MAROT, A. and ZIMINE, S. (1976) Les complexes annulaires de syénites et granites alcalins dans la péninsula Rallier du Baty. Îles Kergulen (TAAF). Thèse 3e cycle. Université Pierre et Marie Currie. Paris.

NOUGIER, J. (1969) Contribution à l'étude géologique et géomorphologique des îles Kerguelen (Territoire des Terres Australes et Antarctiques Françaises). *Com. Natl. Français Recherches Antarctiques*, no. 27, 440 pp.

NOUGIER, J. and LAMEYRE, J. (1973) Les Nordmarkites des îles Kerguelen (TAAF) dans leur cadre structural. Problème de leur origine et celle de certaines roches plutoniques alcalines en domaines océaniques. *Bull. Soc. Géol. Fr.* (7), XV, no. 3–4, pp. 306–311.

POWELL, J. L. and BELL, K. (1974) Isotopic composition of strontium in alkalic rocks. In *The Alkaline Rocks*, H. SORENSEN ed. Wiley–Interscience, pp. 412–421.

SCHLICH, R. and PATRIAT, P. (1971) Anomalies magnétiques de la branche Est de la dorsale médio-indienne entre les îles Amsterdam et Kerguelen. *C.R. Acad. Sci., Paris* **272**, 773–776.

SCHLICH, R. (1975) Campagnes océanographiques dans l'Océan indien. *Courrier du CNRS* no. 17, 14–23.

SHAW, D. M. (1968) A review of K–Rb fractionation trends by covariance analysis. *Geochim. Cosmochim. Acta* **32**, 573–601.

SUN, S. S. and HANSON, G. N. (1975) Evolution of the mantle: geochemical evidence for alkali basalt. *Geology*, June 1975, 297–302.

TOUGARINOV, A. I., KOVALENKO, V. I., ZNAMENSKY-LEGEIDO, V. A., SOBATOVICH, E. V., BRANDT, S. B. and TSYHANSKY, V. D. (1968) Distribution of Pb isotopes, Sn, Nb, Ta, Zr and Hf in granitoids of Nigeria. 687–699. In *Origin and Distribution of the Elements*, L. H. AHRENS, ed., Int. Series Monographs in Earth Sciences, 30, Pergamon Press, Oxford, 1178 pp.

VAN BREEMEN, O., HUTCHINSON, J. and BOWDEN, P. (1975) Age and origin of the Nigerian Mesozoic granites: a Rb–Sr isotopic study. *Contr. Mineral. Petrol.* **50**, 157–172.

VIDAL, Ph., BONIN, B. and LAMEYRE, J. (1978) Teneurs en K et Rb dans certains complexes alcalins anorogéniques. *Bull. Soc. Géol. Fr.* In press.

WATKINS, N. D., GUNN, B. M., NOUGIER, J. and BAKSI, A. K. (1974) Kerguelen: continental fragment or oceanic island? *Geol. Soc. Amer. Bull.* **85**, 201–212.

WILSON, J. T. (1963) Evidence from islands on the spreading of ocean floors. *Nature* **197**, 536–538.

WILSON, J. T. (1965) Evidence from oceanic islands suggesting movement in the Earth. *Phil. Trans. Roy. Soc. Lond.* Ser. A, **258**, 145–167.

ZHIVAGO, A. V. (1965) Géomorphologie et tectonique du fond de l'Océan Antarctique. *Rech. Océano.* no 13. ed. "Science", Moscou. Traduct. in CNFRA no. 17, pp. 19–26.

Aspects of Platinum Elements Distribution in Some Ultramafic and Related Rocks

G. AGIORGITIS, R. BECKER and R. WOLF

Fachbereich Geowissenschaften der Technischen Hochschule, Darmstadt, Germany

Abstract

Results of Pt, Pd, Ir, Os and Ru analyses on samples of mafic and ultramafic rocks and some of their minerals obtained by radio chemical neutron activation analysis are reported.

The Pt abundances in the rocks and some of the minerals are lower than the detection limit. In the rocks of the Troodos-complex, Cyprus, Pd shows a distribution in the range of 0.3–12 ppb. Ir and Os are mainly concentrated in the early differentiates and have a distribution range of 0.01–6 ppb.

The constancy of the ratios Ir/Os, Ru/Os and Ru/Ir especially in chromite suggests that these elements were in thermodynamic equilibrium during the crystallization.

Introduction

Data about the abundances of the platinum elements in terrestrial rocks are rather scarce. Furthermore, some of the early reported results are of doubtful quality. Platinum element abundances in ultramafic and mafic rocks and their minerals have been given by Baedecker and Ehmann (1965), Bate and Huizenga (1962), Razin *et al.* (1965), Crocket and Skippen (1966), Das Sarma *et al.* (1966), Ehmann *et al.* (1970), Gottfried and Greenland (1971), Greenland (1971), Page *et. al.* (1972), Crocket and Chyi (1972) and Gijbels *et al.* (1974). Most authors agree that during magmatic differentiation the platinum elements have a tendency to concentrate in the early differentiates, preferable in chromite.

Platinum and Pd enrich during differentiation in the liquid phase and are finally concentrated in sulfide minerals like pyrrhotite and pentlandite and various metal phases. On this basis one can explain the geochemical distribution of Pt and Pd against Ir, Os and Ru. The distribution of the last three elements in rocks is governed by fractional crystallization. Other processes like partial melting in situ must be also considered.

Analytical Procedures

Platinum element concentrations were measured by wet chemical, carrier-based activation analysis methods.

Irradiation time was 7 h at a thermal neutron flux of 1.6×10^{12}n cm^{-2} s^{-1} for the determination of Pt and Pd and 1 week at a thermal neutron flux of 8×10^{13} for the detèrmination Ir, Os and Ru. Os and Ru were distilled simultaneously, using sodium chlorate and sodium bromate as oxidizing agents. Another two distillations and two precipitations (as sulfide and hydrated oxide respectively) yield pure Os and Ru fractions.

The Ir fraction was obtained by an anion exchange separation, followed by bromate hydrolysis. The chemical yields were determined by reactivation.

The Ni contents were determinated by AAS. The relative standard deviation for duplicate analysis was: Ir ± 5–15%, Os ± 5–15%, Ru ± 10–20%, Pd ± 10–30%, dependent on the sample content.

Discussion

The platinum elements belong to two coherent groups according to their geochemical distribution in rocks and minerals.

The elements Pt and Pd possibly build quadratic-planar complexes with S, As or Sb which are soluble in sulfur-saturated melts. They also show a stronger chalcophilic tendency as compared to Ir, Os and Ru. So Pt and Pd tend to concentrate in residual or late magmatic solutions and build discrete minerals in association with sulfides.

The geochemical distribution of these elements in a series of peridotites, dunites, pyroxenites, gabbros, basalts and diabases, their separated minerals, as well as in some chromites are described below.

The element Pt was determined only in some chromites with higher content than the determination limit of 25 ppb. The rocks have Pd values in the range from 0.3 to 12 ppb, with the exception of some chromites which show higher Pd as well as Pt contents.

The abundance of Pd in periodotites, dunites and gabbros has the ratio 24:1:8. The peridotites from the Troodos complex show a Pd concentration of 8.1 ppb and are therefore comparable with the alpin peridotite from Mt. Albert pluton (Crocket and Chyi, 1972).

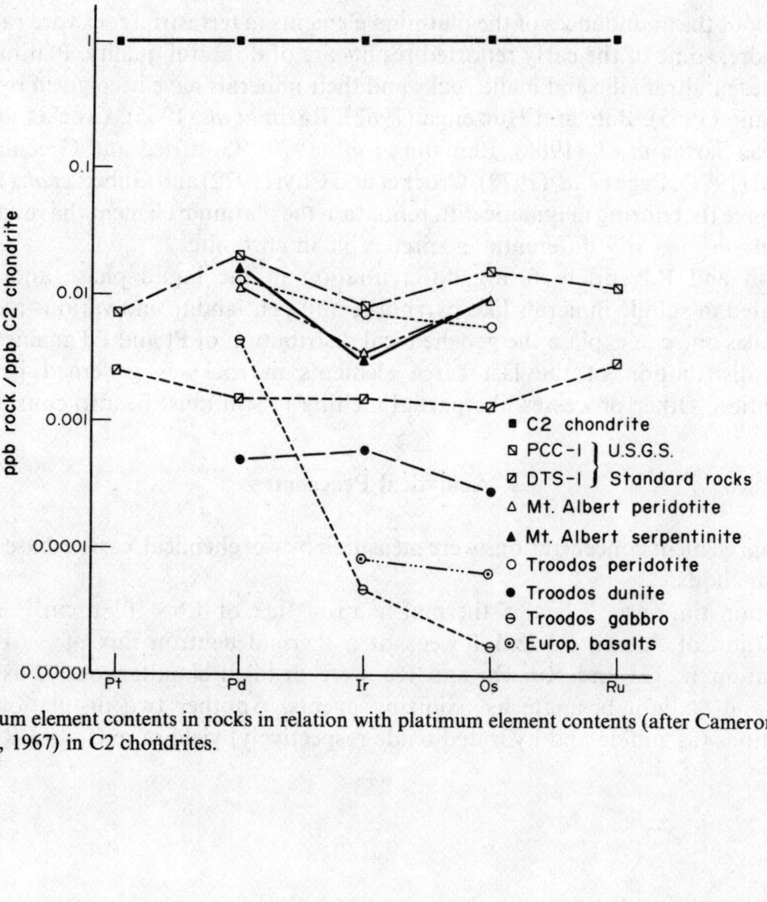

FIG. 1. Platinum element contents in rocks in relation with platimum element contents (after Cameron, 1966 and Crocket *et al.*, 1967) in C2 chondrites.

The lower sulfide content in dunite as compared to other rocks might be an explanation of the lower Pd contents.

Another possibility suggested by Crocket and Chyi for the decrease of Pd in serpentinite of the Mt. Albert pluton is that Pd has higher mobility during the serpentinization process. Probably in this case Pd^{2+} replaces Fe^{2+} in the olivine lattice. This mineral, as shown by recently obtained results from peridotite and dunite, contains 3 ppb Pd.

A substitution of Cr^{3+} by Pd^{4+} in chromite lattice should not be expected due to the high redox potentials Pd^{2+}/Pd^{4+} of 1.26 V. Such a substitution for Pt^{4+} is possible due to the lower redox potentials Pt^{2+}/Pt^{4+} of -0.77V. This was also observed by Grimaldi *et al.* (1969) in chromite from the Stillwater complex. This conclusion is supported by chromite from Skyros Island, Greece, with an average content of 225 ppb Pt against 34.5 ppb Pd (Wolf and Agiorgitis, 1976). Gabbro, with an average of 2.9 ppb Pd, shows an enrichment as compared with dunites. The appearance of pyrrhotite in gabbro is connected with this higher Pd concentration.

The strong siderophilic and weak chalcophilic properties of Ir, Os and Ru result in their enrichment in early differentiates of the orthomagmatic crystallization.

Accurate results for Os, Ir and Ru have become available only within the last few years. Analytical problems are present for Ru due to higher uranium contents in most of the rocks and silicate minerals ($^{235}U(n,f)\ ^{103}Ru$ interfering reaction). The element concentrations in Fig. 1 show that the elements Ir and Os are similarly distributed in peridotites. They decrease from peridotite to basalt, except for Pd in the Troodos gabbros. The enrichment of this element is probably related to sulfide minerals.

The Ir ratios of 1000:100:6:2 in Troodos peridotite, dunite, gabbro and diabase show that this element has a fractionation tendency during the magmatic differentiation. This was also observed in basaltic rocks by Agiorgitis (1977).

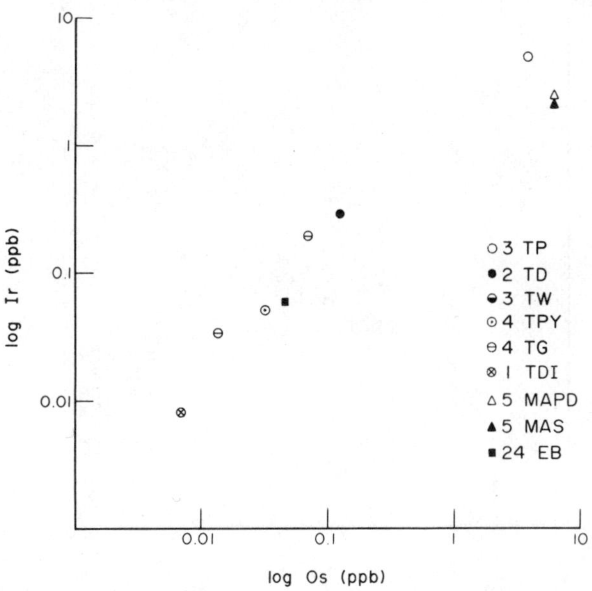

FIG. 2. Correlation of Ir/Os, average values obtained from 5 Mt. Albert peridotites and dunites and 5 serpentinites after Crocket *et al.* (1972), 3 Troodos peridotites, 2 dunites, 3 wehrlites, 4 pyroxenites, 4 gabbros, 1 diabase and 24 European basalts.

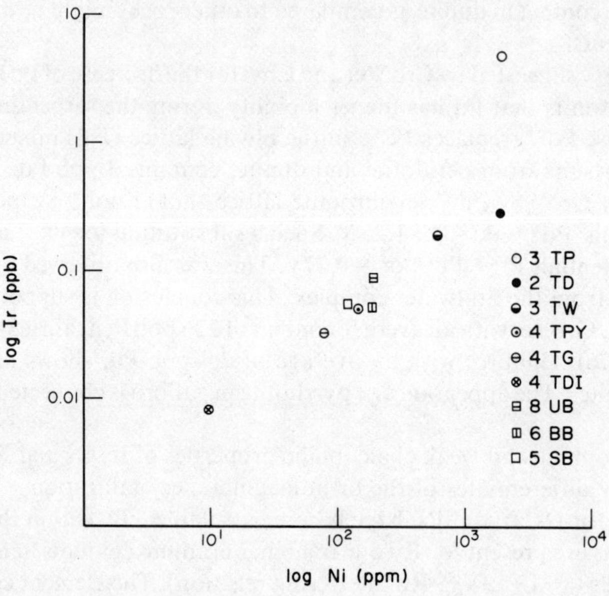

FIG. 3. Correlation of Ir/Ni, average values obtained from Troodos complex rocks and different basalts from central Europe like those from Hungary (8), Burgenland, Austria (6) and Styria, Austria (5).

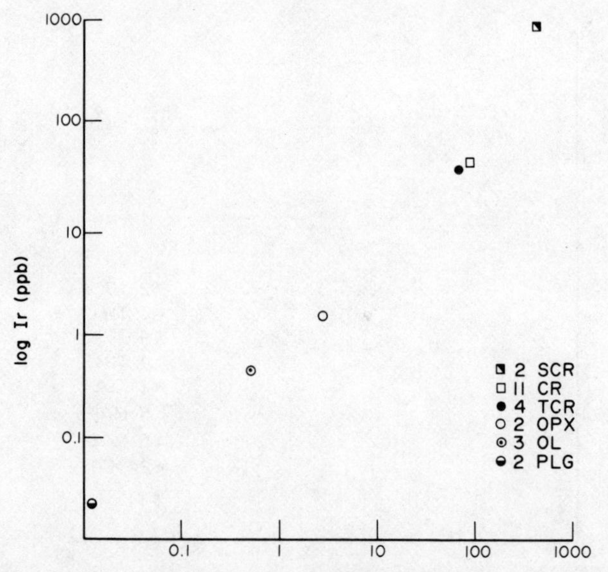

FIG. 4. Correlation of Ir/Os in rock-forming minerals. Average values obtained from 2 Skyros island, Greece, chromites, 11 chromites from other different Greece occurrences, 4 Troodos chromites, 2 orthopyroxenes, 3 olivines and 2 plagioclases.

The ratio Ir to Os is approximately 1:0.8 for ultramafic rocks of the Troodos complex and approximately 1:0.4 according Crocket *et al.* for the rocks of the Mt. Albert pluton. The relation Ir/Os as shown in Fig. 2 is similar in all the Troodos rocks ($r = 0.9996$ with a confidence limit of 99.9%). The log ratio Ir/Ni (Fig. 3) is constant with the exception of peridotites, in which both elements are concentrated, mainly in the chromite.

The correlation coefficient for the Troodos rocks is $r = 0.8745$ and for European basalts $r = 0.485$ with a confidence limit of 99% and 95%, respectively.

The results of all known data are that chromite is the principal concentrator for Ir. The enrichment ratio of this element for the minerals chromite, orthopyroxene, olivine, clinopyroxene, plagioclase is 10,000:300:100:9:4. The homologous element Os behaves similarly, as can be seen from Fig. 4.

An unusually high Ir/Os ratio of approximately 2 is shown by the chromites from Skyros island, whereas the other chromites have an average ratio of about 0.9. In comparison to the chromite, the silicate minerals of Troodos have a ratio of about 0.9. The statement of the constancy of Ir/Os ratio supports the conclusion that this system was in thermodynamic equilibrium during the orthomagmatic crystallization.

References

AGIORGITIS, G. (1977) Distribution of iridium in some basalts. *Tschermaks Miner. Petrogr. Mitt.*, to be published.

BAEDECKER, P. A. and EHMANN, W. (1965) The distribution of some noble metals in meteorites and natural materials. *Geochim. Cosmochim. Acta*, **29**, 329–342.

BATE, G. L. and HUIZENGA, J. R. (1963) Abundances of ruthenium, osmium and uranium in some cosmic and terrestrial sources. *Geochim. Cosmochim. Acta*, **17**, 345–360.

CAMERON, A. G. W. (1966) Abundances of the elements. *Handbook of Physical Constants. Geol. Soc. Amer Mem.* **97**, 8–10.

CROCKET, J. H. and SKIPPEN, G. B. (1966) Radioactivation determination of Pd in basaltic and ultrabasic rocks. *Geochim. Cosmochim. Acta*, **30**, 129–141.

CROCKET, J. H., KEAYS, R. R. and HSIEH, S. (1967) Precious metal abundances in some carbonaceous and enstatite chrondites. *Geochim. Cosmochim. Acta* **31**, 1615–1623.

CROCKET, J. H. (1969) Platinum metals. In *Handbook of Geochemistry*, Vol. II/2, Springer Verlag.

CROCKET, J. H. (1971) Neutron activation analysis for noble metals in geochemistra. In A. O. BRUNFELD, and E. STEINES, *Activation Analysis in Geochemistry and Cosmochemistry*, Universitets Forlaget, 339–351.

CROCKET, J. H. and CHYI, L. L. (1972) Abundances of Pd, Ir, Os and Au in an alpine ultramafic pluton. *Proc. XXIV. Int. Geol. Congress*, Sect. 10, 202–209.

DAS SARMA, B., SEN, B. and MCKOWN, D. M. (1970) Gold and iridium in meteorites and some selected rocks. *Geochim. Cosmochim. Acta* **34**, 493–507.

DAS SARMA, B., SEN, B. and CHOWDHURY, A. (1966) Studies on the geochemistry of platinum. *Econ. Geol.* **61**, 592–597.

GIJBELS, R., MILLARD, H. T., DESBOROUGH, G. A. and BARTEL, A. J. (1974) Osmium, ruthenium, iridium and uranium in silicates and chromite from the eastern Bushveld complex, S. Africa. *Geochim. Cosmochim. Acta* **38**. 319–337.

GOTTFRIED, D. and GREENLAND, L. P. (1971) Variation of iridium and gold in oceanic and continental basalts. *24th Int. Congress*, Sect. 10,. 135–143.

GREENLAND, L. P. (1971) Variation of iridium in a differentiated tholeiitic dolerite. *Geochim. Cosmochim. Acta*, **35**, 319–322.

GRIMALDI, F. S. and SCHNEPFE, M. M. (1969) Mode of occurrences of platinum, palladium and rhodium in chromite. *U.S. Geol. Surv. Prof. Paper*, 650 C. 149–151.

PAGE, N. I., RILEY, L. and HAFTY, J. (1972) Vertical and lateral variation of platinum, palladium and rhodium in the Stillwater complex, Montana. *Econ. Geol.*, **67**, 915–923.

RAZIN, L. V., KHVOSTOV, V. P. and NOVIKOV, V. A. (1965) Platinum metals in the essential and accessory minerals in ultramafic rocks. *Geochim. Int.* **2**, 118–131.

WOLF, R. and AGIORGITIS, G., GOVAERTS, A. and GIJBELS, R. (1975) Abundance of platinum elements in some Greek chromites. *Int. Platin. Symp., Denver*, Col. Proc.

WOLF, R. and AGIORGITIS, G. (1977) An unusual enrichment of platinum elements in chromites from Skyros island, Greece. *N. J.b. Mineral. Mh.*, to be published.

WRIGHT, T. and FLEISCHER, M. (1965) Geochemistry of the platinum metals. *U.S. Geol. Surv. Bull.* 1214–A, A1 with corrections and additions (1967).

YUSHKO-ZAKHAROVA, O. YE., IVANOV, V. V., RAZINA, I. S. and CHERNYAYEV, L. A. (1967) Geochemistry of platinum metals. *Geochemistry Int.* **4**, 1106–1118.

The Distribution of Some Elements in the Direct Contact Aureole of the Krušné hory Mts Granite (Bohemia)

EMIL ROOS

Český Geologický Úřad, Praha, Czechoslovakia

Abstract

Results of a study of distribution of the principal constituents (Li_2O, Na_2O, K_2O, Rb, Be, Zn, B, Sn, TiO_2, V) in the direct contact aureole of a granite body in contact with a phyllite complex are presented. The study was aimed at the recognition of the element distribution in a 16-cm contact zone in both the endogenous and exogenous contact.

The results did not yield conclusive evidence of transfer of elements to considerable distances. The enrichment/impoverishment is interpreted also in terms of postmagmatic processes.

Geological Setting

The studied area belongs to the Erzgebirge–Fichtelgebirge metallogenic subzone within the Saxothuringian metallogenic zone (Sattran, 1966; Tischendorf *et al.*, 1969). The area is chiefly built up by phyllites and subjacent mica schists which border in the west on the Krušné hory Mts, pluton. These metamorphic rocks (probably of Lower Ordovician to Cambrian age) were intruded by granites belonging to the Krušné hory Mts pluton as well as by dikes of quartz porphyrites, abundant Tertiary basalts and to a lesser extent by lamprophyres, aplites.

Magmatic rocks in the studied area represented by autometamorphosed granite of the Variscan post-kinematic Younger intrusive complex—"Erzgebirgsgranite" according to Laube, 1876 (Baumann *et al.*, 1974).

Phyllite Complex

The phyllite complex is formed of epizonally metamorphosed sedimentary pelitic rocks. Phyllites belonging to the green-schist mineral facies (quartz–albite–muscovite–chlorite or quartz–albite–epidote–biotite subfacies) are composed mainly of chlorite and sericite. The thickness of the complex varies from 300 to 600 m. Only weak manifestations of contact metamorphism were observed on the contact with Krušné hory Mts granite. The metamorphism did not substantially change the mineral association typical of the green-schist mineral facies.

Krušné hory Mts Granite

The deep-seated granite body reaches an average depth of 1–3 km in the whole area of the Krušné hory Mts (e.g. Watznauer, 1954). Within this body major elevations (av. length 20–30 km) protrude to the surface as partial plutons (corresponding to the second-order plutons in the sense of Cloos, 1936).

Still smaller local intrusions, plutons of the third order, 100–500 m in diameter, intrude the upper area. The Blatná granite body is the most significant massif of this kind intruding the phyllite complex in the studied area. On the surface, only the apical part of the Blatná granite body is exposed.

An autometamorphosed biotitic, medium-grained granite is the most frequent rock in the body (Stattran, 1966).

Contacts of the Granite with Phyllite Mantle

In the country rocks the effects of the contact metamorphism can be traced to a distance of approx. 400 m from the contact. The usual pyroxenic hornfels contact was not formed—the association of minerals belongs to the low-temperature facies of albite–epidote hornfelses, especially in its pelitic subfacies: quartz–albite–muscovite–biotite (Fyfe *et al.*, 1958). The albite–epidote hornfels also contains some andalusite and cordierite, and in this it differs from the mineral association met in the original phyllite unaffected by contact metamorphism.

The composition of the granite magma and to a great extent also of postmagmatic solution were controlled by palingenesis and assimilation (Tischendorf, 1965).

Sampling Techniques

Samples were taken to a distance of approximately 130 m from the contact both in the granite (endogenous contact or endocontact) and phyllite complex (exogenous contact or exocontact). The sampling intervals were approximately 10 m. Samples were taken from drill cores and in old mining works, where the influence of the autometamorphic changes were a minimum. The petrographical and geochemical results of the study have already been published by the present writer (Roos, 1976 a,b).

TABLE 1. STATISTICAL PARAMETERS FOR THE PHYLLITE (0–8 cm)

	1	2	3	4
Sn	0.0045	0.0000	0.0027	60.85
Rb	0.0725	0.0014	0.0387	53.39
Zn	0.0271	0.0013	0.0364	134.21
V	0.0105	0.0000	0.0063	60.45
B	0.0858	0.0044	0.0666	77.66
Be	0.0008	0.0000	0.0003	42.00
TiO_2	0.4933	0.1440	0.3796	76.92
K_2O	3.7566	2.0440	1.4297	38.05
Na_2O	2.5666	4.4724	2.1148	82.39
Li_2O	0.0960	0.0045	0.0673	70.15

1. Average content in per cent. 2. Dispersion coefficient. 3. Standard deviation. 4. Coefficient of variation.

TABLE 2. STATISTICAL PARAMETERS FOR THE GRANITE (0–8 cm)

	1	2	3	4
Sn	0.0023	0.0000	0.0018	79.51
Rb	0.0685	0.0012	0.0351	51.29
Zn	0.0042	0.0000	0.0024	59.17
V	0.0017	0.0000	0.0022	130.21
B	0.0322	0.0019	0.0439	136.40
Be	0.0009	0.0000	0.0004	47.60
TiO_2	0.1569	0.0440	0.2098	133.64
K_2O	4.1820	3.1275	1.7685	42.28
Na_2O	4.6220	1.7834	1.3354	28.89
Li_2O	0.0582	0.0015	0.0395	67.96

1. Average content in per cent. 2. Dispersion coefficient. 3. Standard deviation. 4. Coefficient of variation.

In this paper the results of a very detailed study of four direct contacts are presented. The rock was cut into approximately 2-cm-thick plates, orientated parallel to the plane of the contact. Altogether twenty-four samples from a contact zone of 16 cm were studied (8 cm from the endocontact, 8 cm from the exocontact).

Analytical Techniques

Samples were analyzed by flame photometry (Li, Na, K), colorimetry (Be) and by X-ray fluorescence (Rb, Zn, Sn, Ti, V), in the laboratories Geoindustria, Brno (apparatus Müller-Micro 111). The spectrometer Qu 24 was used for determination of boron. Several quantitative mineralogical analyses of phyllite and granite were done (see Table 3).

Mathematical Statistical Evaluation

Analytical results were treated by the "simple correlation" method on the Minsk 22 computer. The mean, variance and standard deviation were computed for all sample values (see Tables 1 and 2). In addition the correlation ratios between constituents were counted (see Table 6).

The obtained results were compared with the regional average content (i.e. geochemical background) of the constituents in the rock not affected by contact metamorphism (see Tables 4 and 5).

TABLE 3. QUANTITATIVE MINERALOGICAL ANALYSIS—PHYLLITE

	1	2	3
Quartz	76.5	45.4	5.6
Feldspar	0.3	—	—
Biotite	9.6	8.9	3.1
Sericite muscovite	3.9	21.5	57.9
Chlorite	7.8	21.5	25.7
Ti minerals	2.2	1.8	4.1
Andalusite	0.06	0.16	3.1
Accessories	1.4	0.6	0.2

(Integrator Eltinor, number of grains counted: 3000)
Distance from the contact: 1.2 cm, 2.5 cm, 3.10 cm.

TABLE 4. COMPARATIVE TABLE OF THE CONSTITUENTS IN THE PHYLLITE

	1	2	3	4
Sn	0.0045	0.0028	0.003	0.0002
Rb	0.0725	0.0360	?	0.009
Zn	0.0271	0.0182	0.015	0.007
V	0.0105	0.0781	?	0.013
B	0.0858	0.0430	0.042	0.001
Be	0.0008	0.0008	0.00005–0.0008	0.0002
TiO_2	0.4933	0.7582	0.500	0.570(Ti)
K_2O	3.7566	3.7245	3.000	2.900(K)
Na_2O	2.5666	0.9409	0.980	2.360(Na)
Li_2O	0.0960	0.0704	0.009–0.021	0.002(Li)

1. Average content in per cent (distance interval 0–8 cm from the contact).
2. Average content in per cent (distance interval 10–60 cm from the contact (Roos 1976a)).
3. Regional average content—phyllite (Sattran, 1960; Roos, 1976a).
4. Clark of the element (Taylor 1964).

Results of the Study

For illustration of the results, both graphical and tabular representations were used. Comparison of the content of all studied constituents in both endo-exocontact, with the regional average content has shown that the content of:

(a) 35% of the constituents were approximately equal; no statistically significant changes occurred (*t*-test),
(b) 45% were enriched,
(c) 20% were impoverished,
in the direct contact aureole.

The detailed survey of element distributions is given in Tables 4 and 5 and Figs. 1, 2 and 3.

Analysis of Correlations

The main attention was given to the study of positive correlations between the constituents. Statistically significant correlations between separate constituents in the

TABLE 5. COMPARATIVE TABLE OF THE CONSTITUENTS IN THE GRANITE

	1	2	3	4
Sn	0.0023	0.0039	0.004	0.0004
Rb	0.0685	0.0768	0.086	0.022
Zn	0.0042	0.0092	?	0.004
V	0.0017	0.0053	0.001–0.0004	0.001
B	0.0322	0.0122	0.001–0.005	0.0002
Be	0.0009	0.0004	0.001	0.0003
TiO_2	0.1569	0.1149	0.084	0.150(Ti)
K_2O	4.1820	3.6729	4.770	4.510(K)
Na_2O	4.6220	2.8589	3.440	2.460(Na)
Li_2O	0.0582	0.1102	0.110	0.0024(Li)

1. Average content in per cent (distance interval 0–8 cm from the contact).
2. Average content in per cent (distance interval 10–60 cm from the contact) (Roos, 1976a).
3. Regional average content—granite (Absolonová and Matoulek, 1975).
4. Standard G-1 sample (Mason, 1971).

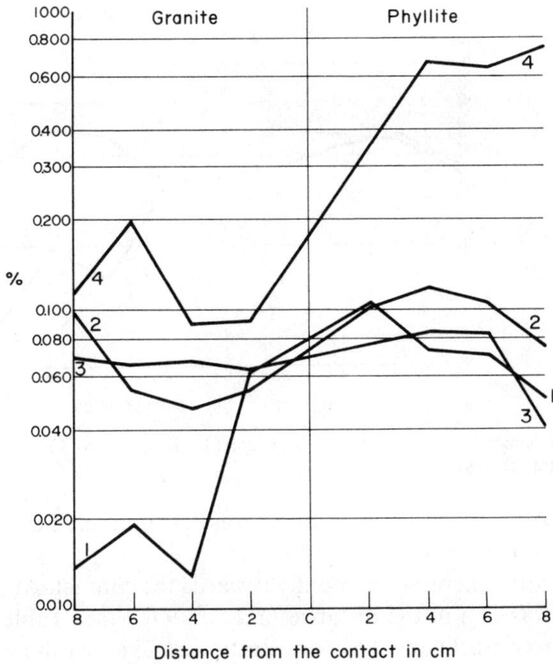

FIG. 1. Distribution of elements in the direct contact aureole: 1—B, 2—Li_2O, 3—Rb, 4—TiO_2 (average of 24 samples–localities: Bystřice, Pila, drill HB 4, JD 55).

distance interval 0–8 cm in both the granite and phyllite are introduced in Table 6. The following statistically significant positive correlations were found:

(a) in the direct endocontact (0–8 cm): K_2O/Rb, TiO_2/Sn, Li_2O/Zn, Zn/Sn,

(b) in the direct exocontact (0–8 cm): K_2O/V, Na_2O/B, Zn/Rb, Rb/Sn.

Quantitative Mineralogical Analysis

Exogenous contact aureole—a strong quarzification was observed in the exogenous contact, along with considerable changes in grain size. In the distance of 20 cm from the

TABLE 6. CORRELATION CHART—PHYLLITE, GRANITE (distance interval 0–8 cm)

	Sn	Rb	Zn	V	B	Be	TiO_2	K_2O	Na_2O	Li_2O
Sn	—									
Rb	+0.75	—							$\underline{-0.92}$	
Zn	$\underline{+0.62}$	+0.76	—						$\underline{-0.66}$	
V				—						
B					—					
Be						—				
TiO_2	$\underline{+0.80}$						—		−0.82	
K_2O		$\underline{+0.90}$		+0.94				—	$\underline{-0.93}$	
Na_2O					+0.78				—	$\underline{-0.58}$
Li_2O			$\underline{+0.64}$							—

Underlined values refer to the granite.
Values not underlined refer to the phyllite.

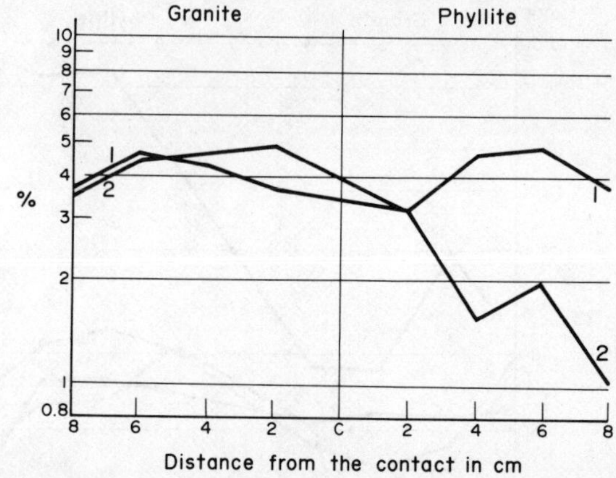

FIG. 2. Distribution of elements in the direct contact aureole: 1—K_2O, 2—Na_2O (average of 24 samples–localities: Bystřice, Pila, drill HB 4, JD 55).

contact plane the average size of quartz grains was 0.25 mm, further from the contact (20–50 cm) the average grain size was 0.11–0.12 mm.

The content of biotite increases markedly towards the contact. At a distance of 20 cm the content of biotite was 2%, in the distance of 2 cm, 9.6% (see Table 3).

Within the 16-cm contact zone the concentration of Ti minerals and andalusite decreases

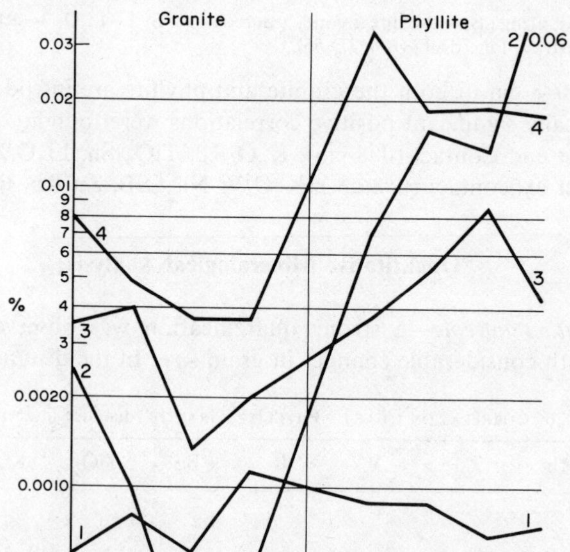

FIG. 3. Distribution of elements in the direct contact aureole: 1—Be, 2—V, 3—Sn, 4—Zn (average of 24 samples–localities Bystřice, Pila, drill HB 4, JD 55).

towards the contact plane. The highest concentration of andalusite, as shown by heavy minerals analysis, is at a distance of 20 cm from the contact.

Endogenous contact aureole—the content of K-feldspar in the granite increases towards the contact and at a distance of 10 cm from the contact it amounts to almost 45%, whereas the content of plagioclase decreases in the direct contact aureole. In the direct contact aureole all mineral grains become coarser. The average grain size at a distance of 2 m from the contact is 0.6 mm, in the direct contact, 1 mm.

Conclusions

1. No significant changes (*t*-test) in the direct contact aureole in comparison with regional average content (goechemical background) were found for:

Rb, Be, K_2O (endogenous contact aureole) and

Sn, Zn, V, B (TiO_2?) (exogenous contact aureole).

2. The following constituents were found enriched within the direct contact aureole as compared with regional average content:

B, TiO_2, Na_2O, V (endogenous contact aureole) and

K_2O, Na_2O, Li_2O, Rb, Be (exogenous contact aureole).

3. The direct contact aureole is impoverished in several constituents in comparison with regional average content:

Sn, Zn, Li_2O (endogenous contact aureole),

(TiO_2?) (exogenous contact aureole).

4. From 1–3 it follows that, compared with the regional average content, 35% of the analyzed constituents in both the exo-endocontact were approximately equal (no statistically significant changes occurred), 45% were enriched, and 20% impoverished.

5. In the contact aureole the elements have usually only slight migrating capacity (e.g. Dennen, 1951; Lipp, 1957; Brookins-Dennen, 1964). The results did not yield conclusive evidence of transfer to considerable distances. The enrichment/impoverishment of the contact aureole can be interpreted also in terms of postmagmatic processes which having affected the distribution of elements make the study of contact-metamorphism effects difficult.

It may be assumed that there is a transfer of elements in the course of contact metamorphism and postmagmatic processes in the area studied which leads to the enrichment of the direct contact, to the lesser extent to its impoverishment.

6. The study of element distributions in the direct contact aureole of the Blatná body (Krušné hory Mts granite) with phyllites has revealed that the elements are distributed unevenly in the rocks, and that no economically significant concentrations of the studied constituents are likely to be found near to the contact zone.

References

Absolonová, E. and Matoulek, M. (1975) Geochemická distribuce prvků v Karlovarském masívu. *Sbor. geol. Věd. Lož. Geol., Mineral.* **17,** 1–32.

Baumann, L., Štemprok, M., Tischendorf, G. and Zoubek, V. (1974) Metallogeny of tin and tungsten in the Krušné hory–Erzgebirge. International geological correlation programme, MAWAM Symposium. Academia, Praha.

BROOKINS, D. G. and DENNEN, W. H. (1964) Trace element variation across igneous contacts. *Trans. Kans. Acad. Sci. (Lawrence)*, **67,** 70–91.

CLOOS, H. (1963) *Einführung in die Geologie,* Verlag Borntraeger, Berlin.

DENNEN, W. H. (1960) Variations in chemical composition across igenous contacts. *Bull. Geol. Soc. Amer.* **62,** 547–558.

FYFE, W. S.,TURNER, F. J. and VERHOOGEN, J, (1958) Metamorphic reactions and metamorphic facies. *Mem. Geol. Soc. Amer.* **73,** 1–259.

LAUBE, G. (1876) *Geologie des Böhmischen Erzgebirges,* Praha.

LIPP, U. (1957) *Stoffverschiebungen im Kontakthof der westerzgebirgischen Granite*, MS Diplomarbeit., Freiberg. i. Sa.

MASON, B. (1971) *Principles of Geochemistry* 3rd ed., J. Wiley, Inc., New York (Russian translation).

ROOS, E. (1976a) Migrace prvků v kontaktní zóně krušnohorských žul s fylity (English summary). *Čas. mineral. geol.* **21,** 3. 285–299.

ROOS, E. (1976b) Preliminary report on the influence of contact metamorphism on the distribution of some metallic elements in regionally metamorphosed rocks of the western Krušné hory (Erzgebirge) Mts. in Bohemia. In *The Current Metallogenic Problems of Central Europe*, edit. Geological Institute Warsaw, 208–215.

SATTRAN, V. (1960) Závěrečná zpráva o komplexním výzkumu Jáchymovska za rok 1957–1959. Not published. MS Geofond, Praha.

SATTRAN, V. (1966) Problémy metalogeneze Českého masivu. (Problems of metallogeny of the Bohemian Massif). *Sbor. geol. Věd,* Praha, Ř. LG, **8,** 7–112.

TAYLOR, S. R. (1964) Abundance of chemical elements in the continental crust. *Geochim. Cosmochim. Acta* **28,** 1273–1285.

TISCHENDORF, G. *et al.* (1965) Zur Lage der Granitoberfläche im Erzgebirge und Vogtland. *Z. Angew. Geol.* **11,** 8, 410–423.

TISCHENDORF, G. *et al.* 1969) Zur Problemen der Metallogenie in der Deutschen Demokratischen Republik. *Z. Angew. Geol.* **15,** 9, 455–472.

WATZNAUER, A. (1954) Die Erzgebirgische Granitintrusionen. *Geologie (Berlin)* **6–7,** 688–706.

The Distribution of Zirconium and Hafnium in Terrestrial Rocks, Meteorites and the Moon

W. D. EHMANN, L. L. CHYI†, A. N. GARG‡ and M. Z. ALI

Department of Chemistry, University of Kentucky, Lexington, Kentucky 40506, U.S.A.

Abstract

We have recently completed radiochemical neutron activation analysis studies for Zr and Hf in a wide variety of natural materials. The Palisade, New Jersey, gravitational differentiation series shows more than a factor of 4 enrichment of Zr and Hf in the latest granophyric diabase as compared to the chilled border, while the Zr/Hf weight ratios exhibit no systematic variation within the sill and fall within the narrow range of 35 to 39. In the pelitic Devonian Littleton, New Hampshire metamorphic formation we find that the Zr/Hf ratio exhibits no systematic trend with degree of regional metamorphism. The abundances of Zr and Hf vary by a factor of nearly 3, but sixteen out of eighteen samples fall in the narrow Zr/Hf ratio range of 38 to 42. Our studies to date do not support the existence of systematic variations of the Zr/Hf ratio in tholeiitic differentiation, or in regional metamorphism. However, for volcanic rocks derived from the alkaline olivine-basalt magma from St. Helena Island we observe a uniform Zr–Hf fractionation trend from Zr/Hf weight ratios of 47 in the primary basalt to 60 in late differentiated phonolite. The increase in the ratio parallels a factor of nearly 4 increase in the abundances of these elements and may be attributed to crystal chemistry considerations.

Zr–Hf fractionation has been shown to occur in lunar basalts. This fractionation may be due, in part, to selective reduction of Zr^{4+} to Zr^{3+} under highly reducing conditions. Lunar Zr/Hf weight ratios range from approximately 21 for anorthosites (Zr <1 ppm) to approximately 46 for both the KREEP-rich basalts (Zr = 400–2000 ppm) and the non-mare KREEP-poor basalts (Zr = 200–340 ppm). The mare basalts exhibit intermediate ratios largely in the range from 32 for the high Ti basalts (Zr = 200–640 ppm) to 36 for the low Ti basalts (Zr = 95–230 ppm). Data on Zr and Hf in co-existing lunar minerals support the suggestion that Zr, possibly as Zr^{3+}, is excluded selectively from early-formed lunar minerals.

Zr–Hf fractionation has also been observed among the achondrites. Eucrites exhibit high Zr abundances and high Zr/Hf ratios, while diogenites have extremely low Zr abundances and low Zr/Hf ratios. The data are consistent with a fractionation resulting, in part, from the presence of Zr^{3+} under extremely low f_{O_2} conditions. The Zr and Hf data for howardites are consistent with a mixing model origin with eucrites and diogenites, as end members. New data for Zr and Hf in enstatite chondrites indicate typical Zr abundances of 5.7 ± 1.5 ppm, while Zr/Hf ratios are somewhat scattered and can not be distinguished clearly from those for the ordinary chondrites.

Introduction

As a consequence of the lanthanide contraction, Zr and Hf have very similar atomic and ionic radii. Both elements are lithophilic, exhibit a stable 4+ oxidation state and are always found together in terrestrial rocks and minerals. Zirconium is believed to be a strongly protective element for hafnium, which is normally present at only ~2% by weight of the total

† Present address: Institute of Mining and Minerals Research, University of Kentucky, Lexington, Kentucky, 40506, U.S.A.

‡ Present address: Department of Chemistry, University of Udaipur, Udaipur (RAJ), 313001, India.

pair abundance. However, Hf-rich minerals are also known (Neves *et al.*, 1974). In this paper we will review previously published data and new data obtained in our laboratory relating to the fractionation of Zr and Hf in terrestrial, lunar and meteoritic materials.

Terrestrial Rocks

Most literature values of the Zr/Hf (mass/mass) ratio in bulk terrestrial rocks fall in the range from 30 to 60. Horn and Adams (1966) using computer-derived geochemical balances, estimated a mean Zr/Hf ratio of 41 for terrestrial igneous rocks. Brooks (1970), based on analyses of forty-seven rocks and minerals, suggests a Zr/Hf ratio of 37 for terrestrial crustal rocks. Kosterin *et al.* (1958, 1960) have reported a regular decrease in the Zr/Hf ratio in zircons from calc-alkali rocks with increasing degree of differentiation. They obtained Zr/Hf ratios from approximately 60 for zircons from early-formed gabbros and diorites to 30–40 for zircons from later-formed granites. Numerous other workers have similarly reported decreases in the Zr/Hf ratio in the course of differentiation leading to the formation of silica-enriched products. Brooks (1969, 1970), however, reported increases in the Zr/Hf ratios with differentiation in the Skaergaard intrusion and also in differentiation products of an alkali olivine–basalt magma from Hawaii. Brooks (1970) also noted that the Hawaiian volcanic rocks he studied had a mean Zr/Hf ratio of 57, a value much higher than the nearly constant value of 37 he found for other rock series from widely different geographical occurrences.

Esson *et al.* (1968) have suggested that a Zr^{4+}–Hf^{4+} fractionation should be observed for calc-alkali granites, based on crystal chemistry arguments. This argument has been supported by data of Condie and Lo (1971) and Gottfried *et al.* (1968) who note a Zr–Hf fractionation in a granitic batholith and two diabase granophyre suites, respectively, although the stated analytical error limits are high. Butler and Thompson (1965) advanced the suggestion that Zr–Hf fractionation in terrestrial rocks could be related to the formation of Zr and Hf complex ions in solution and mass related diffusion rate differences.

In contrast to the previously mentioned variations in the terrestrial Zr/Hf ratio with differentiation, Gulson (1969) found no systematic variation in the Zr/Hf ratio in diorites, granites and gabbros from the Yeoval diorite complex in Australia. Herrera and Heurtebise (1974) reported nearly constant Zr/Hf ratios of ~ 25 for twenty widely scattered sands from Venezuela and concluded that the constancy of the Zr/Hf ratio precluded its use in the identification of the origins and classification of the sands. Finally, Randle (1974) concluded on the basis of published data that the terrestrial Zr/Hf ratio is essentially constant at a value of 50 for a wide variety of rock types. Randle also pointed out the absence of a sufficient number of analytical determinations for which the Zr data have reasonably small standard deviations.

Lunar Rocks

Data for Zr and Hf abundances in lunar samples analyzed by our group have been summarized by Garg and Ehmann (1976) and Garg *et al.* (1977) Compilations are given in Tables 1 and 2. It is clear that Zr and Hf are strongly fractionated in lunar rocks as evidenced by a wide range of Zr/Hf mass ratios from 12 for anorthosite, 60015, to 50 for KREEP breccia, 14321. We have previously suggested (Chyi and Ehmann, 1973) that this fractionation might be explained *in part* by selective reduction of Zr^{4+} (0.79 Å) to Zr^{3+} (0.85 Å) under highly reducing lunar conditions and exclusion of the larger Zr^{3+} from the early

TABLE 1. Zr ABUNDANCES AND Zr/Hf RATIOS IN LUNAR ROCK TYPES

Types	No. of samples	Range of Zr (ppm)	Mean Zr/Hf (w/w)
Anorthosites (ANT)	2	0.1–0.5	21.0±9
Troctolites (ANT)	1	12–16	37.1±0.5
High Ti mare basalts (HTM)	14	200–640	31.7±0.6
Low Ti mare basalts (LTM)	7	95–230	36.4±1.2
KREEP-poor non-mare basalts (KPN)	3	200–340	45.9±1.0
KREEP-rich basalts and breccias (KRN)	8	400–2000	45.1±0.9
Low KREEP breccias (LKB)	20	50–1300	43.6±0.7
Granitic rock 12013			
Bulk sample	1	1810	36.0
Light lithology	1	1290	35.8
Dark lithology	1	1720	35.3

Note: Where only one sample was analyzed, the range given is for splits. Normally, data are based on analyses of three to five splits per sample. All data are from this laboratory (Garg and Ehmann, 1976; Garg *et al.* 1977).

crystallized minerals. Hafnium would remain in the Hf^{4+} (0.78 Å) oxidation state. A selective reduction of Zr has been used commercially to separate Zr and Hf. Zr–Hf fractionation is also observed between the high titanium and low titanium mare basalts which seem to have been derived from deep-lying primitive condensates. It should be noted that all the principal chemically defined lunar rock groups are clearly characterized on the basis of their Zr abundances and Zr/Hf ratios. Laul *et al.* (1974) suggest solid–liquid partition processes invoked to explain REE fractionation (e.g. Nd–Sm) would be adequate to fractionate Zr^{4+}–Hf^{4+} without the necessity of postulating the presence of Zr^{3+}. The extremely similar ionic radii of Zr and Hf, and the greater fractionation found in lunar rocks for Zr and Hf, as compared to the Nd–Sm pair, suggest to us that some process in addition to simple Zr^{4+}–Hf^{4+} solid–liquid partition must be involved. The selective reduction of Zr^{4+} to Zr^{3+} is one possibility (under conditions of very low oxygen fugacity) that would lead to *increased* fractionation of the pair, although simple solid-liquid partition of Zr^{4+} and Hf^{4+} could also certainly contribute.

TABLE 2. Zr ABUNDANCES AND Zr/Hf RATIOS IN LUNAR SOILS

Site	No. of samples	Range of Zr Abd (ppm)	Mean Zr/Hf (w/w)
Apollo 11	1	330	35.5
Apollo 12	3	470–650	40.7±0.7
Apollo 14	2	950–1100	40.4±0.5
Apollo 15	1	210–250	37.1±0.2
Apollo 16	4	170–210	43.8±0.6
Apollo 17	4	190–280	35.4±2.0
Orange soil	1	190–200	32.2±1.0
Rake soil 72501	1	269–270	41.4±1.0

Note: Most samples analyzed in replicate (three to five splits). Where only one sample was analyzed, range given is for splits. All data are from this laboratory.

Meteorites

Work by Ehmann and Rebagay (1970) and Ehmann and Chyi (1974) failed to confirm any trends in the Zr/Hf ratios among the various classes of meteorites, except for a few high Zr/Hf ratios in achondrites and a high ratio of 52 for the Cl chondrite Orgueil. Ehmann and Rebagay reported a Zr/Hf mean value of 44 for all stone meteorites they analyzed and a value of 39 for just the H- and L-group ordinary chondrites. These values closely approximate the various estimates of the terrestrial Zr/Hf ratio. Ganapathy *et al.* (1976) report a Zr abundance of 3.1 ppm and a Zr/Hf ratio (mass/mass) of 28 for the Cl carbonaceous chondrite Orgueil. Ehmann and Chyi and Palme (1974) reported Zr abundances of 5.2 ppm and 4.1 ppm, respectively, for Orgueil. The Hf abundances for Orgueil of Ehmann and Chyi and Ganapathy *et al.* are in good agreement at 0.10 and 0.11 ppm, respectively. Ehmann and Chyi pointed out their Zr/Hf ratio of 52 for Orgueil appeared to be high. The lower Zr abundances of Palme and Ganapathy *et al.* are probably more representative of the Cl group, although Ganapathy *et al.* also reported a high Zr/Hf ratio of 75.9 in the Cl chondrite, Alais, which they attributed to possible contamination. The scattered high Zr/Hf ratios reported for carbonaceous chondrites reflect the the difficulty of precise Zr and Hf determinations at the ppm level and indicate contamination problems may exist in storage or in the laboratory processing.

Experimental

The Zr and Hf abundances of this study are determined simultaneously by a radiochemical neutron activation analysis procedure (Kumar *et al.*, 1977). Primary standards for both Zr and Hf are used in the analyses and various U.S.G.S. standard rocks are always analyzed as secondary standards in each irradiation. Corrections for the U fission contribution to the ^{95}Zr activity have been applied where required. The correction factor was determined by irradiation of four pure U standards under irradiation conditions and with identical separation chemistry used for the Zr–Hf analyses. This experiment indicated that the activity of ^{95}Zr produced by 1 ppm U is equivalent to 11 ± 1 ppm Zr. For most samples the correction to the Zr values was well under 10% relative.

Chemical yields for both Zr and Hf are determined by a reactivation procedure, using Hf carrier and the ^{178}Hf (n,γ) ^{179m}Hf reaction induced with a 1.25 mg ^{252}Cf isotopic neutron source. Tables illustrating the precision and accuracy of the procedure are given in detail by Kumar *et al.* (1977). Typical examples of mean values and standard deviations of the mean for samples exhibiting three different levels of Zr–Hf abundances are:

(1) Johnstown achondrite (eight analyses), $Zr = 1.27 \pm 0.04$ ppm, $Hf = 0.050 \pm 0.001$ ppm, $Zr/Hf = 25.4 \pm 0.7$.
(2) Allende chondrite (eight analyses), $Zr = 6.29 \pm 0.10$ ppm, $Hf = 0.200 \pm 0.005$ ppm, $Zr/Hf = 31.5 \pm 0.5$.
(3) U.S.G.S. Rock BCR-1 (twelve analyses), $Zr = 179 \pm 2$ ppm, $Hf = 4.95 \pm 0.09$ ppm, $Zr/Hf = 36.3 \pm 0.5$.

Agreement of our data with literature values for standard materials is excellent.

TABLE 3. Zr AND Hf ABUNDANCES IN TERRESTRIAL IGNEOUS AND METAMORPHIC ROCKS

Rock description	Zr† (ppm)	Hf (ppm)	Zr/Hf (weight)
Englewood Cliff Section of the Palisade Sill‡			
Chilled lower contact	101	2.86	35
Early dolerite	91	2.62	35
Mg-olivine dolerite	71	1.86	38
Early hypersthene dolerite	94	2.55	37
Late hypersthene dolerite	80	2.16	37
Early pigeonitic dolerite	94	2.61	36
Late pigeonitic dolerite	117	3.17	37
Ferrodolerite	162	4.40	37
Granophyric dolerite	417	10.6	39
Ferrodolerite	216	6.02	36
Upper chilled dolerite	103	2.90	36
St. Helena Island§			
Basalt	243, 247	5.11, 5.02	48
Trachy basalt	572, 576	10.3, 10.7	55
Phonolite	885, 874	15.1, 14.6	59
Littleton Formation in New Hampshire‖			
Low-grade shales and slates	244	5.72	43
	237	5.79	41
	266	6.82	39
	289	6.82	42
	177	4.54	39
	333	7.94	42
	218	5.48	40
Medium-grade schists	252	5.18	49
	303	7.10	43
	257	6.01	43
	286	6.72	43
	296	7.09	42
High-grade schists and gneisses	117	4.62	25
	147	3.92	38
	266	6.27	42
	146	3.75	39
	229	5.58	41
	300	7.10	42

†All Zr contents are corrected for U interference; activity produced by 1.0 μg of U is the equivalent of 11 μg Zr after reactor irradiation.

‡U contents of these rocks are calculated from K contents and a U/K ratio of 1.2×10^{-4} (Heier and Rogers, 1963; Walker, 1969). Ratio is our estimate.

§U contents are estimated from Th and a Th/U = 3.4 which is an average of Heier *et al.* (1964) and Larsen and Gottfried (1960).

‖U contents are estimated from Th and a Th/U = 3.8 of Adams and Weaver (1958); Th contents are calculated from 312 keV photopeak of ^{233}Pa.

Discussion

New Zr and Hf Determinations in Terrestrial Rocks

Zr and Hf have been determined in eleven samples of the Palisade, New Jersey gravitational differentiation series Table 3 and (Fig. 1). There is a factor of more than 4

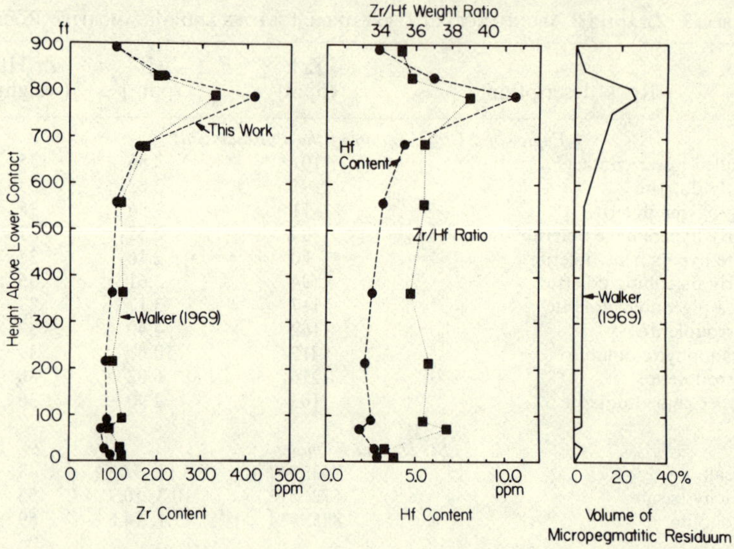

FIG. 1. Zr and Hf in the Palisade, New Jersey gravitational differentiation series.

enrichment of Zr and Hf in the latest granophyric diabase as compared to the chilled border, while the Zr/Hf mass ratios exhibit no systematic variation within the sill and fall within the narrow range of 35 to 39. These results, in contrast to earlier results by some other groups, do not support the existence of any systematic trend for the Zr/Hf ratio as a function of degree of differentiation in a tholeiitic igneous magma. The mean Zr/Hf ratio for the series is 36.3 ± 0.4, in good agreement with the mean value of 37 reported by Brooks (1970) for widely scattered crustal rocks (excluding the Hawaiian volcanic rocks).

We have also determined Zr and Hf in eighteen samples from the pelitic Devonian Littleton metamorphic series in New Hampshire. The results are also summarized in Table 3. The abundances of Zr and Hf vary by a factor of nearly 3 among these samples, but sixteen out of eighteen samples exhibit Zr/Hf ratios in the range of 38 to 42 with a mean value of 40.5 ± 0.4. These data do not offer any evidence for systematic trends in the Zr/Hf ratio with degree of terrestrial regional metamorphism.

Table 3 also presents data for Zr and Hf in rocks derived from the alkali olivine magma at St. Helena Island. Here we do see evidence of a uniform Zr–Hf fractionation trend ranging from Zr/Hf values of 48 in the primary basalt to nearly 60 in the late differentiated phonolite. The increase in the Zr/Hf ratio parallels a factor of nearly 4 increase in the abundances of these elements. Brooks also reported an increase in the Zr/Hf ratio with differentiation among Hawaiian volcanic rocks which were derived from an alkali basalt magma. The abundance of complex stabilizing ions in the Hawaiian and St. Helena magmas and/or crystal chemistry considerations may play a role in this apparent fractionation.

New Data for Zr and Hf in Separated Lunar Minerals

Summaries of our Zr and Hf data for lunar minerals hand picked from crystalline rocks 10047, 10058, 12021, 12035, 15058 and 15065 are presented in Fig. 2 and Table 4. Although extreme care was used during handpicking, the possible presence of minor amounts of mesostasis, especially in the ilmenites, cannot be excluded. Olivine, pyroxene and plagioclase

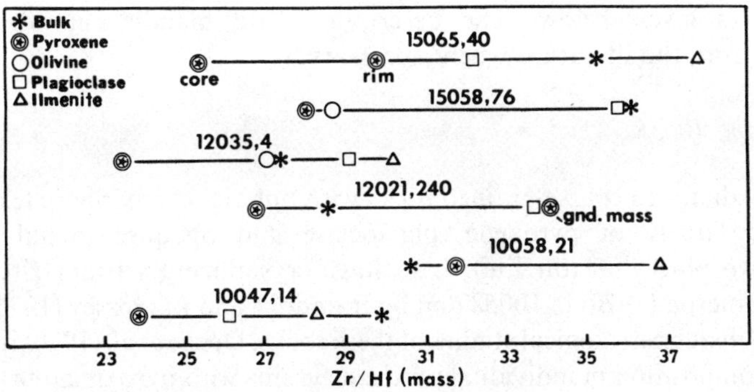

FIG. 2. Zr/Hf ratios in separated lunar minerals.

separates were in general estimated optically to be ~99% pure. The high Zr content of the 10047 plagioclase (158 ppm), while all other plagioclase separates contained much lower Zr abundances, may be indicative of contamination of this separate with mesostasis.

A comprehensive understanding of the Zr–Hf fractionation at the mineralogical level requires a full knowledge of not only the composition and mineralogy of the concerned rocks, but also the paragenesis of the minerals, and the evolution of the parent magma. As not all the information is directly available, several assumptions have to be made. A brief description of rocks and minerals, relevant to our present investigation of Zr–Hf

TABLE 4. ZR AND HF ABUNDANCES AND ZR/HF RATIOS IN SEPARATED LUNAR MINERALS

Mineral		Abundances, ppm 10047	10058	12021	12035	15058	15065	Mean Zr/Hf (w/w)
Pyroxene	Zr	133	186	[32.1, 64.6]†	32.7	82.1	[5.6, 54.5]‡	27.9 ± 1.3
	Hf	5.60	5.86	[0.94, 2.41]†	1.40	2.92	[0.22, 1.83]‡	
Olivine	Zr	—	—	—	10.3	43.8	—	27.9 ± 0.8
	Hf	—	—	—	0.38	1.49	—	
Ilmenite	Zr	462	608	—	550	—	533	33.3 ± 2.4
	Hf	16.3	16.5	—	18.2	—	14.1	
Plagioclase	Zr	(158)	—	42.5	7.9	55.2	6.1	31.4 ± 1.7
	Hf	6.06	—	1.26	0.27	1.54	0.19	
Bulk Rock	Zr	386	272	96.1	47.4	128	92.9	31.3 ± 1.5
	Hf	12.9	8.91	3.36	1.73	3.55	2.63	

† Lower value is for ground mass and higher value is for phenocryst pyroxene.
‡ Lower value is for core and higher value is for rim material.
Note: Only the bulk Zr data are corrected for the U fission contribution to ^{95}Zr. However, reference to the available literature indicates the correction would be negligible as compared with the estimated analytical uncertainty of ±5–10% for Zr in these minerals. Contamination with mesostasis is suspected in the case of the 10047 plagioclase. The ground mass pyroxene in 12021 was misidentified in our previous preliminary report (Garg *et al.*, 1977).

fractionation, is given below. The descriptions are mainly our own observations, substantiated from the literature where necessary.

Rocks 10047 and 10058

These are medium- to coarse-grained rocks with ophitic to subophitic texture. The rocks are composed largely of pyroxene, plagioclase and opaques (mainly ilmenite with inseparable ulvöspinel and tonolite). Late stage crystallizing yttrium–zirconium silicates reported by Cameron (1970) in 10058 and by Ramdohr and El Gorsey (1970) in 10047 were not recognized in our petrological studies of these rocks. Dence *et al.* (1970) reported extreme variations in composition in individual pyroxene grains with pyroxferroite occurring at the margins. In the present study no attempt was made to separate this late stage pyroxene from the main crystals.

The textural relationship of these rocks should suggest a clear-cut crystallization sequence. Ilmenite is a mineral that may crystallize from an early to almost the final stages of crystallization of the igneous rocks. Undoubtedly, some ilmenite crystallized early from the magma which produced these rocks. The absence of olivine from rocks 10047 and 10058 indicates that either olivine was completely resorbed or underwent a complete fractional separation by sinking from this portion of the magma. If the latter is true, then it is possible that the early formed ilmenite also sank with the olivine. Agrell *et al.* (1970) have described a late stage ilmenite in rock 10058 which we take to correspond to our material. Simpson and Bowie (1970) described ilmenite from these rocks as partially molded on feldspar and intergrown with pyroxene. In the light of the above a crystallization sequence can safely be assumed where plagioclase is followed by and crystallizes with pyroxene over a major portion of the crystallization sequence. Ilmenite now found in the rock is no doubt the last to crystallize. The presence of ophitic texture in these rocks where pyroxene crystals enclose plagioclase crystals further supports the suggestion that plagioclase crystallized early and was followed by pyroxene, or the two crystallized simultaneously.

In these rocks ilmenites are enriched in Zr and Hf and also have high Zr/Hf ratios, consistent with late crystallization. The pyroxenes and the plagioclase in 10047 have low and similar Zr/Hf ratios, suggesting their simultaneous early crystallization. As noted previously, however, the 10047 plagioclase we separated does have high Zr–Hf abundances and may be contaminated by mesostasis.

Rocks 12021 and 12035

These crystalline rocks represent the two compositional groups of Goles *et al.* (1971). Rock 12021 is a coarse-grained porphyritic basalt (or micro-gabbro) with pyroxene phenocrysts in a variolitic ground mass of pyroxene and plagioclase. The textural break in the pyroxene might be explained by two stages of crystallization, although two types of texture can also be produced at linear cooling rates involving only a single-stage cooling history (Lofgren *et al.*, 1974). In either case phenocrystic pyroxene would be the first to form. These early phenocrysts show a high Zr content and do have the lowest Zr/Hf ratio, whereas the later groundmass pyroxene and plagioclase have almost identical high Zr/Hf ratios.

Rock 12035 is described by Rhodes *et al.* (1977) as a very slowly cooled olivine basalt. It contains approximately 40% olivine, 45% plagioclase, 15% pyroxene and less than 1% opaques. Like 12021, its crystallization history may be complex. Pyroxene from 12035 has the lowest Zr/Hf ratio and ilmenite the highest ratio. This suggests ilmenite crystallized late along with plagioclase, which has a similar high Zr/Hf ratio in this rock.

Rock 15065

This is a coarse-grained, quartz normative rock from the rim of Elbow Crater. Pyroxene, plagioclase and ilmenite are the major phases. The pyroxene is strongly zoned with a pigeonite core and an augite rim. We have separated all three phases and have also managed to separate the outer augitic pyroxene from the pigeonite core. The separation is optically estimated to be >98% pure. Figure 2 shows the Zr/Hf ratios in different phases from this rock. Ilmenite again has the highest Zr/Hf ratio followed by plagioclase and then pyroxene. This is strongly supported by the crystallization sequence in this rock where pyroxene crystallized earlier than the plagioclase. The textural relationships and phase equilibrium experimental studies by Longhi *et al.* (1970) suggest that plagioclase appeared late in the crystallization sequence. The pigeonite cores in pyroxenes have lower Zr/Hf ratios compared to the late formed margins. Ilmenite which has the highest ratio probably crystallized last. The high Zr and Hf abundances in the outer rim of the pyroxene could be related to small ilmenite inclusions which crystallized simultaneously with the late stage pyroxene.

In summary, for the rocks studied, ilmenite appears to be an excellent concentrator for Zr and Hf. Ilmenite also has the highest Zr/Hf ratio and its apparently late position in the crystallization sequence for these rocks would lend support to our suggestion of Zr–Hf fractionation through exclusion of Zr^{3+} from the earliest formed minerals. It is recognized that lunar ilmenite is difficult to separate with high purity and may contain mesostasis. Therefore, the high Zr/Hf ratios in the ilmenites may partially reflect their content of this final liquid. However, the argument based on the presence Zr^{3+} causing Zr to be enriched with respect to Hf in the final liquids would be equally valid, if this were the case. The argument is further supported by the enrichment in Zr and slightly higher Zr/Hf ratios in the rims of the zoned pyroxene crystals.

Obviously, none of the above proves the presence of Zr^{3+}. However, if oxygen fugacities are extremely low, the presence of some Zr^{3+} would enhance the Zr–Hf fractionation in a manner consistent with data observed for the mineral phases in the rocks examined.

New Data for Zr and Hf in Meteorites

Drake (1975) has suggested that some achondrites (Juvinas and Moore Co.) appear to have been formed under f_{O_2} conditions several orders of magnitude less than for lunar samples. If there is a Zr–Hf fractionation due to the presence of Zr^{3+}, achondrite Zr/Hf ratios could be quite variable. New Zr and Hf data in achondrites are given in Table 5. The data for Nakhla yield a Zr/Hf ratio of 32 which is similar to that for the ordinary chondrites and many terrestrial rocks. Reid and Bunch (1975) have discussed the uniqueness of Nakhla among meteorites, its higher degree of oxidation and its similarity to terrestrial rocks. Our Zr–Hf data for the eucrites, howardites and diogenites, as well as for the unique achondrite Nakhla are plotted in Fig. 3. It may be seen that the eucrites and diogenites appear to define a straight line representing a distinct Zr–Hf fractionation.

McCarthy *et al.* (1972, 1973) and Fukuoka *et al.* (1977) have pointed to chemical evidence in support of a eucrite–diogenite mixing model for howardite origin. The fact that howardites are polymict breccias is also consistent with this model. As may be seen from Fig. 3, our data for Zr and Zr/Hf ratios also support this mode of origin for the howardites.

Our mean values for diogenites are Zr = 0.67 ppm and Zr/Hf = 25. If we combine our

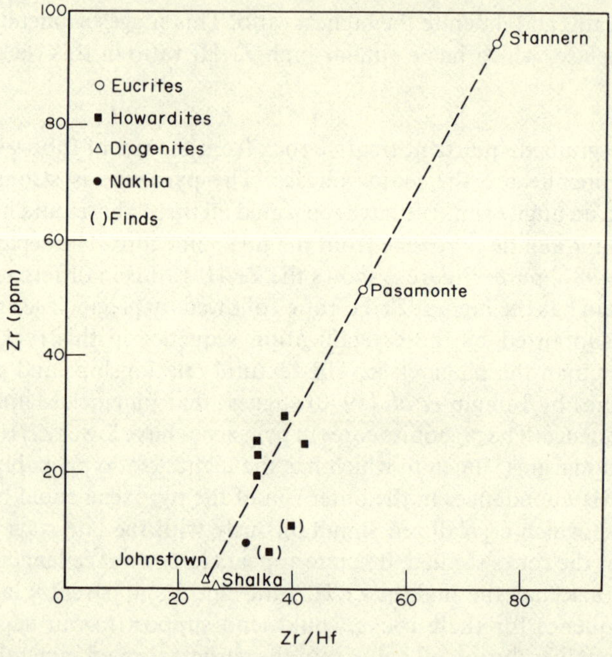

FIG. 3. Zr–Hf relationships in eucrites, diogenites, howardites and nakhlites.

values for Zr and Hf with the data for Zr in eight eucrites analyzed by McCarthy *et al.* (1973), we obtain mean values for eucrites of Zr = 56 ppm and Zr/Hf = 64. Using these values for the end-members in the mixing model, the mean composition of the seven howardites we analyzed would be equivalent to a 35% eucrite and 65% diogenite mixture. Chou *et al.* (1976) have noted that a chondritic component of approximately 2–3% is required to account for siderophile contents in a number of howardites. This minor component would not significantly effect the Zr/Hf ratio in the howardites.

For the howardite, Washougal, our calculated mixture of 44% eucrite and 56% diogenite is in exact agreement with the mixture calculated by Fukuoka *et al.* (1977), based on a mixing model employing sixteen elements.

New data for Zr and Hf in enstatite chondrites are also presented in Table 5. Excluding the high Zr abundance for Blithfield, the mean Zr abundance for enstatite chondrites is 5.7 ± 1.5 ppm and the mean Zr/Hf ratio is 42 ± 6. The Zr/Hf ratios for individual meteorites do exhibit an appreciable scatter, and the data to not permit us to conclude that the enstatite chondrites differ from the ordinary chondrites in their Zr/Hf ratio.

Conclusions

We have shown that Zr and Hf are not always fractionated in terrestrial igneous differentiation or regional metamorphism. Our data do support a Zr–Hf fractionation in the differentiation of the alkali olivine magma at St. Helena Island. A strong Zr–Hf fractionation does exist among lunar rocks and minerals. Ilmenite, an apparently late forming mineral in the lunar rocks we have studied here is an excellent concentrator of Zr and Hf and always exhibits high Zr/Hf ratios. The lunar fractionation patterns would be

TABLE 5. NEW DATA FOR Zr AND Hf AND SEVERAL ADDITIONAL ELEMENTS IN METEORITES

Class/Meteorite	No. of analyses	Zr (ppm)	Hf (ppm)	Zr/Hf (w/w)	Fe (%)	Co (ppm)	Cr (ppm)	Sc (ppm)
Howardites:								
Binda	1	11.0	0.27	40	12.4	16.1	7150	18.9
Kapoeta	1	30	0.77	39	13.4	22.6	4810	21.5
Frankfort	1	23	0.68	34	11.8	17.6	4770	18.4
Washougal	1	25	0.60	42	13.2	22.8	5290	20.2
Yurtuk	1	19.5	0.58	34	11.1	11.4	3770	17.4
Isna	1	6.3	0.174	36	23.1	764	3660	8.7
Bholghati	1	25	0.75	34	13.2	27.2	5960	20.9
Eucrites:								
Pasamonte	2	52	1.01	52	14.2	5.3	1870	29.0
Stannern	1	95	1.26	75	13.9	6.3	1810	29.6
Nakhlite:								
Nakhla	1	8.1	0.25	32	15.4	48.4	1600	54.5
Diogenites:								
Shalka	2	0.11	0.0040	—	12.6	22.9	10200	9.6
Johnstown	10	1.25	0.050	25	12.0	19.5	4900	14.8
Aubrites:								
Cumberland Falls	2	0.25	0.0098	—	0.76	3.8	432	5.6
Shallowater	2	8.0	0.19	42	12.8	376	884	6.0
Norton County	2	0.32	0.0036	—	0.78	4.7	364	7.3
Ureilites:								
Havero	2	0.34	0.0038	—	15.8	105	3810	4.9
Enstatite chondrites:								
Indarch	1	5.1	0.102	50	31.4	988	3290	5.8
Pillistfer	1	4.5	0.112	40	35.8	1300	2900	7.2
Blithfield	1	(16.6)	0.255	—	22.2	369	6690	7.6
Abee	2	4.2	0.096	44	33.2	1090	3730	5.0
Daniel's Kuil	1	7.1	0.162	44	25.5	755	4650	7.5
Atlanta	1	7.7	0.150	51	28.0	1000	3930	6.7
Hvittis	1	4.3	0.127	34	26.0	936	2460	7.2
Jajh deh Kot lalu	1	7.3	0.168	44	22.2	725	2990	7.9
Meteoritic phases:								
Bondoc stony-iron								
Gross silicates	2	2.5	0.077	32	—	—	—	—
Magnetic silicates	1	1.68	0.053	32	39.9	1850	3260	5.1
Non-magnetic silicates	1	3.8	0.097	39	13.2	338	5130	16.1
Sardis octahedrite troilite	2	11.6	0.26	45	—	—	—	—
Antofagasta pallasite olivine	1	0.7	0.02	—	9.6	24.5	9130	1.1

Note: The Zr data for separated meteorite phases are not corrected for the U fission contribution to ^{95}Zr. The data for Fe, Co, Cr and Sc were determined in the same samples by INAA. Zr/Hf ratios are not listed when Zr abundances are below 1 ppm, due to the inferior counting statistics at this low abundance level.

consistent with a model involving the exclusion of the larger Zr^{3+} ion from the earliest formed minerals, if oxygen fugacities are low enough to selectively reduce the Zr^{4+} to Zr^{3+} while Hf remains as Hf^{4+}. Chondritic meteorites exhibit rather uniform Zr abundances and Zr/Hf ratios, but a distinct Zr–Hf fractionation is observed among the eucrite and diogenite achondrites. The Zr and Hf data strongly support a simple eucrite–diogenite mixing model for the genesis of the howardites we have studied.

Acknowledgements

This work was supported in part by N.A.S.A. Grant NGR-18-001-058. We gratefully acknowledge the assistance of the individuals who provided samples for these studies and personnel at the University of Missouri and Georgia Institute of Technology research reactor facilities who aided in irradiations.

Sources of meteorites: Dr. D. Lal, Physical Research Laboratory, India; Dr. K. K. Turekian, Peabody Museum, Yale University; Dr. C. B. Moore, Arizona State University; Dr. E. Jarosewich, Smithsonian Institution; Dr. K. R. Dawson, Geological Survey of Canada; Dr. M. H. Hey, British Museum; Dr. Y. O. Fortier, Geological Survey of Canada; Dr. J. F. Lovering, University of Melbourne; Dr. Roy S. Clarke, Jr., U.S. National Museum; Dr. D. Heymann, Rice University; Dr. H. B. Wiik, Geological Survey of Finland.

Sources of rocks: Dr. A. Poldervaart, Columbia University—Palisade Sill; Dr. P. W. Gast, Columbia University—St. Helena Series; Dr. D. M. Shaw, McMaster University—Littleton Formation.

References

ADAMS, J. A. S. and WEAVER, C. E. (1958) Thorium-to-uranium ratios as indicators of sedimentary processes—an example of geochemical facies. *Bull. Amer. Ass. Petrol. Geol.* **42**, 387.

AGRELL, S. O., SCOON, J. H., MUIR, I. D., LONG, J. V. P., MCCONNELL, J. D. C. and PECKETT, A. (1970) Observations on the chemistry, mineralogy and petrology of some Apollo 11 lunar samples. *Proc. Apollo 11 Lunar Sci. Conf., Geochim. Cosmochim. Acta*, Suppl. 1, **1**, 93–128.

BROOKS, C. K. (1969) On the distribution of zirconium and hafnium in the Skaergaard intrusion, East Greenland. *Geochim. Cosmochim. Acta* **33**, 357–374.

BROOKS, C. K. (1970) The concentrations of zirconium and hafnium in some igneous and metamorphic rocks and minerals. *Geochim. Cosmochim. Acta* **34**, 411–416.

BUTLER, J. R. and THOMPSON, A. J. (1965) Zirconium: hafnium ratios in some igneous rocks. *Geochim. Cosmochim. Acta*, **29**, 167–175.

CAMERON, E. N. (1970) Opaque minerals in certain lunar rocks from Apollo 11. *Proc. Apollo 11 Lunar Sci. Conf., Geochim. Cosmochim. Acta*, Suppl. 1, **1**, 221–245.

CHOU, C-L., BOYNTON, W. V., BILD, R. W., KIMBERLIN, J. and WASSON, J. T. (1976) Trace element evidence regarding a chondritic component in howardite meteorites. *Proc. 7th Lunar Sci. Conf. Geochim. Cosmochim. Acta*, Suppl. 7, **3**, 3501–3518.

CHYI, L. L. and EHMANN, W. D. (1973) Zirconium and hafnium abundances in some lunar materials and implications of their ratios. *Proc. 4th Lunar Sci. Conf., Geochim. Cosmochim. Acta*, Suppl. 4, **2**, 1219–1226.

CONDIE, K. C. and LO, H. (1971) Trace element geochemistry of the Louis Lake batholith of early Precambrian age, Wyoming. *Geochim. Cosmochim. Acta* **35**, 1099–1119.

DENCE, M. R., DOUGLAS, J. A. V., PLANT, A. G. and TRAILL, R. J. (1971) Mineralogy and petrology of some Apollo 12 samples. *Proc. 2nd Lunar Sci. Conf., Geochim. Cosmochim. Acta*, Suppl. 2, **1**, 285–299.

DRAKE, M. J. (1975) The oxidation state of europium as an indicator of oxygen fugacity. *Geochim. Cosmochim. Acta* **39**, 55–64.

EHMANN, W. D. and CHYI, L. L. (1974) Zirconium and hafnium in meteorites. *Earth Planet. Sci. Lett.* **21**, 230–234.

EHMANN, W. D. and REBAGAY, T. V. (1970) Zirconium and hafnium in meteorites by activation analysis. *Geochim. Cosmochim. Acta* **34**, 649–658.

ESSON, J., HAHN-WEINHEIMER, P. and JOHANNING, H. (1968) Determinations of hafnium by neutron activation and variation in the Zr/Hf ratio of some granite masses. *Talanta* **15**, 1111–1118.

FUKUOKA, T., MA, M-S. and SCHMITT, R. A. (1977) Genesis of howardites as mixtures of eucrites and diogenites. In *Lunar Science-VIII*, The Lunar Science Institute, Houston, 340–342

GANAPATHY, R., PAPIA, G. M. and GROSSMAN, L. (1976) The abundances of zirconium and hafnium in the solar system. *Earth Planet. Sci. Lett.* **29**, 302–308.

GARG, A. N. and EHMANN, W. D. (1976) Zirconium-hafnium fractionation in chemically defined lunar rock groups. *Proc. 7th Lunar Sci. Conf., Geochim. Cosmochim. Acta,* Suppl. 7, **3**, 3397–3410.

GARG, A. N., ALI, M. Z. and EHMANN, W. D. (1977) The distribution of Zr and Hf in lunar minerals and rock 12013. In *Lunar Science-VIII,* The Lunar Science Institute, Houston, 346–348.

GOLES, G. G., DUNCAN, A. R., LINDSTROM, D. J. MARTIN, M. R., BEYER, R. L., OSAWA, M., RANDALE, K., MEEK, L. T., STEINBORN, T. L. and MCKAY, S. M. (1971) Analyses of Apollo 12 specimens—compositional variations, differentiation processes, and lunar soil mixing models. *Proc. 2nd Lunar Sci. Conf., Geochim. Cosmochim. Acta,* Suppl. 2, **2**, 1063–1081.

GOTTFRIED, D., GREENLAND, L. P. and CAMPBELL, E. Y. (1968) Variation of Nb–Ta, Zr–Hf, Th–U and K–Cs in two diabase–granophyre suites. *Geochim. Cosmochim. Acta,* **32**, 925–947.

GULSON, B. L. (1969) Electron microprobe determination of Zr/Hf ratios in zircons from the Yeoval diorite complex, N.S.W., Australia. *Lithos* **3**, 17–23.

HEIER, K. S., MCDOUGALL, I. and ADAMS, J. A. S. (1964) Th, U and K in Hawaiian lavas. *Nature,* Lond. **201**, 254.

HEIER, K. S. and ROGERS, J. J. W. (1963) Radiometric determination of thorium, uranium and potassium in basalts and in two magmatic differentiation series. *Geochim. Cosmochim. Acta* **27**, 137–154.

HERRARA, R. and HEURTEBISE, M. (1974) Neutron activation analysis of trace elements in quartz sands: Its possibilities in the assessment of provenance. *Chem. Geol.* **14**, 81–93.

HORN, M. K. and ADAMS, J. A. S. (1966) Computer-derived geochemical balances and element abundances. *Geochim. Cosmochim. Acta* **30**, 279–297.

KOSTERIN, A. V., ZUEV, V. N. and SHEVALEEVSKII, I. D. (1958) Zirconium: hafnium ratio in zircons in some igneous rocks of Northern Kirgizia. *Geochemistry* **1958**, 116–119.

KOSTERIN, A. V., SHEVALEEVSKII, I. D. and RYBALOVA, E. K. (1960) The Zr/Hf ratio in the zircons of some igneous rocks of the north slope of the Karamin Mountain Range. *Geokhimiya* **5**, 541–545.

KUMAR, P. A., GARG, A. N. and EHMANN, W. D. (1977) Precise simultaneous determination of zirconium and hafnium in silicate rocks, meteorites and lunar samples. *J. Radioanal. Chem.* **40**, 51–64.

LARSON, E. S. and GOTTFRIED, D. (1960) Uranium and thorium in selected suites of igneous rocks. *Amer. J. Sci.* **258-A**, 151–169.

LAUL, J. C., HILL, D. W. and SCHMITT, R. A. (1974) Chemical studies of Apollo 16 and 17 samples. *Proc. 5th Lunar Sci. Conf., Geochim. Cosmochim. Acta,* Suppl. 5, **2**, 1047–1066.

LOFGREN, G., DONALDSON, C. H., WILLIAMS, R. J., MULLINS, O. JR. and USSELMAN, T. M. (1974) Experimentally reproduced textures and mineral chemistry of Apollo 15 quartz normative basalts. *Proc. 5th Lunar Sci. Conf., Geochim. Cosmochim. Acta,* Suppl. 5, **1**, 549–567.

LONGHI, J., WALKER, D., GROVE, T. L., STALPER, E. M. and HAYS, J. F. (1974) The petrology of the Apollo 17 mare basalts. *Proc. 5th Lunar Sci. Conf., Geochim. Cosmochim. Acta,* Suppl. 5, **1**, 447–469.

MCCARTHY, T. S., AHRENS, L. H. and ERBANK, A. J. (1972) Further evidence in support of the mixing model for howardite origin. *Earth Planet. Sci. Lett.* **15**, 86–93.

MCCARTHY, T. S., ERBANK, A. J. and WILLIS, J. P. (1973) On the origin of eucrites and diogenites. *Earth Planet. Sci. Lett.* **18**, 433–442.

NEVES, J. M. C. and NUNES, J. E. L. (1974) High hafnium members of the zircon-hafnon series from the granite pegmatites of Zambezia, Mozambique. *Contr. Miner. Petrol.* **48**, 73–80.

PALME, H. (1974) Non-destructive determination of several trace elements in terrestrial and meteoritic samples with 14 MeV neutrons. In *Analyse Extraterrestrischen Materials,* Springer-Verlag, New York, 147–161.

RAMDOHR, P. and EL GORESY, A. (1970) Opaque minerals of the lunar rocks and dust from mare tranquillitatis. *Science* **167**, 615–618.

RANDLE, K. (1974) Some trace element data and their interpretation for several new reference samples obtained by neutron activation analysis. *Chem. Geol.* **13**, 237–256.

REID, A. M. and BUNCH, T. E. (1975) The Nakhlites. Part II: Where, when and how. *Meteoritics,* **10**, 317–324.

RHODES, J. M., BLANCHARD, D. P., BRANNON, J. C., RODGERS, K. V. and DUNGAN, M. A. (1977) Chemistry, classification and petrogenesis of Apollo 12 mare basalts. In *Lunar Science–VIII,* The Lunar Science Institute, Houston, 804–806.

SHIMA, M. and HINTENBERGER, H. (1977) Elemental and isotopic abundances of Ti, Zr, and Hf in stony meteorites. *Meteoritics* **11**, 364–368.

SIMPSON, P. R. and BOWIE, S. H. U. (1970) Quantitative optical and electron-probe studies of opaque phases in Apollo 11 samples. *Proc. Apollo 11 Lunar Sci. Conf., Geochim. Cosmochin. Acta,* Suppl. 1, **1**, 873–889.

WALKER, K. R. (1969) The Palisades Sill, New Jersey, a reinvestigation. *Geol. Soc. Amer., Special Paper III,* Boulder, Colorado, 178 pp.

Distribution and Source of Elements in Gabbroids of Different Geochemical Types

O. M. GLAZUNOV and G. S. PLYUSNIN

Vinogradov Institute of Geochemistry, Academy of Sciences, Irkutsk, U.S.S.R.

Abstract

Such gabbroids show an unequal distribution of elements. The contrast of distribution increases with the degree of differentiation and contamination. Gabbroids can be with a different source of elements. Sources of orebearing gabbroids (Fe–Ti–V and Ti–Zr) are in the deep mantle. The low ratio, the small range of variance of Sr^{87}:Sr^{86} (0.7035 + 0.7056) and also the absence of presence of a direct connection with petrogenic elements indicate a weak contamination of gabbroids of this type by the Earth's crust matter.

The contamination is checked by a quick intrusion of melt. Therefore the content of melt reflects rather the evolution of the deeper stage than the inner chamber one. And the melt there is a product of the pyroxenitic mantle part enriched by Ti, Zr, Nb, ΣTR. Possibly most such melts do originate directly in the mantle. All rocks associations have alkali-ultrabasic features. So the high concentration of Ti and other elements is related to a deep source. Mantle source elements lie with experimental data.

On the other hand (gabbroid Cr–Ti–Au type—stratiform intrusion), the distribution of most rare elements does not go with the mechanism of fractional crystallization of basaltic magma. The character of mineralization and the distribution of elements because of intermediate chambers is defined by the contamination process. The contamination can be taken as an extraction of rare elements from the substratum. The development of such a process belongs to the deep earth's crust. The role of mass-exchange *in situ* is very small, since there is a low rare element content in endocontact gabbroid and mother rocks.

Gabbroids are magmatic but a part of clinopyroxenites and ore magnetites are metasomatic products. The influence of crust factors produces a higher value and dispersion of incoherent elements (Rb, K, Sr). This leads to an increase of the isotope ratio of Sr (at 0.7096 ± 0.0007) and its correlation with K and Sr.

The contamination is wide ranging and eventually produces dioritoids accompanied by, unusual for gabbroids, gold–sulphide mineralization. The new division of gabbroids can explain the difference of the geochemical specialization and the source of ore elements.

In gabbroids of different geochemical types the uneven elements distribution depends on the chamber depth, the degree of selective melting and the contamination process. The distribution contrast also increases with a higher differentiation degree.

In the folded regions there are the following association and geochemical gabbroid types: A—gabbro -pyroxenite, Ti–V, Ti–Zr; B—gabbro-norite, Ni–Au, Ti, Cr–Pt; C—gabbroids together with ultrabasites (Glazunov, 1977)†. Additionally there are gabbroids comagmatic to ultrabasic rocks yet without a distinct geochemical position.

Gabbroids of gabbro-pyroxenite association form steep dike bodies of a great extension. They are characteristic with a gabbro-pyroxene composition: low Mg/Fe (0.8–1.2), anomalous Ti, V, Zr and K (Table 1). They have a low initial ratio of Sr^{87}:Sr^{86} (0.7043–0.7049 ± 0.0007) with slight variations.

† The geochemical type of gabbroids is determined by the content of elements important for prospecting.

TABLE 1. GEOCHEMICAL TYPES OF GABBROIDS

	N	Ti	V	Zr	Cr	Ni	K	Rb	Sr	Ti/Cr	Ni/Co	K/Rb	Sr/Ca	Rb/Sr	$\frac{Sr^{87}/Sr^{86}}{N}$
1	62	31000	320	300	150	120	6200	10	880	207	3.0	620	0.011	0.011	$\frac{0.7043\text{–}0.7054}{0.7047}$ 6
2	24	3300	218	70	550	267	2600	7	240	6	4.2	371	0.0025	0.03	$\frac{0.7030\text{–}0.7050}{0.7040}$ 4
3	84	7300	176	44	98	11	5525	12	770	74	1.3	433	0.01	0.015	$\frac{0.7056\text{–}0.7080}{0.7074}$ 4

1—Ti–V–Zr (gabbro-pyroxene association); 2—gabbroids comagmatic to ultrabasic rocks; 3—Ti (gabbro-norite association). Concentration: $n \times 10^{-4}\%$. N: number of samples.

Gabbro-norites are represented by stratified lopolites. Gabbroids are particular with a great chemical heterogeneity on sections, a high Sr^{87}:Sr^{86} (0.7065–0.7096 ± 0.0007), Sr, Rb and a low V, Zr, Ni:Co.

Gabbroids comagmatic to ultrabasic rocks do differ by a higher Mg (9–10%), Cr, a low Sr^{87}:Sr^{86} (0.7040–0.7050), Ti:Cr, K:Rb and also a small concentration of Sr and Rb. They are compared with basic rocks of the palingenic series with a high Li, Rb, Sr^{87}:Sr^{86} (up to 0.7090), which do also accompany ultrabasites (N 5825, Г-57; Fig. 1—1).

The diverse trace-element composition of gabbroids questions their elements source. Therein a good indicator of genesis and place of rock generation is the ratio Sr^{87}:Sr^{86} (Faure *et al.*, 1972) and series of primary geochemical characteristics of most deep parageneses (Cr:Ti, Ni:Co, etc.).

Different element sources in gabbroids are admitted. Gabbroids of Ti–V and Ti–Zr arise apparently from the deep upper mantle part. This is put down to the anomalous alkalinity and Ti-bearing of rocks, presence of picroilmenite, chromspinelide and low-alumina enstatite. Significant is the direct ratio Cr and Ti in rocks and minerals comparable with that of mantle xenoliths and kimberlites. Low Sr^{87}:Sr^{86} and the absent connection with petrogenic elements explain the weak contamination of this gabbroid type with the earth's crust matter. The contamination is checked by a quick intrusion of melt in the comparatively thin earth's crust (Table 2). Therefore, the magma composition characterizes the evolution of the deep but not the inner-chamber phase. The small variation of Sr^{87}:Sr^{86} witnesses the closed Rb–Sr system and that the primary melt belongs to the direct mantle melt.

TABLE 2. CORRELATION OF GEOCHEMICAL GABBROIDS PARAMETERS WITH THE THICKNESS OF CONSOLIDATED EARTH'S CRUST

	1	2	3
K_2O/Na_2O	0.32	0.35	0.40
Fe_2O_3/FeO	0.40	0.50	0.70
$Rb/Sr \cdot 10^2$	1.13	2.33	8.43
Sr^{87}/Sr^{86}	0.7049 ± 0.0005	0.7067	0.7131
Earth's crust thickness in km†	40	50	60

Geochemical types of gabbroids: 1—Ti–V–Zr (gabbro-pyroxene association); 2—Ti (gabbro-norite association); 3—Fe–V (gabbro-diabase association).

† On Fotiadi *et al.* (1963).

The assumption of a deep origin of the association can be confirmed by known experiments (McGregor, 1969) and thermodynamic calculations (Marakushev *et al.*, 1975). But first of all it is based on the low value and the slight variation of $Sr^{87}:Sr^{86}$. The calculation of the association composition shows a wide spreading of the pyroxene zone in the upper mantle enriched by Ti, Zr and K and comes near the model of the pyroxene pyrolite (Green, Ringwood, 1968).

Gabbro comagmatic to ultrabasics undergo an even lesser contamination (Γ-72, Fig. 1—1). At the formation of palingenic–metasomatic gabbro, as seen from the strong dependence of $Sr^{87}:Sr^{86}$ on Mg (Fig. 1) and on their saturation with K–Li, processes of mass-exchange of gabbroid magmas take part with the surrounding rocks of various composition.

In stratiform gabbroids of Ti, Ni–Au and Cr–Pt the distribution of most trace elements does not agree with the mechanism of fractional crystallization of the basalt magmas. The particularities of element distribution and mineralization character are explained by contamination processes owing to the existing intermediate melt chambers. The development of such a process belongs to the deep lithosphere parts where, in passage the intrusion has overcome the resistance of heavy strata of the earth's crust and extracted series of incoherent elements. The role of mass-exchange *in situ* is very small since there is a low trace-element content in the endocontact. The distinct direct correlation of $Sr^{87}:Sr^{86}$ with Sr, K_2O and also MgO (Figs. 1, 2) can be explained by a supply of acid and partly ultrabasic

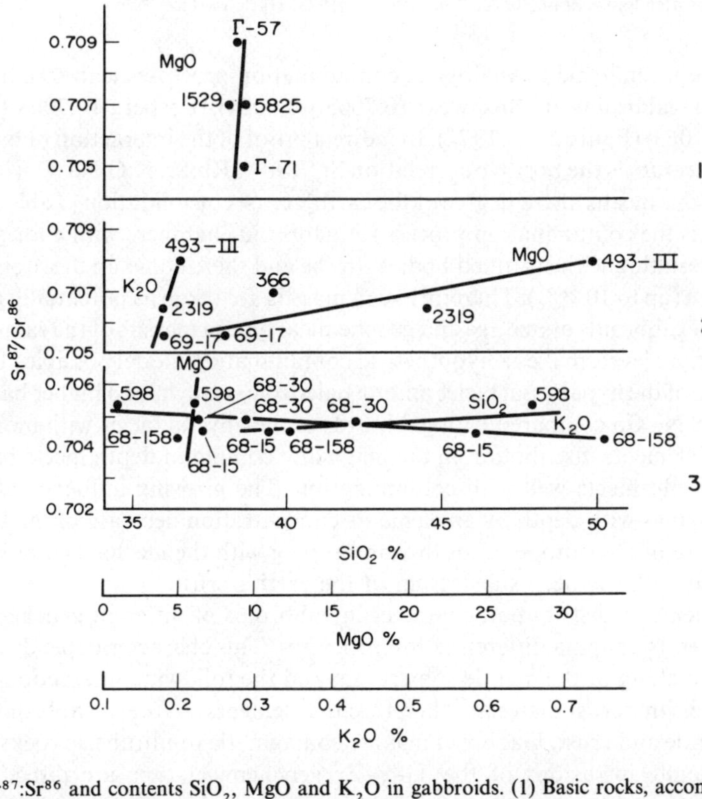

FIG. 1. Ratio $Sr^{87}:Sr^{86}$ and contents SiO_2, MgO and K_2O in gabbroids. (1) Basic rocks, accompany ultrabasic rocks. (2) Ti and Ni-Au geochemical (gabbro-norites type). (3) Ti–V–Zr geochemical type (gabbro-pyroxene).

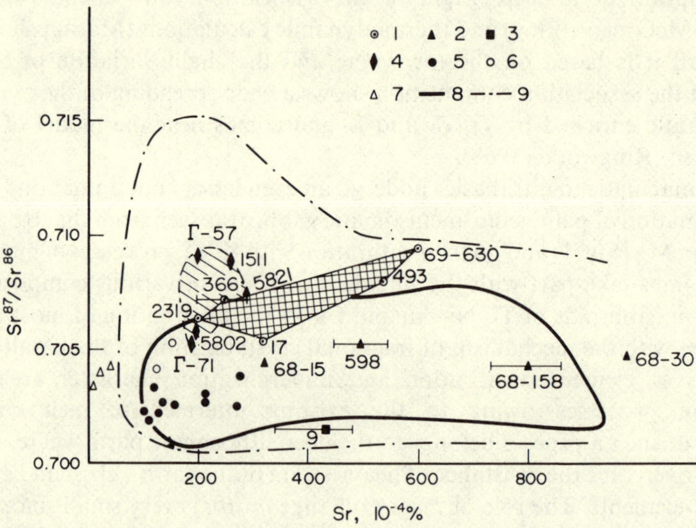

FIG. 2. Ratio Sr^{87}:Sr^{86} and concentration Sr in gabbroids. Associations: (1) Ti and Ni–Au geochemical type (gabbro-norite) and (2) Ti–V–Zr geochemical type (gabbro-pyroxene), (3) gabbro-peridotite, (4) gabbro, accompany ultrabasics, (5) basalts, (6) oceanic amphibolites, (7) oceanic peridotites. Basalts field. (8) Oceans. (9) Continents;(5–9) (after Faure *et al.*, 1972; Bonatti *et al.*, 1973; Hart, 1971).

material in the juvenile melt. Analogous contamination processes lead to a higher isotope ratio of Sr in gabbroids of Bushweld (0.7065 ± 0.0024), Losberg (0.7064 ± 0.0024) and Skaergard (0.7063) (Faure *et al.*, 1972). In indirect proof of the interaction of basaltoids with the crust substratum is the positive correlation Sr^{87}:Sr^{86}, Rb:Sr, K_2O:Na_2O with a thickness variation and this means also a degree of the earth's crust consolidation (Table 2). Especially characteristic is the contamination process for gabbroid chambers with a longer halt in the earth's crust resulting with stratified bodies. In the end there come up dioritoids sometimes enriched by Au (up to 10^{-4} %). Therefore such massifs are taken as potentially gold-bearing.

The study of gabbroids of one age and geochemical type in massifs of the same space but of different structure levels makes a hypothetical contamination model possible. Table 3 shows that gabbroids of the hypabyssal facies among paleozoic strata have a higher basicity and are richer with Cr, Ni, Co compared with gabbroids of the abyssal facies with more K, Zr, Au. The observed elements distribution in the magmatic column in depth meets badly with the differentiation but meets well with contamination. The growing influence of elements of "granitic" magmas with depth by an elements concentration decrease of the Fe group and also the increase of the isotope ratios Sr is in keeping with the mechanism of interaction of the basalt melt with the deep substratum of the earth's crust.

Thus, the elements distribution character in gabbroids of different geochemical types is not only defined by magma differentiation processes. This character depends on the degree of the partial melting of the mantle substratum and the following interaction of the basalt melt with the earth's crust material. Therefore the elements source in gabbroids can also be the upper mantle and crust. Gabbroid massifs comagmatic to ultrabasic rocks and dikelike gabbro-pyroxenite intrusions of the Ti–V–Zr geochemical type are direct mantle melt products. They are without contamination which is checked by a quick melt intrusion along

TABLE 3. GABBROID COMPOSITION CHANGE UNDER CONTAMINATION IN THE HYPOTHETICAL COLUMN (gabbro-norite association)

Massif and rock composition	Size, km^2	Mother rocks, age	Weight %					$n \times 10^{-4}$					10^{-5}	
			TiO_2	FeO	MgO	K_2O	Na_2O	Cr	Ni	Co	Sc	Zr	Au	Sr^{87}/Sr^{86}
Kryzhin (gabbro, gabbro-norite)	70	Carbonate-volcano genous, Cambrian	1.89 (1.57)	7.39 (2.75)	6.69 (3.05)	0.62 (0.50)	3.17 (0.89)	280	140	60	29	42	0.46	0.7056–0.7080
Oseredok (gabbro, gabbro-pyroxenites)	15	Carbonate-schist. Upper Proterozoic	1.98 (1.56)	7.08 (3.04)	6.50 (3.26)	0.72 (0.78)	2.55 (0.71)	125	92	44	18	63	2.6	0.7065–0.7096
Demir-Taiga (gabbro-norites, gabbro-diorites)	6.0	Carbonate. Lower Proterozoic	0.90 —	6.67 (3.50)	6.09 (3.60)	0.92 (0.80)	2.98	99	29	25	20	87	8.2	0.7070–0.7125

In brackets: average square deviation. The emplacement depth of different massifs is determined by geological data.

the tectonic zones. Gabbroid intrusions with longer and repeated halts in the thick earth's crust do form under contamination. It is strong in depth and gives a higher $Sr^{87}:Sr^{86}$ and also a correlation of this index with series of petrogenic elements. Crust factors can change the intrusion's ore profile. The new division of gabbroids reflects various geochemical specialization and elements sources.

References

BONATTI, E., GONNORETS, X. and FERRARA, G. (1973) Peridotite–gabbro–basalt complex of the equatorial of the middle Atlantic ridge. In *Petrology of Igneous and Metamorphic Rocks of the Ocean Floor*, Mir, Moscow, pp. 9–29 (in Russian).

FAURE, G. and POWELL, J. L. (1972) *Strontium Isotope Geology*, Springer Verlag, Berlin, 263 pp.

FOTIADI, E. E. and KOROTAEV, G. I. (1963) Structure of earth's crust of Siberia and Far East on data of regional geophysical studies. *Geology and Geophysics*, no. 10, 5–19 (in Russian).

GLAZUNOV, O. M. (1977) Geochemical types of ultrabasic rocks. In *Geochemical Problems of Endogenic Processes*, Nauka, Novosibirsk, pp. 221–230 (in Russian).

GREEN, D. H., RINGWOOD, A. E. *et al.* (1968) *Petrology of the Upper Mantle*. Mir, Moscow, 335 pp.

HART, S. R. (1971) K, Rb, Cs and Ba: Contents and Sr isotope ratios of ocean floor basalts. *Phil. Trans. Roy. Soc. London*, **268**, no. 1192.

MCGREGOR, I. D. (1969) The system $MgO–SiO_2–TiO_2$ and its bearing on the distribution of TiO_2 in basalts. *Amer. J. Sci.* **267** A, 342–363.

MARAKUSHEV, A. A. and PERCHUK, L. L. (1975) Basic trends of magmatism evolution and development of the earth's crust. In *Crust and Upper Mantle of the Earth*, MGU, Moscow, issue 2, 18–34 (in Russian).

SECTION V

GEOCHEMISTRY OF THE SURFACE

Secular Variations in Chemical Composition of Sediments: A Review

J. VEIZER

Department of Geology, University of Ottawa, Ottawa, Ontario K1N 6N5, Canada

Abstract

The average chemical composition of sedimentary rocks differs from the upper continental crust in (a) enrichment in mafic components; (b) higher oxidation state; and (c) high concentrations of excess volatiles. Their major element composition, except for Ca and Na, is consistent with ~30% basaltic admixture. This requirement is only partially satisfied by direct contribution from the crust of oceanic and intermediate composition. A more mafic composition of source areas, and consequently of sedimentary mass, during early terrestrial history could account for this discrepancy. The strongly cannibalistic nature of sedimentary cycle, combined with fast surficial recycling rates (~0.66% × 10^{-6} y), were instrumental in effective homogenization of sediments and perpetuation of this chemical anomaly. The known secular trends for alkalis, Sr isotopes and incompatibles are consistent with this alternative.

Introduction

At the beginning of this century Daly (1909) observed that, statistically, the Mg/Ca ratio of carbonate rocks increases with their geologic age. However, only the post-war accumulation of analytical data has enabled consideration of sedimentary processes in their geochemical context, as a part of the global exogenic cycle. In addition, the extension of the geological time scale into the Precambrian, complemented by new discoveries of apparent secular trends in chemical composition of sediments, has raised the question of the applicability of the present-day exogenic cycle to the geological past.

Following the pioneering work of Clarke (1924) the major analytical contribution must be credited to three decades of research by Ronov and associates at the Vernadsky Institute of Geochemistry. They delineated secular trends and interpreted them within an essentially evolutionary framework (e.g. Ronov, 1964, 1968, 1972; Ronov and Yaroshevsky, 1969; Ronov and Migdisov, 1971). In contrast, the "American" school emphasized the role of recycling processes in a steady state exogenic cycle (e.g. Garrels and Mackenzie, 1971; Li, 1972) and a basically uniformitarian application to the geological past.

The apparent secular trends, in the "evolutionary" approach, are believed to reflect secular variations in the nature or intensity of exogenic processes, whereas in the "uniformitarian" interpretation the role of post-depositional alterations is emphasized or the existence of the trends is questioned. Stated in this polarized form, the two interpretations are mutually exclusive. However, this is not necessarily the case. Minute deviations from an apparent steady state, although within observational error on a short time scale, may result in measurable differences over a long time span ($\geq 10^7$ y). Consequently, with some modifications, the two approaches may be combined. This review is an attempt at such a unified model.

Due to space limitations the present discussion is based frequently on summary papers and some original sources may not therefore be included in the text. The review is restricted to chemical parameters reflecting the evolution of the detrital component of sediments. The restriction precludes the discussion of variations in stable isotopes, oxidation state of metals, silica and alkaline earth; all associated at least partially with authigenic (bio)chemical phases.

Chemical Composition of the Total Sedimentary Mass

Present-day sedimentary accumulations are derived principally from disintegration of the upper continental crust. The association of ancient cratons with peripheral sedimentary wedges indicates that this was so not only during the Phanerozoic but also during a considerable portion of the Precambrian. The average chemical composition of the total sedimentary mass might be therefore expected to compare with that of the present day upper continental crust. This is not the case as already pointed out by Ronov (1968). Compared with the upper continental crust sediments are typified by the following features (Goldschmidt, 1933; Rubey, 1951; Vinogradov, 1967; Ronov, 1968) (Fig. 1):

(A) more mafic overall chemical composition;
(B) higher oxidation state; and
(C) strong enrichment in excess volatiles.

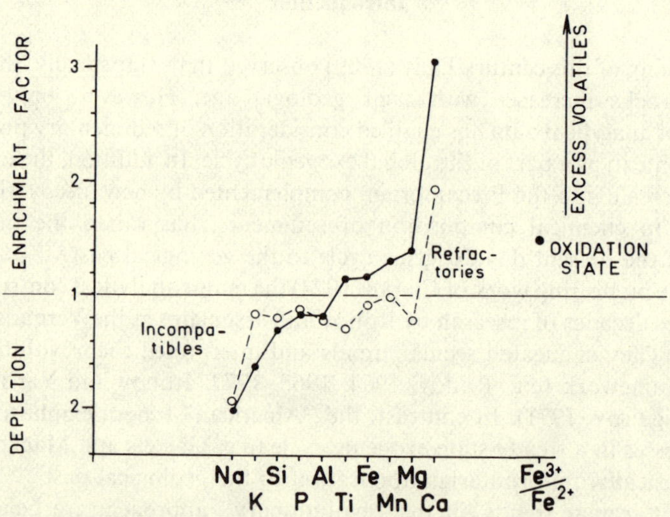

FIG. 1. Chemical features of sediments. Explanations: filled circles—average chemical composition of sediments normalized to average upper continental crust; open circles—average chemical composition of sediments normalized to the mixture of 70% upper continental crust and 30% oceanic crust (~andesite). (Based on the data of Ronov and Yaroshevsky, 1969.) The oceanic layer II is excluded from the sedimentary mass. Its inclusion (*if* sedimentary) would only magnify the depletion–enrichment slope.

Restricting discussion to the first point it is of interest to calculate the proportion of the excess mafic component in the total sedimentary mass. Figure 1 shows that, among major elements, sodium and calcium show the maximal depletion and enrichment respectively with the others falling on the slope between these end members. Trace element concentrations of sediments are not well known but, in general, the "refractories" are enriched and the "incompatibles" slightly depleted or comparable to concentrations in the upper continental crust (cf. Wedepohl, 1969).

Calculations show that to eliminate the depletion–enrichment slope a source area composed of ~70% of upper continental crust (dacite) and 30% oceanic crust (basalt) is required. Normalization of chemical composition of sediments by such a mixture results in a flat major element pattern (Fig. 1), although Ca and Na do not conform with this trend. However, the pattern shows ~10 to 15% overall depletion due to dilution by an as yet unaccounted for phase(s). Since Ca is overabundant, calcite may be indicated as the diluting phase. The directly measured percentage of carbonates in the sedimentary column is indeed in excess, by a similar factor, of the theoretically calculated proportions based on balance considerations during weathering of igneous rocks (Pettijohn, 1975; Ronov and Yaroshevsky, 1969; Wedepohl, 1969; Garrels and Mackenzie, 1971). Exclusion of this anomalous calcite brings the total pattern, except for Na, to the 1.0 level (Fig. 1). Material of this composition is comparable in its major element chemistry to the earth's crust as a whole (~andesite) and not to the upper continental crust.

Causes of the Anomalous Abundances of Calcium and Sodium

Recent developments (Muehlenbachs and Clayton, 1972; Hart, 1973; Spooner and Fyfe, 1973) led to a re-evaluation of the role of direct interaction between sea water and oceanic crust as a source and sink for elements involved in the exogenic cycle. The proposed overall (halmyrolytic and hydrothermal) chemical reactions (e.g. spilitization) can be formulated as follows (Garrels and Mackenzie, 1971; Mackenzie and Wollast, 1977):

(1) basaltic rocks + $2CO_2 + H_2O$ = Na-feldspar + clay minerals + (Ca^{2+}, Fe^{2+}, Mn^{2+}) carbonate + $2HCO_3^-$ + Fe oxides + SiO_2

and

(2) NaCl + Na, Ca-feldspar = Na-feldspar + $CaCl_2$.

These reactions may well account, qualitatively and quantitatively, for the observed $CaCO_3$ (and slight Mn, Fe) enrichment and Na depletion in sea water and hence sediments.

Estimates of the total sedimentary mass, by various authors and techniques, are summarized in Table 1. The lowest estimate, based on geochemical balance calculations, did not take into account sediments, such as greywackes, chemically similar to magmatic rocks (Wedepohl, 1969). Consequently, the estimated limits for total sedimentary mass are 1.7 to 3.2×10^{24} g, with $\sim 2.0 \times 10^{24}$ g being the most recent direct estimate. In this mass the 15% excess of $CaCO_3$ therefore amounts to $10.2–19.2 \times 10^{22}$ g of Ca. Similarly, the enrichment in Mn (by ~0.02% MnO) amounts to $2.6–4.8 \times 10^{20}$ g, and the enrichment in Fe (by ~0.25%) to $4.3–8.0 \times 10^{21}$ g. On the other hand, the Na depletion (by ~1.5% Na_2O) accounts for $1.9–3.6 \times 10^{22}$ g. Correcting for $\sim 0.2 \times 10^{22}$ g Na present in the hydrosphere (Wedepohl, 1969) leaves a total depletion in sediments of $1.7–3.4 \times 10^{22}$ g.

Comparison with total contribution/extraction values (Table 2) shows that, with the

TABLE 1. TOTAL SEDIMENTARY MASS

Author	Mass in 10^{24}g
Goldschmidt (1933)	0.86
Kuenen (1950)	2.7
Poldervaart (1955)	1.7
Horn and Adams (1966)	2.0
Ronov (1968)	2.4
Ronov and Yaroshevsky (1969)	2.0†
Garrels and Mackenzie (1971)	3.2
Li (1972)	2.4

† Excluding the oceanic layer II.

possible exception of Ca, the present-day fluxes are more than adequate to account for the above estimates. Considering the uncertainties involved in calculations, and the probability of higher fluxes during early terrestrial history, the quantitative agreement is acceptable even for Ca. Thus, this direct interaction may conceivably explain the observed Ca, Na, (Mn, Fe) anomalies.

In contrast, it is unlikely that this direct interchange between sea water and oceanic crust produced the overall mafic composition of sediments. If the estimated directions of fluxes for Si, K and Mg (Table 2) are real it appears that such a process will result in felsic contribution to the sediments. Furthermore, it is difficult to see how the net fluxes, for the whole suite of elements, could remain directly proportional to the dacite–basalt mixing trend.

The Mafic Component

The above discussion leaves unresolved the cause of the ~30% "basaltic" admixture in the sedimentary mass. It may be argued that a direct contribution from island arc and ocean floor environments accounts for this admixture. Volcanogenic sediments form ~16–19% of the total sedimentary mass on the continents and the terrigeneous and volcanic sediments of the oceanic layer I can account for an additional ~1% (Ronov and Yaroshevsky, 1969; Ronov, 1976). It is therefore evident that the percentage of volcaniclastic sediments is lower than required. Furthermore, such assemblages are more likely of andesitic than of basaltic provenance (e.g. Jakeš and White, 1971). Thus, chemically, the direct volcanogenic contribution into sedimentary mass can account for ~1/5 of the observed mafic component. The remainder will have to be accounted for by other phenomena.

TABLE 2. FLUXES BETWEEN OCEANIC CRUST AND SEA WATER

	Ca	Na	Mn	Fe	Si	K	Mg
Present-day flux in 10^{13}g y^{-1}†	+1.74	−4.52	+0.91	+5.43	+11.58	+1.48	−4.03
Total contribution/extraction during 4.6 b.y. (in 10^{22}g)	+8.0‡	−20.8	+4.2	+25.0			

† Estimates from Maynard (1976), tables 5–7.
+ indicates addition into sea water;
− subtraction from sea water.

‡ Other estimates (e.g. Mackenzie and Wollast, 1977) are up to one order of magnitude higher.

Secular Variations in Chemical Composition of Sediments

Extensive studies on the Russian Platform (Ronov, 1964, 1968, 1972; Ronov and Migdisov, 1971) corroborated the initial observations of Daly (1909) that, statistically, the chemical composition of sediments varies with their geologic age. Their conclusions were confirmed by North American data (Nanz, 1953; Engel, 1963; Ronov and Migdisov, 1971) and recent studies on Australian sediments (Van Moort, 1972, 1973; Veizer, 1976).

The best defined secular variations are those for alkalis, particularly the K/Na ratio (e.g. Ronov and Migdisov, 1971; Engel *et al.*, 1975). Figure 2 shows that sediments reflect the composition of basement rocks with, in general, a mafic to felsic transition with decreasing geologic age. The major apparent change appears to correlate with the Archean–Proterozoic boundary. A reverse of the trend is indicated by younger sediments, although this is in part due to the kaolinite-montmorillonite mineralogy of Cenozoic rocks. Their diagenetic transformation, with age, into the stable assemblage of illite (chlorite) will increase their K content (cf. Dunoyer de Segonzac, 1970; Weaver and Beck, 1971).

Data for trace elements are rare at present. The increase in incompatibles is demonstrated by the Th/Al trend for the Russian Platform (Fig. 2) and a similar trend has been observed

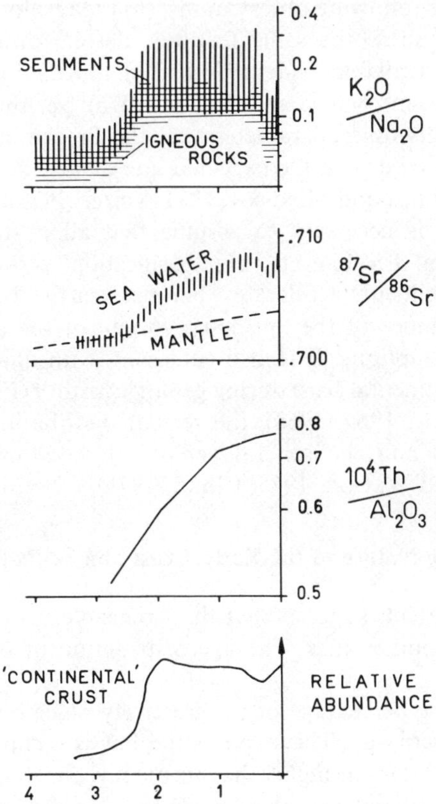

FIG. 2. Secular variations in composition of sediments and crust. The K_2O/Na_2O and "continental crust" curves from Engel *et al.* (1975); $^{87}Sr/^{86}Sr$ from Veizer and Compston (1976); and Th/Al from Ronov and Migdisov (1971).

also for U/Al (Ronov and Migdisov, 1971). Several authors (e.g. Ronov *et al.*, 1974; Nance and Taylor, 1976, 1977) noted that ΣREE also increases in progressively younger sediments. Their relative distribution patterns and the distinct depletion in Eu appear to have been uniform since at least 1.5 b.y. ago. In contrast, the Archean sediments, although variable, are generally less depleted in Eu and show a less pronounced enrichment in light REE. The secular distribution of incompatible trace elements therefore suggests again a mafic to felsic general trend with a major "event" at the Archean–Proterozoic transition.

The existence of a major phase of mafic to felsic crustal transformation at the Archean–Proterozoic boundary is supported also by secular variation of $^{87}Sr/^{86}Sr$ in ancient sea water (Veizer and Compston, 1976). Figure 2 shows that the pre-2.5 b.y. marine water had Sr isotopic ratios similar to that of the contemporary upper mantle, suggesting that stable continental crust played only a subordinate role in the Archean exogenic cycle of Sr. This was followed by a sharp increase in radiogenic ^{87}Sr, from felsic crust, during the early Proterozoic.

All the secular trends discussed show a mafic to felsic progression with decreasing geologic age. Furthermore, they all suggest that a major mafic to felsic transformation occurred during the late Archean–early Proterozoic interval. Are these trends real and representative? If representative, are they "primary" or epigenetic? Their representative nature and "primary" origin is supported by the following observations: (a) the results from three continents are mutually consistent and show the same features, despite differences in approach and instrumentation; (b) the available evidence shows that REE are little affected by either diagenesis or metamorphism (Nance and Taylor, 1976), yet they show the same secular tendencies as the other discussed parameters; (c) the Precambrian portion of the alkali variations is in direct contrast with the expected late diagenetic–metamorphic increase in K/Na ratio (p. 273 and Ronov and Migdisov, 1971; Veizer, 1973); (d) to explain these trends as secondary features it is necessary to assume that all post-depositional phenomena (diagenesis, metamorphism, recycling, etc.) are "basification" processes. Even if true, the rate of such processes must suddenly accelerate when passing the 2.5-b.y. age limit.

In contrast, the acceptance of the "primary" origin of the trends does not require a multitude of ad hoc assumptions. It is also consistent with quantitative estimates of the relative abundance of continental crust during geologic history (Fig. 2), secular variations in sedimentary facies (Ronov, 1964, 1968), the secular distribution of ores of sedimentary affiliation (Veizer, 1976) and the general geological development of the shields (e.g. Goodwin, 1970; Reed and Watson, 1975; Cloud, 1976).

The Nature of the Early Crust and Sediments

The review of secular trends suggests that the Archean crust (and sediments) was more mafic than its younger counterparts. The age distribution of ore deposits supports this observation.

Figure 3 shows that the percentage of progressively older crust, sediments, ores, etc., decreases exponentially with age. This exponential law is a consequence of recycling (cf. Garrels and Mackenzie, 1971). The higher the rate of recycling, the smaller the proportion of the old units preserved. In other words, the slope in Fig. 3 increases with the rate of recycling. For example, sediments are recycled at a faster rate than their basement and surficial sediments faster than the buried ones (cf. curve 1 vs. 2). Figure 3 shows that ~85% of the known Cr (and Ni) reserves are associated with crustal segments older than ~1.7 b.y.

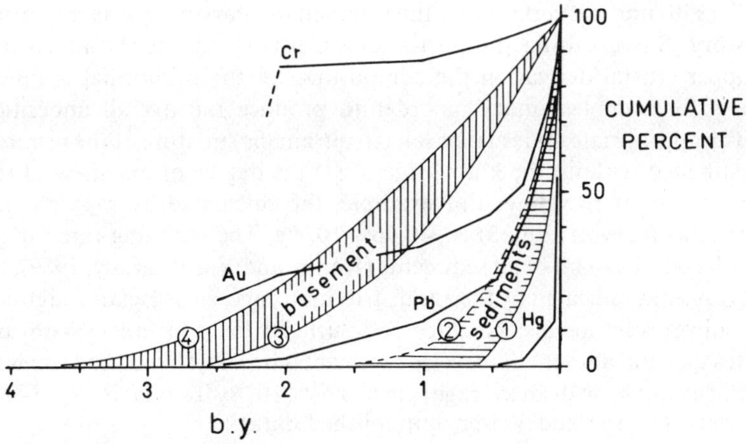

FIG. 3. Present-day cumulative age-distribution pattern of sediments, basement and selected ores. The sedimentary field includes the data for cumulative areal distribution of sediments by Blatt and Jones (1975) (curve 1) and cumulative distribution of thicknesses from Holmes (1965, fig. 257) (curve 2). The basement field is based on cumulative areal distribution of age provinces by Hurley and Rand (1969) and Ronov (1976) (curve 3) and Tugarinov and Bibikova (1976) (curve 4). The cumulative age distribution of the total tonnage of selected ores from Laznicka (1973).

Similarly, old Au accumulations are overabundant. These features cannot be explained by a deeper level of erosion in ancient blocks because the accumulations are in high-level associations (Watson, 1976). Since the deposits could not have been recycled (dispersed) at a rate slower than the recycling rate of their host rocks (basement), the high frequency in early Precambrian must be a relic of an unusually high original abundance. This association of "mafic" ores with old crustal segments is again consistent with a more mafic composition of the early crust (and associated sediments).

As a consequence, a high percentage of older rocks in the total sedimentary mass might conceivably explain the "basaltic" admixture.

The proportion of Archean sediments in the total sedimentary mass is not known. However, an upper limit can be estimated. Extrapolation of sedimentary age curves to 2.5 b.y. ago (Fig. 3) shows that the proportion of Archean sediments, whether by area or thickness (and thus volume), in the total sedimentary mass is negligible. However, the thickness curve is based on Phanerozoic estimates only and may not be representative. Theoretically, the minimal recycling rate of sediments is that of the basement itself, because the destruction of the latter is causing the destruction of the overlying sediments as well. The percentage of basement greater than 2.5 b.y. in age is <1–13% (curves 3 and 4). Thus, even this unrealistically high estimate is not sufficient to account for the remaining four-fifths of the "basaltic" anomaly. In addition, the average composition of Archean sediments was probably less mafic than that of basalt (Baragar and Goodwin, 1969).

Recycling of the Sedimentary Mass

It has been shown that sedimentary recycling is essentially cannibalistic (Gilluly *et al.*, 1970), with "new" sediments forming at the expense of the old ones. A completely closed recycling system of a constant mass will perpetuate the initial chemical composition of primordial sediments. However, the recycling system is neither entirely cannibalistic ($\gtrsim 3/4$;

Gilluly *et al.*, 1970) nor is it certain that the total sedimentary mass was not growing during geologic history. Since, during the post-Archean period, the newly added material was mostly of upper crustal derivation the composition of the primordial sedimentary mass would have to be complementary in order to produce the overall andesitic nature of sediments. The appropriate andesite–basalt (to ultramafic) mixture of the primordial source cannot be estimated without the knowledge of (1) the degree of openness of the recycling system, (2) the rate of recycling. For example, the calculated average recycling rate of surficial sediments (curve 1; Fig. 3) is ~0.66% 10^{-6}y. The recycling rate for geosynclinal settings, which contain ~3/4 of all sediments (Ronov and Yaroshevsky, 1969), may even be considerably faster as indicated by the age distribution curve of associated high-level Hg ores (also other subvolcanic metals, bauxites and surficial Fe, Mn ores). With any of those recycling rates the total amount of recycled sediments considerably exceeds their present-day total mass. For example, with an average rate of 0.66% 10^{-6}y it would by ~ 15–30 times their present-day mass (Jansen and Veizer, unpublished data).

This rapid turnover, of essentially cannibalistic type, causes efficient homogenization of sediments. However, over a long time span, the partially open nature of the recycling system results in a steadily increasing proportion of material derived from the upper continental crust in progressively younger sedimentary rocks. In other words, sediments should show mafic to felsic secular trends in chemical composition with their decreasing geologic age.

Summary

The average chemical composition of sedimentary rocks is comparable to that of the total crust of the earth (~andesite). Furthermore, the sediments contain excess Ca, (Mn, Fe) and are depleted in Na. These Ca, (Mn, Fe) and Na anomalies may result from a direct chemical interaction between sea water and oceanic crust. If the residual sediments were derived mostly from disintegration of the upper continental crust, an ~30% "basaltic" admixture in the total sedimentary mass is required to account for their overall andesitic composition. Contributions from oceanic crust, volcano–sedimentary sequences or preserved Archean sediments appear inadequate to account for this mafic component.

The age-distribution pattern of sedimentary mass shows that sedimentary rocks are subject to fast recycling in an essentially cannibalistic system, which results in perpetuation of their original chemical composition. However, due to the partially open nature of this cycle, the proportion of material derived from the upper continental crust increases in progressively younger sediments. In such a model, the "basaltic" component is a relic feature of a more mafic primordial sedimentary mass (and crust) and the observed mafic–felsic secular trends are a consequence of continuous upper crustal addition, particularly during the post-Archean stage of terrestrial development.

Future Considerations

The Progress in this field depends on further quantification of sedimentological data concerned with (1) volumes, masses and thicknesses of various facies of sedimentary rocks (ores), and (2) establishment of recycling rates for different tectonic settings and rock (ore) types. The major stumbling block for acceleration of the geochemical research is (1) the absence of readily available and representative sample collection(s), covering the known terrestrial time span. In addition, studies of (2) trace element and isotopic composition of sediments; (3) mineralogical control of elemental distribution; and (4) oceanic crust–sea water interaction are essential as a basis for the quantification of theoretical models.

References

BARAGAR, W. R. A. and GOODWIN, A. M. (1969) Andesites and Archean volcanism of the Canadian Shield. *Oregon Dept. Geol. Mining Ind. Bull.* **65**, 121–142.

BLATT, H. and JONES, R. L. (1975) Proportions of exposed igneous, metamorphic, and sedimentary rocks. *Bull. Geol. Soc. Amer.* **86**, 1085–1088.

CLARKE, F. (1924) The data of geochemistry. *U.S. Geol. Surv., Bull.* 770, 841 pp.

CLOUD, P. (1976) *Major Features of Crustal Evolution*, The Geological Society of South Africa, Vol. LXXIX, 1–32.

DALY, R. A. (1909) First calcareous fossils and evolution of limestones. *Bull. Geol. Soc. Amer.* **20**, 153–170.

DUNOYER DE SEGONZAC, G (1970) The transformation of clay minerals during diagenesis and low grade metamorphism: a review. *Sedimentology* **15**, 281–346.

ENGEL, A. E. J. (1963) Geological evolution of North America. *Science* **140**, 143–152.

ENGEL, A. E. J., ITSON, S. P., ENGEL, C. G., STICKNEY, D. M. and CRAY, E. J. (1974) Crustal evolution and global tectonics: a petrogenic view. *Bull. Geol. Soc. Amer.* **85**, 843–858.

GARRELS, R. M. and MACKENZIE, F. T. (1971) *Evolution of Sedimentary Rocks*, Norton, 397 pp.

GILLULY, J., REED, J. C. and CADY, W. M. (1970) Sedimentary volumes and their significance. *Geol. Soc. Amer. Bull.* **81**, 353–375.

GOLDSCHMIDT, V. M. (1933) Grundlagen der quantitativen Geochemie. *Fortschr. Miner. Krist. Petrogr.* **17**, 112 pp.

GOODWIN, A. M. (1970) Metallogenic evolution of the Canadian Shield. In *Geology and Economic Minerals of Canada*, R. J. W. DOUGLAS, ed., Geol. Surv. Can., Econ. Geol. Rep. 1, 156–162.

HART, R. A. (1973) A model for chemical exchange in the basalt–seawater system of oceanic layer II. *Can. J. Earth Sci.* **10**, 799–816.

HOLMES, A. (1965) *Principles of Physical Geology* 2nd ed., Nelson, 1288 pp.

HORN, M. K. and ADAMS, J. A. S. (1966) Computer derived geochemical balances and element abundances. *Geochim. Cosmochim. Acta* **30**, 279–297.

HURLEY, P. M. and RAND, J. R. (1969) Pre-drift continental nuclei. *Science* **164**, 1229–1242.

JAKEŠ, P. and WHITE, A. J. R. (1971) Composition of island arcs and continental growth. *Earth Planet. Sci. Lett.* **12**, 224–230.

KUENEN, PH. H. (1950) *Marine Geology*, Wiley, 568 pp.

LAZNICKA, P. (1973) Development of nonferrous metal deposits in geological time. *Can. J. Earth Sci.* **10**, 18–25.

LI, Y. -H. (1972) Geochemical mass balance among lithosphere, hydrosphere and atmosphere. *Amer. J. Sci.* **272**, 119–137.

MAYNARD, J. B. (1976) The long-term buffering of the oceans. *Geochim. Cosmochim. Acta* **40**, 1523–1532.

MACKENZIE, F. T. and WOLLAST, R. (1977) Sedimentary cycling models of global processes. In *The Sea* (vol. 6). *Marine Modelling*, E. GOLDBERG, A. N. MCCAVE, J. J. O'BRIAN and J. H. STEELE, eds., pp. 739–785. Wiley.

MUEHLENBACHS, K. and CLAYTON, R. N. (1972) Oxygen isotope studies of fresh and weathered submarine basalts. *Can. J. Earth Sci.* **9**, 172–184.

NANCE, W. B. and TAYLOR, S. R. (1976) Rare earth element patterns and crustal evolution—I. Australian post-Archean sedimentary rocks. *Geochim. Cosmochim. Acta* **40**, 1539–1551.

NANCE, W. B. and TAYLOR, S. R. (1977) Rare earth element patterns and crustal evolution—II. Archean sedimentary rocks from Kalgoorlie, Australia. *Geochim. Cosmochim. Acta* **41**, 225–231.

NANZ, R. H. JR. (1953) Chemical composition of Precambrian slates with notes on the geochemical evolution of lutites. *J. Geol.* **61**, 51–64.

PETTIJOHN, F. J. (1975) *Sedimentary Rocks*, 3rd ed., Harper & Row, 628 pp.

POLDERVAART, A. (1955) Chemistry of the earth's crust. *Geol. Soc. Amer.*, Spec. Papers 62, pp. 119–144.

REED, H. H. and WATSON, J. (1975) *Introduction to Geology*, vol. 2, part 1, *Early Stages of Earth History*, Macmillan, 221 pp.

RONOV, A. B. (1964) Common tendencies in the chemical evolution of the earth's crust, ocean and atmosphere. *Geochem. Int.* **1**, 713–737.

RONOV, A. B. (1968) Probable changes in the composition of sea water during the course of geological time. *Sedimentology* **10**, 25–43.

RONOV, A. B. (1972) Evolution of rock composition and geochemical processes in the sedimentary shell of the Earth. *Sedimentology* **19**, 157–172.

RONOV, A. B. (1976) Volcanism, carbonate accumulation, life. (The regularities of the global geochemistry of carbon.) *Geokhimija* **8**, 1252–1277 (in Russian).

RONOV, A. B., BALASHOV, YU. A., GIRIN, YU. P., BRATISHKO, R. KH. and KAZAKOV, S. A. (1974) Regularities of rare-earth element distribution in the sedimentary shell and in the crust of the earth. *Sedimentology* **21**, 171–193.

RONOV, A. B. and MIGDISOV, A. A. (1971) Evolution of the chemical composition of the rocks in the shields and sediment cover of the Russian and North American Platforms. *Sedimentology* **16**, 137–185.

RONOV, A. B. and YAROSHEVSKY, A. A. (1969) Chemical composition of the earth's crust. In *The Earth's Crust and Upper Mantle*, P. J. HART, ed., Amer. Geophys. Union, Geophys. Monogr. **13**, 37–57.

RUBEY, W. W. (1951) Geologic history of sea water. An attempt to state the problem. *Bull. Geol. Soc. Amer.* **62**, 1111–1147.

SPOONER, E. T. C. and FYFE, W. S. (1973) Sub-sea-floor metamorphism, heat and mass transfer. *Contr. Mineral. Petrol.* **42**, 287–304.

TUGARINOV, A. I. and BIBIKOVA, E. V. (1976) Evolution of the chemical composition of the earth crust. *Geokhimija* **8**, 1151–1159 (in Russian).

VAN MOORT, J. C. (1972) The K_2O, CaO, MgO and CO_2 contents of shales and related rocks and their implication for sedimentary evolution since the Proterozoic. *24 Int. Geological Congr.*, Section 10, pp. 427–439.

VAN MOORT, J. C. (1973) The magnesium and calcium contents of sediments, especially pelites, as a function of age and degree of metamorphism. *Chem. Geol.* **12**, 1–37.

VEIZER, J. (1973) Sedimentation in geologic history: recycling vs. evolution or recycling with evolution. *Contr. Mineral. Petrol.* **38**, 261–278.

VEIZER, J. (1976) Evolution of ores of sedimentary affiliation during geologic history: Relations to the general tendencies in evolution of the crust, hydrosphere, atmosphere and biosphere. In: *Handbook of Strata-bound and Stratiform Ore Deposits*, K. H. WOLF, ed., Elsevier. Vol. 3, 1–41.

VEIZER, J. and COMPSTON, W. (1976) $^{87}Sr/^{86}Sr$ in Precambrian carbonates as an index of crustal evolution. *Geochim. Cosmochim. Acta* **40**, 905–914.

VINOGRADOV, A. P. (1967) The formation of the ocean. *Izv. Akad. Nauk U.S.S.R.*, Geol. Ser. **4**, 3–9 (in Russian).

WATSON, J. V. (1976) Vertical movements in Proterozoic structural provinces. *Phil. Trans. Roy. Soc. Lond.* A, **280**, 629–640.

WEAVER, C. E. and BECK, K. C. (1971) Clay water diagenesis during burial: how mud becomes gneiss. *Geol. Soc. America*, Spec. Papers 134, 96 pp.

WEDEPOHL, K. H. (1969) Die Zusammensetzung der Erdkruste. *Fortschr. Miner.* **46**, 145–174.

Kinetics of Nutrient Regeneration in Anoxic Marine Sediments

ROBERT A. BERNER

Department of Geology and Geophysics, Yale University, New Haven, Connecticut 06520, *U.S.A.*

Abstract

Anoxic sediments undergoing bacterial sulfate reduction, because of the buildup in them of dissolved ammonia and phosphate from the decomposition of organic matter, may serve as important sources of these nutrients for the oceans. This paper demonstrates how pore water concentration-vs-depth data, along with laboratory measurements, can be combined with theoretical models to deduce rates and rate laws for nutrient release and bacterial sulfate reduction in sediments. Further application of the models enables determination of the stoichiometric C:N:P composition of the organic matter undergoing decomposition and the fraction of liberated nutrient which is returned to the overlying water.

For an anoxic sediment from Long Island Sound, U.S.A., diagenetic modeling of pore water data has led to the conclusion that rates of organic nitrogen and phosphorus decomposition and of sulfate reduction by micro-organisms, below the top few cm of sediment, is first order with respect to the organic matter being decomposed. Application of the first-order rate law to a variety of other sediments has led to the discovery of a direct proportionality between the rate constant for sulfate reduction and the square of the rate of deposition. Also, decomposition rates calculated from modeling are in good agreement with rates measured in the laboratory using the same sediment. The calculated and measured rates have been used, by means of further modeling, to show that in a Long Island Sound sediment about 70% of the ammonia liberated below 10 cm and about 50% of the sulfide liberated above 10 cm is lost from the sediment by diffusion and mixing.

In the upper portion of anoxic sediments overlain by oxygenated water, higher concentrations of more readily metabolized organic matter are found, and migration of dissolved constituents is enhanced due to irrigation and other modes of mixing by benthic organisms (bioturbation). This results in much faster nutrient regeneration rates than in the underlying sediment.

Introduction

Nutrient regeneration in the marine environment is essential to maintenance of life in the sea. The major source of the phosphorus, nitrogen, etc., which is required for marine photosynthesis is liberation from organic matter by the respiratory activity of marine organisms. Only a small proportion is supplied by rivers. Although nutrient production in oxygenated seawater has drawn considerable attention, the release of phosphorus and nitrogen by microbial activity in anoxic sediments has, by contrast, been largely neglected. In areas where organic matter is rapidly sedimented to the bottom, the rapid buildup of dissolved phosphorus and nitrogen in sediment pore waters via anaerobic microbial activity can cause the sediments to become a nutrient source for the overlying water (e.g. Suess, 1976). Mass transfer is accomplished by diffusion, bioturbation, and wave and current stirring at and near the sediment–water interface. Proof that sediments act as a potentially important source of nutrients is provided by determinations of high concentrations of dissolved P and N in sediment pore waters as compared to overlying seawater (e.g. Rittenberg *et al.*, 1955;

Nissenbaum *et al.*, 1972; Sholkovitz, 1973; Bray, 1973; Hartmann *et al.*, 1973; Berner, 1974; Suess, 1976; Morse and Cook, 1978).

A proper understanding of nutrient regeneration in sediments involves addressing such questions as how, where, how fast, and in what proportions this regeneration occurs. Appropriate answers require detailed study of pore water concentration-vs-depth data, combined with flux determinations (e.g. Hallberg *et al.*, 1972) and laboratory measurements, and apppropriate diagenetic modeling. Also, since much of the nutrient regeneration in anoxic sediments occurs in conjunction with bacterial sulfate reduction, this process must be considered along with the release of dissolved P and N.

The purpose of the present paper is to illustrate how laboratory and field data can be combined with theoretical models to both test the models and to deduce: (1) rates and rate laws for bacterial sulfate reduction and organic N and P release; (2) the stoichiometric composition (C:N:P) of organic matter undergoing decomposition; (3) the role of bioturbation in nutrient regeneration; and (4) the proportion of P and N liberated, and sulfide formed by sulfate reduction, that actually escapes from the sediments into the overlying water.

Rates and Rate Laws for Sulfate Reduction and Nutrient Regeneration

Two basic approaches can be used to deduce rates of early diagenetic chemical reactions. They are calculation, by means of diagenetic models, and direct measurement using natural sediments themselves. Both approaches are used here. The first method employs the following assumptions: steady state diagenesis has been attained (a reasonable assumption—see Lasaga and Holland, 1976); effects due to compaction or externally impressed flow can be ignored; bioturbation and wave and current stirring can be ignored; adsorption is simple, linear, and rapidly reversible; mineral precipitation or dissolution can be ignored; diffusion occurs through the pore water only; and concentration gradients are predominantly vertical. Under these conditions the diagenetic equation of Berner (1976) becomes:

$$D\frac{\partial^2 c}{\partial x^2} - (1+K)\omega\frac{\partial c}{\partial x} + R(x) = 0 \tag{1}$$

where c = concentration of dissolved sulfate, ammonium, or phosphate in the sediment pore water,

x = depth measured positively downward from the sediment–water interface,

D = diffusion coefficient of dissolved species in the sediment pore water which includes the effects of tortuosity but not adsorption,

K = adsorption constant equal to the ratio of rapidly exchangeable adsorbed phosphate or ammonium to dissolved phosphate or ammonium. The concentrations are expressed in terms of the same units so that K is dimensionless,

ω = rate of depositional burial,

R = rate of sulfate reduction, rate of decomposition of organic phosphorus, or rate of decomposition of organic nitrogen.

Now, from a curve fitted to c vs x data, for a sediment which has probably undergone steady state diagenesis, values of $\partial c/\partial x$ and $\partial^2 c/\partial x^2$ at each depth can be determined either graphically or analytically. If independent measurements of D, K, and ω have also been made, then sufficient data exists to calculate the rate of reaction R for each depth x using equation (1). From the functional relation between R and x the rate law for R can be deduced.

Values of R can also be determined by direct measurement at various depths in the sediment. Jorgensen (1977) has used $^{35}SO_4$ tracer to determine *in situ* sulfate reduction rates in anoxic sediments from the Limfjorden, Denmark. Additional measurements have been obtained for anoxic Long Island Sound (U.S.A.) sediments (Goldhaber *et al.*, 1977; Rosenfeld, 1977; Aller, 1977) by incubating in a series of sealed containers (hereinafter referred to as jars) samples of sediment from a given depth zone and sequentially opening the jars and analyzing for dissolved sulfate, ammonium, and reactive phosphate over a period of several months. Results of jar experiments (corrected for adsorption and adjusted to average *in situ* temperatures using measured temperature coefficients for the same or similar sediments) agree with values calculated from equation (1) within a factor of 2 for both sulfate reduction (Goldhaber *et al.*, 1977) and ammonium formation (Rosenfeld, 1977). Factor-of-2 agreement is good considering errors involved in determining D, K, and ω and the temperature coefficient of R for sediments, and in fitting curves to pore water and jar experiment data. The agreement, in fact, is better than previous attempts because D, K, and ω were actually measured on the sediment under study and not estimated as is the usual case. The jar experiment results thus help to place some confidence in theoretical calculations of regeneration rates.

In the course of our studies it has been found that concentration-vs-depth data for dissolved sulfate, ammonium, and phosphate in a number of anoxic sediments can be well fitted by exponential expressions if data within a narrow surficial zone of bioturbation are excluded. If such an exponential fit is, in fact, correct, we can use the resulting empirical expressions to directly deduce the rate laws for bacterial sulfate reduction and ammonium and phosphate production in sediments. An example is provided by Fig. 1 where interstitial water concentration data for a sediment locality in Long Island Sound, U.S.A., are fitted to exponential expressions for sulfate, ammonium, and phosphate. As shown in the Appendix, such expressions, along with equation (1), lead directly to the rate equations:

$$\varepsilon = \varepsilon_0 \exp(-k_S t), \tag{2}$$

$$N = N_0 \exp(-k_N t), \tag{3}$$

$$P = P_0 \exp(-k_P t) \tag{4}$$

where t = time

ε, N, P = concentrations of metabolizable (i.e. decomposable) organic carbon, organic nitrogen, and organic phosphorus respectively,

and

$$k_S = 2R_{S_0}/\varepsilon_0, \qquad k_N = R_{N_0}/N_0, \qquad k_P = R_{P_0}/P_0.$$

The subscripts S, N, and P refer to sulfate reduction, ammonium formation, and phosphate formation respectively and the subscript 0 refers to zero time.

Equations (2), (3), and (4) are the integrated forms for first-order rate expressions with rate constants k. This indicates that if dissolved sulfate, ammonium, and phosphate profiles in sediments can be well fitted by exponential expressions of the form shown in this paper, then the rate of decomposition of metabolizable organic carbon, nitrogen, or phosphorus is first

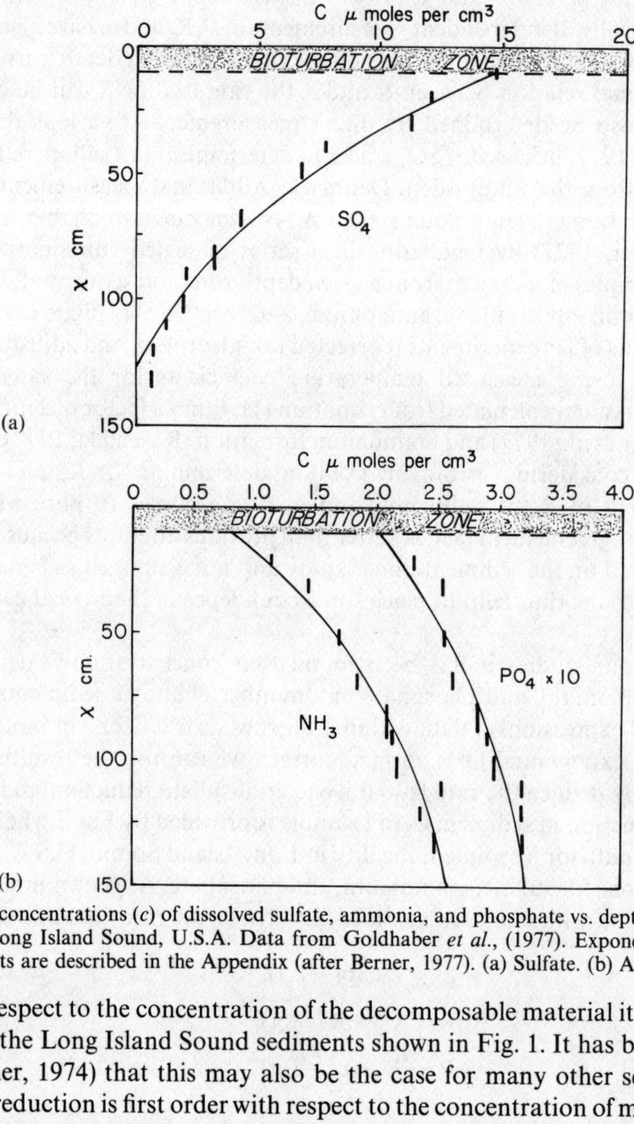

FIG. 1. Plot of concentrations (c) of dissolved sulfate, ammonia, and phosphate vs. depth (x) for sediments of the FOAM site, Long Island Sound, U.S.A. Data from Goldhaber *et al.*, (1977). Exponential curves fitted to the measured points are described in the Appendix (after Berner, 1977). (a) Sulfate. (b) Ammonia and phosphate.

order with respect to the concentration of the decomposable material itself. This seems to be the case for the Long Island Sound sediments shown in Fig. 1. It has been suggested by the writer (Berner, 1974) that this may also be the case for many other sediments.

If sulfate reduction is first order with respect to the concentration of metabolizable organic carbon and if metabolizable carbon is held essentially constant by reacting a large excess with a limited amount of sulfate, then the rate of sulfate reduction should be constant. This is demonstrated in Fig. 2 by a jar experiment where the starting conditions were $c_{S_0} = 20\ \mu\text{mole cm}^{-3}$ and $\varepsilon_0 \approx 2000\ \mu\text{mole cm}^{-3}$. Here the concentration of sulfate decreases linearly with time, as expected. The data do not decrease exponentially as would be expected if sulfate reduction were first order with respect to sulfate. (At low concentrations of sulfate, around 2–3 mmole per liter, sulfate reduction may become first order with respect to sulfate as shown by Ramm and Bella, 1974, but the method of measurement used in determining the data of Fig. 2 was insufficiently precise to detect this effect; it would be expected from consideration of the Michaelis-Menten equation.) Thus, again jar experiments agree with predictions based on diagenetic modeling.

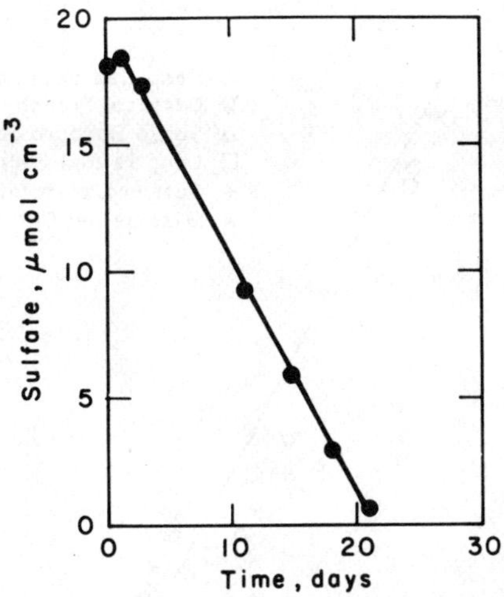

FIG. 2. Jar experiment data for sediment from a tidal marsh panne, Guilford, Conn., U.S.A. After Martens and Berner (1977).

Fitting of pore water concentration data with exponential expressions has enabled the determination of rate constants k, for sulfate reduction and ammonia and phosphate formation, for a variety of different sediments exhibiting a wide range of sedimentation rates (Berner, 1974; Goldhaber, 1974; Goldhaber and Kaplan, 1975; Shokes, 1976; Toth and Lerman, 1977; Rosenfeld, 1977). A log-log plot of k_S, for sulfate reduction, vs sedimentation rate ω, for some sediments where ω is known with a reasonable degree of accuracy (e.g. by radiometric dating), is shown in Fig. 3. Note that the data delineate a straight line with a slope of two; in other words:

$$k_S = A\omega^2 \tag{5}$$

where A is a constant. This fundamental empirical relation has already been pointed out by Toth and Lerman (1977) and Shokes (1976). It was previously hinted at by Goldhaber and Kaplan (1975). Since k_S is a measure of the reactivity or metabolizability of organic matter (Berner, 1974), the proportionality indicates that more rapid deposition entails the addition to the sediment of organic matter which can be more readily decomposed. This is reasonable. If deposition is slow, considerable time is available for aerobic organisms present near, at, and above the sediment–water interface to destroy metabolizable compounds before they can become buried and made available to sulfate-reducing bacteria. Conversely, rapid burial results in the initial preservation of compounds which can subsequently be decomposed by sulfate reducers plus associated microflora.

The reason why the rate constant in equation (5) should vary as the square of the rate of deposition, and not some other power, is not as simple. One explanation can be offered in terms of an equation which is directly derivable from the rate law expression for sulfate (Appendix equation (19′)) when substituted in equation (1). This equation (see also Berner, 1974) is:

$$k_S = \left[\frac{L\varepsilon_0}{D_S(c_{S_0} - c_{S\infty})} - \frac{1}{D_S}\right]\omega^2 \tag{6}$$

(see Appendix for symbol definitions).

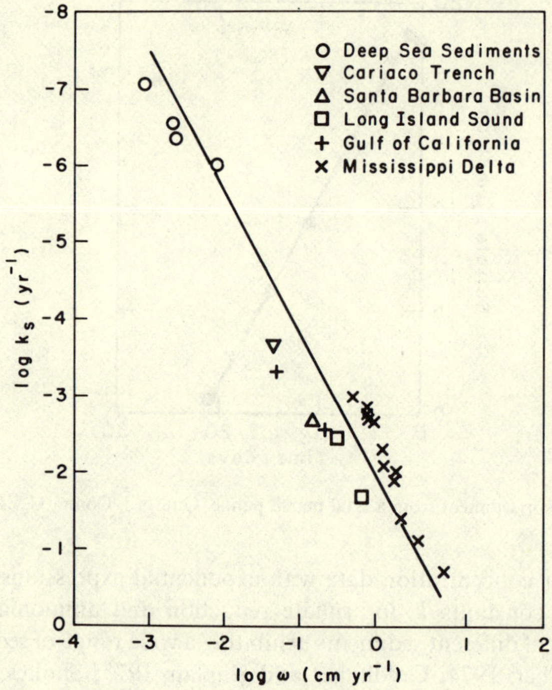

FIG. 3. Log-log plot of k_S, the first-order rate constant for sulfate reduction, vs. ω, the rate of deposition, for anoxic sediments where values of ω are well established. Sources of data: Deep sea sediments and Cariaco Trench, Toth and Lerman (1977); Santa Barbara Basin, Berner (1974); Long Island Sound, Rosenfeld (1977); Gulf of California, Goldhaber (1974); Mississippi Delta, Shokes (1976).

Since L and D_S are essentially constant for all fine-grained sediments and ε_0 and $(c_{S_0} - c_{S\infty})$ vary sympathetically (and each varies over no more than about an order of magnitude), the terms within the parentheses are essentially constant. In this way equation (5), an empirical expression, can be derived from diagenetic kinetic modeling.

Stoichiometry of Nutrient Regeneration

Through the use of diagenetic equations one can deduce the stoichiometric C:N:P ratio of decomposing organic matter from pore water data for dissolved sulfate, ammonium, and phosphate. Such equations are especially useful when plots of ammonium vs. sulfate or phosphate vs. sulfate are linear indicating that the decomposing organic matter maintains a constant stoichiometric C:N:P ratio during decomposition and is attacked mainly by sulfate-reducing bacteria plus associated microflora. In this case the decomposition process can be written as a simple chemical reaction (Redfield, 1958; Richards, 1965):

$$(CH_2O)_x(NH_3)_y(H_3PO_4)_z + \frac{x}{2}SO_4^{--} \rightarrow xCO_2 + yNH_3 + zH_3PO_4 + \frac{x}{2}S^{--} + xH_2O.$$

(Here both reactants and products are highly generalized simply to indicate the average oxidation state of sedimentary organic matter and to represent total dissolved carbonate, sulfide, etc.) Use of linear plots of sulfate, ammonium, and phosphate concentrations vs. one another has been made by several workers (e.g. Sholkovitz, 1973; Hartmann *et al.*, 1973; Suess, 1976) to deduce the stoichiometric coefficients x, y, and z in the above reaction and their change (if any) with depth.

In a recent paper the writer (Berner, 1977) has shown that the stoichiometric C:N:P ratio of organic matter which is decomposed during bacterial sulfate reduction cannot be deduced from pore water data alone. This is because phosphate, ammonium, and sulfate undergo adsorption on sediment solids to different degrees and the three species diffuse in sediments at different rates. The effects of differential adsorption and differential diffusion, as well as release from the decomposition of organic matter, are included in equations which are derived from diagenetic equations similar to equation (1). In the diagenetic equations it is assumed that organic C, N, and P undergoing decomposition are non-diffusable and that decomposition follows first-order kinetics, as discussed above. To describe simple straight-line-plot stoichiometric decomposition, it is also necessary to assume that the constants k for decomposition are equal, or in other words that $k_S = k_N = k_P \equiv k$. In this case one obtains the expressions (Berner, 1977):

$$\frac{dc_N}{dc_S} = -\frac{\alpha_N}{L}\left[\frac{D_S k/\omega^2 + (1+K_S)}{D_N k/\omega^2 + (1+K_N)}\right], \tag{7}$$

$$\frac{dc_P}{dc_S} = -\frac{\alpha_P}{L}\left[\frac{D_S k/\omega^2 + (1+K_S)}{D_P k/\omega^2 + (1+K_P)}\right] \tag{8}$$

where α_N = N:C ratio of the decomposing organic matter,
α_P = P:C ratio of the decomposing organic matter,
L = ratio of number of sulfate ions reduced per atom of organic carbon oxidized (in the reaction given above $L = 1/2$).

Note that since all terms on the right-hand side of each equation are constants, both expressions describe straight lines. Also, note that in general $D_S \neq D_N \neq D_P$ and $K_S \neq K_N \neq K_P$ so that the stoichiometric ratios α_N/L and α_P/L are not simply equal to the slopes dc_N/dc_S and dc_P/dc_S.

The exponential fits to the data of Fig. 1 have been used along with the known rate of deposition ω to obtain values of k_S, k_N, and k_P for a sediment site in Long Island Sound, U.S.A. The three calculated k values were found to be essentially equal as would be expected since the concentration data of Fig. 1 form straight lines when plotted against one another. From independent measurements (or estimates) of D_S, D_N, D_P, K_S, K_N, K_P, and the k value derived from curve fitting, equations (7) and (8) were solved to obtain:

$$\frac{dc_N}{dc_S} = -0.47\frac{\alpha_N}{L}, \tag{9}$$

$$\frac{dc_P}{dc_S} = -0.54\frac{\alpha_P}{L}. \tag{10}$$

Thus, because of differential diffusion and differential adsorption the stoichiometric ratios

α_N/L and α_P/L for the Long Island Sound sediments are distinctly different from slopes of the linear plots of c_N vs. c_S and c_P vs. c_S. In future stoichiometric modeling of nutrient regeneration these factors should be taken into consideration.

Regeneration of Nutrients in the Zone of Bioturbation

During the discussion of rate laws and stoichiometry, pore water data for a Long Island Sound sediment were used to derive various parameters, but only for depths greater than about 10 cm where the effects of bioturbation are negligible. However, much nutrient regeneration occurs in the top 10 cm and here the effects of bioturbation (as well as enhanced microbial activity) cannot be ignored. Bioturbation can be considered to be the movement of sediment grains and pore water as a result of the vital activities of benthic organisms. Because of irrigation, which is flushing of burrow tubes by host infauna, the turnover of pore water generally is faster than that of adjacent particles (Aller, 1977). This necessitates the use of different parameters for describing pore water bioturbation and particle bioturbation.

The simplest model for bioturbation is to assume that, averaged over space and time, the movement of grains and water is random. If so, bioturbation can be considered as a diffusional process and effects described in terms of simple mixing coefficients D_B for particles and for water. The mixing coefficients then take the place of D, the molecular diffusion coefficient, in diagenetic equations. This approach, employing a constant D_B, has been employed by Goldberg and Koide (1962), Guinasso and Schink (1975), and Aller (1977) to describe bioturbation in the upper portions of sediments. A similar approach has been used by Vanderborght *et al.* (1977) to describe wave and current stirring.

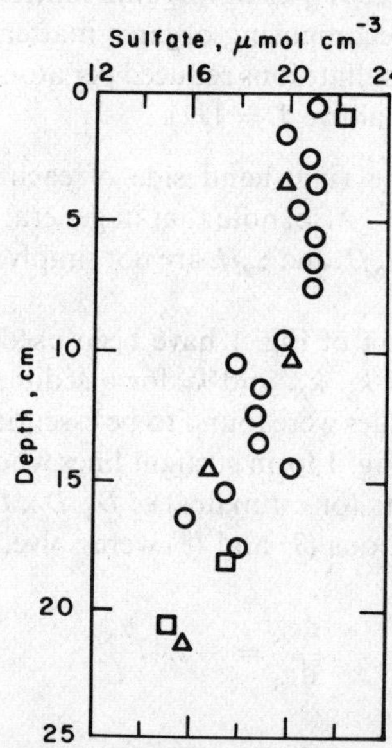

FIG. 4. Sulfate concentration vs. depth for the upper 25 cm of sediments from the FOAM site, Long Island Sound, U.S.A. (summer profile). Data from Goldhaber *et al.*, (1977). Circles represent box core data; other symbols gravity core data.

Some idea of the magnitude of D_B for pore water is provided by the dissolved sulfate data of Goldhaber *et al.* (1977). In Fig. 4 are shown dissolved sulfate concentrations for the top 20 cm of a sediment, obtained by box coring, taken near to that depicted in Fig. 1. The upper region above about 8 cm shows seasonal changes (Aller, 1977) and cannot be assumed to be at steady state. (The data shown in Fig. 4 illustrate a typical late summer distribution.) However, at any given time the value of D_B can be calculated, if it is assumed that there is a discontinuous change in D at 8 cm. With this assumption we have, for conservation of flux using Fick's First Law of diffusion, the relation (Goldhaber *et al.*, 1977):

$$D_B \frac{\partial c_S}{\partial x}\bigg|_{x<8} = D \frac{\partial c_S}{\partial x}\bigg|_{x>8} \tag{11}$$

where D_B = bioturbation diffusion coefficient and D = molecular diffusion coefficient. Since $\partial c_S/\partial x$ at $x > 8$ is over 5 times larger than $\partial c_S/\partial x$ at $x < 8$, by this model D_B for pore water is at least 5 times greater than D during the late summer when bioturbation is at a maximum. Similar ratios of D_B/D ranging from 2–10 have been calculated by Aller (1977) for other sediments of Long Island Sound. Thus, bioturbation appreciably accelerates the rate of migration of nutrient species liberated near the sediment–water interface.

The zone of bioturbation has been excluded when fitting pore water data with simple exponential functions, not only because migration is accelerated there, but also because organic matter decomposition rates are much higher in this region due to the presence of much more readily decomposable organic matter. This has been demonstrated by jar experiments which show, for example, that the average rate of sulfate reduction in the top 10 cm of the Long Island Sound sediment depicted in Fig. 1 is over 12 times higher than rates found at greater depths.

Nutrient Loss to the Overlying Water

An important question, so far unaddressed, is the proportion of ammonium or phosphate regenerated in sediments that is actually returned to the overlying water. In other words, how much is lost from the sediment? Considerable regeneration must occur right at the sediment–water interface because the C:N ratio of dead organic matter in the uppermost portions of sediments is distinctly higher than that of plankton (e.g. Emery, 1960) thus indicating selective regeneration of nitrogen prior to burial. This regenerated material is given off more or less directly to the overlying water. The question considered here is how much additional material is given off from sediment below the top ~2 cm. In addition, it is instructive to determine the proportion of $H_2S + HS^-$ produced by sulfate reduction, which is returned to the overlying water and re-oxidized to sulfate. In this way the openness of sediments, with respect to exchange of nutrients and sulfur with the overlying water, can be examined.

One way of answering the proportional loss question is to measure directly the fluxes of ammonium and phosphate out of a sediment. Direct flux measurements have been made by Hallberg *et al.* (1972), Aller (1977), and Rosenfeld (1977). One precaution in interpreting direct flux measurements, however, is that if a collecting chamber is left for any appreciable time over a sediment (say for more than a few days) and encloses originally oxygenated seawater, it is probable that oxygen consumption within the chamber will result in alteration of bioturbation and microbial decomposition rates which may in turn alter the fluxes. In addition, there are inherent problems for all sediments relating to possible alteration of the structure of the benthic boundary layer by use of a bottom flux sampler (e.g. McCave, 1974).

Another approach is the use of pore water data and diagenetic modeling. The total amount of ammonium and phosphate produced or sulfate reduced (sulfide produced) over a given depth interval $(x_2 - x_1)$ (assuming constant sedimentation rate) is given by:

$$\Sigma = \frac{1}{\omega}\int_{x_1}^{x_2} R(x)\mathrm{d}x. \tag{12}$$

The amount present both in solution and as adsorbed or precipitated solids at any given depth divided by Σ and subtracted from one gives the proportion lost during burial λ. For instance, in the case of nitrogen loss during burial between depths 0 and x_1 (assuming $c_0 = 0$):

$$\lambda_N = 1 - \frac{(1+K_N)c_N(x_1)}{\Sigma_N(x_1)} \tag{13}$$

which, if pore water data can be fitted to the exponential expression, Appendix equation (2′), is:

$$\lambda_N = \frac{D_N\beta_N}{\omega(1+K_N)+D_N\beta_N}. \tag{14}$$

Here λ_N represents the proportion of ammonium lost.

By way of example the data of Fig. 1 can be used to calculate λ_N for depths below the zone of bioturbation. The exponential fit to the data provides the value $\beta_N = 0.012\ \mathrm{cm}^{-1}$. If depth x is measured from the base of the zone of bioturbation which is located at a depth of 10 cm, equation (14) can be solved for λ_N using the additional independently determined values for ω ($0.3\ \mathrm{cm\ y}^{-1}$), D_N ($6 \times 10^{-6}\ \mathrm{cm}^2\ \mathrm{s}^{-1}$), and K_N (1.6 as determined by Rosenfeld, 1977) to be:

$$\lambda_N = 0.71.$$

Thus, in this sediment roughly 70% of the ammonia generated below 10 cm is lost to the overlying water. (To maintain a long-term steady state within the bioturbation zone, no ammonium can accumulate there so that ammonium lost from the underlying sediment must be given off to the overlying water.) This value is in good agreement with the results of Rosenfeld (1977) who has found, by direct determinations of changes in organic nitrogen and dissolved plus adsorbed ammonium with depth, that near-shore sediments in Long Island Sound show λ_N values ranging from 0.47 to 0.98 with a value of 0.74 in the vicinity of the site portrayed in Fig. 1.

For sulfate and phosphate the calculation of λ is more complicated because precipitation of sulfide and phosphate to form authigenic minerals occurs and the precipitation cannot be accounted for simply as an adsorption. Sulfide precipitates to form various FeS minerals and pyrite whereas phosphate may precipitate to form apatite, vivianite, and other minerals. In addition, phosphate is adsorbed strongly by ferric oxides present at the sediment–water interface, and, as a result it may be partially trapped and prevented from diffusing into the overlying water. Goldhaber *et al.* (1977) have measured the concentration of pyrite plus FeS minerals as a function of depth in a sediment from Long Island Sound and, by the use of equation (14) and jar experiment determinations of $R(x)$, have concluded that as much as 50% of the H_2S formed in the top 10 cm by sulfate reduction is lost to the overlying water. Analogous studies of phosphate precipitation and adsorption in marine sediments are needed.

Appendix

Here a derivation is presented to show how rate laws for sedimentary organic matter decomposition can be derived from exponential curves fitted to interstitial water concentration data. In Fig. 1 data for a sediment station in Long Island Sound, U.S.A., are fitted to expressions of the form:

$$c_S = (c_{S_0} - c_{S\,\infty}) \exp(-\beta_S x') + c_{S\,\infty}, \tag{1'}$$

$$c_N = c_{N\,\infty}[1 - \exp(-\beta_N x')], \tag{2'}$$

$$c_P = c_{P\,\infty}[1 - \exp(-\beta_P x')] \tag{3'}$$

where c = concentration in pore water,
x' = depth below the base of the zone of bioturbation (in this case below 10 cm—i.e. $x' = x + 10$),
β = constant derived from curve fitting,
c_0 = concentration at $x = 0$,
c_∞ = asymptotic concentration or concentration where $x' \to \infty$ (For sulfate this can be a negative number for the purpose of curve fitting)

and the subscripts S, N, and P refer to sulfate, ammonium, and phosphate respectively. (At the site shown in Fig. 1 organic nitrogen and phosphorus are released almost entirely as ammonium ion and molybdate-reactive phosphate respectively.) From text equation (1) and (1′), (2′), and (3′) one can solve for R:

$$R_S = (c_{S_0} - c_{S\,\infty})\beta_S[\beta_S D_S + (1 + K_S)\omega] \exp(-\beta_S x'), \tag{4'}$$

$$R_N = c_{N\,\infty}\beta_N[\beta_N D_N + (1 + K_N)\omega] \exp(-\beta_N x'), \tag{5'}$$

$$R_P = c_{P\,\infty}\beta_P[\beta_P D_P + (1 + K_P)\omega] \exp(-\beta_P x'). \tag{6'}$$

(Note—because of negligible sulfate adsorption K_S is usually assumed equal to zero.) Denoting the pre-exponential factors as R_{S_0}, R_{N_0}, and R_{P_0}, equations (4′), (5′), and (6′) can be rewritten as:

$$R_S = R_{S_0} \exp(-\beta_S x'), \tag{7'}$$

$$R_N = R_{N_0} \exp(-\beta_N x'), \tag{8'}$$

$$R_P = R_{P_0} \exp(-\beta_P x'). \tag{9'}$$

Substitution of equations (7′), (8′), and (9′) in equations (1′), (2′), and (3′) to eliminate β_S, β_N, and β_P and solving for R we obtain:

$$R_S = \frac{R_{S_0}}{(c_{S_0} - c_{S\,\infty})}(c_S - c_{S\,\infty}), \tag{10'}$$

$$R_N = \frac{R_{N_0}}{c_{N\infty}}(c_{N\infty} - c_N), \tag{11'}$$

$$R_P = \frac{R_{P_0}}{c_{P\infty}}(c_{P\infty} - c_P). \tag{12'}$$

Equations (10′), (11′), and (12′) can refer to a closed, homogeneous system, such as a jar experiment, as well as to the open system of a sediment undergoing deposition and diffusion. If closed, $R_S = dc_S/dt$, $R_N = (1+K_N)dc_N/dt$, $R_P = (1+K_P)dc_P/dt$. Using these relations and solving the resulting first-order differential equations for the conditions: $t = 0$, $c_S = c_{S_0}$, $c_N = 0$, $c_P = 0$ and $t = \infty$, $c_S = c_{S\infty}$, $c_N = c_{N\infty}$, $c_P = c_{P\infty}$ we obtain for a closed, homogeneous system:

$$\frac{c_S - c_{S\infty}}{c_{S_0} - c_{S\infty}} = \exp\left[-\frac{R_{S_0}}{(c_{S_0} - c_{S\infty})}t\right], \tag{13'}$$

$$\frac{c_{N\infty} - c_N}{c_{N\infty}} = \exp\left[-\frac{R_{N_0}}{(1+K_N)c_{N\infty}}t\right], \tag{14'}$$

$$\frac{c_{P\infty} - c_P}{c_{P\infty}} = \exp\left[-\frac{R_{P_0}}{(1+K_P)c_{P\infty}}t\right]. \tag{15'}$$

From the stoichiometry of the generalized Redfield–Richards reaction for *P* and *N* release during sulfate reduction (given in the text) for a closed homogeneous system:

$$c_{S_0} - c_S = L(\varepsilon_0 - \varepsilon), \tag{16'}$$

$$(1+K_N)c_N = N_0 - N, \tag{17'}$$

$$(1+K_P)c_P = P_0 - P \tag{18'}$$

where ε = concentration of metabolizable organic carbon,
N = concentration of metabolizable organic nitrogen,
P = concentration of metabolizable organic phosphorus

and the subscript 0 refers to time zero (ε, N, and P are expressed in mass per unit volume of pore water). Substituting equations (16′)–(18′) in (13′)–(15′) and noting that when ∞, $c \to c_\infty$ and ε, N, P$\to 0$, we obtain:

$$\varepsilon = \varepsilon_0 \exp(-k_S t), \tag{19'}$$

$$N = N_0 \exp(-k_N t), \tag{20'}$$

$$P = P_0 \exp(-k_P t). \tag{21'}$$

where

$$k_S = R_{S_0}/L_0\varepsilon_0 = R_{\varepsilon_0}/\varepsilon_0,$$

$$k_N = R_{N_0}/N_0,$$

$$k_P = R_{P_0}/P_0.$$

Acknowledgements

Research supported by American Chemical Society Grant PRF 9318 AC–2–C, NSF Grant DES 73–00365, and Yale University. The writer has benefitted from discussions with J. K. Rosenfeld and R. C. Aller. Musical inspiration courtesy of Joseph Szulc, Henri Duparc, and Gabriel Fauré.

References

ALLER, R. C. (1977) The influence of macrobenthos on chemical diagenesis of marine sediments. Ph.D. thesis, Yale University.

BERNER, R. A. (1974) Kinetic models for the early diagenesis of nitrogen, sulfur, phosphorus, and silicon in anoxic marine sediments. In *The Sea*, Vol. 5, E. D. GOLDBERG, ed., Wiley, Interscience, 427–450.

BERNER, R. A. (1976) Inclusion of adsorption in the modeling of early diagenesis. *Earth Planet. Sci. Lett.* **29**, 333–340.

BERNER, R. A. (1977) Stoichiometric models for nutrient regeneration in anoxic sediments. *Limnol. Oceanog.* **22**, 781–786.

BRAY, J. T. (1973) The behaviour of phosphate in the interstitial waters of Chesapeake Bay sediments. Ph.D. thesis, Johns Hopkins University, 149 pp.

EMERY, K. O. (1960) *The Sea Off Southern California*, Wiley, 366 pp.

GOLDBERG, E. D. and KOIDE, M. (1962) Geochronological studies of deep-sea sediments by the Io/Th method. *Geochim. Cosmochim. Acta* **26**, 417–460.

GOLDHABER, M. B. (1974) Equilibrium and dynamic aspects of the marine geochemistry of sulfur. Ph.D. thesis, University of California, Los Angeles, 399 pp.

GOLDHABER, M. B., ALLER, R. C., COCHRAN, J. K., ROSENFELD, J. K., MARTENS, C. S. and BERNER, R. A. (1977) Sulfate reduction, diffusion, and bioturbation in Long Island Sound sediments: report of the FOAM group. *Amer. J. Sci.* **277**, 193–237.

GOLDHABER, M. B. and KAPLAN, I. R. (1975) Controls and consequences of sulfate reduction rates in recent marine sediments. *Soil Sci.* **119**, 42–55.

GUINASSO, N. L. and SCHINK, D. R. (1975) Quantitative estimates of biological mixing rates in abyssal sediments. *J. Geophys. Res.* **80**, 3032–3043.

HALLBERG, R. O., BAGANDER, L. E., ENGVALL, A. G. and SCHIPPEL, F. A. (1972) Method for studying geochemistry of sediment–water interface *Ambio* **1**, 71–72.

HARTMANN, M., MÜLLER, P., SUESS, E. and VAN DER WEIJDEN, C. H. (1973) Oxidation of organic matter in recent marine sediments. *Meteor. Forschungs Ergebhisse, Reihe C,* **12**, 74–86.

JORGENSEN, B. B. (1977) The sulfur cycle of a coastal marine sediment (Limfjorden, Denmark). *Limnol. and Oceanog.* **22**, 814–832.

LASAGA, A. C. and HOLLAND, H. D. (1976) Mathematical aspects of non-steady-state diagenesis. *Geochim. Cosmochim. Acta* **40**, 257–266.

MARTENS, C. S. and BERNER, R. A. (1977) Interstitial water chemistry of anoxic Long Island Sound sediments. I. Dissolved gases. *Limnol. and Oceanog.* **22**, 10–25.

MCCAVE, I. N. (1974) *The Benthic Boundary Layer*, Plenum Press, 323 pp.

MORSE, J. W. and COOK, N. (1978) A preliminary investigation of the distribution of phosphorus in North Atlantic Ocean deep sea and continental slope sediments. *Limnol. and Oceanog.* **23**, 825–830.

NISSENBAUM, A., PRESLEY, B. J. and KAPLAN, I. R. (1972) Early diagenesis in a reducing fjord, Saanich Inlet, British Columbia I. chemical and isotopic changes in major components. *Geochim. Cosmochim. Acta* **36**, 1007–1027.

RAMM, A. E. and BELLA, P. A. (1974) Sulfide production in anaerobic microcosms. *Limnol. and Oceanog.* **19**, 110–118.

REDFIELD, A. C. (1958) The biological control of chemical factors in the environment. *Amer. Scientist* **46**, 205–222.

RICHARDS, F. A. (1965) Anoxic basins and fjords. In *Chemical Oceanography*, Vol. 1, J. P. RILEY and G. SKIRROW, eds., Academic Press, 611–645.

RITTENBERG, S. C., EMERY, K. O. and ORR, W. L. (1955) Regeneration of nutrients in sediments of marine basins. *Deep Sea Res.* **3**, 23–45.

ROSENFELD, J. K. (1977) Nitrogen diagenesis in nearshore anoxic sediments. Ph.D. thesis, Yale University.

SHOKES, R. F. (1976) Rate dependent distributions of lead–210 and interstitial sulfate in sediments of the Mississippi River delta. Ph.D. thesis, Texas A. and M. University.

SHOLKOVITZ, E. (1973) Interstitial water chemistry of the Santa Barbara Basin sediments. *Geochim. Cosmochim. Acta* **37**, 2043–2073.

SUESS, E. (1976) Nutrients near the depositional interface. In *The Benthic Boundary Layer*, I. N. MCCAVE, ed., Plenum Press, 57–79.

TOTH, D. J. and LERMAN, A. (1977) Organic matter reactivity and sedimentation rates in the ocean. *Amer. J. Sci.* **277**, 465–485.

VANDERBORGHT, J. P., WOLLAST, R., and BILLEN, G. (1977) Kinetic models of diagenesis in disturbed sediments, Part I. Mass transfer properties and silica diagenesis. *Limnol. and Oceanog.* **22**, 787–793.

Discussion

J. GONI:

1° Il faut introduire le paramètre de rendement de réaction dans votre formule qui lie *D* avec *R*(*x*), surtout dans le domaine de l'hypocapillaire (~100 Å de ∅).

2° Les phénomènes biochimiques de réduction du $SO_4^=$ sont beaucoup plus complexes; même si l'on travaille avec du $SO_4^=$ entièrement ionique, on n'a pas le droit de simplifier les réactions (précipitation du $CO_3^=$, sulfures colloïdaux, . . .) enfin blocage final de la réaction.

R. A. BERNER:

The approach used is phenomenological. Critical parameters are *measured* and a straightforward mass balance is employed. One can *measure* bulk-diffusion rates and bulk-reaction rates without having to get into detailed mechanistic explanations such as considerations of dead-end micropores or biochemical pathways within each cell. Such mechanistic studies are important but cannot be used, at the present state of knowledge, to predict bulk kinetics of sediments.

Répartition des Eléments Traces dans les Phosphorites Marines

L. Prévôt†, J. Lucas†, Y. Nathan‡ and Y. Shiloni‡

† *Centre de Sédimentologie et de Géochimie de la Surface (C.N.R.S.), Laboratoire de Géologie de l'Université Louis Pasteur 1, rue Blessig, 67000 Strasbourg, France*

‡ *Geological Survey of Israël, 30, Malkei Israël St., Jerusalem, Israël*

Résumé

466 échantillons appartenant à 6 séries phosphatées marines différentes ont été étudiés. Le minéral apatite de ces séries de provenances diverses (Maroc, Sénégal, Israël, plate-forme continentale nord-espagnole) est invariablement une fluorapatite carbonatée dont le rapport F/P_2O_5 est le même partout et dont la quantité de CO_2 varie d'un maximum de 6 % au Maroc et sur la plate-forme à 4 % en Israël et 1 % seulement dans les phosphates légèrement altérés de Taïba. La proportion moyenne de P_2O_5 varie d'une série à l'autre entre 7 % et 32 % environ. 23 à 27 éléments majeurs, mineurs et traces ont été dosés, les coefficients de corrélation entre ces éléments ont été calculés et diverses méthodes statistiques simples ont été utilisées. Certains éléments sont relativement abondants: Sr, V, Ni, Cr, Zn, Ba, U et Y. D'autres sont faibles ou inexistants: B, Co, Ga, Cd, Pb. Il apparaît certains groupements caractéristiques. Sr est toujours étroitement associé au phosphate dans les séries non altérées. D'autres éléments, répartis selon les cas étudiés dans des phases différentes, sont cependant presque toujours intercorrélés: Sr–Na; V–Ni–Cr–Zn. Bien que la qualité de l'apatite varie peu, on constate que les quantités d'éléments traces associés ne sont pas fonction de la quantité de phosphate; elles paraissent plutôt caractériser les différents milieux de sédimentation. Leur présence paraît liée à celle du phosphore, leur abondance à celle des autres composants majeurs, en particulier des argiles. Il est intéressant de remarquer que les argiles, bien que présentes en faible quantité, contiennent une grande partie des éléments traces alors que les carbonates, plus abondants, n'en contiennent pas. L'altération modifie la répartition des éléments traces dans les phosphorites.

Abstract

466 samples belonging to 6 different marine phosphorite series have been studied. The apatite mineral of these series, whether from Morocco, Senegal, Israel, or from the North Spanish continental platform, is always a carbonate-fluorapatite; its F/P_2O_5 ratio is the same everywhere, and its CO_2 content varies from a maximum of 6% in Morocco and on the platform, to 4% in Israël, and 1% only in the slightly weathered phosphates of Taïba (Senegal). Throughout the series, the average proportion of P_2O_5 varies from about 7% to 32%. 23 to 27 major, minor and trace elements have been determined, the correlation coefficients between these elements have been calculated and a number of simple statistical methods have been used. Some elements appear to be relatively abundant: Sr, V, Ni, Cr, Zn, Ba, U and Y. Others are scarce or nonexistent: B, Co, Ga, Cd, Pb. Some characteristic associations become apparent. Strontium is always closely associated with phosphate in the non-altered series. Other elements, although distributed differently in the mineral phases according to the various series studied, are nevertheless intercorrelated: Sr–Na; V–Ni–Cr–Zn. Although the quality of the apatites is almost constant, it is noteworthy that there is no relation between the proportion of these intercorrelated elements and the proportion of phosphate; in fact, they seem to characterize the sedimentary environment. Their presence seems to be linked to that of phosphorus, their abundance to that of the other major components, especially clays. It is interesting to note that most of the traces are contained in clays which constitute only a small fraction of the rock, whereas the more abundant carbonates do not contain any trace elements. Weathering can modify the distribution of traces in phosphorites.

MAROC, SÉNÉGAL, PLATEFORME NORD-ESPAGNOLE

L. PRÉVÔT et J. LUCAS

Dans le cadre d'un programme plus général d'étude des formations phosphatées marines, un premier lot d'échantillons variés provenant de gisements dont les conditions géologiques de mise en place sont connues, a été retenu pour une réflexion préliminaire sur la répartition des éléments traces dans ce type de roche. Plusieurs auteurs qui ont donné des analyses de traces à l'occasion de l'étude de gisements divers (Krauskopf, 1955; Gulbrandsen, 1966; Naumenko, 1966; Bliskovskiy, 1969; Altschuler, 1973) constatent que les quantités absolues d'éléments varient d'un gisement à l'autre, que le milieu phosphaté est un des milieux favorables à l'enrichissement en éléments traces, que la matière organique paraît jouer un rôle certain bien qu'encore mal défini. L'analyse de 281 échantillons nouveaux permet d'apporter quelques précisions supplémentaires.

Définition Sommaire des Lieux de Prélèvement

Deux coupes proviennent du gisement crétacé-éocène des Ganntour (Maroc) dont l'étude géologique a été faite par Boujo (1976). Ce gisement de phosphates en grains peut être considéré comme une formation en place affectée seulement de remaniements locaux. Déposé dans un golfe long et étroit ne communiquant avec la mer que par une ouverture située à l'Ouest, il s'étend sur une longueur de plus de 100 km. La série phosphatée, marno-phosphatée, cherteuse et carbonatée, remarquablement alternée, traduit une sédimentation rythmique qui s'est poursuivie du Maestrichtien à la base du Lutétien. Fortement condensée elle varie d'Est en Ouest aussi bien en épaisseur (de 180 m à Youssoufia à 100 m à la Tessaout) qu'en faciès. Par ailleurs on distingue une frange nord dite "dénoyée" où le phosphate blanc affleure et une zone "noyée" où la série entièrement située dans la nappe phréatique ne contient que du phosphate noir. Les deux coupes choisies sont une tranchée de recherche placée en zone dénoyée vers l'Est du gisement et un sondage recoupant les formations en zone noyée à l'Ouest. Dans la tranchée, des couches de phosphate plus ou moins riches alternent avec des niveaux de carbonate où calcite et dolomite sont généralement séparées; les niveaux cherteux y sont assez nombreux, la silice détritique y est rare; les minéraux argileux peu abondants sont des smectites en bas de la série progressivement remplacées par des attapulgites vers le haut. Dans le sondage, le seul carbonate présent est la dolomite; les cherts sont plus rares mais le quartz détritique plus abondant que dans la tranchée; les minéraux argileux sont semblables.

Une seconde série d'échantillons provient du gisement de Taïba (Sénégal). Daté du Lutétien supérieur (Flicoteaux et Lappartient, 1972) ce gisement a été décrit par Menor (1976). Il est formé d'une couche de phosphate d'environ 8 m d'épaisseur reposant sur des marnes finement feuilletées et recouverte par une épaisse formation sablo-gréseuse. Entre la couche phosphatée et la couverture s'intercale un niveau silico-ferrallitique riche en phosphate d'alumine, témoignant d'un intense épisode d'altération météorique semblable à celle qui est étudiée à Thiès par Flicoteaux (Flicoteaux et Tessier, 1971). Dans la couche phosphatée on reconnait un ensemble hétérogène riche en silice-silex et phosphate silicifié—dans la partie inférieure, et un ensemble homogène à silex beaucoup plus rares dans la partie supérieure. Bien qu'à proximité de formations calcaires, la roche phosphatée est pratiquement dépourvue de carbonates, ce qui paraît provenir d'une décarbonatation secondaire par altération. Ainsi, le gisement de Taïba apparaît comme le résultat d'un dépôt sédimentaire en place légèrement modifié par altération. Les échantillons analysés proviennent soit de la carrière d'exploitation, soit de puits de recherche.

C'est d'un contexte tout à fait différent que proviennent les échantillons du dernier groupe, puisqu'ils ont été dragués en mer sur le plateau continental au Nord et au Nord-Ouest de l'Espagne, près du Cap Ortegal et sur le Banc Le Danois. La situation géographique et géologique de cette région est clairement exposée par Lamboy (1976). Prélevés à une profondeur variant entre 100 et 800 m, les phosphates se présentent en nodules plus ou moins réguliers pouvant atteindre la taille d'un oeuf, en encroûtements superficiels ou en ciment de brêches à éléments variés. Les nodules sont emballés dans un sable généralement dépourvu de phosphate, les encroûtements recouvrent le "bed-rock" en place. Les deux régions étudiées diffèrent, entre autres caractères, par la teneur en fer et surtout par le minéral qui héberge ce fer: glauconie au Cap Ortegal, oxydes au Banc Le Danois. La période de formation des phosphates reste imprécise; elle paraît en tout cas être post-Miocène . . . peut-être se poursuit-elle encore de nos jours.

Méthodes

Les échantillons analysés ne sont pas triés en fonction de leur teneur éventuelle en phosphore; au contraire tous les faciès ont été retenus depuis les silexites et les carbonates jusqu'aux phosphates les plus riches. Les teneurs sont donc très variables d'un échantillon à l'autre, ce qui explique l'importance des écarts-types observés sur les moyennes. Malgré l'imperfection de la méthode, ce sont ces moyennes qui ont été retenues pour les calculs statistiques, chaque série étant représentative d'une région bien déterminée.

Tous les échantillons ont été analysés par diffraction des rayons X. La plupart des éléments majeurs et traces ont été dosés par quantométrie (Ataman et Besnus, 1965; Besnus et Rouault, 1973). Le fluor est dosé à l'aide d'une électrode spécifique. Au-dessous de 1 % le P_2O_5 est dosé par colorimétrie. Les résultats de l'analyse chimique ont été traités pour obtenir la matrice de corrélation des éléments deux à deux, les diagrammes de corrélations et les groupes à intercorrélations positives ont été constitués suivant la méthode décrite par Besnus (1975) et Besnus et Mosser (1976).

Nature du Minéral Apatitique

La teneur moyenne en P_2O_5 des cinq séries étudiées varie entre 7 % au Banc Le Danois et 32 % à Taïba. Comme l'a déjà remarqué Altschuler (1973) dans d'autres gisements, le minéral apatite est sensiblement le même dans tous les cas où roche n'est pas altérée. L'étroite similitude des diagrammes de diffraction des rayons X le suggère, la mesure des teneurs en CO_2 et F le confirme: il s'agit d'une *fluorapatite carbonatée*.

Les teneurs en CO_2, obtenues dans certains cas par analyse chimique, dans d'autres par la méthode des couples de pics de Gulbrandsen (1970), ou enfin à la fois par les deux méthodes qui concordent de façon assez satisfaisante, sont de l'ordre de 6 % au Maroc et sur la plateforme nord-espagnole. Menor (1975) a montré que la faible teneur en CO_2 de l'apatite de Taïba, inférieure à 3 % et souvent voisine de 1 % seulement, est due à une décarbonatation de la roche lors de l'altération.

A première vue les teneurs en fluor des apatites sont toutes semblables, avec un rapport F/P_2O_5 peu différent de 1/10. Dans le détail ce rapport F/P_2O_5 prend des valeurs situées entre 0,099 et 0,121 et ces variations, bien que très faibles, sont cependant significatives. Dans les fluorapatites les plus carbonatées, F/P_2O_5 est le plus élevé. Dans les fluorapatites faiblement carbonatées de Taïba, ce rapport est légèrement inférieur au rapport stoechiométrique proposé par Altschuler *et al.* (1958) pour la fluorapatite carbonatée ($\geq$ 0,10), mais nettement supérieur au rapport idéal de la fluorapatite non carbonatée (0,090). Cette corrélation positive entre les teneurs en CO_2 et les teneurs en fluor des apatites confirme, comme l'admettent maintenant beaucoup d'auteurs, qu'une partie du fluor compense le déficit de charge dû au remplacement de certains ions PO_4^{3-} par des ions Co_3^{2-}. Mais la variation du fluor est faible, car à une perte ou un gain de poids en CO_3^{2-} de 3% correspond une perte ou un gain de poids en fluor de 1% seulement.

Les Eléments Traces Principaux

Quelques caractères communs à toutes ces séries se dégagent de la simple lecture des résultats d'analyse chimique en éléments traces. Les analyses étant trop nombreuses, seules les teneurs moyennes avec leurs écarts-types sont données dans le tableau I pour chacune des séries étudiées. Certains éléments sont partout relativement abondants: Sr, Ba, V, Ni, Cr, Zn. S'ajoutent à cette liste U et Y qui n'ont été dosés que dans une partie des échantillons. D'autres éléments sont rares voire inexistants, tels Co, Ga, Cd, Pb. Le bore, dont la teneur est très variable, se rattache, selon la provenance des échantillons, à l'un ou à l'autre de ces deux groupes.

L'élément majeur spécifique des sédiments étudiés étant le phosphore, les rapports entre les teneurs moyennes de chacun des éléments traces abondants et la teneur moyenne en P_2O_5 ont été calculés pour chaque série, (Tableau 1). Ces rapports sont d'une grande variété; ainsi *les quantités d'éléments traces ne sont pas proportionnelles à la quantité de phosphate.*

L'Exemple de la Tranchée du Maroc

L'examen combiné des coefficients de corrélation et des diagrammes de corrélation fait apparaître des groupements d'éléments intercorrélés. La figure 1 donne le détail des intercorrélations pour la tranchée marocaine; elle visualise les groupements d'éléments intercorrélés tels que:

(1) Si – Al – Mg – Fe – Ti – Ni – Mn
(2) Si – Al – Mg – Ti – Mn – Zn
(3) Ca – P – F – Sr
(4) P – F – Sr – Ba – B – Na
(5) V – Cr – Ni – Zn

Il est aisé de reconnaître dans ces groupes une partie des minéraux connus par l'étude pétrographique et l'analyse aux rayons X, et d'y lire les liens des éléments traces et des éléments mineurs avec ces phases minérales. C'est ainsi que les groupes (1) et (2) dont les éléments majeurs sont ceux des argiles contiennent le nickel, le manganèse et le zinc. A noter

TABLEAU 1. ANALYSES CHIMIQUES MOYENNES

Les éléments majeurs et mineurs sont exprimés en pour cent; les éléments traces en ppm. Les rapports réels à P_2O_5 sont obtenus en multipliant les rapports donnés par 10^{-5}. Les chiffres en italiques sous les moyennes sont les écarts-types.

LIEU (nombre d'échantillons)	SiO_2	Al_2O_3	Fe_2O_3	MgO	CaO	P_2O_5	F	Perte à 1000°C	K_2O	Na_2O	TiO_2	Mn_3O_4
Maroc tranchée	17,79	1,73	0,45	7,02	36,23	13,00	1,58	22,13	0,14	0,21	0,11	<lim. dét.
(65)	*15,34*	*1,55*	*0,55*	*5,70*	*12,33*	*11,22*	*1,36*	*11,39*	*0,14*	*0,14*	*0,10*	
Maroc sondage	41,60	1,82	0,56	5,59	22,69	9,54	1,71	16,20	0,48	0,24	0,14	<lim. dét.
(81)	*21,92*	*1,82*	*0,78*	*4,22*	*11,54*	*7,80*	*0,95*	*10,15*	*0,28*	*0,10*	*0,10*	
Cap Ortegal	17,02	2,33	7,94	1,85	35,35	12,68	†	17,77	0,83	0,65	0,13	0,04
(50)	*17,69*	*1,65*	*6,96*	*2,91*	*11,65*	*8,90*		*10,27*	*0,84*	*0,32*	*0,10*	*0,03*
Banc Le Danois	15,92	2,81	28,64	1,67	23,25	7,17	†	17,60	0,46	0,51	0,19	0,29
(29)	*15,69*	*1,23*	*17,60*	*0,78*	*13,21*	*6,54*		*6,85*	*0,34*	*0,19*	*0,14*	*0,32*
Sénégal Taïba	14,26	2,37	1,37	0,37	43,75	31,84	3,17	3,94	0,13	0,12	0,13	0,06
(56)	*10,82*	*2,44*	*1,13*	*0,41*	*7,87*	*5,65*	*0,66*	*1,87*	*0,12*	*0,09*	*0,12*	*0,11*

† F n'a pas été dosé dans tous les échantillons.

	Sr	Ba	V	Ni	Co	Cr	B	Zn	Ga	Cu	Pb
Maroc tranchée	679	73	207	77	2	280	30	185	1	47	1
	465	*73*	*134*	*48*	*4*	*131*	*26*	*89*	*1*	*24*	*1*
Maroc sondage	602	61	69	39	1	169	6	177	1	35	8
	375	*26*	*80*	*45*	*3*	*122*	*10*	*140*	*2*	*16*	*15*
Cap Ortegal	987	22	44	17	3	49	112	30	1	20	4
	477	*26*	*51*	*17*	*6*	*46*	*83*	*21*	*1*	*16*	*21*
Banc Le Danois	762	120	356	283	97	113	159	162	2	22	39
	621	*160*	*191*	*272*	*178*	*74*	*54*	*95*	*2*	*29*	*42*
Sénégal Taïba	609	66	529	150	4	415	2	496	1	80	1
	335	*100*	*577*	*156*	*15*	*299*	*2*	*232*	*1*	*65*	*0,3*

	Sr/P_2O_5	Sr/CaO	V/P_2O_5	Ni/P_2O_5	Cr/P_2O_5	Zn/P_2O_5
Maroc tranchée	52,25	19,25	15,92	5,92	21,53	14,23
Maroc sondage	69,29	29,15	7,23	4,08	17,71	18,55
Cap Ortegal	77,84	27,92	3,47	1,34	3,86	2,36
Banc Le Danois	106,27	32,77	49,65	39,47	15,76	22,59
Sénégal Taïba	19,12	13,92	16,61	4,71	13,03	15,57

que le Mg est celui des éléments majeurs dont la corrélation positive avec les autres est la moins forte; la bonne corrélation qui le lie par ailleurs à la perte au feu montre bien que le Mg se partage entre la phase argileuse et la phase dolomitique. Les groupes (3) et (4) correspondent à la phase apatitique qui renferme Sr, Ba, B et Na. On peut ajouter à ces éléments le soufre† qui se trouve être en parfaite corrélation avec le phosphore et le fluor, ce qui le situe dans le réseau de l'apatite. Sa teneur moyenne pour l'ensemble des échantillons de la tranchée est de 0,30%. Si on rapproche cette teneur de celle du sodium (0,16% Na), on constate que ces valeurs approximatives sont compatibles avec l'introduction simultanée d'un S^{6+} et d'un Na^+ dans la maille (Gulbrandsen, 1966). L'absence du calcium dans le groupe d'intercorrélations (4), alors qu'il est présent dans le groupe (3) est intéressante. Elle

† Nous remercions M. Vine du U. S. Geological Survey à qui nous devons ces analyses de soufre.

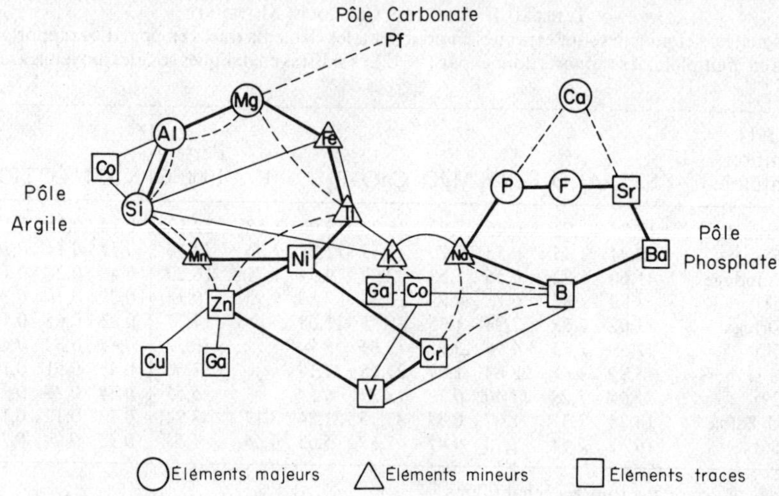

FIG. 1. Intercorrélations entre éléments dans la tranchée du Maroc

s'explique par le partage du calcium entre les carbonates de calcium, calcite et dolomite, et l'apatite. Mais elle permet de préciser que, mis à part le strontium dont la corrélation est positive tant avec le calcium qu'avec le phosphore, les autres éléments traces du groupe apatitique sont bien liés au phosphore. Le groupe (5) mérite l'attention: il n'est constitué que d'éléments traces, plus précisément d'une partie de ceux qui se caractérisent par leur abondance dans toutes les séries étudiées ici. La figure montre que ces traces font en quelque sorte le joint entre la phase argileuse à laquelle se rattachent Zn et Ni et la phase apatitique à laquelle se rattachent indirectement V et Cr par l'intermédiaire du B. Un autre pont est jeté entre ces deux phases par la corrélation positive entre les deux éléments mineurs K et Na. L'examen du nuage de point dans le diagramme de corrélation entre Al et Na confirme d'ailleurs qu'une partie de Na est bien liée aux argiles.

Sur la figure 1 et sur le triangle des intercorrélations positives dont elle découle, tous les éléments se groupent autour de deux phases qui à elles seules expliquent la totalité du sédiment: phosphates et argiles. La silice et les carbonates n'apparaissent pas, bien qu'ils représentent respectivement 18 et 30% environ de l'ensemble du sédiment. Sans doute le quartz détritique est-il pauvre en traces, mais on aurait pu penser que les minéraux néoformés, calcite et dolomite, et surtout opale, piégeaient certains des éléments traces présents; ce n'est pas le cas, *phosphate et argiles se partagent tout, silice et carbonates sont vides*. Pour les éléments majeurs, le camouflage du pôle carbonate s'explique par l'appartenance de Mg^{2+} et Ca^{2+} à plusieurs phases; seule la liaison (Mg - Pf) suggère l'existence de dolomite. Il est certain que l'étude statistique, sortie de son contexte, serait très dangereuse puisqu'elle laisserait ignorer la présence de deux phases importantes.

Comparaison des séries entre elles

Les autres séries d'échantillons ont été étudiées de la même manière. Le tableau 2 résume les principaux résultats sur la répartition des éléments traces.

TABLEAU 2. REPARTITION DES ELEMENTS TRACES DANS LES DIFFERENTES PHASES MINERALES.
Certains groupes d'éléments intercorrélés ne sont pas reliés à une phase minérale.

Maroc Tranchée

Calcite et dolomite —

Apatite: [Na Sr] Ba B

Argiles: Na Ni Zn B Mn

[V—Ni—Cr—Zn; Zn—Cu]

Maroc Sondage

Dolomite: Cu

Apatite: [Na Sr] Zn Cr

Argiles Fe: V Ni Cr Cu

Argiles K: Ba Mn

[V—Ni—Cr—Zn—Cu]

Cap Ortégal

Calcite —

Apatite: Sr

Argiles: Na B Cr

Fer (glauconite et goethite): Na B [Cr V Ni Zn] Mn

[Na—Sr]—Cr—B—Zn

Banc le Danois

Calcite —

Apatite: [Na Sr] [V Cr Zn]

Argiles —

Fer (goethite): B

[V—Ni—Co—Zn]

Sénégal Taïba

Calcite — (Sr?)

Apatite —

Argiles Mg: [V Ni Cr Zn Cu]

Kaolinite: Mn Ba

[Na—Sr]—Pf

Le strontium est le seul élément trace toujours étroitement lié au phosphore de l'apatite dans les phosphates non altérés. Seule la série altérée de Taïba fait exception: là le strontium dont la teneur relative par rapport à P_2O_5 est nettement plus faible qu'ailleurs, aurait quitté l'apatite pendant la décarbonatation. Son affinité avec le sodium, que l'on remarque dans tous les cas étudiés, demeure: le sodium quitterait également l'apatite pendant son altération ménagée.

D'autres corrélations positives existent entre le phosphore et certains éléments traces, mais elles n'ont pas de caractère général. Au contraire chaque lieu paraît caractérisé par un cortège propre. Ainsi, au Cap Ortegal le strontium est le seul élément trace dans l'apatite, alors qu'au Banc Le Danois pourtant proche l'apatite renferme également vanadium, chrome et zinc. Au Maroc, dans le même gisement, l'apatite de la tranchée dénoyée renferme strontium, baryum et bore pro-parte; celle du sondage sous le niveau hydrostatique contient strontium, zinc et chrome.

La pauvreté en traces du carbonate de calcium, observée dans l'exemple de la tranchée du Maroc, est de règle. La seule exception serait peut-être celle du sondage où une partie du cuivre, d'ailleurs peu abondant, paraît associé à la dolomite.

Le cortège d'éléments traces des phosphorites est relativement semblable dans tous les gisements. L'apatite et les argiles se partagent les traces. Mais l'affinité des traces pour les minéraux argileux paraît plus forte puisque les teneurs en éléments traces sont plus fortes lorsqu'il y a plus d'argile. A Taïba où les argiles sont plus abondantes après l'altération, la plupart des éléments traces sont également plus abondants.

Si les raisons qui conduisent à cette partition ne nous sont pas encore connues, on ne peut manquer de remarquer le rôle particulier que semble jouer l'abondance du fer exprimé dans la concentration des traces au Banc Le Danois. Ce rôle est bien souligné par Bliskovskiy (1969).

Enfin on peut constater dans les cinq cas présentés la constance de l'association V–Ni–Zn–Cr. Presque toujours intercorrélés, ces éléments auxquels s'associe quelquefois le cuivre, forment un groupe isolé ou en association avec les argiles ou le fer. Ce groupe qui semble caractériser les sédiments phosphatés n'est qu'exceptionnellement rattaché à l'apatite.

Conclusion

De cette étude, qui mérite d'être complétée par l'analyse systématique de nombreux autres échantillons pour acquérir un caractère de généralité suffisant, on peut cependant conclure que le cortège d'éléments traces confère aux gisements phosphatés marins mis en place dupuis le tertiaire une allure de famille transparaissant à travers leur diversité. Paraissent ainsi caractéristiques et communes à tous ces phosphates:

—l'association constante du strontium et du phosphore;
—l'association constante du strontium et du sodium;
—l'association constante V–Ni–Zn–Cr.

La diversité est introduite par les éléments et minéraux associés, ceux du fer entre autres, provoquant une répartition des éléments traces particulière à chaque gisement, dépendant de chaque étape de leur histoire. Si cette histoire n'est pas encore connue dans le détail, on peut cependant penser que les éléments observés ont été piégés par un même mécanisme, probablement biologique sur la marge continentale (Lucas et Prévôt, 1975). Phosphore, vanadium, chrome, zinc, nickel ont été retenus ensemble dans la même phase organique au cours de la concentration initiale, comme le suggèrent entre autres Gulbrandsen (1966) et Krauskopf (1955); ils ont ensuite été séparés dans les phases différentes pendant la période de diagenèse précoce, c'est–à–dire au cours de la minéralisation. A l'inverse le strontium et le phosphore qui sont actuellement associés au sein du même minéral apatite étaient probablement séparés. Ils ont été rapprochés, l'un dans les carbonates, l'autre dans la matière organique et se sont trouvés réunis au cours de la minéralisation.

Pendant l'altération météorique ménagée, certains éléments peuvent changer de situation, comme le strontium qui abandonne le phosphate de calcium décarbonaté; mais le cortège général reste pratiquement le même. Ainsi les caractères acquis par les mécanismes initiaux de concentration et de piégeage se conservent globalement à travers les transformations successives du matériel. Les changements interviennent à l'échelle du minéral; mais les migrations sont de courte distance et la roche se comporte longtemps comme un système clos.

ISRAEL

Y. NATHAN and Y. SHILONI

Geologic Background

In Israel phosphorites are found throughout the Negev in the same stratigraphic horizon (Upper Mishash–Late Campanian). They overlie and are interbedded with a chert, porcelanite and chalk sequence and are generally overlain by a Maestrichtian marl. The northern part of the Negev is dominated by more or less parallel anticlinal ranges whose general trend is north-east, south-west. These anticlinal ranges subdivide the area into a series of synclinal valleys. Two of these synclinal valleys comprise all the present exploited fields. The Oron-Zef'a syncline and the Zin syncline. The eight sampled fields in this study are within these two synclines. Most of the phosphate fields are either in the middle of the syncline or on the lower flanks. Some are found on the higher flanks. These are the so-called anticlinal fields. Towards the structural highs the Mishash Formation thins out, becoming also relatively more cherty and at times even disappears completely. Because of the topography, which is roughly parallel to the geologic structure, the "anticlinal" fields are much more weathered than the "synclinal" fields. A more detailed geologic background with a detailed bibliography on previous works, together with more complete chemical data, will be found in Nathan *et al.* (1977).

Samples and Methods

Of 185 samples, 164 taken from outcrops (trenches) and boreholes from 8 phosphate fields, and 21 collected sporadically were studied. Most of the samples are composite representing from 40 cm to 1 m section; 27 elements and organic matter were determined in each sample. The major elements were analysed by classical wet chemical methods and most of the trace elements were determined by X-ray fluorescence. The CO_2 content within the apatite lattice was estimated from the Gulbrandsen curve (Gulbrandsen, 1970). The chemical results were treated by an R-mode factor analysis (SPSS). A similar factor analysis has already been successfully used to interpret analysis of phosphorites from Jordan (Reeves and Saadi, 1971). We used a Varimax orthogonal matrix and interpreted the loadings of the factors as the probabilities of occurrence of the various elements in the different minerals, each factor representing a mineral or a phase. There is a similarity between the factors and the groups of positive intercorrelations used by Prévôt and Lucas. More details on the methods and exact location of the samples are given in Nathan *et al.* (1977)

Mineralogy

The main mineral is carbonate-fluorapatite (francolite) with calcite, opal C-T, gypsum, dolomite, clay minerals and iron oxides as gangue minerals. Halite is found in surface exposures only. The francolite is very uniform in all the samples with an F/P_2O_5 ratio of 0.119 ± 0.005 and CO_2 content of 3.9% $\pm$ 0.2. This last value is intermediate between the 6% obtained for the Morrocan phosphorite and the 1% of Taïba.

TABLE 1

	"Synclinal" fields				"Anticlinal" field			
	Yorke'am $n = 23$		Zefa $n = 18$		Hazeva $n = 11$		All samples $n = 185$	
Element	Mean (%)	St. Dev.	Mean (%)	St. Dev.	Mean (%)	St. Dev.	Mean (%)	St. Dev.
CaO	50.72	2.53	50.69	4.41	50.41	1.63	50.88	2.82
P_2O_5	26.80	2.11	25.37	7.93	26.76	2.50	25.71	4.71
F	3.15	0.24	3.00	0.91	3.25	0.35	3.06	0.56
CO_2	9.95	1.81	12.06	6.33	11.05	2.73	11.42	4.95
SiO_2	4.54	3.68	3.48	7.84	3.12	0.63	3.10	3.83
Al_2O_3	0.44	0.17	0.39	0.22	0.69	0.17	0.47	0.45
Fe_2O_3	0.28	0.10	0.41	0.33	1.13	1.17	0.35	0.40
MgO	0.44	0.24	0.31	0.22	0.35	0.09	0.56	0.78
Na_2O	0.67	0.24	0.55	0.22	0.78	0.33	0.56	0.26
Cl	0.46	0.29	0.21	0.21	0.63	0.82	0.50	0.64
SO_3	2.31	1.40	1.80	1.08	1.05	0.30	2.44	1.33
O.M.	0.20	0.08	0.85	1.26	0.14	0.06	0.70	1.17
	ppm		ppm		ppm		ppm	
K	525	202	562	224	604	452	564	389
Ti	263	76	238	107	379	136	246	105
V	214	78	172	85	193	98	155	77
Cr	236	28	199	63	294	177	227	75
Mn	34	14	41	15	103	26	40	25
Co	5	21	5	8	16	17	5	6
Ni	93	21	67	15	99	52	89	29
Cu	33	10	26	22	11	8	20	20
Zn	617	203	431	177	351	154	521	204
Rb	4	2	3	1	3	2	4	2
Sr	2661	172	2132	504	1795	214	2300	564
Y	73	42	83	38	90	48	73	45
Nb	4	2	7	7	4	1	5	3
Mo	46	20	28	19	34	7	41	19
Ba	326	71	502	904	445	262	338	319
U	120	20	147	51	108	14	111	49

	V/P_2O_5 (ppm/%)	Cr/P_2O_5	Zn/P_2O_5	Sr/P_2O_5	Y/P_2O_5	U/P_2O_5
Yorke'am	8.0	8.8	24.9	99	2.7	4.5
Zefa	6.8	7.8	17.3	84	2.5	5.7
Hazeva	7.2	10.9	13.1	67	3.4	4.0

Trace Elements

Table 1 gives the means and standard deviations for two typical "synclinal" fields, one "anticlinal" field and for all the samples. The ratio of the more abundant trace elements to P_2O_5 is also given. We see that similarly to what Prévôt and Lucas observed on a much larger scale, there is a great variation in these ratios and that the trace elements are not proportional to the P_2O_5 found. This is true in very similar fields a few km apart. This leads us to the conclusion that trace-element distribution is a very sensitive parameter to environmental changes or to weathering. To make sense of these differences, we must first know the distribution of the trace elements within the various minerals. This is one of our uses of the factor analysis. Seven significant factors occurred in most of the fields and in the total analysis: (1) the apatite (phosphorous) factor, (2) the clay (alumina) factor, (3) the chert (silica) factor, (4) the iron oxide factor, (5) the sulphate factor, (6) the halite factor, (7) the organic matter factor. It is interesting to note that similarly to what has been found by Prévôt and Lucas no carbonate factor is found, although carbonates form 20 to 30% of the phosphorite. Only three of these factors are relevant to the trace element discussion: (1) apatite, (2) clay and (4) iron oxide. The clay factor is a very important one explaining 20–35% of the variance, in spite of the fact that clay forms only 1–3% of the phosphorite. Sometimes it includes all the iron variance or most of it. In the more weathered fields, it includes only a small part of it and there an iron oxide factor independent of the clay occurs. The trace elements in the apatite (according to the apatite factor in all the fields and in their sum) are: Sr and U, these two elements occurring mainly within the apatite lattice; Zn, Y, V and Cr, these elements occurring in the apatite but also in other minerals.

We see from Table 1, where the ratio of these elements to P_2O_5 is given, that the trace elements in apatite can be divided into two groups: (1) those that are sensible to weathering (Sr and Zn) and (2) those that are not sensible to weathering (U, V and Y). The differences in content of the group (2) elements between the various fields are clear and definite but cannot be explained by weathering. They are due to differences in the environment of deposition (or diagenesis). It is difficult to classify Cr since it occurs in high amounts in other minerals. It is therefore doubtful whether the ratio given in Table 1 is significant.

References

ALTSCHULER, Z. S., CLARKE, R. S., Jr. and YOUNG, E. J. (1958) The geochemistry of uranium in apatite and phosphorite. *U.S. Geol. Surv. Prof. Paper*, 314 D, 45–90.

ALTSCHULER, Z. S. (1973) The weathering of phosphate deposits. Geochemical and environmental aspects. In *Environmental Phosphorous Handbook*, Ed. Wiley, 33–96.

ATAMAN, G. et BESNUS, Y. (1965) Une méthode de dosage des éléments traces dans les roches par spectrométrie à lecture directe. *Bull. Serv. Carte géol. Als. Lorr.* **18**, 179–190.

BESNUS, Y. et ROUAULT, R. (1973) Une méthode d'analyse des roches au spectromètre d'arc à lecture directe par un dispositif d'électrode rotative. *Analusis* **2**, 2, 111–116.

BESNUS, Y. (1975) Etude géochimique comparative de quelques gisements supergènes de fer. Thèse Sci, Strasbourg, 184 pp.

BESNUS, Y. et MOSSER, C. (1976) Les groupes à intercorrélations positives: une méthode statistique pour déterminer la répartition des éléments chimiques entre les phases constitutives d'une roche. *C.R. Acad. Sci., Paris*, Série D, **283**, 587–590.

BLISKOVSKIY, V. Z. (1969) Molybdenum, chromium and vanadium in phosphorites. *Geochemistry Int. U.S.A.*, 878–887, trad. de *Geokhimiya* **9**, 1082–1091.

BOUJO, A. (1976) Contribution à l'étude géologique du gisement de phosphate crétacé-éocène des Ganntour (Maroc Occidental). *N et Mém. Serv. Géol. Maroc* **262**, 227 pp. *Sci. Géol. Mém.*, no. 43, 227 pp.

FLICOTEAUX, R. et TESSIER, F. (1971) Précisions nouvelles sur la stratigraphie des formations du plateau de Thiès (Sénégal occidental) et sur leurs altérations. Conséquences paléogéographiques. *C.R. Acad. Sci., Paris*, **272**, 364–367.

FLICOTEAUX, R. et LAPPARTIENT, J. R. (1972) Le passage de l'Eocène inférieur à l'Eocène moyen sur la bordure orientale de Dôme de N'Dias (Sénégal occidental). *Marseille, Trav. Lab. Sci. Terre* (1), 6, 28 pp.

GULBRANDSEN, R. A. (1966) Chemical composition of phosphorites of the Phosphoria Formation. *Géochim. Cosmochim. Acta* **30**, 769–778.

GULBRANDSEN, R. A. (1970) Relation of carbon dioxide content of apatite of the Phosphoria Formation to regional facies. *U.S. Geol. Survey Prof. paper* 700, B, B9–B13.

KRAUSKOPF, K. B. (1955) Sedimentary deposits of rare metals. *Econ. Geol.*, 50th Ann. Vol., 411–463.

LAMBOY, M. (1976) Géologie marine et sous-marine du plateau continental au Nord-ouest de l'Espagne. Genèse des glauconies et des phosphorites. Thèse Sci. Rouen, 285 pp.

LUCAS, J. et PRÉVÔT, L. (1975) Les marges continentales, pièges géochimiques; l'exemple de las marge atlantique de l'Afrique à la limite Crétacé-Tertiaire. *Bull. Soc. Géol. Fr.*, XVII, 496–501.

MENOR E. de A. (1975) La sédimentation phosphatée. Pétrographie, minéralogie et géochimie des gisements de Taïba (Sénégal) et d'Olinda (Brésil). Thèse Doct. Ing., Strasbourg, 153 pp.

NATHAN, Y., SHILONI, Y., RODED, R., GAL, I and DEUTSCH, Y. (1977) Chemistry and geochemistry of the Northern and Central Negev Phosphorites (southern Israel). *Bull. Geol. Surv. Israel* (à l'impression).

NAUMENKO, V. V. (1966) Répartition des microéléments dans les phosphates. *Dop. Akad. Nauk Ukrain RSR* **2**, 229–231.

REEVES, M. J. and SAADI, T. A. K. (1971) Factors controlling the deposition of some phosphate bearing strata from Jordan. *Econ. Geol.* **66**, 451–465.

The Major Element Geochemistry and the Mineralogical Evolution of Granitic Rocks during weathering

W. CHESWORTH

University of Guelph, Ontario, Canada

Abstract

The geochemical evolution of weathering granite is described in terms of the development of residua that become increasingly enriched in the components SiO_2, Al_2O_3, Fe_2O_3 and H_2O. Mineralogical evolution mirrors this trend in that the residua tend to become dominated by phases such as kaolinite, quartz, gibbsite or goethite/hematite. The principle controlling factor is the behaviour of groundwaters most of which, in granitic terrains, appear to be buffered to compositions in equilibrium with kaolinite. Most physical ("soil-forming") factors of the environment affect the availability, residence-time and composition of groundwaters and thereby determine the geochemical and mineralogical evolution of the weathering granite.

Introduction

The weathering of granite was reviewed by Collier (1961) and by Millot (1964) in his more comprehensive work. In the last decade the major advances that have taken place in our understanding of the processes involved have come about in the study of the chemistry of granite–water interactions at the surface of the earth. This has been promoted by laboratory experiment (e.g. Pedro, 1964) and by developments in thermodynamics, particularly by Helgeson (e.g. 1970), in which the theoretical treatments of open systems developed by Korzhinskii (1959) and Thompson (1955) have been wedded to the graphical techniques of Garrels and Christ (1965).

The approach I take here is to summarize the main trends in terms of a geochemical and mineralogical evolution towards the residua system of weathering (Chesworth, 1973a). Later I examine ways of interpreting these trends.

The Parent Material

The term granite means different things to different workers as Streckeisen (1967) clearly shows. I take the term in the broad sense to mean a plutonic rock with essential quartz and feldspar. The feldspar may be present as one phase (usually a K-feldspar/Na-feldspar solid solution) or as two phases (a K-feldspar solid solution and a plagioclase). The commonest ferromagnesian mineral is biotite and it may or may not be accompanied by muscovite. Hornblende is common in more calcium-rich granitoids. Less common are pyroxene-bearing and fayalite-bearing varieties. Magnetite is virtually ubiquitous as an accessory mineral.

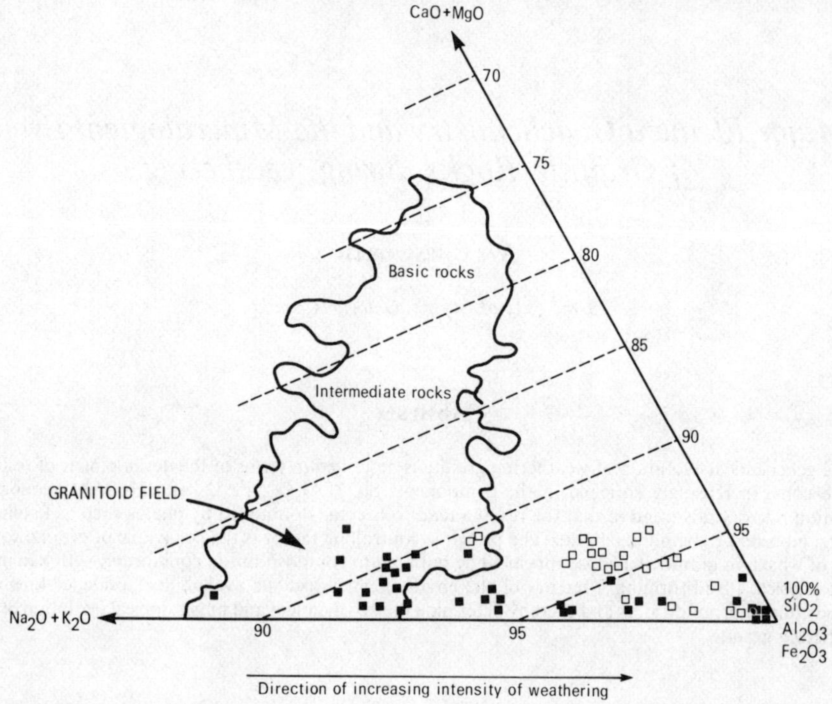

FIG. 1. The trend towards residua rich in one or more of the components SiO_2, Al_2O_3 and Fe_2O_3 shown by weathering granite. The field of the common igneous rocks is taken from Chesworth (1973b), where a rationale for the diagram is given. Analyses plotted are from Stace *et al.* (1968), Dejou *et al.* (1970), Chevalier and Dejou (1972), Rice (1973), Blaxland (1974) and La Hay (1976) (■ bulk analyses of rock, regolith and soil; □ clay fractions of weathered material).

The composition of the parent material dictates the starting-point of the weathering trend (Fig. 1) and the primary minerals determine the initial transformations that are possible. A general finding is that biotite and plagioclase feldspar weather more readily than muscovite, K-feldspar and quartz (e.g. Goldrich, 1938; Harriss and Adams, 1966).

The Trend Towards Geochemical Maturity

Many published schemes of the geochemistry of the major elements at the surface of the earth indicate a tendency for the components SiO_2, Al_2O_3 and Fe_2O_3 to accumulate in the weathering zone at the expense of other components such as CaO, MgO, Na_2O and K_2O, that have a greater tendency to be carried away in aqueous solution. Since all components are in fact leached away the accumulation is relative rather than absolute, however the trend towards residua rich in one or more of the components SiO_2, Al_2O_3 and Fe_2O_3 is well established and is illustrated in Fig. 1. Chemically weathered granites and granitic soils all move towards the SiO_2–Al_2O_3–Fe_2O_3 apex, the trend showing up first in the clay fraction, where the most active weathering proceeds. These three components, together with H_2O, the prerequisite component for weathering, constitute a residua system that marks in the general case, the locus of possible end points of the weathering of silicate rocks.

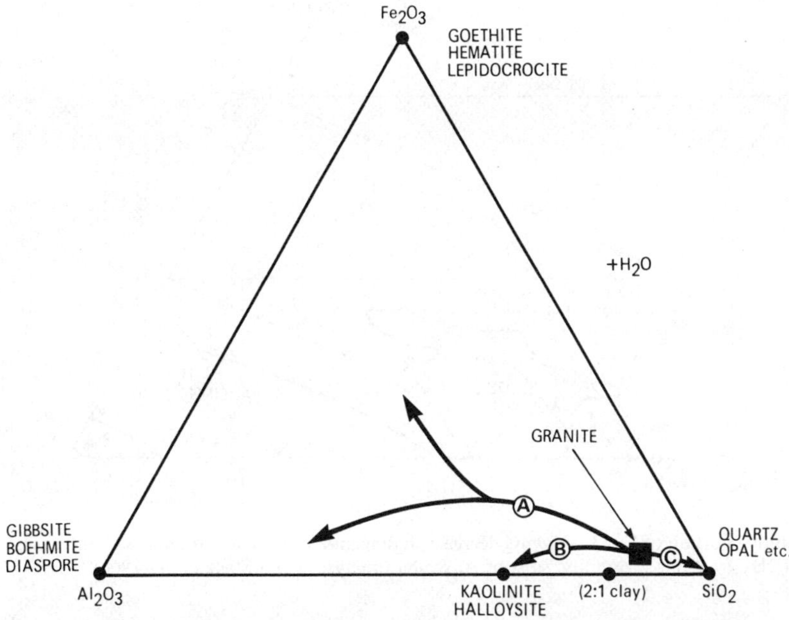

FIG. 2. Schematic diagram to show three common geochemical processes in the weathering of granite. A: ferrallitization; B: monosiallitization; C: podzolization. Common phases found in earth-surface deposits are shown (cf. Chesworth, 1973a, fig. 1). A particular trend will largely be determined by natural variations of Eh and pH (e.g. Garrels and Christ, 1965, chapter 11).

Figure 2 shows a projection of this system that enables three separate trends to be distinguished within the general trend. First and most commonly there is a trend towards kaolinite-rich compositions (kaolinization). Secondly, there is a trend towards SiO_2 enrichment shown in the formation of albic horizons in podzols. (It should be noted that elsewhere in the podzol profile, and particularly in the spodic horizon, kaolinite may still form.) Finally there is the trend towards enrichment in Al, Fe or both, relative to Si, seen in the formation of oxisols, bauxites and laterites. In Fig. 2, and slightly adapting Pedro's (1966) terminology, these processes are called monosiallitization, podzolization and ferrallitization. It is important to stress that kaolinite can and does form in all three processes.

The Trend Towards Mineralogical Maturity

Since the weathering of granite leads to compositions enriched in SiO_2, Al_2O_3, Fe_2O_3 and H_2O, it is not surprising that the clay fraction becomes progressively enriched in phases that are made up of one or more of these components. These include kaolinite, halloysite, quartz, gibbsite, boehmite, goethite and hematite. Clay minerals with 2:1 structure may form at an early stage of weathering but 1:1 clays, hydroxides and oxides become increasingly dominant. Figure 3 shows this point, and also serves to illustrate the fact that kaolinite is the commonest accumulative new phase. Its gradual accumulation can be taken as an indicator of mineralogical maturity in the weathering material.

For those materials in which kaolinization is not the dominant process, the accumulation

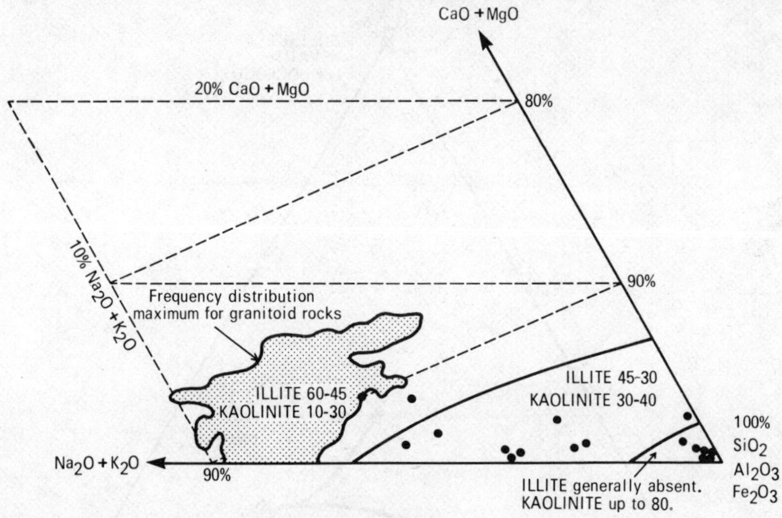

FIG. 3. "Granitic" portion of Fig. 1, showing decrease in illite and increase in kaolinite with increasing weathering intensity, for clay fractions of granitic soils of Australia (analyses from Stace *et al.*, 1968).

of other phases that plot within the residua system of weathering can be taken as an indication of mineralogical maturity as, for example, quartz in the albic horizon of a podzol, or gibbsite or goethite (or both) in a ferrallitic soil. The generally observed mineralogical trends are summarized in Fig. 4.

A General Interpretation of the Weathering System

Water and its chemistry are of paramount importance to the weathering system. The availability of water, its residence time in the weathering deposit, its composition and its

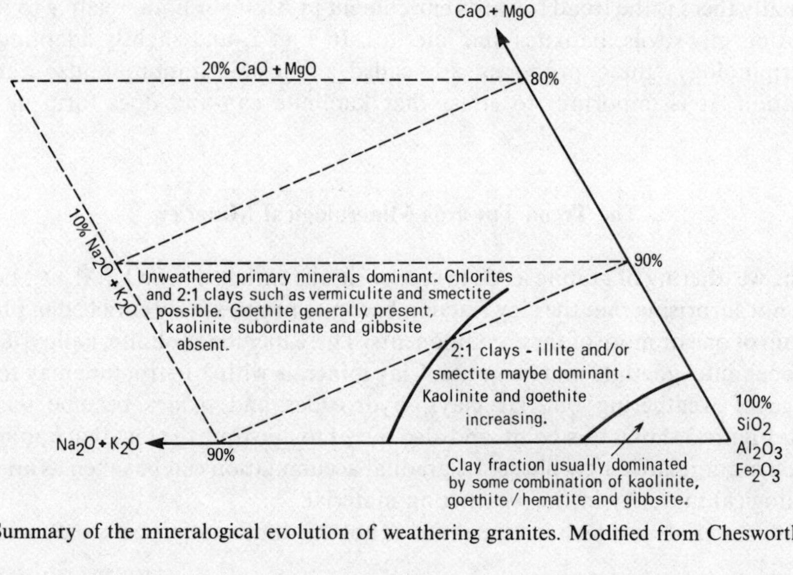

FIG. 4. Summary of the mineralogical evolution of weathering granites. Modified from Chesworth (1977).

vertical mobility in leaching, all in effect dictate the directions of geochemical and mineralogical evolution for the solid residua. The principal chemical action of water is hydrolysis and the hydrolysis of primary phases is an important agent in the promotion of the initial fragmentation of a granite to form an arene (Robert, 1966). As new phases, both amorphous and crystalline, are formed, cations and cationic complexes are released into aqueous solution. The importance of organisms at this stage cannot be overestimated. Not only do they produce organic compounds that accelerate the acid attack on mineral surfaces and act as complexing agents to hold inorganic components in solution (Huang and Keller, 1971), they also produce opaline phytoliths that greatly increase the amount of silica in solution since amorphous silica is much more readily soluble than quartz (Krauskopf, 1956).

The leaching process has been elegantly simulated in a series of experiments by Pedro (e.g. 1964), using various rocks, including granite, and confirming the tendency for solid residua to become enriched in SiO_2, Al_2O_3 and Fe_2O_3. The leachate composition is obviously important in determining the amounts of each component subtracted from the rock and an examination of an average groundwater draining granitic terrain (White *et al.*, 1963) indicates that the major components are present in the following decreasing order:

$$Si > Ca > Na > Mg > K > Fe > Al$$

The analyses of other workers (e.g. Feth *et al.*, 1964; Wolff, 1967) yield different sequences but agree in placing Fe and Al at the low solubility end and Si at the high. Consequently these three elements accumulate in a weathering granite for two different reasons: Fe and Al because they have an inherently low solubility under normal conditions of Eh and pH, and Si, in spite of its high solubility (with respect to amorphous silica), because the potential reservoir of SiO_2 in a rock such as granite is so great that other less mobile components are exhausted before it. This reservoir effect is clearly demonstrated in Fig. 5, which assumes an average granite leached by an average groundwater from granitic terrain. It is an

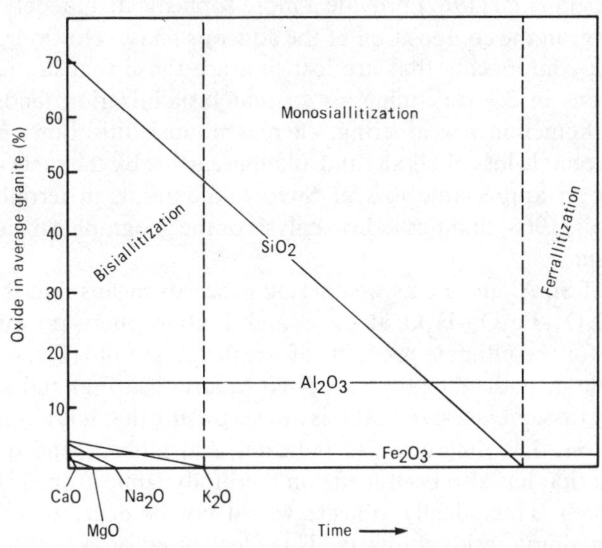

FIG. 5. Subtraction diagram showing hypothetical leaching of an average granite by an average groundwater from granitic terrain (cf. Chesworth, 1973a, fig. 5). The stages of secondary mineral formation are named following Pedro (1966) with the substitution of ferrallitization for allitization.

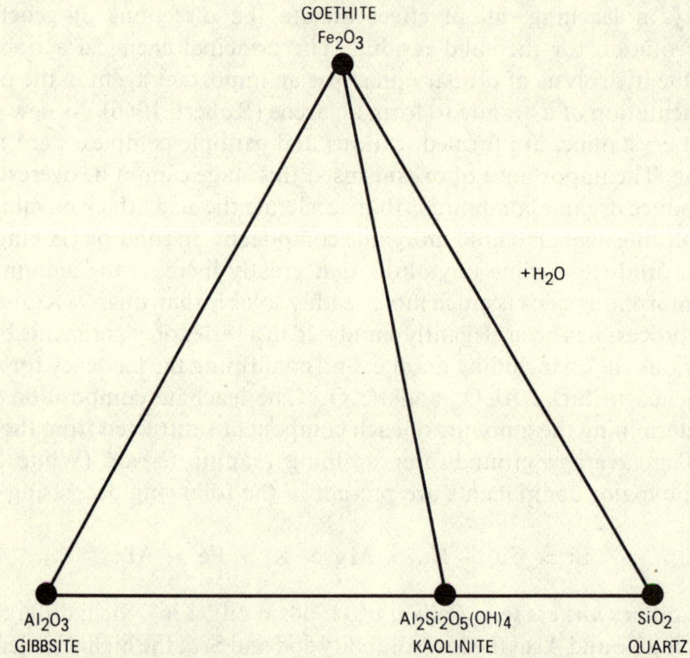

FIG. 6. Projection of the system SiO_2–Al_2O_3–Fe_2O_3–H_2O (Chesworth, 1975) to show phase assemblages stable under earth-surface conditions in the presence of water. This interpretation implies that such phases as halloysite, hematite, lepidocrocite, boehmite and diaspore are unstable or metastable under these conditions. Kittrick (1969) provides an opposed viewpoint.

oversimplification in that the groundwater composition would not normally be constant with time. Helgeson *et al.* (1969) provide a more sophisticated model based on continuous irreversible change in the composition of the aqueous phase. However, Fig. 4 does serve to indicate that the components that are lost first are those that are used in structural or exchange positions in 2:1 clay minerals so that bisiallitization tends to be an early or intermediate phenomenon in weathering, whereas monosiallitization tends to dominate later stages after appreciable loss of alkalis and alkaline earths. By the same argument prolonged leaching leading to appreciable loss of SiO_2 would result in ferrallitization. These are essentially Pedro's (1968) findings in his analysis of the geographical distribution of clays in terrestrial deposits.

The buildup of Si, Al and Fe as weathering proceeds means that the mineralogy of the system SiO_2–Al_2O_3–Fe_2O_3–H_2O at 25°C and 1 atmosphere pressure can serve as an idealized model for the ultimate products of weathered granite (Fig. 6). This indicates that kaolinite is stable in both silica-oversaturated (quartz-bearing) and silica-undersaturated (gibbsite-bearing) assemblages, so that it is not surprising that it is a common mineral of the weathering process. The diagram also indicates that gibbsite and quartz are not stable together, a point that has also been made on thermodynamic grounds (Garrels and Christ, 1965, pp. 352–358). Thus ideally gibbsite would not be expected to form in weathered granites still containing residual quartz. This ideal situation is realized in humid tropical climates where weathering systems are most likely to approach equilibrium (e.g. Delvigne, 1956). Where gibbsite is produced in these silica oversaturated environments it can be explained as an intermediate product of the hydrolysis of feldspar surfaces, by H_2O undersaturated with respect to kaolinite (Helgeson, 1974). However, most analyses of groundwaters from granitic terrain indicate oversaturation with respect to kaolinite (e.g. Feth *et al.*, 1964; Tardy, 1971; Paces, 1972), again confirming that kaolinization should be a common phenomenon in the weathering of granite. Where formation of gibbsite has been important (one of the best documented examples being the granitic arene of Marcoles, Dejou *et al.*, 1970) factors such as abundant rainfall, inclined land surfaces and relatively high porosity have intervened to cause a low residence time for the groundwater in the arene, so that the concentration of SiO_2 in solution remains low. Ferrallitization will also be promoted by lack of vegetative cover, as for example on many tropical laterites, since this would lead to a lack of biogenic opal which would therefore tend to lower the concentration of SiO_2 in solution.

The pathways for mineral transformations in weathering granites have been looked at most recently by Tardy *et al.* (1973), who have tentatively related them to climate. One could equally well propose arguments relating them to any of the other so-called soil-forming factors (e.g. Souchier, 1971; Gradusov, 1974). However, a knowledge of the behaviour of water at the earth's surface is of more fundamental concern and in fact enables one to predict the effect of such physical factors as, for example, climate, organic matter, relief and porosity in terms of how they influence residence time and therefore water chemistry. The longer the residence time, the more concentrated the solution will be, therefore the more likelihood of bisiallitization. Similarly, the shorter the residence time, the more dilute the solution will be and the greater the chance of ferrallitisation.

Conclusion

The reaction of granite with water at the surface of the earth leads to chemical trends such as monosiallitization, podzolization and ferrallitization that produce deposits enriched in the components SiO_2, Al_2O_3, Fe_2O_3 and H_2O. Consequently the resulting weathered deposits tend to become enriched in phases that have compositions made up of one or more of these components and in fact the commonest accumulating phase is kaolinite. These tendencies can be described in terms of the concept of a residua system of weathering and can be explained in terms of the availability, residence time and chemistry of percolating groundwaters.

One of the major areas for which information is still required is on the true nature of the mineralogical reactions that take place. These can involve (Millot, 1964) degradation (or aggradation) of existing solid phases, transformation via an amorphous phase, or direct precipitation from aqueous solution. Direct investigation of parageneses microscopically (e.g. Protz *et al.*, 1974; Meunier and Velde, 1976) and submicroscopically (e.g. Rimsaite, 1975; Stoch and Sikora, 1976) together with the laboratory investigation of reactions (e.g. Siffert and Wey, 1962; Robert and Razzaghe-Karimi, 1974) and *in situ* experiments in the field (e.g. Duchaufour et al., 1968) will produce many advances in this direction.

Acknowledgement

I thank Dr. Les Evans for his helpful comments.

References

BLAXLAND, A. B. (1974) Geochemistry and geochronology of chemical weathering, Butler Hill granite, Missouri. *Geochim. Cosmochim. Acta* **38**, 843–852.

CHESWORTH, W. (1973a) The residua system of chemical weathering: a model for the chemical breakdown of silicate rocks at the surface of the earth. *J. Soil Sci.* **24**, 69–81.

CHESWORTH, W. (1973b) The parent rock effect in the genesis of soil. *Geoderma* **10**, 215–225.

CHESWORTH, W. (1975) The system SiO_2–AlOOH–Fe_2O_3–H_2O and the kaolinitic stage of the geothite facies. *Clays and Clay Min.* **23**, 389–392.

CHESWORTH, W. (1977) Weathering stages of the common igneous rocks, index minerals and mineral assemblages at the surface of the earth. *J. Soil Sci.* **28** (in press).

CHEVALIER, Y. and DEJOU, J. (1972) Evolution géochimique superficielle en climat méditerranéen d'un granite à deux micas doté d'une porosité très élevée. Exemple choisi dans le massif des Maures, au Plan-de-la-Tour (Var). *C.R. Acad. Sci. Fr.* **274**D, 1240–1243.

COLLIER, D. (1961) Mise au point sur les processus de l'altération des granites en pays tempéré. *Ann. Agron.* **12**, 273–332.

DEJOU, J., GUYOT, J. PEDRO, G. CHAUMONT, C. and ANTOINE, H. (1970). Etude sur le problème de l'origine de la gibbsite dans les arènes granitiques des pays tempérés. *Science du Sol.* **2**, 15–25.

DELVIGNE, J. (1965) *Pédogenèse en zone tropicale. La formation des minéraux secondaires en milieu ferrallitique*, Dunod Ed., Paris, 177 p.

DUCHAUFOUR, PH., VEDY, J. C. and BRUCKERT, S. (1968) Contribution à l'étude expérimentale "in situ", de l'évolution chimique et de l'altération d'un grès et d'un granite vosgiens sous l'influence de deux types de litières. *Bull. A.F.E.S.* **1**, 5–18.

FETH, J. H., ROBERSON, C. E. and POLZER, W. L. (1964) Sources of mineral constituents in water from granitic rocks Sierra Nevada, California and Nevada. *U.S. Geol. Surv. Water Supply* Paper 1535–I.

GARRELS, R. M. and CHRIST, C. L. (1965) *Solutions, Minerals and Equilibria*, Harper & Row, New York, 435 pp.

GOLDICH, S. S. (1938) A study in rock-weathering. *J. Geol.* **46**, 17–23.

GRADUSOV, B. P. (1974) A tentative study of clay mineral distribution in soils of the world. *Geoderma* **12**, 49–55.

HARRISS, R. C. and ADAMS, J. A. S. (1966) Geochemical and mineralogical studies on the weathering of granitic rocks. *Amer. J. Sci.* **264**, 146–173.

HELGESON, H. C. (1970) Description and interpretation of phase relations in geochemical processes involving aqueous solutions. *Amer. J. Sci.* **268**, 415–438.

HELGESON, H. C. (1974) Chemical interaction of feldspars and aqueous solutions. In *The Feldspars*, W. S. MACKENZIE and J. ZUSSMAN, eds., 184–217.

HELGESON, H. C., GARRELS, R. M. and MACKENZIE, F. T. (1969) Evaluation of irreversible reactions in geochemical processes involving minerals and aqueous solutions—II. Applications. *Geochim. Cosmochim. Acta* **33**, 455–481.

HUANG, W. H. and KELLER, W. D. (1971) Dissolution of clay minerals in dilute organic acids at room temperature. *Amer. Min.* **56**, 1082–1095.

KITTRICK, J. A. (1969) Soil minerals in the Al_2O_3–SiO_2–H_2O system and a theory of their formation. *Clays and Clay Min.* **17**, 157–167.

KORZHINSKII, D. S. (1959) *Physiochemical Basis of the Analysis of the Paragenesis of Minerals*, Chapman & Hall, London, 142 pp.

KRAUSKOPF, K. B. (1956) Dissolution and precipitation of silica at low temperatures. *Geochim. Cosmochim. Acta* **10**, 1–26.

LA HAY, K. L. (1976) Early stages of the weathering of the Methuen granite. Unpub. M.Sc. thesis, Guelph, 58 pp.

MEUNIER, A. and VELDE, B. (1976) Mineral reactions at grain contacts in early stages of granite weathering. *Clay Min.* **11**, 235–240.

MILLOT, G. (1964) *Géologie des Argiles*, Masson et Cie., Paris, 499 pp.

PACES, T. (1972) Chemical characteristics and equilibration in natural water–felsic rock–CO_2 system. *Geochim. Cosmochim. Acta* **36**, 217–240.

PEDRO, G. (1964) Contribution à l'étude expérimentale de l'altération géochimique des roches cristallines. Thèse Fac. Sci., Paris, 334 pp.

PEDRO, G. (1966) Essai sur la caractérisation des différents processus zonaux résultant de l'altération des roches superficielles (cycle aluminosilicique). *C.R. Acad. Sci., Fr.* **262** D, 1828–1831.

PEDRO, G. (1968) Distribution des principaux types d'altération géochimique à la surface du globe. Préséntation d'une esquisse géographique. *Rev. géog. phys. et geol. dyn.* **10**, 457–470.

PROTZ, R., GILLESPIE, J. E. and BREWER, R. (1974) Micromorphology and genesis of four soils derived from different rocks in Peterborough County, Ontario, Canada. In *Soil Microscopy*, by G. K. RUTHERFORD, ed., 481–497.

RICE, C. M. (1973) Chemical weathering on the Carnmellis granite. *Min. Mag.* **39**, 429–447.

RIMSAITE, J. (1975) Natural alteration of mica and reactions between released ions in mineral deposits. *Clays Clay Miner.* **23**, 247–255.

ROBERT, M. (1966) Le processus d'arénisation des granites: nouvelles données experimentales. *Comptes Rend. 91ᵉ Congrès Nat. des Soc. Savantes, Rennes*, Tome II, 153–161.

ROBERT, M. and RAZZAGHE-KARIMI, M-H. (1974) Evolution des micas trioctaédriques en présence d'acides organiques. *Bull. du Gr. Fr. Argiles*, **26**, 307–317.

SIFFERT, B. and WEY, R. (1962) Synthèse d'une sépiolite à température ordinaire. *C.R. Acad. Sci. Fr.* **254** D, 1460–1463.

SOUCHIER, B. (1971) Evolution des sols sur roches cristallines à l'étage Montagnard (Vosges). *Mém. Serv. Carte. Géol. Alsace Lorraine* **33**, 134 p.

STACE, H. C. T., HUBBLE, G. E., BREWER, R., NORTHCOTE, K. H., SLEEMAN, J. R., MULCAHY, M. J. and HALLSWORTH, E. G. (1968) *A Handbook of Australian Soils*, Rellim, S. Australia, 435 pp.

STOCH, L. and SIKORA, W. (1976) Transformations of micas in the process of kaolinization of granites and gneisses. *Clays Clay Miner.* **24**, 156–162.

STRECKEISEN, A. (1967) Classification and nomenclature of igenous rocks. *N. Jb. Miner. Abh.* **107**, 144–240.

TARDY, Y. (1971) Characterization of the principal weathering types by the geochemistry of waters from some European and African crystalline massifs. *Chem. Geol.* **7**, 253–271.

TARDY, Y., BOCQUIER, G., PAQUET, H. and MILLOT, G. (1973) Formation of clay from granite and its distribution in relation to climate and topography. *Geoderma* **10**, 271–284.

THOMPSON, J. B. (1955) The thermodynamic basis for the mineral facies concept. *Amer. J. Sci.* **253**, 65–103.

WHITE, D. E., HEM, J. D. and WARING, G. A. (1963) *Data of Geochemistry*, 6th ed., Chapter F: Chemical composition of subsurface waters. *U.S. Geol. Surv. Prof. Paper* 440–F.

WOLFF, R. G. (1967) Weathering of Woodstock granite, near Baltimore, Maryland. *Amer. J. Sci.* **265**, 106–117.

Discussion

S. E. HAGGERTY:

In a series of papers this morning we heard of both the mobility and immobility of titanium. Can you throw any light on what Ti does during the weathering of granite?

W. CHESWORTH:

Titanium behaves as one of the accumulating elements at the surface of the earth during weathering. Any extension of the concept of a residua system of weathering to include minor and trace components would naturally include titanium.

J. GONI:

Rôle de la biogéochimie dans les processus d'alteration.

W. CHESWORTH:

I think the full text takes your point into account. Biological processes are obviously important in earth-surface processes particularly in pedogenesis and can alter the mobility of inorganic components by several orders of magnitude.

Eléments Traces dans quelques Argiles des Altérations et des Sédiments

CH. MOSSER

Centre de Sédimentologie et Géochimie de la Surface
1, rue Blessig—67084 Strasbourg Cedex

Abstract

The content of trace elements in kaolinites and smectites formed by alteration under equatorial and tropical conditions is closely similar to that observed for the parent rocks. The content of Ni, Co and Cr increases from clays formed on acid rocks to those formed on basic rocks. The same variation is observed with V, but it is less regular. The content of Ba, Rb, Li, Ga and Pb, on the other hand, decreases in clays developed on progressively more basic rocks.

Two mineral groups, the kaolinites, and iron oxides and hydroxides, are found in the $<2\mu$ fraction of kaolinitic alteration products. Trace elements are shared between these two mineral groups.

Mainly only smectite is found in the $<2\mu$ fraction of smectitic alteration products and trace elements are then localized with the dominant smectite component.

During transport out of the alteration profiles and during subsequent sedimentation, kaolinites and iron compounds can be separated. The trace elements associated with these two minerals are then also separated. This is observed with kaolinites sedimented in the Charentes (France). These kaolinites are nearly free of iron (1.5% Fe_2O_3), most of which occurs as sulfide minerals, in which Pb and Zn are also concentrated.

The trace elements from some sedimentary clay minerals allow us to distinguish between original characteristics of alteration and later sedimentation environments.

Two examples are given: kaolinites sedimented in the Charentes and tertiary illites and smectites from Cormeilles en Parisis. The high level of Ga and the low content of other trace elements in the Charentes kaolinites suggest that these clays are alteration products which may have developed on granite-like acid parent rocks. A similar parent rock type is suggested for the illites and smectites on the basis of the content of V, Ni, Co, Zn, Ga, Cu, Pb and Sn. The very high Li level (500 ppm) of the Charente kaolinites is indicative of the influence of the depositional environment; the content of Sr, Cr, B, Li and Rb of the Cormeille illites and smectites are indicative of the laguno-marine sedimentation environment.

Introduction

Les minéraux argileux se forment dans les conditions thermodynamiques et géochimiques existant à la surface du globe terrestre. Ce sont des minéraux qui du fait de leur petite taille ($<2\ \mu$) offrent une grande surface de réaction au milieu dans lequel ils se trouvent. Ces minéraux peuvent ainsi absorber ou adsorber des cations présents dans leur environment de genèse ou de dépôt. L'analyse et l'étude des éléments traces ainsi adsorbés ou absorbés par les minéraux argileux peut alors devenir une technique pour connaître les milieux de formation et de dépôt des argiles, à condition que les éléments traces analysés soient représentatifs de ces milieux.

Actuellement kaolinites et smectites sont des minéraux argileux qui se néoforment massivement dans les régions à climat tropical et équatorial, dans des profils d'altération établis sur différentes roches mères. Des kaolinites et des smectites ont été analysées pour montrer l'influence des roches mères sur la nature des éléments traces et leurs teneurs dans les deux types de minéraux argileux formés.

Lorsque les profils d'altération sont démantelés, tous leurs produits constitutifs, et en particulier les argiles, sont entrainés dans des zones de sédimentation. Une argile déposée dans un bassin sédimentaire garde-t-elle suffisamment la trace de son origine pour que l'on puisse reconnaître la roche mère qui lui a donné naissance ou l'histoire sédimentaire et diagénétique précoce est-elle marquante au point d'effacer toute empreinte reconnaissable d'une histoire précédente. Deux exemples sont étudiés ici; les kaolinites "sidérolitiques" des Charentes et les illites et smectites tertiaires de Cormeilles en Parisis.

I. Techniques d'Etude

Extraction des Minéraux Argileux

Les diverses roches dont on veut étudier les argiles sont broyées avec ménagement puis mises en suspension dans de l'eau distillée. La fraction granulométrique $<2\ \mu$ est extraite par sédimentation et siphonnage. On évite au maximum tout traitement chimique, mais on ne peut éviter l'utilisation de HCl pour décarbonater certaines roches ou certains sédiments afin de libérer les argiles et débarasser au maximum la fraction $<2\ \mu$ des carbonates.

La fraction $<2\ \mu$ obtenue est souvent presque exclusivement formée de minéraux argileux. Cependant elle renferme parfois aussi des minéraux tels que goethite, hématite, limonite, quartz et quelques minéraux lourds. Il faut donc contrôler systématiquement à l'aide de la diffraction RX la composition minéralogique de la fraction $<2\ \mu$ extraite.

Lorsque la fraction argileuse est accompagnée d'une trop forte proportion d'oxydes et d'hydroxydes de fer elle a été traitée par le réactif de Tam (acide oxalyque, oxalate d'amonium) sous rayonnement UV pour la débarasser de ces composés du fer. Cette méthode de De Endredy a été décrite par Segalen (1970).

Analyses Chimiques

Les fractions inférieures à 2 μ ont été analysées par spectrométrie à lecture directe pour le dosage de SiO_2, Al_2O_3, MgO, CaO, Fe_2O_3, Mn_3O_4, TiO_2, Sr, Ba, V, Ni, Co, Cr, B, Zn, Ga, Cu, Pb, Sn, par spectrométrie d'absorption atomique pour le dosage de Li, Rb et par spectrométrie de flamme pour le dosage de Na_2O et K_2O.

Les argiles qui ont été déferrifiées par le réactif de Tam ont été analysées avant et après traitement afin de pouvoir faire la part de ce qui revient aux argiles et aux oxydes et hydroxydes de fer.

Présentation des Résultats

Lorsque le nombre d'échantillons est suffisant ce sont les moyennes des analyses chimiques des fractions $<2\ \mu$ qui sont présentées. Des analyses de fraction $<2\ \mu$ isolées mais représentatives sont présentées lorsqu'on manque de points de prélèvements. Ce cas se présente surtout lors de l'étude des argiles des altérations car le nombre d'échantillons prélevés par profils d'altération est souvent restreint.

Les éléments traces sont souvent répartis dans les différentes phases minéralogiques constitutives de la fraction $< 2\ \mu$ lorsque celle-ci est composée de plusieurs espèces s'argiles et de minéraux autres que des argiles. Deux méthodes ont été utilisées en fonction du nombre d'échantillons dont on dispose. La méthode statistique par régressions, décrite par Mosser, Weber et Gac (1974) a été pratiquée lorsque le nombre d'échantillons était suffisant. Cette méthode permet de tenir compte de tous les minéraux présents dans $1' < 2\ \mu$. La méthode de déferrification est utilisée lorsque le nombre d'échantillons est très petit. On ne peut alors tenir compte que de la phase argile et de la phase fer.

Les analyses des roches mères sont données pour tous les échantillons issus des profils d'altération. Les roches mères ont été classées d'après un indicè de basicité inspiré de l'indice "an" de Rittman (1963).

$$an = \frac{CaO}{1{,}2(K_2O + 1{,}5Na_2O) + CaO}$$

Les valeurs de cet indice pour les différentes roches mères présentées dans cette étude sont les suivantes:

0,02 Granite de RCA
0,10 Granite de Nayouri Hte Volta
0,15 Granite de Tassamakat Hte Volta
0,17 Granite de Diebiga Hte Volta
0,29 Gneiss de Soffokel Hte Volta
0,46 Diorite de Garango Hte Volta
0,53 Schiste amphibolitique RCA
0,58 Basalte de Madagascar
0,60 Dolérites porphyroblastiques Hte Volta
0,60 Amphibolite de Mene Hte Volta
0,62 Gabbro de Kollel Hte Volta
0,64 Basalte de Taïba Sénégal
0,70 Gabbro de Kollel Hte Volta
0,73 Océanite de la Réunion

II. Kaolinites et Smectites d'Altération

Les analyses chimiques des fractions $<2\ \mu$ à smectites et à kaolinites sont présentées Tableau 1. La répartition des éléments traces dans les phases constitutives des fractions $< 2\ \mu$ est donnée Tableau 3. Les analyses des roches mères correspondantes sont présentées Tableau 2. Les figures 1 et 2 permettent de visualiser les variations de teneurs des éléments traces des fractions $< 2\ \mu$ à smectites et kaolinites, lorsqu'on passe progressivement de profils d'altération établis sur roche acide à des profils établis sur roches basiques. Elles présentent en même temps les teneurs en éléments traces des différentes roches mères.

Les Roches Mères

Les roches mères classées par ordre de basicité croissante présentent des variations de teneurs en éléments traces cohérentes. Les teneurs en Ba, Li, Rb, Ga et Pb diminuent, les teneurs en Ni, Co, Cr augmentent ainsi que celles en V qui ont cependant tendance à décroitre pour les roches mères du pôle le plus basique. Ces variations d'ensembles de teneurs en éléments traces des roches mères sont en excellent accord avec celles données par Rössler et Lange (1972). Dans le détail on observe cependant une rupture de pente importante de la variation des teneurs en Cr et Ni au niveau de l'amphibolite de Mene. Ceci peut être dû au fait que l'on n'a pas classé les roches métamorphiques à part des roches éruptives et intrusives.

Tableau 1. Analyses Chimiques de Fractions $<2\mu$ issued d'Alterations Tropicales et Équatoriales

1. *Analyses de fractions argileuses à smectite dominante > 90%*

Nom de la Roche mère			SiO_2	Al_2O_3	MgO	CaO	Fe_2O_3	Mn_3O_4	TiO_2	Na_2O	K_2O	Perte 1000°C	Sr	Ba	V	Ni	Co	Cr	B	Zn	Ga	Cu	Pb	Sn	Li	Rb
Granite de Tassamakat Haute Volta	N=4	$\bar{x}$	51,5	25,5	1,92	1,38	6,85	0,083	0,68	0,08	0,62	10,94	55	241	52	37	5	63	<5	47	33	20	16	<5	59	46
		σ	1,23	1,24	0,42	0,65	1,12	0,058	0,09	0,04	0,16	0,42	21	78	11	10	3	19		15	3	6	8		22	12
Granite à gros grains Diébiga Haute Volta	N=1		51,0	22,9	1,76	1,0	9,6	0,444	1,18	0,93	1,60	8,67	85	1670	89	39	23	48	<5	165	40	20	36	<5	30	137
Gneiss à biotite et épidote de Soffokel Haute Volta	N=7	$\bar{x}$	47,3	25,4	2,62	1,43	9,9	0,076	0,85	0,10	0,72	11.71	110	355	110	100	12	208	<5	59	25	90	12	<5	83	59
		σ	2,04	1,09	0,22	0,20	1,6	0,020	0,12	0,03	0,08	0,31	40	76	25	19	4	33		17	4	25	5		16	10
Diorite quartzique Garango Hte Volta	N=1		47,5	23,1	2,12	1,8	11,6	0,095	1,18	0.09	0,35	11,35	44	407	109	114	15	231	<5	73	31	87	4	<5	30	44
Dolérite porphyroblastique de Mene Haute Volta	N=1		50,3	14,6	2,45	1,8	21,3	0,149	0,32	0,63	0,77	7,27	43	106	73	82	46	47	<5	88	13	266	<2	<5		
Amphibolite de Mene Haute Volta	N=1		49,8	16,9	2,32	1,8	18,1	0,180	1,41	0,24	0,59	8,64	53	84	164	31	34	18	20	53	22	61	8	<5	10	12
Gabbro de Kollel Haute Volta	N=1		54,1	18,0	4,86	1,7	12,2	0,059	0,36	0,43	0,45	8,14	74	<5	30	122	48	218	22	37	19	61	<2	<5	6	4
Basalte de Taïba Sénégal	N=1		45,6	19,7	2,57	3,8	13,9	0,153	1,58	0,08	0,07	9,51	36	49	271	452	42	216	<5	95	21	86	5	<5		
Océanite de la Réunion	N=7	$\bar{x}$	54,0	1,91	12,77	0	22,14	0,33	0,65	0,07	0,51	7,74	104	75	54	2400	143	433	<5	114	21	104	6	<5	1	8
		σ	1,34	1,54	1,79		1,42	0,04	0,13	0,09	0,10	0,52	37	22	13	297	20	29		9	3	16	3		—	—

Table 1 Continued

2. *Analyses de fractions argileuses à kaolinite + oxydes et hydroxydes de fer.—Kaolinite* > 80%.

Nom de la Roche mère			SiO_2	Al_2O_3	MgO	CaO	Fe_2O_3	Mn_3O_4	TiO_2	Na_2O	K_2O	Perte 1000°C	Sr	Ba	V	Ni	Co	Cr	B	Zn	Ga	Cu	Pb	Sn	Li	Rb
Granite de R.C.A.	N=3	$\bar{x}$	41,8	36,1	0,21	0,15	5,7	0,040	1,01	0,04	0,37	14,56	93	225	62	42	9	<5	5	35	41	21	27	<5	26	56
		σ	2,40	1,52	0,08	0,13	2,13	0,04	0,24		0,13	1,67	50	150	21	10	5	—	6	12	4	7	15	—	12	28
Granite de Nayouri Haute Volta	N=1		47,1	34,7	0,49	0,3	4,6	0,020	0,72	0,12	0,74	12,49	10	179	38	21	<2	39	<5	39	30	23	22	<5	27	45
Migmatite de Madagascar	N=4	$\bar{x}$	44,6	36,6	0,10	<0,2	4,4	0,046	0,31	<0,05	0,23	14,13	48	408	71	40	6	<5	12	15	16	30	41	<5	18	11
		σ	2,7	1,6	0,04		4,4	0,024	0,30		0,12	0,43	17	96	52	26	5		4	8	4	20	51		9	3
Schiste amphibolique R.C.A.	N=17	$\bar{x}$	38,0	29,7	0,39	0,37	15,1	0,080	1,09	<0,05	0,14	15,04	41	141	195	73	21	73	<5	48	29	86	<2	<5	8	19
		σ	2,13	1,68	0,30	0,30	2,47	0,04	0,28	0,02	0,07	2,02	19	76	30	29	12	87		35	4	13			4	9
Basalte de Madagascar	N=21	$\bar{x}$	35,86	30,15	0,34	0,11	17,95	0,140	2,85	0,03	0,13	12,51	12	121	210	116	24	38	<5	24	26	59	<2	<5	<5	<10
		σ	3,04	2,38	0,25	0,03	5.50	0,07	1.05	0,02	0,16	1,21	11	114	94	82	40	66		23	7	48				
Gabbro de Kollel Haute Volta	N=1		43,2	33,2	0,47	0,2	10,0	0,101	0,37	<0,05	0,16	13,20	<10	26	108	185	44	46	<5	31	22	158	<2	<5	11	3

Les teneurs en éléments majeurs sont exprimées en %.
Les éléments traces sont exprimés en ppm.

TABLEAU 2. ANALYSES CHIMIQUES DES ROCHES MÈRES SUR LESQUELLES SE SONT FORMÉS LES SMECTITES ET LES KAOLINITES DONNÉES TABLEAU 1

Nom de la Roche mère	SiO_2	AL_2O_3	MgO	CaO	Fe_2O_3	Mn_3O_4	TiO_2	Na_2O	K_2O	Perte 1000°C	Sr	Ba	V	Ni	Co	Cr	B	Zn	Ga	Cu	Pb	Sn	Li	Rb
Granite de Tassamakat	73,3	15,0	0,49	2,1	1,6	0,041	0,21	4,88	2,75	0,45	408	832	27	6	7	11	<5	49	15	10	29	<5	21	66
Granite à gros grains Dièbiga	70,7	12,8	0,55	2,6	2,4	0,045	0,24	3,59	5,05	0,64														
Gneiss à biotite et épidote de Soffokel	59,3	17,9	3,40	3,1	7,0	0,173	0,70	2,52	2,66	3,41	559	1016	112	60	18	192	<5	60	13	135	23	<5	33	49
Diorite quartzique de Garango	50,9	18,3	4,65	7,3	8,9	0.147	0,96	4,13	1,02	1,68	473	462	125	64	23	119	<5	101	17	80	9	<5	8	18
Dolérite porphyroblastique Mene	49,2	16,7	4,64	9,2	12,2	0,162	1,40	2,87	0,88	1,63	233	116	182	85	48	84	<5	114	19	210	5	<5	13	36
Amphibolite de Mene	53,2	13,5	4,53	7,6	14,4	0,210	0,97	2,46	0,56	2,29	168	0	242	24	49	17	17	92	15	78	5	<5	7	16
Gabbro de Kollel	52,5	18,7	8,10	8,7	8,5	0,142	0,30	2,77	0,26	0,25	217	<5	85	112	44	85	5	36	11	53	<2	<5	5	5
Basalte de Taïba	44,0	10,4	9,48	10,3	9,2	0,130	1,50	2,36	1,35	8,47	1177	152	159	178	27	280	<5	52	2	78	<2	<5		
Océanite de la Réunion	44,7	7,3	23,0	5,6	13,9	0,20	1,33	0,96	0,24	0,79	93	57	130	1310	120	>1100	<5	230	10	153	<2	<5		
Granite de R.C.A.	69,5	17,1	0,45	<0,2	2,5	0.03	0,37	0,58	4,10	4,08	186	1242	44	20	6	<5	18	60	23	41	38	13	40	230
Granite de Nayouri Hte Volta	73,8	14,6	0,43	1,4	1,5	<0,10	0,21	3,87	4,67	0,37	436	1476	25	2	2	13	18	27	13	32	26	<5	9	98
Schiste amphibolique R.C.A.	55,1	15,2	3,48	6,7	11,2	0,20	1,76	3,06	0,36	2,22	254	88	229	41	29	52	<5	97	16	67	<2	<5	<5	<10
Basalte Madagascar	50,05	15,9	5,27	8,8	13,08	0,20	2,10	2,97	0,78	0,96	332	689	254	109	57	<5	11	117	16	54	<2	<5		
Gabbro Hte Volta Kollel	49,0	16,0	7,69	10,0	12.2	0,147	0,70	2,21	0,19	0,17	196	93	320	221	73	132	<5	42	15	132	<2	<5	3	2

Les teneurs en éléments majeurs sont exprimées en %.
Les éléments traces sont exprimés en ppm.

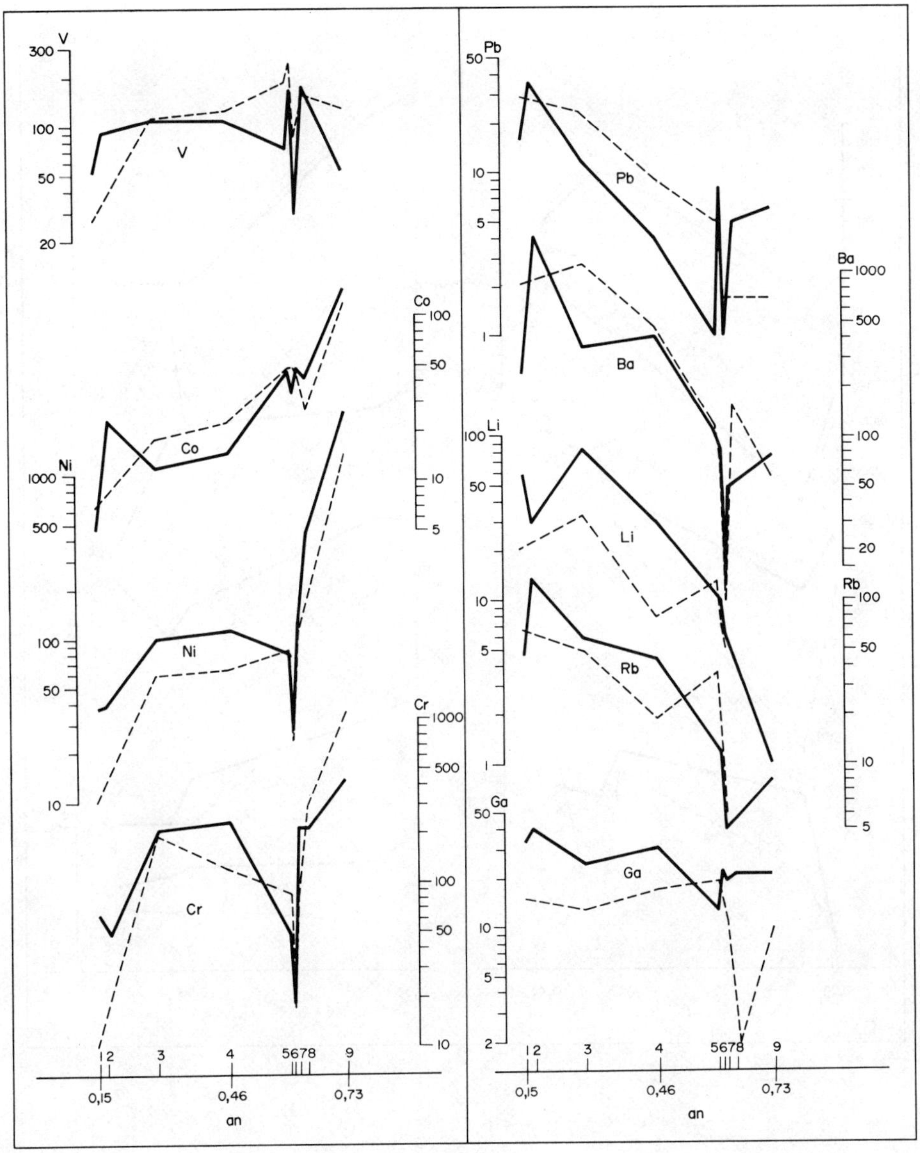

FIG. 1. Teneurs en éléments traces de fractions $<2\,\mu$ à smectites dominantes et teneurs en éléments traces des roches mères correspondantes.

Les roches mères sont classées d'après l'indice de basicité

$$an = \frac{CaO}{1{,}2(1{,}5Na_2O + K_2O) + CaO} \text{ de Rittmann.}$$

1 = Granite de Tassamakat, Hte Volta. 2 = Granite à gros grains, Diébiga, Hte Volta. 3 = Gneiss à biotite et épidote de Soffokel, Hte Volta. 4 = Diorite de Garango, Hte Volta. 5 = Dolérite porphyroblastique, Hte Volta. 6 = Amphibolite de Mene, Hte Volta. 7 = Gabbro de Kollel, Hte Volta. 8 = Basalte de Taïba, Hte Volta. 9 = Océanite de la Réunion. —— teneurs des fractions $<2\,\mu$ à smectite; ---- teneurs des roches mères correspondantes.

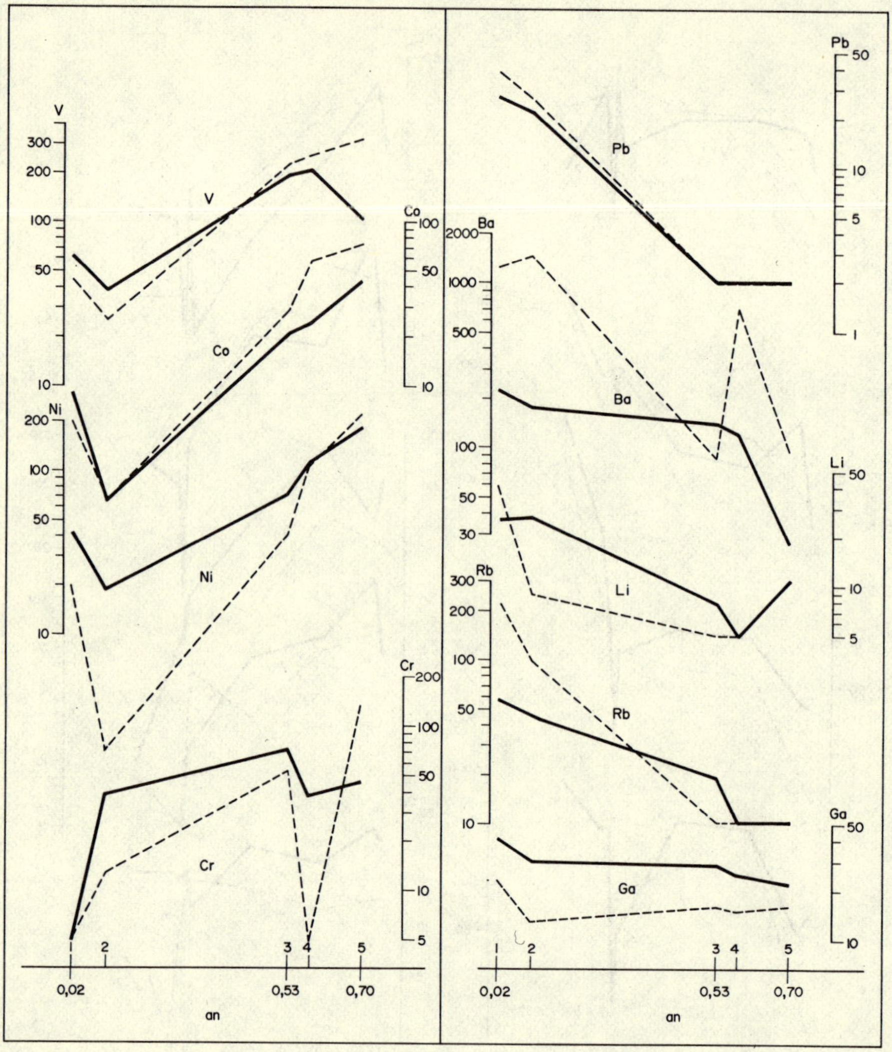

FIG. 2. Teneurs en éléments traces de fractions $< 2\,\mu$ à kaolinites dominantes et teneurs en éléments traces des roches mères correspondantes.

Les roches mères sont classées d'après l'indice de basicité

$$an = \frac{CaO}{1{,}2(1{,}5Na_2O + K_2O) + CaO} \text{ de Rittmann.}$$

1 = Granite de R.C.A. 2 = Granite de Nayouri, Hte Volta. 3 = Schiste amphibolitique R.C.A. 4 = Basalte de Madagascar. 5 = Gabbro de Kollel, Hte Volte. —— teneurs des fractions $< 2\,\mu$ à kaolinite; ---- teneurs des roches mères correspondantes.

Composition Chimique des Fractions <2 μ à Smectites et Kaolinites

Les variations des teneurs en éléments traces des fractions <2 μ à smectites et à kaolinites suivent bien celles des roches mères. Les fractions à smectites et à kaolinites formées sur roches mères basiques ont des teneurs plus faibles en Ba, Li, Rb, Ga, Pb que celles formées sur roches acides, alors que les teneurs en Ni, Co, Cr et V sont plus fortes. La diminution de teneurs en V des roches très basiques se marque aussi pour les fractions <2 μ à smectites et les fractions <2 μ à kaolinites. Les teneurs en B et Sn sont faibles ou nulles.

Ce qu'il est important de noter c'est que les éléments traces des fractions argileuses à smectites et à kaolinites reflètent bien la composition en éléments traces des roches mères sur lesquelles elles se sont formées.

Répartition des Eléments Traces dans les Phases Minéralogiques Constituant les Fractions <2 μ

La répartition des éléments traces dans les phases minérales constituant les fractions <2 μ, par calcul de régressions, permet de tenir compte de toutes les phases minéralogiques mineures présentes dans les échantillons. Lorsqu'on procède par élimination des oxydes et des hydroxydes de fer par le réactif de Tam on est obligé de faire certaines hypothèses. On néglige les teneurs en éléments traces de phases minéralogiques mineures autres que celles des composés du fer et des argiles. On estime de plus que les éléments traces localisés dans les oxydes et hydroxydes de fer sont, lors de la solubilisation de ces derniers par le réactif de Tam, mis eux aussi en solution et éliminés par lavage. On ne tient dans ce cas pas compte de la réadsorption possible de ces éléments par les argiles.

La localisation des éléments traces par régression n'a pû être faite que pour les fractions <2 μ à kaolinites car le nombre d'échantillons de fraction <2 μ à smectites n'était pas suffisant pour pouvoir être traité statistiquement. Le traitement de déferrification a été appliqué aussi bien à des échantillons contenant de la kaolinite qu'à des échantillons contenant de la smectite. Tous les résultats sont donnés Tableau 3.

Les analyses chimiques après déferrification confirment ce que nous avions noté précédemment après analyse RX des échantillons, à savoir que les smectites constituent la plupart du temps la totalité de la fraction <2 μ alors que les kaolinites n'en constituent qu'une partie, le reste étant formé d'oxydes et d'hydroxydes de fer. Ainsi les éléments traces analysés, lorsque la fraction <2 μ est composée de smectite, sont presque tous liés à l'argile alors que les éléments traces se répartissent entre la kaolinite et les oxydes et hydroxydes de fer lorsque la fraction <2 μ est formée de kaolinite.

Il s'avère, lorsqu'on examine les produits d'altération sur gabbro, Tableau 3, que la fraction <2 μ à kaolinite peut présenter des teneurs en éléments traces aussi élevées que la fraction smectitique, mais dans ce cas Sr, Ba, V, Co, et les autres éléments traces dans une proportion moindre, sont liés aux oxydes et aux hydroxydes de fer et non à la kaolinite.

III. Kaolinites, Illites et Smectites Sédimentaires

Les kaolinites présentées ici sont celles du "sidérolitique" des Charentes, prélevées dans des carrières exploitées à Montguyon. Les smectites et les illites ont été prélevées dans l'éocène supérieur du Bassin de Paris à Cormeilles en Parisis au niveau des Marnes bleues d'Argenteuil.

Tableau 3. Répartition des Éléments Traces dans les Phases Minéralogiques Constituant les Fractions $< 2\mu$.

		SiO_2	Al_2O_3	MgO	CaO	Fe_2O_3	Mn_3O_4	TiO_2	Na_2O	K_2O	Perte 1000°C	Sr	Ba	V	Ni	Co	Cr	B	Zn	Ga	Cu	Pb	Sn	Li	Rb
Localisation des éléments traces dans les argiles et dans les oxydes et hydroxydes de fer par déferrification																									
Smectite s/Gabbro	sans traitement	53,8	23,9	3,15	1,1	7,7	0,024	0,42	0,10	0,28	9,58	12	<5	46	156	3	270	11	27	16	55	48	<5	9	4
	apres déferrification	52,7	22,8	2,80	1,1	7,3	0,020	0,43	0,27	0,29	11,48	49	30	47	148	7	338	48	38	17	67	<2	<5	non dosés	
Smectite s/ dolérite porphyrobl.	sans traitement	50,3	14,6	2,45	1,8	21,3	0,149	0,32	0,63	0,77	7,27	43	106	73	82	46	47	<5	88	13	266	<2	<5		
	après déferrification	52,2	14,7	2,01	1,9	17,3	0,077	0,29	0,76	0,88	9,46	50	110	20	80	35	70	<5	92	30	235	<2	<5		
Kaolinite s/Gabbro	sans traitement	43,2	33,2	0,47	0,2	10,0	0,101	0,37	<0,10	0,16	13,20	<10	26	108	185	44	46	<5	31	22	158	<2	<5	11	3
	après déferrification	45,8	35,9	0,41	<0,2	3.9	0,017	0,22	0,22	0,17	14,19	<10	<5	6	137	<2	31	<5	31	17	136	<2	non dosés		
Localisation des éléments traces dans les différents minéraux de l' <2 μ par traitement statistique de type régression																									
Kaolinites sur Granite R.C.A.	<2 μ (80% K+9% I+S+Mo+Q+11% GG) N = 24 x̄	41,8	36,1	0.21	0,15	5,7	0,04	1,04	0,04	0,37	14,56	93	225	62	42	9	<5	5	35	41	21	27	<5	26	56
	σ	2,40	1,52	0.08	0,13	2,13	0,04	0,24	0,02	0,13	1,67	50	150	21	10	5	—	6	12	4	7	15	—	12	28
	Kaolinite											55	43	0	19	0	—	2	9	40	10	30	—	33	18
	I+S+Mo+Q											211	1022	222	156	56	—	33	311	78	67	44	—	0	433
	Goethite Gibbsite											304	964	411	127	36	—	0	3	21	68	0	—	0	0
Kaolinite s/Schiste amphibolique R.C.A.	<2 μ (68% K+15% S+Mo+Q+16% GG) N=17 x̄	38.0	29,7	0,39	0,37	15.1	0,08	1,09	0,03	0,14	15,04	41	141	195	73	21	73	<5	48	29	86	<2	<5	8	19
	σ	2,13	1,68	0,30	0,30	2,47	0,04	0,28	0,02	0,07	2,02	19	76	30	29	12	87	—	35	4	13	—	—	4	9
	Kaolinite											36	142	45	32	0	10	—	21	19	0	—	—	9	20
	S+Mo+Q											110	350	233	160	84	392	—	217	28	132	—	—	15	30
	Goethite Gibbsite											0	0	829	185	79	0	—	11	73	457	—	—	0	0

Note: K = Kaolinite, I = Illite, S = Smectite, Mo = Matière organique, Q = Quartz, GG = Goethite Gibbsite.

Les teneurs en éléments majeurs sont exprimées en %.
Les éléments traces sont exprimés en ppm.

Kaolinites du Sidérolitique des Charentes

De grandes poches de kaolinites blanches sont exploitées dans les Charentes Maritimes. D'après Kulbicki (1953) et Steinberg (1967) on aurait à faire à des kaolinites sidérolitiques. Ce serait des kaolinites sédimentées en Charentes Maritimes après transport depuis les profils d'altération du Massif Central dans lesquels elles se seraient formées.

Dans les exploitations on trouve, associés à ces kaolinites, des nodules de sulfures formés de pyrite et marcassite, ainsi que des passées noires riches en matières organiques.

L'analyse chimique (Tableau 4) et minéralogique de la fraction argileuse révèle des kaolinites très pures, accompagnées seulement de 0,5% de Fe_2O_3 et 0,75% de TiO_2 probablement sous forme d'anatase. Ces kaolinites sont exceptionnellement riches en Li (500 ppm) mais par contre pauvres ou dépourvues en éléments tels que Sr, Ba, Ni, Co, Cr, B, Zn, Cu, Pb, Sn. Dans les nodules de pyrites et de marcassite, à l'aide de la microsonde nous avons trouvé du Zn et Pb. Ba à été trouvé dans certains échantillons sous forme de sulfate.

Les kaolinites formées sous climat d'altération tropical sont accompagnées ou enrobées d'environ 5 à 20% d'oxydes et hydroxydes de fer. Celles prélevées dans les Charentes ne sont accompagnées que de 0,5% de Fe_2O_3. Fer et kaolinite se sont donc séparés, soit lors du transport des particules argileuses depuis le profil d'altération jusqu'au lieu de leur sédimentation soit plus vraisemblablement dans le milieu sédimentaire lui-même où du fer se trouve sous forme de nodules de pyrite et marcassite. Ba et Pb trouvés dans les sulfures et les sulfates se sont également séparés de la kaolinite. Ce sont des éléments qui dans les altérations sont associés aux kaolinites formées sur granites. Leur rayon ionique assez important (Ba^{2+} = 1,44 Å et Pb^{2+} = 1,26 Å) ne leur permet probablement pas d'intégrer les positions octaédriques des kaolinites. Ba et Pb ne sont probablement qu'adsorbés à la surface des kaolinites et peuvent ainsi dans un milieu favorable se séparer assez aisément de ce minéral.

Si avec Steinberg et Kulbicki on considère que ces kaolinites ont une origine d'altération les fortes teneurs en Ga (39 ppm) des kaolinites et les teneurs en Ba et Pb que l'on retrouve dans les sulfures et les sulfates du sédiment plaident en faveur d'une roche mère acide de type granitique.

Des teneurs de 500 ppm de Li n'ont jusqu'à présent jamais été trouvées dans des kaolinites formées par altération, même pas dans celles formées sur granite qui contiennent pourtant les plus fortes teneurs en Li. On est ainsi tenté de les interpréter comme étant la marque du milieu de sédimentation, d'autant plus que Thiry et Trauth (1976) ont déjà signalé des teneurs élevées en Li (125 ppm) pour les kaolinites des argiles plastiques sparnaciennes dans un environnement sédimentaire analogue avec nodules de sulfures et couches ligniteuses.

Illites et Smectites des Marnes Bleues d'Argenteuil (Eocène supérieur) de Cormeilles en Parisis

Le milieu de dépôt des Marnes bleues d'Argenteuil a été décrit par Chateauneuf et Trauth (1972) comme un milieu lagunaire à apports intermittents en eaux salées. Smectites et illites sont les deux minéraux argileux essentiels. Les compositions chimiques des smectites et des illites, aussi bien en éléments majeurs qu'en éléments traces sont très semblables (Tableau 4). Ceci pose d'ailleurs dans ce contexte le problème de la relation génétique smectites illites.

Les teneurs en Ba, Ni, Co, Zn, Ga, Cu, Pb, Sn des smectites et des illites des Marnes bleues d'Argenteuil sont très proches de celles de smectites d'altération formées sur granite (fig. 4). Elles en diffèrent par des teneurs plus élevées en Sr, V, Cr, B, Li et peut être Rb. Seules V et Cr

TABLEAU 4. KAOLINITES, SMECTITES ET ILLITES SÉDIMENTAIRES

			SiO_2	Al_2O_3	MgO	CaO	Fe_2O_4	Mn_3O_4	TiO_2	Na_2O	K_2O	Perte 1000°C	Sr	Ba	V	Ni	Co	Cr	B	Zn	Ga	Cu	Pb	Sr	Li	Rb
Kaolinites du "Sidérolitique"	N=9	x̄	45,3	39,3	0,04	<0,20	0,5	0,018	0,69	0,17	0,07	13,90	<10	<5	42	7	<2	58	<5	4	39	<2	<2	<5	464	<10
Montroux Charentes		σ	0,36	0,39	0,0		0,07	0,003	0,04	0,02	0,01	0,02	—	—	6	3	—	4		4	12	—		—	114	—
Smectites de	N=20	x̄	55,07	21,28	3,93	0,77	6,02	0,04	0,79	0,24	4,84	6,97	231	462	163	42	11	121	226	65	22	17	16	<5	111	149
Cormeilles en Parisis		σ	0,91	0,36	0,32	0,24	0,22	0,006	0,05	0,01	0,15	0,16	157	65	11	6	2	7	13	7	2	4	4		5	4
Illites de	N=19	x̄	53,77	20,53	4,30	0,75	6,32	0,04	0,75	0,25	5,21	7,46	866	640	158	39	9	113	228	63	21	20	14	<5	128	146
Cormeilles en Parisis		σ	0,59	0,79	0,22	0,12	0,21	0,007	0,06	0,02	0,23	0,53	834	182	11	3	2	6	11	9	2	6	4		5	12

Les teneurs en éléments majeurs sont exprimées en %.
Les éléments traces sont exprimés en ppm.

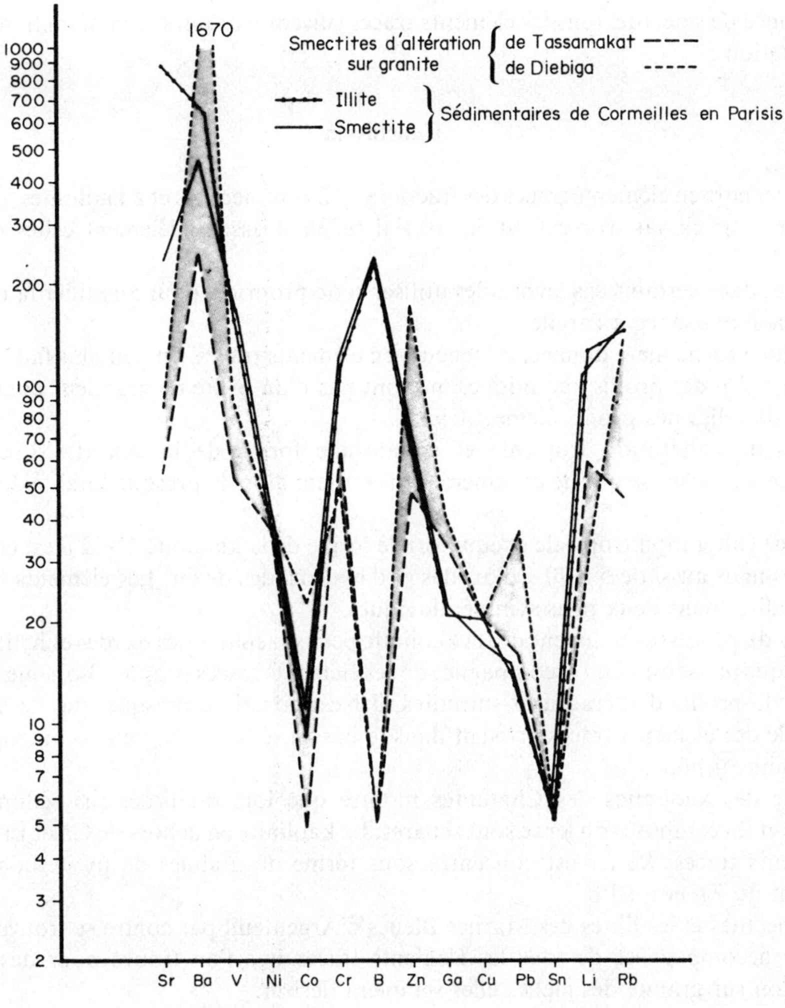

FIG. 3. Eléments traces des illites et smectites sédimentaires comparés à ceux des smectites d'altération sur granite.

ont des teneurs compatibles avec celles de smectites d'altération formées sur roches basiques aussi n'est-il pas raisonnable de considérer cette hypothèse. On est aussi amené à dire que l'enrichissement en Sr, V, Cr, B, Li et peut-être Rb est dû à l'action du milieu de sédimentation sur un matériel smectitique formé par altération de roches granitiques. A ces éléments traces il faut ajouter MgO et K_2O qui semblent eux aussi s'être enrichis dans le milieu de sédimentation.

Contrairement aux kaolinites sédimentaires des Charentes qui étaient plus pauvres en éléments traces que les fractions $<2\,\mu$ des profils s'altération à kaolinite, les smectites et illites des Marnes bleues ont des teneurs en éléments traces équivalentes à celles des fractions $<2\,\mu$ des profils d'altération à smectite. Dans les profils à kaolinite, les éléments traces sont localisés dans deux phases minéralogiques, la kaolinite et les composés du fer oxyde et hydroxyde; les éléments traces se séparent lors de la sédimentation en même temps que leurs minéraux hôtes. Dans le profil d'altération à smectite, la presque totalité de la fraction $<2\ \mu$ étant formée de smectite, tous les éléments traces suivent leur unique minéral hôte lors de la sédimentation.

Conclusion

—Les teneurs en éléments traces des fractions $<2\ \mu$ à smectites et à kaolinites formées par altération sous climat tropical ou équatorial reflètent assez fidèlement celles des roches mères.

On peut dans certains cas favorables utiliser cette propriété pour identifier la roche mère qui a donné naissance à l'argile.

—Sur une roche mère donnée, les teneurs en éléments traces, un peu plus faibles dans les fractions $<2\ \mu$ des profils kaoliniques, ne sont pas d'un ordre de grandeur complètement différent de celles des profils à smectites.

—Lorsque l'altération tropicale et équatoriale forme de la smectite $1'<2\ \mu$ n'est pratiquement composé que de ce minéral qui contient alors la presque totalité des éléments traces.

Lorsque l'altération tropicale et équatoriale forme de la kaolinite $1'<2\ \mu$ est composé de kaolinites mais aussi de 5 à 20% d'oxydes et d'hydroxydes de fer. Les éléments traces sont alors localisés dans deux phases minéralogiques.

—Lors du processus sédimentaire la kaolinite peut se séparer des oxydes et hydroxydes de fer et chaque phase est alors accompagnée de ses éléments traces propres. Lorsque le matériel provient de profils d'altération à smectites, il n'est constitué presque que de smectite et l'ensemble des éléments traces arrivent dans le bassin sédimentaire en accompagnant leur unique minéral hôte.

L'étude des kaolinites des Charentes montre que lors du processus sédimentaire la kaolinite et les composés du fer se sont séparés. La kaolinite en dehors de Ga et Li est pauvre en éléments traces. Le fer est concentré sous forme de nodules de pyrite et marcassite contenant du Zn et du Pb.

Les smectites et les illites des Marnes Bleues d'Argenteuil par contre se trouvent dans le sédiment accompagnées de tous les éléments traces que l'on trouve pour des smectites d'altération sur granite desquelles elles semblent dériver.

—Les kaolinites des Charentes aussi bien que les illites et les smectites des Marnes Bleues d'Argenteuil présentent certaines teneurs en éléments traces incompatibles avec des teneurs héritées des profils d'altération.

Les teneurs en Li pour les trois familles argileuses étudiées et les teneurs en Sr, V, Cr, B et peut-être Rb pour les illites et les smectites du Bassin Parisien sont plus élevées que celles des kaolinites et smectites formées par altération sur des roches granitiques. Ces teneurs plus élevées en certains éléments à affinités basiques associés à des éléments à affinités acides sont interprêtés comme étant dues à l'action du milieu sédimentaire.

Références

Chateauneuf, J. J. et Trauth, N. (1972) Palynologie, composants minéralogiques majeurs et phase argileuses des Marnes Bleues d'Argenteuil. Contribution à la reconstitution du milieu de dépôt. *Mém. B.R.G.M.* n°77, pp. 329–336.

KULBICKI, G. (1953) Constitution et genèse des sédiments argileux sidérolithiques et lacustres du Nord de l'Aquitaine. Thèse Sci. Toulouse *et Sci. Terre*, **4**, 1956, 5–101.

MOSSER, C., WEBER, F. et GAC, J. Y. (1974) Eléments traces dans des kaolinites d'altération formées sur granite et schiste amphiboliteux en République Centrafricaine. *Chem. Geol.* **14**, 95–115.

RITTMANN, A. (1963) *Les Volcans et leur activité*, Masson édit., 461 pp.

RÖSSLER, H. J. et LANGE, H. (1972) *Geochemical Tables*, Elsevier Publishing Company, Amsterdam, 268 pp.

SEGALEN, P. (1970) Extraction du fer libre des sols à sesquioxydes par la méthode de De Endredy par irradiation à l'ultra-violet de solutions oxaliques. *Cah. ORSTOM, Sér. Pédol.* **8**, 483–496.

STEINBERG, M. (1967) *Contribution à l'étude des formations continentales du Poitou* (*Sidérolithique des auteurs*), Thèse Fac. Sci., Orsay, 415 pp.

THIRY, M. et TRAUTH, N. (1976) Les sédiments Paléocènes et Eocènes inférieurs du Bassin de Paris. Rôle des argiles dans la rétention d'éléments traces. *Bull. Sci. Géol.* **29**, 1, 33–43.

Discussion

B. MOINE:

Dans le processus d'évolution des illites et smectites en milieu sédimentaire décrit par Mme Mosser observe-t-on des corrélations entre éléments traces et éléments majeurs, spécialement entre Li et Rb d'une part et K d'autre part?

C. MOSSER:

On observe une corrélation positive entre K_2O et Rb pour les smectites et illites sédimentaires du bassin de Paris depuis les marnes bleues d'Argenteuil (Eocène supérieur) jusqu'aux argiles vertes de Romainville (Oligocène inférieur). Une corrélation positive K_2O–Li s'observe au niveau des illites provenant des argiles vertes de Romainville.

E. N. PENTCHEVA:

Pourquoi le césium qui aurait un comportement intéressant et particulier (surtout comme élément indicateur dans les rapports Rb/Cs et Li/Cs) manque-t-il dans ces études? La même question est posée à propos de Mo et W.

C. MOSSER:

Cs, Mo et W n'ont pas été analysés faute de technique facilement accessible au laboratoire où je travaille pour étudier environ 1 500 échantillons.

Mise en Evidence d'une Certaine Mobilité du Titane dans les Podzols ainsi qu'au Laboratoire sous l'Influence d'Agents Organiques

J. C. DUMON et M. VIGNEAUX

Institut de Géologie du Bassin d'Aquitaine, Université de Bordeaux I, 351, cours de la Libération, 33405 Talence-Cedex

Abstract

In soil science, this element is traditionally considered as particularly stable, inaccessible to the lixiviation downward which is current for most cations.

However, stratification of this element is evident in Aquitain-podzols.

The presumption of the influence of organic acids present in the litters of acidiphil forest led us to investigate the action of certain of these constituents on ilmenite and rutile.

This experimentation, compared to the action of inorganic acids, led us to show a certain mobilization of titanium, particularly by oxalic acid, a very abundant compound in the forest-litters of the "Landes of Gascogne".

A la suite d'une étude particulière (Dumon *et al.*, 1973) destinée à rechercher l'origine des composés ferro-titanés fréquents notamment dans les dunes et sur les plages du littoral aquitain, nous avons été amenés à considérer la distribution du titane dans les sols du contexte régional. Cet élément est en effet représenté dans les podzols qui constituent pour l'essentiel le support de la vaste forêt des Landes de Gascogne. C'est moins son abondance toute relative—nous envisagerons plus loin cet aspect—que son comportement qui présente un certain intérêt.

Bien que rencontré et dosé de manière constante avec les éléments majeurs des roches et des sols, il n'a cependant guère suscité d'intérêt, en particulier chez les agronomes. Nombre de végétaux en renferment pourtant, comme l'a mis en évidence il y a un demi-siècle l'école de Bertrand et notamment Mme Voronca-Spirt (1929) dans un travail volontairement limité aux végétaux comestibles. Plus récemment, Burges (1960) dans une étude consacrée aux végétaux supérieurs, sans avoir pour objectif essentiel le titane, donne des chifres évocateurs. C'est ainsi qu'on peut y observer la place de choix tenue par cet élément. Analysant des cendres de *Pinus sylvestris* sur podzol développé sur sable glaciaire de la région de Liverpool, cet auteur donne les chiffres suivants:

— silice 28%
— oxyde ferrique 18,4%
— oxyde de titane 16,5%

Le titane occupe donc le troisième rang suivi de très loin par Na^+ et Ca^{++}.

Au vu de ces résultats on s'explique assez mal la relative désaffection à l'endroit de cet élément. Cela tient probablement à une mobilité à la fois plus limitée et surtout plus discrète que celle des composés du Fe ou du Mn, éléments pour lesquels les couleurs permettent de suivre le cheminement des composés dans le profil. N'oublions pas que les couleurs propres de ces cations sont à l'origine même de l'idée d'une migration possible de ces éléments per ascensum ou per descensum selon le climat. Rien de semblable ne se manifeste pour le titane et il est sans doute l'un des derniers éléments à se déplacer dans un profil pédologique.

Au plan physiologique, on manque également d'informations, mais sans doute est-il moins indispensable à la vie des végétaux que N, P ou K ou même que les quelques oligo-éléments actuellement largement étudiés sans que les spécialistes soient d'ailleurs totalement unanimes quant à leur rôle dans la hiérarchie des facteurs du milieu.

I. Distribution du Titane dans les Podzols Aquitains

La répartition de cet élément dans les podzols humo-ferrugineux régionaux conduit à formuler des réserves quant à la totale inertie dans les sols qui lui est habituellement attribuée.

Dans le tableau suivant (Tableau 1) figurent en regard les teneurs en Fe et en Ti des divers horizons pédologiques de trois podzols aquitains. On y observera l'existence d'une véritable stratification comparable, à certains égards, à celle connue de longue date pour le Fe et plus récemment pour Al (Juste, 1965).

TABLEAU 1. RÉPARTITION DU TITANE ET DU FER DANS TROIS PODZOLS DU SUD-OUEST AQUITAIN

Localisation des profils	Cestas-le-Puch		Martignas		Montplaisir	
Eléments (ppm)	Fe	Ti	Fe	Ti	Fe	Ti
Horizons						
A_1	750	300			3400	
A_2	580	201	900	380		386
B_1	1900	970	800		4700	1742
B_2	7300	700		803	3000	749
Roche-mère	1440	260	500	202	2400	666

Le maximum de Ti paraît le plus souvent se situer dans l'horizon B1 et non pas, comme pour Fe, dans l'horizon B2. On distingue donc bien horizons lessivés appauvris (A) et horizons d'accumulation, enrichis (B). Outre l'absence de coïncidence étroite et systématique entre maxima du Fe et maxima du Ti—à relier probablement à un décalage dans le temps entre la migration de l'un et de l'autre—on remarquera les teneurs assez limitées en Ti de ces podzols. Elles ne sont en effet en rien comparables à celles observées par example dans les sols Hawaïens, de types latéritiques pour lesquels la richesse en Ti est depuis longtemps connue (Sherman, 1952).

Si le point saillant réside dans la mise en évidence de cette stratification inhérente à la migration de Ti, il est intéressant de noter que sa localisation préférentielle se situe principalement dans l'horizon B1, c'est-à-dire le plus riche en composés organiques. Bien que jamais signalé dans la littérature, ce fait peut cependant y être observé au travers des divers résultats chiffrés consignés par les auteurs (Karim et Khan, 1955). Cette observation s'est en tout cas révélée précieuse pour aborder l'étude des mécanismes mêmes de la migration.

On sait tout le rôle joué par les acides organiques dans la dynamique du Fe dans les sols (Bruckert, 1970). L'acide citrique, par exemple, a un rôle essentiel dans la plupart des sols acides. Ce Fe est en effet complexé par ce composé, tant à l'état Fe^{2+} que Fe^{3+} (Dommergues et Duchaufour, 1965). Kaurichev *et al.* (1963), Schnitzer *et al.* (1964), Muir *et al.* (1964) ont montré, toujours à l'égard du Fe, le pouvoir complexant très élevé d'acides organiques comprenant plusieurs groupements carboxyles et un groupement oxhydrile. D'autre part, leur présence dans les litières et donc dans les végétaux en général est bien connue (Jacquin et Bruckert, 1965). Dans le contexte pédologique régional, Juste (déjà cité), outre la mise en évidence des divers composés organiques présents de manière notable, a montré leur influence sur l'entraînement en profondeur de l'Aluminium.

Pour toutes ces raisons, et surtout du fait de leur prédominance dans les sols (Robert et Razzaghe-Karimi, 1975) ce sont donc des acides organiques simples qui ont été choisis pour tenter une expérimentation de mobilisation de Ti. Leur action a été mise en parallèle avec celle d'acides minéraux et même d'eau permutée sur différents minéraux titanifères.

II. Rôle Respectif d'Acides Organiques et Minéraux dans la Solubilisation de Composés Titanés Naturels de Type Ilménite

(a) *Méthodologie*

Un gramme d'ilménite littorale préalablement porphyrisée a été agité durant 20 heures et maintenu en contact simple pendant 160 heures en présence de 100 ml de chacun des extractants organiques ou minéraux pris à la normalité 0,1 (Tableau 2). Après séparation des phases (centrifugation suivie de filtration) le titane a été dosé sur la phase liquide par spectrophotométrie d'absorption à 405 mμ.

TABLEAU 2. MISE EN SOLUTION DU TITANE D'UNE ILMÉNITE, PAR DIVERS AGENTS ORGANIQUES ET MINÉRAUX

Extractants	ph des divers extractants	Ti extrait (ppm)	Force de l'extractant†
Acides organiques N/10			
oxalique	1,6	17 385	232
ascorbique	—	3 219	43
citrique	2,2	1 271	17
L. tartrique	2,2	767	10
D1. malique	2,2	449	6
D1. aspartique	—	449	6
succinique	—	150	2
acétique	2,9	75	1
Acides minéraux N/10			
sulfurique	1,2	821	11
chlorhydrique	1,1	671	9
nitrique	—	521	7
Eau permutée	7,0	497	6,5

† Le plus mauvais extractant (ac. acétique) s'est vu attribuer l'indice 1.

(b) *Principe du Dosage*

Il s'agit du dosage classique à l'eau oxygénée (Voinovitch *et al.*, 1962). L'H_2O_2 donne avec Ti (IV) en milieu acide (SO_4H_2), un complexe orangé TiO_{2++} dont le spectre d'absorption est situé vers 400 mμ. La coloration jaune-orangé est instantanée et stable indéfiniment.

De l'acide phosphorique est ajouté pour supprimer l'interférence due au Fe^{3+}. Egalement, dans ce même but nous avons mis au point une méthode de dosage par fluorescence X sur la phase liquide (Dumon *et al.* 1975).

(c) *Résultats*

Le Tableau 2 comporte outre les pH initiaux des divers extractants des données chiffrées sur l'aptitude de chacun d'eux à faire passer du titane en solution à partir d'un composé jusqu'ici considéré comme insoluble quelle que soit l'agressivité du milieu.

Le résultat le plus important, en conformité d'ailleurs avec les données rencontrées dans la littérature, mais toujours pour d'autres éléments que Ti, réside dans le fait que la plupart des agents organiques testés excercent sur le sol une dégradation plus intense que les acides forts minéraux, en dépit de pH pourtant plus élevés que ceux des solutions minérales.

En second lieu, on notera que l'acide Oxalique, en extrayant environ 200 fois plus de titane que l'acide acétique (le plus mauvais des agents d'extraction utilisés), constitue le meilleur agent de mise en solution du titane.

Il est suivi par l'acide ascorbique, malgré tout beaucoup moins actif. Egalement des acides minéraux utilisés, c'est l'acide sulfurique qui s'est montré le plus efficace arrivant au 4ème rang après acides oxalique, ascorbique et citrique.

(d) *Discussion*

En considérant les seuls acides organiques, la mise en solution de Ti paraîtrait devoir être attribuée à la seule valeur du pH des solutions puisque le plus efficace, l'acide oxalique est, initialement au moins, à pH 1,6 contre 2,9 à l'extractant le moins efficient (acide acétique). Ceci serait d'ailleurs conforme aux courbes fournies par Loughnan (1969) qui donne entre autres paramètres la solubilité de TiO_2 en fonction du pH. Sur ces courbes, le pH critique marquant l'insolubilité du composé est pH = 2. Seule, d'après ce même auteur, la forme Ti $(OH)_4$ resterait soluble jusqu'aux environs de pH 5,5.

S'il ne faut pas totalement écarter l'action des pH, il est probable qu'elle n'est pas seule en cause puisque les acides minéraux pourtant à pH plus bas (1,1 ou 1,2) se révèlent peu agressifs à l'égard de l'ilménite.

En outre, l'eau désionisée, à pH 7,0, se montre relativement aussi active que des acides organiques tels que: D1 malique, D1 aspartique, succinique ou acétique. Les bas pH ne jouent donc pas semble-t-il de rôle primordial dans la mise en solution du Ti mais ils y participent vraisemblablement.

Si l'action de l'acide oxalique sur les minéraux titanés n'avait pas fait l'objet jusqu'ici de publications, par contre son action sur d'autres éléments (Bruckert, 1966) est bien connue. Cet acide fait partie des composés les plus résistants à la décomposition par la microflore tellurique. Gastuche (1957) l'utilisant sur la kaolinite parallèlement à l'acide chlorhydrique, dans le but de solubiliser l'aluminium, constate que malgré un pH plus élevé que celui de l'acide minéral c'est l'acide organique qui se montre meilleur extractant.

Sur un plan plus général, Leleu et Sarcia (1971) comparant à l'acide sulfurique divers acides organiques au nombre desquels se trouvaient l'acide oxalique et l'acide citrique constatent qu'à pH identiques pour des solutions diluées, il s'ensuit une plus grande altération des minéraux des roches par les molécules organiques.

Selon Juste, déjà cité, il semblerait que "l'agressivité particulière des acides organiques soit à relier à l'action complexante de l'anion organique vis à vis des éléments métalliques des silicates". "Il vient aussi s'y ajouter l'action des protons."

Bruckert, déjà cité, étudiant sur des sols forestiers les quantités respectives d'acides, de Fe et d'Al hydrosolubles constate une distribution parallèle des cations et des acides citrique, oxalique et vanillique laissant supposer leur migration simultanée sous forme de complexes organo-métalliques. Par contre, Ginzburg *et al.* (1963) testant pourtant l'action d'acides organiques variés sur la mise en solution de Al n'évoquent pas l'acide oxalique. Erhart (1973) s'en étonne et écrit: "c'est un des composés organiques des plus importants, produits par la végétation forestière, responsable de la podzolisation."

Pour Bruckert (1970) acides oxalique et citrique constituent les agents complexants des litières les plus puissants. Il rappelle qu'oxalates et citrates constituent les réactifs de base dans l'extraction du Fe et de l'Al libres.

III. Action de l'Acide Oxalique sur le Rutile

Ce minéral constitue le plus résistant des oxydes de titane (Lacroix, 1962). Compte-tenu des résultats précédents, il semblait donc intéressant de suivre son comportement sous l'action de l'acide oxalique.

(a) *Méthodologie*

Un gramme de rutile a été agité en présence de 100 ml d'acide oxalique N/10. Les échantillons ont été multipliés afin de disposer de diverses durées de contact sous agitation. L'expérience n'a pas dépassé 104 heures.

Au terme de la séparation des phases, Ti a été dosé par fluorescence X en phase liquide sous atmosphère d'hélium.

Le principal avantage de cette méthode, réside d'abord dans sa rapidité de mise en oeuvre, mais surtout, non point dans le cas du rutile, mais dans celui des composés ferro-titanés, dans la suppression du problème de l'interférence du Fe, gênante en spectrophotométrie d'absorption.

(b) *Résultats*

Ils sont consignés dans le Tableau 3 et sur la courbe (Fig. 1).

TABLEAU 3. DYNAMIQUE DE LA SOLUBILISATION DU TITANE DU RUTILE PAR L'ACIDE OXALIQUE

Temps d'agitation en heures	8	24	32	48	56	72	80	96	104
TiO_2 de matière sèche (μg/g)	608	1457	1475	2163	2068†	2315	2488	2633	2825
Indice	100	239	242	356	340	381	409	433	464

† Perte lors de la récupération.

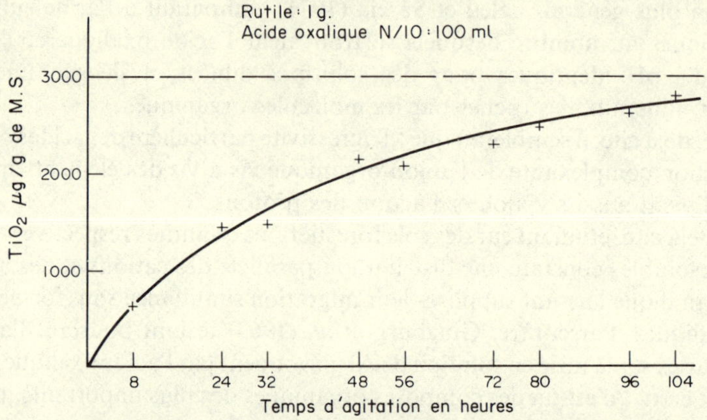

FIG. 1. Extraction de Ti par l'acide oxalique en fonction du temps.

On observera en tout premier lieu que *même le rutile ne résiste pas totalement à l'action solubilisante de l'acide oxalique*.

Afin de rendre plus aisée la lecture du phénomène, l'indice 100 a été affecté à la quantité de TiO_2 passée en solution après seulement 8 heures de traitement. On voit que la quantité de TiO_2 solubilisée augmente jusqu'aux environs de 104 heures d'agitation, Al'issue de l'expérience 0,3% (en TiO_2) du rutile sont passés en solution.

Cette seconde expérience confirme donc tout à fait les résultats de la précédente puisqu'un minéral aussi peu altérable que le rutile subit l'action de composés organiques.

On peut penser raisonnablement que dans la nature les conditions de mobilisation du titane sont plus favorables qu'au laboratoire. En effet, la solution lixiviante issue des litières est constamment renouvelée, de ce fait, il n'y a pas de tendance à l'établissement d'un équilibre entre les différentes phases qui, dans notre expérience au contraire, semble s'établir vers 100 heures. On passe d'un système clos à un système ouvert qui ne peut que favoriser et intensifier le phénomène. D'autre part, si l'acide oxalique est un constituant essentiel des litières, de nombreux autres acides aliphatiques sont également présents ainsi que des acides phénols. On peut donc s'attendre à des phénomènes du même genre de la part d'autres composés.

Bien qu'obtenue au laboratoire, cette solubilisation si limitée soit elle d'un composé réputé stable face aux agressions chimiques les plus variées valait d'être soulignée. Elle permet d'envisager le jeu de phénomènes voisins dans la nature. Elle éclaire ainsi d'une manière inhabituelle le comportement d'un élément assez méconnu en raison de l'inactivité qui lui est trop souvent attribuée. Sans nécessairement constituer l'explication exclusive de la stratification du Ti dans les podzols, cette expérimentation confirme l'intervention probable dans la nature des composés organiques.

Elle confirme l'observation faite sur le terrain de la liaison du titane à l'horizon B1 spécialement enrichi en substances organiques.

Reférences

BRUCKERT, S. (1966) Répartition des acides hydrosolubles dans quelques profils de sols forestiers. *C.R. Acad. Sci., Paris,* **263,** série D, 1572–1575.

BRUCKERT, S. (1970) Influence des composés organiques solubles sur la pédogénèse en milieu acide. *Thèse Doct. Sc. Nat.*, Nancy, 72 pp.

BURGES, A. (1960) Physico-chemical investigations of humic-acid. *Transactions of 7th International Congress of Soil Science, Madison, Wisc., U.S.A., 1960*, Vol. II, Commissions II: *Soil Chemistry* and III: *Soil Biology*, Elsevier.

DOMMERGUES, Y. et DUCHAUFOUR, PH. (1965) Etude comparative de la dégradation biologique des complexes organoferriques dans quelques types de sols. *Science du sol*, **1**, 43–59.

DUMON, J. C., LAPAQUELLERIE, Y. et LATOUCHE, C. (1973) Sur la présence et l'origine possible des ilménites de la zone du Pyla (Gironde). *C.R. Acad. Sci., Paris*, **276**-D, 889–892.

DUMON, J. C. (1975) Action d'acides organiques divers sur un minéral titané. Comparaison de leur pouvoir d'extraction du titane avec celui d'acides minéraux. *Bull. Inst. Géol. Bassin d'Aquitaine*, No. 17, 95–100.

DUMON, J. C. et LAPAQUELLERIE, Y. (1975) Dosage par fluorescence X, sur phase liquide, du titane en traces. *Bull. Inst. Géol. Bassin d'Aquitaine*, No. 18, 133–137.

ERHART, H, (1973) *Itinéraires géochimiques et cycle géologique de l'aluminium*. Paris, Doin édit., 253 pp.

GASTUCHE, M. C. (1957) L'aluminium libre en surface de la kaolinite. *Bull. Gr. fr. Argiles, Paris*, **9**, N.S., No. 4, 49–59.

GINZBURG, I. I., YASHINA, R. S., MATYEYEVA, L. A., BIELETSKI, V. V. et NUZHDENOUSKAYA, T. S. (1963) Decomposition of certain minerals by organic acids. In *Chemistry of the Earth Crust*, trans. of geochemical conference in honour of V. Vernadsky's centenary, Vol. 1.

JACQUIN, F. et BRUCKERT, S. (1965) Identification et évolution des acides hydrosolubles de deux litières forestières. *C.R. Acad. Sci., Paris* **260**-D, 4556–4559.

JUSTE, C. (1965) Contribution à l'étude de la dynamique de l'aluminium dans les sols acides du Sud-Ouest atlantique: application à leur mise en valeur. Thèse Ing. Doct., Nancy, 143 pp., 27 figs., 39 tableaux.

KARIM, A. et KHAN, D. H. (1955) Titania content of some gray brown podzolic soils of East Pakistan. *Soil Sci.*, **80**, 277–280.

KAURICHEV, I. S., IVANOVA, T. N. et NOZDRUNOVA, K. A. (1963) Low molecular organic acid content of water soluble organic matter in soils. *Soviet Soil Sci.* 223–229.

LACROIX, A. (1962) Minéralogie de la France et de ses anciens territoires d'Outre-Mer. *Paris, Libr. Sci. et Tech. Albert Blanchard*, t. III, 816 pp.

LELEU, M. et SARCIA, C. (1971) Méthodes d'études de l'altération supergène des minéraux et des roches. *Bull. B.R.G.M.*, 2e série, sect. II, No. 5.

LOUGHNAN, F. C. (1969) *Chemical Weathering of the Silicate Minerals*. American Elsevier Publishing Co., Inc., New York.

MUIR, J. W., MORRISSON, R. I., BOWN, C. J. et LOGAN, J. (1964) The mobilization of iron by aqueous extracts of plants. I. Composition of the amino-acid and organic acid fractions of a pine needles. *J. Soil. Sci.* **15**, 220–225.

ROBERT, M. et RAZZAGHE KARIMI, M. (1975) Mise en évidence de deux types d'évolution minéralogique des micas trioctaédriques en présence d'acides organiques hydrosolubles. *C.R. Acad. Sci., Paris*, **280**-D, 2175–2178.

SCHNITZER, M. et SKINNER, S. I. M. (1964) Organo-metallic interactions in soils. 3. Properties of iron and aluminium organic matter complexes prepared in the laboratory and extracted from a soil. *Soil Sci.* **98**, 197–203.

SHERMAN, G. D. (1952) The titanium content of Hawaiian soils and its significance. *Soil Sci. Am. Proc.* 15–18.

VOINOVITCH, I. A., DEBRAS-GUEDON, J. et LOUVRIER, J. (1962) *L'Analyse des silicates*, Paris, Hermann, 510 pp.

VORONCA-SPIRT, C. (1929) Recherches sur la présence du titane chez les végétaux et chez les animaux. Thèse Sci., Paris, Imp. de la Cour d'Appel, pp. 1–61, 4 tableaux.

Distribution of Clastophilic Elements in Saline Deposits in their Relation to the Evolution of Saliferous Basins, and their Application for Correlation and Differentiation of Saline Series

D. P. Khrushshov

Institute of Geochemistry and Physics of Minerals of the Academy of Sciences of the Ukrainian SSR, Kiev, U.S.S.R.

Abstract

Clastophilic elements can be considered as indicators of evolution of actively developing saliferous basins; they reflect their relationships with surrounding continental massives and peculiarities of superimposed mineralization. Concentrations of elements and relationships between elements are controlled by the material that is supplied by washout sources; they depend on its transportation, sedimentation environment, and to a lesser degree by geochemical conditions existing in sedimentation basins. The influence of superimposed processes is conditioned by many factors and in the end is lithologically controlled. Changes of concentrations within stratigraphic subdivisions in general, their changes in separate groups and types of rocks, ratios of contents variations, peculiarities of correlation relationships between elements are all indicative of evolutionary processes.

This research work was carried out simultaneously with working out methods for lithological differentiation of Saline Series of different ages within the Dnieper–Donetz depression, the Donbass, the Precarpathian foredeep, the latter being considered as a single geological structure (comprising its Soviet, Rumanian and Polish parts). The data on other saline basins of the Soviet Union and Polish Zechsteins are used for the sake of comparison. With some amendments the resulting universal techniques for lithological differentiation can be applied to any saliferous basin for prospecting for potassium salts and other deposits related to evaporite formations, for geological mapping and for some problems concerning oil geology.

Geochemical researches are an integral part of the techniques for lithological differentiation of saline series referring to different ages. However, our interest is in the studies of clastophilic elements as they are understood by N. M. Strachov (1962). Because of their low abundances and geochemical behaviour they can be considered as minor elements of saline formations.

The water-soluble constituent of saline rocks contains Mn, Ti, Ba, Fe and negligible quantities of Cu, Ti, V, Mo, Pb, La, U, Th and Au. The greater part of clastophilic elements is in the water-insoluble residue of the rocks.

The residue contains clastic and chemogenic material and also some ore minerals, the latter appearing due to sedimentation, diagenesis or superimposed processes.

The Precarpathian foredeep is a most complicated region both from the point of view of its peculiar geological evolution and of the diversity of geochemical processes. In this connection the region studied exemplifies a very wide range of geochemical situations.

Regularities of clastophilic element distribution in general types and groups of rocks have

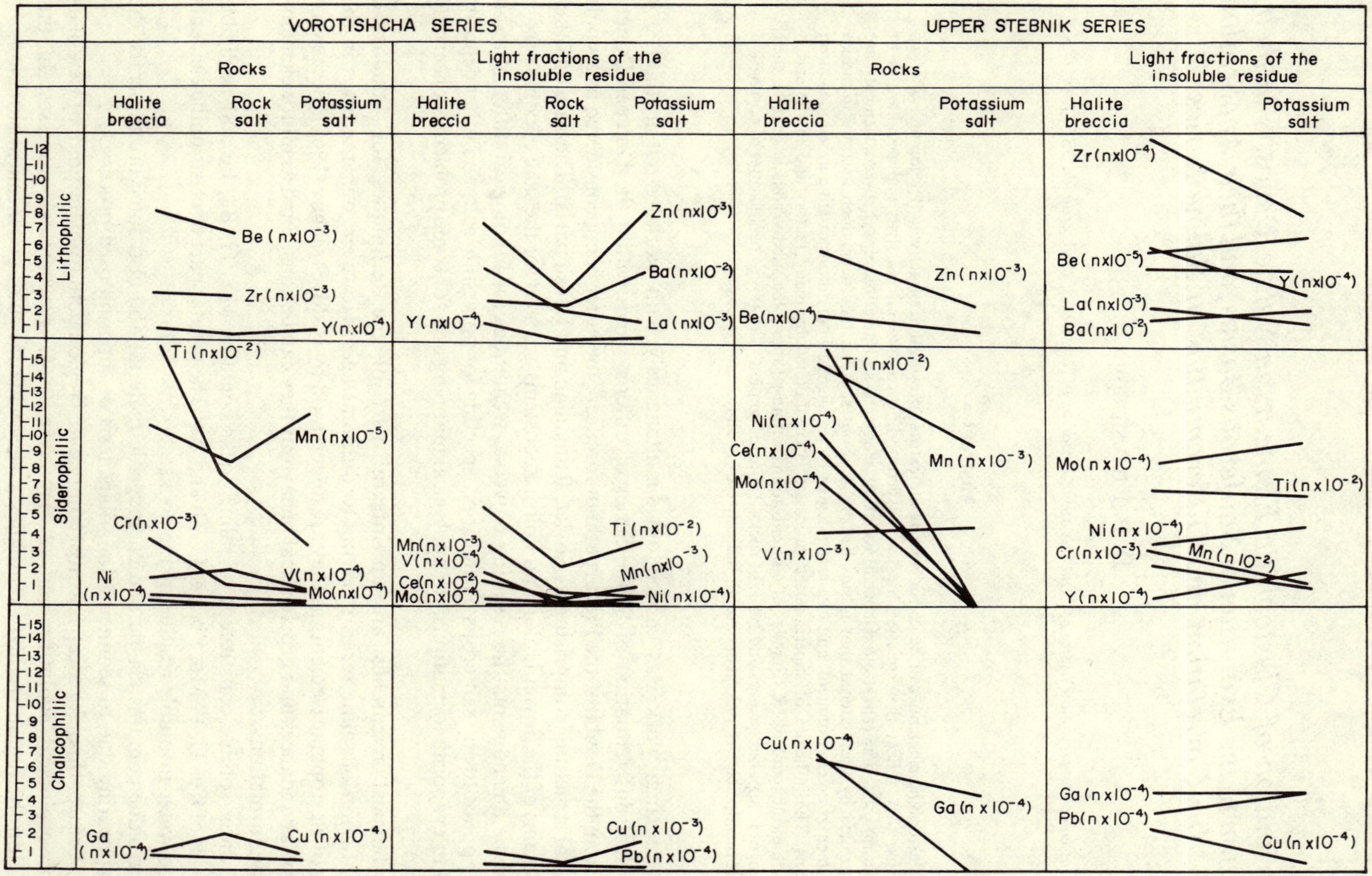

FIG. 1. Distribution of clastophilic elements in the main types of rocks in Precarpathian foredeep.

been considered. In accordance with our nomenclature, salt-bearing rocks (halite breccias), rock salt and potassium salts have been studied. Within these more general groups, representing certain parts of a formational set arranged in accordance with evaporite stages and weakening of ferrigenous sedimentations, the microelemental distribution is rather simple. From the first group and further, the set of contained elements decreases, and it can be due to restricted sets of the constituent rocks and minerals of detrital origin. One can observe a similar decrease in the graphs showing contents of elements (see Fig. 1, here and below we give mean values). Pronounced changes in concentrations are characteristic of siderophilic elements (excepting V and sometimes Mn, Mo, Ni). Changes in lithophilic and chalcophilic element abundances are not so prominent.

Clastophilic element distributions were determined in light fractions of the insoluble residue where there exists a higher concentration of microelements. Siderophilic and chalcophilic elements demonstrate the same tendency (in a weaker form) of decreasing concentrations from saliferous breccias to potassium salts. In some cases increased Ti, Zr and Ba contents are observed in potassium salts. This anomalous tendency could be due to integrated evaporation and sedimentation processes that continued to the highest stages of halogenesis; this phenomenon is characteristic of evaporite formations of foredeeps.

From these data one can conclude that in order to know evolutionary changes of element contents it is necessary to compare some definite groups or types of salt and saliferous rocks. A general scheme of distribution and entry mode of clastophilic elements in different components of insoluble residues has been worked out. The widest range of elements is observed in sandy aleuritic fractions of the insoluble residue. In general they are related to two parts: the clastic and the authigenic ones.

In heavy fractions the general range of minor elements is characterized by prevalence in Ti and Zr; there are Mn, Cr, Cu, Pb (0.n%) and also Ni, Ce, La, Y (0.0n%), etc.

Argillaceous fractions contain Ti, Ba (0.n%); Mn, V, Zn, Cu (0.0n%); Ni, Co, Nb, Pb, Ga, Sc, La, Y, Li (0.00n%), Mo, Sn, Be, Yb (0.000n%), Ag is present sporadically. Co, Zn, Ni, Mn, Cu are found mainly in the acid-soluble residue of argillaceous fractions, and Pb, Cr, V, Zr, Ag, Ti in the acid-insoluble ones (that is, in the aluminosilicate fraction).

The minerals that are the main concentrators for Ti, Zr, Ba, La, Y, Cu, Mn, and Zn have been distinguished. It was only in rare instances that independent minerals containing other elements were identified. It is likely that the presence of other elements usually represents an isomorphous inpurity within the crystal framework of minerals or they are occluded by an argillaceous mycelium (here and below we mean that the sorption of most clastophilic elements takes place at the stages of mineral formation that are followed by halogenesis).

To illustrate geochemical relationships existing between microelements within general groups of rocks of saline formations, schemes of correlation called correlation pleiads have been created (see Fig. 2). The work has been performed on the basis of paired correlation ratios of elements as the formulae of iterative correlation. In the main, both the former and the latter illustrate the same geochemical situations reflecting mainly disorderly connections which are quite natural for the geological environment under consideration. Nevertheless, some regularities can be ruled out that seem to reflect the possible associations of principal geochemical processes. For example, Cu is found only in some definite part of formulae and tends to attract such chalcophilic elements as Pb, Mo, La in varying combinations, sometimes with the elements that belong to other groups. The same relationships are reflected by correlation schemes. The associativity of the chalcophilic elements is supported by their similar variations of contents in the course of time (see below).

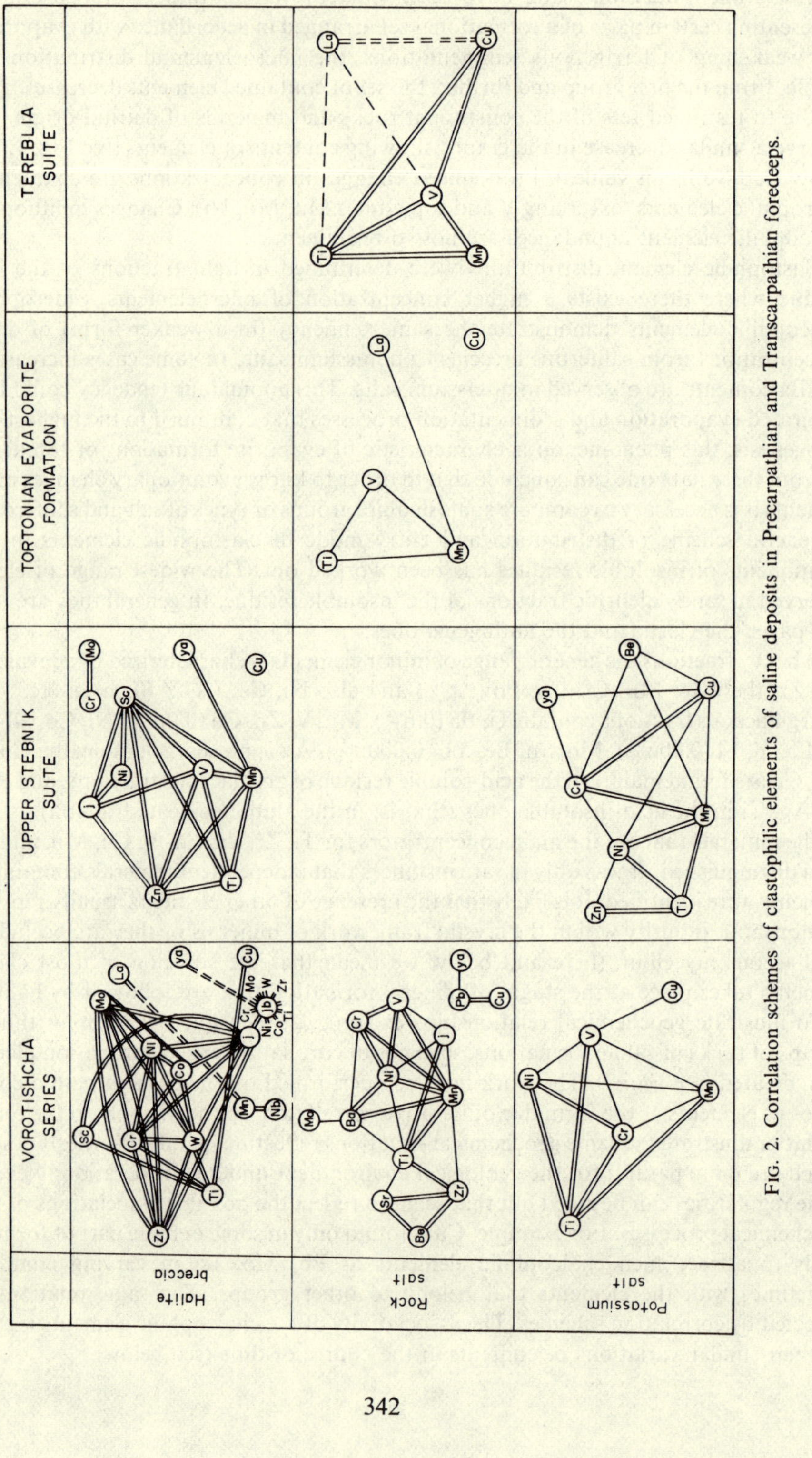

FIG. 2. Correlation schemes of clastophilic elements of saline deposits in Precarpathian and Transcarpathian foredeeps.

Other groups are outlined less clearly. Ba, Sr and also Sc reveal rather stable bonds. These elements of the lithophilic group are also markedly related to the siderophilic ones (Ti, and also Cr, V, Ni, Co) the latter may attract the rare-earth elements of yttrium group (La, etc.). These close relations peculiar to the elements of different groups are explained by their association with the argillaceous fraction and are caused by sorption processes. High values of correlation relationships characterize the elements that are formed in the process of chemogenic sedimentation (Mn, Ba and others). The associativity of Ti and Zr deserves no special comment. Separation of chalcophilic element associations need our special consideration. Not to mention their supply with terrigeneous material there exist two more possibilities, the first being deposition of Cu and Pb within the basin and the second being their supply resulting from superimposed processes (in the last case the term "clastophilic elements" is not altogether right genetically). Mineralizations of both genetic types are present within Ukrainian saliferous basins.

The rock analyses clearly demonstrate the lithological control of their associations that determines participation of one or another process. Thus, within breccias Pb and Ga, Cu and Be group together (it is a chalcophilic group) as well as Co, Sc, Y (less distinctly). In rock salt the associations of Cu, Pb, Ga are quite distinct and those of Ba, Mn and of the elements are not so evident. Thus the data on these two rock types reflect the following: superimposed processes, supply of clastic material with their primary association of chalcophilic elements; combinations that are related to sorption and hemogenic sedimentation processes and, finally, accumulation of heavy stable minerals. In potassium salts Cr and V (sometimes Ni) are closely related; other elements do not reveal their relationship distinctly which means that here we are dealing with clastophilic element supply by argillaceous material.

Saliferous rocks from other saline basins of the Ukraine (the Donnbass, Dnieper–Donets depression, the Transcarpathian depression) demonstrate very weakly expressed correlations of clastophilic elements.

From the viewpoint of genetic interpretation variable concentration characteristics are also of interest for us. As a rule, variation ratios can increase for two reasons: changes of the conditions in which sedimentation occurs or by an additional supply, that is, due to superimposed processes. Siderophilic and lithophilic elements mostly are characterized by moderate variations of ratios which reflect primary conditions of sedimentation. However, anomalously high values of the parameter are observed for some elements (Cu, Mo, Ba, Ni, V). The behaviour of Va and Ni is likely to be dependent on initial sedimentation (may be an irregular distribution of argillaceous or organic materials).

A very high variability of Ba is partly dependent on the sedimentation. But if this were so one should consider the possibility of local barite mineralization associated with brine ore formation (see Khrushshov, 1974). The presence and irregular distribution of those geterogenic type concentrations is likely to influence the behaviour of ratio variations for Cu and Mo.

Our task is aimed at considering regularities of distribution of clastophilic elements and at elucidating their relationships with saline basin evolutions to be applied for solving differentiation and correlation problems. The major proportion of these elements enter the basin during the process of terrigenous sedimentation, the latter superimposing the process of halogenesis, and their changing parameters of distribution reflect (Khrushshov, 1974) the evolution of sedimentation in the course of time and (Strachov, 1962) spatial variations of sediment distributions, i.e. facies changes. Until now many research workers were of the opinion that when one deals with series of close ages the latter (Strachov, 1962) dominates over the influence of the former (Khrushshov, 1974). It is generally true, but if the rock types are clearly distinguished to reflect identical stages of evaporite process, the outline of some tendencies of evolution can be obtained; the latter is also proved statistically. In some cases we have complications due to the processes of superimposed mineralization creating secondary concentrations of elements. We shall be able to reveal time and spatial peculiarities under consideration to the extent that we can differentiate the results of all these phenomena.

Lying within the scope of our interest the groups of rocks of the Vorotishcha series (the Aquitanian and Burdigalian stages), the Upper Stebnik suite (the Helvetian stage) and the Tortonian evaporite formation (the Upper Tortonian) reveal the following tendencies of evolution observed on the background of some deviations, all of which are impossible to consider.

Lithophilic elements in the rocks studied are characterized by small variations in their concentrations, and some of them appear to be regular ones. These are: decreasing concentrations of Ba within the Lower Vorotishcha–Upper Stebnik deposits and their increase (in light fractions of insoluble residues) within Upper Stebnik–Upper Tortonian deposits; increasing zirconium concentrations are observed in all suites successively from the Lower Vorotishcha to the Upper Tortonian. Regularities of the distributions are conditioned by the saline basin migration. The general scheme of evolution for the Precarpatian Miocene basins was given by Strachov (1962) and specified by us.

Appreciable variations of abundances are peculiar to siderophilic elements. It is possible to observe an appreciable increase in concentrations of Cr and V from the Vorotishcha series to the Upper Stebnik suite. These elements have equal concentration within the Tortonian formation. Ti and Mn concentrations decrease or stay at the same level in all groups of rocks from the Upper Vorotishcha to the Upper Stebnik deposits; an abrupt decrease is observed in the Tortonian evaporite formation.

The studied peculiarities of evolutionary dependent Ti, Zr and Ba concentration changes are also supported by mineralogical evidence.

In light fraction of insoluble residues one can observe somewhat different behaviour of siderophilic elements. The changes are conditioned either by successive "purification" of the chemogenic sedimentation or by increased dispersion of the corresponding minerals concentrations and their trapping by chemogenic minerals in the process of their crystallization.

Chalcophilic elements are characterized by most complicated regularities of their distribution. To begin with, quite discrepant tendencies of changing contents are peculiar to different types of rocks and the data suggest that the supply of these elements are lithologically controlled. One should also note that in the Lower Molassa sediments (the Vorotishcha and Stebnik series) mean concentrations of Cu are much higher than those found in their analogous type saliferous rocks of other regions. And, finally, evolutionary characteristics of Cu are opposite to those of most elements related to terrigenous material sources.

The data available imply that Pb and Ga are mostly supplied from wash-down sources. But correlation schemes demonstrate that chalcophilic elements clearly standing apart. It is also possible to observe decreased concentrations of Cu as one goes from breccias to rock salt and further to potassium salts, i.e. to the rocks more resistant to influence of superimposing processes. But in the saline rocks where Cu depends mostly on terrigenous supply, its quantities decrease in the uppermost position of the basin from supply sources (the Late Stebnik age). All these data bear testimony to the heterogeneity of the elements that belong to the group under consideration. In other words, some definite quantities of these elements were separated in the course of sedimentation-diagenesis and also due to superimposed processes. One can observe the behaviour of Cu from this viewpoint. In the salt-bearing breccias the quantity of Cu is more than 8 times increased from the Vorotishcha deposits to the Upper Stebnik ones, and it is significantly lower in the Upper Tortonian formation. Analogous Tortonian rocks of the Transcarpathian foredeep are characterized by lower concentrations of Cu but the sediments belonging to the Lower Permian of the Donbass (they neighbour copper-bearing series) have almost identical Cu contents. Concentrations of Cu increase in those parts of the basin that are mostly remote from supply sources; temporarily they can be referred to the Stebnik age. Most significant are changing ratios of Cu variations from lower values peculiar to moderate levels of concentrations in the Vorotishcha series rocks (normal sedimentary distribution) to an abrupt increase of their variability; the latter, when associated with significant increases of mean concentrations, usually suggest an additional supply of the element in other forms; possibly it suggests significant ore manifestations; and in our case it is represented by the underlying Lower Stebnik red bed series. Similar changes are characteristic of Be and Mo. But these peculiarities are mostly common in saliferous breccias that are, on the one hand, more permeable for ore solutions and, on the other, can include the rocks containing syngenetic copper mineralizations. Low ratios of potassium salts when associated with similar low abundances bear testimony to normal sedimentary genesis of elements.

Evolutionary changes are also demonstrated by schemes of correlation pleiads and also by formulae of iterative correlations for elements (see Fig. 2). Saliferous breccias of the Upper Stebnik suite have Cu concentrations grouping markedly apart as compared to those in the Vorotishcha series. Clearly expressed association between Co, Ni, Sc, Y disappears. Rock salt of the Vorotishcha age is characterized by meaningful relations between Ba and Mn which is likely to reflect peculiarities of the chemogenic sedimentation. Light fractions in the insoluble rock salt residue of the stebnik age are characterized by a disorderly set of their elements. Rock salt of the Tortonian evaporite formations has a restricted set of elements and weak correlation relationships; in light fractions of their insoluble residues one observes large quantities of elements, their relationships being equally obscure. These peculiarities of the Tortonian formation are similar to those of the Tereblia suite of the Transcarpathians.

Potassium salts of the Vorotischcha series are also characterized by well-expressed relationships between Cu, Ni, V (they are supplied by argillaceous materials). Analogous rocks of the Upper Stebnik suite have sets of elements without any regular relationships. Potassium salts of the Lower Permian (the Donbass region) have a very restricted set of elements with extremely poor correlations.

Thus clastophilic elements can be considered as indicators of evolution of actively developing saliferous basins; they reflect their relationships with surrounding continental massives and peculiarities of superimposed mineralization. Concentrations of elements and relationships between them are controlled by the material that is supplied by washout sources, they depend on its transportation, sedimentation environment and to a lesser degree by geochemical conditions existing in sedimentation basins. The influence of superimposed processes is conditioned by many factors and is in the end lithologically controlled. Changes of concentrations within stratigraphic subdivisions in general, their changes in separate groups and types of rocks, ratios of contents variations, peculiarities of correlation relationships between elements are all indicative of evolutionary processes. Practical applications of these indicators are dealt with in other works.

The data considered show that saliferous series possess certain individual parameters that can be applied for correlations of saliferous deposits.

Some of these methods of analysis can be used, and are used by us, for studying ore-bearing sedimentary suites and series of non-saline origin.

Acknowledgements

The author of the article thanks colleagues from the Polish Geological Institute in Warsaw and the Institute of Geology and Geophysics in Bucharest who kindly helped him to collect material from Polish and Rumanian saliferous basins.

References

KRUSHSHOV, D. P. (1974) Osobennosti metallogenii solianocupolnykh oblastej. *Tezisy docladov IV sympoziuma IAGOD*, Varna.

STRACHOV N. M. (1962) *Osnovy teorii litogeneza*, Tome III, Moskwa.

The Behaviour of some Rare Elements in Sedimentation Processes

I. P. ADAMCHUCK, D. N. PACHADZHANOV, N. D. MELNICOVA and YU. YA. VALIJEV

Institute of Chemistry, Tadjikistan Academy of Sciences, Dushanbe, 734030 U.S.S.R.

Abstract

In Tadjikistan Mesozoic deposits are represented by polyfacial complexes of rocks formed under conditions of humid and arid lithogenesis.

The contents of Rb, Cs, Ga, Sn, Zn, Hf and some petrogenic elements (K, Al, Ti, Fe) in 500 average and 300 separate samples have been determined by X-ray fluorescence analysis. Basic regularities of the distribution of the elements in lithologo-facial rocks and weathering products of the feeding areas, and also in granulometric fractions, pelitic fractions and 5% HCl-extract were determined. The balance of the elements in the main lithologic varieties of the deposits of humid and arid types of lithogenesis of the intermountain areas is given. The value of their migrating mobility is given, geochemical criteria of facial and lithogenetical types of the deposits are proposed.

Rare elements geochemistry of sedimentary processes are important because of the necessity to construct the geochemical model of the earth crust, to calculate the clarkes and the substance balances during sedimentogenesis, etc. These data can be used for palaeofacial reconstructions and exploration geology. However, for rare elements such as niobium, tantalum, zirconium, hafnium, bismuth, tungsten and others, the data are poor; moreover, they are not correlated to each other for the following reasons. Element properties in sedimentation processes are affected by: (1) their concentrations and forms in rocks of the source area; (2) intensity of weathering processes in areas of water supply; (3) physico-chemical migration and sedimentation conditions that are greatly affected by climate; (4) regional structure facial peculiarities. In order to attain scientific authenticity in the investigation of the main tendencies in the behaviour of chemical elements, it is necessary that we study a whole sedimentation process in one sedimentation basin and under specific earth crust structures. This process involves substance mobilization in areas of water supply, migration of weathering products, deposition of sedimentary material. For this reason, it is necessary to study geologically homogeneous objects, representing a whole cycle. In spite of this, the accuracy of the analyses being used is of great importance. We agree with the words of D. Shaw that every method gives the best results only in the laboratory where it was worked out. But these words must be supplemented with the necessity of studying homogeneous geological objects.

As an object for investigation we have taken the mesozoic sedimentation basin of the Tadjik depression, where deposition had taken place in a humid and an arid climate from a comparatively constant feeding area. These deposits are represented by a polyfacial complex of the Jurassic humid and early Cretaceous arid formations that were divided into genetic groups by a facial formation analysis worked out by the Geological Institute, Academy of Sciences, U.S.S.R. (Yu. A. Zhemchuzhnickov and P. P. Timofeev). The chemical elements

TABLE 1. CHEMICAL ELEMENTS CONCENTRATION COEFFICIENT IN ANCIENT, HUMID OR ARID AND RECENT ARGILLACEOUS ROCKS

	Ancient deposits			Recent deposits
	Humid	Arid		
Element	C_m/C_c	C_m/C_c	C_l/C_c	C_m/C_c
Li	0.89	1.11	1.42	0.63
K	0.87	—	2.20	—
Cu	0.84	—	1.00	1.35
Rb	0.72	—	—	0.75
Mg	1.86	—	2.76	—
Ca	2.71	—	3.89	1.68
Zn	0.47	—	0.22	1.37
Ba	0.99	—	1.46	1.25
B	2.74	—	—	1.65
Al	0.91	1.10	0.87	—
Ga	0.71	1.06	—	0.98
Y	1.97	1.00	0.68	—
La	0.71	0.94	0.98	—
Yb	1.25	0.95	0.80	—
Th	0.69	1.23	0.88	1.00
U	1.09	—	—	1.53
Pb	0.75	—	2.86	1.12
Cr	0.92	—	1.28	0.97
Mo	2.58	—	1.33	9.30
Mn	2.23	—	1.18	1.23

were determined with an X-ray fluorescence analysator VRA-2 (Firm "Karl Zeiss", Iena, G.D.R.). Detection limits vary from 20 to 0.n ppm for different rare elements, relative reproduction is 1–3.

The sampling of the layers was carried out by making average samples for petrogenetic types of rocks of specific facies. Some simple samples and their fractions were also studied. For the elements migration study, 5%-HCl argillaceous rocks extract (MMF), including salts, some metallic amorphous hydroxides and soluble forms of rare elements (adsorbed, fixed by organic substance, carbonates, etc.) were investigated.

The study of literature on the elements distribution in facial types of sedimentary rocks (more than 150 papers were reviewed) allow us to distinguish essential tendencies in their content being influenced by sedimentation conditions. To establish these tendencies, we have calculated concentration coefficients, i.e. the ratio of element concentration in marine deposits to their concentration in continental deposits (C_m/C_c). If C_m/C_c is ≥ 1.1, an element is accumulated in marine deposits, if C_m/C_c is ≤ 0.9, an element is accumulated in continental deposits.

1. In humid continental deposits the following elements of the Periodical System are accumulated (Table 1): from the first group—lithium, potassium, copper, rubidium; from the second group—zinc; from the third group—gallium, lanthanum, thorium; from the fourth group—boron, yttrium, ytterbium; and from the sixth and seventh groups—molybdenum, manganese. One more group of elements is distinguished, characterized by a steady distribution in facial profile. This group includes barium, aluminium, uranium, chromium.

TABLE 2. ELEMENTS CONTENT IN JURASSIC ARGILLACEOUS ROCKS AND EARLY CRETACEOUS DEPOSITS OF THE TADJIK DEPRESSION (ppm)

	Jurassic				Early Cretaceous		
Element	C	CM	M	I	C	CM	I
Rb	192–218	165–200	148–190	C	150	142	—
Cs	7.4–7.5	6.7–7.0	6.8–6.5	C	6.6	6.6	—
K n 10^4	2.3–2.4	3.2–3.8	2.6–3.6	M	2.2	2.5	CM
Ca n 10^4	0.17–0.19	0.7–0.85	3.46–2.53	M	2.0	2.8	CM
Sr	371–341	273–225	349–302	CM	200	248	CM
Ba	681–689	570–612	493–475	C	548	494	C
Al n 10^4	13.4–13.8	8.2–8.9	8.0–9.2	C	8.3	7.9	—
B	81–100	132–187	163–180	M	77	108	CM
Ga	27–28	25–25	23–23	C	23.7	23.5	—
Y	39–40	36–36	37–37	C	34	35	—
La	73–74	65–64	55–61	C	64	67	—
Ce	109–115	103–95	109–96	—	102	96	—
Yb	4.2–4.3	3.9–3.8	3.5–3.6	C	3.8	3.7	—
In	0.02–0.02	0.02–0.03	0.07–0.03	M	0.17	0.57	CM
Tl	1.7–3.0	1.8–1.8	2.2–2.4	CM	1.2	1.7	CM
Bi	0.24–0.3	0.41–0.38	0.25–0.35	CM	0.28	0.55	CM
Si n 10^4	27.9–25.7	31.6–27–6	21.2–25–2	CM	26.4	24.1	C
Ti n 10^4	0.45–0.45	0.37–0.38	0.25–0.35	C	0.31	0.35	CM
Zr	226–196	179–155	118–133	C	154	161	—
Sn	6.7–6.8	6.0–5.9	5.3–5.8	C	5.9	5.6	—
Hf	6.8–6.3	6.0–4.8	3.8–4.4	C	4.4	4.4	—
As	10.9–10.8	9.9–9.5	9.1–9.2	C	9.8	9.6	—
Mo	2.1–1.7	1.3–0.7	0.65–0.70	C	0.9	0.9	—
W	3.0–2.9	2.4–2.8	2.2–2.4	C	2.2	2.3	—
Fe n 10^4	2.1–1.7	3.4–3.4	1.9–3.4	CM	3.6	3.5	—
Mn	112–148	323–309	472–428	M	2816	904	C

C—continental, CM—coastal-marine, M—marine, I—informativity; 192–218—elements content in alevrolite–argillite.

2. An essential difference in the geochemistry of arid deposits is the inversion of the distribution of some elements in a facial profile. Elements of the first group of the Periodical System tend to concentrate in marine and lagoon deposits; aluminium, thorium, lead, chromium, barium belong to this group. Only for the second group of elements the same tendencies as for humid lithogenesis seem to be conserved.

3. Diagenetic processes affect in some way the elements distribution in deposits. Thus, copper, zinc and lead are accumulated in recent marine deposits, while in ancient deposits these elements tend to be in continental formations.

By studying the process of sedimentation in a specific basin it was stated that those essential tendencies are constant, but some differences may occur as a result of basin specific features.

In the conditions of the humid (Jurassic) sedimentation in the basin of the Tadjik depression (Table 2) rubidium, barium, aluminium, gallium, lanthanum, ytterbium, zirconium, hafnium, molybdenum and tungsten are accumulated in continental argillaceous deposits. Potassium, calcium, boron and manganese tend to concentrate in marine deposits. Moreover, such elements as strontium, silicon and iron are very often met in coastal-marine areas.

In conditions of an arid (early Cretaceous) sedimentation basin, silicon, barium and

TABLE 3. ELEMENTS CONTENT IN PELITIC FRACTION OF JURASSIC DEPOSITS OF THE TADJIK DEPRESSION (ppm)

Element	Continental		Coastal-marine		Marine Arid		Informativity
	KH	HK	KHH	HH	HH	ChH	
Rb	206	238	203	236	210	225	—
Cs	7.6	7.2	6.6	6.9	7.0	6.6	—
K n 10^4	2.2	3.2	3.6	4.6	4.8	4.6	M
Sr	281	274	179	179	239	137	C
Ba	950	855	713	662	638	570	C
Ca n 10^4	0.09	0.09	0.14	0.36	0.62	0.16	M
B	64	72	174	197	262	255	M
Al n 10^4	15.5	15.6	14.0	13.2	12.9	12.4	C
Ga	28.6	27.3	24.2	24.6	23.7	24.5	C
Y	40	40	37	38	38	37	—
La	80	75	67	65	64	61	C
Ce	112	107	104	117	107	106	—
Yb	4.5	4.3	3.9	4.1	4.1	3.9	C
Si n 10^4	24.2	25.1	26.6	25.9	26.3	26.6	M
Ti n 10^4	0.38	0.43	0.46	0.43	0.41	0.46	CM
Zr	150	141	134	137	145	140	—
Sn	7.2	7.1	6.1	6.6	6.1	6.1	C
Hf	4.8	4.5	4.1	4.4	4.0	4.1	C
As	12.1	11.5	10.1	9.9	10.5	9.9	C
Mo	3.0	2.7	1.2	1.2	0.8	0.8	C
W	3.0	2.9	2.5	2.3	2.8	2.2	C
Fe n 10^4	2.0	2.2	4.1	3.9	3.4	5.4	CM
Mn	74	119	300	180	185	242	CM

Association of clay minerals: KH—kaolinite–hydromica, HK—hydromica–kaolinite, KHH—kaolinite–hydrated-hydromica, HH—hydrated-hydromica, ChH—chlorite–hydrated-hydromica.

manganese tend to accumulate in continental deposits, potassium, calcium, strontium, boron and zirconium in coastal-marine ones. It must be said that these tendencies are true for the pelitic fractions (<0.001 mm) of humid terrigenous formations (Table 3). In this case, strontium, barium, aluminium, gallium, ytterbium, tin, hafnium, molybdenum and tungsten are accumulated in continental deposits. The mineralogical composition of the argillaceous fraction affects the distribution of such elements as rubidium, yttrium and cerium more than do the sedimentation conditions.

Some differences in rare element distribution according to the facial profile of argillaceous rocks and in their pelitic fraction may be caused by the existence of different element forms in sedimentary rocks. In order to evaluate the role of migration-mobilizing forms (MMF) in elements balance and its quantity dependence on sedimentation conditions, 5%-HCl extract from argillaceous rocks were studied (Table 4).

In humid rocks the maximum MMF quantity of most chemical elements is found in marine deposits; it varies from 3 to 36% of their total content in rocks. It is noteworthy that elements such as calcium, manganese, iron and strontium are found mainly in MMF—up to 66–99%. In the process of sedimentation many hydroxides (or carbonates) of rare elements are coprecipitated with them. This process also depends on sedimentation conditions, i.e. in the continental conditions, manganese is closely bound to rare-earth elements, tin, etc.; in marine conditions, arsenic, strontium and tungsten are associated with iron and calcium.

A distinctive feature of rocks of arid formation is a higher MMF element content and its

TABLE 4. ELEMENTS CONTENT IN 5% Hcl EXTRACT OF THE TADJIK DEPRESSION ARGILLACEOUS ROCKS, IN RATIO (%)

	Jurassic				Early Cretaceous			
Elements	C	CM	M	L	C	CM	M	FL
Rb	3.3	2.4	9.4	8.4	6.8	19.2	3.4	8.2
Cs	6.0	8.8	30.2	1.7	10.0	27.0	11.5	12.4
K	9.5	9.4	31.7	11.2	5.2	13.9	12.5	8.7
Ca	62	89	99.8	70	88	96	98	90
Sr	7.4	35.4	66.4	8.9	36.0	61.3	83.5	62.9
Ba	13.2	4.6	14.2	8.5	12.3	26.1	16.1	18.4
B	29.2	36.2	35.8	40.6	25.8	32.6	47.7	38.8
Al	6.7	5.6	2.1	8.0	2.0	12.2	7.7	5.1
Ga	0.6	2.1	7.7	3.2	11.4	23.3	16.2	14.7
Y	5.6	6.5	20.0	3.2	5.4	23.4	5.3	12.6
La	5.2	5.1	10.9	2.4	12.0	27.0	9.6	13.8
Ce	1.9	6.4	9.8	5.0	2.0	22.0	2.2	4.9
Yb	2.7	4.5	15.7	4.0	3.7	25.2	9.0	15.2
Si	3.0	4.9	6.0	6.9	6.6	7.8	0.0	4.6
Ti	4.7	10.3	13.3	6.6	2.6	20.8	0.0	2.9
Zr	5.9	0.5	0.0	9.2	18.6	27.0	0.0	4.5
Sn	3.7	1.2	13.6	3.3	0.0	27.6	14.0	12.2
Hf	2.8	2.2	3.8	3.8	8.7	22.3	10.4	8.0
As	0.2	3.5	7.0	2.4	1.2	36.7	10.0	12.0
Mo	3.1	11.3	3.0	0.9	0.0	35.6	10.0	0.0
W	5.6	5.0	15.0	4.7	0.6	24.4	15.0	5.3
Fe	48.7	60.6	78.8	47.3	11.1	33.5	0.0	1.3
Mn	83.0	88.0	98.0	80.0	56.0	92.0	90.0	79.0

C—continental, CM—coastal-marine, M—marine, L—lagoon, FL—freshwater lagoon.

maximum dislocation to coastal-marine area. Data show correlation between MMF elements content and sedimentation conditions.

Such regularities become clear if we bear in mind that the natural surface waters of a hypergenesis area are greatly diluted solutions of most chemical elements. In this case sorption and coprecipitation play a great role in element migration. Radiochemical investigations of I. Ye. Starick and colleagues and many other scientists showed that the intensity of the adsorption depends on the element form in solution, the charge of the sorbent surface, the physical and chemical properties of the element and the pH of the environment. Adsorption is affected also by the presence of ligands in a solution, the carbonate ion in particular. The latter contributes to the formation of coordinate saturated complexes and adsorption maximum dislocation to alkaline marine range.

Experimental data corroborate this statement (Table 5). The elements that in normal conditions are sorbed in acid environment (groups I, III of elements, tungsten, molybdenum, etc.) are sorbed in alkaline environment, if CO_2 is in excess.

As is known, continental formations are characterized by pH ≤ 7, marine formations by $-\geq 7.5$. In surface water during climate aridization the content of carbonate ion is increased. The data described are well correlated with experimental data on sorption (Table 5).

The established regularities of the distribution of chemical elements in different rock types, being formed under specific sedimentation conditions, make it possible to distinguish typomorphic element groups, characterizing continental and marine deposits. In this case we

TABLE 5. EXPERIMENTAL DATA ON ELEMENTS SORPTION AND THEIR DISTRIBUTION IN ARGILLACEOUS DEPOSITS

Elements groups	Experimental data			Geochemistry data			
	Normal conditions		CO_2 excess	Humid deposits		Arid deposits	
	pH <7	pH >7	pH >7	cont.	marine	cont.	marine
I	Li,Na,K, Cu,Rb,Cs	Ag	Li,Na,K,Cu, Ag	Li,Na, K,Cu, Rb,Cs		Rb,Cs	Li,Na, Cu,K
II	Zn,Hg, Sr,Ba	Mg,Ca, Cd,Sr	Mg,Ca,Cs, Ba	Zn,Hg, Ba	Mg,Ca, Sr,Ba, Cd	Zn	Mg,Ca, Sr,Ba
III	Ga,Ce,Th	B,Sc,In, La(?), Tl	B,Sc(?),Y, In,Ce,Th,U,	Ga,La, Ce,Pr, Nd	B,Sc, Y,Sm, Gd,Tl	Er,Yb	B,Sm, Dy,Th
IV	Zr,Sn, Hf,Pb,		Hf(?),Pb(?)	Zr,Sn, Hf,Pb,	Ge,Hf	Ti,Zr, Sn,Hf	C,Pb,Hf
	As, Nb,Sb	As,Bi	V,Nb,Bi	Ta,Sb	P,As, Bi,Nb	P,V(?)	Nb(?)
VI	W	Mo	Cr,Mo,W	Mo	S		S,Cr,Mo
VII		Mn			F,Mn	F,Mn	
VIII			Fe,Co,Ni		Ni		Fe,Ni

can single out a narrow group of elements for detailed description of sedimentation conditions.

According to the facial profile of argillaceous rocks the following ratios of elements are highly informative—Ca:Ba, Mn:Mo, Fe:W, Ca:B, Rb:B, Ba:B, etc. In pelitic fraction the following ratios are informative—Mn:Mo, Fe:Ca, W:Mo, Ca:B, etc. The ratios of elements migrate-mobility forms are significantly changed. In this case all the above-mentioned geochemical ratios are informative of palaeoreconstructions.

Thus, the reviewed literature and experimental data on mesozoic deposits of the Tadjik depression show that many elements have migration mobility in hypergenesis both in humid and arid climate conditions. Their migration is affected by physico-chemical conditions, element existence in natural water, geochemical rock specification of feeding areas, etc.

All these facts are premises for the detailed investigation of the element distribution in sedimentary rocks and their use in paleoreconstruction of sedimentation conditions.

Minor Elements in Clay Rocks, Indicators of the Sedimentation Environment

V. A. KUZNETSOV, K. I. LUKASHOV and V. K. LUKASHOV

Institute of Geochemistry and Geophysics of the B.S.S.R. Academy of Sciences, Minsk, U.S.S.R.

Abstract

At the west of the Russian platform clay lithogenesis is characterized by a predominance of kaolinite in the lower Miocene and of montmorillonite in the middle Miocene–Pliocene. The recent times stream sediments have more kaolinite mixture than the flood plane sediments and gibbsite and halloysite occur in places of the erosion of paleozoic and mesozoic weathering crust. In the Anthropogene clay formations are distinguished by the predominance of hydromica development. Glacial and fluvial erosion of the preanthropogene rocks led to the origin of mixed clay composition, where clay minerals occur in different proportions. The type of clay is interpreted in terms of the physico-chemical and mechanical processes of the geological time. The paleoclimatic and paleogeochemical environments are considered to have had a very big influence on these processes.

The average concentration of Cr, Ni, V, Cu, Ti in pliocene clay sediments is 1.5–2 times more than in anthropogene sediments. The distribution of microelements in the first case have a more differentiate of character in granulometric fractions. The reduction of sorption of copper content in clay rocks in the Pliocene sequence of sediments is marked. Concentration of Ni, V, Cu, Cr, Ti decreases, Zr increases and their ratio changes in clay by the transition to periglacial conditions of the lower Pleistocene and by the transition from warm to cold conditions of interglacials of the Pleistocene. Clay minerals specifics, dispersed organic matters and their components, ferrugination and carbonatization degree of the clay sediments, and the oxidation-reduction environment influence the associations and concentrations of microelements.

In the pore solution of the basement of the Poltav series clays promote the concentration of V and Ni along with Cl, Na and the ratio (Na + K):(Ca + Mg). These trends testify about the lagoon conditions of the sedimentation. The concentration of Mn, Cr, V decreases and Ti and Zr increases in the range-marine–delta-river clays.

Among Pleistocene lake, alluvial and lagoon sediments of the south of the Russian platform the latter are distinguished by a maximum concentration and the specific contents on Mn in the extracts of the exchange, carbonate and organic forms. In marine conditions diagenetic mineral formation accompanied by the sharp concentration of Mn in the carbonate, organic and amorphous contents of clay sediments. Epigenetic processes of deep hydrogenesis and of hypergenesis have an important influence on the clay content.

In contemporary and buried river valleys of the Pleistocene clay sediments are distinguished by the presence of concentrations and ratios of the microelements in near river and inner flood plains, oxbow, lakes, and also in normal, periglacial and glacio-alluvium.

In the Pollessie region alluvial clays, of the southern drainage, formed from the weathering of the Ukraine massif, concentrate Ti, Zr, Cu, Cr and Co more than the clay of the northern drainage, derived from the river erosion of moraine deposits. Similar concentrations of the microelements took place in glacio-alluvial and glacial sediments of the lower Pleistocene in deep valleys in the region of the Byelorussian massif erosion.

Known laws of the microelements distribution in clay sediments are used for the deduction of palaeogeographic and facial environments of the Cenozoic lithogenesis, the correlation of horizons, obtaining an Neogene–Anthropogene boundary, for geological prognoses and for prospecting.

Materials of minor elements used for the partition, correlation and reconstruction of palaeogeographical environments of cenozoic sediments formation in the west of Russian platform are elucidated. This concerned in particular, the division between barren marine and fresh water stratas, the share out of stratigraphic boundaries on the basis of climatic changes and sediments composition; the division and correlation of glacial sediments. The results of these investigations have been published previously by the author (Lukashov V. K., 1970, Lukashov K. I. and Lukashov V. K. 1975). The main generalizations and conclusions on the questions mentioned above are discussed there.

The study of oligocene–miocene sea- and fresh water sediments in the Byelorussian area shows that there are on the average more Mn (in 1.8 times), Cr (in 4 times) (Lukashov K. I. and Lukashov V. K., 1975), Ni (in 5 times) (Lukashov and Derbinskiy, 1974), V (in 3.5 times) and some other elements in the first sediments. These data with data obtained by other methods (analyses of pore solutions, of absorption complexes and others) can be used for division of barren section.

In the sections studied the sea marine sediments are differentiated from freshwater ones by higher boron content (correspondingly 0.019 and 0.008%). This fact is confirmed by the work of other investigators. With that one needs to note that the geochemistry of the element is studied inadequately. At this time many particularities of the boron fixation on clay particles surface are not clear. In order to resolve these questions in the Institute of Geochemistry and Geophysics of the B.S.S.R. Academy of Sciences some experimental works were done (Lukashov and Derbinskiy, 1974; Lukashov *et al.*, 1977). For divisions of the marine and freshwater sediments we also used stage extraction. It is seen, for example, that brackish sediments have more Mn-distribution on fractions. The differences are clearer in the fraction of exchangeable ions and carbonates.

In Byelorussian attention has been drawn to the problem of the Neogene–Antropogene boundary of the international geological correlation programme (IGCP). We have studied the sedimentation environments (climatic, tectonic conditions) and the contents of Neogene and Pleistocene sediments. It is shown that pliocene and pleistocene sediments differ sharply in their composition and minor element distribution. In pliocene clay rocks the average concentration of Cr, Ni, V and Cu are 2 times more than in Antropogene clays. They have also 1.5 times more Ti, but less Zr. There is 6 times more Mn content in the antropogene clays in the Byelorussian area. The curves of the minor elements distribution in the granular fractions (0.01–0.005; 0.005–0.001 and less than 0.001 mm) of the pliocene sediments have a more differentiated character, which is a result of significant weathering and resedimentation. In some sections of the western part of the Russian platform the higher Ba and Sr contents are marked in the boundary pliocene–pleistocene parts of the sections. Possibly the main role of Ba and Sr distribution are the particularities of carbonate content. It is well known that the intensive accumulation of carbonates in sediments took place in many regions of the world. The middle pliocene sediments have 6 times more Cu than the sediments of the end of the pliocene when the climate changes to being cool and arid. The Cu content is reduced as a result of the reduction in biological activity and weathering processes. Higher Co contents in the upper part of the pliocene section indicate the onset of oxidizing processes with uplift of the region.

There are very large variations of glacial facies and formations in the Byelorussian region (as well as in other regions in the U.S.S.R., Europe, Canada and the U.S.A.). To understand the genesis of these variations many geochemical investigations of the composition of these sediments have been carried out, in particular the distribution of minor and rock-forming elements in the different facies of the sediments and the density of the minerals formed in the weathering and diagenetic processes. It is known that mechanical mode of the material sedimentation plays the major role in glacial lithogenesis.

TABLE 1. AVERAGE CONTENT AND RATIO OF TRACE ELEMENTS (0.001%) IN DIFFERENT SIZE FRACTIONS

Colour	Size fractions (mm)	Ti	Zr	Cr	V	Ni	Mn	Ti/Zr	V/Cr
Grey	0.001	450	18	10	22	1.8	31	25	2.2
Red	0.001	475	20	7.3	16	1.7	30	23.8	2.1
Grey	0.005–0.001	600	19	9.0	18.8	1.4	27	31.6	2.1
Red	0.005–0.001	740	23	7.9	18	1.2	54	32.1	2.27
Grey	0.01–0.005	650	31	6.9	13.2	<1.0	25	21	1.9
Red	0.01–0.005	890	33	6.4	11.9	<1.0	45	26.9	1.86
Grey	0.075–0.01	470	50	3.9	3.7	≪1.0	24	9.4	0.95
Red	0.075–0.01	450	53	3.2	2.6	≪1.0	37	8.5	0.81

The latest investigations in the U.S.S.R., Canada and other countries show beyond doubt that some chemical reactions and processes of hydration, hydrolysis, ions exchange took place at low temperatures. Accordingly there was a migration of Fe, Mn, S and other elements. Our investigations show that with periodical freezing and melting the clay minerals were transformed with the formation of different mixed-layered minerals, and that the carbonates and salts were redistributed in the cold arid environments resulting in isotopic differentiation, especially C and O.

Below are given some data of the lithogeochemical investigation of the moraine facies.

There are two moraine horizons in many outcrops in the Dnieper basin which are shown on some hundreds of kilometres: lower (in many cases grey) and upper (red). Most geologists consider them to be of different ages: the red moraine—Moscow—(or Sozian) and the grey moraine—Dnieper—(or Oksian). Others consider that the big differences in the contents of carbonate, Mn and in some cases of iron are grounds for suggesting that the colour differences are due to reduction processes altering the character of the clay formations. As to the content of resistant elements (Ti, Cr, V, Zr) there are practically no differences in the grey and red tills (Table 1).

We propose that the till horizons mentioned above, occasionally separated by fluvio-glacial and cold lacustrine and/or alluvial deposits, often represents different facies of one and the same glacial complex. These tills differ not only geochemically but also by some mechanical composition. Perhaps reduction processes took place in the lower till even during the sedimentation process.

When studying glacial deposits it is necessary to pay attention to the peculiarities of source of rock material and sedimentation process.

For example, the succession of weathering of minerals in moraines is broken. This is explained by the fact that many minerals sensitive to weathering, e.g. hornblende, have greater resistance to shearing than the feldspar minerals. Therefore, in the very fine fractions we find feldspars more intensively weathered than the ferro-magnesium minerals, which are present in slightly coarser fractions. Because of this we have to be careful to estimate the degree of leaching in glacial lithogenesis. According to our data such elements typically resistant to weathering, such as Cr and Zr, are found in the mineral form in rocks in moraines of Byelorussia, which are not stable both to mechanical transport (Cr in the limonitizated grains) and chemical weathering (Zr in biotite). In the environment of moraine lithogenesis they could be very mobile, although the weathering processes as a whole were in general not intense enough to mobilize resistant elements.

As a result of intensive aqueous leaching of rocks, dispersion aureoles of elements are formed which very often reach the surface. This makes it possible to carry out geochemical

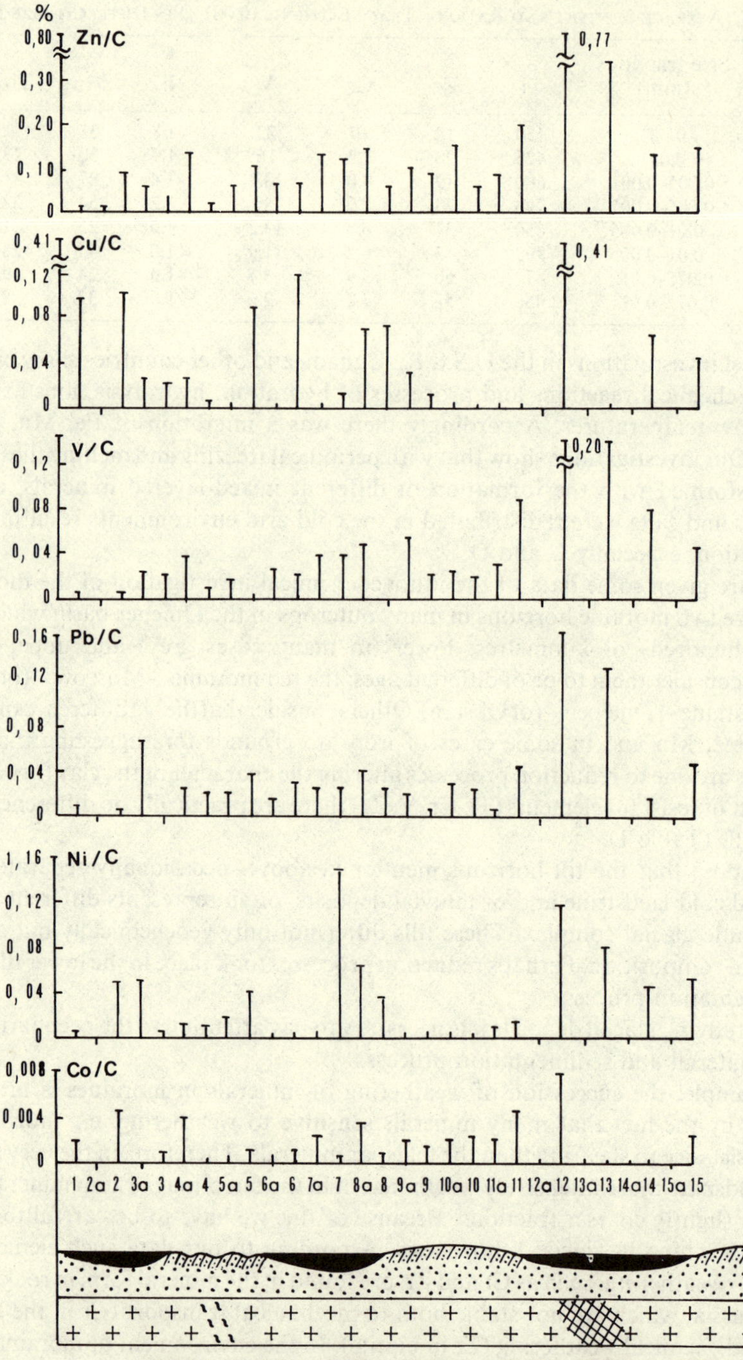

FIG. 1. Content of Zn, Cu, V, Pb, Ni, Co in metallo-organic complexes in peat and soil cover. 1—peat, 2—sand, 3—sand with humus, 4—bog hillocks, 5—rock of crystalline base, 6—orebearing metasomatite, 7—zone mylonitization and cataclasm.

prospecting and mapping of basement crystalline rocks. From the authors' investigations (Lukashov K. I. and Lukashov V. K. and Matruntchik, 1977), made in some regions of the humid zone, the most effective method is the analysis of the mattallo-organic complexes with, for example, the help of sodium pyrophosphate extraction. For example, a zone of mineralization at the depth of 20 m is distinctly noticeable in peat and soil cover by all the elements (Fig. 1). Less distinct data were obtained with the help of cold HCl extraction and analysis of the rock itself. As a conclusion it is necessary to mention that trace elements are useful indicators of environment conditions when we know about the mode of their occurrence and relation to chemical and geological processes.

References

KUZNETSOV, V. A. (1973) *Géochimie de la lithogenèse des alluvions*, Minsk.

LUKASHOV, V. K. (1970) *Géochimie de la lithogenèse du quaternaire*, Minsk.

LUKASHOV, V. K. (1972) *Les Indicateurs géochimiques des processus d'hypergenèse et de sédimentation*, Minsk.

LUKASHOV, V. K. and DERBINSKIY, I. A. (1974) *Etude appliquée et expérimentale de la géochimie du bore*. Voprosy geologiy Belorussiy, Minsk.

LUKASHOV, V. K. and DERBINSKIY, I. A. (1976) Sur le mécanisme de fixation du bore dans les minéraux argileaux. *Materialy XI Vsesoiuznogo sovechtchania poizutcheniu i ispolzovaniu glin i glinistyh Mineralov*, Moscou.

LUKASHOV, K. I. and LUKASHOV, V. K. (1975) *Géochimie de la zone d'hypergenèse*, Minsk.

LUKASHOV, K. I. and LUKASHOV, V. K. (ensemble d'auteurs sous la direction de) (1977) *Etudes expérimentales des formes et des processus de migration hypergène des éléments*, Minsk.

LUKASHOV, K. I., LUKASHOV, V. K. and MATRUNTCHIK, L. I. (1977) Recherches écologico-géochimiques expérimentales des gisements de métaux rares. *DAN. BSSR*, No. 2.

SECTION VI

GEOCHEMISTRY OF THE OCEANIC AND CONTINENTAL CRUST

Geochemistry of Effusive Rocks of Rift Zones

V. I. GERASIMOVSKY

V. I. Vermadsky Institute of Geochemistry and Analytical Chemistry, U.S.S.R. Academy of Sciences, Moscow, U.S.S.R.

ABSTRACT

The following effusive rock series within the rift zones are distinguished: tholeiitic basalts–andesites–rhyolites; basalts–trachytes and phonolites; Na-alkaline series: olivine melanonephelinites–nephelinites and phonolites and K-alkaline one: katungites–ugandites–leucitites–leucitic basanites. The carbonatites are genetically related with two latter series and their separation from alkaline melts is due to liquation. The formation of the effusive rock series is considered as the result of evolution of the initial (mantle) melt which was probably formed within the upper mantle during riftogenesis. The effusive rock series (excluding K-alkaline) are characterized by increase of the rare lithophylic element content and by decrease of content of the mafic typomorphic elements from the initial to the end members of magmatic differentiation. Alkaline varieties of basalts and other rock types are enriched in the rare lithophylic elements, fluorine and some mafic typomorphic elements (Sr,Ti,P) in comparison with less alkaline varieties.

Iceland is located within the region of the Mid-Atlantic ridge. The iceland is mostly comprised of basalts. Only 10% of rocks is represented with acidic effusives, intrusive rocks (gabbro, granophyres) and sedimentary–pyroclastic series (Palmason and Saemundsson, 1974). The most ancient rocks in Iceland are dated approximately about 16 m.y. and are located commonly within the marginal zones of the Iceland. The Quaternary rocks (basic and acidic effusives) are localized mostly within the submeridional (NNE–SSW) rift zone. The effusive paroxysms were observed recently and several times (1970—volcano Hücla; 1971—iceland Heimaey; 1975, 1977—northern part of the rift). The effusive series of rocks contain all varieties from basalts to rhyolites. Two rock series are distinctly identified: tholeiitic basalts–andesitic basalts–andesites–rhyodacites–rhyolites and alkaline basalts–oversaturated alkaline effusives (rhyodacites and rhyolites) (Gerasimovsky *et al.*, 1974). The chemical composition of effusive rocks of Iceland is markedly variable (Jakobsson, 1974; Sigvaldason, 1974; Bailey and Noe-Nygaard, 1974; Wood *et al.*, 1976).

In East Africa, east rift zone and West rift zone are distinguished. The East zone (rift Gregory) is localized in Kenya, Northern Tanzania and Eastern Uganda. Western rift zone (Tanganyika zone) is coincident with the chain of lakes (Alberta, Edward, Kivu and Tanganyika). The formation of the rift zones was accompanied by the Neogene-Quarternary volcanism. The associations of effusive rocks within these zones are extremely variable in their composition and chemistry. Three rock series are distinguished (Gerasimovsky and Polyakov, 1974). The alkaline basaltic rock series (picritic basalts, basalts, olivine basalts, trachy-basalts, andesitic basalts, trachytes and phonolites) is abundant in the Eastern rift and in minor quantities within the Western rift. The Na–alkaline rock series (olivine melilitites, olivine melanephelinites, nephelinites and phonolites) are widely represented within the Eastern rift whereas the K–alkaline rock series are frequent members of the Western rift zone rock series (katungites, mafurites, ugandites, leucitites, leucite basanites, etc.). The alkaline rocks of the Eastern rift are characterized by predominant role of sodium in relation to potassium; the Western rift is characterized by the inverse relations. The main rock-forming mineral of the alkaline series of the Eastern rift is represented by nepheline while the leucite is the typomorphic mineral of the Western rift alkaline series. The effusive carbonatites are observed in the association with both alkaline rock series. The over-saturated per alkaline effusives (comendites, pantellerites) are found in subordinate quantities within the Eastern rift (near Lake Naivasha, Kenya). Furthermore, the intrusive alkaline rocks (nepheline syenites, urtite–ijolite association, carbonatites) are identified among the effusive rocks of the Eastern rift zone (Eastern Uganda, etc.).

All effusive rock series (excluding K–alkaline) are characterized by increase of Na, K, SiO_2 content, rare lithophylic elements (Li, Be, REE, Zr, Nb, Th, etc.), F and by decrease of mafic typomorphic elements (Ca, Mg, Fe, Ti, V, Cr, Ni, Co, Sc, P) from the initial to the end members of magmatic differentiation. Tables 1 and 2 show these concentrations in two effusive rock series: basalt–andesite–rhyolitic (Iceland) and Na–alkaline (Eastern rift of East Africa). Table 3 represents the regular change of ratio values of coherent elements for

TABLE 1. ELEMENT CONTENT IN THE EFFUSIVE ROCKS OF ICELAND IN THE ALKALINE BASALT OF THE EASTERN RIFT OF EAST AFRICA

	Iceland			Africa
Element	Basalt	Andesite	Rhyolite	Basalt
Na,%	1.97	2.96	3.35	2.41
K,%	0.39	1.59	2.85	1.12
Li ppm	7.4	15	24	7.9
Rb ppm	8.7	40	77	21
Ba ppm	134	380	730	440
Y ppm	33	71	109	26
Zr ppm	166	420	620	300
Nb ppm	17.5	59	73	57
F ppm	545	1170	1340	940
Ca,%	8.16	3.83	0.85	6.60
Mg,%	4.02	0.90	0.06	4.82
Fe,%	9.87	6.87	1.82	9.77
Ti,%	1.49	0.92	0.16	1.55
V ppm	374	113	22	263
Cr ppm	166	46	16	322
Ni ppm	82	25	11	135
Co ppm	49	22	8	51
Sc ppm	32	19	5	23
Sr ppm	255	160	80	735
P ppm	1040	1100	170	2100

The following methods were used for the determination of the elements concentration: spectrochemical analysis (Cr, Ni, Co, V, Cu, Sr, Ba, F, P), flame photometry (Na, K, Li, Rb), gamma spectrometry (Th, U, K), X-ray spectrometry (REE, Zr), mass spectrometry (oxygen, carbon isotopic analysis), wet chemistry silicate analysis (petrogenic elements). The determinations has been made mainly in the laboratories of V.I. Vernadsky Institute of Geochemistry and Analytical Chemistry of U.S.S.R., Academy of Sciences, Moscow.

Iceland effusive rocks series (basalts–andesite–basalts–andesites–dacites–rhyodacites–rhyolites).

Geochemical history of minor rare elements is determined by the history of petrogenic elements primarily of Na, K, Ca, Mg, Fe and Si. Usually the content of rare lithophylic elements is found to increase with the increase of alkalies (Na and K) and SiO_2 in certain series of rocks (excluding K-alkaline one). The best correlation is found between the rare lithophylic elements and potassium. Abundances of titanium, phosphorous and other mafic typomorphic elements (V, Cr, Ni, Co and Cs) decreases with the depletion in calcium, magnesium and iron. The mafic typomorphic elements are concentrated predominantly in olivine and pyroxenes substituting isomorphously iron and magnesium.

The rocks of each effusive rock series are genetically interdependent. This feature is resulted in gradual increase or decrease of several element concentrations (Tables 1 and 2) from the initial to the end derivatives (from basalt to rhyolite or from melilitite to phonolite). The analogous regularities are displayed by chemical composition and crystallization temperature of phenocrysts (Polyakov *et al.*, 1976). For example, Iceland effusive rock series (basalts–andesites–rhyolites) is characterized by depletion in anorthite component of the plagioclases (70–80 An for basalt, 60–62 An for andesite = dacite, 37–38 An for rhyolite), by depletion of enstatite and enrichment in ferrosilite in clinopyroxenes (54 En and 6 Fs for basalt; 43 En and 14 Fs for andesite–dacite; 15 En and 45 for rhyolite) and by depletion in fayalite component (10–13 Fa for basalt, 84 Fa for rhyolite) with increase of SiO_2 content in the mentioned rocks.

The most probable crystallization temperature of plagioclase phenocrysts is about 1150–1250°C for basalts, 1000–1150°C for andesites and 920–980°C for dacites. Isotopic composition of helium (Mamyrin *et al.*, 1974), oxygen (Muehlenbachs, 1973), strontium and lead (Brooks and Jakobsson, 1974) reveals mantle source of melts, which give rise to Iceland

TABLE 2. ELEMENT CONTENT IN THE Na–ALKALINE ROCK SERIES OF THE EASTERN RIFT AND IN K–ALKALINE ROCK SERIES OF THE WESTERN RIFT OF EAST AFRICA

	Na–alkaline rocks				K–alkaline rocks
Element	Melilite	Melano-nephelinite	Nephelinite	Phonolite	
Na,%	2.04	3.82	6.43	6.53	1.03–2.49
K,%	1.54	2.03	3.39	3.86	3.26–7.59
Li ppm	17	21	26	20	8.5–11.3
Rb ppm	48	63	116	130	124–485
REE ppm	340	322	414	390	410–955
Zr ppm	285	270	710	860	320–410
Nb ppm	120	110	200	180	120–190
Th ppm	11.0	8.2	17.3	24.6	11.0–27.9
F ppm	1650	1200	2600	1530	800–1400
Ca,%	9.95	8.0	5.67	2.20	6.32–10.70
Mg,%	6.72	3.67	1.39	0.63	3.38–10.22
Fe,%	10.77	9.06	6.85	4.59	8.29–9.82
Ti,%	1.77	1.63	0.89	0.64	1.99–2.79
V ppm	212	250	140	42	201–290
Cr ppm	150	50	16	22	307–1180
Ni ppm	75	38	18	12	47–326
Sc ppm	23	16	3.5	3.5	20–45
P ppm	1900	1400	1450	1100	1200–2300

TABLE 3. RATIO VALUES OF COHERENT ELEMENTS IN ICELAND EFFUSIVE ROCKS SERIES

	Basalt	Andesite–basalt	Andesite	Dacite	Rhyolite–dacite	Rhyolite
Na/K	5.05	2.35	1.86	1.43	1.25	1.18
K/Rb	448	446	397	383	372	370
Ca/Sr	316	223	192	146	124	114
Sr/Y	7.8	5.0	3.0	2.2	1.08	0.73
Sr/Ba	1.90	0.72	0.53	0.29	0.29	0.11
Ti/Nb	851	290	156	67	32	22
Ni/Co	1.64	1.26	1.23	1.22	1.20	0.80
Ca/Mg	2.01	2.43	4.26	3.65	11.2	14.2
Fe/Mg	2.50	4.90	7.80	7.20	27.5	30.3
F/P	0.52	0.59	1.06	3.30	4.98	5.18

effusive rocks. High values of He^3/He^4 ratio for basalt (1×10^{-5}) and rhyolite–dacite (1.5×10^{-5}) are characteristic of basalt layer of the earth's crust. Iceland basalt are closely related to the basalts of the-Mid Atlantic ridge as concerning their value of Sr^{87}/Sr^{86} ratio (0.703040 for the former and 0.702650 for the latter) and average values of lead isotopic ratios (value of Pb^{206}/Pb^{204} ratio is equal to 18.82 and 18.672 correspondingly, that of Pb^{207}/Pb^{204} to 15.59 and 15.543, Pb^{208}/Pb^{206} to 38.68 and 38.307). The regularities of element distribution in both the effusive rock series of the East Africa rift zones (Na–alkaline and K–alkaline) are not the same. The latter in comparison with the Na–alkaline series (Table 2) is characterized by considerable predominance of potassium over sodium (Na/K = 0.20–0.44), high concentration of a number of mafic typomorphic elements (Ca, Mg, Fe, V, Cr, Ni, Co, Sc, Ti, P), some elements (Cr, Ni, Sr, Ti) being more abundant than in alkaline basalts of the East Africa rift zones (Table 1). At the same time the rocks of K–alkaline series are enriched in many rare lithophylic elements: Pb (124–385 ppm), Ba (1300–2380 ppm), REE (410–955 ppm), Nb (120–190 ppm), Ta (9.2–18 ppm) and Th (11–27.9 ppm).

It is difficult to explain the genesis of the potassium alkaline rocks of the Western rift by means of assimilation hypothesis taking into account the concentration of mafic typomorphic elements in these rocks. Assimilation of sialic material of the earth's crust process by magmatic melt would rather promote depletion in these rock typomorphic elements. The effusive rocks of the K–alkaline series are presumably originated from the mantle melts initially enriched in potassium.

It is also worth noting the homogenization temperature of phenocryst inclusions from the rocks of the K–alkaline series varies about 1080–1400°C. This is also a contradiction with the assimilation hypothesis because homogenization temperatures of phenocryst inclusions from other rock types (Na–alkaline series and alkaline basalts) is somewhat lower (Naumov *et al.*, 1972).

These data support the assumption that the primary melts of the K–alkaline rock series are considered as the derivatives of the initial (juvenile) ultrabasic potassium alkaline magma.

Alkaline rock varieties in comparison with less alkaline varieties are enriched in rare lithophylic elements, fluorine and some basic rock typomorphic elements: titanium, strontium and phosphorous and are depleted in mafic typomorphic elements (excluding Ti, Sr and P). The data concerned with the concentration of some elements in Iceland basalts: toleiitic varieties of the Terstareykyre area (northern part of the rift zone) (1) and alkaline varieties from the island Heimaey (2) are predicted:

	Na	K	Cr	Ni	Sc	Ti	P
1	1.44%	0.09%	413 ppm	133 ppm	36 ppm	0.54%	0.03%
2	3.59	0.85	29	20.4	15.5	1.87	0.176

	Sr	Ba	Y	Nb	Zr	Be	F
1	65 ppm	30 ppm	16 ppm	3.1 ppm	58 ppm	1.1 ppm	100 ppm
2	323	625	44	30	287	5.9	720

Alkaline basalts of the East Africa (Eastern rift) are also enriched in rare lithophylic elements: Ba (440 ppm), REE (185 ppm), Zr (300 ppm), Nb (67 ppm), Ta (3.2 ppm), in F (940 ppm) and some mafic typomorphic elements: Ti (1.55%), Sr (735 ppm) and P (2100 ppm) (Gerasimovsky and Polyakov, 1974).

The carbonatites of the East Africa rift zones are very different in their geochemical features. Effusives carbonatites of the Eastern rift zone are subdivided into sodium–calcium and calcium ones (Dawson, 1966). The latter concerning their composition are analogous to the intrusive carbonatites of the same zone. Lavas of carbonate–silicate composition are found at the Western rift zone (Knorring and Bois, 1961). The carbonatites of the Eastern rift zone are spatially related to the Na–alkaline series of effusive rocks whereas the carbonatites of the Western zone to the K–alkaline effusive series.

Typomorphic elements of the carbonatites are the following: calcium, carbon, fluorine, phosphorus, strontium, barium, rare-earth elements, titanium, zirconium and niobium. Many of them (Cl, F, P and TR) have the same geochemical history as calcium being the components of the same minerals (calcite, apatite).

Data on isotopic composition of strontium (values of Sr^{87}/Sr^{86} ratio are about 0.7034 for carbonatite lavas; Bell and Powell, 1970), carbon and oxygen (Vinogradov *et al.*, 1970, 1971), homogenization temperature of inclusions in carbonatite calcite (640–690°C; Romanchev, 1972) and geological observations (repeated sequence of eruptions of carbonatite lava) reveal juvenile (mantle) source of carbonatite matter.

According to data of Romanchev (1971) three-phase equilibrium: two liquids and gas was found in nepheline inclusions from effusive rocks (nephelinites). During cooling of one of the liquids (carbonate melt) calcite crystallized and another (silicate melt) was quenched into glass. These investigations imply that separation of carbonatite melt from the alkaline silicate one is due to liquation.

Data on isotopic composition, geochemical and thermobarometric investigations showed that formation of the above-mentioned rock series could be considered as the result of differentiation of separate initial melts, which were originated within the upper mantle during riftogenesis. Melt composition was determined by the depth and relative melt volumes of upper mantle material. Differentiation of the initial melts occurred in the intermediate chambers.

The deposits of rare and ore-forming elements are absent in the effusive rocks of the rift zones. The deposits of pyrochlore and apatite are related to the carbonatites of the Eastern rift zone of East Africa.

References

Bailey, J. C. and Noe-Nygaard, A. (1976) Chemistry of Miocene plume tholeiites from northwest Iceland. *Lithos* **9**, No. 3, 185–201.

Bell, K. and Powell, J. L. (1974) Strontium isotopic studies of alkalic rocks: the alkalic complexes of Eastern Uganda. *Bull. Geol. Soc. Amer.* **81**, 3481–3490.

BROOKS, C. K. and JAKOBSSON, S. (1974) Petrochemistry of the volcanic rocks of the North Atlantic Ridge. In *Geodynamics of Iceland and North Atlantic Area,* L. KRISTIANSSON, ed.

DAWSÒN, J. B. (1966) Oldonye Lengai, an active volcano with sodium carbonatite lava flows. In *Carbonatites,* O. F. TUTTLE and J. GITTINS, eds., 155–168.

GERASIMOVSKY, V. I. and POLYAKOV, A. I. (1974) Geochemistry of rift zone volcanic rocks of East Africa. *East Africa Rift System,* Vol. III, Moscow, Izd-vo Nauka, 5–194.

GERASIMOVSKY, V. I. and POLYAKOV, A. I. (1977) Geochemistry and genesis of volcanic rocks series of East Africa and Iceland rift zones. In *The Main Problem of Riftogenesis,* N. A. LOGATCHEV, ed., Novosibirsk, Izd-vo Nauka, 149–155.

GERASIMOVSKY, V. I., POLYAKOV, A. I., KRIGMAN, L. V. and NESMEYANOVA, L. I. (1974) Alkali–silica relation in Iceland rocks. *Geockimiya* No. 7, 978–986.

KNORRING, O. VON and DU BOIS, C. G. B. (1961) Carbonatitic lava from Fort Portalarea in Western Uganda. *Nature* **192** No. 4807, 1064–1065.

MAMYRIN, B. A., GERASIMOVSKY, V. I. and KHABARJN, L. V. (1974) Helium isotopes in the rift zone rocks of East Africa and Iceland. *Geockimiya* No. 5, 663–670.

MUEHLENBACHS, K. (1973) The oxygen isotope geochemistry of siliceous volcanic rocks from Iceland. *Carnegie Inst. Year Book* **72**, 593–597.

NAUMOV, V. B., POLYAKOV, A. I. and ROMANCHEV, B. P. (1972) Crystallization conditions of volcanic rocks of East Africa zones according to data of thermobarometric investigations. *Geockimiya* No. 6, 663–668.

PALMASON, G. and SAEMUNDSSON, K. (1974) Iceland in relation to the Mid-Atlantic ridge. *Ann. Rev. Earth Planet. Sci.* **2**, 25–50.

POLYAKOV, A. I., 'ILYIN, N. P. and MURAVIOVA, N. S. (1976) Crystallization conditions of Iceland rocks of rhyolite–basalt association (according to data on composition of phenocryst minerals and distribution coefficients). *Geockimiya* No. 7, 963–982.

ROMANCHEV, B. P. (1972) Rock formation condition of some carbonatite complexes of East Africa according to data on inclusion thermometry. *Geockimiya* No. 2, 172–179.

SIGVALDASON, G. E. (1974) Basalt from the centre of assumed Icelandic mantle plume. *J. Petrology* **15**, No. 3, 497–524.

VINOGRADOV, A. P., DONTSOVA, E. I., GERASIMOVSKY, V. I. and KUZNETSOVA, L. D. (1971) Oxygen isotopic composition of rift zone carbonatites of East Africa. *Geockimiya* No. 5, 507–514.

VINOGRADOV, A. P., KROPOTOVA, O. I. and GERASIMOVSKY, V. I. (1970) Isotopic composition of carbonatite carbon of East Africa. *Geockimiya* No. 6, 643–646.

WOOD, D. A., GIBSON, J. L. and THOMPSON, R. N. (1976) Elemental mobility during zeolite facies metamorphism of the tertiary basalts of Eastern Iceland. *Contr. Mineral. Petrol.* **55**, No. 3, 241–254.

Review of the Geochemistry of Indian and other Oceanic Rocks†

K. V. SUBBARAO,‡ D. R. C. KEMPE,§ V. V. REDDY, § G. R. REDDY†† and R. HEKINIAN‡‡

‡ Indian Institute of Technology, Powai, Bombay-400 076, India.
§ British Museum (Natural History), Cromwell Road, London SW7 5BD, England.
†† Bhabha Atomic Research Center, Trombay, Bombay-400 073, India.
‡‡ Centre Océanologique de Bretagne, 29.273, Brest, France.

Abstract

A review is attempted of the geochemistry of dredged and drilled (DSDP) volcanic rocks from different tectonic environments in the Indian Ocean. These basalts range from quartz tholeiites to nepheline–normative alkali basals with the main concentration lying in the olivine tholeiite field of the Di–Hy–Ol–Qz–Ne tetrahedron. In general, basalts from the spreading ridges are depleted in large ion lithophile (LIL) and rare earth elements (REE). However, between the three main oceanic ridges, the Mid-Indian Oceanic Ridge (MIOR) basalts are enriched in LIL elements and $^{87}Sr/^{86}Sr$ ratios, suggesting that the source of the MIOR was depleted at a much later time and also to a lesser degree than the sources of the Mid-Atlantic Ridge (MAR) and the East Pacific Rise (EPR). Geochemical differences exist between the MIOR and the Carlsberg Ridge (CR). In contrast to the tholeiitic volcanism encountered on the spreading MIOR, the aseismic Ninetyeast Ridge shows alkali volcanism and a wide range of differentiation, illustrated by the presence of pyroxene basalts, picritic basalts, and oceanic andesites, and characterized by the presence of higher Fe_2O_3+FeO ($>11\%$), TiO_2(2–3%), K_2O (2–3%), LIL and light rare-earth elements (LREE), and $^{87}Sr/^{86}Sr$ ratios than are found in mid-oceanic ridge basalt (MORB). The aseismic Ninetyeast Ridge shows geochemical similarities with some of the nearby plateau and islands. It is possible that the Ninetyeast Ridge, Broken Ridge, Naturaliste Plateau, Kergulenen Plateau, and St. Paul–Amsterdam Islands were originally joined together, forming a single ridge system produced from two fixed hot spots. The distribution of $^{87}Sr/^{86}Sr$ ratios for the volcanic rocks from different tectonic environments in the Indian Ocean supports a "zoning-depletion model" with systematically arranged alkali-poor (spreading ridges) and alkali-rich (aseismic ridges, seamounts, islands, and abyssal hills) zones in the mantle beneath the Indian Ocean. Out of the twenty DSDP sites from the Indian Ocean which are included in this review, eight sites with enriched LIL concentrations and LREE patterns may derive from hot spot activity, nine sites with depleted patterns probably formed from spreading mid-ocean ridge sea-floor, and the remaining three sites appear to have been formed from both hot spot and spreading activity. Inconsistencies between LIL, including REE, geochemistry and tectonic models, exist in about a third of the DSDP sites from the Indian Ocean.

INTRODUCTION

In this paper we attempt a review of the existing geochemical data on volcanic rocks dredged and drilled (from Deep Sea Drilling Project (DSDP) sites) from the Indian Ocean, from different tectonic environments. Some new data are also reported. The major objectives are (1) to present major, REE and selected LIL elements and Sr isotopic compositions of volcanic rocks from the spreading Mid-Indian Oceanic Ridge as well as the Carlsberg Ridge,

† Contribution no. 557 of the Scientific Department, Centre Océanologique de Bretagne.

the aseismic Ninetyeast Ridge, and possible hot spot sites, as well as ocean basins from the eastern and western Indian Ocean; (2) to examine relationships between tectonic setting and geochemistry of the Indian oceanic volcanic rocks; and (3) to compare the geochemistry of possible hot spots from the Indan Ocean with those of the Atlantic and Pacific oceans.

The Spreading Mid-Indian Oceanic Ridge and the Aseismic Ninetyeast Ridge

Major Elements

The MIOR has the form of an inverted "Y"-shaped structure in the centre of the Indian Ocean (Fig. 1); its northern extension is the CR. Tholeiitic basalts predominate on these ridges, and their petrology and chemistry have been described by Balashov *et al.*, 1970; Bezrukov *et al.*, 1966; Cann and Vine, 1966; Dmitriev, 1974; Engel, 1965; Engel and Fisher, 1975; Hekinian, 1968; Kempe, 1974; Wiseman and Poole, 1937. However, various types of basaltic rocks, spillites, gabbros, anorthosites, serpentinites, and dunites have also been reported from the fracture zones intersecting the MIOR by Reinisch, 1908; Korzhinskiy, 1962; Gladkikh and Chernysheva, 1966; Heezen and Nafe, 1964; Udintsev and Chernysheva, 1965; Chernysheva and Bezrukov, 1966; Cann and Vine, 1966; Hekinian, 1968, 1970; Engel and Fisher, 1969, 1975.

Normatively, when plotted in Yoder and Tilley's (1962) Di–Qz–Ol–Ne tetrahedron, ocean-floor basalts display variation from quartz normative tholeiites to nepheline normative alkali basalts, but mostly cluster in the olivine tholeiite field (Figs. 2, 3 and 4). The average major element compositions of ocean floor basalts are presented in Table 1; minor

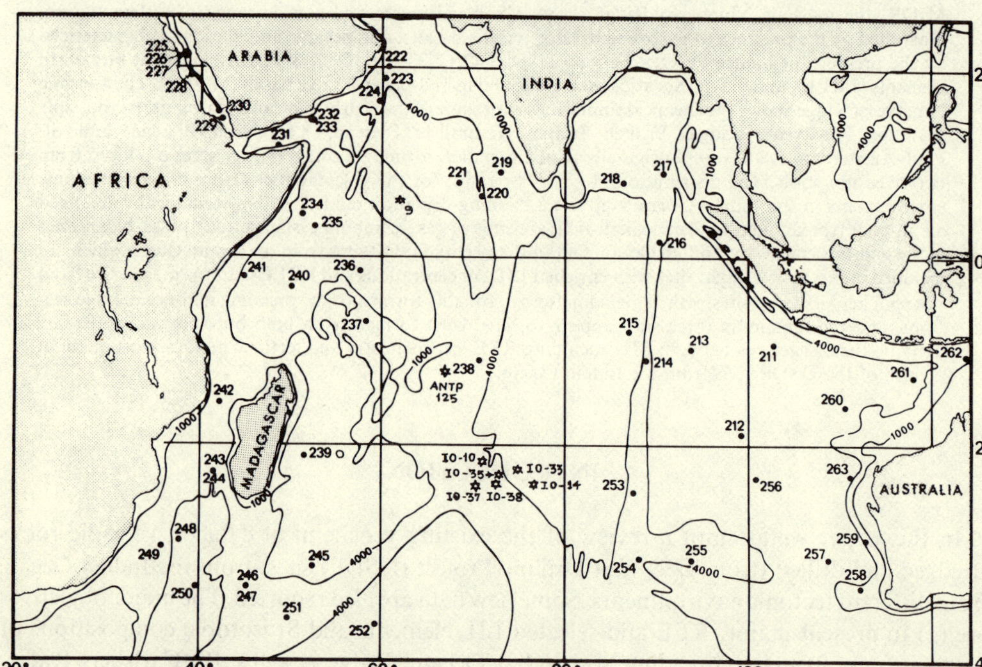

FIG. 1. Location of dredged and drilled (DSDP) samples from the Indian Ocean.

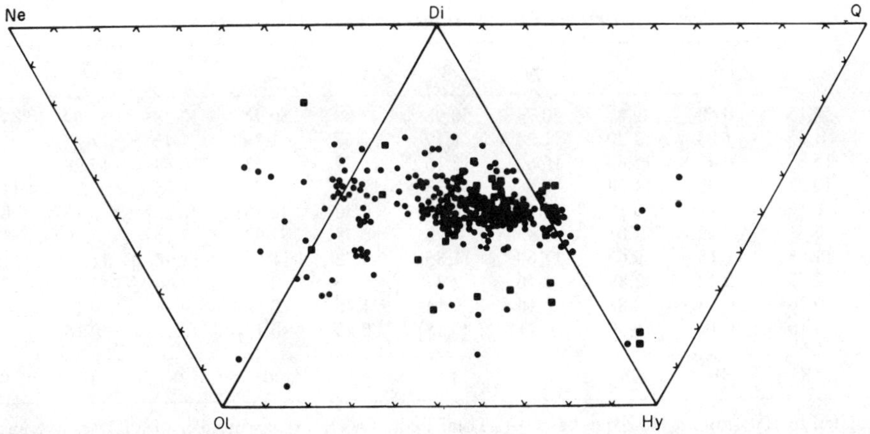

FIG. 2. Normative plot (Yoder and Tilley, 1962; normalized to $Fe_2O_3 = 1.50\%$, *cc*-free) of basaltic rocks from the Atlantic Ocean. Circles, dredged rocks; squares, drilled. (Data sources as in Kempe, 1975; up to late 1974, i.e. excluding FAMOUS and DSDP/IPOD Leg 37 and onwards.)

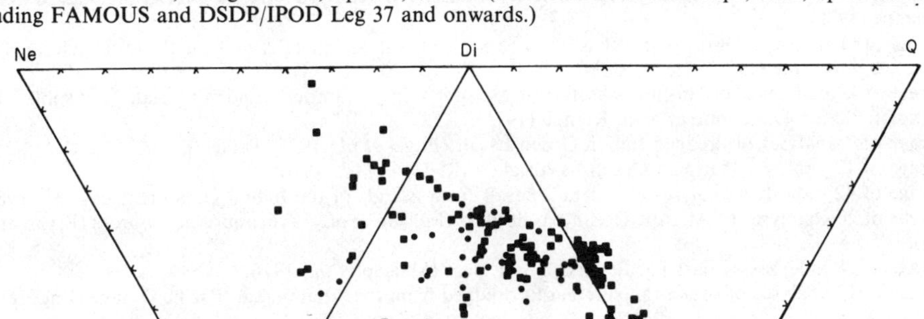

FIG. 3. Normative plot of basaltic rocks from the Indian Ocean. As in Fig. 2 (Data sources as in Kempe 1975; up to late 1974.)

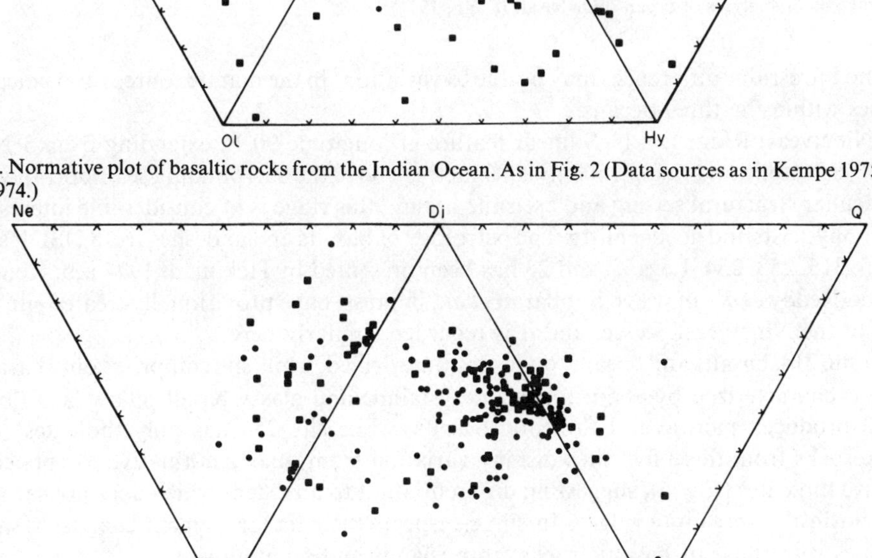

FIG. 4. Normative plot of basaltic rocks from the Pacific Ocean. As in Fig. 2. (Data source as in Kempe, 1975; up to late 1974.)

TABLE 1. AVERAGE COMPOSITIONS OF OCEAN-FLOOR BASALTS

	1	2	3	4	5	6	7	8	9	10
SiO_2	50.15	50.58	49.87	50.74	50.98	47.90	50.74	50.28	50.04	49.11
TiO_2	1.35	1.03	2.20	1.24	1.19	2.67	1.36	1.77	1.51	1.42
Al_2O_3	15.57	16.49	15.69	16.62	15.16	15.79	15.53	14.88	17.28	15.85
FeO*	10.28	9.32	11.34	8.58	10.33	11.97	9.76	11.35	8.73	11.38
MnO	0.16	0.19	0.19	0.16	[0.18]	0.30	[0.18]	[0.18]	0.17	0.18
MgO	7.51	7.89	7.01	8.03	7.70	6.78	7.92	7.11	7.29	7.76
CaO	11.85	11.13	9.63	11.60	11.85	9.72	11.63	11.46	11.88	11.21
Na_2O	2.77	2.42	2.89	2.76	2.32	3.06	2.57	2.66	2.78	2.73
K_2O	0.20	0.83	0.88	0.16	0.14	1.36	0.16	0.16	0.16	0.22
P_2O_5	0.16	0.12	0.30	0.11	[0.15]	0.45	[0.15]	[0.15]	0.16	0.14
n	68	51	48	7	12	42	206	38	10	94

Recalculated to 100% on an H_2O-free basis. Total Fe as FeO*. *n* = number of analyses averaged.

1. Average of 68 analyses of spreading ridge ocean-floor basalt (MORB) from the Indian Ocean. Data sources as in Kempe (1975).
2. Average of 51 analyses of intra-plate ("basin") ocean-floor basalt and metabasalt from the Indian Ocean. Data sources as in Kempe (1975).
3. Average of 48 analyses of ocean-floor basalt from aseismic ridges, seamounts and intra-plate "hot spots" from the Indian Ocean. Data sources as in Kempe (1975).
4. Average of 7 analyses of dredged Indian Ocean basalt (Engel *et al.*, 1975, Table 7).
5. Average of 12 analyses of Indian Ocean basalt glass (Melson *et al.*, 1976).
6. Average of 42 published analyses of "alkali" basalt from islands of the Indian Ocean (Engel *et al.*, 1965).
7. Average of 206 analyses of Atlantic Ocean basalt glass (Melson *et al.*, 1976), and other sources (Bryan *et al.*, 1976).
8. Average of 38 analyses of East Pacific Rise basalt glass (Melson *et al.*, 1976).
9. Average of 10 analyses of ocean tholeiite basalt dredged from the Atlantic and Pacific Oceans (Engel *et al.*, 1965).
10. Average of 94 analyses of ocean-floor basalt (Cann, 1971).

inter- and intra-ridge differences may be due to variations in the mantle sources and volcanic processes within the three oceans.

The Ninetyeast Ridge is a N–S linear feature at longitude 90°E, extending from 5°N to about 30°S latitude and rising 1500 to 2000 metres above the surrounding sea floor. In view of its peculiar structural setting and aseismic nature, this ridge is of considerable interest to both geophysicists and geochemists. The petrology of basalts and andesites from DSDP sites 214, 215, 216, 253, 254, Legs 22 and 26 has been presented by Hekinian, 1974 a,b; Kempe, 1974; and Reddy *et al.*, in press. Subbarao *et al.*, in press, have provisionally treated site 215 as part of the Ninetyeast Ridge, and it is regarded similarly here.

From site 214, basalts and oceanic andesites were drilled, while site comprises only basalts; site 215 is characterized by about fourteen crystalline and glassy basalt pillow lava flows. Site 253 produced picritic and tholeiitic basalts, while site 254 has only tholeiites. The volcanic rocks from these five sites display variation from quartz normative to nepheline normative tholeiites (Fig. 5), suggesting differentiation to an extent which does not seem to be present in the spreading ridges. In some respects their major element chemistry show similarities with those of basaltic rocks from islands and seamounts.

Kempe (1975) showed that there are significant systematic differences between the normative (normalized to $Fe_2O_3 = 1.50\%$, cc free) mineralogies of drilled and dredged volcanic rocks from the Indian, Pacific and Atlantic Oceans, dredged basalts peaking at $di_{47}hy_{36}ol_{17}$ and drilled basalts at $di_{53}hy_{43}ol_4$. Amongst the drilled rocks, the Indian Ocean

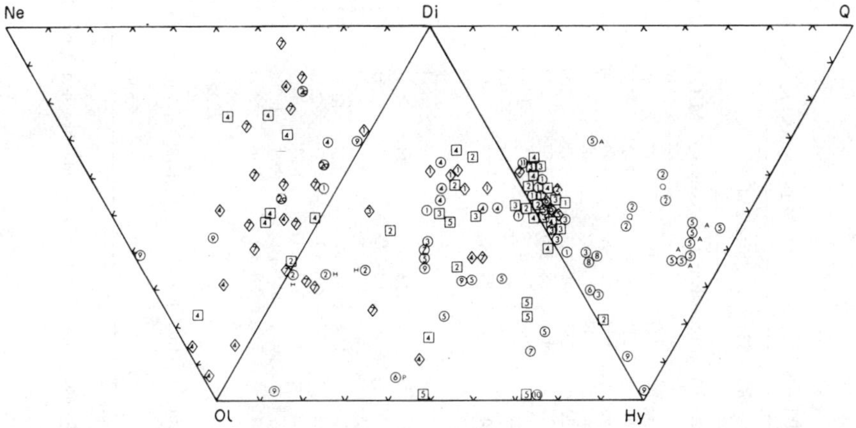

FIG. 5. Normative (normalized) plot (Yoder and Tilley, 1962) of basaltic rocks from aseismic ridges, seamounts, and intra-plate "hot spots" from the Atlantic, Indian and Pacific Oceans. Key is given below. (Data sources as in Kempe, 1975; Dickey *et al.*, 1974; Frey *et al.*, 1974; Marshall, 1975; Kharin, 1976; Ridley *et al.*, 1976; Jackson *et al.*, 1976; Schilling *et al.*, 1976; Bonatti *et al.*, 1977; Foder *et al.*, 1977.)

	Aseismic ridges	Seamounts	Intra-plate "hot spots"
Atlantic Ocean	Iceland–Faroes Ridge, including site 336 ① Walvis Ridge: quartz-enriched rocks ②Q hematite-enriched rocks ②H Rio Grande Rise (2a)	Discovery Tablemount [1] near Bouvet Island [2]	DSDP site 10 ◇1
Indian Ocean	Ninetyeast Ridge, DSDP site 216 ③ site 215 ④ site 214 ⑤ site 214 andesites ⑤A site 253 ⑥ site 253 picrite ⑥P site 254 ⑦ Mozambique Ridge, site 249 ⑧		DSDP site 248 ◇2 DSDP site 250 ◇3 DSDP site 211 ◇4 DSDP site 256 ◇5 DSDP site 257, core 11 ◇6
Pacific Ocean	Cocos Ridge, including site 158 ⑨ Coiba Ridge, site 155 ⑩ Carnegie Ridge, site 157 ⑪	Galapagos centre [3] Abyssal hills and seamounts in the eastern and central Pacific Ocean, including sites 313 and 315A [4] Easter line [5]	DSDP sites 165A; 167; 169, core 5; 170; and 171 ◇7

appears to have a high incidence of quartz tholeiites in the Ninetyeast Ridge and the Wharton Basin, and of *ne*-normative basalts in the north-eastern region. In contrast, the basalts from the southern Indian Ocean are the more normal oceanic (olivine) tholeiites (MORB). In the Pacific Ocean rocks from the eastern Pacific region are generally MORB tholeiites, while those from the central region include a high proportion of *ne*-normative alkali basalt; tholeiitic MORB predominates in the Atlantic Ocean (Figs. 2 to 5).

Table 2. Ranges and Averages of Trace Element Analyses of Volcanic Rocks from the Ninetyeast Ridge (DSDP Legs 22 and 26) and Mid-Indian Oceanic Ridges

	Site 214 basalts		Site 214 andesites		Site 215		Site 216		Site 253		Site 254			
	Range	Average	Range	Average	Range	Average	Range	Average	Range	Average	Range	Average	MIOR	CR
Sc		45	15–17	16.5					46–55	51	39–45	42	49	
V	192–680	476	22–70	41	245–310	262	249–470	419	165–255	215	240–400	342	250	
Cr	30–40	37	5–10	<5	240–325	271	20–165	42	325–890	630	83–690	469	360	315
Co	49–78	65	25–41	37	43–52	48	50–62	52	42–56	49	26–65	46	35	75
Ni	33–60	47	5–15	<5	95–120	104	20–65	40	205–520	297	98–265	193	109	226
Cu	50–185	140	1–5	2.42	54–70	60	55–210	89	12–205	116	38–200	139	93	95
K	1490–6720	2488			6640–6970	6847	4730–10,700	6983	1570–3320	2596	1660–6140	2675	1700	1730
Sr	150–370	265	580–1500	789	330–515	375	115–610	235	30–77	52	62–185	114	108	142
Ba	22–70	45	530–780	612	405–495	437	95–230	140	6–30	14	29–95	56	17	
Rb			35.1–48.7	41.9	19.5–26.8	23				5			2.54	2.1
Cs			0.55–0.79	0.67	0.16–0.42	0.29							0.128	0.018
U	50–185	140	0.4–1.0	0.68	0.29–0.4	0.30	0.1–1.0	0.31					0.51	
Y	13–40	26	62–85	68.66	35–44	38	27–37	30			35–60	49	35	48.75
Zr	95–160	121	235–380	273	155–175	163	150–180	159			80–195	155	71	133.75
La	6.6–31.1	12.28	30.8–38.7	34.38	14.7–17.8	16.3	13.2–16.6	14.86	2.3–9.2	4.6	7.2–12.0	9.9	4.4	2.94
Ce	12.5–81.8	34.76	64.2–85	73.4	30–61.1	37.7	30.2–30.8	30.50	6.2–22	11.6	17–30	25	13.84	13.13
Sm	2.72–5.1	4.73	10.4–11.8	11.1	3.94–5.15	4.5	5.0–5.72	5.39	1.66–3.2	2.32	4.1–5.3	4.77	2.66	3.92
Eu	1.66–1.8	1.55	3.1–3.35	3.10	1.42–1.79	1.57	1.45–1.62	1.55	1.06–1.2	0.88	1.2–1.87	1.63	1.19	1.39
Tb	0.76–1.15	0.83	1.43–2.0	1.77	0.69–1.19	0.89	0.91–1.13	1.02	0.39–0.78	0.57	0.81–1.1	0.93	0.782	1.03
Yb	1.76–3.66	2.60	3.66–5.15	4.6	2.5–3.0	2.73	3.6–3.85	3.69	2.3–3.1	2.66	3.0–3.9	3.34	2.44	3.88
K/Rb	561–670	630	266–369	317	256–341	298	308–420	364		664			831	938
K/Cs	3908–7263	6044	16,390–23,542	19,966	16,586–41,500	29,043	23,304–52,992	38,148					20,882	62,000
$^{86}Sr/^{87}Sr$	.7043–.7048	.7045		0.7049		.7045		.7055					.7034	.7028
$[La/Sm]_{e.f.}$	0.73–2.78	1.36	1.54–1.93	1.69	1.86–2.16	1.97	1.41–1.66	1.50	0.58–1.61	1.03	0.95–1.51	1.13	1.06	0.52
References	1,2		1,2		1,2,3,4		1,2		4,5		4,6		7,8,9	7,10,11

MIOR: Mid-Indian Oceanic Ridge, CR: Carlsberg Ridge

1. Thompson *et al.*, 1974; 2. Reddy *et al.*, 1977; 3. Bougault, 1974; 4. Frey *et al.*, 1977; 5. Kempe, 1974; 6. Frey and Sung, 1974; 7. Subbarao *et al.*, 1977; 8. Reddy, unpublished data; 9. Engel and Fisher, 1975; 10. Cann, 1969, 11. Schilling, 1971.

Large Ion Lithophile Elements

In order to illustrate the chemical variations between and within spreading and aseismic ridge volcanic rocks we have compiled (in Table 2 and Fig. 6) published and unpublished data for certain LIL elements.

The variation in LIL elements (K, RB, Sr, Cs, U) in rocks from the MIOR, CR, Mid-Atlantic Ridge (MAR), East Pacific Rise (EPR) and Ninetyeast Ridge are shown in Fig. 6. The MIOR rocks are enriched in LIL elements relative to MAR and EPR; further, the chemical differences between these three ridges are apparent in the K/Cs–K, K/Rb–K and K/U-K variation diagrams (Figs. 7, 8 and 9). The rocks from MAR and EPR have higher K/Cs and K/Rb ratios and lower K concentrations than the MIOR rocks. These data suggest that the MIOR rocks are less depleted in LIL elements than those from the MAR and EPR

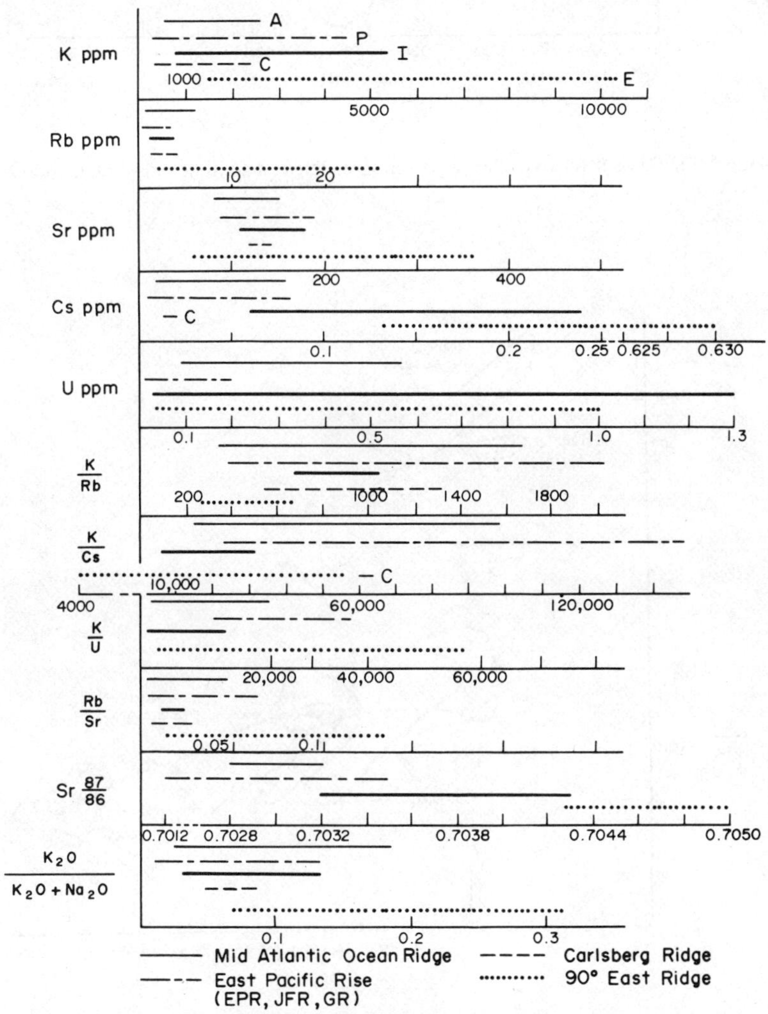

FIG. 6. Variation of LIL elements their corresponding ratios and $^{87}Sr/^{86}Sr$ for spreading ridges (MIOR, MAR, EPR) and the aseismic Ninetyeast Ridge. Data courses as in Subbarao *et al.* (in press.)

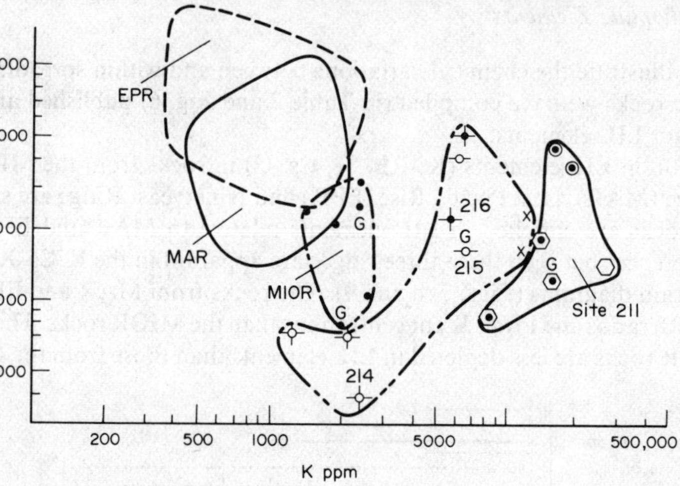

FIG. 7. Variation of K/Cs vs. K for volcanic rocks from spreading and aseismic ridges. Data sources as in Subbarao *et al.* (in press.)

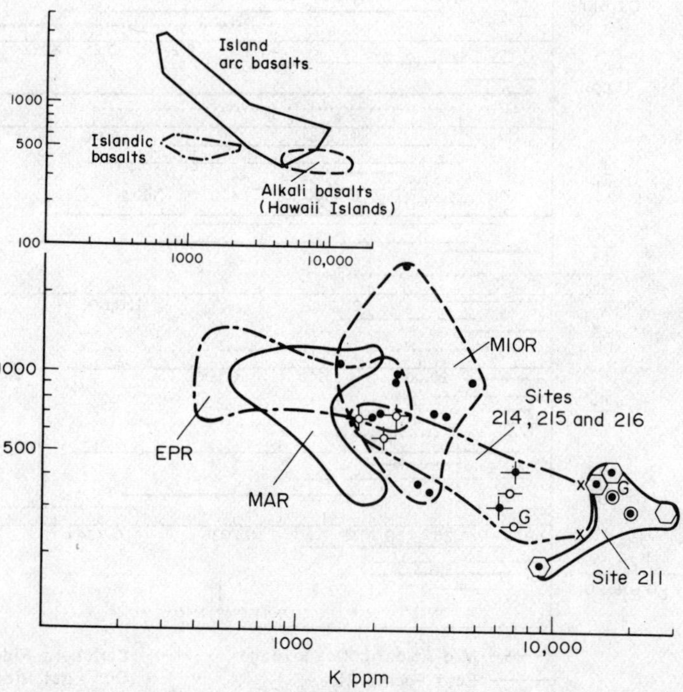

FIG. 8. Variation of K/Rb vs. K for volcanic rocks from spreading and aseismic ridges, island arcs, and islands. Data sources as in Reddy *et al.* (in press.)

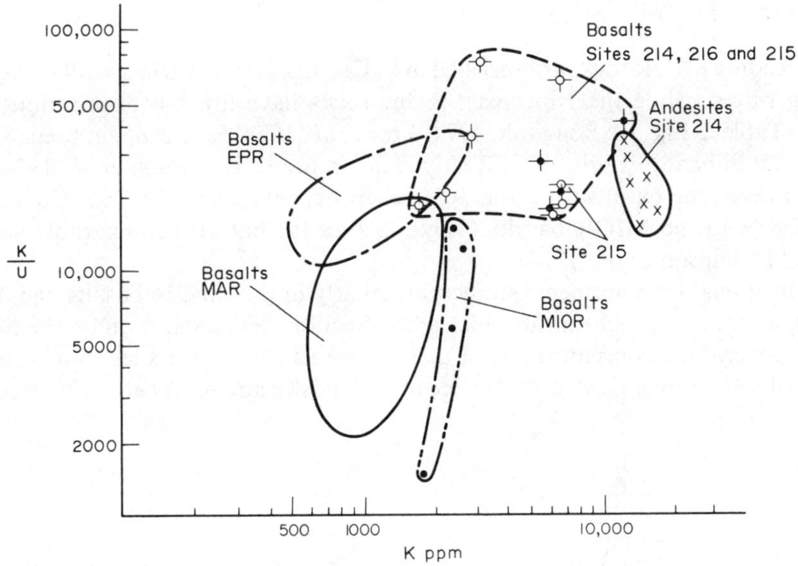

FIG. 9. Variation of K/U vs. K for volcanic rocks from spreading and aseismic ridges. Data sources as in Subbarao *et al.* (in press.)

and it is likely that the underlying mantle might have been subjected to different degrees of depletion, probably at different times in the geological past (Subbarao *et al.*, in press; Reddy *et al.*, in press; Reddy, V. V., unpublished data).

The basalts and andesites from the Ninetyeast Ridge are highly enriched in LIL elements relative to spreading ridges (Table 2, Fig. 6) and the chemical identity of the ridge is well retained in both the elemental concentrations and various ratios. For example, these Ninetyeast Ridge rocks, while following the general trend of decreasing K/Rb with increasing K, do in fact have lower K/Rb ratios, ranging from 250 to 700, than most of the spreading ridge rocks (Table 2, Fig. 8). The Ninetyeast Ridge samples from sites 214, 216 and 215 show two geochemical groupings, with K/Rb ratios around 250 and 700. The lower ratios are similar to those of the alkali island basalts, while the higher ratios overlap the fields of the EPR and MIOR. We interpret these relations to mean a heterogeneous mantle source(s)—perhaps coupled with varying degrees of partial melting.

The basalts and andesites from the Ninetyeast Ridge are also enriched in K, Ba, Sr and U relative to spreading ridges, to a degree similar to alkali island basalts (Table 2, Fig. 6). The K/Cs–K and K/U–K relationships suggest again that differentiation in the Ninetyeast Ridge has progressed further than in the spreading ridges, illustrated by the presence of a well-differentiated sequence of picritic basalts, basalts and andesites in the Ninetyeast Ridge.

We believe that the geochemical trends displayed by the K/Rb–K ratios are more reliable than K/Cs–K, as Cs is more mobile and sensitive to sea water alteration than Rb. Therefore we have relied more on the K/Rb/K variations in speculating on the nature of the source areas. However, we have used the Cs concentrations and K/Cs ratios of fresh crystalline or glassy basalts when drawing some critical conclusions on the source rock chemistry.

Transitional Metal Trace Elements

The spreading MIOR rocks are enriched in V, CR, Co, Ni and Cu in a similar way to other spreading ridges, while the Ninetyeast Ridge rocks have lower concentrations of these elements (Table 2, Fig. 10; Bougault, 1974; Frey *et al.*, 1974; Frey *et al.*, in press; Engel and Fisher, 1975; Subbarao *et al.*, 1976; Reddy *et al.*, in press; Thompson *et al.*, 1974; Kempe, 1974). However, the basalts from site 215 are strongly enriched in Ni, Cr, Cu and V, to a similar degree to the MIOR basalts (Table 2, Fig. 10) but also comparable with island tholeiites (Thompson *et al.*, 1974).

The transitional trace-elements study, particularly in the site 214 basalts and andesites, provides an excellent geochemical evidence for fractional crystallization on the Ninetyeast Ridge. In general the concentrations of these elements from other sites show similarity to alkali basalts and tholeiitic suites from oceanic islands (Ramanan *et al.*, 1976; Reddy *et al.*, 1977).

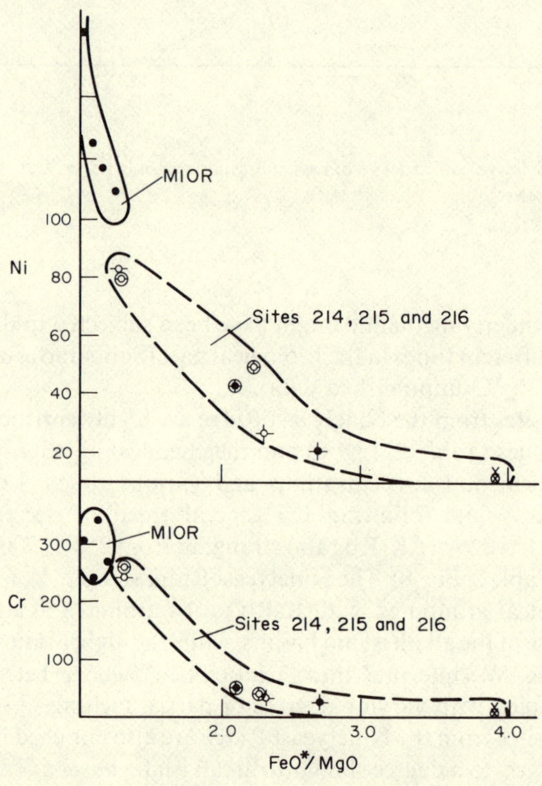

FIG. 10. FeO*/MgO vs. Ni and Cr variation diagram for volcanic rocks from the MIOR and Ninetyeast Ridge. Data sources as in Reddy *et al.* (in press).

Rare Earth Elements

The REE contents of the spreading MIOR and CR display depleted chondrite normalized patterns similar to other spreading ridges like the MAR and EPR (Fig. 11; Frey and Haskin,

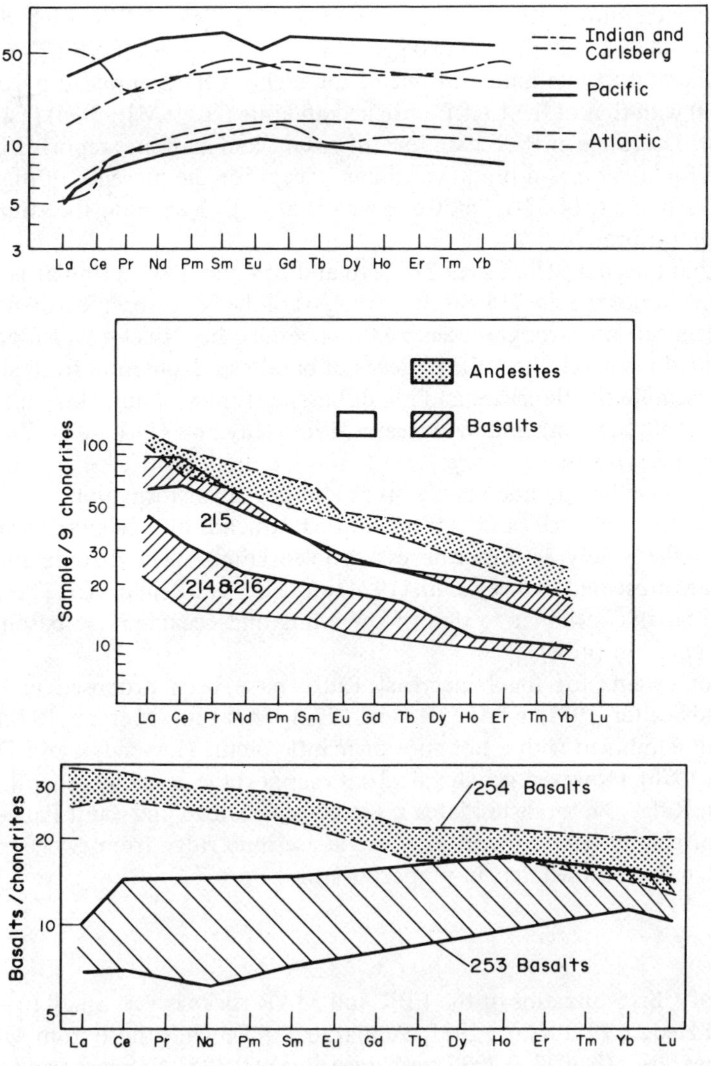

FIG. 11. REE distribution patterns for MIOR, CR, MAR, EPR and Ninetyeast Ridge (sites 214, 215, 216, 253, 254). Data sources: as in Reddy *et al.* (in press); Frey *et al.* (in press); Frey and Sung (1974); Thompson *et al.* (1974).

1964; Kay *et al.*, 1970; Schilling, 1971; Subbarao *et al.*, 1975; Fleet *et al.*, 1976; Reddy, V.V., unpublished data). However, the MIOR rocks have higher $[La/Sm]_{e.f.}$ ratios (about 1.03) than the CR rocks (0.52) (Schilling, 1971; Reddy *et al.*, in press) and these variations may reflect the chemical differences within the mantle source areas of the two ridges.

By contrast, the Ninetyeast Ridge (sites 214, 215, 216, 254) basalts and andesites show marked enrichment in light rare-earth elements (LREE) (Frey and Sung, 1974; Frey *et al.*, in press; Subbarao *et al.*, 1975, in press; Reddy *et al.*, in press, Reddy, V .V., unpublished data) similar to that in island basalts (Fig. 11), while site 253 basalts are depleted in LIL and LREE and show characters similar to spreading ridges (Fig. 11). The LREE-enriched patterns for site 214 basalts and andesites display two geochemical groups with $[La/Sm]_{e.f.}$ ratios varying from 0.97 to 2.8 and a close match of one of the highly LREE-enriched pyroxene basalts ($[La/Sm]_{e.f.}$ 2.8) with that of the LREE-enriched andesite ($[La/SM]_{e.f.}$ 1.94) (Table 2, Reddy *et al.*, in press). The range of REE data for 214 basalts and andesites reported by Frey *et al.* (in press) and Reddy *et al.* (in press) is similar, except for the presence of highly enriched LREE pyroxene basalt (214-53-1, 26-30 cm) with $[La/Sm]_{e.f.}$ 2.8 among the samples reported by Reddy *et al.* (in press).

It is likely that the rocks from sites 214, 216 and 254 may have a similar type of mantle source. Whether or not the site 215 basalts form part of the Ninetyeast Ridge, it is significant that the geochemical data strongly resemble those of the other Ninetyeast Ridge sites as well as alkali basalts. In general, the REE patterns of basalts and andesites from sites 214, 216, 254 and 215 resemble the tholeiites and alkali basalts from St. Paul, Hawaii and Reunion Islands (Gast, 1968; Schilling and Winchester, 1969; Kay and Gast, 1973; Zielinski, 1974; Frey *et al.*, in press) possibly suggesting a similar mode and type of volcanism and differentiation, as well as mantle source areas and depths of formation.

However, the LIL-depleted basalts from site 253, which is located close to site 254 in the southern half of the Ninetyeast Ridge poses a problem (Fig. 1). The REE data for six basalts from site 253 are presented by Frey *et al.* (1974) and Frey *et al.* (in press). The REE pattern for one of the basalts is similar to that of spreading mid-oceanic ridge basalts, structural models (Frey *et al.*, in press).

A variety of origins for the Ninetyeast Ridge have been proposed including uplift (McKenzie and Scalter, 1971), a horst (Bowin, 1973), a hot spot (Morgan, 1972) and a leaky transform fault combined with a hot spot or mantle plume (Luyendyk and Davies, 1974; Johnson *et al.*, 1976). However, geochemical data support the possibility that the Ninetyeast Ridge, Broken Ridge, Naturaliste Plateau, Kerguelen Plateau and Saint Paul–Amsterdam Islands may originally have formed as a single aseismic ridge from two fixed hot spots (Luyendyk, Rennick and Sclater, in preparation).

$^{87}Sr/^{86}Sr$ Ratios

The range of $^{87}Sr/^{86}Sr$ ratios of the EPR and MAR rocks is very small (0.7026–0.7028; Peterman and Hedge, 1971; Hart, 1971; Subbarao, 1972), while those from MIOR display distinctly higher ratios (0.7032–0.7035, with one value of 0.7043; Subbarao and Hedge, 1973; Engel and Fisher, 1975), suggesting that the source of the MIOR was depleted in LIL elements more recently and/or to a lesser degree than the basalts from the EPR and MAR (Subbarao and Hedge, 1973). However, the two analyses of CR rocks have lower $^{87}Sr/^{86}Sr$ ratios (0.7028, 0.7029) than the MIOR (Table 2). These meagre data probably indicate that the source of the CR is more depleted in LIL elements than that of the MIOR.

In contrast to the spreading ridges, the aseismic Ninetyeast Ridge volcanic rocks (sites 214 and 215) have much higher ratios ranging from 0.7043 to 0.7049 (Reddy *et al.*, in press), similar to alkali island basalts (0.7039–0.7066; Peterman and Hedge, 1971), and seamount basalts (0.7030–0.7053; Subbarao *et al.*, 1973; Subbarao and Hekinian, in press) suggesting similar mantle source areas and volcanic processes (Reddy *et al.*, in press) (Table 2).

The Sr isotopic composition of volcanic rocks from the Ninetyeast Ridge display certain interesting features. The $^{87}Sr/^{86}Sr$ ratios for basalts vary from 0.7043 to 0.7048, while the andesites nearly all have the higher ratio of 0.7049 (Table 2). The data sets of higher

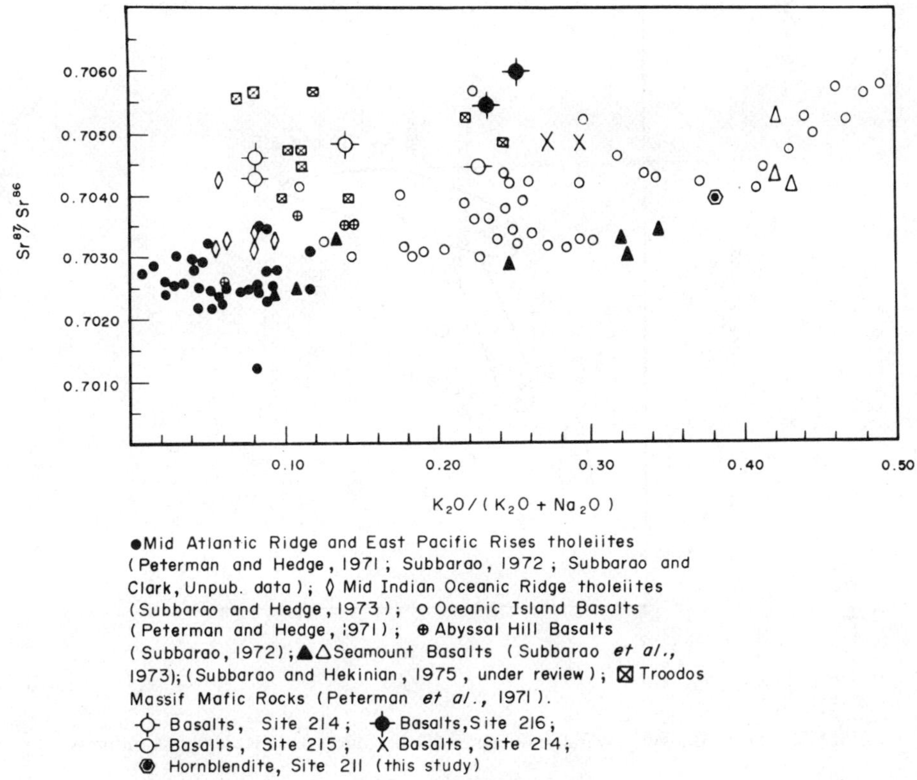

FIG. 12. Variation of $K_2O/(K_2O + Na_2O)$ vs. $^{87}Sr/^{86}Sr$ ratios for oceanic rocks from different tectonic environments. Data sources: as in Subbarao *et al.* (in press); Reddy *et al.* (in press).

$^{87}Sr/^{86}Sr$ ratios for basalts and andesites probably suggest a genetic relationship between the two rock types, and is compatible with a model involving fractional crystallization of a basaltic magma or variable degrees of partial melting of a single mantle source (Reddy *et al.*, in press). The former view is consistent with the major and minor element data, particularly the fractionation trends of Ni, Cr, V and Cu (Ramanan *et al.*, 1976).

In order to obtain information on the mantle source areas, we have plotted (K_2O/K_2O+Na_2O) vs. $^{87}Sr/^{86}Sr$ (Fig. 12) and also $^{87}Sr/^{86}Sr$ vs. $[La/Sm]_{e.f.}$ (Fig. 13). Various degrees of depletion of LIL elements would be reflected in Rb/Sr and K_2O/K_2O+Na_2O, whereas different stages of depletion would be reflected only in the $^{87}Sr/^{86}Sr$ ratios. Similarly, $[La/Sm]_{e.f.}$ is also considered to be a good indicator of fractionation of the light rare earths and little affected by shallow-depth fractional crystallization during magma ascent (Schilling, 1973). We believe this ratio to offer clues on the nature of various chemical domains in the earth's mantle.

Volcanic rocks from different tectonic environments—oceanic ridges, seamounts, abyssal hills—display a positive correlation in the $K_2O/(K_2O+Na_2O)$ vs. $^{87}Sr/^{86}Sr$ diagram (Fig. 12; Peterman and Hedge, 1971; Subbarao, 1972; Subbarao and Hedge, 1973; Subbarao *et al.*, 1973; Subbarao and Hekinian, in press; Reddy *et al.*, in press). This correlation is interpreted as involving multiple episodes of magma extraction with preservation of resultant isotopic chemical heterogeneities.

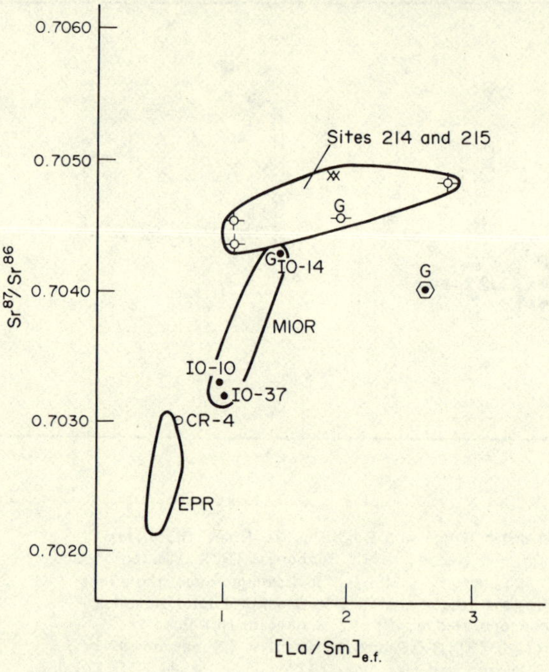

FIG. 13. $^{87}Sr/^{86}Sr$ vs. $[La/Sm]_{e.f.}$ variation diagram. Data sources as in Reddy *et al.* (in press.)

Furthermore, a positive correlation between $^{87}Sr/^{86}Sr$ and $K_2/(K_2O+Na_2O)$ for the rocks from the spreading EPR and MIOR and the aseismic Ninetyeast Ridge is also evident from Fig. 12. This systematic chemical trend may suggest that the mantle beneath the EPR is relatively more depleted in LIL elements than that beneath MIOR. It seems likely that the mantle source of the Ninetyeast Ridge rocks is relatively less depleted in LIL elements than the recent spreading ridges such as the EPR and MIOR.

Distribution of $^{87}Sr/^{86}Sr$ Ratios in the Indian Ocean

The distribution of $^{87}Sr/^{86}Sr$ ratios of volcanic rocks from different tectonic environments from the Indian Ocean is shown in Fig. 14. The rocks from the spreading MIOR and Cr have lower $^{87}Sr/^{86}Sr$ ratios (0.7028 to 0.7035; one value of 0.7043) in contrast to the aseismic Ninetyeast Ridge (0.7043 to 0.7049). However, there is a distinct similarity between the aseismic Ninetyeast Ridge and the associated islands (0.7039–0.7052), and is considered to be significant in terms of tectonic–geochemical models. Subbarao *et al.* (in press) and Reddy *et al.* (in press) interpret these relationships in terms of "zoning-depletion model" with the presence of alkali-poor and alkali-rich zones in the mantle. This means that a mantle zone having a higher $^{87}Sr/^{86}Sr$ ratio of 0.7045 may probably feed aseimic ridges, abyssal hills, seamounts and islands, while a mantle zone with a lower $^{87}Sr/^{86}Sr$ ratio of about 0.7030 may generate the spreading mid-oceanic ridges (Subbarao, 1972).

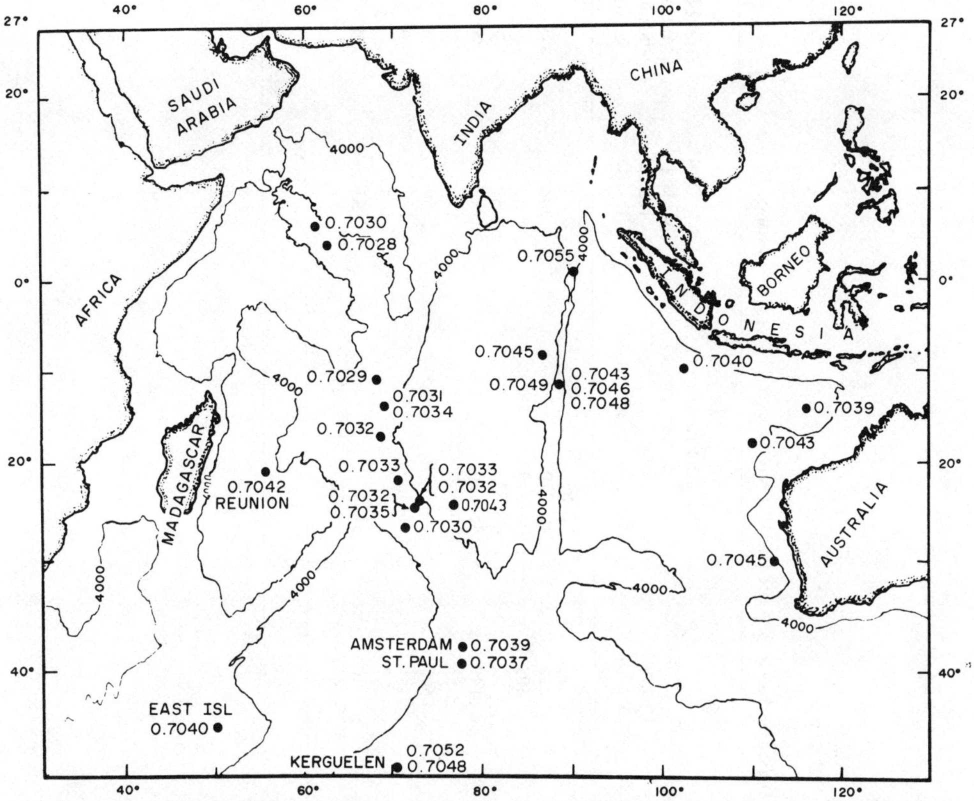

FIG. 14. Distribution of $^{87}Sr/^{86}Sr$ ratios in the volcanic rocks from different tectonic environments from the Indian Ocean. Data sources as in Reddy *et al.* (in press).

Western Indian Ocean

Somali Basin—Sites 235, 236 and 240

These three sites from the Somali basin yielded porphyritic plagioclase-rich basalts, porphyritic augite-rich basalts and fine-grained grey basalts with fresh glass, respectively. These rocks have higher concentrations of transitional metal elements, but lower amounts of Ba, Zr and Y than oceanic island basalts but resemble MORB (Table 3; Engel *et al.*, 1974; Erlank and Reid, 1974).

TABLE 3. RANGES AND AVERAGES OF TRACE ELEMENT ANALYSES OF BASALTS FROM DSDP LEGS 24 AND 25, WESTERN INDIAN OCEAN

	Site 235 (Somali Basin)		Site 236 (Somali Basin)		Site 240 (Somali Basin)	Site 238 (Argo Fracture Zone)		Site 239 (Mascarene Basin)	Site 245 (Madagascar Basin)
	Range	Average	Range	Average		Range	Average		
Sc	33–39	36	43–53	48		52–59	55		
V	170–180	175	180–210	196	220	210–270	240	294	338
Cr	430–450	452	250–280	140	403	300–310	303	53	161
Co	37–54	46	43–75	54	52	47–53	49	52	54
Ni	130–330	227	72–300	157	149	85–160	114	48	70
Cu	48–71	64			65	110–120	113	110	84
K	2150–100,290	27,350	1410–3810	2540	2860	490–1820	933	3050	340
Sr	150–180	170	100–180	130	76	130–170	156	122	79
Rb					7.4			9.1	<0.5
Y	24–29	26		20	20	36–47	41	33	28
Zr	36–45	40	27–32	29	45	40–54	47	87	65
References	1		1		2	1		2	2

1. Engel *et al.*, 1974; 2. Erlank and Reid, 1974.

Argo Fracture Zone—Site 238

Site 238 is located in the sediment-filled north-east part of the Argo fracture zone and contains fine-grained slightly porphyritic basalt. The trace element composition of these basalts is also similar to MORB (Table 3; Engel *et al.*, 1974).

Mascarene Basin—Site 239

Site 239 contained fine-grained basalt with dark-brown altered glass. The minor element contents, particularly Ba and Nb, are similar to those of MORB (Table 3); the higher concentrations of K and Rb may be due to sea-water alteration. Erlank and Reid (1974) remark "an unusual feature is the presence of equivalent amounts of Co, Cr and Ni, as abyssal tholeiites generally contain much higher concentrations of Cr and Ni. It is not possible to assess the significance of this feature at this stage, but it may be noted that the concentrations of Co, Cr and Ni in the site 239 basalt are similar to low K-tholeiites found in island area" (p. 543).

Madagascar Basin—Site 245

Site 245 contains fresh, intrusive fine- to medium-grained and glassy diabasic basalts. The minor element chemistry is similar to that of MORB, with low K and Rb concentrations (Table 3; Erlank and Reid, 1974).

Mozambique Basin, South-western Indian Ocean

Sites 248 and 249

These two sites are characterized by the presence of quartz tholeiite basalts (Figs. 3 and 4; Erlank and Reid, 1974). The trace-element chemistry shows that site 248 basalts are enriched in LIL and LREE (higher $[La/Sm]_{e.f.}$ of 1.26) and resembles other "hot spot" sites (Table 4; Erlank and Reid, 1974; Kempe, unpublished data). In contrast, site 249 basalts are depleted in LIL elements similar only to MORB (Table 4, Erlank and Reid, 1974; Erlank, A. J., personal communication, 1975).

Site 250

Site 250 has yielded olivine-rich basalt with titanaugitic salite, which contains higher K_2O (0.57%), Rb (> 5 ppm), P_2O_5 (0.25%), but normal TiO_2 (1.46%), Ba (110 ppm), U (0.5 ppm) and marked enrichment in LREE relative to MORB (Table 4; Fig. 15; Kempe, 1974; Fleet *et al.*,1976). Fleet *et al.* (1976) regard is as probably a "transitional" or "off-ridge axis" basalt closely resembling the titaniferous augite-bearing basalt from Leg 2, site 10 (Frey *et al.*, 1974). Fleet *et al.* (1976) remark "site 250 and possibly site 248 basalts may derive from hot spot activity that is manifest chemically but not physically" (p. 4258).

Site 251

This site is located from the south-west branch of MIOR and yielded typical MORB. These basalts are depleted in LIL element concentrations and show characteristic depleted LREE patterns (Table 4; Fig. 15; Kempe, 1974; Fleet *et al.*, 1976).

TABLE 4. RANGES AND AVERAGES OF TRACE ELEMENT ANALYSES OF BASALTS FROM DSDP LEGS 25 AND 26, SOUTH-WESTERN INDIAN OCEAN

	Site 248 (Mozambique Basin)		Site 250 (Mozambique Basin)	Site 249 (Mozambique Basin)		Site 251 (SW branch of MIOR)		Mid-Ocean Ridge basalt Average
	Range	Average		Range	Average	Range	Average	
V	293–318	309	250	187–191	189	285–370	313	289
Cr	140–165	153	215	228–249	238	220–245	236	296
Co	57–71	66		57–71	64			41
Ni	89–97	93	110	167–178	172	110–170	148	123
Cu	95–193	156	12	68–115	91	12–70	51	87
K	2240–9880	5220	4730	3090–3640	3365	990–4980	2820	1110
Sr	219–254	241	210	103–107	105	80–85	83	116
Ba		104	110			25–200	118	12.6
Rb			<5			5–9	6.6	1.35
Cs								0.023
U			0.5					
Y	28–34	30	25	19–22	20		50	43
Zr	139–165	154	200	62–63	62		200	100
La		13	20.6			4.0–4.5	4.2	2.67
Ce			30.3			6.4–11.1	8.7	9.99
Sm		5.6	4.0			3.6–4.0	3.8	3.09
Eu			1.3			0.8–1.4	1.1	1.15
Tb			0.6			0.8–1.2	1.0	0.79
Yb			2.0			2.4–3.8	3.1	2.94
K/Rb	139–722	484	946		687		415	822
K/Cs								48,260
$^{87}Sr/^{86}Sr$								0.7027
$[La/Sm]_{e.f.}$		1.26	2.81			0.54–0.68	0.61	0.47
References	1,2		3,4	1		3,4		5–13

1. Erlank and Reid, 1974; 2. Kempe, unpublished REE data; 3. Kempe, 1974; 4. Fleet *et al.*, 1976; 5. Melson and Thompson, 1971; 6. O'Nions and Pankhurst, 1974; 7. Dasch *et al.*, 1973; 8. Subbarao and Hedge, 1973; 9. Subbarao, 1972; 10 Hedge and Peterman, 1970; 11. Hart *et al.*, 1973; 12. Kay *et al.*, 1970; 13. Schilling, 1971.

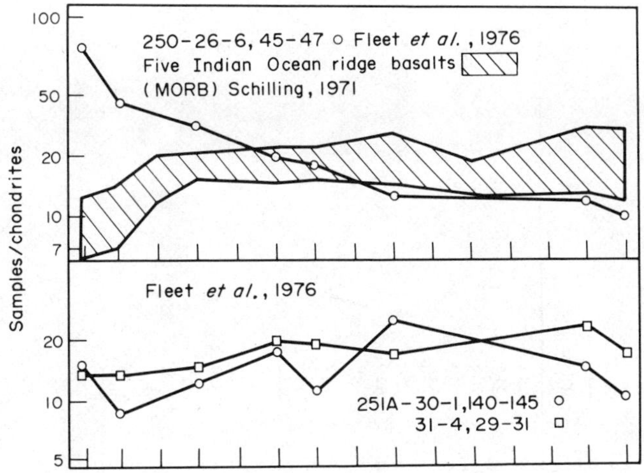

FIG. 15. REE distribution patterns for sites 250, 251. Data sources: Fleet *et al.* (1976).

Wharton Basin

The volcanic rocks drilled from Legs 22 and 26 in the Wharton basin at sites 211, 212, 213, 256 and 257 (Fig. 1) are predominantly quartz or olivine tholeiites (Figs. 2 and 3).

Site 213

Site 213 (western portion of the basin) yielded mainly altered olivine pillow basalts and at least eleven successive pillow lavas with upper and lower chilled margins (Hekinian, 1974a). The crystalline rocks show a high Fe_2O_3/FeO (>1) and H_2O ($>2\%$) with variable K_2O (0.06 to 1.07%) and low MgO ($<5.6\%$) (Hekinian, 1974a), whereas the glassy samples show low K_2O (0.06%) and P_2O_5 (0.09%) (Melson *et al.*, 1976).

The low Zr, Ba, Sr, Ni and $[La/Sm]_{e.f.}$ and slight depletion in LREE relative to heavy REE (Fig. 14) in site 213 basalts are similar to the spreading MORB (Table 5; Frey *et al.*, 1977). This interpretation is also supported by Sclater and Fisher (1974) based on the structural and magnetic profiles.

Site 212

This site lies in the centre and also in the deepest portion of the Wharton basin. Hekinian (1974a) recognized at least seventeen altered olivine pillow metabasalts. These are characterized by low K_2O (0.61%) and P_2O_5 (0.03%) (Melson *et al.*, 1976), also by high Cr, moderately high Ni, and low Zr and REE abundances (Table 5; Fig. 16; Frey *et al.*, in press; Hekinian and Subbarao, 1976; Reddy, V. V., unpublished data).

Frey *et al.* (1977) suggested that these rocks "experienced very little fractional crystallization after formation by partial melting", similar to some of the DSDP Atlantic site 3 (cores and 14 and 18). In general the geochemical characteristics of rocks from sites 212 and 213 are similar, compatible with their formation at a spreading ridge axis as suggested by Sclater and Fisher (1974).

TABLE 5. RANGES AND AVERAGES OF TRACE ELEMENT ANALYSES OF IGNEOUS ROCKS FROM WHARTON BASIN (DSDP LEGS 22 AND 26), SOUTH-EASTERN INDIAN OCEAN

	Site 211 diabase		Site 211 amphibole bearing basalts and hornblendite		Site 212		Site 213	
	Range	Average	Range	Average	Range	Average	Range	Average
Sc		21		18		46		43
V	140–205	173	115–200	163	205–225	215	200–235	217
Cr	270–335	293	40–75	63		1200	290–345	310
Co	24–69	41	16–36	22	33–45	39	30–68	45
Ni	90–135	117	67–95	83	130–150	140	75–87	82
Cu	31–47	37	21–29	25	52–60	56	80–95	86
K	11,700–21,410	16,950	10,040–18,590	14,315	21,490–31,790	26,640	3070–9210	4790
Sr	490–540	515	230–1100	570	65–100	88	87–115	96
Ba	395–450	427	140–800	530	23–135	79	10–16	13
Rb					16.2–65	47	23–35	28
Cs	0.38–0.52	0.45	0.81–1.95	1.28		0.22		
Y	26–27	26	34–39	38	12–24	18	22–32	27
Zr	150–165	157	255–330	300	21–23	22	53–62	57
La	44.3–47.5	45.5	32.8–65.6	49.3	1.52–2.30	1.9		2.2
Ce	72.5–74.0	73.2	82–116	93.6	3.69–6.00	4.89		7.6
Sm		6.12		10.33	0.95–1.29	1.10		2.18
Eu	1.95–2.74	2.32	1.92–3.68	2.58	0.38–0.47	0.45		0.85
Tb	1.0–1.29	1.16	0.58–1.78	1.08	0.28–0.38	0.32		0.64
Yb	2.29–2.5	2.47	2.30–3.38	2.82	1.08–1.75	1.33		2.55
K/Rb	308–349	324		200	358–489	423	148–329	238
K/Cs	41,173–47,605	44,389		9747				
$^{87}Sr/^{86}Sr$				0.7040				
$[La/Sm]_{e.f.}$		4.23	2.57–2.64	2.60	0.96–0.97	0.96		0.55
References	1,2,3,4		1,2,3,4		1,2,3,4		1,2	

TABLE 5. CONTINUED

	Site 256 LREE-enriched basalts		Site 256 LREE-depleted basalts		Site 257 LREE-enriched basalt	Site 257 LREE-depleted basalt		Mid-Ocean ridge basalt
	Range	Average	Range	Average		Range	Average	Average
Sc			41–49	46		44–47	46	
V	450–490	470	395–535	443	300	235–330	290	289
Cr	105–110	108	85–130	108	390	160–670	307	296
Co			32–40	36		27–43	40	41
Ni	100–130	113	78–100	85	120	95–160	127	123
Cu	12–17	13.6	215–2000	444	17	12–140	62	87
K		2070	1660–2400	2027	990	330–9040	2396	1110
Sr	160–165	161	120–160	137	170	65–100	79	116
Ba	25–40	30	35–53	43	25	10–150	35	12.6
Rb								1.35
Cs								0.023
Y		50	38–46	43	40	23–50	34	43
Zr		200	115–175	143	150	30–100	71	100
La	12.2–12.7	12.5	9.0–10.4	9.82	6.9	1.2–2.7	1.85	2.67
Ce	15.1–24.2	21	29–33	30.5	13.8	2.3–6.8	4.20	9.99
Sm	4.0–5.0	4.66	4.31–4.85	4.68	3.4	1.4–2.09	1.77	3.09
Eu	1.2–1.8	1.6	1.37–1.59	1.52	1.2	0.5–0.79	0.65	1.15
Tb	0.8–0.9	0.83	0.8–1.1	1.03	0.6	0.2–0.60	0.41	0.79
Yb	2.4–3.2	2.9	3.0–3.4	3.27	2.1	1.6–2.9	2.13	2.94
K/Rb								822
K/Cs								48,250
$^{87}Sr/^{86}Sr$								0.7027
$[La/Sm]_{e.f.}$	1.37–1.66	1.47	1.10–1.20	1.14	1.10	0.36–0.78	0.53	0.47
References	5,6		1		5,6	1,5,6		7,8,9,10,11 12,13,14,15

1. Frey *et al.*, 1977; 2. Bougault, 1974; 3. Reddy, unpublished data; 4. Subbarao *et al.*, 1976; 6. Kempe, 1974; 7. Melson and Thompson, 1971; 8. O'Nions and Pankhurst, 1974; 9. Dasch *et al.*, 1973; 10. Subbarao and Hedge, 1973; 11. Subbarao, 1972; 12. Hedge and Peterman, 1970; 13. Hart, *et al.*, 1973; 14. Kay *et al.*, 1970; 15. Schilling, 1971.

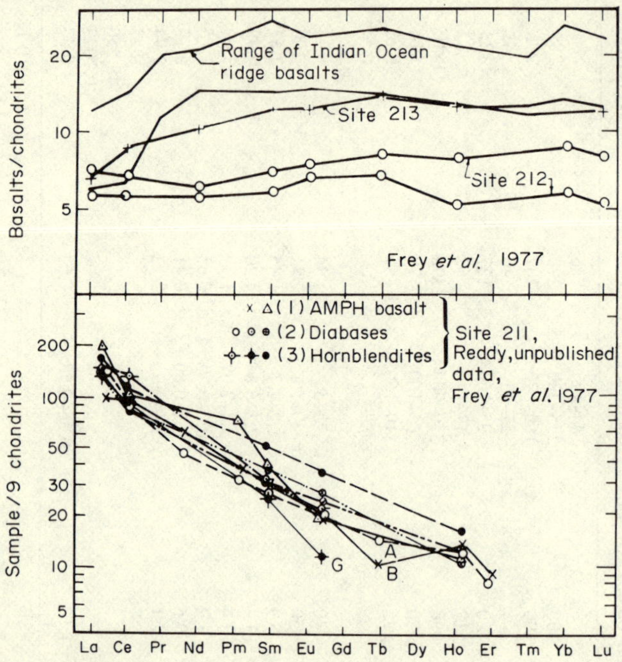

FIG. 16. REE distribution patterns for sites 213, 212, 211. Data sources: Frey *et al.* (in press); Reddy, V. V., unpublished data; Subbarao *et al.* (1976); Hekinian and Subbarao (1976).

Site 211

This site is located in the north-eastern part of the Wharton basin, close to the Java Trench in the north and Christmas Island in the east, and yielded a largely silica undersaturated suite (Fig. 5) consisting of a diabase sill followed by flows of amphibole-bearing basalts and hornblendite with glassy rims (Hekinian, 1974 a,b). The diabase sill is nepheline normative, alkali rich ($K_2O > 1\%$, $Na_2O > 3\%$) and shows high TiO_2 ($>2\%$), marked enrichment in LREE (Fig. 16) and higher $[La/Sm]_{e.f.}$ ratio (4.23). Frey *et al.* (in press), recognized two geochemically distinct units in the sill. The nepheline-normative amphibole-bearing basalts and the hornblendites have lower than usual amounts of transitional elements, but exhibit higher Sr, Ba and Zr concentrations and marked enrichment in LREE, similar to the diabase sill (Table 5). Thus, geochemically, these characters are unlike those of typical MORB but similar to alkali island basalts (Subbarao *et al.*, 1976; Frey *et al.*, in press; Reddy, V. V., unpublished data). The mineralogy of the amphibole-bearing basalts and hornblendites (the first recorded occurrences from the floor of the Indian Ocean) suggest that these rocks were probably formed from an undersaturated hydrous magma (Hekinian, 1974 a,b). It is likely that the basalts from this site are related to the Cocos and Keeling islands and Christmas Island, but are definitely unrelated to spreading mid-oceanic ridge system (Hekinian, 1974 a,b; Sclater and Fisher, 1974; Hekinian and Subbarao, 1976; Subbarao *et al.*, 1976; Reddy, V. V., unpublished data).

Sites 256 and 257

These two sites are located in the central and south-eastern end of the Wharton basin repectively and yielded fractionated Fe- and Ti-rich quartz tholeiites (Fig. 5; Kempe, 1974). The enrichment in V, P and Zn accompanies increases in Fe and Ti and depletion in Al, Ca, Cr, Ni and Cr (Table 5; Kempe, 1974; Fleet *et al.*, 1976).

Basalts from site 256 and the uppermost barrel (core 11) at site 257 show LREE-enriched patterns (Fig. 17) and higher concentrations of Ti, Sr, Zr, Ba, P_2O_5, similar to rocks from the Ninetyeast Ridge sites 214 and 216 (Table 5; Fleet *et al.*, 1976). However, the apparently much older lower rocks from site 257 display depleted LREE patterns (Fig. 17) and LIL concentrations similar to MORB (Table 5; Fleet *et al.*, 1976; Frey *et al.*, in press) Kempe (1975, 1976) also reported a similar pair of sites (320, 321) with extreme compositions from the Nazca plate.

Based on tectonic models Luyendyk (1974) regarded the basement at site 257 as part of the earliest crust formed from the ridge between India and Antartica during the Early Crestaceous. Continuous sea floor spreading at this ridge led to the formation of crust at site

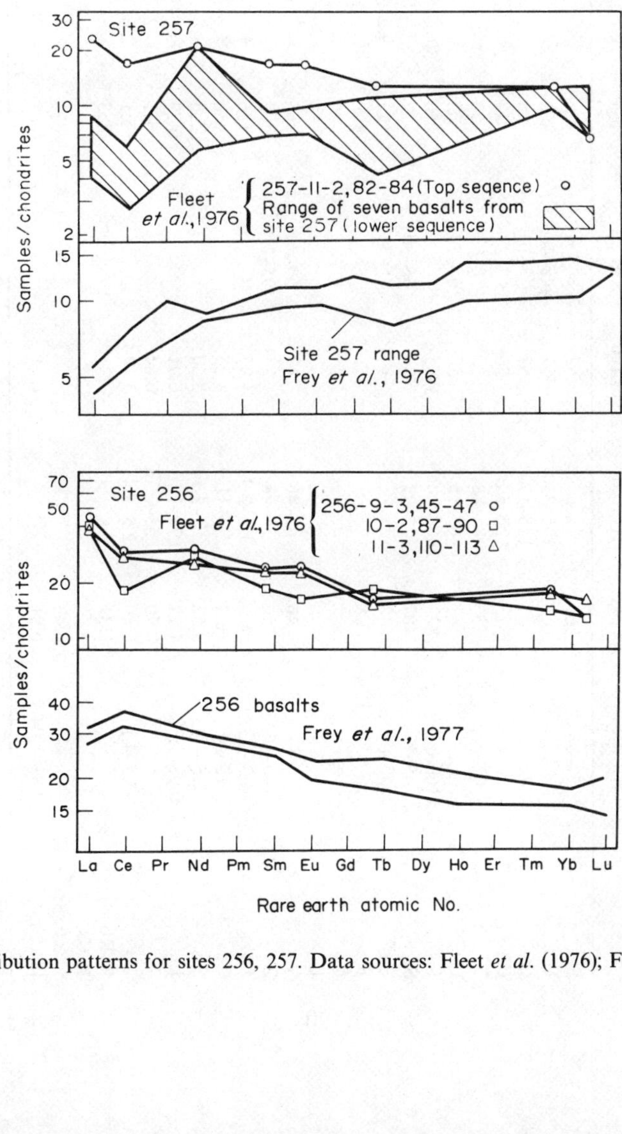

FIG. 17. REE distribution patterns for sites 256, 257. Data sources: Fleet *et al.* (1976); Frey *et al.* (in press.)

TABLE 6. RANGES AND AVERAGES OF TRACE ELEMENT ANALYSES OF DSDP LEG 27 BASALTS FROM THE SOUTH-EASTERN INDIAN OCEAN ABYSSAL PLAINS

	260 (Gascoyne Plain)		261 Unit A basalts (Argo Plain)		261 Unit B basalts (Argo Plain)		261 Unit C basalts (Argo Plain)		259 (Perth Plain)		Mid Ocean Ridge basalt
	Range	Average	Range	Average	Range	Average	Range	Average	Range	Average	Average
Sc	57–66	61		52				52			
V			305–330	317	445–480	462		335		265	289
Cr	72–90	81	145–170	157	7–10	8.5		150	125–170	147	296
Co	36–110	73	22–50	36	13–22	17		41	15–52	33	41
Ni	67–265	166	85–88	86	50–63	56		83	48–68	58	123
Cu			65–95	80	55–70	62			40–65	52.5	87
K	1160–3150	1823	2690–4480	3585		7250	3070–5310	4190		2810	1110
Sr	85–130	107	72–97	84	65–90	77		115	73–100	86	116
Ba	30–60	45			15–35	23	9.2–35.0	21	6.1–9.3	7.6	12.6
Rb		3.5				8.5				2.9	1.35
Cs											0.023
U						0.86				0.07	
Y		46	41–46	43	60–70	65		48	34–46	40	43
Zr	140–160	150	78–88	83	195–210	202		85	58–80	77	100
La	4.3–7.7	6.0	1.6–2.34	1.97		6	1.3–2.45	1.87		1.8	2.67
Ce	12–20	16.3	5.5–11.0	8.25		18	4.2–9	6.6		6.7	9.99
Sm	2.8–4.03	3.52	2.1–3.11	2.60		5.3	1.6–3.16	2.38		2.0	3.09
Eu	0.98–1.32	1.17	0.79–1.03	0.91		0.79	0.61–1.17	0.89		0.76	1.15
Tb		0.70	0.55–0.73	0.64		1.1	0.42–0.79	0.60		0.53	0.79
Yb	3.7–4.1	3.80	2.8–3.4	3.10		5.1	2.0–3.9	2.95		2.80	2.94
K/Rb		331				853				969	822
K/Cs											48,260
$^{87}Sr/^{86}Sr$		0.7043				0.7039				0.7045	0.7027
$[La/Sm]_{e.f.}$	0.83–1.04	0.91		0.41		0.61	0.42–0.44	0.43		0.49	0.47
References	1,2		1,2		1,2		1,2		1,2		3–11

1. Frey *et al.*, 1977; 2. Robinson and Whitford, 1974; 3. Melson and Thompson, 1971; 4. O'Nions and Pankhurst, 1974; 5. Dasch *et al.*, 1973; 6. Subbarao and Hedge, 1973; 7. Subbarao, 1972; 8. Hedge and Peterman, 1970; 9. Hart *et al.*, 1973; 10. Kay *et al.*, 1970; 11. Schilling, 1971.

256 in late Early Cretaceous times. Between 110 and 80 m.y. incipient rifting between Australia and Antarctica led to volcanic upwelling and the formation of the Naturaliste Plateau, the Broken Ridge–Kerguelen Plateau, and the oldest part of the Ninetyeast Ridge (Luyendyk and Davies, 1974). It is possible that site 256 and uppermost site 257 basalts were formed as part of this LIL and LREE-enriched volcanism.

The geochemical difference between various sites from the Wharton basin possibly suggest the presence of at least two magnetic source with considerable fractionation and minor element variations within each main type of basalt (Kempe, 1974; Fleet *et al.*, 1976).

Abyssal Plain Sites

During Leg 27, basalts were drilled from three plain sites nearby 260, 261 and 259 on the edge of the Wharton basin. Site 259 yielded quartz tholeiites showing characteristic depletion in LIL elements, inclusion of Zr, and relative depletion in LREE (Table 6, Fig. 18). Site 260 tholeiites display relatively high Ba and Zr concentrations, a nearby flat chondrite-normalized REE pattern (Fig. 18) and a variable chemical composition including $[La/Sm]_{e.f.}$ (0.83–1.04) (Table 6). Geochemically, these rocks lie between the MORB and island basalts. At site 261, the uppermost and lower units A and C have closely similar REE abundances with LREE-depleted patterns which resemble site 259 basalts. Unit B is enriched in LIL, Ti, Zr, Ba, P and REE, but shows slight depletion in LREE (Table 6, Fig. 6; Robinson and Whitford, 1974; Frey *et al.*, in press). The higher $^{87}Sr/^{86}Sr$ ratios of 0.7039 to 0.7043 for basalts from these three sites are probably due to some degree of low temperature alteration (Robinson and Whitford, 1974; Whitford, 1975). Whitford (1975) remarks "If these ratios are a widespread feature of the oceanic crust which is being subducted to the north, then they

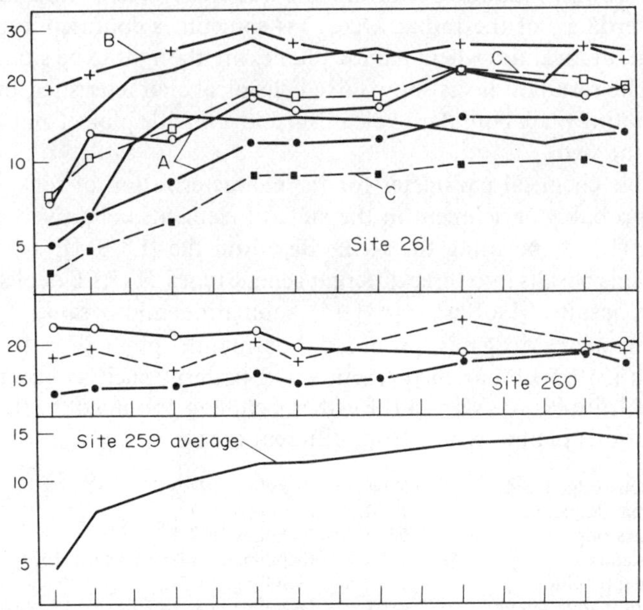

FIG 18. REE distribution patterns for sites 259, 260, 261. Data sources: Frey *et al.* (in press); Robinson and Whitford (1974); Schilling (1971).

may account for the Sr isotopic character of at least the arc tholeiites in Java" (p. 1299). It is, however, significant that these higher $^{87}Sr/^{86}Sr$ ratios fall within the range of MIOR (0.7032 to 0.7043; Subbarao and Hedge, 1973) probably indicating similiar mantle sources for ridge and basin basalts.

Basaltic Rocks from Aseismic Ridges, Seamounts and other Intra-Plate "Areas"

Much attention has been paid recently to the chemical characterization of basaltic rocks from aseismic ridges, seamounts and other intra-plate volcanic sources, as distinct from rocks formed at spreading ocean ridges. Perhaps the most popular model—although by no means universally accepted—that has been suggested for the origin of these rocks is the fixed mantle plume or "hot spot" (cf. Wilson, 1965; Morgan, 1971, 1972; Schilling, 1973). The model proposes a fixed volcanic source at a relatively deep level in the mantle from which volcanism takes place continuously or intermittently. Assuming a constant and fairly low rate of eruption, the product will be a seamount if the overlying crustal plate is stationary at the time, or an (aseismic) ridge, with a structure related to the rate of crustal spreading, if the plate is moving. Given a faster rate of eruption, the result will be an island or a string of islands, probably joined below the sea surface into an island ridge, again controlled also by the spreading rate. An incipient hot spot will not be apparent from bathymetry, tectonic feature, or seismic profile, but will only be detected if drilling penetrates the basement rocks. This type of feature is only now becoming recognized, as intra-plate drilling recovers basalt that is atypical of mid-ocean ridge basalt (MORB). The concentration between the various types of intra-plate volcanism, including that well known and fully described from oceanic islands (e.g. Hawaiian Islands and the linked Emperor seamount chain; Iceland, the Azores and the islands of the South Atlantic Ocean; and the scattered islands—Reunion, Kerguelen, St. Paul and Amsterdam—of the Indian Ocean) is sometimes confused, especially in the Mid-Atlantic Ridge, at localities where mantle plumes are thought to be situated below the mid-ocean ridge. The resultant lavas show mixed chemical characteristics, including some typical of both accreting plate boundary volcanism and a mantle plume, and of gradational mixtures between the two.

The most valuable chemical parameter for the characterization of rocks derived from mantle plumes is probably enrichment in the suite of elements collectively referred to as "incompatible" or (LIL), including the REE. Based on the $[La/SM]_{e.f.}$, Schilling (1976) subdivided the oceanic basalts into three different genetic types: (1) REE depleted submarine mid-oceanic ridge basalts: $[La/Sm]_{e.f.} < 1$; (2) submarine mid-oceanic ridge tholeiites generally occurring near Morgan's proposed hot mantle plumes or triple junctions: $[La/SM]_{e.f.} \simeq 1$; and (3) LREE-enriched fractionated patterns such as occurring in alkali basalts or island tholeiites: $[La/SM]_{e.f.} > 1$. Further, Schilling presented the $[La/Sm]_{e.f.}$ ratios for MORB and various plume basalts from different areas:

Mid oceanic ridge basalts	0.47–0.72	(Schilling, 1971)
Iceland mantle plume	1.1–1.38	(Schilling, 1973)
Rekykjanes ridge	0.9–1.30	(Schilling, 1973)
Faeroe islands	1.0–2.4	(Schilling and Noe-Nygard, 1974)
Azores mantle plume	1.1–3.6	(Schilling, 1975)
Galapagos mantle plume	1.1	(Schilling *et al.*, 1976)
Discovery Tablemound	2.6	(Kempe and Schilling, 1974)

As well as being enriched in LIL and RE elements and having higher $[La/Sm]_{e.f.}$ ratios than

TABLE 7. RANGE OF MAJOR ELEMENTS IN BASALTS FROM ASEISMIC RIDGES AND "HOT SPOTS"

SiO_2	40.77–52.81	(DSDP site 211 diabase; dredged near Bouvet Island)
TiO_2	0.69–4.92	(DSDP site 253 picrite; DSDP site 165)
Al_2O_3	11.51–22.80	(Galapagos spreading centre; Cocos Ridge)
FeO*	6.41–18.36	(Cocos Ridge; Galapagos spreading centre)
MnO	0.05–0.48	(DSDP site 171; Discovery Tablemount)
MgO	2.04–13.85	(Walvis Ridge; DSDP site 253 picrite)
CaO	4.78–12.61	(DSDP site 170; Galapagos spreading centre)
Na_2O	1.39–5.78	(DSDP site 253 picrite; DSDP site 170)
K_2O	0.06–6.05	(Galapagos spreading centre; DSDP site 313)
P_2O_5	0.07–1.28	(Mozambique Ridge; DSDP site 165)

Note: FeO* = total Fe as FeO: DSDP sites 165, 170 and 171: Leg 17; site 313: Leg 32, both central Pacific Ocean; sites 211 and 253; see Fig. 1.

MORB plume-derived basalts show a general tendency towards enrichment in K_2O, TiO_2, P_2O_5 and often in total Fe (FeO*); they also show a wider range in major element chemistry than MORB.

Detrich *et al.* (1977) list the thirteen ridges which they consider to form the major aseismic features in the three main oceans, showing that more than half probably formed close to sea level and subsequently underwent subsidence: a consequence of the crustal uplift associated with mantle plume activity. All known analyses of basalt dredged or drilled from these ridges, together with many others from seamounts and intra-plate "hot spots", have been computed. No islands, for which abundant data are available, nor their slopes, nor mixed areas of the MAR—Reykjanes Ridge, 45°N, the Azores, 37°N (FAMOUS; Leg 37)—are included, the last because of the difficulty in selecting genuine mantle plume samples uncontaminated by MORB. In the Indian Ocean, site 215 has been provisionally treated as part of the Ninetyeast Ridge. Normative compositions have been plotted on the Yoder and Tilley (1962) projection in Fig. 5 (and also, as appropriate, in Figs. 2 to 4), and their elemental ranges given in Table 7. For each element the sources of the lowest and the greatest values, respectively, have been given in brackets alongside (Table 7). Undoubtedly some extreme values are the result of alteration but the overall ranges are greater than and different from those of MORB.

The normative plot (FIg. 5), in which analyses are plotted normalized to $Fe_2O_3 = 1.50\%$, *cc*-free, shows a wide distribution, notable for a mode in the quartz tholeiite field at $Q_5 di_{50} hy_{45}$, a wide scatter in the olivine tholeiite and olivine basalt fields, and for the frequencies of distribution of points with in three basalt fields. If the site 214 oceanic andesites are excluded, there are 27% alkali olivine basalts, 40% olivine tholeiites, and 33% quartz tholeiites. This differs strongly from the distribution obtained for all ocean-floor basalts, where the mass of data derives from analyses of MORB, and which shows a single mode in the olivine tholeiite field at approximately $di_{50} hy_{40} ol_{10}$ (Kempe, 1975). The quartz tholeiite mode in Fig. 5 corresponds almost exactly with that quoted by Chayes' mode for undersaturated oceanic island rocks at $ne_{20} di_{50} ol_{30}$ falls at the centre of the *ne*-normative field scatter in Fig. 5. The simple matrix given below shows the number of ridge sites, seamounts, or other "hot spot" localities to contain one or more analysed rocks of the three types of basalt:

	Alkali olivine basalt	Olivine tholeiite	Quartz tholeiite
Aseismic ridge	4	8	8
Seamount	4	5	4
Intra-plate "hot-spot"	8	4	3

Drilling and dredging of non-MORB sources, then, shows a normative distribution comparable with island and continental basalts, and unlike that for spreading ridge basalts. The matrix above suggests that there is a tendency for aseismic ridge basalts to be quartz tholeiites, intra-plate "hot spots" to be alkali olivine basalts, and seamounts to be any of the three types. Large islands are well known enough to show an association of quartz- and nepheline-normative basalts (e.g. Hawaii, Iceland); smaller islands may contain olivine tholeiite differentiating into nepheline-bearing rocks (Gough, Tristan, St. Helena, Cape Verde islands) or quartz tholeiite differentiating into silicic alkali rocks (Jan Mayen, Azores, Bouvet, Ascension). The strong diaschistic tendency, therefore, for islands and intra-plate "hot spots" to contain both extreme types of basalt is shown less by seamounts but least of all by aseismic ridges, which notably consist of quartz tholeiite rocks. This height or structure-related variation may also be time-related, giving the sequence: islands > seamounts > aseismic ridges in terms of length of time of volcanism, height (locally modified by erosion) and differentiation (Kempe, 1975).

Other aspects of the major element geochemistry of these three types of basalt link them to a greater or lesser degree with each other and with island rocks, and again distinguish them from MORB. Hekinian (1974b) used the differentiation and crystallization indices and the AFM diagram to demonstrate that island and non-MORB ocean-floor basalts generally have similar differentiation trends, which are more extended than the less fractionated MORB. He showed that aseismic ridges in general, with particular reference to the very well-studied Ninetyeast and Walvis Ridges, are enriched in pyroxene, and chemically in (total) FeO* (> 11%) and TiO_2 (1.5–4%), both TiO_2 and K_2O (0.2–3%) being more variable than in MORB. He noted that Fe, Cu and S mineralization is associated with the basalts at these two ridges. Iron-enrichment is apparent particularly in the ferrobasalts from sites 214, 216 and 254 in the Ninetyeast Ridge (Hekinian, 1974a; Kempe, 1974; Frey *et al.*, in press), where they are accompanied by oceanic andesites (icelandites), and Walvis Ridge (Hekinian, 1972). Frey *et al.* (in press) note the similarities of the ferrobasalts with those from oceanic islands such as Galapagos, Iceland, the Faroes, Amsterdam and St. Paul, which include FeO*/MgO ratio, TiO_2, K_2O and the LIL and RE elements.

Chemical characterization of the other types of non-MORB oceanic basalt, from seamounts and intra-plate "hot spots", is more difficult. Their resemblance to island-type alkali olivine basalts rocks has been noted; for example Frey *et al.* (in press) compare site 211 rocks with those of (Indian Ocean) Christmas Island, and at sites 10 (Atlantic) (Bryan *et al.*, 1976) and 250 (Indian Ocean) the basalts contain purplish titanaugite. The basalts from sites 250 and 248 (Indian Ocean) are enriched in LIL and REE (Fleet *et al.*, 1976; Kempe, unpublished data). The ferrbasalt (FETI type) at site 256, Wharton Basin, is notable for its remarkable chemical resemblance to similar rocks from site 321, Nazca Plate, except that the latter is not enriched in LIL and RE elements (Kempe, 1976; Fleet *et al.*, 1976) but is highly fractionated MORB, as also is the extreme FETI basalt from site 32, Delgada Fan, again in the eastern Pacific Ocean (Melson *et al.*, 1976). Remarkable chemical convergence is also known from the uppermost part of core 11, site 257, also in the Wharton Basin, and the remainder of the site 257 core. The top sample has a very similar composition to that of the remaining analysed rocks except that it alone is enriched in LIL and RE elements; it is separated from the lower lavas by a considerable K–Ar age difference and a thin layer of limestone (Fleet *et al.*, 1976), while the sites 10 and 250 basalts can perhaps be regarded as incipient "hot spots".

It is probable that out of twenty DSDP sites from the Indian Ocean which are included in this review, basalts from eight sites (214, 216, 215, 254, 256, 211, 248, 250) with enriched LIL and LREE patterns may derive from intra-plate or "hot spot" activity, basalts from nine sites (212, 213, 261, 259, 249, 251, 235, 236, 240) with depleted patterns belong to the spreading MORB type, while basalts from three sites (253, 257, 260) combine some "hot spot" and MORB characteristics. Further, there is only partial agreement between chemical data and the tectonic models proposed for the DSDP sites in the Indian Ocean. Comparative data for the Atlantic and Pacific Oceans are also presented.

Summary and Conclusion

A review of the geochemistry of volcanics from the mid-oceanic ridges around the world with particular emphasis on the various tectonic provinces in the Indian Ocean suggests that the different types of basaltic rocks encountered are mainly the product of deep-seated processes occurring within an heterogenous mantle. The data gathered show that only partial (about 60%) agreement exists between the geochemical interpretation and the structural models proposed for the DSDP sites from the Indian Ocean. The model repartition of the different types of basalts with respect to their setting indicates that the aseismic ridges consist mostly of quartz–tholeiite, the intra-plate areas are more of alkali–olivine–tholeiite types, and the seamounts are made up of a variety of rock types such as alkali–olivine basalt, olivine–tholeiite and quartz–tholeiite.

The recent spreading ridges from the Indian, Pacific and Atlantic Oceans display variation from quartz-normative tholeiites to nepheline-normative alkali basalts, with the greatest concentration in the olivine tholeiite field. The Mid-Indian Oceanic Ridge (MIOR) is relatively less depleted in LIL elements and enriched in Sr isotopes than the East Pacific Rise (EPR) and the Mid-Atlantic Ridge (MAR). However, these three ridges are all characterized by depleted REE patterns. In general the volcanic rocks from these three spreading ridges show a negative correlation in the K–K/Rb and K–K/Cs variation diagrams, and a positive correlation in the K–K/U variation diagram, suggesting continuous depletion in the mantle.

In sharp contrast to the spreading ridges, *the aseismic Ninetyeast Ridge* (DSDP sites 214, 216, 215, 253 and 254) is characterized by the presence of picritic to quartz-normative basalts and oceanic andesites. These rocks usually have a higher Fe_2O_3+FeO ($>10\%$), TiO_2 ($>2\%$) and K_2O (0.2 to 1.5%) content, and are also enriched in LIL and REE and $^{87}Sr/^{86}Sr$ ratios (0.7053 to 0.7049) and depleted in Ni, Cr, V and Cu when compared to mid-oceanic ridge basalts.

The crustal composition of the *abyssal plains regions* from the Indian Ocean consists of volcanics which resemble both recent mid-oceanic ridge and island types of basalts. The basaltic rocks from the Wharton Basin at sites 212, 213 and 257 (bottom cores) have depeleted LIL and REE patterns similar to mid-oceanic ridge basalts (MORB).

The sites (260, 261, 259) from abyssal plain located between the Wharton Basin and the Australian continental shelf region yielded basalts with relatively depleted REE patterns but variable $[La/Sm]_{e.f.}$ (0.85–1.04) and $^{87}Sr/^{86}Sr$ (0.7039–0.7043) ratios. Geochemically these rocks lie between MORB and island-type tholeiite and therefore are compatible with the seafloor-spreading history for these areas as suggested by tectonic models. Other sites (256 and top cores of 257) from the abyssal plain regions show LIL- and LREE-enriched patterns which differentiate them from normal MORB. In particular, at site 211, the alkali-rich diabase and the hornblendite intrusives show marked enrichment in LIL and REE and appear to be related to the volcanism producing the nearby Cocos, Keeling and Christmas Islands.

The basement composition in the Western Indian Ocean (DSDP sites 235, 236, 240, 238, 239 and 245) mostly resembles the Mid-Indian Oceanic Ridge volcanics with low LIL and higher transitional metal elements, consistent with a spreading tectonic model. However, basalts at sites 248 and 250 from the Mozambique basin are enriched in LIL and LREE relative to MORB, and are similar to basalts from Amsterdam and Kerguelen areas which suggest that the basement at these Mozambique basin sites was produced by similar processes.

Volcanic rocks from different tectonic environments—recent spreading mid-oceanic ridges, aseismic ridges, seamounts, abyssal hills and islands—display positive correlations in the $K_2O/(K_2O+Na_2O)$ vs. $^{87}Sr/^{86}Sr$ and $^{87}Sr/^{86}Sr$ vs. $[La/Sm]_{e.f.}$ variation diagrams, probably suggesting multiple episodes of magma extrusion reflecting isotopic and chemical heterogeneities in the mantle. The distribution of $^{87}Sr/^{86}Sr$ ratios from the Indian Ocean are interpreted in terms of "zoning-depletion model" with the presence of alkali-poor and alkali-rich zones in the mantle. It is significant that the MIOR is relatively depleted in LIL, LREE and Sr isotopes in contrast to seamounts, abyssal hills, islands and aseismic ridges which are probably derived from hot spot activity, suggesting a definite relationship between tectonics and geochemical provinces. In other words, a region of the mantle having a relatively lower $^{87}Sr/^{86}Sr$ ratio may feed a spreading MIOR, while a zone with higher $^{87}Sr/^{86}Sr$ ratios may yield aseismic ridges, abyssal hills, islands and seamounts (Subbarao, 1972; Subbarao *et al.*, in press; Reddy *et al.*, in press). In terms of this model it is possible that the Ninetyeast Ridge, Naturaliste Plateau, the Broken Ridge–Kerguelen Plateau, Cocos-Keeling and Christmas islands chain may have been formed from a region of the mantle enriched with LIL and LREE.

Acknowledgements

One of us (K.V.S.) is grateful to the organizers of the IAGC–UNESCO Symposium held in Paris in May 1977 for providing part of his travel expenses. Also to the British Council and British Museum (Natural History), London, for providing a subsistence grant and office space, respectively, during May-June 1977, thus helping in preparing the background for this paper through extensive discussion with the staff of the Department of Mineralogy. Grateful thanks are also due to Dr. P. Henderson for reviewing the manuscript.

References

Balashov, Yu. A., Dmitriev, L. V. and Sharaskin, A. Ya. (1970) Distribution of rare earths and yttrium in the bedrock of the ocean floor. *Geokhimiya,* **6**, 647–660 (In Russian). Translated in *Geochem. Int.* **7**, 456–468.

Bezrukov, P. L., Krylov, A. Ya. and Chernysheva, V. I. (1960) Petrography and absolute age of basalts from the floor of the Indian Ocean. *Oceanology* **6**, 210–214.

Bonatti, E., Harrison, C. G. A., Fisher, D. E., Honnorez, J., Schilling, J. -G., Stipp, J. J. and Zentilli, M. (1977) Easter volcanic chain (south-east Pacific): a mantle hot line. *J. Geophy. Res.* **82**, 2457–2478.

Bougault, H. (1974) Distribution of first series transition metals in rocks recovered during DSDP Leg 22 in the northeastern Indian Ocean. In *Initial Reports of the Deep Sea Drilling Project,* C. C. von der Borch, J. G. Sclater *et al.*, Vol. 22. Washington: U.S. Government Printing Office, pp. 449–457.

Bowin, C. V. (1973) Origin of the Ninetyeast Ridge from studies near the Equator. *J. Geophys. Res.* **78**, 6029–6043.

BRYAN, W. B., THOMPSON, G., FREY, F. A. and DICKEY, J. S. (1976) Inferred geologic settings and differentiation in basalts from the Deep-Sea Drilling Project. *J. Geophys. Res.* **81**, 4285–4304.

CANN, J. R. (1969) Spilites from the Carlsberg Ridge, Indian Ocean. *J. Petrol.* **10**, 1–19.

CANN, J. R. (1971) Major element variations in ocean-floor basalts. *Philos. Trans. Roy. Soc., Lond.*, A, **269**, 495–505.

CANN, J. R. and VINE, F. J. (1966) An area on the crest of the Carlsberg Ridge: petrology and magnetic survey. *Philos. Trans. Roy. Soc., Lond.*, A, **259**, 198–217.

CHAYES, F. (1972) Silica saturation in Cenozoic basalt. *Philos. Trans. Roy. Soc., Lond.*, A, **271**, 285–295.

CHERNYSHEVA, V. Y. and BEZRUKOV, P. L. (1966) Serpentinite from the crest of the Indo-Arabian Ridge. *Dokl. Akad. Sci.* **166**, 207–210.

DASCH, E. J., HEDGE, C. E. and DYMOND, J. (1973) Effect of seawater interaction on strontium isotope composition of deep-sea basalt. *Earth Planet. Sci. Lett.* **19**, 177–183.

DETRICH, R. S., SCLATER, J. G. and THIEDE, J. (1977) The subsidence of aseismic ridges. *Earth Planet. Sci. Lett.*, **34**, 185–196.

DICKEY, J. S., FREY, F. A. and WATSON, E. B. (1974) Basalts from the South Atlantic triple junction. Abstracts with Programs, Geol. Soc. Amer. Ann. Mtg. 6(7).

DMITRIEV, V. (1974) Petrochemical study of the basaltic basement of the Mid-Indian Ridge: Leg 24, Djibouti to Mauritius. In *Initial Reports of the Deep Sea Drilling Project,* R. L. FISHER, E. T. BUNCE *et al.*, Vol. 24, Washington: U.S. Government Printing Office, 767–779.

ENGEL, A. E. J., ENGEL, C. G. and HAVENS, R. G. (1965) Chemical characteristics of oceanic basalts and the upper mantle. Bull. Geol. Soc. Amer. **76**, 719–734.

ENGEL, C. G. and FISHER, R. L. (1969) Lherzolite, anorthosite, gabbro, and basalt dredged from the Mid-Indian Ocean Ridge. *Science,* **166**, 1136–1141.

ENGEL, C. G. and FISHER, R. L. (1975) Granitic to ultramafic rock complexes of the Indian Ocean Ridge system, western Indian Ocean. *Bull. Geol. Soc. Amer.* **86**, 1553–1578.

ENGEL, C. G., BINGHAM, E. and FISHER, R. L. (1974) Trace element compositions of Leg 24 basalts and one diabase. In *Initial Reports of the Deep Sea Drilling Project,* R. L. FISHER, E. T. BUNCE, *et al.*, Vol. 24, Washington: U.S. Government Printing Office, 543–551.

ENGEL, C. G., FISHER, R. L. and ENGEL, A. E. J. (1965) Igneous rocks of the Indian Ocean floor. *Science* **150**, 605–610.

ERLANK, A. J. and REID, D. L. (1974) Geochemistry, minerology and petrology of basalts, Leg 25, Deep Sea Drilling Project. In *Initial Reports of the Deep Sea Drilling Project,* E. S. W. SIMPSON, R. SCHLICH *et al.*, Vol. 25, Washington: U. S. Government Printing Office, 543–551.

FLEET, A. J., HENDERSON, P. and KEMPE, D. R. C. (1976) Rare earth element and related chemistry of some drilled southern Indian Ocean basalts and volcanogenic sediments. *J. Geophys. Res.* **81**, 4257–4268.

FODER, R. V., HUSLER, J. W. and KUMAR, N. (1977) Petrology of volcanic rocks from an aseismic rise: Implications for the origin of the Rio Grande Rise, South Atlantic Ocean. *Earth Planet. Sci. Lett.*, **35**, 225–233.

FREY, F. A., BRYAN, W. B. and THOMPSON, G. (1974) Atlantic Ocean floor: geochemistry and petrology of basalts from Legs 2 and 3 of the Deep Sea Drilling Project. *J. Geophys. Res.* **79**, 5507–5527.

FREY, F. A., DICKEY, J. S., THOMPSON, G. and BRYAN, W. B. Eastern Indian Ocean DSDP sites: correlations between petrography, geochemistry, and tectonic setting. *Deep-Sea Drilling in the Indian Ocean,* A.G.U. Monograph (in press).

FREY, F. A. and HASKIN, L. A. (1964) Rare earths in oceanic basalts. *J. Geophys. Res.* **69**, 775–780.

FREY, F. A. and SUNG, C. M. (1974) Geochemical results for basalts from sites 253 and 254. In *Initial Reports of the Deep Sea Drilling Project,* T. A. DAVIES, B. P. LUYENDYK *et al.*, Vol. 26, Washington: U.S. Government Printing Office, 567–571.

GAST, P. W. (1968) Trace element fractionation and the origin of tholeiitic and alkaline magma types. *Geochim. Cosmochim. Acta,* **32**, 1057–1086.

GLADKIKH, V. S. and CHERNYSHEVA, V. I. (1966) Rare earth elements in sub-oceanic mafic extrusives. *Geochim. Cosmochim. Acta,* **3**, 786–789.

HART, S. R. (1971) K, Rb, Cs, Sr and Ba contents and Sr isotope ratios of ocean floor basalts. *Philos. Trans. Roy. Soc., Lond.*, A, **268**, 573–587.

HART, S. R. SCHILLING, J. G. and POWELL, J. L. (1973) Basalts from Iceland and along the Reykjanes Ridge: Sr isotope geochemistry. *Nature (Phys. Sci.)* **246**, 104–107.

HEDGE, C. E. and PETERMAN, Z. E. (1970) The strontium isotopic composition of basalts from the Gorda and Juan de Fuca Rises, northeastern Pacific Ocean. *Contr. Mineral Petrol.* **27**, 114–120.

HEKINIAN, R. (1968) Rocks from the Mid-Oceanic Ridge in the Indian Ocean. *Deep Sea Res.* **15**, 195–213.

HEKINIAN, R. (1970) Gabbro and pyroxenite from a deep-sea core in the Indian Ocean. *Marine Geol.* **9**, 287–294.

HEKINIAN, R. (1972) Volcanics from the Walvis Ridge. *Nature (Phys. Sci.)* **239**, 91–93.

HEKINIAN, R. (1974a) Petrology of igneous rocks from Leg 22 in the north-eastern Indian Ocean. In *Initial Reports of the Deep Sea Drilling Project*, C. C. VON DER BORCH, J. G. SCLATER *et al.*, Vol. 22, Washington: U.S. Government Printing Office, 413–447.

HEKINIAN, R. (1974b) Petrology of the Ninetyeast Ridge (Indian Ocean) compared to other aseismic ridges. *Contr. Mineral Petrol.* **43**, 125–147.

HEKINIAN, R. and SUBBARAO, K. V. (1976) Synthesis of geochemical results on rocks obtained from the Deep Sea Drilling Project in the Indian Ocean. Int. Geol. Congress, Sydney.

HEEZEN, B. C. and NAFE, J. A. (1964) Vema trench: Western Indian Ocean. *Deep Sea Res.* **118**, 79–84.

JACKSON, E. D., BARGAR, K. E., FABBI, B. P. and HEROPOULOS, C. (1976) Petrology of the basaltic rocks drilled on Leg 33 of the Deep Sea Drilling Project. In *Initial Reports of the Deep Sea Drilling Project*, S. O. SCHLANGER, E. D. JACKSON *et al.*, Vol. 33, Washington: U.S. Government Printing Office, 571–630.

JOHNSON, B. D., POWELL, C. MCA. and VEEVERS, J. J. (1976) Spreading history of the eastern Indian Ocean and greater India's northward flight from Antarctica and Australia. *Geol. Soc. Amer. Bull.* **87**, 1560–1566.

KAY, R. and GAST, P. W. (1973) The rare earth content and origin of alkali-rich basalts. *J. Geol.* **81**, 653–682.

KAY, R., HUBBARD, N. J. and GAST, P. W. (1970) Chemical characteristics and origin of oceanic ridge volcanic rocks. *J. Geophys. Res.* **75**, 1585–1613.

KEMPE, D. R. C. (1974) The petrology of the basalts, Leg 26. In *Initial Reports of the Deep Sea Drilling Project*, T. A. DAVIES, LUYENDYK *et al.*, Vol. 26, Washington: U.S. Government Printing Office, 465–503.

KEMPE, D. R. C. (1975) Normative mineralogy and differentiation patterns of some drilled and dredged oceanic basalts. *Contr. Mineral. Petrol.* **50**, 305–320.

KEMPE, D. R. C. (1976) Petrological studies on DSDP Leg 34 basalts: Nazca plate, eastern Pacific Ocean. In *Initial Report of the Deep Sea Drilling Project*, R. S. YEATS, S. R. HART *et al.*, Vol. 34, Washington: U.S. Government Printing Office, 189–213.

KEMPE, D. R. C. and SCHILLING, J. G. (1974) Discovery Tablemount basalt: petrology and geochemistry. *Contr. Mineral. Petrol.* **44**, 101–115.

KHARIN, G. N. (1976) The petrology of magmatic rocks, DSDP Leg 38, In *Initial Reports of the Deep Sea Drilling Project*, M. TALWANI, G. UDINTSEV, *et al.*, Vol. 38, Washington: U.S. Government Printing Office, 685–715.

KORZHINSKIY, D. S. (1962) Spilites and the transvaporization hypothesis in the light of new oceanological and volcanological data. *Izv. Acad. Sci. USSR* **9**, 13–19.

LUYENDYK, B. P. (1974) Gondwanaland dispersal and the early formation of the Indian Ocean. In *Initial Reports of the Deep Sea Drilling Project*, T. A. DAVIES, B. P. LUYENDYK, *et al.*, Vol. 26, Washington: U.S. Government Printing Office, 945–952.

LUYENDYK, B. P. and DAVIES, T. A. (1974) Results of DSDP Leg 26 and the geologic history of the southern Indian Ocean. In *Initial Reports of the Deep Sea Drilling Project*, T. A. DAVIES, B. P. LUYENDYK *et al.*, Vol. 26, Washington: U.S. Government Printing Office, 909–943.

MARSHALL, M. C. (1975) Petrology and chemical composition of basaltic rocks recovered on Leg 32, Deep Sea Drilling Project. In *Initial Reports of the Deep Sea Drilling Project*, R. L. LARGON, R. MOBERLY *et al.*, Vol. 32, Washington: U.S. Government Printing Office, 536–570.

MCKENZIE, D. P. and SCLATER, J. G. (1971) The evolution of the Indian Ocean since the Late Cretaceous. *Geophys. J. R. Astron. Soc.* **25**, 437–528.

MELSON, W. G. and THOMPSON, G. (1971) Petrology of a transform fault zone and adjacent ridge segments. *Philos. Trans. Roy. Soc., Lond.*, **268**, 423–441.

MELSON, W. G., VALLIER, T. L., WRIGHT, T. L., BYERLY, G. and NELEN, J. (1976) Chemical diversity of abyssal volcanic glass erupted along Pacific, Atlantic, and Indian Ocean sea-floor spreading centers. In *The Geophysics of the Pacific Ocean Basin and its Margin*, Geophys. Monograph 19, Washington: A.G.U. 351–368.

MORGAN, W. J. (1971) Convection plumes in the lower mantle. *Nature, Lond.*, **230**, 42–43.

MORGAN, W. J. (1972) Deep mantle convection plumes and plate motions. *Bull. Amer. Ass. Petrol. Geol.* **56**, 203–213.

O'NIONS, R. K. and PANKHURST, R. J. (1974) Petrogenetic significance of isotope and trace element variations in volcanic rocks from the Mid-Atlantic, *J. Petrol.* **15**, 603–634.

PETERMAN, Z. E. and HEDGE, C. E. (1971) Related strontium isotopic and chemical variations in oceanic basalts. *Geol. Soc. Amer. Bull.*, **82**, 493–500.

RAMANAN, S., REDDY, V. V., SUBBARAO, K. V. and HEKINIAN, R. (1976) Trace element distributions in igneous rocks from the Indian Ocean. Unpublished Report, I.I.t., Bombay.

REDDY, V. V., SUBBARAO, K. V., REDDY, G. R., MATSUDA, J. and HEKINIAN, R. Geochemistry of volcanics from the Ninetyeast Ridge and its vicinities in the Indian Ocean. *Marine Geology* (in press).

REINISCH, R. (1908) Petrographic: II, Gesteine von der Bouvet—Insel, Kerguelen, St. Paul and New-Amsterdam. In *Wissenschaftliche Ergebnisse der Deutschen Tilfsee Expidition 1898–1899*, **10**, 47–75.

RIDLEY, W. I., PERFIT, M. R. and ADAMS, M. L. (1976) Petrology of basalts from Deep Sea Drilling Project, Leg 38. In *Initial Reports of the Deep Sea Drilling Project*, M. TALWANI, G. UDINTSEV *et al.*, Vol. 38, Washington: U.S. Government Printing Office, 731–739.

ROBINSON, P. T. and WHITFORD, D. J. (1974) Basalts from the eastern Indian Ocean, DSDP Leg 27. In *Initial Reports of the Deep Sea Drilling Project*, J. J. VEEVERS, J. R. HEIRTZLER *et al.*, Vol. 27, Washington: U.S. Government Printing Office, 551–559.

SCHILLING, J. -G. (1971) Sea-floor evolution: rare-earth evidence. *Philos. Trans. Roy. Soc. Lond.*, A, **268**, 663–706.

SCHILLING, J. -G. (1973) Iceland mantle plume: geochemical study of Reykjanes Ridge. *Nature, Lond.*, **242**, 565–571.

SCHILLING, J. -G. (1975) Azores mantle blob: rare-earth evidence. *Earth Planet. Sci. Lett.*, **25**, 103–115.

SCHILLING, J. -G. (1975) Rare-earth variations across "normal segments" of the Reykjanes Ridge, 60°–53°N, Mid-Atlantic Ridge, 29°S, and East Pacific Rise, 2°–19°S, and evidence on the composition of the underlying low-velocity layer. *J. Geophy. Res.* **80**, 1459–1473.

SCHILLING, J. -G., ANDERSON, R. N. and VOGT, P. (1976) Rare earth, Fe and Ti variations along the Galapagos spreading centre, and their relationship to the Galapagos mantle plume. *Nature, Lond., Phys. Sci.* **261**, 108–113.

SCHILLING, J. -G. and NOE-NYGAARD, A. (1974) Faeroe-Iceland plume: rare-earth evidence. *Earth Planet. Sci. Lett.*, **24**, 1–14.

SCHILLING, J. -G. and WINCHESTER, J. W. (1969) Rare-earth contribution to the origin of Hawaiian lavas. *Contr. Mineral. Petrol.* **23**, 27–37.

SCLATER, J. G. and FISHER, R. L. (1974) Evolution of the east Central Indian Ocean with emphasis on the tectonic setting of the Ninetyeast Ridge. *Geol. Soc. Amer. Bull.* **85**, 683–702.

SUBBARAO, K. V. (1972) The strontium isotopic composition of basalts from the East Pacific and Chile Rises and abyssal hills in the eastern Pacific Ocean. *Contr. Mineral. Petrol.* **37**, 111–120.

SUBBARAO, K. V., CLARK, G. S. and FORBES, R. B. (1973) Strontium isotopes in some seamount basalts from the northeastern Pacific Ocean. *Can. J. Earth Sci.* **10**, 1479–1484.

SUBBARAO, K. V. and HEDGE, C. E. (1973) K, Rb, Sr and $^{87}Sr/^{86}Sr$ in rocks from the Mid-Indian Oceanic Ridge. *Earth Planet. Sci. Lett.*, **18**, 223–228.

SUBBARAO, K. V. and HEKINIAN, R. Alkali enriched rocks from the central eastern Pacific Ocean. *Marine Geol.* (in press).

SUBBARAO, K. V., HEKINIAN, R. and REDDY, V. V. (1976) Geochemistry of oceanic rocks. Intern. Geol. Congress, Sydney.

SUBBARAO, K. V., REDDY, V. V. and HEKINIAN, R. (1975) Geochemistry and fission track geochronology of rocks from the Mid-Indian Oceanic Ridge (MIOR) and Leg 22 of the Deep Sea Drilling Project. IUGG XVI General Assembly, Grenoble.

SUBBARAO, K. V., REDDY, V. V. HEKINIAN, R. and CHANDRASEKARAM, D. *Large Ion Lithophile Elements and Sr and Pb Isotopic Variations in Volcanic Rocks from the Indian Ocean*, AGU. Monograph In Indian Ocean geology and Biostratigraphy, J. R. HEIRTZLER, H. M. BOLLI, T. A. DAVIES, J. B. SAUNDERS and J. G. SCLATER, ed., 259–278.

THOMPSON, G., BRYAN, W. B., FREY, F. A. and SUNG, C. M. (1974) Petrology and geochemistry of basalts and related rocks from sites 214, 215, 216, DSDP Leg 22, Indian Ocean. In *Initial Reports of the Deep Sea Project*, C. C. VON DER BORCH, J. G. SCLATER *et al.*, Vol. 22, Washington: U.S. Government Printing Office, 459–467.

UDINTSEV, G. B. and CHERNYSHEVA, V. I. (1965) Rock samples from the upper mantle of the Indian Ocean rift zone. *Dokl. Akad. Sci. U.S.S.R., Earth Sci. Sect.*, **165**, 85–88.

WHITFORD, D. J. (1975) Strontium isotopic studies of the volcanic rocks of the Sunda arc, Indonesia, and their petrogenetic implications. *Geochim. Cosmochim. Acta.* **39**, 1287–1302.

WISEMAN, J. D. H. and POOLE, J. H. I. (1937) Geological and mineralogical investigations of basalt from the Carlsberg Ridge, Indian Ocean. *Report of J. Murray Expedition 1933–1934*, British Museum Nat. History, **3**, 1–28.

WILSON, J. T. (1965) Evidence from oceanic islands suggesting movement in the earth. *Philos. Trans. Roy. Soc., Lond.*, A, **258**, 145–167.

YODER, H. S., JR. and TILLEY, C. E. (1962) Origin of basalt magmas: an experimental study of natural and synthetic rock systems. *J. Petrol.* **3**, 342–532.

ZIELINSKI, R. A. (1975) Trace element evaluation of a suite of rocks from Reunion Island, Indian Ocean. *Geochim. Cosmochim. Acta* **39**, 713–734.

The Lower Continental Crust of the Massif Central (Bournac, France)—With Special References to REE, U and Th Composition, Evolution, Heat-flow Production

C. DUPUY,† A. LEYRELOUP‡ and J. VERNIERES†

† *Centre Géologique et Géophysique, U.S.T.L. 34060 Montpellier Cedex, France*
‡ *Laboratoire de Pétrologie, U.S.T.L. 34060 Montpellier Cedex, France*

Abstract

From the analyses of xenoliths scavenged by the Bournac pipe (Massif Central) an estimated average composition of the lower crust is presented for major and twenty-four trace elements. This composition, relatively basic and close to the average of some andesites, suggests that the lower crust is not always depleted in LILE; if a residual character may be attributed to some rocks, it seems unlikely to consider all the lower crust as a restite. The heat-flow production (1.6–1.7 H.P.U.) is relatively high for lower crustal granulite facies rocks.

Introduction

In the present paper, the chemical data on the granulite facies xenoliths (Leyreloup *et al.*, 1977) scavenged by the Plio-quaternary of the southern and central part of French Massif Central, especially in the Bournac pipe (Velay, fig. 1) are completed by the determination of the following elements: La, Ce, Sm, Eu, Tb, Yb, Lu, Cs, Th, U, Sc, Ta, Zr and Hf. New and previous results are now integrated in order to discuss more precisely the origin, the nature and the average composition of the lower crust beneath Bournac locality and its heat-flow production.

General Considerations and Previous Studies

The studied xenoliths belong to a portion of a stable continental crust as suggested by recent geotectonic models (Mattauer, 1974; Mattauer and Etchecopar, 1976; Burg, 1977; Burg and Matte, 1977). Petrology and chemistry of these xenoliths (Leyreloup, 1973; 1974; Leyreloup *et al.*, 1977) define a metasedimentary sequence and a metaigneous suite with tholeiitic affinities. Both suites, affected by the same tectono-metamorphic events, have undergone a high-pressure granulite facies metamorphism, probably of Precambrian age (Allegre *et al.*, 1975; Hamet *et al.*, 1977) accompanied by an incipient protomobilization most obvious in the acidic rocks. As observed in the xenolithic samples, this primary metamorphic episode was followed by a widespread granulite retrometamorphism (intermediate to low-pressure granulite facies) due to a Variscan and Tertiary post-tectonic uplift beneath the French Massif Central. Among the successive mineralogical assemblages generated during this complex petrologic history, the latest ones and their fluid inclusion

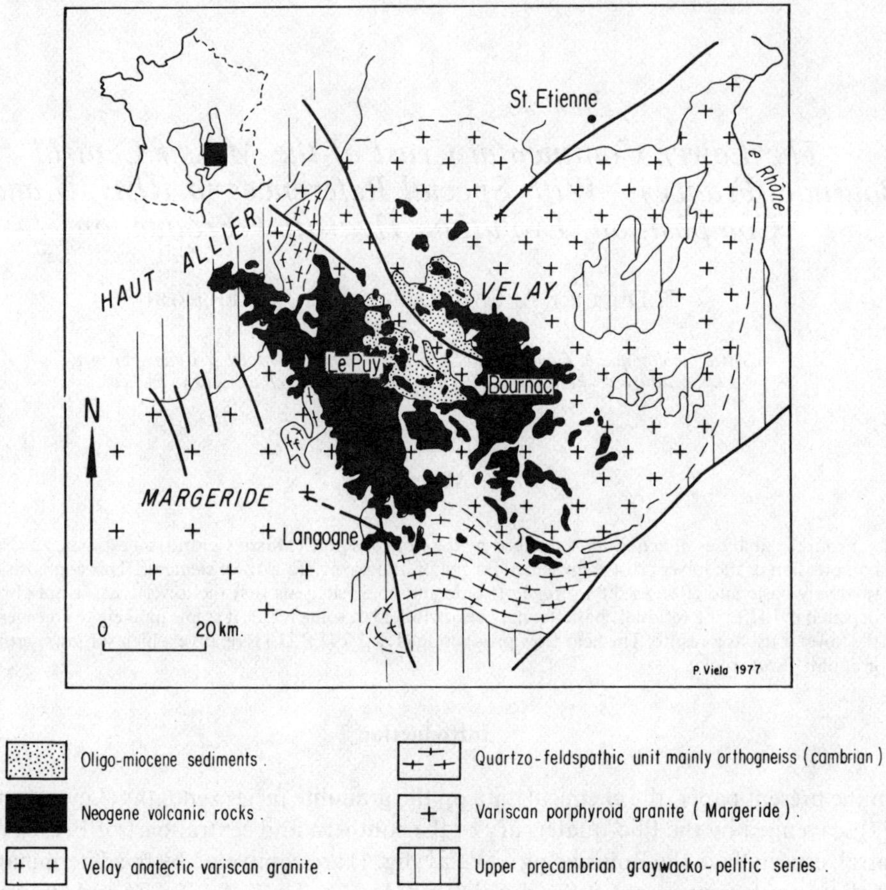

FIG. 1. Geological sketch map of the Velay region (Eastern Massif Central).

studies (Leyreloup, 1974; Bilal, 1976; Leyreloup and Bilal, 1977) locate this granulite facies basement approximately between 25–26 km and 12–15 km depth. The core of xenoliths analyzed is not significantly affected by the thermal metamorphism induced by the basaltic breccia during its ascent and thus furnish available informations about the chemistry of the lower crust beneath Bournac locality during the Neogene epoch. Furthermore, an average composition has been calculated from the analyzed xenoliths and their respective proportions (Table 1). The type rock proportions are calculated assuming that the xenoliths provide a representative sampling of the whole lower crust beneath Bournac and that the volumetric percentage of the various kind of rocks at depth is proportional to their frequency occurrence in the outcropping basaltic breccias. The frequency occurrence has been estimated after petrographic study of more than 2000 xenoliths. The results on major and some trace elements (Li, Rb, Sr, Ba, V, Cr, Co, Ni, Cu, Zn) recently discussed in detail elsewhere (Leyreloup *et al.*, 1977) may be summarized as follows:

—the average established composition differs from various continental crust estimates, by lower SiO_2, Na_2O and K_2O and higher Fe_2O_3 and MgO content. Approximately, the same differences appear when the comparison is extended to the average of surface shields.

TABLE 1. AVERAGE COMPOSITION OF EACH ROCK GROUPS ENCOUNTERED AMONG THE XENOLITHS.

Reference	1	2	3	4	5	6	7	8	9
n	6	25	5	2	5	4	7	9	3
Major elements (%)									
SiO_2	48.8	49.8	63.2	46.7	57.5	65.3	56.0	61.8	69.6
Al_2O_3	8.5	17.5	16.7	23.8	20.5	16.1	19.1	16.9	14.4
Fe_2O_3	4.8	4.8	2.3	7.8	3.3	2.3	3.7	3.8	2.0
FeO	6.8	5.0	3.4	4.9	6.0	4.8	4.3	3.0	2.0
MnO	0.13	0.13	0.07	0.15	0.15	0.07	0.08	0.08	0.06
MgO	18.4	6.9	2.1	3.9	3.2	2.7	3.2	2.6	1.3
CaO	8.6	8.8	4.3	2.7	1.1	1.1	4.3	3.0	1.8
Na_2O	0.5	2.1	3.4	3.4	1.3	1.3	2.8	2.7	2.2
K_2O	0.27	0.36	1.31	1.62	2.85	2.80	2.36	2.10	4.27
TiO_2	0.6	1.4	0.7	0.7	1.1	1.0	0.9	0.9	0.6
P_2O_5	0.09	0.21	0.20	0.17	0.09	0.16	0.16	0.17	0.12
H_2O^+	0.9	1.3	0.5	2.0	0.4	0.3	1.2	0.7	0.5
H_2O^-	1.6	1.5	1.4	1.2	1.3	0.8	1.1	1.3	0.5
Trace elements (ppm)									
Li	7.0	9.0	5.4	3.5	4.6	3.0	9.7	6.4	3.3
Rb	13	7	20	28	70	45	32	32	96
Cs	1.4	2.0	1.1	1.4	0.3	0.2	2.4	0.9	0.2
Sr	125	327	289	249	162	182	355	265	179
Ba	40	297	626	325	773	900	723	792	770
La	6.5	11.3	33.0	42.1	48.9	46.0	30.0	37.6	34.1
Ce	20.8	27.6	69.6	71.0	109.4	89.0	56.1	84.1	65.6
Sm	3.8	4.1	6.4	6.4	9.1	9.5	5.8	6.6	5.6
Eu	1.14	1.30	1.71	1.75	1.75	1.83	1.87	1.81	1.43
Tb	0.37	0.57	0.81	1.64	1.24	1.01	0.80	0.89	1.09
Yb	1.20	2.00	2.96	13.50	4.60	3.72	2.91	3.50	4.65
Lu	0.14	0.34	0.49	2.32	0.78	0.60	0.47	0.54	0.72
U	0.28	0.27	0.43	1.12	1.26	1.22	0.52	0.71	0.53
Th	0.69	1.0	1.5	1.8	20.8	17.0	4.4	7.2	6.3
Zr	40	89	233	807	266	266	262	254	196
Hf	1.1	2.5	6.3	20.2	6.8	6.8	6.6	6.7	5.5
Ta	2.5	0.5	0.4	1.8	1.5	1.7	0.7	0.7	1.1
Sc	32	30	15	47	27	22	25	18	13
V	268	211	107	173	148	130	162	116	67
Cr	1016	324	75	116	103	105	103	104	45
Co	71	39	15	51	23	12	17	18	31
Ni	319	76	21	47	44	18	32	37	69
Cu	40	21	15	34	20	36	20	26	14
Zn	91	85	60	167	115	102	77	89	43
Weighting factors (%)	3.5	44	10	0.5	10	10	2	15	5

n: number of samples.

Meta-igneous rocks: charnockitic ultra-basic cumulates (1); basic charnockites (2); intermediate and acidic charnockites (3).

Metasedimentary rocks: per-aluminous granulites (4); khondalito-kinzigitic metapelitic rocks (5 and 6); hypersthene meta-graywackes or meta-arkose (7); Al_2SiO_5-free granulites (8 and 9).

TABLE 2. COMPARISON OF VARIOUS ESTIMATE COMPOSITIONS (MAJOR ELEMENTS)

	1	2	3	4	5	6	7	8	9	10
SiO_2%	56.3	60.2	61.9	62.1	66.7	60.7	54.8	57.9	57.6	54.2
Al_2O_3	17.1	15.6	15.6	15.5	15.0	15.8	15.5	15.6	15.4	17.2
Fe_2O_3†	8.8	7.9	6.9	6.8	4.5	1.7	11.3	7.0	8.4	8.4
MnO	0.108	0.1	0.1	0.11	0.07	0.11	0.2		0.15	0.2
MgO	5.0	3.9	3.1	3.8	2.3	3.0	4.4	3.0	3.7	4.4
CaO	5.5	5.8	5.7	6.6	4.2	5.3	7.4	6.9	5.6	7.9
Na_2O	2.1	3.2	3.1	3.4	3.6	3.5	2.1	4.2	3.7	3.7
K_2O	1.42	2.5	2.9	1.0	3.2	2.3	1.7	1.3	0.8	1.1
TiO_2	1.1	1.0	0.8	0.6	0.5	0.9	1.3	1.2	1.0	1.3
P_2O_5	0.16	0.2	0.3				0.3		0.2	0.3
H_2O	2.2					0.77		2.4	2.2	0.9

† Total Fe as Fe_2O_3.
1. Lower crust beneath the Bournac pipe (this paper).
2. Continental crust (Taylor, 1964).
3. Continental crust (Ronov and Yaroshevskiy, 1969).
4. Scourie assemblage (Holland and Lambert, 1972).
5. Canadian shield (Shaw *et al.*, 1967, 1976).
6. Granulitic basement, Bahia, Brazil (Sighinolfi, 1969; Sighinolfi and Sakai, 1977).
7. Fraser Range granulites, Australia (Lambert and Heier, 1968b).
8. Archean andesites from South Pass (Condie and Harrison, 1976).
9. Archean andesites from the Canadian Shield (Baragar and Goodwin, 1969; Condie and Harrison, 1976).
10. Average andesite (Wedepohl, 1969).

However, the values we propose are close to those of the Fraser Range granulitic formation especially for SiO_2, MgO, Na_2O, K_2O and TiO_2. This similarity results from the proportion of basic terms which are important in both cases. In the studied rocks, the calculation of the average implies 44% of basic charnockites. This proportion is not surprising since several authors (Wilkinson, 1973, 1974; Irving, 1974; Rogers and Nixon, 1975) have already noticed the predominance of basic rocks among localities where xenoliths have been described. Furthermore, if we accept that these granulitic xenoliths represent the lower crust beneath the Bournac pipe, this assessment agrees with the more basic character generally inferred for the crust at depth. Finally, the mean composition calculated (Na_2O expected) is also similar to the major element content of andesites especially those of Archean age (Table 2);

—the estimated contents of Li, Rb are lower than the crustal average proposed by several authors (e.g. Table 3, columns 2–3). The Li concentration is close to those of Bahia granulitic basement (Sighinolfi, 1969) and andesitic rocks (Dupuy and Lefèvre, 1974; Coulon, 1977). The concentration of Rb attains the values encountered in island arc and Archean andesites (Table 3, columns 8, 9, 11). The K/Rb ratio (318) is intermediate between K/Rb values of surface continental crust (230–260) and usual granulite facies rocks values. The calculated grand mean of Sr is slightly lower than the crustal values obtained by Eade and Fahrig (1971) of Shaw *et al.* (1976) but close to Canadian Archean andesites (Table 3, column 9). The Rb/Sr ratio (0,1) is more than half the average crustal values;

—some elements like V, Co, Ni and Zn display similar content to those of Taylor crustal average. Our data (Cr and Cu excepted) are close to the andesite average reported on the Table 2 (columns 9, 10, 11). The high Cr content is not different from the average presented by Baragar and Goodwin (1969) for the Canadian Archean volcanic rocks. The enrichment of Cr in our average might reflect the important proportion of basic rocks. On the other hand, the Cu depletion might result from the metamorphism. This low Cu content has been already enregistered by Shaw *et al.* (1967) for the Canadian shield.

TABLE 3. COMPARISON OF VARIOUS ESTIMATE COMPOSITIONS (TRACE ELEMENTS)

Reference	1	2	5	6	7	8	9	10	11	12	13	14
Li	7	20	22	10	—	—	—	13	—	—	—	13
Rb	27	90	110	46	45	20	30	64	31	—	—	21
Cs	1.3	3.0	—	—	—	1.1	0.4	—	0.5	3.2	12	1.0
Sr	266	375	316	543	170	408	263	638	385	—	—	192
Ba	545	425	1076	1207	420	420	248	964	270	523	390	177
La	26	30	32	—	—	37	26	37	12	15	38	10.5
Ce	57	60	66	—	—	72	46	76	24	31	80	22.3
Sm	5.8	6	4.5	—	—	7	6.2	5.6	2.9	2.7	5.6	3.2
Eu	1.5	1.2	0.94	—	—	1.8	1.5	1.5	1.0	1.1	1.1	1.0
Tb	0.78	0.9	0.48	—	—	1.1	1.5	0.65	0.68	0.44	0.77	0.59
Yb	2.98	3.0	1.5	—	—	2.7	4.0	2.0	1.9	1.73	2.8	2.3
Lu	0.48	0.5	0.23	—	—	0.62	0.63	0.35	0.4	—	0.50	0.31
U	0.57	2.7	2.5	0.70	—	—	—	0.76	0.69	1.4	2.9	0.52
Th	5.8	9.6	10	13.34	—	—	—	—	2.2	4.3	14	2.2
Zr	174	165	240	—	—	—	160	—	110	—	—	100
Hf	4.6	3.0	5.8	—	—	—	—	—	2.3	2.9	4.8	2.2
Ta	0.8	2.0	—	—	—	—	—	—	—	—	—	0.5
Sc	25	22	7	—	—	—	—	—	30	—	—	31
V	167	135	53	—	—	—	201	150	175	—	—	257
Cr	220	100	35	71	—	100	103	56	56	—	—	121
Co	29	25	12	—	—	32	18	21	24	—	—	47
Ni	61	75	19	—	—	98	63	37	18	—	—	77
Cu	22	55	14	39	—	—	60	53	54	—	—	115
Zn	85	70	52	94	—	—	86	91	—	—	—	85

1 to 9. See legend Table 2.
10. Modern calc–alkali andesites from Peru compiled by the senior author from various sources.
11. Modern calc–alkali andesites from Peru (Condie and Harrison, 1976; Taylor *et al.*, 1969).
12. Archean sediments, Australia (Nance and Taylor, 1977).
13. Post-Archean sediments, Australia (Nance and Taylor, 1976).
14. Values obtained on the U.S. Geol. Standard sample W_1.

Analytical Methods

Rare elements (REE), Cs, Th, U, Sc, Ta, Zr and Hf have been analyzed by instrumental neutronic activation on composite powders obtained by mixing the samples according to their nature and SiO_2 content. A method adapted from that of Gordon *et al.* (1968) was used. Furthermore, U and Th determinations have been checked by isotopic dilution. Analytical precision varies with the element and its abundance; usually it was range of 1–12% of amount present (coefficient of variation). Accuracy may be estimated from the analysis of the standard reference rocks. Values obtained on W_1 are reported in Table 3.

Presentation of Data

The data of each rock group presented with previous results (Leyreloup *et al.*, 1977) in Table 1 are obtained from a weighted average of individual values for the elements determined by A.A.S. (previous study) and from the analyses of composites for the others. There is a good agreement between determinations obtained in the two different ways.

Cs, Zr, Hf, Sc, Ta

The metaigneous tholeiitic suite is characterized by lower Zr, Hf content than the metasedimentary sequence. Cs, Ta and Sc show no obvious difference between the two groups. However, the Cs content is generally higher than the values proposed for the high-grade metamorphic rocks (Heier and Brunfelt, 1970). The corresponding K/Cs ratios are quite low in metaigneous rocks but high in metasedimentary samples which are thus well discriminated. Furthermore, Cs follows Li rather than Rb. This anomalous relationship is not presently understood in such rocks. Within the metasedimentary group, difference appear from one to another type of rocks: for instance, a particular increase of Zr and Hf content may be observed in the aluminous metasedimentary rocks (Table 1, column 4).

U, Th

The foregoing observations are corroborated by U, Th which vary through a large range in accordance with the major element content but also with the rocks types: U concentrate around 0.27–1.26 ppm, Th 0.69–20.8 ppm. The range of variations is lower for U. This observation was already noticed by Sighinolfi and Sakai (1977) for the Bahia granulite facies rocks. The averaged U content reported in the Table 1 are near to those published for other granulite facies terrains (Heier and Adams, 1965; Lambert and Heier, 1967, 1968a; Roelandts, 1973; Del Fiore *et al.*, 1974; Killeen and Heier, 1975; Scharbert *et al.*, 1976) but lower than the values for Canadian granulite facies rocks (4,16 ppm, Burwash and Cumming, 1976). The previous remarks may be extended to Th if the khondalito-kinzigitic paragneisses (KKP) (Table 1, columns 5, 6) are expected. The Th content and Th/U ratios of these paragneisses are among the highest values reported elsewhere (Sighinolfi and Sakai, 1977). Although U and Th tend to increase with K_2O and SiO_2 (Table 1) there is no systematic relationship between U and Th. The U/Th ratios vary in a large range and discriminate nicely the metasedimentary rocks from the metaigneous ones (Table 4), thus reflecting the nature of the premetamorphic material (Petrov *et al.*, 1972). Furthermore, U is

TABLE 4. HEAT-FLOW PRODUCTION VALUES ACCORDING TO XENOLITH TYPES. (Nos. 1 to 9 refer to Table 1)
H.P. (heat production) in units of 10^{-13} cal cm^{-3} s^{-1}. All values computed from given U (ppm), Th (ppm), K (per cent) and d (g cm^{-3}) data, using HPU = 0.317d (0.73U + 0.20 Th + 0.27 K) (after Birch *et al.*, 1968).

Reference	1	2	3	4	5	6	7	8	9
d(measured)	3.01(0.13)	2.74(0.09)	2.62(0.13)	2.90(0.1)	2.72(0.08)	2.61(0.09)	2.65(0.07)	2.65(0.02)	2.54(0.08)
Th/U	2.46	3.70	3.49	1.61	16.51	13.93	8.46	10.1	11.89
$Th/K \times 10^4$	3.1	3.3	1.4	1.3	8.8	7.3	2.2	4.1	1.8
$U/K \times 10^4$	1.3	0.9	0.39	0.84	0.53	0.53	0.26	0.41	0.15
H.P.U.	0.38	0.42	0.76	1.42	4.95	4.09	1.51	2.02	2.11

( ) Standard deviation;

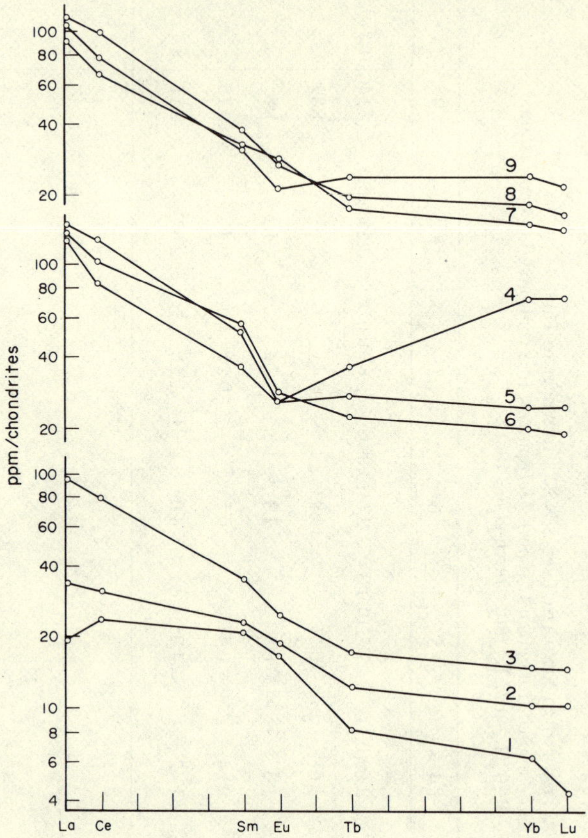

FIG. 2. REE abundances normalized to chondrites in composite samples. The numbers refer to the columns 1 to 9 of the Table 1.

more likely related to K suggesting similar behaviour during the complex petrogenetic history of these xenoliths (Leyreloup, 1973, 1974). However, in this study, implying composite samples analysis, it is difficult to argue in more detail about the relative mobility of these radioactive elements during metamorphic processes.

REE

The lowest REE contents occur in the basic members of the metaigneous suite which display a continental tholeiitic pattern (e.g. see Philpotts and Schnetzler, 1968). The peraluminous granulites (Table 1, column 4) display a particular pattern with an increase of HREE (Fig. 2) reflecting the presence of refractory minerals (garnet, Al_2SiO_5, zircon). All other samples have quite similar REE content and display an enrichment of LREE and no HREE fractionation. Their patterns, characteristic of calc-alkaline suites, may suggest some affinities with such rocks, although this hypothesis is not supported by other major or trace elements (e.g. Ca, Na, K, Sr, Rb). Such patterns are also observed in metasedimentary rocks from other areas (Nance and Taylor, 1976, 1977; Wildeman and Haskins, 1973). If there is little or no change in either the absolute abundances or the relative REE pattern during sedimentary processes, diagenesis and metamorphism (Nance and Taylor, 1976), the behavior or these elements may put some contraints in the relationship between igneous and metasedimentary rocks. In this respect, the similarity of an REE pattern for some rock groups reported (Table 1) may suggest that, among the metasedimentary rocks, the metagraywackes and the Al_2SiO_5–free granulites may be generated by the destruction of the intermediate and acidic charnockites.

Estimated Composition of the Lower Crust Beneath the Bournac Pipe

The calculated averages for newly determinated elements are presented together with the previous results (Leyreloup *et al.*, 1977) on Tables 2 and 3.

Zr, Hf, Cs, Sc, Ta

The concentration of Zr is similar to the average proposed by Taylor (Table 3, column 2), but also to the various estimates of continental crust (160 ppm: Horn and Adams, 1966; 170 ppm: Vinogradov, 1962, see also values reported by Shaw *et al.*, 1976, Table 5). The Zr/Hf ratio equal to 38 is not significantly different from that proposed for the continental surface shield of Canada (Shaw *et al.*, 1976). The estimated Cs content is lower than the crustal average proposed by several authors (e.g. Table 3, column 2) but comparatively higher than the values encountered in modern island arc and Archean andesites (Table 3, columns 8, 9 and 11). The estimated Sc grand average is also notably higher than the values reported by Shaw *et al.* (1976) and not very different from continental crust and modern calc-alkaline andesites (Table 3, columns 2 and 11). Because of the lack of available data, no comment accompanies Ta.

U, Th

The U content is lower than the average crust and shield estimates (Shaw 1967; Eade and Fahrig, 1971.) but quite similar to those of recent andesites (Table 2, columns 10–11). The Th concentration is intermediate and the Th/U ratio (around 8–9) is very similar to that of New Quebec rocks (Eade and Fahrig, 1971) but more than double the common accepted one of 3 to 4 for average continental crustal rocks.

REE

The mean REE content is enriched in LREE relative to chondrites; the pattern display no significant Eu anomaly and no HREE fractionation (Fig. 3). The estimated average is not very different from the values (Eu excepted) of post-Archean sedimentary rocks from Australia and for elements La through Sm similar to the continental surface Precambrian shield of Canada (Table 3). Furthermore, the calculated average shows a striking similarity to the NASC and to the average crust of Taylor. Compared to the andesites, if the LREE are close to those of Archean and modern continental margin andesites, there are some discrepancies with HREE; their contents are intermediate between those of ancient and modern andesites.

The chemistry of xenoliths scavenged by the Neogene volcanoes from other parts of the French Massif Central (Leyreloup *et al.*, 1977) suggest that the chemical properties of the lower crust beneath Bournac might be approximately extended to a larger area.

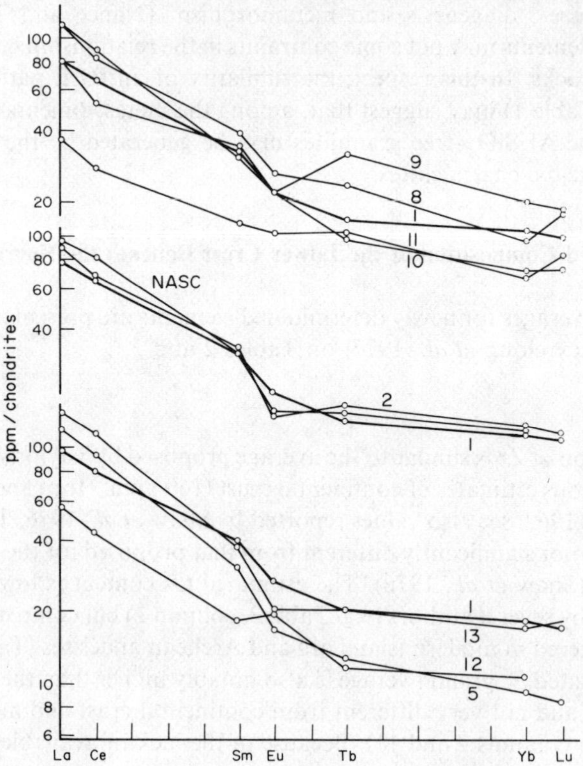

FIG. 3. Average REE abundances normalized to chondrites of lower crust beneath the Bournac pipe compared to various estimates. The numbers refer to columns 1, 2, 5, 8 and 9 of Table 2 and to columns 10, 11, 12 and 13 of Table 3. (NASC: North American shale composite.)

Heat Production

The heat production of the lower crust beneath the Bournac pipe may be estimated from the individual K, U and Th content of each rock types reported in Table 1. The results (Table 4) indicate large range variations; the lower HPU values belong to the metaigneous suite, the highest to the KKP. These variations are related to the relative Th and K content of the different type rocks and also to their different specific gravities. The high heat-flow production of the KKP, quite similar to some values proposed for granite gneisses, migmatites (Husmann, 1956; Lachenbruch and Bunker, 1971; Smithson and Decker, 1974), micaschists (Pliler, 1956, Lachenbruch and Smith, 1968; Smithson and Decker, 1974), contradict the usually accepted hypothesis suggesting that the high-grade metamorphic rocks present the lowest heat-production rate. The HP mean value for the lower crust is approximately $1.6–1.7 \times 10^{-13}$ cal cm^{-3} s^{-1} with averaged specific gravities respectively of 2.7 and 2.8 g cm^{-3}, the last value taking into account the porosity corrections (Leyreloup, 1978). These HPU values are near the highest values admitted for the lower crust by Smithson and Decker (1974). The slight discrepancy between our calculated values directly established from xenoliths scavenged from the lower crust and the theoretical values (in

Smithson and Decker, 1974) cannot be inferred to the different specific gravities but result more likely from the composition of the lower crustal rocks in the Bournac region. It may be also suggested that the xenoliths are not representative of the lowest part of the crust but are scavenged from some intermediate levels in the range of 12–15 km to 25–26 km. In fact, the measured seismic P wave velocities beneath the Velay region between 20 and 26 km depth (Perrier and Ruegg, 1973) and those extrapolated according to Christensen and Fountain (1975) from the specific gravities of the xenoliths at pressure representative of their depth before volcanic eruption (Leyreloup, 1978), are quite similar. Furthermore the depth of the Moho discontinuity is approximated around 27–30 km (Perrier and Ruegg, 1973; Hirn, 1976). This imprecision results in this zone from the presence of 7.7–8.35 km/s P wave velocities related to anomalous upper mantle (Perrier and Ruegg, 1973) or may suggest banded and disrupted structures (Smithson and Brown, 1977). Thus in all cases, these xenoliths, and consequently the fraction of the studied lower crust, seem well located around 20 to 26 km depth quite near the Moho discontinuity; it results that the calculated high heat-flow production in the lower crust at Bournac is only related to the relatively high radioactive elements content of some xenolith types (KKP).

Discussion and Conclusion

The different remarks gathered on major and twenty-two trace elements favor the following comments:

The similar composition of the lower crust beneath the Bournac pipe and the andesites confirms the hypotheses of a relatively basic character of the continental crust at depth. This conclusion does not signify that this crust was derived from andesitic parent rocks. Indeed, among the xenoliths there are two main components, metaigneous xenoliths belonging to a tholeiitic suite and metasedimentary xenoliths which likewise show no petrological or chemical (REE excepted) affinities with andesitic rocks. Likewise it is possible that some of the metasedimentary rocks of Bournac might have been derived from the mataigneous tholeiitic suite. As the REE patterns are similar to those of calc-alkaline suites and, in the absence of more complete geochronological data, we cannot absolutely exclude an origin from the destruction of some older calc-alkali complex. However, the REE patterns are not in this case a definite proof. For instance, the continental crust REE average of Taylor (1964) displaying typical calc-alkali pattern very close to our own estimate was established by the mixing in equal proportion of basaltic and granitic rocks.

The estimated concentrations of various LILE (especially K, Rb, Cs, U, Th) are not lower in the crust at depth beneath the Bournac pipe when compared with igneous rocks of similar major element content from other regions. As already suggested by Moine *et al.* (1972), this argues against the overall LILE depletion advocated by several authors (Brown and Fyfe, 1970; Fyfe, 1973; Heier, 1973; Moorbath, 1975) and consequently against the idea of thick residue of depleted granulite facies rocks at great depth, at least in this region. Our results do not exclude chemical zonation of the crust as a whole. Such zonation between the lower and upper levels of the crust is evidenced by comparison with overall crustal averages (e.g. in the latter K, Li, Rb, Cs, U and Th are higher and some transition elements tend to be lower). Some other elements (Zr, Zn and LREE) do not show this kind of variation. The vertical zonation displayed by some elements is related to the corresponding major element rock composition and is in accord with the generally accepted basic character of the lower crust

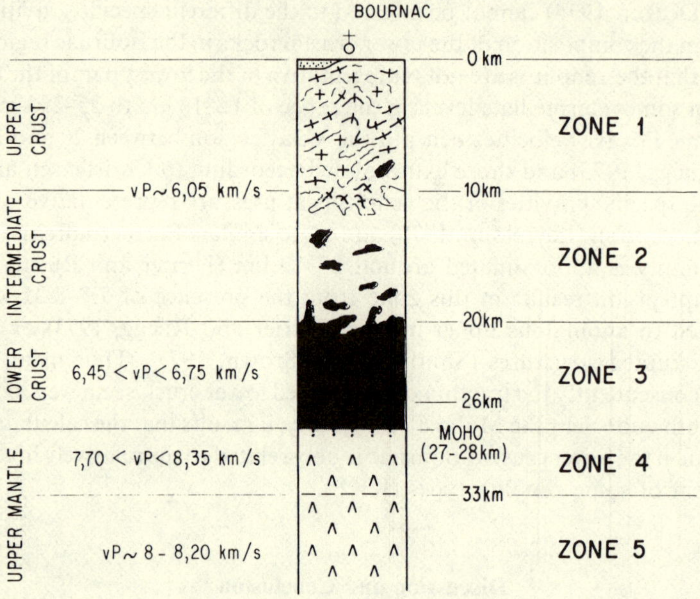

FIG. 4. Hypothetical cross-section of the crust beneath the Bournac pipe. Geophysical data are from Perrier and Ruegg (1973) and Hirn (1976).
Zone 1. (dotted area) Sannoisien green shales; (crosses) Anatectic cordierite, bi, ± garnet, Variscan Velay leucogranite dome; (deformed crosses) Cambrian orthogneisses paleosome.
Zone 2. Intermediate low-pressure granulite facies metapelitic rocks possibly highly anatectic with great proportions of relictual high-pressure granulite facies rocks and low-pressure restites.
Zone 3. Undifferentiated granulite and charnockite basement.
Zone 4. Anomalous upper mantle.
Zone 5. Normal upper mantle.
(vP) P wave velocities.

(Berckhemer, 1969). It does not imply a systematic geochemical differentiation with upward preferential migration of the mobile elements. Although this hypothesis may have local validity, anatectic processes may produce the same effect. Several types of evidence are needed to determine the relative importance of the two processes, in particular, the proportion of rocks with residual character in the whole crust is determined by respective proportion of basic and acidic rocks and also by the origin of granites. If some granites are undoubtedly generated in the crust, the origin of some others is questionable. In the region under consideration, the proportion of restites is around 1%, the other rocks essentially representing an unmelted granulitic paleosome. Consequently, the residual character cannot be applied to the whole lower crust beneath the Bournac pipe. In the highest levels, the presence of ortho-gneiss (Cambrian monzogranite, Kornprobst, pers. comm.) and Variscan anatectic granite (Velay) (Figs. 1 and 4), rich in K and related trace elements, indicate a vertical zonation. The thickness of this granitic formation is estimated from field geology at 8–10 km (Burg and Matte, pers. comm). Between these acidic rocks in the highest levels and the granulitic basement (20–26 km depth) the nature of the crust is unknown. However, by comparison with the Haut–Allier metamorphic belt (50 km to the northwest), where the upper part of the lower catazonal complex (anatectic KKP with high-pressure relictual mineralogical assemblages is largely exposed (Forestier *et al.*, 1973; Marchand, 1974;

Lasnier, 1977) with great proportion of restites, we suggest that similar rocks may exist in the intermediate zone beneath the Bournac locality.

It is noteworthy to emphasize that the cross-section (Fig. 4) proposed for the stable crust beneath Bournac is not obligatory representative for the Massif Central crust as a whole, especially in the overthrusting zones (e.g. Burg, 1977). The specificity of Velay lower crust is the high LILE content associated with relatively high heat-flow production. These characteristics may be primary features, thus conflicting the classical LILE-depletion hypothesis; but it may be also due to some secondary enrichment processes associated to very small hydration during the successive Variscan and Tertiary post-tectonic uplift of the Massif Central granulite facies basement.

Finally, as shown by the present study, the lower crust beneath Bournac is composed of heterogenous banded granulite facies material, thus confirming the main conclusions of Smithson and Brown (1977). However, the type rock proportions may be extremely variable from place to place and a lower granulite facies crust, only composed of metasedimentary paragneisses or at the opposite of basic and/or acidic igneous rocks, cannot be locally excluded inducing particular heat-flow data.

Acknowledgements

The authors are greatly indebted to Prof. D. Strong (St Johns University, Newfoundland) and Prof. J. Touret (Paris VII University) for his helpful comments on the manuscript and to Dr J. Lancelot (Montpellier University) for some U, Th determinations by mass spectrometry. They extend thanks to Misses H. Dubuc and H. Zanca for typing the manuscript and Mr P. Viela for his assistance in the preparation of the figures.

Financial support was provided by the Centre Géologique Géophysique, Montpellier, and Geodynamic A.T.P. (C.N.R.S).

References

ALLEGRE, C. J., HAMET, J. and LEYRELOUP, A. (1975) Etude $^{87}Rb/^{87}Sr$ des enclaves catazonales remontées par les volcans néogènes du Velay. Présence d'un socle antécadomien sous le Massif Central. Abstract. *3ème R.A.S.T.*, Montpellier p. 8.

BARAGAR, W. R. and GOODWIN, A. M (1969) Andesites and archean volcanism of the Canadian shield. *Proc. Andesite Conf.*, A. R. MCBIRNEY, ed., *Bull. Oreg. State Dep. Geol. Mineral., Ind.* **65**, 121–142.

BERCKHEMER, H. (1969) Direct evidence for the composition of the lower crust and the Moho. *Tectonophysics*. **8**, 17–105.

BILAL, A. (1976) Les inclusions fluides dans les enclaves de Bournac (Massif Central français). Implications sur les structures de la croûte continentale Thèse 3ème cycle, Nancy, 151 pp.

BIRCH, F., ROY, F. and DECKER, E. R. (1968) Heat flow and thermal history in New England and New York, in studies of Appalachian Geology (Northern and Maritime). E. ZEN, W. S. WHITE, J. B. HADLEY and J. B. THOMSON, Jr., eds., Ch. 33, Interscience, 475 pp.

BROWN, C. G. and FYFE, W. S. (1970) The production of granitic melts during ultrametamorphism. *Contr. Mineral. Petrol.* **28**, 310–318.

BURG, J. P. (1977) Tectonique et microtectonique des séries cristallophylliennes du Haut-Allier et de la vallée de la Truyère. Contribution du microscope électronique à transmission à l'étude de la déformation des minéraux dans les zones profondes. Thèse 3ème cycle., Montpellier, 79 pp.

BURG, J. P. and MATTE, Ph. (1977) A cross section through the French Massif Central in the scope of its Variscan geodynamic evolution. *Geotagung* **77**, Gottingen.

BURWASH, R. A. and CUMMING, G. L. (1976) Uranium and thorium in the Precambrian basement of Western Canada. I. Abundance and distribution. *Can. J. Earth Sci.* **13**, 284–293.

CHRISTENSEN, N. I. and FOUNTAIN, D. M. (1975) Constitution of the lower continental crust based on experimental studies of seismic velocities in granulites. *Geol. Soc. Amer. Bull.* **86**, 227–236.

CONDIE, K. E. and HARRISON, N. M. (1976) Geochemistry of the Archean Bulawayan group Midlands greenstone belt, Rhodesia. *Precambrian Res.* **3**, 253–271.

COULON, Ch. (1977) Pétrologie et géochimie des roches calco-alcalines de Sardaigne (Italie). Thèse Doct., Univ. Marseille.

DEL FIORE, G., PETERS, J. and ROELANDTS, I. (1974) Precise determination of uranium in rocks and minerals by neutron activation analysis using ^{133}I fission products. *Chem. Geol.* **13**, 309–319.

DUPUY, C. and LEFEVRE, C. (1974) Fractionnement des éléments en trace Li, Rb, Ba, Sr, dans les séries andésitiques et shoshonitiques du Pérou. Comparaison avec d'autres zones orogéniques. *Contr. Mineral. Petrol.* **46**, 147–157.

EADE, K. E. and FAHRIG, W. F. (1971) Chemical evolutionary trends of continental plates—a preliminary study of the Canadian shield. *Bull. Geol. Surv. Can.* **179**, 51 pp.

FORESTIER, F. H., LASNIER, B., LEYRELOUP, A. and MARCHAND, J. (1973) Vues nouvelles sur la catazone dans le Massif Central français es le Massif Armoricain, de l'affleurement au Moho. *Bull. Soc. Géol. France* **XV**, 5–6. 562–578.

FYFE, W. S. (1973) The granulite facies, partial melting and the Archean crust. *Phil. Trans. Roy. Soc., Lond*, **A, 273**, 457–461.

GORDON, G. E., RANDLE, K., GOLES, G. G., CORLISS, J. B., BEESON, M. H. and OXLEY, S. S., (1968) Instrumental activation analysis of standard rocks with high-resolution gamma-rays detectors. *Geochim. Cosmochim. Acta* **32**, 369–396.

HAMET, J., VITRAC, A. and LEYRELOUP, A. (1977) Etude géochronologique Rb/Sr et U/Pb de la croûte inférieure du Massif Central français d'aprés les enclaves catazonales de Bournac (en préparation).

HEIER, K. S. (1973) Geochemistry of granulite facies rocks and problems of their origin. *Phil. Trans. Roy. Soc., Lond*, A, **273**, 429–442.

HEIER, K. S. and ADAMS, S. A. S. (1965) Concentration of radioactive elements in deep crustal material. *Geochim. Cosmochim. Acta* **29**, 53–61.

HEIER, K. S. and BRUNFELT, A. O. (1970) Concentration of Cs in high grade metamorphic rocks. *Earth Planet. Sci. Lett.* **9**, 416–420.

HIRN, A. (1976) Sondages sismiques profonds en France. *Bull. Soc. Géol, France* **5**, 1065–1071.

HOLLAND, J. G. and LAMBERT, R St J. (1972) Major element chemical composition of shields and the continental crust. *Geochim. Cosmochim. Acta* **36**, 673–683.

HORN, M. K. and ADAMS, J. A. S. (1966) Computer-derived geochemical balances and element abundances. *Geochim. Cosmochim. Acta* **30**, 279–297.

HUSMANN, G. (1956) Determination of the thorium and uranium contents of gneisses and anatectic rocks in the Central Black Forest by means of the coincidence method. *Neues Jahrb. Mineral. Monatsh*, pp. 108–120.

IRVING, A. J. (1974) Geochemical and high pressure studies of garnet pyroxenite and pyroxene granulite xenoliths form the Delegate basaltic pipes. *Australia. J. Petrol.* **15**, 1–40.

KILLEEN, P. G. and HEIER, K. S. (1975) Trend surface analysis of Th, U, and K, and heat production in three related granitic plutons. Farsund area, South Norway. *Chem. Geol.* **15**, 163–176.

LACHENBRUCH, A. H. and BUNKER, C. M. (1971) Vertical gradients of heat production in the continental crust. 2. Estimates from borehole data. *J. Geophys. Res*, **76**, 3852–3860.

LACHENBRUCH, A. H. and SMITH, A. R. (1968) Radiogeologic studies in the Central part of the Sierra Nevada batholith, California. *J. Geophys. Res.* **73**, 1481–1496.

LAMBERT, I. B. and HEIER, K. S. (1967) The vertical distribution of uranium, thorium and potassium in the continental crust. *Geochim. Cosmochim. Acta* **31**, 377–390.

LAMBERT, I. B. and HEIER, K. S. (1968a) Estimates of crustal abundances of thorium, uranium and potassium. *Chem. Geol.* **3**, 233–238.

LAMBERT, I. B. and HEIER, K. S. (1968b) Geochemical investigations of deep-seated rocks in Australian shield. *Lithos* **1**, 30–53.

LASNIER, B. (1977) Persistance d'une série granulitique au coeur du Massif Central français (Haut-Allier); les termes basiques, ultrabasiques et carbonatés. Thèse Doctorat. Sci. Nat., Univ. Nantes.

LEYRELOUP, A. (1973) Le socle profond en Velay d'après les enclaves remontées par les volcans néogènes. Son thermo-métamorphisme et sa lithologie: granites et série charnockitique (Massif Central français). Thèse 3ème cycle, Nantes., 356 pp.

LEYRELOUP, A. (1974) Les enclaves catazonales remontées par les éruptions néogènes de France, nature de la croûte inférieure. Partie 1: Lithologie et structurologie. *Contr. Mineral. Petrol.* **46**, 17–27.

LEYRELOUP, A. (1978) Etude de la croûte inférieure continentale sous le Massif Central français d'après les enclaves remontées par les éruptions volcaniques plio-quaternaires. Modèle de formation et d'évolution d'une croûte "primitive". Comparaisons avec d'autres segments crustaux de même type. Thèse d'Etat. (en préparation).

LEYRELOUP, A. and BILAL, A. (1977) Les enclaves catazonales remontées par les éruptions néogènes de France. Nature de la croûte inférieure. Part 3, Pétrologie (Phases solides et inclusions fluides) (en préparation).

LEYRELOUP, A., DUPUY, C. and ANDRIAMBOLOLONA, R. (1977) Catazonal xenoliths in French neogene volcanic rocks; constitution of the lower crust. Part 2, Chemical composition and consequences on the evolution of the French Massif Central Precambrian crust. *Contr. Mineral. Petrol.* **62**, 283–300.

MATTAUER, M. (1974) Existe-t-il des chevauchements de type himalayen dans la chaîne hercynienne du Sud de la France? 2ème R.A.S.T., Pont à Mousson, p. 279.

MATTAUER, M. and ETCHECOPAR, A. (1976) Argument en faveur des chevauchements de type himalayen dans la chaîne hercynienne du Massif Central français. Coll. int. C.N.R.S. n° 268. *Ecologie et Géologie de l'Himalaya*, 261–267.

MOINE, B., LA ROCHE, H. de and TOURET, J. (1972) Structures géochimiques et zonéographie métamorphique dans le Précambrien catazonal du Sud de la Norvège (région d'Arendal). *Sciences de la Terre*, t. XVII, 1–2, 131–164.

MOORBATH, S. (1975) Age and isotope constraints for the evolution of Archean crust. In *The Early History of the Earth*, B. F. WINDLEY, ed., 351–373.

MARCHAND, J. (1974) Persistance d'une série granulitique au coeur du Massif Central français (Haut-Allier): les termes acides, Thèse 3ème cycle, Univ. Nantes.

NANCE, W. B. and TAYLOR, S. R. (1976) Rare earth element patterns and crustal evolution—I. Australian post-Archean sedimentary rocks. *Geochim. Cosmochim. Acta* **40**, 1539–1551.

NANCE, W. B. and TAYLOR, S. R. (1977) Rare earth element patterns and crustal evolution II. Archean sedimentary rocks from Kalgoorlie, Australia. *Geochim. Cosmochim. Acta* **41**, 225–231.

PERRIER, G. and RUEGG, J. C. (1973) Structure profonde du Massif Centrale français. *Ann. Geophys.*, **29**, fasc. 4, 435–502.

PETROV, B. V., KRENDELEV, F. P., BOBROV, V. and TSIMBALIST, V. G. (1972a) Behaviour of radioactive elements and gold during metamorphism of the Patmosk highland sedimentary rocks. *Geokhimiya* **8**, 947–955.

PHILPOTTS, S. A. and SCHNETZLER, C. G. (1968) Genesis of continental diabases and oceanic tholeiites considered in light of rare earth and barium abundances and partition coefficients. In *Origin and Distribution of the Elements*, L. H. AHRENS, ed., **30**, 940–947.

PLILER, R. (1956) The distribution of thorium and uranium in sedimentary rocks and the oxygen content of the Precambrian atmosphere. Unpublished M.A. Thesis. The Rice Institute.

ROELANDTS, I. (1973)—Neutron activation analysis and distribution of rare earths and trace elements in rocks and coexisting minerals from the complex of South Rogaland (Norway). Thesis, Université de Liège.

ROGERS, N. and NIXON, P. H. (1975) Geochemistry of lower crustal granulite xenoliths from lesotho kimberlites, *Ann. Report* **19**, Univ. Leeds.

RONOV, A. B. and YAROSHEVSKIY, A. A. (1969) Chemical composition of the Earth's crust. In *The Earth's Crust and Upper Mantle*, Geophys. Monograph. 13: Pembroke, J. HART, ed., Amer. Geophys. Union, 35–57.

SCHARBERT, H. G., KORKISCH, J. and STEFFAN, I. (1976) Uranium, thorium and potassium in granulite facies rocks, Bohemian Massif, Lower Austria, Austria. *Tschermaks Min. Petr. Mitt.* **23**, 223–232.

SIGHINOLFI, G. P. (1969) K/Rb ratios in high grade metamorphism: a confirmation of the hypothesis of a continual crustal evolution. *Contr. Mineral. Petrol.* **21**, 346–356.

SIGHINOLFI, G. P. and SAKAI, T. (1977) Uranium and thorium in Archean granulite facies terrains of Bahia (Brazil). *Geochem. J.* **11**, 33–39.

SHAW, D. M. (1967) U, Th and K in the Canadian Precambrian shield and possible mantle compositions. *Geochim. Cosmochim. Acta* **31**, 1111–1113.

SHAW, D. M., REYLLY, G. A., MUYSSON, J. R., PATTENDEN, G. E. and CAMPBELL, F. E. (1967) The chemical composition of the Canadian Precambrian shield. *Can. J. Earth Sci.* **4**, 829–854.

SHAW, D. M., DOSTAL, J. and KEAYS, R. R. (1976) Additional estimates of continental surface Precambrian shield composition in Canada. *Geochim. Cosmochim. Acta* **40**, 73–83.

SMITHSON, S. B. and DECKER, E. R. (1974) A continental crustal model and its geothermal implications. *Earth Planet. Sci. Lett.* **22**, 215–225.

SMITHSON, S. B. and BROWN, S. K. (1977) A model for lower continental crust. *Earth Planet. Sci. Lett.* **35**, 134–144.

TAYLOR, S. R. (1964) Abundance of chemical elements in the continental crust: a new table. *Geochim. Cosmochim. Acta* **28**, 1273–1285.

TAYLOR, S. R., CAPP, A. C. and GRAHAM, A. L. (1969) Trace element abundances in andesites. *Contr. Mineral. Petrol.* **23**, 1–28.

VINOGRADOV, A. P. (1962) Average contents of chemical elements in the principal types of igneous rocks of the earth's crust. *Geochem. USSR* (English. Trans.) 641–644.

WEDEPOHL, K. H. (1969) Composition and abundance of common igneous rocks, composition and abundance of common sedimentary rocks, In *Handbook of Geochemistry*, Vol 1, K. H. WEDEPOHL, ed., Springer, New York, 227–271.

WILDEMAN, T. R. and HASKIN, L. A. (1973) Rare earths in Precambrian sediments. *Geochim. Cosmochim. Acta* **37**, 419–438.

WILKINSON, J. F. G. (1973) Pyroxenite xenoliths from an alkali trachybasalt in the Olen Innes area, New South Wales. *Contr. Mineral. Petrol.* **43**, 15–32.

WILKINSON, J. F. G. (1974) Garnet clinopyroxenite inclusions from diatremes in the Gloucester area, New South Wales, Australia. *Contr. Mineral. Petrol.* **46**, 275–299.

Trace Element Variations and Isotopic Composition of Charnockitic Acidic Rocks related to Anorthosites (Rogaland, S.W. Norway)

DANIEL DEMAIFFE,† JEAN-CLAIR DUCHESNE‡ and JAN HERTOGEN§

† Minéralogie-Pétrologie, Université de Bruxelles and Musée Royal de l'Afrique Centrale, Tervuren, Belgium.
‡ Géologie–Pétrologie–Géochimie, Université de Liège, Belgium.
§ Instituut voor Nucleaire Wetenschappen, Rijks Universiteit Gent, Belgium. "Aangesteld Navorser" of the Belgium NFWO.

Abstract

Charnockitic acidic rocks related to anorthosites are well represented in the Rogaland anorthositic province. Quartz mangerites constitute the upper level of the Bjerkrem–Sogndal layered lopolith and are related to anorthosites through a complete sequence of differentiated products which includes leuconorites, norites and mangerites. In the Hidra body, charnockitic rocks occur as a stockwork of dykes cutting across anorthosite and leuconorite. The southwest end of the magmatic province consists of two distinct bodies of acidic rocks: the Farsund charnockite and the Lyngdal hornblende granodiorite. They are considered to be related to the neighboring anorthosite massifs.

Some trace and major elements (REE, Rb, Sr, U, Th, Hf, Zr, K_2O, SiO_2) of these rocks as well as the strontium and oxygen isotopic ratios are investigated.

Quantitative modeling of the behavior of trace elements using the Rayleigh equation indicates in agreement with the isotopic data that direct derivation of the acidic rocks by fractional crystallization of a parental magma is a possible mechanism in Hidra and in the Lyngdal granodiorite. In Bjerkrem–Sogndal, however, the REE evolution in the acidic rocks and the high value of the $^{87}Sr/^{86}Sr$ initial ratio (0.7105) compared to that of the anorthosites (0.705–0.706) cannot be explained by a simple fractional crystallization process. Contamination by supracrustal material relatively poor in REE and with a high Sr isotopic ratio must also be invoked. In the Farsund charnockite REE distribution and Sr and O isotopic ratios suggest an origin by anatexis of gneisses of sedimentary origin.

Introduction

Genesis of anorthosites is one of the most debated problems of the contemporary petrology. The magmatic character of these rocks has long been recognized but the nature of the relations with spatially associated intermediate and acidic rocks,†† mostly of charnockitic type, has not been clearly established. Two hypotheses have been put forward: for some geologists (Buddington, 1939, 1972) anorthosites and charnockites belong to two different magmatic series and thus are not comagmatic; for others (P. Michot, 1956, 1965; Philpotts, 1966; De Waard, 1969; Emslie, 1969) charnockites are supposed to represent an end-product of magmatic differentiation of the anorthositic suite. Kranck (1969) even considers Rapakivi granites as the ultimate term of the differentiation sequence emplaced at a higher level in the crust.

Some other geologists, however, have adopted an intermediate position (J. Michot and

†† Here called for short: acidic rocks.

Pasteels, 1969; Pasteels *et al.*, 1970; Martignole, 1974): charnockites should correspond to a residual liquid of the anorthositic differentiation contaminated at various degrees by supracrustal material.

The present work discusses the petrogenetic relations between the Rogaland anorthosites and their related charnockites from a geochemical (REE, K, Rb, Sr, U, Th, Zr, Hf) and isotopic ($^{87}Sr/^{86}Sr$ and $^{18}O/^{16}O$) point of view.

Previous Geochemical and Isotopic Work

Preliminary data of Philpotts *et al.* (1966) on a mangerite† from Quebec were followed by the work of Green *et al.* (1974) on intrusive mangerites from the Lofoten Islands. In the first case, the chondrite-normalized REE pattern is characterized by a lack of Eu anomaly while, in the Lofoten mangerites, a positive Eu anomaly is observed. These rocks cannot be considered as residual liquids from anorthosite fractionation; if so, the mangerites should present a REE distribution complementary to the anorthositic one, which is characterized by a high positive Eu anomaly.

In the Pickerel meta-anorthositic complex (Simmons *et al.*, 1974) the negative Eu anomaly and the REE enrichment of the quartz monzonites are in agreement with the hypothesis of a comagmatic relation between these rocks and the anorthosites.

In the Hopen charnockite (Ormaasen, 1977) a similar relation with anorthosite has been demonstrated as well as an increase of the REE content and of the Eu anomaly with differentiation indexes.

Data on the Sr isotopic composition of acidic rocks related to anorthosites yield ambiguous conclusions: Heath and Fairbairn (1969) show that, in some cases (Adirondacks), anorthosites and syenites have the same $^{87}Sr/^{86}Sr$ initial ratios and are thus considered comagmatic while, in other cases (Nain), the initial ratio of the acidic rocks is much higher. Barton and Doig (1972) found that rocks of the anorthosite–charnockite series in the Lac Croche complex lie on an isochron. However, this cannot be taken as true a criterion for comagmatic origin, since Schrijver (1975) demonstrated that a high-grade metamorphic event could have caused isotopic homogenization on a regional scale. In the Morin complex, Barton and Doig (1977) have shown that the mangerites related to the anorthosites do not define an isochron, apparently precluding a comagmatic origin.

The Rogaland Igneous Complex

The present study deals with the Rogaland anorthositic province which has been studied from both geological and petrological points of view by P. Michot and J. Michot and their coworkers. The description and the geological evolution of the different units of this complex have been published previously (P. Michot, 1960; J. and P. Michot, 1969; Demaiffe *et al.*, 1973). Field and petrological relations between acidic charnockites and anorthosites–norites are particularly well defined in two cases: the Bjerkrem–Sogndal lopolith and the Hidra body.

† In the nomenclature used here, quartz mangerite is adopted indistinctly for hypersthene granite (charnockite *sensu stricto*) and hypersthene quartz monzonite. Mangerite and jotunite stand for mesoperthite–hypersthene monzonite and hypersthene monzodiorite, respectively, in full agreement with Streckeisen's nomenclature on charnockitic rocks (1974).

In the Bjerkrem–Sogndal (Bk–Sg) layered lopolith (P. Michot, 1960, 1965) anorthosites, leuconorites and norites are associated, through a complete sequence, with mangerites and quartz mangerites which occur in the upper level of the lopolith and correspond to the last two stages of a gravity differentiation process. Layered structures and igneous lamination are present in the lower anorthositic–noritic levels of the lopolith but disappear in the mangerites. The mineralogical evolution throughout the whole sequence has been established by P. Michot (1965) and Duchesne (1972 a, b). Several petrographic varieties of quartz mangerites are studied here: a hypersthene type (73776), a fayalite type (Pa 66N, 66261, 6687) and a pegmatitic hololeucocratic type (64120).

In the eastern part of the complex, the *anorthositic–leuconoritic bodies of Hidra and Garsaknatt* (Demaiffe *et al.*, 1973; Demaiffe, 1977) were emplaced in the surrounding gneisses post-tectonically. The inner border of these massifs is outlined by a fine-grained (300–500 μm) jotunitic rock, locally porphyritic, whose transition towards the central leuconorite is very progressive. This jotunite (see footnote p. 418), interpreted as a chilled margin, corresponds to the initial magmatic liquid as shown by the REE geochemistry (Duchesne *et al.*, 1974; Demaiffe, 1977). A stockwork of charnockitic dykes and pegmatitic lenses containing heavy REE minerals (Adamson, 1942) intrude the leuconorite of the Hidra body.

The southeastern end of the Rogaland province is constituted by two separate acidic intrusions, the *Farsund charnockite* and the *Lyngdal hornblende granodiorite* which have been previously considered as two facies of a single magmatic intrusion, the so-called "farsundite". Available geochemical data (Falkum *et al.*, 1972) have demonstrated the distinct nature of these two intrusions. The genetic relationships with the large anorthosite plutons of the province (Egersund–Ogna, Håland–Helleren, Åna–Sira) have not yet been clearly defined (Pasteels *et al.*, 1970). The present paper contributes to clarify this question.

Analytical Methods

Major elements have been determined by a combination of X-ray fluorescence spectrometry and wet chemical methods; REE, U, Th, Hf by instrumental neutron activation analysis (Hertogen and Gijbels, 1971); Rb, Sr and Zr by X-ray fluorescence spectrometry and by isotopic dilution for Rb- and Sr-poor rocks.

The Sr isotopic compositions have been measured on a VARIAN TH5 mass spectrometer of the Belgian Centre for Geochronology. Ten determinations of the isotopic composition of the Eimer and Amend M.I.T. Standard yield as a mean $^{87}Sr/^{86}Sr = 0.70807 \pm 0.00010$ (1σ) when corrected for $^{86}Sr/^{88}Sr = 0.1194$. Oxygen was extracted from the rocks by reacting with BrF_5 at 600°C in Ni tubes and $^{18}O/^{16}O$ ratios have been determined on a double collecting mass spectrometer at the Laboratoire de Géochimie des Isotopes Stables, Université de Paris VII.

Chemical and Isotopic Data

Table 1 presents some major and trace element data of the various acidic rocks studied here, together with their Larsen differentiation index. Normative compositions (in weight %) are plotted in a Qz–Ab–Or diagram (Fig. 1). Mangerites and quartz mangerites from Bk–Sg show an evolutionary trend characterized by a progressive increase of the Larsen index and

TABLE 1. TRACE ELEMENT CONCENTRATIONS (ppm) IN ROGALAND CHARNOCKITIC ACIDIC ROCKS AND IN A GRANITIC GNEISS

	Bjerkrem–Sogndal lopolith							Hidra charnock. dykes		Lyngdal granodiorite		Farsund charnock.	Granitic gneiss
	Mangerite		Quartz mangerite										
	Pa 66P	TII[1]	73776	Pa 66N	66261	6687	64120	443–4/1	283–2/2	134–1/1	137–1/1	Pa 70A	134/72[2]
Rb	61.5	71.5	78.2[3]	100	112.7[3]	138[3]	198.5[3]	84.8[3]	148	87	101		166
Sr	281	170	207[3]	142	98.4[3]	88.4[3]	71.4[3]	420[3]	299	503	465		130
Rb/Sr	0.219	0.421	0.377	0.703	1.148	1.567	2.78	0.212	0.495	0.173	0.218		1.28
K_2O (%)		5.19	4.53	5.04	5.56	5.54	5.54	3.06	4.80	3.76	3.69	4.08	4.42
K/Rb		624	481	418	409	333	232	300	270	359	303		221
U	0.33		0.97	0.81	0.66	1.29	2.68	3.01	0.63	1.09	1.29	0.72	0.74
Th	1.10		2.34	1.92	0.94	2.54	9.84	9.81	2.17	5.89	5.68	2.66	6.8
Th/U	3.33		2.41	2.37	1.42	1.97	3.67	3.26	3.44	5.4	4.40	3.69	9.2
Zr	109	289	1040	932	690	740	530		744	612	486		
Hf	1.83	5.3	2.86	19.1	19.1	21.2	17.6	33.7	18.4	15.8	12.7	15.1	4.3
Zr/Hf	60	55	37	49	36	35	31		40	39	38		
SiO_2		57.93	59.44	59.08	64.45	67.75	72.54	55.2	65.5	62.1	67.28	64.2	74.8
Larsen index		12	10	10	17	20	26	5	17	14		13	26
$\delta^{18}O$(‰)	6.44		6.47		4.02		6.09	6.82	7.43	6.98	4.38	6.44	

U, Th, Hf: instrumental neutron activation analysis (J. Hertogen, Gent).
Rb, Sr, Zr: X-ray fluorescence spectrometry (M. Delvigne and F. Durez, Tervuren; J. C. Duchesne, Liège).
$\delta^{18}O$: Laboratoire de Géochimie des Isotopes stables, Université de Paris VII.

(1) Analyst: I. Roelandts (Liège).
(2) From Franssen and Deutsch (1973), analyst: J. Hertogen (INAA, Gent).
(3) Analyst: D. Demaiffe (Istopic dilution, Bruxelles).

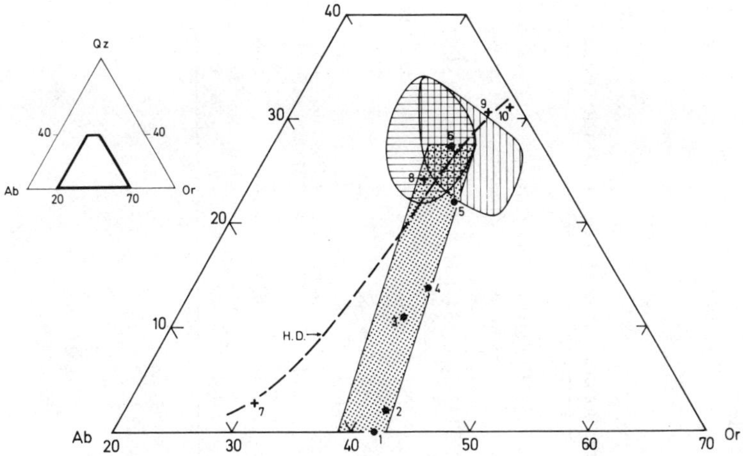

FIG. 1. Normative compositions (in weight %) of the acidic rocks related to anorthosites in the Qz–Ab–Or diagram.

Stripped area: Evolution trend of the Bjerkrem–Sogndal rocks (after Duchesne, unpublished data) on which the rocks studied here are reported. (1: TII; 2: Pa 66N; 3: 73776; 4: 66261; 5: 6687; 6: 64120.)

H.D.: Evolution trend of the Hidra charnockitic dykes (7: 443–4/1; 8: 283–2/2; 9: 309–2/1; 10: 298–2/1). The analyses of the last two samples are not reported in this paper.

Vertically and horizontally hatched: Farsund charnockite and Lyngdal hornblende granodiorite (after Falkum *et al*., 1972; Demaiffe and J. Michot, unpublished data).

of the Qz proportion, the Ab/Or ratio remaining approximately constant. The charnockitic dykes of the Hidra body grossly display a similar evolution. The Lyngdal granodiorite and the Farsund charnockite belong to two restricted and overlapping domains of the Qz–Ab–Or diagram. As a rule in Bk–Sg, the Rb contents and the Rb/Sr ratios increase progressively with the differentiation index, whereas the Zr/Hf and K/Rb ratios decrease. U and Th markedly increase in the late pegmatitic rock (64120). Concentrations of these elements in the Hidra dykes and the Lyngdal granodiorite are approximately in the same range of values.

The REE concentrations and the La/Yb and Eu/Eu* ratios are listed in Table 2. The chondrite-normalized REE patterns are drawn in Figs. 2 to 5 and are compared to some other rocks, namely the average value for thirty-six Precambrian sediments (Wildeman and Haskin, 1973) and a granitic gneiss (134/72) from the metamorphic envelope of the Rogaland anorthosites (Franssen, pers. comm.).

In Bk–Sg, the mangerites (Pa 66P and TII) and the quartz mangerite 6621 are characterized by a positive Eu anomaly (Eu/Eu*: 2.42 to 1.20). In the other rocks the REE content and the Eu/Eu* values show a negative correlation with the differentiation index. The total REE content (ΣREE) varies between 400 and 250 and the Eu/Eu* ratio from 0.85 to 0.50.

In the Hidra dykes, the REE content (ΣREE $\simeq$400 ppm) and the Eu anomaly (Eu/Eu* $\simeq$0.50) are at the higher side of the Bk–Sg range. The Lyngdal granodiorite also plots within the Bk–Sg range but has a smaller negative Eu anomaly (Eu/Eu* $\simeq$0.80–0.85). The Farsund charnockite (ΣREE $\simeq$250 ppm) does not show any Eu anomaly at all.

On the whole, the REE content of the acidic charnockitic rocks is high relatively to average granite (ΣREE $\simeq$250 ppm, Herrmann, 1970) but comparable to the Rapakivi

TABLE 2. REE CONCENTRATIONS (ppm) IN ROGALAND CHARNOCKITIC ACIDIC ROCKS, IN THE HIDRA JOTUNITIC BORDER FACIES AND IN A GRANITIC GNEISS

	Bjerkrem–Sogndal massif							Hidra body						
	Mangerite		Quartz mangerite					Charnock. dyke		Jotunite	Lyngdal granodiorite		Farsund charnock.	Granitic gneiss
	Pa 66P	TII	66261	Pa 66N	64120	6687	73776	443–4/1	283–2/2	200–2/2	137–1/1	134–1/1	Pa 70A	134/72
La	26.1	27.8	31.3	57	51.7	52.2	74.9	71.6	72.6	21.3	75.1	81	45	24.4
Ce	62.7	67	74.5	141	109	122	172	166	169	48.6	160	177	104	44.4
Nd	46.2	42.3	53.3	100	59.0	77.5	109	108	107	29.1	87.7	96	64.4	20.0
Sm	9.95	10.54	11.3	20.7	11.2	15.8	21.8	21.9	20.3	6.60	14.8	17	13.5	3.42
Eu	7.51	6.15	4.36	5.64	1.83	3.13	5.03	3.82	2.94	2.27	3.64	4.16	4.23	0.86
Tb	1.49	1.24	1.76	3.33	1.70	2.43	3.20	3.57	2.90	1.04	2.05	2.42	2.31	0.25
Yb	2.31	2.66	4.67	7.3	4.14	5.42	7.05	10.3	6.28	3.56	5.33	6.12	6.35	0.51
Lu	0.39	0.46	0.78	1.2	0.55	0.87	1.09	1.64	0.92	0.63	0.83	0.94	0.98	0.088
La/Yb	11.3	10.45	6.65	7.8	12.6	9.65	10.7	6.95	11.6	6.0	14.1	13.2	7.1	48
Eu/Eu*	2.42	1.85	1.19	0.85	0.50	0.61	0.73	0.55	0.47	1.0	0.82	0.85	1.0	1.0

Analysts: J. Hertogen (INAA) except sample TII by I. Roelandts (NAA).
Eu* is determined by graphical interpolation between Sm and Tb on the chondrite-normalized diagram.

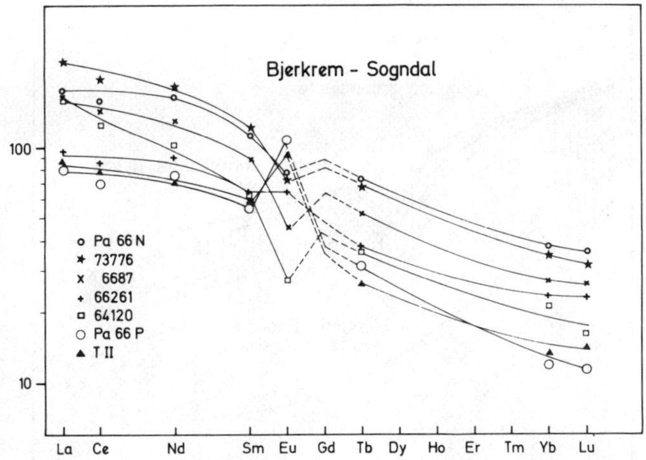

FIG. 2. Chondrite-normalized REE patterns in Bjerkrem–Sogndal mangerites (TII and Pa 66P) and quartz mangerites (Pa 66N, 73776, 66261, 6687, 64120).

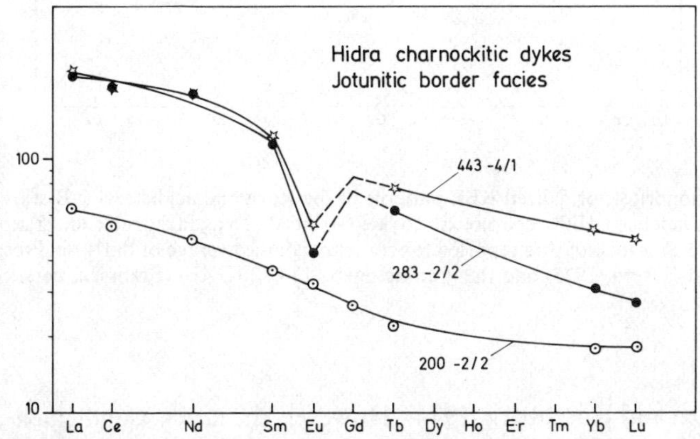

FIG. 3. Chondrite-normalized REE patterns in two Hidra charnockitic dykes (283–2/2 and 443–4/1) and in the jotunitic border facies (200–2/2) corresponding to the parental magma.

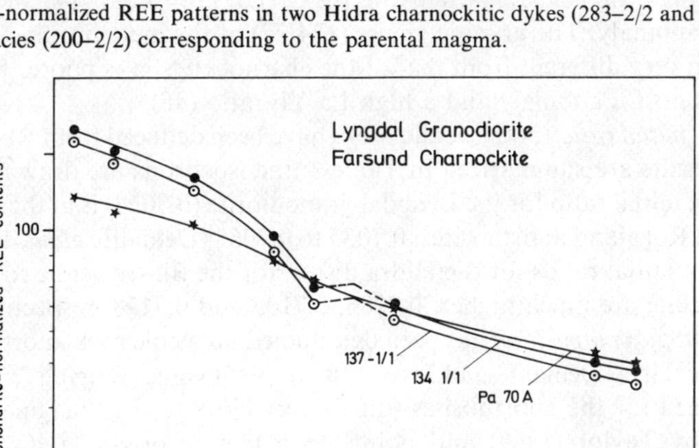

FIG. 4. Chondrite-normalized REE patterns in the Lyngdal hornblende granodiorite (134–1/1 and 137–1/1) and in the Farsund charnockite (Pa 70A).

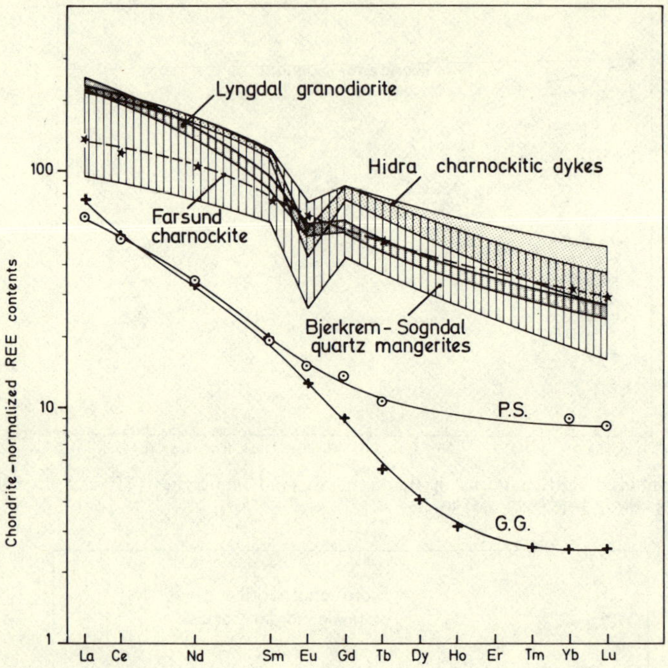

FIG. 5. Ranges of chondrite–normalized REE patterns in the Rogaland acidic rocks: Bjerkrem–Sogndal quartz mangerites (vertically hatched), Hidra charnockitic dykes (stripped), Lyngdal hornblende granodiorite (diagonally hatched) and Farsund charnockite. Are reported for comparison: the average of thirty-six Precambrian sediments (P.S.) (Wildeman and Haskin, 1973) and the granitic gneiss 134/72 (G.G.) (Franssen, pers. comm.).

granites (Koljonen and Rosenberg, 1974). However, the latter usually have a higher La/Yb ratio (15 to 23) than the acidic charnockitic rocks (La/Yb = 6.5 to 14) and show a more pronounced Eu anomaly. The granitic gneiss (134/72) of the metamorphic envelope has a REE distribution very different from that of the charnockites: it is poorer in REE (ΣREE $\simeq$ 100 ppm), has no Eu anomaly and a high La/Yb ratio (48).

The Sr isotopic initial ratios of the acidic rocks have been deduced from Rb–Sr whole rock isochrons; the results are summarized in Table 3; the isochrons are drawn in Fig. 6.

The Sr isotopic initial ratio for the Lyngdal granodiorite (0.7054) is in the range of values measured for the Rogaland anorthosites: 0.7035 to 0.7065 (Demaiffe *et al.*, 1974; Demaiffe, 1977) whereas the initial ratios for the Hidra dykes, for the Bk–Sg acidic rocks and for the Farsund charnockite are much higher: 0.7086, 0.7105 and 0.7128, respectively.

The oxygen isotopic composition has been determined on whole rock anorthosites, norites and acidic charnockites (Demaiffe and Javoy, 1976). $\delta^{18}O$ values ranging from 5.2 to 6.6‰ have been obtained for the anorthosites, this is very close to the "magmatic anorthosite range" defined by Taylor (1969) and points to a mantle origin. $\delta^{18}O$ values for the charnockites (Table 2) fall also in a restricted interval (6.0 to 7.5‰) which is slightly higher than but overlaps the anorthosite range. A few low $\delta^{18}O$ values, around 4‰, are to be pointed out in some quartz mangerites.

TABLE 3. $^{87}Sr/^{86}Sr$ ISOTOPIC INITIAL RATIOS DEDUCED FROM WHOLE-ROCK ISOCHRONS

	n†	$(^{87}Sr/^{86}Sr)_o \pm 2\sigma$	t(m.y.) $\pm 2\sigma$	Reference
Bk–Sg mangerites and quartz mangerites	12	0.7105 ± 0.0003	873 ± 18	Demaiffe (1977)
Hidra charnockitic dykes and pegmatites	6	0.7086 ± 0.0006	909 ± 25	
Lyngdal granodiorite	6	0.7054 ± 0.0005	932 ± 38	Pederson and Falkum (1975)
Farsund charnockite	5	0.7128 ± 0.0009	852 ± 41	

† = number of points of the isochron.

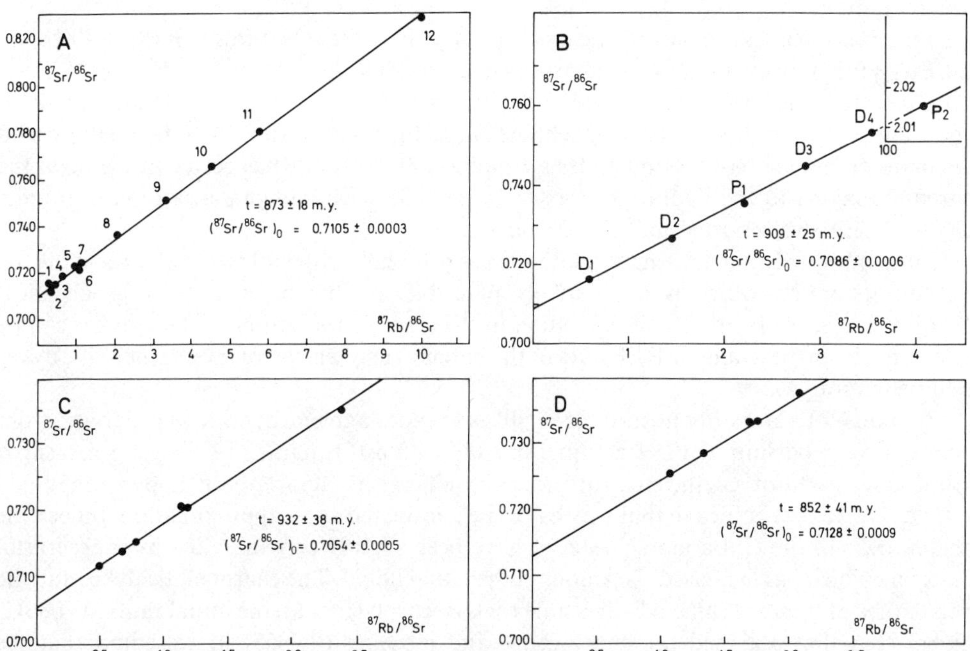

FIG. 6. Rb–Sr whole rock isochrons for the acidic rocks related to anorthosites: A and B from Demaiffe (1977), C and D from Pedersen and Falkum (1975). A: Bjerkrem–Sogndal mangerites and quartz mangerites. (1: 15B/72; 2: 130/72; 3: 133/72; 4: Pa 66P; 5: Pa 66Q; 6: 73776; 7: Pa 66M; 8: Pa 66N; 9: 66261; 10: 6687; 11: 101′/73; 12: 64120.) B: Hidra charnockitic dykes (D) and pegmatites (P) (D1: 443–4/1; D2: 283–2/2; P1: 501–2; D3: 377–2/1; D4: 298–2/1; P2: 501–1.) C: Lyngdal hornblende granodiorite. D: Farsund charnockite.

Interpretation and discussion

Existence of plagioclase cumulates and geochemical evidence (Duchesne and Demaiffe, 1977) indicate that the magmatic evolution of the anorthosite series is mainly controlled by fractional crystallization involving plagioclase.

Under these conditions the REE (except Eu) are expected to behave largely as incompatible or hygromagmatophile (Treuil and Varet, 1973) elements; i.e. the distribution coefficients are much smaller than unity. The Rayleigh equation then simplifies to

$$C_i/C_{i,o} = 1/f \quad (1)$$

where $C_{i,o}$ and C_i are the concentrations of element i in the parental magma and in the residual liquid respectively; F is the mass fraction of residual magma.

The validity of equation (1) for the present purpose is confirmed in the Bk–Sg lopolith by Roelandts and Duchesne (this book) who measured the REE contents in plagioclase and apatite and were able to relate them to an estimation of F.

The hygromagmatophile character of the REE implies that their concentrations in the liquids (and in the fractionated solids) continuously increase with fractionation. Moreover, in anorthositic massifs, extensive crystallization of plagioclase with an important positive Eu anomaly imposes a negative Eu anomaly on the residual liquids. Thus a high REE content and a negative Eu anomaly would characterize genuine residual liquids of anorthosite differentiation.

In the Hidra body the charnockitic dykes clearly show these features (ΣREE $\simeq$400 ppm; Eu/Eu* = 0.50) and can thus be interpreted as residual liquids. It has been previously demonstrated that the fine-grained jotunitic border facies displays a smooth REE distribution with no Eu anomaly (Duchesne *et al.*, 1974; Demaiffe, 1977). By analogy with the undifferentiated basaltic and andesitic magmas, this jotunite has been considered as the parental magma to the Hidra anorthositic suite. The REE contents in a typical jotunite (200–2/2) are also reported in Table 2 and Fig. 3.

If it is admitted that the charnockitic dykes are really cogmatic with the rocks of the anorthosite series, which is qualitatively plausible, on the basis of their geochemical characteristics, the factor F corresponding to the stage of formation of these dykes can be calculated from the available REE data of the parental magma and of the charnockitic dykes using equation (1).

The range of values obtained (0.25–0.30) is in close agreement with that derived from quantitative modeling of the fractional crystallization (Demaiffe, 1977) using measured solid/liquid partition coefficients for the main minerals (plagioclase, orthopyroxene).

The F value determined in that way has to be considered as an approximation. Indeed the acidic rocks of the Hidra massif seem to have been slightly contaminated by supracrustal material, which has increased the amount of residual liquid. The charnockitic dykes and the pegmatites plot on the same Rb–Sr whole rock isochron (Fig. 6); the initial ratio (0.7086) is higher that that obtained for anorthosites and jotunites (0.7055). If we admit that the contaminating supracrustal material is rich in radiogenic strontium (^{87}Sr/^{86}Sr$\simeq$0.720–0.725; Versteeve, 1975), the amount of contamination has to be small. The oxygen isotopic composition of the charnockitic dykes (δ^{18}O: 6.7–7.4‰) is also compatible with this hypothesis.

In Bjerkrem–Sogndal, mangerites display a REE pattern with a positive Eu anomaly. Accumulation of mesoperthitic feldspars can explain this feature. Mangerites (and also the quartz mangerite 66261) would not represent liquid completely separated from their cumulate minerals. As for the other quartz mangerites, the evolution trend in the Qz–Ab–Or diagram as well as the K/Rb trend (Duchesne, 1977) can be explained by continuous substraction of a mangeritic cumulate and the quartz mangerites can represent successive residual liquids. However, the REE evolution cannot be explained by simple fractional crystallization, because the REE contents decrease with increasing differentiation index. Two hypotheses can be invoked to explain this behaviour:

(a) *Apatite crystallization.* In Bk–Sg, apatites are rich in REE, especially in the final stages, e.g. the La content in apatite increases from 980 ppm in TII to 3000 ppm in 66261 and to 4180 ppm in 6687 (Roelandts, pers. comm.). It is conceivable that subtraction of minute quantities of apatite would drastically decrease the REE content in successive liquids. This hypothesis, in which the apatite continues to play the role of a cumulate mineral till the last terms of the series, must be rejected. Indeed, if it were the case and if apatite–liquid partition coefficients could be considered constant (Roelandts and Duchesne, 1977), the REE content in the successive apatites of the quartz mangerites would also decrease.

(b) *Supracrustal contamination.* Progressive contamination of successive mangeritic liquids by supracrustal material relatively poor in REE could explain the decrease of the REE in the quartz mangerites. For instance, such low REE might be similar to the granitic gneiss 134/72 or to the Precambrian sediments (Fig. 5). This contamination hypothesis is corroborated by the high value of the Sr isotopic initial ratio of these rocks (0.7105) compared to the measured values in the Bk–Sg anorthosites and norites (0.705–0.706: Demaiffe, 1977). Indeed, this high value cannot entirely be due to late closure of the Rb–Sr system relative to the U–Pb system, as discussed by Duchesne and Demaiffe (1977).

The mixing equation of Pushkar (1968) gives estimates of 15–30 wt.% for the amount of supracrustal contamination of the residual liquids. Input data for the calculations were the Sr contents and initial isotopic ratios of the migmatitic rocks of the envolope (Versteeve, 1975) and of the anorthosite–norite part of the Bk–Sg (Duchesne, 1977; Demaiffe, 1977). This supracrustal contamination is also in agreement with the K and Rb data (Duchesne and Demaiffe, 1977).

The low $\delta^{18}O$ values (4‰) of some quartz mangerites can be explained by a supracrustal contamination through a late fluid phase rather than by a bulk contamination (assimilation).

The Lyngdal hornblende granodiorite is rich in REE and characterized by a negative Eu anomaly after interpolation of the Gd content from the heavy REE trend; the Sr isotopic initial ratio (0.7054) is in the range of the values obtained on the Rogaland anorthosites. In contrast with those of the Farsund charnockite (see below), the geochemical features of the Lyngdal granodiorite are compatible with a comagmatic link between this rock and the anorthositic complex, although it is not yet possible to decide which anorthositic massif the granodiorite is to be linked with.

However, the magnitude of the Eu anomaly strongly depends, all things being equal, on the plagioclase–magma partition coefficient of Eu (D^{Eu}_{Plag}) which, in turn, is strongly dependent on the redox conditions (Philpotts, 1970). In the Bk–Sg lopolith, Roelandts and Duchesne (this book, Fig. 1) have shown that D^{Eu}_{Plag} may be small enough to give a small Eu anomaly in liquid of approximately equivalent composition to the Lyngdal granodiorite. This indicates that an extensive plagioclase fractionation can be needed to produce a visible negative Eu anomaly in residual liquids.

The Farsund charnockite is poorer in REE (ΣREE $\simeq$250 ppm) than the granodiorite; it shows no Eu anomaly and has a high Sr isotopic initial ratio (0.7128). These data rather favor an independent origin of this charnockite, probably through crustal anatexis and without any chemical relation with the anorthosites.

Conclusions

The combined use of major and trace element data, especially the REE, and of strontium

($^{87}Sr/^{86}Sr$) and oxygen ($^{18}O/^{16}O$) isotopic compositions proves to be a powerful methodological approach in deciphering the petrogenesis of the acidic charnockites related to anorthosites.

Direct derivation of acidic rocks by fractional crystallization of anorthosites appears to be a viable mechanism in some cases only. In most cases, contamination of the residual liquids by supracrustal material has to be invoked to explain some geochemical features of the charnockites.

More data are needed on the trace element geochemistry of high-grade metamorphic terranes to arrive at a better understanding of the contamination process.

Acknowledgements

Determination of the strontium isotopic composition was part of the program of the Belgian Centre for Geochronology. Professor M. Javoy is thanked for providing laboratory facilities during a period of work of one of us (D. Demaiffe) at the Laboratoire de Géochimie des Isotope stables, Université de Paris VII, and for discussing the isotopic aspect of this work. I. Roelandts has allowed us to use some unpublished results. This research has been partly supported by the Belgian FNRS–NFWO.

D. Demaiffe is a former "Aspirant" of this Foundation.

References

ADAMSON, O. J. (1942) The granite pegmatites of Hitterö, S.W. Norway. *Geol. Fören. Forhand, Stockholm* **64**, 97–116.

BARTON, J. M. and DOIG, R. (1972) Rb–Sr isotopic studies of the Lac Croche Complex, Grenville Province, Quebec. *Can. J. Earth Sci.* **9**, 1180–1186.

BARTON, J. M. and DOIG, R. (1977) Sr isotopic studies of the origin of the Morin anorthosite complex, Quebec, Canada. *Contr. Miner. Petrol.* **61**, 219–230.

BUDDINGTON, A. F. (1939) Adirondack igneous rocks and their metamorphism. *Geol. Soc. Amer. Mem.* **7**, 1–354.

BUDDINGTON, A. F. (1972) Differentiation trends and parental magmas for anorthositic and quartz mangerite series, Adirondacks, New York. *Geol. Soc. Amer. Mem.* **132**, 477–488.

DEMAIFFE, D. (1977) De l'origine des anorthosites.Pétrologie, géochimie et géochimie isotopique des massifs anorthositiques d'Hidra et de Garsaknatt (Rogaland–Norvège méridionale). Doctorat Thesis, Université Libre de Bruxelles, 363 pp.

DEMAIFFE, D., DUCHESNE, J. C., MICHOT, J. and PASTEELS, P. (1973) Le massif anorthosito–leuconoritique d'Hidra et son faciès de bordure. *C. R. Acad Sci., Paris* **277**, D17–D20.

DEMAIFFE, D. and JAVOY, M. (1976) Oxygen isotope geochemistry of anorthosites and related rocks of South Norway. Abstract in: *Fourth Eur. Colloq. Geochron.*, E.C.O.G. IV, Amsterdam.

DEMAIFFE, D., MICHOT, J. and PASTEELS, P. (1974) Time relationship and strontium isotopic evolution in the magma of the anorthosite–charnockite suite of South Norway. Abstract in: *Int. Meet. for Geochron. Cosmochron. and Isot. Geology*, Paris.

DE WAARD, D. (1969) The anorthosite problem: the problem of the anorthosite–charnockite suite of rocks. In *Origin of Anorthosites and Related Rocks*, Y. W. ISACHSEN, ed., N.Y. St. Mus. Sci. Serv., Mem. 18, 71–91.

DUCHESNE, J. C. (1972a) Iron–titanium oxide minerals in the Bjerkrem–Sogndal massif, S.W. Norway. *J. Petrol.* **13**, 57–81.

DUCHESNE, J. C. (1972b) Pyroxènes et olivines dans le massif de Bjerkrem–Sogndal (Norvège méridionale). Contribution à *l'Etude de la série anorthosite–mangérite*, 24th I.G.C., Montreal, sect. 2, 320–328.

DUCHESNE, J. C. (1978) Quantitative modeling of Sr, Ca, Rb and K in the Bjerkrem–Sogndal layered lopolith (S.W. Norway). *Contr. Mineral. Petrol.* **66**, 175–184.

DUCHESNE, J. C. and DEMAIFFE, D. (1978) Trace elements and anorthosite genesis. *Earth Planet. Sci. Lett.* **38**, 249–272.

DUCHESNE, J. C., ROELANDTS, I., DEMAIFFE, D., HERTOGEN, J., GIJBELS, R. and DE WINTER, J. (1974) Rare earth data on monzonoritic rocks related to anorthosites and their bearing on the nature of the parental magma of the anorthositic series. *Earth Planet. Sci. Lett.* **24**, 325–335.

Emslie, R. F. (1969) Crystallization and differentiation of the Michikamau intrusion. In *Origin of Anorthosites and Related Rocks,* Y. W. Isachsen, ed., N.Y. St. Mus. Sci. Serv., Mem. 18, 163–174.

Falkum T., Wilson, J. R., Annis, M. P., Fregerslev, S. and Zimmermann, H. D. (1972) The intrusive granites of the Farsund area, South Norway. *Norsk Geol. Tidsskr.* **52**, 463–466.

Franssen, L. and Deutsch, S. (1973) Un cas d'anorthosite résiduelle par ultramétamorphisme de roches supracrustales (Vikesaa, Norvège méridionale). *Ann. Soc. Géol. Belg.* **96**, 373–385.

Green, T. H., Brunfelt, A. O. and Heier, K. S. (1969) REE distribution in anorthosites and associated high grade metamorphic rocks. *Earth Planet. Sci. Lett.* **7**, 93–98.

Green, T. H., Brunfelt, A. O. and Heier, K. S. (1972) Rare earth element distributions and K/Rb ratios in granulites mangerites and anorthosites, Lofoten–Vesteraalen, Norway. *Geochim. Cosmochim. Acta* **36**, 241–257.

Heath, S. A. and Fairbairn, H. W. (1969) Sr^{87}/Sr^{86} ratios in anorthosites and some associated rocks. In *Origin of Anorthosites and Related Rocks,* Y. W. Isachsen, ed., N.Y. St. Mus. Sci. Serv., Mem. 18. 99–110.

Herrmann, A. G. (1970) Yttrium and lanthanides. In *Handbook of Geochemistry,* Wedepohl *et al.*, eds., Springer Verlag.

Hertogen, J. and Gijbels, R. (1971) Instrumental neutron activation analysis of rocks with a low-energy photon detector. *Anal. Chim. Acta* **56**, 61–70.

Koljonen, T. and Rosenberg, R. (1974) Rare earth elements in granitic rocks. *Lithos.* **7**, 249–261.

Kranck, E. H. (1969) Anorthosites and Rapakivi, magmas from the lower crust. In *Origin of Anorthosites and Related Rocks,* Y. W. Isachsen, ed., N.Y. St. Mus. Sci. Serv., Mem. 18, 93–97.

Martignole, J. (1974) L'évolution magmatique du complexe de Morin et son apport au problème des anorthosites. *Contr. Miner. Petrol.* **44**, 117–137.

Michot, J. and Michot, P. (1969) The problem of anorthosites: The South Rogaland igneous complex, southwestern Norway. In *Origin of Anorthosites and Related Rocks,* Y. W. Isachsen, ed., N.Y. St. Mus. Sci. Serv., Mem. 18, 399–410.

Michot, J. and Pasteels, P. (1969) La variation du rapport $(^{87}Sr/^{86}Sr)_0$ dans les roches génétiquement associées au magma plagioclasique. *Ann. Soc. Géol. Belg.* **92**, 255–262.

Michot, P. (1956) La géologie des zones profondes de l'écorce terrestre. *Ann. Soc. Géol. Belg.* **80**, 19–73.

Michot, P. (1960) La géologie de la catazone: le problème des anorthosites, la palingenèse basique et la tectonique catazonale dans le Rogaland méridional. *Norges Geol. Unders.* **212**, 1–54.

Michot, P. (1965) Le magma plagioclasique. *Geol. Rundsch.* **55**, 956–976.

Ormaasen, D. E. (1977) Petrology of the Hopen mangerite–charnockite intrusion, Lofoten, North Norway. Manuscript submitted to *Lithos.*

Pasteels, P., Michot, J. and Lavreau, J. (1970) Le complexe éruptif du Rogaland méridional (Norvège). Signification pétrogénétique de la farsundite et de la mangérite quartzique des unités orientales; arguments géochronologiques et isotopiques. *Ann. Soc. Géol. Belg.* **93**, 453–476.

Pedersen, S. and Falkum, T. (1975) Rb–Sr isochrons for the granitic plutons around Farsund, Southern Norway. *Chem. Geol.* **15**, 97–101.

Philpotts, A. R. (1966) Origin of the anorthosite–mangerite rocks of Southern Quebec. *J. Petrol.* **7**, 1–64.

Philpotts, J. A. (1970) Redox estimation from a calculation of Eu^{2+} and Eu^{3+} concentrations in natural phases. *Earth Planet. Sci. Lett.* **9**, 257–268.

Philpotts, J. A., Schnetzler, E. C. and Thomas, H. H. (1966) RE abundances in an anorthosite and a mangerite. *Nature* **212**, 805–806.

Pushkar, P. (1968) Sr isotope ratios in volcanic rocks of three island arc areas. *J. Geophys. Res.* **73**, 2701–2714.

Roelandts, I. and Duchesne, J. C. (1977) REE in apatite from layered norites and iron–titanium oxide ore-bodies related to anorthosites (Rogaland, S.W. Norway), this volume, 199–212.

Schrijver, K. (1975) Deformed root of a composite diapir in granulite facies. *Geotekt. Forsch.* **49**, 1–118.

Simmons, E. C., Hanson, G. N. and Lumbers, S. B. (1974) Comagmatic andesine anorthosites and quartz monzonites—Products of differentiation? *Ann. Meet. Geol. Soc. Amer.*, Abstract, Florida, 956–957.

Streckeisen, A. (1974) How should charnockitic rocks be named? In *Géologie des domaines cristallins,* J. C. Duchesne and J. Belliere, eds., Centenaire Soc. Géol. Belgique, Liège, 349–360.

Taylor, H. P. (1969) Oxygen isotope studies of anorthosites with particular reference to the origin of bodies in the Adirondack Mountains, New York. In *Origin of Anorthosites and Related Rocks,* Y. W. Isachsen, ed., N.Y. St. Mus. Sci. Serv., Mem. 18, 111–134.

Treuil, M. and Varet, J. (1973) Critères volcanologiques, pétrologiques et géochimiques de la genèse et de la différenciation des magmas basaltiques: exemple de l'Afar. *Bull. Soc. Géol. France* **40**, 506–540.

Versteeve, A. J. (1975) Isotope geochronology in the high grade metamorphic Precambrian of Southwestern Norway. *Norges Geol. Unders.* **318**, 1–50.

Wildeman, T. R. and Haskin, L. A. (1973) Rare earths in Precambrian sediments. *Geochim. Cosmochim. Acta* **37**, 419–438.

Comparative Petrogenesis of Archaean and Modern Low-K Tholeiites. A Critical Review of some Geochemical Aspects

R. C. O. Gill

Department of Geology and Mineralogy, Parks Road, Oxford, England

Abstract

Low-K tholeiites are the most abundant component of both Archaean greenstone belt volcanism and present-day volcanism. This paper compares the geochemistry of Archaean and modern low-K tholeiite associations, paying attention to elements which resist secondary modification during low-grade metamorphism. Allowance is made for differing degrees of fractionation among the suites considered by the use of variation diagrams versus Mg number. The Archaean basalts may be distinguished from modern mid-ocean ridge basalts and marginal basin basalts in having contents of Ti, Zr and P which are about half those in the modern suites, the comparison being referred to a common Mg number. Higher contents of Cr, Ni and Co and lower concentrations of Al discriminate the Archaean suite from modern island-arc tholeiites. When referred to a common Mg number, the Archaean tholeiites are found to have Ni contents at least twice as high as any modern low-K tholeiite.

These comparisons indicate that no universally satisfactory analogue of the Archaean tholeiite association exists among modern low-K tholeiites of any type. It is therefore inappropriate to seek information on the tectonic environment of greenstone belt volcanics by the use of geochemical discriminants developed from compilations of analyses of Phanerozoic basalts from known tectonic associations.

Semi-quantitative calculations suggest that the greater concentrations of Ni in the Archaean tholeiites cannot be explained in terms purely of higher melting temperatures and greater degrees of melting. Melting at higher pressures is a possible explanation, but since this alternative requires the existence of lower geothermal gradients in the Archaean source region than obtain at modern ocean ridges, it is regarded as unattractive. It is concluded that the Archaean source region must have possessed higher concentrations of Ni than in the modern mantle, possibly in the form of disseminated metallic particles reflecting incomplete core separation.

Introduction

In the ten years that have passed since the first Paris Symposium on the Origin and Distribution of the Elements, there have been exciting developments connected with both ends of the geological time-scale. The basalts of the ocean floor must be considered among the youngest of igneous rocks, and through the use of modern oceanographic sampling techniques the ocean bottom is losing much of its mystery. The geochemical evolution of the oceanic crust has now been documented in great detail, and the igneous processes by which these very young volcanic rocks differentiate are now understood in quantitative terms. A lot still remains to be learned about the nature of the source region, its inhomogeneity, and the manner in which different source domains contribute to the variability of basalt chemistry seen at the surface, but these problems would not now be contemplated were it not for the progress that has been made in recent years, particularly with regard to isotope and REE geochemistry.

During the same decade, research on the oldest available terrestial rocks, some of which bear a striking resemblance to the young ocean crust, has been just as productive, although the emphasis has been rather different. The Archaean record has been shown by isotopic methods to extend as far back as 3800 m.y. Isotopic studies have been most successfully applied to the deep-seated granitic and tonalitic rocks that dominate the Archaean shields, and here they have proved to be particularly informative on the processes of crustal accretion. Unfortunately, the application of isotope geochemistry to the volcanic associations that occur in the intervening greenstone belts has been hindered by the widespread low-grade metamorphism to which they have been subjected. Moreover, the distribution of several incompatible elements that have proved their usefulness in relation to the petrogenesis of modern ocean-floor basalts is also too susceptible to alteration and metamorphism to justify complete confidence in their application to Archaean basalt genesis.

While there are now signs of progress being made in this difficult field, many authors working in Archaean geochemistry have relied for petrogenetic information on the distribution of other elements which appear less liable to secondary mobility. A lot of data is now available on the geochemistry of REE and a range of more prosaic elements in greenstone belt volcanics, and the merits and the pitfalls of comparing these early volcanic suites with possible modern analogues have been discussed by numerous writers. In this contribution I shall consider the geochemistry of some of these less mobile elements in Archaean basalts, paying special attention to the significance of drawing comparisons with younger basalt types.

The highly magnesian komatiites, being more or less confined to the Archaean, have naturally captured a great deal of attention. Considered from the point of view of volume, however, the dominant rock type in most greenstone belts is a magnesium-poor tholeiitic basalt. Such Archaean tholeiities, when well preserved, are generally found to be of low-potassium type, possessing all of the chemical features by which one distinguishes present-day mid-ocean ridge and island-arc tholeiites from other modern basalt types. It seems that low-K tholeiitic basalt has the distinction of being the most abundant lava type to be erupted both during Archaean times and at the present day, since it now covers two-thirds of the earth's surface. For this reason it can be used for purposes which are not served by the more exotic members of the greenstone belt successions. If secular changes in the composition of the mantle, or its thermal structure, have been reflected in the geochemistry of mantle-derived magmas through the course of geological time, one might expect to read the record of such changes most clearly in rock types which are common to both Archaean and modern environments, since one thus minimizes the influence of other petrogenetic variables. I hope that this survey of the comparative geochemistry of Archaean and modern low-K tholeiites will illustrate some of the possibilities of this approach.

Archaean Tholeiites and Related Modern Magma Types

Greenstone belts most commonly occur as synclinal structures surrounded and deformed by the later emplacement of extensive tonalite and granodiorite plutons. Basalts are common throughout the typical greenstone succession. A number of authors have pointed out a tendency for komatiitic rocks to be found most commonly near the base of the volcanic sequence, whereas intermediate and acid volcanics of calc-alkaline affinity may become progressively more important in the upper parts of the succession. It should be emphasized, however, that basalts are almost universally the most abundant rock type, even in belts in which andesites are thought to be common, such as those of the Canadian shield (Baragar and Goodwin, 1970; Hallberg, 1972).

The possibility exists that the tholeiitic basalt magma may have been more abundant than present exposures suggest. O'Nions and Pankhurst (1978) present detailed arguments in favour of a two-stage model for the generation of the abundant tonalitic plutonism that succeeds and envelopes the greenstone belts. Trace element data are interpreted in terms of an immediate (second-stage) source material of basaltic composition, having the flat REE and low Rb/Sr characteristics of the Archaean tholeiites but adopting a plagioclase–amphibole–garnet mineralogy. This viewpoint implies that the abundance of low-K tholeiites may be under-represented by an order of magnitude or more in areas where tonalitic plutons or gneisses predominate, the balance presumably having been either kept at depth or consumed by some process of subduction.

It is unusual to see the basement upon which a given greenstone belt was laid down, the unconformity being in nearly every case obliterated by deformation or the intrusion of later plutons. For this reason the original tectonic environment of the early Precambrian greenstones has been the subject of vigorous controversy. The chemical distinctiveness of the ultramafic and mafic lavas and their common occurrence in the form of pillows have given rise to the view that they represent a primordial oceanic crust. The recent identification of a sialic basement underlying one or two greenstone belts, and the occurrence of chemically similar low-K tholeiitic dykes cutting sialic crust in the interior of the Greenland craton, indicate that this interpretation cannot be applied universally, though it may still be applicable to the majority of greenstone belts. Another view, based on the calc-alkaline affinities of the intermediate and acid rocks associated with many belts, is that they represent developing island-arcs. Gill and Bridgwater (1976) and others have, however, questioned the use of chemical parallels with modern volcanic suites to identify Archaean crustal environments, since the nature of modern volcanism is rarely determined by crustal type but reflects the mantle substructure with which the crustal environment is linked geographically. In the discussion given here, magma chemistry will be viewed as reflecting the attributes of the source region, not those of the crustal environment in which the lavas are now preserved.

Volcanic rocks resembling the Archaean low-K tholeiite suite (hereafter abbreviated to "Archaean tholeiite") are found in three modern volcanic associations:

(i) Mid-ocean ridge basalts (MORB).

(ii) Basalts erupted in marginal basins at centres of "behind-arc" ocean-floor spreading. The chemical difference between the marginal basin (MB) association and MORB is slight (see Figs. 5 and 6), but certain structural parallels between this tectonic environment and the Archaean greenstone belts (Tarney *et al.*, 1976) justify a separate category.

(iii) The island-arc tholeiite series (IAT) found in the outer and lower parts of several arc structures.

The comparisons discussed in the following pages will consider ocean-floor basalts from the north and south Atlantic and eastern Pacific oceans and the Caribbean Sea, marginal basin basalts from the Mariana and Scotia arcs, and island-arc tholeiites from Tonga, Fiji, the South Sandwich Islands and the West Indies. The data for Archaean basic rocks are drawn from the literature dealing with southern Africa, Australia, Minnesota and West Greenland. Individual data sources are given in the figure captions.

In assembling the data for Archaean basalts, I have attempted to minimize the influence of alteration where possible by omitting analyses with high CO_2 or abnormal oxidation ratios, and those which are identified in the original reference as representing significantly altered rocks. Most of the analyses selected have K_2O contents less than 0.3%.

Comparative Geochemistry—Fractionation and Covariance

Much has been written on the geochemistry of the Archaean tholeiite suite, comparing its composition with that of present-day basalt suites (Hart *et al.*, 1970; Glikson, 1971; White *et al.*, 1971; Hallberg and Williams, 1972; Jahn *et al.*, 1974; Condie, 1976; Nesbitt and Sun, 1976; O'Nions and Pankhurst, 1978; Hawkesworth and O'Nions, 1977; Gill and Bridgwater, in press). There is little consensus in the literature on the generalizations that may be drawn from the available data, and, of these, which are the most important. But the differences distinguished between the two populations depend to a large degree on the manner in which the comparison is made. Distinctions are often drawn simply by comparing the means, a technique which has two obvious shortcomings. In the first place there is the elementary statistical limitation that the difference between two means cannot be regarded as significant unless the variances of the two populations are also considered. The second, and geochemically more important, factor concerns the degree of fractionation which is developed in each of the basalt populations.

Figure 1 shows the distribution of published analyses of Archaean tholeiites and modern mid-ocean ridge basalts with respect to Mg number†. It is clear from this diagram that the Archaean and MORB populations exhibit somewhat different degrees of fractionation, and the difference is statistically resolvable*. The more advanced degree of fractionation apparent among the Archaean basalts, whatever the cause may have been, must be accompanied by geochemical changes additional to that in Mg number. Any comparison that is made between mean Archaean tholeiite and MORB compositions, which does not take account of such superficial changes in chemistry, will conceal a systematic error introduced by the more extended fractionation in the Archaean suite, particularly when the discussion is framed in terms of the parent magmas of each suite.

The point is illustrated by Al_2O_3 in Fig. 2. It has been suggested (Glikson, 1971; Condie, 1976) that Archaean tholeiites are consistently poorer in Al_2O_3 than recent ocean tholeiites. Considering the wide range of Al_2O_3 contents in each association, one would in fact require statistical tests to establish such an inference even when the Al_2O_3 data is examined in isolation. When considered in conjunction with Mg number, the distinction between the Al_2O_3 values of the two populations is, in one sense, eliminated. The MORB data, though scattered, lie on a clear trend, presumably representing low-pressure differentiation, and the Archaean tholeiites are found to fall on the same converging trend. When comparison is made between samples of each population having the same range of Mg number, the two Al_2O_3 distributions have indistinguishable means. The ill-defined difference in Al_2O_3 contents between Archaean tholeiites and MORB tells us nothing that was not already

† Mg number=100 Mg/(Mg+Fe^2) in cation units. In calculating Fe^2, Fe_2O_3 has been set arbitrarily at 0.15ΣFeO regardless of the measured Fe_2O_3/FeO ratio.

* In selecting analyses for the MORB histogram, ferrobasalts from such places as the Juan fe Fuca and Gorda ridges have been included when they appear in general accounts of ocean-floor basalts. Data from papers dealing specifically with these unusually fractionated MORB suites have been excluded, to avoid introducing a bias into the MORB distribution.

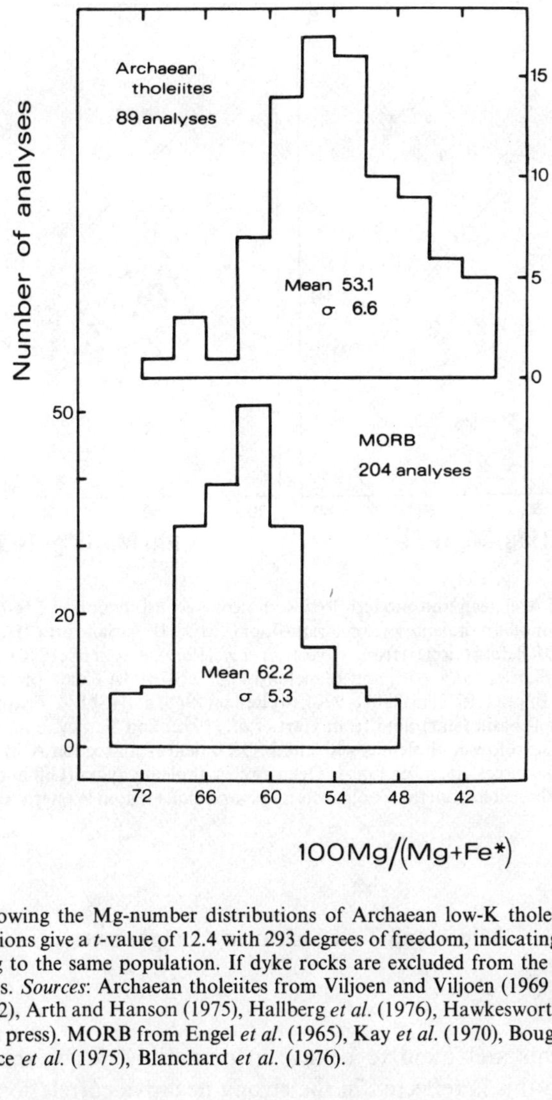

FIG. 1. Histograms showing the Mg-number distributions of Archaean low-K tholeiites and modern MORB tholeiites. The distributions give a *t*-value of 12.4 with 293 degrees of freedom, indicating a probability of less than 0.05% that they belong to the same population. If dyke rocks are excluded from the Archaean suite, the same probability limit applies. *Sources*: Archaean tholeiites from Viljoen and Viljoen (1969 and other papers in same volume), Hallberg (1972), Arth and Hanson (1975), Hallberg *et al.* (1976), Hawkesworth and O'Nions (1978) and Gill and Bridgwater (in press). MORB from Engel *et al.* (1965), Kay *et al.* (1970), Bougault and Hekinian (1974), Frey *et al.* (1974), Bence *et al.* (1975), Blanchard *et al.* (1976).

apparent from the differing degrees of Mg/Fe fractionation. One cannot maintain, as one might conclude from the literature, that the primary magmas (those in equilibrium with residual mantle mineral assemblages, assuming these to have constant Mg number) which produced the two observed magma sequences differed with respect to Al_2O_3.

Marginal basin (MB) basalts, though scattered, overlap with the MORB trend, and it is not possible to draw a consistent distinction between them and the Archaean tholeiites. The Al_2O_3 distribution of the island-arc tholeiite series, on the other hand, is quite distinct from all of the other basalt types shown in Fig. 2. This major geochemical departure sets the island-arc tholeiites apart from other low-K basalts, regardless of age, and constitutes a major objection to any proposed affinity between the Archaean basalts and modern island-arc associations. With respect to aluminium, the Archaean suite shows much closer identity with MORB.

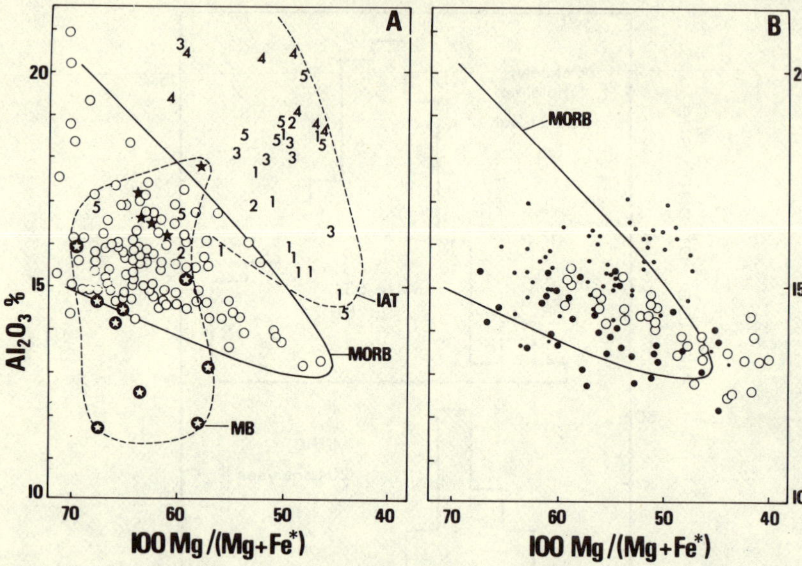

FIG. 2. Al_2O_3 contents of Archaean and modern low-K tholeiites as a function of Mg-number:
A. Composition fields of modern tholeiites, from ocean-floor ("MORB"), island arc ("IAT") and marginal basin ("MB") environments. MORB data (circles) from Miyashiro *et al.* (1969), Kay *et al.* (1970), Bougault and Hekinian (1974), Frey *et al.* (1974), Bence *et al.* (1975) and Blanchard *et al.* (1976). IAT data (numbered) from (1) Ewart *et al.* (1973), (2) Ewart and Bryan (1972), (3) Gill (1970), (4) Baker (1968) and (5) J. A. Pearce (unpublished data; S. Sandwich Islands). Marginal basin (star) data from Hart *et al.* (1972) and Tarney *et al.* (in press).
B. Distribution of Archaean low-K tholeiites with the MORB field from section A for comparison.
Data from (solid circles) sources given for Fig. 1. Open circles: tholeiite dykes (Gill and Bridgwater, in press). Small points: 80 Archaean tholeiites from the Coolgardie and Norseman areas in Western Australia (Hallberg, 1972 and unpublished data).

Minor and Trace Element Comparisons

Ti, Zr and P

All of these elements are found to behave incompatibly during the differentiation of MORB magmas, and this is reflected in the strong negative correlation with Mg number (Figs. 3 and 4). The four magma associations shown in Figs. 3 and 4 overlap appreciably when differing degrees of fractionation are disregarded, but divergent trends become apparent when the data are plotted against Mg number. The differences resolved in this way can be expressed in semi-quantitative terms. By reference to a given Mg number, one may see that in the Archaean basalts Ti and P are only 0.5–0.7 times as abundant as in MORB (and MB basalts), whilst for Zr the same factor is less than 0.5. If data from individual provinces were to be considered separately, these distinctions might be still further refined. Most of the P_2O_5 data for the East Pacific ferrobasalts, for example, plot in the lower part of the MORB envelope (data from Clague and Bunch, 1976). Given the assumption of incompatible behaviour and a constant relationship between Mg number and volume fractionation, the above abundance ratios extend also to the primary magmas (in the sense used above) from which the MORB and Archaean tholeiite trends have developed.

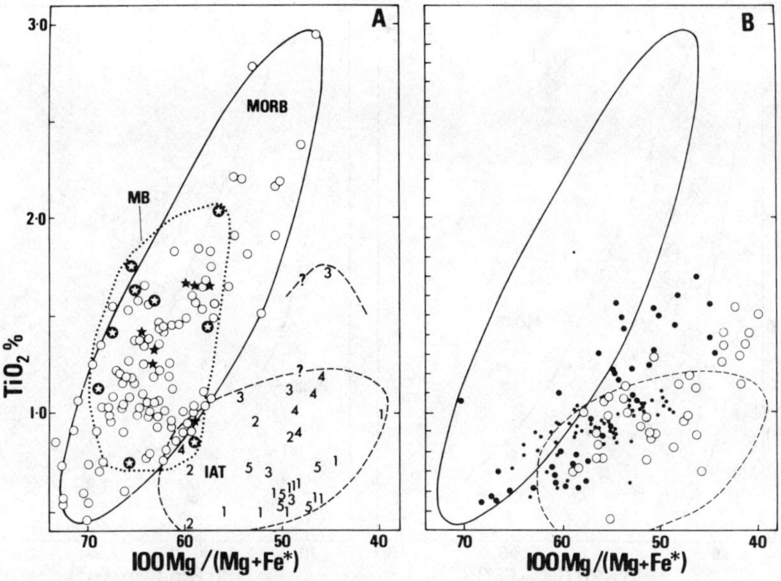

FIG. 3. TiO_2 contents of modern and Archaean low-K tholeiites plotted against Mg-number.
A. Modern tholeiites. Sources, symbols and abbreviations as in Fig. 2. The upper boundary shown for the IAT field ("?") is ambiguous on account of a single (altered oxidized) analysis from Fiji.
B. Archaean tholeiites (sources and symbols as in Fig. 2.).

There is a nearly perfect overlap between the Archaean tholeiites and the island-arc tholeiites with respect to Ti, Zr and P in Figs 3 and 4. This parallel with the island-arc association is thought to extend to other incompatible elements such as K and Rb (O'Nions and Pankhurst, 1978; Jahn and Sun, this volume), although with these elements the

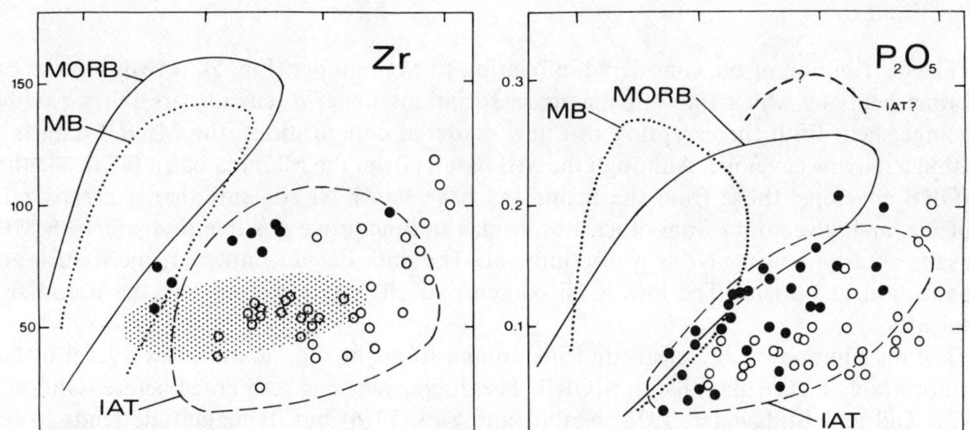

FIG. 4. Zr and P_2O_5 contents. MORB field based on data from Engel *et al.* (1965), Thompson *et al.* (1972), Clague and Bunch (1976) and references given in Fig. 2. IAT and marginal basin fields from references already cited. The upper boundary marked "IAT?" takes account of two unrepresentative analysis from Fiji. Archaean compositions from sources in Fig. 2. The stippled field indicates the Coolgardie–Norseman suite (Hallberg, unpubl. data)

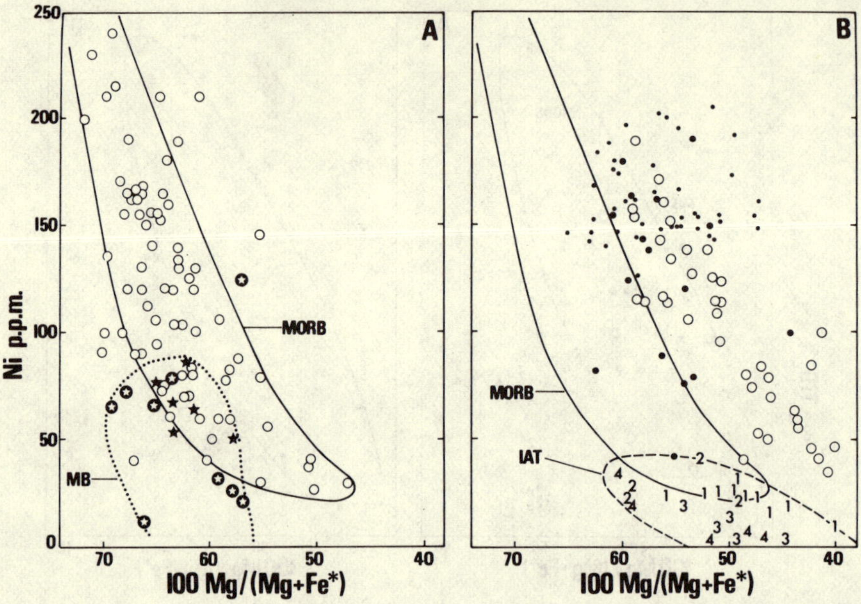

FIG. 5. Ni contents of modern and Archaean tholeiites. A. MORB and marginal basin data from sources in preceding diagrams. B. Archaean tholeiites and island-arc tholeiites from sources cited.

departure from MORB is in the opposite direction, towards relative enrichment in the IAT and Archaean tholeiite suites. The magnitude of this enrichment relative to MORB is uncertain, particularly in the Archaean basalts, because of the susceptibility of K and Rb to secondary migration.

Ni, Cr and Co

The distribution of Ni, considered in relation to Mg number (Fig. 5), reveals a series of distinct fields by which the four magma associations under discussion are fairly readily distinguished. With the exception of a few scattered compositions, the MORB data fall within a narrow envelope. Although the MB basalts from the Mariana basin fall inside the MORB envelope, those from the Scotia sea have lower Ni contents that generally fall outside, and, given data from other arcs, it may in time prove possible to distinguish MB basalts as a type using Ni as a discriminant. The data do not, however, merit such an application at present. The lowest Ni contents of all types are found in the island-arc tholeiites.

The distribution of Archaean tholeiite compositions in Fig. 5b indicates a level of Ni demonstrably higher than that in MORB. The discrepancy has been noted before (Glikson, 1971; Gill and Bridgwater, 1976; Nesbitt and Sun, 1976) but its magnitude tends to be underestimated. Nesbitt and Sun (1976) regard Ni contents in the Archaean tholeiites as being 20% higher than in MORB. When the comparison is referred to a common Mg number, however, the Ni distributions of the two suites can be seen to differ by a factor of 2 or 3 times. To the extent that it applies to the primary liquids parental to the two lines of descent (an assumption which is less satisfactory than for the incompatible elements), this large departure is petrogenetically very significant, and its implications are discussed in some

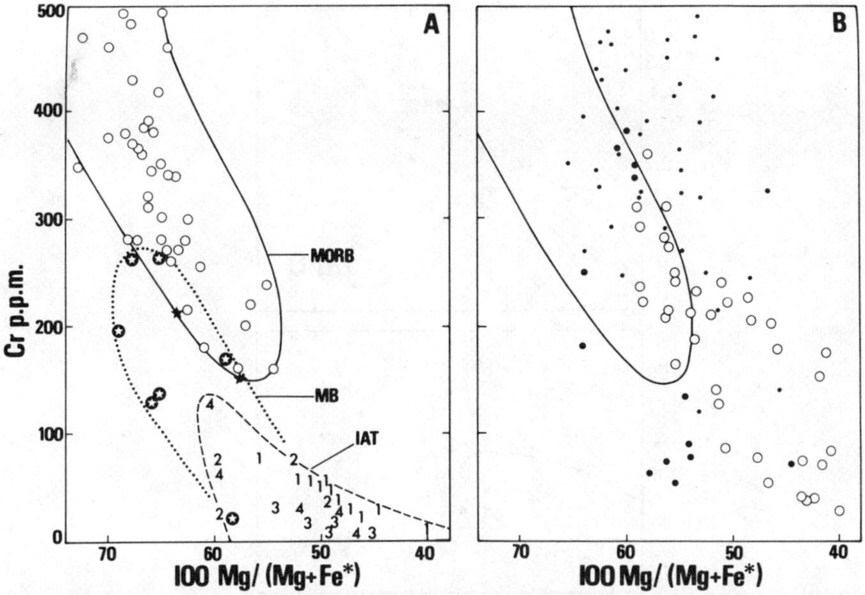

FIG. 6. Cr contents of modern and Archaean tholeiites. A. MORB, IAT and marginal basin compositions from sources cited. B. Archaean compositions from sources cited.

detail later. The Archaean tholeiites clearly do not share the Ni-poor tendency of the island-arc tholeiites, nor does Fig. 5 lend any geochemical support to a relationship between the Archaean basalts and the marginal basin association.

A similar pattern appears, though less distinctly, in the behaviour of Cr (Fig. 6). Because Cr data do not seem to be available for the more fractionated of ocean-ridge basalts, the MORB envelope cannot be extended as far as for Ni. This difficulty, in conjunction with the wide scatter of Cr data for the Archaean basalts†, causes the possible slight divergence between the MORB and Archaean distributions to be a subjective matter. The MB and IAT fields are nonetheless easily distinguished from that of MORB, and one may reiterate that neither of these types can be regarded as a complete modern analogue of the Archaean basalt association.

The direction of the MORB fractionation curve for Co, unlike those for Ni and Cr, is not clearly defined by the available data (Fig. 7). Calculated evolution vectors based on quantitative least-squares modelling are shown as arrows in Fig. 7 to indicate the expected direction of development of magma composition. The arrows show that the relationship between the Co concentration in MORB and the Archaean tholeiite suite could be attributed solely to routine magma fractionation. It would be inappropriate therefore to attach more fundamental petrogenetic significance to the departure in mean Co concentrations. The same remark applies to the marginal basin tholeiites, whose contents fall within the MORB field.

The island-arc tholeiite association (not shown) falls mainly below the MORB Co distribution. Co therefore reinforces the indication given by Ni and Cr that the island-arc suite fails to provide a satisfactory modern parallel for greenstone belt tholeiites.

† There is circumstantial evidence for interlaboratory discrepancies in the analysis of Cr and, until they are eliminated, one cannot determine whether interprovince differences exist, nor can one develop a quantitative interpretation of Archaean–MORB relationships.

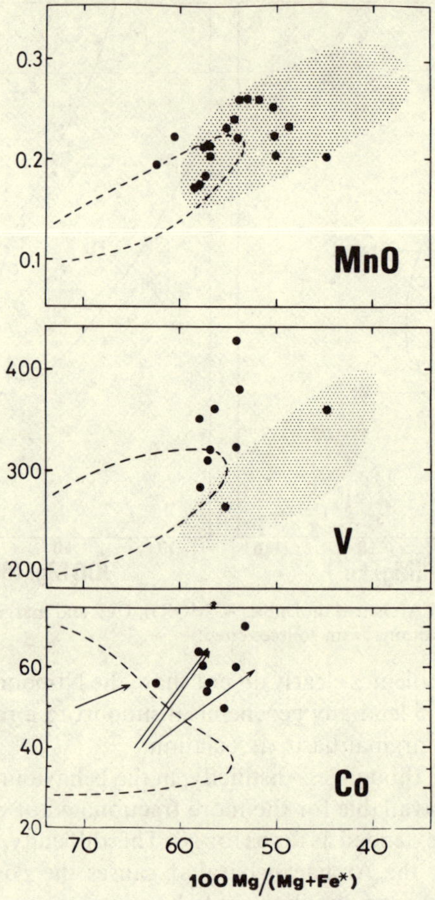

FIG. 7. MnO, V and Co contents in Archaean and modern populations. The MORB field is indicated by a dashed line (sources as cited previously). Averages of Archaean suites are indicated by filled circles (mainly from Hallberg, 1972). The stippled area indicates the distribution of Archaean tholeiite dykes from West Greenland (Gill and Bridgwater, in press). In the case of Co, the MORB fractionation trend is not clear from the shape of the field. The arrows represent the calculated Co trends based on the fractionation models of Frey *et al.* (1974), using the observed initial Co content of 50 ppm, and of Clague and Bunch (1976), based on an arbitrary initial content of 40 ppm Co. The olivine/liquid partition coefficient for Co was taken as 4 (Frey *et al.*, 1974; Leeman, 1973), those for other phases being neglected.

Other Transition Metals

The Mn contents of the Archaean tholeiites lie on a common trend with the MORB distribution, and the same may well be true of V, although the data are too scattered to give a clear picture. Limited data for Sc (not shown) also indicate colinearity between MORB and Archaean tholeiites when plotted against Mg number. The remarks made concerning Co probably apply to each of these elements. Cu is believed to be too susceptible to secondary mobility to merit comparison.

Rare-earth Elements

An abundance of REE data now exists on the Archaean volcanic suites, but only a small proportion of it is presented with accompanying major element analyses, and few of the REE fractionation model calculations that have appeared in the literature on Archaean basalts have been subjected to major element fractionation constraints. Consequently little of the REE data available is amenable to the kind of treatment discussed in the paragraphs above.

La/Sm ratios are higher in Archaean tholeiites than in modern ocean-ridge basalts (Sun and Nesbitt, 1977) and this, together with the apparently higher K, Rb and Ba contents and lower K/Rb ratios in Archaean tholeiites, favours a parallel with modern island-arc tholeiites. However, Hawkesworth and O'Nions (1977) conclude that the Archaean suite has higher Ce/Yb and Rb/Sr ratios than IAT basalts (of similar Mg content) and prefer an analogy with marginal basin basalts. In view of the scarcity of data at present, and the apparent heterogeneity of the MB basalt association in the case of such elements as Al and Ni (Figs. 2 and 5), one should treat this comparison cautiously.

Mg Number and Alteration

By reference to Figs. 3, 4, 5, 6 and 7, it will be seen that the departures between MORB and the Archaean tholeiites with respect to Ti, Zr, P, Ni and Cr could be eliminated by shifting the Archaean distribution to higher values of Mg number. Since Mg and Fe contents can suffer modification during severe alteration of basalts, it is natural to ask whether some of the differences between these two associations may simply reflect a systematic shift to lower Mg numbers accompanying the observed alteration and metamorphism of the Archaean basalts.

This interpretation cannot be conclusively disproved, but several lines of evidence make it seem unlikely. Hydrothermal experiments on the sea-water alteration of *crystalline* basalt suggest that if Mg and Fe contents are affected at all, the change is most probably towards higher Mg numbers (Mottl *et al.*, 1974; Humphris, 1975; Hajash, 1975; Bischoff and Dickson, 1975), the reverse of the change required to reproduce the Archaean tholeiite trend from the MORB trend. It is harder to predict the changes that would occur during regional metamorphism, but it is not plausible to suggest that it introduces a large shift in Mg/Fe ratio uniformly over a worldwide sample of Archaean basalts without having a significant influence on more mobile elements such as potassium. The retention of K_2O contents less than 0.3% in most of the Archaean analyses used in this work is not consistent with such a universal post-magmatic shift in Mg number. The incidence of this hypothetical alteration process would have to be universal among Archaean tholeiites to account for the observed lack of overlap between the Archaean suite and MORB in Figs. 3, 4, 5 and 6; most natural processes would on the contrary leave an even scatter of intermediate compositions representing less complete reaction.

Petrogenetic Application—the Significance of the Ni Anomaly

Figures 2 to 7 depict a series of elements by means of which the magma types being discussed are most readily distinguished from one another. When these diagrams are considered together, one is forced to the conclusion that no universally satisfactory analogue of the Archaean tholeiite association can be found among modern low-K tholeiites. Whilst noting, for example, that the Archaean suite most resembles modern island-arc tholeiites with regard to Ti, Zr, P and light REE (and other incompatible elements), one must not lose sight of the divergent behaviour of the compatible elements Al, Ni, Cr and Co, regarding which the Archaean tholeiites have much closer affinity with mid-ocean ridge basalts. It is therefore no simple matter to identify a single modern petrogenetic standard against which an informative comparison of Archaean tholeiite geochemistry can be made.

Resolution of this fundamental ambiguity must await a more complete understanding of the relationships between the several modern types of low-K tholeiite. The only aspects of the comparative approach to Archaean tholeiite geochemistry that can prove petrogenetically informative at present are the ones in which a consistent disparity exists between the Archaean tholeiite and *all* modern types of low-K basalt. The foremost example is Ni, to which the remainder of this section will be devoted. The compositions of the Archaean basalts in Fig. 5 define a trend that lies to the high-Ni side of the MORB envelope and overlaps with it very little. Referring to an arbitrary Mg number of 60, the centres of the two trends occur at about 200 and 80 ppm Ni respectively. When expressed in such terms Ni contents in the greenstone belt rocks therefore exceed those in MORB by an average of 2.5 times, and by a still larger factor in the case of the MB and IAT associations.

An explanation of the Ni anomaly may be sought either through the influence of different conditions of melting, or in terms of higher Ni concentrations in the Archaean source region than in the sources of modern basalts. The first suggestion is easier to test, and several such models will be discussed below.

Nesbitt and Sun (1976) interpret the high-Ni trend as being a consequence of higher melting temperatures during the production of the Archaean tholeiites than are applicable for modern low-K basalts. Their case rests on the temperature dependence of the olivine–liquid partition coefficient of Ni. $K_{Ni}^{ol\text{-}L}$ fall off rapidly with increasing temperature above 1200°C, and it is suggested that melting at higher temperatures will generate a series of basalt liquids with higher Ni contents, assuming the Ni content of the mantle before melting remains constant. The proposal is not as simple as it may seem, because in most circumstances elevating the temperature of melting will change the composition of the liquid produced (to more magnesian and olivine-normative compositions), so that a compensating stage of magma fractionation must be introduced into the model if the final liquid is still to have a major element composition comparable to MORB. The additional fractionation step must influence the Ni content of the liquid.

A test of the hypothesis may be formulated in the simplest quantitative terms by considering two alternative melting models: (a) basalt is produced by 20% melting of four-phase lherzolite at 10 kb† and (b) a higher-temperature liquid is produced by 50% melting‡ of the same material at 10 kb, that liquid undergoing subsequent partial crystallization to yield a Mg number comparable to the liquid produced directly in (a). The results of these calculations are summarized in Table 1. The basis of the calculation is as follows.

† 10 kb is consistent with the range of melting depths inferred from MORB eruption temperatures (Scheidegger, 1973; Hekinian *et al.*, 1976). It can be shown that the choice of pressure does not significantly affect the result of the calculation.

‡ An upper of limit of 50% melting has been assumed. Green (1975) infers that a South African peridotitic komatiite may represent as much as 70% melting, but doubt exists as to whether such an extreme composition can be parental to the more prolific Archaean tholeiites (Nesbitt and Sun, 1976; O'Nions and Pankhurst, 1978). It happens that the conclusions drawn from these calculations are not particularly sensitive to the chosen upper limit of melting.

TABLE 1. NICKEL IN THOLEIITE AS A FUNCTION OF MELTING TEMPERATURE

	(1)	(2)		
(a) *20% melting at 1370°C, 10 kb,* $K_{\text{Ni}}^{\text{ol/L}} = 5.2$				
(i) Modal melting	0.349	700		
(ii) Non-modal melting All cpx and ga/sp in liquid All ol and opx in residue	0.283	566		
(iii) Mean of (i) and (ii)	0.316	633†		
(b) *50% melting at 1500°C, 10 kb,* $K_{\text{Ni}}^{\text{ol/L}} = 2.4$			*Fractional crystallization* Two 16% stages, K = 2.4 and 4.0	
(iv) Modal melting	0.76	1520	A	B
(v) Non-modal melting A All cpx and ga/sp in liquid Modal props. of ol and opx in residue	0.66	1320	ol only	$ol_{50}\,opx_{50}$
(vi) Non-modal melting B All cpx + ga/sp + opx in liquid. Residue of ol alone	0.59	1176†	→ 917 ppm →	635 ppm

• (1) Concentration in liquid/concentration in initial solid assemblage.
(2) Ppm in liquid if initial mantle Ni content = 2000 ppm.
† Preferred model.

In order to develop a numerical model of this kind, it is necessary to know the relationship between the degree of partial melting and temperature. The dry melting behaviour of a garnet peridotite nodule (1611) reported by Mysen and Kushiro (1976) will be used for this purpose (this nodule aproximates to the compositon of pyrolite). At 20 kb the assemblage is 20% molten at about 1500°C (ol + opx + Ca–poor–cpx + liq) and 50% molten at about 1630° (ol + liq). The partial melting interval (20–50%) is 130°C. Translating these figures to 10 kb one gets temperatures of 1370°C and 1500°C and an interval of about 130°C respectively. The model is insensitive to the lower percentage-melting limit chosen, since the melting-curve slope is slight between 5 and 30% melting.

The calculations have been carried out using the equilibrium-melting equation of Shaw (1970). The partition coefficient for Ni in olivine/liquid partition has been calculated using the temperature dependence of Leeman (1973), which agrees well with that of Irvine (1974). Both functions conceal a compositional term, but this compound dependence is appropriate to the present basalt liquid application. The orthopyroxene/liquid coefficient is assumed to approximate to unity (Mysen, 1976) and the coefficients for the other phases are assumed to be low. The model mantle composition is ol_{60}–opx_{20}–cpx_{10}–ga/sp_{10} (mica assumed absent).

Each of the two melting models can be formulated several ways depending on the residual assemblage, as shown in Table 1. Modal melting is thought to be an unrealistic model particularly for the 50% melting case, since the experiments of Mysen and Kushiro (1976) indicate olivine as the sole residual phase at these temperatures.

In extending the degree of partial melting from 20% to 50%, one is altering the parameters of what is generally assumed to approximate to an *equilibrium* process. The partial crystallization step which has been introduced to compensate for the more advanced partial melting in model (b) would on the other hand approximate to a *fractional* crystallization process in nature, particularly if it takes place at low pressure. The chemical changes (in the concentrations of the compatible elements) during fractional crystallization will be significantly greater for a given temperature interval than in equilibrium melting. Such differences are difficult to quantify accurately, but are of great importance in determining the Mg/Fe and Ni behaviour of the derived liquids.

The mass fraction of liquid at 20% melting is only four-tenths of that at 50% melting. By reversing the process precisely, 60% crystallization of the higher temperature liquid would be required in order to reproduce the Mg number of the other magma (case (a)). This assumption cannot be applied in the case of fractional crystallization, and it has been assumed instead that only 30% crystallization will be required to achieve parity of Mg number.† This is a lower limit most favourable to the hypothesis of Nesbitt and Sun (see later). Since the crystallization will take place over a continuous temperature range between 1500° and 1370°C, the fractionation of Ni has been approximated by two 16%-crystallization stages, using K values appropriate to 1500° and 1410°C. Details are given in the lower part of Table 1.

The final results, shown in boxes in Table 1, suggest that the two melting paths would produce two liquids with similar Ni contents for a given Mg number. Many uncertainties enter into this calculation, and there is no doubt that future developments will allow the model to be formulated more accurately‡. But most of the uncertainty relates to absolute values, whereas the utility of the model calculation depends on a comparison, from which (to a first approximation) the errors will have cancelled out. The existing data suggest that increasing the percentage of melting in the Archaean source region does not provide a satisfactory explanation of the large (more than 2-fold) discrepancy in Ni content in the Archaean tholeiite suite relative to MORB. This conclusion would only break down if extremely small degrees of subsequent crystallization were sufficient to bring the higher temperature liquid into major-element conformity with the lower-temperature liquid (representing MORB). The proportion of normative olivine in analyses of high-Mg basalts and komatiites from greenstone belts, together with their low Al_2O_3 contents, indicate that the above assumptions in fact under-estimate the degree of olivine fractionation required to reproduce low-Mg basalt. It seems likely that a natural high-percentage-melting history would in fact have imparted a low-Ni trend, not a high-Ni trend, to the Archaean basalt suite. This may be an argument against a simple fractional crystallization relationship between komatiites and tholeiites in greenstone belts.

An alternative means of increasing melting temperature, but without changing the percentage of melting, would be to produce Archaean tholeiites at greater depths than MORB. An Archaean source at 20 kb would yield 20% melt at 1500°C, using the data given above. Such a change in the depth of melting would again introduce a temperature increment of 130°C. Calculations analogous to model (a), version (iii) in Table 1 show that a 20% melt at 1500°C will contain 1200 ppm Ni (same initial mantle composition as before) compared to 630 ppm at 1370°C and 10 kb. Preliminary data (Mysen, 1976) suggest that the Ni partition between olivine and liquid is slightly pressure-dependent, in a direction that would reinforce this difference. One must assume that the change in major element composition accompanying the deeper depth of generation would be slight. Since the two liquids (at 10 and 20 kb) represent comparable degrees of melting, no subsequent fractionation stage will be required in this model to adjust for differences in Mg number.

Existing data therefore do provide a basis for deriving the observed high-Ni trend in

† Either liquid may subsequently fractionate to lower values of Mg number, but it is assumed that the effects of this stage will cancel out when the two "melting paths" are compared. The figure of 30% applies only to that degree of crystallization required to compensate the extended degree of melting in model (b).

‡ The calculation might be improved by using an exchange distribution coefficient $K^{ol\text{-}L}_{Ni\text{-}Mg}$, as suggested by Sato (1977). This approach would, however, require a more precise knowledge of Mg concentrations at each stage of the melting process.

Archaean tholeiites by manipulating the physical conditions of melting alone. The adoption of such a model, however, introduces the necessity for a lower geothermal gradient during the Archaean melting episode than obtains at the mid-ocean ridges today (regardless of the absolute values chosen). Archaean heat-flow is widely thought to have exceeded present-day values, (e.g. McKenzie and Weiss, 1975) and this requirement must be seen as a severe weakness of the model.

It therefore seems inescapable from the calculations given here that the source regions which fed the copious greenstone belt tholeiite association in Archaean times possessed higher Ni concentrations than the source regions of all modern low-K tholeiites. Discounting the possibility that the Archaean volcanism systematically sampled high-Ni domains in a heterogenous mantle that are no longer available, this observation leads one to consider a secular change in source region chemistry between Archaean times and the present. The chalcophile character of Fe and Ni have led to the suggestion (Nesbitt and Sun, 1976) that sulphides may have been more abundant in the Archaean source region than at the present day. Each of these elements has an equally strong siderophile aspect, so that the presence of small iron particles in the Archaean upper mantle, possibly as a result of imperfect separation of the metallic phase into the core, seems an equally valid interpretation. The presence either of sulphide or metallic phases must influence the composition of the melts produced, probably in the observed direction of Ni and Fe enrichment. One must note, however, that the addition of sulphides or a metallic phase to the source regions of modern mid-ocean ridge basalts is unlikely to explain the discrepancies in the behaviour of the lithophile elements Cr, Ti, Zr and P that are seen in the Archaean tholeiites.

Conclusions

In this paper I have considered some of the ways in which basalts erupted during the first half of the earth's history differed from those being produced today. The bulk of Archaean basalts are found to be relatively magnesium-poor, low-K tholeiites, broadly similar to modern low-K basalts from oceanic and circum-oceanic regions. More detailed examination shows, however, that none of the individual low-K basalt types that can be distinguished today provides in all respects a satisfactory analogy for the Archaean tholeiite suite, whose geochemistry seems to be an ambiguous composite of mid-ocean ridge tholeiite, island-arc tholeiite and marginal basin basalt trends. It seems therefore that little will be learned from comparisons which are confined to one of these modern associations, until the petrogenetic differences between these modern types themselves are understood more fully. The failure to find an appropriate modern analogue of the Archaean tholeiite suite casts considerable doubt on the applicability of most geochemical discriminant functions, developed from compilations of Phanerozoic basalt analyses, to determine the magmatic or tectonic affinity of greenstone belt basalts.

There are some respects in which the Archaean tholeiites differ geochemically from all of the modern low-K basalt types. Ni is the best defined example, but the same is probably true of Cr. Calculations given here suggest that the 2- or 3-fold enrichment of Ni in Archaean tholeiites, compared to the most Ni-enriched modern basalt trend, is unlikely to be due solely to differences in the conditions of melting, unless one is prepared to accept the notion of a lower geothermal gradient in the Archaean source region than is found at ocean ridges today. If this view is discounted, there seem to be good grounds for suspecting the presence of additional amounts of Ni (and by analogy Cr) in the Archaean source region.

Acknowledgements

Dr. J. A. Hallberg and Dr. J. A. Pearce have kindly provided compilations of published and unpublished data, and I thank them and Dr. J. Tarney for allowing me to make use of unpublished material. I have benefited from the comments of Drs. M. J. Norry, R. J. Pankhurst and D. Bridgwater. I warmly thank Mrs. K. Webb and Mr. P. Deussen for last-minute preparation of typescript and diagrams respectively.

References

ARTH, J. G. and HANSON, G. L. (1975) Geochemistry and origin of the early Precambrian crust of northeastern Minnesota. *Geochim. Cosmochim. Acta* **39**, 325–362.

BAKER, P. E. (1968) Petrology of Mt. Misery volcano, St. Kitts, West Indies. *Lithos* **1**, 124–150.

BARAGAR, W. R. A. and GOODWIN, A. M. (1969) Andesites and Archaean volcanism of the Canadian shield. In *Proceedings of the Andesite Conference*, A. R. MCBIRNEY, ed., Bulletin 65. Oregon Dept. of Geology and Mineral Industries, 121–142.

BENCE, A. E., PAPIKE, J. J. and AYUSO, R. A. (1975) Petrology of submarine basalts from the central Caribbean: DSDP Leg 15. *J. Geophys. Res.* **80**, 4775–4804.

BISCHOFF, J. L. and DICKSON, F. W. (1975) Sea water-basalt interaction at 200°C and 500 bars: implications for origin of sea-floor heavy metal deposits and regulation of sea water chemistry. *Earth Planet. Sci. Lett.* **25**, 385–387.

BLANCHARD, D. P., RHODES, J. M., DUNGAN, M. A., RODGERS, K. V., DONALDSON, C. H., BRANNON, J. C., JACOBS, J. W. and GIBSON, E. K. (1976) The chemistry and petrology of basalts from Leg 37 of the Deep-Sea Drilling Project. *J. Geophys. Res.* **81**, 4231–4246.

BOUGAULT, H. and HEKINIAN, R. (1974) Rift valley in the Atlantic Ocean near 36°50′N: petrology and geochemistry of basaltic rocks. *Earth Planet. Sci. Lett.* **24**, 249–261.

CLAGUE, D. A. and BUNCH, T. E. (1976) Formation of ferrobasalt at East Pacific midocean spreading centres. *J. Geophys Res.* **81**, 4247–4256.

CONDIE, K. C. (1976) Trace element geochemistry of Archaean greenstone belts. *Earth Sci. Rev.* **12**, 393–417.

CONDIE, K. C. and HARRISON, N. M. (1976) Geochemistry of the Archaean Bulawayan Group, Midlands greenstone belt, Rhodesia. *Precamb. Res.* **3**, 253–271.

ENGEL, A. E. J., ENGEL, C. G. and HAVENS, R. G. (1965) Chemical characteristics of oceanic basalts and the upper mantle. *Geol. Soc. Amer. Bull.* **76**, 719–734.

EWART, A. and BRYAN, W. B. (1972) Petrography and geochemistry of the igneous rocks from Eua, Tongan Islands. *Geol. Soc. Amer. Bull.* **83**, 3281–3298.

EWART, A., BRYAN, W. B. and GILL, J. B. (1973) Mineralogy and geochemistry of the younger volcanic islands of Tonga, S. W. Pacific. *J. Petrol.* **14**, 429–465.

FREY, F. A., BRYAN, W. A. and THOMPSON, G. (1974) Atlantic ocean floor: geochemistry and petrology of basalts from legs 2 and 3 of the Deep-Sea Drilling Project. *J. Geophys. Res.* **79**, 5507–5527.

GILL, J. B. (1970) Geochemistry of Viti Levu, Fiji, and its evolution as an island arc. *Contr. Mineral. Petrol.* **27**, 179–203.

GILL, R. C. O. and BRIDGWATER, D. (1976) The Ameralik dykes of West Greenland, the earliest known basaltic rocks intruding stable continental crust. *Earth Planet. Sci. Lett.* **29**, 276–282.

GILL, R. C. O. and BRIDGWATER, D. Early Archaean basic magmatism in West Greenland: the geochemistry of the Ameralik dykes. *J. Petrol*, (in press).

GLIKSON, A. Y. (1971). Primitive Archaean element distribution patterns: chemical evidence and geotectonic significance. *Earth Planet. Sci. Lett.* **12**, 309–320.

GREEN, D. H. (1975) Genesis of Archaean peridotitic magmas and constraints on Archaean gradients and tectonics. *Geology* **3**, 15–18.

HAJASH, A. (1975) Hydrothermal processes along mid-ocean ridges: an experimental investigation. *Contr. Mineral. Petrol.* **53**, 205–226.

HALLBERG, J. A. (1972) Geochemistry of Archaean volcanic belts in the Eastern Goldfields region of western Australia *J. Petrol.* **13**, 45–56.

HALLBERG, J. A. and WILLIAMS, D. A. C. (1972) Archaean mafic and ultramafic rock associations in the Eastern Goldfields region, Western Australia. *Earth Planet. Sci. Lett.* **15**, 191–200.

HALLBERG, J. A., CARTER, D. N. and WEST, K. N. (1976) Archaean volcanism and sedimentation near Meekatharra, Western Australia. *Precambrian Res.* **3**, 577–595.

HART, S. R., BROOKS, C., KROGH, T. E., DAVIS, G. L. and NAVA, D. (1970) Ancient and modern volcanic rocks: a trace element model. *Earth Planet. Sci. Lett.* **10**, 17–28.

HART, S. R., GLASSLEY, W. E. and KARIG, D. E. (1972) Basalts and sea floor spreading behind the Mariana Island arc. *Earth Planet. Sci. Lett.* **15**, 12–18.

HAWKESWORTH, C. J. and O'NIONS, R. K. (1977). The petrogenesis of some Archaean volcanic rocks from southern Africa. *J. Petrology* **18**, 487–520.

HEKINIAN, R., MOORE, J. G. and BRYAN, W. B. (1976) Volcanic rocks and processes of the Mid-Atlantic ridge rift valley near 36°49′N. *Contr. Mineral. Petrol.* **58**, 83–110.

HUMPHRIS, S. E. (1975) Elemental fluxes during hydrothermal alteration of basalts (abstract). *Trans. Amer. Geophys. Union* **56**, 469.

IRVINE, T. N. (1973) Chromitite layers in stratiform instrusions. *Carnegie Inst. Wash. Yb.* **73**, 300–316.

JAHN, B.-M., SHIH, C.-Y. and MURTHY, V. R. (1974) Trace element geochemistry of Archaean volcanic rocks. *Geochim. Cosmochim. Acta* **38**, 611–627.

KAY, R., HUBBARD, N. J. and GAST, P. W. (1970). Chemical characteristics and origin of oceanic ridge volcanic rocks. *J. Geophys. Res.* **75**, 1585–1613.

LEEMAN, W. P. (1973) Partitioning of Ni and Co between olivine and basaltic liquid: an experimental study (abstract). *Trans. Amer. Geophys. Union* **54**, 1222–1223.

MCKENZIE, D. and WEISS, N. (1975) Speculations on the thermal and tectonic history of the earth. *Geophys. J. R. Astr. Soc.* **42**, 131–174.

MIYASHIRO, A., SHIDO, F. and EWING, M. (1969) Diversity and origin of abyssal tholeiite from the Mid-Atlantic Ridge near 24° and 30° North latitude. *Contr. Mineral. Petrol.* **23**, 38–52.

MOTTL, M. J., CORR, R. F. and HOLLAND, H. D. (1974) Chemical exchange between sea water and mid-ocean ridge basalt during alteration: an experimental study (abstract). *Geol. Soc. Amer. Abstracts with Programs* **6**, 879.

MYSEN, B. O. (1976) Nickel partitioning between upper mantle crystals and partial melts as a function of pressure, temperature and nickel concentration. *Carnegie. Inst. Wash. Yb.* **75**, 662–668.

MYSEN, B. O. and KUSHIRO, I. (1976) Compositional variation of coexisting phases with degree of melting of peridotite under upper mantle conditions. *Carnegie Inst. Wash. Yb.* **75**, 546–555.

NESBITT, R. W. and SUN, S.-S. (1976) Geochemistry of Archaean spinifex-textured peridotites and magnesian and low-magnesian tholeiites. *Earth Planet. Sci. Lett.* **31**, 433–453.

O'NIONS, R. K. and PANKHURST, R. J. (1978) Early Archaean rocks and geochemical evolution of the Earth's crust. *Earth Planet. Sci. Lett.* **38**, 211–236.

SATO, H. (1977) Nickel content of basaltic magmas: identification of primary magmas and a measure of the degree of olivine fractionation. *Lithos* **10**, 113–120.

SCHEIDEGGER, K. F. (1973) Temperatures and compositions of magmas ascending along mid-ocean ridges. *J. Geophys. Res.* **78**, 3340–3355.

SHAW, D. M. (1970) Trace element fractionation during anatexis. *Geochim. Cosmochim. Acta* **34**, 237–243.

SUN, S.-S. and NESBITT, R. W. (1971) Chemical heterogeneity of the Archaean mantle, composition of the earth and mantle evolution. *Earth Planet. Sci. Lett.* **35**, 429–448.

TARNEY, J., DALZIEL, I. W. D. and de Wit, M. J. (1976) Marginal basin "Rocas Verdes" complex from S. Chile: a model for Archaean greenstone belt formation. In *The Early History of the Earth*, B. F. WINDLEY, ed., London, J. Wiley & Sons, 131–146.

TARNEY, J., SAUNDERS, A. D. and WEAVER, S. D. (1977) Geochemistry of volcanic rocks from the island-arcs and marginal basins of the Scotia Arc region. In *Ewing Mem. volume* of A.G.U., 367–377.

THOMPSON, G., SHIDO, F. and MIYASHIRO, A. (1972) Trace element distributions in fractionated oceanic basalts. *Chem. Geol.* **9**, 89–97.

VILJOEN, R. P. and VILJOEN, M. J. (1969) An introduction to the geology of the Barberton granite–greenstone terrain. *Geol. Soc. S. Africa* Spec. Publ. **2**, 9–28.

WHITE, A. J. R., JAKES, P. and CHRISTIE, D. M. (1971) Composition of greenstones and the hypothesis of sea-floor spreading in the Archaean. *Geol. Soc. Austr.* Spec. Publ. **3**, 47–56.

Rare-earth Element Geochemistry of Regional Metamorphic Rocks

G. K. MUECKE, C. PRIDE and P. SARKAR

Department of Geology, Dalhousie University, Halifax, Nova Scotia, Canada B3H 3J5

Abstract

A considerable body of data on the distribution and behaviour of rare-earth elements (REE) in igneous and sedimentary rocks has been accumulated over the past decade. Metamorphosed and metasomatized rocks have by comparison received scant attention. Some investigators have reported REE mobility, but most of the available data suggests that REE levels and patterns are essentially unaffected by metamorphism. However, the hypothesis of REE "immobility" during metamorphism has not been rigorously tested. We present new evidence from amphibolite and granulite facies terrains which bears upon the mobility of rare-earth elements during high-grade metamorphism.

Metasomatic zones provide an excellent test of element mobility. Examples are described from an amphibolite facies terrain in which these zones have retained their original REE chemistry, as well as their Ti, P, Y, Zr, Hf, Nb and Ta contents. In the associated metavolcanic suite strong linear correlations exist between the REEs and the other "immobile" elements. Such correlations are frequently explained in cogenetic volcanic suites as being produced by fractional crystallization. It seems unlikely that element migration during metamorphism would preserve such coherent behaviour of elements with different chemical properties. We conclude that in these cases metamorphism and metasomatism has affected neither the REE abundances and patterns of this metavolcanic suite, nor its Ti, P, Y, Zr, Hf, Nb or Ta content. Only these elements provide a reliable guide to the parentage of the metasomatized volcanics. By contrast, our work on Archean granulite-facies gneisses from the Lewisian of N.W. Scotland shows that these deep crustal rocks are depleted in REEs (except Eu) by a factor of 2 or 3 relative to upper crustal values. A distinct positive europium anomaly supports the idea of preferential retention of Eu in the deep crust proposed by Jakeš and Taylor (1974) and Nance and Taylor (1976). Mineral REE patterns in the granulites and their distribution coefficients closely resemble those reported for phenocrysts in dacites, suggesting that the minerals equilibrated with a granodioritic liquid. The trace and major element chemistry of this terrain is compatible with the model of a granulite facies residuum, left after the generation of a granodioritic upper crust.

Rare-earth elements provide a promising new tool in the determination of the parentage of metamorphosed and metasomatized rocks up to the upper amphibolite facies of regional metamorphism and may prove valuable in assessing the importance of anatexis in the evolution of the lower continental crust.

Introduction

In unaltered igneous suites REE geochemistry has been used to elucidate the chemical and mineralogical nature of the magma source, the degree of melting of the parental material, the nature and sequence of fractionating phases, and the genetic relationships between members of the suite (e.g. Zielinski and Frey, 1970; Kay and Gast, 1973; Sun and Hanson, 1975). In detrital sedimentary rocks REEs can yield information on the provenance, age and composition of the source region, as well as indications of the physico-chemical weathering conditions (e.g. Ronov *et al.*, 1967; Wildeman and Haskin, 1973; Ronov *et al.*, 1974; Nance and Taylor, 1976). REE patterns in chemical sediments are thought to reflect the REE chemistry of sea water, the nature of precipitating phases, as well as redox conditions in the atmosphere and at the site of deposition (e.g. Piper, 1974; Fryer, 1977). Most Precambrian shields and orogenic belts, and to some extent Paleozoic, Mesozoic and Cenozoic mobile belts, consist of rocks which have been modified by metamorphism. Their present mineralogy and major element chemistries are frequently no longer reliable guides to the processes active during their original formation. Any group of elements, such as the REEs, which has been shown to be a valuable environmental and petrogenetic indicator in unaltered rocks therefore has the potential of substantially extending our understanding of metamorphic terrains, provided that their distribution is not affected by metamorphism.

However, at the present time opinion is still divided among those investigators who cite evidence indicating REE mobility during metamorphism and a majority who consider this group of elements as essentially immobile. The hypothesis of rare-earth immobility has, however, rarely been rigorously tested, particularly in the higher metamorphic grades. In this paper we review the current status of REE geochemistry of metamorphic rocks and present new evidence from amphibolite and granulite facies terrains which relates to the mobility of the REEs during high-grade metamorphism.

Summary of Previous Work

(i) Low-grade Metamorphism

Much of the work on the possible effects of low-grade metamorphism on the abundance and relative distribution of REEs has concentrated on oceanic volcanic rocks. Frey *et al.* (1968) and Philpotts *et al.* (1969) discerned no distinct differences in the REE abundances of altered and fresh submarine basalts and concluded that initial alteration by seawater does not affect the REEs. Later work by Frey *et al.* (1974) detected changes in abundances of REE and their fractionation during the alteration of basaltic glasses and crystalline basalts from the Atlantic ocean floor. Palagonitization of glasses was found to result in La and heavy REE (HREE) depletion, whereas alteration of crystalline basalts produced light REE (LREE) enrichment. Similarly Wood *et al.* (1976) note that in the partly zeolitized Tertiary lavas of eastern Iceland the LREEs, particularly La, show a less well-defined correlation with elements which are thought to have remained immobile during low-grade metamorphism (e.g. Zr) than do the HREEs. Furthermore, they note that within a single flow LREE variability is much more pronounced in zeolitized flows than in their fresh equivalents. From these observations they conclude that the LREE concentrations in the lavas were substantially modified during zeolitization. By analogy with Frey *et al.*'s observations on the effects of glass alteration, it may be postulated that the breakdown of residual interstitial glass enriched in REE accounts for REE mobility during zeolitization.

Metabasalts from ophiolite complexes frequently show the LREE depleted character and low absolute concentrations of REE found in fresh oceanic basalts (Montigny *et al.*, 1973; Kay and Senechal, 1976; Smewing and Potts, 1976). Herrmann and Wedepohl (1970) and Herrmann *et al.* (1974) investigated spilites from the continents and the Mid-Atlantic Ridge and found their REE abundance and fractionation comparable to those of unaltered continental and ocean-floor tholeiites. These similarities led to the conclusion that the REE are not substantially affected by zeolite and lower greenschist facies metamorphism or spilitization. However, recent work by Hellman and Henderson (1977) and Floyd (1977) provides evidence for REE mobility during these processes.

(ii) Medium- and High-grade Metamorphism

Haskin *et al.* (1968) have reported REE abundances for a composite of metamorphosed shales from North America, and found them to be essentially similar to a North American shale composite. In a more detailed study Cullers *et al.* (1974) studied progressively metamorphosed pelitic schists ranging from the greenschist to upper amphibolite facies and concluded that they show no change in REE content as a function of metamorphic grade. Although this data indicates the absence of any gross changes, the limited number of samples taken from each metamorphic zone and the substantial range of abundances within lithologic units does not exclude the possibility of considerable REE redistribution. Kay *et al.* (1970) and Masuda *et al.* (1971) noted the similarity between REE abundances and patterns in amphibolites and foliated metagabbros from the ocean floor and apparently equivalent fresh samples. Similarly Garmann *et al.* (1975) recognized that REE patterns from eclogites in Norwegian gneisses resemble those of unaltered tholeiites. Van de Kamp (1969) and Balashov *et al.* (1974) have used REE abundances and patterns to differentiate para-amphibolites from ortho-ampibolites.

REE studies of gneisses from the amphibolite and pyroxene granulite facies have been reported by Green *et al.* (1969, 1972), O'Nions and Pankhurst (1974), Lausch *et al.* (1974), Hanson (1975), Dostal (1975) and Pride and Muecke (1976). The similarity between REE abundances and patterns in amphibolite and granulite facies gneisses from Norway led Green *et al.* (1969, 1972) to conclude that granulite facies metamorphism did not substantially affect the REEs. In view of the small sample number (two amphibolite facies and four granulite facies gneisses) this conclusion must be regarded as tentative.

(iii) REE in Minerals in Metamorphic Rocks

Very little information is currently available pertaining to REE distribution in the rock-forming metamorphic minerals. Roaldset (1975) has reported the REE patterns of some metamorphic micas and Reitan and Roelandts (1973) analyzed coexisting pyroxenes from pyroxene–granulite facies rocks. Zayats and Kuts (1964), Orsa *et al.* (1967) and Puchelt and Emmermann (1976) have reported REE abundances in accessory minerals of metamorphic rocks.

Few of the studies published so far have been designed to determine the mobility of REE during medium- to high-grade metamorphism. Similarities between REE patterns of metamorphic rocks and unaltered rocks of equivalent composition are not a reliable test of immobility since REE abundances in unmetamorphosed sedimentary and igneous rocks show considerable variability which may mask any effects of metamorphism.

REE Geochemistry of an Amphibolite Facies Terrain

The White Rock Formation of southwestern Nova Scotia, Canada, consists of a thick sequence of early Paleozoic clastic sediments and silicic, intermediate and basaltic volcanics which has undergone Acadian regional metamorphism (Taylor, 1967). Most of the rocks attained conditions of the upper garnet zone, although locally they may reach the staurolite zone. Temperatures and lithostatic pressures during metamorphism have been estimated to have been 500–580°C and 3–4 kb respectively (Sarkar, 1977). In order to determine the extent of element migration during metamorphism, some well-developed local metasomatic zones in this region have been investigated. Ovoid epidosite lenses, often 1–2 meters in length, occur within thick homogeneous amphibolite units. Their random distribution, lack of correlation with distinct structural features, and transitional contact with the metabasites suggests an *in situ* origin (Harrigan and MacLean. 1976). The mineralogy changes from a hornblende–plagioclase–epidote assemblage in the metabasite, through a thin transition zone, to an assemblage dominated by epidote ($\sim 80\%$ of mode). As may be expected from the mineralogical changes, concentrations of some of the major elements in the epidosites differ drastically from the host metabasite (Table 1, analyses 1–3). Particularly striking are the changes from $\simeq 10$ wt.% CaO in the metabasites to $\simeq 20\%$ in the epidosite, a marked increase

TABLE 1. AVERAGE MAJOR ELEMENT (wt. %) AND TRACE ELEMENT (ppm) CHEMISTRY OF METASOMATIC ZONES FROM THE WHITE ROCK FORMATION, NOVIA SCOTIA

	1	2	3	4	5	6	7
SiO_2	40.84	47.23	46.27	39.44	43.30	57.04	57.04
TiO_2	2.19	2.35	2.38	3.78	3.70	1.24	1.88
Al_2O_3	16.28	15.27	16.01	18.02	15.96	21.83	22.61
Fe_2O_3	8.19	3.84	3.63	1.64	2.69	0.11	0.43
FeO	3.81	4.82	7.32	15.72	17.95	6.15	6.40
MnO	0.14	0.15	0.16	0.53	0.33	0.05	0.04
MgO	4.52	6.19	7.48	5.19	5.61	1.58	1.62
CaO	19.72	14.32	10.76	10.11	3.45	3.45	0.52
Na_2O	0.87	3.09	3.23	0.88	0.79	5.15	1.46
K_2O	0.08	0.16	0.42	0.44	0.20	0.87	3.78
P_2O_5	0.30	0.40	0.30	0.96	0.94	0.12	0.11
CO_2	1.59	1.12	0.19	0.24	0.12	0.11	0.19
H_2O^+	0.88	0.84	1.60	2.76	4.90	2.25	3.68
Total	99.41	99.78	99.75	99.71	99.94	99.95	99.76
La	16.4	16.7	14.3	49.0	52.2	—	7.44
Ce	36.2	37.9	33.8	98.2	85.7	—	15.7
Sm	5.43	5.89	5.15	12.3	13.4	—	2.79
Eu	1.84	1.88	1.89	3.82	3.04	—	0.79
Tb	0.73	0.83	0.95	1.37	1.31	—	0.73
Yb	1.54	1.47	1.54	2.79	2.69	—	3.41
Lu	0.25	0.24	0.22	0.39	0.36	—	0.55
Y	17	23	22	46	47	33	38
Li	8	13	52.5	86	134	65	80
Rb	1	3	8	12	4	37	176
Sr	2012	742	409	152	148	702	255
Zr	133	155	148	370	372	223	208
Hf	4.08	4.31	3.72	11.2	7.6	6.3	6.0
Nb	23	25	24.5	57	60	23	27
Ta	1.46	1.51	1.57	4.76	3.40	—	1.92
Th	1.60	1.37	1.49	7.23	4.40	15.0	14.3

1. Average of epidosite lens.
2. Thin transition zone between epidosite and metabasite.
3. Average of metabasite surrounding epidosite lens.
4. Average metabasite sill.
5. Transition zone between metabasite sill and schists.
6. Poikiloblastic schists.
7. Pelitic schists.

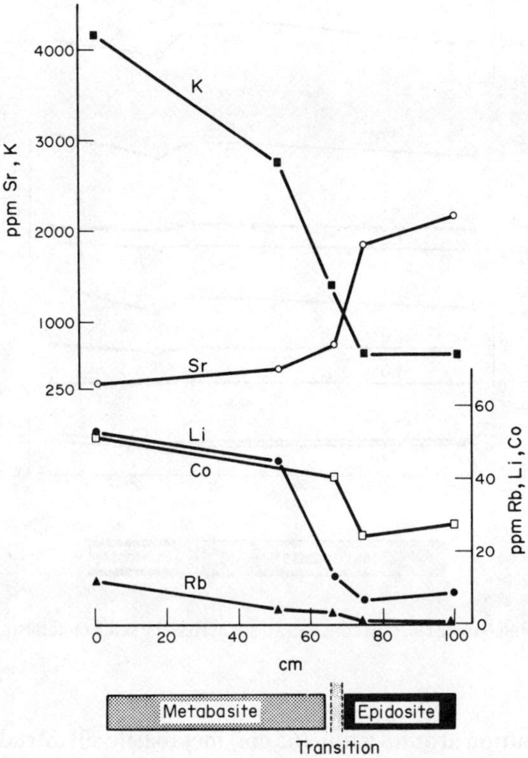

FIG. 1. Variation of mobile trace element concentrations across metabasite-epidosite contact, White Rock Formation, N.S. Sample at 100 cm located in the center of the epidosite lens.

in ferric iron, and the decrease in the magnesium and sodium concentrations. Other elements, notably aluminum, titanium and phosphorus, remain essentially constant. Among the minor and trace elements potassium, rubidium and lithium show strong compositional gradients and are distinctly less abundant in the metabasite near the lens and very low in the epidosite (Table 1, Fig. 1). Strontium shows a concomitant increase from $\simeq 250$ ppm in the metabasite to over 2000 ppm in the epidosite.

However, a suite of trace elements including Zr, Hf, Nb, Ta, Th and the REE shows approximately equal concentrations in the metabasite and the epidosite (Table 1, Fig. 2). Any deviations from constancy fall within the range of expected sampling and analytical errors. It follows that normalized rare-earth patterns (using the average chondrite of Frey *et al.*, 1968) in the metabasite and epidosite are indistinguishable from each other within the margins of error (Fig. 3a). This seems even more remarkable if one considers the affinity of the epidote mineral structure for REE (Lyakhovich and Barinskii, 1961; Fleischer, 1965). We suggest that the epidosite lenses are the product of metasomatism which has strongly affected concentrations of the alkali and alkaline earth elements. Despite these major compositional changes the original Al, Ti, P, Zr, Hf, Nb, Ta, Th and REE content of the metabasite has been preserved in the epidosites.

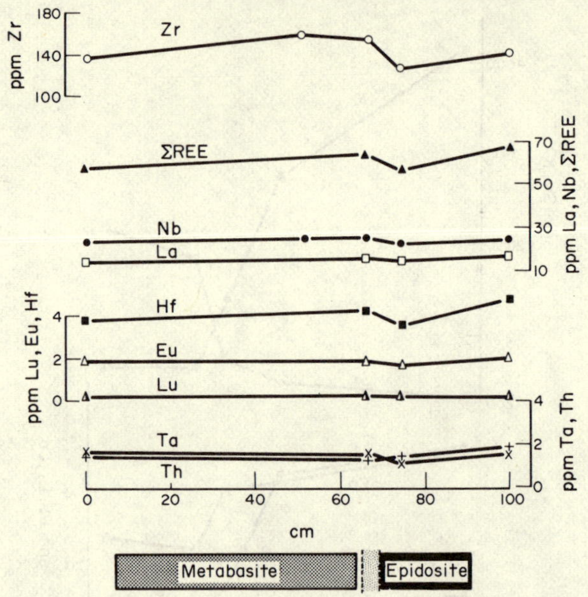

FIG. 2. Variation of "immobile" trace element concentrations across the same metabasite-epidosite contact depicted in Fig. 1.

The element distribution around a thin (65 cm) metabasite sill intruded into metapelite has also been investigated. Both contacts are marked by distinct zones in which the mineralogy of the metabasite and pelite have been strongly modified. The hornblende-rich central amphibolite is rimmed by a zone in which chlorite, quartz and iron oxide predominate. The pelite near the contact consists of $\simeq 80\%$ albite porphyroblasts and very little muscovite. With increasing distance from the contact the albite content is sharply reduced and muscovite shows a concomitant increase. Chemically the marginal zones are marked by strong gradients in CaO (Table 1, analyses 4–7) and also in ferrous iron, silicon, sodium and potassium. Similar to the previous example, phosphorus and titanium appear to have remained essentially unaffected by the metasomatism and serve to identify the chlorite–quartz zone as part of the metabasite and the poikiloblastic schist as metapelite. Among the trace elements Sr, Rb and Li show very strong compositional gradients (Table 1). Zr, Nb and Y, on the other hand, maintain nearly constant concentrations in the metabasite and metapelite and their respective reaction zones. The REE fractionation and abundances in the pelitic schist and metabasite differ considerably, particularly for the LREE (Fig. 3b). However, the REE abundances and pattern in the quartz–chlorite transition zone are essentially indistinguishable from the remainder of the metabasite, except for a possible slight reduction in Ce and Eu. This data confirms our earlier observation that Ti, P, Zr, Nb, Y and REEs have remained essentially undisturbed by the metasomatism which substantially changed the alkali and alkaline earth content of the rocks.

The question then remains as to whether the redistribution of the elements affected the rocks outside the documented metasomatic zonations and whether REE immobility can be demonstrated in them. A representative range of acid to basic rocks of the White Rock metavolcanic suite which are not visibly affected by metasomatism have been analyzed for their trace element content. Plots of REE versus the other previously defined "immobile" trace elements yield excellent linear correlations (correlation coefficients >0.96). High correlations between these element pairs are characteristic of fresh cogenetic volcanic suites in which the members are allegedly related by fractional crystallization (e.g. Weaver *et al.*, 1972; Barbieri *et al.*, 1975; Ferrara and Treuil, 1975; Wood *et al.*, 1976). It is highly unlikely that these good correlations could have been maintained in the metavolcanics if the elements had undergone significant redistribution during metamorphism. Their chemical properties differ substantially and lead one to expect a scatter of data points, except in the improbable event that the elements were added or lost in such a systematic fashion as to preserve their correlations. In contrast, a plot of La versus Rb shows only a very low degree of correlation, whereas in unaltered cogenetic volcanic suites these two elements exhibit a strong covariance (Weaver *et al.*, 1972; Ferrara and Treuil, 1975). This confirms the previous findings from the metasomatized zones that the alkalis in this terrain underwent major redistribution during metamorphism, but that this redistribution did not affect the REE.

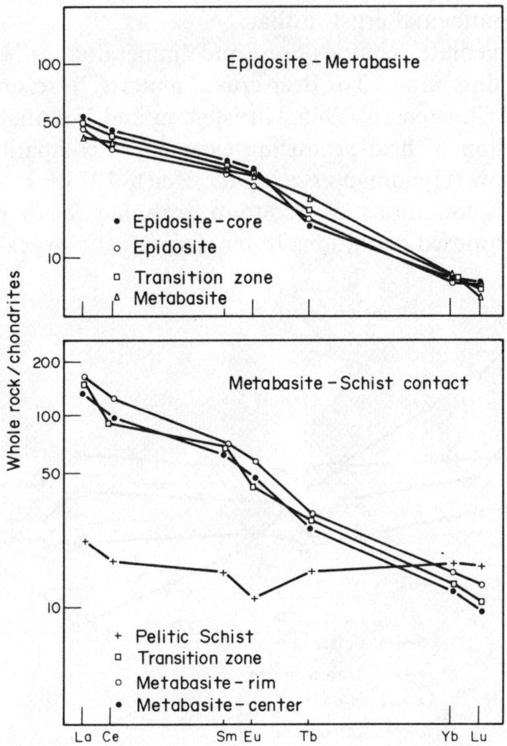

FIG. 3. Chondrite-normalized rare earth patterns of rocks associated with metasomatic zones. (a) *Top*: Rare earth patterns of epidosite lens, transition zone and metabasite. (b) *Bottom*: Rare earth patterns of metabasite sill, marginal transition zone, and adjacent pelitic schist.

REE Geochemistry of a Granulite Facies Terrain

The Archean Lewisian complex of mainland N.W. Scotland represents one of the most intensely studied sections of deep crust and, as such, is particularly conducive to detailed geochemical investigations (Holland and Lambert, 1975). The Scourian assemblage, the oldest portion of this complex, formed and was metamorphosed between 2.6–2.9 b.y. ago (Moorbath *et al.*, 1969). The rocks underwent a granulite facies metamorphism (medium- to high-pressure type of Green and Ringwood, 1967) at which time they became depleted in incompatible elements (Moorbath *et al.*, 1969). Very low concentrations of uranium and rubidium and very high K/Rb ratios up to 4300 (Muecke, 1969; Holland and Lambert, 1975) are characteristic of the Scourian gneisses; features not unique to the Scourian assemblage, but also encountered in other medium- to high-pressure granulite facies terrains (Lambert and Heier, 1968; Sighinolfi, 1969; Green *et al.*, 1972; Heier, 1973; Lewis and Spooner, 1973). The complex, as a whole, is of intermediate composition with an estimated 80% by volume consisting of tonalitic pyroxene gneisses (Muecke, 1969; Holland and Lambert, 1975). The dominant phases present are plagioclase, quartz, aluminous orthopyroxene and clinopyroxene, and lesser amounts of pyrope-rich garnet and pargasitic hornblende—an essentially anhydrous assemblage. The Scourian terrain thus closely conforms to models proposed for the lower continental crust in that:

(a) the overall intermediate composition and mineralogy are compatible with geophysical constraints imposed on deep crustal material by seismic evidence (Green, 1970; Heier, 1973; Christensen, 1965; Christensen and Fountain, 1975);

(b) the extreme depletion of heat-producing elements is compatible with models of continental heat flow (Hyndman *et al.*, 1968; Heier, 1973);

(c) the estimated P–T conditions of Scourian granulite facies metamorphism are compatible with proposed conditions in the deep crust (Muecke, 1969).

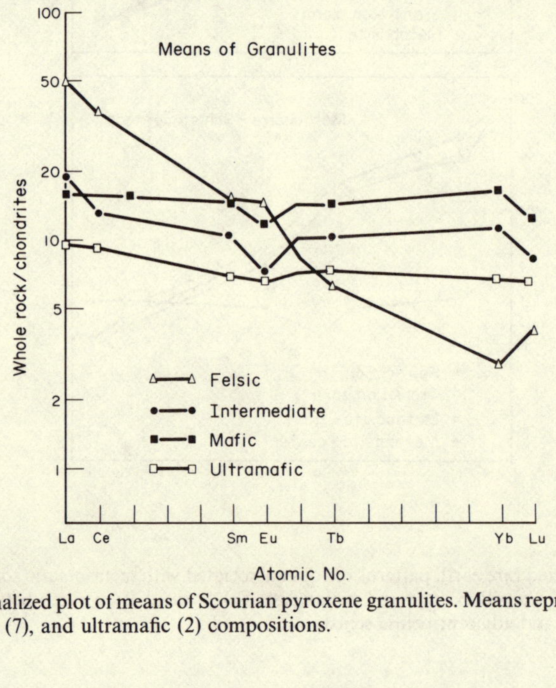

FIG. 4. Chondrite-normalized plot of means of Scourian pyroxene granulites. Means represent felsic (nine samples), intermediate (1), mafic (7), and ultramafic (2) compositions.

TABLE 2. AVERAGE REE CONTENT (ppm) OF SCOURIAN GNEISSES AND MINERALS, LEWISIAN, N.W. SCOTLAND

Rock type or mineral	No. of analyses	La	Ce	Sm	Eu	Tb	Yb	Lu
Felsic pyroxene granulites	9	16.8	32.7	2.75	1.02	0.297	0.568	0.137
Intermediate pyroxene granulite	1	6.48	11.8	1.90	0.494	0.491	2.25	0.278
Mafic pyroxene granulites	7	5.24	13.8	2.71	0.808	0.672	3.30	0.411
Ultramafic pyroxene granulites	2	3.22	8.39	1.26	0.456	0.348	1.48	0.221
Average Scourian terrain†		14.4	28.7	2.67	0.957	0.391	1.02	0.182
Orthopyroxenes	4	1.50	4.81	0.356	0.082	0.090	0.927	0.308
Clinopyroxenes	7	4.67	20.2	6.24	1.04	1.17	3.32	0.751
Plagioclases	8	5.53	7.35	0.427	0.590	0.088	0.087	0.013
Garnets	5	0.636	4.03	1.89	0.828	1.77	20.4	3.24
Hornblendes	2	14.1	48.9	8.04	1.78	1.55	5.35	0.62

† Calculated using an estimated abundance of 80 parts felsic; 15 parts mafic; 5 parts ultramafic gneisses.

Average REE abundances for Scourian pyroxene granulites of felsic, intermediate, basic and ultramafic composition are presented in Table 2 and their chondrite-normalized patterns in Fig. 4.

The tonalitic or felsic gneisses all have LREE-enriched patterns with small, but distinct, positive Eu anomalies. The other rock types are all lower in absolute concentration of LREE with small positive or negative Eu anomalies. The relative abundances of the various rock types have been estimated (Müecke, 1969) and weighted estimate can be made of the REE

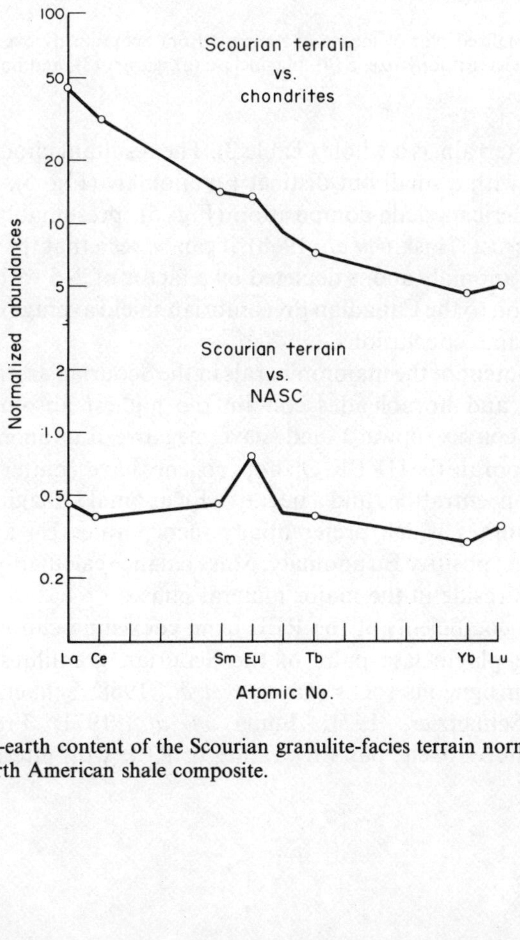

FIG. 5. Estimated average rare-earth content of the Scourian granulite-facies terrain normalized with respect to average chondrite and the North American shale composite.

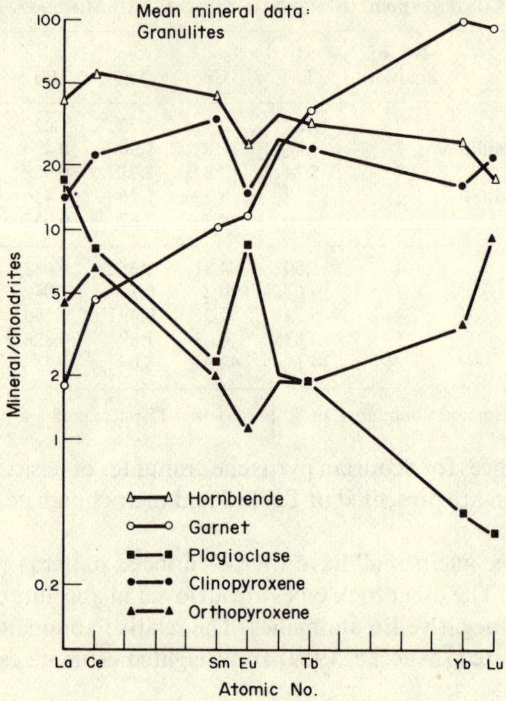

FIG. 6. Chondrite-normalized plot of means of minerals from Scourian pyroxene granulites. Means represent clinopyroxene (7 samples), orthopyroxene (4), plagioclase (8), garnet (5), and hornblende (2).

composition of the terrain as a whole (Table 2). The resultant chondrite-normalized pattern is LREE enriched with a small but distinct Eu anomaly (Fig. 5). When normalized to the average North American shale composition (Fig. 5), presumably a good estimate of the upper continental crust (Haskin *et al.*, 1968), it can be seen that the Scourian assemblage has a large positive Eu anomaly and is depleted by a factor of 2–3 with respect to upper crustal rocks. Normalization to the Canadian Precambrian shield average of Shaw *et al.* (1976) leads to essentially the same conclusions.

Average REE content of the major minerals in the Scourian assemblage are given in Table 2. Clinopyroxenes and hornblendes contain the highest absolute values of REE, with patterns that are convex upward and have negative Eu anomalies (Fig. 6). Garnets preferentially incorporate the HREE. Orthopyroxenes have a pattern that is concave upward, has low absolute concentrations and a negative Eu anomaly. Plagioclase is the only phase of the pyroxene granulites which preferentially incorporates Eu and the resulting pattern always shows a large, positive Eu anomaly. Mass balance calculations indicate that generally $>80\%$ of the REE reside in the major mineral phases.

The distribution coefficients of the REE from coexisting clinopyroxene–orthopyroxene and clinopyroxene–plagioclase pairs of the Scourian granulites can be compared with values derived from igneous rocks (Onuma *et al.*, 1968; Schnetzler and Philpotts, 1970; Nagasawa and Schnetzler, 1971; Buma *et al.*, 1971; Frey *et al.*, 1974). For clinopyroxene–orthopyroxene pairs K_D values of REE with intermediate atomic numbers

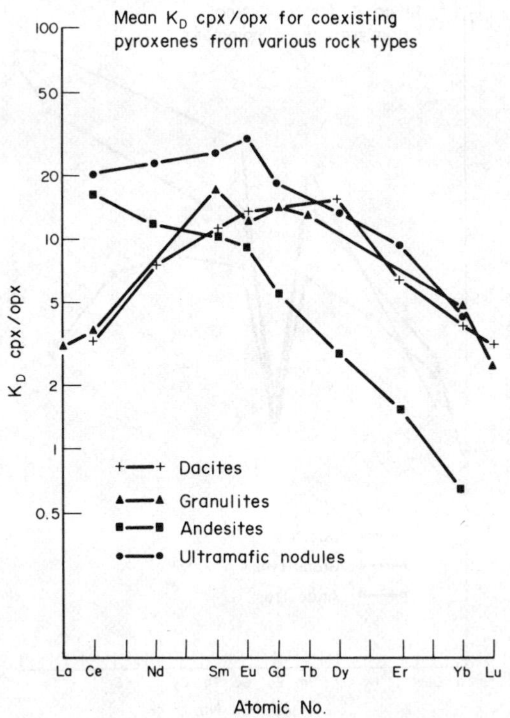

FIG. 7. Variation of distribution coefficients of rare-earth elements between clinopyroxene and orthopyroxene. Values from Scourian pyroxene granulites are compared to those of other rock types (for sources see text).

(Sm, Eu, Tb) are distinctly higher than those of either LREE or HREE (Fig. 7). Both the pattern of variation of K_D with atomic number and absolute value strongly resemble those encountered in dacites. This similarity extends to the absolute concentrations and fractionation in these minerals (Pride, 1977). In Fig. 8 coexisting clinopyroxene–plagioclase pairs of the granulites show increasing K_D with higher atomic number (except Eu). Again the pattern of change and the absolute values of K_D closely resemble those of dacites. One possible explanation for the close resemblance between distribution coefficients in the pyroxene granulites and dacites is that the granulite minerals equilibrated with a melt resembling dacite (i.e. granodiorite) in composition. Such a conclusion can only be tentative since the influence of pressure and temperature on K_D of REE is essentially unknown. REE distribution coefficients (K_D) for liquidus phases in equilibrium with a melt have been shown to be dependent on the composition of the melt, and possibly the temperature and pressure (Nagasawa and Schnetzler, 1971; Mysen, 1976; Wood, 1976).

From geophysical, petrological and geochemical considerations it has been concluded that the Scourian assemblage exhibits the features expected in the lower continental crust (Muecke, 1969; Holland and Lambert, 1975; Pride, 1977). Furthermore, both the REE depletion of the Scourian assemblage and the large positive Eu anomaly relative to the upper crust conform to predictions by Jakeš and Taylor (1974) and Nance and Taylor (1976, 1977) regarding the REE geochemistry of lower continental crust. Their model is based on the well-documented decrease of Eu in clastic sediments from the Precambrian to the Recent

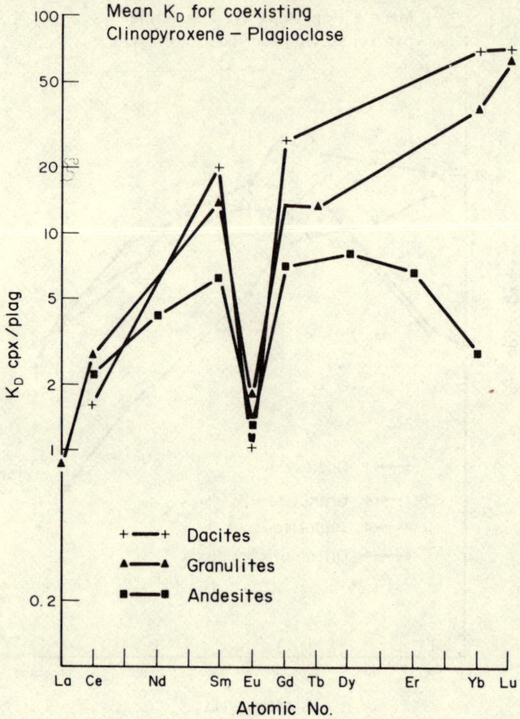

FIG. 8. Variation of distribution coefficients of rare-earth elements between clinopyroxene and plagioclase. Values from Scourian pyroxene granulites are compared to those of other rock types (for sources see text).

(Wildeman and Haskin, 1973; Wildeman and Condie, 1973). The very strong incompatible element depletion, high K/Rb ratios, low REE content and relative enrichment of Eu in the Scourian gneisses presupposes either parental material of unusual composition or the action of a crustal fractionation process. We favour the latter interpretation in which a Scourian terrain of "average" crustal composition underwent partial melting in the deep crust, leading to the generation and separation of granodioritic magmas. The strong affinity of silicic magmas for the incompatible elements, REE and water can account for the depleted, anhydrous character of the Scourian assemblage. Progressive melting would be accompanied by an increase in modal plagioclase content of the residuum and the formation of the tonalitic gneisses. The preferential retention of Eu by the plagioclase (Nance and Taylor, 1976) and its exclusion of Rb (McCarthy, 1976) best explain the positive Eu anomaly and the high K/Rb ratios. Furthermore, REE distribution coefficients in the pyroxene granulites suggest that the minerals equilibrated with melt of granodioritic composition. We conclude that the Scourian gneisses probably represent the residuum after extraction of a "granodioritic" melt. It is therefore likely that the continental crust is stratified with respect to REE as has already been documented for other incompatible elements (Heier, 1965; Heier and Brunfelt, 1970; Sighinolfi, 1971; Heier and Thoresen, 1971; Heier, 1973; Lewis and Spooner, 1973).

Conclusions

The work of Frey *et al.* (1974) and Wood *et al.* (1976) has demonstrated that REE can become mobile during early stages of sea water–rock interaction and zeolite-facies metamorphism. Igneous rocks in which glass has been altered may be particularly prone to changes in REE chemistry. From the available data it is less clear how REEs behave at higher metamorphic grades. REE studies on metasomatic zonations constitute one possible test of mobility. In cogenetic metavolcanic suites the preservation of the strong coherences between REE and other LIL elements provides a further test. The White Rock metavolcanic suite, Nova Scotia, has undergone amphibolite facies metamorphism and both the above criteria indicate that the rocks have retained their pre-metamorphic REE chemistry. It therefore becomes possible to use these elements to analyze the crystal-magma fractionation and paleotectonic setting of this suite despite the advanced metamorphism (Sarkar, 1977).

Granulite-facies gneisses from the Lewisian of N.W. Scotland show that these rocks are depleted in REE by a factor of 2 or 3 relative to upper crustal values. Together with the REE data on the minerals, these results suggest that the Lewisian granulites represent the residuum after partial melting and removal of a granodioritic magma.

Rare-earth elements promise to provide a valuable new tool to the study of regional metamorphic terrains; however, it is clearly desirable to obtain considerably more information from other terrains before our conclusions can be regarded as having general validity. In metamorphic terrains where REE immobility can be demonstrated, Nd isotopic techniques (DePaolo and Wasserburg, 1976) may provide valuable petrogenetic information not otherwise obtainable.

Acknowledgements

We thank Drs. P. Jagam and Ian Gibson for their assistance with the analytical work. Zr, Nb, Y, Sr and Rb analyses were performed for the authors at the Department of Geology, Bedford College, London. Jan Aumento is thanked for the drafting. The comments of Anne Muecke, Ian Gibson and Jarda Dostal greatly improved the manuscript. C. P. acknowledges the receipt of a National Council Post-graduate Scholarship and P. S. of a Killam Post-graduate Scholarship from Dalhousie University. The work was supported by the National Research Council of Canada.

References

Balashov, Yu. A., Kremenetskiy, A. A. and Shvets, V. M. (1974) Geochemical criteria of the nature of Precambrian amphibolites. *Geochem. Int.* **1973**, 918–931.

Barbieri, F., Ferrara, G., Santacroce, R., Treuil, M. and Varet, J. (1975) A transitional basalt-pantellerite sequence of fractional crystallization, the Boina centre (Afar Rift, Ethiopia). *J. Petrol.* **16**, 22–56.

Buma, G., Frey, F. A. and Wones, D. R. (1971) New England granites: trace element evidence regarding their origin and differentiation. *Contr. Mineral. Petrol.* **31**, 300–320.

Christensen, N. I. (1965) Compressional wave velocities in metamorphic rocks at pressures to 10 kilobars. *J. Geophys. Res.* **70**, 6147–6164.

Christensen, N. I. and Fountain, D. M. (1975) Constitution of the lower continental crust based on experimental studies of seismic velocities in granulite. *Geol. Soc. Amer. Bull.*, **86**, 227–236.

CULLERS, R. L., YEH, L-T., CHAUDHURI, S. and GUIDOTTI, C. V. (1974) Rare earth elements in Silurian pelitic schists from N.W. Maine. *Geochim. Cosmochim. Acta* **38**, 389–400.

DEPAOLO, D. J. and WASSERBURG, G. J. (1976) Nd isotopic variations and petrogenetic models. *Geophys. Res. Lett.* **3**, 249–252.

DOSTAL, J. (1975) The origin of garnet–cordierite–sillimanite bearing rocks from Chandos Township, Ontario. *Contr. Mineral. Petrol.* **49**, 163–175.

FERRARA, G. and TREUIL, M. (1975) Petrological implications of trace element and Sr isotope distributions in basalt–pantellerite series. *Bull. Volcan.* **38**, 548–574.

FLEISCHER, M. (1965) Some aspects of the geochemistry of yttrium and the lanthanides. *Geochim. Cosmochim. Acta* **29**, 755–772.

FLOYD, P. A. (1977) Rare earth element mobility and geochemical characterisation of spilitic rocks. *Nature* **269**, 134–137.

FREY, F., HASKIN, M. A., POETZ, J. A. and HASKIN, L. A. (1968) Rare earth abundances in some basic rocks. *J. Geophys. Res.* **73**, 6085–6098.

FREY, F. A., BRYAN, W. B. and THOMPSON, G. (1974) Atlantic ocean floor: geochemistry and petrology of basalts from Legs 2 and 3 of the Deep Sea Drilling Project. *J. Geophys. Res.* **79**, 5507–5527.

FRYER, B. J. (1977) Rare earth evidence in iron-formations for changing Precambrian oxidation states. *Geochim. Cosmochim. Acta* **41**, 361–367.

GARMANN, L. B., BRUNFELT, A. O., FINSTAD, K. G. and HEIER, K. S. (1975) Rare-earth element distribution in basic and ultra-basic rocks from West Norway. *Chem. Geol.* **15**, 103–116.

GREEN, D. H. and RINGWOOD, A. E. (1967) An experimental investigation of the gabbro to eclogite transformation and its petrological applications. *Geochim. Cosmochim. Acta* **31**, 767–833.

GREEN, T. H., BRUNFELT, A. O. and HEIER, K. S. (1969) Rare earth element distribution in anorthosites and associated high grade metamorphic rocks, Lofoten-Vesteraalen, Norway. *Earth Planet. Sci. Lett.* **7**, 93–98.

GREEN, T. H. (1970) High pressure experimental studies on the mineralogical constitution of the lower crust. *Phys. Earth Planet. Interiors* **3**, 441–450.

GREEN, T. H., BRUNFELT, A. O. and HEIER, K. S. (1972) Rare earth element distribution and K/Rb ratios in granulites, mangerites and anorthosites, Lofoten-Vesteraalen, Norway. *Geochim. Cosmochim. Acta* **36**, 241–257.

HANSON, G. N. (1975) REE analyses of the Morton and Montevideo gneisses from the Minnesota River valley. *Geol. Soc. Amer. Abstr.* **7**, 1099.

HARRIGAN, D. B. and MACLEAN, W. H. (1976) Petrography and geochemistry of epidote alteration patches in gabbro dykes at Matagami, Quebec. *Can. J. Earth Sci.* **13**, 500–511.

HASKIN, L. A., FREY, F. A. and WILDEMAN, T. R. (1968) Relative and absolute terrestrial abundances of the rare earths. In *Origin and Distribution of the Elements*, Int. Ser. Monographs Earth Sci., **30**, L. H. AHRENS, ed., Pergamon Press, Oxford.

HEIER, K. S. (1965) Metamorphism and chemical differentiation of the crust. *Geol. Forens.* **87**, 249–256.

HEIER, K. S. and BRUNFELT, A. O. (1970) Concentration of cesium in high grade metamorphic rocks. *Earth Planet. Sci. Lett.* **9**, 416–420.

HEIER, K. S. and THORESEN, K. (1971) Geochemistry of high grade metamorphic rocks, Lofoten-Vesteraalen, N. Norway. *Geochim. Cosmochim. Acta* **35**, 89–99.

HEIER, K. S. (1973) Geochemistry of granulite facies rocks and problems of their origin. *Phil. Trans. Roy. Soc., Lond., A*, **23**, 429–442.

HELLMAN, P. L. and HENDERSON, P. (1977) Are rare earth elements mobile during spilitisation? *Nature* **267**, 38–40.

HERRMANN, A. G. and WEDEPOHL, K. H. (1970) Untersuchungen an spilitischen Gesteinen der variskischen Geosynkline in Nordwestdeutschland. *Contr. Mineral. Petrol.* **29**, 255–274.

HERRMANN, A. G., POTTS, M. J. and KNAKE, D. (1974) Geochemistry of the rare earth elements in spilites from the oceanic and continental crust. *Contr. Mineral. Petrol.* **44**, 1–16.

HOLLAND, J. G. and LAMBERT, R. St. J. (1975) The chemistry and origin of the Lewisian gneisses of the Scottish mainland: the Scourie and Inver assemblages and sub-crustal accretion. *Precambrian Res.* **2**, 161–188.

HYNDMAN, R. D., LAMBERT, I. B., HEIER, K. S., JAEGER, J. C. and RINGWOOD, A. E. (1968) Heat flow and surface radioactivity measurements in the Precambrian shield of Western Australia. *Phys. Earth Planet. Interiors* **1**, 129–135.

JAKEŠ, P. and TAYLOR, S. R. (1974) Excess europium content in Precambrian sedimentary rocks and continental evolution. *Geochim. Cosmochim. Acta* **38**, 739–745.

KAY, R., HUBBARD, N. J. and GAST, P. W. (1970) Chemical characteristics and origin of oceanic ridge volcanic rocks. *J. Geophys. Res.* **75**, 1585–1613.

KAY, R. and GAST, P. W. (1973) The rare earth content and origin of alkali-rich basalts. *J. Geol.* **81**, 653–682.

KAY, R. W. and SENECHAL, R. G. (1976) The rare earth geochemistry of the Troodos ophiolite complex. *J. Geophys. Res.* **81**, 964–970.

LAMBERT, I. B. and HEIER, K. S. (1968) Geochemical investigations of deep-seated rocks in the Australian shield. *Lithos* **1**, 30–53.

LAUSCH, J., MÖLLER, P. and MORTEANI, G. (1974) Die Verteilung der seltenen Erden in den Karbonaten und penninischen Gneisen der Zillertaler Alpen (Tirol, Österreich). *Neues Jahrb. Mineral. Monatsh.* 490–507.

LEWIS, J. D. and SPOONER, C. M. (1973) K/Rb ratios in Precambrian granulite terrains. *Geochim. Cosmochim. Acta* **37**, 1111–1118.

LYAKHOVICH, V. V. and BARINSKII, R. L. (1961) Characteristics of the rare earth assemblages in the accessory minerals of granitoids. *Geochem.* **1961**, 495–509.

MASUDA, A., NAKAMURA, N. and TANAKA, T. (1971) Rare earth elements in metagabbros from the Mid-Atlantic Ridge and their possible implications for the genesis of alkali olivine basalts as well as the Lizard peridotite. *Contr. Mineral. Petrol.* **32**, 295–306.

MCCARTHY, T. S. (1976) Chemical interrelationships in a low-pressure granulite terrain in Namaqualand, South Africa, and their bearing on granite genesis and the composition of the lower crust. *Geochim. Cosmochim. Acta* **40**, 1057–1068.

MONTIGNY, R., BOUGAULT, H., BOTTINGA, Y. and ALLÈGRE, C. J. (1973) Trace element geochemistry and genesis of the Pindos ophiolite suit. *Geochim. Cosmochim. Acta* **37**, 2135–2147.

MOORBATH, S., WELKE, H. and GALE, N. W. (1969) The significance of lead isotope studies in ancient high-grade metamorphic basement complexes, as exemplified by the Lewisian rocks of northwest Scotland. *Earth Planet. Sci. Lett.* **6**, 245–256.

MUECKE, G. K. (1969) Petrogenesis of granulite-facies rocks from the Lewisian of North-west Scotland. Unpubl. Ph.D. Thesis, Oxford University.

MYSEN, B. O. (1976) Rare earth partitioning between crystals and liquid in the upper mantle. *Carn. Inst. Wash. Yr. Bk.* **75**, 656–659.

NAGASAWA, H. and SCHNETZLER, C. C. (1971) Partitioning of rare earth, alkali and alkaline earth elements between phenocrysts and acidic igneous magma. *Geochim. Cosmochim. Acta* **35**, 953–968.

NANCE, W. B. and TAYLOR, S. R. (1976) Rare earth element patterns and crustal evolution–I. Australian post-Archean sedimentary rocks. *Geochim. Cosmochim. Acta* **40**, 1539–1551.

NANCE, W. B. and TAYLOR, S. R. (1977) Rare earth element patterns and crustal evolution–II. Archean sedimentary rocks from Kalgoorlie, Australia. *Geochim. Cosmochim. Acta* **41**, 225–232.

O'NIONS, R. K. and PANKHURST, R. J. (1974) Rare-earth element distribution in Archean gneisses and anorthosites, Godthaab area, West Greenland. *Earth Planet. Sci. Lett.* **22**, 328–338.

ONUMA, N., HIGUCHI, H., WAKITA, H. and NAGASAWA, H. (1968) Trace element partition between two pyroxenes and the host volcanic rocks. *Earth Planet. Sci. Lett.* **5**, 47–51.

ORSA, V. I., YELISEYEVA, G. D. and KAZANTSEVA, A. I. (1967) Rare earth assemblages in the accessory minerals of the ancient crystalline rocks of the Middle Dnepr region. *Geochim.* **1967**, 170–173.

PHILPOTTS, J. A., SCHNETZLER, C. C. and HART, S. R. (1969) Submarine basalts: Some K, Rb, Sr, Ba, rare-earth, H_2O and CO_2 data bearing on their alteration, modification by plagioclase, and possible source materials. *Earth Planet. Sci. Lett.* **7**, 293–299.

PIPER, D. Z. (1974) Rare earth elements in the sedimentary cycle: a summary. *Chem. Geol.* **14**, 285–304.

PRIDE, C. R. and MUECKE, G. K. (1976) Trace element geochemistry of granulite-facies granites and their host from the Lewisian of N.W. Scotland. *Geol. Ass. Canada, Program with Abstr.* **1**, 43.

PRIDE, C. (1977) Rare earth element studies of a granulite facies terrain: the Lewisian of North-west Scotland. Ph.D. thesis, Dalhousie University.

PUCHELT, H. and EMMERMANN, R. (1976) Bearing of rare earth patterns of apatites from igneous and metamorphic rocks. *Earth Planet. Sci. Lett.* **31**, 279–286.

REITAN, P. H. and ROELANDTS, I. (1973) Rare earth partitioning: coexisting metamorphic pyroxenes. *Geol. Soc. Amer. Abstr.* **5**, 778–779.

ROALDSET, E. (1975) Rare earth element distributions in some Precambrian rocks and their phyllosilicates, Numedal, Norway. *Geochim. Cosmochim. Acta* **39**, 455–469.

RONOV, A. B., BALASHOV, Y. A. and MIGDISOV, A. A. (1967) Geochemistry of the rare earths in the sedimentary cycle. *Geochem. Int.* **4**, 1–17.

RONOV, A. B., BALASHOV, Y. A. and GIRIN, Y. P. (1974) Regularities of rare earth element distribution in the sedimentary shell and in the crust of the earth. *Sedimentology* **21**, 171–193.

SARKAR, P. (1977) Petrology and geochemistry of White Rock meta-volcanics, Yarmouth, Nova Scotia. Ph.D. thesis, Dalhousie University.

SCHNETZLER, C. C. and PHILPOTTS, J. A. (1970) Partition coefficients of rare-earth elements between igneous matrix material and rock-forming mineral phenocrysts—II. *Geochim. Cosmochim. Acta* **34**, 331–340.

SHAW, D. M., DOSTAL, J. and KEAYS, R. R. (1976) Additional estimates of continental surface Precambrian shield composition in Canada. *Geochim. Cosmochim. Acta* **40**, 73–83.

SIGHINOLFI, G. P. (1969) K/Rb ratio in high-grade metamorphism: a confirmation of the hypothesis of a continental crustal evolution. *Contr. Mineral. Petrol.* **21**, 346–356.

SIGHINOLFI, G. P. (1971) Investigations into deep crustal levels: fractionating effects and geochemical trends related to high-grade metamorphism. *Geochim. Cosmochim. Acta* **35**, 1005–1021.

SMEWING, J. D. and POTTS, P. J. (1976) Rare-earth abundances in basalts and metabasalts from the Troodos massif, Cyprus. *Contr. Mineral. Petrol.* **57**, 245–258.

SUN, S. S. and HANSON, G. N. (1975) Origin of Ross Island basanitoids and limitations upon the heterogeneity of mantle sources for alkali basalts and nephelinites. *Contr. Mineral. Petrol.* **52**, 77–106.

TAYLOR, F. C. (1967) Reconnaissance geology of Shelburne map-area, Queens, Shelburne, and Yarmouth counties, Nova Scotia. *Geol. Surv. Canada Memoir* 349.

VAN DE KAMP, P. C. (1969) Origin of amphibolites in the Beartooth Mountains, Wyoming and Montana: new data and interpretation. *Geol. Soc. Amer. Bull.* **80**, 1127–1135.

WEAVER, S. D., SCEAL, J. S. C. and GIBSON, I. L. (1972) Trace element data relevant to the origin of trachytic and pantelleritic lavas in the East African Rift System. *Contr. Mineral. Petrol.* **36**, 181–194.

WILDEMAN, T. R. and HASKIN, L. A. (1973) Rare earths in Precambrian sediments. *Geochim. Cosmochim. Acta* **37**, 419–438.

WILDEMAN, T. R. and CONDIE, K. C. (1973) Rare earths in Archean graywackes from Wyoming and from the Fig Tree Group, South Africa. *Geochim. Cosmochim. Acta* **37**, 439–453.

WOOD, J. B. (1976) Samarium distribution between garnet and liquid at high pressure. *Carn. Inst. Wash. Yr. Bk.* **75**, 659–662.

WOOD, D. A., GIBSON, I. L. and THOMPSON, R. N. (1976) Elemental mobility during zeolite facies metamorphism of the Tertiary basalts of eastern Iceland. *Contr. Mineral. Petrol.* **55**, 241–254.

ZAYATS, A. P. and KUTS, V. P. (1964) Rare earth elements in the accessory minerals of gneisses of the Ukranian crystalline shield. *Geochem.* **1964**, 1126–1127.

ZIELINSKI, R. A. and FREY, F. A. (1970) Gough Island: evaluation of a fractional crystallization model. *Contr. Mineral. Petrol.* **29**, 242–254.

Abundance of Lithium in Spilites and its Implications for the Spilitization Process

NICOLE VATIN-PERIGNON,† DENIS M. SHAW‡ and JOHN R. MUYSSON‡

†*Institut Dolomieu, Géologie et Minéralogie, Rue Maurice Gignoux, 38031, Grenoble-Cedex, France*
‡*Department of Geology, McMaster University, Hamilton, Ontario, Canada L8S 4M1*

Abstract

Recent discoveries on sea-floor basalts, and new studies on old basaltic rocks interbedded with clay sediments and shales of various ages, show that spilitization results from the action of hydrothermal solutions on basic volcanic rocks during, or immediately following, the late stage of cooling.

Geochemically, spilites differ from basalts in many aspects. Among the minor constituents, Li is a good indication of spilitization. The Li contents are always *abnormally* high (average 80 ppm); their wide range (20–200 ppm) is connected with increasing H_2O contents, itself a function of the chlorite content. There is also a concomitant increase of Na and depletion of Ca. Moreover, the basaltic rocks are more deeply transformed when the hydrothermal solutions are enriched in CO_2. Such alteration requires waters rich in both Na and Li with a 10^5 Li/Na ratio of about 750; the ratio for seawater is 1,8 but most subsurface waters have higher ratios (250–450). This evidence confirms that spilitization is not a low-temperature submarine weathering process (halmyrolysis).

Spilitization of basaltic rocks necessarily involves the combined action of heat (volcanic activity) and of fluids (water). But the process can be significantly more complex according to the composition of the fluids (late magmatic fluids, connate waters in sediments and, perhaps, seawater, or various mixtures of these fluids), and to the textures of volcanic rocks. The most complete transformation occurs in either vitreous or porous lavas in an argillaceous sedimentary environment, and a fluid consisting of Li-, Na-, and CO_2-enriched hot water (400-500° or more) with a low Ca/Na ratio.

Résumé

Les récentes découvertes sur les basaltes des fonds océaniques et les études nouvelles sur les séries basaltiques anciennes inclues dans des sédiments argileux ou des schistes de différents âges montrent que la spilitisation est dûe à l'action de solutions hydrothermales agissant sur les roches basiques à la fin ou immédiatement aprés leur cristallisation.

Les anomalies géochimiques des spilites par rapport aux basaltes sont nombreuses. Parmi les éléments mineurs, le Li est un bon traceur de la spilitisation. Les teneurs en Li sont, en effet, toujours *anormalement* élevées (moyenne 80 ppm) et la distribution de cet élément fait apparaître une forte dispersion (20–200 pmm) liée à une corrélation positive avec H_2O, d'autant plus forte que la roche sera riche en chlorites. Il s'agit donc bien de l'eau des chlorites. Ces variations s'accompagnent d'une augmentation de Na et d'une perte de Ca. De plus, la roche basique, des modifications plus poussées que lorsque le fluide est composé principalement d'H_2O. Une altération de la sorte nécessite des eaux enrichies en Na et Li avec un rapport 10^5 Li/Na voisin de 750. Ce rapport est de 1,8 pour l'eau de mer, mais beaucoup d'eaux souterraines ont des rapports plus élevés (250–450). Cette évidence confirme que la spilitisation n'est pas une altération sous-marine à basse température (halmyrolyse).

La spilitisation d'une roche basique implique, *obligatoirement*, l'action combinée de la chaleur (activité volcanique) et d'un fluide (eau). Elle peut être módifiée par la nature du fluide: fluides magmatiques de fin de cristallisation, eaux connées de sédiments, voire même eau de mer ou le mélange de ces fluides, et par la structure de la roche, mais elle est la plus complète lorsque la sédimentation environnante est argileuse, la lave vitreuse ou alors poreuse et perméable et le fluide une eau chaude (400-500° ou plus), riche en Li, Na et CO_2 et à rapport Ca/Na bas.

L'extension des investigations sur les fonds océaniques permet d'exclure pour les spilites draguées sur la Ride de Carlsberg (Cann, 1969) toute origine liée à un métamorphisme orogénique. Par ailleurs, de nouvelles données concernant les roches vertes spilitiques de Ligurie (Spooner et Fyfe, 1973) indiquent que la spilitisation de ces laves est le produit d'un hydrothermalisme lié à un gradient géothermique apparent trés élevé. Les éxpériences d'Hajash (1975), quant à elles, montrent qu'une hydrothermalisation est suffisante pour produire la quasi totalité des caractères tant minéralogiques que géochimiques des spilites. Ces données, jointes à toutes celles déjà acquises sur les spilites, concourrent à démontrer que les paragénèses spilitiques se développent précocement dans les roches basiques sous l'action des fluides hydrothermaux. Un enrichissement en Li ayant été noté poûr des spilites alpines (Vatin-Pérignon et Shaw, 1972), il paraît interessant de connaître si ce phénomène se produit également dans d'autres séries spilitiques et s'il peut apporter des éléments nouveaux dans l'étude du processus de spilitisation.

Echantillonnage et Environnement Géologique

Les 240 échantillons étudiés ici proviennent presque tous d'Europe Occidentale, à l'exception de quelques laves du Canada et de la Jamaïque.

Une distinction est établie entre les spilites à paragenèse BT développée et les dolérites (ou diabases des auteurs de langue anglaise ou allemande) où le processus d'altération hydrothermale n'est qu'amorcé; ceci en raison de la structure même de la roche en voie de spilitisation. En effet, à chimisme initial identique, une roche basique dont la plus grande partie des éléments est intégrée aux reseaux cristallins des mineraux (cas des dolérites intrusives filoniennes) sera moins atteinte par la spilitisation qu'une roche d'épanchement ne sont pas liés à une structure capable de les retenir. De même une distinction est introduite entre basaltes et spilites de même origine magmatique lorsque la paragénèse minérale HT (labrador, augite) est encore conservée et que seul le verre est hydraté (chlorites) avec des amygdales de calcite.

Ces laves appartiennent à des formations étalées du Précambrien à l'Eocène (tableau 1) et correspondent à des *associations* laves-sédiments-tufs *mises en place sous faible profondeur d'eau.*

Les coulées précambriennes de l'*Ontario*, intercalées dans des conglomérats, sont recouvertes par des grés arkosiques et des schistes du Keweenawan supérieur. Dans les *Vosges* septentrionales (Schirmeck, Rabodeau) les spilites et keratophyres dévono-dinantiens s'insèrent dans une série sédimentaire détritique schisto-gréseuse à lentilles calcaires. A l'extrémité SW du faisceau synclinal du *Morvan* les basaltes, spilites et keratophyres sont liés à une sédimentation dévono-dinantienne détritique siliceuse fine (schistes, psammites) ou grossière (arkoses, poudingues). Dans la province de *Huelva* (Espagne) les formations volcano-sédimentaires faméniennes et viséennes (spilites, keratophyres et tuffites détritiques) s'intercalent entre deux faciés flysch. Le bassin de *Chateaulin* (Bretagne) est, au Viséen inf., un bassin subsident à mer très peu profonde où se déposent des conglomérats et schistes à intercalations de spilites et keratophyres. Les vases argileuses initiales sont à dominante illito-chloritique. La sédimentation du Trias dans le *Massif des Ecrins-Pelvoux* (Alpes) est gréseuse et calcaro-dolomitique à intercalations de coulées et tufs spilitiques. Au *Montgenèvre* (Hautes Alpes) les dolérites traversent les ophiolites des schistes lustrés, surmontés par des calcaires, schistes et radiolaires. Dans le

TABLEAU 1. AGE, PROVENANCE DES BASALTES, DOLÉRITES ET SPILITES

Age	Nature			Provenance Pays et localités	Collecteurs et Références
	Basalte	Dolérite	Spilite		
Précambrien		24		Canada, Mamanse Pt., District d'Algoma Ontario	N. W. D. Massey (non publié) R. N. Annells, 1973
Dévonien (Givetien)		24		France, Vosges (Bas-Rhin) massif du Schirmeck	Th. Juteau et G. Rocci, 1965, 1966
Dévono–Diantien		26	2	France, Vosges (Bas-Rhin) massif du Rabodeau	Th. Juteau, 1971
Dévonien (Faménnien)	11		11	France, Morvan (Nièvre) extrémité SW du faisceau	J. Bébien, 1970 et 1971
Dinantien (Viséen)	8		7	France, Morvan (Nièvre) extrémité SW du faisceau	J. Bébien, 1970 et 1971
Devono–Diantien		2	5	Espagne, prov. de Huelva (Rio Tinto)	E. Soler, 1973
Dinantien		6	3	France, Bretagne, Chateaulin (Cotes-du-Nord)	J. P. Sagon, 1969 et 1976
Trias sup.		4	74	France, massif des Ecrins (Hautes Alpes)	R. Aumaître et G. Buffet, 1973; N. Vatin-Pérignon *et al.*, 1974
Jurassique sup. Crétacé inf.		1		France, Montgenèvre (Hautes-Alpes)	M. Vuagnat (non publié) C. Mevel, 1975
Crétacé moyen Eocène inf.		1	3	France, Thônes (Haute-Savoie), Val d'Illiez	M. Delaloye et G. Sawatzki, 1973
Crétacé sup.		34	7	France-Italie, Bourg St-Maurice (Savoie) et la Thuile, prov. d'Aoste	J. L. Lasserre et C. Laverne, 1976
Eocène			5	Jamaïque	T. E. Smith et T. A. Jackson, 1974

synclinal de *Thônes* (Haute Savoie) les dolérites des grauwackes du flysch helvétique se rattacheraient au volcanisme océanique piémontais-ligure. Les dolérites et laves en coussins crétacées du *Versoyen* (Savoie) appartiennent à un système sills-coulées épanchées au cours d'une phase de distension au début de la période de subsidence du bassin pélitique. Enfin, les spilites et kératophyres éocènes de *Jamaïque* sont interstratifiés avec des tufs lithiques déposés dans une eau peu profonde.

Par comparaison, tout un ensemble de roches volcaniques basiques (cf. Shaw *et al.*, 1974a et b, 1975) constitue le groupe A des figs. 1, 2 et 3 et du tableau 3. Il comprend des basaltes tholéiitiques du plancher océanique (75), des îles océaniques (19), des plateaux (22) et des coulées et dykes continentaux variés (79); ainsi que des basaltes alcalins d'îles (4), d'arcs (6) et continentaux (46).

Pour toutes ces roches basaltiques ainsi que pour celles des séquences basaltes-spilites et dolérites-spilites, il a été également déterminé soit H_2O+, H_2O-, CO_2, soit perte au feu et CO_2, soit perte au feu seule. Le Li est dosé par absorption atomique à McMaster University.

Résultats

Les anomalies géochimiques des laves des séquences basaltes ou dolérites–spilites reflètent le plus ou moins grand développement des assemblages minéralogiques de Basse Température au cours du processus de spilitisation. Si l'on prend comme terme de comparaison le basalte no. 27 et la dolérite no. 25 de Le Maître (1976), on constate les variations suivantes (tableau 2):

$\bar{X}$ des basaltes par rapport au basalte no. 27 =
+ Li, H_2O, Na, K, Si, Al
− Ca, Fe, Mg, Ti;

$\bar{X}$ des dolérites par rapport à la dolérite no. 25 =
+ Li, H_2O, Na, CO_2
− Ca, Fe, (K)
peu variants: Al, Mg, Ti, Si;

Tableau 2. Comparaison des Basaltes, Dolérites et Spilites

		Basaltes 1		Basalte no. 27		Dolérites 2		Dolérite no. 25		Spilites 3		Spilite S
%	*n*	$\bar{X}$	σ	$\bar{X}$	n	$\bar{X}$	σ	$\bar{X}$	n	$\bar{X}$	σ	$\bar{X}$
SiO_2		51,82	2,36	49,20		50,48	7,13	50,18	116	47,46	5,09	49,65
TiO_2		1,07	0,18	1,84	96	1,51	0,62	1,14		1,84	0,66	1,57
Al_2O_3		17,52	1,35	15,74		16,09	1,70	15,26		15,66	2,26	16,00
Fe_2O_3		9,43(a)	,67	3,79	99	10,31(a)	2,19	2,86	117	10,66(a)	2,28	3,85
FeO				7,13				8,05				6,08
MnO		0,18	,04	,20	93	,16	,06	,19	115	,15	,08	,15
MgO		4,81	1,52	6,73		6,39	2,68	6,78		6,42	2,12	5,10
CaO		8,80	1,37	9,47		6,18	3,11	9,24		5,26	3,43	6,62
Na_2O		3,06	,39	2,91		3,78	1,08	2,56		3,95	1,38	4,29
K_2O		1,21	,48	1,10		,84	,87	1,04		1,42	1,33	1,28
P_2O_5		n.d.		,35	6	,27	,13	,27	12	,29	,18	,26
H_2O+		1,64(b)	,54	,95		3,08(b)	1,42	1,46	96	3,86(b)	1,52	3,49
H_2O-				,43				,43				
CO_2		n.d.			43	,94	1,03	,18	97	3,01	2,89	1,63
Li ppm	12	48(c)	17	10(d)	98	52(c)	32		119	80,5(c)	42	

1, 2, et 3: échantillons traités dans cet article.
n: nombre d'échantillons (19 pour les basaltes, 100 pour les dolérites, 118 pour les spilites sauf lorsqu'un autre chiffre est indiqué).
$\bar{X}$: moyenne arithmétique.
σ: écart-type.
(a): Fe_2O_3 total de l'analyse quantométrique.
(b): H_2O calculé = perte au feu − CO_2.
(c): Li ppm déterminé par absorption atomique à McMaster University.
no. 27; moyenne des basaltes (à caractères intermédiaires entre alcalins et tholéiitiques) et no. 25: moyenne des dolérites (à caractères tholéiitiques dominants)—Le Maitre, 1976.
(d): valeur médiane de tous les basaltes (Prinz, 1968, tableau II, no. 8).
S: moyenne des spilites (Vallance, 1960).

$\bar{X}$ des dolérites par rapport aux spilites =
+ Li, H_2O, Na, CO_2, K
− Ca, Si
peu variants: Al, Fe, Mg, Ti.

En exceptant les basaltes devono-dinantiens à distribution perturbée au niveau de tous les éléments, y compris Al, les dolérites apparaissent proches d'une souche originelle à caractère tholéiitique et les éléments les plus affectés par des processus post-magmatiques sont les mêmes pour les trois catégories de roches envisagées ici (basaltes, dolérites et spilites): Li, H_2O et Na ont des teneurs toujours croissantes, Ca est sans cesse en diminution. Si paraît également impliqué dans les migrations d'élements alors que *Al*, dont le comportement est assez controversé, présenterait une relative invariance (de la Roche *et al.*, 1974) confirmée ici au cours de la transition dolérite–spilite. Sa distribution reflète la quantité de plagioclase initial dans les dolérites et de mineraux de déstabilisation de celui-ci (albite, épidote) dans les spilites. Cet élement paraît donc particulièrement interessant puisque sa distribution semble peu perturbée par les altérations hydrothermales.

Parmi les éléments mobiles, *Ca* est le seul en diminution constante. Sa distribution est caractérisée par une extrème dispersion qui reflète le transfert de cet élément à l'extérieur de la roche, lié à la circulation de fluides chauds. Les études expérimentales à 0,5–1,0kb, en présence de vapeur d'eau et à des températures supérieure à 500° (Hajash, 1974, 1975; Maurin, 1974; Mottl *et al.*, 1974) montrent en effet que Ca est trés facilement relaché par les basaltes dans une ambiance hydrothermale et que sa migration entraine celle de Si, Na, K

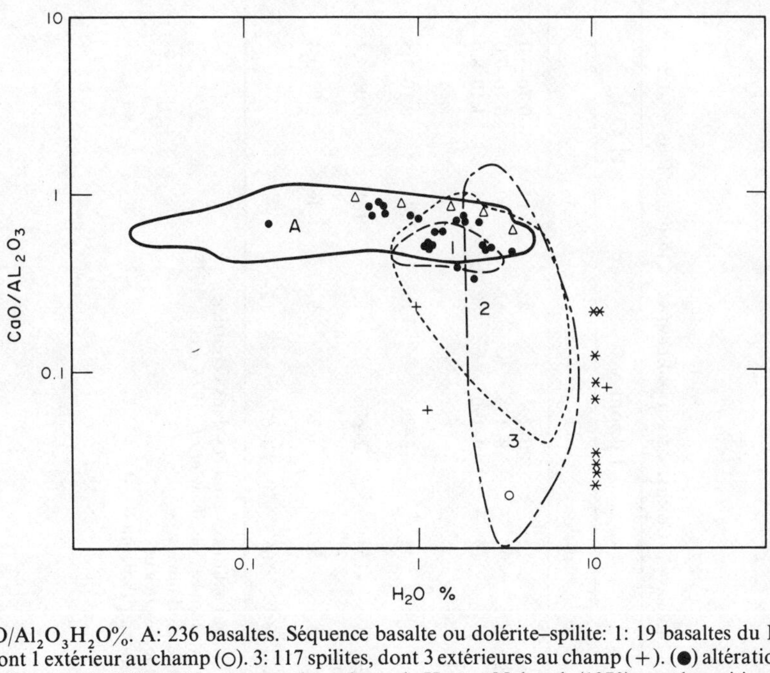

FIG. 1. $CaO/Al_2O_3H_2O\%$. A: 236 basaltes. Séquence basalte ou dolérite–spilite: 1: 19 basaltes du Morvan. 2: 98 dolérites, dont 1 extérieur au champ (○). 3: 117 spilites, dont 3 extérieures au champ (+). (●) altération des basaltes océaniques (Bass *et al.*, 1973). (△) moyenne des valeurs de Hart et Nalwark (1970). ⁎ palagonitisation des verres basaltiques (Frey *et al.*, 1974).

Pour les figs. 1 et 2: H_2O = P.F. —CO_2 pour les échantillons provenant des Alpes, d'Espagne et de Jamaïque; H_2O = P.F. pour les échantillons provenant de Bretagne, des Vosges et du Morvan.

Tableau 3. Comparaison des Élements Chimiques Principaux des Basaltes et des Laves de la Séquence Basalte ou Dolérite–spilite

	Li ppm			H_2O %		CaO/Al_2O_3		Na_2O %			
	n	G	$\bar{X}$	n	G	n	G	n	G	$\bar{X}$	10^5 Li/Na
1–Basaltes	12	20–72	48	19	0,70–3,0	19	0,40–0,70	19	2,49–4,02	3,06	211
2–Dolérites	98	8–155	52	100	0,70–7,0	100	0,04–0,97	100	,63–6,36	3,78	185
3–Spilites	119	14–190	80	96	1,10–8,6	118	0,01–1,10	118	,22–7,25	3,95	274
Groupe A:											
Tholéiites océaniques	48	3–22	8,3	47	0,10–3,5	68	0,57–0,90	75	1,49–4,00	2,44	46
Toutes les tholéiites	154	2–67	13	161	0,07–4,4	116	0,41–0,90	195	1,36–4,52	2,54	69
Tous les basaltes alcalins	52	5–26	9,4	46	0,02–3,4	52	0,45–1,13	56	1,21–5,09	3,57	35
Tous les basaltes	206	3–67	12					251	1,21–5,09	2,77	58

1, 2 et 3: laves spilitisées étudiées dans cet article.
Groupe A: données de Shaw *et al.*, 1974 a et b, 1975.
n: nombre de données.
G: gamme des teneurs.
$\bar{X}$: moyenne arithmétique.

(Maurin) ou de K, Fe, Mn (autres auteurs). La composition chimique des saumures naturelles d'Islande (Mottl *et al.*, 1975) est d'ailleurs voisine de celle des fluides obtenus expérimentalement.

Le rapport CaO/Al_2O_3 est donc un bon indicateur de l'altération hydrothermale. Il varie indépendamment de H_2O à basse concentration, mais une corrélation apparaît aux environs de $H_2O = 2\%$ à la fois pour les phénomènes d'halmyrolyse (Bass *et al.*, 1973; Hart et Nalwalk, 1970) et de spilitisation. Sur la Fig. 1 la position des domaines 1, 2 et 3 des laves de la séquence basaltes ou dolérites–spilites permet de suivre également la destruction du clinopyroxène en même temps que l'expulsion de Ca et l'hydratation de la roche, c'est à dire la fixation d'eau dans le réseau des mineraux hydratés néoformés (chlorite, épidote). La plus ou moins grande intensité du phénomène spilitique apparaît ainsi lié à la plus ou moins grande perméabilité de la roche aux fluides hydrothermaux, donc à sa fissuration et à la possibilité de circulation entre grains. Si la roche constitue un système ouvert, la circulation des fluides chauds autorise les transferts de matière et en particulier la perte de Ca et la fixation d'H_2O, mais si elle constitue un système fermé ou avec échange de matière limité, seule une déstabilisation des paragénèses magmatiques intervient = destruction du plagioclase.

La répartition de *Li* dans les laves en cours de spilitisation montre toujours une grande dispersion. Les teneurs moyennes sont 4 à 6 fois supérieures à celles des basaltes (tableau 3) et pour certaines spilites, des valeurs de 200 et 250 ppm peuvent être atteintes (Aumaître et Buffet, 1973).

Sur la fig. 2, H_2O des basaltes du groupe A est trés variable alors que Li reste bas: seules 10 valeurs sur 205 sont supérieures à 30 ppm. Les faibles teneurs en Li (10–20 ppm) des basaltes sont bien connues (Civetta *et al.*, 1974) et certaines laves de l'Océan Indien (Engel et Fischer, 1975) n'en contiennent que 5,4 ppm en moyenne; cependant, dés que ces roches sont altérées,

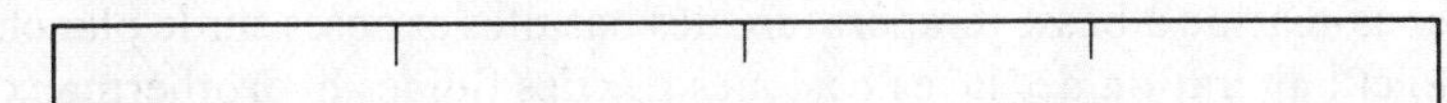

FIG.2. Li ppm–H_2O%. A: 205 basaltes, Séquence basalte ou dolérite–spilite: 1:12 basaltes du Morvan. 2:98 dolérites, dont 2 extérieures au champ (○). 3: 117 spilites, dont 4 extérieures au champ (+). (●) palagonitisation des verres des basaltes océaniques (Frey *et al.*, 1974).

on constate une augmentation trés nette de Li, pouvant dépasser 75 ppm (Thompson et Melson, 1970; Thompson *et al.*, 1974; Henrich et Thompson, 1976). Cette altération correspond à une palagonitisation du verre (Frey *et al.*, 1974) ou à une argilisation du basalte sous l'action de l'eau de mer, entrainant une corrélation positive $\delta^{18}O$/log ppm Li (Muehlenbachs et Clayton, 1972). La formation de smectite à partir des basaltes océaniques est d'ailleurs bien connue (Anderson *et al.*, 1976) mais peu d'auteurs, hormis Cann (1969) rapportent l'altération des basaltes océaniques à un processus de spilitisation.

Quant aux valeurs de Li des spilites, elles sont rarement données dans la littérature: 8 et 10 ppm sont signalés dans les basaltes spililisés du Deccan (Vallance, 1974), entre 50 et 100 ppm dans certains pillow-lavas du Baer Bassit, environ 70 ppm dans les spilites triasiques et jusqu'a 217 ppm dans les terres d'ombre les accompagnant (Parrot et Vatin-Pérignon, 1974; Parrot et Delaune-Mayere, 1974).

Dans la présente étude, les spilites ont été choisies dans des environnements géologiques d'âges différents mais la plupart proviennent de bassins à sédimentation argileuse dominante. Elles ont en moyenne 48 (basaltes), 52 (dolérites) et 80 (spilites) ppm de Li et sont riches en fluides. Leurs concentrations en H_2O sont toujours élevées (jusqu'à 8,6%) surtout en comparaison de celles des basaltes dont la moyenne est de 0,9% d'H_2O (Le Maitre, 1976).

Les laves vitreuses apparaissent systématiquement plus hydratées que les dolérites bien cristallisées et, de ce fait, *plus spilitisées* c'est à dire enrichies en un minéral typique de la spilitisation qui est *la chlorite*. La tendance dispersant les teneurs en Li apparaît donc normalement liée à H_2O et ne peut être attribuée à un métamorphisme de faciés schistes verts car un phénomène thermique est incapable d'enrichir les roches en Li. Seuls des processus d'altération hydrothermale peuvent s'accompagner d'un accroissement de Li et ceux-ci sont au moins au nombre de 3: l'altération deutérique des verres basaltiques au cours du refroidissement (*palagonitisation*, Melson, 1973; Melson et Thompson, 1973; Moore, 1966); l'altération sous-marine à basse température des basaltes exposés sur le plancher océanique (*halmyrolyse*) et l'altération des laves basiques par des fluides hydrothermaux agissant à la fin ou immédiatement aprés leur cristallisation (*spilitisation*). Dans ce dernier cas la corrélation positive Li-H_2O est en fait établie entre Li et H_2O des chlorites car ce minéral est le seul qui contienne suffisamment d'eau et soit, en volume, assez important, dans la transition basalte ou dolérite–spilite. La localisation de Li dans le réseau des chlorites néoformées au cours de la spilitisation semble donc la plus plausible et pourrait se confirmer par les dosages de cet élement dans les chlorites par spectrométrie gamma sous irradiation (travaux actuellement en cours).

Il apparaît néanmoins que l'abondance de Li liée à H_2O et le rapport CaO/Al_2O_3 sont de bons indicateurs de la spilitisation et la discrimination s'établit très nettement (fig. 3) entre le domaine des laves spilitiques et celui des basaltes.

Au cours du processus de spilitisation, outre H_2O et Li, *Na* augmente également (tableau 3) et cet enrichissement est lié à la transformation anorthite-albite. Les expériences d'hydrothermalisme sur les basaltes (Hajash, 1975; Mottl *et al.*, 1974) à partir de l'eau de mer montrent que les teneurs en Na décroissent dans le fluide, alors que l'albite eristallise dans le basalte. Il y a donc incorporation de Na dans la roche, mais *seulement* pour des températures de 400–500° et plus. L'altération des basaltes à basse température (eau de mer froide) libère au contraire Na qui passe dans l'eau de mer (Hart, 1970). On peut donc penser que dans une ambiance hydrothermale chaude (400–500° ou plus), le fluide circulant à travers les microfractures ou entre les grains de la roche provoquera la destruction du plagioclase, la cristallisation d'albite à partir du Na qu'il contient et l'évacuation du Ca Liberé. Mais ce

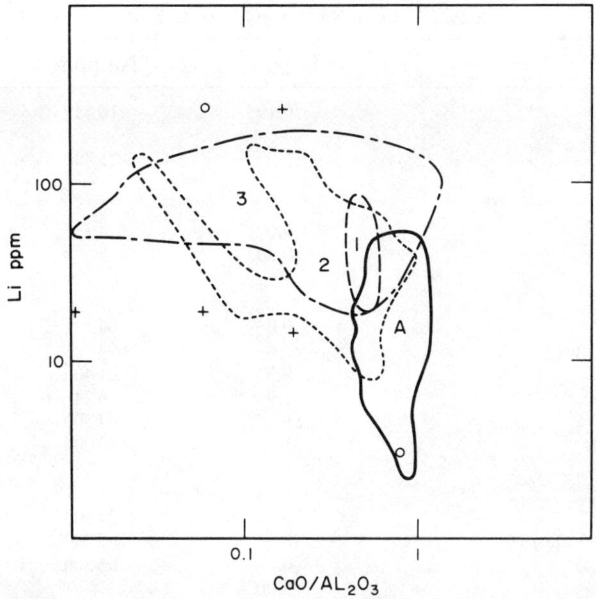

FIG. 3. Li ppm–CaO/Al_2O_3. A: 205 basaltes. Séquence basalte ou dolérite–spilite: 1: 12 basaltes du Morvan. 2: 100 dolérites, dont 2 extérieures au champ (○). 3: 118 spilites, dont 4 extérieures au champ (+).

processus de spilitisation pourrait également être amorcé dans une ambiance hydrothermale plus froide (dés 200–250°C) et sans implication de CO_2 comme l'indiquent les travaux de Jehl (1975), Jehl *et al.* (1976) sur le developpement d'associations de type spilitique résultant d'une convection d'eau de mer dans les basites et ultrabasites de l'Océan Nord Atlantique.

Source de Li dans les Spilites

Na et Li étant tous deux incorporés à la lave au cours de la spilitisation, leur origine peut donc être recherchée ensemble.

Si l'on considère l'abondance de ces éléments dans les eaux naturelles (tableau 4) il est clair que l'eau de mer peut, dans certaines conditions, fournir du Li au basalte (bien que sa teneur soit faible), mais il est non moins clair que pratiquemment toutes les eaux juvéniles, de lacs ou souterraines en fourniront encore plus.

Ce n'est pas le cas pour Na et le rapport Li/Na est donc beaucoup plus haut dans les eaux terrestres que dans l'eau de mer. Les données du tableau 3 indiquent que les rapports 10^5 Li/Na des moyennes des basaltes et des spilites sont respectivement 58 et 274, ce qui requière pour la transformation basalte–spilite un rapport 10^5 Li/Na voisin de 750 pour le matériel incorporé. Cette valeur n'est pas necessairement celle des fluides hydrothermaux responsables de la spilitisation mais bien que les coefficients de partage de Li soient mal connus (Volfinger, 1970) il semble cependant necessaire que la valeur du rapport Li/Na soit élevée dans ces solutions hydrothermales.

Il ne manque donc pas d'eaux assez riches en Li et Na pour alterer les roches basiques en un assemblage de mineraux de basse température mais la question reste de savoir pourquoi la spilitisation provoque une éxagération des teneurs en Li des roches basiques?

TABLEAU 4. LI ET Na DANS QUELQUES EAUX

	Li ppm	Na ppm	10^5 Li/Na
Seawater (1)	0,19	10 555	1,8
Formation waters (1)	26	25 800	101
Rivers (2)			
Japanese	0,0011	4 15	27
North American	0,0033	5	66
Lakes			
Lake Tanganyika (2)	0,8	64 2	1250
Laguna Escandato, Peru (2)	7,6	694	1100
Mono lake, Calif. (2)	8,5	21 400	40
Salton Sea, Calif. (2)	1,9	6 249	30
Great Lake, Utah (3)	56	92 200	61
Searles Lake, Calif. (4)	81	110 000	74
Subsurface brine			
Michigan basin (4)	70	22 500	310
Niland Well, Imperial Valley (4)	18	7 280	250
Clayton valley, Nev. (5)	300	66 200	450
Paris basin (6)	0,0074	—	—
Non-volcanic hot springs (7 et 8)	0,24–8	120–6 100	88–960
	0,13–14	328–9140	16–150
Petroleum waters (8, 9 et 10)	2,0–2,7	3 770–40 200	44–53
	0,09–2,21		1–40
	0,8–3,6		
Red Sea bottom brines (11)	0,21–4,72	12 600–119 000	3,6
Red Sea core interstitial water (11)	0,36–7,92	10 400–115 000	3,0–7,2
Volcanic springs (7, 8, 9 et 12)	0,1–196	27–1 398	400–1 900
	0,21–0,7		12
	2,7–14,8	280–1 320	248–1 430
	0,05–8,4	4.5–526	

(1) Heier et Billings, 1970, chap. 3-Li, chap. 11-Na; (2) Livingstone, 1963; (3) Cohenour, 1966, in Vine, 1975; (4) White, Hem et Waring, 1963; (5) Kunasz, 1970, in Vine, 1975; (6) Fruchart et Pinta, 1975; (7) Ellis et Mahon, 1964; (8) White, 1957 a et b, 1963; (9) Khod'kova et Gemp, 1970; (10) Merino, 1975; (11) Brooks *et al.*, 1969; (12) Rowe *et al.*, 1973.

L'enrichissement en Li des eaux souterraines s'explique mal et a été discuté par Vine (1975). Il paraît résulter en partie de la grande solubilité des composés du Li., aussi y a-t-il corrélativement peu de sédiments riches en Li. On note, toutefois, une accumulation notable de Li dans les argiles formées sous climat désertique et dans les évaporites (Tardy *et al.*, 1972; Vine, 1975; tableau 5).

Les teneurs en Li des pélites encaissantes de certaines séries spilitiques alpines (Lasserre et Laverne, 1976) s'étalent sur une gamme identique à celle des spilites: 30–100 ppm. Roches vertes et pélites sont, ici, étroitement imbriquées et les schistes les plus éloignés des intrusions ont des teneurs en Li avoisinant 100 ppm, alors que ces valeurs sont moindres (65 ppm) et identiques à celles des sills, au contact des deux formations. De même, les enveloppes classiquement chloriteuses des laves en coussins et leurs matrices d'origine volcanique ou

TABLEAU 5. ABONDANCE DE LI DANS LES ROCHES

	Li ppm
Ultramafic rock (1)	tr–4,0
Serpentinite (1)	tr–19
Basalt	2–40
Mud-volcano Sediment (2)	11–190
Shale (3)	
glacial climate	65
aride climate	83
humid climate	100
saline lagoon	112–340
Evaporite clay minerals (4)	620–3290
Detrital clay minerals (4)	50–80

(1) Maksimovic et Maksimovic, 1968,
(2) Khod'koda et Gemp, 1970,
(3) Ronov *et al.*, 1970,
(4) Tardy *et al.*, 1972.

sédimentaire sont jusqu'a 4 fois plus riches en Li que les coeurs de coussins correspondants (Li = 20 ppm). Par ailleurs, les différents gisements spilitiques envisagés ici appartiennent à des ensembles lithologiques contenant tous des pélites; on peut donc se demander si l'origine du Li des spilites n'est pas à rechercher dans les pélites mêmes: les teneurs élevées en Li des sédiments sont certainement primaires puisque ceux-ci sont essentiellement constitués à l'origine de mineraux argileux (illite surtout) comptant parmi ceux qui accumulent le plus de Li ($\bar{X}$ = 66 ppm pour 105 sédiments argileux marins; gamme: 17–207 ppm; Hirst, 1962). Il est donc concevable q'au cours de l'altération hydrothermale de roches magmatiques en cours de refroidissement, incluses dans des séries pélitiques, la teneur en Li de ces roches augmente alors que celle des pélites diminue légèrement; il s'établirait ainsi une sorte de compensation du gradient de teneur en Li entre un basalte ou une dolérite initialement à 10 ppm de Li et un encaissant pélitique à 80–100 ppm de Li au moyen d'un transfert de matière dont le vecteur serait le fluide hydrothermal (principalement H_2O).

Lors des processus tardi et post magmatiques impliquant la circulation de fluides, le Li des argiles peut être mobilisé et concentré dans ces fluides. On sait que la teneur maximale en Li des eaux hydrothermales est une fonction directe de la température (Ellis et Mahon, 1964), de ce fait la teneur en Li des fluides chauds doit pouvoir être importante. Or, ce sont de tels fluides chauds qui, percolant à travers les masses basiques, sont responsables des altérations propres à la spilitisation et sont également impliqués dans la cristallisation de la chlorite. Les teneurs en Li des spilites étant liées à celles de H_2O des chlorites, ceci peut expliquer l'enrichissement en Li des chlorites corrélativement à celui des roches qui les contiennent.

En conclusion, des teneurs en Li élevées seront une indication sur la réalisation la plus complète du processus de spilitisation. Celui-ci depend de la température, de la structure de la roche et de la composition du fluide.

Le fluide doit être chaud (400–500° ou plus): la chaleur provenant de la roche basique elle-même en cours de refroidissement et devant être suffisante pour amorcer dans les pélites encaissantes le système hydrothermal responsable de la spilitisation.

Le fluide doit pouvoir circuler à travers la roche: les verres (assemblages silicatés non organisés) sont plus sensibles à l'hydratation à haute température que les roches bien cristallisées (dolérites) où la circulation des fluides se fait par les microfractures ou dans le film intergranulaire.

Le fluide de base est H_2O, que ce soit les fluides magmatiques de fin de cristallisation, les eaux connées des sédiments, l'eau de mer ou un mélange de ces fluides. La présence de CO_2 d'origine exogène (provenant du milieu de sédimentation saturé en carbonates) ou endogène (d'origine profonde relaché par le magma et se dissolvant dans le fluide aqueux lorsque celui-ci est au sein de la roche basique) ou mixte, augmentant l'action déstabilisatrice du fluide et favorisant la destruction de la paragenèse HT (plagioclase–pyroxène–olivine) et les recristallisations BT (albite–chlorite–calcite).

Les conditions les plus favorables pour spilitiser une roche basique sont donc réunies lorsqu'une lave vitreuse (ou poreuse et perméable) s'épanchant dans un bassin de sédimentation peu profond, riches en vases argileuses, est traversée, au cours de son refroidissement, par des fluides hydrothermaux chauds (400–500° ou plus), riches en Li, Na et CO_2, pauvres en Ca et à rapport Li/Na élevé.

References

ANDERSON, T. F., DONNELLY, T. W., DREVER, J. I., ESLINGER, E., GIESKES, J. M., KASTNER, M., LAWRENCE, J. R. and PERRY, E. A. (1976) Geochemistry and diagenesis of deep-sea sediments from leg 35 of the Deep Sea Drilling Project. *Nature* **261**, 473–475.

ANNELLS, R. N. (1973) Proterozoic flood basalts of eastern lake Superior: the Keweenawan volcanic rocks of the Mamainse Point area, Ont. *Geol. Survey Canada* Paper 72–10, 51 p.

AUMAITRE, R. et BUFFET, G. (1973) Minéralogie, Pétrographie et Géochimie des laves spilitiques et des filons basiques associés du Massif des Ecrins–Pelvoux (Alpes françaises occidentales). Thèse 3 ème cycle, Grenoble, 301 pp.

BASS, M. N., MOBERLY, R., RHODES, J. M., SHIH, C. and CHURCH, S. E. (1973) Volcanic rocks cored in the Central Pacific, Leg 17, Deep Sea Drilling Project. *Eos* **54**, 991–995.

BEBIEN, J. (1970) Etude pétrographique et géochimique des formations volcaniques devono-dinantiennes de l'extrémité SW du faisceau synclinal du Morvan. Thèse 3ème cycle, Nancy, 110 pp.

BEBIEN, J. (1971) Eléments nouveaux sur le volcanisme dévono-dinantien de l'extrémité SW du faisceau synclinal du Morvan. *C.R. Acad. Sci., Paris* **273** D, 466–468.

BROOKS, R. R., KAPLAN, I. R. and PETERSON, M. N. A. (1969) Trace element composition of Red Sea geothermal brine and interstitial water. In *Hot Brines and Recent Heavy Metal Deposits in the Red Sea*, DEGENS et ROSS, eds., Springer-Verlag, N.Y., pp. 180–203.

CANN, J. R. (1969) Spilites from the Carlsberg ridge, Indian Ocean. *J. Petrol.* **10**, 1–19.

CIVETTA, L., DE FINO, M., LA VOLPE, L. and LIRER, L. (1974) Geochemical trends in the alkali basaltic suites of the Assab range (Ethiopia). *Chem. Geol.* **13**, 149–163.

DELALOYE, M. et SAWATZKI, G. (1973) Géochronométrie des éléments volcaniques du flysch helvétique du synclinal de Thônes (Haute-Savoie, France). *Arch. Sci.* **28**, 1, 95–99.

ELLIS, A. J. and MAHON, W. A. J. (1964) Natural hydrothermal systems and experimental hot water: rock interactions. *Geochim. Cosmochim. Acta* **28**, 1323–1358.

ENGEL, C. G. and FISHER, R. L. (1975) Granitic to ultramafic rock complexes of the Indian Ocean ridge system; western Indian Ocean. *Geol. Soc. Amer. Bull.* **86**, 1553–1578.

FREY, F. A., BRYAN, W. B. and THOMPSON, G. (1974) Atlantic Ocean floor: Geochemistry and petrology of basalts from Legs 2 and 3 of the Deep-sea Drilling Project. *J. Geophys. Res.* **79**, 5507–5527.

FRUCHART, A. and PINTA, M. (1975) Geochemical occurrence of lithium in the subterranean waters of the Parisian Basin. *C.R. Acad. Sci., Paris* **281**, B, 2–3, 93–96.

HAJASH, A. (1974) An experimental investigation of high temperature seawater–basalt interactions. *Geol. Soc. Amer.*, abstracts with programs, **6**, 7, 771.

HAJASH, A. (1975) Hydrothermal processes along Mid-ocean Ridges: an experimental investigation. *Contr. Mineral Petrol.* **53/3**, 205–226.

HART, S. R. (1970) Chemical exchange between sea water and deep ocean basalts. *Earth Planet. Sci. Lett.* **9**, 269–279.

HART, S, R. and NALWALK, A. J. (1970) K. Rb, Cs and Sr relationships in submarine basalts from the Puerto Rice trench. *Geochim. Cosmochim. Acta* **34**, 145–155.

HEIER, K. S. and BILLINGS, G. K. (1970) Geochemistry of lithium, Chap. 3; Geochemistry of sodium; Chap. 11, *Handbook of Geochemistry*, WEDEPOHL, ed., Vol. 11–1; Springer-Verlag.

HENRICHS, S. and THOMPSON, G. (1976) The low-temperature weathering of oceanic basalts by seawater: 2 Trace element fluxes. *Geol. Soc. Amer., Abstr. with Programs,* **8**, 913.

HIRST, D. M. (1962) The geochemistry of modern sediments from the Gulf of Paria. II: The location and distribution of trace elements. *Geochim. Cosmochim. Acta* **26,** 1147–1188.

JEHL, V. (1975) Le métamorphisme et les fluides associés des roches océaniques de l'Atlantique Nord. Thèse de docteur-ingénieur, Univ. Nancy I, 242 pp.

JEHL, V., POTY, B. and WEISBROD, A. (1976) Hydrothermal metamorphism of the oceanic crust in the north atlantic ocean. *Amer. Geophys. Union,* Spring annual meeting program, p. 37.

JUTEAU, Th. (1971) Nouvelles données cartographiques, pétrographiques et chimiques sur le massif dévono–dinantien du Rabodeau (Vosges septentrionales). *Sci. de la Terre,* Fr., **16**, 45–106.

JUTEAU, Th. et ROCCI, G. (1965) Contribution à l'étude pétrographique du massif volcanique dévonien de Schirmeck (Bas-Rhin). *Bull. Serv. carte géol. Als. Lorr.* **18**, 3, 145–176.

JUTEAU, Th. et ROCCI, G. (1966) Etude chimique du massif volcanique dévonien de Schirmeck (Vosges septentrionales). *Sci. de la Terre* **11**, 68–104.

KHOD'KOVA, I. A. and GEMP, S. D. (1970) Occurrence of lithium, rubidium and cesium in the water and solid ejecta from mud volcanoes in the Kerch'-Tamam region. *Geochem. Int.,* **6**, 1040–1047.

LA ROCHE, H. de, ROCCI, G. et JUTEAU, Th. (1974) Essai de caractérisation chimique des associations spilitiques. In *Spilites and Spilitic Rocks,* AMSTUTZ, ed., Springer-Verlag, pp. 39–50.

LASSERRE, J. L. et LAVERNE, Ch. (1976) Le volcanisme tholéiitique de la zone du Versoyen (Alpes franco-italiennes): minéralogie, pétrographie et géochimie. Thèse 3ème cycle, Grenoble, 252 pp.

LE MAITRE, R. W. (1976) The chemical variability of some common igneous rocks. *J. Petrol.,* **17**, 4, 589–598.

LIVINGSTONE, D. A. (1963) Chemical composition of rivers and lakes. In *Data of Geochemistry,* FLEISCHER, ed., *U.S. Geol. Survey, Prof. Paper* 440–G.

MAKSIMOVIC, Z. and MAKSIMOVIC, S. (1968) Lithium, sodium and potassium in Alpine ultramafic rocks. In *Origin and Distribution of the Elements,* AHRENS, éd., Pergamon Press, pp. 761–776.

MAURIN, N. (1974) Principaux résultats expérimentaux sur la mise en solution d'éléments chimiques constitutifs d'un basalte pour différentes conditions de pression de vapeur d'eau et de température *Géol. alpine* **50**, 107–110.

MELSON, W. G. (1973) Basaltic glasses from the Deep Sea Drilling Project. Chemical characteristics, compositions of alteration products and fission track "ages". *Eos* **54**, 1011–1014.

MELSON, W. G. and THOMPSON, G. (1973) Glassy abyssal basalts. Atlantic sea floor near St. Paul's rocks: petrography and composition of secondary clay minerals. *Geol. Soc. Amer. Bull.* **84,** 703–716.

MERINO, E. (1975) Diagenesis in Tertiary sandstones from Kettleman North Dome, California. II: Interstitial solutions: distribution of aqueous species at 100°C and chemical relation to the diagenetic mineralogy. *Geochim. Cosmochim. Acta* **39**, 1629–1646.

MEVEL, C. (1975) Les "pillow-lavas" spilitiques des Massifs ophiolitiques du Chenaillet et des Gêts (Alpes françaises). Thèse 3ème cycle, Paris, 223 pp.

MOORE, J. G. (1966) Rate of palagonitization of submarine basalt adjacent to Hawaii. *U.S. Geol. Surv. Prof. Paper* 550–D, pp. 163–171.

MOTTL, M. J., CORR, R. F. and HOLLAND, H. D. (1974) Chemical exchange between sea water and mid-ocean ridge basalt during hydrothermal alteration; an experimental study. *Geol. Soc. Amer., Abstr. with Programs* **6**, 879–880.

MOTTL, M. J., CORR, R. F. and HOLLAND, H. D. (1975) Trace element content of the Reykjanes and Svartsengi thermal brines, Iceland. *Geol. Soc. Amer., Abstr. with Programs,* **7**, 1206–1207.

MUEHLENBACKS, .and CLAYTON, R. N. (1972) Oxygen isotope geochemistry of submarine greenstones. *Can. J. Earth Sci.* **9**, 471–478.

PARROT, J. F. et DELAUNE-MAYERE, M. (1974) Les terres d'ombre du Bassit (NW syrien) comparaison avec les termes similaires du Troodos (Chypre). *ORSTOM, ser. Géol.* **6**, 2, 147–160.

PARROT, J. F. et VATIN-PÉRIGNON, N. (1974) Répartition de quelques élements en trace dans les différentes roches éffusives de la région ophiolitique du NW syrien. *ORSTOM, Sér. Géol.* **6**, 2, 185–226.

PRINZ, M. (1968) Geochemistry of basaltic rocks: trace elements. In *Basalts,* POLDERVAART et HESS, eds., Wiley-Interscience, 271–323.

RONOV, A. B., MIGDISOV, A. A., VOSKRESENSKAYA, N. T. and KORZINA, G. A. (1970) Geochemistry of lithium in the sedimentary cycle. *Geochem. Int.* **1–2**, 75–102.

ROWE, J. J., FOURNIER, R. O. and MOREY, G. W. (1973) Chemical analysis of thermal waters in Yellowstone National Park, Wyoming, 1960–65. *U.S. Geol. Surv. Bull.,* **1303**, 1–31.

SAGON, J. P. (1969) Le volcanisme dinantien dans la partie orientale du bassin de Châteaulin: région de Saint-Martin-des-Prés, Allineuc (Côtes-du-Nord). *Bull. Soc. géol. France* **7**, 11, 561–573.

SAGON, J. P. (1976) Contribution à l'étude géologique de la partie orientale du bassin de Châteaulin (Massif Armoricain): stratigraphie, volcanisme, métamorphisme, tectonique. Thèse Doc. Etat Sci. Nat., Univ. Paris VI, 671 pp.

SHAW, D. M., DUPUY, C., FRATTA, M. and HELSEN, J. (1974) An application of factor analysis to basic volcanic rock geochemistry. *Bull. Volc.* **37–4**, 1070–1089.

SHAW, D. M., FRATTA, M., DUPUY, C. et VATIN–PÉRIGNON, N. (1974) Comportement de quelques éléments incompatibles dans les roches volcaniques à caractère tholéiitique, alcalin et spilitique. *Rev. Haute Auvergne*, **44**, 415–440.

SHAW, D. M. and MUYSSON, J. R. (in press) Granitophile trace elements and alteration in Leg 37 drill-core. *Initial Reports of the Deep Sea Drilling Project*, XXXVII.

SMITH, T. E. and JACKSON, T. A. (1974) Tertiary spilites and quartz keratophyres of the Wagwater Belt, Jamaica, West Indies. *Bull. Volc.* **37–4**, 870–890.

SOLER, E. (1973) L'association spilites–quartz keratophyres du SW de la péninsule ibérique. *Geol. en Mijnbouw* **52**, 277–288.

SPOONER, E. T. C. and FYFE, W. S. (1973) Sub-seafloor metamorphism, heat and mass transfer. *Contr. Mineral. Petrol.* **42**, 287–304.

TARDY, Y., KREMPP, G. et TRAUTH, N. (1972) Le lithium dans les minéraux argileux des sédiments et des sols. *Geochim. Cosmochim. Acta* **36**, 397–412.

THOMPSON, G. and MELSON, W. G. (1970) Boron contents of serpentinites and metabasalts in the oceanic crust: implications for the boron cycle in the oceans. *Earth Planet. Sci. Lett.* **8**, 61–65.

THOMPSON, G., BRYAN, W. B., FREY, F. A. and SUNG, C. M. (1974) Petrology and geochemistry of basalts and related rocks from sites 214, 215, 216, DSDP leg 22, Indian Ocean. *Initial Reports of the Deep Sea Drilling Projects*, VON DER BORCH and SCLATTER, eds., 459–468.

VALLANCE, T. G. (1960) Concerning spilites. *Proc. Linnean Soc. New South Wales* **85**, 1, 8–52.

VALLANCE, T. G. (1974) Spilitic degradation of a tholeiitic basalt. *J. Petrol.* **15**, 79–96.

VATIN-PÉRIGNON, N. et SHAW, D. M. (1972) Comportement du lithium et d'autres éléments en trace dans quelques coulées triasiques de l'Aiguille de Morges (Alpes françaises méridionales). Une contribution au problème de l'origine des roches spilitiques. *Géol. Alpine* **48**, 201–206.

VATIN-PÉRIGNON, N., AUMAITRE, R. et BUFFET, G. (1974) La spilitisation dans le Massif des Ecrins–Pelvoux: un cortège intrusif et éffusif dolérito–spilitique. *Géol. Alpine* **50**, 153–194.

VINE, J. D. (1975) Lithium in sediments and brines—how, why, and where to search? *U.S. Geol. Surv.* Open File Report 75–86, 1–14.

VOLFINGER, M. (1970) Partage de Na et Li entre sanidine, muscovite et solution hydrothermale à 600°C et 1000 bars. *C.R. Acad. Sci. Paris*, **271** D, 1345–1347.

WHITE, D. E. (1957a) Thermal waters of volcanic origin. *Geol. Soc. Amer. Bull.* **68**, 1637–1658.

WHITE, D. E, (1957b) Magmatic, connate and metamorphic waters. *Geol. Soc. Amer. Bull.* **68**, 1659–1682.

WHITE, D. E. (1963) Geothermal brine well: mile-deep drill hole may tap ore-bearing magmatic water and rocks undergoing metamorphism. *Science*, **139**, 919–922.

WHITE, D. E., HEM, J. D. and WARING, G. A. (1963) Data of geochemistry. Chap. F, Chemical composition of subsurface waters *U.S. Geol. Surv., Prof. Paper* 440–F.

Rare Earths in Nigerian Mesozoic Granites and Related Rocks

P. BOWDEN†, J. N. BENNETT†, J. E. WHITLEY‡ and A. B. MOYES‡

†*Department of Geology, University of St. Andrews, Fife, Scotland.*
‡*Scottish Universities Research and Reactor Centre, East Kilbride, Glasgow, Scotland.*

Abstract

Nine rare earths have been determined by Instrumental Neutron Activation Analysis (INAA) in syenites and related granitic rocks from the Nigerian younger granite (Mesozoic) Province. The majority of the granites contain dispersed U–Nb and vein-controlled Zn–Sn mineralization. It is inferred from a study of the chondrite-normalized rare-earth patterns that europium depletion in alkaline and sub-alkaline mineralized granites appears to be not just related to magmatic processes but also to hydrothermal changes induced during sub-solidus recrystallization.

Introduction

Until recently little attention has been paid to the possible modifications that can occur in granites during and after their cooling history. The importance of sub-solidus changes has been generally neglected although stable isotope studies have indicated the need for detailed re-examination of petrogenetic hypotheses (Forester and Taylor, 1977). The Nigerian "younger granite" province has long been regarded as a classical example of the development and evolution of magmatic granites but it is also well known that the same granites shed abundant cassiterite into the nearby alluvial deposits. In fact the Nigerian anorogenic granitic province is significantly mineralized with not only cassiterite, but also there are substantial quantities of zinc as well as minor lead, copper, rare earths, thorium, uranium, and niobium, etc. Mineralization is both dispersed and vein controlled.

In this paper we present the results of the determination of rare earths in a representative suite of rocks from some of the anorogenic ring complexes in Nigeria, and show that the derived chondrite-normalized rare-earth patterns for augite syenites can be simply explained by normal magmatic processes. However, we believe that the chondrite normalized rare-earth patterns for granites associated with mineralization can be interpreted as representing original magmatic abundances which have been modified by hydrothermal processes during sub-solidus cooling. Of particular interest to us is the magnitude and significance of the europium depletion in mineralized granites.

Rare-earth abundances

The absolute abundances of the rare earths in granites and related rocks, and all natural materials, as a function of atomic number show a characteristic saw-toothed pattern which is attributable to the favoured formation of even-numbered nuclides at the time of nucleosynthesis (Haskin and Haskin, 1968). Modifications of this pattern by natural processes are best seen by dividing the observed abundances by abundances representative of primordial matter. It is now common practice to discuss rare-earth abundances in terms of their chondrite-normalized patterns. In this paper the values used for chondritic abundances are those provided for the Leedey chondrite by Nakamura (1974).

The rare earths are a group of trace elements that are found both dispersed in common rock-forming minerals and as accessory minerals in acid igneous rocks. The fourteen naturally occurring rare earths are grouped together in the periodic table because of their similar chemical properties, but it is recognized that rare earths do not necessarily behave as a coherent group in geochemical processes (Herrmann, 1970). In granitic rocks the absolute abundances are greater than in basalts with some added variations depending on the tectonic setting (orogenic vs, anorogenic) and geochemistry (calc-alkaline vs, alkaline). Whilst in general chondrite-normalized rare-earth patterns generally exhibit smooth trends from the light to heavy group of rare earths (within experimental error), this trend may be interrupted by an enrichment or depletion of europium. This anomaly has been attributed to the ease with which europium can be reduced to its divalent form and subsequently removed from a magma by the separation of plagioclase feldspar (Philpotts and Schnetzler, 1968). Drake (1975) has suggested that the oxygen fugacity of the rock at the time of crystallization controls the relative distribution of Eu^{2+} and Eu^{3+} and therefore the size of the europium anomaly. Drake further argues that because of this control the magnitude of the europium anomaly provides a measure of the oxygen fugacity. More recently geochemical modelling using distribution coefficients between crystal and liquid have been introduced by Haskin (see this volume) although the credibility, basis, interpretation and actual values of the distribution coefficients assuming a simple crystal–liquid relationship have been questioned by Shaw (this volume). Furthermore, it has been assumed that the distribution of rare earths in igneous rocks is strictly controlled by magmatic processes such as crystal fractionation, partial melting, mixing of magmas, etc., and that the rare earths are inert to any mobilization at sub-solidus temperatures. Bowden and Whitley (1974) first hinted that rare-earth patterns in granites may be modified by hydrothermal overprinting. Thorpe *et al.* (1977) believe that any hydrothermal mobility of rare earths in the Skye granites was insignificant despite the overwhelming stable isotope evidence Forester and Taylor (1977) that these granites had extensively interacted with heated meteoric groundwaters. Some metasomatic mobility of light rare earths has been postulated by Wood *et al.* (1976) during zeolite facies metamorphism of some Icelandic Tertiary basalts. Recently Martin *et al.* (1978) have shown that all rare earths are mobile during metasomatic fenitization of quartzites. This means that hydrothermal transport of rare earths is a distinct possibility under certain circumstances. It also means that another factor causing variations in chondrite-normalized rare-earth patterns is rock–fluid interaction. Bearing these facts in mind it is hardly surprising that the interpretation of chondrite–normalized rare-earth patterns is currently an area of considerable activity and controversy.

Nigerian Mesozoic Granites

Over forty individual ring complexes, emplaced into Pan-African basement rocks, have been recognized in Nigeria (Fig. 1). They constitute the southern part of an extensive zone along the ninth meridian extending from the Niger Republic and have been attributed to successive periods of mid-plate magmatism and mineralization which migrated southwards from Ordovician to late Jurassic times (Bowden *et al.* 1976). The Triassic-Jurassic (younger granite) ring-complexes in Nigeria are circular or elliptical in outline, 10–25 km in diameter

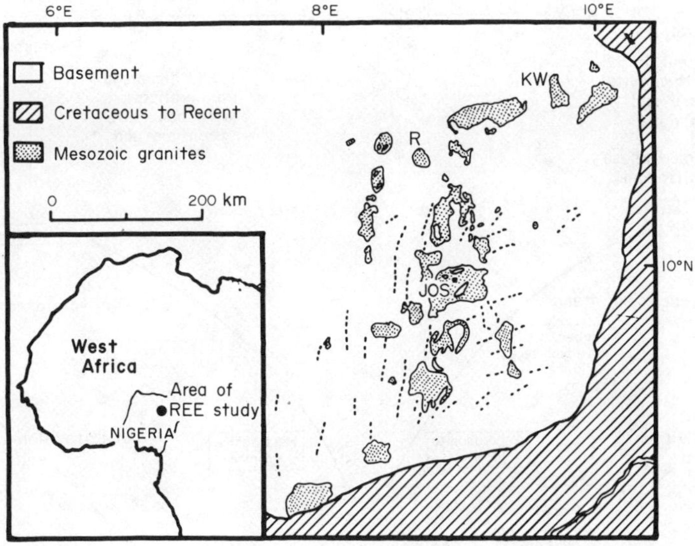

FIG. 1. Geological sketch map of the Nigerian Mesozoic Granite Province. Individual ring complexes sampled in this study are identified as follows: KW–Kila Warji, R–Ririwai (Liruei), and Jos.

with concentrically arranged ring-dyke intrusions of augite syenites, fayalite granites and peralkaline granites. Hastingsite granites and related biotite granites may also form an associated series of concentric intrusions, but some biotite granites may exhibit shallower dipping contacts resembling late-stage stocks or cupolas (Bowden and Turner, 1974). It is primarily within the biotite granites that dispersed deposits of columbite, xenotime, monazite, thorite and some cassiterite occurs; and vein deposits of sphalerite, cassiterite, base metal sulphides, monazite, uraninite are found (Kinnaird, 1977). Some minor ore mineralization has also been recorded in all other rock types including the "basement" which overlie cupolas of biotite granite. As well as this dominant phase of sub-alkaline granite-related mineralization the peralkaline granites show a dispersed mineralization associated with recrystallization and formation of sub-solidus albite, microcline, quartz and alkali amphibole (riebeckite). They may contain pyrochlore, cryolite and fluorite, and in addition an enrichment in many other LIL elements including the so-called "incompatibles".

Although the origin of the Nigerian Younger Granites is still in dispute, research by Martin and Bowden (in preparation) on the alkali feldspars indicates that a common magmatic liquid may have been one which crystallized fayalite, iron-rich CPX, alkali feldspar and some quartz at low pressure. The products of crystallization of such a liquid are the oversaturated syenites and related fayalite granites. The other rock types, in particular the riebeckite granites and the biotite granites, exhibit textures and mineralogy which suggest that they may have recrystallized at sub-solidus temperatures. One cause of these features could be the interaction of the granites with hydrothermal fluids which were responsible for the development of U–Nb and zinc-enriched tin mineralization. A schematic petrogenetic evolution diagram for the Nigerian anorogenic province is presented in Fig. 2.

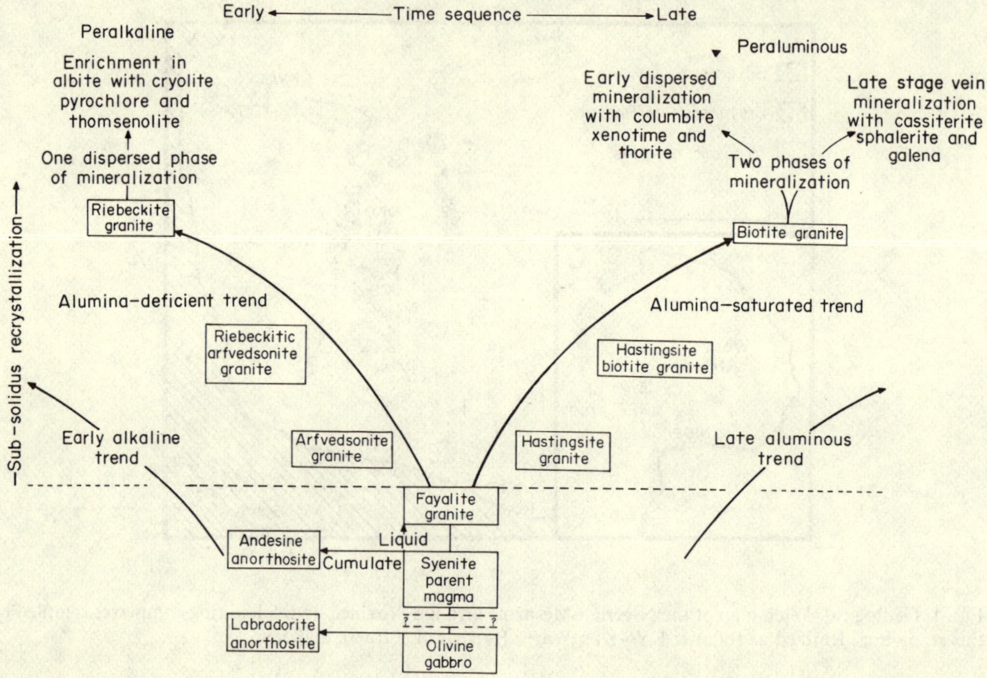

FIG. 2. Schematic petrogenetic evolution diagram for the origin of Nigerian Mesozoic granites (after Bowden and Kinnaird, 1978).

Rare Earths in Nigerian Granites and Related Rocks

In this paper we present the results of the determination of rare earths on a syenite and related quartz syenite porphyry: riebeckite granites, biotite granites and a mineralized replacement vein with its associated wall rock alteration. Although some of the samples were collected from differently located ring complexes in Nigeria it is accepted that the overall magmatic and post-magmatic processes were similar and repetitive at each centre. These samples are believed to be representative of the petrological and geochemical variations in the Nigerian Mesozoic anorogenic subvolcanic province (Bowden and Turner, 1974).

Method and Results

Instrumental neutron activation analysis of powdered whole-rock samples was carried out at the S.U.R.R.C., East Kilbride, by a technique similar to that described by Gordon *et al.* (1968). Whilst full details of the non-destructive method which determines nine rare earths can be obtained from the third author (Whitley, 1974), salient features relevant to its application to the Nigerian granites are discussed below. Acquisition of detectors of improved resolution and a small computer for data processing have led to improved precision in a continually developing rare-earth analysis programme. The main feature of results presented are improved figures for the riebeckite granites where interferences due to relatively high tantalum contents were previously not so well resolved.

Inconvenient half-lives or excessive interferences limit the number of rare earths that can be conveniently determined by instrumental activation analysis to nine or ten elements but this is sufficient to provide information about the relative levels of light, middle and heavy rare earths, and includes europium which is particularly important because of the behaviour of this element in its reduced state. A number of other trace elements (Cs, Hf, Ta, Th, Sc and Co) can also readily be determined from the gamma-ray spectra that are produced.

Samples and standards (prepared from solutions of spectrographic grade metals or oxides previously ignited to 1000°C) were irradiated for 3 hours in a flux of 3.6×10^{12} n cm^{-2} s^{-1} in the central vertical stringer of the reactor at East Kilbride. The location of irradiation is situated between twin fuel tanks and there is no significant variation in flux density over the volume in which the samples were contained.

Samples were counted several times between 5 days and 2 months after irradiation, the optimum times for the determination of each element being suggested by Gordon *et al.* (1968). Whilst all samples were analysed using a 25-cm^3 Ge (Li) detector with a resolution of 4.2 keV and a 0.5-cm^3 planar detector with a resolution of 520 eV, analysis of the riebeckite granites was repeated with the improved detectors: an 80-cm^3 Ge(Li) detector with a resolution of 2.2 keV and 0.5-cm^3 planar detector of 520 eV resolution. This repeat analysis confirmed results for most of the elements determined, and eliminated irregularities attributable to incomplete resolution of interferences at low counting rates, such as ^{160}Tb at 962.6 and 966.4 keV and ^{152}Eu at 964 keV and ^{141}Ce at 145.4 keV and ^{59}Fe at 143 keV. The use of a computer simplified the calculation of results with reference to more than one photopeak per element where this was feasible. The computer was also used for calculation of absolute concentrations and calcomp plotting of chondrite normalized abundances, as described by Moyes and Whitley (1977).

The accuracy of the method was assessed by processing duplicate samples of the standard rock G2 at the same time as the Nigerian samples. The results obtained for La, Ce, Sm and Eu were within 5% of the 1972 values for this rock (Flanagan, 1973). The results for Nd and Gd were within 30% of the 1972 values, the difference being associated with the poor counting statistics due to the low levels of these elements in the standard rock. The remaining heavy rare earths were below the limit of detection for G2, so for these elements comment can only be offered on the precision attained in the Nigerian rocks.

For the Nigerian rocks the precision of the determination of La, Ce, Sm and Eu, derived from counting statistics, was always better than ±5% and frequently approached ±1%. For

TABLE 1. RARE-EARTH VALUES FOR SYENITES FROM KILA WARJI, AND A MINERALIZED REPLACEMENT VEIN FROM RIRIWAI (LIRUEI), NIGERIAN MESOZOIC PROVINCE

Rare-earth elements	Augite syenite KW6	Quartz-syenite porphyry KW15	Reddened wall rock (B)	Central vein + greisen (C)
La	131±2.9	257±5.6	55.2±0.7	89.8±0.9
Ce	202±14.7	450±53.2	93.3±4.0	190±8
Nd	63.8±16.0	304±37.9	33.2±2.8	49.9±4.2
Sm	18.9±0.21	66.5±1.3	8.0±1.0	15.2±0.4
Eu	4.84±0.13	7.10±0.11	0.18±0.04	0.24±0.04
Gd	13.1±1.4	54.8±2.3	7.5±2.1	17.6±4.0
Tb	2.35±0.1	6.0±0.1	1.6±0.1	3.4±0.2
Yb	4.79±0.42	14.4±1.2	6.1±0.7	17.3±1.6
Lu	0.86±0.02	2.55±0.1	1.01±0.12	2.84±0.13

Analysts: A. B. Moyes and J. E. Whitley

the remaining elements the precision was generally $\pm 20\%$ or less, except for Nd in samples where the concentrations were particularly low, and the precision decreased to 40%. The rare-earth concentrations found in the Nigerian syenites and granites are given in Table 1, and chondrite-normalized abundances are those provided for the average of ten chondrites by Nakamura (1974).

Discussion

The chondrite-normalized rare-earth patterns indicate there is considerable rare-earth enrichment, both in the light and heavy end of the spectra. Compared to the oversaturated syenite (KW6) from Kila Warji there are significant Eu anomalies in all the granites, the magnitude of which appears to increase in the mineralized granite and replacement vein from Ririwai (Liruei).

Europium Depletion and Crystal Fractionation

It is common to assume for basic rocks that the crystallization of plagioclase effectively causes europium depletion in the fractionating liquids. Because there is a marked europium depletion in the Nigerian riebeckite granites and to a lesser extent in the biotite granites we must seriously consider whether the Nigerian granites are evolved from a more basic parent whose plagioclase removal cause the development of peralkalinity trends in later fractionates. This concept known as the "plagioclase effect" was first discussed by Bowen (1917) and effectively demands that plagioclase cumulates should be prevalent as xenoliths in the Nigerian Mesozoic granites. A recent discovery by Wright (1975) of a plagioclase xenolith (An 58–65) from the Rafin Jaki porphyry, a final phase composite xenolithic ring dyke in the Jos–Bukuru cycle lends some support to the "plagioclase effect" although it might be expected that examples of fractionating liquids of intermediate composition should occur within the younger granite province. The porphyritic syenite (KW6) and the related quartz–syenite porphyry provides a possible insight into the magmatic processes operating in the Nigerian anorogenic province. This rock type (KW6) was chosen because it may provide an example of the parental magma for the province, and because it is distinctive through its content of labradorite–andesine crystals. The syenite has a characteristic porphyritic texture in which well-formed, partly rounded large (5 mm) plagioclase crystals (near An_{50}) rimmed by turbid alkali feldspar occur scattered throughout the rock (up to 9% vol). The intervening areas consists of well-formed alkali feldspars (65% by volume, 0.1–3 mm) as clusters or individuals with small regions of turbid anhedral alkali feldspars, minor olivine and more abundant pale-brown ferro-augite which is often rimmed by brown hornblende and itself rimmed by blue-green amphibole. A little granophyre may fill the interstices between the larger alkali feldspars but sometimes adjacent crystals have sutured interfaces or smooth concave–convex junctions. The crystals, often rhombic in section parallel to 010, are sometimes zoned with a perthite core rich in potash feldspar; veins and rims of turbidity are common. The core sometimes contains inclusions of augite, iron ore and apatite which may be arranged along the edges of perthite lamellae. Accessory minerals are iron oxides and apatite and occasionally zircon which forms needle-like crystals within the interstitial granophyre.

The large-sized plagioclase crystals and often rounded shape suggest that they formed from a slowly crystallizing magma and may have been carried to high levels of the crust by

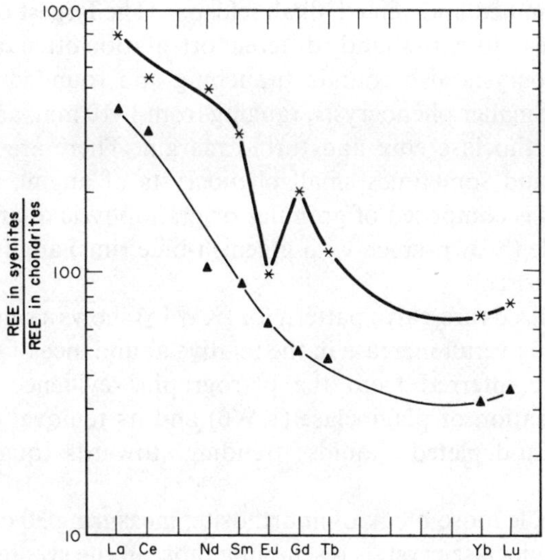

FIG. 3. Chondrite-normalized rare-earth patterns for syenite KW6 (closed triangles) and for quartz-syenite porphyry KW15 (closed stars) from the Kila Warji complex. Data presented in Table 1.

the magma. The smaller alkali feldspar phenocrysts probably formed subsequently in high-level magma reservoirs. The large plagioclase crystals could have originated in several different ways.

1. Crystallization from a hypothetical basaltic magma of which the syenite would be a differentiate.
2. The plagioclase crystals could be unmelted relicts of a source rock which is dioritic or monzonitic in composition, from which the syenite magma was formed through partial melting.
3. Crystallization from a syenite or monzonite–syenite magma.

From the chondrite-normalized rare-earth patterns for the augite syenite (KW6) it can be seen that there is no negative or positive Eu anomaly (Fig. 3). This pattern is a feature which is consistent with syenites from other ring complexes in Nigeria (Aleksiyev, 1970) and recorded in other alkaline provinces (Balashov, 1971; Khomyakov, 1972). It indicates on a magmatic model that the syenite cannot have been fractionated to any great extent from a more basic parent otherwise a negative Eu anomaly might be expected. A positive Eu anomaly could have indicated that the syenites represent cumulates developed from a fayalite granite magma. From the chondrite-normalized rare-earth patterns alone one is led to the assumption that syenitic liquids containing fayalite, ferro-augite–hedenbergite alkali feldspar and minor quartz are the parental unfractionated syenite or monzonite–syenite magmas in the Nigerian Province.

The presence under the microscope of a small plagioclase phenocryst in the chilled margins to the porphyritic syenite shows that the syenite liquid was still able to precipitate early plagioclase. The rare-earth evidence and the petrographic evidence supports the concept of a parental syenitic magma.

The quartz–syenite porphyry (KW15) is intermediate between granite and syenite in composition, and is conspicuous by its lack of plagioclase, but contains abundant phenocrysts of turbid rimmed and veined alkali feldspar. The largest of these are 1.5 cm in length and are composite units with slightly different orientation often separated by vein-like quartz. The large phenocrysts also contain branching and rounded inclusions of zoned augite and hornblende. Smaller phenocrysts, ranging from 1–10 mm, are rounded and often zoned with a clear anorthoclase core and turbid margin. There are occasional rounded phenocrysts of quartz and sometimes small phenocrysts of augite, rimmed by aegirine augite. The groundmass is composed of granular or granophyric quartz and cloudy alkali feldspar with hornblende (brown-green with greenish-blue rims) augite and a little biotite, iron oxide, allanite and zircon.

The chondrite-normalized rare-earth pattern for (KW15) shows a small but significant Eu depletion as well as some overall increase in the relative abundance of the other rare earths (Fig. 3). It is therefore inferred from the petrographic evidence that the effects of fractionational crystallization of plagioclase (KW6) and its removal (KW15) could yield rare-earth enriched, Eu-depleted liquids trending towards quartz syenite–granite compositions.

The discovery of a single loose block of anorthosite, measuring 30 cm across in the Kila Warji complex, with plagioclase crystals resembling those in the syenite is significant, since sub-volcanic intrusions of anorthosite occur in association with syenite and peralkaline granites in Niger (Karche and Moreau, 1977).

Although the "plagioclase effect" depletes the fractionating liquids in aluminium and calcium, the generation of peralkaline silicic sodic liquids needs the removal of potash in the crystallizing phenocrysts. Bailey and Schairer (1964) suggested an "orthoclase effect" for the process, whereby separation of alkali feldspar from a slightly alumina-deficient liquid may fractionate alumina and potash producing strongly peralkaline sodic residual liquids. Since experimental evidence (Thompson and MacKenzie, 1967) fully justifies this concept, it means that peralkaline granites may be derived by fractional crystallization of a parent magma which was slightly peralkaline or became peralkaline through the operation of a plagioclase effect. It is possible that europium depletion in fractionating granitic liquids could be induced by europium removal in crystallizing alkali feldspars from syenite parental liquids. This means that europium depletion in alkaline and sub-alkaline granitic rocks may not just represent a historical record of plagioclase removal from a more basic liquid but indicates the operation of the orthoclase effect in intermediate to silicic parents.

Alkali feldspars are not the only minerals which affect rare-earth partitioning between crystals and liquid. For example, Nagasawa and Schnetzler (1971) found that crystallization of hornblende, zircon and apatite in acidic magmas could also remove substantial quantities of the rare earths from the liquid. Balashov (1972) was able to show that partition coefficients for the light rare earths were considerably lower than for the heavy rare earths. Fractionation of mafic minerals would therefore result in a concentration of light rare earth in the remaining liquid, which would build up to such an extent that crystallization of light REE accessory minerals would occur restoring the balance between the rare-earth spectrum. Further fractionation could then result in a build up of heavy rare earths. Some effects of these processes may be evident in the chondrite-normalized rare-earth patterns for the Nigerian province.

Europium Depletion and Hydrothermal Processes

An alternative interpretation of chondrite-normalized rare-earth patterns in mineralized granitic rocks is that the distribution of the rare earths is in part controlled by migrating hydrothermal fluidş derived either from the same magmatic source or from outside the magmatic system. Generally speaking, research groups in America and Europe are still sceptical about the validity of metasomatic (re)distribution of rare earths. In contrast some workers in East European countries and Russia remain unconvinced that crystal–liquid equilibria can be attributed as the sole control of rare-earth variations. They attach considerable significance to the possibility of the transport of rare earths as alkali fluoride complexes (Mineyev, 1963; Aleksiyev, 1970). It is pertinent at this stage to consider granites associated with mineralisation discussed earlier and referred to in Fig. 2.

In Nigeria two styles of mineralization have been recognized (Bowden and Kinnaird, 1978). One, a dispersed mineralization in which the ore minerals are found scattered throughout the host granite. Under the microscope there is substantial evidence for sub-solidus recrystallization of felsic and mafic minerals. Two alkali feldspars show a greater degree of ordering, and texturally the granites are very variable, but often saccharoidal. Such textured granites have often been referred to as "apogranites" (Tugarinov *et al.*, 1968). Both the peralkaline riebeckite granites and the sub-alkaline biotite granites show evidence to different degrees of the influence of dispersed U–Nb mineralization and also contain somewhat different ore minerals dependent on the alkalinity of the granite and the composition of the hydrothermal fluid. Biotite granites which show evidence of dispersed mineralization also frequently contain later vein-controlled zinc–tin mineralization with minor tungsten. These veins sometimes coalesce as lodes and are concentrated near the roof zones of the intrusions. The detailed petrography mineralogy and major element geochemistry of these rocks has been presented elsewhere (Buchanan *et al.*, 1971; Kinnaird, 1977; Bowden and Turner, 1974). In the Ririwai (Liruei) complex the variably-textured biotite (apo)granite shows good evidence for dispersed mineralization and is cut by a later east–west group of replacement veins coalescing to form the Ririwai lode. Chondrite-

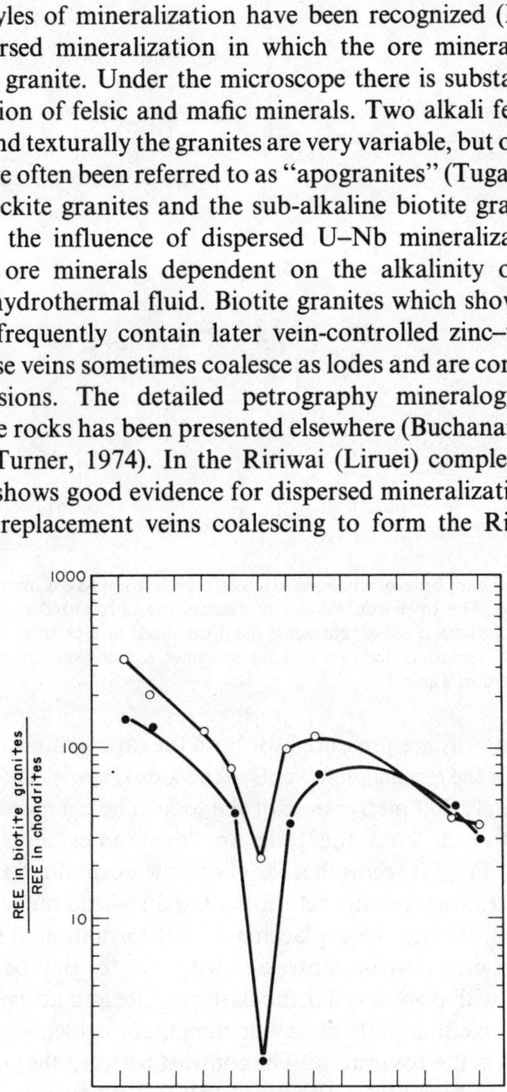

FIG. 4. Chondrite-normalized rare-earth patterns for the variably textured biotite (apo)granite (closed circles) from the Ririwai (Liruei) complex, and for the coarse-grained biotite granite (open circles) from the Jos complex. Data taken from Bowden *et al.* (1976b, Table 4, analyses 4 and 6).

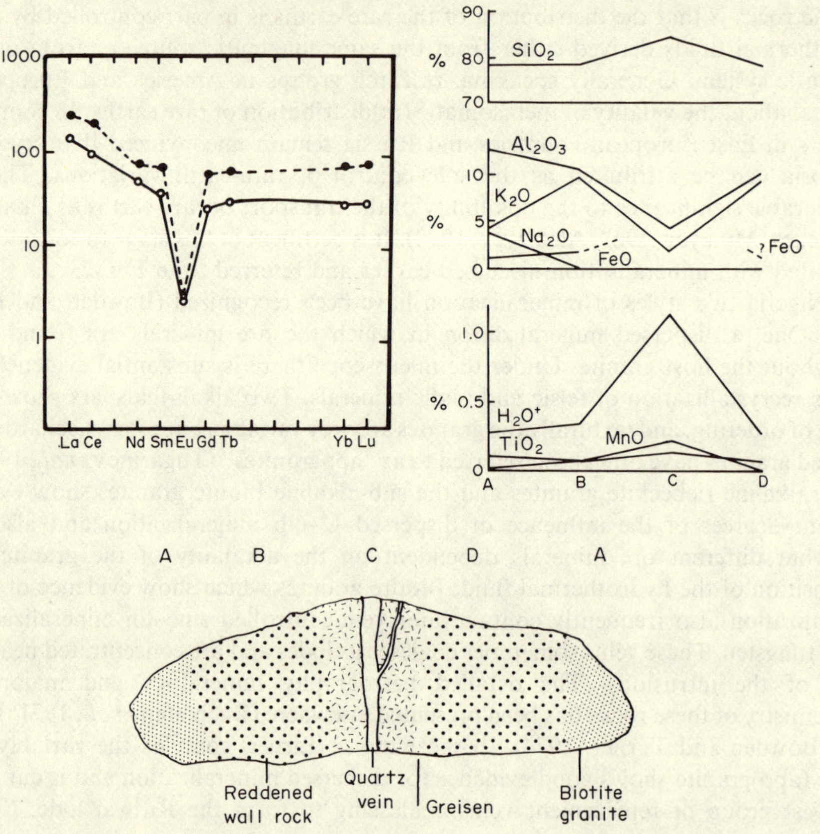

FIG. 5. Chemical variations and chondrite-normalized rare-earth patterns across a mineralized replacement vein from the Ririwai (Liruei) lode. The open-circle curve represents the normalized rare-earth variations in the reddened wall rock (B). The closed-circle curve represents the normalized rare-earth variations in the central vein plus greisen halo (C). Chemical variation diagram and the specimen sketch taken from Bowden and Kinnaird (1978); rare-earth data presented in Table 1.

normalized rare-earth patterns are presented for both the (apo)granite (Fig. 4) and a north-south section across one of the replacement veins in the lode (Fig. 5). The biotite (apo)granite was collected approximately 200 metres east of the lode. The chemical variations across a mineralized replacement vein from the lode are also indicated in Fig. 5. From an examination of Fig. 4 and Fig. 5 it seems that the chondrite normalized rare earth patterns in mineralized biotite granites were mainly set during the dispersed mineralization phase and enhanced, apart from Eu, during the replacement vein formation. These two chondrite-normalized rare-earth patterns may be contrasted with that for the Jos biotite granite (Fig. 4), a coarse-grained rock with disordered orthoclase perthite and no evidence of apogranite textural recrystallization. Neither is there any documentary evidence of dispersed or vein-controlled mineralization in the Jos granite. The contrast between the chondrite-normalized rare-earth patterns is most dramatic with regard to the Eu anomaly. In our opinion the increased Eu anomaly in the mineralized granite and its replacement vein is due to hydrothermal overprinting of an original rare-earth magmatic distribution. The chondrite-normalized rare-earth pattern for the Jos biotite granite may represent an approximation to

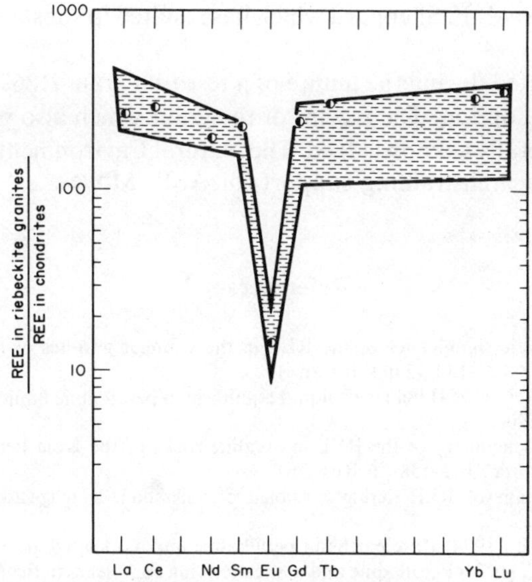

FIG. 6. Chondrite-normalized rare-earth patterns for riebeckite (apo)granites. Shaded region covers the variations in normalized rare-earth abundances for the Ririwai (Liruei) complex. Half-closed circles are the individual normalized rare-earth values for a riebeckite (apo)granite from the Ririwai complex. Data taken from Bowden *et al.* (1976b, Table 4, analyses 1–3).

the magmatic rare-earth distribution achieved through crystal fractionation from a syeno-monzonite parental magma.

In Fig. 6 we present the chondrite-normalized rare-earth patterns for riebeckite (apo)granites that show good textural and mineralogical evidence for sub-solidus recrystallization and mineralization. We are well aware that the peralkaline granites have been regarded as the end products of crystal fractionation from a syenitic parental magma. Earlier we demonstrated evidence to show a characteristic magmatic chondrite-normalized rare-earth abundance patterns in syenite, quartz-syenite and related rocks. We believe that the dispersed mineralization phase overprinted the earlier magmatic rare-earth abundances. Thus for mineralized granites the principal difference between magmatic and hydrothermal patterns is the extent of the Eu anomaly. Some modification to the proportion of other rare earths is also evident.

We conclude that europium depletion in alkaline and sub-alkaline mineralized granitic rocks appears to be not just related to magmatic processes, especially crystal fractionation, but may also be due to hydrothermal changes induced during sub-solidus recrystallization at successively lower temperatures. The important inference is that most rare-earth elements, except for example Eu, are mobile at sub-solidus temperatures in hydrothermal mineralizing fluids.

Acknowledgements

The authors thank Professor E. K. Walton at St. Andrews and Professor H. W. Wilson at East Kilbride for their interest and encouragement. Valuable assistance and discussions have been provided by Professor R. F. Martin of McGill University. Important constructive criticisms were given by S. M. F. Sheppard which has resulted in substantial improvement of this paper.

This work was completed during the tenure of a research grant R2679 from the Overseas Development Ministry, London. The receipt of the grant which also provided support for one of us (J.N.B.) is gratefully acknowledged. The Natural Environmental Research Council are thanked for their research training support for A. B. Moyes.

References

ALEKSIYEV, E. K. (1970) Genetic significance of the REE in the younger granites of northern Nigeria and the Cameroons. *Geochem. Int.* 127–132. (English trans.)

BAILEY, D. K. and SCHAIRER, J. F. (1964) Feldspar–liquid equilibria in peralkaline liquids— the orthoclase effect. *Amer. J. Sci.* **262**, 1198–1206.

BALASHOV, YU. A. (1971) Geochemistry of the REE in alkaline rocks of the Kola Peninsula. *Geokhim. Petrol. Mineral. shchelochnykh Porod.* 142–158 (in Russian).

BALASHOV, Y. A. (1972) Partition of REE during fractional crystallisation of magmatic melts. *Geochem. Int.* **9**, 305–320.

BOWDEN, P. and WHITLEY, J. E. (1974) Rare earths in peralkaline and associated granites. *Lithos,* **7**, 9–21.

BOWDEN, P. and TURNER, D. C. (1974) Peralkaline and associated ring complexes in the Nigerian–Niger Province, West Africa. In *The Alkaline Rocks* H. SØRENSEN, ed., Wiley, 330–351.

BOWDEN, P., VAN BREEMEN, O., HUTCHINSON, J, and TURNER, D. C. (1976a). Palaeozoic and Mesozoic age trends for some ring complexes in Niger and Nigeria. *Nature (Lond.)* **259**, 297–299.

BOWDEN, P., WHITLEY, J. E. and VAN BREEMEN, O. (1976b) Geochemical studies of the younger granites of northern Nigeria. In *Geology of Nigeria,* C. A. KOGBE, ed., 177–193.

BOWDEN, P. and KINNAIRD, J. A. (1978) The Younger Granites of Nigeria—a zinc-rich tin province. *Trans. Instn. Min. Metall.* (Sect. B Appl. Earth Sci.) (in press).

BOWEN, N. L. (1917) The problem of the anorthosites. *J. Geol.* **25**, 209–243.

BUCHANAN, M. S., MACLEOD, W. N., TURNER, D. C., BERRIDGE, N. G. and BLACK, R. (1971) *The Geology of the Jos Plateau,* Vol. 2, *Younger Granite Complexes. Bull. Geol. Surv. Nigeria,* **32**, 160 pp.

DRAKE, M. J. (1975) The oxidation of Eu as an indicator of oxygen fugacity. *Geochem. Cosmochim. Acta* **39**, 55–64.

FORESTER, R. W. and TAYLOR, H. P., Jr. (1977) $^{18}O/^{16}O$, D/H, and $^{13}C/^{12}C$ studies of the Tertiary Igneous Complex of Skye, Scotland. *Amer. J. Sci.* **277**, 136–177.

GORDON, G. E., RANDLE, K., GOLES, G. G., CORLISS, J. B., BEESON, M. H. and OXLEY, S. (1968) Instrumental activation analysis of standard rocks with high resolution γ-ray detectors. *Geochim. Cosmochim. Acta* **32**, 369–396.

HASKIN, L. A. and HASKIN, M. A. (1968) Relative and absolute terrestrial abundances of the REE. In *Origin and Distribution of the Elements* L. H. AHRENS, ed., Pergamon Press, 889–912.

HASKIN, L. A. (this volume) On rare earth element behaviour in igneous rocks. In *Origin and Distribution of the Elements.* Pergamon Press, 175–189.

HERRMANN, A. G. (1970) Yttrium and the lanthanides. In *Handbook of Geochemistry* K. H. WEDEPOHL, ed., Springer, Vol. II–2, chap. 39.

KARCHE, J. P. and MOREAU, C. (1977) Note préliminaire sur le massif subvolcanique à structure annulaire d'Abontorok (Aïr-Niger). *C.R. Acad. Sci., Paris* **284**, 1259–1262.

KHOMYAKOV, A. P. (1972) REE in rocks and accessory minerals of the Burpala alkaline complex. *Izv. Akad. Nauk SSSR. Ser. Geol.,* 94–101 (in Russian).

KINNAIRD, J. A. (1977) A geological investigation of some Nigerian Jurassic granites and their mineralisation. Thesis for M.Sc. (University of St. Andrews), Vol. 1, 172 pp.

MARTIN, R. F., WHITLEY, J. E. and WOOLLEY, A. R. (in press) An investigation of rare earth mobility: fenitized quartzites, Borralan complex, NW Scotland. *Contr. Mineral. Petrol.*

MINEYEV, D. A. (1963) Geochemical Differentiation of the rare earth elements. *Geochemistry* **12**, 1129–1150.

MOYES, A. B. and WHITLEY, J. E. (1977) Determination of rare earth elements in igneous rocks by neutron activation analysis with group separation. Scottish Universities Research and Reactor Centre, Report no. 59/77.

NAGASAWA, H. and SCHNETZLER, C. C. (1971) Partitioning of rare earths, alkali and alkaline-earth elements between phenocrysts and acidic magma. *Geochim. Cosmochim. Acta* **37**, 2227–2242.

NAKAMURA, N. (1974) Determination of REE, Ba, Fe, Mg, Na and K in carbonaceous and ordinary chondrites. *Geochim. Cosmochim. Acta* **38**, 757–775.

PHILPOTTS, J. A. and SCHNETZLER, C. C. (1968) Eu anomalies and the genesis of basalt. *Chem. Geol.* **3**, 5–13.

SHAW, D. M. (this volume) Trace element melting modes. In *Origin and Distribution of the Elements,* Pergamon Press, 577–586.

THOMPSON, R. N. and MACKENZIE, W. S. (1967) Feldspar-liquid equilibria in peralkaline acid liquids: an experimental study. *Amer. J. Sci.* **265**, 714–734.

THORPE, R. S., POTTS, P. J. and SARRE, M. B. (1977) Rare earth evidence concerning the origin of granites of the Isle of Skye, northwest Scotland. *Earth Planet. Sci. Lett.* **36**, 111–120.

TUGARINOV, A. I., PAVLENKO, A. S. and KOVALENKO, V. I. (1968) Geochemical evidence on the origin of apogranites. *Geochem. Int.* **12**, 1156–1171.

WHITLEY, J. E. (1974) Determination of rare earths and other trace elements in igneous rocks by instrumental activation analysis. Scottish Universities Research and Reactor Centre, Report no. 51/74.

WOOD, D. A., GIBSON, I. L. and THOMPSON, R. N. (1976) Element mobility during zeolite facies metamorphism of the Tertiary basalts of eastern Iceland. *Contr. Mineral. Petrol.* **55**, 241–254.

WRIGHT, J. B. (1975) Anorthosite—first occurrence in Nigeria and relevance to the younger granites. *Mineral. Mag.* **40**, 193–196.

Apport de la Géochimie—Majeurs—À la Reconstitution Antémétamorphique de la Série du Bas-Limousin (Massif Central, France)

P. L. GUILLOT,† M. TEGYEY,‡ C. ALSAC,‡ J. P. FLOC'H,§ J. GROLIER† and D. SANTALLIER§

†Laboratoire de Pétrologie, Université d'Orléans, 45045–Orléans Cédex, France.
‡B. R. G. M. Service géologique national, 45018–Orléans Cédex, France.
§Laboratoire de Géologie, Université de Limoges, France.

Abstract

Within metamorphic formations, geochemistry is often the only method of determining the origin of rocks whose initial characters have more or less completely disappeared through recrystallization. In the South-Limousin metamorphic basement, a geochemical study has provided us with new data for the discussion of this problem for common metamorphic rocks such as leptinites, gneiss, eclogites.

Three main groups have been defined in the series: Dronne's Group at the bottom (phyllites with basic bodies), Bas-Limousin Group (leptinites, gneiss, metagraywackes and phyllites, basic rocks mainly eclogites and amphibolites), Genis Group at the top (phyllites, volcanic acid and basic rocks).

A poliphased metamorphism, with an important Barrovian phase, took place during the Devonian.

For this geochemical study, diagrammatic representations—de la Roche and Miyashiro diagrams—have been used. The origins of basic rocks, leptinites and para-rocks (gneiss, metagraywackes, phyllites) are discussed successively.

Basic rocks as a whole have an abyssal tholeiitic chemism; most of them are very depleted in potassium; some with calco-magnesian composition represent anorthite and/or olivine cumulates; some are sodium-rich and this tendency is specially important in basic formations within the chlorite zone.

Para-rocks are derived from: volcanoclastic deposits of rhyo-dacitic composition (Grès de Thiviers and Quartzites de Payzac formations), detrital sediments mainly of graywacke compositions (gneiss) and shales.

Leptinites, in general, show igneous-rock compositions with, however, a higher Si content than in the calco-alkalin suite: rhyolitic to kèratophyric (Vergonzac), granitic with potassic tendency (Aubazine) or arkosic tendency (Dronne), granitic to grano-dioritic (St-Yrieix), trondhjemitic (Albussac).

It should be emphasized that this reconstitution is in complete harmony with geological and petrographical data.

Dans l'étude des socles anciens, une étape essentielle est la mise en évidence de la lithologie prémétamorphique. Hormis de rares cas où des textures reliques indiquent de façon claire la nature originelle des roches, seule l'étude géochimique apporte des éléments de solution à ce problème. Elle mène ainsi à une meilleure connaissance de la croûte (océanique ou continentale) à une époque donnée.

Dans cette optique, nous présentons les résultats d'une étude géochimique sur la série métamorphique du Bas-Limousin, l'une des mieux connues du Massif Central Français et pour laquelle on dispose à l'heure actuelle de données cartographiques, structurales et pétrographiques multiples.

1. Le Cadre Géologique d'Ensemble

Le Bas-Limousin métamorphique s'étend sur environ 200 km à la bordure sud-ouest du

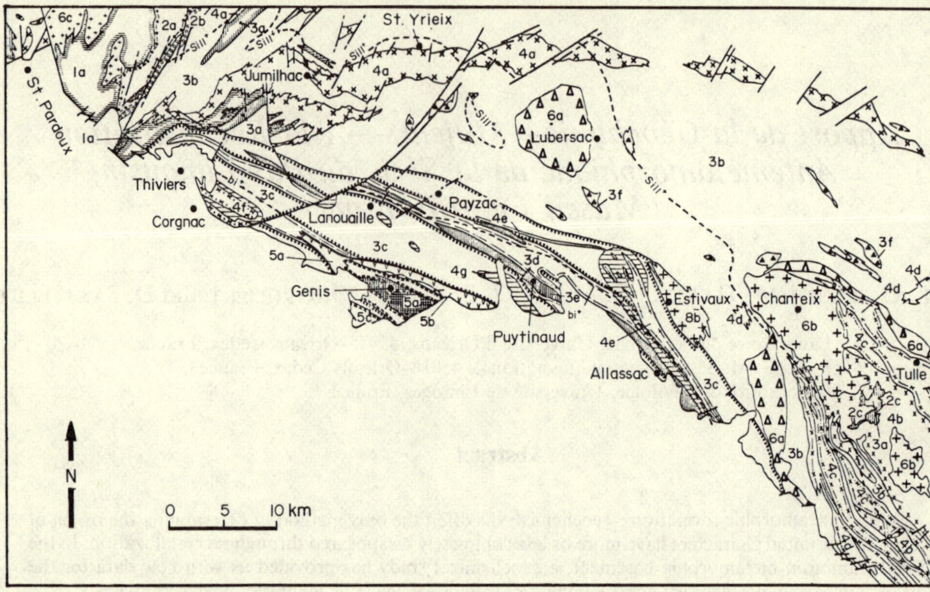

FIG. 1. Carte litho-stratigraphique du Bas-Limousin (d'après les levés de Boissonnas, Floc'h, Grolier, Guillot, Santallier et Talbert).

1. (a). Micaschistes de la Dronne.
2. (a). Métagranites 2(b) Méta-arkoses (Arc de la Dronne), 2(c) Orthogneiss de Chameyrat et Mulatet.
3. Groupe Bas-Limousin. 3(a) Gneiss pélitiques micacés, 3(b) Gneiss plagioclasiques, 3(c) "Grès de Thiviers–Quartzites de Payzac", 3(d) Schistes sériciteux et métagraywackes, 3(e)Tufs basiques (Formation de Puytinaud), 3(f): Eclogites et amphibolites.
4. (a). Leptynites de l'Arc de St Yrieix. 4(b) Leptynites d'Albussac 4(c) Leptynites d'Aubazine 4(d) Leptynites de Vergonzac 4(e) Granite du Saut du Saumon, 4(f) Granite de Corgnac.
5. Groupe de Genis 5(a) Méta-ignimbrites, 5(b) Arkoses, 5(c) Schistes à acritarches, 5(d) Roches basiques de Génis.
6. Granitoïdes post-métamorphes—en noir: Massifs ultra-basiques, hachures horizontales: Métamorphisme de contact. bi^+ Isograde de la Biotite. $sill^+$ Isograde de la Sillimanite.

Massif Central français. Le secteur étudié se situe entre les villes de Tulle et de St Pardoux (fig. 1). On rappelle ici quelques données géologiques récemment acquises, concernant la lithologie et le métamorphisme.

1. La Lithologie

La série comprend trois groupes lithostratigraphiques (Floc'h, 1977), de bas en haut:

A. *Le groupe de la Dronne*. Son âge est incertain: Cambrien ou Infra-Cambrien. Il est représenté par des micaschistes mésozonaux, à rares niveaux quartziteux intercalés.

B. *Le groupe Bas-Limousin*: d'âge cambrien moyen–supérieur (Floc'h, 1977; Autran *et al.*,

1977) il a un cachet d'ensemble volcanique et terrigène. La succession est schématiquement la suivante:

— gneiss plagioclasiques plus ou moins micacés;
— formations volcano-détritiques: "grès de Thiviers", "quartzites de Payzac";
— métabasites à caractère tuffacé;
— schistes à fréquents niveaux graphiteux, graywackes verdâtres en alternance.

D'importantes formations leptynitiques et de nombreux corps amphibolitiques s'intercalent dans ces deux groupes.

C. *Le groupe ordovico-silurien de Génis*: il comprend trois termes superposés: méta-ignimbrites et arkoses associées, schistes à Acritarches avec lentille de calcaire à Conodontes, formation basique enfin.

2. *Le Métamorphisme*

Il débute par une phase à haute pression dont les nombreux massifs éclogitiques recensés (Santallier, 1976a) sont le témoin. La phase majeure est barrovienne: la succession classique des zones a été reconnue, épizone dans le groupe de Génis et les termes élevés du groupe Bas-Limousin, mésozone dans le domaine des gneiss plagioclasiques et des micaschistes de la Dronne, catazone enfin localement atteinte aux environs de St-Pardoux.

II. Présentation de l'Etude Géochimique

Cette étude géochimique s'appuie sur un support de 192 analyses dont 130 nouvelles, les autres provenant d'une compilation au cours de laquelle on a été amené à rejeter un nombre restreint d'analyses difficiles à interpréter au seul vu des renseignements bibliographiques.

Parmi les nombreuses représentations possibles, on a retenu deux des diagrammes de de La Roche:

—diagramme (Al/3—K) en fonction de (Al/3—Na) qui sépare nettement les domaines des roches sédimentaires et ignées et, pour ces dernières, fait apparaître clairement les tendances spilitiques et kératophyriques (de la Roche, 1968; Miyashiro, 1975); ce diagramme a été utilisé récemment en Haut-Limousin dans un contexte voisin (de La Roche *et al.*, 1974);

—diagramme (Si/3—(K + Na + 2/3Ca)) en fonction de (K—(Na + Ca)) qui donne une grille de référence des roches ignées se basant sur la nature des feldspaths et sur l'excès ou le déficit en silice libre.

Pour les formations basiques, nous avons utilisé, en plus, trois diagrammes proposés par Miyashiro—respectivement SiO_2, FeO^+, TiO_2, en fonction du rapport FeO^+/MgO,—qui séparent clairement les roches magmatiques des séries calco-alcaline, tholéitique d'arc et tholéitique abyssale (Miyashiro, 1975).

En utilisant ces diagrammes, nous admettons implicitement, qu'au moins en première approximation, le métamorphisme n'a pas modifié sensiblement la composition des roches originelles.

III. Les Formations Basiques

L'échantillonnage a porté sur l'ensemble des roches basiques intercalées dans la série.

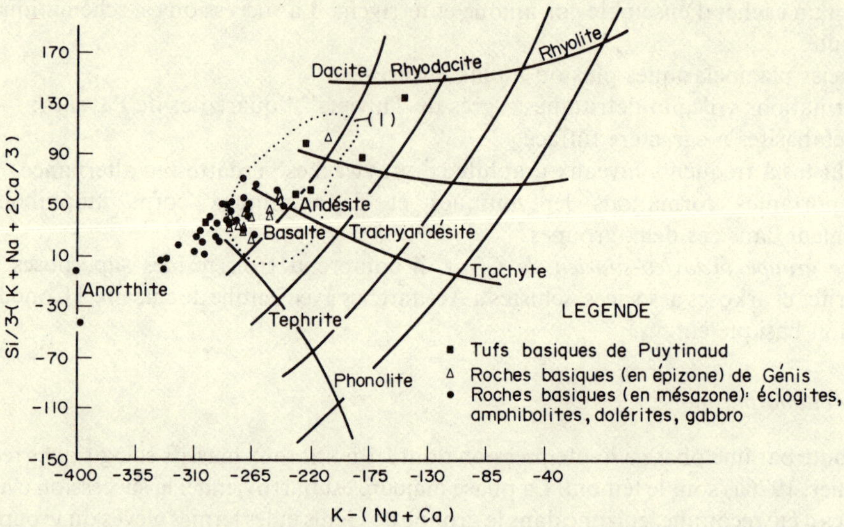

FIG. 2. Les formations basiques dans le diagramme de H. de La Roche (1966). (1) Enveloppe des basaltes d'après M. Tegyey (1974).

Parmi les nombreux corps basiques affleurant dans les gneiss plagioclasiques et les micaschistes de la Dronne, on distingue d'anciennes éclogites avec ou sans disthène et zoïsite primaires, d'anciens gabbros, d'anciennes dolérites, enfin des amphibolo-pyroxénites et amphibolites dont toute l'histoire ancienne a été effacée par le métamorphisme (Santallier, 1976a, b).

La formation basique de Génis est un complexe volcano-plutonique où des termes microlitiques voisinent avec des dolérites et des gabbros. Les textures primaires sont parfois conservées, mais en général ces roches ne montrent qu'un assemblage granoblastique fin à paragenèse "schistes verts".

Parmi les niveaux tuffacés lenticulaires interstratifiés dans la partie supérieure du groupe Bas-Limousin, seul celui de Puytinaud a fait l'objet d'analyses (Roques, 1941). Il s'agit de tufs basiques, associés à d'anciennes laves basiques avec quelques termes acides, dans un contexte de métagrauwackes et de chlorito-schistes.

Les tendances géochimiques sont illustrées par trois séries de diagrammes:

1. Le Diagramme (Si3—(K + Na + 2 Ca/3))/ (K—(Na + Ca))—fig. 2

L'ensemble des points est réparti selon une bande qui, à partir du champ des basaltes, s'étend vers le pôle anorthite: cette tendance calcique est mieux marquée dans les éclogites à disthène. Quelques termes, au contraire sont décalés vers le domaine des andésites: ils traduisent, à l'exception de deux points correspondant à des tufs plus ou moins remaniés de la formation de Puytinaud, une amorce de différenciation tholéitique ou calco-alcaline.

2. Le Diagramme (Al/3—K)/(Al/3—Na)—fig. 3

Les échantillons analysés occupent trois champs bien distincts:

(a) Une partie des éclogites dépourvues de disthène, les amphibolites, les métadolérites et l'amphibolo-pyroxénite ont des compositions proches de celles des basaltes. Les

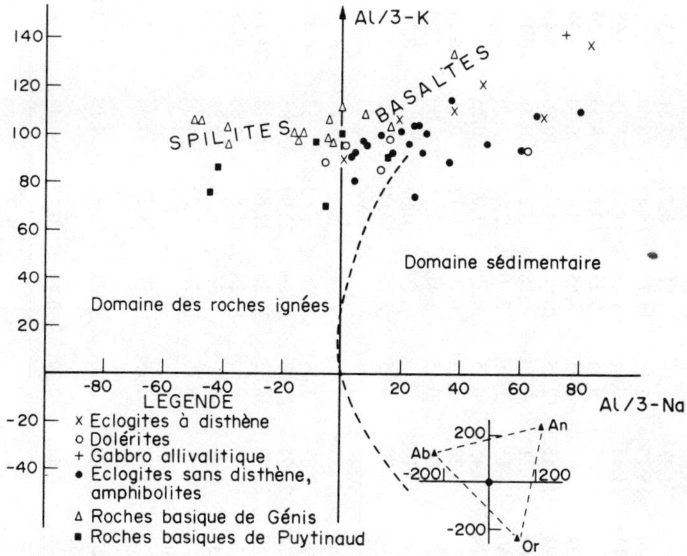

FIG. 3. Les formations basiques dans le diagramme de H. de La Roche (1968).

éclogites à disthène et quelques éclogites sans disthène présentent, au contraire, un étalement en direction du pôle anorthite traduisant une composition alumineuse qui, jointe à des teneurs élevées en calcium (cf. tableau 1) évoque des cumulats à plagioclase basique. Une amphibolite interprétée par l'un de nous (D. S.) comme un métagabbro allivalitique, se situe au voisinage de ces points;

(b) les roches basiques de Génis montrent, au contraire, un étalement marqué vers la gauche du diagramme, du fait des variations importantes de leurs teneurs en sodium; elles se placent dans le champ spilitique (de La Roche *et al.*, 1974);

(c) les tufs basiques de Puytinaud occupent, au voisinage, une position plus basse en raison de leurs teneurs en potassium plus élevées.

3. *Les Diagrammes de Miyashiro*—fig. 4

Outre la tendance spilitique soulignée précédemment, trois traits essentiels se dégagent de la composition de ces roches:

— teneur en potassium, généralement inférieure à 0,4 %, y compris dans les termes non spilitiques;

— rapport FeO + MgO peu élevé (fréquemment inférieur à 2);

— Olivine normative dans plus de 80 % des échantillons analysés.

Ces particularités géochimiques soulignent le faible degré de différenciation de ces roches basiques. En outre, elles sont celles retenues par Miyashiro (1975) dans sa définition des tholéites abyssales.

Dans les représentations graphiques présentées ici et prenant en compte, le titane, le fer et le magnésium, il y a une excellente corrélation entre les courbes de référence "tholéites abyssales" et les points représentatifs des diverses formations étudiées. La corrélation est

Tableau 1. Compositions Chimique et Normative des Principaux Facies

	1	2	3	4	5	6	7	8	9	10	11	12	13	14	15	16
SiO_2	66,00	66,60	66,66	76,00	70,80	76,60	75,80	74,00	65,45	70,61	49,40	47,40	47,20	49,00	43,80	53,00
TiO_2	0,79	0,68	0,75	0,17	0,30	0,11	0,23	0,20	0,96	0,56	1,78	1,18	1,71	0,74	2,21	1,59
Al_2O_3	15,00	14,55	14,32	12,10	14,20	12,45	12,65	13,05	15,47	14,44	15,75	17.35	15,75	15,40	16,00	16,50
Fe_2O_3	1,20	0,90	1,27	0,60	0,80	0,60	0,90	1,15	6,56†	4,11†	1,35	0,65	1,15	1,05	5,25	2,65
FeO	4,55	3,80	4,20	0,80	2,65	0,65	1,20	2,15			9,05	9,35	9,55	10,12	8,35	6,30
MgO	2,55	2,25	2,82	0,35	0,90	0,20	0,50	0,60	2,24	1,34	7,05	8,05	8,65	8,07	6,70	5,20
CaO	3,05	3,10	1,51	0,75	3,00	0,50	0,50	1,50	2,26	0,93	10,45	12,25	10,06	9,65	10,00	5,45
MnO	0,10	0,12	0,10	0,01	0,05	0,03	0,03	0,04	0,11	0,05	0,21	0,20	0,18	0,21	0,21	0,15
Na_2O	3,35	4,60	3,84	3,20	3,35	3,05	2,40	3,00	3,55	2,60	3,15	2,35	2,80	2,66	2,70	4,65
K_2O	1,80	1,60	1,94	4,95	2,50	4,60	4,95	3,30	2,29	4,14	0,45	0,25	0,29	1,00	0,25	1,05
P_2O_5	0,16	0,14	0,17	0,03	0,06	0,03	0,16	0,03			0,20	0,13	0,19	0,12	0,21	0,16
H_2O+	1,80	1,60	1,40	0,65	0,75	0,50	0,70	1,05			1,00	0,45	1,41	1,11	4,00	2,70
H_2O-		0,10	0,05	0,02		0,05	0,10				0,10			0,09	0,05	0,20
P.F.	1,35					0,40	0,90	0,80			0,70	0,95	2,25	2,00		
Qz	27,26	22,10	28,03	36,36	32,88	40,19	41,26	39,04	28,17	36,09						0,84
Or	10,80	9,62	11,76	29,59	15,00	27,53	29,48	19,71	13,70	24,79	2,69	1,49	1,76	6,03	1,55	6,42
Ab	28,73	39,53	33,26	27,33	28,71	26,09	20,42	25,61	30,34	22,24	26,94	20,03	24,26	22,94	23,85	40,64
An	14,41	14,54	6,66	3,58	14,75	2,33	1,55	7,34	11,34	4,67	27,79	36,32	30,26	27,64	32,15	21,73
Di		0,23									19,23	19,84	16,14	16,86	14,88	4,27
Hy	12,79	11,02	12,96	1,60	6,13	1,08	2,40	4,28	5,66	3,39	6,69	0,07	4,36	6,26	5,26	18,63
Ol											10,82	18,76	17,76	17,03	9,50	
Cor	2,36		3,60	0,16	0,65	1,61	2,78	1,88	3,06	4,03						
Ma	1,77	1,33	1,89	0,88	1,18	0,88	1,31	1,68			1,98	0,95	1,71	1,55	7,96	3,97
Hm									6,63	4,16						
Ilm	1,52	1,31	1,46	0,33	0,58	0,21	0,44	0,38	0,26	0,13	3,42	2,26	3,33	1,43	4,39	3,12
Rut									0,83	0,50						
Ap	0,35	0,31	0,38	0,07	0,13	0,07	0,35	0,57			0,44	0,29	0,43	0,27	0,48	0,36

† Fer total

1—Grès de Thiviers
2—Grès de Thiviers
3—Quartzite de Payzac
4—Leptynite d'Aubazine
5—Leptynite d'Albussac
6—Leptynite de l'arc de St Yrieix
7—Leptynite de l'arc de la Dronne
8—Leptynite de Vergonzac
9—Gneiss plagioclasique
10—Gneiss plagioclasique
11—Métadolérite
12—Eclogite à disthène
13—Eclogite sans disthène
14—Amphibolite banale
15—Roche basique de Gènis
16—Tuf basique de Puytinaud

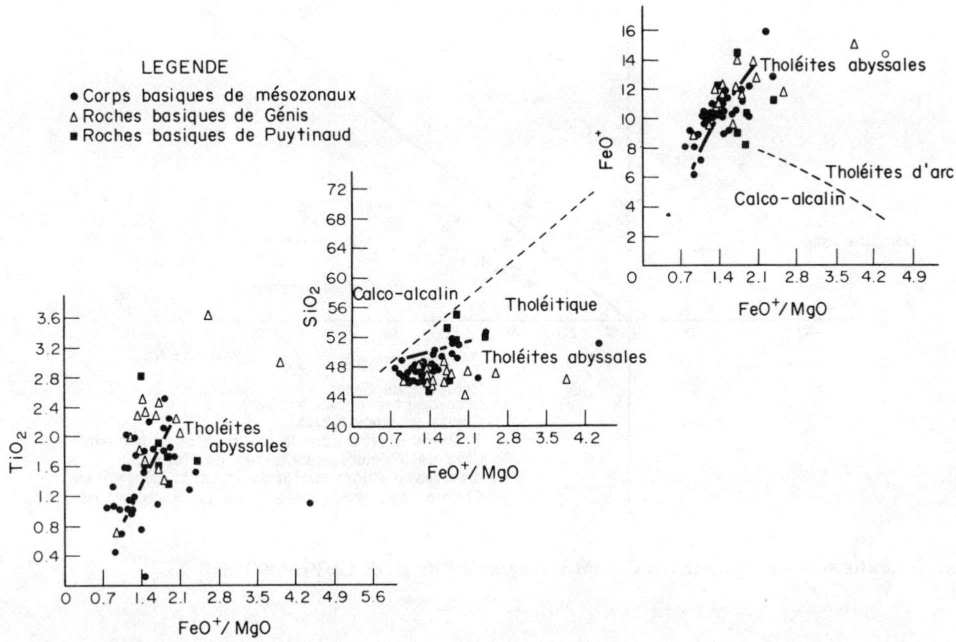

FIG. 4. Les formations basiques dans les diagrammes de Miyashiro (1975).

moins bonne pour la silice; ceci peut être dû en partie à la mobilité de SiO_2 au cours du métamorphisme (Fonteilles, 1976), en partie à la nature tuffacée de certains échantillons (tufs de Puytinaud se disposant au dessus de la courbe "tholéites abyssales").

Par ailleurs, ces compositions sont très proches de celles données récemment par Middlemost (1975) pour les "basaltes océaniques profonds".

Le caractère tholéitique avait déjà été mis en évidence dans le secteur d'Uzerche essentiellement (Coffrant et Piboule, 1975). Nos résultats précisent ce point de vue et font apparaître une parenté magmatique entre les formations basiques du Bas-Limousin.†

IV. Les Formations "Para-Dérivées"

Elles comprennent:

—dans le groupe Bas-Limousin, les gneiss plagioclasiques, les "Grès de Thiviers" et "Quartzites de Payzac" ainsi que des schistes prélevés dans les différents termes stratigraphiques;

— dans les groupes de la Dronne et de Génis, des schistes et micaschistes.

Reportées dans le diagramme (A1/3—K) (A1/3—Na) (Fig. 5), ces roches montrent une forte dispersion horizontale, traduisant une variation importante du paramètre (A1/3—Na). Les points se répartissent entre le domaine des roches ignées et celui des roches sédimentaires chimiquement très dégradées (shales). Quatre champs s'individualisent néanmoins:

†Les premiers résultats d'une étude des éléments traces menés avec M. Treuil, conduisent à la même interprétation.

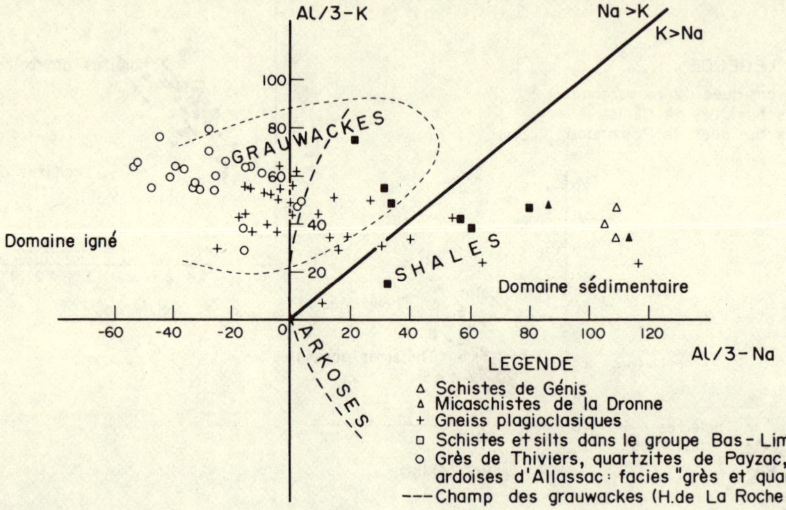

FIG. 5. Les formations "para-dérivées" dans le diagramme de H. de La Roche (1968).

1. *Les "Grès de Thiviers"— "Quartzites de Payzac"*: leurs points figuratifs sont groupés dans une zone restreinte située dans le domaine igné, avec un large débordement dans le champ des grauwackes. La forte dispersion horizontale des points traduit une variation importante des teneurs en sodium; celles-ci sont particulièrement élevées dans quelques échantillons qui se situent à la limite du domaine des spilites. Les caractéristiques pétrochimiques de ces formations ont fait l'objet d'une étude récente (Guillot *et al.*, 1977) qui a permis de préciser la nature rhyo-dacitique du matériel volcanoclastique originel.

2. *Les gneiss plagioclasiques* s'étendent du champ des shales jusqu'aux confins du domaine sédimentaire et igné, où ils se superposent partiellement aux roches précédentes. Cette grande dispersion est liée à la diversité de cette formation, où se succèdent des gneiss micacés à tendance micaschisteuse, puis des gneiss plagioclasiques à deux micas ou biotite seule, avec dans le détail de nombreuses variations de grain; l'ensemble de ces caractères confère à cette formation un incontestable cachet sédimentaire. Néanmoins quelques-uns des échantillons analysés (analyse n°8) ont des compositions comparables à celles des "Grès de Thiviers" et des "Quartzites de Payzac" et sont probablement aussi d'anciens tufs.

L'étalement des compositions entre le domaine igné dacitique et le domaine des shales, avec une plus forte densité de points dans le champ intermédiaire entre arkoses et grauwackes, nous suggère que ces roches représentent des produits plus ou moins remaniés d'un matériel tuffacé initial de composition relativement homogène.

3. *Les schistes et micaschistes des groupes de la Dronne et de Génis* dérivent de sédiments à composition de shales.

4. *Les schistes et micaschistes du groupe Bas-Limousin* s'étalent largement, depuis des compositions de shales jusqu'à des compositions de sédiments moins évolués, proches de celles des gneiss. Cette tendance s'accorde avec les observations pétrographiques qui conduisent à voir dans la plupart de ces roches d'anciennes siltites à débris quartzo-plagioclasiques.

V. Les Formations Leptynitiques

1. Rappels Sommaires sur le Problème des Leptynites en Bas-Limousin

Dans le Massif Central français, le terme de leptynite s'applique à des roches gneissiques claires, essentiellement "quartzo-feldspathiques"; l'usage en a été abusivement étendu à des termes mésocrates.

Les leptynites constituent, en Bas-Limousin, les arcs de St-Yrieix et de la Dronne, ainsi que le coeur du domaine de l'anticlinal de Tulle avec les "leptynites de Vergonzac", les "leptynites roses d'Aubazine", et les "leptynites grises d'Albussac"—dénommées de façon plus rigoureuse "gneiss mésocrates" par Boissonnas et Talbert (1976). L'origine de ces roches est débattue depuis fort lontemps (Roques, 1941; Giraudon, 1956; Alinat, 1975); des observations récentes apportent de nouveaux arguments, essentiels pour aborder ce problème.

Les données cartographiques montrent clairement l'obliquité des leptynites des arcs de St Yrieix et de la Dronne (Floc'h, 1977), des leptynites d'Aubazine (feuille 1/50.000e Tulle) sur les formations encaissantes, évoquant des masses plutoniques à caractère initialement intrusif.

Les données structurales et pétrographiques: dans chacune des formations suivantes: leptynites de St Yrieix, de la Dronne, d'Albussac et d'Aubazine, une partie au moins des observations suivantes a pu être faite: absence de litage prémétamorphique à caractère sédimentaire, présence de "noyaux" peu ou pas déformés à texture grenue originelle conservée, persistance de microcline relique en phénoclastes, existence d'enclaves de l'encaissant gneissique ainsi que d'aplites et de pegmatites foliées dont la composition rappelle celle des leptynites-hôtes. Ces arguments suggèrent fortement que ces leptynites proviennent de la recristallisation de roches plutoniques.

Les leptynites de Vergonzac, par contre, se singularisent par leur aspect lité—bancs décimétriques à métriques, avec par places, des intercalaires amphiboliques nombreux—et

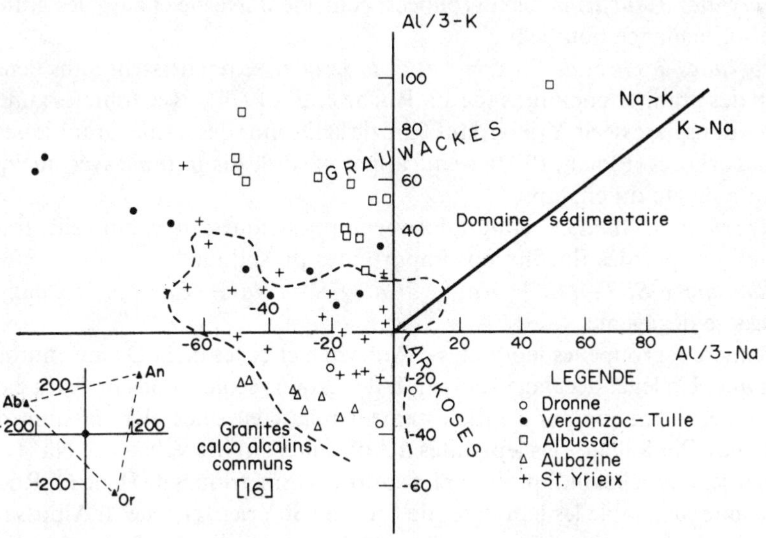

FIG. 6. Les formations leptyniques dans le diagramme de H. de La Roche (1968).

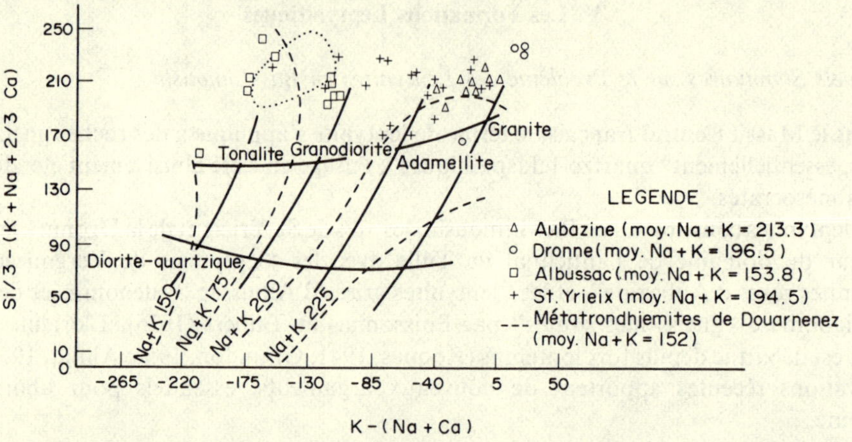

FIG. 7. Les formations leptyniques dans le diagramme de H. de La Roche (1966).

par un passage progressif aux gneiss plagioclasiques qui les surmontent. Ce mode de gisement est compatible avec une mise en place effusive ou sédimentaire: en l'absence de texture relique, la géochimie apporte des arguments pour lever cette ambiguité.

2. *Apports de l'Etude Géochimique*

(a) *Le diagramme (Al/3—K)/(Al/3—Na)—fig. 6.* La quasi totalité des échantillons analysés se situe dans le domaine igné:

- *Les leptynites de Vergonzac* sont largement étalées horizontalement, en direction du pôle albite, ce qui traduit le caractère très sodique de certaines d'entre elles dont les compositions se rapprochent de celles de kératophyres quartziques. L'origine effusive—volcanites acides rhyolitiques à kératophyriques—semble la plus probable.
- *Les leptynites d'Aubazine,* bien groupées, coïncident avec le champ des granites calco-alcalins à tendance potassique.
- *Les leptynites des arcs de St-Yrieix et de la Dronne* se répartissent dans l'ensemble du champ des granites communs (de La Roche *et al.*, 1974), avec toutefois une tendance sodique dans l'arc de St-Yrieix; dans l'arc de la Dronne des termes granitiques voisinent avec des arkoses (Floc'h, 1977), ce qui apparaît sur le diagramme avec quelques points sortant à droite du champ.
- *Les leptynites d'Albussac* sont faiblement potassiques; une nouvelle fois, elles se caractérisent par des fluctuations importantes du sodium.

(b) *Le diagramme Si/3—(K + Na + 2Ca/3)/K—(Na + Ca) fig. 7.* Deux tendances géochimiques se dégagent:

- La première regroupe les leptynites d'Aubazine et celles de la Dronne autour du pôle granitique. Un léger décalage vers le NE de certains échantillons prélevés dans l'arc de la Dronne, s'explique par l'enrichissement relatif en silice et potassium des termes arkosiques. Par ailleurs, les leptynites d'Aubazine, ont des valeurs de Na + K, un peu plus élevées que celles indiquées par les courbes altimétriques de H. de La Roche (1966).
- La seconde rassemble les leptynites de l'Arc de St Yrieix et celles d'Albussac, dont on peut remarquer que les points représentatifs sont fortement décalés vers le Nord par rapport à la ligne tonalite-granite.

Les leptynites de l'Arc de St Yrieix s'étalent largement de part et d'autre du pôle adamellitique, ce qui reflète la grande variété des faciès reconnus pétrographiquement (Floc'h, 1977). Les compositions évoluent de granites à des grano-diorites, ce que traduit le paramètre Na + K oscillant entre 178 et 220 (moyenne 194), la tendance sodique relevée dans quelques échantillons étant responsable des valeurs les plus élevées.

Les leptynites d'Albussac se distinguent par:

. un caractère silico-sodique remarquable;

. des teneurs en potassium faibles;

. des teneurs en calcium assez constantes et voisines de 3%.

Ces particularités rendent compte de leur position dans la grille de référence. Ces compositions s'apparentent à celles de trondjhemites s.1. Remarquons à ce propos les analogies chimiques de ces roches avec les métatronjhemites ordoviciennes, probablement contemporaines, de la baie de Douarnenez, Massif Armoricain (Barrière, 1972), en particulier les valeurs très voisines du paramètre Na + K (voir figure).

Conclusions

Aux compositions des roches examinées dans cette étude, on a pu rapporter celles d'équivalents non métamorphiques; ces rapprochements sont, par ailleurs, parfaitement compatibles avec les données cartographiques et pétrographiques. Dans l'ensemble, le métamorphisme peut donc être considéré comme topochimique. Néanmoins, on a relevé deux dérives fréquentes: tendance silico-sodique de certaines leptynites, tendance spilitique dans les métabasites, sans liaison directe avec le gradient du métamorphisme régional.

En définitive, les résultats suivants doivent être soulignés:

— *caractère tholéitique abyssal des métabasites*, à tous les niveaux de la série; tenant compte des travaux récents (Mouthier, 1976; Coffrant, Piboule, 1975), il semble que l'on puisse généraliser cette interprétation à l'ensemble du Limousin;

— *grand développement des termes tuffacés et volcano-sédimentaires* dans les formations du groupe Bas-Limousin; cette tendance paraît générale dans le Massif Central occidental (de La Roche *et al.*, 1974; Chenevoy, 1958; Bouloton, 1974; Mouthier, 1976).

— *origine "ortho" des formations leptynitiques*, prouvée par les observations cartographiques, pétrographiques et géochimiques. Ces importantes manifestations magmatiques cambro-ordoviciennes sont essentiellement plutoniques, avec la mise en place d'un cortège de granitoïdes variés (granites calco-alcalins et trondhjemites). Les termes volcaniques sont essentiellement d'anciens kératophyres quartziques.

Annexe

Analyses Nouvelles

—B.R.G.M.—Orléans—1974—1975—1976 (voie chimique classique).

—C.R.P.G.—Nancy—1964—1975 (quantométrie).

—Départment des sciences de la Terre, Université d'Orléans—1976 (absorption atomique).

Analyses tirées de la Bibliographie:

—Roques (1941),
—Fourches (1959),
—Alinat (1975).

Références

ALINAT, M. (1975) Le zircon dans les roches de la série métamorphique de l'anticlinal de Tulle, Massif Central français. Applications pétrogénétiques. Thèse 3ème cycle, Univ. de Nice, 196 pp.

AUTRAN, A., FLOC'H, J. P., GROLIER, J., GUILLOT, P. L., PROST, A. et SANTALLIER, D. (1977) Essai de corrélations lithostratigraphiques dans le socle de l'Ouest Massif Central. Evolution tectonique et métamorphique. *Communic. à la 5ème réunion annuelle des Sciences de la Terre, Rennes.*

BARRIERE, M. (1972) Origine et mise en place de la trondhjémites gneissique de Douarnenez. *Bull. BRGM*, 2ème série, n°2.

BOISSONNAS, J. et TALBERT, J. C. (1976) *Notice de la feuille Brive la Gaillarde de la carte géologique de France au 1/50.000*. BRGM ed.

BOULOTON, J. (1974) Etude géologique de la région d'Aigurande (NO du Massif Central français). Lithostratigraphie, structure et pétrographie de la série métamorphique. Thèse 3ème cycle, mention pétrographie, Clermont-Ferrand.

CHENEVOY, M. (1958) Sur l'importance et les caractères du paléovolcanisme en Limousin septentrional (Massif Central français). *B.S.G.F.*, 6ème série, VIII, n°2, 113–122.

COFFRANT, P. et PIBOULE, M. (1975) Les métavulcanites basiques du Bas-Limousin (Massif Central français): étude géochimique des éclogites et des dolérites. *B.S.G.F.*, 7ème série, XVII, 620–628.

FLOC'H, J. P. (1977) Age Cambrian moyen-supérieur des formations métamorphiques de groupe Bas-Limousin. *C.R. Acad. Sci., Paris* (à paraître).

FONTEILLES, M. (1976) Essai d'interprétation des compositions chimiques des roches d'origine métamorphique et magmatique du massif hercynien de l'Agly (Pyrénées orientales). Thèse Doct. Etat, Paris.

GIRAUDON, R. (1956) Etude géologique des terrains cristallophylliens de la feuille de Brive situés à l'Ouest de la faille d'Argentat. Diplôme géol. pétro. Fac. Sci. Clermont, 93 pp.

GUILLOT, P. L., BOYER, C. et TEGYEY, M. (1977) Grès de Thiviers, ardoises d'Allassac et quartzites de Payzac: un complexe volcano–détritique rhyo-dacitique dans la série métamorphique du Bas-Limousin. *Bull. BRGM*, sect. 1, n°3.

LA ROCHE, H. de (1966) Sur l'usage du concept d'association minérale dans l'étude chimique des roches: modèles chimiques, statistiques, représentations graphiques, classification chimico-minéralogique. *C.R. Acad. Sci., Paris* **262**, 1665–1668.

LA ROCHE, H. de (1968) Comportement géochimique différentiel de Na, K et Al dans les formations volcaniques et sédimentaires: un guide pour l'etude des formation métamorphiques et plutoniques. *C.R. Acad. Sci., Paris,* **267**, série D, 39–42.

LA ROCHE, H. de, AUTRAN, A., CHANTRAINE, J. et MOINE, B. (1974) Etudes géochimiques associées à la cartographie géologique. *Bull. BRGM*, 2ème série, sect, IV, n°2, 109–124.

LA ROCHE, H. de, ROCCI, G. et JUTEAU, Th. (1974) Essai de caractérisation chimique des associations spilitiques. In *Spilites and Spilitic Rocks*, IUGS, ser. A, n°4, 39–58.

MIDDLEMOST, E. (1975) The basalt clan. *Earth Sci. Rev.* **11**, 341.

MIYASHIRO, A. (1975) Volcanic rock series and tectonic setting. *Ann. Rev. Earth and Planetary Sci.* **3**, 251–269.

MOUTHIER, B. (1976) Lithostratigraphie et métamorphisme des formations cristallophylliennes de la région de Chateauneuf–La-Forêt. Haut-Limousin. Massif Central français. Thèse 3ème cycle. 2 vols., 161 pp.

ROQUES, M. (1941) Les schistes cristallins de la partie Sud-Ouest du Massif Central français. *Mém. Serv. Cart. Géol. fr.*, Imp. Nat., Paris.

SANTALLIER, D. (1976a) Métamorphisme polyphasé en Bas-Limousin: les phases précoces. *C. R. Acad. Sci., Paris,* série D, **283**, 737–740.

SANTALLIER, D. (1976b) Quelques minéraux des éclogites et amphibolites dérivées en Bas-Limousin. Etude chimique préliminaire. *Bull. BRGM*, sect. 1, n°3, 153–183.

TEGYEY, M. (1974) Essai d'informatique pétrographique: constitution d'un fichier et application à l'étude des roches volcaniques des zones orogéniques. Thèse 3ème cycle, Univ. Orléans, 146 pp.

Chemical Evolution in an Old Crustal Area (Sistema Central Espanol)

A. Aparicio, J. L. Brändle and F. Bellido

Departamento de Petrolog a y Geoqu mica, C.S.I.C., Madrid, Spain

Abstract

This area is constituted by metamorphic rocks, of Lower Paleozoic age, affected by the Hercynian orogeny. The pre-metamorphic rocks, mainly pelitic rocks with some quartzites and limestone layers, underwent regional metamorphism and later intrusion of plutonic rocks generating a contact metamorphism. According to the metamorphic intensity, three groups are established in the metamorphic rocks: low (slates), medium (schists), and high metamorphic grade (gneisses and migmatites).

The chemical evolution, during metamorphism, shows an increase of some elements (SiO_2, CaO, Na_2O, K_2O, Rb) and a decrease of some others (Al_2O_3, Fe_2O_3, MgO, H_2O, TiO_2, Th). If a comparison is made, there is almost a chemical identity between slates and schists, which can be clearly seen on the Si–FeK–Mag triangle. The same triangle also shows a sharp change for the gneissic rocks, as some of them even fall into the field of plutonic rocks; a theoretical line of separation has also been drawn between the range of maximum probability of appearance of plutonic and metamorphic rocks.

The Q-mode analysis delimit some relations between the two types of rocks. The statistical analysis was separately made for major and minor elements, and the results were plotted on triangular diagrams of three main components (90% of communality). The low–medium-grade metamorphic rocks present a maximum of concentration in the first component (Al_2O_3, TiO, H_2O, MgO) and a weak dispersion towards the second (SiO_2, Na_2O, K_2O). The gneissic rocks, extend up to the second component. In the same triangle the plutonic rocks show a trend from the third component (CaO, MgO, FeO, P_2O_5) to the second. Both trends (plutonic and metamorphic) meet at the second component, where the most differentiated plutonic rocks are also gathered.

Minor elements show a similar trend; both groups meet at the second component (Rb, Th). The overlapping allows us to regard the possibility of an influence of the plutonic materials in the compositional variability of the high-grade metamorphic rocks (gneisses and migmatites) while the chemical transformation and mobilization is weak in low–medium grade of metamorphism, the migration and chemical change is strong between the gneissic–migmatitic rocks and the plutonic materials.

Introduction

The Spanish Central System is a massive made of Lower Paleozoic rocks (mainly Cambro–Ordovician) that were affected during the Hercynian orogeny by plutonic and metamorphic processes, which produced a strong transformation of the primitive rocks.

The stratigraphic series is formed by pelitic materials, some layers of limestone (in the Cambrian series) and quartzites (mainly in the Ordovician series). As a result of the metamorphic processes, the initial materials were transformed into slates, schists, augen- and branded gneisses, marbles, amphibolites and metaquartzites.

Two types of regional matamorphism have been distinguished (Fúster *et al.*, 1974; López Ruiz *et al.*, 1975)†. The eastern sector developed an intermediate pressure metamorphism of

† Exhaustive petrographic and bibliographic information of this area is avilable from López Ruiz *et al.* (1975) and Aparicio *et al.* (1975).

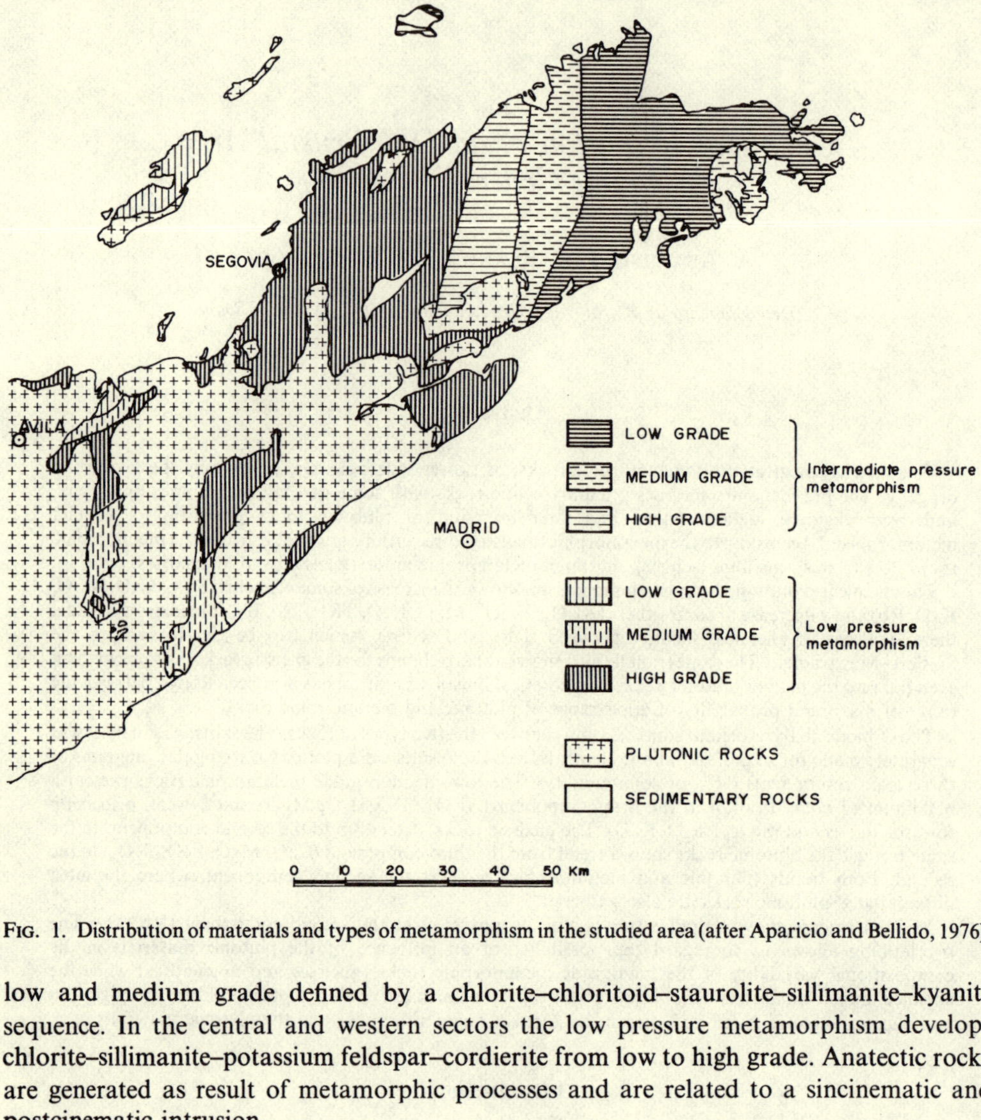

FIG. 1. Distribution of materials and types of metamorphism in the studied area (after Aparicio and Bellido, 1976).

low and medium grade defined by a chlorite–chloritoid–staurolite–sillimanite–kyanite sequence. In the central and western sectors the low pressure metamorphism develops chlorite–sillimanite–potassium feldspar–cordierite from low to high grade. Anatectic rocks are generated as result of metamorphic processes and are related to a sincinematic and postcinematic intrusion.

Sometimes a contact metamorphism is superimposed on the regional metamorphism. The plutonic rocks show mainly adamellitic–granodioritic composition and a calcoalkaline trend (Aparicio *et al.*, 1975).

This paper describes the geochemical evolution and the interchange of some major and minor elements during the metamorphic–plutonic processes as well as the possible genetic interactions between both type of rocks.

In order to establish this relationship, we have collected already published data (López Ruiz *et al.*, 1975; Aparicio *et al.*, 1975; Aparicio and Bellido, 1976; Aparicio and Galan, 1977) and new analyses elaborated in the Department of Petrology and Geochemistry (C.S.I.C., Madrid), the available analyses (106) were selected and reduced to a smaller number (85), according to the following accounts: it has been observed that samples

TABLE 1. AVERAGES, STANDARD DEVIATIONS, MINIMUM AND MAXIMUM VALUE OF MAJOR ELEMENTS (wt. %) AND MINOR ELEMENTS (ppm) FOR FOUR SETS OF DATA EXAMINED

	Slates				Schists			
	Average	St. Dev.	Max. V	Min. V	Average	St. Dev.	Max. V.	Min. V
SiO_2	58.13	2.17	63.30	55.50	60.09	4.08	71.20	54.20
Al_2O_3	20.92	1.39	22.84	18.35	20.75	2.29	24.84	14.52
Fe_2O_3	2.74	1.25	5.61	1.47	2.25	0.96	3.90	0.80
FeO	4.06	1.55	6.13	0.64	4.55	1.04	6.41	2.37
MgO	2.35	0.94	4.07	0.71	2.04	0.85	3.46	0.60
CaO	0.52	0.45	1.98	0.14	0.60	0.41	1.70	0.14
Na_2O	0.90	0.50	1.72	0.13	1.07	0.57	2.59	0.46
K_2O	3.75	0.67	5.35	2.46	3.86	0.62	5.13	2.77
MnO	0.07	0.04	0.18	0.01	0.12	0.11	0.50	0.03
TiO_2	1.02	0.10	1.23	0.88	0.94	0.28	1.26	0.19
P_2O_5	0.13	0.03	0.19	0.06	0.16	0.09	0.39	0.07
H_2O	4.54	1.24	5.56	0.90	3.25	0.75	4.35	1.84
Ba	843	200	1349	636	989	289	1766	626
Rb	149	22	198	113	167	30	209	112
Pb	53	43	145	13	50	52	249	24
Sr	165	46	260	101	128	51	212	34
La	55	21	91	17	56	11	78	36
Th	17	4	24	8	16	5	25	6
Cu	25	18	70	3	55	42	185	7
Ni	44	15	72	18	49	19	90	19
Zr	172	39	260	112	213	77	419	121
SiO_2	6889	271	7482	6567	7060	350	7981	6515
FEK	1022	141	1320	859	1090	165	1402	815
MAG	1066	207	1371	721	1017	236	1410	577

	Gneisses				Plutonic rocks			
	Average	St. Dev.	Max. V	Min. V	Average	St. Dev.	Max. V	Min. V
SiO_2	64.54	6.93	74.65	52.60	69.44	4.08	76.25	60.90
Al_2O_3	17.66	3.51	24.98	12.61	14.62	1.22	17.04	12.49
Fe_2O_3	1.23	0.69	2.60	0.19	0.71	0.40	1.57	0.17
FeO	4.00	2.20	8.22	1.07	2.32	1.09	4.49	0.60
MgO	1.95	0.91	3.46	0.41	1.20	0.64	2.52	0.20
CaO	1.30	0.64	2.55	0.42	2.47	1.02	4.49	0.84
Na_2O	2.06	0.93	3.81	0.54	3.55	0.27	4.10	3.07
K_2O	4.21	0.89	6.16	2.89	3.83	0.47	4.62	2.79
MnO	0.07	0.10	0.50	0.01	0.07	0.18	0.95	0.01
TiO_2	0.71	0.35	1.35	0.08	0.44	0.25	0.91	0.03
P_2O_5	0.11	0.07	0.37	0.02	0.20	0.10	0.41	0.02
H_2O	2.09	1.13	4.54	0.30	0.99	0.30	1.75	0.46
Ba	771	249	1284	112	446	194	842	26
Rb	179	45	313	92	209	52	335	142
Pb	42	20	88	9	28	7	44	14
Sr	155	58	291	49	125	48	200	24
La	42	18	79	12	34	2	42	31
Th	16	6	37	4	26	5	36	11
Cu	34	44	194	1	10	7	24	4
Ni	38	20	69	8	8	1	12	6
Zr	199	63	359	82	149	63	246	18
SiO_2	7286	600	8250	6181	7635	398	8268	6767
FEK	1448	324	2236	987	1824	093	2115	1687
MAG	827	385	1464	217	469	229	1026	107

separated from the main distribution of elements, by two empty-classes, are related to not representative types, either by their scarcity or because of analytical or transcription errors. For instance, plutonic rocks with a SiO_2 content lower than 60% were discarded since basic rocks are scarcely represented in this area; as for the metamorphic rocks, those containing $SiO_2 > 80\%$, $K_2O > 8\%$, $P_2O_5 > 2\%$ were also discarded since they show no relationship to pelitic terms.

The metamorphic rocks were separated in three groups on the basis of textural and paragenetic characteristics. The first group was delimited by the presence of chlorite (slates), the second includes biotite (schists), and the third, K-feldspar (gneisses and migmatites). Plutonic rocks from the fourth group.

In spite of the different mineralogical features, it was not possible to establish a separation between slates and schists affected by low- and intermediate-pressure metamorphism due to their similar chemical composition.

Geochemical Evolution

The average values of major and minor elements are shown in Table 1. An increase in SiO_2, CaO, Na_2O and Rb concentrations can be observed in line with the increasing metamorphic intensity reaching maximum values in plutonic rocks, wheras Al_2O_3, Fe_2O_3, MgO, TiO_2, H_2O and Pb (Fig. 2) follow an inverse trend. K_2O increases with metamorphic grade, without reaching a maximum in plutonic rocks. Other elements (FeO, Ba, La, Cu, Ni and Zn) present higher content in metamorphic than in plutonic rocks.

The majority of the major and minor elements tested show a similar concentration in both slates and schists; but conversely the gneissic–migmatitic rocks and plutonic rocks show a signficant increase or decrease in concentration.

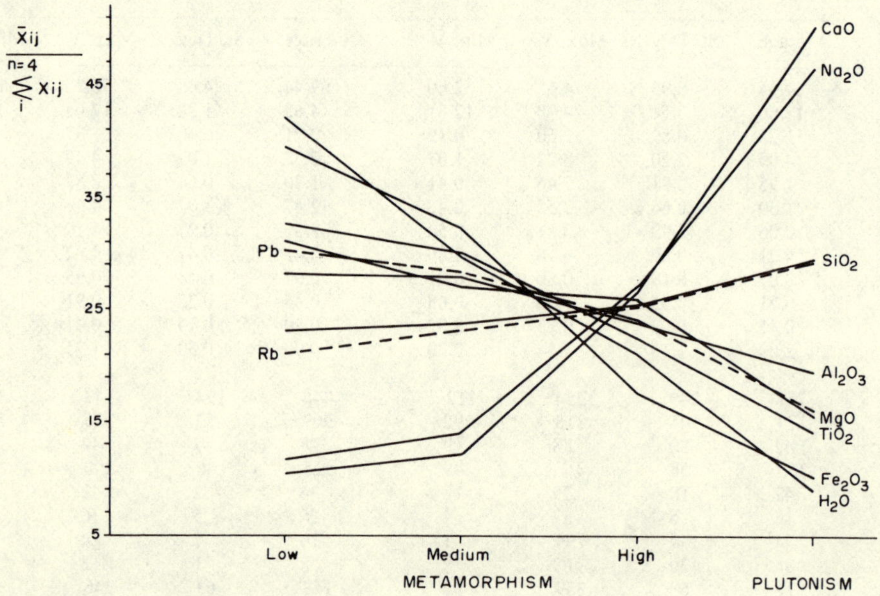

FIG. 2. Relationships between averages of elements of different rocks groups.

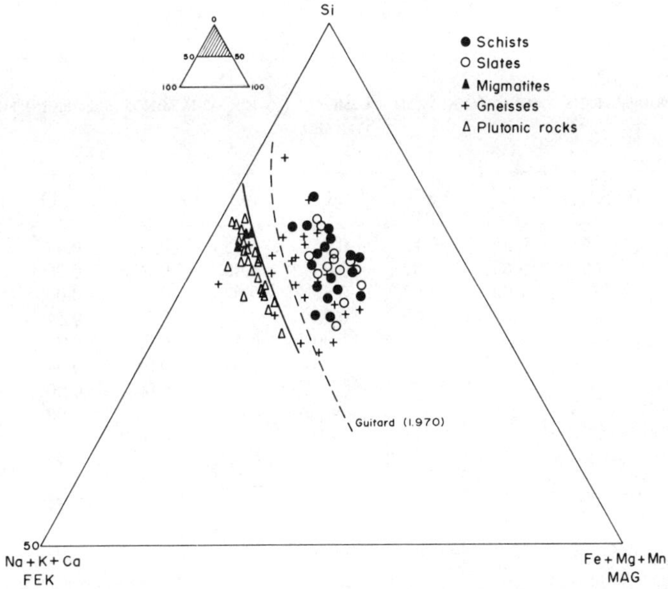

FIG. 3. Si–Fe+Mg+Mn–Na+K+Ca diagram (after Guitard, 1970) showing ortho–para line (solid) in Central System and the same line (dashed) for Pyrenees.

This fact is noted in the Si–Na+Ca+K– Fe+Mg+Mn triangle projection (Guitard, 1970) which separates the ortho–para materials where the projection field for slates and schists are superimposed (Fig. 3). At the same time, gneisses and migmatites show a wider field, which sometimes overlaps the field of plutonic rocks.

On the same triangle is plotted the theoretical ortho–para line of separation between plutonic–metamorphic rocks in this geological area, which deviates from a similar line established in the Pyrenees by Guitard (1970).

Some petrogenetic considerations and data about the evolution in these materials were determined separately in each group of rocks. These values allow us to observe a strong influence of detrital parental minerals in low–medium-grade metamorphic rocks and a large geochemical interchange and mobility in gneissic and plutonic rocks.

We can thus observe in the slates low values of the correlation coefficients among the different elements (Table 2) which is, in fact, a logical result of their inherited sedimentary characteristics which have only been affected by slight transformations. The similar behavior between detrital quartz and zircons may contribute to the positive correlation between SiO_2 and Zr (although the former element is located in the silicates framework).

The negative relation SiO_2–Al_2O_3 is interpreted as the inversion between the sand and argillaceous fraction in the sedimentary materials. CaO and Na_2O show a high positive mutual correlation, but the correlation of both elements against SiO_2 is negative, showing dependence on plagioclase and the antagonism of this mineral with detrital quartz.

Table 2 ("slates") shows a high positive correlation between CaO, Na_2O and P_2O_5 (and negative of these elements against SiO_2) that would indicate the inverse relation between the detrital influence and the amount of phosphatic materials.

TABLE 2. INTERCORRELATIONS AMONG ANALYZED ELEMENTS (MAJOR AND MINOR ELEMENTS) FOR EVERY GROUP STUDIED

Slates

	SiO_2	Al_2O_3	Fe_2O_3	FeO	MgO	CaO	MnO	Na_2O	K_2O	TiO_2
SiO_2	1.00	−0.76	−0.35	−0.15	−0.05	−0.45	0.39	−0.47	0.10	0.09
Al_2O_3	—	1.00	0.63	−0.32	−0.39	−0.08	−0.59	0.00	0.12	0.44
Fe_2O_3	—	—	1.00	−0.82	−0.62	−0.02	−0.29	−0.00	0.12	0.60
FeO	—	—	—	1.00	−0.75	0.30	0.25	0.24	−0.23	−0.72
MgO	—	—	—	—	1.00	0.30	0.58	0.24	−0.47	−0.68
CaO	—	—	—	—	—	1.00	0.05	0.94	−0.49	−0.68
MnO	—	—	—	—	—	—	1.00	−0.00	−0.50	−0.35
Na_2O	—	—	—	—	—	—	—	1.00	−0.56	−0.68
K_2O	—	—	—	—	—	—	—	—	1.00	0.61
TiO_2	—	—	—	—	—	—	—	—	—	1.00
P_2O_5	—	—	—	—	—	—	—	—	—	—
H_2O	—	—	—	—	—	—	—	—	—	—
Ba	—	—	—	—	—	—	—	—	—	—
Rb	—	—	—	—	—	—	—	—	—	—
Pb	—	—	—	—	—	—	—	—	—	—
Sr	—	—	—	—	—	—	—	—	—	—
La	—	—	—	—	—	—	—	—	—	—
Th	—	—	—	—	—	—	—	—	—	—
Cu	—	—	—	—	—	—	—	—	—	—
Ni	—	—	—	—	—	—	—	—	—	—
Zr	—	—	—	—	—	—	—	—	—	—

Schists

	SiO_2	Al_2O_3	Fe_2O_3	FeO	MgO	CaO	MnO	Na_2O	K_2O	TiO_2
SiO_2	1.00	−0.85	−0.40	−0.65	−0.27	0.12	−0.32	0.14	−0.40	−0.22
Al_2O_3	—	1.00	0.16	0.46	0.01	−0.41	0.00	−0.26	0.63	0.27
Fe_2O_3	—	—	1.00	0.00	0.25	−0.07	0.26	−0.11	−0.03	0.11
FeO	—	—	—	1.00	0.15	−0.05	0.26	−0.35	0.16	0.23
MgO	—	—	—	—	1.00	0.08	0.34	0.55	−0.48	−0.51
CaO	—	—	—	—	—	1.00	0.37	0.46	−0.48	−0.31
MnO	—	—	—	—	—	—	1.00	0.23	0.00	−0.10
Na_2O	—	—	—	—	—	—	—	1.00	−0.45	−0.58
K_2O	—	—	—	—	—	—	—	—	1.00	0.35
TiO_2	—	—	—	—	—	—	—	—	—	1.00
P_2O_5	—	—	—	—	—	—	—	—	—	—
H_2O	—	—	—	—	—	—	—	—	—	—
Ba	—	—	—	—	—	—	—	—	—	—
Rb	—	—	—	—	—	—	—	—	—	—
Pb	—	—	—	—	—	—	—	—	—	—
Sr	—	—	—	—	—	—	—	—	—	—
La	—	—	—	—	—	—	—	—	—	—
Th	—	—	—	—	—	—	—	—	—	—
Cu	—	—	—	—	—	—	—	—	—	—
Ni	—	—	—	—	—	—	—	—	—	—
Zr	—	—	—	—	—	—	—	—	—	—

TABLE 2. CONTINUED.

Slates

P_2O_5	H_2O	Ba	Rb	Pb	Sr	La	Th	Cu	Ni	Zr
−0.59	−0.08	0.30	0.06	0.03	0.30	−0.26	0.24	−0.17	0.27	0.52
0.11	0.43	−0.00	0.32	0.04	−0.09	0.39	0.10	−0.27	−0.51	−0.42
0.87	0.27	0.39	0.45	0.24	−0.03	0.46	−0.04	−0.01	−0.17	0.07
0.25	−0.23	−0.55	−0.56	−0.32	−0.26	−0.49	−0.26	0.18	0.19	−0.32
0.10	0.05	−0.15	−0.60	−0.17	−0.35	−0.14	−0.34	0.43	0.52	0.12
0.90	−0.50	−0.39	−0.37	0.01	−0.17	0.11	−0.26	0.79	0.16	−0.12
−0.18	0.06	0.19	−0.58	0.10	−0.02	−0.26	−0.44	0.47	0.62	0.45
0.91	−0.40	−0.35	−0.32	−0.14	−0.13	0.18	−0.23	0.68	0.01	−0.09
−0.33	−0.17	0.17	0.58	0.15	0.05	−0.02	0.26	−0.59	−0.17	−0.17
−0.56	0.39	0.54	0.58	0.23	0.22	0.26	0.37	−0.55	−0.34	0.05
1.00	−0.44	−0.39	−0.14	−0.20	−0.30	0.06	−0.28	0.58	−0.06	−0.19
—	1.00	0.27	0.15	−0.07	−0.20	0.21	−0.05	−0.37	0.00	0.00
—	—	1.00	0.47	−0.13	−0.15	0.37	0.19	−0.06	0.09	0.71
—	—	—	1.00	−0.18	−0.33	0.23	0.35	−0.49	−0.18	0.22
—	—	—	—	1.00	0.44	0.11	−0.21	0.15	0.26	−0.12
—	—	—	—	—	1.00	0.04	0.38	−0.10	−0.41	−0.14
—	—	—	—	—	—	1.00	0.03	−0.04	−0.06	0.15
—	—	—	—	—	—	—	1.00	−0.19	−0.53	−0.00
—	—	—	—	—	—	—	—	1.00	0.30	0.16
—	—	—	—	—	—	—	—	—	1.00	0.28
—	—	—	—	—	—	—	—	—	—	1.00

Schists

P_2O_5	H_2O	Ba	Rb	Pb	Sr	La	Th	Cu	Ni	Zr
−0.19	−0.60	0.41	−0.27	−0.09	−0.00	−0.24	0.04	−0.24	−0.29	0.76
0.27	0.37	−0.26	0.58	−0.14	0.06	0.27	0.12	0.06	0.03	−0.56
−0.07	0.21	−0.14	0.05	0.16	−0.27	0.00	−0.04	0.36	0.29	−0.19
0.20	0.57	−0.32	−0.02	0.01	−0.30	0.48	0.02	0.26	0.27	−0.53
−0.31	−0.33	−0.65	−0.37	0.15	0.47	−0.22	−0.25	0.44	0.59	−0.44
−0.12	0.14	0.07	−0.63	0.38	0.40	−0.39	−0.57	0.05	0.03	−0.18
−0.12	0.12	0.04	−0.28	0.82	0.01	−0.13	−0.26	0.31	0.44	−0.38
−0.26	−0.41	−0.27	−0.46	0.24	0.56	−0.64	−0.45	−0.13	0.02	−0.24
0.29	0.14	0.39	0.79	0.00	−0.12	0.26	0.18	0.13	−0.18	−0.07
0.08	0.43	0.05	0.30	−0.07	−0.15	0.51	0.50	0.11	−0.16	0.13
1.00	0.10	−0.06	0.11	−0.13	−0.09	0.11	−0.14	0.17	−0.28	−0.31
—	1.00	−0.25	0.00	0.05	−0.12	0.41	0.11	0.12	0.29	−0.34
—	—	1.00	0.24	0.18	−0.20	0.10	0.00	−0.03	−0.18	0.57
—	—	—	1.00	−0.23	−0.25	0.21	0.47	−0.04	−0.07	0.16
—	—	—	—	1.00	0.16	−0.30	−0.20	0.25	0.45	−0.09
—	—	—	—	—	1.00	−0.43	−0.21	−0.19	−0.29	−0.27
—	—	—	—	—	—	1.00	0.58	−0.06	0.18	0.13
—	—	—	—	—	—	—	1.00	−0.19	0.10	0.43
—	—	—	—	—	—	—	—	1.00	−0.38	−0.46
—	—	—	—	—	—	—	—	—	1.00	−0.00
—	—	—	—	—	—	—	—	—	—	1.00

TABLE 2. CONTINUED.

Gneisses

	SiO_2	Al_2O_3	Fe_2O_3	FeO	MgO	MnO	CaO	Na_2O	K_2O	TiO_2
SiO_2	—	−0.94	−0.46	−0.91	−0.69	−0.46	0.01	0.45	−0.05	−0.80
Al_2O_3	—	—	0.33	0.86	0.51	0.51	−0.10	−0.50	0.06	−0.68
Fe_2O_3	—	—	—	0.26	0.57	−0.17	−0.10	−0.33	−0.19	0.51
FeO	—	—	—	—	0.70	0.55	−0.06	−0.47	−0.15	0.83
MgO	—	—	—	—	—	0.14	−0.07	−0.35	−0.22	0.73
MnO	—	—	—	—	—	—	0.19	−0.37	−0.03	0.33
CaO	—	—	—	—	—	—	—	−0.04	0.15	0.03
Na_2O	—	—	—	—	—	—	—	—	0.37	−0.58
K_2O	—	—	—	—	—	—	—	—	—	−0.24
TiO_2	—	—	—	—	—	—	—	—	—	—
P_2O_5	—	—	—	—	—	—	—	—	—	—
H_2O	—	—	—	—	—	—	—	—	—	—
Ba	—	—	—	—	—	—	—	—	—	—
La	—	—	—	—	—	—	—	—	—	—
Rb	—	—	—	—	—	—	—	—	—	—
Pb	—	—	—	—	—	—	—	—	—	—
Sr	—	—	—	—	—	—	—	—	—	—
Th	—	—	—	—	—	—	—	—	—	—
Zr	—	—	—	—	—	—	—	—	—	—
Cu	—	—	—	—	—	—	—	—	—	—
Ni	—	—	—	—	—	—	—	—	—	—

Plutonic rocks

	SiO_2	Al_2O_3	Fe_2O_3	FeO	MgO	MnO	CaO	Na_2O	K_2O	TiO_2
SiO_2	—	−0.97	−0.64	−0.94	−0.94	−0.48	−0.93	0.25	0.80	−0.68
Al_2O_3	—	—	0.60	0.93	0.92	0.46	0.88	−0.36	−0.75	0.65
Fe_2O_3	—	—	—	0.40	0.65	0.30	0.55	0.02	−0.62	−0.06
FeO	—	—	—	—	0.90	0.45	0.90	−0.36	−0.73	0.67
MgO	—	—	—	—	—	0.48	0.83	−0.32	−0.79	0.68
MnO	—	—	—	—	—	—	0.44	0.01	−0.50	−0.27
CaO	—	—	—	—	—	—	—	−0.19	−0.83	0.62
Na_2O	—	—	—	—	—	—	—	—	0.00	−0.30
K_2O	—	—	—	—	—	—	—	—	—	−0.47
TiO_2	—	—	—	—	—	—	—	—	—	—
P_2O_5	—	—	—	—	—	—	—	—	—	—
H_2O	—	—	—	—	—	—	—	—	—	—
Ba	—	—	—	—	—	—	—	—	—	—
La	—	—	—	—	—	—	—	—	—	—
Rb	—	—	—	—	—	—	—	—	—	—
Pb	—	—	—	—	—	—	—	—	—	—
Sr	—	—	—	—	—	—	—	—	—	—
Th	—	—	—	—	—	—	—	—	—	—
Zr	—	—	—	—	—	—	—	—	—	—
Cu	—	—	—	—	—	—	—	—	—	—
Ni	—	—	—	—	—	—	—	—	—	—

TABLE 2. CONTINUED.

Gneisses

P_2O_5	H_2O	Ba	La	Rb	Pb	Sr	Th	Zr	Cu	Ni
0.34	−0.64	−0.54	−0.83	0.20	0.45	0.07	−0.09	−0.10	−0.21	−0.85
−0.36	0.59	0.55	0.74	−0.15	−0.42	−0.07	0.09	−0.05	0.23	0.77
−0.34	0.71	0.46	0.37	−0.22	−0.27	0.11	−0.32	0.30	0.00	0.51
−0.23	0.49	0.40	0.86	−0.32	−0.40	−0.15	0.15	0.17	0.32	0.86
−0.29	0.49	0.18	0.60	−0.26	−0.39	−0.19	−0.05	0.25	0.17	0.74
−0.08	0.34	0.15	0.35	−0.03	−0.10	−0.31	0.11	0.01	0.39	0.33
0.58	−0.24	0.14	0.04	0.16	−0.36	0.35	−0.21	0.15	−0.24	0.05
0.06	−0.57	−0.59	−0.39	0.36	0.56	0.21	0.28	−0.21	−0.21	−0.54
−0.17	−0.31	0.08	−0.04	0.52	0.50	0.00	0.31	−0.22	−0.42	−0.27
−0.19	0.60	0.55	0.85	−0.46	−0.43	−0.01	0.11	0.51	0.33	0.84
—	−0.29	−0.06	−0.12	−0.04	0.12	0.00	−0.13	0.21	0.14	−0.19
—	—	0.54	0.48	−0.40	−0.24	−0.11	−0.21	0.24	0.34	0.60
—	—	—	0.55	−0.42	−0.44	0.13	−0.12	0.30	0.03	0.47
—	—	—	—	−0.40	−0.29	0.05	0.15	0.41	0.39	0.86
—	—	—	—	—	0.02	−0.01	−0.06	−0.48	−0.28	−0.40
—	—	—	—	—	—	−0.10	0.27	0.05	0.19	−0.42
—	—	—	—	—	—	—	−0.15	0.16	−0.03	0.07
—	—	—	—	—	—	—	—	0.01	−0.03	−0.08
—	—	—	—	—	—	—	—	—	0.37	0.28
—	—	—	—	—	—	—	—	—	—	0.29
—	—	—	—	—	—	—	—	—	—	—

Plutonic rocks

P_2O_5	H_2O	Ba	La	Rb	Pb	Sr	Th	Zr	Cu	Ni
−0.85	−0.53	−0.86	0.67	0.87	0.73	−0.93	0.61	−0.94	−0.17	0.33
0.86	0.47	0.87	−0.70	−0.83	−0.76	0.91	−0.61	0.92	0.09	−0.30
0.44	0.76	0.49	−0.55	−0.59	−0.53	0.54	−0.26	0.57	0.26	−0.05
0.88	0.29	0.87	−0.59	−0.83	−0.67	0.92	−0.63	0.92	0.10	−0.41
0.80	0.49	0.83	−0.65	−0.82	−0.81	0.90	−0.49	0.89	0.02	−0.21
0.40	0.24	0.17	−0.20	−0.28	−0.39	0.32	−0.43	0.33	0.07	−0.05
0.74	0.43	0.79	−0.60	−0.90	−0.56	0.91	−0.65	0.86	0.32	−0.41
−0.41	0.10	−0.54	0.26	0.36	0.31	−0.35	0.18	−0.37	0.00	0.09
−0.60	−0.55	−0.53	0.57	0.73	0.65	−0.75	0.56	−0.69	−0.12	0.22
0.60	0.28	0.80	−0.54	−0.72	−0.51	0.77	−0.27	0.76	0.01	−0.35
—	0.29	0.81	−0.54	−0.76	−0.69	0.85	−0.67	0.89	−0.04	−0.44
—	—	0.32	−0.49	−0.36	−0.41	0.34	−0.21	0.41	0.32	0.22
—	—	—	−0.69	−0.87	−0.66	0.90	−0.45	0.92	0.08	−0.46
—	—	—	—	0.70	0.64	−0.67	0.33	−0.70	−0.01	0.17
—	—	—	—	—	0.64	−0.94	0.63	−0.89	−0.16	0.53
—	—	—	—	—	—	−0.72	0.45	−0.73	0.36	0.17
—	—	—	—	—	—	—	−0.62	0.93	0.01	−0.48
—	—	—	—	—	—	—	—	−0.53	−0.10	0.36
—	—	—	—	—	—	—	—	—	0.10	−0.48
—	—	—	—	—	—	—	—	—	—	0.02
—	—	—	—	—	—	—	—	—	—	—

In schists (Table 2) SiO_2 also maintains a negative correlation with the rest of major and minor elements related to ferromagnesian minerals, except Zr, which presents the same correlation against SiO_2 as exactly as it happened in the slates. However, the schists begin to show a weak positive correlation between SiO_2–CaO, SiO_2–Na_2O that indicates an increase in the plagioclase content with a parallel increase in silicification.

All these data show that a recrystallization process is taking place in these rocks (slates and schists) with a weak contribution in SiO_2, maintaining clearly the inherited sedimentary characteristics of pelitic rocks. The gneissic and migmatitic rocks show more significant differences. With silicification the total FeO, MgO, TiO_2, H_2O, Al_2O_3, Ba, La and Ni decreases and Na_2O and Pb increases.

In mineralogical terms these results represent an increase in quartz and albitic plagioclase and a decrease in biotite and Al-minerals. The K_2O-content is not related to the variation in SiO_2, and biotite is the major mineral which controls Ba, and Ni evolution.

If we admit that metamorphism at this level is accompanied by an enrichment in SiO_2 in the primitive pelitic composition, this enrichment could be parallel to the increase in Na_2O (albitic plagioclase), Pb and to a lesser extent Rb, that will be furnished to the system. The weak correlation between K_2O and SiO_2 would be explained by the decrease in the amount of biotite simultaneous to the increase of feldspar.

This fact explains why the strong decrease in elements, which form part of micas, is not accompanied by a decrease in K_2O in the same degree within the silicification process. If, in addition, we take into account the weak increase in Rb in this process, it may point out that Rb was added to the system in the quartz–feldspar phase, since if Rb were mainly controlled by the mica minerals, it would decrease in the same measure as these minerals.

The plutonic rocks show a strong correlation among the different elements. Thus, in accordance with the sign of correlation coefficient we can make two groups. The first group

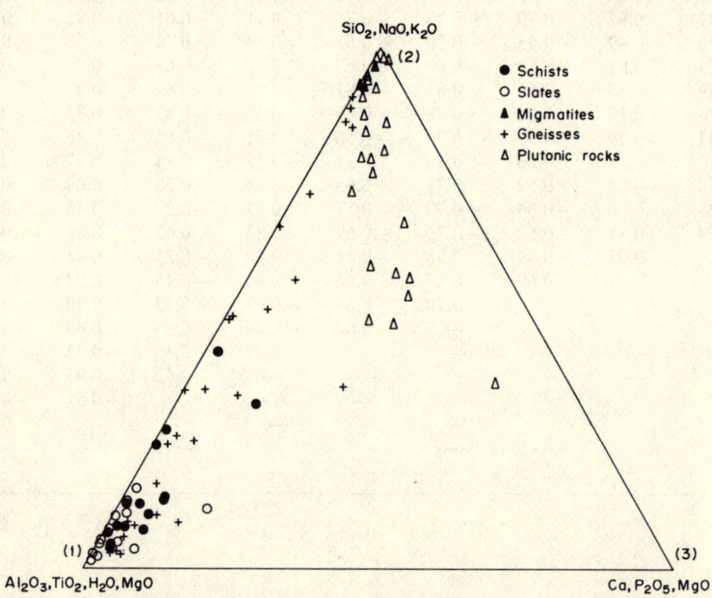

FIG. 4. Q-mode diagram of major elements that have communalities in excess of 0.9 in the three-factor Varimax solution

(A) is composed of elements with positive correlation among them (SiO_2, Na_2O, K_2O, La, Rb, Pb, Th). The second group (B) includes every element with a negative correlation against the A group elements (Al_2O_3, Fe_2O_3, FeO, MgO, CaO, TiO_2, P_2O_5, H_2O, Ba, Sr, Zr, Cu, Ni).

It is possible to observe that the A group is formed by elements which are in residual liquids, whereas the B group includes the minerals of early crystallization. The low correlation coefficient between Na_2O–K_2O is due to the influence in the change of distribution between the liquid and the forming feldspar. In this sense it is necessary to note that while Na_2O is principally included in feldspar minerals, K_2O is distributed mainly in both biotite and K-feldspar. The high values of correlation coefficients show the influence of the crystallization process, which follows a constant rule.

Q-Mode Analysis

Factor analysis is a multivariant method used to study statistical relationships. In this paper we follow the Q-mode method since it was suggested (Davis, 1973) as the most suitable method when dealing with variables with constant sums.

We have adopted programs from Klovan and Imbrie (1971) and Klovan and Miesch (1976). Data were treated in three different ways allowed in the program, with transformed data to be plotted (Figs. 4 and 5). To get a better separation and to diminish the influence of the more abundant elements we selected the

$$X_{ij} = \frac{X_{ij} - X_{i\,\min}}{X_{i\,\max} - X_{i\,\min}}$$

transformation so that the values for all elements and oxides are limited between 0 and 1. Two different Q-mode analyses have been developed, for major and minor elements. In both analyses, we can see the three principal components have a communality higher than 90% but we must distinguish a variation in the influence of three components, since in the Q-mode of the major elements, the variances of each component are: first 48,22, second 39,0, third 5,5 while in the Q-mode minor elements the variances are first 42,11, second 29,84, third 21,33. The minor elements present a similar influence in the second and third components.

In Fig. 4 the major elements are represented by the Q-mode in the three factor Varimax, the first component is affected mainly by the Al_2O_3, TiO_2, H_2O and MgO; the second by SiO_2, Na_2O and K_2O and the third by P_2O_5, CaO, MgO and FeO. In this figure slates and schists define a field at the vertix of the first component, although the schists spread on a band from the 10% of the third component to the 40% of the second component, while the gneissic rocks reach the top of this component (SiO_2 + alkalis), where the most differentiated plutonic rocks and the metamorphic materials possibly affected by infiltrations of the residual liquids are plotted.

On the other hand, the evolution line of plutonic rocks varies from the second to the third component, which is defined by the mafic elements in the crystallization process, by a decrease in SiO_2–K_2O and also by an increase of CaO, MgO, FeO and P_2O_5. Thus, we can discern two trends that are joined at the second component, one trend constituted by plutonic rocks and the other one by metamorphic rocks. The joining takes place when the crystallization process is finishing and the residual liquid is richer in SiO_2, Na_2O and K_2O.

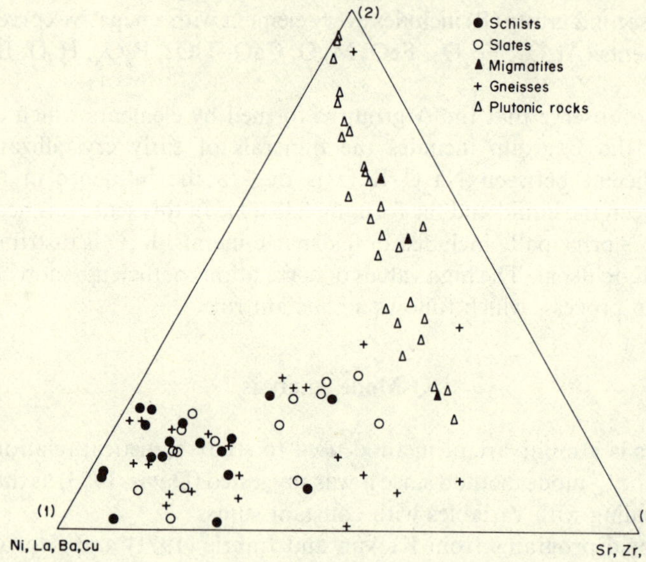

FIG. 5. Q—mode diagram of minor elements that have communalities in excess of 0.9 in the three-factor Varimax solution.

The same representation for minor elements (Fig. 5) shows that the first component is mainly affected by Ni, La and Ba, the second by Rb and Th, and the third by Sr, Zr and Ba. This figure also shows two trends. One of them mainly for slates and schists, from the first to the third component with a weak inflexion towards the second. The hard inflexion to this second component is marked by gneissic and migmatitic rocks. The other trend is conditioned by plutonic rocks which drift in a band from the second to the third component. This line marks the variation during the crystallization process. So, the crystallization of a calcic plagioclase, which includes Sr and Ba, evolves into a plagioclase with a lower CaO content, where Rb is more easily introduced, since this element and also Th remain in the residual liquids.

Thus, some metamorphic rocks which undergo as strong an infiltration as migmatites suffer an enrichment in Rb and Th, giving a similar composition to the most differentiated plutonic rocks and some gneissic and mainly migmatitic rocks.

Conclusions

The characteristics of a metamorphic series in the Spanish Central System allow us to establish the chemical identity between the series of low and medium metamorphic grades. The evolution toward banded and augen gneisses and migmatites is conditioned by a strong transformation. The evolution into this crustal zone shows a strong influence from the inherited sedimentary characteristics of the rocks belonging to the low and medium metamorphic grades.

In spite of the presence of two types of metamorphism (low and intermediate pressure) with different paragenesis in this area, both metamorphic series point to similar chemical features.

The chemical change is perceptible when the rocks reach the highest metamorphic grade. In these levels, SiO_2, CaO, Na_2O, K_2O, Rb and Th show a great diffusion and mobility with a progressive decrease towards a more superficial zone.

The geochemical behavior of associated plutonic rocks marks a positive trend of increase for the same elements (exception of K_2O and Th), a normal variation in the crystallization process at calcoalkaline series. The statistical study shows a convergence between the most differentiated plutonic rocks and the high-grade metamorphic rocks. This joining takes place on the field of influence of Rb, Th, SiO_2, Na_2O and K_2O and suggests the possibility of transfer of these elements to a high-grade metamorphic rocks.

Acknowledgements

We wish to thank J. L. Casaseca, F. Hernan and V. Calleja for their assistance with the preparation of the manuscript.

References

Aparicio, A., Barrera, J. L., Caraballo, J. M., Peinado, M. and Tinao, J. M. (1975) Los materiales graníticos del Sistema Central Español. *Mem. Inst. Geol. Min.* **88**, 145 pp.

Aparicio, A. and Bellido, F. (1976) Geochemical features of the metamorphism in the Sistema Central (Spain). *Chem. Geol.* **17**, 281–293.

Aparicio, A. and Galan, E. (1977) El metamorfismo de bajo grado en el área central del Macizo Hespérico (Sistema Central Español, Toledo). *Bol. Geol. Min. España* (in press).

Davis, J. C. (1973) *Statistics and Data Analysis in Geology,* J. Wiley, London, 550 pp.

Fuster, J. M., Aparicio, A., Casquet, C., Garcia Cacho, L., Mora, A. and Peinado, M. (1974) Interacciones entre los metamorfismo plurifaciales y polifásicos del Sistema Central Español. *Bol. Geol. Min.* **85**, 595–600.

Guitard, G. (1970) Le métamorphisme hercynien mésozonal et les gneiss oeillés du massif du Canigou (Pyrénées orientales). *Mém. B.R.G.M.* **63**, 350 pp.

Klovan, J. E. and Imbrie, J. (1971) An alogorithm and Fortran-IV program for large-scale Q-mode factor analysis and calculation of factor scores. *J. Int. Ass. Math. Geol.,* **3**, No. 1, 61–77.

Klovan, J. E. and Miesch, A. T. (1976) Extended CABFAC and QMODEL computer programs for Q-mode factor analysis of compositional data. *Computers and Geosciences* **1**, No. 3, 161–178.

Lopez Ruiz, J., Aparicio, A. and Garcia Cacho, L. (1975) El Memorfismo de la Sierra de Guadarrama (Sistema Central Español). *Mem. Inst. Geol. Min. de España* **86**, 127 pp.

The Chemical Composition and Origin of the Primeval Continental Crust

ALEXEI A. BEUS

Centre for Natural Resources, Energy and Transport, United Nations, New York, U.S.A.

Abstract

Existing geological and geochemical evidence permits the judgment to be made that the primeval crust was formed through fractionation of magmatic melts from the mantle during early stages of the earth's evolution. The basalt-andesitic composition of the primeval continental lithosphere explains the formation of andesitic "proto-continents" floating above the hot basaltic bed. The consequent evolution of the proto-continents (mainly during the Precambrian time) involves, according to the scheme discussed earlier, an interaction of rocks composing the primeval continental crust with ascending flows of mantle-borne superheated aqueous solutions carrying potassium and silicon from the mantle. The development of metasomatic granitization and subsequent fusion of granitic magma from the metasomatically altered rocks are considered the main effects of such interactions which resulted, finally, in the formation of the present continental lithosphere.

Average compositions of the present continental lithosphere, orthoamphibolite and andesite of the island arcs have been used to reconstitute the composition of the primeval continental crust, using the equations of geochemical balance. Only a basalt–andesitic primeval continental crust fits the requirements of the geochemical balance. The calculated model of this consists of $15,337 \times 10^{24}$ g andesites and of $5,508 \times 10^{24}$ g basalts and has the following average composition (in %): O—46.24; Si—26.22; Al—8.86; Fe—6.32; Mg—2.63; Ca—5.40; Na—2.32; K—1.04; Ti—0.63.

According to the geochemical balance the rock-forming chemical elements composing the lithosphere can be divided into three groups: (1) elements which, after magmatic fractionation of the primeval crust, were not brought into the lithosphere by the mantle-borne aqueous solutions, or emigrated with the residual solutions onto the surface of the crust (Al, Mg); (2) elements which, after magmatic fractionation of the primeval crust, were additionally introduced into the continental lithosphere by aqueous solutions from the mantle (Si, K, O and, in minor quantities, Na); (3) elements of the primeval crust which were partially freed during the process of granitization and were transported in large amounts by the residual endogenous solutions onto the surface of the crust (Ca, in lesser degree Fe).

The problem of chemical composition and origin of the primeval continental crust of the globe is of great significance for the solution of some important theoretical and practical questions, for instance those related to the Precambrian geology and metallogeny.

It seems that the opinion is nearly unanimous that the primeval crust was formed by means of fractionations of the magmatic melts from the mantle during the early stages of the earth's evolution. As shown in many works (Yoder and Tilley, 1962; Green and Ringwood, 1967; Green, 1971; and others) basalts and also andesites of the island arcs type are the most typical primary products fractionating from the mantle. In the last case the strontium isotopes ratio (Faure and Powell, 1972) leaves no doubt regarding the mantle origin of the andesitic magma related to volcanoes of the island arcs. Taking these data into account the composition of the primeval continental lithosphere most probably could be: (a) basaltic; (b) andesitic; or (c) basalt–andesitic. The second and third cases are the most attractive from the general geological viewpoint because a fractionation from the mantle of andesites, which are less dense than basalts, allows one to explain the formation of the andesitic proto-continents "floating" above the hot basaltic bed.

Basalts are the predominant products of fractionation from the mantle at the present time. However, there is no conclusive evidence that the situation was the same during the initial stages of development of the earth's crust when the upper mantle was not yet depleted in the easily fusible components. Moreover, existing data on the composition of the Precambrian metapelitic rocks of the Russian and North American platforms, thoroughly analysed by Ronov and Migdisov (1970), suggest andesitic contents of iron in Archean metapilitic sediments and gradual increase of iron contents from Archean to Proterozoic meta-sediments with the highest iron values in them (corresponding to basaltic source) between Pt_{1-2} and Pt_3. For instance, the average abundance of iron in early Archean metapelitic sediments of the Russian platform was estimated at 4.5% which is close to the iron average of andesites (5.0%). The average rises in the Proterozoic metamorphosed clayey sediments to 6.8% which can be taken as evidence of sufficient increase in quantity of basic volcanics on the surface of continents in the early Proterozoic time in comparison with the early Archeozoic time when the share of rocks having an andesitic composition could have been much more significant.

Andesite, which is the more easily fusible component of this pair of primary products fractionated from the mantle, had to be fractionated at the very beginning, composing the "nuclei" of proto-continents.

The possibility of development of several andesitic, or even more siliceous nuclei or proto-continents, in the early Precambrian has also been discussed by Engel *et al.* (1974) Goodwin (1974), and Lowman (1976).

A consequent evolution of the proto-continents (mainly during the Precambrian time) involves, according to a scheme discussed earlier (Beus, 1972, 1973, 1975), migration of potassium and silicon from the mantle with ascending flows of the super heated aqueous solutions and interaction of these solutions with the rocks composing the primeval

TABLE 1. AVERAGE CHEMICAL COMPOSITION OF THE CONTINENTAL LITHOSPHERE, ANDESITE OF ISLAND ARCS, AMPHIBOLITE AND DIORITE

Elements	Continental lithosphere (without sedimentary layer) $d = 2.80$		Andesite of island arcs (150 analyses) $d = 2.84$		Amphibolite (390 analyses)† $d = 3.00$		Diorite (678 analyses, 48 regions)
	%	g/cm³	%	g/cm³	%	g/cm³	(%)
O	46.6	1.3048	46.8	1.3299	44.6	1.3375	46.00
Si	27.7	0.7756	27.4	0.7789	22.9	0.6876	26.20
Al	8.1	0.2268	9.07	0.2573	8.35	0.2500	8.90
Fe	5.7	0.1596	5.42	0.1534	8.88	0.2655	6.13
Mg	2.4	0.0672	2.00	0.0570	4.31	0.1300	2.61
Ca	4.3	0.1204	4.76	0.1353	7.14	0.2143	5.03
Na	2.3	0.0644	2.49	0.0707	1.85	0.0556	2.50
K	1.9	0.0532	1.17	0.0330	0.71	0.0212	1.41
Ti	0.6	0.0168	0.52	0.0145	0.96	0.0288	0.60
P	0.1	0.0028	0.07	0.0019	0.07	0.0021	0.81
Mn	0.09	0.0023	0.11	0.0031	0.12	0.0037	0.12
H	0.10	0.0028	0.11	0.0031	0.11	0.0033	0.12

† Recalculated using the average H_2O content in basalt.

continental crust. The development of metasomatic granitization of the primeval crust and a subsequent fusion of the siliceous magmas having a granitic composition are considered the main effects of such interactions which resulted finally in formation of the present continental lithosphere of the globe.

The average chemical composition of the continental lithosphere which was used for computing the model composition of the primeval crust is represented in Table 1 (Beus, 1972). More representative average compositions of andesite of the island arcs and amphibolite are used in this study in comparison with the previous works of the author (Beus, 1975, 1976).

Average orthoamphibolite is considered to be more representative of the ancient basaltic rocks than any type of younger basalts. Average andesite of the islands arc certainly differs to some extent from the precambrian andesitic rocks. However, this type of andesite, according to existing data on the strontium isotopes ratio, undoubtedly was fractionated from the mantle.

The physical parameters for calculation (volume and mass of the lithosphere) were taken from works of Mr. Horn and J. Adams, (1964, 1966) and also A. B. Ronov and A. A. Yaroshevsky (1969). The volume of eroded rocks calculated by M. Horn (their mass is 2.04×10^{24} g) was included in the whole volume of the lithosphere. An idea concerning the participation of the mantle-originated aqueous solutions in modification of the primary continental crust accepts the progressing expansion of its volume and mass in the process of modification. This is in accordance with conclusions drawn by many investigators. In order to estimate this expansion in volume, a couple of rock-forming elements should be selected, which presumably were not brought to the crust from the mantle by aqueous solutions and also were not carried out in any significant scale onto the surface of the crust with the residual solutions in the process of crustal geochemical modification. Aluminum is the most probable member of this group of elements; its average contents in "granitic" and "basaltic" layers and also in main types of igneous rocks are subjected only to very minor deviations. The possibility of magnesium emigration from the mantle with aqueous solutions seems also negligible because of strong ties which keep this element under mantle conditions in structures of olivine and pyroxene. Investigators who studied the geochemical balance between eroded primary rocks of the lithosphere and sedimentary rocks (Horn, 1964; Horn and Adams, 1966; Ronov and Yaroshevsky, 1969) found no difficulties in balancing these elements. Their studies supported the assumption that any significant transfer of aluminum and magnesium with the deep-originated residual aqueous solutions into the sedimentary layer did not take place.† The geochemical model suggested in this work is based on the assumptions discussed above.

It is necessary to emphasize that hypotheses of basaltic or entire andesitic compositions of the primeval continental crust do not fit the requirements of a geochemical balance, and therefore must be abandoned.‡

A hypothesis considering the basaltic composition of the primeval continental crust cannot be accepted because the huge masses of iron (0.66×10^{24} g) and magnesium (0.41×10^{24} g) which would have to be freed in the process of the primeval crust modification

† The possibility of such a transfer of some elements in the process of transformation of the substance under an influence of the mantle-originated aqueous solutions (especially at early stages of the crust evolution) should be taken into consideration in any modelling of this kind.

‡ Volume of the primeval crust in these cases can be estimated on the basis of the mass of Al existing in the continental lithosphere assuming that this element was not brought in or emigrated out of the crust during its evolution with aqueous solutions.

into the present lithosphere do not exist in the earth's crust. In the case of the entirely andesitic primary crust (of the island arcs type) the deficit of these metals (0.20×10^{24} g Fe and 0.14×10^{24} g Mg) takes place. Therefore it would be necessary to accept unlikely transportation of large amounts of Fe and Mg into the andesitic primeval crust by the mantle-originated aqueous solutions or also to abandon this hypothesis.

The third assumption of the basalt-andesitic primeval continental crust seems to be the most realistic from the geological viewpoint. The following equations based on the presumption of constant Al and Mg masses in the terrestrial crust after magmatic fractionation of basalts and andesites from the mantle were used in calculations:

$$a.\mathrm{Al}_a + b.\mathrm{Al}_b = \mathrm{Al}_{\mathrm{lm}},$$

$$a.\mathrm{Mg}_a + b.\mathrm{Mg}_b = \mathrm{Mg}_{\mathrm{lm}},$$

where a and b are volumes of the andesitic and basaltic layers in the primeval continental crust accordingly, in cm^3; Al_a and Al_b—amounts of Al in the andesitic and basaltic layers in g/cm^3 (see Table 2); Mg_a and Mg_b—the same for Mg; Al_{lm} and Mg_{lm}—masses of Al and Mg in the present lithosphere, in g.

From the system of equations it is possible to estimate the volumes of andesitic and basaltic layers as 5.337×10^{24} cm^3 and 1.836×10^{24} cm^3 accordingly (Table 2).

The compositional characteristics of the basalt-andesitic primeval continental crust are shown in Table 3. On its basis the rock-forming elements composing the lithosphere in regard to their origin can be separated for three groups: (1) chemical elements which, after magmatic fractionation of basalts and andesites composing the primeval crust were not transported by the mantle-originated aqueous solutions into the lithosphere and were not removed with the residual solutions out of it onto the surface of the crust (aluminum, magnesium); (2) chemical elements which, after magmatic fractionation of basalts and andesites are carried into the continental lithosphere by the mantle-originated aqueous solutions (silicon, potassium, oxygen and, in very minor quantities, sodium); (3) chemical elements of the primeval crust, which are freed in the process of granitization and are transported in large amounts by the residual endogenous solutions onto the surface of the crust directly into the sedimentary layer (calcium and, to a lesser degree, iron).

Quantitative relationships between potassium, silicon and oxygen supposedly carried out of the mantle by aqueous solutions allow one to conclude that potassium hydrosilicate exists in these solutions. The solubility of some potassium hydrosilicates, according to Morey and Penner, (cited from Eitel, 1954), increases sharply at temperatures exceeding 420°C. The latter explains the possibility of mutual potassium and silicon emigration with superheated aqueous solutions in the process of dehydration of the upper mantle.

TABLE 2. PHYSICAL PARAMETERS OF THE PRIMEVAL CONTINENTAL CRUST (MODEL) AND OF THE PRESENT CONTINENTAL LITHOSPHERE

Parameters	Primeval continental crust: Andesitic layer	Basaltic layer	The whole crust	Continental lithosphere (without sedimentary layer)
Volume (10^{24} cm^3)	5.337	1.836	7.173	8.079†
Density (g/cm^3)	2.84	3.00	2.88	2.80
Mass (10^{24} g)	15.157	5.508	20.665	22.621†

† Taking into account eroded crystalline rocks (2.04×10^{24} g).

TABLE 3. CHEMICAL COMPOSITION OF THE PRIMEVAL CONTINENTAL CRUST (MODEL)

Rock-forming elements	Primeval continental crust			Continental lithosphere (without sedimentary layer (10^{24} g)	Brought into primeval crust with solutions from the mantle (10^{24} g)	Carried onto the surface (10^{24} g)	Average composition of the primeval continental crust (%)
	Andesitic layer (10^{24} g)	Basaltic layer (10^{24} g)	The whole crust (10^{24} g)				
O	7.098	2.457	9.555	10.542	0.987	—	46.24
Si	4.157	1.262	5.419	6.266	0.847	—	26.22
Al	1.373	0.459	1.832	1.832	—	—	8.86
Fe	0.819	0.487	1.306	1.289	—	0.017	6.32
Mg	0.304	0.239	0.543	0.543	—	—	2.63
Ca	0.722	0.393	1.115	0.973	—	0.142	5.40
Na	0.377	0.102	0.479	0.520	0.041	—	2.32
K	0.176	0.039	0.215	0.430	0.215	—	1.04
Ti	0.077	0.053	0.130	0.136	(0.006)	—	0.63
Others	0.054	0.017	0.071	0.090	0.019	0.159	0.34
Total	15.157	5.508	20.665	22.621	2.115	0.159	100.00

The different behavior of sodium should be emphasized. The main amount of sodium in the continental crust is related to the fractionation of andesitic and basaltic magmas composing the primeval crust.

Calcium and iron, as mentioned before, compose a group of elements which not only were freed from the basic and intermediate rocks in the process of Precambrian granitization, but also were partly removed by the residual volcanic solutions onto the surface of the primeval crust. The large excess of calcium in the sedimentary layer of the continental crust is well known. It appeared impossible to balance an amount of calcium existing in the sedimentary rocks with that of the eroded upper part of the lithosphere (Horn, 1964; Ronov and Yaroshevsky, 1969).† The probable explanation of this excess supported by calculations of the geochemical balance consists of partial calcium emigration with aqueous solutions in the process of modification of the primeval continental crust onto its surface.

The calculation (Table 3) also supports the possibility of partial iron emigration with the endogenous aqueous solutions onto the Precambrian surface of the lithosphere which was followed by formation of volcano–sedimentary iron concentrations, which later on were strongly metamorphosed and transformed into ironstones and metamorphic iron ore of an iron formation which were so typical of the Precambrian time.

It is seen from Table 3 that the hypotheses considering the basalt–andesitic composition of the primeval continental crust is in good agreement with the geological and geochemical data. On the basis of this model, two main stages in formation of the present continental lithosphere can be suggested: (1) fractionations of andesitic and basaltic magmas from the mantle and formation of the proto-continents; (2) modifications of primary rocks of the primeval continental crust under an influence of aqueous solutions carrying silicon and potassium from the upper mantle.

The similarity between average compositions of the primeval continental crust and that of diorite should be noted. Therefore the primeval continental crust composed of the more basic andesite (analogous in composition to diorite) could be regarded as an alternative to the two-layered basalt–andesitic primeval continental crust modelled here.

It is necessary to point out that the figures describing the composition of the primeval continental crust, which are presented in the above tables, could vary depending on changes in average compositions of andesite and basalt types selected as a basis for calculations. They also depend on reliability of accepted parameters characterizing the present lithosphere. However, this will not change the main conclusions discussed in this paper concerning the predominant role of andesitic volcanics in the composition of the primeval continental crust and of its geochemical modification under the influence of the mantle-borne superheated water solutions.

References

BEUS, A. A. (1972) *Geochemistry of the Lithosphere*, Nedra, Moscow. (English translation of this book, Mir, Moscow, 1976.)

BEUS, A. A. (1973) Average chemical composition and origin of "granitic" and "basaltic" shells in lithosphere. *Int. Geol. Rev.* **15**, 1241.

BEUS, A. A. (1975) Origin and geochemical evolution of the earth's crust. In *Problemy mineralnogo Syriya*, Nauka, Moscow, pp. 116 (in Russian).

† Excess of calcium determined by Horn (1964) in the process of computations varied from 0.0195 to 0.0226×10^{24} g. According to Horn it could be eliminated only by using some arithmetical manipulations.

BEUS, A. A. (1975) Model of the chemical composition of the continental primeval crust of the globe. *Dokl. Akad. Sci. U.S.S.R.* **222**, 44–46, 950–953.

BEUS, A. A. and GOLUBEV, V. S. (1977) Geochemical model of evolution of the continental lithosphere (in print) (in Russian).

EITEL, W. (1954) *The Physical Chemistry of the Silicates,* University of Chicago Press, Chicago, U.S.A.

ENGEL, A. E. J., ITSON, S. P., ENGEL, C. G., STICKNEY, D. N. and CRAY, E. J., Jr. (1974) Crustal evolution and global tectonics: a petrogenetic view. *Geol. Soc. Amer. Bull.* **85**, 843–858.

FAURE, G. and POWELL, J. L. (1972) *Strontium Isotope Geology,* Springer Verlag, New York.

GOODWIN, A. M. (1974) Precambrian belts, plumes and shield development. *Amer. J. Sci.* **274**, 987–1028.

GREEN, D. H. (1971) Composition of basaltic magmas as indicators of origin: application to oceanic volcanism. *Phil. Trans. Roy. Soc. London,* 268–307.

GREEN, D. H. and RINGWOOD, A. E. (1966) The origin of calc–alcaline igneous rocks suite. *Earth Planet. Sci. Lett.* **1**, 307.

GREEN, D. H. and RINGWOOD, A. E. (1967) The genesis of basaltic magmas. *Contr. Mineral. Petrol.* **15**, 103–190.

HORN, M. K. (1964) A computer method for the geochemical balance of the elements. Thesis, Houston, Texas.

HORN, M. K. and ADAMS, J. A. S. (1966) Computer-derived geochemical balances and element abundances. *Geochim. Cosmochim. Acta* **30**, 279–297.

LOWMAN, P. D., Jr. (1976) Crustal evolution in silicate planets: implications for the origin of continents. *J. Geol.* **84**, 1–26.

RONOV, A. B. and YAROSHEVSKY, A. A. (1969) Chemical composition of the earth's crust, p. 37. In *The Earth's Crust and Upper Mantle,* P. J. HART, ed., Amer. Geophys. Union, Washington, 736 pp.

RONOV, A. B. and MIGDISOV, A. A. (1970) Evolution of the chemical composition of the rocks in the shields and sediment cover of the Russian and North-American platforms. *Geochem. Int.* **7**, 294–325.

YODER, H. S. and TILLEY, C. E. (1962) Origin of basalt magmas: an experimental study of natural and synthetic rock systems. *J. Petrol.* **3**, 342–532.

Geochemical Investigation of Permian Andesites from Central Europe

F. J. ECKHARDT

Bundesanstalt für Geowissenschaften und Rohstoffe, Stilleweg 2, 3000 Hannover 51, West Germany

Abstract

In the Upper Carboniferous and the Permian volcanic rocks are more abundant than one expects from surface outcrops in Central Europe. About 500 chemical analyses always show similar compositions over the whole region. The chemical composition of this calc-alkaline rock suite covers the range from basalts to rhyolites—including andesites and dacites. This composition can easily be compared with that of the rocks in the young andesite regions (island arcs, South America, Turkey). The concentrations of the stable trace elements (average from 45 andesites) are close to the known data from young andesites too. Applying Ringwood's model one may conclude that the permocarbonian volcanic rocks have assimilated crustal material too.

Arguments for plate-tectonics during the Carboniferous and the Permian in Europe have been published by several authors in the last years. The application of these geochemical investigations is proving this assumption.

Subsequent volcanism (as defined by Stille) produced large amounts of eruptive rocks during the Permo-Carboniferous of Central Europe. These rocks have been described as "quartzphorphyries, porphyries and only a few melaphyres" by the authors of geological maps and geological textbooks.

The intensive oil and gas exploration during the last 25 years has provided proof of Permian volcanic rocks in several areas of the Central European Basin. These rocks have been modified by low-grade metamorphism of the zeolite and greenschist facies (spilitization).

The eruptive rocks of the Central European Basin can be related to the tectonic lineaments known in this area (Dvorak and Paproth, 1969). There should be no difficulty in connecting the rocks of the other areas of Permo-Carboniferous volcanic activity to tectonic elements, since Central Europe experienced intensive tectonic partitioning during the Variscian orogenesis.

Different volcanologic and morphologic types of volcanism have been described (Schwab, 1968): ignimbrites, different types of volcanoes, subvolcanic rocks, dome structures, lava flows, glowing avalanches. These phenomena are significant for the morphology of the young andesitic provinces known in Central America and Turkey (Anatolia).

When collecting petrological and geochemical data from all areas of Central Europe with Permo-Carboniferous volcanic rocks one will always find the association basalt–porphyrite–(=andesite)=rhyolithe. Very often the basic rocks are the lowest part of the sequence. This fact does not exclude that they are also present in geologically younger periods of this age.

About 500 chemical analyses, some taken from literature, some XRF-analyses carried out at the B.G.R., have been compiled (Eckhardt, 1977). The quality of the analyses has been controlled as closely as possible. Still, nearly all of the analyses could be used for the following discussion.

Calculations and different composition diagrams (K_2O/SiO_2, Alk/SiO_2, AFM, etc., Eckhardt, 1977) show that:

1. all the volcanic rocks are very similar in the range of composition of the main elements;
2. these volcanic rocks represent a calc-alkaline rock association, as is well known in late andesitic provinces.

The values for concentrations of K_2O, Na_2O and CaO show some scattering at constant SiO_2 values. This can be explained by the mobility of these elements under the influence of low-grade metamorphism as stated already (Eckhardt, 1968). In spite of this scattering one can demonstrate the calc-alkaline character of the rock suite (Fig. 1). Following the recommendation of Pearce and Cann (1973) the triangle Ti–Zr–Y is proving this too.

Besides the major elements eighteen trace elements have been determined in ninety samples from the Central European Basin by XRF in the B.G.R.

The LIL elements (large-ion-lithophil elements: K, Na, Rb, Sr, Ba, Pb, La) could not be used for geochemical correlations due to their sensitivity to metamorphism (Pearce and

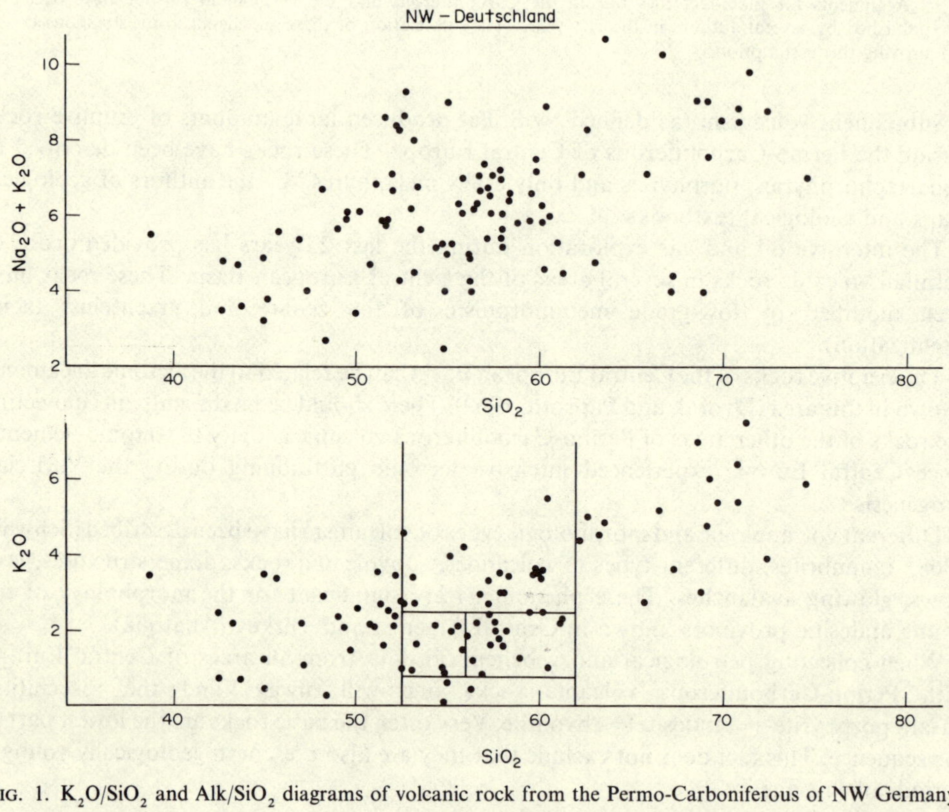

FIG. 1. K_2O/SiO_2 and Alk/SiO_2 diagrams of volcanic rock from the Permo-Carboniferous of NW Germany.

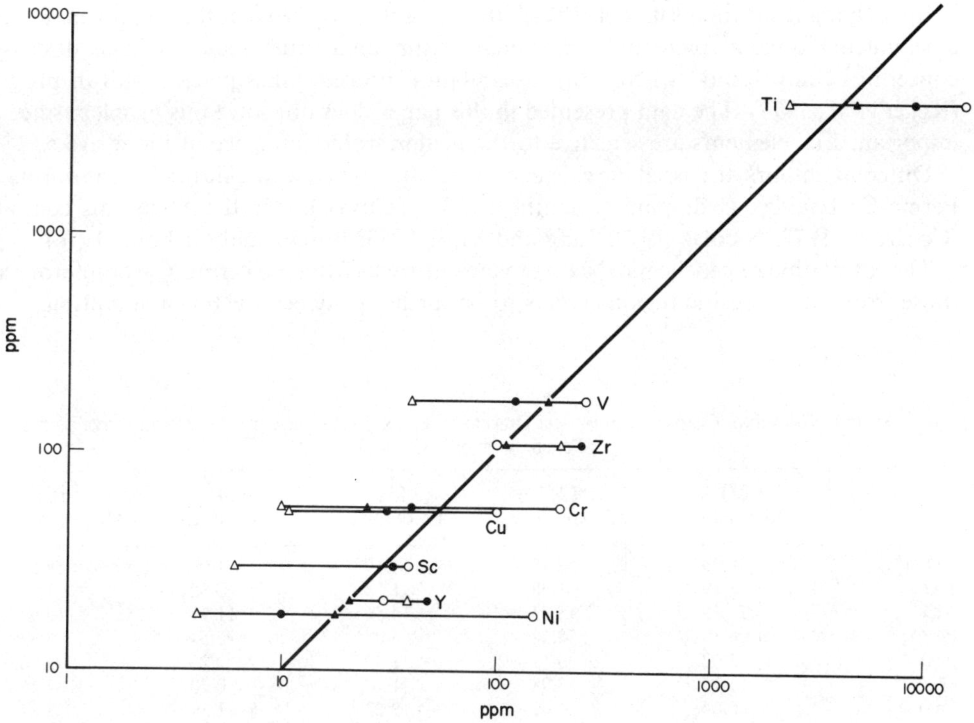

FIG. 2. Trace element abundances of island-arc andesites (▲) compared with those of granites (△), basalts (○) and NW-German andesites (●). (Data from Taylor, 1968; Jakeš and White, 1972.)

Cann, 1973; Pearce, 1975). Nevertheless it is possible to draw conclusions by comparing other trace elements (Fig. 2, Tables 1–3).

1. The samples show a maximum SiO_2 concentration between 55–60% SiO_2 with a range of 45–72%.
2. K_2O–60 is about 3%, a high value.
3. The concentrations of Ti, Y, Zr, Th and Cr are higher than the data published for island–arc andesites (Taylor, 1968). The concentrations of these "stable" elements (Pearce, 1975) show a distinct trend to the values known for granites and the continental crust (Fig. 2, Table 2) (Taylor, 1968; Taylor and White, 1966).
4. An analogous trend can be shown for other trace elements (V, Nb, Cu, Ce, La), although the mean values are scattered (Eckhardt, 1977) (i.e. a certain amount of sensitivity to metamorphism is observed).

The geological environment and the geochemical data of the major and the trace elements lead to the conclusion that volcanic rocks were produced in Central Europe during the Permo-Carboniferous which are analogous to the volcanic rocks present in the late regions of the Andes and of Anatolia. These are andesites of the "continental margin type".

In applying Ringwood's model (1974) different authors agree that the melting material of a subducted oceanic plate has additional, assimilated crust material. The discussion concerning "how" and "where" this assimilation process takes place is still in progress (Keller *et al.*, 1977). The data presented in this paper does not solve this problem since the important LIL elements are sensitive to the metamorphic influence of these rocks.

Different authors discussed the existence of subduction zone in Central Europe during the Permo-Carboniferous dipping from the south (Tethys) under the European continent (Cosgrove, 1972; Nicolas, 1972; Jung and Vinx, 1973; Lorenz and Nicholls, 1976).

The correlation of geochemical data of volcanic rocks from the Permo-Carboniferous and those from late andesitic regions seems to be further evidence for this assumption.

TABLE 1. CHEMICAL COMPOSITION OF THE DIFFERENT ROCK TYPES FROM THE ANDESITE SUITE OF THE PERMIAN OF NW GERMANY

	(1) 13 Basalts	(2) 58 Andesites	(3) 11 Dacites	(4) 8 Rhyolithes	(5) (All samples 1-4)
SiO_2 (%)	48.76	57.79	64.10	71.90	58.51
TiO_2	1.19	1.38	1.14	0.38	1.23
Al_2O_3	15.39	15.33	14.56	11.91	14.94
Fe_2O_3	6.89	5.58	5.08	3.51	5.52
FeO	3.62	3.12	1.80	0.25	2.76
MgO	5.17	3.98	1.78	0.20	3.55
CaO	6.64	4.01	2.22	1.39	3.94
K_2O	2.41	2.41	3.86	6.15	2.92
Na_2O	3.10	4.17	3.42	2.58	3.78
P_2O_5	0.21	0.23	0.22	0.04	0.21
K_2O/Na_2O	0.78	0.58	1.13	2.38	0.77
K/Rb	167	194	294	199	189
Sr (ppm)	194	218	124	78	191
Rb	120	103	177	256	128
Nb	14	19	28	31	20
Zr	171	203	323	434	234
Y	32	39	54	72	43
Pb	23	38	31	14	33
Th	8	10	17	23	12
Ni	25	15	11	6	15
Co	42	29	27	0	28
Mn	1444	1198	842	220	1103
V	150	124	68	7	111
Zn	129	180	124	29	152
Cu	20	28	27	48	29
Cr	124	86	25	2	77
Ce	61	72	114	131	81
La	28	37	69	80	43
Ba	419	542	631	1006	576
Sc	35	33	25	9	30

Table 2. CHEMICAL COMPOSITION OF ANDESITIC ROCK SUITES OF THE ISLAND-ARC TYPE (Taylor, 1968) FROM CENTRAL ANATOLIA (TURKEY) (Keller *et al.*, 1977) AND THE NW GERMAN PERMIAN COMPARED WITH DATA FROM THE CONTINENTAL CRUST (Taylor, 1967; Taylor and White, 1966)

	(1) Andesites (Island arc)	(2) Andesites (Turkey)	(3) Permian (NW Germany)	(4) Continental crust
SiO_2 (%)	59.50	61.80	58.51	60.3
TiO_2	0.7	0.8	1.23	1.0
Al_2O_3	17.2	17.17	14.94	15.6
Fe_2O_3	6.7	5.5	5.52	7.9
FeO			2.76	
MgO	3.42	2.6	3.55	3.9
CuO	7.03	5.58	3.94	5.8
K_2O	1.60	2.96	2.92	2.5
Na_2O	3.68	3.47	3.78	3.2
P_2O_5	—	0.21	0.21	0.24
K/Rb	430	242	189	240
K_2O/Na_2O	0.43	0.85	0.77	0.78
Sr (ppm)	385	782	191	375
Rb	31	102	128	85
Nb	4	—	20	20
Zr	110	164	234	145
Y	21	23	43	33
Pb	7	—	33	17.5
Th	2	—	12	9.6
Ni	18	9.9	15	77
Co	24	7.3	28	25
Mn	1200	697	1103	950
V	175	—	111	145
Zn	—	57.6	152	—
Cu	54	39.5	29	55
Cr	56	28.7	77	105
Ce	24	88	81	60
La	12	38	43	30
Ba	270	1010	576	425
Sc	30	—	30	22

TABLE 3. GEOCHEMICAL CHARACTERISTICS OF DIFFERENT ANDESITE-TYPES COMPARED WITH THE DATA OF PERMIAN ANDESITES (Taylor, 1968; Jakeš and White, 1972; Keller *et al.*, 1976)

	Island arc	Continental margin (Andes type)	Konya (Turkey)	Permian (NW Germany)
SiO_2-range	50–66%	56–70%	56–68%	45–72%
K_2O/Na_2O	0.8	0.6–1.1	(0.45)–0.6–1.3	0.4–2.0
K_2O-60	1.6%	2.2%	2.9%	3.0%
Trace elements (at SiO_2-60)	Low Rb, Ba, Sr, Zr, Th	High Rb, Ba, Sr, Zr, Th, U	High Rb, Ba, Sr, Zr	High Y, Zr, Th
Rb	25–30 ppm	80 ppm	100 ppm	140 ppm
Ba	270 ppm	680 ppm	1000 ppm	600 ppm
Sr	400 ppm	700 ppm	800 ppm	160 ppm
K/Rb	400	230	240	230

References

COSGROVE, M. E. (1972) The geochemistry of the potassium-rich Permian volcanic rocks of Devonshire, England. *Contr. Mineral. Petrol.* **36**, 155–170.

DVORAK, J. and PAPROTH, E. (1969) Über die Position und die Tektogenese des Rhenoherzynikums und des Sudetikums in den mitteleuropäischen Varisziden. *Neues Jb. Geol. Paläontol., Monatsh.* 65–88.

ECKHARDT, F.-J. (1968) Vorkommen und Petrogenese spilitisierter Diabase des Rotliegenden im Weser–Ems–Gebiet. *Geol. Jb.* **85**, 227–264.

ECKHARDT, F. -J. (1977) Dep permische Vulkanismus Mitteleuropas. *Geol. Jb.*, D (in press).

JAKEŠ, P. and WHITE, A. J. R. (1972) Major and trace element abundances in volcanic rocks of Orogenic areas. *Geol. Soc. Amer. Bull.* **83**, 29–40.

JUNG, D. and VINX R. (1973) Einige Bemerkungen zur Geochemie der Magmatite des Saar–Nahe–Pfalz–Gebietes. *Ann. Sci. Univ. Besançon, Géol.* 3° Sér., n°18, 197–202.

KELLER, J., JUNG, D., BURGATH, K. and WOLFF, F. (1977) Geologie und Petrologie des neogenen Kalkalkali-Vulkanismus von Konya (Erenler Dağ–Alaca Dağ Massiv, Zentralanatolien). *Geol. Jb.*, B, **25** (in press).

LORENZ, V. and NICHOLLS, I. A. (1976) The Permo-Carboniferous basin and range province of Europe. An application of plate tectonics. In *The Continental Permian in Central, West, and South Europe*, W. FALKE, ed., D. Reidel Publ. Co., 313–342.

NICOLAS, A. (1972) Was the Hercynian Orogenic belt of Europe of the Andean type? *Nature* **236**, 221–223.

PEARCE, J. A. (1975) Basalt geochemistry used to investigate past tectonic environments on Cyprus. *Tectonophysics* **25**, 41–67.

PEARCE, J. A. and CANN, J. R. (1973) Tectonic setting of basic volcanic rocks determined using trace element analyses. *Earth Planet. Sci. Lett.* **19**, 290–300.

RINGWOOD, A. E. (1974) The Petrological evolution of island arc systems. *J. Geol. Soc. Lond.* **130**, 183–204.

SCHWAB, M. (1968) Zur Vulkanotektonik im Permokarbon. *Freiberger Forschungshefte* **C 219**, 109–125.

TAYLOR, S. R. (1967) The origin of growth of continents. *Tectonophysics* **4**, 17–34.

TAYLOR, S. R. (1968) Geochemistry of andesites. In *Origin and Distribution of the Elements*, L. H. AHRENS, ed., Pergamon Press, Oxford, 559–583.

TAYLOR, S. R. and WHITE, A. J. R. (1966) Trace element abundances in andesites. *Bull. Volcanol.* **29**, 177–194.

Geochemistry of the Carbonatite Complexes in East Africa

ISAAC O. NYAMBOK

Present address: *Department of Mineralogy and Petrology, Institute of Geology, University of Uppsala, Box 555, S–751 22, Uppsala, Sweden*

Permanent address: *Department of Geology, University of Nairobi, Box 30197, Nairobi, Kenya*

Abstract

Carbonatite centres are characterized by the common occurrence of carbonatites and alkalic igneous rocks, both as volcanic and as intrusive rocks. In the eastern rift zone, the carbonatite complexes are sodium-rich while in the western rift zone they are potassium-rich. The complexes are undersaturated in silica content and enriched in alkalis. The two suites of alkaline rocks have variable K/Rb ratios as well as variable K and Rb contents. The K/Rb ratio (approximately 287) of lavas from the western rift zone is very close to estimates for the average continental crust while lavas from the eastern rift zone have higher K/Rb ratios.

The ratios $^{87}Sr/^{86}Sr$ of the carbonatite complexes show a distinct variation between those of the eastern rift zone and those of the western rift zone. The carbonatite complexes of the eastern rift zone have lower $^{87}Sr/^{86}Sr$ ratios than those of the western rift zone. The $^{87}Sr/^{86}Sr$ ratios of the potassium-rich alkaline rocks of the western rift zone vary considerably and are thought to have resulted from mixing of crustal and upper mantle material.

The carbonatite complexes are greatly enriched in trace elements–Sr, Ba, Nb and REE. Most of these trace elements are concentrated in the carbonatites rather than the associated alkaline rocks.

Introduction

There are a number of carbonatite centres in the eastern and western rift zones in East Africa. In the carbonatite centres, there is typically an association of carbonatites and alkaline rocks. Commonly two types of carbonatites occur, namely dolomitic and calcitic carbonatites. On the other hand, the composition of the alkaline rocks of the rift zones varies considerably. The eastern rift zone is characterized by sodium alkaline rocks while the western rift zone is dominated by potassium alkaline rocks. The rock series of the eastern rift zone is composed of alkaline picrites, nephelinites, meliltites, phonolites, ijolites, nepheline syenites and carbonatites and the rock series of the western rift zone is composed of katungites, ugandites, leucitites, basanites, melilitites, nephelinites, mafurites, trachybasalts, trachytes, latites and carbonatites. Comparison of the contents of SiO_2, K_2O and Na_2O and the ratios of K/Rb, Rb/Sr and $^{87}Sr/^{86}Sr$ of the eastern and western rift zones is made in order to unravel the petrogenetic variations of the two zones.

Geological Setting of the Carbonatite Complexes

The carbonatite centres in East Africa are closely associated with the Rift Valley system. In general, they are localized individual occurrences, although they are distributed over a large area. The centres in the eastern rift zone can be separated into two groups; the older intrusive complexes and the younger volcanic and intrusive complexes (King *et al.*, 1972). The older complexes occur at Budeda, Panda, Sekululu, Sukulu and Tororo while the younger ones occur at Elgon, Moroto, Toror, Homa and Wasaki. These complexes intrude the Precambrian basement rocks. The geology of some of the complexes is summarized below.

One of the small occurrences is Budeda. The *Budeda* complex is composed of melteigites–ijolites which are surrounded by fenitized basement granite. Carbonatites occur only in small masses (King and Sutherland, 1966). The *Panda Mbeya* complex occurs in the south-west part of Tanzania. It intruded the Precambrian Ubendian System and consists of carbonatites, volcanic ash and fenitized basement rocks. The complex is roughly circular (Suwa *et al.*, 1975). The *Tororo* complex consists of two bodies of carbonatite, separated from one another by fenitized rocks; the northern complex consists of a series of steeply inclined concentric rings of carbonatites (Williams, 1952) and the southern consists of nepheline syenites and ijolites. On the other hand, the *Sukulu* complex is composed almost entirely of carbonatite with the exception of a narrow ring of alkaline silicate rocks.

At *Toror*, the carbonatite complex intruded fenitized granitic gneiss. The whole complex is also intruded by phonolitic and trachytic dykes. Erosion exposed a carbonatite core which is surrounded by an ijolite outer ring. The *Homa* complex, which intruded andesitic lava of the Nyanzian system, is composed of carbonatite and ijolite. It is a multi-centred complex. Plugs and dykes of nephelinites and melilitites occur (Saggerson, 1952). The *Wasaki* complex consists of ijolites surrounded by fenites. Carbonatite masses occur at Sokolo, and phonolitic lavas at Nyamaji (Pulfrey, 1954).

In the western rift zone, the rocks are predominantly volcanic. The volcanic activity in this region is thought to date from Upper Pliocene to Present. The Birunga field is composed of eight major volcanoes (Bell and Powell, 1969) and has many plagioclase-bearing rocks (Higàzy, 1954). In the south of Birunga, nepheline–leucite–melilite rocks are characteristic. In the Bunyaruguru field there is a greater variety and the rock types are predominantly ugandite, mafurite and katungite.

Chemistry of the Carbonatite Complexes

The chemical compositions of the carbonatites and the associated alkaline rocks of the rift zones vary considerably (Table 1). Data on complete chemical analyses of the carbonatite complexes are well documented (Bell and Powell, 1969; King, 1965; Saggerson, 1952). The common feature of the chemistry of carbonatite complexes of both the eastern and western rift zones is the high content of alkalies. The alkalies show significant variation in sodium and potassium. The alkali content and variation forms a starting point for a comparative study of the complexes of eastern and western rift zones, which results in the division of the complexes into two series: the sodium alkaline rocks and the potassium alkaline rocks. The sodium alkaline rocks are found in the eastern rift zone and the potassium alkaline rocks in the western rift zone (Table 1 and Fig. 1A). In general, the potassium-rich rocks of the western rift zone are dominated by melanocratic varieties and the sodium-rich rocks of the eastern rift zone by leucocratic varieties.

The alkaline rocks are enriched in titanium. The series of potassium alkaline rocks is more enriched in titanium than that of sodium alkaline rocks. Maximum content of titanium is

TABLE 1. SOME CHEMICAL DATA ON THE CARBONATITE COMPLEXES OF EASTERN AND WESTERN RIFT ZONES

Locality	Rock type	SiO_2 (wt.%)	CaO (wt.%)	K_2O (wt.%)	Na_2O (wt.%)	Rb (ppm)	Sr (ppm)	La (ppm)	Ce (ppm)	K/Rb (wt.)	Rb/Sr (wt.)	$^{87}Sr/^{86}Sr$ corrected atomic
Eastern Rift												
Budeda	carbonatite					0.0	2992				0.0	0.7032
	nepheline syenite	56.71	5.38	5.33	7.71	170.0	865			314	0.197	0.7036
	ijolite	45.68	12.32	3.62	5.45	71.3	437			508	0.163	0.7037
Homa	carbonatite	1.30	52.51		0.12	0.0	2185	349	568		0.0	
Panda-Mbeya	carbonatite	3.52	50.19	0.0	0.20	0.0	5100	510	490		0.0	
Toror	phonolite	52.3	3.18	3.84	8.80	88.0	330			436	0.267	0.7042
	nephelinite	46.2	14.30	3.12	9.32	79.0	1308			395	0.060	0.7036
	carbonatite					0.0	2031				0.0	0.7033
Tororo	carbonatite					0.0	5075				0.0	0.7037
	nepheline syenite			4.89		74.1	528			660	0.140	0.7043
	ijolite			2.59		66.7	2081			388	0.032	0.7040
Sukulu	carbonatite					0.0	4628				0.0	0.7036
Wasaki	carbonatite	0.93			0.38	0.0	1275	322	957			
Western Rift												
Birunga	nephelinite	38.48	13.01	5.70	4.96	147.0	2265			387	0.065	0.7049
	melilitite	35.35	15.72	5.08	4.95	114.0	2882			445	0.040	0.7048
	leucitite	43.37	10.67	4.86	3.10	200.0	1002			243	0.150	0.7062
	basanite	44.74	9.70	3.68	3.05	145.0	1529			254	0.095	0.7075
	ugandite			2.35		98.0	862			240	0.114	0.7059
Bunyaruguru	mafurite	39.22	12.47	5.41	1.28	176.0	2122			307	0.083	0.7055
	katungite	33.72	16.10	4.73	1.34	200.0	2114	62		237	0.095	0.7048
	ugandite	40.47	8.06	3.46	0.68	163.0	1054	35		212	0.152	0.7055
Katunga	katungite	36.21	16.79	4.09	1.32	118.0	2560			346	0.046	0.7051
	carbonated lava					99.0	1605				0.062	0.7046
Katwe-Kikorongo	katungite	35.51	16.00	2.11	1.25	83.0	2108			254	0.039	0.7044
	carbonated lava					0.0	1149				0.0	0.7055

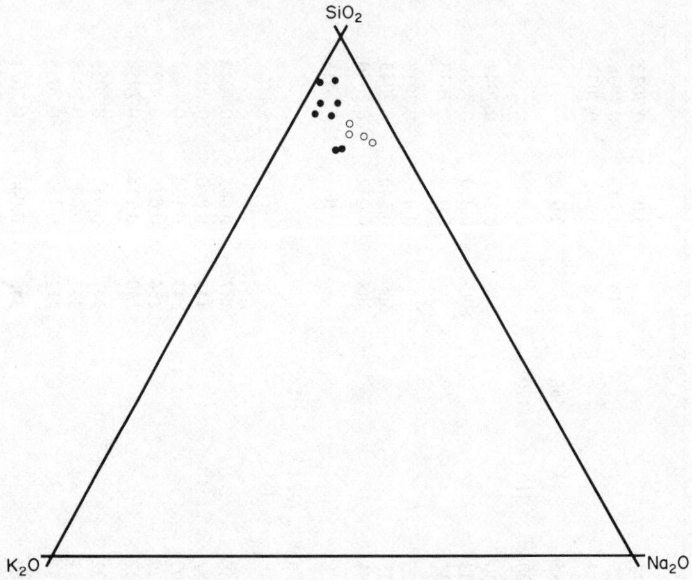

FIG. 1A. Ternary plot of SiO_2, K_2O and Na_2O wt.-% from the carbonatite centres. ○ = alkaline rocks from the eastern rift; ● = alkaline rocks from the western rift.

found in the rocks which are depleted in silica and minimum content in rocks with high silica.

The two rock series have been analysed for the content of K and Rb (Bell and Doyle, 1971). The results summarized in Table 1 show that the lavas from the western rift zone are more enriched in K and Rb. However, the eastern rift zone seems to have higher K/Rb ratios than the western rift zone. Bell and Doyle (1971) found that the average K/Rb ratio for the carbonatite complexes in the eastern rift zone is close to 377 while in the western rift zone it is about 287. This is an interesting variation and it focuses on the possible genetic variation of the two rock series. It is important to note that the carbonatites in these complexes have values of K and Rb too low to be detected except in rare cases of effusive carbonatites.

Higàzy (1954) first investigated the content of trace elements in the alkaline rocks of the western rift zone and gave data concerning the enrichment of trace elements in these rocks. Bell and Powell (1969) followed with the study of the Sr isotopic composition which threw more light on the genesis of the rocks. The content of Sr and the initial $^{87}Sr/^{86}Sr$ ratio can give indications in the case of plutonic and volcanic igneous rocks as to the source of a particular magma and whether it has been subjected to interaction with other rocks. From Table 1, it is evident that the isotopic ratio of Sr is higher in the western rift zone than in the eastern rift zone. The carbonatite complexes are extremely enriched in strontium, with carbonatites having the highest Sr content in the eastern rift zone. In general, the rocks of the eastern rift zone have higher Sr contents than those of the western rift zone (Table 1). However, the initial $^{87}Sr/^{86}Sr$ ratios in the carbonatites and the associated alkaline rocks show no significant variation within individual complexes, indicating that the carbonatites and the associated alkaline rocks are most likely comagmatic. Nonetheless, the carbonatites show slightly lower initial $^{87}Sr/^{86}Sr$ ratios with those of the eastern rift zone having the lowest ratios. The carbonatites of the western rift zone do not show a distinctly low initial $^{87}Sr/^{86}Sr$ ratio and vary significantly. The sodium-rich alkaline rocks of the eastern rift zone

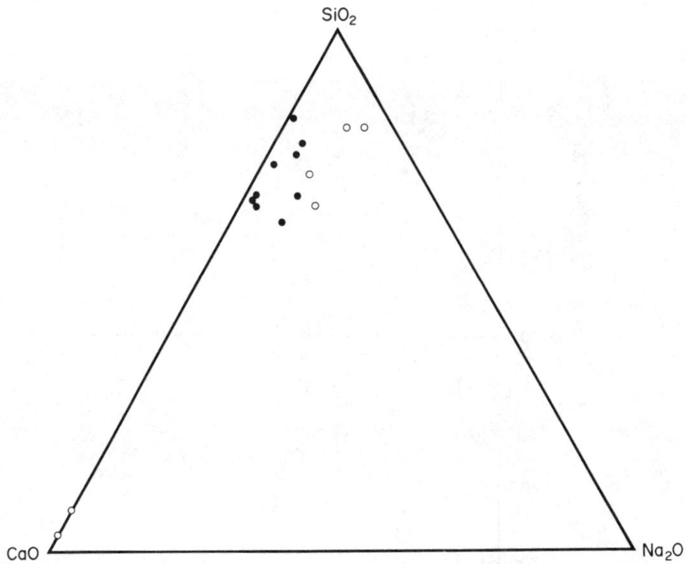

FIG. 1B. Ternary plot of SiO_2, Na_2O and CaO from the carbonatite centres. ○ = rock samples from the eastern rift; ● = rock samples from the western rift.

have initial $^{87}Sr/^{86}Sr$ ratios in the range of 0.7036–0.7045 and on the other hand the potassium–rich alkaline rocks of the western rift zone have higher initial $^{87}Sr/^{86}Sr$ ratios in the range of 0.7046–0.7055. A comparison is made between the carbonatite complexes of the eastern and western rift zones based on plots of K/Rb against initial $^{87}Sr/^{86}Sr$ ratios (Fig. 2A); a trend of distinct grouping emerges.

As shown in Fig. 2B the initial $^{87}Sr/^{86}Sr$ ratios of the potassic rocks of the western rift zone and those of the sodic rocks of the eastern rift zone plotted against Rb/Sr ratios, though variable, show a pattern of grouping. This grouping demonstrates the variation between the carbonatite complexes of the eastern and western rift zones. The Rb/Sr ratio of the alkaline rocks varies considerably. However, on the average it appears to be higher in the eastern rift zone than in the western one. Some values of the Rb/Sr ratio of the eastern rift zone are $\geq$ 0.196 while in the western rift zone the highest values are approximately 0.165.

Usually strontium, barium, rare-earths, niobium and rubidium are characteristic elements of the carbonatite complexes. It is notable that the complexes in East Africa are very enriched in these trace elements. The Sr content is generally greater than 2000 ppm in carbonatites of the eastern rift zone, but somewhat lower in carbonatites of the western rift zone. Another characteristic feature of the carbonatite complexes is the exceptionally high concentrations of REE. There have been studies on the concentrations of REE in the carbonatites and the associated alkaline rocks. However, a detailed study on the distribution and concentration of the rare-earth elements is still needed. Such a study would further throw more light on the genesis and source magma of the carbonatite complexes. However, the available data show REE to be more highly concentrated in carbonatites than in the associated alkaline rocks. The carbonatites are particularly rich in Ce-earth elements (Mwinyihamisi, unpublished data; Barber, 1974).

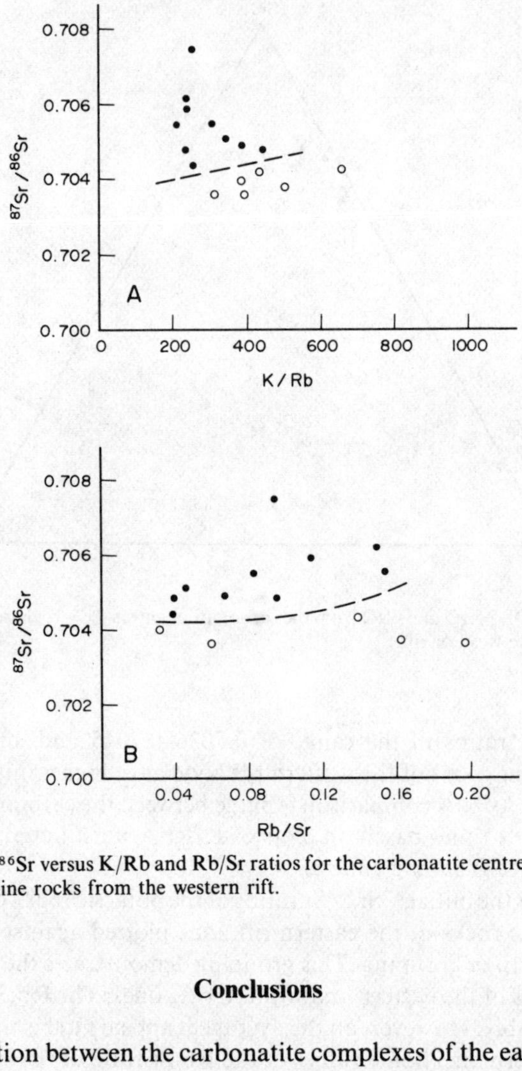

FIG. 2. Plots of initial $^{87}Sr/^{86}Sr$ versus K/Rb and Rb/Sr ratios for the carbonatite centres. ○ = alkaline rocks from the eastern rift; ● = alkaline rocks from the western rift.

Conclusions

The chemical variation between the carbonatite complexes of the eastern and western rift zones leads to the question: Is there a clear difference in the genesis of these rocks? Clearly the carbonatites and the associated alkaline rocks at individual complexes are comagmatic. Furthermore, there is a clear correlation between the carbonatite complexes within each rift zone. On the other hand there is a clear distinction between the carbonatite complexes of the eastern rift zone and those of the western rift zone. The K/Rb ratios of lavas from the western rift zone are very close to estimates of the average continental crust. In addition, the $^{87}Sr/^{86}Sr$ ratios of the potassic alkaline rocks of the western rift zone vary considerably and perhaps resulted from contamination by crustal material. This variation can be attributed to the depth of the magma chamber and possibly implies a distinctive temperature difference between melts of the eastern rift zone and the western one.

The lower $^{87}Sr/^{86}Sr$ ratios of carbonatite complexes of the eastern rift zone suggest that these rocks may be derived from deeper magma chambers, possibly below the upper mantle region.

Acknowledgements

I wish to thank Dr. O. A. Mwinyihamisi for allowing me the use of his unpublished data. Doc. B. Lindqvist kindly read the manuscript and made helpful suggestions. The Swedish Natural Science Research Council provided financial support which is greatly acknowledged.

References

BARBER, C. (1974) The geochemistry of carbonatites and related rocks from two carbonatite complexes, south Nyanza, Kenya. *Lithos* **7**, 53–63.

BELL, K. and DOYLE, R. G. (1971) K–Rb relationships in some continental alkalic rocks associated with the East African Rift Valley System. *Geochim. Cosmochim. Acta* **35**, 903–915.

BELL, K. and POWELL, J. L. (1969) Strontium isotopic studies of alkalic rocks: The potassium-rich lavas of the Birunga and Toro–Ankole regions, East and Central Equatorial Africa. *J. Petrol.* **10**, 536–572.

HIGÀZY, R. A. (1954) Trace elements of volcanic ultrabasic-potassium rocks of south-west Uganda and adjoining part of Belgian Congo. *Bull. Geol. Soc. Amer.* **65**, 39–70.

KING, B. C. (1965) Petrogenesis of the alkaline igneous rock suites of the volcanic and intrusive centres of eastern Uganda. *J. Petrol.* **6**, 67–100.

KING, B. C., LE BAS, M. J. and SUTHERLAND, D. S. (1972) The history of the alkaline volcanoes and intrusive complexes of eastern Uganda and western Kenya. *J. Geol. Soc.* **128**, 173–205.

KING, B. C. and SUTHERLAND, D. S. (1966) The carbonatite-bearing complexes of eastern Uganda. In *Carbonatites,* London, Wiley & Sons, 73–126.

PULFREY, W. (1954) Alkaline syenites at Ruri, South Nyanza, Kenya. *Geol. Mag.* **91**, 209–219.

SAGGERSON, E. P. (1952) Geology of the Kisumu district. *Kenya Geol. Surv.*, Rep. no. 21.

SUWA, K., OANA, S., WADA, H. and OSAKI, S. (1975) Isotope geochemistry and petrology of African carbonatites. *Phys. Chem. Earth* **9**, 735–745.

WILLIAMS, C. E. (1952) Carbonatite structure: Tororo Hills, eastern Uganda. *Geol. Mag.* **89**, 286–292.

Hydrogen, Carbon, Nitrogen and Helium in Gases of Igneous Rocks

I. A. PETERSILJE and W. A. PRIPACHKIN

Geological Institute of Kola Branch of the U.S.S.R. Academy of Sciences, 184200 APATITY, U.S.S.R.

Abstract

The gases of closed pores of igneous rocks were investigated. A research of such kind offers a new geochemical method that can be used for the elucidation of complicated petrogenetic problems of gaseous regime of the earth's crust and for practical geological problems.

In the igneous rocks of earth–alkaline series the content of gases varies from 0.4 to 9.5 cm^3/kg of rock. Granitoids contain a minimum quantity of gases. In the composition of gaseous phase hydrogen evidently prevails (up to 60–88% of gaseous mixture); hydrocarbon gases, nitrogen, helium, argon and sometimes carbon dioxide are the rest of the species.

Agpaitic alkaline rocks (Khibiny, Lovosero, Illimaussak, Middle-Tatarien and other massifs) in their gaseous phase amounts and in composition differ sharply. The predominating component of the composition of these gases is not hydrogen, as in the previous case, but methane which amounts to 92% of the whole volume of gases, maximal quantity of gases in rocks reaches up to 150 cm^3/kg of rock. Agpaitic alkaline rocks by gaseous phase composition resemble the rocks of sedimentary formation, but methane of alkaline rocks differ from methane of sedimentary rocks in the carbon isotopic composition. The average isotopic composition of methane carbon, expressed in per cent of δC^{13}, is 0.85 for the Khibini massif, 0.53 for the Lovosero massif and 0.77 for Illimaussak (PDB etalon).

The carbon isotope distribution between different fractions of hydrocarbon gases (methane–ethane–propane) is the reverse of that by sedimentary rock gases. The former allows the confirmation with a sufficient degree of certainty the abiogenetic origin of hydrocarbons in agpaitic alkaline rocks.

The prevailing part of non-carbonic carbon in the igneous rocks is concentrated not in hydrocarbon gases, but in so-called dispersed carbon (C_{disp}), which represents, presumably, carbon–graphite matter, uniformly disseminated in the rocks. The content of C_{disp} varies from 0.008 to 0.09 wt.% on the rock.

In the fracture zones the filtration of helium is revealed. The content of helium in water-dissolved gases of bore-holes in these zones is up to 4%. Such differences in helium content permit us to reveal and trace the fracture zones by superficial gaseous and water–helium mapping.

The results of investigations of volcanic gases give us some approximate ideas about the composition of gases, which accompany the initial stage of magmatic process. When magmatic melts intrude and crystallize their gaseous phase becomes poor in the more active elements which are lost by passing to the solid state and accumulates gaseous components which are the most stable under present thermodynamic conditions. Gases reserved in magmatic rocks after their consolidation are located in cavities of minerals, in intergranular pores and microfissures of igneous rocks. They are more sensitive to the changes of thermodynamic conditions of environment than solid minerals. The investigation of gases inside the rocks of the agpaitic alkaline massifs is of special interest because they differ greatly from other geologic formations by amount and composition of their gaseous phase.

We call by the term "gaseous constituent" the gases which are contained in closed pores of igneous rocks and are extracted by their grinding in vacuum without heating. The contents of gases in rocks is given in 1 cm^3 per 1 kg of rock through all the text (Petersilje, 1964). Carbon's isotopic composition is given in term δC^{13} % (International standard PDS). The main object of our investigation is the Kola Peninsula, where Precambrian, crystalline rocks mainly occur and magmatism is widely spread. Besides that, the data obtained in some Siberian massifs (Goryachegorsky and Sredne-Tatarsky), in the Urals (Vishnjovogorsky) and in Greenland (Illimaussak) are used for discussion.

For investigating and analysing rocks of the upper mantle large unchanged xenoliths of garnet peridotites and mantle eclogites from kimberlite tube the "Obrajornaya, Yakutiya" were chosen. A small amount of gases which vary from 0.53 to 1.33 is characteristic of peridotites and eclogites. It has been found that the gaseous constituent of the upper mantle rocks on the level of garnet peridotites (i.e. at the depth over 70 km) has essentially hydrogenic composition, with some admixture of methane, nitrogen, carbon dioxide and helium. The composition of the gaseous constituent in the vertical section of the upper mantle varies with depth. At a depth of over 70 km it has essentially a hydrogenic composition and has a hydrogen-methanic one at the very top of the mantle (Lutz *et al.*, 1976).

With the exception of agpaitic alkaline rocks, intrusive igneous rocks contain comparatively small amounts of gases which vary in the range 0.5–15.0. In the composition of gases H_2 prevails greatly over other components (60–88%). There are small quantities of hydrocarbonic gases which are represented mainly by methane, nitrogen, helium, argon and sometimes carbon dioxide.

The contents of hydrocarbonic gases inside the analysed rocks average tenth parts of 1 cm^3 and the maximum quantity of methane reaches 3.4, ethane 0.9 and propane 0.05. The igneous rocks differ one from another by the amount of gases inside them. A minimum quantity of gases is observed in granitoids.

It seems that processes leading to the formation of hydrogen (H_2) inside the earth's crust are widely enough spread. The most probable way of its formation is the interaction of water-vapour with the ferrous oxides compounds.

These reactions are isothermic and reversible. Isolation of the hydrogen occurs intensively enough under 500°C and the higher the temperature the more intensive the reaction flows.

Matsuo's thermodynamic calculations (1960) show that in a gaseous phase of basaltic magma the hydrogen is a main component, if water is excluded.

In 1970 and 1974 there was fixed helium coming from the depth by fissure zones of the Monchegorsky massifs (Petersilje *et al.*, 1970) and Pechenga (Petersilje *et al.*, 1974).

From the bore-holes of the fissure zone cutting the basic–ultrabasic rocks of the Monchegorsky massif, the pure samples of the gases containing helium were taken by passing them through water. Helium content in airless mixture varied from 0.83 to 3.74%, while in the rocks enclosing fissures its content did not exceed 0.17%. In the Pechenga region some observations were done and some measurements were performed in three bore-holes, having disclosed the gas-conducting fissure zones. The content of He in the sample of gas picked out during a gas-display 1.86%. The fissure zones were well investigated and outlined by a water-soluble gas-carotage method. In the area of one such bore-hole some profiles of superficial gaseous mapping were carried out at a depth of 1–1.2 m. The sampling had proved the possibility of revealing the presence of such zones when they approach the surface. As filtration of gases is always accompanied by their diffusion an attempt was made to discover its trace. For this purpose, we chose one of the deep bore-holes, drilled in volcanic subaqueous deposits of Pechenga, in which many covers of diabasis were present. We studied representative samples of diabasis which had been subjected to the regional metamorphism influence at a depth of 0 to 6000 m. In the most part of the rocks the contents of uranium, thorium and K^{40} were determined and it was found their contents were low and without any essential variations. The average content of gases given for the thousand metres intervals is shown in Table 1. The investigations showed that as the depth increases the total content of hydrogen and methane gases decreases while the helium content increases.

TABLE 1.

Depth intervals m	Total sum of gases	He	H_2	N_2	CH_4
0–1000	5.45	0.004	4.48	0.42	0.55
2000–3000	4.02	0.024	3.58	0.34	0.074
5000–6000	1.05	0.186	0.16	0.70	0.009

The data obtained show us that helium is produced in the rocks not only as a result of the decay of radioactive elements, but helium (He) can be accumulated in the rocks by means of diffusion from subcrustal depth. The foregoing data show us also that there is not any diffusion flow of deep-seated hydrogen and hydrocarbonic gases in the investigated region.

Agpaitic alkaline rocks differ greatly from all other igneous rocks by the amount and the composition of their gases. Here methane, but not hydrogen, is a prevailing component composing up to 70–90% of the total content of gases. Besides methane, some quantities of ethane, propane, butane, hydrogen and helium are present in a natural gaseous mixture. Sometimes carbon dioxide is present there. The maximum quantity of hydrocarbonic gases fixed in the Khibini massif rocks reaches 200 cm^3.

The high content of hydrocarbonic gases is a typical feature for all the agpaitic alkaline rocks studied, without any dependence on their age and geological conditions. When the rocks of different intrusive complexes composing the given massifs are similar in their chemical and mineral compositions there is also a resemblance of quantities and compositions of their gases.

The Khibini massif was studied most completely. Its hydrocarbonic gases are contained in the rocks and are absent in the fracture fissure zones and vein deposits, testifying to the completion of the gas-producing process and to the absence of deep-seated hydrocarbonic gases flow. The investigations of massif contacts show that hydrocarbonic gases came in it, not from the massif enclosing rocks, but on the contrary, they penetrated into the contacts from the massif and formed narrow haloes in the exo-contact zone in the area of the contact metasomatic alterations. From this evidence we suggest that methane was produced as a result of a synthesis in cooling alkaline melt.

The rocks and rock-forming minerals differ one from another in their contents of hydrocarbonic gases. The analysis of seventy-four samples of monomineral fraction of the Khibini massif rocks gave the results shown in Table 2.

The gases within intrusive complexes occur unevenly and are distributed in a concentric zonal way in accordance with the general annular structure of the massif (Ikorsky and Pripachkin, 1971).

The main peculiarity of the results is an essential impoverishment of gases in light isotopes in comparison with other hydrocarbon generations previously studied.

The average isotopic composition of methane for the Khibini massif's rocks is -0.85, Lovoservsky -0.53, Illimaussak -0.77. Let us remember that the carbon isotopic

TABLE 2.

Mineral	Ch_4	C_2H_6	C_3H_8
Nepheline	54.02	1.44	0.24
Eudialite	14.62	0.82	0.12
Sphene	13.24	0.23	0.05
Aegyrine	8.23	0.13	0.07
Apatite	1.27	0.02	0.01

composition of methane of sedimentary depositions varies as a rule from -3.2 to -5.0.

The distribution of carbon's isotopic composition among fractions of hydrocarbonic gases of alkaline rocks is the reverse to that of sedimentary rocks, testifying to their different genesis. The following regularity of carbon isotopic composition variation in methane and its homologues is typical for the alkaline rocks of the Khibini massif:

$$CH_4 = -0.32 \rightarrow C_2H_6 = -0.91 \rightarrow C_3H_8 = -1.62$$

Since the fractionation of isotopes in hypogene conditions is restricted it is logical to suppose that the components which are the products of differentiations or degassing of magmatic melts inherit the isotopic composition of primary carbon. We suppose that in this case the primary carbon is the graphite carbon of the mantle, the origin of which is connected with the carbon of carbonaceous chondrites, the latter's isotopic composition is found to be in the same range of values as in the rocks mentioned above (from -0.97 to -1.57) (Galimov and Petersilje, 1967).

In all studied samples of magmatic igneous rocks, besides methane there are dispersed liquid and solid organic compounds which are represented by bitumen and insoluble organic matter.

The amount of bituminous components is not so great and the content of chloroformic extract in them varies from 0.0006 to 0.00016 wt. %. Chloroformic bitumen is oily, has a smell typical for aromatic compounds and a colour which varies from yellow to brown. The bitumen from igneous rocks differs from biogenic bitumen of sedimentary depositions by the higher oil content and as a rule by the absence in them (with rare exceptions) of a methane–naphthene fraction of oils. The naphthene aromatic bicyclaromatic and policyclaromatic fractions are prevailing in the compositions of oil of the most part of igneous rocks. The paraffin compositions in bitumen of the igneous rocks, with the exception of the alkaline rocks, is presented by a very narrow interval of compounds of standard construction C_5–C_{30}. The composition of bitumen in the igneous rocks depends on the conditions of rock-forming processes, of the intensity of metamorphism and of the character of the post-magmatic processes.

The average content of C_{org} in the rocks is 0.02 wt. % and the main part of non-carbonate carbon is accumulated in them. The composition C_{org} is not clear until now.

The results of our investigations of endogenous deposit gases were used for the solution of some practical problems. For example, on the basis of the investigation of gaseous compositions in quartzes of the Kola Peninsula we know that granitic pegmatites have the following parameters (CO_2/CH_4 and CH_4/C_2H_6 ratios) which are typical for pegmatites of different types and which allow us to evaluate the character of ore mineralization in the process of researching. Oreless, ceramic, mica-bearing, rare-metalic and intergranitic pegmatites can be easily differed by this method (Voloshin *et al.*, 1975).

The detections of H_2 and CO_2 contents in quartzes of different stages of hydrothermal processes in wolfram ore deposits were revealed and it revealed the possibility of the application of this indicator for searching ore veins in these districts. There were stated the distinctions in quantities and composition of apatite gases from different deposits and that allows us to use this indicator for revealing typomorphic features of this mineral (Karjavin *et al.*, 1975). In Tuva, where a dozen autonomic alkaline massifs were observed within the limits of single magmatic province, the investigation of gaseous constituent of the rocks allowed us to observe that different complexes of rocks differ in their gases content and the thermal conditions of their crystallization are well fixed by the H_2 and CH_4 ratio. The regular variation of gases was seen during the process of formation of the alkaline massifs.

So the foregoing results of the investigation of gaseous constituent can give us important information for a solution of the complex problems of petrology, of the gaseous regime of the earth's crust and practical geology.

References

GALIMOV, E. M. and PETERSILJE, I. A. (1967) On hydrocarbonic composition of gases and CO_2 contained in igneous rocks of Khibini, Lovosero and Illimaussak. *Dokl. AN SSSR* **173**, No. 4, 914–917.

IKORSKY, S. V. and PRIPACHKIN, W. A. (1971) On some peculiarities of distribution of hydrocarbonic gases in intrusive complexes of Khibini massif. V. sb. "Probl. Magm. Balt. S.", Nauka, L., 278–288.

KARJAVIN, V. K., DUDKIN, O. B. and PRIPACHKIN, W. A. (1975) The gaseous phase of some natural samples apatite. *Dokl. AN SSSR* **222**, No. 4, 942–945.

LUTZ, W. G., PETERSILJE, I. A. and KARJAWIN, W. A. (1976) The composition of gaseous matters in rocks of the earth's mantle. *Dokl. AN SSSR* **226**, No. 2, 440–443.

MATSUO, S. (1960) On the origin of volcanic gases. I. *Earth Sci. Nagoya Univ.* **8**, 222–245.

PETERSILJE, I. A. (1964) *The Geology and Geochemistry of the natural Gases and Dispersed Bitumens of Some Geological Formations of Kola Peninsula,* Nauka, M., L.

PETERSILJE, I. A., ACHMEDOV, A. M. and UVADIEV, L. N. (1974) Hydrocarbonic gases and organic carbon in proterozoic deposits of the Kola Peninsula. *Dokl. AN SSSR* **215**, No. 4, 961–964.

PETERSILJE, I. A., KOZLOV, E. K. and BELIAIEV, K. D. (1970) The display of nitrogen–hydrocarbonic gases in ultrabasic rocks of Monchegorsky pluton's Sopcha massif. *Dokl. AN SSSR* **194**, No. 5, 1190–1193.

VOLOSHIN, A. V., IKORSKY, S. V. and DAVIDENKO, I. V. (1975) The gluctic bond of carbon compounds with processes forming of granitic pegmatites. Tesisi dokl. na sov. 2 "Uglerod i ego soed. v. endog. proc. mineraloobr.", Lvov, pp. 13.

Distribution Regularities of Trace Elements in Rocks and Minerals of the Ultrametamorphism Zone

Z. I. PETROVA

Vinogradov Institute of Geochemistry, Academy of Sciences, Irkutsk, U.S.S.R.

Abstract

The distribution of petrogenic and trace elements: Co, Ni, Cr, V, Cu, Zr, Zn, Pb, Sn, Ba, Sr, Be, Li, Rb, F, Ce, La, Nd, Y, Yb—in rocks and minerals of deep crustal horizons—in the ultrametamorphism zone—is investigated in detail for the example of the Precambrian Aldan phlogopite-bearing region. The most early formations are rocks of the granulitic facies. Their composition changes from ultra-acid to ultrabasic. Rocks of the middle and basic composition do prevail. The distribution regularities of chemical elements therein are analogous to those in rocks of younger and less metamorphosed strata. As to element content level and their ratios: Co:Ni, V:Cr, Ba:Sr, etc., there the studied stratum is found to be primary effusive-sedimentary.

The ultrametamorphic transformation of the granulitic facies rocks is accompanied by an intensive loss of most trace elements and leads to a wide development of various granitoids. Lost are mafic elements: Co, Ni, Cr, Sn, Sr, etc., and granitic ones: F, Be, Li. There the forming granites have a higher alkalinity and a specific trace-element composition, e.g. alaskite has (ppm): 12.6 V, 7.0 Li, 0.6 Be, 150 F, 156 Rb, 2940 Ba, 360 Sr.

Intermediate products of ultrametamorphic alteration of the basic and carbonate rocks are basificates and skarns. Therein do partially concentrate those elements lost from areas of massive granitoid development such as Ca, Mg, Fe, Co, Sn, etc., and do crystallize such minerals as phlogopite, wollastonite and spinel.

Ultrametamorphism leads to formation of more contrasted rocks as to composition and trace element content than the initial ones. Rock areas enriched by a certain set of petrogenic and trace elements become a convenient medium for the formation of ore concentrations at a succeeding superposition of metasomatic processes.

These zones are open to observation in regions where deeply eroded Precambrian strata are exposed. In rocks and minerals of such a zone—within the Aldan crystalline shield (Aldan phlogopite province)—has been studied the distribution of many trace elements and regularities of their behavior in ultramorphic processes have been found. Quantitative values are given in Tables 1 and 2.

The most early rocks there are schists and gneisses of the granulitic facies belonging to the Fedorov range. They have no relicts and are a substratum for the following geological formations. There prevail schists of the average and basic composition (Table 1). This particularity is common for ancient granulitic complexes depending on the specific formation conditions of their initial rocks in the oldest geosynclinals lying on the primary basic crust. Following their composition and geological character, these rocks have analogs among non-metamorphosed magmagenic and sedimentary rocks. This gives an isochemical character to the regional metamorphism and the primary composition of the deep-metamorphosed stratum. Using petrochemical and geochemical diagrams of para- and orthorocks (Predovsky, 1970: de La Roche, 1972), indicators of trace and petrogenic elements (Table 2), and also correlative connections of trace elements with petrogenic ones—

TABLE 1. TRACE-ELEMENTS CONTENTS (ppm)

Phase	Rocks	n	Co	Ni	Cr	V	Cu	Zn
Regional metamorphism (granulitic facies)	Biotite–hypersthene plagioschists	8	15 / 4–24	22 / 9–40	64 / 29–100	188 / 96–330	47 / 18–83	79 / 25–130
	Hypersthene, bi-pyroxene and diopside–hornblende (with biotite) plagioschists	45	20 / 7–40	20 / 6–50	37 / 3–130	214 / 83–450	66 / 13–175	71 / 20–125
	Biotite–diopside and diopside plagioschists	12	7 / 3–9	18 / 6–20	70 / 11–137	69 / 27–159	12 / 4–27	36 / 16–80
	Quartz–diopside plagioschists and diopside plagiogneisses	10	14 / 10–17	38 / 26–50	100 / 50–180	116 / 79–160	25 / 6–74	34 / 16–80
Ultrametamorphism (amphibolite facies)	Plagiomigmatites on biotite–hypersthene plagioschists	6	16 / 13–22	57 / 29–95	126 / 26–230	151 / 52–250	34 / 10–60	77 / 48–110
	Plagiomigmatites on substantial hornblende plagioschists	16	9 / 3–16	9 / 5–13	13 / 3–36	99 / 13–250	22 / 12–37	56 / 16–115
	Quartz–diopside–scapolite with rock sphene	7	5 / 2–13	13 / 3–29	—	36 / 8–98	14 / 2–21	49 / 40–63
	Shadow K-feldspar migmatites	25 28	4 / 1–13	5,5 / 2–14	8 / 2–29	32 / 7–60	12 / 4–50	50 / 19–85
	Alyaskites	15 17	2.5 / 0.9–5.5	3.5 / 1.6–6.0	3.5 / 1.5–7.0	12 / 3–18	17 / 5–50	43 / 14–110
	Orthotectites	14	2 / 0.9–6	3 / 1.8–6.5	5 / 1.5–14	12 / 2–25	20 / 3.5–53	36 / 10–83
	Basificates	10	23 / 10–57	100 / 6–350	186 / 8.5–690	221 / 20–760	60 / 3–280	53 / 22–118
	Spinel–pyroxene skarns	17	15 / 5–30	25 / 10–66	43 / 9–184	49 / 20–93	1–500	25–1200
	Pargasite–pyroxene skarns	12	8 / 4–18	22 / 10–37	37 / 7–70	52 / 31–100	8 / 2–18	27–600
	Pyroxene–scapolite skarns	12	11 / 2–25	18 / 9–50	45 / 18–100	51 / 31–100	12 / 1–21	64 / 25–100
	Phlogophite–pyroxene skarns	11	8 / 4–15	17 / 9–35	23 / 5–62	32 / 14–60	13 / 1–78	74 / 25–135

Note: n—sample quantity. Numerator: average; denominator: minimal and maximal element content in choice samples. Analyses: Vinogradov Institute of Geochemistry, Irkutsk.

particularly, trace elements of iron Co and Ni with Mg and Fe (Petrova *et al.*, 1975), together with a detailed analysis of the geological structure, the rhythmic-layer stratum in the Aldan province is reconstructed as sedimentary-effusive. The primary composition is mostly of magmagenic rocks of the andesite–basalt type. With metamorphism they become hypersthenic (with biotite and hornblende), or bi-pyroxenic and hornblende (with biotite and diopside), plagioschists. Sedimentary rocks were converted into pyroxenic schists and gneisses and forsterite calciphyres and marbles.

The ultrametamorphic rock transformation of the granulitic facies has an imposed character. They are separated in time from the regional metamorphism (Volkova and Brandt, 1970) and have a distinct tectonic control, confined to deep faults, but not chaotically distributed within the areas of the granulitic facies denudation. The metamorphism proceeds in T, P conditions of the amphibolite facies or between the granulitic and amphibolite facies. Characteristic is that in these conditions ultrametamorphic processes proceed mainly as gradual replacements of the initial rocks by a new parageneses through the metasomatic zone column. The character of the processes strongly depends on the chemical composition of the initial rocks. Acid rocks differently migmatized become shadow migmatites, compositionally near granites. In the heart of large fields of such migmatites develop processes of melting and crystallization of alaskites and leucocratic granites. The analogous changes of basic rocks together with migmatites on the periphery of the migmatite fields develop basificates, specific melanocratic rocks of the

IN ROCKS OF THE ALDANSKY PHLOGOPITE PROVINCE

Pb	Sn	Zr	Be	Sr	Ba	Li	Rb	La	Ce	Nd	Yb	Y
12	4.6	184	2.0	403	1209	33	183	58	99	48	3.0	15
6–20	2.6–6.3	83–480	1.1–2.8	210–1300	660–3200	28–38	100–315	41–66	71–120	41–55	2.1–3.7	5129
12	7.2	130	1.7	841	600	21	24	58	100	55	4.8	26
12–53	5–13	55–120	0.2–4.0	350–1600	270–1350	7–39	5–96	21–120	46–160	10–82	3.6–6.8	13–50
11	9.6	183	1.9	484	599	24	19	66	116	55	4.0	23
7–80	4.5–17.5	63–300	0.7–3.7	350–650	13–650	12–38	%/vol–50	25–155	47–280	33–145	2.9–7.2	13–54
7.5	8.2	178	2.9	357	86	31	8	44	88	45	4.2	24
4–10	4.5–19.0	100–300	1.6–6.3	240–460	51–220	28–52	5–12	35–60	44–150	32–63	3.5–6.5	19–28
12	4.9	171	2.1	461	951	34	101	62	93	50	3.6	11
5–24	2.4–7.1	118–260	1.0–2.9	330–1050	480–1050	18–58	86–108	45–120	72–130	32–74	2.5–5.9	10–19
9	5.3	315	1.9	944	1177	20	33	74	129	73	4.5	27
4–22	3.7–7.0	67–720	1.1–3.3	690–1200	430–1650	15–30	16–84	54–130	86–180	52–100	3.3–8.3	15–43
11	7.0	121	2.3	335	236	25	10	62	95	39	3.8	23
6–15	5.0–12.0	40–270	1.0–3.9	96–540	35–390	14–40	5–14	22–140	54–190	19–49	2.5–6.2	14–39
31	4.9	158	1.2	291	1711	10	124	59	96	42	2.8	17
7–62	1.7–9.0	15–400	0.2–2.0	82–540	690–3900	2–30	34–232	22–140	42–200	22–100	1.4–5.9	4–38
14	2.6	120	0.6	363	2939	7	156	45	75	33	2.0	14
5–36	1.0–4.7	32–230	0.2–1.0	130–830	660–5500	tr–17	100–232	11–120	30–180	6–66	1.1–4.0	3–60
14	4.0	99	1.8	393	3962	5	224	73	115	51	2.8	19
5–37	2.0–6.5	15–360	0.8–4.5	250–670	2200–5700	tr–8	102–360	32–135	51–190	18–95	1.4–5.8	6–39
18	13	15	3.3	312	429	28	59	67	197	79	5.4	38
7–40	4–26	89–208	1.1–4.4	66–930	10–1800	14–58	tr–201	5–190	35–500	24–170	3.7–7.2	18–55
13	24.5	137	2.3	72	105	10	5	49	122	62	3.8	27
5–25	7.2–80	37–350	1.0–3.7	23–170	19–310	4–23	tr–27	24–91	47–250	38–130	2.5–5.3	17–80
10	11.7	91	1.8	141	346	10	21	55	130	56	3.5	26
5–20	4–30	48–140	0.5–4.0	60–260	64–1000	tr–14	7–34	24–100	47–220	24–95	2.3–4.5	10–54
94	8.0	126	1.7	251	301	20	18	49	87	47	3.3	19
5–18	4.5–16.0	53–260	0.5–5.5	110–1200	34–1200	10–39	tr–50	29–85	37–160	17–86	2.5–3.8	10–41
14	9.3	65	1.4	102	1211	12	225	28	72	34	2.5	14
6–25	6.0–11.0	22–96	0.5–2.0	40–260	220–4800	3–21	42–423	13–56	30–160	23–59	1.7–3.9	10–21

amphibole composition, but in the central migmatite fields are formed granitoids of a higher alkalinity—leucocratic granosyenites with amphibole and pyroxene. The transformation of ultrabasic schists and dolomite marbles is accomplished through the magnesian skarn column. There granitoids are represented by yet higher alkaline varieties, mainly pyroxenic syenites. Granitoid magmas, melting in the central parts of the migmatite fields, crystallize *in situ* or after a slight displacement.

Table 1 shows that rocks forming from ultrametamorphism have a different trace-elements content than initial schists. Following their behavior in the transformation series: schist–plagiomigmatite–shadow K-feldspar migmatite–granite, the trace elements fall into three groups:

1. elements with a gradually decreasing content: Co, Ni, Cr, V, Cu. Therewith go Zn, Sn, Sr, Li and Be. Their content in plagiomigmatites is equal to that in schists but decreases in K-feldspar migmatites and granites;
2. elements with a gradually and regularly increasing content: Rb and Ba. There belongs also Pb with a slight increase only in granitoids;
3. elements with a constant content through all phases as in the initial schists, slightly increasing only in plagiomigmatites: Zr, Ce, La, Nd, Yb and Y.

The trace-elements content of basificates are the contrary of shadow K-feldspar migmatites. They have a higher content of Co, Ni, Cr, V, Cu, Zn, Sn, Be, Li, also La, Ce, Nd, Yb, Y and a lower content of Zr, Sr, Ba, Rb. Compared with initial plagioschists the

TABLE 2. ELEMENT RATIOS IN THE FEDOROV ROCKS RANGE AND IN SOME MAGMATITE ONES

	Rocks	$\frac{Li}{Mg} \times 1000$	K:Rb	Co:Ni	V:Cr	Ratios Ba:Sr	$\frac{Ba\dagger}{Rb}$	$\frac{K\ \dagger}{Ba}$	$\frac{K+Ca\dagger}{Ba}$	Ca:Sr
Fedorovsky range	Hypersthene, bi-pyroxene- and diopside–hornblende plagioschists	0,73	653	1,12	13,7	0,65	22,2	16,7	125	100
	Biotite–diopside plagioschists	0,53	490	0,50	0,75	1,0	32,1	12,0	135,5	234
	Pyroxene schists	0,17	125	0,56	1,98	0,50	3,8	45,4	5076	3000
	Dolomite marbles	0,24	27	0,07	0,56	0,41	11,7	2,3	3331	1370
	Pyroxene plagioschists	0,37	667	0,73	1,77	0,05	16,5	57,0	2609	277
	Quartz–pyroxene plagioschists and pyroxene plagiogneisses	1,85	545	0,36	1,31	0,16	10,8	48,8	1166	270
Data by Taylor (1968)	Andesites (average)	0,48	430	1,33	3,13	0,70	8,7	49,2	235	130
	Average Saipan andesite	—	590	1,33	4,34	0,54	11,0	54,5	615	286
	Average granodiorite	2,63	230	0,67	2,5	1,14	4,5	50,8	101	58
	Average basalt	0,19	425	0,32	1,25	0,54	12,5	33,2	313	165
	Average oceanic tholeiite	0,21	1170	0,32	1,00	0,12	11,7	100	6064	725

† *Note*: Ratios are average elements contents.

basificates are enriched by those elements lost through migmatite formation. The content of Rb, Pb, Be, La, Ce, Nd, Yb, Y is higher and Zr lower than in schists. Chemically, the basificates are opposite to the migmatites.

Skarns are intermediate products of replacement of ultrabasic schists and dolomite marbles by granitoids and have a very complex trace-elements distribution. Generally the outer zones of skarns, represented by spinel–pyroxenic and pargasite–pyroxenic parageneses, show a basificate tendency but those of inner zones a migmatite one.

Table 3 reports the contents of elements in dark-colored rock-forming minerals of the rocks. Their particular distribution is in keeping with the particularities for rocks.

The data show an intensive redistribution and matter exchange in ultrametamorphic processes. From formation areas of granitoid parageneses a large group of trace elements related to Mg, Ca and Fe is lost geochemically, since at the granitization of meso- and melanocratic rocks of the granulitic facies the dark-colored minerals—pyroxenes, amphiboles and micas (being the main concentrators of these elements: Co, Ni, Cr, V, Cu, Zn, Sn, Sr, Li)—are destroyed. These elements are partly supplied and deposited in areas of basificate and skarn formation. But only partly and because basificates are 50–90% made up by their mineral concentrators—amphiboles and pyroxenes. These elements contents in minerals of basificates are the same as in those of schists or even lower (Table 2). Also the development of basificates is much smaller than that of migmatites.

In the granitoid formation area there are intensively supplied only Ba and Rb. Pb, like its basic petrogenic analog K, is less changed. An active concentrator and carrier of these

TABLE 3. TRACE ELEMENTS CONTENTS (ppm) IN THE MAIN ROCK-FORMING MINERALS OF THE ALDAN PHLOGOPITE-BEARING PROVINCE

Minerals	n	Co	Ni	Cr	V	Cu	Sn
Clinopyroxenes:							
from plagioschists	12–9	37	128	171	300	22	9
		14–80	24–460	83–330	71–850	5–85	4–23
from plagiomigmatites	3	52	76	133	286	15	—
		50–56	40–110	30–220	170–440	10–20	
from shadow K–feldspar magmatites	6	22	18	22	148	18	10
		8–39	4–43	10–38	66–370	3–38	7–14
from orthotectites	2	23	11	16	200	10	6
from basificates	1	9	4	10	36	4	10
from skarns on ultrabasic schists	7	8	16	43	54	4	7
		5–14	7–25	6–68	37–100	3–8	3–14
from skarns on dolomite marbles	8	3	10	34	59	5	9
		1.5–10	4–20	20–48	42–70	3–8	4–15
Amphiboles							
from plagioschists	10	47	57	135	386	26	16
		36–71	13–200	19–335	300–550	7–56	7–45
from plagiomigmatites	5	31	24	48	264	40	21
		22–42	18–29	27–60	150–360	13–55	11–28
from shadow K-feldspar migmatites	6	52	24	34	312	44	18
		24–110	10–55	14–101	120–750	16–70	14–25
from basificates	4	25	43	155	297	15	46
		13–36	9–130	39–530	73–760	10–21	35–57
from skarns on ultrabasic schists	5	19	32	182	125	10	22
		11–31	13–61	26–600	78–220	6–16	14–50
from skarns on dolomite marbles	6	5	19	79	167	11	—
		3–10	10–35	32–110	110–230	1–30	
Micas (biotites and phlogopites)							
from plagioschists	7	78	107		416		6.8
		28–120	35–230	22–1000	85–800	7–260	6.3–7.0
from plagiomigmatites	3	52	116	224	225	19	—
		46–62	49–240	16–610	180–255	14–22	
from shadow K-feldspar migmatites	1	53	75	130	440	22	—
from basificates	2	26	160	320	325	12	—
from skarns on ultrabasic schists	2	15	46	47	41	11	2.5
from skarns on dolomite marbles	3	9	17	42	52	7	2
		4–14	10–25	40–45	49–55	4–10	1.6–2.5

elements is the crystallizing K-feldspar. The rock analysis of the ultrametamorphism zone shows that the element sources lie beyond the zone. It is not excluded that they are juvenile solutions.

It seems that the indifferent behavior in the ultrametamorphic processes of Zr, La, Ce, Nd, Yb and Y comes from position change. Thus Zr in schists is dispersed in dark-colored minerals, but in granites it forms a mineral phase—accessory zircon. REE in dark-colored minerals and plagioclase schists are dispersed, but in higher concentrations, in the accessory granitoid minerals apatite, sphene and allanite.

The confinement of ultrametamorphism processes to deep faults conditions the zonal structure of the ultrametamorphism zones. There are three: outer, middle and inner.

The outer zone is characteristic of a least intensive ultrametamorphic process. Here processes of recrystallization, plagiomigmatization and basification of initial schists form bodies of slightly displaced granites.

The middle zone goes with the greatest mass of transition products of granitization of the initial rocks: differently migmatized plagioschists, basificates, various skarns, bodies of displaced granites rooting in the inner zone. This zone is the most contrasted as to composition and trace-elements contents. Areas enriched by certain petrogenic and trace elements become a good medium for ore-concentration formation in subsequent metasomatic processes. Thus areas of high-magnesian and alumina skarns and basificates in the following post-magmatic stage get phlogopite ores.

The inner zone has an intensive development of ultrametamorphic processes. They go with a loss of most of the trace elements and bring a wide development of shadow migmatites and granites crystallizing *in situ*. Lost are mafic elements and "granitic" ones: Sn, Be, Li. The forming granites have a higher alkalinity and a specific trace-element composition (Table 1). Interestingly, compared with Clarkes for analogous low-calcic granites (Turekian and Wedepohl, 1961), the contents of Ba are high and of Be, Li and V are low.

References

La Roche, H. de (1972) Geochemical characteristic of metamorphism areas, signs and proofs of their pre-metamorphic history. In *International Geochemistry Congress, Metamorphism and Metasomatose,* Moscow, Vol. 3, Book 1, 263–290 (in Russian).

Petrova, Z. I., Požharitskaya, L. K., Roisenman, V. M., Ronenson, B. M. and Utenkov, V. A. (1975) *Metamorphic Complex of Aldan Phlogopite Deposits,* Nauka, 150 pp. (in Russian).

Predovsky, A. A. (1970) *Geochemical Reconstruction of the Primary Composition of Metamorphosed Volcanogenous–Sedimentary Precambrian Formations,* Apatity, 120 pp. (in Russian).

Taylor, S. R. (1968) Geochemistry of andesites. In *Origin and Distribution of the Elements,* Pergamon Press, Oxford, 559–583.

Turekian, K. K. and Wedepohl, K. H. (1961) Distribution of the elements in some major units of the earth's crust. *Geol. Soc. Amer. Bull.* **72**, 175–191.

Volkova, N. V. and Brandt, S. B. (1970) Using K–Ar isochrons for absolute rock age determination of the Fedorov group of phlogopite deposits (Aldan). In *Institute of Geochemistry Yearbook,* Irkutsk, 213–217 (in Russian).

Evolution of the Ore Formation in the History of the Earth

A. I. TUGARINOV

Institute of Geochemistry, Moscow, U.S.S.R.

Abstract

The formation of the earth's crust proceeded unevenly, mainly during the first 2 milliards of years of the earth (4.6–2.6 mld.).

This has led to the inclusion into the early protocrust of the first undifferentiated blocks of the oceanic crust containing effusives of ultrabasic composition of the type of comatiites, which were attended by unique nickel deposits. The presence of large gold ore deposits (Yellowknife, Colar) is characteristic of the same old protocrust.

The further geological history of the earth's crust is connected with an intensive erosion of the oldest crust which occurred under conditions of a changing earth atmosphere, of an increase of free oxygen in it and of new conditions of sedimentation. One of the most striking examples of this was the appearance and huge development during the lower Proterozoic (2.6–1.9 mld. of years ago) of big deposits of ferruginous quartzites, jaspilites (Hammersli, Krivoi Rog, KMA, etc.). From this time dates the appearance of the first large uranium deposits (2.6–1.9 mld. of years ago) (Witwatersrand, Atabaska, Singhbum, Alligator River).

The following middle and upper Proterozoic (1.9–0.6 mld. of years ago) was characterized by a similar global appearance of large polymetallic deposits being mainly of a stratiform type. The genesis of the deposits was connected with the accumulation of thick carbonate strata enriched in C_{org} (MacArthur, Mount-Isa, Zavar, Sullivan, Sardana, Kholodnenskoye) appearing at that time.

The geological history of the earth's crust during the Phanerozoic is distinguished by a further complication of rock composition. Owing to the appearance of numerous linear structures and connected with the latter sediments having been subject to unprecedented sedimentary differentiation, enriched in organic materials, which in their turn were distinguished by accumulation, sorption of rare elements and metals—metallogenic provinces of most different type arose in the earth's crust: tin-bearing provinces (Pacific Ocean zone), tungsten provinces (Middle Asia), germanium provinces (Donbass), uranium ones (Hercynides of Europe), etc. Much favourable to this was the locally displayed magmatism in the same mobile zones which favoured the redeposition of the ore load of sediments and the creation of vein deposits.

The irreversible and strictly directed character of the evolution of chemical distribution of elements in the structures of the earth's crust, its relatively slight relation to mantle sources of metals and submission to life development and the laws of organic geochemistry—such are the features which determine the contemporaneous distribution of elements in the earth's crust.

Before speaking about ore formation in the history of the earth we must consider the major events which are the basis for the present divisions of geologic time. The age of the earth has been calculated to be 4.6 b.y. within an error of $\pm$ 50 m.y.

At the present time we know nothing about the first 600 million years. It has been established recently that the first most ancient relicts of proto-continents appeared in Greenland, the eastern coast of Labrador, in Minnesota and South Africa, about 4 billion years ago.

On the given geochronological scale (see Fig. 1) the triangles denote the main tectono-magmatic cycles in the history of the earth dated by the methods of absolute geochronology.

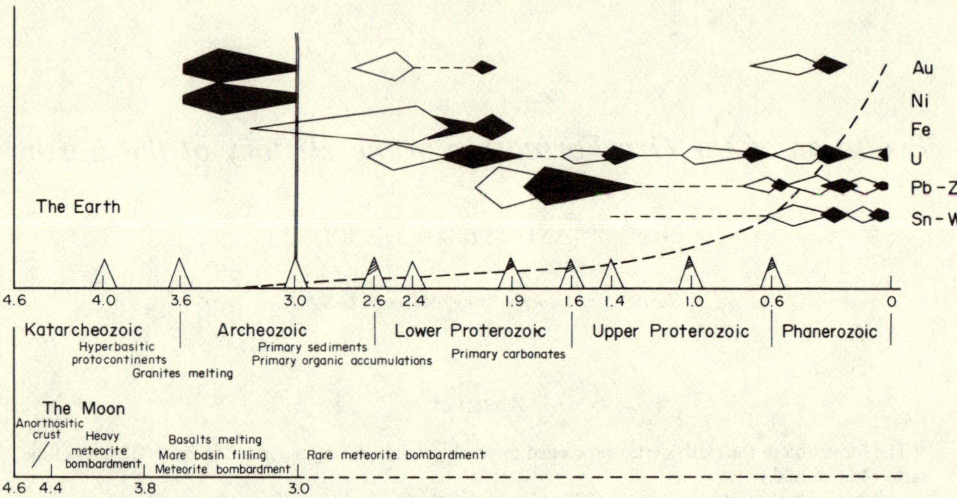

FIG. 1. Epochs of maximum development of ore deposits in the history of the Earth: filled diamonds—endogenous deposits; open diamonds—exogenic deposits; interrupted line—sedimentation rate (mass of sediments in time unit); triangle—epoch of magmatogenesis established in 1956; shaded triangle—epoch of magmatogenesis established in 1976; parallel line—time of emergence of proto-continent above the ocean level.

Dashed triangles designate epochs established for the first geochronological time-scale in 1966. The open triangles designate the epochs fixed recently on the basis of extensive work of geochronologists (Tugarinov, 1956, 1972).

The observed character of the ancient proto-continents leaves no doubt about the fact that their thickness (Sial) was much less than have the recent continents (40 km), what is most important is that the study of the composition of rocks and their interrelations indicate that the most ancient crust, the so-called greenstone formation, represents areas of the ancient oceanic crust into which intrusions of plagiogranites, anorthosites, etc., were intruded about 3.6 billion years ago (Tugarinov and Bibikova, 1976).

Rock interrelations of such a kind which for a long time caused animated discussion in the world on the question of "what was first, the chicken or the egg?" are today settled rather unambiguously (Glikson, 1972). Thus, for example, a detailed investigation proved the well-known Morton "gneisses" (Minnesota) to be pyroxen–amphibole relicts of 4.0-b.y.-old strata representing remnants of plastically deformed rocks of the ancient basement injected and disrupted with granite material about 2.8 billion years ago. It is characteristic that in granite apophysae corroding this ancient rock even orthite grains are met which indicate the considerable radioactivity increase of this multiphase stratum of gneisses (Sims and Morey, 1972).

In the region of Greenland the oldest stratum is 3.8 billion years old and is known to be represented by shales, graywackes, ferruginous quartzites which dip at a relatively low angle, squeezed by Amitsok gneisses of quartz–feldspathic composition (Sutton and Windly, 1974; Bridgewater *et al.*, 1973; McGregor, 1973).

In the Barberton (South Africa) the series Onvervakht are the most ancient formation being obviously intruded by tonalitic diorites 3.3 billion years old and possibly being of a still more ancient age (according to Hunter and Condie). It is represented by basic and ultrabasic effusives including comatiites (Condie and Hunter, 1976; Dungworth and Schwartz, 1974).

One of the important peculiarities of these ancient rocks is the absence of a continental or of a coastal facies among them. All the above-mentioned rocks have without any doubt a submarine origin. Facies connected with the erosion of continents first appear only approximately 3 billion years old. The early aleurolite intercalations in greenstone strata of corresponding age in the Dnepro region is an example of such an event.

Further on about 2.7 billion years ago on all continents of the world a massive accumulation of conglomerates is noted. This accumulation indicates that the denudation of a continent of an already substantial area had reached an enormous extent and that the sedimentation processes had acquired a quite different character. Beginning from this time sedimentation became the major process leading to an extreme differentiation of terrestrial matter. There is indeed no other factor like sedimentation which leads to such a complete and extreme decomposition and isolation of the initial material of crystalline rocks: quartzites and sandstones, limestones and dolomites, clays. . . A new type of sediments—of C_{org}—must be added to this. Why is it that about 2 billion years after the moment of the earth's origin such an intensive process of disintegration of the intially uniform matter began?

The most important reasons for this is that about 3 billion years ago the floating to the surface of the first proto-continents for the first time occurred. This favoured early formation of sediments and subsequently also the increasing sedimentation rate, as is well shown by the diagram (see Fig. 2) adopted from Hargraves (Tugarinov, 1956; Tugarinov and Bibikova, 1976).

But things were not limited to this factor. The origin and growth of the ancient proto-continents, which had already begun under submarine conditions, consisted at early stages (as it is deduced from very fragmental observations in Greenland and South Africa) of the intrusion into the most ancient strata of magmas of more acid composition, However, these magmas only approached granites in composition as tonalites, plagiogranites, anorthosites, etc. (Fig. 3).

Fusion of magmatic masses of such composition at a sufficiently shallow depth (10–8 km) under the primary protocrust could occur at an early stage of our planet existence because of the high thermal regime at that time. It is possible to say that due to the extinction of

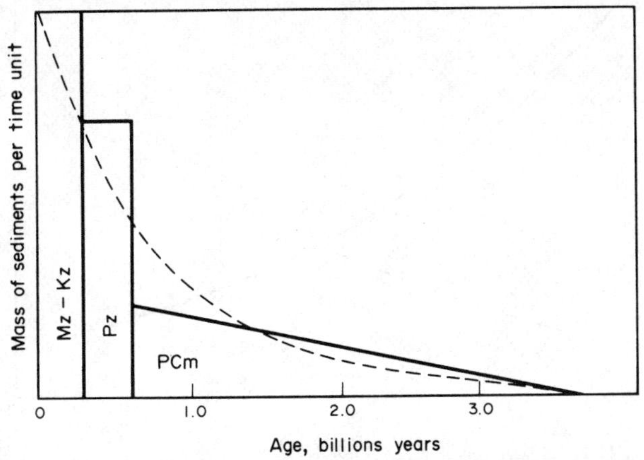

FIG. 2. Increase of sedimentation rate in the history of the earth (according to Hargraves, 1976).

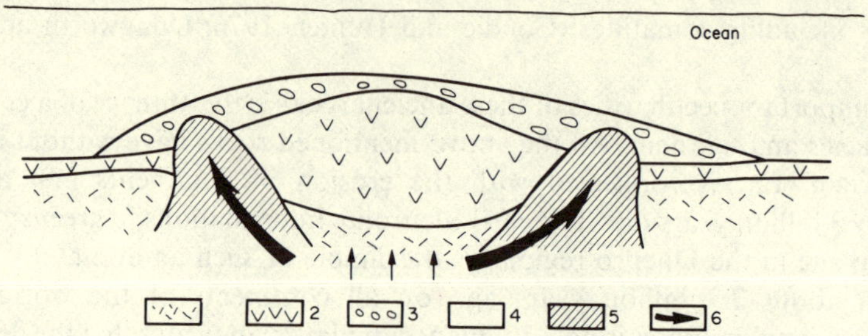

FIG. 3. Scheme of the rise of proto-continents of the earth: 1—mantle; 2—basaltic crust; 3—sediments; 4—ocean; 5—intrusions of the tonalite–plagiogranite type; 6—direction of the movements of radiogenic heat and the fusion of rocks.

radioactive elements 4.6 billion years ago the thermal gradient was not 30° per 1 km as at present, but in the order of 100° per 1 km. In accordance with this, about 4 billion years ago the 750° isotherm occurred at a depth of approximately 10 km (see Fig. 4) (Hargraves, 1976; Heier, 1973).

All this favoured not only the animation of the early magmatic activity, but also its less differentiated character.

Undoubtedly the initial areas of the ancient crust consolidation "cores" served as screens below which the radiogenic heat was trapped and which in its turn favoured the formation of fusion chambers.

The scheme of the early continent formation is given in Fig. 3. Among the geologists of the world a tendency has recently arisen to assume that the formation of sialic crustal blocks occurred almost simultaneously and that the whole subsequent geological history is the history of cleavages, crushings and motions of separate fragments of these blocks (Anhaeusser, 1973; Kröner *et al.*, 1973). Today, however, we have at our disposal a number of striking facts confirming the old point of view about the gradual growth of continents at

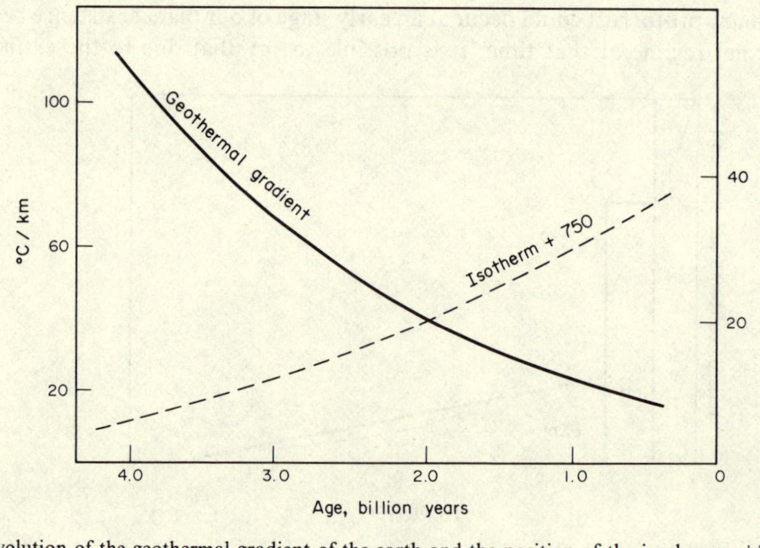

FIG. 4. Evolution of the geothermal gradient of the earth and the position of the isochrone +750°C.

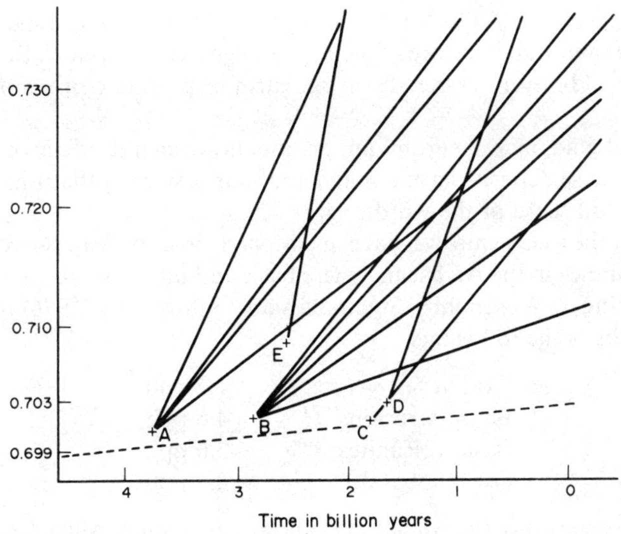

FIG. 5. Origin of blocks of different ages of the Greenland shield (according to Moorbath, 1976). Interrupted line—evolution of Sr^{87}/Sr^{86} in the mantle; line with crosses—Sr^{87}/Sr^{86} in sources of the fusion of rock blocks of the Greenland shield. A—Amitsoq gneiss. B—Nuk gneiss. C–D—Ketilidian gneisses. E—Amiralik dykes.

the expense of the fusion of their material from the mantle. Thus, on Fig. 5, a diagram by Moorbath is shown which gives hundreds Sr^{87}/Sr^{86} age measurements of separate blocks of the Greenland shield and of the Sr^{87}/Sr^{86} ratio in rocks of the fusion source established on the basis of these measurements (Moorbath, 1976). This diagram clearly shows that each younger block of an age from 3.5 to 2.0 billion years old was fused from the mantle which always varied the Sr^{87}/Sr^{86} ratio at the expense of its slight increase owing to small rubidium concentrations in its mass.

It follows from Fig. 5 that the isotopic composition of strontium in the mantle changed from 0.699 to 0.703. But in the case of crustal rock refusion, as we see for the Amiralik dikes, this strontium isotope ratio jumps to 0.707. This jump indicates that in this case the question is not about the fusion of corresponding magma from the mantle, but about its genesis at the expense of crustal rock refusion.

When we now return to the peculiarities of sediment formation, we must also note another aspect of this process. More than 3 billion years ago life is fixed on the earth only in the form of separate relicts (for example, according to Dangvort's and Schwarz's data, in the formation Onvervakht intercolations with C_{org} content from 0.02 to 1.58% were met). The core consisted of kerogen and aliphatic hydrocarbons with sufficiently light carbon; $\delta C^{13} = -25$ to -35%. However, after 3 billion years, it becomes a constant factor of sediment formation.

We can affirm today that the moment of the emergence of continents was not only characterized by a strong development of weathering and sediment formation, but also by an increase of life, by the generation of the primary biosphere. This leads immediately to a strong development of photosynthesis and to emission of molecular oxygen into the atmosphere. From this time on the atmosphere, at first represented by gasses of the type H_3, CH_4, CO_2, H_2, etc., approached more and more the present composition, i.e. that consisting of N_2, O_2, CO_2, Ar, etc.

Of what importance had all this for ore formation? At the present time is is frequently heard that "the Precambrian is a store of metals, of huge treasures, etc.". But is it true? Can the 3.5 billion-year-old history of the life on our earth be piled into one concept? Probably not!

Large deposits of gold, nickel, chromium, platinoids are characteristic of the first half of the Archaean. All these deposits have a magmatic source since at that time there were no precipitates in the full sense of the word.

Gold, like the other elements we have mentioned, is a typical siderophile element. Therefore its appearance in the Archaean, where basic and ultrabasic rocks prevail, was not accidental. According to Wedepohl's, Sighinolfi's and Santos data (1976) the present gold distribution in rocks is the following:

Granites	1.7 ppb
Basic intrusions	4.6 ppb
Basic volcanites	3.6 ppb
Oceanic basalts	0.5 ppb

All known gold ore deposits of the Archaean (Yellowknife, Grange Mine, Colar, Kalgoorlie, and others) occur in the basic greenstone rocks of various composition and are distinguished by a large size and substantial extension with depth and along the strike are associated with considerable alterations of enclosing rocks. In the Archaean there are no gold deposits associated with granites such as are found in the Phanerozoic. This suggests that gold-bearing granites of younger epochs derive their gold from those numerous placers or basic rocks which contaminate these granites had arisen (note the sequence of the non-painted and painted lenses on the main scheme (Fig. 1)).

Still more striking is the fate of nickel of which the largest deposits were relatively recently discovered in the Archaean. To them belong Mount Vindarra (Australia) and two deposits in the region of Kalgoorlie–Kambalda Dom (Fig. 6) (Brooks and Hart, 1974; Lusk, 1976). These deposits represent ribbon ores of the exhalation type and are characterized by pentlandite, and millerite scattered among ultrabasic effusives of the comatiite type. Comatiites are rocks represented by olivine plates in the skeleton-like forms which are cemented by pyroxenes. The magnesium content amounts to 10 to 20%. The texture of the rock resulting from the peculiar mineral composition is named spinifex (after the Australian thorny plant). Despite the fact that rocks similar in chemistry have been found also in the Phanerozoic, comatiites are known only in Archaean. In 1976 trough fragments of a rock similar to comatiite and named marianite were dredged from the bottom of the Mariansk trough. The fact is that, as we have already noted, the isotherms at that time passed much closer to the surface than during later epochs. Owing to this fact appeared comatiites which had arisen from a magma heated to 1450°C and were extruded and reacted with sea water. The skeleton growth forms of rock-forming minerals are in particular connected with this fact (Nesbitt and Shen-Su Sun, 1976).

After their discovery in Australia such deposits were also found in Canada and Finland.

This epoch of metallogeny was followed by a very productive epoch of already typical sedimentary deposits of the lower Proterozoic. Uranium–iron deposits are noted in it, as well as strata enriched in lead and zinc. Uranium is the main element which forms its largest sedimentary and residual deposits in this epoch. Endogenous uranium deposits were absent at that time. One may name among these deposits the Witwatersrand, Elliot Lake (Blind River), a number of Arnkhim deposits in Australia, and others. The age of these deposits is

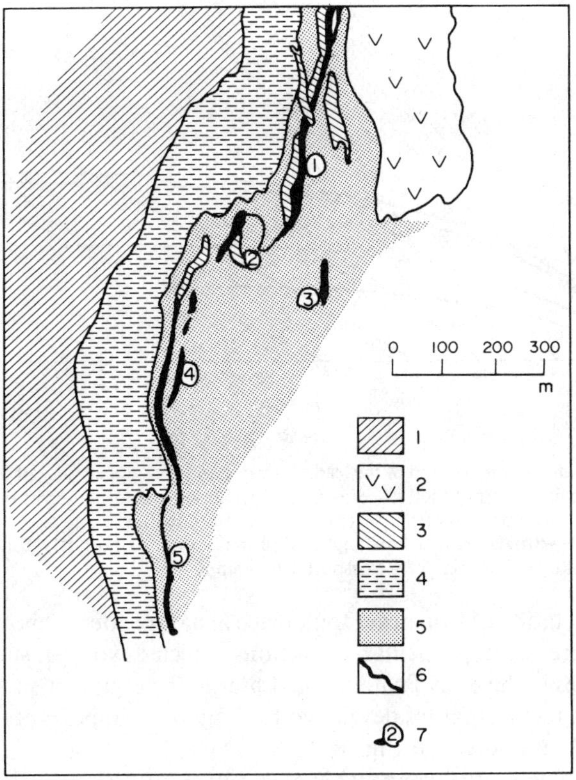

FIG.6. Geological map of the deposit Mount Vindarra: 1—granites and granitogneisses; 2—metabasalts; 3—intrusions of the main composition; 4—iron ore formation; 5—rocks of ultrabasic composition; 6—shale horizons; 7—nickel deposits.

about 2.8 billion years. Uranium is accumulated in conglomerates in which, together with pyrite and C_{org}, it cements the quartz gravel. These deposits have been studied in great detail but we cannot give a complete description here. Let us only note that together with uranium in some of them substantial accumulations of gold and platinoids have been found for the first time. It is very typical that uranium as well as pyrite in some cases is found in the form of detrital grains as well as which survived for a long time in the alluvium. This fact suggests that the conditions were more reducing than at present (Tugarinov, 1972, 1973).

About that time occur many ferruginous quartzites on different continents. To this type of deposit belongs the deposits of Hammersly, of the Upper Lake, of Singkhbkhum, Krivoi Rog and the Kursk magnetic anomaly, and others. The synchroneity of the accumulation of huge amounts of iron in this epoch is striking. This suggests that the causes of such a mass precipitation of iron were of a global character. It is considered that the best explanation of this is that of A. A. Saukov's. He proposed that the appearance of free oxygen in the atmosphere at that time led to mass conversion of divalent iron which in the form of bicarbonate was soluble in the ocean, to the trivalent form which was rapidly precipitated in the form of hydroxide (Bailey and James, 1973; Goodwin *et al.*, 1976).

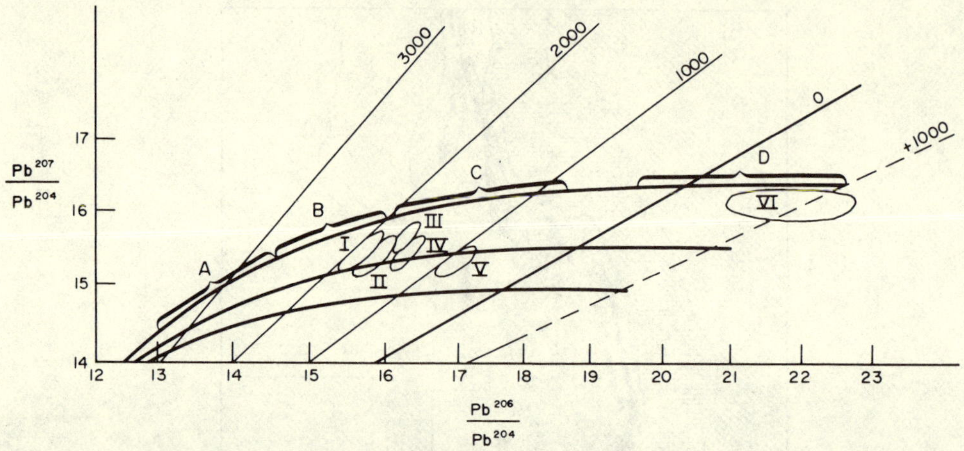

FIG. 7. Evolution of the isotopic composition of the lead of ore deposits: A—appearance of the first ore showings of lead, B—generation of the first lead-productive sedimentary formations, C—formation of the first commercial lead deposits, D—formation of lead deposits from several sources of ore matter.
Deposits: I—Vikhanti; II—Broken Hill, Ram-Jangle, Mount Isa, MacArthur; III—Sullivan; IV—Zava; V—Gorevskoye, Kholodnenskoye, Sardinia; VI—Missouri–Mississippi.

Such a mass precipitation of iron never took place in nature later. This obviously indicates once again that there were particular conditions which favoured such a sedimentary formation of iron ores in the early Proterozoic. Later in the region of such deposits (Krivoi Rog, KMA) alkaline metasomatites developed leading to the appearance of rich epigenetic magmatic ores which are noted on Fig. 1.

On Fig. 7 lead–zinc deposits form a stable maximum of sedimentary lead accumulations which subsequently developed into an equally stable maximum of endogenous deposits. What is the matter? At the end of the lower Proterozoic and the beginning of the upper Proterozoic an important epoch of the formation of the largest lead–zinc deposits is known. Most of these deposits are of the so-called stratiform type: Sullivan in Canada, Vikhanti in Finland, Zavar in India, Mount-Iza, MacArthur and Broken Hill in Australia, and others with ages from 1.5 to 2 billion years. A peculiarity of these deposits is their confinement to carbonate rocks containing a large amount of organic matter. These deposits are partly metasomatic deposits which, as a rule, pass along strike into conformable deposits. Despite the difference in the initial notion about their age, according to the isotopic age data they have turned out to be synchronous with the enclosing strata. More than this, the lead of the deposits proved to be identical with the lead of the enclosing rocks after correcting for radiogenic lead which formed after their accumulation in the enclosing strata. In other words, for each of the above-mentioned deposits the enclosing complex of rocks proved to be a productive formation bearing scattered lead concentrations, which in the course of the subsequent metasomatosis had formed ore deposits (see Fig. 7).

Later on in geological history analogous situations arose more than once during which large lead–zinc deposits were formed at the expense of rocks initially enriched in these elements (particularly carbonate rocks). Thus in the east of the U.S.S.R. a whole province has been found: the deposits Ozernoye, Kholodneuskoye and Sardinia. In Kazakhstan has been found in another province: Zhairem, Achisai, Mirgalimsai, and others. The first province is attributed to the late Precambrian and the second to Variscan time (Tugarinov, 1973).

It is very striking that lead deposits are practically unknown at a more ancient time. As it is seen in Fig. 7, during this early period of the Archaean and the beginning of the Proterozoic lead is only met in the form of small mineralogical finds and ore outcrops. Moreover, in ancient uranium deposits (1.8 billion years) lead as an accompanying element is practically unknown (common lead and not the radiogenic one is referred to here while during the Phanerozoic, lead and uranium paragenesis is commonly found).

We connect the lack of lead, as well as of uranium and iron ore deposits in the Archaean, first of all with the fact that in this epoch there were yet no facies which could become productive formations for these elements. In particular carbonate facies which are so characteristic of lead deposits appeared in the history of the earth only about 2.3 billion years ago.

In striking contrast with the earlier mentioned deposits are the tin-tungsten deposits. They are in contrast first of all because they are very young. The maximum age determined for them is 300–400 m.y. The main region of occurrence is the Pacific belt in which the deposits are dated at 100–150 m.y. What is the reason of such a restriction in time? It is true that small outcrops of tin are known, such as Pitkäranta, which are dated at 1500 m.y. However, this is an evident exception.

Quite recently in the G.D.R. tin deposits were found in Silurian stratum near the contact with a granite intrusion of Variscan age. Skarns formed at the contact contained a considerable amount of cassiterite, and the whole Silurian rock mass for some hundreds of kilometres proved to be contaminated with tin which was concentrated in biotite. In other words, the genesis of the deposits was somewhat different, tin was not introduced into the skarns by granites. The productive packet interacting with the granite is considered to be the tin source.

We have noted a similar picture in the Far East, where tin deposits are scattered in the stratum of the lower Cretaceous represented by shales forming substantial concentrations in the so-called biotitic hornfelses or are met in exocontacts of numerous intrusions breaking through this stratum. However, during injection of the same intrusions into the overlying effusive stratum of the upper Cretaceous, the cassiterite mineralization disappears. The question is again not about the tin content in the intrusions, but about the same productive stratum on the lower. Cretaceous being distinguished by high tin content. How has it formed? As the study of marine cassiterite placers in the same region has shown, the greatest cassiterite concentrations are in variegately stratified sediments where sands alternate with clays sufficiently frequently. It is known that cassiterite is not preserved in placers farther than 50 km from the source. In short, accumulations of cassiterite during erosion of some massif will occur at a relatively small distance from it in a sand–shale stratum. To the west of the shale denudations of the lower Cretaceous along the break extends a series of intrusions of Permian age distinguished by poor tin content at the expense of whose erosion the formation of the productive stratum of the lower Cretaceous had occurred. Thus it is concluded that in distinction to lead and uranium whose accumulation in strata containing organic matter occurred at once in sufficiently great amounts, tin had to pass through a series of subsequent enrichments before its productive formation could arise.

For a number of minor elements, such as, for example, germanium and vanadium a change occurs in the course of geological history of the form of their concentrations as evidenced by the appearance of new types of deposits.

Thus, for example, at the beginning of the Proterozoic there were no deposits of only germanium, because the whole mass of germanium was scattered among ferruginous quartzites in which germanium isomorphously entered the magnetite in the form of its divalent oxides instead of iron. Only in the Phanerozoic appeared biogenic germanium concentrations in the form of germanium containing coals and germanium accumulations in the form of the proper mineral germanite in sulphide deposits (Tugarinov, 1973).

An analogous fate was suffered by vanadium which existed in the early Precambrian only in the form of isomorphous substitutions in titanomagnetites of magmatic origin and which during the subsequent development of organic life began to accumulate in coals and oils in which it is a constant associated element.

In short, the history of ore formation in the crust of the earth obeys the general laws of the development of the crust itself. By virtue of the irreversibility and the trend of its development the process of ore formation in its turn obeys the general unalterable laws of geological development. This is displayed by the fact that at different epochs deposits of certain metals are dominant over others. There is no constancy in ore formation. At the same time it is of great importance to know the causes of the appearance of some particular epochs of metallogeny for the purpose of correctly conducting exploration programs.

In this article have been given examples of some of the major epochs of metallogeny and their causes have been discussed.

References

ANHAEUSSER, C. R. (1973) *The Geology and Geochemistry of the Archaean Granites and Gneisses of the Johannesburg–Pretoria Dome,* Spec. Publ. Geol. Soc. S. Afr., 3.

BAYLEY, R. W. and JAMES, H. L. (1973) Precambrian iron formations of the United States. *Econ. Geol.*, **68**, 934–959.

BRIDGEWATER, S. D., ESHER, A., JACKSON, G. D., TAYLOR, F. C. and WINDLEY, B. F. (1973) Development of the Precambian Shield in West Greenland, Labrador and Baffiin Island. In *Geology of the Arctic*, Amer. Ass. Pet. Geol., mem. 19, pp. 99.

BROOKS, C. and HART, S. (1974) On the significance of kematiite. *Geology,* Feb.

CONDIE, K. C. and HUNTER, D. R. (1976) Trace element geochemistry of Archaean granitic rocks from the Barberton Region, South Africa. *Earth Planet. Sci. Lett.* **29**, 389–400.

DUNGWORTH, G. and SCHWARTZ, A. W. (1974) Organic matter and trace elements in Precambrian rocks from South Africa. *Chem. Geol.* **14**, 167–172.

GLIKSON, A. (1972) Early Precambrian evidence of a primitive Ocean crust and Island nuclei of sodic granite *Geol. Soc. Amer. Bull.* **83**, 3323–3344.

GOODWIN, A. M., MONSTOR, J. and THODE, H. G. (1976) Carbon and sulfur isotope abundances in Archaean iron-formations and early Precambrian life. *Econ. Geol.* **71**, 870–891.

HARGRAVES, R. B. (1976) Precambrian geologic history. *Science,* **193**, No. 4251.

HEIER, K. S. (1973) A model for the composition of the deep continental crust. *Fortschr. Miner.* **50**, 174.

KRÖNER, A., ANHAEUSSER, C. R. and VAJNER, V. (1973) Neue Ergebnisse zur evolution der Prekambrischen Kruste im südlichen Afrika. *Geol. Rundsch.* **62**, 218–309.

LUSK, J. (1976) A possible volcanic–exhalative origin for lenticular nickel sulfide deposits of volcanic association, with special reference to those in Western Australia. *Can. J. Earth Sci.* **13**, 451–458.

MCGREGOR, V. R. (1973) The early Precambrian gneisses of the Godthab district, West Greenland. *Philos. Trans.* Roy. Soc. Lond. A, **273**.

MOORBATH, S. (1976) Age and isotope constraints for the evolution of Archaean crust. In *The Early History of the Earth*, J. Wiley, New York.

NESBITT, R. W. and SHEN-SU SUN (1976) Geochemistry of Archaean spinifex-textured peridotites and magnesian and low-magnesian tholeiites. *Earth Planet. Sci. Lett.* **31**, 433–453.

SIGHINOLFI, P. and SANTOS, A. M. (1976) Geochemistry of gold in Archaean granulite facies terrains. *Chem. Geol.* **17**, 113–123.

SIMS, P. K. and MOREY, G. B. (1972) *Geology of Minnesota,* Minnesota Geol. Surv.

SUTTON, J. and WINDLEY, B. (1974) The Precambrian. *Sci. Prog. Oxf.* **61**, 401–420.

TUGARINOV, A. I. (1956) Epoques métallogéniques dans le Précambrien. *Izv. AN SSSR,* sér. géol., No. 9.

TUGARINOV, A. I. (1972) On evolution of ore formation in history of the earth. *M.G.U. XXIV session, Dokl. Sov. Geol., Geokhimiya.*

TUGARINOV, A. I. (1973) General geochemistry. *Izd. Atomizdat.*

TUGARINOV, A. I. and BIBIKOVA, E. V. (1976) Evolution of the chemical composition of the earth crust. *Geokhimiya*, No. 8.

TUGARINOV, A. L. and VOÏTKEVITCH, G. V. (1972) Precambrian geochronology of continents. *Izd. Nedra.*

VOÏTKEVITCH, G. V. and LEBEDKO, G. I. (1975) Ore deposits and Precambrian metallogeny. *Izd. Nedra.*

Ore Element Accumulation in the Earth's Crust Evolution

YU. K. BURKOV and D. V. RUNDQUIST

Scientific Research Geological Institute (VSEGEI), Sredny Prospect, 74, Leningrad V–26, U.S.S.R.

Abstract

Among many problems of the evolution of the earth's crust the phenomenon of the accumulation of the ore elements during various epochs of the history of the geological evolution is much less studied. The data accumulated during the last years allow us to examine the given problem in a new way.

The study of statistical connections between contents of the chemical elements in a great number of geological objects of various composition and origin leads to the conclusion on the possibility of establishing two leading systems of lithosphere based on geochemical associations: the first uniting the magmatic formations and sedimentary rocks—products of their mechanical breaking—and the second including the sedimentary products of chemical weathering, biological objects and biogenic formations. For the first of these systems, having an endogenous nature, typical is the differentiation of elements by their femaphilic–felsiphilic features: antagonism of Fe, Ni, Co, etc., concerning K, Pb, Ba, etc. For the second (exogenous), typical is the differentiation in the first place in accordance with the properties of inertia–mobility of the chemical elements in the zone of hypergenesis: antagonism of Al, Ti, Ga, etc., concerning Cl, K, Ca, etc.

Simultaneously the existence was established of a provisional and causal relation between endogenous and exogenous ore formation and the phenomenon of transfer of conditions favourable for the concentration of ore formation in the process of the geological history from an "endogenous" geochemical system to an "exogenous" one. The analysis of regularities of change of types of associations of ore-forming elements during the transition between named systems allowed the establishment of the main principles of evolution of geochemical associations in rocks and ores during the geological history connected with the fundamental properties of the chemical elements: the greater and greater spreading of felsiphilic associations, the gradual decomposition of femaphilic associations and the increase of the number and complexity of the combined femaphilic–felsiphilic associations.

Among many problems of ore formation, the time problem, including such questions as the evolution of the deposit genetic types and different metal accumulation during various epochs of the geological development history, remains insufficiently studied. Recently data have been obtained that enable the consideration of many aspects of the mentioned problem from a new standpoint. Firstly, as compared to the existing ones, principally different notions on the location and character of the boundary dividing the leading geochemical systems of the lithosphere were formed; one of these might be conventionally termed "endogenous" and the second "exogenous"† (Burkov, 1973); secondly, the time and causal relationship between endogenous and exogenous ore formation was determined and, thirdly, some general regularities of the primary distribution of different type ore formation relative to the mentioned boundary were revealed.

†The conventionality of the terms is defined by the fact that the types of geochemical associations typical of these varieties are assumed as a basis of distinguishing the systems but not the aspects of the conditions of the formation of the rock types of which they are comprised.

1. On Two Basic Geochemical Systems of the Lithosphere

To characterize the leading geochemical systems of the lithosphere and to make more precise the position of the boundary dividing them, the data on the distribution of the statistic relations between chemical elements in the main varieties of rocks and ores of various composition and origin have been defined and interpreted: intrusive volcanic, metasomatic, metamorphic, crusts of weathering as well as sedimentary ones: clastogene, chemogenic and biogenic. †When evaluating the mentioned relations, algorithms and programs were used that were specially worked out based on the correlation statistic analysis and that have been realized at the computers (Burkov, 1971).

The geochemical study of magmatic formations differing in the age of intrusions and volcanic bodies of the basic, intermediate and acid compositions has shown that a great variety of the relation types between the elements is characteristic of them. At the same time, the generalization of the evaluations of the mentioned relations enabled us to determine stable tendencies to the formation of two concrete geochemical associations in all the magmatic formations. As Table 1 shows, summarizing the results of the analysis of geochemical associations of magmatogene rocks, the formation of two antagonistic groups is characteristic of their varieties basic in composition (Table 1, a), the "nucleus" of one of which is composed of cobalt, nickel, iron, rhodium, etc., that are closely related between themselves. The greatest antagonism towards these elements is exhibited by potassium, rubidium, chlorine and barium that are joined by high positive correlations. Chromium, copper, manganese and silver are "joined" to cobalt, nickel, etc., by weaker relations and sulphur, selenium, phosphorous and arsenic to potassium, rubidium, etc. It is interesting that these eight elements form analogous (weaker) relations and between themselves within the groups which they comprise. Even lower positive correlations join vanadium, cadmium, zinc and niobium with the "cobalt group" and yttrium, lanthanum, silicon, etc. with the "potassium group". Finally, aluminium, titanium and gallium occupy an "intermediate position" between the mentioned antagonistic associations forming insignificantly weak relations with the elements of which they are comprised.

According to the character of geochemical associations the rocks of the intermediate composition, as follows from Table 1 (b), are almost analogous to the formations of the basic composition. They differ from the latter only in the titanium behaviour which forms the composition of the "cobalt group".

Acid magmatic formations (Table 1, c) differ from the considered types only in the position of zirconium which "leaves" the "potassium group" in them.

Thus, in spite of the differences in the compostion (basic–intermediate–acid), magmatic formations are characterized by the high stability of the character of the relations between the elements and they are very similar between themselves in this indication. As a result, a middle type of geochemical associations might be outlined for these formations as presented in Table 1.

It is interesting to compare the obtained generalized geochemical characteristic of magmatic formations based on the study of the types of relations between the elements with the existing evaluations of the basic regularities of the average content distribution of the

†More than 5000 endogenous and 9000 exogenous objects have been studied, each of which was characterized by not less than seventy samples analysed by spectral and chemical methods. A sufficiently large amount of the rock types studied enables us to consider the obtained average characteristics of the chemical element relations as some kind of constants like clarks of elements.

concrete chemical elements in a set basic rocks–granitoids. For the first time these regularities were revealed and described in detail by A. E. Fersman (1953), who stressed the existence of two leading geochemical associations of magmatic formations: a group of elements increasing their contents in the set basic rocks–granitoids and a group of elements decreasing their contents in the mentioned direction. D. M. Shaw (1963) suggested that the elements increasing contents in basic rocks should be termed femaphiles, those increasing contents in granitoids, felsiphiles, and elements having no distinct tendencies in the distribution, homeophiles. The associations characteristic of each of the studied types of magmatic formations that are revealed on the basis of the study of the relations between the elements coincide in many aspects with those distinguished before—they are, on the whole, conformable to the groups of femaphiles (iron, cobalt, nickel, etc.) of felsiphiles (silicon, potassium, etc.) and of homeophiles (aluminium, gallium, etc.).

In sedimentary formations, which are products of mechanical destruction of magmatic formations of the feeding areas (beds of arkoses, greywackes, etc.), geochemical associations have been defined slightly differing or not differing at all from the main groups of elements typical of magmatogene formations. Joining up of elements by the indications of their femaphile, felsiphile or homeophile features is also characteristic of them. Thus, products of mechanical weathering inherit element associations characteristic of parent rocks. (The generalized geochemical characteristic of this type of rocks is not cited as it is fully conformable to the geochemical formula of magmatic rocks, Table 1.)

TABLE 1. AVERAGE CHARACTERISTICS OF GEOCHEMICAL ASSOCIATIONS OF MAGMATOGENE FORMATIONS OF THE BASIC (a), INTERMEDIATE (b) AND ACID (c) COMPOSITIONS AND A GENERAL CHARACTERISTIC OF MAGMATOGENE, FORMATIONS†

a	Co Fe Pd Ni Rh Pt	Cr Mn Cu Ag	V Cd	Zn Ti Nb	Al Zr Ga	Si Ge	Y La	P S As Se	Cl K Ba Rb
b	Co Fe Pd Ni Rh Pt	Cr Mn Cu Ag	V Cd	Zn Ti Nb	Al Ga	Si Zr Ge	Y La	P S As Se	Cl K Ba Rb
c	Co Fe Pd Ni Rh Pt	Cr Mn Cu Ag	V Zn Cd Nb		Al Ti Ga	Si Zr Ge	Y La	P S As Se	Cl K Ba Rb
d	Co Fe Pd Ni Rh Pt	Cr Mn Cu Ag	V Cd	Zn Ti Nb	Al Ga	Si Zr Ge	Y La	P S As Se	Cl K Ba Rb

†In Tables 1 to 3 the elements forming the highest positive correlation relations are joined by thick frames and those weaker related by thin ones. The elements divided by the highest negative correlations are arranged on the opposite sides of the cited geochemical formulas.

TABLE 2. AVERAGE CHARACTERISTICS OF GEOCHEMICAL ASSOCIATIONS OF SEDIMENTARY ROCKS OF THE BIOGENIC ORIGIN (a), SEDIMENTARY ROCKS PRODUCTS OF CHEMICAL WEATHERING OF THE HYPERGENESIS ZONE (b) AND A SUMMARY CHARACTERISTIC OF THE MENTIONED ROCK TYPES (c)

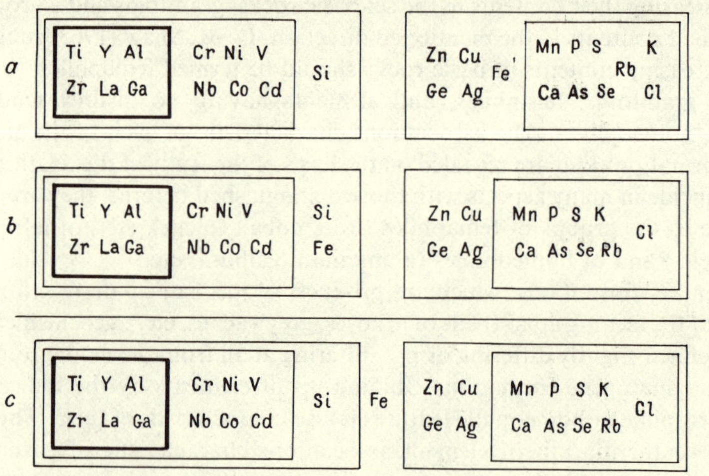

In sedimentary formations which are products of chemical weathering of catchment area rocks (some bauxites, kaolins, a great many rock varieties of clay and carbonate composition, etc.), as well as in the magmatogene formations, at the background of a considerable variety of relation types a tendency is established to the formation of two main geochemical associations between elements. However, as Table 2 (b) shows, the composition of the latter differs sharply from the composition of the main group elements of magmatic rocks. Titanium, yttrium, zirconium, lanthanum, etc., closely related between themselves are a part of one of these "sedimentary" associations. Chromium, niobium, nickel, cobalt, etc., "join" them, and with weaker relations—silicon. The association, of which potassium, chlorine, rubidium, etc., form the composition, occupies an antagonistic position with respect to the mentioned elements. Copper, silver, zinc, etc., form weaker positive relations with the latter. It is interesting that the parallel study of geochemical features of biogenic deposits—peats, coals, etc. (Table 2, a)—and some biological objects—plants, fishes, warm-blooded animals—has shown that they are very similar in relation characteristics between elements. At the same time it was found that the biogenic formations contain element associations practically analogous to the associations in the considered rocks products of chemical weathering. As is seen, the differences between the two mentioned types of formations (Table 2, a and b) are defined only by some changes of iron behaviour. This allows the averaging of their characteristics (Table 2, c).

It is interesting to interpret the revealed regularities of the relation distribution between elements in chemogenic and biogenic sedimentary formations from the positions of modern notions on the character of the behaviour of individual elements under the earth's surface conditions. From these positions the teaching on the sets of the element geochemical mobility in the hypergenesis zone proposed by N. M. Strakhov (1960–1962) seems the most important. He paid attention to the fact that under these conditions the element migrations are defined not only by chemical features of their associations but by the forms in which these associations migrate. The analysis of these forms enabled N. M. Strakhov to reveal the set "the beginning" of which is made up of inert elements almost not passing into a solution composition (for instance, titanium, zirconium, etc.), the middle part, elements with complex forms of migration (the majority of so-called "minor" elements, for instance, vanadium, zinc, copper, etc.), and the "end" elements, the main forms of migration of which are solutions (for instance, calcium, manganese, potassium, etc.). As can be seen, the geochemical characteristics obtained on the basis of the analysis of the relations between elements are well conformable with N. M. Strakhov's data that enable us to interpret the revealed regularities of association of elements from the positions of properties of their inertness–mobility in the hypergenesis zone. The groups shown in Table 2 exhibiting the highest antagonism are made up of the most inert (titanium, zirconium, lanthanum, etc.) and the most mobile (chlorine, potassium, rubidium) components, and the elements found between these groups are characterized, evidently, by complex forms of migration.

Summing up some preliminary results, it is possible to come to a conclusion that the accumulated actual material enables us to outline the following process of "rebuilding" of geochemical associations at the "boundary" between endo- and exogenous conditions. Chemical weathering leads to the destruction of parent rock minerals and the passing of elements contained in them into the composition of solutions and suspensions. The geochemical associations belonging to parent formations are broken and further element migration occurs in conformity with their individual physico-chemical features exhibited under the conditions of transportation and sedimentation. In the process of migration and settling in the territory of the terminal basin run-off the elements, the considerable mass of which is capable of migrating in the form of ionic or colloid solutions (lithium, fluorine, sodium, potassium, calcium, strontium, etc.), form positive relations between themselves and separate from the elements, the associations of which are poorly dissolved under these conditions (beryllium, aluminium, gallium, hafnium, etc.).

Thus, when passing mechanogenic to chemogenic and biogenic sedimentary formations the transition is observed of element associations similar to the associations in parent rocks to the associations reflecting sets of element mobility under the sedimentation conditions, i.e.

TABLE 3. AVERAGE CHARACTERISTICS OF GEOCHEMICAL ASSOCIATIONS OF MAGMATOGENE FORMATIONS (c), SEDIMENTARY ROCKS PRODUCTS OF MECHANICAL WEATHERING OF MAGMATIC FORMATIONS (b) AND THE JOINED GEOCHEMICAL FORMULA OF CHEMOGENIC AND BIOGENIC SEDIMENTARY FORMATIONS (a)

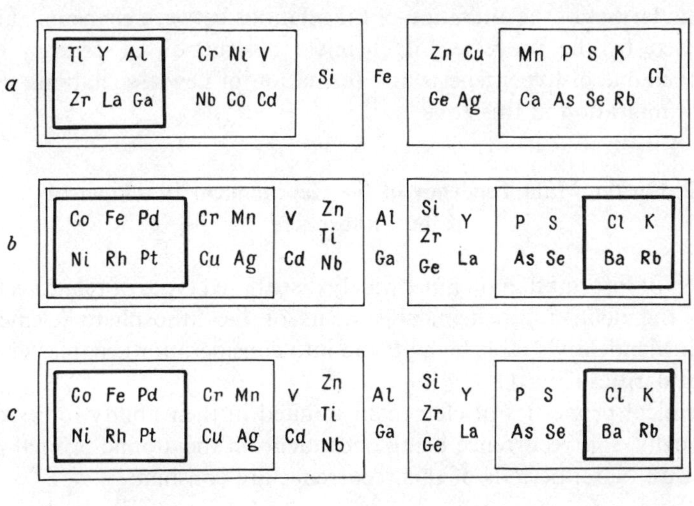

the alternation of element groups differentiated according to their femaphile–felsiphile features with the groups differentiated according to their inertness–mobility in the hypergenesis zone.

The data generalization on the regularities of the distribution of statistic relations between chemical elements in the investigated objects enables us to come to the following conclusions: in sedimentary formations, which are products of mechanical destruction of magmatic rocks of feeding areas element associations similar to those in magmatogene (and, in particular, volcanic) rocks are preserved (Table 3, b, c); in sedimentary rocks, which are products of chemical weathering of the formations of feeding areas, the associations defined by the forms of element migration under the hypergenesis conditions are formed. The analogous associations are typical of biogenic rocks and bios (Table 3, a).

Thus, from the positions of the distribution of the relations between elements within the lithosphere two common geochemical systems are outlined: "endogenous", which joins magmatic formations and products of their mechanical breaking, and "exogenous", that includes products of chemical weathering, bios and biogenic formations. Hence, the distribution of the associations in the lithosphere according to the degree of their community might be presented in the form of the following scheme:

Magmatogen formations	Products of mechanical weathering of magmatogene formations	Zone of rebuilding of the character of relations between chemical elements	Products of chemical weathering of rocks of feeding areas	Bios	Biogenic formations

The zone of the gradual transition between the two defined systems might be applied evidently to one of the main geochemical boundaries of the lithosphere. It does not contradict the geochemical boundaries between magmatic and sedimentary rocks (Saukov, 1951) established before or between living and lifeless matter (Vernadsky, 1977). Being defined by the regularities of the alteration of the relations between elements, this boundary is fully determined by the processes of chemical destruction of primary endogenous associations in the zone of hypergenesis and formation of new associations reflecting the forms of element migration in this zone.

2. On the Main Tendency of the Geochemical Development of the Lithosphere

The summation of information on most widely distributed types of relations between the elements within the defined geochemical systems of the lithosphere (element features reflected by D. I. Mendeleev's table being taken into consideration) enables us to establish the following regularities (Fig. 1):

1. The geochemical properties of elements exhibited in their ability for associating are repeated periodically, this recurrence being dependent on the atomic ordinal number.

2. Beginning with N 13, periods of this recurrence are constant.

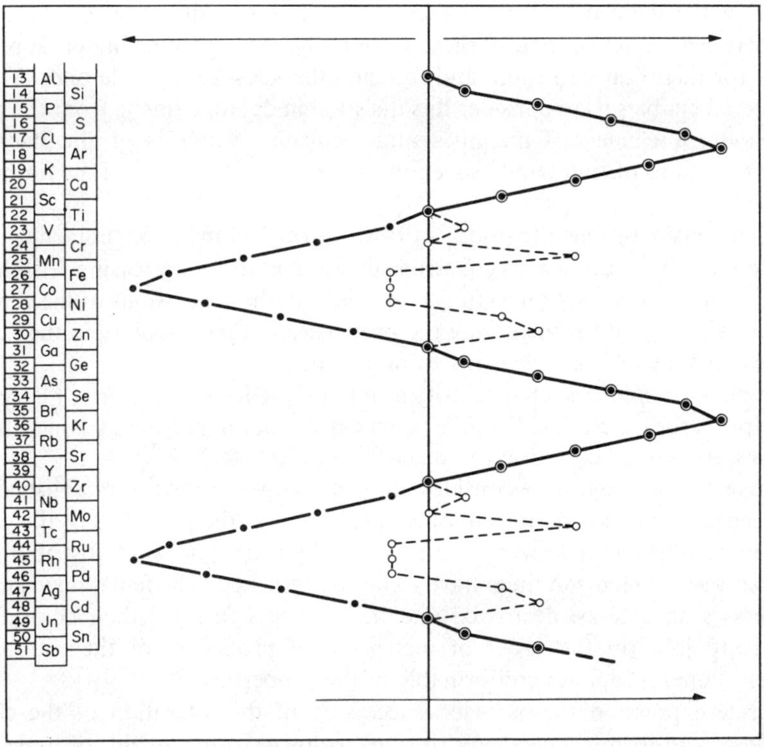

Fig. 1. The behaviour of elements in "endogenous" (dots in the scheme) and "exogenous" (circles) geochemical systems. Arrows in the upper part of the figure show the increase of femaphile property (left) and felsiphile property (right) of chemical elements in the "endogenous" system. The arrow in the lower part shows the direction of increasing of element mobility in the "exogenous" system. (The position of Tc is shown presumably.)

3. In the succession of elements constructed in conformity with the increasing of their ordinal numbers, beginning with N 13, groups alternate regularly, the neighbour members of which form positive relations in magmatic rocks and products of their mechanical weathering. The alternation of these groups is subordinated to the following simple rule: one homeophile and eight felsiphiles, one homeophile and eight femaphiles, one homeophile and eight felsiphiles, etc. (Fig. 1), the properties of femaphile or felsiphile features as a rule being regularly strengthened in the central parts of the distinguished groups and being weakened to their "ends" in the zones of joining to homeophiles (this regularity is shown in Fig. 1 by a solid line).

4. Element pairs with difference of numbers that equals 18 form positive relations in the formations of the zone of hypergenesis and biological objects.

The defined leading features of geochemical associations of the two most contrasting systems—"endogenous" and "exogenous"—open up additional possibilities for the analysis of the main tendency of the lithosphere evolution on the whole. As was stated by the works (Fersman, 1953; Vernadsky, 1977), the process of rock and ore formation in geological history is a result of repeating cycles: magmatism—metasomatism—crust formation—sedimentation—metamorphism—magmatism, etc. In the course of these repeated cycles it is natural that some geochemical associations are destroyed and others appear. Taking into consideration the defined and considered features of the alteration of the character of relations between elements, some of these elements have to be found under more favourable conditions for their concentration, and for the others, less favourable ones. The obtained actual material enables us to consider this question in detail, namely, from the positions of the behaviour of elements–femaphiles and elements–felsiphiles at the background of repeated circulations of endo- and exogenous processes in the course of the formation of the earth's crust.

With such considerations attention is drawn to the following phenomena:

—for felsiphiles, their successions constructed with regard for increasing felsiphile features (of weak felsiphiles to strong ones) fully coincide with the successions of the same elements constructed with regard for increasing the properties of their mobility in the "exogenous" geochemical system (of inert elements to mobile ones);

—for elements–femaphiles such regularity is not observed, some of them which possess high femaphile properties are related to inert elements of the hypergenesis zone, the others to mobile ones, the same being true for weak femaphiles, etc.†

The above being taken into consideration, it follows from the established facts that felsiphile geochemical associations in ores (and rocks) at the processes of the earth's crust evolution have to be characterized on the average by wider and wider distribution, and the relations between elements in them have to be strengthened. The femaphile associations at these processes have to be destroyed and the elements forming their composition have successively to join (in the order of increasing of properties of their mobility in the hypergenesis zone) felsiphiles conformable in the properties.

In concrete expression the assumed succession of the alteration of the character of associating of ore-forming elements during geological time might be reflected in the successive alternation of the following groups, for instance: cobalt, iron, nickel, vanadium, chromium, rhodium, ruthenium, etc. (a great many of the most inert femaphiles); copper, zinc, lead, molybdenum, tungsten, tin, etc. (inert ones and femaphiles and felsiphiles characterized by intermediate mobility); arsenic, silver, cadmium, antimony, gold, mercury, bismuth, etc. (the most mobile femaphiles and felsiphiles conformable to them in mobility). The comparison of the composition of the enumerated associations with the character of alteration in time and space (as the "endogenous" geochemical system to the "exogenous" one) of really existing groups of ore-forming elements (Rundquist, 1969, 1975) is indicative of their considerable similarity, which might be evidence of the truth of all the above-mentioned assumptions.

Thus, the study of some most common geochemical properties of elements (femaphile–felsiphile features, inertness–mobility in the hypergenesis zone) enables us to construct the evolution models of their associations in rocks and ores during geological time. The main indications of this evolution are: wider and wider distribution of felsiphile associations, the successive disintegration of femaphile associations and, finally, the increase of the number and complication of combined femaphile–felsiphile associations. From this background one more regularity becomes apparent: when the ordinal numbers of the D. I. Mendeleev table are increased (and, hence, on the average of their "rarity" in the composition of rocks and ores) the properties of femaphile features of femaphile elements are decreased and those of felsiphiles are increased. With the general tendency to "felsiphilization" of the lithosphere in the course of geological history it cannot help but lead

†The changes of the positions of elements–femaphiles revealed empirically in the hypergenesis zone are shown in Fig. 1 by a dotted line.

to the common trend of behaviour for two mentioned element groups—constant increase of the role and contents of elements in ores with higher ordinal numbers and lower clarks of the lithosphere (and the "weakest" femaphiles and "strongest" felsiphiles are related to the latter).

The observed regularities defined by fundamental properties of chemical elements themselves (atoms) are apparently the basis of such phenomena established empirically as the increase of the role of acid and the decrease of basic (in composition) geological formations in the process of the lithosphere development, exponential increase of the variety of these formations and drawing of a greater number of "rare" elements with low clarks and high numbers of D. I. Mendeleev's table into the composition of ores (and, it is possible, of rocks) during later epochs.

References

BURKOV, Yu. K. (1971) Study of the conditions of the sedimentary bed formation by the methods of the statistic treatment of geochemical data. *Trudy VSEGEI*, new ser., **158**, 346–356.

BURKOV, Yu. K. (1973) Chemical element associations as indicators of the primary nature of metamorphosed sedimentary rocks. In *Lithology and Precambrian Sedimentary Geology* (The materials of the Xth All-Union Lithological Conference), Isd-vo AN SSSR, M., 197–199.

FERSMAN, A. A. (1953) *Selected Works*, Isd-vo AN SSSR, M., **2**, 768 pp.

RUNDQUIST, D. V. (1969) The ore formation evolution in time. In *Geological Structure of the USSR*, Nedra, M., **5**, 303–332.

RUNDQUIST, D. V. and NEZHENSKY, I. A. (1975) *Zoning of Endogenous Ore Deposits*, Nedra, L., 224 pp.

SAUKOV, A. A. (1951) *Geochemistry*, Gosgeolisdat, M., 382 pp.

SHAW, D. M. (1963) *Geochemistry of Crystalline Rock Microelements*, Nauka, L., 207 pp.

STRAKHOV, N. M. (1960–1962) *The Bases of the Lithogenesis Theory*, M., **1–3**, 760 pp.

VERNADSKY, V. I. (1977) *Naturalist's Reflections (A scientific thought as a planet phenomenon)*, Nauka, M., 192 pp.

SECTION VII

GEOCHEMISTRY OF THE UPPER MANTLE

Trace Element Melting Models†

DENIS M. SHAW

Department of Geology, McMaster University, Hamilton, Ontario, Canada L8S 4M1

Abstract

Application of the Berthelot–Nernst partition coefficient permits development of a general theory for behaviour of trace elements during anatexis. In addition to constraints imposed by conditions of modal melting, non-modal melting and incongruent melting, it is shown that the presence of a fluid phase may profoundly modify the behaviour of some elements.

Applications of the theory to petrological problems are impeded by inadequate knowledge of D-values, of phase petrology and of mantle chemistry.

Spectacular progress has been made in the application of trace-element studies to petrological problems during the ten years which have elapsed since the first Symposium on Origin and Distribution of the Elements of the International Association for Geochemistry and Cosmochemistry.

This paper is concerned with trace-element behaviour during rock melting or anatexis, the theory of which lies at the heart of many studies published during this ten-year period.

Melting Theory for Trace Elements

The simplest expressions governing trace-element behaviour during melting of a rock are:

$$\frac{c^l}{c_0} = \frac{1}{D}(1-F)^{((1/D)-1)}, \tag{1}$$

$$\frac{\bar{c}^l}{c_0} = \frac{1}{F}[1-(1-F)^{1/D}], \tag{2}$$

$$\frac{c^L}{c_0} = \frac{1}{D+F(1-D)}. \tag{3}$$

Based on expressions for step-wise melting developed by Schilling and Winchester (1967) and Gast (1968), and on derivations for fractionation first obtained by Rayleigh (1896), these equations were given by the author (1970) and have since been developed by several authors, notably Hertogen and Gijbels (1976; see also Consolmagno and Drake, 1976). The first equation gives the concentration c^l of the trace element in the infinitesimal mass of liquid separating from the solids (whose aggregate concentration is $c^s = Dc^l$) after a fraction F has already melted, assuming that all previous liquids have been unable to re-equilibrate with the residual solids: this may be called *Rayleigh fractionation*. The aggregate concentration in all liquids produced by this process, if they were to accumulate and mix together, is $\bar{c}^l$ in equation (2): this is the *accumulated Rayleigh fractionation*. If, on the other hand, the liquid

†Contribution number 62 of the McMaster Isotopic, Nuclear and Geochemical Studies Group.

TABLE 1. LA CONCENTRATIONS DURING MODAL MELTING OF GARNET LHERZOLITE (ppm)

F	c^l	$\bar{c}^l$	c^s	c^L	c^S
0.001	39.0	39.5	1.3	39.0	1.3
0.01	29.7	34.7	0.96	30.9	1.0
0.1	1.73	12.5	0.056	10.1	0.33
0.2	0.051	6.5	0.002	5.8	0.19
0.7	0	1.9	0	1.8	0.059

$D = 0.0324$. $C_0 = 1.3$ ppm.

remains continually in equilibrium with residual solids, its concentration after a fraction F has melted is c^L (equation (3): this is *equilibrium* or *batch melting*.

Equations (1) to (3) are referred to above as the simplest expressions because they are subject to only two constraints. The first of these is that the modal proportions X^i of the mineral phases i in the rock are the proportions also going into the liquid, i.e. *modal melting*, and are therefore constant. In practice it is unrealistic to assume that this condition could remain applicable throughout melting. The second constraint is that the partition coefficient D^{i-l} of the trace element between each mineral i and liquid is constant, so that the bulk distribution coefficient $D(=\Sigma_i X^i D^{i-l})$ is also constant.

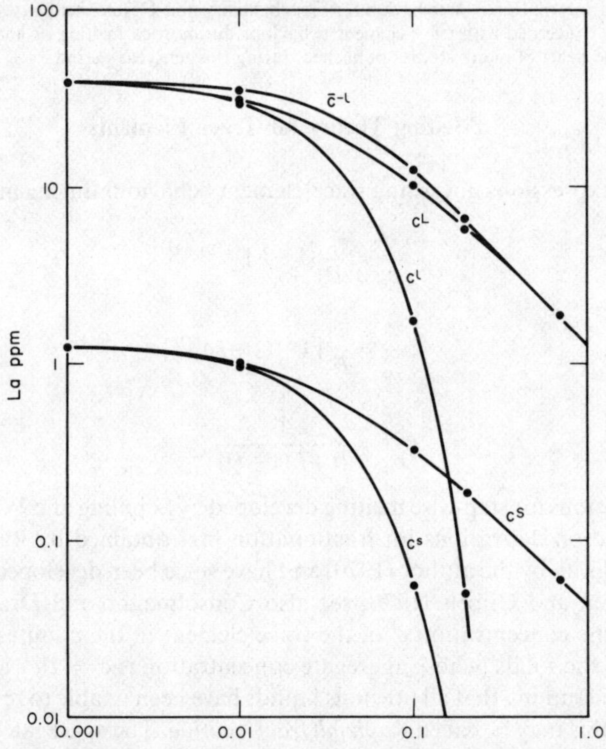

FIG. 1A

FIG. 1. La concentrations in melting of garnet lherzolite, $C_0 = 1.3$ ppm. (A) Modal melting; $D = 0.0324$. (B) Non-modal melting; $D_0 = 0.0324$, $P = 0.05$. (C) With incongruent melting of clinopyroxene; $Q = 0.19$.

Suppose, for example, that a lherzolite consists of olivine, orthopyroxene, clinopyroxene and garnet in the mass proportions 0.3, 0.2, 0.3 and 0.2, and that the respective partition coefficients for La are 0.008, 0.01, 0.08 and 0.02, then $D = 0.0324$. If we assume that the initial rock contains 1.3 ppm La then the concentrations in the liquid and solid at various stages of melting are shown in Table 1 and Fig. 1A.

Modal melting is unrealistic since rocks will begin to melt at an invariant point, where the proportions p^i of phases entering the liquid are different from their initial proportions X_0^i in the rock. After a fraction F has melted the residual phases are now in proportions X^i, where for each mineral we have

$$X_0^i = (1-F)X^i + p^i F \tag{4}$$

and

$$\sum_i X_0^i = \sum_i X^i = \sum_i p^i = 1.$$

This is *non-modal melting* and the melting stage covered by this stage will end when one phase, say α, is exhausted, which occurs when

$$F_{\max} = X_0^\alpha / p^\alpha. \tag{5}$$

The bulk distribution coefficient is no longer constant and is given by

$$D = \frac{D_0 - FP}{1-F} \tag{6}$$

$$\text{where } D_0 = \sum_i X_0^i D^{i-l} \text{ and } P = \sum_i p^i D^{i-l} \text{ (Shaw, 1970).} \tag{7}$$

The trace-element concentration is now given by the following expressions:

$$\frac{c^l}{c_0} = \frac{1}{D_0}\left(1 - \frac{PF}{D_0}\right)^{((1/P)-1)}, \tag{8}$$

$$\frac{\bar{c}^l}{c_0} = \frac{1}{F}\left[1 - \left(1 - \frac{PF}{D_0}\right)^{1/P}\right], \tag{9}$$

$$\frac{c^L}{c_0} = \frac{1}{D_0 + F(1-P)}. \tag{10}$$

If we continue with the same example as before, but assume that only clinopyroxene and garnet melt, each in the proportion 0.5, then melting will proceed until all the garnet is used up, which occurs (equation 5) when $F_{\max} = 0.2/0.5 = 0.4$. The concentrations obtained in this model are given in Table 2 and Fig. 1B. The value of D_0 is 0.0324, the same as D in the previous case, and $P = 0.05$.

It should be noted that equation (8) is limited to values of $PF/D_0 < 1$ (and similarly equation (9)), which may not apply where $P > D_0$. This is the case in the present example and the limiting value of F is D_0/P or 0.64; this may, however, be disregarded since F is already restricted to values <0.4.

Another constraint which might arise is that of *incongruent melting*. This has been considered by Hertogen and Gijbels (1976) and the following statement uses their terminology in part. Suppose phase α melts incongruently to produce phases β, γ, δ (already present in the rock) plus liquid, in proportions t^β, t^γ, t^δ, t^l respectively, where $\Sigma t^j = 1$. Suppose also that the rock also contains phases ε to ν and that the proportion of any phase in

TABLE 2. La CONCENTRATIONS DURING NON-MODAL MELTING OF GARNET LHERZOLITE (ppm)

F	c^l	$\bar{c}^l$	c^s	c^L	c^S
0.001	39.0	39.5	1.3	39.0	1.3
0.01	30.0	34.8	0.96	31.0	1.0
0.1	1.7	12.5	0.051	10.2	0.31
0.2	0.036	6.5	0.001	5.8	0.16
0.64	0	2.0	0	2.0	0.002

$D_0 = 0.0324$. $P = 0.05$. $C_0 = 1.3$ ppm

the liquid is q^i, where $\Sigma q^i = 1$ (if one or more of the q^i equals zero then we have non-modal melting). It may then be shown that

$$D = \frac{D_0 - FQ}{1-F} \tag{11}$$

where

$$Q = \sum_{i=\alpha}^{\nu} q^i D^{i-l} + \frac{q^\alpha}{t^l}\,(D^{\alpha-l} - t^l D^{\alpha-l} - t^\beta D^{\beta-l} - t^\gamma D^{\gamma-l} - t^\delta D^{\delta-l}). \tag{12}$$

Note that $\sum_{i=\alpha}^{\nu} q^i D^{i-l} = P$ as defined above. If phase α produces fewer than or more than three solid phases during melting then equation (12) may be modified appropriately.

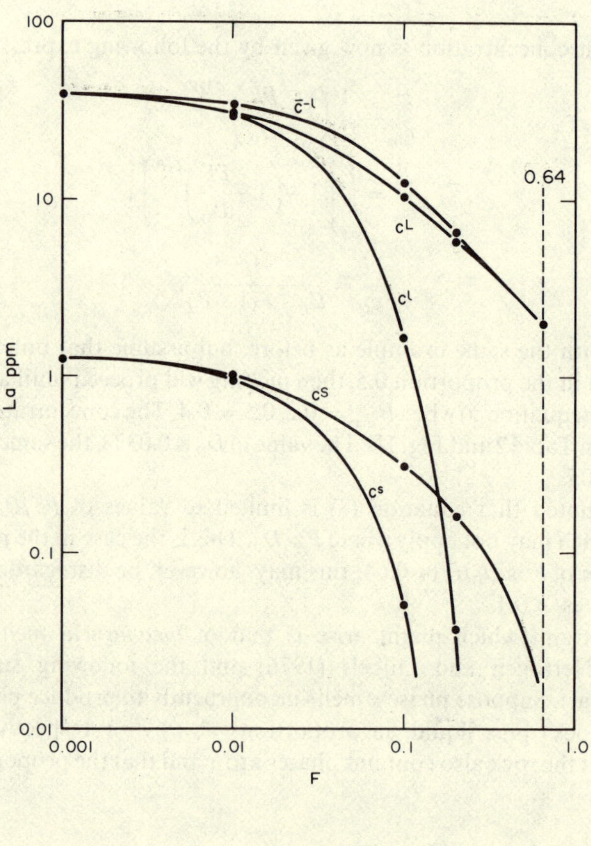

FIG.1B

TABLE 3. La CONCENTRATIONS DURING NON-MODAL MELTING OF GARNET LHERZOLITE, ASSUMING INCONGRUENT MELTING OF CLINOPYROXENE (ppm)

F	c^l	$\bar{c}^l$	c^s	c^L	$\bar{c}^S$
0.001	39.1	39.6	1.3	39.1	1.26
0.01	31.0	35.4	0.96	32.1	0.99
0.1	0.93	12.9	0.014	11.5	0.17
0.12	0.22	10.8	0	10.0	0.11
0.17	0	7.6	0	7.6	0

$D_0 = 0.0324. \quad Q = 0.19. \quad C_0 = 1.3$ ppm

Equation (12) appears to differ from the expression obtained by Hertogen and Gijbels. Expressions governing the trace element concentration are identical to equations (8), (9) and (10) if Q is substituted for P:

$$\frac{c^l}{c_0} = \frac{1}{D_0}\left(1 - \frac{QF}{D_0}\right)^{((1/Q-1)}, \tag{13}$$

$$\frac{\bar{c}^l}{c_0} = \frac{1}{F}\left[1 - \left(1 - \frac{QF}{D_0}\right)^{1/Q}\right], \tag{14}$$

$$\frac{c^L}{c_0} = \frac{1}{D_0 + F(1-Q)}. \tag{15}$$

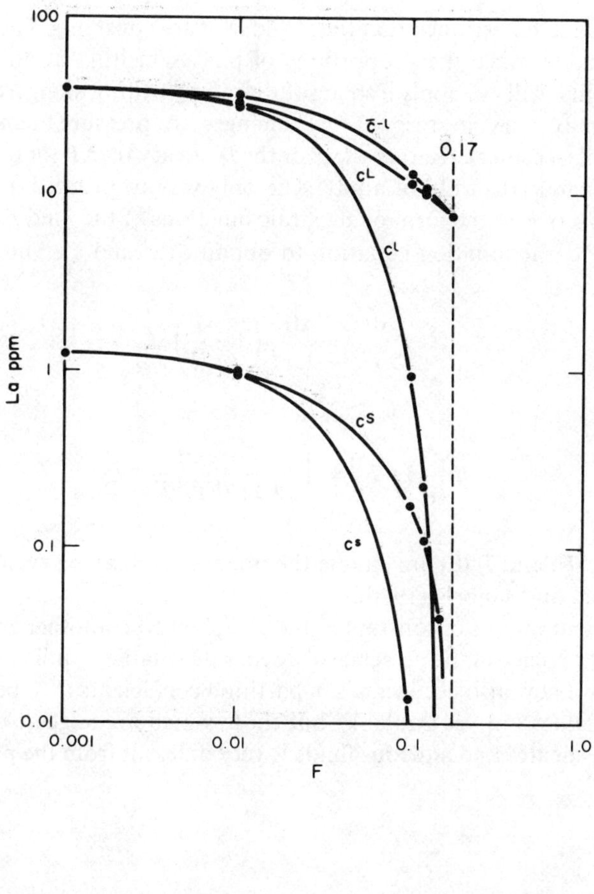

FIG. 1C

These expressions may be evaluated if the proportions t^j are assumed to be constant. Thus, in the same example, we may accept as a first approximation (Yoder, 1976, Fig. 8–7, p. 143) that clinopyroxene melts incongruently at 30 kb, 1620°C to give orthopyroxene + liquid, and we will let $t^l = 0.2$, $t^{opx} = 0.8$, so that

$$Q = P + \frac{q^{cpx}}{t^l}\,(D^{cpx-l} - t^l D^{cpx-l} - t^{opx} D^{cpx-l}) = 0.05 + 0.14 = 0.19.$$

In this case, although $F_{max} = 0.4$ once again, the constraint that $QF/D_0 < 1$ means that melting will end at $F = D_0/Q = 0.17$. Values of the calculated concentrations are given in Table 3 and Fig. 1C.

It is seen from Fig. 1 that there are significant differences in the concentrations of La depending on which of c^l, $\bar{c}^l$ or c^L is considered to give the most realistic model. On the other hand, there is very little difference between the variations of any one of these as a function of the choice of modal melting, non-modal melting or incongruent melting. The reason for this is that the partition coefficients for individual minerals are all small, and do not differ greatly. If, however, the D-value for one mineral differs greatly from the others, then the melting behaviour of that mineral can influence strongly the course of liquid evolution: the effect of different minerals for different trace elements has been summarized by Arth and Hanson (1975, p. 326), and the role of phlogopite in this connection has been analysed in detail by Beswick (1976).

It follows that element ratios will often be more sensitive to variations between different melting models than absolute concentrations, as pointed out by numerous authors (e.g. McCarthy, 1976).

The analysis has so far assumed that the p^i and D^{i-l} are constants. The former assumption is usually reasonable, since the proportions of phases melting at an invariant point are constant; although it will not apply if solid solutions are melting along a cotectic line. But the values of D^{i-l} may vary in response to changes in pressure, temperature and bulk composition or all (see later). If either the p^i or the D^{i-l} vary then P (or Q) is not constant and equations (8), (9) and (10) will not apply. The only way to proceed is then to express the variations in terms of F in the form of algebraic functions of the kind $p^i(F)$ and $D^i(F)$ and to integrate the basic fractionation equation to obtain c^s, c^l and $\bar{c}^l$. This equation is

$$\frac{\mathrm{d}c^s}{c^s} = \frac{\mathrm{d}F}{1-F}\left(1 - \frac{1}{D}\right) \tag{16}$$

or

$$\ln\frac{c^s}{c_0}(1-F) = \int_0^F \frac{\mathrm{d}F}{F\Sigma p^i(F)D^i(F) - D_0}\,. \tag{17}$$

If the functions $p^i(F)$ and $D^i(F)$ are simple the integration may be evaluated: examples are given by Hertogen and Gijbels (1976).

Returning to assumptions of constant p^i and D^{i-l}, however, another important situation is where melting takes place in the presence of a separate volatile or fluid (f) phase. In addition to values of D^{i-l} we now also need values of partition coefficients D^{i-f} between minerals and fluid. Few such values are yet available, but those which are suggest that the partition of alkalies between silicates and aqueous fluids is very different from the partition of the REE.

TABLE 4. SELECTED D-VALUES FOR SILICATE–MAGMA AND SILICATE–AQUEOUS FLUID

Mineral	D^{i-l}	Source	D^{i-f}	Source
		Rb		
q	0	assumed	0	assumed
or	0.5	Philpotts and Schnetzler, 1970; Arth, 1976	0.5	Beswick, 1973 and pers. comm.
pl	0.03	Griffin and Murthy, 1969; Arth, 1976	0.03	assumed same as D^{i-l}
bi	1.5	Dupuy, 1968; Arth, 1976	1.5	Beswick, 1973 and pers. comm.
hb	0.03	Philpotts and Schnetzler, 1970; Arth, 1976	0.03	assumed same as D^{i-l}
		Ce		
q	0	assumed	0	assumed
or	0.044	Arth, 1976	0.44	see text
pl	0.27	Arth, 1976	2.7	modified from Cullers *et al.*, 1973
bi	0.32	Arth, 1976	3.2	see text
hb	1.52	Arth, 1976	15.2	see text

q quartz
or K-feldspar
pl plagioclase
bi biotite
hb hornblende

Values from Arth (1976) are for melt of rhyolitic composition

Thus, whereas $D_{\mathrm{Rb}}^{\mathrm{or}-l}$ is similar to $D_{\mathrm{Rb}}^{\mathrm{or}-f}$ (0.5, see Table 4), Cullers *et al.* (1973) found that $D_{\mathrm{Ce}}^{\mathrm{pl}-f}$ is approximately 25 (value obtained by interpolation on their Fig. 3), whereas $D_{\mathrm{Ce}}^{\mathrm{pl}-l}$ is given as 0.27 for rhyolitic rocks by Arth (1976). The values obtained by Cullers *et al.*, using pure water as the fluid, may be too high according to similar experiments by Zielinski and Frey (1974). In the values assembled in Table 4, the value of $D_{\mathrm{Ce}}^{\mathrm{pl}-f}$ has been decreased by a factor of about 10 from the figure given by Cullers *et al.*, and values for other minerals have been interpolated, assuming them to be proportional to the ratio $D_{\mathrm{Ce}}^{\mathrm{pl}-l}/D_{\mathrm{Ce}}^{\mathrm{pl}-f}$. This approach may be considered overly conservative and is admittedly arbitrary, but should be sufficiently realistic for the present purposes, until more secure values are available.

If now we consider melting of a granodiorite, consisting of q, or, pl, bi, hb in proportions 0.2, 0.15, 0.5, 0.075, 0.075, to a melt containing q, or, pl in proportions 0.3, 0.2, 0.5 respectively, the *D*-values in Table 4 give the parameters shown in Table 5. In the last two columns are given the (calculated) partition coefficients between melt and aqueous fluid: it should be

TABLE 5. PARAMETERS FOR Rb AND Ce CONCENTRATIONS DURING GRANODIORITE MELTING

Element	D_0^{s-l}	D_0^{s-f}	P^{s-l}	P^{s-f}	$D_0^{l-f}f$	P^{l-f}
Rb	0.205	0.205	0.115	0.115	1	1
Ce	0.280	2.80	0.142	1.42	10	10

noted that, since

$$D^{l-f} = \frac{D_0^{s-l} - PF^{s-l}}{1-F} \cdot \frac{1-F}{D_0^{s-f} - FP^{s-f}}$$

then D^{l-f} is constant (= 10). The example is based on the melting experiments of Robertson and Wyllie (1971) on granodiorite (rock sample 766–H_2O) in the presence of excess H_2O, ignoring for simplicity the contributions of bio and hb to the initial melt, which forms at a temperature just above 700°C at 2 kb pressure. The end of this melt step is reached at $F_{max} = 0.20/0.30 = 0.67$, when *qtz* is exhausted.

The appropriate theory may be developed in several ways (see Shaw, in press, for a different approach) and simplifying assumptions are needed. Here it will be assumed that the rock and melt react with a constant mass of fluid and v will define the mass-ratio of fluid/initial rock in the (closed) system. The equations obtained are:

$$\frac{c^l}{c_0} = \frac{D^{l-f}}{D_0^{s-f}+v}\left[1 - \frac{P^{s-f}F}{D_0^{s-f}+v}\right]^{((1/P^{-l})-1)} \tag{18}$$

$$\frac{\bar{c}^l}{c_0} = \frac{1}{F}\left[1 - \left(1 - \frac{P^{s-f}F}{D_0^{s-f}+v}\right)^{1/P^{s-l}}\right] \tag{19}$$

$$\frac{c^L}{c_0} = \frac{1}{F(1-P^{s-l}) + D_0^{s-l} + v/D^{l-f}}. \tag{20}$$

It is of course highly artificial to treat v as a constant, as is necessary in these equations. But by evaluating the concentrations using both large and small values of v it is possible to establish approximate limiting conditions (they are approximate because small values of v will correspond to systems undersaturated in H_2O, which may change the mineral stabilities: see Wyllie *et al.*, 1976).

The granodiorite melting parameters summarized in Tables 4 and 5 have been used to evaluate equations (18), (19) and (20), taking the initial concentrations of Rb and Ce as equal.

Results for c^{-l} are shown as the ratio Ce/Rb in Table 6 and Fig. 2, including the dry melting case ($v = 0$). It is clear from this example that, if the *D*-values chosen are of the correct order of magnitude, the presence of an aqueous fluid during melting of granodioritic or average upper continental crustal material can produce very significant variations in the relative behaviour of alkalies and REE, both for small and large degrees of melting.

TABLE 6. RATIOS Rb/Ce CALCULATED FROM $\bar{c}^l/c_0$ FOR NON-MODAL GRANODIORITE FRACTIONAL MELTING

	v			
F	0	0.1	1	100
0.01	1.36	0.95	0.32	0.10
0.03	1.34	0.95	0.32	0.10
0.1	1.29	0.96	0.34	0.10
0.3	1.16	0.96	0.39	0.10
0.6	1.05	0.98	0.48	0.10

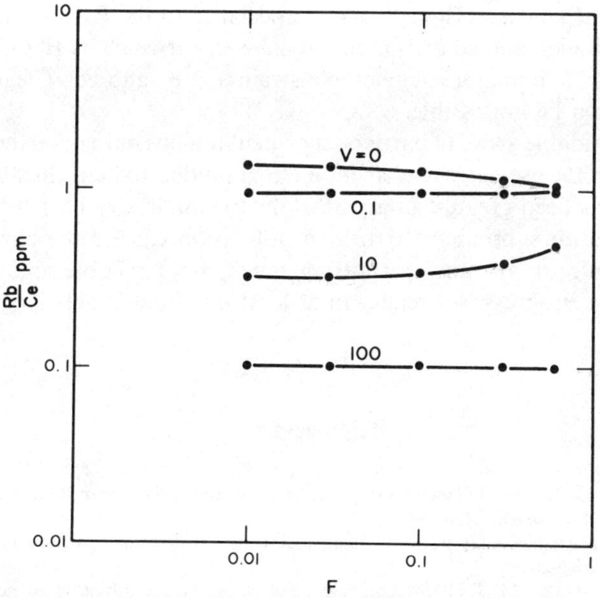

FIG.2. Effect of the ratio υ = volatile mass/silicate mass on the ratio Rb/Ce ($\bar{c}^l$) during fractional non-modal melting of granodiorite.

Other Factors influencing Melting Models

Application of the foregoing theory to questions of magma origin is fraught with difficulty. Three interrelated problems are dominant.

The first is the present imperfect knowledge of *D*-values. In addition to difficulties of measurement, either by experiment or from natural rock systems, *D*-values for most elements vary significantly with pressure, with temperature and above all with bulk composition. A rapidly growing literature on this subject exists and has been reviewed in some detail elsewhere (Shaw, in press).

The second problem is that the phase petrology of melting is much more complex than the simplified models used in the last section. Out of the vast literature of experimental and phenomenological petrology relevant to this matter, reference may be made to significant contributions by Presnall (1969), Yoder (1976).

The third problem is concerned with the movement of magma from its environment of origin, at depth, to its extrusion or intrusion near the earth's surface. A magma chamber will be fed by assorted products of melting, produced at varying temperatures, probably from different parent rock compositions, moving under little-understood tectonic controls. Moreover, the ascent of magma towards the surface will result in changes in its composition, by crystallization, by reaction with wall-rocks or by further melting, and in changes in the liquidus phases resulting from pressure and temperature decrease. Many aspects of these problems have been examined by Yoder (1976) in relation to basalt genesis.

A final problem relates to logistics of research. Because of the incompleteness of most laboratory facilities a problem is often studied only in terms of a few trace elements, perhaps without major element analyses. Some workers specialize in the REE, some on transition elements, some on alkalies, and so on. Often critical elements such as H_2O, CO_2, P_2O_5, F_2 are not considered; often major element constraints are ignored. Clearly a complete interpretation will then be impossible.

In spite of this formidable array of barriers, spectacular gains have nevertheless been made in the last ten years in the use of trace-element melting models to help elucidate problems of petrology. Success has been greatest in applications to simple, dry basic rocks which have undergone little alteration subsequent to their consolidation, e.g. the moon, young terrestrial volcanics. New techniques are now permitting much more reliable analyses and better estimates of *D*-values. Progress is a reality in at least one human activity.

References

ARTH, J. G. and HANSON, G. N. (1975) Geochemistry and origin of the early Precambrian crust of northeastern Minnesota. *Geochim. Cosmochim. Acta* **39**, 325–362.

BESWICK, A. E. (1973) An experimental study of alkali metal distributions in feldspars and micas. *Geochim. Cosmochim. Acta* **37**, 183–208.

CONSOLMAGNO, G. J. and DRAKE, M. J. (1976) Equivalence of equations describing trace element distributions during equilibrium partial melting. *Geochim. Cosmochim. Acta* **40**, 1421–1422.

CULLERS, R. L., MEDARIS, L. G. and HASKIN, L. A. (1973) Experimental studies of the distribution of rare earths as trace elements among silicate minerals and liquids and water. *Geochim. Cosmochim. Acta* **37**, 1499–1512.

GAST, P. W. (1968) Trace element fractionation and the origin of tholeiitic and alkaline magma types. *Geochim. Cosmochim. Acta* **32**, 1057–1086.

HERTOGEN, J. and GIJBELS, R. (1976) Calculation of trace element fractionation during partial melting. *Geochim. Cosmochim. Acta* **40**, 313–322.

MCCARTHY, T. S. (1976) Chemical relationships in a low-pressure granulite terrain in Namaqualand, South Africa, and their bearing on granite genesis and the composition of the lower crust. *Geochim. Cosmochim. Acta* **40**, 1057–1068.

PRESNALL, D. C. (1969) The geometric analysis of partial fusion. *Amer. J. Sci.* **267**, 1178–1194.

RAYLEIGH, J. W. S. (1896) Theoretical considerations respecting the separation of gases by diffusion and similar processes. *Philos. Mag.* **42**, 493–499.

ROBERTSON, J. K. and WYLLIE, P. J. (1971) Experimental studies on rocks from the Deboullie Stock, Northern Maine, including melting relations in the water-deficient environment. *J. Geol.* **79**, 549–571.

SCHILLING, J. G. and WINCHESTER, J. W. (1967) Rare-earth fractionation and magmatic processes. In *Mantles of the Earth and Terrestrial Planets*, S. K. RUNCORN, ed., Interscience, 267–283.

SHAW, D. M. (1970) Trace element fractionation during anatexis. *Geochim. Cosmochim. Acta* **34**, 237–242.

SHAW, D. M. (1977) Trace element behaviour during anatexis. Magm. Genesis. *Proc. Chapman Conf., Partial Melting in the Earth's Upper Mantle*, Bull. 96, Dept. Geol. Mineral. Industr., Portland, Oregon, 189–213.

WYLLIE, P. J., HUANG, W.-L., STERN, C. R. and MAALØE, S. (1976) Granitic magmas: possible and impossible sources, water contents, and crystallization sequences. *Can. J. Earth Sci.*, **13**, 1007–1019.

YODER, H. S., Jr. (1976) *Generation of Basaltic Magma*, Nat. Acad. Sci., Washington, D.C., 265 pp.

ZIELINSKI, R. A. and FREY, F. A. (1974) An experimental study of the partitioning of a rare earth element (GD) in the system diopside–aqueous vapour. *Geochim. Cosmochim. Acta* **38**, 545–566.

Chemical Differentiation of the Lower Lithosphere as Represented by the Ronda Ultramafic Massif, Southern Spain

JOHN S. DICKEY, Jr., MASAAKI OBATA and C. JOHN SUEN

Department of Earth & Planetary Sciences, Massachusetts Institute of Technology, Cambridge MA, U.S.A. 02139

Abstract

The Ronda ultramafic massif of southern Spain was differentiated into peridotite (95%) with mafic segregations (5%) before it was elevated to the crust from a depth of 70 km. No igneous textures have survived the metamorphism which occurred during the rise and emplacement of the massif, but certain primary features can be identified which indicate that the process of differentiation involved partial fusion of peridotite followed by segregation and modification of the melts by wallrock reactions, crystal fractionation and, possibly, remelting. The partial melts are now represented by mafic layers which are termed *orthomagmatic* if they truly represent the compositions of their parent melts and *depleted* if they do not.

Introduction

The Ronda ultramafic massif, which outcrops over 300 km^2 near the south coast of Spain between Marbella and Estepona (Fig. 1), is a 1.5-km-thick slab of peridotite and related rocks which has been thrust northwards from high density roots in the Alboran Sea (Bonini *et al.*, 1973). The massif is dominantly (95%) peridotite with mafic segregations (5%) ranging in mineralogy from garnet pyroxenites to olivine gabbros. There are also minor intrusions and inclusions of felsic rocks which were introduced during the elevation and emplacement of the massif. The massif is surrounded by a variety of contact metamorphic rocks, and petrological studies of these metamorphic rocks (Loomis, 1972; Dickey and Obata, 1974) indicate that the Ronda massif entered the outer crust at approximately 800°C.

Petrological studies (Dickey, 1970; Obata and Dickey, 1976; Obata, 1977) have established that the mafic segregations, which occur as layers 1 cm to 3 m thick, were formed when the massif lay at great depth (70 km). During its rise to the earth's surface the massif was more or less metamorphosed along a decreasing P–T gradient so that a spectrum of mineral assemblages are present, ranging from high P and T varieties to lower P and T varieties (Fig. 2). No primary igneous textures survived this metamorphism.

In order to learn how the Ronda massif became differentiated it is necessary to look beyond the effects of metamorphism and identify features which were established by the initial differentiation process. In this report we discuss such primary features in terms of fractional crystallization and partial fusion. Our data show that no single-step process can produce all of the primary features. These features may have been produced by partial melting of lherzolite followed by fractional crystallization of the melts as they migrated through the mantle.

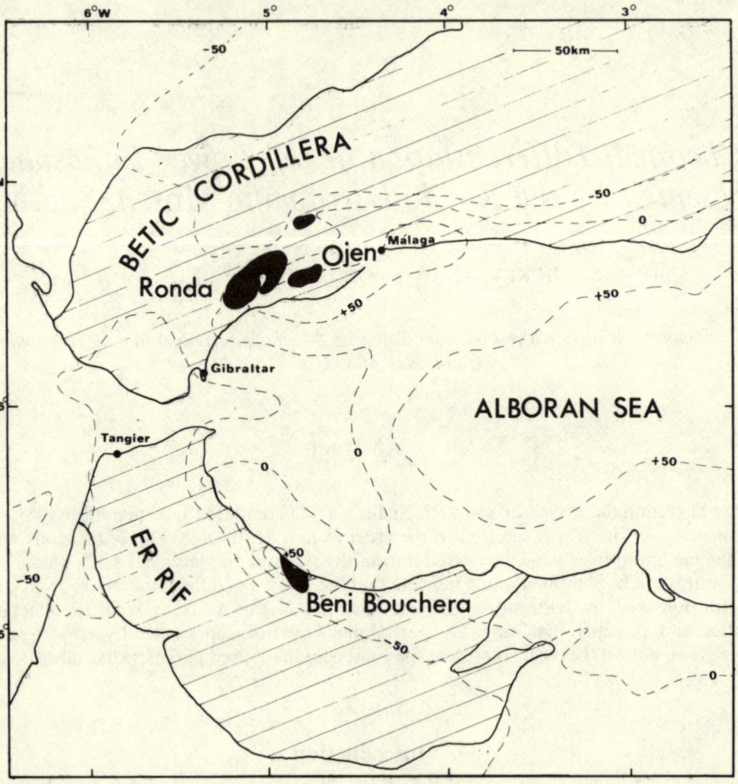

FIG. 1. Locations of the major peridotite massifs in the Betic-Rif orocline. Bouger gravity anomalies (in mgal), shown as dashed lines, are from Bonini *et al.* (1973). Positive anomalies as high as +110 mgal occur within the +50 mgal closure south-east of the Ronda massif. Petrological descriptions of the Ronda, Ojen and Beni Bouchera massifs can be found in Dickey (1970), Hernandez-Pacheco (1967) and Kornprobst (1969) respectively.

Primary Characteristics of the Ronda Massif

Megascopic Relationships

The massif contains 95% peridotite and 5% mafic layers, but the distribution is uneven, ranging from 100% peridotite to 70% peridotite and 30% mafic layers (Fig. 3). The southwestern half of the massif, which contains the highest concentration of mafic layers, also contains the highest pressure mineral assemblages (Fig. 2). Mafic layers are also abundant in rocks of the plagioclase lherzolite facies, however, so there is no close correlation between metamorphic grade and abundance of mafic layers.

The mafic layers typically have sharp contacts with the host peridotites. There is no evidence which indicates that primary megascopic gradations existed between the mafic layers and the peridotites. The interlayering of peridotite and mafic layers exists on a variety of scales, from a few centimeters to meters. Most common is interlayering of units 5 to 10 cm thick.

Although the present textures and mineral assemblages are due to metamorphism there are modal variations within some mafic layers which may represent primary variations.

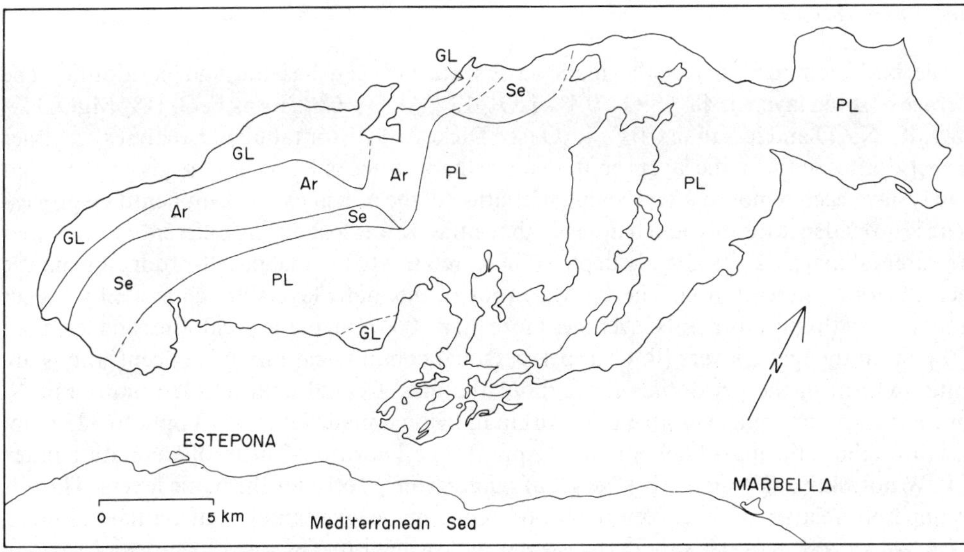

FIG. 2. Distribution of mineral facies in the Ronda peridotite massif. GL, garnet lherzolite facies. AR ariegite subfacies of the spinel lherzolite facies. SE, seiland subfacies of the spinel lherzolite subfacies. PL plagioclase lherzolite facies. These facies are defined in the sense of O'Hara (1967).

Many layers have pyroxene-rich margins. Some garnet pyroxenite layers have spinel pyroxenite margins. A few layers have marginal bands of chromian websterite. Isolated layers of chromian websterite are common throughout the massif.

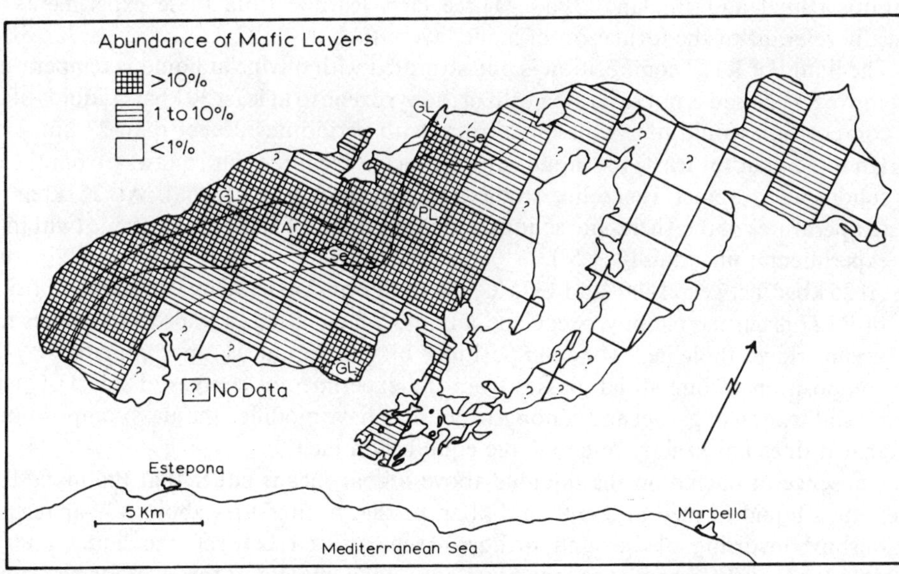

FIG. 3. Distribution of mafic layers in the Ronda peridotite massif.

Bulk Compositions

The bulk compositions of the mafic layers are between basaltic and peridotitic. The average of seven layers is 48% SiO_2, 0.4% TiO_2, 13% Al_2O_3, 6.6% Fe as FeO, 18% MgO, 11% CaO, 1% Na_2O and less than 0.01% K_2O (see Dickey, 1970, for tabulated analyses). Neither the peridotites or the mafic layers contain more than a trace of K. Although it is possible that K may have been removed after the differentiation of the massif by some unidentified process there is no evidence for this, and it appears that either K was lost during differentiation or that the parental material was already depleted in K, relative to normal basalt-producing mantle rocks, before differentiation. The Fe:Mg ratios of the mafic layers are consistently higher than those of the peridotites. Ni averages more than 2000 ppm in the peridotites and less than 300 ppm in the mafic layers (B. Chappel, personal communication). Ni concentrations are quite uniform in the peridotites throughout the massif (1920 ppm to 2380 ppm), but Ni concentrations are quite variable both within individual mafic layers (113 ppm to 325 ppm) and throughout the massif (89 ppm to 810 ppm). The Thornton–Tuttle differentiation index (CIPW normative Q + Or + Ab + Ne + Lc) ranges from 1 to 16 for the mafic layers. There is no apparent relationship between the location of a mafic layer and its differentiation index.

Chemically the mafic layers and the peridotites are separated by a wide compositional gap. The mafic layers have many magmatic characteristics, relatively high concentration of magmaphile elements and Fe-Mg ratios, whereas the peridotites are refractory in nature. No rocks in the massif bridge the compositional gap between the peridotites and the mafic layers.

Phase Equilibria

Experimentally determined phase relations for one mafic layer, plagioclase–garnet–clinopyroxenite R127, are shown in Fig. 4. The experiments (Obata and Dickey, 1976) were made under dry conditions in graphite crucibles in a Boyd–England solid media apparatus (Boyd and England, 1960). Three facts learned from these experiments are especially relevant to the formation of mafic layers such as R127:

1. The liquid of R127 composition is not saturated with olivine at liquidus temperatures above 8 kbar and is not saturated with orthopyroxene to at least 30 kbar. Liquids of this composition would not be in equilibrium with peridotites deeper than 27 km.
2. Up to 30 kbar (at least) the liquids of R127 is never more than 15° away from the dry solidus of a spinel lherzolite studied by Kushiro *et al.* (1968). At 25 kbar the temperatures of the lherzolite solidus and the liquidus of R127 are the same (within the experimental uncertainty ±5°).
3. At 25 kbar, between 1400° and 1375°C, the equilibrium assemblage for the composition of R127 is garnet, clinopyroxene and a liquid which in many respects resembles mid-ocean ridge tholeiite. The composition of the liquid is approximated by the composition of interstitial glasses from the experimental charges (Table 1). Quench crystallization of garnet and clinopyroxene may have modified the glass composition so that it does not exactly resemble the equilibrium melt.

The absence of olivine on the liquidus above 8 kbar means either that the mafic layer represents a liquid formed at less than 8 kbar, or that at pressures above 8 kbar reaction relationships involving olivine and orthopyroxene existed between the liquid and the peridotite, or that the mafic layer represents a crystal cumulate. If the layer formed from a liquid generated by melting peridotite at less than 8 kbar the massif must have been later

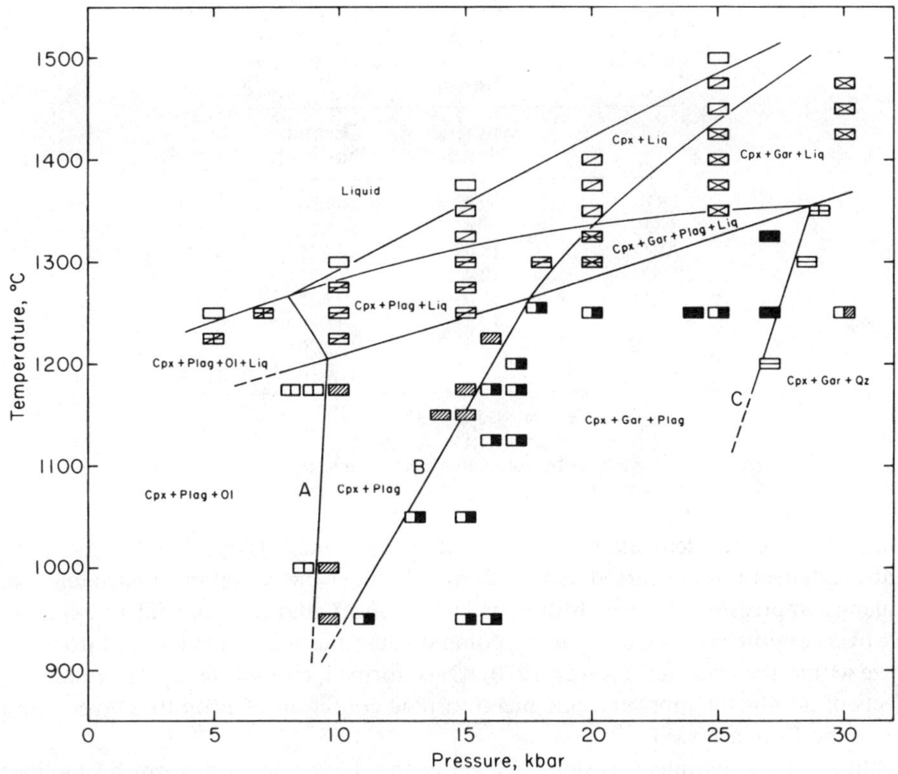

FIG. 4. Pressure–temperature diagram for the composition of plagioclase–garnet–clinopyroxenite R127. Experiments run in graphite crucibles under dry conditions. Subsolidus runs indicated by open squares. Runs above the solidus indicated by circles (Obata and Dickey, 1976).

metamorphosed at higher pressures (above 20 kbar). This is possible, especially in the context of subduction zone tectonics, but there is no evidence for compressional metamorphism in the Ronda massif. The existence of reaction relationships involving olivine, orthopyroxene and mafic liquids at high pressures has been demonstrated by O'Hara and Yoder (1967) and Kushiro and Yoder (1974). At 20 kbar, for example, Kushiro and Yoder have shown that a liquid generated by melting peridotite might react with olivine as it cools producing a different liquid plus clinopyroxene and spinel. Such reactions may be prominent in basalt genesis, for most basalt liquids are not saturated with olivine under mantle conditions (Yoder, 1976). If the mafic layers originated as partial melts in the mantle, therefore, they might have reacted with the enclosing peridotites producing wallrock reaction zones rich in clinopyroxene and spinel. Such zones may be represented by the pyroxene-rich layer margins, especially the margins of chromian websterite which are in some respect (such as Cr + Al content and Fe:Mg ratio) more like peridotite than like the typical mafic layers. The isolated chromian websterite layers may also represent reaction products.

If the interlayered peridotites and mafic rocks originated as rhythmically layered magmatic cumulates in a large (or growing) magma chamber the correspondence between the mafic layer liquidus and the lherzolite solidus, over a range of at least 30 kbar, is a remarkable coincidence. If the chamber was deep (70 km) in the earth it is difficult to

TABLE 1. COMPARISON OF GLASS FROM EXPERIMENTAL RUN PRODUCTS (R127 at 25 kbar, 1375° and 1400°C) WITH AVERAGE OCEANIC THOLEIITE

	Synthetic glass§	Oceanic tholeiite‡
SiO_2	50.9	49.94
TiO_2	1.4	1.51
Al_2O_3	18.0	17.25
FeO†	10.4	8.71
MgO	5.0	7.28
CaO	10.3	11.68
Na_2O	2.5	2.76
K_2O	?	0.16

† Total Fe expressed as FeO.
‡ Average from Engel *et al.* (1965).
§ Analysis by automated electron microprobe. Average of 7-point analyses.

understand the fine scale (cm) alternation of peridotite and mafic layers. On the other hand, if magmatic sedimentation occurred in a shallower chamber the massif must have undergone subsequent compression (for which there is no evidence), and it is difficult to explain the absence of chromitite layers. Chromite is a common cumulus mineral in mafic plutons which crystalize within the crust (cf. Thayer, 1970). Once formed, chromitite layers would survive the effects of all but the most extreme metamorphic conditions. Chromitite layers are not present in the Ronda massif.

In summary, the experimental evidence suggests that the mafic layers formed from partial melts at pressures above 20 kbar. As they migrated through cooler peridotite the melts were modified by wallrock reactions and crystal fractionation. The close correspondence of the mafic layer liquidus with the lherzolite solidus suggests, however, that the compositions of layers such as R127 have not undergone drastic changes in major element concentrations. The composition of the interstitial glass quenched from 1325° and 1400°C at 25 kbar suggests that R127 could have formed by crystal fractionation of oceanic tholeiite magma precipitating garnet and clinopyroxene.

Rare-earth Element Distribution

Rare-earth element concentrations (determined by instrumental neutron activation analysis and normalized to chondritic abundances) are higher in the mafic layers than in the peridotites (Fig. 5). All of the samples are depleted in light rare-earth elements. One sample shows a slight positive Eu anomaly. The depletion of the peridotites in light rare earths is compatible with an origin as residues of partial melting. The absence of a Eu anomaly in the peridotite patterns precludes the possibility of partial melting of plagioclase-bearing peridotite. Many mantle-derived peridotites have similar patterns (e.g. Frey, 1969; Frey *et al.*, 1971; Loubet *et al.*, 1975). The rare-earth element patterns of the mafic layers are more varied than those of the peridotites. Some patterns resemble those of ocean tholeiites although the overall concentration levels are lower in the mafic layers. None of the analyzed mafic layers is as enriched in light rare-earth elements as would be expected if the mafic layers were simply the unmodified products of small degrees of partial melting of the Ronda peridotites.

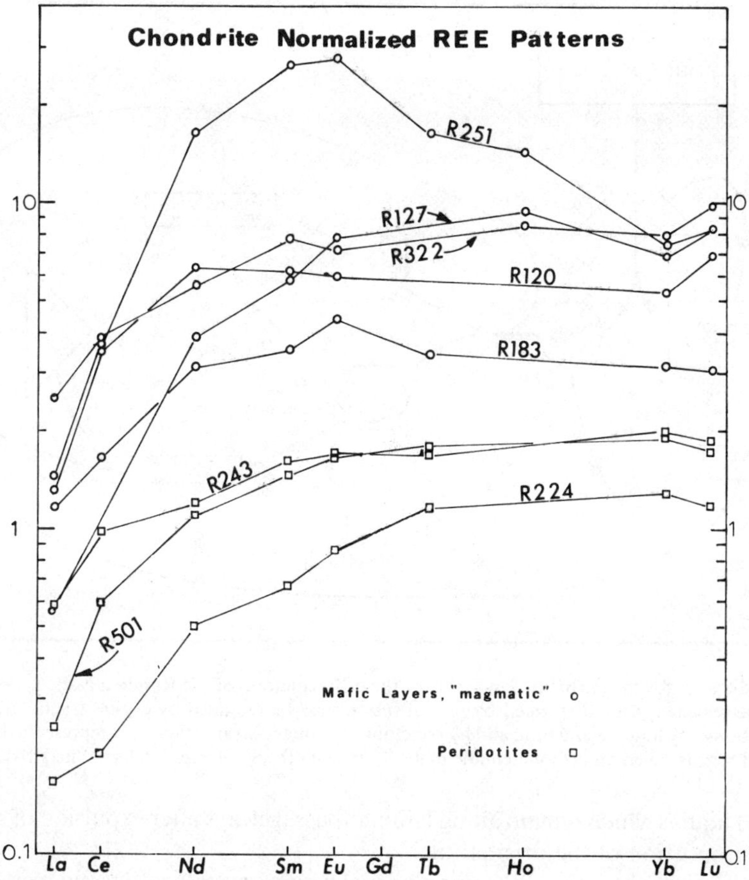

FIG. 5. Concentrations of rare-earth elements, relative to chondritic meteorites, in mafic layers and peridotites of the Ronda massif. R251, spinel pyroxenite. R127, plagioclase–garnet–clinopyroxenite. R322, olivine gabbro. R120, olivine gabbro. R183, olivine gabbro. R243, spinel lherzolite. R224, spinel harzburgite. R501, garnet lherzolite.

No single-step process is capable of explaining the rare-earth element distributions in the peridotites and mafic layers. The rare-earth data suggest that the mafic layers might be crystal fractionates from partial melt segregations or residual material from repeated partial fusion of segregations which were originally formed by partial melting or by fractional crystallization.

Conclusions

The primary characteristics of the Ronda massif, described and discussed above, appear to have resulted from differentiation of solid peridotite by partial melting at high pressure ($\sim$25 kbar) followed by segregation and modification of the melts by wallrock reactions and crystal fractionation as the melts migrated through cooler peridotite. There is no evidence which supports any model based on large-scale fractional crystallization of any generally molten, magmatic source. The mafic layers are combinations of cumulus materials reaction products

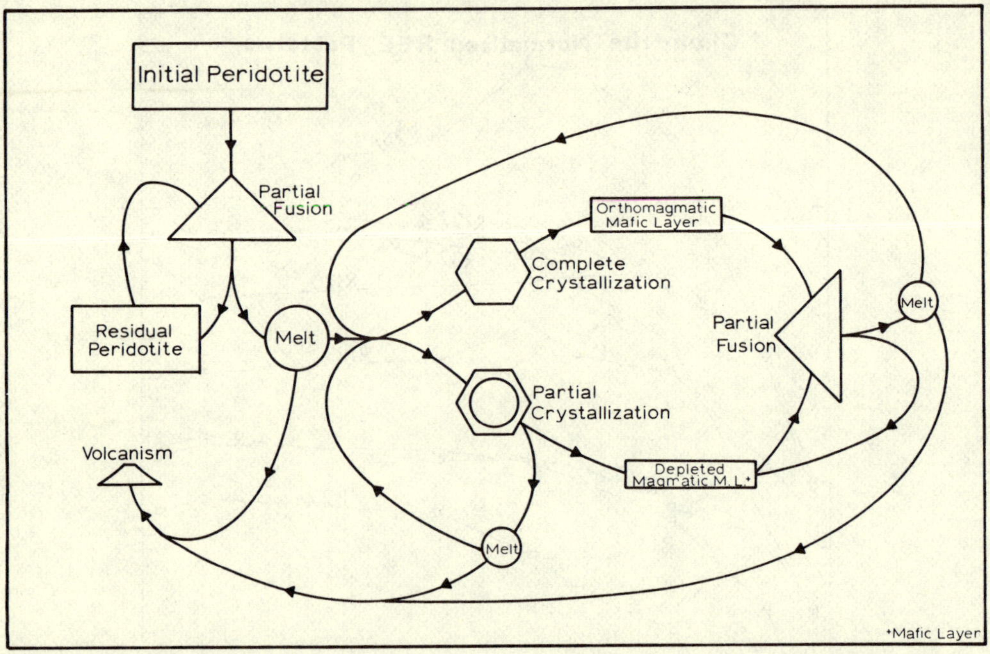

FIG. 6. Flow diagram for the partial fusion model of the differentiation of the Ronda massif. Solid residues from fusion may be remelted. Liquids formed by partial fusion may be modified by crystal fractionation (including wallrock reactions), by liquid extraction, and by remelting. Any mafic layer which truly represents the composition of its parental melt is called an orthomagmatic mafic layer. All others are called depleted magmatic mafic layers.

and trapped liquids which remain along laminar passageways after expulsion of some (as yet unknown) proportion of the original melt.

Figure 6 illustrates how the processes of partial fusion, crystallization and melt segregation could interact and how complicated cycles might develop. The mafic segregations formed in this model are divided into two classes: those which represent the compositions of their parent melts (orthomagmatic) and those which lack some part of their parent melts (depleted). Several generations of melts may be created by partial fusion of residual peridotite or of mafic layers. Orthomagmatic segregations may form from *any* of these melts. Their definitive characteristic is that their bulk compositions are the same as the compositions of their parent melts. Depleted segregations can also form during any cycle. Furthermore, depleted segregations can form in two ways: as crystal cumulates lacking some part of the parent melt or as unmelted residues formed by partial fusion of mafic layers. In the absence of clear textural evidence these two kinds of depleted segregations are probably indistinguishable.

Acknowledgements

Research on the Ronda ultramafic massif is supported by the Earth Sciences Division of the National Science Foundation of the United States (grant numbers DES73–00264, DES75–02900 and EAR76–84067). The study of rare-earth element distribution was guided by Professor F. A. Frey. Technical assistance was provided by Dr. A. S. Parkes. Assistance in the field was provided by Mr. Dorian Nicol. Suen's field expenses were partially paid by the Geological Society of America. The Carnegie Institution of Washington provided materials and facilities for the experimental studies which were supervised by Dr. Ikuo Kushiro of the Geophysical Laboratory.

References

BONINI, W. E., LOOMIS, T. P. and ROBERTSON, J. D. (1973) Gravity anomalies, ultramafic intrusions, and the tectonics of the region around the Strait of Gibraltar. *J. Geophys. Res.* **78**, 1372–1382.

DICKEY, J. S., Jr. (1970) Partial fusion products in alpine-type peridotites: Serrania de la Ronda and other examples. *Min. Soc. Amer. Special Paper,* **3**, 33–49.

DICKEY, J. S., Jr. and OBATA, M. (1974) Graphitic hornfels dikes in the Ronda high-temperature peridotite massif. *Amer. Miner.* **59**, 1183–1189.

ENGEL, A. E. J., ENGEL, C. G. and HAVENS, R. G. (1965) Chemical characteristics of oceanic basalts and the upper mantle. *Geol. Soc. Amer. Bull.* **76**, 719–734.

FREY, F. A. (1969) Rare earth abundances in a high-temperature peridotite intrusion. *Geochim. Cosmochim. Acta* **33**, 1429–1447.

FREY, F. A., HASKIN, L. A. and HASKIN, M. A. (1971) Rare earth abundances in some ultramafic rocks. *J. Geophys. Res.* **76**, 2057–2070.

HERNANDEZ-PACHECO, A. (1967) Estudio petrografico y geochimico del macizo ultramafico de Ojen (Malaga). *Estudios Geologicos* **23**, 85–143.

KORNPROBST, J. (1969) Le massif ultrabasique des Beni Bouchera (Rif Interne, Maroc): étude des péridotites de haute température et de haute pression, et des pyroxénolites à grenat qui leur sont associées. *Contr. Mineral. Petrol.* **23**, 283–322.

KUSHIRO, I. and YODER, H. S. Jr. (1974) Formation of eclogite from garnet lherzolite: liquidus relations in a portion of the system $MgSiO_3$–$CaSiO_3$–Al_2O_3 at high pressure. *Carnegie Inst. Washington Year Book,* **73**, 266–269.

LOOMIS, T. P. (1972) Contact metamorphism of pelitic rock by the Ronda ultramafic intrusion, southern Spain. *Geol. Soc. Amer. Bull.* **83**, 2449–2474.

LOUBET, M., SHIMIZU, N. and ALLÈGRE, C. J. (1975) Rare earth elements in alpine peridotites. *Contr. Mineral. Petrol.* **53**, 1–12.

OBATA, M. (1977) Petrology and petrogenesis of a high-temperature peridotite intrusion: Serrania de la Ronda, southern Spain. Ph.D. Thesis, Massachusetts Institute of Technology.

OBATA, M. and DICKEY, J. S. Jr. (1976) Phase relations of mafic layers in the Ronda peridotite. *Carnegie Inst. Washington Year Book* **75**, 562–566.

O'HARA, M. J. (1967) Mineral paragenesis in ultrabasic rocks. In *Ultramafic and Related Rocks*, P. J. WYLLIE, ed., John Wiley & Sons Inc., New York, 393–401.

O'HARA, M. J. and YODER, H. S. Jr. (1967) Formation and fractionation of basic magmas at high pressures. *Scot. J. Geol.* **3**, 67–117.

THAYER, T. P. (1970) Chromite segregations as petrogenetic indicators. *Geol. Soc. S. Africa Special Publication* **1**, 380–390.

YODER, H. S., Jr. (1976) *Generation of Basaltic Magma*, National Academy of Sciences, Washington, D.C., 265 pp.

Trace Element Distribution and Isotopic Composition of Archean Greenstones

BOR-MING JAHN† and SHEN-SU SUN‡

†*Université de Rennes-Institut de Géologie, Avenue du Général-Leclerc, B.P. 25A,* 35031 *Rennes-Cedex, France*
‡*Department of Geology and Mineralogy, The University of Adelaide, G.P.O. Box* 498, *Adelaide, Australia*

Abstract

Trace element abundances and Sr isotopic compositions in Archean and modern volcanic rocks are reviewed. K, Rb, Ba and perhaps Sr are more mobile than rare earth and some transition elements in alteration and low-grade metamorphism. Data of K, Rb, Sr and Ba from individual greenstone samples usually show significant variations. However, average values of these elements from a large set of samples tend to show a rather consistent feature. Geochemical arguments derived from these elements are also in general agreement with those from more refractory REE. Alteration effects on REE patterns are observed, but most Archean volcanic rocks appear to possess their original magmatic patterns.

Archean low-K tholeiites and high-Mg komatiites are characterized by their generally flat REE patterns of about 2.5 to 15 times chondritic abundances. Their $(La/Sm)_N$ ratios range from 0.7 to 1.3 and are significantly higher than those of typical MORB (0.4 to 0.7). The data of peridotitic komatiites have been used to estimate the REE abundances in the Archean upper mantle. The two times chondritic abundances thus derived is similar to those estimated for the modern upper mantle.

For andesitic rocks, most Archean REE patterns can find their modern analogues, but in some cases Archean patterns (with severe HREE depletion) have rare modern analogues. Archean siliceous volcanic rocks commonly show severe HREE depletion. This cannot be explained as due to fractional crystallization of basaltic magmas, rather, it suggests a separate melting event in which garnet plays an important role in the source region. The source is hence likely an eclogite or garnet amphibolite in the mantle due to lithosphere subduction or that converted from a thick pile of basaltic rocks of the same greenstone belt.

The application of trace element abundances to identification of the tectonic settings for Archean volcanic rocks is not very successful. This is mainly due to the following factors: (1) insufficient understanding of geochemical characteristics of trace elements or simply due to indiscriminate LIL and REE patterns in modern volcanic rocks from different tectonic settings, (2) insufficient understanding of geochemical behaviors of some transition elements during partial melting and crystal fractionation under different P, T, X conditions, (3) evolutionary change of the upper mantle composition through time, (4) existence of heterogeneities in both the Archean and the modern upper mantle, and finally, (5) insufficient understanding in tectonic styles during the Archean.

Available Sr isotopic data for Archean volcanic rocks show that they are essentially evolved along a path with Rb/Sr ratio = 0.026–0.034. Most modern oceanic island volcanic rocks and continental basalts also have Sr isotopic composition evolved along the same path. Some Archean basalts, e.g. 2.7 b.y. Minnesota, show an early depletion of Rb relative to Sr. Likewise, the modern ocean ridge basalts show a significant depletion of Rb relative to Sr. This depletion probably took place about 2 b.y. ago. Evidence from both Nd and Sr isotopes suggests that, counting MORB sources apart, the upper mantle has evolved essentially with a constant Rb/Sr and Sm/Nd ratios. Trace element abundances also show a gross constancy in Archean and modern volcanic rocks. This seems to militate against the consequences of partial melting processes operating in the upper mantle of non-infinite trace element reservoir. It may, however, be explained as due to the interplay of (1) replenishment of trace elements from lower part of the mantle, and (2) recycling of trace elements in the crust-upper mantle system.

I. Introduction

Ten years ago, at the first Symposium on the origin and distribution of the elements in Paris, there was not a single article which dealt with trace element distribution and isotopic compositions of Archean greenstones. Since then large bodies of data have been accumulated on these rocks. These data are extremely valuable and are frequently used in delineating the temporal sequence of Archean volcanisms, the origin of Archean volcanic and sedimentary rocks (e.g. greenstones), the tectonic setting of greenstone belts, and finally, the upper mantle composition of the Archean time and its possible evolutionary change throughout geologic history. We shall present in the following sections a brief review of the current state of knowledge concerning the distribution of some trace elements and isotopes in the Archean greenstones. Many other studies related to the same topic have been published in recent years, for examples, Hart *et al.* (1970); Glikson (1971); Arth and Hanson (1972, 1975); Jahn *et al.* (1974); Condie (1976); Condie and Baragar (1974); Jahn and Nyquist (1976); Nesbitt and Sun (1976); Jahn (1977); Hawkesworth and O'Nions (1977); Sun and Nesbitt (1977a, b).

Before the discussion of trace element distribution in Archean greenstones, a general feature of greenstone belts is described below. Windley (1973, 1976) has classified the Archean terrains into two types: (1) Low-grade greenstone-granite terrains, in which elongate belts of volcanic rocks and volcanogenic and chemical sediments are metamorphosed to a low-grade, commonly in the greenschist facies. These belts are invariably surrounded and partly intruded by vast areas of granitic rocks. Examples: Canadian Shield, including the Vermilion district of NE Minnesota, the Rhodesian craton, the Kaapvaal craton, the Yilgan and Pilbara Blocks of W. Australia, India, and E. Finland. (2) High-grade terrain, in which high-grade metamorphic rocks of tonalite-granodiorite composition predominate ($\sim$80–85%). Rocks are generally metamorphosed in the amphibolite and granulite facies, and the primary igneous or sedimentary features are generally not preserved. Examples: Morton and Montevideo gneisses of the Minnesota River Valley, Amitsoq gneisses of W. Greenland, Limpopo Belt of Southern Africa, etc. These two types of terrains may or may not have a genetic relationship (see Shackleton, 1976, for review of genetic relationships).

In greenstone belts, volcanic rocks of tholeiitic composition usually predominate. Andesitic and siliceous volcanic rocks occur, but commonly in subordinate amounts. Locally, high-Mg basaltic and peridotitic komatiites occur, such as in the Onverwacht Group, South Africa (Viljoen and Viljoen, 1969a, b), in the Rhodesian craton (Hawkesworth and O'Nions, 1977; Bickle *et al.*, 1976), in the Superior Province (Pykes *et al.*, 1973; Brooks and Hart, 1972), in western Australia (Nesbitt, 1971; Nesbitt and Sun, 1976) and in Finland (Blais *et al.*, 1977). Because of the similarity in rock association, depositional environments and general geochemical characteristics (with the exception of komatiitic varieties) between greenstone belts and modern volcanic sequences occurring in various tectonic environments (ridges, island arcs, marginal basins, continental margins), it would seem possible to infer some analogous tectonic setting for Archean greenstone belts. However, it is our opinion that the application of trace element abundances in identifying the ancient tectonic environments is not that successful. We shall give a critical review on this subject.

The trace elements of interest are K, Rb, Sr, Ba, rare earths and some transition metals, such as Ni, Co, Cr, etc. Isotopic composition of Sr will be used together with trace-element data to place constraints in modelling crustal evolution.

II. Alteration and Low-grade Metamorphic Effects on Trace Element Abundances

A. *General*

The effects of alteration on trace element abundances in basalts have been studied (e.g. Hart, 1969, 1971; Hart *et al.*, 1974; R. A. Hart, 1970, 1973; Frey *et al.*, 1974). Frey *et al.* (1974) made a detailed study on the elemental migration patterns in both glassy and crystalline basalts of DSDP cores as due to sea-water alteration. They summarized: (1) The conversion of glass to palagonite results in a much greater loss of Na, Ca, Mg, Mn and Si and larger enrichment in K, H_2O, total Fe and Fe^{3+} than the alteration of crystalline rocks. Larger change of trace element abundances were also found in glasses than in crystalline rocks. (2) Sea-water alteration causes enrichment in K, Pb, Cu, Ba, Sr, B, Li and Rb; Cr, Ni and V tend to be slightly depleted; but Co, Y, Zr, Hf and Sc remain essentially unchanged. (3) HREE (heavy rare-earth elements, i.e. Gd–Lu) tend to be depleted by alteration, but there is no significant internal fractionation among HREE. LREE (light rare-earth elements, i.e. La–Sm) are fractionated by glass alteration in some cases, particularly in palagonites. Even in crystalline portions, LREE (especially La and Ce) are subjected to change up to more than a factor of 2, thus alteration processes could have changed the REE distribution from a LREE-depleted pattern to a LREE-enriched pattern and thus tholeiitic basalts would appear to be more alkalic. According to these observations, except for a few elements like Co, Zr, Hf and Sc, *almost all* trace elements would be subject to various degrees of change as a result of sea-water alteration. This makes it difficult to assess precisely the original magma composition for volcanic rocks erupted in oceanic environments and experienced sea-water alteration, regardless of geologic age. The prevailing low-grade metamorphism in Archean greenstone belts poses another problem for trace element redistribution. In contrast with sea-water alteration, metamorphism is essentially a process of dehydration. How trace elements would behave in low-grade metamorphism after sea-water alteration is a long-standing and not yet well-understood problem. Because the study of alteration effect on transition metal abundances is scanty at present, we shall tentatively consider that trace transition metals remain intact in the alteration and low-grade metamorphic processes. In the following we examine these two specific groups of elements:

B. *On* K, Rb, Sr *and* Ba *abundances*

K, Rb, Ba and, to a lesser extent, Sr are demonstrably more mobile than some transition metals and rare-earth elements (REE) during the processes of sea-water alteration and low-grade metamorphism (Hart, R. A., 1970, 1973; Hart, 1969, 1971; Hart *et al.*, 1974; Frey *et al.*, 1974). In order to assess the geochemical characteristics of Archean greenstone belts using these four elements, Hart *et al.* (1970) and Jahn *et al.* (1974) have made an attempt to overcome the mobility problem by sampling a geographically large area within a given greenstone belt. This is under a premise that although these elements are mobile, the effect of enrichment in one sample will be cancelled out by the depletion in another. That is, it is assumed that a greenstone belt as a whole remains a closed system and the combined sea-water reaction and low-grade metamorphism is a two-way balanced reaction. This assumption may not be strictly valid, but is perhaps justified by the close agreement between the measured average values and the theoretically estimated values (Sun and Nesbitt, 1977a). The conclusions derived from the distribution of these four elements are also consistent with

TABLE 1. K, Rb, Sr AND Ba ABUNDANCES AND RATIOS OF VOLCANIC ROCKS FROM DIFFERENT TECTONIC ENVIRONMENTS

	K (ppm)	Rb (ppm)	Sr (ppm)	Ba (ppm)	K/Rb	Rb/Sr	Sr/Ba	K/Ba	Ni (ppm)	Cr (ppm)	Ni/Cr	References
Archean basalts (A)	2006	5.84	161	69	343	0.036	2.33	29.1				1
Archean basalts (B)	2330	6.38	152	96	365	0.042	1.58	24.3	120	320	0.38	2
MORB	1160	1.10	136	10.5	1050	0.008	12.95	110	140	480	0.29	3
Oceanic alkali basalt	12,500	36.0	900	450	400	0.040	2.0	28.0				4
Low-k tholeiites of Island arcs	2170	3.2	250	100	680	0.012	2.5	21.7				3
Basalt of Island arc tholeiite series (IATS)	5000	5	200	75	1000	0.025	2.67	66.7	30†	50†	0.60	5
Marginal Basin basalts	2660	3.1	163	34	858	0.019	4.79	78.2	121	309	0.39	6
Andesite of calc-alkaline series	12,900	30	385	270	430	0.078	1.43	47.8	18†	25†	0.72	5
Archean andesites + felsic rocks	9942	33.1	187.3	270	300	0.177	0.694	36.8				7

References:
1. Weighted average from "Grand basic" composite of seventy samples of Hart *et al.* (1970) and Table 1 of Jahn (1977). All values determined by the isotopic dilution method.
2. Average of 525 Archean basalts from Canadian shield, W. Australian shield and Rhodesian craton; Ni and Cr values estimated by Tarney *et al.* (1976).
3. Hart *et al.* (1970); Ni and Cr values from Frey *et al.* (1974).
4. Sun and Hanson (1975).
5. Jakeš and White (1972).
6. Average of Mariana (Hart *et al.*, 1972) and Lau Basin (Hawkins, 1976). Ba value taken from Hart *et al.* (1972) only.
7. From (B) of Table 1 in Jahn (1977).

† See text for explanation.

those from the more refractory REE and transition metals (for more details, see Jahn, 1977).

The abundances of K, Rb, Sr and Ba in Archean and some modern volcanic rocks of different tectonic environments are shown in Table 1 and Figs. 1 and 2, it is seen that the average Archean basalts generally distinguish themselves from MORB by having significantly higher Rb and Ba contents. Archean basalts are chemically more similar to the low-K tholeiites of island arc or marginal basin basalts. Also, Archean andesitic and felsic rocks are geochemically similar to modern calc-alkaline series rocks. However, inference of Archean tectonic settings based on these similarities cannot be warranted. This point will be elaborated in a later section.

C. *On REE abundances*

Although Frey *et al.* (1974) have shown the important effect of alteration on the redistribution of REE, most workers still regard them as immobile. Indeed, the alteration effect on REE is considerably smaller than on the alkali elements. Furthermore, the small change of REE is often masked by insufficient precision of analysis, thus the effect of alteration has been commonly overlooked. Our experience in REE analysis by the isotopic dilution method (with a precision of 2 to 5%) allows us to reach the following observations in the process of alteration/low-grade metamorphism: (1) Eu is more mobile than other REE, perhaps due to the different behavior of divalent Eu from other trivalent REE (Sun and Nesbitt, 1977b). This is exemplified by the inconsistent and unpredicted Eu anomalies (both positive and negative) in many REE patterns of Archean basalts (see Fig. 3); (2) likewise, due

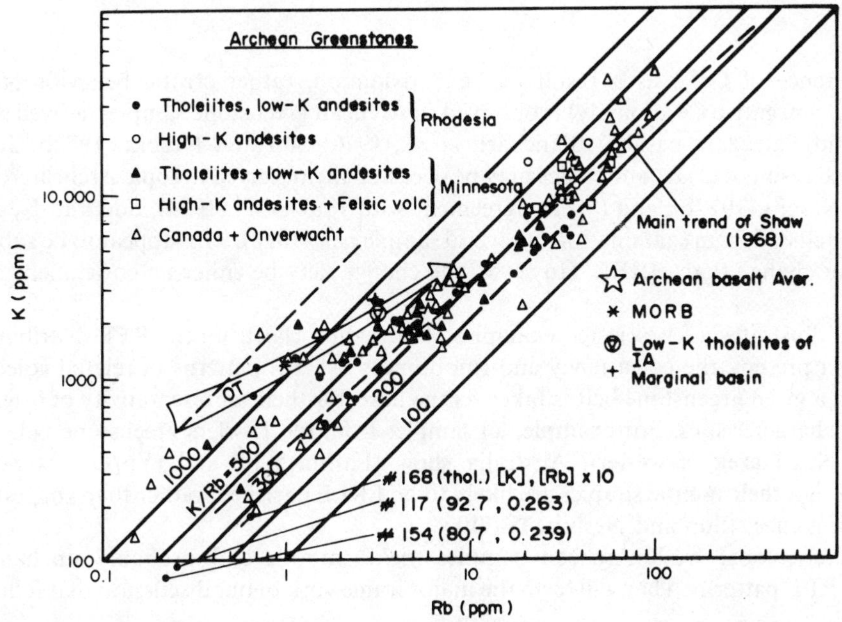

FIG. 1. K vs. Rb variation diagram for Archean volcanic rocks from Rhodesia, Minnesota, Canada, and Onverwacht, S. Africa. All values are determined by the isotopic dilution method (uncertainties <3%). The data clearly show severe depletion of K and Rb in some rocks (lower left end of the diagram). Enrichment is not easily detected, but must exist in some rocks. Data sources: Hart *et al.* (1970), Jahn *et al.* (1974), Jahn and Condie (1976), Jahn (unpublished). Data sources for average values, see Table 1. Abbreviations: OT = oceanic trend, LKT = low-K tholeiites of island arcs (I.A), MORB = mid-ocean ridge basalts.

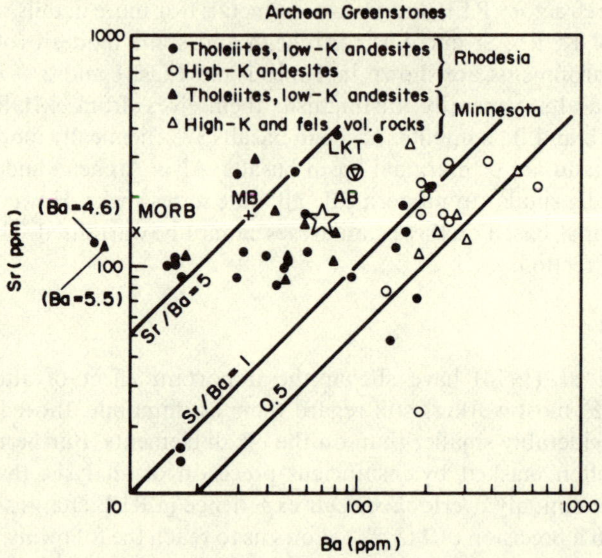

FIG. 2. Sr vs. Ba variation diagram for Archean volcanic rocks from Rhodesia and Minnesota. Like in the case of K and Rb variation, depletion of Sr and Ba is clearly seen in some cases; enrichment is again not easily detected but must exist. Abbreviations: MB = marginal basin basalts, AB = Archean basalts; others, see Fig. 1. Data sources: Jahn *et al.* (1974), Jahn and Condie (1976), Jahn (unpublished). For average values, see Table 1.

to the presence of Ce^{4+} as a result of Ce^{3+} oxidation, rather erratic behavior of Ce (commonly a negative Ce anomaly) is observed in Archean greenstone samples, as well as in modern and Paleozoic basalts (Blanchard *et al.*, 1976; Sun and Nesbitt, 1977b; Jahn, unpublished results); (3) La anomalies may be occasionally observed in some Archean rocks (Sun and Nesbitt, 1977b); and (4). In agreement with Frey *et al.* (1974), Sun and Nesbitt (1977b) concluded from examination of several sample pairs that LREE appear to be subject to a greater change than HREE. However, the change may be either an enrichment or a depletion.

Figure 3 illustrates a few selected examples of alteration effects on the REE distribution patterns. At present, the consistency and smoothness of REE patterns of related volcanic rocks from a given greenstone belt is taken as the basis for their representativity of original magmatic characteristics. For example, all samples from the Lawlers greenstone belt and from the Negri area of western Australia show (La/Sm) > 1.0 and (Yb/Gd) 0.8–0.9, suggesting that their mantle sources are likely to be LREE enriched, rather than suggesting an alteration effect (Sun and Nesbitt, 1977b).

Fortunately, most fresh Archean samples yield smooth and probably unchanged magmatic REE patterns. They will form the major framework of our discussion that follows.

III. REE Abundances in Archean Greenstones

Condie (1976) recently made an important review on the trace-element geochemistry of Archean greenstone belts. His classification of Archean volcanic rocks based mainly on REE

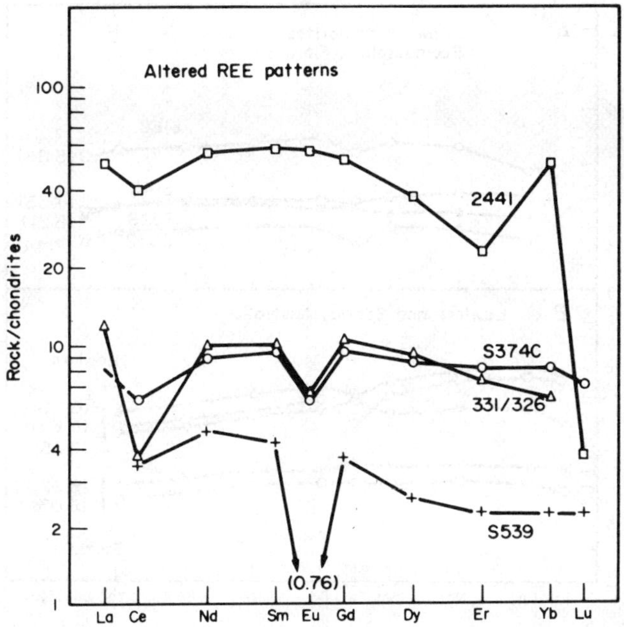

FIG. 3. Altered REE patterns found in some Archean and Paleozoic (2441) metavolcanic rocks. Data sources: Jahn (unpublished), Sun and Nesbitt (1977b). Chondrite normalization values (Masuda values divided by 1.2) used are (in ppm) = La (0.315), Ce (0.813), Nd (0.597), Sm (0.192), Eu (0.0722), Gd (0.259), Dy (0.325), Er (0.213), Yb (0.208), Lu (0.0323). The Yb anomaly in a metamorphic rock (2441) is highly questionable.

abundances is generally valid. However, in this review only the range and the average REE abundances are reported; thus many characteristics such as the regularity or smoothness of individual REE patterns, the (La/Sm) ratio and Eu anomaly, are not well expressed. In the following we choose to present a group of representative individual analyses. We understand that each greenstone belt may have its own individual characteristic REE patterns inherited from its own source region (see Sun and Nesbitt, 1977b), so it is not possible to cover all the spectrum in this review. Only a selected few that represent the majority of rock types are presented.

A. *Low-K Tholeiites*

In most greenstone belts tholeiitic rocks (including komatiitic varieties) predominate (Condie, 1976). In the belts where there are no komatiitic rocks, low-K tholeiites generally predominate. Figures 4a to 4d illustrate the REE distribution patterns in some low-K tholeiites from various greenstone belts. It is clear that most of them are characterized by very flat chondrite-normalized pattern with about 6 to 20 times chondritic abundances. The $(La/Sm)_N$ ratios range from about 0.7 to 1.3, slightly higher than those of typical modern mid-ocean ridge basalts (MORB, see Fig. 9). The Eu anomalies are not consistent; some are negative and some positive. This inconsistency is not attributable to plagioclase fractionation. Rather, it is more likely to be induced by a post-crystallization alteration (Sun and Nesbitt, 1977b, and see the argument in the following section on komatiites).

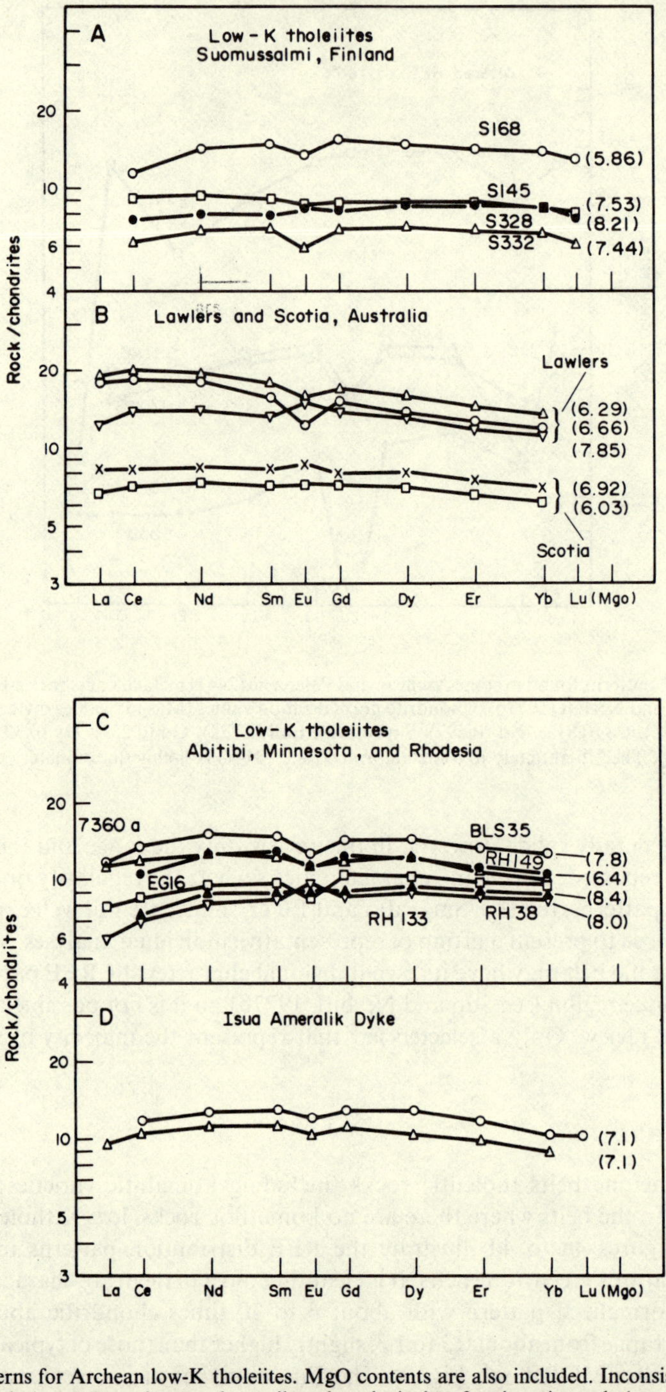

FIG. 4. REE patterns for Archean low-K tholeiites. MgO contents are also included. Inconsistent Eu anomalies (positive and negative) are seen and cannot be attributed to plagioclase fractionation only (see text for arguments). Fractionated HREE patterns are seen in Lawlers and Isura rocks (4B and 4D) but are flat in 4A and 4C. Data sources: Sun and Nesbitt (1977b), Jahn *et al.* (1974), Jahn *et al.* (1977) and Hawkesworth and O'Nions (1977). For internal consistency, all values are normalized by the same chondrite values (see Fig. 3 caption).

B. *Peridotitic and Basaltic Komatiites*

The definition of komatiite is currently in a state of confusion. In this paper we shall loosely define komatiite with the following restrictions: (1) komatiites usually occur in a suite including an ultramafic–mafic assemblage; (2) the peridotitic variety must not be a cumulate rock, and generally have quench texture (spinifex); and (3) they must have $MgO \geq 9\%$, and $CaO/Al_2O_3 \geq 1.0$. The texture evidence is of secondary importance, but spinifex textures are often observed in komatiitic rocks.

Figures 5a to 5d show the REE distribution patterns in some komatiitic rocks from different localities. Some important observations include: (1) in a given greenstone belt, the REE abundances show a inverse correlation with MgO contents; (2) basaltic komatiites (generally having MgO = 9–18%) have REE of 6 to 20 times chondritic abundances; these values are similar to those of low-K tholeiites, despite their higher MgO contents; (3) peridotitic komatiites with $MgO > 28\%$ show REE about 2 to 3 times chondritic abundances, normally exhibiting flat HREE but fractionated LREE with $(La/Sm)_N < 1$; (4) rocks of intermediate composition ($MgO = 18-28\%$) also show REE abundances intermediate between the basaltic and peridotitic komatiites. In general, the REE distribution patterns of both low-K and komatiitic basalts are quite similar to those of modern MORB, island arc or maginal basin tholeiites. The difference lies at the slightly higher $(La/Sm)_N$ ratios in the Archean basalts. This may be indicative of an evolutionary feature in the upper mantle source(s).

Like in low-K tholeiites, the Eu anomalies in these high-Mg rocks are not consistent. Sun and Nesbitt (1977b) have argued that in a high-degree melting event, plagioclase is not likely to be a residual phase after extraction of a high-Mg melt. It is not a liquidus phase in this melt either. The effects of oxygen fugacity and temperature on Eu anomalies have been determined experimentally (Sun *et al.*, 1974). At 1200°C and $f_{O_2} = 10^{-9}$ atm for MORB, a 5% Eu anomaly would require a 30% plagioclase removal. A lower temperature results in lower f_{O_2} and higher D_{Eu} value and thus leads to more efficient Eu removal. Consequently lower temperature differentiates (with lower MgO contents, see Figs. 4 and 5) are expected to have larger negative Eu anomalies if plagioclase separation is involved. The REE patterns in Figs. 4 and 5 fail to demonstrate this point. We therefore conclude that the inconsistency of Eu anomalies is probably caused by alteration effect for which the mechanism is not yet clear. Plagioclase fractionation may occur in some cases, but certainly cannot be the major factor for the Eu anomalies observed.

C. *Andesites*

Condie (1976) classified Archean andesites into three categories: (1) DAA (depleted Archean andesite) showing flat REE patterns of about 20 to 40 times chondritic abundances with negative Eu anomalies; (2) LAA (low-alkali Archean andesite) showing small LREE enrichment (La = 25–50 ×, Lu = 4–12 ×); and (3) HAA (high-alkali Archean andesite) showing significant LREE enrichment (La = 60–120 ×, Lu = 5–12 ×). Condie (1976) also observed that LAA is grossly similar to modern calc-alkaline andesites; for example, andesite from Lau Volcanics (Gill, 1976), but HAA and DAA do not appear to have modern counterparts. In Fig. 6a we present some Archean results. Two Rhodesian samples apparently fall in the category of LAA, while andesites from Marda, Australia, fall in the HAA category. The andesite from Finland exhibits a strong LREE enrichment (La = 190 ×) exceeding Condie's range of HAA. Figure 6b shows the REE data of modern island arc

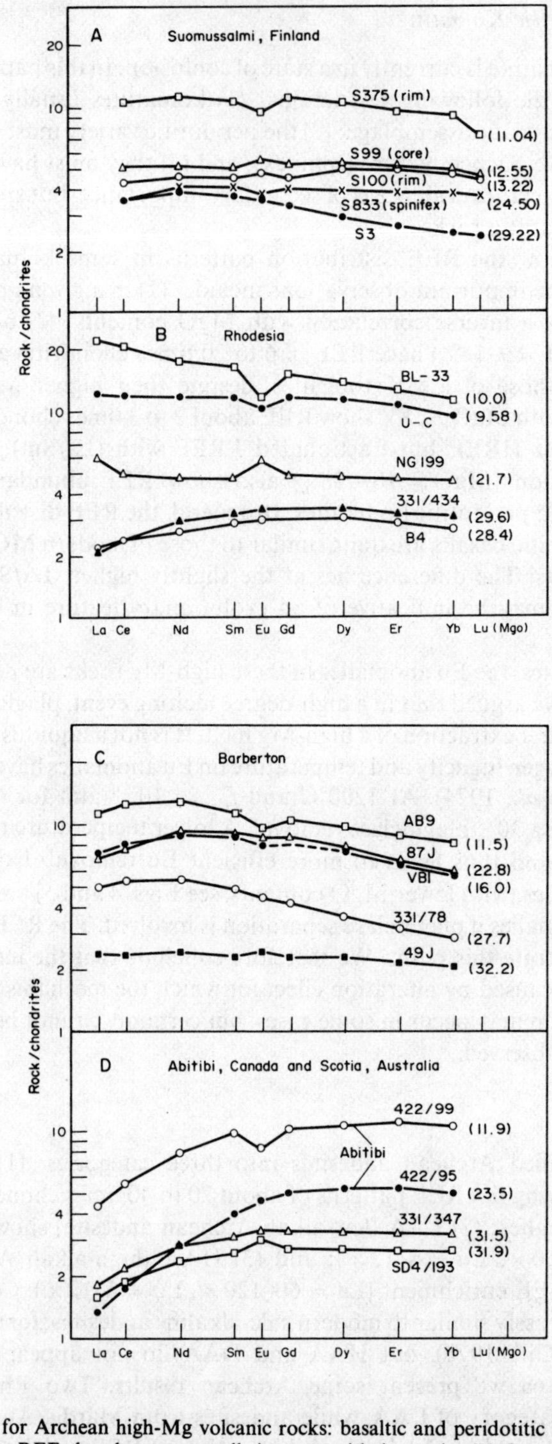

FIG. 5. REE patterns for Archean high-Mg volcanic rocks: basaltic and peridotitic komatiites. Inconsistent Eu anomalies are also seen. REE abundances generally increase with decreasing MgO contents. Data sources: Jahn *et al.* (1977, for 4a), Sun and Nesbitt (1977b, for 4b, 4c, and 4d).

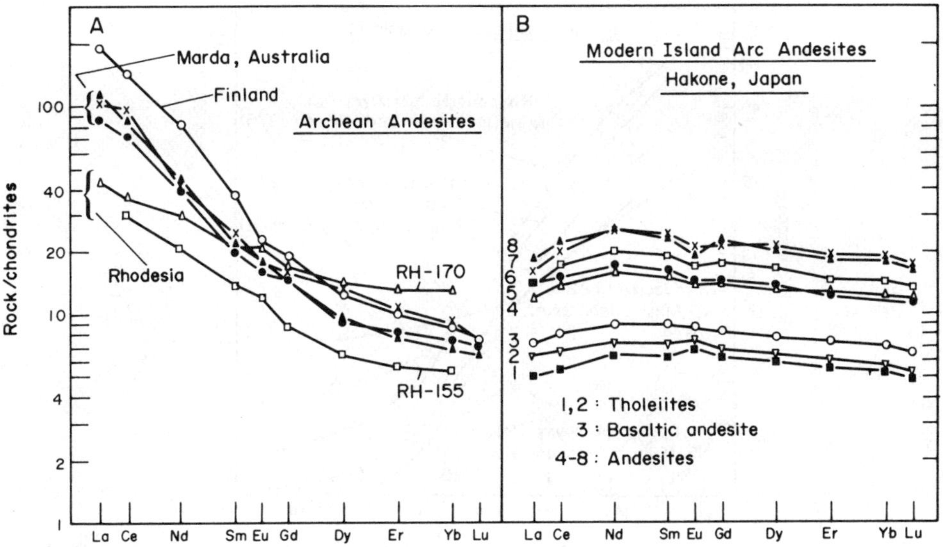

FIG. 6. REE patterns for Archean andesitic rocks (6A) and some modern island arc andesites from Japan (6B). Condie's HAA (Condie, 1976) is represented by the data of Finland and Marda, Australia (6A), while LAA is by those of Rhodesia (6A). Condie's third type, DAA, is not shown in 6A, but it has its modern analogues found in Japan (samples 6, 7, 8 of 6B). In contrast, some modern andesites (e.g. sample 3 of 6B) are characterized by tholeiitic REE patterns, and they have no Archean equivalents. Data sources: 6A: Jahn *et al.* (1977), Taylor and Hallberg (1977), Sun and Nesbitt (1977b), Hawkesworth and O'Nions (1977). 6B: Fujimaki (1975).

volcanic rocks from the Hakone volcano, Japan (Fujimaki, 1975). This is to emphasize that DAA do have their modern counterparts, as shown by samples 6, 7 and 8. In addition, andesites with very low and nearly flat REE abundances of 7 to 15 × chondrites (samples 3, 4, 5) occur in the modern island arc environment but are not found in Archean greenstone belts. The andesites from the Marda igneous complex of western Australia have been suggested to have an affinity to the Andean-type volcanism but apparently they were deposited in a stable environment (Taylor and Hallberg, 1977). In addition Archean andesites are often found to have severe HREE depletions, e.g. Rhodesian andesites (Hawkesworth and O'Nions, 1977), a characteristic not commonly observed in modern andesites.

D. *Siliceous Volcanic Rocks—Rhyolites and Rhyodacites*

In many Archean terrains, a bimodal volcanic association, namely, tholeiites and overlying rhyolites/rhyodacites, occur (Barker and Peterman, 1974). Figure 7 illustrates some typical REE patterns for Archean siliceous volcanic rocks. These patterns invariably show strong overall fractionation with severe HREE depletion. Examples: NE Minnesota (Arth and Hanson, 1972, 1975; Jahn *et al.*, 1974), Finland (Jahn *et al.*, 1977), Rhodesia (Condie and Harrison, 1976; Hawkesworth and O'Nions, 1977), and Barberton (Glikson, 1976).

Another type of siliceous volcanic rock has been found to have LREE enriched but HREE relatively flat patterns (La = 80–200 ×, HREE ~ 10–30 ×, Condie, 1976). This type of rock does not occur so often as the first type. Examples: Barberton, South Pass, Yellow Knife and western Kenya (Condie and Baragar, 1977).

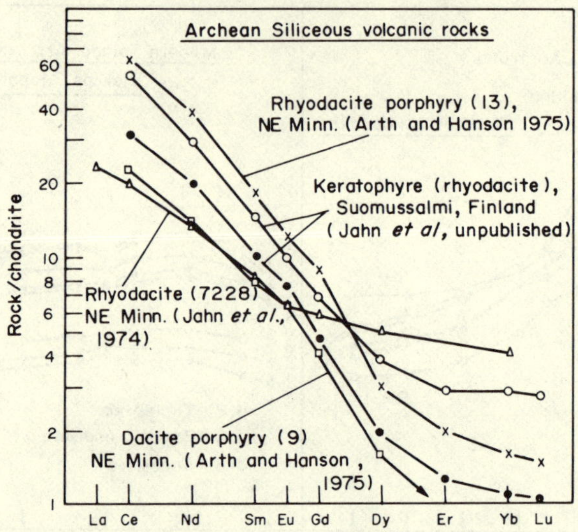

FIG. 7. REE patterns for Archean siliceous volcanic rocks. They are shown to be characterized by HREE depletion.

E. *Sedimentary Rocks*

The REE distribution in Archean and post-Archean sediments has been studied by many workers in recent years (Wildeman and Haskin, 1973; Wildeman and Condie, 1973; Jakeš and Taylor, 1974; Nance and Taylor, 1976, 1977). A major contrasting feature appears in the REE distribution of Archean compared with post-Archean sediments. All the post-Archean sediments (greywackes, shales, carbonates, arkoses, quartzites, etc.) are characterized by remarkably parallel REE patterns with (a) strongly fractionated LREE, (b) relatively flat HREE, (c) almost constant negative Eu anomalies (average $Eu/Eu^* = 0.67$), and (d) $\Sigma LREE/\Sigma HREE = 9.7 \pm 1.8$ (Nance and Taylor, 1976; Wildeman and Haskin, 1973; Wildeman and Condie, 1973). Figure 8a illustrates these characteristics. All the distribution patterns are parallel to that of the North American Shale composite sample (NAS).

In contrast, most Archean sedimentary rocks show either no or positive Eu anomalies, despite the fact that overall distribution patterns are similar to the post-Archean sediments (Fig. 8b). That is, when compared with NAS data, all Archean sediments, with few exceptions, show excess Eu abundances (Wildeman and Haskin, 1973; Wildeman and Condie, 1973; Jakeš and Taylor, 1974; Nance and Taylor, 1977).

Does this Eu excess serve as a Eu "spike" and imply the existence of possible early anorthositic crust? Or is it simply a local effect due to selective sedimentary differentiation leading to feldspar accumulation? A test from Devonian sediments led Nance and Taylor (1977) to conclude that the local effect of sedimentary differentiation is probably the major cause of such excess Eu in the Archean sediments.

Another important fact is that most Archean sedimentary rocks have REE patterns similar to those of Archean andesites (Fig. 6a) and modern calc-alkaline rocks (Jakeš and Taylor, 1974; Nance and Taylor, 1977). Similar comparisons also extend to most other elements. Because Archean sedimentary rocks in principle could provide information about the overall composition of the Archean crust, Taylor (1978) believes that the average Archean crustal

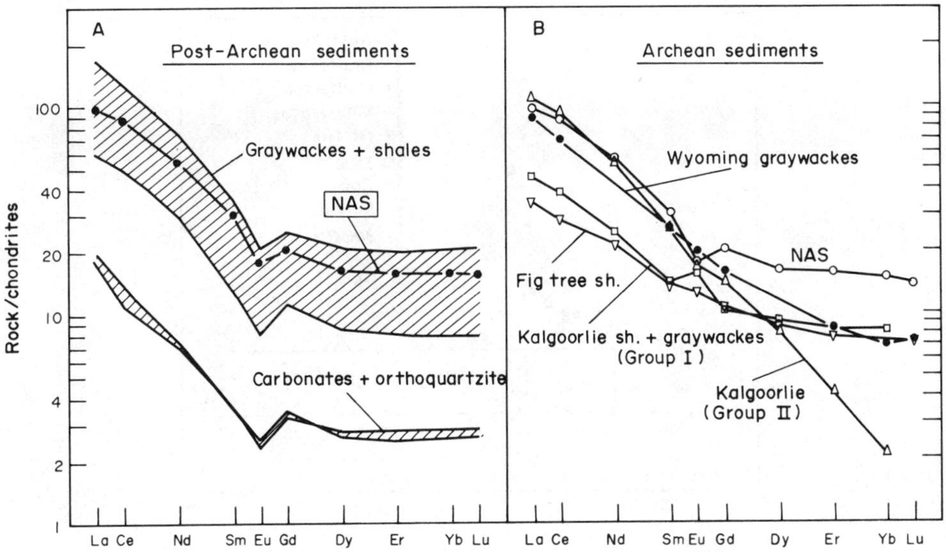

FIG. 8. REE patterns for Archean and post-Archean sediments. The post-Archean sediments are characterized by negative Eu anomalies, fractionated LREE and relatively flat HREE (8A). They may be exemplified by the North American Shale composite (NAS). In contrast, Archean sediments show either no or positive Eu anomalies. Both LREE and HREE are fractionated. Compared to NAS, they show "Eu excess". Data sources: Wildeman and Haskin (1973), Wildeman and Condie (1973), Haskin and Haskin (1966), Nance and Taylor (1976, 1977).

composition may be approximated by the modern calc-alkaline volcanic rocks. However, in many greenstone belts of bimodal distribution, andesitic rocks are generally lacking. The source for the "calc-alkaline" Archean sediments must then be various mixtures of tholeiites and siliceous volcanic rocks. In fact, a mass balance calculation shows that the results are consistent with such a mixing hypothesis (Nance and Taylor, 1977).

For the almost constant negative Eu anomalies in post-Archean sedimentary rocks, Jakeš and Taylor (1974) hypothesized that such Eu depletion may be a consequence of progressive growth of the upper sialic crust by partial melting of the lower crust which selectively retains Eu in residual plagioclase. The relatively flat HREE observed in some Archean and many post-Archean rocks (Figs. 8a and 8b) suggests that garnet is not an important residue in the melting episodes which lead to the formation of ultimate sources for sedimentary rocks.

F. *The* $(La/Sm)_N$ *Ratios of Archean Basalts*

We have shown earlier in Figs. 4 and 5 that most Archean basalts (both low-K and high-Mg komatiitic varieties) exhibit rather flat REE patterns. The majority of their chondrite-normalized $(La/Sm)_N$ ratios falls in the range of 0.7 to 1.3 (total range = 0.4 to 3.5; Sun and Nesbitt, 1977a). This suggests that the composition of their sources are close to primary chondritic mantle material. In contrast, the typical modern mid-ocean ridge basalts (MORB) have $(La/Sm)_N$ ratios ranging from 0.4 to 0.7 (Fig. 9; Sun and Nesbitt, 1977a). It is now well known that the sources for MORB have been depleted in many LIL elements (e.g. K, Rb and Ba) and light REE. Such depletion is also reflected by its very high Rb–Sr model age ($\sim$8.2 b.y.) and low Sr^{87}/Sr^{86} ratios (see Sun and Hanson, 1975, and Jahn and Nyquist, 1976, for

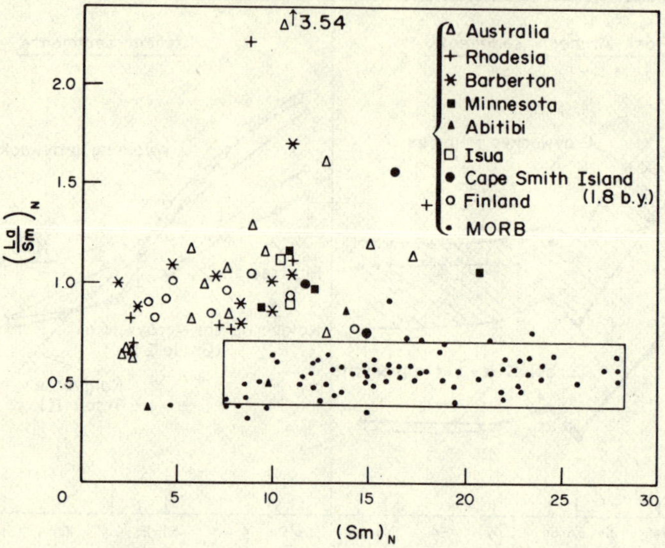

FIG. 9. $(La/Sm)_N$ vs. $(Sm)_N$ variation diagram for Archean volcanic rocks and MORB (after Sun and Nesbitt, 1977a). La values of Finland volcanic rocks are obtained by extrapolation. $(La/Sm)_N$ ratios in Archean rocks are clearly higher than in MORB. Data sources: see references given by Sun and Nesbitt (1977a) for Barberton, sources are also from Herrmann *et al.* (1976).

discussion). Assuming that Archean basalts have randomly sampled the Archean upper mantle, the gross difference in the $(La/Sm)_N$ ratios of the Archean basalts and the MORB may be indicative of an evolutionary feature in terms of the upper mantle chemistry.

It has been argued by Sun and Nesbitt (1977a) that during the generation of Archean high-Mg komatiites only olivine and orthopyroxene would be left in the source as residual minerals. Small amount of garnet may be left in some cases to explain the slightly fractionated HREE with $(Yb/Gd)_N < 1$ as observed in the Barberton samples (Fig. 5c). Because most Archean basalts possess nearly flat HREE, it follows that garnet does not usually play an important role as a residual phase. Furthermore, due to the extremely small mineral/melt K_D values for REE in olivine and orthopyroxene (Schnetzler and Philpotts, 1970), the REE abundances in the melts will be enriched according to the reciprocal of the degree of partial melting (i.e. $1/F$). This leads to an important conclusion that the REE patterns of the melt will be similar to that of the source and that the overall abundance of the REE can be used to estimate the percentage of melting (Sun and Nesbitt, 1977b).

A more detailed examination of the REE patterns in Figs. 4 and 5 reveals that some rocks are slightly LREE-enriched (e.g. Lawlers, Fig. 4b) and some LREE-depleted (e.g. Abitibi, Fig. 5d). The source for the Abitibi rocks is probably an example of depleted Archean mantle, like that for the MORB. While the Minnesota low-K tholeiites (Fig. 4c) do not show conspicuous LREE depletion in its mantle source, the isotopic evidence (initial Sr^{87}/Sr^{86} ratios < 0.7010) suggests a Rb depletion (relative to Sr) in its mantle source (see Jahn and Nyquist, 1976, for discussion). The above observations merely indicate that LREE or alkali element depletion in some upper mantle sources had already taken place during the Archean.

On the other hand, the LREE-enriched Archean basalts are likely to be generated from upper mantle sources with compositions characterized by original LREE enrichment.

Otherwise, a two-stage model may apply as follows. In modern marginal basins, such as the Mariana and the Lau Basins, basalts are found to have LREE-enriched patterns (Hart *et al.*, 1972; Gill, 1976). Marginal basins are above subduction zones and are different from mid-ocean ridges in tectonic setting. The subducted lithosphere may undergo partial melting producing melt with fractionated (LREE-enriched) REE by garnet control. The melt could be subsequently incorporated into the overlying mantle, thus producing the light LREE enrichment in this part of mantle. Later partial melting induced by the basin rifting may result in slightly LREE-enriched basalts. Archean greenstone belts have been compared with present-day marginal basins (e.g. Tarney *et al.*, 1976). Perhaps this two-stage model may explain the REE characteristics in some Archean basalts as well as in the modern marginal basin basalts.

In terms of HREE, Figs. 4 and 5 show that in a given area or greenstone belt, rocks with higher MgO content (peridotitic) commonly show flatter HREE than rocks with lower MgO (basaltic). That is, $(Yb/Gd)_N$ values decrease with decreasing MgO contents. This difference may be due to crystal fractionation involving garnet at depth or due to the residual mineralogy containing garnet. Sun and Nesbitt (1977b) estimated that 5% garnet settling in the partial melt may produce a $(Yb/Gd)_N$ ratio of 0.85, as observed in Rhodesia and Australian rocks (Figs. 5b and 4b).

G. *REE Abundances in the Archean Mantle*

Sun and Nesbitt (1977a) have recently used the chemical regularities in Archean high-Mg rocks (spinifex-textured peridotites, or STP, and basaltic komatiites) to estimate an average composition of the Archean mantle. In these rocks, MgO and many other oxides or elements show rather systematic variations. For example, TiO_2 and MgO show an inverse correlation which intersects the MgO axis at about 50% MgO when $TiO_2 = 0$. Based on the information from meteorites and terrestrial ultramafic rocks, a value of MgO = 38% was chosen to represent the mantle value and the TiO_2 abundance (0.16–0.21%) was thus estimated.

Yb abundances show a similar variation like TiO_2 against MgO contents. An estimation at 38% MgO yields an Yb value of 0.30 to 0.46 ppm (Sun and Nesbitt, 1977a). Overall, the refractory lithophile elements, including REE, Ti and Zr, were estimated to have about 1.8 to 2.4 times chondritic abundances. Their model composition for the Archean mantle agrees with a new pyrolite model composition of Ringwood (1975).

Bickle *et al.* (1976) have done similar excercise in estimating the mantle composition. Their results show some substantial differences from Sun and Nesbitt (1977a) in Al_2O_3, CaO, MnO, Na_2O, P_2O_5 and TiO_2. This is partly resulted from their different assumed MgO value (41%). However, if 41% Mgo is used in Sun and Nesbitt (1977a), the differences, such as TiO_2, may even be magnified. In addition, 41% of MgO seems too high for the "primary" upper mantle composition (see any standard textbook for comparison).

IV. Application of Trace Element Geochemistry to Identification of Archean Tectonic Settings

Trace-element abundances have been used by several workers to identify the tectonic environments of Archean greenstone belts (e.g. Hart *et al.*, 1970; White *et al.*, 1971; Jahn *et al.*, 1974; Condie, 1975, 1976; Condie and Harrison, 1976; Condie and Baragar, 1974). Since 1970 a great deal of major and trace element data have been reported for both modern and

Archean volcanic rocks. These data allow us to reevaluate the application of trace-element geochemistry to the problems of Archean tectonic settings. We now conclude that the identification of Archean Tectonic settings by geochemical means is not very successful. The basic reasons are as follows:

1. Insufficient understanding of trace-element characterization in modern volcanic rocks from different tectonic environments which are taken as the frames of reference for Archean volcanic rocks. That is, many trace-element patterns are indiscriminant in rocks from different tectonic environments. For example, the Troodos Ophiolite Complex has given rise to a controversy as to whether it represents a slice of oceanic crust (e.g. Gass, 1968; Gass *et al.*, 1975; Moores *et al.*, 1971; Moores, 1975; Hynes, 1975) or whether it was created in a young island arc environment (Miyashiro, 1973, 1975a, b), or a marginal basin (Smewing *et al.*, 1975). Kay and Senechal (1976) tried to resolve this problem by using the abundances of REE, Ti and Zr in the basaltic rocks and found that this geochemical tool yielded no clear distinction between the rocks from different tectonic environments.

On the other hand, although some distinct geochemical differences may exist between, for example, modern island arc tholeiites (IAT) and MORB (Jakes and Gill, 1970), we cannot be sure that the same differences would also exist in the Archean analogues.

2. Insufficient understanding of geochemical behavior of the transition elements during partial melting and crystal fractionation under different P, T and X (composition) conditions. Using K, Rb, Cs, Sr, Ba and REE abundances, Hart *et al.* (1970) and Jahn *et al.* (1974) proposed that Archean greenstone belts were formed in ancient island arc environments. However, some refractory transition metal data (Ni, Co, Cr and V, etc.) show that Archean volcanic rocks generally have higher contents of these elements than modern island arc rocks (see Table 1). This same evidence has been used by some to dismiss the island arc model for Archean greenstone belts (e.g. Hawkesworth and O'Nions, 1977). In fact, the distribution coefficients of transition metals are sensitive to the melt composition, temperature and sometimes redox condition (e.g. for Cr, Schreiber and Haskin, 1976; for Ni, Hart *et al.*, 1976).

Hart *et al.* (1976) have determined the compositional and temperature effects on the distribution coefficient of Ni between olivine and silicate melts. They found that the compositional effect is much greater than the temperature effect, and that most of the observed range of D_{Ni} is due to compositional variation following the equation: $\ln D_{Ni} = 3.325 - 0.0885\,(MgO)$. That is, the higher the MgO content, the lower the D_{Ni} value. To a lesser degree, D_{Ni} decreases slightly with increasing temperature of equilibrium. The pressure effect has not been evaluated. Thus, if primary liquid is MgO rich, such as in pricritic or komatiitic basalts, considerable olivine fractionation (D_{Ni} will be as low as 4) would still not drastically deplete Ni content of the liquid. Consequently, olivine fractionation could have occurred in Archean island arc environments, but there would have been no substantial Ni, and possibly Co, depletion. In other words, the difference of transition metal abundances between Archean greenstones and modern IATs cannot be used to argue strongly against an island arc model for Archean rocks. Besides, olivine fractionation may not necessarily have occurred in Archean island arc environments if they ever existed.

It should also be noted that although IAT rocks generally have low Ni, Cr and Co contents compared with MORB or Archean volcanics, many other IAT rocks contain "undepleted" abundances of the transition metals (Miyashiro and Shido, 1975). Miyashiro and Shido (1975) have shown that the fields of Ni and Cr in MORB are completely enveloped by those in island arc and active continental margin volcanic rocks.

3. Evolutionary change of the upper mantle composition. In all the trace element models

proposed so far, a basic assumption has been implicitly made: the Archean upper mantle is not different in composition from the modern upper mantle. If the mantle composition has undergone a continuous (or episodic) temporal change due to extraction of volcanic liquids throughout geologic times, the application of trace element geochemistry to the identification of Archean tectonic settings would be totally unacceptable. In terms of REE, the $(La/Sm)_N$ ratios in most modern IAT, MORB and marginal basin basalts are similar but are significantly lower than those of Archean basalts (see Section III F). This suggests a temporal variation in REE abundances in the upper mantle.

In addition, some geochemical differences in modern rocks of different tectonic environments may indeed be important (e.g. Ba content), but, due to the evolving nature of the mantle composition, the same differences in Archean rocks may not be so apparent as in their modern analogues.

4. Existence of heterogeneities in both the Archean and the modern upper mantle. The compositional heterogeneity is well known for the modern upper mantle. This accounts for the compositional differences in MORB (source in LVZ?), oceanic island volcanic rocks (source in deeper mantle as mantle plumes? or in the top of the LVZ?), continental basalts (source in subcontinental undepleted mantle?, see Brooks *et al.*, 1976), marginal basin, island arc and continental margin basalts (source in subducted lithosphere? or overlying mantle?). The Archean upper mantle also shows demonstratable heterogeneity from the evidence of $(La/Sm)_N$ ratio (see previous discussion) and initial Sr^{87}/Sr^{86} ratios (e.g. Jahn and Nyquist, 1976). The heterogeneities have introduced another degree of complexity and uncertainty into the application of trace-element modeling.

5. Insufficient understanding of tectonic styles operating in the Archean time. For example, for the evolution of global tectonics, Burke and Dewey (1973) have proposed three tectonic régimes: (a) permobile régime (>2.7 b.y.), (b) transitional régime (2.7–2.0 b.y.) and (c) plate tectonics régime (<2.0 b.y.). In the Archean time, due to a higher thermal gradient, the global tectonic régime appears to have been characterized by great mobility and absence of rigid features as of today. Burke and Dewey (1973) believe that there were no "continents or large cratonic areas". Individual granodioritic areas rarely exceed 10^5 km^2 and are bounded by greenstone belts. Archean greenstone belts are generally about 10 times as long as they are wide, in contrast with the rocks of comparable composition preserved in the suture zones of later continental collision orogenies (with length/width ratio about 100).

Because of a higher heat production in the Archean than today (2–3 times), but the surface temperature of Archean crust cannot have been much different from today (<100°C), a more efficient heat-dissipating process is needed. There were probably greater average spreading rates, greater total ridge lengths, or both (Burke *et al.*, 1976).

The above serves to point out the difference in tectonic style between the Archean and present times. It is thus unlikely that we can precisely equate any Archean greenstone belt with modern tectonic settings. If the modern volcanic rocks can be formed in different tectonic settings, it is needless to hypothesize that Archean volcanic rocks were formed in any unique tectonic environment; they could also be formed in many tectonic settings as of today.

V. Isotopic Compositions of Archean Volcanic Rocks

A substantial amount of work on isotopic dating of Archean rocks in recent years has yielded valuable information about temporal sequences and modes of crustal development.

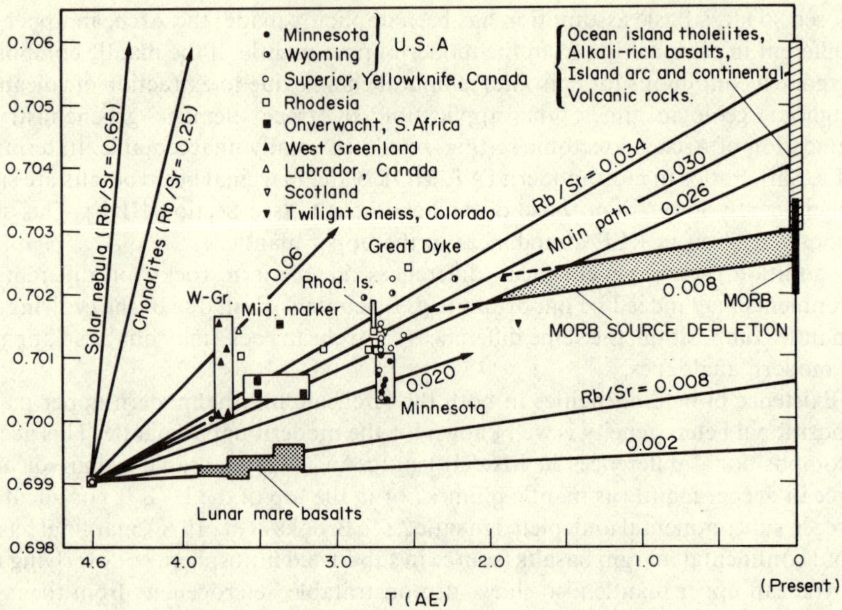

FIG. 10. Sr isotopic evolution (*I, T*) diagram for terrestrial rocks. A field of lunar mare basalts is shown for comparison. I = initial Sr^{87}/Sr^{86} ratios, T = time in aeon (A.E. = 10^9 yrs). Data sources: see references given by Jahn and Nyquist (1976) and Moorbath (1976). Additional sources: Hurst *et al.* (1975), and Stueber *et al.* (1976).

Available Sr isotopic data have been used to construct a Sr isotope evolution (I,T) diagram for the upper mantle (e.g. Jahn and Nyquist, 1976). Figure 10 summarizes the currently available Sr isotopic data for Archean rocks, including those from low-grade greenstone belts and high-grade gneissis terrains (data sources, see figure caption). Also included are the isotopic composition of the modern upper mantle. For the Archean data, the following observations can be made:

1. The I values ($\equiv$ initial Sr^{87}/Sr^{86} ratios) are scattered but restricted in the range of 0.700 to 0.702. Not all of them have been interpreted as representing their respective magmatic I values. For example, the I values of Middle Marker (0.7015) and Rhodesian Bulawayan limestones (0.7009–0.7014) are more likely to represent the isotopic compositions in equilibrium with Archean sea water (for details, see Jahn and Nyquist, 1976). Nevertheless, the large range of isotopic composition for 2.7 b.y. terrains (Minnesota, Superior Province of Canada, Rhodesia) strongly indicate an isotopic heterogeneity in the Archean upper mantle. This is consistent with our previous discussion on the $(La/Sm)_N$ ratios.

2. Some I values for the 3.7 b.y. rocks from West Greenland are relatively high (0.7010 $\pm$0.0010–0.0020) for rocks of that age. We suspect that they may represent "metamorphic initial ratios". However, the large statistical errors (common for I values obtained from acidic rocks) extend the ratios to less than 0.7000. Then I value of 0.701 for the 3.6 b.y. Uivak gneisses of Labrador was regarded as a metamorphic initial ratio by Hurst *et al.* (1975). The Minnesota values are lower than those of the contemporaneous Superior Province and Rhodesian Craton. We believe that the mantle source for the Minnesota rocks had been previously depleted in Rb relative to Sr, probably a billion years or more beforehand. Except

for the cases mentioned above, the majority of I values seem to have evolved along a Main Path with Rb/Sr ratios = 0.026 to 0.034. This Main Path is thought to represent the isotopic composition of the undepleted upper mantle.

3. For modern volcanic rocks, MORB are not only characterized by depletion of some LIL elements and LREE, but also a rather restricted range of low I values (0.7025–0.7035, see Brooks *et al.*, 1976, for data compilation). Ocean island basalts, island arc and continental basalts have slightly higher I values ranging from 0.7030 to 0.7060. These values have been considered to be primary (i.e. mantle values) and not due to crustal contamination (see Brooks *et al.*, 1976, for detailed arguments). A possible source of ocean island basalts is thought to be below the now convecting asthenosphere (Sun and Hanson, 1975; Brooks *et al.*, 1976). Common Pb and Rb–Sr isotopic data suggest that a major event of mantle differentiation leading to the separation of MORB sources (LVZ or asthenosphere) from the rest of the mantle probably took place about 1.5 to 2.0 b.y. ago (Sun and Hanson, 1975; Brooks *et al.*, 1976). The array of Archean data and those of ocean island basalts and many island arc and continental volcanic rocks (for some deviations, see Brooks *et al.*, 1976) in Fig. 10 suggests that they may have evolved from a presumably closed mantle system throughout geologic history. The MORB sources probably have a slightly lower or similar Rb/Sr ratio to average MORB (0.008, Hart, 1971). A backward extrapolation from MORB results in an intercept with the Main Path at about 1.5 to 2.0 b.y.; a value which coincides remarkably with the Pb secondary isochron age (Sun and Hanson, 1975) or the so-called Rb–Sr mantle isochron age (Sun and Hanson, 1975; Brooks *et al.*, 1976).

O'Nions *et al.* (1977) have made a systematic study on both Nd and Sr isotopic compositions in oceanic basalts. They found that the Nd^{143}/Nd^{144} ratios display a negative correlation with the Sr^{87}/Sr^{86} ratios. DePaolo and Wasserburg (1976) have measured Nd isotopic compositions in some rocks of different ages. They found that modern alkali basalts and a continental basalt (BCR–1) have evolved along with Precambrian rocks, essentially from a mantle source of nearly constant chondritic Sm/Nd ratio. Richard *et al.* (1976) have arrived at an identical conclusion. Using the correlation of Nd and Sr isotopic ratios, O'Nions *et al.* (1977) suggest that the bulk Sr^{87}/Sr^{86} ratio for the mantle might be close to 0.705, and the bulk Rb/Sr ratio equal to 0.032. Again, this is another independent way of estimating the I value or Rb/Sr ratio for the modern "undepleted mantle" assuming a single-stage mantle evolution model.

4. Although both MORB and the Minnesota data (Fig. 10) form a linear trend and thus argue for the existence of depleted Archean mantle sources, MORB cannot have the same parentage as the Minnesota rocks. If MORB have evolved from a single-stage isolated mantle source having the primordial Sr composition ($\simeq 0.699$) 4.6 b.y. ago, the calculated Rb/Sr ratio for this source is 0.02, a value too high for the inferred value of about ≤ 0.008 for the source of MORB. On the other hand, if a MORB source, with Rb/Sr = 0.008, had evolved from the Minnesota mantle (<0.701 at 2.7 b.y.), then the present Sr^{87}/Sr^{86} ratio for the MORB would be less than 0.7018, a value too low and never found for modern rocks.

In conclusion, the data of trace element abundances and isotopic compositions indicate that some evolutionary features of the earth's mantle are evident, e.g. depletion of LIL and LREE in the MORB source. The compositional heterogeneity of the mantle has existed ever since the Archean time, and apparently it is enhanced even more today. If the continents were not formed all at once, but rather, as advocated by Moorbath (1975, 1976), by gradual addition of sial over a period of time, we may expect a concomitant gradual depletion of "continental elements" (alkalis and other lithophile trace elements) in the upper mantle.

However, counting MORB apart, most Archean and younger basaltic volcanic rocks tend to show a gross compositional constancy (constant trace element abundances and ratios, Jahn, 1977). This may be explained as due to the interplay of (1) replenishment of trace elements from lower parts of the mantle and (2) some recycling of trace elements in the crust-upper mantle system (for more details, see Jahn, 1977). For the evolution of oceanic mantle, see Sun and Hanson (1975) and Sun and Nesbitt (1977a).

Acknowledgements

We thank our colleagues at Rennes and at Adelaide for helpful discussions on Archean problems. Technical assistance from C. Dalibad and M. Lautram of Rennes in the preparation of this manuscript is greatly appreciated.

References

ARTH, J. G. and HANSON, G. N. (1972) Quartz diorites derived by partial melting of eclogite or amphibolite at mantle depths. *Contr. Mineral Petrol.* **37**, 161–174.

ARTH, J. G. and HANSON, G. N. (1975) Geochemistry and origin of the early Precambrian crust of northeastern Minnesota. *Geochim. Cosmochim. Acta* **39**, 325–362.

BARTON, J. M., Jr. (1975) Rb–Sr isotopic characteristics and chemistry of the 3.6–b.y. Hebron gneiss, Labrador. *Earth Planet. Sci. Lett.* **27**, 427–435.

BICKLE, M. J., HAWKESWORTH, C. J., MARTIN, A., NISBET, E. G. and O'NIONS, R. K. (1976) Mantle composition derived from the chemistry of ultramafic lavas. *Nature* **263**, 577–580.

BLAIS, S., AUVRAY, B., JAHN, B. M., CAPDEVILA, R. and HAMEURT, J. (1977) Petrography and geochemistry of the Suomussalmi greenstone belt, eastern Finland (in preparation).

BLANCHARD, D., MENZIES, M. and JACOBS, J. (1976) Rare earth geochemistry of the Point Sal ophiolite—evidence of basalt seawater interaction. *EOS, Trans. Amer. Geophys. Union* **57**, 1027.

BROOKS, C. and HART, S. (1972) An extrusive basaltic komatiite from a Canadian Archean metavolcanic belt. *Can. J. Earth Sci.* **9**, 1250–1253.

BROOKS, C., JAMES, D. E., HART, S. R. and HOFMANN, A. W. (1976) Rb–Sr mantle isochrons. *Ann. Rept. Dept. Terr. Mag. (1975–1976)* 176–207.

BURKE, K. C. A. and DEWEY, J. F. (1973) An outline of Precambrian plate development. In *Implications of Continental Drift to the Earth Sciences*, Vol.2. D. H. TARLING and S. K. RUNCORN, eds., Academic Press, London, 1035–1045.

BURKE, K., DEWEY, J. F. and KIDD, W. S. F. (1976) Dominance of horizontal movements, arc and microcontinental collisions during the later Permobile regime. In *The Early History of the Earth*, B. F. WINDLEY, ed., John Wiley & Sons, London, 113–129.

CONDIE, K. C. (1975) Mantle-plume model for origin of Archean greenstone belts based on trace element distributions. *Nature* **258**, 413–414.

CONDIE, K. C. (1976) Trace element geochemistry of Archean greenstone belts. *Earth Sci. Rev.* **12**, 393–417.

CONDIE, K. C. and BARAGAR, W. R. A. (1974) Rare-earth element distributions in volcanic rocks from Archean greenstone belts. *Contr. Mineral. Petrol.* **45**, 237–246.

CONDIE, K. C. and HARRISON, N. M. (1976) Geochemistry of the Archean Balawayan Group, Midlands greenstone belts, Rhodesia. *Precam. Res.* **3**, 253–271.

DE PAOLO, D. J. and WASSERBURG, G. J. (1976) Nd isotopic variations and petrogenetic models. *Geophys. Res. Lett.* **3**, 249–252.

FREY, F. A., BRYAN, W. B. and THOMPSON, G. (1974) Atlantic ocean floor: geochemistry and petrology of basalts from Legs 2 and 3 of the deep-sea drilling project. *J. Geophys. Res.* **79**, 5507–5527.

FUJIMAKI, H. (1975) Rare earth elements in volcanic rocks from Hakone volcano and northern Izu Peninsula, Japan. *J. Fac. Sci. Univ. of Tokyo* **19**, 81–93.

GASS, I. G. (1968) Is the Troodos Massif of Cyprus a fragment of Mesozoic ocean floor? *Nature* **220**, 39–42.

GASS, I. G., NEARY, C. R., PLANT, J., ROBERTSON, A. H. F., SIMONIAN, K. O., SMEWING, J. D., SPOONER, E. T. C. and WILSON, R. A. M. (1975) Comments on "The Troodos ophiolitic complex was probably formed in an island arc" by A. Miyashiro and subsequent correspondence by A. Hynes and A. Miyashiro. *Earth. Planet. Sci. Lett.* **25**, 236–238.

GILL, J. B. (1976) Composition and age of Lau Basin and Ridge volcanic rocks: Implications for evolution of an interarc basin and remnant arc. *Bull. Geol. Soc. Amer.* **87**, 1384–1395.

GLIKSON, A. Y. (1971) Primitive Archean element distribution patterns: chemical evidence and geotectonic significance. *Earth Planet. Sci. Lett.* **12**, 309–320.

GLIKSON, A. Y. (1976) Trace element geochemistry and origin of early Precambrian acid igneous series, Barberton Mountain Land, Transvaal. *Geochim. Cosmochim. Acta* **40**, 1261–1280.

GREEN, D. H., NICHOLLS, I. A., VILJOEN, M. and VILJOEN, R. (1975) Experimental demonstration of the existence of peridotitic liquids in earliest Archean magmatism. *Geology*, **3**, 11–15.

HART, R. A. (1970) Chemical exchange between seawater and deep ocean basalts. *Earth Planet. Sci. Lett.* **9**, 269–279.

HART, S. R. (1969) K, Rb, Cs contents and K/Rb, K/Cs ratios of fresh and altered submarine basalts. *Earth Planet. Sci. Lett.* **6**, 295–303.

HART, R. A. (1973) A model for chemical exchange in the basalt–seawater system of oceanic layer 2. *Can. J. Earth Sci.* **10**, 799–816.

HART, S. R. (1971) K, Rb, Cs, Sr and Ba contents and Sr isotope ratios of ocean floor basalts. *Phil. Trans. Roy. Soc. Lond.* A, **268**, 573–587.

HART, S. R., BROOKS, C., KROGH, T. E., DAVIS, G. L. and NAVA, D. (1970) Ancient and modern volcanic rocks: A trace element model. *Earth Planet. Sci. Lett.* **10**, 17–28.

HART, S. R., GLASSLEY, W. E. and KARIG, D. E. (1972) Basalts and sea floor spreading behind the Mariana island arc. *Earth Planet. Sci. Lett.* **15**, 12–18.

HART, S. R., ERLANK, A. J. and KABLE, E. J. D. (1974) Sea floor alteration: some chemical and Sr isotopic effects. *Contr. Min. Petrol.* **44**, 219–230.

HART, S. R., DAVIS, K. E., KUSHIRO, I. and BRUCE, E. (1976) Partitioning of nickel between olivine and silicate liquid (abs). *Annual Meeting Geol. Soc. America* **8**, 906–907.

HASKIN, L. A. and HASKIN, M. A. (1966) Rare earths in European shales: a redermination. *Sciences* **154**, 507–509.

HAWKESWORTH, C. J. and O'NIONS, R. K. (1977) The petrogenesis of some Archean volcanic rocks from southern Africa. *J. Petrol.* (in press).

HAWKINS, J. W. (1976) Petrology and geochemistry of basaltic rocks of the Lau Basin. *Earth Planet. Sci. Lett.* **28**, 283–297.

HERRMANN, A. G., BLANCHARD, D. P., HASKIN, L. A., JACOBS, J. W., KNABE, D., KOROTEV, R. L. and BRANNON, J. C. (1976) Major, minor and trace element compositions of peridotitic and basaltic komatiites from the Precambrian crust of southern Africa. *Contr. Mineral. Petrol.* **59**, 1–12.

HURST, R. W., BRIDGWATER, D., COLLERSON, K. D. and WETHERILL, G. W. (1975) 3600 m.y. Rb–Sr ages from very early Archean gneisses from Saglek Bag, Labrador. *Earth Planet. Sci. Lett.* **27**, 393–403.

HYNES, A. (1975) Comment on "The Troodos ophiolite complex was probably formed in an island arc" by A. Miyashiro. *Earth Planet. Sci. Lett.* **25**, 213–216.

JAHN, B. (1977) Trace element geochemistry of Archean volcanic rocks and its implications for the chemical evolution of the upper mantle. *Bull. Soc. Géol. France* (in press).

JAHN, B. M., SHIH, C. Y. and MURTHY, V. R. (1974) Trace element geochemistry of Archean volcanic rocks. *Geochim. Cosmochim. Acta* **38**, 611–627.

JAHN, B. M. and NYQUIST, L. E. (1976) Crustal evolution in the early earth-moon system: Constraints from Rb–Sr studies. In *The Early History of the Earth*, B. F. WINDLEY, ed., John Wiley & Sons, London, pp. 55–76.

JAHN, B. M. and CONDIE, K. C. (1976) On the age of Rhodesian greenstone belts. *Contr. Mineral. Petrol.* **57**, 317–330.

JAHN, B. M., CORNICHET, J., BLAIS, S. and AUVRAY, B. (1977) REE distribution in volcanic rocks of the Suomussalmi greenstone belt, Finland (in preparation).

JAKEŠ, P. and GILL, J. B. (1970) Rare earth elements and the island arc tholeiitic series. *Earth Planet. Sci. Lett.* **9**, 17–28.

JAKEŠ, P. and WHITE, A. J. R. (1972) Major and trace element abundances in volcanic rocks of orogenic areas. *Bull. Geol. Soc. Amer.* **83**, 29–40.

JAKEŠ, P. and TAYLOR, S. R. (1974) Excess europium content in Precambrian sedimentary rocks and continental evolution. *Geochim. Cosmochim. Acta* **38**, 739–745.

KAY, R. W. and SENECHAL, R. G. (1976) The rare earth geochemistry of the Troodos Ophiolite complex. *J. Geophys. Res.* **81**, 964–970.

MIYASHIRO, A. (1973) The Troodos ophiolite complex was probably formed in an island arc. *Earth Planet. Sci. Lett.* **19**, 218–224.

MIYASHIRO, A. (1975a) Origin of the Troodos and other ophiolites: A reply to Hynes. *Earth Planet. Sci. Lett.* **25**, 217–222.

MIYASHIRO, A. (1975b) Origin of the Troodos and other ophiolites: A reply to Moores. *Earth Planet. Sci. Lett.* **25**, 227–235.

MIYASHIRO, A. and SHIDO, F. (1975) Tholeiitic and calc-alkalic series in relation to the behavior of titanium, vanadium, chromium and nickel, *Amer. J. Sci.* **275**, 265–277.

MOORBATH, S. (1975) Evolution of Precambrian crust from Sr isotopic evidence. *Nature* **254**, 395–398.

MOORBATH, S. (1976) Age and isotope constraints for the evolution of Archean crust. In *The Early History of the Earth*, B. F. WINDLEY, ed., John Wiley & Sons, London, pp. 351–360.

MOORES, E. M. (1975) Discussion of "Origin of Troodos and other ophiolites: A discussion; A reply to Hynes" by A. Miyashiro. *Earth Planet. Sci. Lett.* **25**, 223–226.

MOORES, E. M. and VINE, F. J. (1971) Troodos Massif, Cyprus and other ophiolites as oceanic crust: evaluation and implications. *Phil. Trans. Roy. Soc. Lond.* A, **268**, 443–466.

NANCE, W. B. and TAYLOR, S. R. (1976) Rare earth element patterns and crustal evolution—I. Australian post-Archean sedimentary rocks. *Geochim. Cosmochim. Acta* **40**, 1539–1551.

NANCE, W. B. and TAYLOR, S. R. (1977) Rare earth element patterns and crustal evolution—II. Archean sedimentary rocks from Kalgoorlie, Australia. *Geochim. Cosmochim. Acta* **41**, 225–231.

NESBITT, R. W. (1971) Skeletal crystal forms in the ultramafic rocks of the Yilgarn Block, western Australia: Evidence for an Archean ultramafic liquid. *Geol. Soc. Aus. Spec. Publ.* **3**, 331–347.

NESBITT, R. W. and SUN, S. S. (1976) Geochemistry of Archean spinifex-textured peridotites and magnesian and low-magnesian tholeiites. *Earth Planet. Sci. Lett.* **31**, 433–453.

O'NIONS, R. K., HAMILTON, P. J. and EVENSEN, N. M. (1977) Variations in $^{143}Nd/^{144}Nd$ and $^{87}Sr/^{86}Sr$ ratios in oceanic basalts. *Earth Planet. Sci. Lett.* **34**, 13–22.

PYKES, D. R., NALDRETT, A. J. and ECKSTRAND, O. R. (1973) Archean ultramafic flows in Munro Township, Ontario. *Bull. Geol. Soc. Amer.* **84**, 955–978.

RICHARD, P., SHIMIZU, N. and ALLEGRE, C. J. (1976) $^{143}Nd/^{146}Nd$, a natural tracer: an applications to oceanic basalts. *Earth Planet Sci. Lett.* **31**, 269–278.

SCHACKELTON, R. M. (1976) Shallow and deep-level exposures of Archean crust in India and Africa. In *The Early History of the Earth*, B. F. WINDLEY, ed., John Wiley & Sons, London, 317–321.

SCHNETZLER, C. C. and PHILPOTTS, J. A. (1970) Partition coefficients of rare earth elements between igneous matrix material and rock forming mineral phenocrysts—II. *Geochim. Cosmochim. Acta* **34**, 331–340.

SCHREIBER, H. D. and HASKIN, L. A. (1976) Chromium in basalts: experimental determination of redox states and partitioning among synthetic silicate phases. *Proc. Lunar Sci. Conf. 7th* **2**. 1221–1259.

SMEWING, J. D., SIMONIAN, K. O. and GASS, I. G. (1975) Metabasalts from Troodos Massif Cyprus: Genetic implications deduced from petrography and trace element geochemistry. *Contr. Mineral Petrol.* **51**, 49–64.

STUEBER, A. M., HEIMLICH, R. A. and IKRAMUDDIN, M. (1976) Rb–Sr ages of Precambrian mafic dikes, Bighorn Mountains, Wyoming. *Bull. Geol. Soc. Amer.* **87**, 909–914.

SUN, C. O., WILLIAMS, R. J. and SUN, S. S. (1974) Distribution coefficients of Eu and Sr for plagioclase–liquid and clinopyroxene–liquid equilibria in oceanic ridge basalt: an experimental study. *Geochim. Cosmochim. Acta* **38**, 1415–1433.

SUN, S. S. and HANSON, G. N. (1975) Evolution of the mantle: geochemical evidence from alkali basalt. *Geology* **3**, 297–302.

SUN, S. S. and NESBITT, R. W. (1977a) Chemical heterogeneity of the Archean mantle, composition of the bulk earth and mantle evolution. *Earth Planet. Sci. Lett.* (in press).

SUN, S. S. and NESBITT, R. W. (1977b) Petrogenesis of Archean ultrabasic and basic volcanics: Evidence from rare earth elements. *Contr. Mineral. Petrol.* (in press).

TARNEY, J., DALZIEL, I. W. D. and de WIT, M. J. (1976) Marginal basin "Rocas Verdes" complex from S. Chile: a model for Archean greenstone belt formation. In *The Early History of the Earth,* B. F. WINDLEY, ed., John Wiley & Sons, London, pp. 131–146.

TAYLOR, S. R. (1978) Island arc models and the composition of the continental crust. *Amer. Geophys. Union Monogr.* (Ewing Symp.) (in press).

TAYLOR, S. R. and HALLBERG, J. A. (1977) Rare earth elements in the Marda calc-alkaline suite: an Archean geochemical analogue of Andean-type volcanism. *Geochim. Cosmochim. Acta* (in press).

VILJOEN, M. J. and VILJOEN, R. P. (1969a) The geology and geochemistry of the Lower Ultramafic Unit of the Onverwacht Group and a proposed new class of igneous rock. *Geol. Soc. S. Africa Spec. Publ.* **2**, 55–85.

VILJOEN, M. J. and VILJOEN, R. P. (1969b) Evidence for the existence of a mobile extrusive peridotitic magma from the Komati Formation of the Onverwacht Group. *Geol. Soc. S. Africa Spec. Publ.* **2**, 87–112.

WHITE, A. J. R., JAKES, P. and CHRISTIE, D. M. (1971) Composition of greenstones and the hypothesis of sea-floor spreading in the Archean. *Spec. Publ. Geol. Soc. Aust.* **3**, 47–56.

WILDEMAN, T. R. and HASKIN, L. A. (1973) Rare earths in Precambrian sediments. *Geochim. Cosmochim. Acta* **37**, 419–438.

WILDEMAN, T. R. and CONDIE, K. C. (1973) Rare earths in Archean greywackes from Wyoming and from the Fig Tree Group, South Africa. *Geochim. Cosmochim. Acta* **37**, 439–453.

WINDLEY, B. F. (1973) Crustal development in the Precambrian. *Phil. Trans. Roy. Soc. Lond.* A, **273**, 321–341.

WINDLEY, B. F. (1976) New tectonic models for the evolution of Archean continents and oceans. In *The Early History of the Earth*, B. F. WINDLEY, ed., John Wiley & Sons, London, pp. 105–111.

The Role on Partial Melting of Mantle Diapirism, CO_2 and H_2O from the Study of Lherzolite Nodules of Intracontinental Alkali Basalts: Example of the French Massif Central

E. BERGER

Ecole Nationale Supérieure des Mines, 60, boulevard Saint-Michel, 75272 Paris Cédex 06

Abstract

The origin of ultrabasic xenoliths of intracontinental alkali basalts may be enlightened by the study of several parameters such as major, minor and trace elements, nature of fluid inclusions. Ultrabasic xenoliths may be originated from either depleted or undepleted mantle which in turn may or may not be metasomatized, or by crystallization from magmatic liquids.

Paleogeotherms may be obtained from the temperatures and pressures deduced from lherzolites' ortho and clinopyroxene compositions. Two end-types of pyroxene-geotherms are observed: a lower temperature ("continental") South African type, and an upper temperature ("oceanic") French Massif Central type.

In the case of the French Massif Central mantle diapirism may be inferred from available paleogeotherms. This interpretation is in good agreement with geophysical records and with the distensive crustal evolution of the area during Tertiary times. The lowering of pressure induced by diapirism of upper mantle would favour decarbonization reactions which in turn would facilitate the genesis of alkali basalts. A "hot-oceanic" type paleogeotherm would allow the formation of alkali basalts (Massif Central type) whereas a "continental" paleogeotherm (South African Shield) would favour the production, among other magmas, of kimberlitic or carbonatitic magmas.

Continental alkali basalts frequently contain a great number of ultrabasic xenoliths (Wilshire and Binns, 1961; Forbes and Kuno, 1967; Brousse, R., 1967, etc.). With the exception of some websterites most xenoliths are upper mantle fragments. More precisely, three types are to be distinguished: (1) undepleted mantle; (2) residues of partial melting; (3) directly crystallized products within the upper mantle (i.e. the websterites). These three groups (Mercier, 1976) may have been metasomatized within the mantle or not. In fact the metasomatism is difficult to prove, e.g. the discrimination between primary and secondary phlogopites (Dawson and Smith, 1975) is not satisfactorily solved and the significance of some pargasites remains partially controversial. In some cases indeed ratios between heavy and light REE (Frey, 1974) suggest a metasomatism within the mantle a process that cannot be generalized in the present state of our knowledge. As a matter of fact REE analyses are still scarce and of difficult interpretation. It is henceforth possible to define upper mantle equilibrium conditions, thus appreciating coexisting phases and their stability field. New paths of investigation can now be followed: (1) nature (Roedder, 1965; Chemla *et al.*, 1968) and localization (Green and Radcliffe, 1975) of fluid inclusions; (2) geothermometers and geobarometers (Mercier, 1976); (3) partial melting experiments (Ito and Kennedy, 1967); at various pressures (Kushiro *et al.*, 1972; Green, 1973a, b) both in dry conditions (Holloway and Eggler) and with CO_2 and/or H_2O present (Kushiro, 1968; Mysen, 1973; Nicholls, 1974; Mysen *et al.*, 1975); (4) carbonation–decarbonation experiments (Newton and Sharp, 1975; Brey and Green, 1975, 1976; Wyllie and Huang, 1976; Eggler *et al.*, 1976). On these bases, the conditions of formation of continental alkali basalts may be discussed using as an example the French Massif Central.

The French Massif Central (Symposium Jung, J., 1971) is a large outcrop of the European Variscides. From the Oligocene it was subjected to distensional intra-plate tectonics generating a system of rifts and grabens (mostly N–S oriented and later filled by lacustrine and brackish Tertiary sediments) and a widespread and important volcanic activity (Brousse, 1971) till less than 10,000 years. Three systems of volcano–tectonic alignments (e.g. the "sillon houiller" NNE–SSW; the NW–SE Velay, Coiron and Ardèche alignments; the N–S "chaine des Puys", etc.) are present. The Villafranchian volcanic paroxysm can be correlated with subsequent uplift of Massif Central.

During the last ten years four models have been presented for the Tertiary–Quaternary volcanism of the Massif Central: (1) subduction-related volcanism in connection to Alpine orogeny (Roever, de, 1975); (2) drift of the European plate above a fixed hot-spot (Duncan *et al.*, 1972; Burke and Dewey, 1973; Wilson, 1973); (3) hot-spot, without relative displacement between the European plate and the spot (Froidevaux *et al.*, 1973); (4) upper mantle diapirism facilitating an adiabatic partial melting (Berger, 1977b, c).

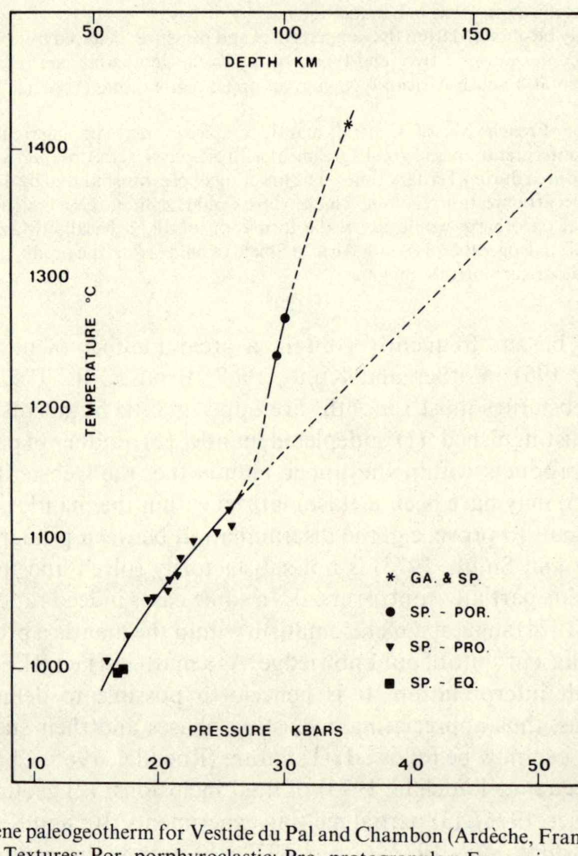

FIG. 1. Inflected pyroxene paleogeotherm for Vestide du Pal and Chambon (Ardèche, France) from Berger (1977b). Ga. garnet; Sp. spinel Textures; Por. porphyroclastic; Pro, protogranular; Eq. equigranular.

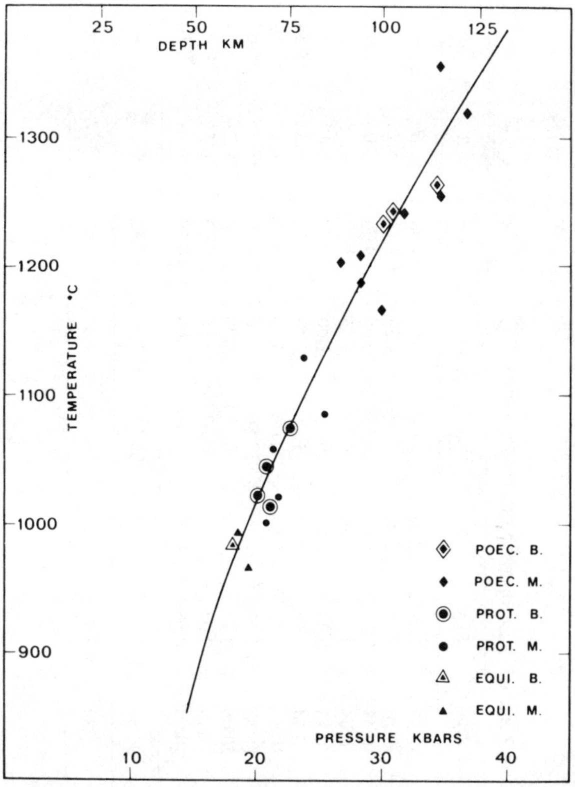

FIG.2. Noninflected pyroxene paleogeotherm for Borée (Ardèche, France) from Mercier (1976) and Berger (1977c). Poec. poikiloblastic; B. Berger; M. Mercier.

The first model, originally suggested in order to explain the local leucite-basanites, cannot be taken into consideration for the following two reasons: the leucite-basanites are chemically similar to basic alkaline rocks associated to intracontinental distensional areas (Rhinegraben, East African Rift, etc.) and have no similarities with deep subduction-related and weakly potassic basalts (shoshonites).

The second model, although being in agreement with numerous and "up-to-date" models, cannot explain the time and space relationships of the Massif Central volcanic activity.

The third one can be accepted only if neglecting some volcanic series or imagining a rather complex mechanism such as a triple plate junction a branch of which would be subsequently broken into two parts (Brousse, 1974).

Finally, the fourth model accounts for: (1) the deformations observed in the peridotite xenoliths, (2) their equilibrium conditions, and (3) the chemical composition of the parental basalts. In agreement with geophysical data (Perrier and Ruegg, 1973) is also the inflexion of the paleogeotherm curve at least in one case (Berger, 1977b: fig. 1) and possible in a second one the study of which is in progress.

The paleogeotherm curves here presented (Mercier, 1976; Berger, 1977b, c: Figs. 1 and 2) have been determined in considering the spinel–lherzolites (the most frequent rocks among

Table 1. Microprobe Analyses of Some Typical Lherzolites Nodules Pyroxenes

	1d	1e	2d	2e	3d	3e	4d	4e	5d	5e	6d	6e	7d	7e
SiO_2	52.61	55.09	52.51	54.74	52.74	55.28	52.32	55.49	52.63	54.98	52.86	54.96	52.69	53.99
Al_2O_3	5.46	3.79	5.08	3.74	5.54	3.89	5.26	3.98	5.39	4.10	4.33	4.11	5.74	5.19
TiO_2	0.39	0.07	0.41	0.10	0.51	0.07	0.48	0.32	0.49	0.18	0.05	0.06	0.07	0.02
Cr_2O_3	0.90	0.25	0.53	0.25	1.05	0.27	0.96	0.22	1.07	0.27	1.35	1.23	0.98	0.76
Fe_2O_3	0.34	1.14	0.76	1.41	0.00	0.78	0.33	0.33	0.06	0.85	1.00	0.00	0.17	1.21
FeO	1.93	5.74	2.91	6.44	2.65	5.66	2.54	5.67	2.62	5.51	2.47	5.68	3.37	4.84
MnO	0.06	0.13	0.12	0.18	0.01	0.17	0.07	0.12	0.03	0.06	0.10	0.12	0.11	0.14
MgO	15.35	33.11	15.18	32.31	15.57	33.08	16.38	33.23	16.09	32.86	19.00	31.59	18.75	31.67
CaO	21.57	0.39	21.29	0.55	20.21	0.57	20.06	0.79	20.25	0.79	18.70	1.88	18.21	1.94
Na_2O	1.35	0.03	1.21	0.05	1.49	0.05	1.18	0.04	1.29	0.07	0.56	0.09	0.56	0.05
NiO	0.02	0.06	0.03	0.06	0.04	0.06	0.03	0.08	0.04	0.08	0.06	0.10	0.04	0.10
Σ	99.98	99.80	100.03	99.83	99.81	99.88	99.61	100.27	99.96	99.75	100.48	99.82	100.69	99.91
FeO*	2.23	6.77	3.60	7.71	2.65	6.36	2.84	5.97	2.67	6.28	3.37	5.68	3.52	5.92

Cations to 6 oxygens

	1d	1e	2d	2e	3d	3e	4d	4e	5d	5e	6d	6e	7d	7e
Si	1.901	1.904	1.905	1.901	1.907	1.907	1.896	1.905	1.900	1.900	1.894	1.905	1.881	1.869
Al^{IV}	0.099	0.096	0.095	0.099	0.093	0.093	0.104	0.095	0.100	0.100	0.106	0.095	0.119	0.131
Al^{VI}	0.133	0.058	0.122	0.054	0.143	0.065	0.120	0.066	0.129	0.067	0.077	0.073	0.122	0.080
Ti	0.011	0.002	0.011	0.003	0.014	0.002	0.014	0.008	0.013	0.005	0.001	0.002	0.002	0.001
Cr	0.026	0.007	0.015	0.007	0.030	0.007	0.027	0.006	0.031	0.007	0.038	0.034	0.028	0.021
Fe^{3+}	0.009	0.030	0.021	0.037	0.000	0.020	0.008	0.009	0.002	0.022	0.027	0.000	0.005	0.031
Fe^{2+}	0.058	0.165	0.088	0.186	0.080	0.163	0.076	0.162	0.079	0.159	0.074	0.164	0.100	0.140
Mn	0.002	0.004	0.004	0.005	0.000	0.005	0.002	0.003	0.001	0.002	0.003	0.004	0.003	0.004
Mg	0.832	1.716	0.826	1.683	0.844	1.712	0.890	1.712	0.871	1.703	1.021	1.642	1.004	1.645
Ca	0.835	0.014	0.828	0.020	0.784	0.021	0.779	0.029	0.783	0.029	0.718	0.070	0.697	0.072
Na	0.094	0.002	0.085	0.003	0.104	0.003	0.083	0.003	0.090	0.005	0.039	0.006	0.039	0.003
Ni	0.001	0.002	0.001	0.002	0.001	0.002	0.001	0.002	0.001	0.002	0.002	0.003	0.001	0.003
Σ	4.001	4.000	4.001	4.000	4.000	4.000	4.000	4.000	4.000	4.001	4.000	3.998	4.001	4.000
Wo	48.35	0.74	47.53	1.06	45.90	1.11	44.64	1.52	45.18	1.53	39.60	3.73	38.70	3.88
En	48.18	90.55	47.42	89.09	49.42	90.29	51.00	89.97	50.26	90.06	56.32	87.53	55.75	88.58
Fs	3.47	8.71	5.05	9.85	4.68	8.60	4.36	8.51	4.56	8.41	4.08	8.74	5.55	7.54
$\frac{100Mg}{Mg+Fe^{2+}}$	93.48	91.22	90.37	90.05	91.34	91.30	92.13	91.35	91.68	91.46	93.24	90.92	90.94	92.15

Samples origins: 1, 3, 5, 7—Vestide du Pal maar (Ardèche, France); 2, 4, 6 Borée maar (Ardèche, France).
Structures: 1, 2, equigranular—3, 4, protogranular—5, 7, porphyroclastic (blastomylonitic)—6, poikiloblastic.
d, clinopyroxene; e, orthopyroxene—FeO, total Fe as FeO–Fe^{2+} and Fe^{3+} are calculated from stoichiometric formules.
Analysts: E. Berger, G. Frot, M. C. Forette. (G.G.M., E.N.S.M.P.).

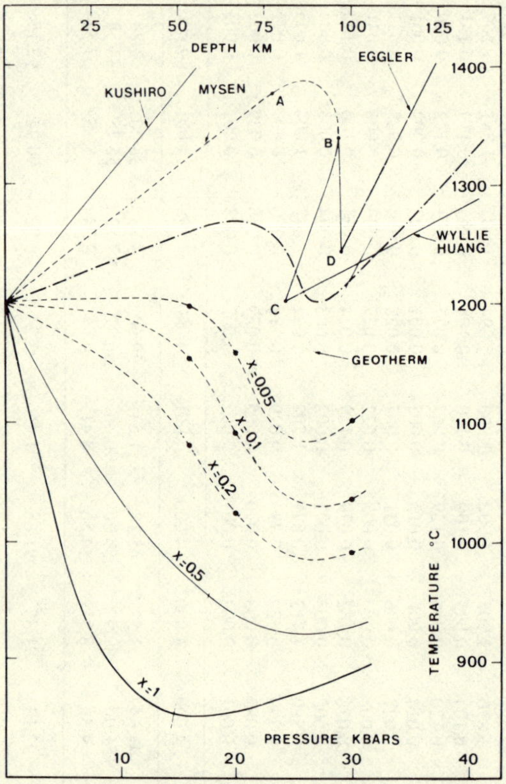

FIG. 3. Lherzolite solidus curves with H_2O and CO_2 present. Dry solidus (solid line) from Kushiro (1968). Solidi, CO_2 present only—at pressures from 0 to 25 kbars (dashed line)—and H_2O present only are from Mysen and Boettcher (1975). $X = H_2O/(CO_2 + H_2O)$. BC and BD segments are interpolated from data for the CaO–MgO–SiO_2–CO_2–H_2O system (Eggler, 1976; Wyllie and Huang, 1976). Dot–dashed line: Massif Central hypothetical lherzolite solidus (Berger, 1977c).

the xenoliths), and some scarce spinel–garnet lherzolites (Albert *et al.*, 1967; Mercier, 1976; Berger and Brousse, 1976; Berger 1977a). Some characteristic microprobe analyses are given in Tables 1 and 2. Three different geothermometers (Mysen, 1973; Wood and Banno, 1973; Mercier, 1976) gave concordant results. Pressures were calculated from the Mercier's single-pyroxene geobarometer (1976). For all volcanic areas of the Massif Central paleogeotherms of "hot-oceanic type" are obtained. Such paleogeotherms disagree with a continental-type paleogeotherm which occurs in the South African Shield.

Our fourth model which has been recently developed in the light of a possible role of both CO_2 and H_2O (Berger, 1977c) will now be discussed.

During the last ten years experiments on partial melting (particularly by Carnegie and Canberra groups) were carried out showing the important roles of CO_2 and/or H_2O besides these of temperature and pressure. The lowering of solidus by H_2O is evident from Fig. 3. As for CO_2 is concerned it has very little effect below 25 kb whereas between 25 and 30 kb it dramatically lowers the solidus curve.

In the light of the above considerations, the knowledge of CO_2 and H_2O contents in peridotites is of utmost importance.

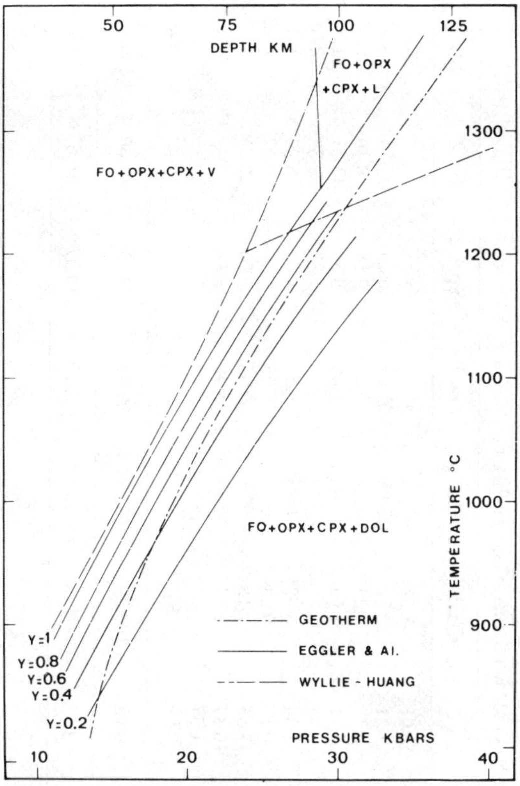

FIG. 4. Hypothetical intersection of French Massif Central geotherm and carbonation–decarbonation curves from Eggler *et al.* (1976) for the system $CaO–MgO–SiO_2–CO_2–H_2O$; $Y = CO_2/(CO_2 + H_2O)$.

The basalts of the Massif Central are of alkaline-type (Velde, 1967; Brousse, 1971); consequently their origin could not have occurred with water present only, otherwise this melting would have led to the genesis of silica-oversaturated magmas (Kushiro, 1968). Fluid inclusions containing H_2O and CO_2 are actually known in minerals from the Massif Central lherzolites (Chemla *et al.*, 1968), especially in xenoliths from the Ardèche area. It must be noted that CO_2 is the main phase in these inclusions (the CO_2/H_2O ratio being always superior to 5).

Green and Radcliffe (1975) studied the distribution of fluid inclusions in some xenoliths comparable in our opinion to those from the Massif Central. They have shown that fluid inclusions are located on crystal defects induced by a deformation which is subsequent to the first plastic deformations of the rock and prior to the incorporation of the xenoliths into the magma. Wyllie and Huang (1976) have shown that in the system $CaO–MgO–SiO_2–CO_2$ a pressure decrease or a temperature increase may induce a decarbonation reaction. As shown in Fig. 4, the carbonation–decarbonation curve may be either very close to the French Massif Central paleogeotherm (when using Wyllie and Huang results) or can intersect it (using results of Eggler *et al.*, 1976). We have observed in a previous paper (Berger, 1977c) that the $CO_2–H_2O$ present solidus intersects the paleogeotherm near by 1200°C and 28 kb (using results of Mysen and Boettcher, 1975).

TABLE 2. MICROPROBE ANALYSES OF TWO SPINEL–GARNET PHASES

	Ol.1	Ol.2	En.1	En.2	Di.1	Di.2	Ga.1	Ga.2	Ga.2	Sp.1	Sp.2	Sp.2
SiO_2	40.74	40.99	53.87	54.27	51.39	52.01	41.52	41.23	41.45	—	—	—
Al_2O_3	0.10	0.07	6.26	6.20	7.41	7.31	23.31	22.98	23.04	56.73	55.94	56.37
TiO_2	tr.	tr.	0.11	0.13	0.31	0.28	0.25	0.22	0.22	0.18	0.16	0.13
Cr_2O_3	0.05	tr.	0.57	0.52	0.97	0.71	1.65	1.46	1.46	10.84	11.31	11.53
Fe_2O_3	0.00	0.00	0.72	0.00	0.98	0.37	—	—	—	1.97	2.39	2.38
FeO	9.05	9.31	5.37	5.47	2.99	3.38	6.98*	6.85*	6.53*	8.60	8.43	8.21
MnO	0.14	0.12	0.07	0.13	0.08	0.11	0.28	0.26	0.26	0.11	0.12	0.12
MgO	49.03	49.56	31.63	31.10	17.53	17.56	21.16	20.78	20.65	20.76	20.86	21.12
CaO	0.14	0.13	1.61	1.51	17.72	16.84	5.14	5.39	5.40	—	—	—
Na_2O	0.00	0.00	0.08	0.15	1.02	1.37	—	—	—	—	—	—
NiO	0.32	0.35	0.10	0.10	—	0.04	—	—	—	0.31	0.29	0.29
Σ	99.57	100.53	100.39	99.57	100.30	99.98	100.29	99.17	99.17	99.50	99.60	100.15
FeO*	9.05	9.31	6.02	5.47	3.87	3.72	6.98	6.85	6.53	10.37	10.67	10.35

Numbers of ions on the basis of n oxygens (n)

	Ol.1	Ol.2	En.1	En.2	Di.1	Di.2	Ga.1	Ga.2	Ga.2	Sp.1	Sp.2	Sp.2
n	4	4	6	6	6	6	24	24	24	32	32	32
Si	0.998	0.995	1.854	1.878	1.841	1.865	5.868	5.892	5.920	—	—	—
Al^{IV}	0.002	0.002	0.146	0.122	0.159	0.135	—	—	—	—	—	—
Al^{VI}	0.001	0.000	0.108	0.130	0.154	0.173	3.876	3.864	3.871	13.861	13.692	13.704
Ti	0.000	0.000	0.003	0.004	0.008	0.007	0.026	0.023	0.023	0.028	0.024	0.020
Cr	0.001	0.000	0.015	0.014	0.027	0.020	0.184	0.164	0.164	1.777	1.857	1.881
Fe^{3+}	0.000	0.000	0.018	0.000	0.027	0.009	—	—	—	0.307	0.387	0.368
Fe^{2+}	0.185	0.188	0.154	0.158	0.089	0.101	0.822*	0.815*	0.777*	1.488	1.461	1.413
Mn	0.003	0.002	0.002	0.004	0.002	0.003	0.033	0.031	0.031	0.019	0.021	0.020
Mg	1.801	1.805	1.633	1.614	0.943	0.944	4.486	4.455	4.423	6.467	6.510	6.546
Ca	0.004	0.003	0.059	0.056	0.680	0.647	0.778	0.825	0.826	—	—	—
Na	0.000	0.000	0.005	0.010	0.071	0.095	—	—	—	—	—	—
Ni	0.006	0.006	0.003	0.003	—	0.001	—	—	—	0.052	0.048	0.048
Σ	3.001	3.001	4.000	3.993	4.001	4.000	16.073	16.069	16.035	23.999	24.000	24.000
$\frac{100Mg}{Mg+Fe^{2+}}$	90.68	90.57	91.08	91.08	91.37	90.33	84.51	84.54	85.05	81.29	77.89	78.61

1—Spinel–garnet lherzolite from Eglazines pipe (Causses, France): in E. Berger and R. Brousse (1976); En.1: new analyse.
2—Spinel–garnet lherzolite from Vestide du Pal maar (Ardèche, France): in E. Berger (1977a).
Ol., olivine; En., orthopyroxene; Di.: clinopyroxene; Ga., garnet; Sp., spinel.
FeO*, total Fe as FeO–Fe^{2+} and Fe^{3+} are calculated from stoichiometric formules.
Analysts: E. Berger, G. Frot, M. C. Forette (G.G.M., E.N.S.M.P.).

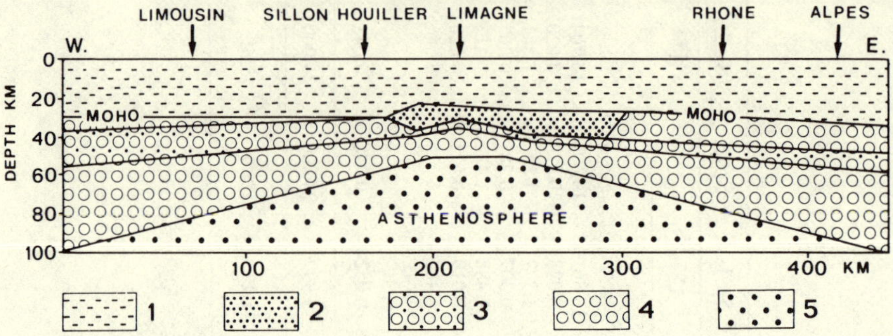

FIG. 5. Cross-section of French Massif Central from geophysical data (Perrier and Ruegg, 1973). 1—Continental crust; 2—"Anomalous Mantle" (P waves speed: Vp = 7.2 to 7.4); 3—Lesser velocity section (Vp = 7.7 to 8.2); 4—"Normal" mantle (Vp = 8.2 to 8.4); 5—Asthenosphere.

The lowering of pressure induced by diapirism of the upper mantle would favour decarbonation reactions which in turn would facilitate the genesis of alkali basalts. The interpretation we suggest is consistent with both geophysical (Fig. 5, where can be seen the local dome structure of the asthenosphere) and structural data (distensional tectonics and graben generation since the Oligocene; general uplift during Villafranchian time).

In our opinion it is significant that our whole interpretation is based on the up to now unproved hypothesis that spinel-lherzolites are the source of basic magmas. With such a reservation in mind, the nature of the geotherm of a given area may help to understand the different kinds of parental magmas locally present. A "hot-oceanic" type paleogeotherm would allow the formation of alkali basalts (Massif Central type) whereas a "continental" paleogeotherm (South African Shield) would favour the production, among other magmas, of kimberlitic or carbonatitic magmas (Koster van Groos, 1975; Wyllie and Huang, 1976). Further studies on different areas are necessary to test the soundness of these hypotheses.

References

ALBERT, D., ALBERT, R. and BROUSSE, R. (1967) Enclaves de péridotites à pyrope chromifère dans les pipes de la région de Bédarieux (Hérault). *C.R. Acad. Sci.* D, **265**, 657–660.

BERGER, È. and BROUSSE, R. (1976) Une lherzolite à grenats du pipe d'Eglazines (Lozère, France) stabilisée à 1400°C et 30 kbars. *C.R. Acad. Sci.* D, **282**, 1477–1480.

BERGER, E. (1977a) Sur la présence d'une lherzolite à grenats en enclave dans le basalte alcalin de la Vestide du Pal (Ardèche): Conditions d'équilibre, implications pétrogénétiques et géotectoniques. *C.R. Acad. Sci.* D, **284**, 709–712.

BERGER, E. (1977b) Le diapirisme du manteau supérieur générateur du volcanisme du Massif Central français. *C. R. Acad. Sci.*, D, **284**, 791–794.

BERGER, E. (1977c) Rôle de CO_2 et H_2O dans la genèse des basaltes alcalins du Massif Central français. *C. R. Acad. Sci.* D, **284**, 1859–1862.

BOYD, F. R. (1976) Inflected and noninflected geotherms. *Carn. Inst. Wash. Year Book* **75**, 521–523.

BREY, G. and GREEN, D. H. (1975) The role of CO_2 in the genesis of olivine melilite. *Contr. Mineral. Petrol.* **49**, 93–103.

BREY, G. and GREEN, D. H. (1976) Solubility of CO_2 in olivine melilitite at high pressures and role of CO_2 in the Earth's upper mantle. *Contr. Mineral. Petrol.* **55**, 217–230.

BROUSSE, R. (1967) La place des ultrabasites en France. *Geol. Runds.* **57**, 621–655.

BROUSSE, R. (1971) Magmatologie du volcanisme néogène et quaternaire du Massif Central. In *Symposium J. Jung*, Plein Air Service Edit., Clermont-Ferrand, 337–478.

BROUSSE, R. (1974) Le volcanisme en France. In *Volcanisme et phénomènes associés, Revue Haute-Auvergne* **44**, 221–242.

BURKE, K. and DEWEY, J. F. (1973) Plume generated triple junctions: Key indicators in applying plate tectonics to old rocks. *J. Geol.* **81**, 406–433.

CHEMLA, M., BROUSSE, R., TOURAY, J. C. and CAUSSE, C. (1968) Abondance relative des gaz inclus dans quelques minéraux magmatiques d'origine profonde. In *Origin and Distribution of the Elements*, L. H. AHRENS, ed., Pergamon Press, 971–979.

DAWSON, J. B. and SMITH, J. V. (1975) Chemistry and origin of phlogopite megacrysts in kimberlite. *Nature* **253**, 5490, 336–338.

DUNCAN, R. A. PETERSON, N. and HARGRAVES, R. B. (1972) Mantle plumes, movement of the European plate, and polar wandering. *Nature* **239**, 82–86.

EGGLER, D. H. (1976) Does CO_2 cause partial melting in the low-velocity layer of the mantle? *Geology* **4**, 2, 69–72.

EGGLER, D. H., KUSHIRO, I. and HOLLOWAY, J. R. (1976) Stability of carbonate minerals in a hydrous mantle. *Carn. Inst. Wash. Year Book* **75**, 631–636.

FORBES, R. B. and KUNO, H. (1967) Peridotite inclusions and basaltic host rocks. In *Ultramafic and Related Rocks*, P. J. WYLLIE, ed., J. Wiley & Sons, Inc., 328–337.

FREY, F. A. (1974) The mineralogy, geochemistry and origin of lherzolite inclusions in Victorian basanites. *Geochim. Cosmochim. Acta* **38**, 1023–1059.

FROIDEVAUX, C., BROUSSE, R. and BELLON, H. (1973) Hot spot in France? *Nature* **248**, 5451, 749–751.

GREEN, D. H. (1973a) Conditions of melting of basanite magma from garnet peridotite. *E.P.S.L.* **17**, 2, 456–465.

GREEN, D. H. (1973b) Experimental melting studies on a model upper mantle composition at high pressure under water-saturated and water-undersaturated conditions. *E.P.S.L.* **19**, 1, 37–53.

GREEN, H. W. and GUEGUEN, Y. (1974) Origin of kimberlite pipe by diapiric upwelling in the mantle. *Nature* **249**, 5458, 617–620.

GREEN, H. W. and RADCLIFFE, S. V. (1975) Fluid precipitates in rocks from the Earth's mantle. *Geol. Soc. Amer. Bull.* **86**, 846–852.

HOLLOWAY, J. R. and EGGLER, D. H. (1976) Fluid-absent melting of peridotite containing phlogopite and dolomite. *Carn. Inst. Wash. Year Book* **75**, 636–639.

IRVING, A. T. (1976) On the validity of paleogeotherms determined from xenolith suites in basalts and kimberlites. *Amer. Miner.* **61**, 638–642.

ITO, K. and KENNEDY, G. C. (1967) Melting and phase relations in a natural peridotite to 40 kilobars. *Amer. J. Sci.* **265**, 519–538.

KOSTER VAN GROOS, A. F. (1975) The effect of high CO_2 pressures on alkalic rock and its bearing on the formation of alkalic ultrabasic rocks and the associated carbonatites. *Amer. J. Sci.* **275**, 163–185.

KUSHIRO, I. (1968) Melting of a peridotite nodule at high pressures and high water pressures. *J. Geoph. Res.* **73**, 18, 6023–6029.

KUSHIRO, I., SHIMIZU, N., NAKAMURA, Y. and AKIMOTO, S. (1972) Compositions of coexisting liquid and solid phases formed upon melting of natural garnet and spinel lherzolites at high pressures: A preliminary report. *E.P.S.L.* **14**, 19–25.

MERCIER, J. C. (1976) Single-pyroxene geothermometry and geobarometry. *Amer. Miner.* **61**, 7–8, 603–615.

MERCIER, J. C. (1977) Hétérogénéïté chimique et rhéologique du manteau supérieur. Thèse, Nantes (in the press).

MERCIER, J. C. (1977) Criteria for a petrochemical and genetic classification of peridotites (in the press).

MYSEN, B. (1973) Melting in a hydrous mantle: phase relations of mantle peridotite with controlled water and oxygen fugacities. *Carn. Inst. Wash. Year Book*, **72**, 467–478.

MYSEN, B. and BOETTCHER, A. L. (1975) Melting of a hydrous mantle: I. Phase relations of natural peridotite at high pressure with controlled activities of water, carbon dioxide and hydrogen. *J. Petrol.* **16**, 520–548.

NEWTON, R. C. and SHARP, W. E. (1975) Stability of forsterite + CO_2 and its bearing on the role of CO_2 in the mantle. *E.P.S.L.* **26**, 239–244.

NICHOLLS, I. A. (1974) Liquids in equilibrium with peridotite mineral assemblages at high water pressures. *Contr. Mineral. Petrol.* **45**, 289–316.

PERRIER, G., RUEGG, J. C. (1973) Structure profonde du Massif Central français. *An. Geophys.* **29**, 4, 435–502.

ROEDDER, E. (1965) Liquid CO_2 inclusions in olivine-bearing nodules and phenocrysts from basalts. *Amer. Miner.* **50**, 1746–1782.

ROEVER DE, W. P. (1975) Belts with scattered leucite occurrences indicate direction of dip of coeval subduction zones. *Pétrologie* **I**, 1, 37–42.

VELDE, D. (1967) Sur le caractère alcalin des basaltes tertiaires et quaternaires de France. *C.R. Acad. Sci.* **264**, 1141–1144.

WILSHIRE, H. G. and BINNS, R. A. (1961) Basic and ultrabasic xenoliths from volcanic rocks of New South Wales. *J. Petrol.* **2**, 2, 185–208.

WILSON, J. T. (1973) Mantle plumes and plate motions. *Tectonophysics* **19**, 149–164.

WOOD, B. J. and BANNO, S. (1973) Garnet–orthopyroxene and orthopyroxene–clinopyroxene relationships in simple and complex systems. *Contr. Miner. Petrol.* **42**, 109–124.

WYLLIE, P. J. and HUANG, W. L. (1976) Carbonation and melting reactions in the system CaO–MgO–SiO_2–CO_2 at mantle pressures with geophysical and petrological applications. *Contr. Mineral. Petrol.* **54**, 79–107.

Inclusions in Diamond and the Mineral Chemistry of the Upper Mantle

HENRY O. A. MEYER and HSIAO-MING TSAI

Department of Geosciences, Purdue University, West Lafayette, Indiana 47907, U.S.A.

Abstract

During the last decade significant progress regarding our understanding of the mineralogy and geochemistry of the earth's upper mantle has been achieved through studies of the ultramafic and eclogitic xenoliths in kimberlite, and of mineral inclusions in diamond.

The ultramafic and eclogitic xenoliths in kimberlite represent fragments of various horizons of the mantle that have been rapidly transported to the surface by the mechanism of kimberlite eruption. Textural differences in these mineralogically simple ultramafic xenoliths (lherzolite, harzburgite, pyroxenite) are in part reflected in chemical differences between the constituent minerals. Thus many xenoliths which display a "fluidal" or "mosaic" texture indicative of intense deformation are much less depleted in Ti, Al, Ca, Fe and Na than xenoliths in which granular or coarse tabular textures predominate.

Interestingly, there is generally a contrast in chemistry between all the mineral constituents of the xenoliths and similar ones that occur as inclusions in diamonds. Typically, inclusions in diamond are monominerallic and homogeneous. Olivine and orthopyroxene inclusions have very small compositional differences in spite of the host diamonds being markedly different in geologic age and geographic locality. Characteristically, two distinct, but always separate, suites of minerals occur as inclusions. One suite resembles the minerals that are present in ultramafic xenoliths, whereas the second is similar to minerals in eclogitic xenoliths. Thus it is likely that natural diamond can form in at least two distinct chemical environments in the mantle.

Much of the above data has been combined with the results of experimental phase equilibria and theoretical studies with the conclusion that it is now possible to suggest conditions of pressure and temperature for the genesis of these xenoliths as well as diamond. All the pressure and temperature values are consistent with final equilibration at depths of the order of 150–200 km in the mantle.

Introduction

A minor revolution has occurred during the last 10 or so years in our knowledge and understanding of the earth's upper mantle. This revolution perhaps best considered as a major increase in endeavor has been particularly noticeable in the fields of mineralogy, petrology and geochemistry. Undoubtedly the major stimulus for this increase in mantle research has been the realization that many of the xenoliths that occur in kimberlite and basalt are of mantle origin. Detailed examination of these xenoliths thus provides information that has direct bearing on the mineralogy and chemistry of the upper mantle. Another source of information is provided by the mineral inclusions in natural diamond. These inclusions have been locked within the unreactive diamond host since their inception and thus represent pristine mantle material.

Prior to 1967 no analyses of mineral inclusions in diamond were available and relatively few papers had been published on the mineralogy and chemistry of the mantle xenoliths in kimberlite and basalt. The marked change in research effort in the mid-sixties with regard to mantle mineralogy can be attributed in part to the development of the electron microprobe for routine mineral analysis. The availability of the microprobe and its capability to analyze *in situ* grains opened up to many workers exciting fields of research that were hitherto closed. The many thousands of mineral analyses now available on mantle xenoliths are almost entirely the product of microprobe study.

In spite of the increase in knowledge our understanding of mantle chemistry and processes is still in part speculative. It is likely this state of affairs will remain for some time because of the impossibility of direct *in situ* observation. Furthermore, it is possible that what information we have is biased since we are at the whim of nature to supply us with samples. The samples we have are from localized occurrences in basalts and kimberlites and thus represent a meager sampling of the upper mantle. Nevertheless, the importance of these mantle fragments must not be underestimated. Studies of these rocks and associated minerals during the last decade have resulted in petrologic models for the mantle which in turn, although constraining geophysical interpretations, must also not be contrary to known physical laws.

The purpose of this review is to present the results of investigations into the mineralogy and petrology of the upper mantle as evidenced by the xenoliths in kimberlite and associated mineral inclusions in diamond. It should be remembered, however, that the interpretation of such data is often only accomplished with reference to the results of experimental investigations upon systems of comparable, albeit simplified, mineralogy and chemistry.

Occurrence of Mantle Xenoliths and Diamond

Xenoliths, consisting of ultramafic and eclogitic rocks, often occur in basalts (Jackson and Wright, 1970; Wilshire and Shervais, 1975; Aoki, 1968) and in kimberlite (Williams, 1932; O'Hara and Mercy, 1963; Boyd and Nixon, 1975). Kimberlite is also the primary source rock for natural diamond; thus the mineral inclusions in diamond and the mineral constituents of the xenoliths are spatially, although not necessarily genetically, associated. It is therefore our intention to devote this review to the studies of mantle rocks from kimberlite, and to the mineral inclusions in diamond.

The occurrence of xenoliths in kimberlite has been known since the discovery of kimberlite itself (Lewis, 1887). Prior to the last decade most of the studies were confined to xenoliths from Southern Africa (Wagner, 1914; Williams, 1932). However, the discovery of kimberlites in Siberia as well as interest and discovery of new kimberlites in North America (Brookins, 1970; McCallum and Eggler, 1971; Meyer, 1976) has considerably extended the geographic coverage.

In the first instance many authors (e.g. Williams, 1932) referred to the xenoliths of ultramafic and ultrabasic rocks as cognate inclusions, believing them to be genetically related to kimberlite. The term cognate is no longer used although the problem of a genetic link between kimberlite and the xenoliths has never been adequately solved. Undoubtedly, knowledge of the genesis of kimberlite would help this problem. Several authors (Wyllie, 1967; Dawson, 1971; Mitchell, 1970) have discussed the origin of kimberlite but no unanimous opinion has been formulated. Nevertheless, it is generally accepted that the mechanism of emplacement of kimberlite in the crust is one of fluidization and final explosive activity (Holmes, 1965; Dawson, 1971; Clement, 1975; Hawthorne, 1975). It is also recognized that kimberlite may have ascended very rapidly from the upper mantle

(McGetchin and Ullrich, 1973) and undoubtedly this rapid ascent is the main reason for the preservation of diamond and the relatively unaltered state of the xenoliths.

Mineralogical Nature of the Xenoliths

The xenoliths are relatively simple in mineralogy, having for the most part less than four major phases. These phases are usually olivine, clinopyroxene, orthopyroxene and garnet. To a lesser extent spinel, ilmenite and mica may be present. However, there is an abundance of petrographic and mineralogical varieties among the xenoliths. For example, garnet lherzolite, lherzolite, websterite, harzburgite may all be present in a single kimberlite pipe, or one may find complex ilmenite–pyroxene intergrowths or large single crystals up to 17 cm in size of one of the above minerals (xenocryst). Also occurring are xenoliths of eclogite (garnet and omphacitic pyroxene) but interestingly in South Africa these eclogite (or griquaite) xenoliths, on average, are much less common than the ultramafic ones.

Texturally, the xenoliths are coarse grained and for the most part are unaltered, except near their outer margins. With regard to the ultramafic examples three distinct petrographic types have been recognized.

1. Large, single-phase xenoliths, consisting either of one crystal (xenocryst, megacryst) or an aggregate of several crystals (discrete nodule, Boyd, 1973).
2. Xenoliths with coarse granular or coarse tabular textures with polyphase assemblages (Boullier and Nicolas, 1975; Boyd, 1973).
3. Polyphase xenoliths whose texture is indicative of intense deformation or recrystallization. These have been referred to as "sheared" (Boyd, 1973) or mosaic and porphyroclastic (Boullier and Nicolas, 1975; Harte, 1977).

Apart from some work by MacGregor and Carter (1970) little attempt has been made to characterize the textural relationships of the eclogite xenoliths.

Although the mineralogy of the granular and mosaic (sheared) xenoliths is similar there are in detail differences in mineral chemistry and in overall bulk chemistry between the two groups. For example, the granular xenoliths have in general less Ti, Al, Ca, Na and iron compared to the mosaic types. This apparent depletion has led to the terms sterile and fertile (Boyd and McCallister, 1976) respectively for these xenoliths. Ideally a fertile xenolith represents mantle that if partially molten would produce basaltic material. In contrast a sterile xenolith represents mantle that could not form such a magma, and is itself possibly the residuum of an episode of partial melting.

It should be noted that these two textural types represent extremes, perhaps best exemplified in the studies of Boyd (1973) and Nixon and Boyd (1973 a, b) on the xenoliths from Lesotho. Recent work has shown that gradation between these extremes is common. Furthermore, mosaic or sheared xenoliths from some localities (e.g. Thaba Putsoa, Premier) indicate consistently higher temperatures of equilibration than those from other kimberlites (e.g. Matsoku, Bultfontein).

Significant and characteristic differences in mineral chemistry occur between the types of ultramafic xenoliths and xenocrysts. For example, the minerals that are present as large xenocrysts or as discrete nodules generally are more Fe-rich than minerals in the mosaic and granular xenoliths. Olivine and orthopyroxene in the xenoliths are magnesia-rich; however, if both minerals occur in the same rock the orthopyroxene is more Mg-rich than the olivine (Fig. 1) (Gurney *et al.*, 1975). Olivine is usually more Fe-rich in the mosaic or sheared

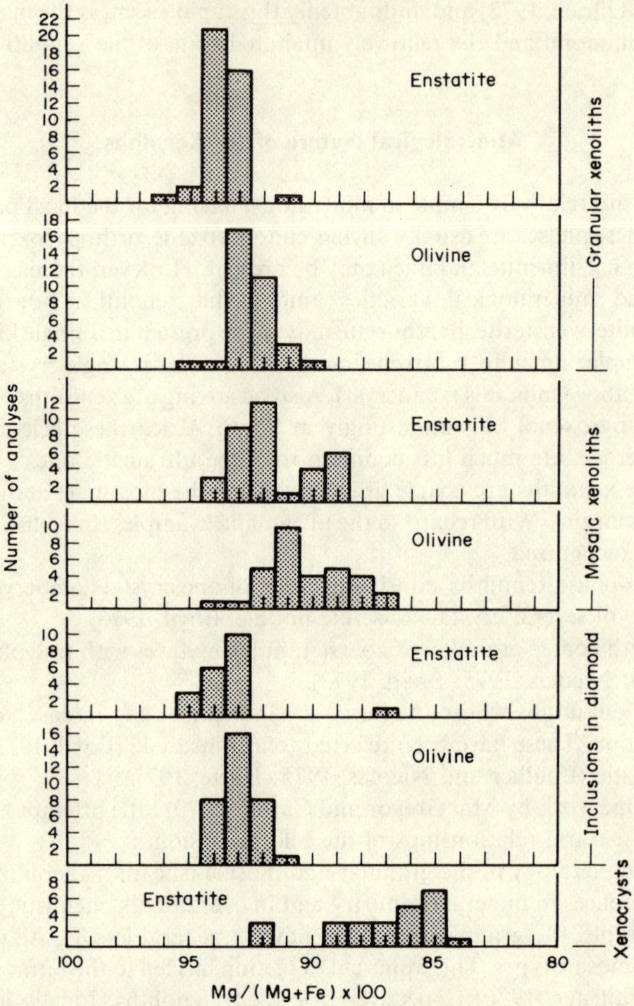

FIG. 1. Forsterite and enstatite (Mg/Mg + Fe) contents of olivines and orthopyroxenes from granular and mosaic xenoliths, xenocrysts (discrete xenoliths) and inclusions in diamond.

xenoliths than in the granular types (Fig. 1). Also the olivine and enstatite in the mosaic xenoliths usually contain more Al_2O_3 and Cr_2O_3 than the same minerals in the granular xenoliths. Furthermore, the orthopyroxenes in the mosaic xenoliths generally show more solid solution towards diopside (CaO) and pyrope (Al_2O_3) than those in the granular xenoliths.

Clinopyroxenes from each group (mosaic, granular, xenocryst) are distinct (Fig. 2). Although the clinopyroxenes from both the mosaic and discrete xenoliths (or xenocrysts) show a similar large range in Ca/Ca + Mg ratios, the xenocryst clinopyroxenes are more Fe-rich than the mosaic ones. This variation in Ca/Ca + Mg indicates a greater range in equilibration temperatures (Davis and Boyd, 1966; Boyd, 1973) than for the clinopyroxenes in the granular xenoliths. These latter clinopyroxenes have a much more restricted range in Ca/Ca + Mg and generally have equilibrated at lower temperatures.

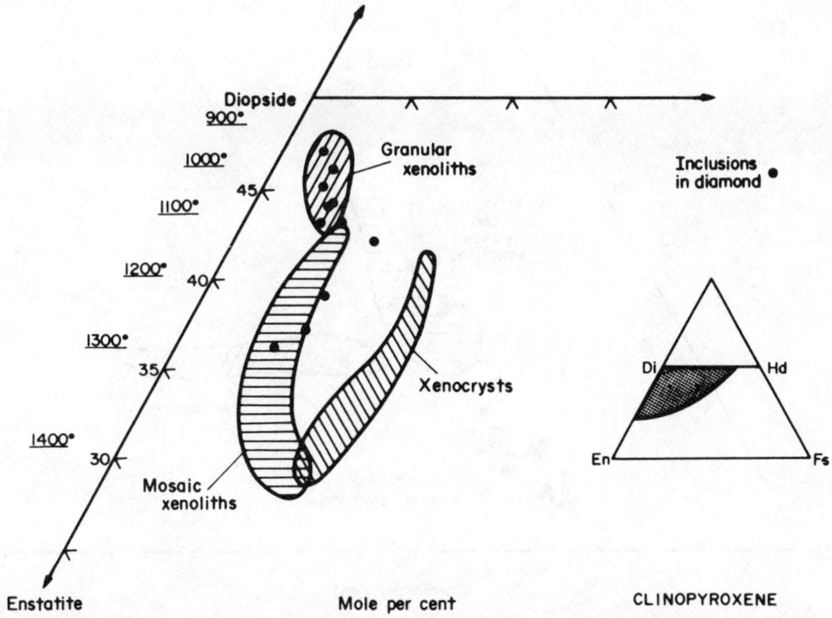

FIG. 2. Clinopyroxenes from mosaic and granular xenoliths, xenocrysts and inclusions in diamond in terms of Ca–Mg–Fe. Temperature scale for Ca/Ca + Mg is from Davis and Boyd (1966). Omphacitic-type inclusions in diamond are not shown.

Garnets from each xenolith type also form distinct groups, although in terms of Ca–Mg–Fe both granular and sheared varieties plot within the same region (Fig. 3). However, other elements (e.g. Ti and Cr) may be used to differentiate the various types. For example, Nixon and Boyd (1973a) noted that garnets in mosaic xenoliths from Lesotho had higher Ti and lower Cr contents than the granular garnets. Interestingly, in the case of garnets from the Frank Smith pipe, South Afric, Boyd (1974) observed the reverse trend with regard to Cr.

Several other minerals are to be found in the xenoliths, e.g. ilmenite, mica, spinel, sulfides. The ilmenite is characterized by having up to 15 wt.% MgO (Fig. 4) and usually occurs as large single crystals or as bimineralic complexes, sometimes as symplectic intergrowths, with clinopyroxene, enstatite or garnet (Ringwood and Lovering, 1970; Dawson and Reid, 1970; Boyd, 1971; Gurney *et al.*, 1973; McCallister *et al.*, 1975). Ilmenite is not a common constituent of the mosaic or granular xenoliths. It has been suggested (Boyd and Nixon, 1975) that slight variations occur in the average MgO contents of ilmenite between different kimberlites. Some of the ilmenites may contain up to 4 wt.% Cr_2O_3 (Haggerty, 1975).

Spinels occurring in the xenoliths are more common in the granular rather than the mosaic types. Overall the spinels contain Fe, Mg, Al and Cr and are thus chromites, or Ti-poor spinels. Smith and Dawson (1975) have reviewed much of the literature on spinel-bearing xenoliths and have shown that $Fe^{2+}/(Fe^{2+}+Mg)$ varies much less than Al/(Al + Cr).

Hydrous phases in the xenoliths are confined to mica and amphibole. Phlogopite mica is a primary phase in many xenoliths (Carswell, 1975; Erlank, 1976; Gurney *et al.*, 1975) although a secondary Ti-rich phlogopite may occur. Smith *et al.* (1975) believe that two subgroups of phlogopite based on Na–K relationships may be present.

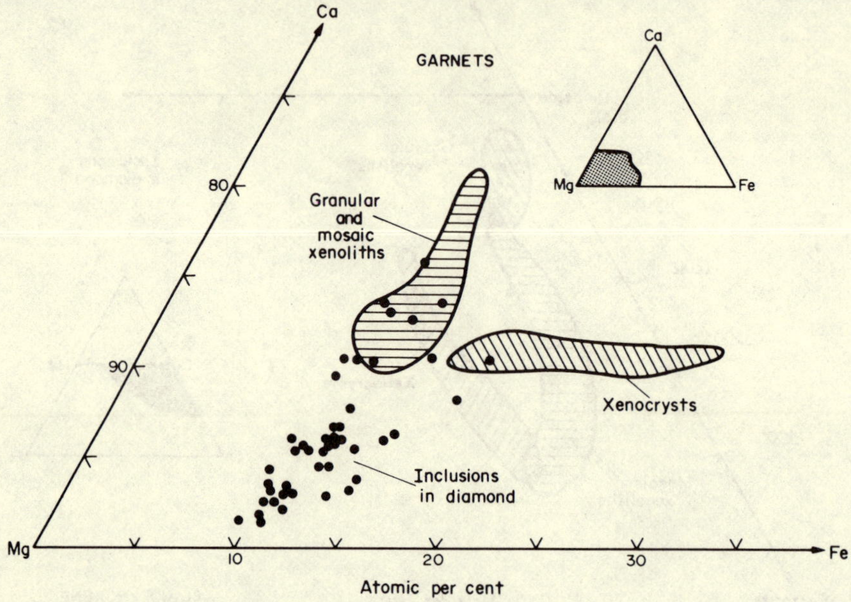

FIG. 3. Garnets from mosaic and granular xenoliths, xenocrysts and inclusions in diamond shown in terms of Ca–Mg–Fe. Note the sub-group of garnet inclusions that fall within or near to the region of ultramafic xenoliths. These inclusions do not include the eclogitic-type garnet inclusions.

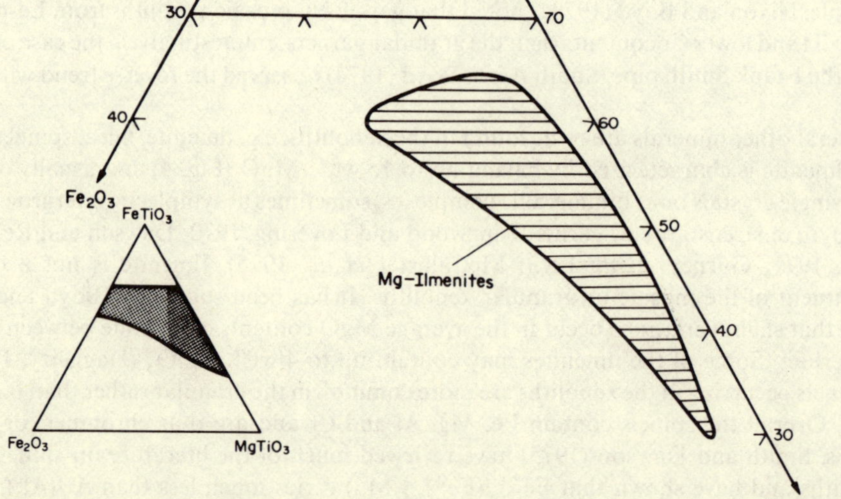

FIG. 4. General region in which kimberlitic and xenolithic ilmenites plot in the system $FeTiO_3$–$MgTiO_3$–Fe_2O_3.

Amphibole is much less common in xenoliths than most other phases and appears to be predominantly a potassic richterite ($KCaNaMg_5Si_8O_{22}(OH,F)_2$) (Erlank, 1976). This amphibole appears to be confined to granular xenoliths. Dawson and Smith (1975) have examined amphibole and mica in "glimmerite nodules" (mica-rich xenoliths) and suggest the amphiboles and other phases are chemically distinct from these occurring in the other xenoliths. It is likely these glimmerite nodules have equilibrated at higher levels in the mantle than the granular or mosaic xenoliths.

Sulfides are not too common in ultramafic xenoliths but when present are usually complex aggregates of pyrrhotite, pentlandite and chalcopyrite. The latter mineral usually occurs as a rim to the complex, whereas the pentlandite has the appearance of having exsolved from the pyrrhotite.

Eclogite Xenoliths

Xenoliths of eclogitic type that consist primarily of pyrope-almandine garnet and jadeitic-rich clinopyroxene (omphacite) are not as common as the ultramafic xenoliths. Nevertheless, it is usual to find such xenoliths as a minor portion of the total xenolith population and in special instances, notably the Roberts Victor kimberlite, South Africa, eclogite is predominant. Originally these clinopyroxene–garnet xenoliths were referred to as griquaite, and this name is once again finding favor to describe these mantle eclogitic rocks.

For a short period in the mid-1960s eclogite xenoliths received much attention but more recently investigations of these rocks have been superseded by studies on the ultramafic xenoliths. However, MacGregor and Carter (1970) believe two distinct groups of eclogite occur based on textural and mineralogical grounds. Group I eclogites are characterized by cumulate textures which may perhaps be equated with the granular type ultramafic xenoliths. Group II eclogites have textures somewhat analagous to the mosaic or fluidal textures of the mosaic or sheared ultramafic xenoliths. Chemically, it appears that the clinopyroxenes in the group II eclogites are poorer in the enstatite molecules than group I, thus indicating group II clinopyroxenes are overall more calcic than group I.

Recently, Reid *et al.* (1977) have described the mineralogy and chemistry of diamond-bearing eclogites and Smyth and Hatton (1977) have described the first occurrence of coesite (other than in natural diamond) in an eclogite. At present Tsai *et al.* (1977) are studying the sulfide minerals and sulfur isotopes in eclogitic xenoliths.

Mineral Inclusions in Diamond

The occurrence of minerals inside natural diamond is a relatively common feature. Usually the inclusions are small ($\sim 100\ \mu m$ size on average), have extremely well-formed crystal faces that are sometimes pseudomorph after the diamond host. Almost all silicate and oxide inclusions are monominerallic although it is not uncommon for different or similar minerals to be isolated from each other within the diamond.

Three periods of inclusion formation relative to diamond are known. Some inclusions are of primary origin in that they either formed contemporaneously with the host diamond or prior to the host and were subsequently incorporated during growth. Other inclusions obviously post-date the formation of diamond and appear as alteration products of pre-existing phases or have entered the diamond along cracks and cleavage planes. Recently,

TABLE 1. MINERAL INCLUSIONS IN DIAMOND
Identification by X-ray diffraction and/or electron microprobe analysis

Pre- and/or Syngenetic	Epigenetic
Olivine	Serpentine
Chrome–pyrope garnet	Goethite
Pyrope–almandine garnet	Graphite
Enstatite	Hematite
Diopside	Sellaite
Coesite	Kaolinite
Zircon	Amphibole
Rutile	Sanidine
Ilmenite	Xenotime
Chrome–spinel	
Diamond	
Pyrrhotite	
Pentlandite	

Origin uncertain: mica, quartz, amphibole, magnetite, kyanite

Meyer and Tsai (1976a) have summarized and discussed most of the literature of mineral inclusions in diamond.

The significance of the inclusions lies in the fact that for the pre- or syngenetic mineral the host diamond has acted as an armor against chemical reaction or re-equilibration with external silicate or oxide phases. Thus the inclusions have retained their primary compositions since inception and thus represent pristine mantle minerals. Typical mineral inclusions are listed in Table 1.

For the most part the chemistry of the inclusions is comparable to similar minerals present in kimberlite and in the ultramafic and eclogitic xenoliths. Nevertheless, in detail significant and characteristic differences are present. Furthermore, Meyer and Boyd (1968, 1972) noted two major characteristics of the inclusion:

1. Two major suites of minerals are present; one suite consists of minerals that resemble those in eclogitic xenoliths, and the other contains minerals comparable to those in the ultramafic xenoliths. Minerals in these suites are mutually exclusive.
2. In spite of major differences in geographic locality and geological time most inclusions of a specific type have relatively limited compositional ranges.

Olivine occurring as inclusions is forsterite rich, on average Fo_{93} and thus resembles that from granular xenoliths (Fig. 1). In contrast the Cr_2O_3 contents are relatively high compared to most terrestrial olivines (Meyer, 1975) but do approximate those in sheared ultramafic xenoliths. Enstatite inclusions have $Mg/Mg+Fe$ resembling those of granular xenoliths, although the Ca content of enstatites coexisting with diopside inclusions are generally higher than those existing alone.

Two types of pyroxene occur as inclusions. One type resembles those in the ultramafic xenoliths (Fig. 2), especially the granular ones, whereas the second type is somewhat akin to the omphacitic pyroxene of eclogite. Extremely Cr-rich pyroxenes which also contain fairly high contents of Na_2O have been noted in Siberian diamonds by Sobolev *et al.* (1971). Meyer and Tsai (1976a), commenting on the occurrence of the clinopyroxene inclusions, have observed that this mineral is most common in diamonds from the Precambrian Premier kimberlite, South Africa.

Garnet inclusions in diamond also occur in two distinct groups (Meyer, 1968). Like the

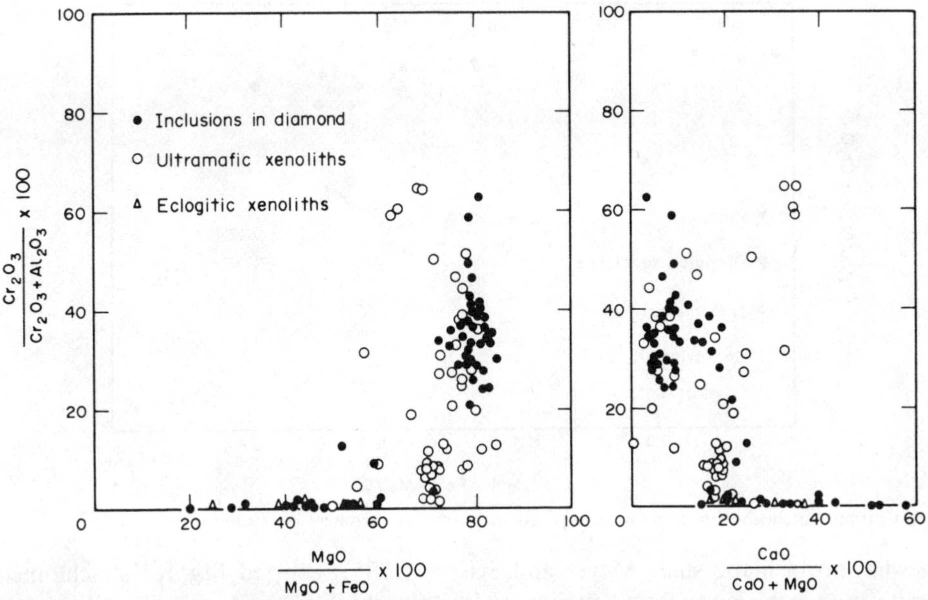

FIG. 5. Garnet inclusions in diamond. Two major groups are obvious: (1) Cr-pyrope garnets with high MgO/MgO+FeO and low CaO/CaO+MgO. (2) Eclogitic-type garnets with very low $Cr_2O_3/Cr_2O+Al_2O_3$ and wide ranges of MgO/MgO+FeO and CaO/CaO+MgO. Note 3 inclusions that span the gap between the two groups.

pyroxenes, the garnets may be divided into an ultramafic suite and an eclogitic one. The ultramafic garnets are generally low in CaO (<4 wt.%) (Fig. 5) and contain considerable quantities of Cr_2O_3 (up to 15 wt.%). These garnets contain the highest recorded MgO and since there is insufficient Al to balance the Mg it is necessary to utilize the rare molecule $Mg_3Cr_2Si_3O_{12}$–chrome pyrope (Meyer, 1968) or knorringite (Nixon and Hornung, 1968). The chrome-pyrope inclusions are claret colored whereas the eclogitic types are orange-brown. Sobolev *et al.* (1972) have examined garnet and coexisting clinopyroxene inclusions in a diamond that was contained within an eclogite, and found that the mineral phases in the eclogite are comparable in chemistry to those inside the diamond.

Although there is some variation in mineral chemistry of both clinopyroxenes and garnets between different diamonds it is usual to find that within a single host the inclusions of a specific type are identical in chemistry. Thus eclogitic and ultramafic suite inclusions have never been found together in the same diamond.

Spinel occurring as inclusions in diamond is virtually always Cr-rich and can be referred to as chromite. Compositions lie close to the join $MgCr_2O_4$–$FeCr_2O_4$ with Cr_2O_3 contents of up to 62 wt.%. The majority of chromite inclusions lie in the region MgO/(MgO+FeO) of 0.4 to 0.8 but some rare cases have been found of much lower values (Meyer and Boyd, 1972; Meyer and Tsai, 1976a) (Fig. 6).

Several other minerals occur as primary inclusions, including coesite, ilmenite zircon, rutile and sulfides. Coesite (Milledge, 1961) is important as until recently (Smyth and Hatton, 1977) its presence in diamond was the only known terrestrial occurrence other than in meteorite impact craters. Rutile has been found by several authors (Gnevushev and Nikolaeva, 1961; Harris, 1968; Meyer and Svisero, 1975; Prinz *et al.*, 1975). Gurney *et al.* (1969) also described rutile as an inclusion in diamond which itself was found within an eclogite. Ilmenite is

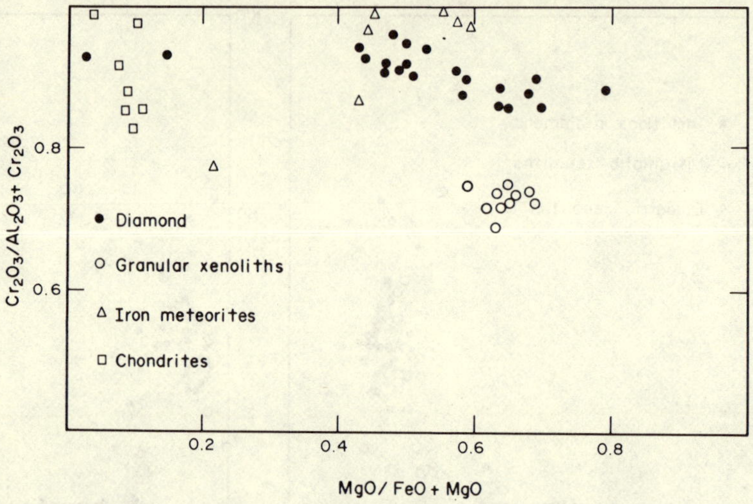

FIG. 6. Chromite inclusions in diamond compared with other occurrences of chromite.

somewhat problematic since Meyer and Svisero (1975) reported Mg-free stoichiometric ilmenite as an inclusion in Brazilian diamonds. Recently, Tsai and Meyer have discovered a Mg-rich ilmenite, comparable in composition to ilmenite occurring as xenocrysts, in a diamond from the Finsch kimberlite, South Africa.

The problem of several other mineral phases, such as kyanite, quartz, mica and amphibole, occurring within diamond has not been resolved as yet. If diamond forms stably in the mantle some of these phases should not occur as *primary* inclusions and at the present time there is often difficulty in actually assigning any inclusion to a primary or secondary source based on visual observation of the inclusion–host assemblage. This problem together with the mineralogy of these and other secondary inclusions has been reviewed by Meyer and Tsai (1976a).

Discussion

Virtually all the data presented above for the mineral chemistry of xenoliths and inclusions in diamond are the result of investigations during the last 10 to 12 years. These data plus the results of phase equilibria studies at high pressures have been combined to produce petrologic models of the upper mantle (Boyd, 1973; MacGregor and Basu, 1976; O'Hara *et al*., 1975). For example, Chatterjee (1969), Boyd (1973) and MacGregor (1974) have attempted to determine temperatures and pressures of equilibration for the minerals of the xenoliths. These physical parameters are obtained by utilizing the experimental data on the enstatite–diopside solvus (Davis and Boyd, 1966; Nehru and Wyllie, 1974; Mori and Green, 1976; Lindsley and Dixon, 1975) to determine a temperature, and the Al_2O_3 content of the enstatite to obtain the pressure (Boyd and England, 1964; MacGregor, 1974).

Utilizing these methods it has been found that the mosaic or sheared ultramafic xenoliths for certain localities have generally equilibrated at higher temperatures and pressures than the granular xenoliths. Descrete xenoliths or xenocrysts have somewhat intermediate ranges. The trend of points on the P–T plot were first interpreted by Boyd (1973) and MacGregor (1974) as representing a fossil geotherm which had been perturbed by frictional heating at the

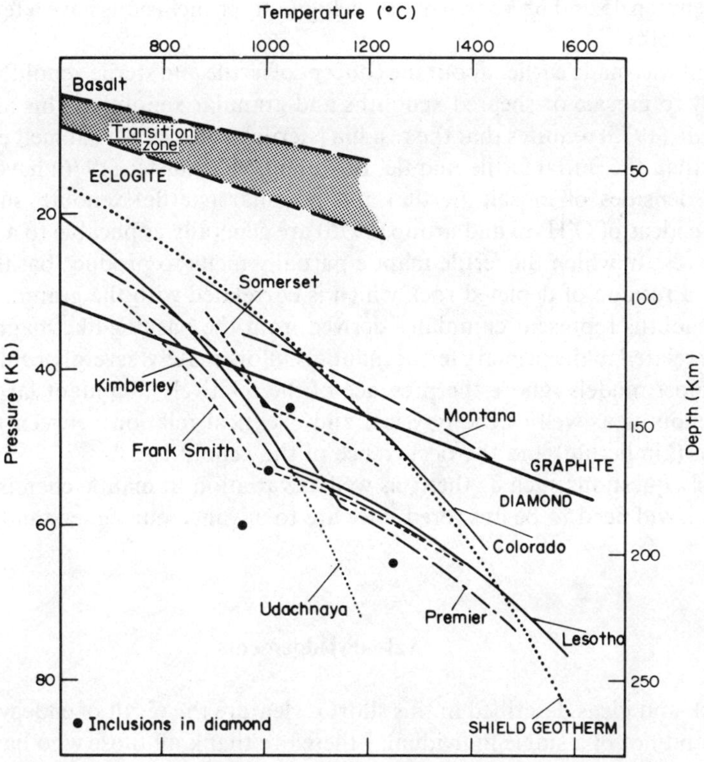

FIG. 7. Geotherms based on studies of ultramafic xenoliths in kimberlite, and possible pressures and temperature of equilibration for four diamonds based on their inclusions. Lesotho, Kimberley, Frank Smith, Premier represent the South African continent. Montana, Colorado, Somerset (N.W.T.–Canada) are localities in the North American continent, and Udachnaya represents the Siberian continent.

base of the lithosphere. This unique and exciting model received much criticism (Wilshire and Jackson, 1975; Mercier and Carter, 1975; Thomsen, 1975; Mori and Green, 1976) on the basis that the inflection was an artifact of the experimental data and choice of xenoliths. However, recently using the same techniques and experimental data palaeogeotherm for other mantle regions (e.g. Kimberley, South Africa; Western U.S.; Siberia; Northern Canada) have been obtained (Fig. 7). Some of these show inflections and others do not (Boyd, 1976; Boyd *et al*., 1976; Mitchell, 1976; Hearn and Boyd, 1975; Eggler and McCallum, 1976; Danchin and Boyd, 1976) (Fig. 7). It is thus necessary to explain the absence or presence of the inflection in the geotherm. Currently, it is believed that the deformation textures and thermal effects are not consistent with stress-heating in the LVZ (low velocity zone) (Mercier and Carter, 1975; Goetze, 1975). Accordingly, other models at present under consideration to explain the various geotherms include convective movement of partially molten mantle, thermal diapirs or combinations of both. These ideas are well summarized in Boyd (1976).

Nevertheless, in spite of the above discussion and controversy it is important to remember that the pressures and temperatures derived by these techniques are all compatible with an upper mantle origin for the xenoliths. Similar, but more scattered pressures and temperatures have been determined by Meyer and Tsai (1976b) for inclusions in diamond (Fig. 7). The mineral inclusions of the ultramafic suite have equilibrated in the range 1000° to 1300°C at pressure between 45 and 65 kbar, whereas eclogitic-type inclusions have a temperature range of 850° to 1250°C.

Comment was made earlier about the concept of fertile and sterile xenoliths corresponding respectively to mosaic or sheared xenoliths and granular xenoliths. This idea expressed by O'Hara *et al.* (1975) requires that the residua (sterile) from the partial melt episode would be less dense than the initial fertile mantle. Boyd and McCallister (1976) have shown that the calculated densities of mosaic (fertile) and granular (sterile) xenolith support this idea. Overall, the ideas of O'Hara and group (1975) are generally applicable to a simplified upper mantle process in which the fertile mantle partially melts to produce basaltic-like material and leaves a residua of depleted rock which is correlated with the granular xenoliths. The eclogite xenoliths represent cumulates derived from the basaltic-like magma and are thus genetically related to the primary fertile mantle. Unfortunately, several problems remain. For example, most models ignore the presence of the relatively abundant large xenocrysts or discrete xenoliths as well as the physical and chemical relation between the transporting mechanism (kimberlite) and the occurrence of the xenoliths.

Obviously questions such as these, as well as variation in mantle chemistry, both lateral and vertical, will need to be answered if we are to advance our understanding of the upper mantle.

Acknowledgements

The work and ideas described in this short review are the result of endeavor by numerous scientists, and not of a single individual. I therefore thank all those who have increased our knowledge of mantle mineralogy and geochemistry through the last decade and hope their contributions in the future will also provide much stimulation. The research of which this forms a part is aided by grants from the National Science Foundation, Earth Sciences Section, DES72–01659 and GA–43990. HMT also acknowledges support as a David Ross Fellow from Purdue Research Foundation.

References

Aoki, K. (1968) Petrogenesis of ultrabasic and basic inclusions in alkali basalts, Iki Island, Japan. *Amer. Mineralogist* **53**, 241–256.

Boullier, A. M. and Nicolas, A. (1975) Classification of textures and fabrics of peridotite xenoliths from South African kimberlites. *Phys. Chem. Earth* **9**, 467–476.

Boyd, F. R. (1971) Enstatite–ilmenite and diopside–ilmenite intergrowths from Monastery Mine. *Carnegie Inst. Wash. Yearbook* **70**, 134–138.

Boyd, F. R. (1973) A pyroxene geotherm. *Geochim. Cosmochim. Acta* **37**, 2533–2546.

Boyd, F. R. (1974) Ultramafic nodules from the Frank Smith kimberlite pipe, South Africa. *Carnegie Inst. Wash. Yearbook* **73**, 285–294.

Boyd, F. R. (1976) Inflected and non-inflected geotherms. *Carnegie Inst. Wash. Yearbook* **75**, 521–523.

Boyd, F. R. and England, J. L. (1964) The system enstatite–pyrope. *Carnegie Inst. Wash. Yearbook* **63**, 157–161.

Boyd, F. R. and McCallister, R. H. (1976) Densities of fertile and sterile garnet peridotites. *Geophys. Res. Lett.* **3**, 509–512.

Boyd, F. R. and Nixon, P. H. (1975) Origins of the ultramafic nodules from some kimberlites of Northern Lesotho and the Monastery Mine, South Africa. *Phys. Chem. Earth* **9**, 431–454.

Boyd, F. R., Fujii, T. and Danchin, R. V. (1976) Noninflected geotherm for the Udachnaya Kimberlite Pipe, USSR. *Carnegie Inst. Wash., Yearbook* **75**, 523–531.

BROOKINS, D. G. (1970) The kimberlites of Riley County, Kansas. *Kans. Geol. Surv. Bull.* **200**, 1–32.

CARSWELL, D. A. (1975) Primary and secondary phlogopites and clinopyroxenes in garnet–lherzolite xenoliths. *Phys. Chem. Earth*, **9**, 417–429.

CHATTERJEE, N. D. (1969) Aus welchen Erdtiefen stammen die diamantfuhrenden kimberlite? *Neues Jb. Mineral.* **7**, 289–305.

CLEMENT, C. R. (1975) The emplacement of some diatremefacies kimberlite. *Phys. Chem. Earth* **9**, 51–60.

DANCHIN, R. V. and BOYD, F. R. (1976) Ultramafic nodules from the Premier kimberlite pipe, South Africa. *Carnegie Inst. Wash. Yearbook* **75**, 531–538.

DAVIS, B. T. C. and BOYD, F. R. (1966) The join $Mg_2Si_2O_6$–$CaMgSi_2O_6$ at 30 kb pressure and its application to pyroxenes from kimberlite. *J. Geophys. Res.* **71**, 3567–3576.

DAWSON, J. B. (1971) Advances in kimberlite geology. *Earth Sci. Rev.* **7**, 187–214.

DAWSON, J. B. and REID, A. R. (1970) A pyroxene–ilmenite intergrowth from the Monastery Mine, South Africa. *Contr. Mineral. Petrol.* **26**, 296–301.

DAWSON, J. B. and SMITH, J. V. (1975) Mineral chemistry of amphibole-diopside glimmerite nodules from some South African kimberlites. (Abstr.) *Kimberlite Symposium,* Cambridge, England.

EGGLER, D. H. and MCCALLUM, M. E. (1976) A geotherm from megacrysts in the Sloan kimberlite pipes, Colorado. *Carnegie Inst. Wash. Yearbook* **75**, 538–541.

ERLANK, A. J. (1976) Upper mantle metasomatism as revealed by potassic richterite bearing peridotite xenoliths from kimberlite. (Abstr.) *Amer. Geophys. Union meeting, Wash.* 37.

GNEVUSHEV, M. A. and NIKOLAEVA, E. S. (1961) Solid inclusions in diamonds from the deposits of Yakutia. *Tr. Yak. Filiala Sibirsk Otd. Akad. Nauk SSSR., Ser. Geol.* No. 6, 97–105 (in Russian).

GOETZE, C. (1975) Sheared lherzolites: from the point of view of rock mechanics. *Geology* **3**, 172–173.

GURNEY, J. J., FESQ, H. W. and KABLE, E. J. D. (1973) Clinopyroxene–ilmenite intergrowths from kimberlite: a reappraisal. In *Lesotho Kimberlites*, P. H. NIXON, ed., Lesotho Nat. Dev. Corp., Maseru, 238–253.

GURNEY, J. J., HARTE, B. and COX, K. G. (1975) Mantle xenoliths in the Matsoku kimberlite pipe. *Phys. Chem. Earth* **9**, 507–523.

GURNEY, J. J., SIEBERT, J. C. and WHITFIELD, G. G. (1969) A diamondiferous eclogite from the Roberts Victor Mine. *Trans. Geol. Soc. S. Afr., Spec. Pub.* No. 2, 351–359.

HAGGERTY, S. E. (1975) The chemistry and genesis of opaque minerals in kimberlites. *Phys. Chem. Earth* **9**, 295–307.

HARRIS, J. W. (1968) The recognition of diamond inclusions. Pt. 1: Syngenetic mineral inclusions. Pt. 2: Epigenetic mineral inclusions. *Ind. Diam. Rev.*, 402–410, 458–461.

HARTE, B. (1977) Rock nomenclature with particular relation to deformation and recrystallization textures in olivine-bearing xenoliths. *J. Geol.* **85**, 279–288.

HAWTHORNE, J. B. (1975) Model of a kimberlite pipe. *Phys. Chem. Earth* **9**, 1–16.

HEARN, B. C. and BOYD, F. R. (1975) Garnet peridotite xenoliths in a Montana, U.S.A., kimberlite. *Phys. Chem. Earth* **9**, 247–256.

HOLMES, A. (1965) *Principles of Physical Geology*, Nelson, London, 1288 pp.

JACKSON, E. D. and WRIGHT, T. L. (1970) Xenoliths in the Honolulu Volcanic Series, Hawaii. *J. Petrol.* **11**, 405–430.

LEWIS, C. (1887) On a diamondiferous peridotite and the genesis of diamond. *Geol. Mag.* **4**, 22–24.

LINDSLEY, D. and DIXON, S. A. (1975) Coexisting diopside and enstatite at 20 kbar and 900°–1200°C. (Abstr.) *Geol. Soc. Amer. Prog. Abstr.* **7**, 1171.

MACGREGOR, I. D. (1974) The system MgO–Al_2O_3–SiO_2: Solubility of Al_2O_3 in enstatite for spinel and garnet peridotite compositions. *Amer. Mineralogist* **59**, 110–119.

MACGREGOR, I. D. and BASU, A. R. (1976) Geological problems in estimating mantle geothermal gradients. *Amer. Mineralogist* **61**, 715–724.

MACGREGOR, I. D. and CARTER, J. L. (1970) The chemistry of clinopyroxenes and garnets of eclogite and peridotite xenoliths from the Roberts Victor Mine, South Africa. *Phys. Earth Planet. Int.* **3**, 391–397.

MCCALLISTER, R. H., MEYER, H. O. A. and BROOKINS, D. G. (1975) "Pyroxene"–ilmenite xenoliths from the Stockdale pipe, Kansas: Chemistry, crystallography, and origin. *Phys. Chem. Earth* **9**, 287–293.

MCCALLUM, M. E. and EGGLER, D. H. (1971) Mineralogy of the Sloan diatreme, a kimberlite pipe in northern Larimer County, Colorado, *Amer. Mineralogist,* **56**, 1735–1749.

MCGETCHIN, T. R. and ULLRICH, G. W. (1973) Xenoliths in maars and diatremes with inferences for the Moon, Mars, and Venus. *J. Geophys. Res.* **78**, 1833–1853.

MERCIER, J. C. C. and CARTER, N. L. (1975) Pyroxene geotherms. *J. Geophys. Res.* **80**, 3349–3362.

MEYER, H. O. A. (1968) Chrome pyrope: an inclusion in natural diamond. *Science* **160**, 1446–1447.

MEYER, H. O. A. (1975) Chromium and the genesis of diamond *Geochim. Cosmochim. Acta* **39**, 929–936.

MEYER, H. O. A. (1976) Kimberlites of the continental United States: a review. *J. Geol.* **84**, 377–403.

MEYER, H. O. A. and BOYD, F. R. (1968) Mineral inclusions in diamond. *Carnegie Inst. Wash. Yearbook* **67**, 130–135.

MEYER, H. O. A. and BOYD, F. R. (1972) Composition and origin of crystalline inclusions in natural diamonds. *Geochim. Cosmochim. Acta* **36**, 1255–1273.

MEYER, H. O. A. and SVISERO, D. P. (1975) Mineral inclusions in Brazilian diamonds. *Phys. Chem. Earth* **9**, 785–795.

MEYER, H. O. A. and TSAI, H.-M. (1976a) Mineral inclusions in natural diamond—their nature and significance: A review. *Min. Sci. Eng.* **8**, 242–261.

MEYER, H. O. A. and TSAI, H.-M. (1976b) Mineral inclusions in diamond: Temperature and pressure of equilibration. *Science* **191**, 849–851.

MILLEDGE, H. J. (1961) Coesite as an inclusion in G.E.C. synthetic diamonds. *Nature* **190**, 1181.

MITCHELL, R. H. (1970) Kimberlite and related rocks—a critical reappraisal. *J. Geol.* **78**, 271–278.

MITCHELL, R. H. (1976) Ultramafic xenoliths from the Elwin Bay kimberlite: the first Canadian palaeogeotherm. *Canad. J. Earth. Sci.* **14**, 1202–1210.

MORI, T. and GREEN, D. H. (1976) Subsolidus equilibria between pyroxenes in the CaO–MgO–SiO_2 system at high pressures and temperatures. *Amer. Mineralogist* **61**, 616–625.

NEHRU, C. E. and WYLLIE, P. J. (1974) Electron microprobe measurement of pyroxenes coexisting with H_2O-undersaturated liquid in the join $CaMgSi_2O_6$–$Mg_2Si_2O_6$ at 30 kilobars, with applications to geothermometry. *Contr. Mineral. Petrol.* **48**, 221–228.

NIXON, P. H. and BOYD, F. R. (1973a) Petrogenesis of the granular and sheared ultrabasic nodule suite in kimberlites. In *Lesotho Kimberlites,* P. H. NIXON, ed., Lesotho Nat. Dev. Corp., Maseru, 48–56.

NIXON, P. H. and BOYD, F. R. (1973b) The discrete nodule association in kimberlites from northern Lesotho. In *Lesotho Kimberlites,* P. H. NIXON, ed., Lesotho Nat. Dev. Corp., Maseru, 67–75.

NIXON, P. H. and HORNUNG, G. (1968) A new chromium garnet end member, knorringite, from kimberlite. *Amer. Mineralogist* **53**, 1833–1840.

O'HARA, M. J. and MERCY, E. L. P. (1963) Petrology and petrogenesis of some garnetiferous peridotites. *Trans. Roy. Soc. Edin.* **65**, 251–314.

O'HARA, M. J., SAUNDERS, M. J. and MERCY, E. L. P. (1975) Garnet peridotite, primary ultrabasic magma and eclogite; interpretation of upper mantle processes in kimberlite. *Phys. Chem. Earth* **9**, 571–604.

PRINZ, M., MANSON, D. V., HLAVA, P. F. and KEIL, K. (1975) Inclusions in diamonds. Garnet lherzolite and eclogite assemblages. *Phys. Chem. Earth* **9**, 797–815.

REID, A. M., BROWN, R. W., DAWSON, J. B., WHITFIELD, G. G. and SIEBERT, J. C. (1976) Garnet and pyroxene compositions in some diamondiferous eclogites. *Contr. Mineral. Petrol.* **58**, 203–220.

RINGWOOD, A. E. and LOVERING, J. F. (1970) Significance of pyroxene–ilmenite intergrowths among kimberlite xenoliths. *Earth Planet. Sci. Lett.* **7**, 371–375.

SMITH, J. V. and DAWSON, J. B. (1975) Chemistry of Ti-poor spinels, ilmenites and rutiles from peridotite and eclogite xenoliths. *Phys. Chem. Earth* **9**, 309–322.

SMITH, J. V., STEELE, I. M., HRUBEC, J. and DAWSON, J. B. (1975) Chemistry of mantle minerals. (Abstr.) *Kimberlite Symposium*, Cambridge, England.

SMYTH, J. R. and HATTON, C. J. (1977) A coesite–sanidine grospydite from the Roberts Victor kimberlite. *Earth Planet. Sci. Lett.* **34**, 284–290.

SOBOLEV, N. V., BOTKUNOV, A. I., LAURENT'YEV, Yu. G. and POSPELOVA, L. N. (1971) Peculiarities of the composition of minerals coexisting with diamond from Mir pipe, Yakutia. *Zap. Vses. Mineral. Obshch.* **100**, 558–564.

SOBOLEV, V. S., SOBOLEV, N. V. and LAURENT'YEV, Yu. G. (1972) Inclusions in diamond extracted from a diamondiferous eclogite. *Dokl. Akad. Nauk* **207**, 164–167.

THOMSEN, L. (1975) The lower "Petrologic Geotherm": a transitory state. *Science* **188**, 1130–1131.

TSAI, H. M., MEYER, H. O. A. and SHIEH, Y. N. (1977) Sulfide minerals in eclogite xenoliths, in diamond, and sulfur isotope abundances. (Submitted to 2nd Int. Kimb. Conf., Santa Fe, U.S.A.)

WAGNER, P. A. (1914) "The Diamond Fields of Southern Africa". *Transvaal Leader*, Johannesburg. Reprinted 1971, C. Struik, Cape Town, 355 pp.

WILLIAMS, A. F. (1932) *The Genesis of Diamond,* 2 vols., E. Benn, London.

WILSHIRE, H. G. and JACKSON, E. D. (1975) Problems in determining mantle geotherms from pyroxene compositions of ultramafic rocks. *J. Geol.* **83**, 313–329.

WILSHIRE, H. G. and SHERVAIS, J. W. (1975) Al-augite and Cr-diopside ultramafic xenoliths in basaltic rocks from western United States. *Phys. Chem. Earth* **9**, 257–272.

WYLLIE, P. J. (1967) *Ultramafic and Related Rocks*, Wiley, New York, 464 pp.

Discussion

P. HENDERSON:

Do you find inclusions in diamond that are not monomineralic?

H. O. A. MEYER:

Polymineralic inclusions do occur; for example, garnet, olivine, enstatite; however, these inclusions are rare. For geothermometry and barometry such specimens are most desirable.

SECTION VIII

GEOCHEMISTRY AND DETECTION OF MINERAL DEPOSITS

Introductory Remarks on Geochemistry in Discovering Ore Deposits with Particular Reference to Uranium

S. H. U. BOWIE and J. PLANT

Institute of Geological Sciences, Geochemical Division, 64–78 Gray's Inn Road, London WC1X 8NG, England

Large tonnages of uranium are likely to be required between now and the end of the century. On the evidence of the best estimates of demand, annual tonnages of over 100,000 will have to be added to presently known reserves. Previously, annual discovery rate has been about 50,000 tU/y for <\$30/lb U_3O_8 or 80,000 tU/y for <\$50/lb U_3O_8. This was during a period when serious prospecting was in its infancy and the cream of deposits with surface manifestations were awaiting discovery. Most such deposits were identified by their gamma radiation which was detected and measured at, or just above the surface, or from aircraft.

Many of the deposits that have to be developed in the future have no such surface expression and must be found by other methods. One of the most important of these is by the application of geochemistry.

Of immediate importance are the regional geochemical methods based on systematic analyses of sediment and water samples from tributary drainage or lakes. Although the cost of surveys based on these methods should be carefully assessed in relation to aeroradiometric surveys for the terrain investigated, some advantages of geochemistry are:

1. It provides a direct measurement of U.
2. Many other elements which may indicate U mineralization or help to interpret the type of occurrence can be determined on the samples collected.
3. Measurements such as the pH and conductivity of surface waters which can critically influence the dispersion of U in surficial materials can also be determined.

In spite of the widespread application of geochemical methods more work is needed to enable accurate and reliable interpretation to be made. For example, the interrelationship between the organic geochemistry of U and the chemistry of surface waters should be further investigated.

Different sample types should be evaluated. For example, water samples may provide a more sensitive indication of U mineralization than stream sediments in some environments while for other elements heavy mineral concentrate samples may be preferable. Biogeochemical methods also require more study. For all these methods case histories particularly examples where techniques have failed are required.

Because of its natural radioactivity, uranium was one of the first metals for which exploration was possible using on site determination by physical techniques. Such methods still have important application, for example mapping of the continental shelf and gamma spectrometric methods of measuring eU, eTh, and K have recently been developed in the United Kingdom for underwater surveying. Developments of XRF and neutron-activation methods, particularly in the application of mini computers which enable quantitative results to be obtained in the field, are likely to extend the application of on-site methods for U and for other elements. Methods to detect hidden uranium ore deposits (for which a fundamental understanding of groundwater geochemistry is essential) are of increasing importance. Quantitative data on the dispersion of radon and helium, and more particularly of ^3He:^4He ratios, are needed.

In some parts of the world anomalous concentrations of radiogenic lead isotopes occur in non-radioactive minerals such as galena and pyrite and this method should be evaluated from the point of view of discovering hidden ore deposits.

Statistical analysis of data obtained in uranium exploration may help to identify interelement relationships indicative of mineralization; the statistical techniques which enable interrelationships between geochemical and geophysical data recently described by Sinding-Larsen *et al.* (1978) may provide a tool particularly relevant to uranium exploration.

Of very great importance is the need to improve our understanding of the precise physical and chemical conditions under which uranium deposits form and are *preserved*. This requires the applications of theoretical data and models in addition to observations of the dispersion of uranium.

Such observations may result from applied investigations. A good example of this is the information being obtained on the leaching and migration of uranium in sedimentary basins in the western U.S.A. using high-resolution, low-energy spectrometry to examine disequilibrium between U and its daughter products which have very different chemical mobilities.

Finally, uranium may be particularly useful in improving our understanding of ore deposition because of the possibility of examining its distribution quantitatively at low concentration (for example, Lexan—1 ppm) in thin section using fission track methods so that physicochemical models can be evaluated using textural information.

International co-operation is clearly essential and this must include studies between fundamental and applied geochemists.

Reference

SINDING-LARSEN, R., BOTBOL, J. M. and MCCAMMON, R. B. (1978) The use of weighted characteristic analysis as a tool in resource assessment. In *Proceedings of the I.A.E.A.* Rome meeting 1976, (in the press).

Discussion

U. ASWATHANARAYANA:

How useful is prospecting for U using radon in soil gases employing the technique of track registration in lexan-type plastic?

J. PLANT:

We have not had very much success possibly because of the importance of groundwater movements and disequilibrium of U and its daughter products.

R. SINDING-LARSEN:

How does variations in pH complicate the interpretation of uranium results in geochemical exploration?

J. PLANT:

pH can vary very much in regional surveys. For example, if it is high because of high concentrations of bicarbonate U is partitioned into water and stream sediments fail. Therefore pH is necessary for interpretation.

The Redox Potential Field of the Earth

BJØRN BØLVIKEN

Geological Survey of Norway, P.O. Box 3006, N-7001 Trondheim, Norway

Abstract

It may be assumed that the redox potential within the earth generally decreases with decreasing distance from the centre of the earth, forming a field of geochemical redox potentials in many ways analogous to other potential fields. Large- and small-scale variations in the electrical properties of the crust—for example, differences in conductivity between ocean floor and continents or between massive sulphides and country rock—will cause deviations from the primary field of redox potential and production of electromotive forces. Secondary redox potential fields are formed, and cause natural electrical currents, which may influence the distribution of chemical elements. Self-potentials as recorded in mineral exploration are supposed to be a result of the redox potential field of the earth. Empirical data from literature, which support these theories, are summarized and comments are given about their possible consequences in various fields of geology and geochemistry such as weathering of sulphides, mineral and oil-exploration methods, element distribution around folded rocks and composition of seawater.

Primary Redox Potential Field

In the earth there must exist a vertical redox potential field, in many ways analogous to other potential fields, e.g. those of gravity and magnetism. This redox potential field may have important implications which have been overlooked.

The field is vizualized in Fig. 1, which is a hypothetical and idealized homogenous vertical section in the crust. Terrestrial waters constitute an electrolyte with H^+ and OH^- ions as well as redox pairs (e.g. Fe^{3+}/Fe^{2+}) and other ions (e.g. Na^+ and Cl^-). The bedrock could be considered as a system of interconnected porous membranes in the electrolyte. The redox potential of the field will be relatively high near the surface and low at depth. Correspondingly, pH will be relatively low near the surface and high at depth. The potential gradients are probably steepest near the surface. The vertical distance between isopotential surfaces thus increases with depth, but this trend could be upset by various heterogeneities at depth. In the present paper only sections with constant gradients are considered.

The idealized system described above could conveniently be called the primary redox potential field of the earth. If ground water flow is excluded, diffusion becomes an important way by which matter can move in solution in the primary redox-potential field. Oxidizing species (e.g. Fe^{3+}) will diffuse downwards and reduced species (e.g. Fe^{2+}) upwards. Protons, hydroxyl and other ions will be distributed so that electrical neutrality is maintained everywhere in the electrolyte. The oxygen of the atmosphere and material of low oxidation potential in the interior of the earth would serve as reservoirs for oxidizing and reducing species, respectively.

The lithosphere is, of course not homogeneous, topography, permeability, pore volume,

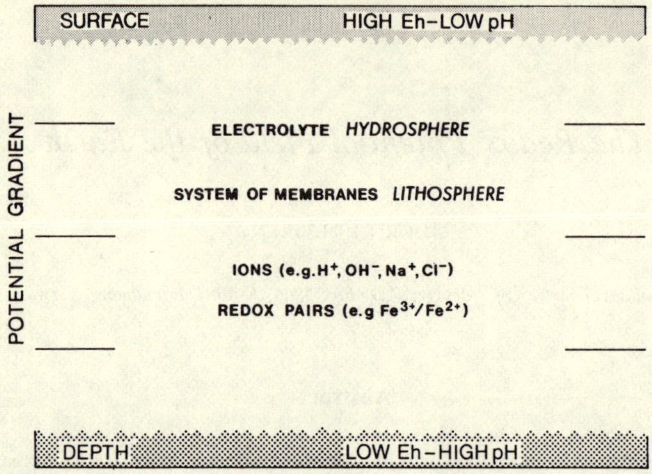

FIG. 1. Diagrammatic representation of the primary redox potential of a vertical section through a homogenous part of the lithosphere. "Hydrosphere" indicates terrestrial waters.

electric conductivity, etc., vary. The influence of some of these factors will be discussed below, but first some data will be given to substantiate the existence of the primary redox potential field.

Indications of the Redox Potential Field

Data indicating the existence of the primary redox-potential field of the earth can be grouped into (1) redox properties of earth material, (2) results of Eh measurements in natural environments, and (3) occurrence of natural self-potentials.

1. The earth's atmosphere is highly oxidizing, while the interior of the earth is strongly reducing. The first statement is an empirical fact, the other is a consequence of widely accepted theories, which suggest that the core consists of metals such as Fe and Ni in oxidation state zero. A simple explanation of this variation in redox properties is provided by sulphur, which at the surface dominantly occurs as SO_4^{2-} (oxidation state +6), while sulphur and sulphides (oxidation states 0 and −2 respectively) prevail in volcanic emanations. Another example is provided by Dodge (1972), who has shown a systematic decrease in ferrous–ferric ratios of granitic rocks across a batholith, the lowest ratio presumably indicating the deepest level of magma generation.
2. The direct measurement of redox potential in natural environments is a difficult task, and the reliability of the reported figures can in some cases be questioned (Sillén, 1965). Redox (Eh) data have been summarized by, for example, Baas Becking *et al.* (1960), Garrels and Christ (1965) and Krauskopf (1967), and they all point in the same direction indicating that terrestrial waters near the surface normally have higher Eh and lower pH than those at depth.
3. When two unpolarizable (e.g. calomel) electrodes are placed some distance apart at the earth's surface or in drill holes (Fig. 2), potential differences of the order of 100–1000 mV are normally found around highly conductive ore bodies. This is the self-potential (SP) method used in exploration. The origin of the SP is not completely understood,

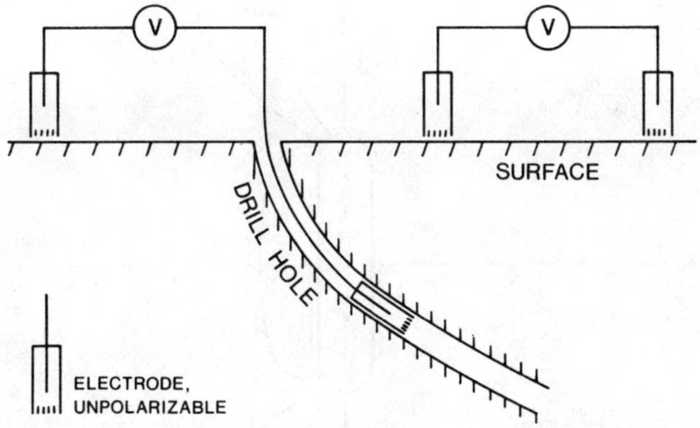

FIG. 2. Principles of measurements of self-potentials in drill holes and at the surface.

but the explanation given by Sato and Mooney (1960) seems now to be generally accepted (Becker and Telford, 1965; Corwin, 1976; Dyck, 1975; Parasnis, 1966). According to this, SP is due to electrochemical processes caused by Eh differences in the upper part of the lithosphere. The vast amount of SP data from mineral exploration can therefore be considered as an indication of the existence of a primary redox potential field of the earth. As suggested on page 662, the SP recorded in oil-well logging could also be interpreted in the same way.

Secondary Redox Potential Fields

Let us consider the effect on the primary redox potential field of a vertical body of an inert material with negligible electrical resistance. This represents an idealized model of an ore body such as, for example, graphite, magnetite or massive sulphides in the ground water of the country rock. The ore body could be regarded as an electrode in an electrolyte, the composition of which varies between the ends of the electrode. The electrode takes on the character of a dipole, and since the redox potential of the electrolyte generally decreases with depth, the lower end of the electrode becomes an anode and the upper a cathode. (Fig. 3). (This electrode differs somewhat from common electrodes which are either cathodes or anodes.)

Secondary currents will flow between the anode and the cathode. Within the electrode (ore body) electrons will move upwards from the anode to the cathode in order to compensate for the electrons produced by the electrolyte at the anode, or consumed by the electrolyte at the cathode. In the electrolyte (country rock ground water) near the electrode (ore body) cations will move upwards from the anode to the cathode, and anions downwards from the cathode to the anode.

The local redox potentials around the electrode could conveniently be termed a secondary redox potential field.

Protons have the highest equivalent conductance of all ions and will most probably be the main carrier of current in an acidic electrolyte. The type of cathode reaction could be illustrated by the equation:

$$2H^+ + 2e^- = H_2$$

or alternatively, by

$$O_2 + 4H^+ + 4e^- = 2H_2O.$$

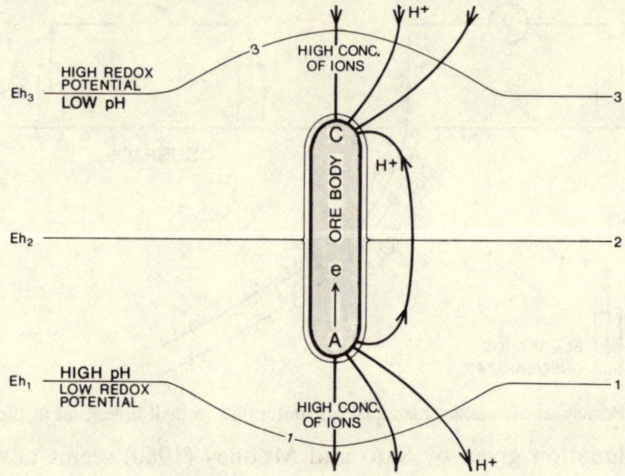

FIG. 3. Schematic model of an ore body as electrode in a primary redox potential field. The lower end develops into an anode (A) and the upper into a cathode (C). Direction of positive electricity (cations) in the electrolyte are indicated with arrows. Arrow inside ore body shows the direction of electron flow in the ore. Thin lines: equipotential surfaces. Deviation from linear isopotential curves show the local secondary redox potential field causing increased potentials around the anode and decreased potentials around the cathode. Species oxidized at the anode will migrate outwards from the anode and be reduced by species moving towards the anode. Species reduced at the cathode will migrate outwards from the cathode and be oxidized by species moving towards the cathode.

Hydroxyl ions have the highest equivalent conductance of all anions and will be the main current carrier in a basic electrolyte. The type of anode reaction could be illustrated by the equation:

$$2OH^- = H_2O + \tfrac{1}{2}O_2 + 2e^-$$

or alternatively, by

$$2OH^- + H_2 = 2H_2O + 2e^-.$$

The products of the cathodic reactions will be transported away (by diffusion of convection) from the cathode, and may gradually become oxidized by oxidizing agents being transported towards the cathode from its surroundings. Correspondingly, the products of the anodic reactions will be transported away and may gradually become reduced by agents being transported towards the anode.

Ground-water flow will promote these processes and thereby sustain the flow of electric currents. Assuming negligible ground water flow, a stationary, nearly static state will evolve, where the only currents flowing are those which compensate for the displacement from this state caused by diffusion. The system will then tend to approach the situation shown schematically in Fig. 3.

A local zone of relatively low redox potentials and high pH will develop around the cathode, and a zone of relatively high redox potentials and low pH around the anode. In the electrolyte the density of electric currents will presumably be highest near the ends of the electrode.

Heterogeneities in the environment of the electrode make the model presented above more complicated. Of particular interest is the contact between overburden and bedrock. Owing to the relatively low electrical resistivity (ρ) in loose material current directions will bend

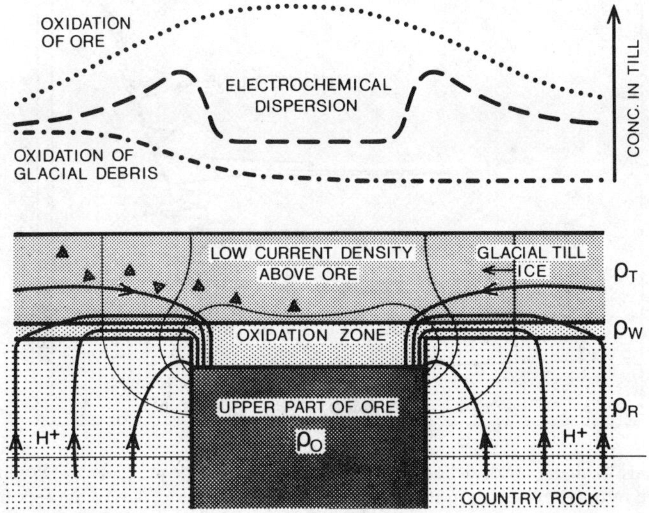

FIG. 4. Schematic model of the upper part of a sulphide ore body outcropping under glacial till. Thin lines: equipotential surfaces. Heavy lines with arrows: current direction of positive electricity (cations) in the electrolyte. ρR, ρw, ρT, ρo: electrical resistivities, being lowest in the ore and highest in the bedrock. Upper curves indicate paths to be expected of ore element from various dispersion mechanisms (After Bølviken and Logn, 1975).

abruptly at the boundaries between bedrock and overburden. In the overburden the resulting current density will be high in the vicinity of the outer edges of the ore, and low above the ore, see Fig. 4.

The above considerations deal with current distribution and chemical conditions in ground water only, without taking into account any impact on rock material. The two phases ground water and rock will tend to approach chemical equilibrium. Consequently, if electrochemical processes cause gradients of chemical potential in the ground water, then these gradients may also influence the local mineralogical composition of the bedrock (providing that favourable reaction speeds exist in relation to available time). Normally not all the ore minerals are stable in the local environment, and in reality the electrode itself (ore body) will take part in chemical reactions, of which oxidation and formation of secondary enrichment zones are of major importance.

Field Indications of Secondary Redox Potential Fields Around Ore Bodies

Natural self-potentials as recorded in mineral exploration (Fig. 2) provide strong indications of the existence of secondary redox potential fields around ore deposits. The distribution of SP at the Joma massive pyrite deposit, Norway (Logn and Bølviken, 1974), is used here as an example.

The SP curve in a typical drill hole (Fig. 5) through the Joma deposit consists of two distinct superimposed patterns: (1) abruptly alternating potentials with minima of −800 mV and (2) a generally smooth potential trend, the variation of which is within approximately 200 mV. The abrupt potential changes occur at the boundaries between sulphide horizons and country rock giving low potentials inside the sulphides. The smooth trend shows the potentials in the country rock.

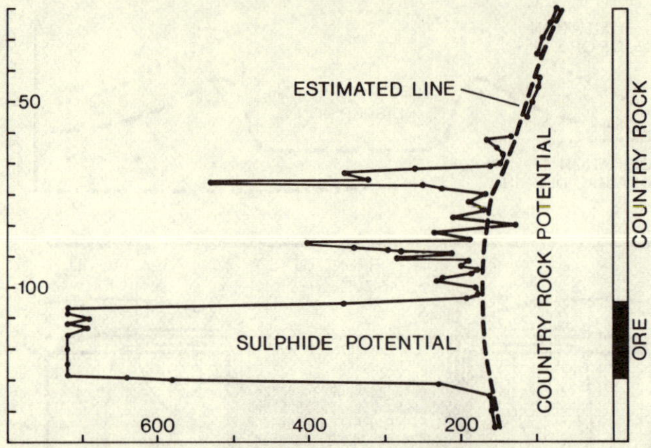

FIG. 5. Self-potentials in diamond drill hole No. 119 Joma pyrite deposit, Norway. Potential pattern can be graphically separated into: (1) potentials in the sulphides (abruptly changing part) and (2) potential in the country rock (smooth trend). (After Bølviken and Logn, 1975.)

The two patterns can be separated graphically as indicated in Fig. 5. When this procedure is used on rather complicated drill hole SP-data taken from a section through the ore, a simple country rock potential pattern appears, see Fig. 6.

The potentials in the country rock are at a minimum around the upper part of the ore body and at maximum around the lower part. The potential distribution indicates that a positive electric current flows upwards in the country rock and consequently downwards in the ore.

These results agree well with the concept of an ore body as a dipole electrode (the lower end being the anode and the upper the cathode) in a galvanic cell, in which cations migrate upwards and anions downwards in the country rock, and electrons upwards in the ore body.

Potential patterns of the same type as those found in the country rock at Joma have been demonstrated for a number of sulphide and iron ore deposits (Becker and Telford, 1965; Blain and Brotherton, 1975; Ivankin and Moiseyev, 1972; Logn and Bølviken, 1975; Parasnis, 1974; Semenov and Kalashnikov, 1967; Sveshnikov, 1967; Sveshnikov and Ryss, 1963) and the dipole character of conducting ore bodies seems to be a general rule.

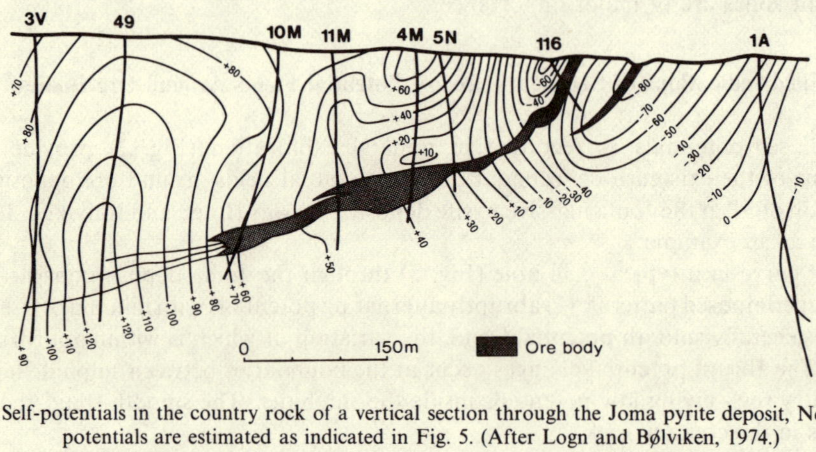

FIG. 6. Self-potentials in the country rock of a vertical section through the Joma pyrite deposit, Norway. The potentials are estimated as indicated in Fig. 5. (After Logn and Bølviken, 1974.)

Natural Currents and Element Distribution at Ore Bodies

According to the model presented above, the natural electrical currents in the country rock are carried by ions and it has been suggested by Govett (1973), Bølviken and Logn (1975) and several Russian authors (e.g. Sveshnikov, 1967; Shvartsev, 1976) that natural currents might influence the distribution of chemical elements.

Webber (1975) doubts that this would be the case to any practical extent, because ionic transport would be a relatively slow and, therefore, weak mechanism compared to diffusion and convection. Webber bases his views on calculations of the velocity at which Ag ions flow in an electrical field of 5 mV/m, which is the order of magnitude of the gradients normally encountered in SP surveys. What Webber estimates is in fact the likelihood for a specific ion released at the anode to reach the cathode. On that point he may be correct; solving such a problem involves consideration of absolute speeds of material transport. However, the formation of geochemical anomalies is not necessarily controlled by speeds of particle movement. More important factors are the transportation fluxes (i.e. the current densities of the transport) and the retaining mechanisms (concentration traps). Furthermore, the net transportation of material would in any case be the vectorial sum of the effect of the various submechanisms. Consequently, one transportation mechanism may dominate another totally only in cases when their directions are more or less parallel. When acting at more or less right angles the various mechanisms will not counteract each other. On the contrary, convection and diffusion may promote the flow of electric currents. Webber's calculations, though interesting, do not take these factors into account, and provide no reason for eliminating the possibility of electrochemical formation of geochemical anomalies.

Moreover, empirical data seem to indicate that the distribution of chemical elements in nature can be influenced by natural electric currents. These data include (1) results of laboratory experiments and (2) field observations.

Laboratory Experiments Indicating Electrochemical Processes in Connection with Sulphides

Two types of laboratory investigations are referred to here: (1) measurements of sulphide potentials (Buehler and Gottschalk, 1910; Gottschalk and Buehler, 1912; Sveshnikov, 1967), and (2) model experiments (Govett and Whitehead, 1974; Govett *et al.*, 1976; Thornber, 1975b).

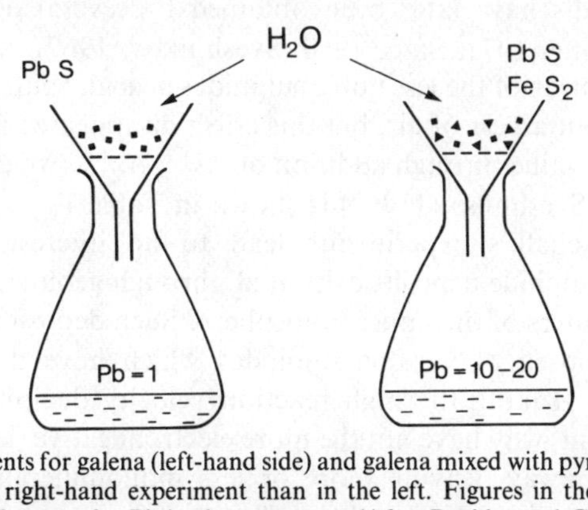

FIG. 7. Weathering experiments for galena (left-hand side) and galena mixed with pyrite (right-hand side). Galena weathers much faster in the right-hand experiment than in the left. Figures in the collected solution indicate concentration differences for Pb in the two cases. (After Buehler and Gottschalk, 1910.)

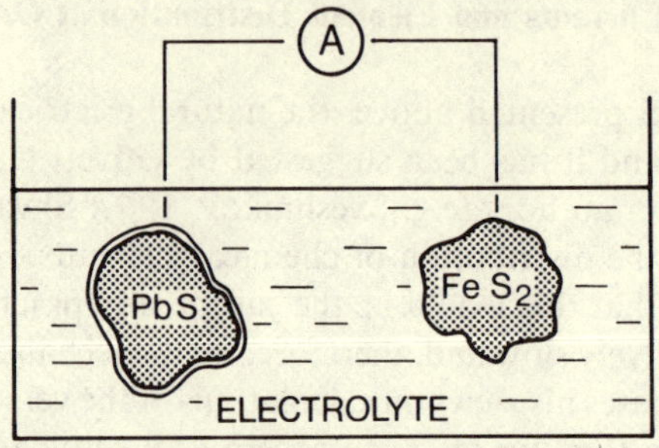

FIG. 8. Experiment showing that galena and pyrite form a galvanic cell when soaked in an electrolyte and connected through an amperemeter. The PbS crystal will gradually be coated by $PbSO_4$ contemporary with a decrease in the electric current measured. (After Gottschalk and Buehler, 1912.)

1. *Sulphide Potentials*

Buehler and Gottschalk (1910) performed weathering experiments in which crushed sulphides such as sphalerite, galena and covellite were washed repeatedly with water over a period of about three months. The various sulphides were treated alone or mixed with equal portions of pyrite or marcasite (Fig. 7). The wash water was collected and analysed after the end of the period. Buehler and Gottschalk observed that galena oxidized from 8 to 20 times faster in the presence of an equal amount of pyrite than when treated alone under the same conditions. They also found that the amount of iron transferred when galena and pyrite were treated together was but a small fraction of the amount transferred when sulphides of iron alone were treated.

Gottschalk and Buehler (1912) also placed one piece each of pyrite and galena in distilled water and connected them with a copper wire through an amperemeter (Fig. 8). They then observed a small electric current which decreased after a few days. In the course of this time there was no observed effect on the pyrite, while the galena became coated with lead sulphate. A similar piece of galena placed in the same solution, but not connected with pyrite, remained perfectly bright. It was concluded that there is a difference of potential between the various sulphides, which can be arranged in a series analogous to the electrochemical potential series for metals.

Corresponding results have later been obtained by several others (see, for example, Bitterlich and Woebking, 1972; Sato, 1960; Sveshnikov, 1967). Sveshnikov (1967) found rather intensive dissolution of the less noble sulphides in acid, neutral and basic solutions, as well as with and without access of air, but this effect disappeared if the Eh of the solutions were buffered at a low value through addition of As^{3+}/As^{5+}. An electrochemical series for sulphides taken from Sveshnikov (1967) is shown in Table 1.

Buehler and Gottschalk's experiments lead to an interesting question: how can polymetallic massive sulphide deposits exist at all through geological periods of time in the oxygen-rich ground waters of the upper lithosphere? Such deposits are in fact good electric conductors and composed of various sulphides which have different positions in the electrochemical series. Taking the high reaction velocity demonstrated by Buehler and Gottschalk into account, why have not the more electronegative (less noble) sulphides been dissolved and moved away, leaving more or less monomineralic deposits of the more electropositive (more noble) sulphides behind?

TABLE 1. ELECTROCHEMICAL SERIES OF SULPHIDE POTENTIALS (After Sveshnikov, 1967)

		V
Chalcocite	Cu_2S	0.64
Argentite	Ag_2S	0.49
Pyrite	FeS_2	0.49
Cinnabar	HgS	0.46
Covellite	CuS	0.45
Galena	PbS	0.22
Pyrrhotite	FeS	0.2
Millerite	NiS	0.11
Sphalerite	ZnS	0.16

There are two possible answers to this question: (1) either the grains of the less noble sulphide must all be coated with more noble (and/or less conducting) minerals, or (2) there must be some sort of cathodic protection.

Coatings would show up in polished sections under the microscope; in cases where no such coatings are observed, alternative (1) could not be the case.

Cathodic protection by the self-potential could, however, be the explanation. The configuration of the various sulphides in some polymetallic ore bodies is perhaps such that they match the distribution of self potentials. The cathodic protection would be most effective in massive steep dipping ore bodies with pronounced self-potential anomalies. Disseminated or flat-lying bodies would produce weaker SP anomalies, and hence less cathodic protection, a reason why the individual grains of disseminated mineralizations would tend to be monomineralic. Also, flat-lying massive sulphide deposits would probably tend to be more monomineralic than steep dipping massive deposits. The validity of these statements could easily be tested by treatment of existing data from known ore deposits.

2 Model Experiments

Govett *et al.* (1976) simulated the dispersion of elements in rocks around sulphide deposits through long period laboratory experiments, where glass or plastic tanks were filled with crushed rock or silica sand in and aqueous solution covering the rock. Crushed Cu-, Pb-, Zn-, Fe-sulphides contained in a nylon mesh bag were placed within the rock matrix to simulate a sulphide body. The tanks were covered, but not sealed from contact with the atmosphere. The potentials at various locations in the tanks were measured with platinum electrodes using a calomel electrode as a reference. At the end of the experiment the tanks were sampled for chemical analysis.

The potential patterns obtained were in some cases difficult to interpret, but generally the Eh measurements showed a relative positive centre below the sulphides and a relative negative centre above the sulphides, as can be predicted from models of a secondary redox potential field. The distribution of elements around the sulphide columns showed an "intriguing similarity to a theoretical electrochemical model derived from the hypothesis proposed bu Sato and Mooney (1960) to explain the generation of self-potential currents".

Thornber (1975) did model experiments to simulate ore alteration. Small slabs of massive pyrrhotite pentlandite primary ore from the Kambalda nickel deposit, Australia, were placed in agar gel, so that the upper end of the slab was exposed to air. The agar gel contained colour indicators for pH, Fe^{2+} and Ni^{2+} in order to visualize the chemical reactions taking place. Self-potentials were also measured in the system. The experiments showed cathodic reactions producing high pH at the upper end of the sulphide slabs, and anodic reactions with release of Fe^{2+} and Ni^{2+} at the lower. Minimum SP was recorded near the "outcrop" of the slab, while a positive maximum occurred just above its lower/middle part.

Thornber's results agree well with what should be expected from the concept of secondary redox potential field around conducting sulphide bodies.

Field Observations Indicating Electrochemical Processes in Connection with Sulphide Deposits

Several authors support the idea that weathering of sulphide deposits includes cathodic reactions near the surface and anodic reactions at depth (e.g. Habashi, 1966; Thornber, 1975a). Many observations and measurements have been made disclosing geochemical distribution patterns, which are related to the SP or in other respects indicate effects of electrochemical processes. The research cited here is grouped into (1) investigations of element distribution *within* sulphide bodies (Karasev *et al.*, 1970; Ivankin and Moiseyev, 1972; Sveshnikov, 1967; Thornber, 1975a) and (2) investigations of element distribution *around* sulphide bodies (Sveshnikov, 1967; Govett, 1973, 1974, 1975, 1976; Juve, 1978).

1 Element Distribution within Sulphide Bodies

Karasev *et al.* (1970) and Ivankin and Moiseyev (1972) found that the crystal structure and electrical properties of individual minerals of steeply dipping ore bodies may vary systematically from the lower (anode) to the upper (cathode) end. At great depths pyrites with *n*-type conductivity prevail, while *p*-type conductivity dominate close to the surface.

Sveshnikov (1967) and others observed a correlation between intensity of self-potential anomalies and formation of secondary enrichment zones. The stronger the natural self-potential anomalies, the more pronounced formation of secondary enrichment zones would be. Sveshnikov commented on the question of causes and effects, and concluded that the enrichment zones are the effects of the natural potentials, and not vice versa.

Based on data from several sulphide deposits Sveshnikov (1967) also stated that self-potential anomalies normally are weak at deposits which have well developed zones of oxidation. Because of the reducing effect of the more pronounced cathodic reactions sulphide bodies with strong self-potential anomalies are more protected from oxidation than bodies with weak self-potential anomalies.

Modifying a technique devised by Lisitsin (1967) Thornber (1975) placed some samples of sulphides taken of various depths from the Kamalda nickel deposit, Australia, in sealed jars without access to air, and shook and ground them in an electrolyte similar to the local ground water. Eh and pH were then measured in the sludge. The results were compared with the "supergene index" which is a parameter related to the depth location of the samples, ranging from low (0) for primary ore to high (15) for gossan. It was shown that the pH decreases with increasing supergene index, and Eh increases with increasing supergene index, and Thornber's results support the hypothesis of galvanic corrosion of the ore.

2 Element Distribution around Sulphide Bodies

Sveshnikov (1967) and Shvartsev (1976) describe "electro-chemical diffusion auroles", which are thought to exist around many Russian deposits. This type of geochemical anomaly is supposed to be due to dissolution of the less noble mineral in microgalvanic cells formed by sulphide couples. The local cathodic cell reactions would normally produce metals of low oxidation state as well as sulphide ions and the local anodic reactions metal ions as well as elementary sulphur.

Govett (1973, 1976) found characteristic distribution patterns of various elements in residual soils, transported soils and post-mineralization rocks near known sulphide bodies in Cyprus and Canada. The complex metal distribution were obviously due to secondary processes, but simple solution and movement by ground water did not offer completely adequate explanation of the detailed patterns. He concluded that there could be significant differential secondary rearrangement of metal in soils and rocks due to electrical potential related to sulphide bodies.

Govett (1975, 1976) also reported data from conductivity measurements in soil slurries. Based on his results from traverses of closely spaced soil sampling at the White Lake massive sulphide deposit, Manitoba, Canada, Govett suggested that the location of deeply buried or blind sulphide deposits may be found by (1) A "rabbit ear" H^+ anomaly of two lateral peaks on the rocks of a central trough (deep deposit); or an H^+ anomaly of three peaks separated by troughs (moderately deep deposit); (2) conductivity anomalies which are the inverse of the H^+ anomalies; and (3) distribution patterns of metals generally similar to those of conductivity (moderately deep deposits) or similar to those of H^+ (very deep deposits) (Fig. 9). Govett also states that wherever there is a detectable SP anomaly at the surface there will be an anomalous distribution pattern of H^+ in the soils. These patterns are in agreement with those expected from electrochemical mechanisms modified by oxidation processes. In Norway (Bølviken and Logn, 1975) results have been obtained which show trends similar to those reported by Govett.

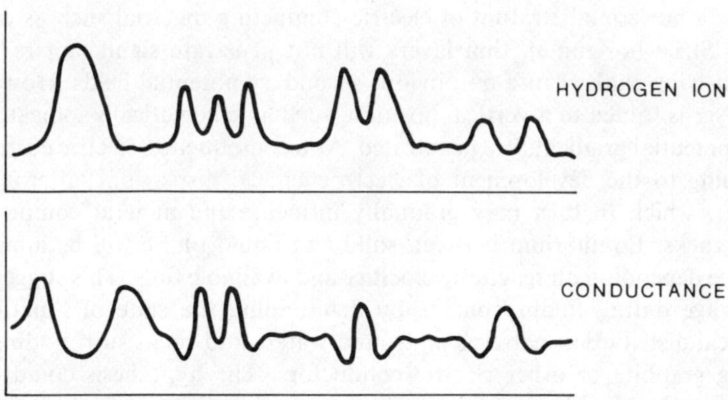

FIG. 9. Schematic illustration of hypothetical distribution of H^+ and conductance (corrected for H^+) in surface soils overlying sulphide and graphite conductors at various depth. (After Govett, 1976.)

A recent observation indicating cathodic reduction of copper above a sulphide deposit is described by Juve (1978). During preparation for open-pit mining of the Stekenjokk massive pyrite deposit in Sweden the overburden was stripped off. In the glacial till just above the suboutcrop of the ore several tons of native copper were found occurring as a post-glacial cement precipitated between the rock fragments and mineral grains of the till. One may wonder why native metals such as Cu, which is unstable under normal atmospheric conditions, can form in glacial soil above a sulphide deposit which is an environment normally expected to be acidic and oxidizing. The primary mineralization of Stekenjokk consists of banded, massive to disseminated iron, zinc and copper sulphides lying conformably on the transition between acid pyroclastic rocks and black shales (Juve, 1974). Both ores and shales are highly conductive. The mineralization reaches great depths (700 m below the surface), thus penetrating relatively large gradients of the earth redox potential field. A well-defined surface SP anomaly was found above the deposit prior to the mining activity (Ø. Logn, pers. comm.). It might then be that the cathodic potential of this long sulphide carbon electrode is low enough to discharge copper ions from ground water percolating in the zone of the suboutcrop of the ore.

The general conclusion about relation between element distribution and natural electric currents is: it is well documented that electrically conducting sulphide, graphite or oxide deposits can cause secondary redox potential fields, which may influence the geochemical dispersion patterns in the conducting body as well as in the surrounding ground water, bedrock and overburden.

Some Speculative Consequences of Redox Potential Fields

In addition to the effects treated above there are many other interesting aspects of the redox potential field of the earth, of which three are suggested here as subjects for discussion.

I. Consider a horizontal stratum of electric conducting material such as carboniferous black shales. Since horizontal, thin layers will not penetrate significant redox potential gradients, flat-lying shales cause no obvious secondary potential fields. However, when a horizontal layer is folded to a vertical position, such as schematically indicated in Fig. 10, much larger potential gradients are penetrated. As a consequence, electric currents will start to flow, leading to the development of electrochemical dispersion patterns in the local ground water, which in turn may gradually influence the mineral composition of the surrounding rocks. Equilibrium between solid and liquid phase will be approached to a varying degree depending on reaction velocities and available time. This suggests a possible new way of age dating folding phases by determining the state of "maturity" of the electrochemical distribution patterns in ground waters and rocks surrounding deposits of steep dipping graphite or other electric conductors. The hypothesis could be tested by geochemical investigations at a variety of steep dipping electrically conductive zones of already known ages.

The model is probably exemplified by the Bidjovagge deposit, Northern Norway. In this area where the rocks have been tectonized during the Precambrian, large deposits of albite carbonate rocks of unknown origin occur in close connection with outcrops of strongly folded graphitic rocks of regional extent. Could these albite carbonate rocks have developed as a result of sodium enrichment in the environment of graphite cathodes?

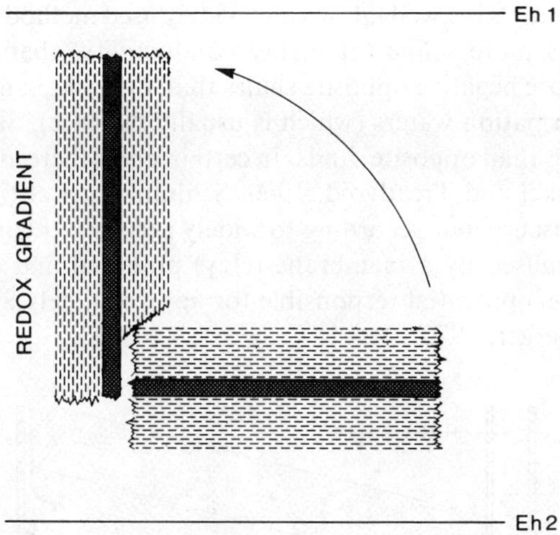

FIG. 10. Schematical illustration of a hypothetical model of a horizontal thin layer of highly conductive material being raised by folding to a vertical position, thus penetrating a large gradient of redox potentials.

II. Consider a section through the upper lithosphere with a vertical zone of relatively high ionic conductivity. The horizontal conductance variations will cause deviations of the primary redox potential field and a flow of electric currents through the system (Fig. 11). Due to this effect positive ions (e.g. H^+) will flow downwards in the zone of the higher conductivity and upwards in the zone of the lower conductivity. Negative ions (e.g. OH^-) will flow in the opposite directions.

This may represent a simplified model of the system of rocks commonly encountered in oil-well drilling: the zone of relatively high conductivity would represent the shales, and the zone of relatively low conductivity would represent the sands.

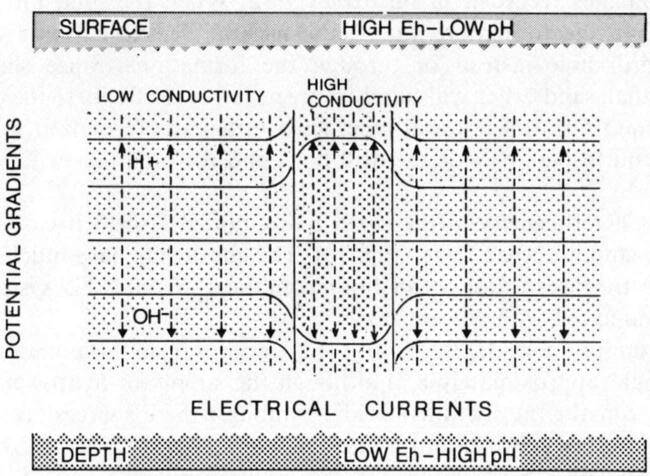

FIG. 11. Schematical representation of a vertical rock formation of high conductivity in a country rock of relatively low conductivity. Horizontal lines: isopotential curves. Vertical lines: current direction. The current directions are relative and do not show the absolute direction flow. Potentials will be relatively low and pH high in the upper part of the higher conducting zone.

Logging of self-potentials is a well known and widely used method in oil exploration. If the drilling mud filtrate is more saline (of higher conductivity) than the formation waters, SP-readings will be more negative opposite shales than opposite sands. If the mud filtrate is less saline than the formation waters (which is usually the case), SP readings will be more positive opposite shales than opposite sands. In certain areas there are occasional exceptions from these rules (Gaskel and Treadgold, 1960; Schlumberger, 1972). The origin of these effects is somewhat obscure, but according to widely accepted theories the SP anomalies in oil wells are mainly caused by a membrane (clay) potential and a liquid junction (mud filtrate/formation water) potential responsible for approximately 80 and 20% of the effect respectively (Schlumberger, 1972).

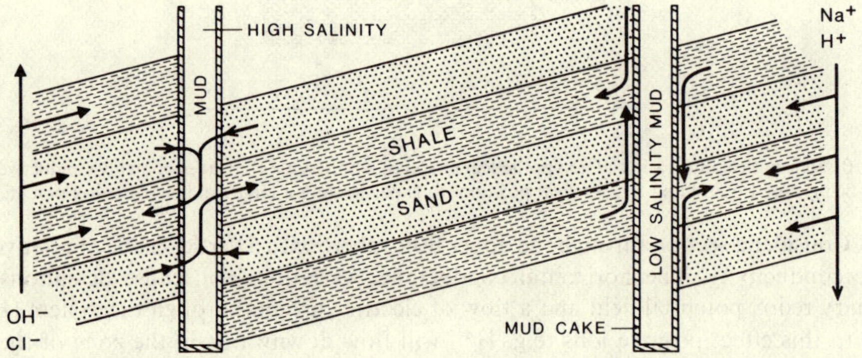

FIG. 12. Schematical illustration of natural currents causing SP anomalies in oil wells. If the mud filtrate has higher conductivity than the formation waters, the electric currents will tend to pass through the drill hole. If the mud filtrate has lower conductivity than the formation waters, the electric currents will pass outside the drill hole.

As indicated in Fig. 12, the redox potential field of the earth could be an additional source of the SP anomalies recorded in oil exploration. When the mud filtrate has a higher conductivity than the formation waters, the natural electric currents will tend to pass through the drill hole instead of through the formations. Since shales have higher conductivities than sands, they will supply more negative electricity to the drill hole than the shales. If the mud filtrate has lower conductivity that the formation waters, the natural currents will tend to pass outside the drill hole, and the polarities will be reversed.

III. A major heterogeneity in the earth redox potential field is caused by the system continents—ocean and ocean floor (Fig. 13). The continents have much greater electrical resistivity than the ocean and ocean floor combined. Would this system then form an enormous galvanic cell with current flow?

Let us, for example, consider the acidity of the oceans from this point of view. The pH of sea water is high (approximately 8.5) although the supply of hydrogen ions from many sources ought to make the sea water acidic. Amongst these sources are:

acid river and other continental waters;

carbon dioxide and other volatile acids of the atmosphere;

organic matter produced in the sea and its acidic decay products;

submarine volcanic emanations;

hydrolysis during precipitation of Fe and Mn oxides;

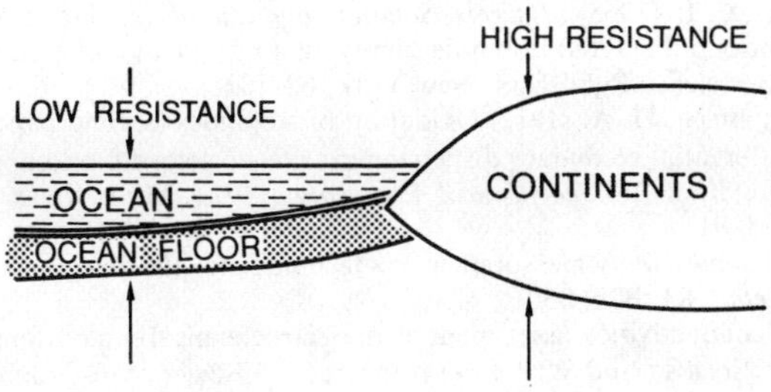

FIG. 13. Schematical section through the ocean, ocean floor (low resistance) and continents (high resistance) illustrating a possible major galvanic cell, the energy source of which is oxygen of the atmosphere and material at low oxidation state in the interior of the earth.

shorter residence time in the atmosphere for Cl^- than for Na^+ (Låg, 1968; Rösler and Lange, 1972);

longer residence times in the oceans for Cl^- than for Na^+ (Rösler and Lange, 1972).

Calcium carbonate equilibria are supposed to regulate the pH of the oceans, but sea water is not everywhere saturated with $CaCO_3$. Why is not the pH lower? Could electrochemical processes in galvanic cells formed by the system oceans–continents cause OH^- ions to migrate from the interior of the earth through the ocean floor into the sea? Measurements of self-potentials along regional profiles crossing continental margins and/or midocean ridges in a similar way as done by Anderson and Johnson (1976) over a geothermal area would be of great interest in this connection.

Acknowledgements

I want to thank H. Hovdan and W. E. Sharp for fruitful discussions and many valuable comments.

References

ANDERSON, L. A. and JOHNSON, G. R. (1976) Application of the self potential method to geothermal exploration in Long Valley California. *J. Geophys. Res.* **81**, 1527–1532.

BAAS BECKING, L. G. M., KAPLAN, J. R. and MOORE, D. (1960) Limits of the natural environment in terms of pH and oxidation-reduction potentials. *J. Geol.* **68**, 243–284.

BECKER, A. and TELFORD, W. M. (1965) Spontaneous polarization studies. *Geophysical Prospecting* **13**, 173–188.

BITTERLICH, W. and WÖBKING, H. (1972) *Geoelektronik*, Springer-Verlag, Wien, 349 pp.

BLAIN, C. F. and BROTHERTON, R. L. (1975) Self-potentials in relation to oxidation of nickel sulphide bodies. *Trans. Inst. Mining Metall.* **84**, B123–127.

BØLVIKEN, B. and LOGN, O. (1975) An electrochemical model for element distribution around sulphide bodies. In: *Geochemical Exploration* I. L. ELLIOT and W. K. FLETCHER, eds., 1974, Elsevier, Amsterdam 631–650.

BUEHLER, H. A. and GOTTSCHALK, V. H. (1910) Oxidation of sulphides. *Econ. Geol.* **5**, 28–35.

CORWIN, R. F. (1976) Offshore use of the self-potential method. *Geophys. Prospecting* **24**, 79–90.

DODGE, F. C. (1972) Variation of ferrous-ferric ratios in the Central Sierra Nevada batholith, U.S.A. 24th International Geological Congress, Canada 1972, Section 10, *Geochemistry* 12–19.

DYCK, A. V. (1975) Electrical borehole methods applied to metallic mineral prospecting. *Geological Survey Canada Paper* 75–31, 13–29.

GARRELS, R. M. and CHRIST, C. L. (1965) *Minerals, Solutions and Equilibria*, Harper & Row, New York, 450 pp.

GASKELL, T. F. and THREADGOLD, P. (1960) Borehole surveying. In *Methods and Techniques in Geophysics*, Vol. I, S. K. RUNCORN ed. Interscience Publishers, New York, 62–103.

GOTTSCHALK, V. H. and BUEHLER, H. A. (1912) Oxidation of sulphides (second paper). *Econ. Geol.* 7, 15–34.

GOVETT, G. J. S. (1973) Differential secondary dispersion in transported soils and post-mineralization rocks: an electrochemical interpretation. In *Geochemical Exploration 1972*. M. J. JONES, ed. Institution of Mining Metallurgy, London, 81–91.

GOVETT, G. J. S. (1974) Soil conductivity measurements: a technique for the detection of buried sulphide deposits, *Trans. Inst. Mining Metall.* **83**, B29–B30.

GOVETT, G. J. S. (1975) Soil conductivities: assessment of an electrochemical exploration technique. In *Geochemical Exploration 1974*, I. L. ELLIOTT and W. K. FLETCHER, eds., Elsevier, Amsterdam, 101–118.

GOVETT, G. J. S. (1976) Detection of deeply buried and blind sulphide deposits by measurement of H^+ and conductivity of closely spaced surface soil samples. *J. Geochem. Explor.* **6**, 359–382.

GOVETT, G. J. S. and WHITEHEAD, R. E. S. (1974) Origin of metal zoning in stratiform sulphides: a hypothesis. *Econ. Geol.*, **69**, 551–556.

GOVETT, G. J. S., GOODFELOW, W. D. and WHITEHEAD, R. E. S. (1976) Experimental aqueous dispersion of elements around sulfides. *Econ. Geol.*, **71**, 925–940.

HABASHI, F. (1966) The mechanism of oxidation of sulfide ores in nature. *Econ. Geol.* **61**, 587–589.

IVANKIN, P. F. and MOISEYEV, V. S. (1972) The nature and structure of natural electrical fields of sulfide deposits. *Dokl. Akad. Sci. U.S.S.R.*, **206**, 9–10.

JUVE, G. (1974) Ore mineralogy and ore types of the Stekenjokk deposit, central Scandinavian Caledonides, Sweden. *Sveriges geologiska Undersökning* C706, 162 pp.

JUVE, G. (1978) Formation of native copper in glacial overburden above the Stekenjokk ore body. In preparation.

KARASEV, A. P., KNYAZEV, G. I., KRASNIKOV, V. I. and SEYFULLIN, R. S. (1970) Influence of electrophysical properties of sulphides and electrochemical phenomena on hydrothermal ore formation. *Int. Geol. Rev.* **12**, 952–958.

KRAUSKOPF, K. B. (1967) *Introduction to Geochemistry*, McGraw-Hill, New York, 721pp.

LISITSIN, A. K., (1967) Method for determination of Eh and pH of equilibrium between aqueous solution, rocks and minerals. *Geokhimiya* **8**, 994–1002.

LOGN, Ø. and BØLVIKEN, B. (1974) Self potentials at the Joma pyrite deposit, Norway. *Geoexploration* **12**, 11–28.

LOGN, Ø. and BØLVIKEN, B. (1975) Self potentials in ore evaluation. Abstract. European Association of Exploration Geophysicists Meeting, Bergen June 1975. To be printed.

LÅG, J. (1968) Relationships between the chemical composition of the precipitation and the contents of exchangeable ions in the humus layer of natural soils. *Acta. Agric. Scand.* **18**, 148–152.

PARASNIS, D. S. (1966) *Mining Geophysics*, Elsevier, Amsterdam, 356 pp.

PARASNIS, D. S. (1974) Some present-day problems and possibilities in mining geophysics, *Geoexploration* **12**, 97–120.

RÖSLER, H. J. and LANGE, H. (1972) *Geochemical Tables*, Elsevier Publishing Company, Amsterdam, 311–317.

SATO, M. (1960a) Oxidation of sulfide ore bodies, I. Geochemical environments in terms of Eh and pH. *Econ. Geol.* **55**, 928–961.

SATO, M. (1960b) Oxidation of sulfide ore bodies, II. Oxidation mechanisms of sulfide minerals at 25°C. *Econ. Geol.* **55**, 1201–1231.

SATO, M. and MOONEY, H. M. (1960) The electrochemical mechanisms of sulfide self-potentials. *Geophys.* **25**, 226–249.

SCHLUMBERGER, (1972) *Log Interpretation*, Vol. I *Principles*. Schlumberger Ltd., 277 Park Avenue, New York.

SEMENOV, M. V. and KALASHNIKOV, JU. A. (1967) Natural electric fields of some polymetallic deposits at Rudny Altai (In Russian). *Geologia i geofizika* **7**, 131–143.

SHVARTSEV, S. L. (1976) Electrochemical dissolution of sulphide ores (translated from Russian). *J. Geochem. Explor.* **5**, 71–72.

SILLÉN, L. G. (1965) Oxidation state of earth's ocean and atmosphere: a model calculation on earlier states: the myth of the probiotil soup. *Arkiv för kemi*, 24 pp.

SVESHNIKOV, G. B. (1967) *Electrochemical Processes in Suphide Deposits*, Leningrad University, 158 pp. (in Russian).

SVESHNIKOV, G. B. and RYSS, JU. S. (1963) Electrochemical processes in sulfide deposits and their geochemical significance. *Geochem.* **8**, 198–204.

THORNBER, M. R. (1975a) Supergene alteration of sulphides, I. A chemical model based on massive nickel sulphide deposits at Kambalda, Western Australia. *Chem. Geol.* **15**, 1–14.

THORNBER, M. R. (1975b) Supergene alteration of sulphides, II. A chemical study of the Kambalda nickel deposit. *Chem. Geol.* **15**, 117–144.

WEBBER, G. R. (1975) Efficacity of electrochemical mechanisms for ion transport in the formation of geological anomalies. *J. Geochem. Explor.* **4**, 231–233.

Discussion

A. BEUS:

Is there any interdependence between the dipole effect and the vertical dimension of a given ore body?

B. BØLVIKEN:

The greater the vertical penetration of the ore, the larger the differences in redox potential between the lower and upper parts of the ore body, and consequently the larger the electric potential differences available for the dipole formation. This, in addition to the fact that deeper penetration may lead to a larger anodic area, will increase the polarization of the ore body and thereby the intensity of the dipole effect.

These features suggest future possibilities for discovering relations between strength of geochemical anomaly and size of deposit.

D. T. RICKARD:

I would like to ask exactly what do you mean by redox potential in this case? If you mean *measured Eh* then, as you know, this has no direct relation usually with the chemical state of the system (due to incomplete redox equilibrium as, for example, in oxygenated surface waters where the measureed Eh is lower than the theoretical because the oxygen couple is not reversible). If you mean the *theoretical Eh* or electron activity then this must be calculated for all couples in the system, which is difficult in natural environments.

B. BØLVIKEN:

When I use the term "Redox potential field of the Earth", I mean the theoretical Eh (defined by Nernst's formula), a quantity which I agree, is very difficult to measure directly in nature. However, this fact should not prevent a meaningful discussion about the consequences of its expected variation with depth.

S. E. HAGGERTY:

If the redox potential of the earth's crust is as large as you suggest, how do you explain the fact that current data show that differences in oxidation states between deep seated (e.g. gabbros) and extrusive (e.g. basalts) suites is negligible, and that the extremes in oxidation are more directly related to SiO_2 contents?

B. BØLVIKEN:

I would tentatively suggest that the basalts have attained equilibrium with the redox potential of their environment at the place of solidification, to a less extent than gabbros have, such that they both reflect the conditions at some depth. On the other hand, our knowledge of the redox potentials variance with depth is rather incomplete. The redox gradients are probably not constant as assumed for the sake of simplicity in this paper. In reality they could be large at some places and small at others. Rock types originating from different depths would not necessarily always show corresponding significant differences in oxidation properties, even in cases where they were chemically in equilibrium with their surroundings at the time of formation. The Fe^{3+}/Fe^{2+} ratio in igneous rocks is not governed by the redox potential of the corresponding melt only, but also by the solubility of the various minerals. Some correlation between contents of Fe^{3+}/Fe^{2+} and SiO_2 should therefore be expected.

C. GRANIER:

Avez-vous étudié les anomalies de fuite dans le même esprit?

B. BØLVIKEN:

Regardless of any fault or fracture zones in the vicinity of the ore, it still remains true that the total amount of current leaving the ore must enter it again at some other site. Faults and fracture zones passing the mineralization would be possible pathways for these natural electric currents, and convection and diffusion might promote or hinder the electrochemical migration along the zones depending on the circumstances. The author is, however, not aware of any empirical data elucidating this situation.

Relative Behavior of Uranium and Lead in some Acidic and Chelating Environments

C. GRANIER†, G. MONTEIL‡ and J. TRICHET§

†*Géochimiste-Conseil, Groupe Imetal, 1, Boulevard de Vaugirard, 75751 Paris Cedex 15, France.*
‡ *Géochimiste, Amok Ltd., PO Box 9204, Saskatoon, Sask., Canada.*
§*Professeur-Directeur du Laboratoire de Géochimie, Minéralogie et Géologie Appliquées, Domaine Universitaire, 45045 Orléans Cedex, France.*

Abstract

Cluff Lake uranium deposits, located in northern Saskatchewan, Canada, are overlaid by various glacial sediments. In addition to uranium, several other metals are present in the mineralized zones. Among these elements, lead is one of the most abundant (partially as a product of uranium decay). The quantity represents approximately 8% of the total uranium content.

During the glacial stages, mineralized particles have been dislocated and dispersed in the overburden. Uranium and lead are therefore both subject to the action of surficial biochemistry related to the nature of organic compounds. These organic materials, which are produced in both podzolic soils and swampy areas, have acidic and chelating properties.

In the two types of environment (soils and swamps), uranium and lead are both affected by organic compounds which result in some similarities in their behaviour. However, due to the substantial differences in their chemical characteristics, lead has less mobility than uranium. A study carried out on some swamps and their margins shows how these differences could be used to allow the localization of their common origin.

Characteristics of Mineralizations—The Glacial Causes

Presently known uranium deposits in Cluff Lake contain several elements in addition to uranium. Lead is one of the most abundant (approximately 8% of the total uranium content, partially as a product of uranium decay).

During the glacial stages, mineralized particles have been dislocated and dispersed in the overburden; they are now concealed in the glacial till. Other mineralized fragments were reworked during fluvial events and spread away from their origins.

As far as these types of mechanical dispersions were involved, lead and uranium had, therefore, a common destiny and an identical scattering.

Glacial Morphology—Present Geochemical Conditions

A smooth topography corresponds to the unconsolidated overburden which covers most of the area. Glacial formations as drumlins, eskers and various other glacial structures are displayed over a landscape of lakes, swamps and muskegs.

Dry lands are mainly covered with evergreen trees (pines) of which the decay produces acidifying and chelating organic materials. Porous sandy soils then are forming an easily

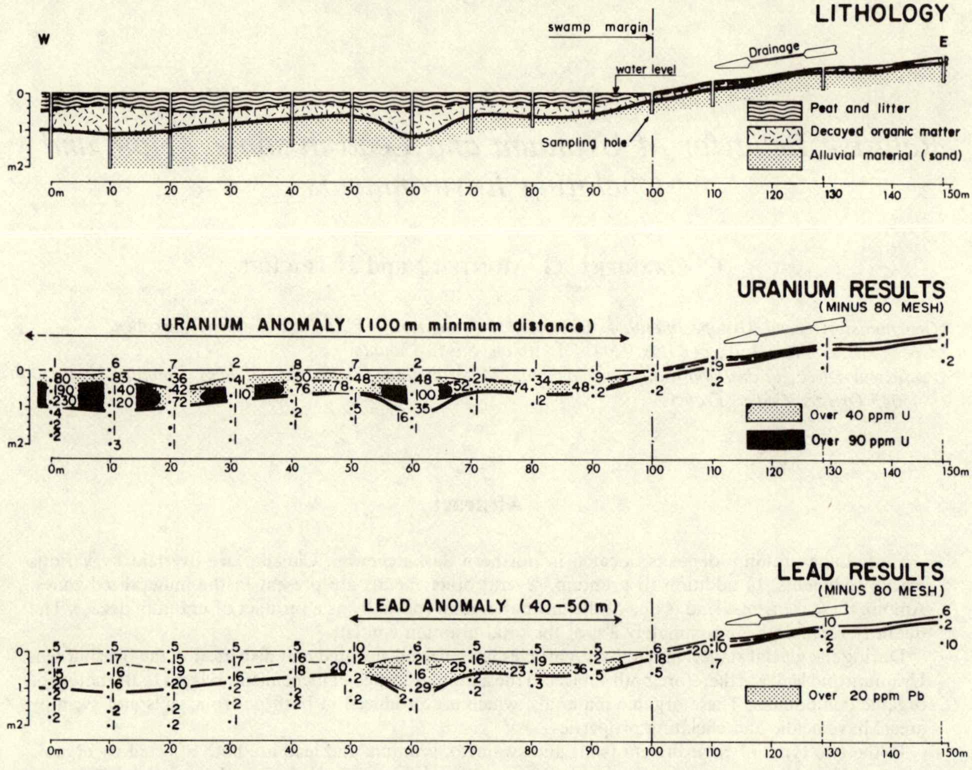

FIG. 1. Sections showing the nature of sediments in the peat swamp (lithology) and the extension of uranium and lead from the swamp border towards the swamp's heart.

leachable and oxidizing media. As a matter of fact, pedologic alteration frequently leads to a podzolic soil.

Lower lands (ponds, peat-bogs) are acting as geochemical traps where chelating organic sediments hang on the elements leached from the environment.

Purpose of the Study

Under such oxydizing and acidic conditions, the uranium is easily dissolved from the glacial overburden (or from the mineralization *in situ*) and carried down the possible collectors. At the same time, lead, which has less mobility than uranium, is likely to stay near its source. Accordingly, this study was intended to show how such a difference in the chemical characteristics could apply to a prospecting purpose.

The experiment was carried out on a swamp where the organic sediments are thick enough to allow vertical sections (Figs. 1 and 2) and also on some temporary channels and the collector in which they flow (Fig. 3).

1. Study of a Swampy Area

Several sections were set through a swamp which was of already known anomalous content. Lead and uranium were analyzed, both after dry ashing.

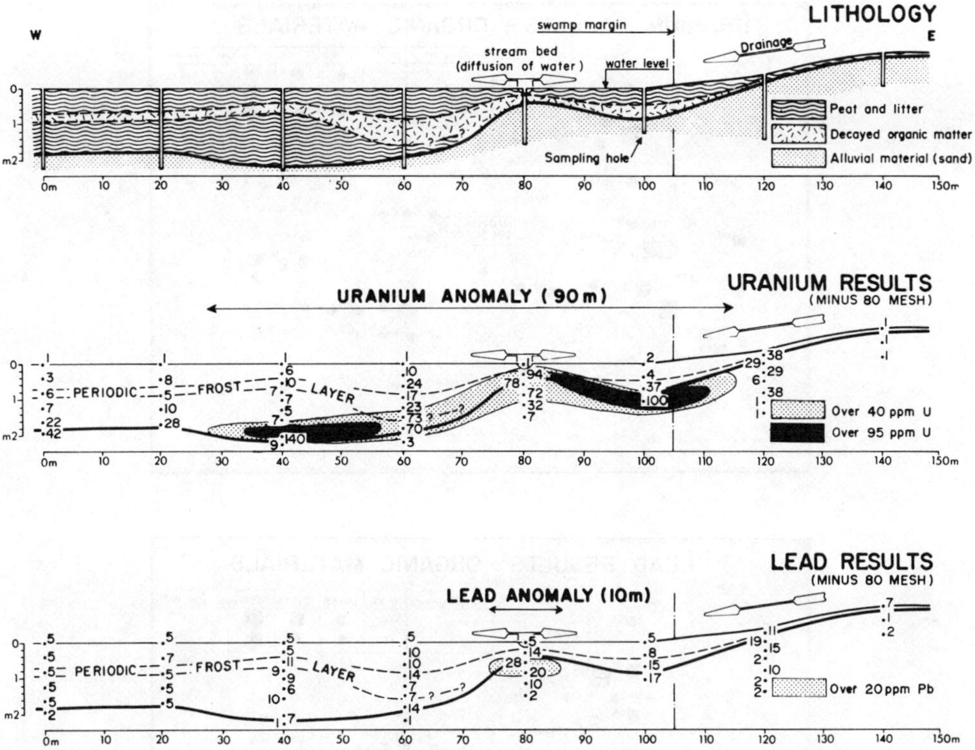

FIG. 2. Sections showing the uranium and lead movements inside the swamp when the metal source is provided by a lateral and an inner drainage.

The contouring of the results shows that, from the contributions of lateral and diffuse drainages (Fig. 1) or, on the contrary, of a central and linear channel (Fig. 2), the uranium is distributed over most of the length of organic sediments whereas the lead is restricted to the edges of the contributing area (side of the swamp or bed of the channel). The patterns illustrate the great mobility of uranium and its concentration in the deepest levels, i.e. the more decayed sediments, and the not very mobile character of lead as well as its great affinity for organic compounds. The content in both elements drops when reaching the alluvial sandy sediments.

The existence of a periodic frozen level (Fig. 2) is a common feature for many swamps of this type; this is a frequent obstruction to sample collection early in the summer.

2. *Study of Channels and their Swampy Outlets*

In order to ascertain the behaviour of uranium and lead between their source (which can be either a mineralization *in situ* or a glacial formation which contains mineralized particles) and the swampy collector where they concentrate, a detailed sampling was carried out in the vicinity of a previously discovered showing. This mineralization is a few meters wide, with an average content of 0.76% U_3O_8.

Three sampling stages were performed in order to follow up the original anomaly. Figure 3 takes these successive steps into consideration:

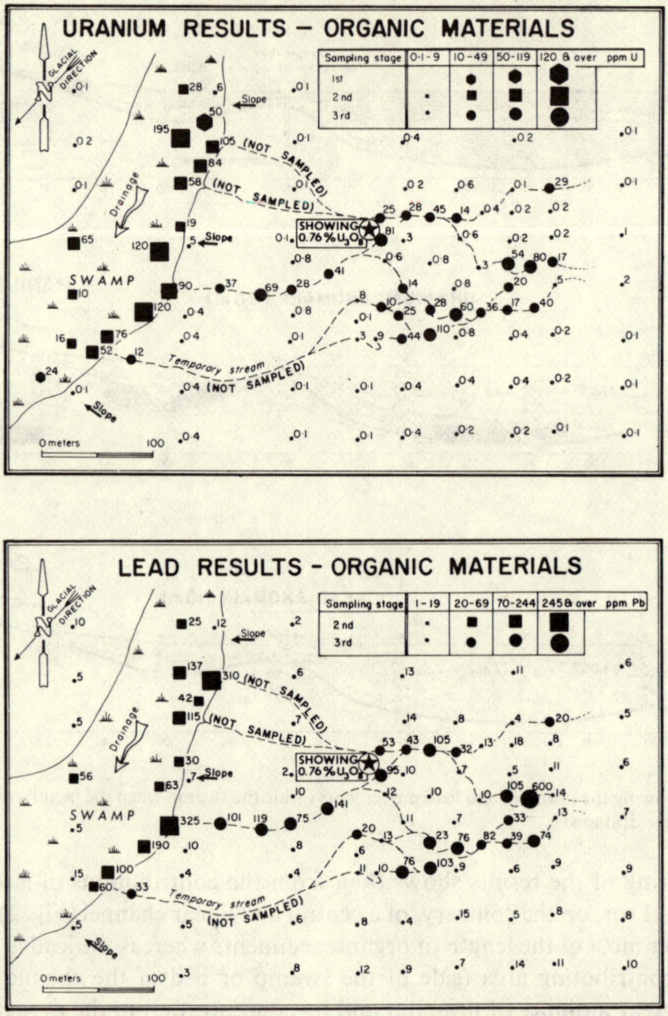

FIG. 3. Geochemical prospecting sketch for uranium and lead in Cluff Lake area.

First stage. Sampling, with a 300-metre spacing, of organic matters in the swamp downslope of the showing. The uranium results ratify the expected existence of an anomaly.

Second stage. Collection of samples, with a 40-meter spacing, along the edge of the swamp. The uranium and lead analyses enhance local concentration near the outlets of temporary channels.

Third stage. Samples are collected on a portion of the channels, upstream of the showing and further, until they disappear in the general morphology; at the same time organic materials from the surrounding soils are collected (A_0–A_1 horizons). This sampling disclosed a continuous chemical dispersion along the channels, and also the possibility of a new discovery beyond the showing. The very low values related to samples collected outside the channels indicated their usefulness as favourable sampling media.

Some observations can be pointed out:

Whenever the uranium content is very high within the channels, the maximum value appears to be in the swamp itself. On the opposite, the maximum lead value is located far upstream.

At the level of outlets, the lead is sharply trapped, whereas the uranium has a wider diffusion over the collector.

Conclusion

These characteristics make possible a methodology adapted to surficial geochemical prospecting.

The sampling on a large grid of organic sediments located at the bottom of the swamps would allow an outline of anomalous zones among the wide swampy areas. The analysis of uranium alone could be sufficient. Collection of more samples along the edge of anomalous swamps and the sampling of channels flowing into them would allow determination of the directions of supply by observation of lead concentrations; thus, followup work could be undertaken with more reliability.

Discussion

B. BØLVIKEN:

When both U and Pb are enriched in humus, how do you explain the different patterns of the two in your bog profiles?

G. MONTEIL:

Cette différence de comportement est à rattacher au caractère extrêmement mobile de l'uranium (ce qui le conduit à aller plus loin à travers le marécage) et à la très grande insolubilité du plomb lié à la matière organique (ce qui explique le blocage de cet élément dès son arrivée dans le marécage).

R. SINDING-LARSEN:

Vous ne considérez pas le pH dans votre exposé alors que l'orateur précédent Jane Plant le signale comme un facteur déterminant.

G. MONTEIL:

A l'échelle de recherche qui correspond, somme toute, déjà à celle d'une prospection semitactique, l'homogénéisation générale résultant de l'action glaciaire autorise à considérer l'influence du pH comme constante à travers la région. Cependant il serait faux d'affirmer que les variations de pH n'ont pas un rôle dans le niveau de teneur des anomalies. En fait, lorque la maille d'échantillonnage est choisie de manière à ce qu'une anomalie soit définie par un nombre suffisant de points, il n'est pas nécessaire de tenir compte d'un tel facteur, puisque ce qui intéresse le prospecteur n'est pas le niveau de teneur dans l'absolu, mais bien plutôt la localisation et la forme de l'anomalie.

Eléments en Traces dans les Pyrites de la Province de Huelva (Espagne)

A. M. DE KERSABIEC ET G. ROGER

Equipe de recherche du C.N.R.S. n° 06 0194, Laboratoire de Géologie appliquée, Université Pierre et Marie Curie, 4, place Jussieu, 75230 Paris Cedex 05

Résumé

Dans 55 échantillons de pyrite triée provenant de la province à amas sulfurés volcano-sédimentaires de Huelva les éléments suivants ont été dosés par spectrométrie d'absorption atomique: Co, Ni, V, Cr, Mn, Pb, Zn, Cu et Sb. Se a été dosé par colorimétrie. Ces dosages ont été entrepris à la suite d'une étude géologique et métallogénique détaillée de la partie centrale de cette province, qui a mis en évidence à la fois le milieu de dépôt sédimentaire des amas et leur liaison spatiale plus ou moins étroite avec le toit de deux épisodes volcaniques acides. Le modèle métallogénétique comporte une indécison quant à la source de la minéralisation: faut-il conclure à une liaison très directe avec des fluides tardi-volcaniques (dépôt près de griffons), ou au contraire à une intervention prépondérante du cycle sédimentaire (amas distaux par rapport aux griffons)? Les rapports Co/Ni et S/Se nous fournissent les principaux éléments de réponse. Le rapport Co/Ni est très élevé dans les amas (19,2 en moyenne) et décroît fortement dans la pyrite disséminée dans les roches de l'encaissant immédiat des amas (Co/Ni = 4,2) et plus encore dans les pyrites disséminées dans des roches métasédimentaires loin de tout amas connu (Co/Ni = 0,8). Le rapport S/Se est beaucoup plus faible dans les amas (moyenne = 3403) que dans la pyrite disséminée (7852). Les variations de ces deux rapports étayent l'hypothèse d'une liaison directe entre les amas et des fluides tardi-volcaniques.

Abstract

The following trace elements: Co, Ni, V, Cr, Mn, Pb, Zn, Cu and Sb were measured by atomic absorption spectrometry in 55 samples of selected pyrite from the volcanic sedimentary massive deposits of Huelva's province. Se was measured by colorimetry. These analyses were started after a geological and metallogenic study of the province's central area. This study has proved the sedimentary origin of the deposit and more or less close spatial relationship with the top of two acid volcanic sequences. The metallogenetic model is uncertain about the mineralizations origin: it could be concluded in favor of either a very close relationship with post-volcanic fluids (close feeder-pipes deposition) or the sedimentary sequence's predominance (massive deposits distant of the feeder-pipes). The answer is mainly given by the Co/Ni and S/Se ratios. The Co/Ni ratio is very high in the massive sulphide deposits (about 19.2) and strongly decreases in the pyrite disseminated in the close massive deposit's host rocks (Co/Ni = 4.2). This ratio is much lower in the pyrite of metasedimentary rocks which are distant of any known massive deposits (Co/Ni = 0.8). The S/Se ratio is very much lower in the massive deposits (mean = 3403) than in the disseminated pyrite (7852). The evolution of these two ratios' values proves the assumption of the direct relationship between the massive deposits and post-volcanic fluids.

Les amas pyriteux à Cu, Pb, Zn et les lentilles de jaspes à manganèse de la ceinture sud-ibérique sont encaissés dans des terrains volcano-sédimentaires dévono-dinantiens plissés et faiblement métamorphisés. Dans le cadre d'une Action Concertée de la D.G.R.S.T., † nous

†Convention de recherche n° 74.7.1181.

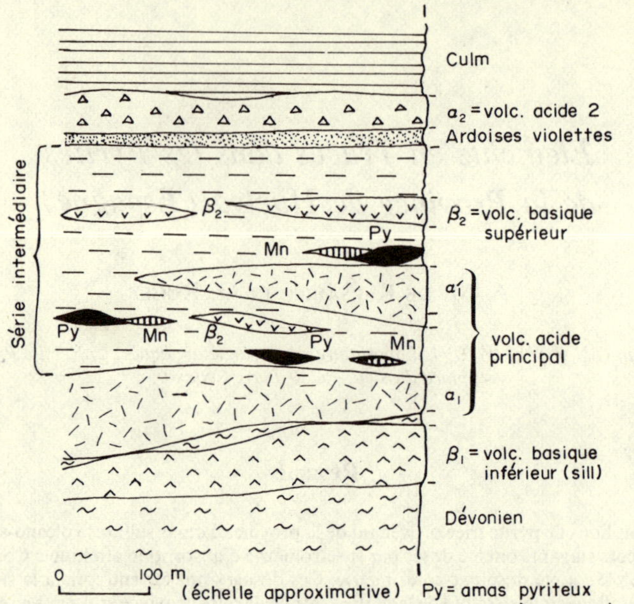

FIG. 1. Sucession lithologique simplifiée.

avons étudié une large transversale dans le partie centrale de la province de Huelva (P. Routhier *et al.*, 1977). Ce travail cartographique et cette étude géologique s.l. et métallogénique nous ont permis d'établir une succession lithologique détaillée et d'ébaucher de grands traits paléogéographiques. Les amas sulfurés, de même que les jaspes à manganèse, se localisent tous au toit stratigraphique de deux épisodes volcaniques acides 1 et 1′, dans la "série intermédiaire" (fig. 1). Certains amas se trouvent en contact direct avec les volcanites sous-jacentes, mais la plupart d'entre eux en sont séparés par une intercalation de shales, de jaspes ou de tuffites parfois très mince (quelques mètres), parfois puissante (jusqu'à 80 m à Sotiel). L'encaissant immédiat des amas est souvent représenté par des shales sombres. La relation spatiale amas-toit du volcanisme acide n'apparaît donc pas douteuse, mais la nature sédimentaire de l'environnement immédiat de la plupart des amas amène à s'interroger sur le bien-fondé de la liaison génétique qui est supposée unir les corps minéralisés au volcanisme sous-jacent par l'intermédiaire d'un hydrothermalisme tardi-volcanique. Le lien génétique volcanisme—amas ne serait-il pas plus distant, une place prépondérante revenant aux facteurs de concentration jouant dans le milieu de dépôt sédimentaire? Il faut tenter de répondre à cette question si l'on veut préciser les particularités de la province sud-ibérique dans le modèle métallogénique volcano-sédimentaire.

Des éléments de réponse sont apportés par l'étude de la distribution d'éléments en traces dans les roches (Co, Ni, V, Cu, Pb, Zn, Cr et Mn), qui montre que les corps minéralisés apparaissent comme nettement "étrangers" à leur environnement sédimentaire, et ne s'inscrivent pas dans un épisode d'enrichissement en métaux du bassin à l'échelle régionale (A. M. de Kersabiec et G. Roger, 1976). Ce résultat étaye l'hypothèse de concentrations métallifères issues de sources relativement ponctuelles, pouvant correspondre aux émergences d'un hydrothermalisme tardi-volcanique dont les traces minéralogiques sont encore assez rarement saisies dans la province (Rio Tinto, San Miguel . . .). Une étude de la

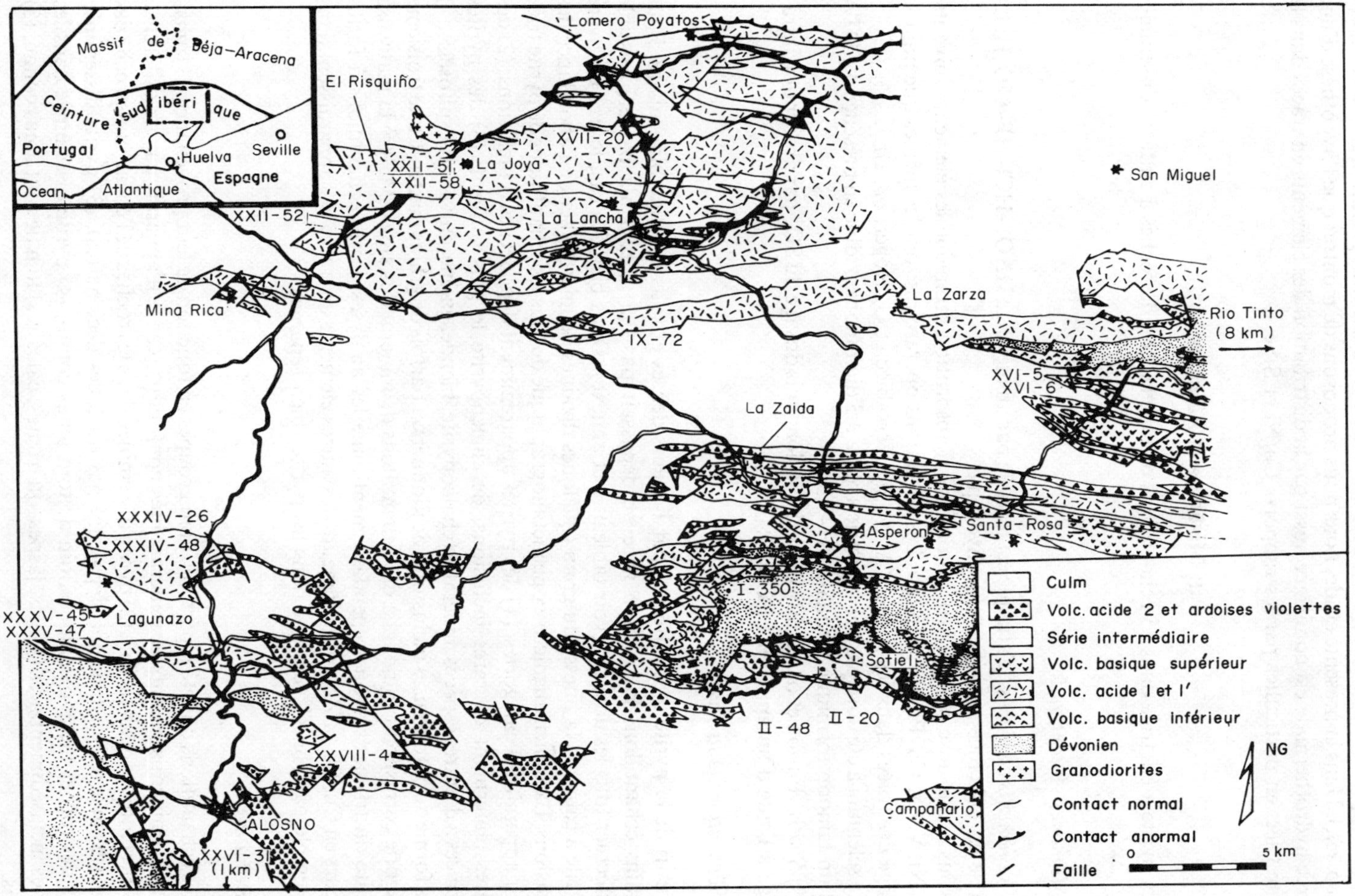

FIG. 2. Carte de localisation des échantillons sur fond géologique simplifié.

composition isotopique du soufre de la pyrite montre une évolution depuis les stockwerks jusqu'à la pyrite disséminée, également cohérente avec cette interprétation (P. Routhier *et al.*, *op. cit.*). Dans la présente étude, nous nous proposons de montrer que l'hypothèse d'une source hydrothermale est confirmée aussi par la distribution des éléments en traces dans la pyrite triée, en particulier par les rapports Co/Ni et S/Se.

I. Méthodes et Résultats

La pyrite a été triée dans 55 échantillons de minerais ou de roches à sulfures disséminés (Fig. 2).

A. Méthodes Analytiques

1°) *Eléments en traces.* L'attaque est réalisée par un mélange HNO_3–HCl–HF–ClO_4H. La reprise est effectuée en milieu chlorhydrique.

Nous avons dosé par spectrophotométrie d'absorption atomique les éléments suivants: Co, Ni, V, Cr, Cu, Pb, Zn, Mn, Sb. Les conditions opératoires sont les mêmes que pour le dosage des traces dans les roches (A. M. de Kersabiec et G. Roger, *op. cit.)*.

Le sélénium a été dosé par colorimétrie avec la 3.3 diaminobenzidine en présence d'EDTA et d'un tampon formique-formiate pH–2,5.

2°) *Soufre.* Fusion et oxydation de l'échantillon, puis dosage du SO_2 par absorption infrarouge à l'aide d'un appareil LECO IR 32.

B. Séparation de la Pyrite

Le tri de la pyrite a été effectué par liqueurs denses et sous la loupe binoculaire. Pour certains échantillons la fraction triée ne dépassait pas 0,5 g, et il n'a pas été possible d'effectuer tous les dosages (voir tableau 1 et annexe). Dans bien des cas, le tri s'est avéré difficile à cause de la finesse du grain (quelques dizaines de microns) et de l'abondance des inclusions. L'examen au microscope métallographique de tous les échantillons de pyrite triée a fait apparaître 3 catégories: (1) des pyrites apparemment dépourvues d'inclusions; (2) des pyrites renfermant de rares inclusions, de chalcopyrite le plus souvent; (3) des pyrites associées à d'autres espèces minérales (chalcopyrite, blende, galène, cuivre gris, sulfosels . . .) sous forme de nombreuses inclusions, de filonnets et parfois de grains mixtes. Cette troisième catégorie correspond toujours à des échantillons de minerai des amas. Malgré la mauvaise qualité du tri, nous avons cependant mentionné les analyses de ces échantillons à fortes teneurs en Cu, Pb, et Zn, car les teneurs des autres éléments en traces n'y apparaissent dans l'ensemble pas plus élevées que dans les pyrites bien séparées.

C. Résultats

Les échantillons ont été répartis en cinq groupes en fonction de leur provenance: (1) amas sulfurés (minerais massifs et minerais lités confondus, car il n'apparaît pas de différence sensible entre ces deux faciès); (2) pyrite disséminée dans les roches de l'encaissant proche des amas (jusqu'à une distance ne dépassant pas quelques décamètres); (3) pyrite disséminée dans des roches métavolcaniques loin de tout amas connu; (4) pyrite disséminée dans les roches métasédimentaires (shales, jaspes, quartzites, calcaires) loin de tout amas connu; (5) filons.

Dans l'hypothèse d'une liaison génétique entre les amas sulfurés et un hydrothermalisme tardi-volcanique, la liaison sulfures-hydrothermalisme la plus étroite doit apparaîte dans le groupe I, le groupe IV représentant à l'opposé le pôle sédimentaire exempt d'influence volcanique. Ce groupement guidera notre interprétation. Le groupe V (filons) ne comprend que trois mesures, et il ne représente actuellement qu'un point de comparaison très peu significatif.

Le tableau 1 indique dans chacun des cinq groupes les teneurs moyennes des éléments dosés, les minimums et les maximums. Le tableau des résultats analytiques détaillés figure en annexe. L'interprétation statistique n'a pas été appliquée à cause du trop petit nombre de mesures dans certains groupes.

II. Discussion

A. Remarques Générales

Le tableau montre que la variance est grande dans chaque groupe pour la plupart des éléments. Cela est particulièrement accentué pour Cu, Pb et Zn dans les groupes I, II et III; même pour des éléments tels Cr, V, Co, Ni dont les moyennes ne sont pas très différentes d'un groupe à l'autre, la variance peut être relativement élevée.

Les teneurs en cuivre, plomb et zinc des pyrites des amas, et à und degré moindre, des pyrites disséminées dans leur enveloppe et dans les roches volcaniques (groupes I à III) sont considérablement plus élevées que celles des pyrites disséminées dans les roches métasédimentaires (gr. IV). Nous avons vu que ces fortes teneurs ne représentent pas seulement des "traces" dans la pyrite, mais qu'elles expriment en grande partie la présence d'espèces minérales autres que la pyrite, que nous n'avons pas pu séparer de façon satisfaisante.

Les teneurs en Cr et en Sb sont assez uniformes; les écarts entre les teneurs moyennes des différents groupes sont peu considérables, et ils n'évoluent pas de façon ordonnée du groupe I au groupe IV. Cette dernière remarque est également valable pour Mn, dont il faut toutefois souligner la forte teneur moyenne dans le groupe IV "pyrite disséminée dans des roches métasedimentaires loin d'amas connus".

Nous examinerons maintenant les éléments dont la distribution dans la pyrite se prête à une interprétation génétique: Co et Ni, puis Se et V.

B. Cobalt et Nickel

Les teneurs en Co et en Ni de la pyrite semblent varier, de façon plus ou moins nette, en fonction du type de gisement, ainsi que le montrent les résultats dans l'ensemble assez convergents d'un certain nombre d'études géochimiques: F Hegemann (1943), G. Berg et F. Friedensburg (1944), C. F. Davidson (1962), B. Cambel et J. Jarkovsky (1967), G. Loftus-Hills et M. Solomon (1967).

Sans entrer dans le détail, il ressort des travaux cités que les pyrites d'origine "chaude", hydrothermale plutonique ou volcanique, sont caractérisées par des valeurs du rapport Co/Ni supérieures à 1, pouvant atteindre plusieurs dizaines et même parfois plusieurs centaines (jusqu'à 830 selon G. Berg et F. Friedensburg, *op. cit.*), alors que dans les pyrites sédimentaires les valeurs de ce rapport sont faibles (0,1 selon F. Hegemann, *op cit.*, moins de 0,35 selon G. Loftus-Hills et M. Solomon, *op. cit.*).

Tableau 1. Teneurs en Éléments Traces et en Soufre de 55 Échantillons de Pyrite Triée Répartis en Cinq Groupes: moyennes arithmétiques, minimums et maximums

	Teneurs moyennes μg/g minimum–maximum												
	Co	Ni	V	Cr	Cu	Pb	Zn	Mn	Sb	Se	S%	S/Se	Co/Ni
I. Pyrite d'amas, minerais massifs et lités (27 éch.)	173	≤9	≤19	11	3 522	4 777	7 830	266	175	134	46,1	3 403	19,2
(Mn, Sb et S: 26 éch.)	10–	4–	≤10–	3–	24–	20–	49–	4–	10–	8–	30,0–	(45,6%)†	
(Se: 24 éch.)	616	48	115	30	14 600	41 000	92 000	1 786	1 650	1 280	53,3		
II. Pyrite disséminée dans l'encaissant proche des amas	151	36	≤23	16	1 269	805	1 191	144	60	55	43,0	7 852	4,2
(r. volcaniques acides, shales, jaspes). (10 éch.)	18–	3–	≤10–	4–	23–	30–	60–	13–	18–	24–	28,1–	(43, 3%)†	
(Cr, Sb et S: 9 éch.)	354	115	49	48	8 750	3 250	4 300	440	80	116	50,2		
(Mn: 8 éch.‡)–(Se: 3 éch.)													
III. Pyrite disséminée dans des r. volcaniques, loin d'amas	242	≤69	≤34	45	502	2 385	4 057	230	136		39,4		
connus (3 éch.)	102–	≤5–	≤10–	11–	88–	75–	57–	31–	95–	—	34,0–	—	3,5
(S: 2 éch.)	394	182	82	102	1 180	6 940	12 000	610	180		44,8		
IV. Pyrite disséminée dans jaspes, shales, quartzite,	63	76	≤16	18	192	98	83	805	49	60	47,2	8 067	0,8
calcaire, loin d'amas connus (12 éch.)	12–	≤5–	≤10–	4–	19–	20–	23–	5–	20–	10–	43,0–	(48,4%)†	
(Cr, Mn, Sb et S: 10 éch.)	184	290	43	71	720	435	165	3 200	115	140	53,5		
(Se: 6 éch.)													
V. Pyrite de filons (3 éch.)	46	≤5	≤12	8	1 329	1 753	1 792	284	283	126	46,4	3 754	9,2
(Se: 2 éch.)	18–		≤10–	5–	20–	≤30–	66–	6–	130–	48–	44,5–	(47,3%)†	
	92		16	12	3 410	4 200	5 100	770	560	204	49,0		

Tous les éléments n'ont pas pu être dosés dans certains échantillons, à cause d'une trop faible quantité de pyrite triée.

† Teneur moyenne en S des échantillons dans lesquels Se a été dosé.

‡ Une très forte valeur (10,08%) a été éliminée; elle correspond à des cristaux de pyrite dans les jaspes hématitiques à manganèse de La Zaïda, criblés d'inclusions d'oxydes de fer et de manganèse (éch. EA 191).

ANNEXE. RÉSULTATS ANALYTIQUES

No. éch.	Provenance	Co	Ni	V	Cr	Cu	Pb	Zn	Mn	Sb	Se	S%
		Eléments en traces ($\mu g/g$ ou %)										
I. *Py d'amas*												
I 87	Santa-Rosa	616	<5	<10	10	996	370	228	n.d.	n.d.	n.d.	n.d.
I 281	Sotiel	116	<5	<10	13	257	295	58	12	40	50	48,3±1,6
I 286	Sotiel	12	14	31	18	810	4,10%	3,70%	940	415	480	39,7±2,6
XVII 7	La Lancha	310	<5	<10	6	3 940	75	173	77	70	48	48,8±1,7
XXII 19	La Joya	10	5	<10	10	1 800	7 600	2,40%	350	150	70	47,2±1,0
XXIV 21	Mina Rica	18	<5	14	12	1 640	1,82%	1,25%	46	340	8	48,7±1,0
XXXIV 16	Lagunazo	18	48	<10	9	1 720	1,12%	165	532	580	16	48 0±1,0
XXXIV 17	Lagunazo	32	14	11	9	3 660	2 520	590	87	150	18	45,3±1,0
E.A. 246	Sotiel	162	12	48	30	4 010	9 350	3 160	330	1 650	202	45,7±1,0
E.A. 320	Lagunazo	22	8	<10	8	24	20	49	5	55	n.d.	52,5±1,0
E.A. 323	Lagunazo	56	<5	<10	5	6 060	275	3 510	1 786	66	48	42,5±3,0
E.A. 326	Lagunazo	162	10	22	9	8 990	2 380	3 200	550	35	96	36,6±2,0
E.A. 387	Sotiel	222	<5	<10	10	181	80	63	4	50	16	53,3±1,0
L.P. 19	Lomero-Poyatos	117	5	12	17	132	420	195	83	23	180	45,8±1,0
L.P. 20	Lomero-Poyatos	86	5	20	20	5 400	870	2 000	370	85	120	48,4±1,0
L.P. 22	Lomero-Poyatos	30	5	115	9	3 100	530	1 600	165	160	40	46,6±1,0
	Lomero-Poyatos	160	5	16	14	1,22%	390	3 500	134	55	192	44,2±1,0
L.P. 32	Lomero-Poyatos	18	8	33	18	3 300	2,25%	9,20%	310	170	1 280	47,8±1,0
R.T. 18	Rio Tinto (Atalaya)	206	<5	<10	7	58	20	51	30	20	n.d.	51,8±2,0
SM A	San Miguel	112	7	<10	5	53	62	68	23	25	128	44,9±1,0
SM 5	San Miguel	144	4	<10	5	57	56	117	19	25	200	49,5±2,0
SOT 57	Sotiel	260	6	13	7	1,46%	270	230	10	35	56	50,3±1,0
SOT 58	Sotiel	460	5	<10	3	2 600	680	150	58	65	74	43,9±2,5
SOT 59	Sotiel	272	14	22	14	3 200	816	190	106	10	50	45,8±1,0
SOT 60	Sotiel	520	34	27	24	8 100	1 100	420	400	40	70	30,0±1,5
Z 15	La Zarza	504	4	11	8	4 400	390	700	57	65	152	46,6±1,0
Z 19	La Zarza	20	5	<10	11	3 800	7 500	2,55%	440	180	26	46,5±1,0
II. *Py. disséminée proche amas*												
II 49	Jaspes, Asperon	138	115	28	n.d.	150	76	60	n.d.	n.d.	n.d.	n.d.
E.A. 191	Jaspes, La Zaïda	134	24	21	48	8 750	750	525	10,08%	68	n.d.	46,6±1,0
EA 319	Shales, Lagunazo	146	21	30	12	110	70	579	202	18	26	32,1±2,0
EA 396	Jaspes, Lagunazo	22	66	<10	6	23	30	169	13	35	24	47,6±1,4
LP 16	Tufs, Lomero-Poyatos	308	15	49	34	761	3 250	3 320	440	80	n.d.	48,1±1,2
LP 18	Tufs, Lomero-Poyatos	354	8	42	9	696	1 210	323	172	61	n.d.	45,0±1,0
LP 29	Tufs, Lomero-Poyatos	240	96	13	14	300	230	510	70	35	n.d.	45,5±2,5
PM 12	Volc. acide, Campanario	46	5	<10	8	387	120	1 290	33	80	n.d.	28,1±1,4
SM3	Rhyolite, San Miguel	104	6	<10	4	310	810	830	110	80	n.d.	43,5±2,0
Z 16	Tufs, La Zarza	18	3	14	10	1 200	1 500	4 300	15	80	116	50,2±1,0
III. *Py. disséminée dans r. volc. loin d'amas.*												
XXVI-31	Tufs, Alosno	394	<5	82	102	238	75	57	48	95	n.d.	34,0±1,1
XXVIII-4	V. acide 2, W Mina Chaparral	102	182	<10	21	1 180	6 940	1,20%	610	180	n.d.	44,8±1,0
XXXIV-48	V. basique, N. Lagunazo	230	19	<10	11	88	140	115	31	134	n.d.	n.d.
IV. *Py. disséminée dans r. sédimentaires, loin d'amas*												
I 350	Shales, S Calañas	156	29	<10	4	439	435	165	92	25	n.d.	47,1±1,0
II 20	Shales, Sotiel	40	56	17	20	48	92	23	1 650	40	n.d.	43,0±2,0
II 48	Calcaires, Sotiel	85	290	23	n.d.	52	125	27	n.d.	n.d.	n.d.	n.d.
III 17	Shales, S Calañas	82	38	19	n.d.	40	88	42	n.d.	n.d.	n.d.	n.d.
III 29	Quartzites, S Calañas	33	110	20	71	125	107	96	3 200	115	72	45,6±2,9
IX 72	Jaspes, SW La Zarza	12	31	43	39	209	64	141	342	39	10	48,3±2,0
XVI 5	Jaspes, W El Villar	18	85	<10	8	38	<20	65	112	29	140	46,7±1,0
XVI 6	Jaspes, W El Villar	22	144	<10	5	197	75	56	2 550	20	n.d.	48,2±1,0
XXII 52	Shales, N Cabezas Rubias	184	94	<10	7	372	67	92	127	70	34	47,8±2,8
XXXIV 26	Jaspes, NE Lagunazo	48	14	<10	12	720	57	162	28	20	n.d.	43,5±2,4
XXXV 45	Shales, W Tharsis	30	21	<10	6	41	20	76	5	65	80	53.5±1,1
XXXV 47	Shales, W Tharsis	48	<5	11	13	19	30	59	43	70	22	48,7±1,0
V. *Py. de filons*												
XVII 20	Sud San Telmo	92	5	<10	5	20	<30	66	6	160	48	49.0±1.2
XXII 51	El Risquiño	18	<5	<10	6	556	1 030	210	75	130	204	45,6±2,0
XXII 58	El Risquiño	28	<5	16	12	3 410	4 200	5 100	770	560	n.d.	44,5±1,9

n.d. = non dosé.

Les moyennes obtenues dans la province de Huelva (tableau) montrent une décroissance ordonnée du rapport Co/Ni depuis le groupe I "amas" (19,2) jusqu'au groupe IV "pyrite disséminée dans les roches métasédimentaires loin d'amas connus" (0,8), avec des moyennes intermédiaires dans les groupes II "pyrite disséminée dans l'encaissant proche des amas" (4,2) et III "pyrite disséminée dans les roches volcaniques loin d'amas connus" (3,5). La moyenne élevée du groupe I s'accorde bien avec l'hypothèse d'une liaison assez directe entre les amas et un hydrothermalisme tardi-volcanique. La décroissance du rapport Co/Ni dès la bordure de l'amas sulfuré traduit bien le caractère assez ponctuel du phénomène minéralisant. La valeur moyenne inférieure à 1 du rapport Co/Ni dans les pyrites du groupe IV représentant le pôle sédimentaire est conforme aux données de la littérature.

Les teneurs en cobalt et en nickel ne varient pas symétriquement: les teneurs en nickel croissent régulièrement du groupe I au groupe IV, alors que les teneurs en cobalt demeurent presque constantes dans les trois premiers groupes puis diminuent brutalement dans le groupe IV.

Quant aux filons, le très petit nombre d'échantillons analysés (trois) ne permet d'aboutir à aucune conclusion. Enregistrons simplement les teneurs particulièrement basses en Co, Ni, V et Cr, la forte teneur en Sb, et le caractère erratique des quelques teneurs en Cu, Pb et Zn, avec des moyennes assez élevées.

C. *Sélénium et Vanadium*

Depuis les travaux de V. M. Goldschmidt et L. Strock (1935) et de C. W. Carstens (1941), le rapport S/Se dans les sulfures et en particulier dans la pyrite est considéré comme un critère possible de distinction entre les sulfures "chauds" d'origine hydrothermale et les sulfures "froids" d'origine sédimentaire ou supergène. Un certain nombre de travaux ont été entrepris dans le but d'éprouver la validité de ce critère dans différents types de gisements et dans différentes régions; citons par exemple ceux de A. B. Edwards et G. C. Carlos (1954), J. E. Hawley et I. Nichol (1961), B. Cambel et J. Jarkovsky (1967), G. Loftus-Hills et M. Solomon (1967), C. A. Anderson (1969). Nous ne nous attacherons pas aux écarts parfois importants entre les valeurs obtenues dans ces différentes études pour les teneurs en Se et pour le rapport S/Se. Nous en retiendrons la tendance essentielle: lorsque la conclusion est en faveur de la validité de ce critère, ce sont toujours les fortes teneurs en Se et donc les faibles valeurs du rapport S/Se qui caractérisent les sulfures "chauds", et à l'inverse les fortes valeurs du rapport S/Se qui apparaissent dans les sulfures "froids".

Les teneurs en vanadium sont également utilisées comme critère de distinction génétique des sulfures (C. W. Carstens, 1943; J. E. Hawley, 1952; B. Cambel et J. Jarkovsky, *op. cit.*). Les variations du vanadium seraient inverses de celles du sélénium: les pyrites "froides" seraient plus riches en vanadium que les pyrites "chaudes".

Le tableau montre que dans la province de Huelva les teneurs en sélénium sont nettement plus élevées dans les pyrites des amas que dans les pyrites disséminées des groupes II et IV (aucun dosage de Se n'a pu être effectué sur les échantillons du groupe III). La chute des teneurs moyennes intervient dès le passage des amas à la pyrite disséminée dans leur enveloppe, mais ensuite les teneurs en Se ne diminuent pas dans les pyrites disséminées loin des amas. La variation du rapport S/Se est donc cohérente avec l'hypothèse d'un pôle "hydrothermal" (amas) et d'un pôle "froid" (pyrite disséminée dans les roches sédimentaires). Il faut toutefois souligner les faibles valeurs du rapport S/Se dans la province de Huelva, par rapport aux chiffres obtenus dans d'autres régions: en Norvège, Se = 1 ppm dans des pyrites sédimentaires (C. W. Carstens, *op. cit.*); en Australie, Se = 30 à 50 ppm et S/Se = 9000 à 13 000 dans des pyrites hydrothermales, S/Se = 500 000 à 38 500 dans des pyrites en terrains sédimentaires (A. B. Edwards et G. C. Carlos, *op. cit.*).

Quelle est la signification de cette richesse en sélénium? Deux hypothèses au moins peuvent être envisagées: soit un facteur régional, soit une influence du métamorphisme.

Les teneurs en vanadium sont au contraire très uniformes; les variations du groupe I au groupe IV sont faibles et non ordonnées.

III. Conclusions

Les variations de la distribution de certains éléments en traces dans la pyrite montrent que dans les amas ce minéral présente des caractères correspondant à une origine "hydrothermale", alors que les caractères des pyrites disséminées se rapprochent de ceux de sulfures "froids". La coupure la plus nette intervient dès la bordure des amas, aussi bien pour le rapport Co/Ni que pour le rapport S/Se. La distribution du vanadium est au contraire uniforme, et cet élément ne peut ici être retenu comme ayant valeur de critère de distinction génétique. Ces résultats confirment le caractère nettement "étranger" des amas sulfurés par rapport aux roches de leur milieu de dépôt sédimentaire, déjà étayé par l'étude de la distribution d'éléments en traces dans les roches (A. M. de Kersabiec et G. Roger, *op. cit.*). L'hypothèse d'une liaison génétique assez directe entre les amas et une source hydrothermale tardi-volcanique véhiculant les cations métalliques apparaît bien fondée. Pour le soufre, une hypothèse génétique semblable est compatible avec la composition isotopique de la pyrite des amas (P. Routhier *et al.*, *op. cit.*).

Il faut enfin remarquer la grande variance enregistrée dans les différents groupes pour la plupart des éléments. Un très petit nombre de mesures ne serait donc guère significatif.

Références

ANDERSON, C. A. (1969) Massive sulfide deposits and volcanism. *Econ. Geol.* **64**, n° 2, 129–146.

BERG, G. et FRIEDENSBURG, F. (1944) *Nickel und Kobalt. Die metallischen Rohstoffe, ihre Lagerungsverhältnisse und ihre Wirtschaftliche Bedeutung*, Stuggart, Enke, 280 pp.

CAMBEL, B. et JARKOVSKY, J. (1967) *Geochemie der Pyrite einiger Lagerstätten der Tschechoslowakei*, Bratislava, 493 pp.

CARSTENS, C. W. (1941) Zur Geochemie einiger norwegischen Kiesvorkommen. *Kongl. Norske Vidensk. Selsk., Forh.* **14**, n° 10, 36.

CARSTENS, C. W. (1943) Über den Vanadiumgehalt norwegischer sedimentärer Eisenoxyd und Eisensulfiderze. *Kongl. Norske Vidensk. Selsk., Forh.* **16**, 1–4.

DAVIDSON, C. F. (1962) On the cobalt: nickel ratio in ore deposits. *Mineral. Mag.* n° 106, 78–85.

EDWARDS, A. B. et CARLOS, G. C. (1954) The selenium content of some Australian sulphide deposits. *Proc. Australasian Inst. Min. Met.* **172**, 31–63.

GOLDSCHMIDT, V. M et STROCK, L. (1935) Zur Geochemie der Selen. II. *Nachr. Ges. Wiss. Göttingen, math-nat. Kl.* **1**, 123–142.

HAWLEY, J. E. (1952) Spectrographic studies of pyrite in some Eastern Canadian gold mines. *Econ. Geol.* **47**, 260–403.

HAWLEY, J. E. et NICHOL, I. (1961) Trace elements in pyrite, pyrrhotite and chalcopyrite of different ores. *Econ. Geol.* **56**, 467–487.

HEGEMANN, F. (1943) Die geochemische Bedeutung von Kobalt und Nickel im Pyrit. *Z. Angew. Mineral* **4**, 121–239.

KERSABIEC, A. M. de et ROGER, G. (1976) Sur la distribution de quelques éléments en traces dans l'encaissant des amas pyriteux de la partie centrale de la province de Huelva (Espagne). *Bull. BRGM*, 2e sér., sect. II, n° 5, 439–485.

LOFTUS-HILLS, G. et SOLOMON, M. (1967) Cobalt, nickel and selenium in sulphides as indicators of ore genesis. *Miner. Deposita* **1**, 228–242.

ROUTHIER, P., LECOLLE, M., ROGER, G., AYE, F., MOLIERE, P., BOYER, C. et PICOT, P. (1977) Amas sulfurés volcano-sédimentaires. La ceinture sud-ibérique à amas sulfurés dans sa partie espagnole médiane. Tableau géologique et métallogénique. Compte-rendu final d'une Action Concertée D.G.R.S.T. *Valorisation des Ressources du Sous-sol*, décision d'aide n° 74.7.1181, 387 pp.

Discussion

C. GRANIER:
Avez-vous utilisé l'Ag et l'As qui sont d'excellents indicateurs?

G. ROGER:
Nous avons pensé à ces deux éléments, mais nous n'avions pas la possibilité de les doser au moment de l'étude. Pour le dosage de As et d'autres éléments volatils (Se, Te, Sb . . .), Mme A. M. de Kersabiec poursuit actuellement la mise au point d'une méthode par absorption atomique. Quant à Ag, nous pouvons maintenant le doser, également par absorption atomique. Nous envisageons donc de compléter ce travail par l'étude d'une plus large gamme d'éléments en traces.

J. C. DUCHESNE:
What about using thallium as a tracer element in your particular problem?

G. ROGER:
Nous n'avons pas étudié cet élément; nous vous remercions de cette information et serions intéressés par des publications à ce sujet.

D. T. RICKARD:
How did you check that the trace elements were in solid solution in the pyrite?

G. ROGER:
La localisation de Co et Ni en traces dans les cristaux de pyrite n'a pu être étudiée. Pour cela, il serait sans doute préférable de faire appel au micro-analyseur à sonde ionique plutôt qu'à la microsonde électronique, dont les seuils de détection sont vraisemblablement trop élevés.

D. T. RICKARD:
What are the effects of metamorphism and alteration on the trace element distribution?

G. ROGER:
Il n'a pas été possible d'évaluer les effets du métamorphisme, faute de points de comparaison soit avec une région exempte de métamorphisme, soit avec une région à métamorphisme plus intense. Toutefois, dans la province de Huelva, le métamorphisme s'est appliqué sur toutes les pyrites étudiées, et *a priori* il n'a donc pas pu introduire de différences importantes entre les divers groupes.

J. PLANT:
Avez-vous envisagé la possibilité d'une contamination par les roches?

G. ROGER:
Nous vous remercions de nous signaler ces intéressantes études. Cette possibilité de contamination n'a pas été envisagée dans le cadre de notre travail.

C. GRANIER:
Avez-vous utilisé l'approche de Stanton pour l'étude des teneurs des pyrites dans les matériaux siliceux. L'étude des gradients permet quelquefois de découvrir les sources ponctuelles ou leur direction?

G. ROGER:
Ce problème ne peut être traité que dans le cadre d'études détaillées de gisements en exploitation. Au cours du travail présenté ici, nous n'avions pas eu cette possibilité. Récemment, nous avons obtenu l'autorisation d'étudier les mines de La Zarza et d'Aznalcollar, et nous pourrions envisager de vous apporter une réponse ultérieurement.

L. GUILLOUX:
Avez-vous étudié à la microsonde électronique la distribution de ces éléments dans la pyrite?

G. ROGER:
Nous n'avons pas eu la possibilité d'étudier la distribution des éléments en traces à l'échelle du cristal. Nous envisageons d'aborder cette étude ultérieurement.

Problèmes Posés par l'Insertion de la Géochimie en Méthodologie des Recherches minières

C. GRANIER

Géochimiste Conseil, Imetal, 1, boulevard de Vaugirard, 75751 Paris Cédex 15

Abstract

The evolution of the geochemical methods, applied to mining exploration, has lead to a generalization of the geochemical approach and has created, and continues to create, for these comparatively new techniques, some problems of adaptation to the existing habits. The correct use of these techniques is hampered at different stages of the exploration programme: planning, execution, interpretation of results.

We give a critical view of the problems involved, such as:

inertia in decision-making and their inadequate knowledge of the exact conditions of application of the geochemical methods;

the failure to properly appreciate the technical requirements, sometimes facilitated by the lack of esoterism of the method;

the statistical ghetto in which some geochemists are confined;

the lack of rigour which often leads to the use of semi-quantitative analytical information for multiple elements, instead of more tested methods;

the inadequacy of geochemical maps with a lack of the necessary indications for an interpretation more connected with a good knowledge of geology, pedology, etc. This interpretation remains for the author the main approach.

L'introduction de la géochimie en prospection minière a bouleversé les méthodologies de recherche. A ce jour, près du tiers de la motivation pour le choix d'une zone de prospection ou pour la sélection d'indices dépend de l'interprétation de documents de géochimie.

Ainsi, la méthode est utilisée aussi bien au niveau stratégique qu'au niveau de l'étude de détails associée étroitement aux approches traditionnelles géologique et géophysique.

Après les premiers tests du nouvel outil de prospection, il a fallu près de 25 ans pour que les méthodes soient parfaitement généralisées. Cependant, les difficultés rencontrées pour leur bonne utilisation ont été nombreuses et subsistent encore parfois.

La première de ces difficultés provient de ce que nous pourrions appeler une pesanteur sociologique.

Il est difficile d'intégrer parfaitement et immédiatement une technologie nouvelle, au réseau de réflexes et d'habitudes motivant le comportement de spécialistes chevronnés lorsque cette technologie implique une approche psychologique inhabituelle; par exemple au plan scientifique:

— nous trouvons encore des pétrographes considérant qu'en milieu résiduel le rapport sol roche (effet pédogénétique) présente trop de fantaisie pour être considéré comme significatif et utilisable;
— la plupart des pétrographes d'ailleurs sont surpris des différenciations qu'apportent les analyses de traces dans les roches. Il faut noter leur étonnement quand on leur indique

que tel massif de granit intrusif est à 30 ppm de plomb et tel autre apparemment de même nature pétrographique est à 800 ppm, ou que telle série volcano sédimentaire ou tel épisode de volcanisme peut être cartographié facilement par la teneur en cuivre d'échantillons de sol résiduel ou d'esquilles de roches superficielles (chip sampling).

Cette approche par la géochimie des traces des problèmes pétrographiques sera sûrement fructueuse. Elle est cependant freinée pour des raisons de méthodes analytiques souvent trop inadaptées dont nous parlerons plus loin.

Un autre exemple: au plan économique. En France, près de 10 ans ont été perdus après la découverte par Sakowitsch d'anomalies géochimiques Pb, Zn en Bretagne (indices de Bodenec et autres) qui n'était même pas mentionnée dans certaines synthèses de travaux en Bretagne! en contrepartie, aujourd'hui, nous voyons certains spécialistes passer sur ces sujets abruptement de la plus totale indifférence à la crédulité la plus béate, ceci alors que certains secteurs ont déjà donné lieu à des géochimies stratégiques à près de 12 points km^2 sans résultat.

Ces erreurs deviennent rares—mais dans de nombreux cas, l'absence d'enseignement sérieux a laissé persister même chez les jeunes géologues ou responsables d'explorations certaine inadaptation, à vrai dire, facilitée quelque peu par le manque d'ésotérisme de la méthode géochimique.

Le temps, la vulgarisation et la diffusion des résultats typiques par les Congrès et la publication du *Journal of Geochimical Exploration* jouent parallèlement pour conduire à l'élimination de ces défauts de jeunesse.

La deuxième de ces difficultés soulevée par l'application des méthodes de prospection géochimique est liée au fossé qui sépare de plus en plus le laboratoire du terrain.

A ce jour, l'échec des méthodes d'analyse sérieuse in situ n'est plus à démontrer. Aussi, l'homme de terrain envoie au laboratoire ses échantillons, mais trop souvent il ne précise pas exactement la qualité du traitement qu'ils doivent subir. Trop souvent on analyse n'importe quoi, n'importe comment!

Il faut préciser, car c'est déjà une interprétation, les conditions de broyage, de bris des mottes, de séparation et de choix de fraction. Les conditions de sélection du type d'attaque de l'échantillon, totale ou ménagée, suivant les problèmes à traiter, peuvent mettre en évidence des différences de contraste supérieures à une puissance de dix.

Au niveau de l'analyse proprement dite, la connaissance du niveau de teneur, où il est nécessaire d'avoir le maximum de différenciation, est capitale pour choisir la méthode d'analyse.

Exemple, pour les sols et alluvions:

0,5—16 ppm pour Mo
0,2— 5 ppm pour Ag
2 —50 ppm pour As
0,1— 3 ppm pour Au
3 —30 ppm pour Sb, etc.

Ces intervalles assez étroits où il est nécessaire de bien différencier les résultats impliquent qu'il faut choisir la méthode d'analyse.

Trop souvent encore, pour des raisons fallacieuses, les impératifs de la recherche sont subordonnés à des questions de rendement de laboratoire ou simplement liés à son équipement. Ces calculs économiques subtils! négligent de prendre en compte les erreurs coûteuses causées par l'implantation de campagnes de recherche et de sondages, sur des informations géochimiques inadaptées; dépenses qui sont d'un ordre de grandeur sans comparaison possible avec les premières!

Un exemple: l'analyse polymétallique est couramment employée en recherche géochimique. Associés à l'élément dominant recherché, les élément mineurs de la paragénèse peuvent conforter la discrimination et permettre un meilleur classement des anomalies significatives, voire même pour l'Ag, l'Au et le Mo apporter l'indication essentielle.

Dans ces conditions, la vogue de l'analyse multiélément par quantométrie en spectro d'émission ou de la fluo X est dangereuse et parfois ridicule.

Les effets de matrice au niveau des éléments en traces n'étant pas maîtrisés, cela conduit souvent à des analyses semi-quantitatives.

Je donnerai un exemple pour la prospection de gisements sulfurés: on nous propose, sans rire, d'étudier 20 à 30 éléments par ces méthodes, l'argument invoqué étant leur bas prix! Or quels sont les éléments les plus utiles et significatifs? Ce sont Pb, Zn, Cu, Ag, As, Au.

Avec les méthodes en question, l'analyse du Zn est assez fantaisiste au niveau de teneur qui nous intéresse 20 à 800 ppm; Pb subit les interférences de As ce qui n'assure pas toujours l'indépendance essentielle des deux informations!

Il y a des difficultés sérieuses pour Au et As au niveau des teneurs utiles pour le géochimiste. Aussi, les méthodes à utiliser sont atomique pour Zn, Pb, Cu, Ag et méthode colorimétrique pour As. C'est toujours assez difficile à faire admettre au chef de laboratoire fier de son quantomètre, qui donne le résultat sur imprimante en une seule opération au lieu de cinq.

Nous aurions pu prendre des exemples similaires pour Mo, Au, Sn, W, Sb! qui ne peuvent être traités correctement par spectro d'émission.

Le géochimiste doit, à mon sens, exiger les analyses les plus adaptées à ses besoins. L'inadaptation d'un prodécé est souvent masquée par des traitements statistiques, indice de corrélation—analyse multivariable, etc.—dont nous constatons à ce jour qu'ils ne sont appliqués qu'à des problèmes déjà résolus. Mais, la découverte de gisements a posteriori ne nous intéresse pas.

Incidemment, nous préciserons que dans l'approche géochimique les populations géochimiques significatives ne sont pas identiques pour chaque élément et que les corrélations au niveau d'une vaste population ne sont pas toujours significatives à tous les niveaux de teneur.

Pour reprendre l'exemple des sulfures Pb, Zn, Cu, les anomalies élevées et ponctuelles d'Ag sont utilisées efficacement comme indicateurs, alors que, globalement, la corrélation (Pb, Zn, Cu) Ag sur l'ensemble du prospect est souvent très mauvaise!

Ces observations sur la nécessité de la parfaite adaptation des méthodes de préparation et d'analyse sont importantes et dans le futur le seront plus encore à notre avis. L'évolution des méthodes de prospection géochimique nécessitera une finesse et une reproductibilité des résultats toujours accrues.

Finie en effet, ou presque, la prospection générale en milieux résiduels pour trouver des anomalies liées à des gisements ayant eu dans un paléoreleif des manifestations dont on trouve d'abondantes reliques, soit chapeaux de fer bien visibles, soit en auréoles de dispersion mécanique, puis chimique dans les sols, voire en puissants chenaux de dispersion dans les alluvions. Maintenant, la plupart de ces indices sont sondés; vérifiés ou en voie de l'être.

Le Futur, c'est la recherche de manifestations polymétalliques infimes liées à l'environnement des minéralisations qui n'ont pas affleuré.

C'est l'étude des dispersions primaires, c'est l'étude des fuites, des dispersions ioniques, des effets biochimiques sous les recouvrements allochtones.

C'est l'étude des différenciations globales du chimisme des inclusions fluides, des formations quartzeuses.

Toutes études étroitement liées à la connaissance naturaliste la plus complète de l'environnement du gîte supposé.

Certaines difficultés sont masquées par l'utilisation de méthodes statistiques, comme nous l'avons vu.

Certains résultats étudiés par des méthodes statistiques raffinées permettent de plaisantes constatations et montrent un tel mépris du résultat brut associé à l'information de terrain qu'ils paraissent des caricatures et qu'il est charitable de ne pas citer leurs auteurs.

Telle étude sur les chapeaux de fer qui ne mentionne même pas les teneurs brutes (qui à vrai dire permettent un classement immédiat aussi efficace par n'importe quel géochimiste de bon sens).

Telle étude sur les courbes de tendance ne situant ni les teneurs, ni les points du réseau de prélèvement mais faisant apparaître l'intérêt d'une courbe de tendance (dérivée 4e) sur un environnement géologique assez hétérogène.

Le problème n'est pas d'optimiser une approche géochimique pour chercher les gisements du passé mais de chercher les gisements futurs nécessaires à nos économies Ce problème se pose avec acuité pour la réalisation des inventaires géochimique ou pour certaines recherches stratégiques réalisées par de grandes compagnies minières. C'est du laxisme que de faire l'analyse semi-quantitative multiélément, en lieu et place d'analyses précises. C'est imprudent et souvent inefficace de rechercher des gisements bien dissimulés avec des mailles de prélèvement de l'ordre de 2 à 3 points km^2 adaptées à la recherche de gisements ayant subi peu ou prou les effets de l'altération météorique, avec l'espoir de découvrir beaucoup d'indices nouveaux dans des zones anciennement prospectées.

Le traitement statistique, aussi élaboré soit-il, d'informations insuffisantes n'est qu'une notation nouvelle trompeuse qui n'améliore en rien la discrimination du résultat.

A ce sujet, il est nécessaire que les statisticiens sortent de leur ghetto. Il est probable que dans le futur, il sera mieux tenu compte, dans l'introduction des données, de tous les facteurs importants conduisant à des populations différentes, qu'il est imprudent de confondre et essentiel de traiter séparément.

Ainsi l'intégration des informations géologiques est rarement assez poussée; il n'est pas tenu compte des informations sur les pollutions accidentelles qui engendrent des populations surimposées (vieux travaux, engrais, etc.), et de la séparation fondamentale entre zones résiduelles du corps minéralisé cherché et zones non résiduelles oblitérées par des recouvrements de tous types.

On sait que dans le premier cas, l'interprétation est assez facile, la filiation de teneurs étant d'abord d'origine mécanique. Il y a une certaine relation entre le niveau de teneur, l'importance spatiale de l'auréole de dispersion et celle du corps minéralisé dans un paléorelief.

Dans le deuxième cas, où les dispersions chimiques sont prédominantes avec passage du niveau de teneur du % (corps minéralisé) au niveau du ppb (dispersion chimique) et retour au ppm suivant les conditions de fixation dans les sols ou les alluvions, les filiations sont aléatoires et souvent plus ténues. C'est alors, plus que le niveau de teneur absolue, les formes des courbes qu'il faut prendre en compte.

Alors, il est impératif que les cartes et le traitement des données tiennent compte de ces différentes populations.

La nécessité de toutes ces précisions pour l'interpretation correcte nous fait douter de l'intérêt des études de courbes de tendance et de tous les procédés qui tendent à homogénéiser les informations plutôt qu'à les différencier.

En conclusion, les découvertes, dans le futur, de gisements toujours plus cachés, demanderont une adaptation plus parfaite qu'aujourd'hui des méthodes géochimiques.

La qualité des analyses devra être affinée, les mailles de prélèvement adaptées à des occurrences plus ponctuelles, plus serrées, les recouvrements plus continus ou plus épais reconnus en géochimie tridimensionnelle par utilisation systématique de sondage ou de prélèvement à la tarière. Les travaux seront toujours plus coûteux.

Aussi l'étroite liaison du géochimiste, du géophysicien, du pédologue, du géomorphologue et du géologue sera plus nécessaire encore qu'aujourd'hui pour saisir et interpréter les plus subtiles modifications de l'environnement du corps minéralisé cherché, épiphénomènes géochimiques différenciés d'un fond local dont la variance devra être circonscrite.

C'est cette approche essentiellement naturaliste qui donnera à notre avis les meilleurs résultats dans le futur.

Discussion

J. Plant:

To what extent do you think published geochemical maps for trace elements are affected by major element interferences. Work on interferences in emission spectroscopy at the I.G.S. London suggests that some maps for Sn are badly affected by interference from iron.

C. Granier:

Nous ne sommes pas très favorables à la spectrographie d'émission systématique pour la cartographie géochimique appliquée à la recherche minière.

Les interférences sont de deux types. Elles sont difficiles à éviter, sauf si l'on fait une attaque spéciale et un traitement systématique de l'échantillon pour chaque population pétrographique, ou pédologique, homogène, ce qui, bien entendu, enlève beaucoup d'intérêt à la méthode qui est alors alourdie. Ce traitement suppose de plus que les populations soient faciles à circonscrire et qu'elles ne soient pas interpénétrées.

1. Il y a des interférences qui sont dues à des éléments majeurs, c'est le classique effet de matrice.
2. Il y a des interférences qui sont dues à l'élément analysé lui-même, dont la réponse peut être différente s'il s'agit d'éléments dispersés dans le réseau ou du même élément individualisé cristallographiquement.
3. De plus, pour une même excitation en spectrographie d'émission, il faut ajouter les difficultés que cause, pour un minéral identique, dans des conditions de dispersion égales, le problème des écarts causés parfois par la température de formation du minéral porteur.

Le problème de l'étain que vous signalez est une illustration de ces difficultés, le comportement de l'élément Sn dispersé dans un réseau cristallin ou dans un environnement altéré ferrugineux ne peut se comparer facilement à celui appartenant à la cassitérite.

J. Goni:

Je m'inscris en faux sur la plupart des affirmations de M. Granier sur l'état actuel de la prospection géochimique. Nous (B.R.G.M.) avons conçu l'inventaire des possibilités minières du sous-sol métropolitain et guyannais, comme un effort coordonné entre analystes, géochimiques et statisticiens, comme un vrai travail d'équipe et non seulement comme une approche du seul "géochimiste prospecteur".

De plus, les méthodes d'analyse multiélément, surtout grâce à l'utilisation des sources au plasma à l'Ar, ont fait d'énormes progrès et ces méthodes ne sont pas "semi-quantitatives"—comme vous l'affirmez—mais quantitatives, reproductibles, sensibles et fidèles (voir communication antérieures, de Barbier, Sakowitsch, Wilhelm sur

"Geochemistry and ore deposits prospecting") pour une conception nettement différente de l'approche géochimique actuelle dans la recherche des anomalies métallifères.

C. Granier:

M. Goni, l'intérêt de votre contestation peut-être un peu globale, est qu'elle révèle la nécessité de notre exposé et les dangers encourus par les non spécialistes de prospection géochimique.

Il est certain que vous mêlez les problèmes d'utilisation rationnelle des laboratoires avec ceux, fondamentaux pour la prospection géochimique minière. C'est une tendance qui ne vous est pas spéciale et dont nous souffrons, hélas, aussi dans l'industrie.

Cependant, nul ne peut contester que les effets de matrice sont difficiles à éliminer pour un échantillon inconnu en spectrographie d'émission et en fluorescence X.

De plus, pour les éléments comme As, Sb, Sn, W, Mo, entre autres, dès qu'il s'agit de faire plus que de déterminer des anomalies à teneur très élevée plus ou moins erratiques, mais les parties marginales significatives des anomalies, on est le plus souvent contraint d'utiliser des méthodes:

—colorimétriques pour Sb, Sn, W, Mo; —ou confined spot pour As, etc.

Nous ne signalons que pour mémoire l'analyse du Zn par spectro d'émission!

Il n'est pas discutable que l'absorption atomique est dans ce cas plus sûre au niveau de teneur 50 à 1 000 ppm qui nous intéresse. Votre remarque, M. Goni, indique simplement que dans vos services la priorité est donnée à la rentabilité et à l'automatisme de l'analyse plutôt qu'à l'efficacité de la méthode de prospection géochimique.

Il nous faut signaler un autre point essentiel de l'approche que vous définissez. La priorité est donnée à une démarche, à notre avis périmée, qui consiste à mettre en évidence des anomalies à teneur élevée et étendues.

La recherche minière actuelle demande plutôt, ceci en particulier en France dans les zones déjà fort étudiées: analyse de précision, grande densité d'echantillonnage, ceci pour discerner les zones anomales par les methodes additives ou multiplicatives.

En ce qui concerne les programmes d'inventaire que vous soulevez, si le but en est de faire des cartes géochimiques, pour découvrir, émerveillé, que le rapport Sn/Ni permet de séparer des massifs ultrabasiques de massifs granitiques, je suis d'accord avec vous et ceci peut être fait avec n'importe quelle méthode analytique.

Mais, en ce qui concerne la recherche minière, il faut rechercher maintenant des minéralisations que l'on ne peut atteindre que par un fin chevelu ou par les auréoles primaires syngénétiques, toutes manifestations impliquant une faible dispersion des éléments constitutifs du gisement dans un paléo-relief.

Nous répèterons donc que dans ces conditions il est nécessaire d'avoir une analyse polymétallique extrêmement précise, permettant de définir la partie marginale des anomalies pour obtenir une anomalie faiblement différenciée certes, mais permettant de travailler à des densités raisonnables.

Certains diront qu'en prenant seulement la fraction lourde de l'échantillon (en sol) on augmente à ce point le niveau de teneur, que les méthodes que vous préconisez, sont suffiisamment efficaces. C'est une tentation fallacieuse. Nous objecterons avec les spécialistes de prospection que l'anomalie élevée et erratique obtenue par ce procédé est fondamentalement moins utilisable que l'anomalie moins élevée, mais plus continue, relevée dans la fraction inférieure à 80 mesh, sans séparation de la fraction lourde.

Cette méthode demande moins d'échantillons. Elle est plus efficace, mais elle implique de pouvoir différencier parfaitement le niveau de teneur qui dépasse le fond.

Nous vous remercions, M. Goni, d'avoir suscité cette très utile controverse, car c'est un sujet important qui mérite réflexion pour les responsables du financement des inventaires, car il convient de leur donner le maximum d'efficacité et il serait stupide et économiquement malsain qu'ils ne servent qu'à redécouvrir ce que l'on connaît déjà.

The Distribution of Elements in the Geochemical Provinces and Ore Deposits

YURI G. SHCHERBAKOV

Institute of Geology and Geophysics, Siberian Branch of the Academy of Sciences, Novosibirsk, 630090, U.S.S.R.

Abstract

The concentration coefficients of elements (CC) in ultrabasites (σ), basaltoids (ν), granitoids (γ), clay shales (c), etc., with respect to primary protoplanetary (meteoritic μ) material show both nonlinear periodical dependence on the nuclear charge and tendencies of radial distribution of elements within the Earth. The degree of element accumulation in the peripheral spheres of the planet, i.e. their centrifugality increases with the increase of the chemical activity of the natural mobile compounds as well as with the decrease of their density and abundance.

Four groups of elements are distinguished on the bases of the ratios of their mean contents (ν/μ and c/ν): centripetal (C_1) with ν/μ and $c/\nu<1$, minimal-centrifugal (C_2) with $\nu/\mu>1$ and $c/\nu<1$, deficiency-centrifugal (C_3) with $\nu/\mu<1$ and $c/\nu>1$, and centrifugal (C_4) with ν/μ and $c/\nu>1$. The fifth group consists of maximum-centrifugal elements whose concentrations in the light envelopes of the planet (atmosphere, hydrosphere and organisms) are higher than those in μ, ν, c. The geochemical evolution of the continental crust is characterized by the relative accumulation of more and more centrifugal elements together with total and general differentiation of the earth material.

The associations and the quantitative ratios of elements concentrating in endogenic ore deposits are determined first of all by the degree of the evolutional development of substratum at the expense of which the ore deposits are formed. It is only ore deposits with the ultrabasites that elements C_1 are connected and it is only ore deposits with gabbroids that elements C_{1+2} and partly C_3 are connected.

Finally, in the ore deposits connected with granitoid complexes the C_3+C_4 elements, Li, Ta, Nb, W, Ba, Pb, etc., are of great importance.

The dependence in endogenic ore deposits between geochemical element ratios and the initial rock composition as well as the degree of substance redistribution during the ore-forming processes is shown by representative examples.

In order to discover the main tendencies in the chemical development of the oceanic and continental earthcrust from protoplanetary material, the relative concentration of elements in these systems is compared with some of their properties: abundance, density, melting point and relative affinity with O, S and Fe, i.e. their lithophility, chalcophility and siderophility (Fig. 1). The composition of the primary protoplanetary material has in this paper been identified with the composition of meteorites (μ) in the proportion 94% stone and 6% metallic ones in accordance with their impact on the surface of the earth (Mason, 1962). The composition of the oceancrust has been identified with basaltoids (ν). The clay shales (c) are of most widespread specific in its composition type of sedimentary rocks of the continental crust. Therefore at first approach the concentration coefficients ν/μ and c/ν of elements illustrate the tendencies in the development of the earth material from a primary composition to the oceanic crust and from the oceanic to the continental one. The degree of change of the concentration of elements in the derivative systems with respect to the initial

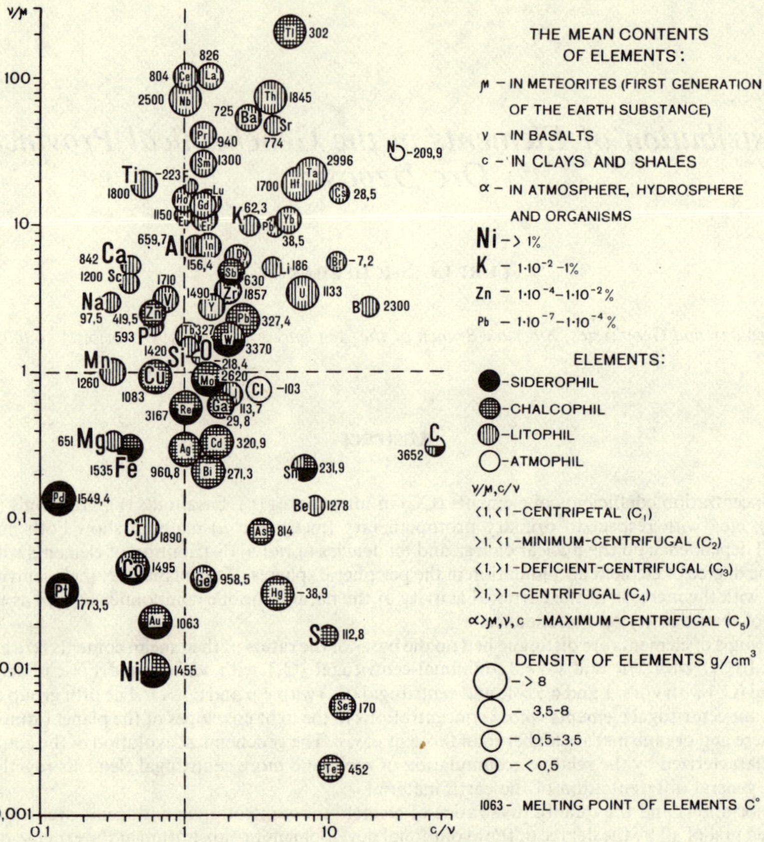

FIG. 1. On elements distribution in the earth and some of their properties.

systems is not a regular function of the above properties of the elements. The degree of accumulation of the elements in the peripheral spheres of the planet, i.e. their centrifugality $C=v/\mu \cdot c/v=c/\mu$, increases with the increase of their chemical activity (*a*) as well as with the decrease of the density of their mobile compounds (*d*) and of their abundance

$$(u): C = f\frac{a}{d \cdot u}.$$

Thus relative increase of centrifugality of odd and (or) rare elements is connected with their low abundance. In Fig. 1 four groups of elements are distinguished on the bases of the ratios of their mean contents (v/μ and c/v): centripetal (C_1) with v/μ and $c/v<1$, minimal-centrifugal (C_2) with $v/\mu>1$ and $c/v<1$, deficiency-centrifugal (C_3) with $v/\mu<1$ and $c/v>1$, and centrifugal (C_4) with v/μ and $c/v>1$. The fifth group consists of maximum-centrifugal elements whose concentrations in the light envelope of the planet (atmosphere, hydrosphere and organisms) are higher than those in μ, v, c. The comparison of concentration coefficients for the main rock types with respect to the protoplanetary material show nonlinear periodical dependence on the nuclear charge, and is more complicated than the dependence on chemical and physical properties (Fig. 2).

The geochemical evolution of the earthcrust from the oceanic type to the continental one from the eugeosynclines to the miogeosynclines, and further on to the orogene and to the platform is characterized by the relative and absolute accumulation of more and more

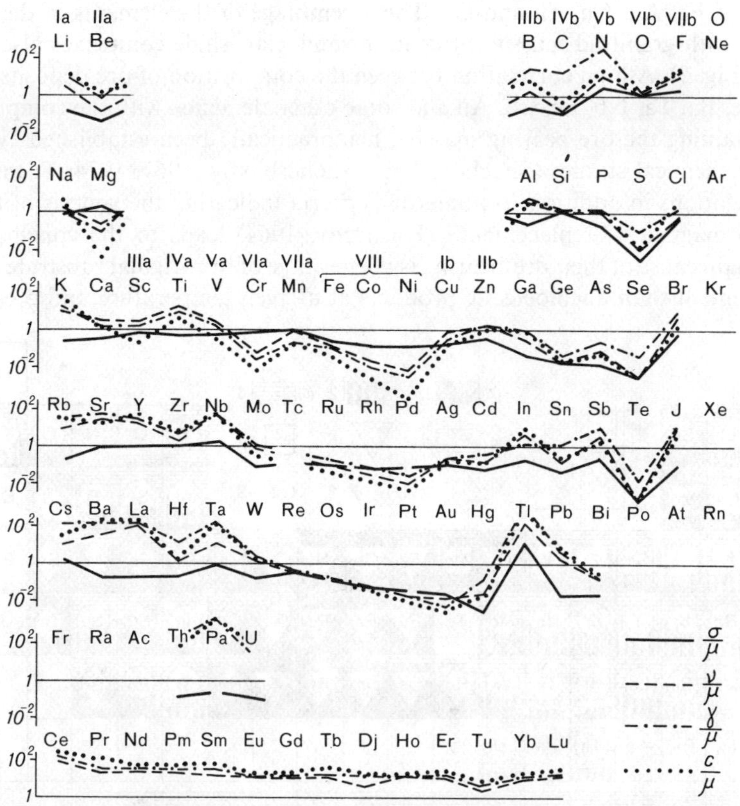

FIG.2. The concentration coefficients of elements in the ultrabasites (σ), basaltoids (ν), granitoids (γ) and clay shales (c) with respect to primary protoplanetary (meteoritic μ) material.

centrifugal elements together with total and general differentiation of the earth material (Shcherbakov, 1976 a,b). As a result of each geochemical process and of the chemical evolution of the earth as a whole, the dispersion and contrast of the distribution of various elements in new geochemical systems increase in proportion to the centrifugality of the elements. The difference between concentration coefficients of elements in the common subgroups of the Periodical system increases from the ultrabasites (σ) to the basaltoids (ν), clay shales (c), granitoids (γ) and finally to the hydrothermalites among the above-mentioned rocks.

The associations and the quantitative ratios of elements concentrating in endogenic ore deposits are primarily determined by the degree of the evolutional development of substratum at the expense of which the ore deposits are formed, i.e. by the concentration and ratios of elements in them.

Accordingly it is only ore deposits of C_1 elements that are connected with ultrabasites. With gabbroids only C_{1-2} elements can be concentrated and the alkaline gabbroids are accompanied only by ore deposits of C_{1-2-3} elements. Granitoids within basic volcanic-sedimentary rocks, independently of their age, are generally characterized by essentially granodioritic composition. They are accompanied by ore deposits only of those chalcophil and siderophil elements whose contents in the basaltoids are higher than in clay shales, i.e.

C_{1-2}, such as Fe, Au, Cu, Zn and Co. The assemblage of the elements in the ore deposits connected with granitoid massifs situated in sand–clay–shale complexes also include C_{3-4} elements (Fig. 3). A high correlation between the composition of ore deposits of B, Sn, U, Mo, W, Be, Ba, Ta, Nb, Fe, Co, Au and some other elements, with the composition of the rocks containing the ore-bearing massifs, has practically been established everywhere by special geochemical studies (Shachov, 1960; Shcherbakov, 1967, 1974; Tugarinov, 1968). Such correlations in addition to numerous criteria indicating the genesis of the granitoid massifs as magmatic "replacement" (Kusnezov, 1964) leads to the conclusion that the first and main cause of their ore-bearing is the heritage of the original substrate with a higher element content than anchioetectic products at a given temperature and pressure.

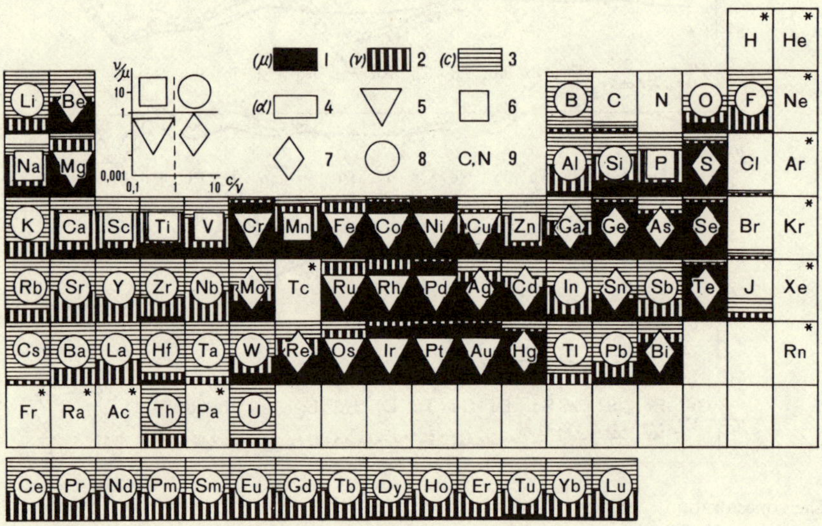

FIG. 3. Geochemical classification of elements according to their radial distribution in the earth and the earth's crust. Relative contents of elements: 1 = in meteorites—μ, 2 = basaltoids, 3 = in clay shales, 4 = in atmosphere, hydrosphere and organisms. Group of elements: 5 = centripetal (C_1), 6 = minimal-centrifugal (C_2), 7 = deficiency-centrifugal (C_3), 8 = centrifugal (C_4), 9 = maximum-centrifugal (C_5). (*) No facts.

The second ore-bearing condition of granitoids is a high degree of magma fractionation responsible for the concentration of emanational ore-forming elements (L. V. Tauson, 1973). Relatively closed systems and long duration of the magmatic precrystallizational state result in higher volatile content in the magma which can concentrate in apical part of the massifs, where ore deposits often are found. The mean content of the elements in the ore-bearing massifs approaching their apical part usually increases approximately one order of magnitude and the elements ratios change in favour of the centrifugal ones (Table 1) Deep zones of the ore-bearing granitoid massifs are accordingly often a little impoverished by the same elements (Shcherbakov *et al.*, 1976; Kosals, 1976).

The third condition to be satisfied by ore deposit formation is the development of hydrothermal processes within the ore-rich component area such as granitoid massifs. Thus the formation of hydrothermal deposits with respect to granitoids may be expressed as the final result of a repeated fractionation of earth material, first by sedimentation then by the remelting of sedimentary or magmatic rocks during the precrystallization differentiation and finally by the hydrothermal metamorphism and redistribution of the metals (Table 1).

TABLE 1. TYPICAL AVERAGE CONTENTS AND RATIOS OF SOME ELEMENTS IN THE ROCKS AND ORES

Elements	Contents of the elements, ppm						Ratios of elements in the rocks and ore deposits from various parts of the hydrothermal systems: root (r) and apical (a) both among basic (v) and acidic ($c-\gamma$) rocks							Examples of provinces
	Basites (v)	Shales (a), clays (c)	Depth parts of granite massifs	Zones of magmatic enrichment	Zones of preores leaching	Hydrotherm. ores	Ratios	v	$c-\gamma$	r in v(1)	a in v(2)	r in $c-\gamma$(3)	r in $c-\gamma$(4)	
Ta	0,48	2,5	2,5	15	0,8	2000	Ta:Nb / Taμ:Nbμ	0,32	2,4	0,1–0,3	0,05–1	0,3–2,6	0,1–5	1–2. Transbaikal, Altai 3–4. Kola Peninsula
Nb	20	20	15	120	5	1000								
W	1	2	1	15	0,5	1000	W:Mo / Wμ:Moμ	1,7	2,5	—	—	0,01–4,0	0,05–200	3–4. Transbaikal, Altai, Caucasis
Mo	1,4	2	0,7	6	0,5	1000								
Pb	8	20	10	30	10	10,000	Pb:Zn / Pbμ:Znμ	0,9	3,8	1,5–5	0,1–10	3–25	1–300	1. Ural, Caucasis 2. Altai 3. Canada, Sikhote Alin 4. Utah, Transcaucasis
Zn	30	80	30	100	50	20,000								
Co	45	20	3	50	2	3000	Co:Ni / Coμ:Niμ	4,6	3–10	0,2–2	30–100	—	16–800	1. Canada, Kola 2. Finland, Morocco 4. Tuva, Transcaucasis
Ni	160	95	3	20	5	40,000								
Ag	0,1	0,1	0,02	0,07	0,01	500	Ag:Au / Agμ:Auμ	15	30–50	0,2–20	0,3–60	0,3–20	3–1000	1. Canada, W. Australia, Ural 2. Japan, Altai-Sajan 3. E. Australia 4. Utah, Colorado, Mexico
Au	0,004	0,001	0,001	0,01	0,0008	10								

The effectiveness of this process which frequently involves the rocks enclosing the pluton depends on the concentration of metals, activity of solutions as well as on the conditions of their discharge.

The broadening of the element associations in the groups of the successively forming ore formations in the folding zones occur together with the enlargement of the thickness of the earthcrust and its compositional complexity (V. A. Kusnezov, 1972). In ore formations this tendency is accompanied by a regular change in the ratios of the geochemically related metals to more centrifugal ones. The geochemical cycle of the continental development is completed by the decrease of the ore element associations mainly at the expense of the centrifugal metals. This tendency develops during the thinning of the continental crust as a result of its deepbasification and eclogitization. At this stage the mono- and few-metallic ore deposits are formed (Au–Ag, Au–Te, Ag–Co, Hg) and the share of C_3 elements in the ores increases.

Table 1 gives the ratios of the element contents to each other in ores and rocks as compared with the corresponding ratios in meteorites. These ratios are minimal for the polyvalent and lithophil elements (Ta/Nb) and maximal for univalent and few valent chalcophil and noble metals (Ag/Au).

The elements ratios discussed increase with the increase in the difference of their centrifugality. Moreover, these ratios and their variations increase from the root-located metasomatic ores to the apical-located veins and epithermal deposits (Table 1). All this shows the role of element ratios as possible indicators of the evolution level of the earth's crust and geochemical provinces as well as indicating the source of metals in the endogenic deposits.

1. The concentration coefficients (CC) of elements in the rocks and ores with respect to meteorites are important geochemical parameters.
2. Using "CC" five groups of elements are distinguished according to the various tendencies of their radial distribution in the earth and earthcrust.
3 The geochemical evolution is characterized by a common tendency of increasing "CC" deviation in each group of the periodical system from initial to derivative geological formations in the course of the fractionation of earth material.
4. The comparison of ratios of "CC" of geochemical-related elements in the rocks and ores makes it possible to consider the level of their geochemical development, the question of the source of metals and even to approach the understanding of redistribution distances.
5. The fact that the composition and localization of epigenetic ore deposits depend in a regular manner on the earth's chemical evolution makes regional geochemical investigations reasonable since their results can be used in the prospecting for new mineral resources.

References

KOSALS, YA. A. (1976) *Principal Features of Geochemistry Rare Elements in the Granitoids Molten and Solutions*, "Nauka", Novosibirsk, 239 pp.

KUSNEZOV, V. A. (1972) Ores formations. *Geol. and Geophys.* No. 6, pp. 7–14, Novosibirsk.

KUSNEZOV, Y. A. JU. (1964) Main types of magmatic formation and their distribution in the fundamental structures of the Earth's crust. *Report of the XXII Session of Int. Geol. Congr.*, Part 16, 271–282, New Delhi.

MASON, B. (1962) *Meteorites*, J. Wiley & Sons, New York, 274 pp.

SHAKHOV, F. N. (1960) About the origin of the granitic magmas and ore deposits. In *Magmatism and Ore Deposits Connection with it,* "Nauka", Moscow.

SHCHERBAKOV, YU. G. (1967) *Distribution and Conditions of Gold Concentration in the Ore Provinces,* "Nauka", Moscow, 267 pp.

SHCHERBAKOV, YU. G. (1974) *Geochemistry of Gold Deposits in Kusnezky Alatay and Gorni Altai,* "Nauka", Novosibirsk, 227 pp.

SHCHERBAKOV, YU. G. (1976a) Chemical evolution of the earth's material and endogene ore formations. In *Gold and Rare Elements in Geochemical Processes,* "Nauka", Novosibirsk, 14–33.

SHCHERBAKOV, YU. G. (1976b) Geochemical evolution and ore formations. In *Problems of Endogene Ore-formation and Metallogeny,* "Nauka", Novosibirsk, 217–229.

SHCHERBAKOV, YU. G., KOSALS YA. A. *et al.* (1976) In *Ores Formations and Geochemistry of the Ore-forming Processes,* Novosibirsk, 119–176.

TAUSON, L. V. (1973) Potential ore-bearing and ore-productivity of the granitoids. In *Problems of Magmatic Geology,* "Nauka", Moscow, p. 330.

TUGARINOV, A. JV. (1968) On the reasons of ore provinces's forming. In *Chemistry of the Earth's Crust,* Vol. 1, "Nauka", Moscow.

Discussion

R. D. SCHUILING:

It is not confusing to introduce a new classification scheme for the elements, like centripetal or centrifugal, implying that the earth's gravity field has acted as the major cause for the distribution of the elements, whereas in reality their geochemical character determines their fate during the differentiation of the earth's materials?

Y. G. SHCHERBAKOV:

The affinity of elements determines only the types of minerals. But the migration and precrystallizational differentiation of mobile ions and their compounds really take place in the earth's gravity field. Therefore the gravitation in a great degree determines the radial distribution of the elements in the earth. I think neither classification is excluding the other but complementing it.

R. SINDING-LARSEN:

Is this method used in the Soviet Union for the production of metallogenetic maps?

Y. G. SHCHERBAKOV:

We have only lately started to use the above-mentioned geochemical principles for the composition of metallogenetic maps.

Prospection Géochimique en France: Evolution et Tendances de la dernière Décennie

J. BARBIER, E. WILHELM et W. SAKOWITSCH

B.R.G.M., B.P. 6009, 45018 Orleans Cédex, France

Abstract

There has been a noticeable evolution within the last ten years in the geochemical prospecting methods, particularly with increased use of emission spectrometry; several factors favoured this evolution:

In France as well as in many countries, large geochemical surveys are carried out in order to promote exploration and/or draw up an inventory of the potential mining resources of the country; such an exhaustive inventory requires the analysis of a great number of elements, and leads to the use of emission spectrometry.

Geochemistry gradually proves itself as one of the most efficient tools in exploration (even the most efficient in some cases), if used with a relatively high sample density, in order to locate directly the deposits sought; the preliminary selection of favourable areas through low samples density reconnaissance surveys seems a hazardous and hardly efficient procedure.

Fortunately the utilization of the computer, which geologists have become more familiar with, allows to make better interpretation: the computer makes it possible to overcome the difficulties which inevitably would have occurred in the processing of large amounts of analytical data.

An important technical effort has been made to provide sufficiently accurate analyses for use in geochemical prospecting, at as low a cost as possible.

The obtained data have caused a reconsideration of the usual models of secondary dispersion around ore deposits; in France and in the regions with similar climate (and paleoclimate) there is mainly a mechanical migration from weathered rocks, related to former permafrost phenomena in a tundra-type environment. The old idea of the metal fixation on clays and organic compounds must, in most cases, be given up, at least with regard to soil or stream sediment anomalies related to ore deposits.

L'objet du présent article est de présenter brièvement l'évolution observée en France ces dix dernières années, dans le domaine de l'exploration géochimique, tant sur le plan des concepts que sur celui des techniques. Le point de vue exposé est celui de géochimistes appartenant à un Service Géologique National qui, en prospection géochimique, est plus spécifiquement orienté vers des travaux systématiques à caractère d'Inventaire minier. Cette optique ne sera pas forcément celle de géochimistes plus spécialement intéressés par la recherche d'une ou plusieurs substances particulières.

Introduction—La Prospection Géochimique en France vers les années 1960

D'abord localisées autour de gîtes ou d'indices de façon à rechercher l'extension de minéralisations connues, les prospections ont rapidement pris un caractère systématique dès les premiers succès obtenus. Elles ont été faites à maille serrée (15 à 20 prélèvements/km^2) de manière à ne laisser échapper aucun gîte même petit, en échantillonnant, le long du réseau hydrographique, le sol recouvrant les alluvions d'une part, les versants des deux côtés du cours d'eau d'autre part.

Ces prospections n'ont généralement concerné qu'un ou deux éléments au maximum, analysés par colorimétrie la plupart du temps; c'est ainsi que les districts les plus prometteurs ont été prospectés surtout pour plomb et zinc d'autres pour antimoine ou tungstène, certains pour cuivre.

L'interprétation des résultats était simple. Elle consistait à retenir pour des prospections détaillées en sol des zones caractérisées par un groupement d'échantillons à teneurs anomales; ces dernières étaient déterminées à partir de l'examen d'histogrammes, le mode étant censé représenter le fond géochimique régional moyen.

I. Evolutions Intervenues lors de la Dernière Décennie

1. *Les Progrés Analytiques Réalisés*

Vers 1965, les *progrès effectués dans le domaine analytique,* tant en absorption atomique qu'en spectrométrie optique d'émission à lecture directe, ont permis l'utilisation progressive de ces techniques pour la recherche des gîtes minéraux. Est très vite apparu aux géochimistes l'avantage d'exploiter au maximum les possibilités des appareils en effectuant non plus le dosage d'un ou deux éléments, mais l'analyse simultanée de 6, 8, 10 voire 20 éléments sur le même échantillon, le gain d'information compensant largement le surcroît de dépenses.

2. *Les Avantages de l'Analyse Multi-éléments*

(a) *Une meilleure définition de la notion d'anomalie.* Les premières études méthodologiques réalisées autour de gîtes connus ont montré qu'une minéralisation se manifeste en général par des anomalies de plusieurs éléments, et qu'il était possible en conséquence de connaître l'origine d'une anomalie d'après la répartition et le niveau de teneur de ces éléments: ainsi des valeurs anomales en cuivre ou zinc peuvent être considérées comme traduisant la présence de sulfures ou bien de roches basiques, selon les teneurs en argent, chrome, nickel, etc., associées.

Les dosages multi-éléments réalisés sur chaque échantillon ont permis de ce fait une excellente sélection des anomalies, en considérant non seulement les principaux métaux Cu, Pb, Zn, Sn, W, mais aussi les indicateurs que sont As, Ag, Sb, Cd, Bi, Mo, . . . ; d'autres éléments tels que Cr, Mn, V, Fe, Ni, etc., ont servi à écarter les anomalies liées non pas à des minéralisations mais à la composition lithologique d'une formation.

(b) *Le caractère exhaustif des prospections réalisées.* La seule analyse du cuivre par exemple, n'est pas en mesure de localiser un gîte d'antimoine, ou inversement. Or il est souvent impossible, dans une région donnée, de prédire ou bien d'exclure la présence d'une concentration économique pour tel ou tel élément. En conséquence, l'analyse multi-éléments s'impose si l'on souhaite effectuer un inventaire *complet* du potentiel minéral d'un secteur déterminé.

(c) *Une réduction possible de la maille d'échantillonnage.* De façon classique, une anomalie isolée n'est pas considérée a priori comme significative, et ne doivent faire l'objet de travaux plus détaillés que des groupements de points anomaux; un tel principe implique nécessairement une densité d'échantillonnage élevée. En réalité un échantillon isolé, qui présente des anomalies de teneurs pour plusieurs métaux évoquant une association minérale rencontrée dans un gîte métallifère, par exemple Ag–Pb–Cu ou Pb–As et Sb, constitue bel et bien une anomalie digne d'intérêt. L'information obtenue par l'analyse multi-éléments permet dans une certaine mesure, d'éviter des échantillonnages trop serrés, donc trop coûteux, surtout dans des régions d'accès difficile; la densité de prélèvements préconisée a ainsi pu être ramenée à 3–5 échantillons par kilomètre carré.

3. *Le Rôle de l'Informatique*

Il s'est avéré primordial. En effet, une prospection faite avec la densité indiquée plus haut, sur une superficie de 1 000 km^2 (20 × 50 km par exemple) avec un dosage de 20 éléments, conduit à un volume de données numériques imposant, de l'ordre de 100 000. Cette abondance de chiffres pose quelques problèmes, ne serait-ce que d'ordre matériel, dans la mesure où les opérations habituelles de report des teneurs sur cartes, calcul des seuils d'anomalies, tracé d'histogrammes etc., se trouvent multipliées par le nombre d'éléments dosés. Seule l'utilisation de l'*ordinateur*, couplé à un moyen de visualisation (table traçante) peut permettre de résoudre convenablement les difficultés. En plus d'être un instrument de calcul, l'ordinateur s'est révélé un moyen de gestion, complément indispensable des analyses multi-éléments.

II. Etat Actuel des Techniques et des idées

1. *Les Types de Prélèvements*

Aucun changement n'a été apporté dans les techniques d'échantillonnage: le matériau prélevé reste toujours, soit du *sol en prospection détaillée,* soit des alluvions ou du *sol sur alluvions* en prospection régionale; l'analyse des eaux n'a pas débouché sur une utilisation intensive en prospection des gîtes minéraux, sauf le dosage du fluor par ionométrie pour la recherche de gîtes à fluorine; le dosage des gaz n'en est encore qu'à ses premiers essais (émanométrie du radon mise à part) et il faudra de longues années probablement avant que cette méthode ne puisse relayer les autres techniques de prospection au niveau de la recherche des gîtes cachés.

2. *Maille d'Echantillonnage*

La conception générale ne s'est guère modifiée durant cette dernière décennie. Le principe fondamental reste la recherche direct d'anomalies liées aux minéralisations, en utilisant un échantillonnage à maille relativement serrée (3 à 5 éch./km^2), à l'inverse des méthodes qui préconisent, en première phase, une sélection de zones favorables basée sur les résultats d'un échantillonnage à large maille (1 éch. pour plusieurs km^2). Cette dernière approche ne peut être efficace que pour des objectifs de grande taille, et ne sera donc légitime que si la concentration minérale, elle même de faibles dimensions, se trouve associée à une formation géologique plus étendue caractérisée par un fond géochimique élevé pour les éléments présents dans la minéralisation, ou encore si la minéralisation est elle même de très grande dimension (gîtes à Cu disséminé, concentration de nickel latéritique, etc .).

Ce cas n'est malheureusement pas général, et l'on connait un grand nombre de gisements qui, au contraire, ne sont pas associés à des unités géologiques à fort "background": ainsi par exemple les amas sulfurés volcano-sédimentaires, liés à des émissions de tufs ou de laves acides, qui ne sont pas spécialement riches en Fe, Cu ou Zn; beaucoup de gîtes filoniens à fluorine se trouvent encaissés dans des roches ayant un fond géochimique en fluor de l'ordre de 700 ppm, ce qui est très banal. Cette liste est loin d'être exhaustive.

3. *Les Techniques Analytiques*

Les méthodes classiques de colorimétrie ou d'absorption atomique sont toujours utilisées, mais au niveau des prospections régionales, elles sont progressivement remplacées par les techniques spectrométriques.

(a) *Spectrométrie d'emission à lecture directe (quantomètre)*. Il s'agit d'un appareillage maintenant classique, qui peut doser simultanément de 20 à 30 éléments-traces et fournir également une bonne estimation du niveau de teneurs des éléments majeurs.

Les impératifs de rendement compatible avec le très grand nombre d'échantillons prélevés, ont pu être surmontés grâce à l'utilisation d'une alimentation par ruban (tape-machine). Dans le même temps, les interférences spectrales et les effets de matrice ont été corrigés à l'aide de programmes statistiques prévisionnels (Moal, J. Y. et Pinault, J. L., BRGM, inédit). La précision des analyses est variable selon les éléments; elle est de l'ordre de 15 à 30 % relatif.

(b) *Spectrométrie d'émission à plasma*. Elle permet le dosage d'une vingtaine d'éléments† préalablement mis en solution. Les interférences sont minimisées par la qualité du système dispersif et par un système inédit de réglage des fentes; elles sont en outre corrigées par voie chimique (solutions tampon) et par calculs sur mini-ordinateur relié au spectromètre. La qualité des résultats analytiques fournie par cette technique est comparable à celle des dosages obtenus par absorption atomique (Boucetta, M. et Fritsche, J., BRGM, inédit).

4. *L'Utilisation de l'Ordinateur*

Deux problèmes majeurs se posent au géochimiste: présenter les résultats analytiques, et sélectionner les anomalies intéressantes.

Le premier point est du ressort de la cartographie, c'est-à-dire de procédés de *visualisation*. Aucune difficulté majeure n'a été rencontrée de ce côté: le traceur de courbes commandé par l'ordinateur permet le *report sur plan* des teneurs à n'importe quelle échelle; les anomalies peuvent être mises en relief par un figuré de taille croissante en fonction de l'importance de l'anomalie, et les variations du fond géochimique représentées simultanément par exemple, au moyen de courbes d'isoteneurs. Il est toutfois difficile à l'ordinateur de prendre en compte puis de restituer les divers éléments géographiques: aussi dessinera-t-on à part un fond topographique, qui sera ultérieurement superposé sur les cartes tracées automatiquement.

Au contraire, pour la sélection des anomalies, aucun programme classique pris individuellement n'est entièrement satisfaisant à l'heure actuelle; il est tout à fait illusoire de baser cette sélection sur les seuls paramètres statistiques, moyenne et écart-type, ce pour de multiples raisons: d'une part ce serait admettre implicitement que, quelle que soit la zone prospectée, la proportion d'anomalies intéressantes est toujours la même ou presque; d'autre part, l'expérience enseigne que les anomalies les plus intenses ne sont pas forcément en relation avec des minéralisations économiques et sont parfois trompeuses.

Les divers modes d'*analyse factorielle* sont également peu efficaces pour la sélection des anomalies. Ces dernières ne peuvent en aucun cas constituer une population suffisamment importante, dans un échantillonnage régional, pour être isolées par analyse multi-variable.

† Mn, P, Ti, Li, Be, B, V, Cr, Co, Ni, Cu, Zn, As, Cd, Sr, Y, Zr, Nb, Mo, Ag, Sn, Sb, Ba, La, W, Pb, Bi.

Les méthodes de *régression* permettent souvent de s'affranchir d'une partie du fond géochimique et de ne garder que certaines valeurs résiduelles facilitant ainsi la distinction entre anomalie formationnelle et anomalie à caractère métallogénique.

Les techniques de *classification* facilitent le regroupement des échantillons présentant des spectres géochimiques voisins; elles favorisent, par analogie avec l'information multi-éléments obtenue au cours de reconnaissance antérieures d'indices ou de gîtes, la sélection des associations géochimiques significatives d'origine métallogénique.

5. *La Conception Actuelle sur la Dispersion Superficielle des Métaux autour des Gîtes*

L'obtention de très nombreuses données analytiques sur des gîtes variés ainsi que leur traitement multivariable (par analyse factorielle, qui donne d'excellents résultats à échelle détaillée), ont permis d'observer l'existence d'un certain nombre de règles constantes dans la répartition superficielle des métaux autour des gîtes. Il y a, d'une manière très générale, une grande similarité entre les associations géochimiques mises en évidence dans les sols et alluvions et celles des produits d'oxydation des minéralisations sous-jacentes (minéraux supergènes, "fragments ferrugineux" du type chapeaux de fer, débris de roches plus ou moins minéralisés.

On est ainsi conduit à admettre, contrairement aux schémas classiques, que la dispersion superficielle des métaux est essentiellement d'orde mécanique. Une telle migration est favorisée par les phénomènes de gélifluction produits lors des épisodes de glaciations pléistocènes. L'idée ancienne d'une fixation des métaux sur des argiles ou des matières organiques doit être abandonnée dans bien des cas, au moins en ce qui concerne les anomalies trouvées autour de gîtes métallifères. Cette nouvelle interprétation serait valable non seulement pour la France, mais pour une ceinture comprenant notamment les Etats-Unis et l'Europe moyenne. En d'autres termes, la géochimie du sol autour des gîtes serait, pour ces régions, beaucoup plus sous la dépendance de *paléoclimats*, qu'influencée par le climat actuel (Barbier J. et Wilhelm E., publication en cours).

Il faut souligner que l'élaboration d'un tel concept n'a été possible que grâce aux très nombreuses analyses réalisées, et à la synthèse de cette information par traitement sur ordinateur. Ce résultat est, en quelque sorte une *retombée scientifique obtenue après des efforts accomplis sur le plan technologique*.

III. Les Problèmes en Suspens et les Perspectives d'Evolution des Techniques Géochimiques.

Les recherches méthodologiques actuellement en cours permettent d'entrevoir les changements susceptibles d'intervenir dans les années à venir dans le domaine des techniques d'exploration géochimique. Deux directions principales se dessinent: tout d'abord la mise au point de programmes de traitement mieux adaptés aux données géochimiques; en effet, comme cela a été dit plus haut, aucun traitement statistique n'est à même, à l'heure actuelle, de réaliser de façon entièrement satisfaisante la sélection des anomalies. Une amélioration sensible dans ce domaine peut être attendue avec la mise au point, par des géochimistes, de programmes faisant avant tout références à l'expérience acquise, ce que ne peuvent faire les calculs multivariables actuellement utilisés.

Ensuite, une meilleure sélection, au niveau de l'échantillonnage, des matériaux sur lesquels devront porter les analyses: l'accent mis sur la migration mécanique a en effet conduit à fonder beaucoup d'espoir dans la séparation et l'analyse des principles phases supports des métaux dans les sols ou alluvions (oxydés, débris de "chapeaux de fer", etc.). Les premiers résultats obtenus à ce sujet laissent entrevoir une possibilité d'amélioration des réponses géochimiques (contraste, dispersion) par rapport à l'analyse du tout venant. Cette idée diffère de celle préconisant l'analyse des "coatings" à hydroxydes de fer et manganèse d'origine essentiellement chimique.

Après les développements enregistrés ces dernières années, dans le domaine de l'analyse et de l'interprétation multi-éléments, il semble donc que les progrès futurs soient à rechercher principalement au niveau du prélèvement géochimique, et en particulier de la sélection des phases à analyser.

Discussion

A: A. Beus:

What is the time log between sampling in the field and having the results plotted on the map in your case?

J. Barbier:

Le report sur carte de l'ensemble des résultats peut être fait dans la journée qui suit la réception des données analytiques. Pour des raisons d'encombrement de l'ordinateur, ce délai est en général étalé sur une semaine. Le temps qui s'écoule entre l'échantillonnage sur le terrain et la production de cartes diverses dépend avant tout du délai entre échantillonnage et analyse en laboratoire.

Quelques Applications de la Géochimie des Eléments en Traces en Exploration Pétrolière et Minière

B. PORTHAULT et P. SAUVAN

S.N.E.A.(P.), Département Laboratoire de Géologie de Pau, Avenue du Président P. Angot, 64000 Pau

Abstract

The types of geological information supplied by trace elements are substantiated by four case-histories in petroleum and mining geology.

In the first one, relationships between trace elements, organic matter and depositional environment are emphasized.

The second case displays the behaviour of some trace elements resulting from effects of ground water circulation.

The third one shows how geological information can be obtained by ascribing concentrations of a given element to its preferential mineralogical bearer.

The fourth one is an example of chemical–mineralogical changes in a hydrothermal system.

Résumé

Quatre exemples illustrent le type d'informations géologiques apportées par les éléments en traces dans le cadre d'études pétrolières et minières.

Le premier cas étudié met en évidence les relations qui peuvent exister entre les éléments-traces, la matière organique et le milieu de dépôt.

Le second cas montre le comportement de certains éléments en traces sous l'effet de circulations de fluides dans des roches poreuses et perméables.

On montre dans le troisième cas comment des informations géologiques non décelables directement apparaissent si l'on rapporte par un calcul les teneurs d'un élément à sa phase porteuse privilégiée.

Le quatrième est un cas homologue du second dans un système hydrothermal.

Généralités

La géochimie minérale a été introduite au Laboratoire de Géologie de Pau de la SNEA(P) il y a une quinzaine d'années. Le premier objectif poursuivi était la recherche d'indicateurs de milieux de dépôt dans les séries sédimentaires. Le développement de cette discipline s'appuyait sur l'apparition relativement récente du spectromètre d'émission U. V. permettant d'obtenir rapidement et à peu de frais le dosage simultané d'un grand nombre d'éléments en traces (Kulbicki G. *et al.*, 1967). Le développement des études de géochimie sédimentaire, puis d'une activité minière, a conduit à sélectionner, en fonction des seuils de détection et de la précision de cet appareillage, 24 éléments en traces:

B, V, Mo, Pb, Zn, Cu, Cr, Ni,
Co, Sr, Ba, Ga, Mn, As, Sb, Ag,
Ge. P, Sn, W, Zr, Be, Bi, Sc.

En outre, le dosage de six éléments majeurs (SiO_2, Al_2O_3, Fe_2O_3, CaO, MgO, TiO_2) est fourni simultanément avec une précision moindre, suffisante toutefois pour être utilisé dans des calculs de routine.

La mise en oeuvre de cette technique permet de disposer d'un grand nombre de données, dont l'interprétation n'est généralement pas aisée si l'on se borne à l'examen des teneurs brutes, car la concentration d'un élément est le plus souvent le résultat global de plusieurs causes géologiques. Ces difficultés d'interprétation sont aggravées par le fait que les matériaux analysés sont presque toujours polyphasés: dans un même type de sédiments par exemple, le nickel peut se trouver en association avec:

—des chlorites; Ni^{2+} y remplace Mg^{2+} par substitution diadochique;
—de la matière organique sapropélique; celle-ci fixe certains métaux dans des conditions réductrices (Leriche, 1959; Delfaud et Gauthier, 1968; Kulbicki et Rumeau, 1969);
—des oxydes de fer et de manganèse sur lesquels il est aisément adsorbé;
—des sulfures;
—etc.

Ces différents comportements d'un même élément dans un lot d'échantillons peuvent être mis en évidence par une analyse statistique multivariable: l'analyse factorielle en mode R groupe les éléments covariants dans un même facteur (Kulbicki *et al.*, 1968); elle fait apparaître dans l'exemple précédent quatre associations impliquant le nickel:

—Mg–Zn–V–Ni (chlorite)
—Carbone–Mo–Cu–Ni–V (matière organique)
—Mn–Fe–Ni–Co (oxydes)
—Zn–Cu–As–Ni–Co (sulfures).

Au vu de ces groupements, le géochimiste formule une hypothèse sur la nature des phases porteuses préférentielles de chaque élément: dans les cas favorables, cette hypothèse est vérifiée grâce à des recoupements avec des résultats d'autres techniques (minéralogie, microfaciès, physique des roches, chimie organique, etc.).

Une telle identification des phases porteuses n'est possible que si la variance analytique est faible par rapport à la variance géologique. En conséquence, on ne peut prendre en compte que les éléments dont les teneurs sont nettement supérieures au seuil de détection. Dans le domaine des sédiments, les dosages sur roches totales présentent un inconvénient majeur: certaines phases, dont les variations de proportions sont prépondérantes, jouent un rôle de diluant par rapport aux autres; tel est le cas du quartz, qui ne supporte aucun élément en traces en proportion décelable, ou des carbonates qui n'en fixent qu'un petit nombre (Sr, Mn, accidentellement Ba, Zn . . .): seules les variations de proportions de ces diluants apparaissent clairement, car seules elles sont supérieures à la variance analytique des éléments dilués. Pour pallier cet inconvénient, on est souvent amené à éliminer ces phases au cours de la préparation (décarbonatation, désulfatation, décantation fractionnée); grâce à ces traitements préliminaires, on amplifie les variations liées à la fraction fine des sédiments, qui est la plus sensible aux conditions physico-chimiques des milieux de dépôt et aux modifications diagénétiques.

Les méthodes ainsi mises au point pour l'étude géochimique des sédiments ont pu être utilisées en géologie pétrolière autant qu'en géologie minière. Dans ce dernier domaine, où l'on traite des matériaux d'une plus grande diversité, les résultats acquis par cette méthode de travail sont souvent recoupés et complétés par des analyses sur fractions séparées par voie physique. Or ces separations sont toujours dispendieuses, parfois impossibles sans modifier la composition chimique de certaines fractions. L'analyse factorielle est un moyen commode de discriminer les diverses phases de manière indirecte.

1. Etudes Pétrolières

Dans le cadre des études pétrolières, la géochimie minérale participe à l'obtention d'informations géologiques variées: nature des apports détritiques, chimisme des milieux de dépot, modifications diagénétiques, détermination des roches-mères d'hydrocarbures, etc.

Nous sélectionnerons deux exemples: le premier montre le rôle des éléments en traces en tant qu'indicateurs de roches-mères; le second illustre leur comportement sous l'effet de mobilisations postérieures à la sédimentation.

1.1. *Exemple de Comportement "Organophile"*

Dans le domaine de la recherche des hydrocarbures, on observe souvent un parallélisme entre les propriétés de la matière organique et le comportement de certains éléments en traces. Cette convergence, lorsqu'elle se manifeste, traduit l'existence de conditions réductrices au moment du dépôt. En outre, la fixation d'éléments en traces par la matière organique est fonction de la présence de composés hétéroatomiques, comme par exemple les porphyrines.

La mise en évidence d'associations d'éléments "organophiles" reflète donc une composition et des conditions de conservation de la matière organique favorables à la génération d'hydrocarbures. En ce sens, la géochimie minérale apporte une contribution à la diagnose des roches-mères.

L'exemple choisi concerne un ensemble de trois formations paléozoïques d'un Bassin des Etats-Unis, recoupé par plusieurs sondages. La série, dont la puissance varie entre 100 et 400 m, est principalement constituée par une alternance d'argilites et de calcaires légèrement dolomitiques à grain fin.

Les éléments en traces ont été dosés sur la fraction argileuse de chacun des échantillons; l'analyse factorielle les groupe en quatre associations dominantes (Fig. 1).

—Le facteur 1 oppose la silice au cortège des éléments liés aux minéraux argileux les plus riches en alumine. L'évolution verticale de cette association fait apparaître un mécanisme séquentiel de la sédimentation. La confrontation avec l'étude des microfaciès permet d'identifier sans ambiguïté les phases porteuses des éléments: le cortège alumineux, pôle posifif du facteur, est lié à des apports terrigènes fins qui envahissent le bassin au début de chaque séquence; chacun de ces épisodes correspond à une tendance à la fermeture du bassin, donc à un certain confinement. Les périodes d'ouverture se marquent en sommet des séquences par une prolifération de Spongiaires dont les spicules occasionnent des concentrations de silice très fine. Cette augmentation de la silice va de pair avec une décroissance des proportions de minéraux argileux.

—Le facteur 2 groupe les éléments fixés par la matière organique; dans le cas étudié, ce sont le molybdène, le vanadium, le cuivre, le nickel et le chrome.

—Le facteur 3 rassemble des éléments sidérophiles. L'étude des lames minces montre que cette association est principalement liée aux oxydes et hydroxydes, concentrés dans les

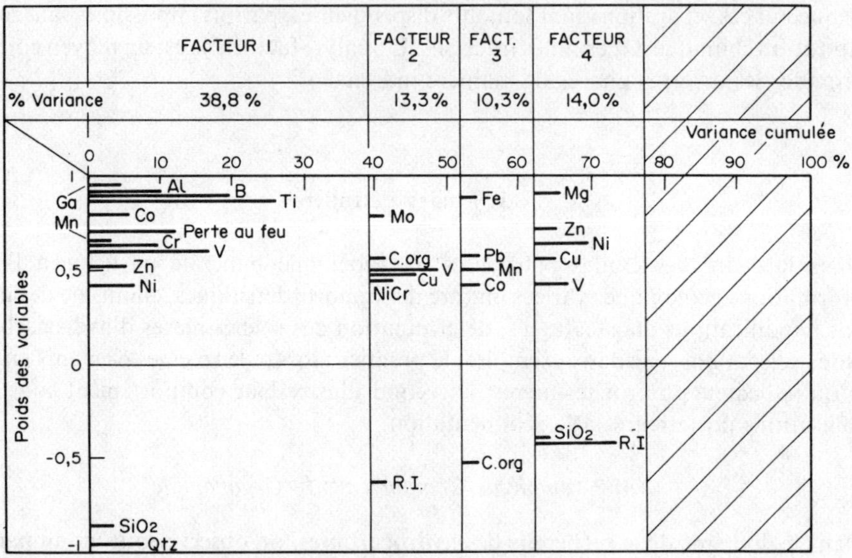

FIG. 1. Exemple n°1: Principales associations géochimiques d'après l'analyse factorielle. Paléozoïque, Etats-Unis. Matrice Variemax. 106 échantillons, 20 variables.

horizons à forte agitation (calcarénites); le fait qu'une part de la variance de la matière organique s'oppose à ces éléments s'explique par l'oxydation de cette dernière.

—Le facteur 4 est une association classique d'éléménts liés à la chlorite. Cette chlorite est préférentiellement néoformée dans les niveaux carbonatés, d'où l'opposition avec le résidu insoluble à l'acide chlorhydrique (R. I. sur la figure 1).

La convergence entre les variations verticales du facteur "organophile" et les diverses analyses effectuées sur la matière organique ressort clairement sur la figure 2 relative à un sondage particulièrement représentatif de ce bassin. On constate que la matière organique se concentre particulièrement dans les bases de séquences, c'est-à-dire lors des périodes durant lesquelles le confinement est le plus accentué. L'étude microscopique des éléments figurés de la matière organique fait voir une nette prédominance des constituants d'origine végétale. Cette caractéristique est en harmonie avec l'afflux de matériaux terrigènes dans ces mêmes niveaux. La qualité de cette matière organique n'est donc pas excellente pour la génération d'hydrocarbures. On constate néanmoins une nette augmentation de la fraction extractible au chloroforme (fraction qui inclut les hydrocarbures) dans les niveaux où cette matière organique est plus abondante.

Les variations du facteur "organophile" reflètent donc assez fidèlement celles de la phase organique, tant en ce qui concerne la quantité que la qualité de celle-ci. Dans le cas présent, on localise à l'aide des paramètres minéraux et organiques trois niveaux de roches-mères potentielles dont l'existence est étroitement contrôlée par les conditions de dépôt.

1.2. *Exemple d'Influence de Phénomènes Diagénétiques*

La série sédimentaire argilo-sableuse que nous allons considérer est d'âge tertiaire et se situe dans un bassin côtier du Pacifique. Elle a été étudiée à l'aide de 19 coupes de terrain représentant 141 échantillons au total. Comme dans l'exemple précédent, l'analyse

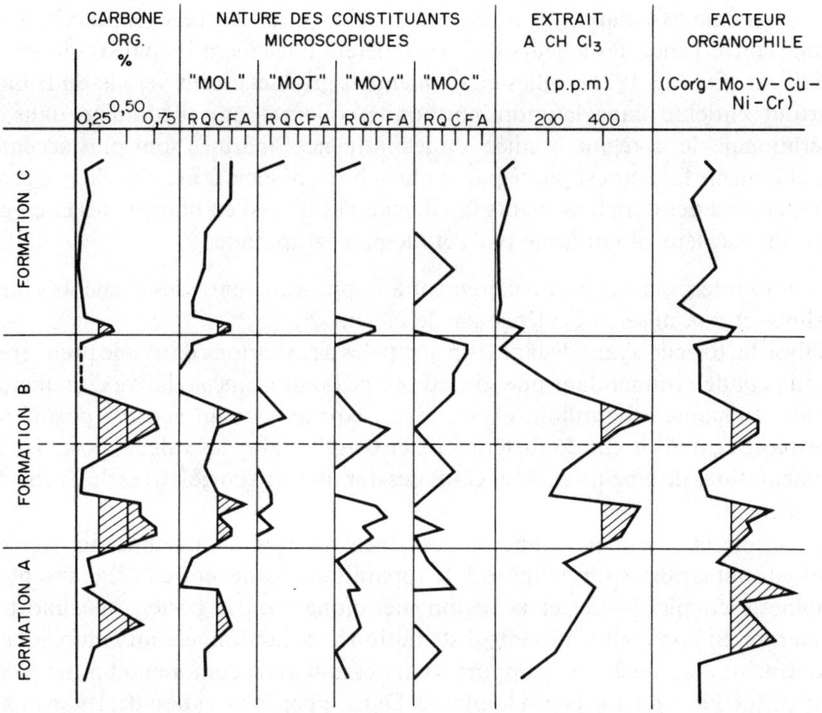

FIG. 2. Exemple n°1: Comparaison entre diverses données géochimiques organiques et minérales de l'un des sondages. Paléozoïque, Etats-Unis.
—Carbone organique: expression en % du poids de la roche totale.
—Nature des constituants microscopiques (sur préparation palynologique).

MOL: matière organique ligneuse,
MOT: matière organique trachéidique;
MOV: matière organique cellulaire végétale,
MOC: matière organique colloïdale.
R: rare; Q: quelques; C: commun; F: fréquent; A: abondant
(indices d'abondance des différents constituants).

— Extrait chloroformique: expression en ppm, par rapport au poids de la roche totale.
— Facteur organophile: "factor-scores" du facteur 2 de la figure 1, exprimés en valeurs relatives par rapport à la moyenne des échantillons (aires hachurées positives).

factorielle a porté sur les dosages d'éléments majeurs et en traces effectués sur la fraction argileuse. Elle met en évidence cinq associations:

—Une association d'éléments ferromagnésiens, qui correspond essentiellement à des chlorites dont les teneurs élevées en Cr et Ni évoquent une origine volcanique basique:
Cr—V—Ni—Co—Mg—Zn—Fe—Mn

—Le deuxième facteur oppose Al et Ga à l'ensemble Ca–Sr–Mg–Ba. Sa signification sera étudiée en détail par la suite.

—Le troisième facteur est piloté par la matière organique, de nature charbonneuse, dont les concentrations évoluent en opposition avec celles des éléments Fe–Mn–Mg–Co, principalement associés dans cette étude à des interstratifiés chlorite–montmorillonite.

—La quatrième association groupe bore et silice. Cette corrélation inhabituelle résulte d'une convergence: les teneurs en bore reflètent fidèlement les paléosalinités du milieu de dépôt (Couch, 1971); celles-ci augmentent graduellement vers le Sud. La silice est surtout enrichie dans la montmorillonite, minéral qui prédomine dans la partie méridionale de la région étudiée, où les influences marines sont plus accentuées.

—Le cinquième facteur est piloté par le phosphore, associé à Fe, Cu, Zn. Ce groupement fait penser à des enrichissements (au niveau des traces) en phosphates et en glauconie. Ceci est localement confirmé par l'étude pétrographique.

Nous nous intéresserons particulièrement à l'opposition entre les éléments alumineux et les alcalino-terreux mise en évidence par le facteur 2.

L'analyse factorielle ayant dégagé les principales associations dans une première étape du calcul, on peut déterminer dans une seconde étape l'importance relative d'un facteur donné au niveau de chaque échantillon. Ce résultat s'exprime par un nombre positif ou négatif ("factor-score"); dans le cas du facteur 2 de l'exemple traité, les valeurs positives indiquent des augmentations de teneurs en Al et Ga et des diminutions corrélatives de Ca, Sr, Mg, Ba et réciproquement.

On a calculé la valeur moyenne de ces "factor-scores" pour chacune des coupes; les résultats en sont reportés sur la figure 3. Il apparaît une limite entre la région septentrionale où domine le couple Al–Ga et la région méridionale où le cortège alcalino-terreux est prépondérant. Si l'on confronte cette distribution avec les données minéralogiques, on voit que les éléments alumineux sont supportés par des minéraux comprenant principalement des feldspaths, des illites-micas et de la kaolinite. Dans la partie méridionale, les alcalino-terreux sont surtout supportés par de la montmorillonite.

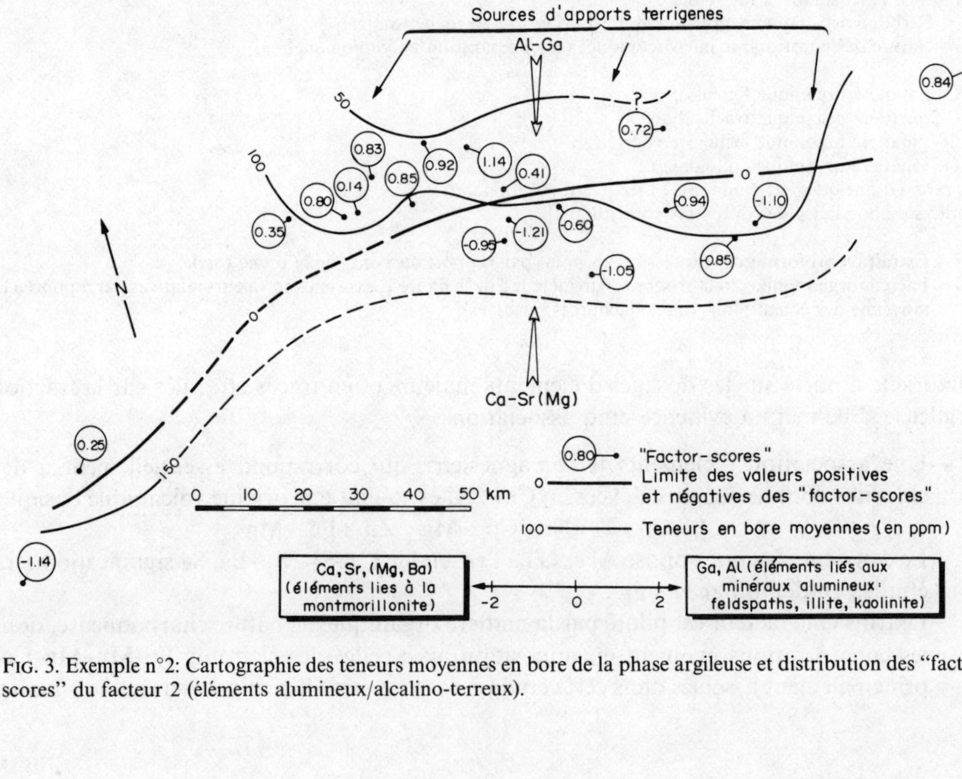

FIG. 3. Exemple n°2: Cartographie des teneurs moyennes en bore de la phase argileuse et distribution des "factor-scores" du facteur 2 (éléments alumineux/alcalino-terreux).

A première vue, cette distribution géographique pourrait suggérer une répartition zonale des apports terrigènes fins. L'examen attentif des lames minces montre cependant que les minéraux alumineux se sont principalement développés dans les niveaux poreux par un processus diagénétique: leurs habitus impliquent en effet qu'ils se sont formés après le dépôt, oblitérant une grande partie de l'espace poreux disponible. Ce processus est beaucoup moins développé dans la zone à montmorillonite, ce qui explique que les cations solubles n'y ont pas été expulsés de leur support originel.

Par ailleurs, on note une certaine concordance entre la limite de ces deux provinces et la carte des teneurs moyennes en bore (fig. 3). Dans le commentaire du facteur 4, nous avons dit que cet élément peut être interprété sans ambiguïté comme un indicateur de la paléosalinité du milieu de dépôt. En première approximation, on peut considérer que la limite entre le domaine continental et le domaine marin dans ce cas particulier correspond à des teneurs moyennes en bore de l'ordre de 100 ppm. Il s'agit en fait de valeurs moyennes qui intègrent des variations liées aux fluctuations de la ligne de rivage au cours de l'intervalle de temps considéré.

La frontière entre les deux provinces minéralogiques est très voisine de la limite entre influences marine et continentale. Ceci évoque une transformation des minéraux par lessivage lié à des circulations d'eau douce dans la région septentrionale. Cette hypothèse permet d'expliquer la recristallisation de minéraux authigènes dans les niveaux poreux de cette région, alors que dans la province méridionale, qui demeure en permanence sous influence marine, les minéraux originels ne sont pas modifiés.

Les incidences pétrolières concernent principalement les propriétés des réservoirs gréseux potentiels du secteur étudié. Dans la partie septentrionale, les grès sont en effet très peu perméables, par suite de l'obturation de l'espace poreux par les recristallisations diagénétiques. Au Sud par contre les caractéristiques primaires des réservoirs sont mieux conservées.

2. Etudes Minières

Dans ce domaine, les matériaux soumis à l'analyse sont sensiblement plus variés:

—roches de tous types;
—sols, sédiments de rivières, batées;
—concentrés, mixtes et stériles dans le traitement des minerais.

Les méthodes couramment utilisées en géochimie pétrolière peuvent aider à la compréhension des minéralisations à chacun des stades de leur recherche. Deux exemples ont été retenus: le premier montre comment des informations géologiques peuvent être tirées de l'exploitation raisonnée d'analyses chimiques et minéralogiques de sols. Le second est relatif à des sondages de reconnaissance dans un système hydrothermal; il illustre l'apport des éléments en traces dans la recherche d'une polarité sur ce type d'objectifs.

2.1 *Informations Géologiques Tirées d'une Etude de Sols*

Les matériaux étudiés sont des sols prélevés au cours d'une campagne de prospection géochimique de faible extension dans une région montagneuse où le substratum n'affleure pratiquement pas. Cette campagne était destinée à délimiter un indice de tungstène; la maille de prélèvements est de 50 × 100 m. Outre les dosages par spectrométrie d'émission U. V., on dispose des teneurs en fluor et en lithium. Les analyses ont été effectuées sur la fraction de granulométrie inférieure à 0,15 mm.

L'analyse factorielle fait ressortir quatre associations principales:

—Le facteur 1 comprend les éléments de la minéralisation: W–F–B–Cu–Bi–As. C'est l'expression d'une paragenèse classique; bore et fluor indiquent la présence de tourmaline. La grille d'échantillonnage ne fait apparaître aucune zonalité des différents éléments entrant dans ce facteur; ceux-ci sont tout-à-fait covariants.

—Le facteur 2 groupe Mg–Zn–Li, association caractéristique des minéraux silicatés magnésiens. L'analyse diffractométrique des fractions argileuses permet de vérifier que les valeurs positives élevées de ce facteur correspondent à des sols riches en talc et en chlorite magnésienne, localisés dans la partie occidentale du secteur. Les valeurs négatives par contre caractérisent des sols entièrement dépourvus de ces minéraux.

—Le facteur 3 rassemble Al–Ti–Ga–Ba–B. Il s'agit d'illite alumineuse ainsi qu'on peut le vérifier par diffractométrie de rayons X. La cartographie de l'association montre que cette illite est principalement cantonnée dans la partie orientale du secteur.

—Le facteur 4 fait intervenir les éléments suivants: Mo–Ge–Ni–Pb–Co–Ga–Cr–Fe. Cette association évoque soit des oxydes, soit des sulfures. Un contrôle à la tarière sur deux points fortement anomaliques en ces éléments montre que le "bed-rock" y est constitué par des quartzites à nombreux granules d'oxydes de fer. Ceux-ci peuvent provenir d'une

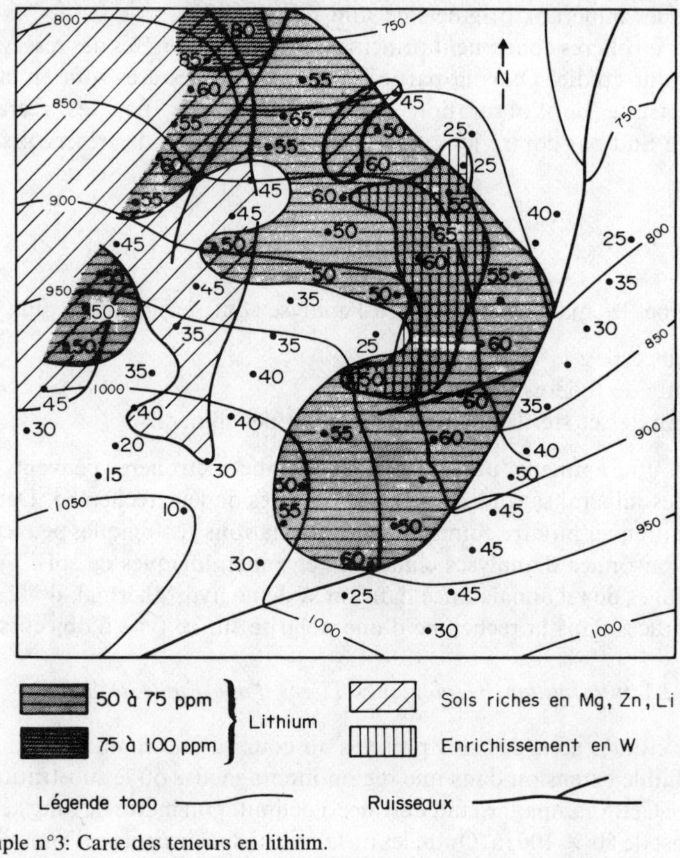

FIG. 4a. Exemple n°3: Carte des teneurs en lithiim.

altération météorique de sulfures originels. La cartographie de ce facteur met en évidence des bandes alternativement riches et pauvres en ces éléments. L'hypothèse d'un contrôle stratigraphique est assez probable pour expliquer cette particularité.

Nous considérerons tout particulièrement le comportement du lithium dont le facteur 2 nous a permis d'identifier les phases porteuses privilégiées.

La cartographie des teneurs brutes en lithium (fig. 4a) ne suggère pas de relation évidente entre la distribution de cet élément et la minéralisation d'une part, la composition minéralogique des sols d'autre part.

Sachant que le lithium est préférentiellement supporté par le talc et la chlorite, seuls minéraux riches en magnésium de ce secteur, le report sur carte du rapport Li/Mg permet d'éliminer l'effet de dilution du lithium occasionné par les phases minérales non porteuses. La figure 4b montre l'aspect de l'anomalie résiduelle résultant de ce calcul. Cette carte fait apparaître des relations qui n'étaient pas discernables sur la précédente:

—Un lien beaucoup plus net se manifeste avec le pointement granitique du secteur Centre-Est.
—La forme des courbes se rapproche beaucoup plus du contour de la zone minéralisée.
—Par ailleurs le contraste entre les valeurs extrêmes est nettement amplifié par comparaison avec la figure 4a et l'anomalie résiduelle est indépendante des teneurs en lithium les plus élevées.

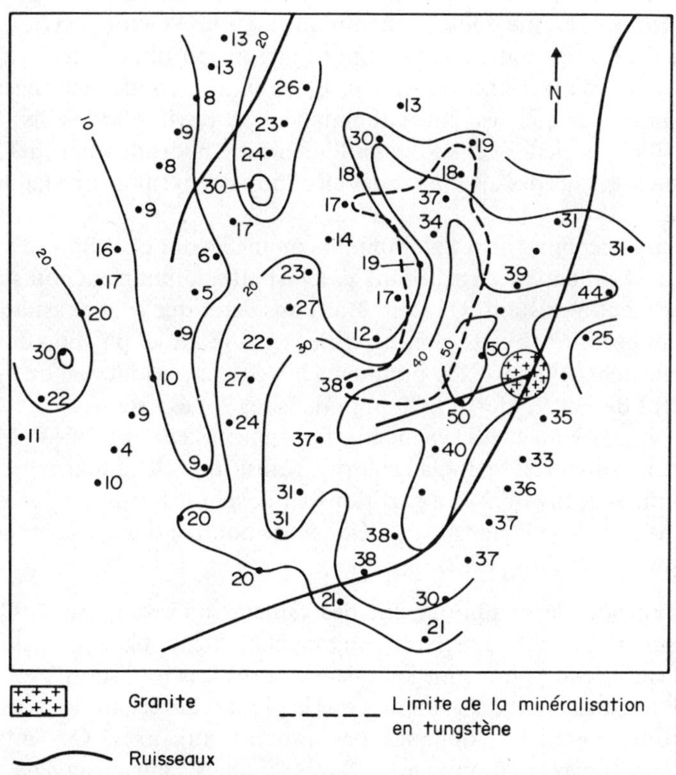

FIG. 4b. Exemple n°3: Carte du rapport Li/Mg.

Au stade actuel des travaux sur ce prospect, l'information obtenue par ce traitement fournit une logique du comportement du lithium. La relation qui semble apparaître entre cet élément et la minéralisation pourra constituer un indicateur de proximité lors des travaux ultérieurs.

2.2 *Evolution Imputable à une Altération Hydrothermale*

Cet exemple concerne des roches traversées par des sondages carottés implantés dans la zone externe d'un système hydrothermal de type "porphyre cuprifère". Les roches étudiées sont des microgranites, laves, tufs de diverses générations. L'étude pétrographique montre que les feldspaths sont plus ou moins intensément transformés; les ferromagnésiens, originellement des biotites, sont presque toujours chloritisés. L'étude des minéraux d'altération a été effectuée par diffractométrie de rayons X sur la fraction fine; elle montre une grande variété de phyllosilicates: montmorillonites, chlorite alumineuse, séricite, kaolinite, interstratifiés illite–montmorillonite et chlorite–montmorillonite (plus précisément sudoïte–beidellite).

Nous nous bornerons ici à considérer le comportement des éléments en traces dosés sur cette même fraction inférieure à 4 microns, plus représentative des phénomènes liés à l'altération hydrothermale.

L'analyse factorielle fait apparaître en particulier les groupements suivants.

—Le facteur 1 oppose les variations de Al, Ga, B, et de la perte au feu à celles de Sr, Ba, Ca, P. Cet antagonisme reflète l'altération hydrothermale des feldspaths: l'analyse diffractométrique confirme que les échantillons dont les "factor-scores" sont négatifs, riches en Ba, Sr, Ca, P, contiennent d'importantes proportions de feldspaths alors que les anomaliques positifs, riches en Al, Ga, B, H_2O, n'en contiennent pas; ces derniers sont par contre caractérisés par l'abondance de phyllosilicates hyperalumineux, notamment illite ("séricite") et kaolinite. L'altération hydrothermale des feldspaths se traduit par un lessivage des alcalins et des alcalino-terreux, alors que l'aluminium reste inerte.

Bore et gallium accompagnent l'aluminium comme ils ont coutume de le faire le plus souvent. Or, si le gallium est enrichi dans les feldspaths, notamment potassiques, le bore y est pratiquement inexistant. On doit donc considérer que le bore est apporté par les fluides qui ont occasionné cette altération. Ses teneurs sur les phyllites de ces sondages peuvent localement atteindre 200 ppm dans la zone la plus interne du système, alors qu'elles restent de l'ordre de 10 à 30 ppm dans les zones externes.

—Le facteur 2 rassemble les éléments ferromagnésiens Cr–Ni–V–Mg–Fe–Ti–P– essentiellement supportés par la chlorite, ainsi que le confirme l'examen des diffractogrammes de rayons X. La partition du phosphore au niveau des traces entre les ferromagnésiens et les plagioclases a déjà été démontrée dans les roches granitiques d'autres régions (Vistelius, 1962).

Pour tenter d'élucider le comportement des minéraux ferromagnésiens au cours de l'altération hydrothermale, chaque échantillon est défini dans le plan des facteurs 1 et 2 par ses coordonnées cartésiennes, qui sont les valeurs de ses "factor-scores".

La figure 5, relative au sondage n°2, montre que la plupart des échantillons s'ordonnent au voisinage de droites parallèles, obliques par rapport aux axes. Ces intervalles sont indépendants de la nature pétrographique des faciès originels. Dans chacun des sondages, les niveaux situés dans la partie inférieure gauche du diagramme renferment exclusivement

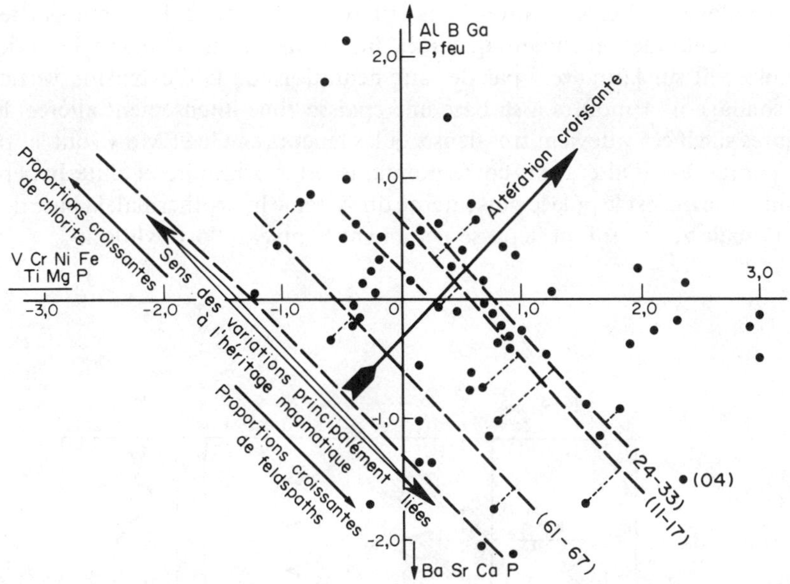

FIG. 5. Exemple n°4: Caractérisation chimique de l'amplitude de l'altération hydrothermale.

feldspaths et chlorites; on définit ainsi une "ligne de base" correspondant aux échantillons n'ayant subi aucune altération; le déplacement le long de cette ligne traduit le changement de proportions de feldspaths et de chlorite. Plus un échantillon est distant de la ligne de base, plus il contient de minéraux phylliteux d'altération. Ceci encore est confirmé par l'estimation semi-quantitative des divers minéraux à partir des diffractogrammes de rayons X: la proportion de minéraux originels décroît dans le sens de la flèche de la figure 5. Les lignées observées doivent donc être interprétées comme correspondant à diverses "amplitudes" d'altération: par "*amplitude*" de l'altération hydrothermale, nous entendons la proportion de minéraux originels transformés, indépendamment de la nature des phyllites qu'elle engendre: cette dernière est fonction du *niveau* d'altération, c'est-à-dire de la température, de la pression et des concentrations dans les fluides.

Le passage d'une lignée (échantillons 61 à 67) à une autre (échantillons 11 à 17 de la figure 5) traduit l'appauvrissement concomitant des feldspaths en Ba, Sr, Ca, P et des ferromagnésiens en Cr, Ni, V, Fe, Mg, Ti, P. Ce passage se fait souvent par saut, ce qui peut s'interpréter de diverses façons: contacts par failles (localement prouvés par la présence de mylonite au niveau de cette rupture), différences de propriétés physiques des roches originelles, etc.

On peut ainsi caractériser l'amplitude de l'altération hydrothermale subie par chaque échantillon par la distance D entre son point représentatif et la ligne de base: cette distance fournit une estimation des quantités de minéraux originels transformés. La figure 6 représente les variations verticales de ce paramètre dans les 4 sondages considérés. Nous n'en commenterons que quelques points:

—Le sondage n°3 est peu altéré: il est à considérer comme le plus externe dans le système hydrothermal. Ceci est confirmé par le fait que la paragenèse d'altération y est dominée par la montmorillonite.

—Les sondages n°2 et 4 montrent des critères d'extériorité dans leur partie inférieure. Celle-ci renferme cependant quelques filons métriques à quartz et séricite, qui se manifestent sur la figure 6 par des augmentations de D d'extension verticale limitée.

—Le sondage n° 1 montre à sa base une épaisse zone intensément altérée: le réseau de fissures silicifiées y devient très dense, et les teneurs en Cu et Mo y sont les plus élevées. La paragenèse d'altération correspondante est à kaolinite et illite hyperalumineuse dominantes. C'est le pôle le plus interne du système hydrothermal; la base de ce sondage est la seule zone où l'on dépasse nettement le niveau "propylitique".

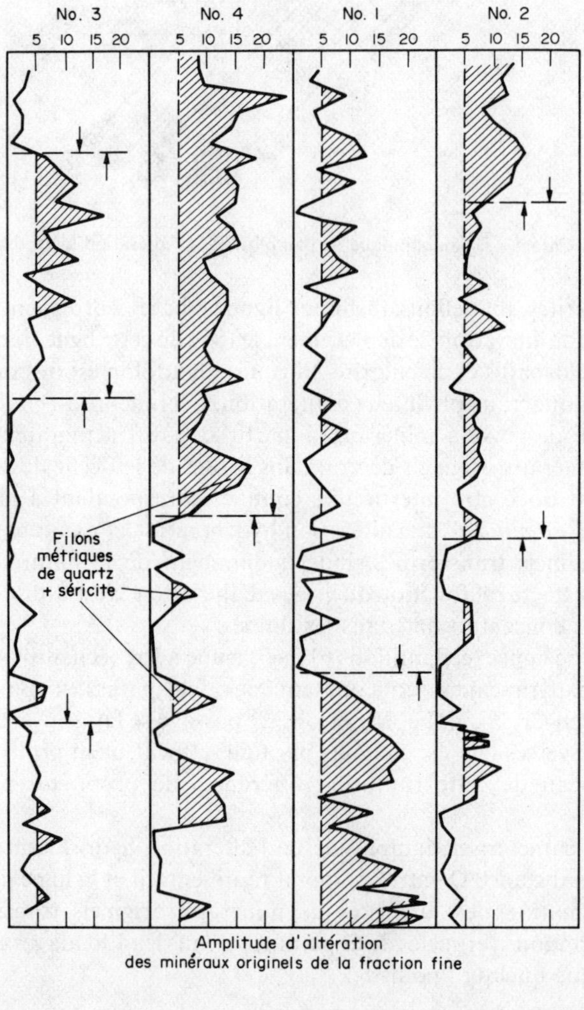

FIG. 6. Exemple n°4: Report des variations verticales de la distance *D* (amplitude de l'altération) dans les sondages n° 1 à 4. Les traits interrompus séparent des lignées différentes telles que représentées sur la fig. 5.

Nombre de paramètres auxiliaires ont été étudiés Nous n'en retiendrons que trois:

—Le rapport B/Al constitue un excellent indicateur du *niveau* d'altération hydrothermale.

—Le rapport Sr/Ba est sous la double dépendance de l'héritage magmatique (Sr/Ba est influencé principalement par les proportions relatives de plagioclases et d'orthose) et de l'altération hydrothermale. Le baryum est en effet plus aisément accepté que le strontium par le réseau des illites et autres minéraux résultant de la destruction des feldspaths.

—Le plus intéressant est le rapport P/(Ca + Mg), dont la considération est dictée par la partition du phosphore dans les minéraux originels des fractions fines de ces roches, plagioclases et ferromagnésiens. Ce rapport décroît d'autant plus que l'amplitude de l'altération est importante, et ceci quelle que soit la position dans le système hydrothermal. En d'autres termes, l'expulsion du phosphore à partir de ses minéraux-hôtes s'effectue indépendamment des conditions thermodynamiques dans l'ensemble traversé par les quatre sondages étudiés. C'est vraisemblablement ce qui explique la répartition du phosphore dans les sols de la campagne de prospection sur cet objectif: il présente une anomalie négative de teneurs sur une vaste surface centrée sur le foyer du système enrichi en cuivre et molybdène; il est par contre enrichi le long d'une ceinture très éloignée de ce foyer, localisée dans la zone propylitique la plus externe. Le phosphore est donc, dans des provinces de lithologie comparable, l'un des indicateurs les plus distaux de ce type de cible.

3. Conclusions

Les quelques exemples présentés sont loin de constituer une revue exhaustive des applications de la géochimie des éléments en traces en exploration pétrolière et minière. Dans certains cas, on peut seulement émettre des hypothèses sur les causes des variations observées. Parfois cependant, il est possible d'établir un lien avec le support minéral ou organique des éléments considérés: ceci permet de dépasser le stade purement descriptif et d'appréhender une logique de leur comportement. On constate que la qualité des informations fournies par les éléments en traces est valorisée au maximum lorsque l'on peut recouper ces informations à l'aide d'autres données géologiques: microfaciès, biofaciès, minéralogie, pétrographie, physique des roches, analyses de phases enrichies . . .

La contribution de la géochimie minérale à la définition des milieux de dépôt a déjà été abordée dans de nombreuses études antérieures. Les applications de cette discipline à la compréhension de processus diagénétiques en est encore à ses débuts. Ces processus, qui sont liés à des déplacements de fluides minéraux ou organiques, présentent autant d'intérêt en géologie pétrolière qu'en géologie minière. Les similitudes de comportement de certains éléments mobiles apparaissent particulièrement, au travers des exemples 2 et 4 de cet article, dans des contextes pétrographiques bien différents. Ce mode d'approche ouvre des perspectives intéressantes dans la recherche d'accumulations de substances utiles liées à des déplacements de fluides, car les éléments en traces sont des indicateurs possibles de ces mouvements.

L'étroite coopération entre géochimistes spécialisés dans des problèmes miniers et pétroliers trouve ici sa pleine justification.

Références

Couch, E. L. (1971) Calculation of paleosalinities from boron and clay mineral data. *Bull. Amer. Ass. Petrol. Geol.* **55**, n°10, 1829–1837.

Delfaud, J. et Gauthier, J. (1968) Contribution à la connaissance de l'environnement de dépôts carbonatés de plate-forme. Caractères géochimiques et sédimentologiques du Jurassique nord-aquitain. *Bull. Centre Rech. Pau—SNPA* **2**, n°2, 347–383.

Kulbicki, G., Sourisse, C. et Baradat, J. (1967) Détermination des teneurs en éléments-traces dans les silicates par spectrométrie U.V. d'émission à lecture directe. *Bull. Centre Rech. Pau—SNPA* **1**, n°2, 437–452.

Kulbicki, G., Sourisse, C. et Cazabat, V. (1968) Aspects actuels du traitement sur ordinateur des données géochimiques au CRP. *Bull. Centre Rech. Pau—SNPA* **2**, n°1, 191–212.

Kulbicki G. et Rumeau J. L. (1969) Caractérisation géochimique des conditions d'oxydo-réduction dans une sédimentation marine carbonatée. Exemple de la Keg River Formation (NW Alberta, Canada). *Bull. Centre Rech. Pau—SNPA* **3**, n°1, 155–166.

Leriche H. H. (1959) The distribution of certain trace-elements in the Lower Lias of Southern England. *Geochim. Cosmochim. Acta* **16**, 101–122.

Vistelius A. B. (1962) Phosphorus in the granitic rocks of Central Tien Shan. *Geochemistry n*°2, 131–151 (trad. de Geokhimiya).

Discussion

J. Lucas:

Vous avez parlé de chlorite née aux dépens des carbonates. Pouvez-vous me dire de quelle chlorite il s'agit et comment vous envisagez le passage du carbonate à la chlorite?

B. Porthault:

L'observation microscopique des échantillons en lames minces montre que la chlorite se développe au contact des grains de dolomie. En outre, l'association géochimique représentée par le facteur groupant Mg, V, Zn, Ni est tout à fait caractéristique des chlorites magnésiennes. Or cette association présente une répartition relativement isotrope dans le bassin, avec une augmentation sensible des "scores" vers le centre. La chlorite n'est donc pas liée à des apports venant des bordures et elle a plus vraisemblablement une origine authigène. Dans cette hypothèse, il est logique de considérer qu'elle puise son magnésium de la dolomie à laquelle elle est étroitement associée.

B. Kubler:

L'analyse factorielle présentée ici passe par les coefficients de corrélation simples et linéaires. Il n'est donc pas étonnant que de bonnes covariations visibles déjà sur les valeurs brutes se traduisent par des facteurs élevés et dans ceux-ci par des poids élevés.

B. Porthault:

Il est certain que l'analyse factorielle en mode R, qui passe par la matrice des coefficients de corrélation entre variables mesurées sur un certain nombre d'échantillons, ne peut que restituer des corrélations mathématiques entre ces variables. Ce procédé de calcul ne constitue donc qu'un outil, mais un outil particulièrement précieux lorsque, comme dans les cas traités, on se trouve en présence d'un grand nombre de valeurs numériques (ici plus d'une vingtaine de variables mesurées sur plus d'une centaine d'échantillons) dont il faut rechercher les causes de variations. Dans de tels cas, l'analyse factorielle constitue un procédé particulièrement rapide et efficace pour réduire les variables considérées en un petit nombre de groupements covariants, pour lesquels il est alors beaucoup plus facile de découvrir une signification géochimique, que si l'on considérait simplement les valeurs brutes.

Distribution du Manganèse dans les Seriés Volcaniques: un Guide de la Genése des Gisements Manganésifères Volcanogènes

HUBERT GUÉRIN

Groupe d'Etudes Géologiques de l'Université de Reims, Faculté des Sciences, B.P. 347, 51062 Reims Cédex et Laboratoire de Pétrographie-Volcanologie, Université de Paris XI, 91405 Orsay

Abstract

Fe, Ti and Mn were dosed from 600 volcanic rocks (Massif Central, Eifel, Tahiti and Marquises Islands) taking into account the rules proper to sampling, crushing and dissolving. Mn was determined through an original chemical method. The optical density of the manganese–formaldoxime complex is measured spectrophotometrically at 490 mμ. The accuracy is $\pm 0.002\%$ MnO. The minerals were analysed with an electron microprobe by R. Brousse and R. C. Maury.

Data show that Mn has a different behaviour in the basalt–rhyolite and basanite–phonolite series. In the latter series, the Mn distribution is not correlated with the Fe one, so is generally accepted, since Mn is constant or even increased from the basanites to the phonolites. Mn is more easily washed than Fe, especially in alkaline volcanic rocks.

In alkaline series, Mn distribution ratios, Mn-minerals/Mn-liquid, show that the fractional crystallization mechanisms explain satisfactorily the behaviour of Mn. Whereas the Mn contents in the suite of rocks decrease on the whole, following the incorporation of this element in easily fractionating minerals (olivines, clinopyroxenes) the ratios Mn/Fe in all the femic phases increase with the evolution of the series because of the intense fractionation of Fe due to the separation of opaque minerals.

Mn, which is very little incorporated in all the phenocrysts (0.01–0.05% MnO), concentrates in the femic and opaque microlites of the groundmass.

From a metallogenic point of view manganesiferous occurrences are located in rocks with a very low rate of MnO (0.02–0.07%), such as the rhyolites in the Massif Central, whereas trachytes or phonolites which have 0.17% MnO contain very few or no manganese ore deposits. Thus, manganese ore deposits will be rather found in the volcanic rocks with a very low rate of MnO. This scheme has been confirmed in Sardinia and in Morocco (Tiouine) where volcanic and volcanic–sedimentary ore deposits are linked with ignimbrites and rhyolites with low MnO rate (0.03–0.08%).

Introduction

Si la grande majorité des gisements de manganèse est d'origine sédimentaire, le rôle du volcanisme est souvent évoqué pour en expliquer l'origine. Les rapports entre le comportement du manganèse dans des séries volcaniques différenciées et les gîtes manganésifères à environnement volcanique d'Auvergne, de Sardaigne et de l'Anti-Atlas sur lesquels j'ai travaillé seront dégagés.

I. Géochimie

1. Techniques

La teneur en MnO des roches bien que donnée dans la plupart des analyses chimiques est très imprécise. "Cette observation s'applique à la quasi totalité des analyses publiées" (Letolle, 1967).

Les intensités de couleur de Mn^{7+} obtenues par oxydation par le périodate, procédé le plus communément employé ou par le peroxydisulfate ne sont pas très stables (Abdullah, 1968). Ici, Mn a été déterminé spectrophotométriquement en mesurant à 490 mμ la densité optique du complexe Mn–formaldoxime. La précision $\pm 0{,}002\%$ MnO est largement suffisante car les valeurs de MnO varient d'environ $\pm 0{,}005\%$ d'un échantillon à l'autre dans une même coulée comme par exemple pour un basalte pris en 4 lieux différents dans la vallée de la Veyre (Puy de Dôme); 0,200–0, 200–0, 193–0, 191%. Sur le même essai le complexe Fe-tiron est dosé volumétriquement par E.D.T.A. (Guérin, 1975).

Le poids d'un échantillon broyé se rapproche de 10 000 fois celui du cristal le plus gros (Laffitte, 1957), sauf quand les phénocristaux atteignent de très grande taille (5 à 6 cm). Les surfaces altérées ont soigneusement été éliminées.

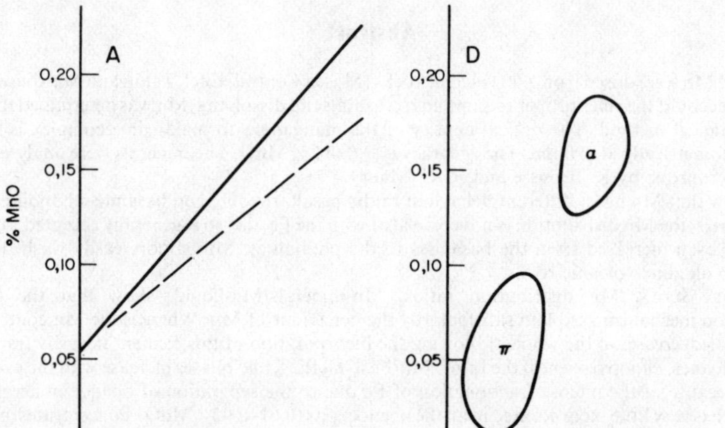

FIG. 1. Courbes de variation des teneurs en MnO en fonction de celles de Fe_2O_3 (fer total) dans les séries volcaniques: A—Mont-Dore basalte–rhyolite (trait plein) et Anti-Atlas andésite–rhyolite (traits discontinus); B—Mont-Dore basanite–phonolite; C—Chaîne des Puys basalte–trachyte; D—Sulcis α-andésite–π-rhyolite; E—Eifel basanite–phonolite; F—Tahiti océanite, basalte, tahitite, phonolite.

2. Répartition de Mn dans Diverses Séries Volcaniques (Fig. 1)

Dans le Mont-Dore, Mn décroît comme Fe dans la suite: basalte, doréite, sancyite, rhyolite. Ceci traduit la substitution bien connue de Fe par Mn dans les oxydes et silicates. La proportion de ceux-ci diminuant au cours de la différenciation, Mn diminue corrélativement.

Par contre dans la suite: basalte alcalin-phonolite, si MnO décroît des basaltes à olivine (0,21%) aux téphrites (0,17%), MnO reste constant des téphrites aux phonolites alors que Fe_2O_3 passe de 8 à 4%. Cette forte teneur en MnO des phonolites présente quelques difficultés à mettre en évidence, car Mn est évacué lors de l'altération alors que Fe n'est pas encore mobilisé. Les phonolites gris-vert prennent une teinte de plus en plus blanche et la teneur en MnO peut être très faible: 0,03% à Fosse (Cantal).

Dans le Cantal, les phonolites inaltérées de la région de Menet ont également une forte teneur en MnO qui peut atteindre 0,24% (Auteroche, Roucoule).

Cet enrichissement n'est pas dû à une contamination sialique quelconque puisqu'il existe aussi dans la série basalte-phonolite de Tahiti (Brousse et Guérin, 1974).

Les basanites, téphrites, phonolites du district de Laach (Eifel) présentent une remarquable constance de la teneur en MnO. Cette non diminution de Mn par rapport à Fe se retrouve dans la suite basalte, trachyandésite, trachyte de la Chaîne des Puys où MnO atteint 0,21% dans les dômites. Notons que les dômites rouges ou jaunes modifiées par les influences fumerolliennes, ne contiennent que 0,07% MnO.

Si le manganèse se comporte comme le fer dans les séries basaltes-rhyolites, de façon inattendue sa teneur reste constante ou augmente dans les séries basaltes-trachytes ou basaltes-phonolites.

3. Répartition de Mn dans les Minéraux

Dès 1936 (Tab. 22, p. 125), Otto montrait que l'augite, la hornblende et la biotite des granites ou des syénites néphéliniques contenaient plus de manganèse que celles des gabbros ou des théralites. D'autres auteurs confirmèrent ce fait (in Berger, 1968). Des études particulières montrèrent que Mn s'enrichissait dans les magnétites tardives au cours de la cristallisation d'une hawaïte (Prévot et Mergoil, 1973) et qu'il s'incorporait plus dans les ilménites que dans les magnétites (Czamanske et Mihalik, 1972; Neumann, 1974).

Récemment Brousse et Maury analysèrent systématiquement à la microsonde tous les minéraux des séries volcaniques alcalines de la Chaîne des Puys, du Mont-Dore et de Nuku-Hiva (Pacifique Central). Des valeurs de MnO que Maury (1976) a déterminé sur les minéraux de la Chaîne des Puys, il apparaît que (Tab. I):

—les microlites contiennent plus de Mn que les phénocristaux correspondants;
—les oxydes absorbent plus de Mn que les silicates correspondants;
—la teneur en Mn des minéraux fémiques augmente vers les termes les plus différenciés.

Ces constatations sont également valables à Nuku-Hiva et dans le Mont-Dore, la rhyolite hyperalcaline de Charlannes contient des minéraux exceptionnellement riches en MnO: magnétite à 14%, pyrophanite à 34%, arfvedsonite à 13% (Brousse et Maury, 1976).

Quantitativement, l'ensemble des phénocristaux de silicates ferro-magnésiens et d'oxydes d'une roche incorpore peu de manganèse: 0,0I à 0,03% MnO à Nuku-Hiva et 0,03 à 0,06% dans la Chaîne des Puys. Ce sont les microlites qui intègrent principalement Mn. Ainsi dans

TABLEAU 1. TENEUR EN MnO DES MINÉRAUX DES ROCHES DE LA CHAÎNE DES PUYS.
Valeurs tirées de la thèse de Maury (1976).

	Magnétite	Ilménite	Olivine	Pyroxène	Amphibole ou biotite	% MnO incorporé dans les phénocristaux
Basalte	0,93†			0,13†		
Saint-Saturnin	0,93		0,61	0,22		0,03
K-Hawaiite	0,07†		0,36†	0,09†		
Pont des Soupirs	1,77†		1,07	0,15		0,02
K-Mugéarite	0,91†			0,12†		
Fonfreyde	1,15	1,35	1,04	0,32		0,06
K-Benmoréite	3,19†				0,77†	
Puy de Clierzou	3,77			0,10	(Amphibole)	
Trachyte	3,31†			2,13†	1,47	
Puy de Dôme	5,48	11,29		2,77	(Biotite)	0,04 à 0,05

† Phénocristaux. Estimation quantitative des minéraux, Guérin.

l'hawaïte du Pont des Soupirs, 0,02% MnO sont incorporés dans les phénocristaux alors que 0,10% le sont dans les magnétites de la pâte. Au cours de la différenciation, si les minéraux sombres et en particulier les oxydes s'enrichissent en Mn, leurs quantités tendent à diminuer comme l'a montré Tsusue (1973) pour les ilménites des granites.

Le rôle du fractionnement des oxydes, des olivines ou des pyroxènes et celui des feldspaths ne sera que brièvement évoqué (Maury et al., 1977). En schématisant, le fractionnement des olivines et surtout des oxydes prépondérant au stade de formation des basaltes-trachyandésites appauvrit le liquide magmatique dans lequel ces minéraux cristallisent. Par contre, en fin de différenciation du magma, le fractionnement des feldspaths qui n'absorbent pas de Mn, enrichit en Mn le liquide résiduel.

A ce stade, si Mn trouve une quantité suffisante de minéraux hôtes, la teneur en MnO de la roche est élevée: 0,17% à 0,24% dans les phonolites et 0,21% dans certains trachytes comme les dômites.

Ces roches, à teneur élevée en Mn, ne recèlent pas de gîtes manganésifères. On peut donc penser que tout le Mn du magma qui leur a donné naissance s'est intégré aux minéraux de ces roches.

A l'opposé, si Mn ne trouve pas assez de minéraux hôtes, la teneur en MnO de la roche est faible: 0,02 à 0,07% comme dans les rhyolites. Ces roches recèlent alors des gîtes d'imprégnation ou des filons manganésifères. Ainsi on peut donc envisager, que Mn qui n'arrive pas à s'intégrer complètement dans les minéraux des rhyolites reste dans la phase pneumatolytique ou hydrothermale et formera des minéralisations.

Si les gîtes de manganèse sont liés aux roches à faible teneur en MnO, ils seront d'autant plus importants que le volume de ces roches le sera également.

II. Implications Métallogéniques

1. Massif-Central

Les volcanites alcalines du Mont-Dore et du Cantal recèlent des occurrences manganésifères alors que les basaltes, trachy-andésites et téphrites en sont exempts.

Dans le Cantal, les trachytes de la région de Menet, présentent des minéralisations noires d'oxydes de manganèse dans leurs plans de diaclases. A Haut-Alliés, les dépôts ont quelques mm d'épaisseur, à Meynoyre ils recouvrent des grenats andratites. Dans le Mont-Dore, les trachytes du Bazet, de Montmie et du Barrie sont aussi minéralisés. Ces modestes indices sont encaissés dans des roches à teneur un peu moins élevée en MnO que celle des basaltes, 0,14% à 0,15% au lieu de 0,18 à 0,20%.

Les roches rhyolitiques, offrent de plus belles minéralisations (Guérin, 1970). Dans le Mont-Dore, des filonnets de psilomélane assez nombreux mais minces sillonnent la rhyolite de Lusclade, dans celle de Voissière les filonnets s'anastomosent et créent un réticulum. A Voissieux, Le Barrie, Rochefort-Montagne, des filons de psilomélane puissants d'une dizaine de cm traversent les nappes de ponces.

Ces rhyolites ont un faible taux en MnO (0,07%). Cette pauvreté en manganèse et le volume plus grand de ces roches expliquent que les minéralisations soient plus importantes que dans les trachytes.

Les minéralisations dans les phonolites sont rares. Dans le Cantal, un filon large de quelques cm imprègne la phonolite de Brocq. Je rapprocherai cette minéralisation du stockwerk de psilomélane associé aux phonolites du Siroua à Asfzimmer.

Ces deux phonolites ont une teneur sensiblement inférieure à la normale: 0,14% à Brocq et 0.10% à Asfzimmer.

2. Sardaigne

L'ouest de la Sardaigne a été le siège d'un important volcanisme cénozoïque.

Dans le Sulcis, les andésites sont surmontées par une importante masse de laves, d'ignimbrites et de tufs dont un niveau est constitué de comendites. Des plaquages, des filonnets, des veinules et des rognons noirs d'oxydes de manganèse sont intimement liés aux roches rhyo-trachytiques alors que les andésites en sont dépourvues. Ces minéralisations ont été exploitées en de nombreux chantiers artisanaux voire industriels (Macchione, île de San Pietro). Des bandes de psilomélane de 5 m de puissance ont été reconnues (Cala Lunga, île de San Antioco).

Ces minéralisations ne sont pas un "produit d'éluvion continental des tufs" (Burckhardt et Falini, 1956) mais résultent de l'activité fumerollienne qui suit la mise en place de ces roches qui sont inaltérées (Guérin, 1973).

La figure 1 montre une séparation entre les deux termes volcaniques du Sulcis et une liaison nette des minéralisations avec les roches à faible teneur en MnO.

Le gisement de Capo-Becco, dans l'île de San Pietro, est le plus important de la Sardaigne avec 45 000 t Mn métal reconnues et 90 000 t Mn métal possibles (Uras, 1965).

Des bancs d'ocre, de jaspe, d'argile et une couche de wad d'un m d'épaisseur sont interstratifiés dans des liparites. Le tout visible sur 2 km a 8 m de puissance.

On peut concevoir un bassin confiné (aucun fossile n'a été découvert dans ces couches) en environnement volcanique influencé par des apports fumerolliens. Ce gisement n'est donc pas d'origine éluviale continentale d'autant que les roches environnantes ont une faible teneur initiale en MnO (0,04% en moyenne). Il peut être défini comme un gisement volcano-sédimentaire de type proximal au sens défini par Bouladon, Picot et Laforêt (1976).

Dans le N–W de la Sardaigne, le volcanisme étudié par Coulon (1977) comprend sommairement 5 formations calco-alcalines (carte 1/50 000 géologique Bonorva):

α_1: andésites inférieures,
τ_1: trachytes inférieurs,
α_2: andésites supérieures,
τ_2: trachytes supérieurs,
α_3: intrusions finales d'andésites.

Ces laves "anté-helvétiennes" sont recouvertes par des conglomérats et des calcaires marins transgressifs du Langhien. Un volcanisme basaltique alcalin post-tortonien s'est poursuivi jusqu'au pléistocène (Coulon *et al.*, 1974).

Les minéralisations décrites par Burckhardt et Falini (1956) puis par Deriu (1964) dans la région de Bosa se localisent dans les volcanites anté-helvétiennes et dans les conglomérats.

Les nappes ignimbritiques τ_1 et τ_2 recèlent les principales minéralisations de psilomélane et de pyrolusite. Celles-ci sont semblables à celles du Sulcis: imprégnations des roches, veinules les traversant, filons dans failles comme à Monte-Minerva près de Montresta. Près de ce village, à Punta Casteddu, de beaux rognons de romanéchite ont été découverts. A Punta Capuzzu, le psilomélane est accompagné de barytine.

Si la plupart des minéralisations sont liées à des rhyo-trachytes à faible teneur en MnO (0,03 à 0,06%), quelques filons de pyrolusite à gangue de jaspes rouges ou jaunes traversent les andésites comme à Monte-Pramma à l'E de Bosa. A Porto Baosu, les andésites α_1 très altérées contiennent des filons de manganèse, mais aussi de galène, de chalcopyrite et de pyrite. Les minéralisations sulfurées dont la genese demande à être précisée sont antérieures à celles de manganèse (Deriu, 1964)

Les plus grosses minéralisations se situent dans les conglomérats. Les gros éléments arrondis ou à peine usés de trachyte rose sont cimentés par de la pyrolusite et accessoirement de la manganite (Monte Crasta d'Ittiri). La minéralisation se prolonge dans les calcaires par des taches noires. A Sos Aghedos, le trachyte altéré et minéralisé est recouvert d'une couche de pyrolusite qui passe à un conglomérat et à un falun d'huîtres à ciment manganésifère.

Le gisement de Sos Aghedos est estimé à 50 000 t de pyrolusite, mais en d'autres gîtes les lentilles minéralisées sont irrégulières. Ces gîtes volcano-sédimentaires sont liés à l'activité finale du volcanisme acide.

3. Maroc

Le Précambrien III de l'Anti-Atlas est constitué d'énormes formations volcaniques et volcano-sédimentaires.

Les minéralisations qui sont soit en filons dans les volcanites soit en couches dans d'anciens bassins sédimentaires, ont été décrites par Bouladon et Jouravsky (1955) dans le secteur de Tiouine et ont fait l'objet d'une étude de synthèse de Pouit (1976).

Les filons de manganèse qui sont très nombreux dans l'Anti-Atlas (Choubert et Faure-Muret, 1970) ont les mêmes caractères que ceux de Sardaigne: imprégnations des rhyolites, veinules qui s'anastomosent, plus gros filons avec ou sans jaspe. La barytine est plus abondante et épigénéïse les rhyolites jusqu'à former des géodes. La silice est aussi fréquente. Le caractère pénécontemporain des phénomènes volcaniques, tectoniques et minéralisateurs a été démontré par Pouit (1966).

Le gisement de Tiouine, où les couches de braunite et de psilomélane associées à des jaspes et des pélites se sont formées dans un bassin sédimentaire subsident et faillé en environnment volcanique, rappelle celui de Capo Becco mais en plus important (réserves estimées à 700 000 t Mn métal).

Le chimisme du volcanisme de l'Anti-Atlas est perturbé par rapport à celui de la Sardaigne. Les analyses de 36 échantillons prélevés dans les régions: de Tiouine où les termes identifiés par Bouladon et Jouravsky (1955) ont été repris, du Djebel Bachkoun, d'Ouarzazate et d'Ait-Saoun, confirment l'existence des deux groupes de roches rhyolitiques (SiO_2:65 à 81%) et andésitiques (SiO_2: 45 à 57%), que Dupuy et Leblanc (1976) avaient identifies dans le secteur de Bou-Azzer. Ils décelaient un enrichissement irrégulier en K_2O dans les ignimbrites et en Na_2O dans les andésites.

Nos analyses montrent:

—dans les rhyolites un enrichissement fréquent de K_2O pouvant atteindre 15% et un appauvrissement jusque 0,10% en Na_2O.

—dans les andésites soit un enrichissement en Na_2O jusque 6%, mais aussi un enrichissement en K_2O et un appauvrissement en Na_2O comme pour les rhyolites.

Dans le quartier de Tindaf, les rhyolites comme les andésites sont enrichies en K_2O et appauvries en Na_2O. Ceci confirme les idées de Dupuy et Leblanc selon lesquelles les circulations hydrothermales altérèrent ces roches et celles de Bouladon et Jouravsky selon lesquelles ces circulations sont la source manganésifère.

L'enrichissement sporadique en MnO (0,40% à Tindaf) de certains échantillons est secondaire.

Malgré ces transformations, les valeurs de Mn décroissent comme celles de Fe. Cette évolution est tout à fait comparable à celle du Mont-Dore (Fig. 1). Or ici aussi, la plupart des gîtes manganésifères sont en liaison avec les volcanites pauvres en MnO. Toutefois au Djebel Bachkoun, les andésites prédominent sur les produits acides qui ont été néanmoins repérés (analyse nouvelle de l'auteur).

Signalons enfin que dans l'Ougarta algérien, le gîte manganésifère du Djebel Guettara dont les réserves sont estimées à 690 000 t Mn métal (Byramjee et Meindre, 1956) est situé dans un environnement identique à celui de Tiouine.

Ces gisements illustrent le rapport entre l'abondance de produits volcaniques émis et l'importance des minéralisations.

Conclusion

La distribution du manganèse au cours de la cristallisation d'un magma se différenciant montre que cet élément dont la teneur dans les magmas basaltiques initiaux est de l'ordre de 0,20% s'intègre dans les minéraux fémiques qu'il trouve en quantité suffisante dans les termes basiques. En fin de cristallisation, la proportion de feldspath et de quartz augmentant, Mn enrichit les fluides résiduels. Les minéraux fémiques cristallisant à ce stade ont une plus forte teneur en Mn que ceux des roches basiques. S'ils sont en quantité suffisante, ils absorbent tout le manganèse, les roches ont alors une teneur égale ou supérieure à 0,16%. Cette valeur peut être définie comme un "seuil de minéralisation" au-delà de laquelle les minéralisations en oxydes de manganèse n'apparaîtront pas.

Par contre si Mn ne trouve pas assez de minéraux hôtes, il donnera naissance à des dépôts au sein même des roches ou alimentera des bassins sédimentaires. Les gîtes volcaniques ou volcano-sédimentaires ainsi formés seront d'autant plus importants que la teneur des roches qui leur sont génétiquement liées sera basse et que leur volume sera grand.

Références

ABDULLAH, M. I. (1968) The automatic determination of manganese in silicate rocks and sediments. *Anal. Chim. Acta* **40**, 526–530.

BROUSSE, R. et GUÉRIN, H. (1974) Distribution du manganèse comparée à celle du fer et du titane dans les laves des îles de la Société et des Gambier. *Cahiers du Pacifique* n°18, 1, 283–289.

BROUSSE, R. et MAURY, R. (1976) Paragenèse manganésifère d'une rhyolite hyperalcaline du Mont-Dore. *Bull. Soc. fr. Minéral. Cristallogr.* **99**, 5, 300–303.

BURCKHARDT, C. et FALINI, F. (1956) Memoria sui Giamenti Italioni di Manganese, Symposium del Manganeso. *20ᵉ Cong. géol. inter.* **5**, 221–272.

BOULADON, J. et JOURAVSKY, G. (1955) Les gisements de manganèse volcanogènes de Tiouine (Infracambrien du Sud Marocain). *Serv. géol. Maroc. Notes et Mémoires* **127**, 180.

BOULADON, J., PICOT, P. et LAFORÊT, C. (1976) Les gisements sulfurés stratiformes liés au volcanisme. Quelques réflexions sur l'environnement et la minéralogie des minerais. *Mém. h. sér. Soc. géol. France* **7**, 149–154.

BERGER, A. (1968) Zur Geochimie und Lagerstättenkunde des Mangans. Clausthaler Hefte Lagerst. *Geoch. mineral. Rohstoffe* **7**, 70.

BYRAMJEE, R. et MEINDRE, M. (1956) Le gisement de manganèse de Guettara. Symposium del Manganeso. *20ᵉ Cong. géol. inter.* **2**, 176–196.

CHOUBERT, G. et FAURE-MURET, A. (1970) The Precambrian iron and manganese deposits of the Anti-Atlas. UNESCO Genesis Precambrian iron and manganese deposits. *Proc. Kiev. Symp. 1970*, **9**, 115–124.

COULON, C. (1977) Le volcanisme calco-alcalin cénozoïque de Sardaigne: pétrographie, géochimie et genèse des laves andésitiques et des ignimbrites. Signification géodynamique. Thèse doct. Etat, Marseille.

COULON, C., DEMANT, A. et BELLON, H. (1974) Premières datations par la méthode K/Ar de quelques laves cénozoïques et quaternaires de Sardaigne Nord-Occidentale. *Tectonophysics* **22**, 41–57.

CZAMANSKE, G. K. et MIHALIK, P. (1972) Oxidation during magmatic differenciation, Finnmarka complex, Oslo Area, Norway: Part I, The opaque oxides. *J. Petrol.* **13**, 493–509.

DERIU, M. (1964) Notizie sulla costituzione geologica del Bosano della Planargia e del Montiferro settentrionale e occidentale. *Monogr. Reg. Bosano, Ass. Comm. Bosa e Cuglieri* 80.

DUPUY, C. et LEBLANC, M. (1976) Etude préliminaire des ignimbrites et andésites du Précambrien III de Bou Azzer (Anti-Atlas, Maroc). *4è Réunion Sci. Terre* 154.

GUÉRIN, H. (1975) Géochimie du manganèse dans les roches volcaniques du Mont-Dore (Auvergne). *Actes 95è Cong. nat. Soc. Sav. Reims 1970* **I**, 373–383.

GUÉRIN, H. (1973) Dispersion et concentration du manganèse dans les roches volcaniques. *Mineral. Deposita* **8**, 192–199.

GUÉRIN, H. (1975) Analyse par voie humide des roches et minéraux silicatés. *Ann. Sci. Univ. Reims et A.R.E.R.S.* **13**, 66–72.

LA FFITTE, P. (1957) Introduction à l'étude des roches métamorphiques et des gîtes métallifères. Masson, 97–119.

LETOLLE, R. (1967) Les laves du Mont-Dore. Recherches sur la géochimie et la genèse des magmas. Thèse doct. Etat Paris, 185.

MAURY, R. (1976) Contamination (par l'encaissant et les enclaves) et cristallisation fractionnée de séries volcaniques alcalines, continentales (Massif Central français) et océaniques (Pacifique Central): l'origine des laves acides. Thèse doct. Etat, Orsay, 455.

MAURY, R., GUÉRIN, H. et BROUSSE, R. (1977) Comportement du manganèse au cours de la cristallisation fractionnée de séries volcaniques, (à paraître *Chim. Geol.*)

NEUMANN, E–R. (1974) The distribution of Mn^{2+} and Fe^{2+} between ilmenites and magnetites in igneous rocks *Amer. J. Sci.* **274**, 1074–1088.

OTTO, H. (1936) Die Rolle des Mangans in den Mineralien. *Min. Petr. Mitt.* **47**, 89–140.

POUIT, G. (1966) Mise en place pénécontemporaine du manganèse au Maroc par rapport aux formations qui le contiennent ou aux surfaces topographiques au moment de son dépôt. *C.R. Acad. Sci., Paris* **262**, 623–625.

POUIT, G. (1976) Gisements et volcanisme. *Mém. h. sér. Soc. géol. France* **7**, 115–126.

PREVOT, M. et MERGOIL, J. (1973) Crystallization trend of titanomagnetites in an alkali basalt from Saint-Clément (Massif Central, France). *Min. Mag.* **39**, 474–481.

TSUSUE, A. (1973) The distribution of manganese and iron between ilmenite and granite magma in the Osumi Peninsula, Japan. *Contr. Mineral. Petrol.* **40**, 305–314.

URAS, I. (1965) Depositi manganesiferi in Sardegna. *Resoconti dell' Ass. min. sarda* **60**, 3, 1–32.

Geochemical Distribution of Uranium in Soils and Vegetation of the "Fe 3" Mine, Saelices, Salamanca, Spain

A. ARRIBAS† and J. HERRERO-PAYO‡

†*Catedrático de Mineralogía y Cristalografía de la Universidad de Salamanca, España*

‡*Jefe de Laboratorio, Empresa Nacional del Uranio S.A. (E.N.U.S.A.), Saelices, Salamanca, España*

Abstract

A detailed geochemical study to find out the relationship between, on the one hand, the surface radiometric anomalies and the uranium content in the soils and vegetation and, on the other hand, the real distribution of the element as shown in mining studies and in numerous drill-cores—mesh 50 × 25 m—and wagon-drill probes—mesh 10 × 10 m—has been made in the "Fe 3" area, situated about 10 km to the north of Ciudad Rodrigo, in the district of Saelices el Chico, in the province of Salamanca (Spain).

The biogeochemical analyses were carried out on the leaves of the live oak (*Quercus ilex* L.), since this species is the most abundant and the most regularly distributed in the area. The conclusions reached in the study were as follows:

1. There exists a correspondence between the radiometric anomalies and the uranium content in the soil.
2. In general, there is a statistical correspondence between the anomalies of the vegetation and the corresponding values of the wagon-drill probes between 1 and 10 m depth.
3. All the prospecting methods clearly reveal the existence of four mineralized areas in a N–S direction. The two central areas, which were discovered by surface radiometry, coincide quite closely with positive anomalies in the soil and vegetation. The other two areas, situated on both sides, had not been discovered by surface radiometry, and were revealed by positive anomalies in the soils and vegetation. For this reason, a detailed study of these two areas has been recommended.

1. Introduction

One of the uranium areas in Spain which has been most exhaustively studied geologically and investigated in most detail by mining studies and wagon drill and diamond drill is the "Fe 3" area, situated about 10 km to the north of Ciudad Rodrigo, in the district of Saelices el Chico, in the province of Salamanca (Spain). A detailed geochemical study was carried out here to find out the relationship which exists between, on the one hand, the surface radiometric anomalies and the uranium content in the soils and vegetation and, on the other hand, the real distribution of the element as shown in mining studies and numerous drill-cores—mesh 50 × 25 m—and wagon-drill probes—mesh 10 × 10m—carried out systematically for the evaluation of the deposits.

For the biogeochemical study an analysis of one particular plant rather than of all the plant species in the mineralized area was preferred. In this case, the live-oak* (*Quercus ilex* L.) was selected because of its abundance and regular distribution in the area.

The area chosen has a surface of approximately 1000 × 400 m, in which about 300 live-

* *encina*, in Spanish.

oaks and 285 soils were sampled, in accordance with the 50×25 m mesh used in diamond drilling.

2. Geology of the Area

The "Fe 3" deposit is situated in the Cambrian schists which extend to the south-east of Saelices el Chico, about 10 km to the north of Ciudad Rodrigo. The area is a peneplain excavated by the Quaternary fluvial network. The valley of the river Agueda, which passes by the mine, is the most important topographical accident in the area.

The host rocks are sericitic and chloritic schists, formed by metamorphism of a pelitic sediment, with abundant carbonaceous material, in the green schists facies. The schistosity cleavage coincides generally with the bedding, but the cleavage is sometimes deformed by a crenulation which gave place to the microfracturing of the rock and the recrystallization of pyrite in the fissures.

The main minerals are chlorite, sericite, quartz, carbonaceous material and pyrite. Among the accessory minerals, zircon, tourmaline, detritic feldspars, apatite, pyrrhotine, chalcopyrite and rutile are the most frequent.

The schists are very fine grained, and at times correspond to real ampelites because of the abundance of carbonaceous material. However, the distribution of the mineralization has no relation with the mineralogical composition of the host rock.

Among the tectonic deformations, the most interesting from a metallogenic point of view are the late fracturing and brecciation since the concentration of the primary and secondary uranium minerals which form the deposit takes place in the resulting structures.

3. Mineralogy of the Deposit

Regardless of their ascendent or descendent origin, the hypogene uranium minerals are hydrothermal. It was the weathering of the area which made the mineralization workable, as it gave place to the oxidation of the primary uranium minerals and to the redistribution of the hexavalent uranium in the fractures and cleavage of the slates. Both the hypogenic and supergenic minerals are summarized in Table 1.

TABLE 1

Host rocks	Primary uranium minerals	Secondary uranium minerals	Gangues
Sericitic and chloritic schists	Pitchblende Coffinite	Gummite Autunite-uranotile Saleite Torbernite Kasolite Uranopilite Ianthinite Sooty pitchblende	Pyrite Marcasite Quartz Calcite Siderite Pyrrhotite Chalcopyrite
Essential minerals Quartz, chlorite, sericite, carbonaceous material, pyrite			
Accessory minerals Zircon, tourmaline, apatite, rutile, detrital feldspars			

4. Origin of the Mineralization

The uranium ores of the peribatholithic slates, both in the Fe mine and in other numerous occurrences existing in the Iberian Peninsula, are workable because of a process of supergenic concentration. From this point of view the deposits are epigenetic, as the hexavalent uranium minerals were formed by oxidization of other pre-existing minerals, namely coffinite and pitchblende, of which traces still remain below the oxidization zone.

However, the origin of these primary minerals is under discussion. For some geologists, these mineralizations correspond to concentrations originated in Tertiary times by descending solutions containing the uranium leached from neighboring granites. The uranium was first set free by the weathering process which accompanied the development of the peneplains, and then deposited in subsidence basins, within the diaclases and brecciated zones next to the granitic bodies.

However, due to the lack of geochronological dating, and taking into account the metallogenic character—the low-temperature paragenesis and the vein-like structure of the primary ores—and, above all, the existence of analogous mineralizations in the metasediments located around several granites—as occurs in those of the Pedroches (Córdoba), D. Benito (Badajoz), and Nisa (Portugal)—other geologists think that, independently of their ascendant or descendant nature, these mineralizations are vein type and epithermal.

In Europe, ores with similar characteristics are quite rare. Some of them correspond to the autunite impregnations of Le Bleymard, in the schists surrounding the Cévennes granites, namely that of Mont Lozère, where small veins with coffinite and siderite—Bondons deposit—have been traversed by drilling below the oxidization zone.

From a genetic point of view, the most significant feature of this type of deposit is the weak metamorphic character of the rocks which surround the granite plutons. For this reason, due to a lack of more precise data, and taking into account the fact that the tectonic structures in which the primary minerals are found in the Spanish and Portuguese deposits, as well as their ages, may be either Alpine or Hercynian, it might be considered that this type of uranium occurrence represents an extrabatholithic example of the same metallogenic processes which led to the concentration of the uranium in some intragranitic deposits, for instance, Los Ratones in the Caceres province.

5. Radiometry

Figure 1 shows the values and the distribution of the surface radiometric anomalies in the area studied.

6. Probes

The probes carried out in the area were of two types: wagon-drill and diamond-drill.

Figure 2 shows the situation of the wagon-drill probes. These were carried out on a mesh of 10 × 10 m and at an average depth of 25 m. A.G.M.T. 14 and a CERE-type probe were used for radiometric testing.

Figure 3 shows the distribution of the diamond-drills. They were made on a mesh of 50 × 25 m, and at an average depth of 50 m. A.G.M.T. 14 and a CERE probe were also used here.

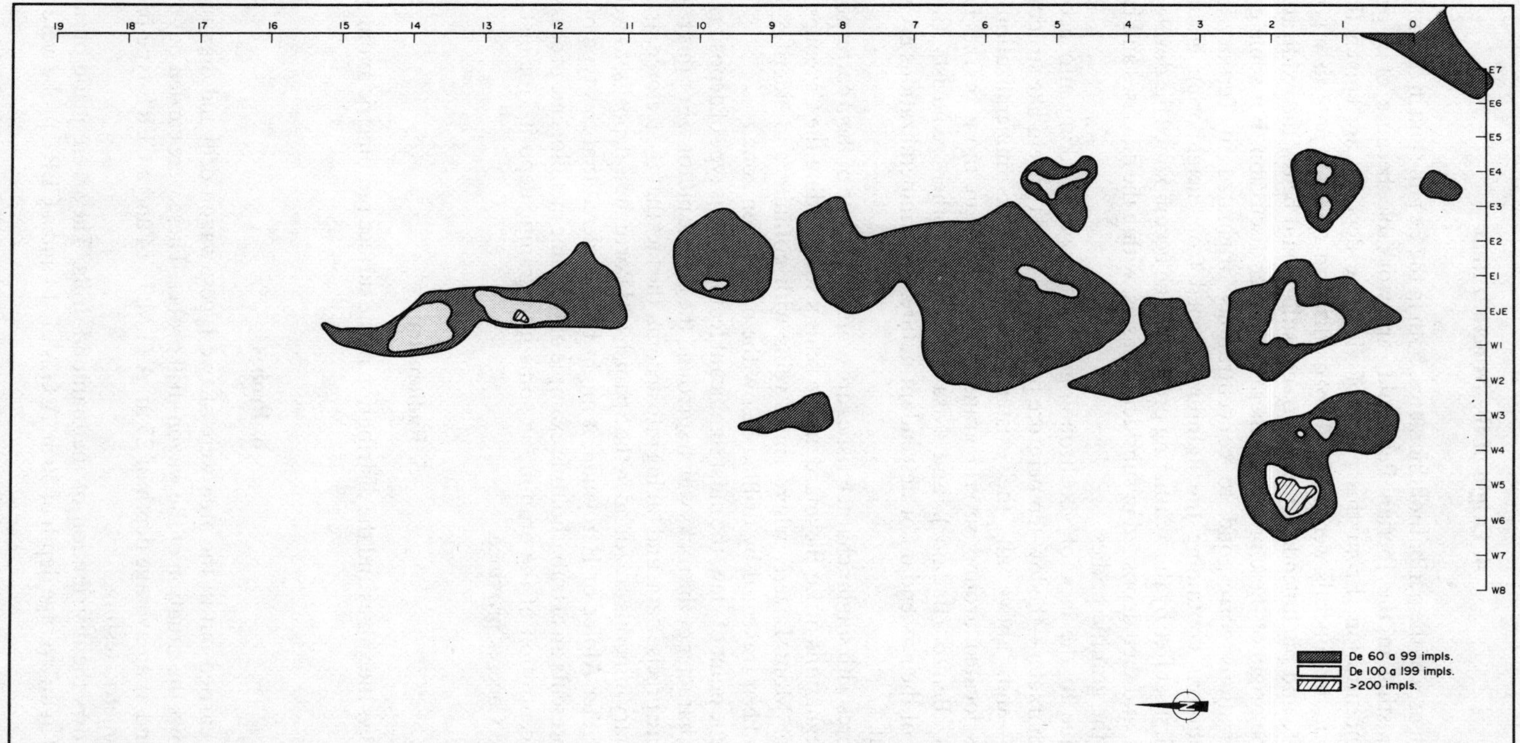

FIG. 1. Distribution of the surface radiometric anomalies in the area studied.

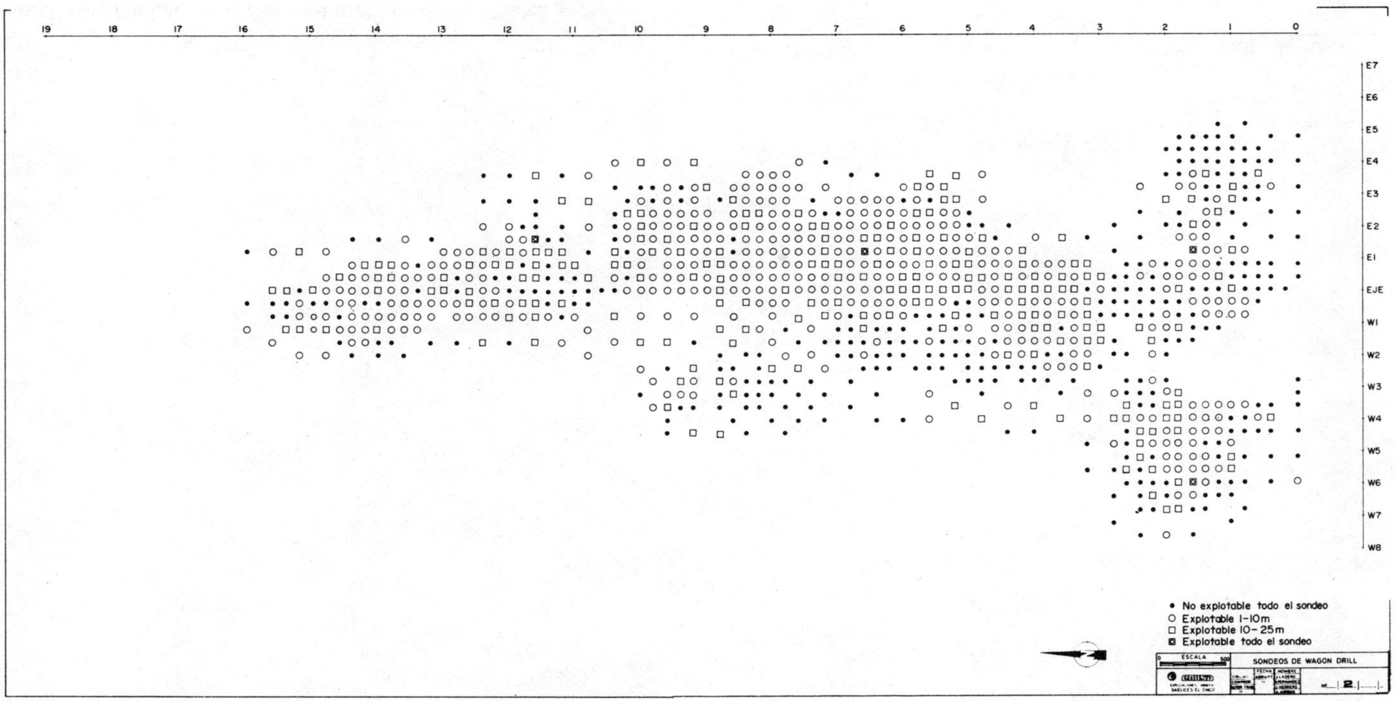

FIG. 2. Situation of the wagon-drill probes in the "Fe 3" area.

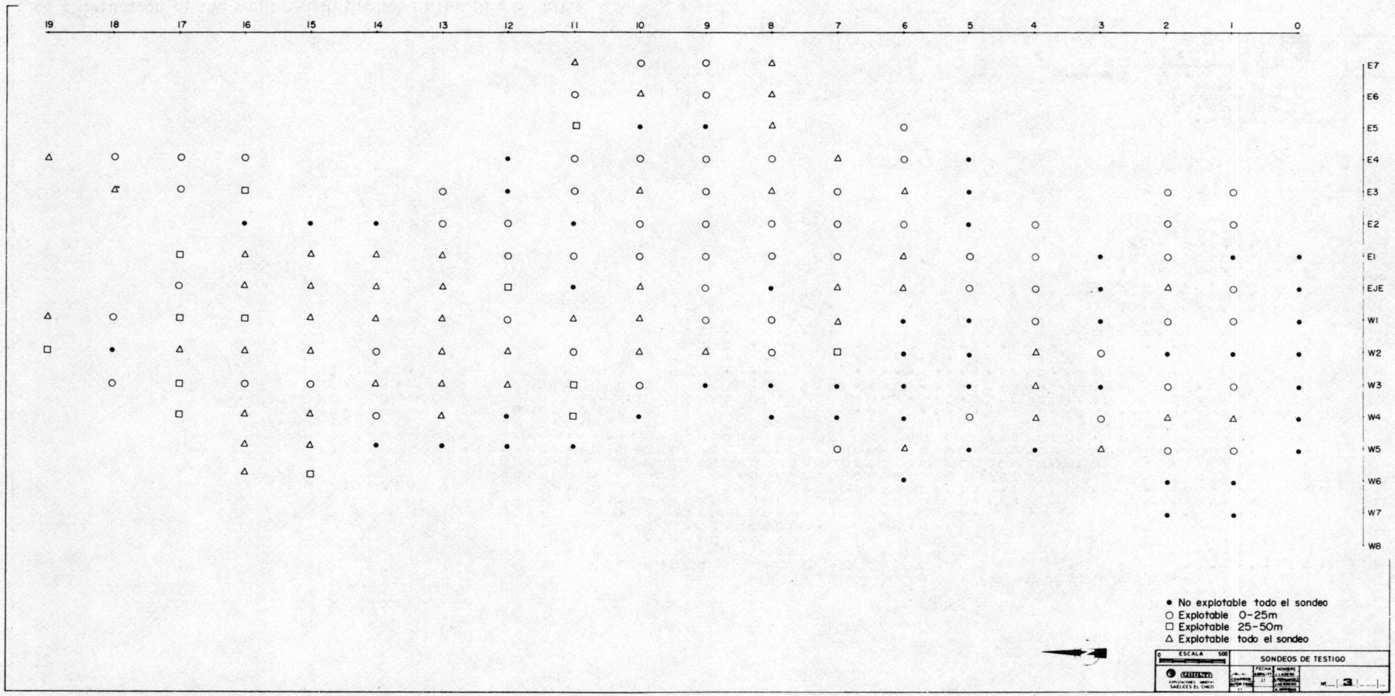

FIG. 3. Distribution of the diamond-drill in the "Fe 3" area.

7. Geochemistry

The geochemical analysis were carried out on samples of soil (B horizon) and vegetation (*Q. ilex* leaves). Soil sampling was made on the central points of the diamond-drill mesh, whereas the biogeochemical samples were taken from the live-oak tree nearest to the knots of the mesh.

(a) *Experimental Methods*

Once dried and ground, the soil samples were calcinated at 450°C. The residue was attacked with nitric acid 1:1 on a heating plate, and the uranium in the solution was determined by extraction with ethyl acetate and spectrophotometry with dibenzoilmethane.

The plants were dried at approximately 90–100°C, and later ground and calcinated at 500°C. The residue was attacked with nitric acid 1:1 on a heating plate. The uranium in the solution was determined by extraction with trioctylphosphine-benzene oxide and spectrophotometry with arsenazo III.

In both cases, a digital spectrophotometer, a Bausch and Lomb "Spectronic" model, was used.

(b) *Plants*

The predominant vegetation in the area is the species *Quercus ilex* L. Among shrubs, the following can be pointed out for their abundance: *Cytisus albus* L., *Lavándula pedunculata* L., *Lavándula stoechas* L. and *Cistus salviaefolius* L.

The species *Quercus ilex* L. was selected because of its good distribution, not only in the area but also in all the west of the Iberian Peninsula. This allowed the study to adapt a particular mesh and, if positive results were obtained, to apply this prospecting method in other areas of the country.

Following this, tests were carried out to find out which part of the live-oak was most suitable for the sampling because of its greatest amount in uranium. For this purpose samples were taken from live-oaks in mineralized areas and each part was analysed separately (leaves, trunk, roots and fruits). The lowest uranium content was found in the fruits, while similar values were obtained from the leaves, trunks and roots. Therefore, the leaves were chosen, as sampling was easier in this way.

The same mesh of 50 × 25 m used in diamond-drill was considered for the sampling of the live-oak leaves.

The samples were taken from the live-oaks nearest to the knots of the mesh, which gave a total of 300 trees (Fig. 4).

(c) *Soils*

The soil sampling of the B horizon was carried out with the same mesh of 50 × 25 m. Due to the shallow depth of the covering, the samples were taken at an average depth of 20 cm. Sampling was carried out on the central points of the rectangles of the mesh, giving a total of 285 soils (Fig. 5).

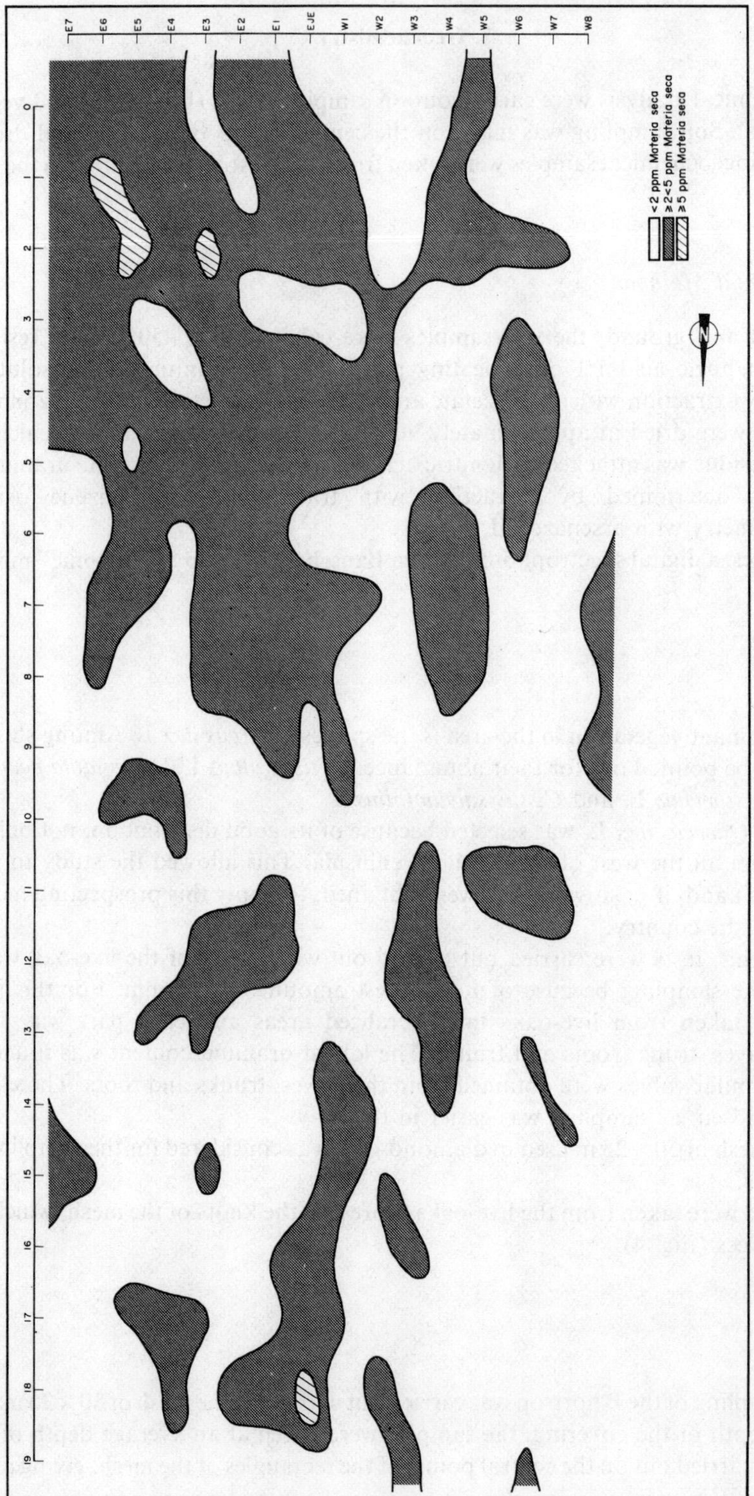

FIG. 4. Distribution of the uranium content in "Fe 3" soils.

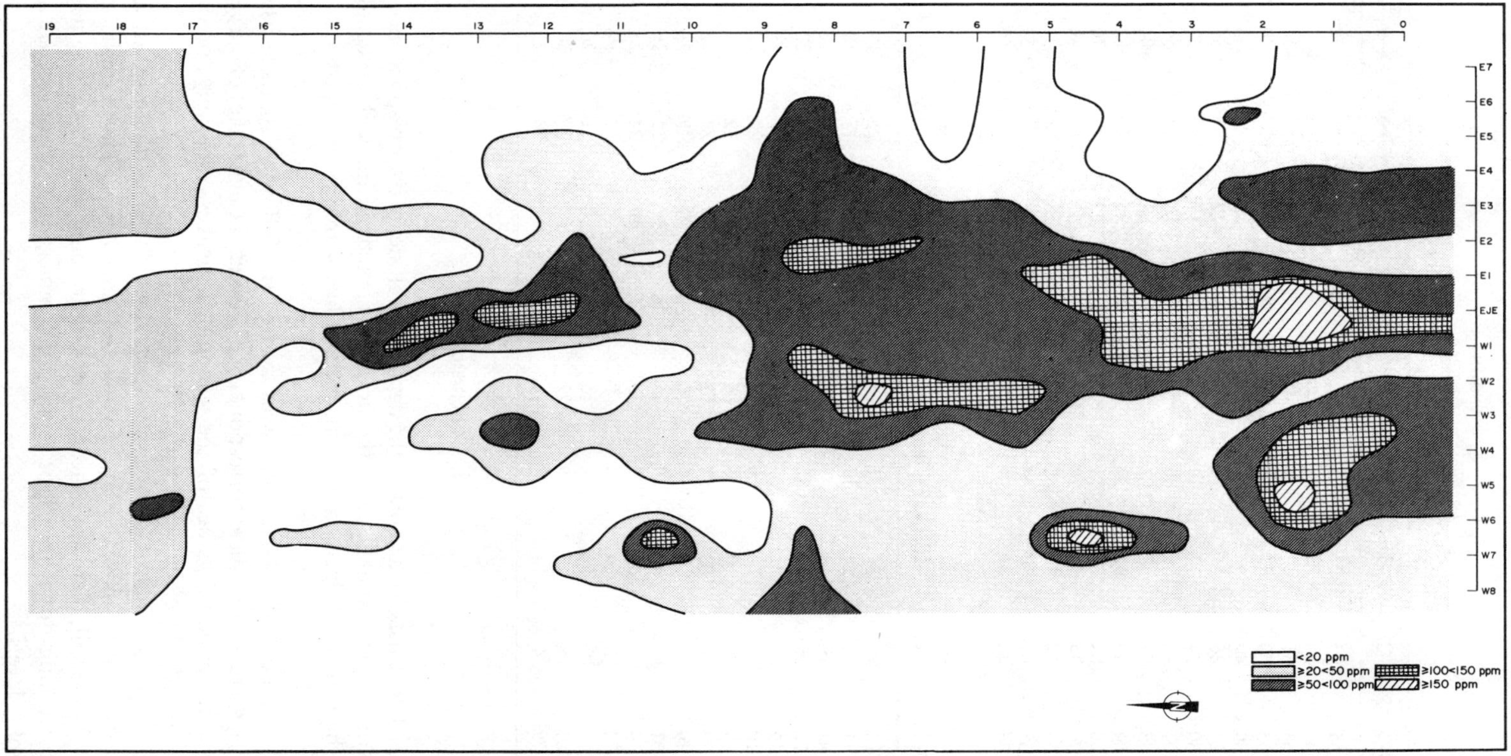

FIG. 5. Distribution of the uranium content in ilex leaves (*Quercus ilex* L.) in the "Fe 3" area.

8. Results

TABLE 2. U_3O_8 CONTENT IN "Fe 3" SOILS (ppm)

Samples	0–1	1–2	2–3	3–4	4–5	5–6	6–7	7–8	8–9	9–10
Eje–E1	55	192	120	66	102	80	52	50	82	54
E1–E2	40	30	52	38	100	71	55	102	103	50
E2–E3	63	51	40	27	41	65	68	52	50	57
E3–E4	55	50	53	16	40	46	21	80·	50	41
E4–E5	40	40	37	15	14	25	9	40	63	24
E5–E6	26	36	51	13	36	21	12	28	53	18
E6–E7	30	28	17	19	11	22	17	25	37	19
Eje–W1	107	296	101	100	54	81	80	61	56	34
W1–W2	45	91	58	105	67	56	64	53	104	33
W2–W3	64	40	62	34	28	122	125	201	82	23
W3–W4	100	102	38	21	28	60	32	55	70	51
W4–W5	79	124	50	20	26	23	22	37	34	24
W5–W6	52	214	37	45	23	25	28	30	30	8
W6–W7	38	61	30	64	268	29	28	23	53	9
W7–W8	34	40	44	37	27	26	25	25	50	22

Samples	10–11	11–12	12–13	13–14	14–15	15–16	16–17	17–18	18–19
Eje–E1	24	69	34	31	16	5	21	25	19
E1–E2	11	78	26	17	13	17	15	17	18
E2–E3	41	43	18	25	34	14	12	31	22
E3–E4	25	35	46	15	11	23	15	24	22
E4–E5	15	24	23	14	9	11	25	20	21
E5–E6	10	19	10	10	8	12	13	20	22
E6–E7	19	6	10	8	6	18	16	22	21
Eje–W1	30	51	102	138	53	30	29	22	22
W1–W2	19	9	9	15	70	18	22	29	22
W2–W3	31	10	45	10	15	38	13	23	26
W3–W4	34	36	62	38	15	15	10	24	23
W4–W5	24	16	20	14	8	12	15	22	17
W5–W6	9	19	16	8	15	8	10	51	28
W6–W7	101	15	15	9	22	20	13	20	20
W7–W8	79	24	14	8	11	16	11	20	21

9. Conclusions

1. As expected, there exists in the area of the deposit a well-defined correspondence between the surface radiometric anomalies and the uranium content in the soil, especially at the contact with the bed-rock, since, in general, the soil has scarcely developed.

2. The uranium content of the live-oak leaves showed the following characteristics:
   (a) In general, there is a statistical correspondence between the anomalies of the vegetation and the corresponding values of the wagon-drill probes, especially for the values obtained at a depth of between 1 and 10 m.
   (b) In some cases, in which this correspondence does not exist, there is a relationship between the positive anomalies of the vegetation and the soil. This is explained by the fact that, in the soil, there are concentrations of uranium which were originated from solutions coming down the slope from other mineralized areas located at higher topographic levels.

TABLE 3. U_3O_8 CONTENT IN ILEX LEAVES (*Quercus ilex* L.) "Fe 3"
(ppm dry material)

Samples	0	1	2	3	4	5	6	7	8	9
Eje	—	2.3	3.3	<2	2.5	2.4	<2	2.0	2.0	<2
E1	3.7	3.2	2.0	<2	2.2	2.5	2.1	2.1	2.0	2.3
E2	3.8	<2	2.0	2.3	2.2	2.0	2.0	2.9	2.5	<2
E3	<2	2.8	11.2	2.1	<2	2.1	3.0	2.0	2.3	<2
E4	4.3	4.3	2.0	<2	<2	2.3	<2	<2	<2	<2
E5	2.6	2.3	12.6	<2	2.7	<2	2.1	2.5	<2	<2
E6	3.7	8.3	3.7	2.5	2.0	2.2	<2	2.6	2.1	<2
E7	3.4	3.2	3.0	2.1	3.1	<2	<2	<2	<2	<2
W1	—	—	2.6	<2	2.2	2.1	<2	2.3	<2	2.1
W2	—	—	<2	2.1	2.8	3.6	<2	2.0	—	<2
W3	<2	<2	—	—	—	—	<2	—	<2	<2
W4	2.4	4.0	2.0	—	<2	<2	2.2	—	2.0	<2
W5	2.1	<2	2.0	<2	<2	<2	<2	2.3	3.4	<2
W6	<2	<2	2.9	<2	2.1	2.2	2.2	<2	<2	<2
W7	—	<2	2.4	<2	—	<2	<2	<2	<2	<2
W8	<2	<2	<2	<2	<2	<2	<2	2.0	<2	4.7

Samples	10	11	12	13	14	15	16	17	18	19
Eje	<2	<2	<2	<2	<2	<2	2.1	2.6	10.0	<2
E1	<2	2.1	2.5	<2	<2	<2	2.3	<2	2.1	<2
E2	<2	<2	2.1	<2	<2	<2	<2	<2	2.3	<2
E3	<2	<2	3.5	2.3	<2	2.9	<2	2.1	<2	<2
E4	2.8	<2	<2	<2	<2	<2	<2	2.9	3.0	<2
E5	<2	<2	<2	3.4	<2	<2	<2	2.3	<2	<2
E6	2.0	<2	<2	<2	<2	<2	<2	<2	<2	<2
E7	<2	<2	<2	<2	<2	2.1	<2	<2	<2	<2
W1	2.2	<2	<2	<2	2.4	2.1	<2	2.0	<2	<2
W2	<2	<2	<2	<2	2.3	2.9	<2	<2	2.0	<2
W3	<2	2.0	2.2	<2	—	<2	2.2	<2	<2	2.0
W4	<2	<2	<2	4.8	<2	—	<2	<2	<2	<2
W5	<2	2.6	<2	<2	<2	<2	<2	<2	<2	<2
W6	<2	2.5	<2	2.0	<2	<2	<2	<2	<2	3.1
W7	<2	2.1	<2	—	2.0	<2	<2	<2	<2	<2
W8	<2	<2	<2	<2	<2	<2	<2	<2	<2	<2

(c) On the contrary, in one particular area, the positive anomalies of the vegetation coincide with negative anomalies in the soil. In this case, due to the marked inclination of the terrain and to the fact that there is hardly any soil, it can be assumed that the plants have taken the uranium from deeper levels, since it would have been leached from the surface. However, this idea is only a hypothesis as no drilling has been made in the area.

3. In any case, all the prospecting methods—surface radiometry, geochemistry of soils and plants, surface mining works, and the study with wagon and diamond-drilling—clearly indicate the existence of four mineralized areas in a north to south direction. The two central areas, which were discovered by surface radiometry, coincide quite closely with positive anomalies in the soil and vegetation. Here, the existence of the underlying deposit was proved by systematic wagon and diamond-drill exploration. The other two areas, situated on

either side of the former, but not so radiometrically distinct, also show appreciable positive anomalies in the uranium content of the soils and vegetation. For this reason, a detailed study of these two areas has been recommended.

References

ARRIBAS, A. (1970) Las pizarras uraníferas de la provincia de Salamanca. *Studia Geologica* **1**, 7–45.

GEFFROY, J. (1971) Les gîtes uranifères dans le Massif Central, pp. 541–579. In *Symp. J. Jung: Géologie, géomorphologie et structure profonde du Massif Central français*. Plein Air Service Edition Scientifique, Clermont-Ferrand, 610 pp.

PILAR, L. (1966) Condições de formação do fazigo uranífero de Nisa. *J.E.N., Lisboa* 51–83.

RODRIGUEZ, B. (1972) Métodos para determinar uranio en minerales, disoluciones y concentrados de uranio. *J.E.N., Madrid* 1–16.

Fe and Zn Partitioning between Stannite and Sphalerite and its Application in Geothermometry.

I. J. NEKRASOV, V. I. SOROKIN and E. G. OSADCHII

Institute of Experimental Mineralogy, Academy of Sciences of the U.S.S.R., Moscow

Abstract

Quantitative evidence of Zn, Fe partitioning between sphalerite (Zn,Fe)S and stannite $Cu_2(Fe,Zn)SnS_4$ at 300–500°C and $P = 1$ and 5 kbars has been obtained. A new geothermometer has been derived. The partitioning constant of Zn and Fe between sphalerite and stannite is independent of pressure and is given by $\lg K = -1274/T + 1.174$.

One of the main objectives in studying ore genesis is obtaining quantitative evidence on the past physico-chemical parameters of mineralization, particularly with regard to temperature and pressure. Analyses of component fractionation among coexisting phases have been widely used recently in geothermometry and barometry, and a great number of geothermometers and geobarometers, that are of great importance for the petrology of silicates, have been derived over the past years (Barth, 1951; Perchuk, 1969; Saxena, 1969, 1975). The authors of these papers also treated the problem from the theoretical aspect.

However, only a limited number of papers discuss the temperature and pressure at which many sulfide associations have been formed (Bethke and Barton, 1971; Besmen *et al.*, 1975). These studies were primarily concerned with the fractionations of minor components among ore minerals: Cd between PbS and (Zn,Fe)S; Co, Ni and Sn between FeS_2 and $Fe_{1-x}S$, etc. Owing to very small concentrations of the trace elements in sulfides (thousandth, hundredth fractions of %), the sulfide formation temperatures could not be determined accurately.

We studied Fe and Zn fractionation between stannite and sphalerite at 300, 400 and 500°C and pressure of 1–5 kbars. This mineral pair occurs in many tin, pyrite and polymetallic deposits and is extremely suitable for experiments due to the ability to form solid solutions. Stannite is shown (Springer, 1972) to form wide regions of solid solutions in which Cu_2FeSnS_4 (stannite proper) and Cu_2ZnSnS_4 (kesterite) are end-members. Many natural stannites contain considerable Zn (to 12 wt.%) which replaces iron isomorphously (atom per atom) (Petruk, 1973). Similarly, in the sphalerite solid solution FeS replaces ZnS up to more than 50 mol.% FeS (Barton and Toulmin, 1966; Sorokin and Chichagov, 1974). Iron and zinc partitioning between sphalerite and stannite therefore can be described by an exchange reaction

$$Cu_2FeSnS_4 + ZnS = Cu_2ZnSnS_4 + FeS. \qquad (1)$$

This reaction is believed to be valid for the entire group of stannites.

Runs were done in sealed 0.7–1.2 cm³ gold ampules under isothermal conditions. The ampules were heated either in autoclaves or in high-pressure vessels where the pressure was measured with a Bourdon pressure-gauge. The ampules were placed into resistance furnaces;

the temperature was regulated within 1.5°C and was recorded by a chromel–alumel thermocouple.

The ampules were charged with ZnS (semiconductor purity grade), Fe (carbonylic, high purity), Cu, Sn and S (high purity) and Cu_2FeSnS_4, Cu_2ZnSnS_4 and (Zn,Fe)S which were synthesized from pure elements in vacuum quartz ampules. The synthesis was done in two stages with grinding. The total weight of the charge was 20 to 70 mg.

The reaction media were aqueous HCL solution (up to 0,1 M HCl) with some oxalic acid. The reaction products were examined by X-ray powder diffraction and in reflected light. The compositions of sphalerite and stannite were analysed on a JXA–5A X-ray microanalyzer. Iron and zinc in coexisting stannites and sphalerites were analyzed on two spectrometers simultaneously. Standards were homogeneous stannites of the series Cu_2FeSnS_4–Cu_2ZnSnS_4 and ZnS–FeS, synthesized in vacuum quartz ampules.

In some experiments stannite–sphalerite equilibria were approached from two directions, which means that in one of the two parallel series of runs the Fe content of original sphalerite was decreased, while in the other series it was increased. Further evidence showed that the phases have equilibrated in the immediate vicinity of the contact of minerals. Therefore, we have restricted our studies to the near-contact zones of the minerals.

Figure 1 summarizes the results for coexisting sphalerite and stannite. The results for 1 and 5 kbars agree well within the experimental error, and consequently the volume effect of reaction (1) approaches zero for this pressure range and is negligible. The same result is obtained by calculating the volume effect of reaction (1) from the known molar volumes of the components of reaction (Clark, 1966). We found that stannite and kesterite molar volumes are 95.037 and 98.986 cm^3/mole, respectively, at $T = 25$°C and $P = 1$ atm. The volume effect of reaction (1) is 1,685 cm^3/mole.

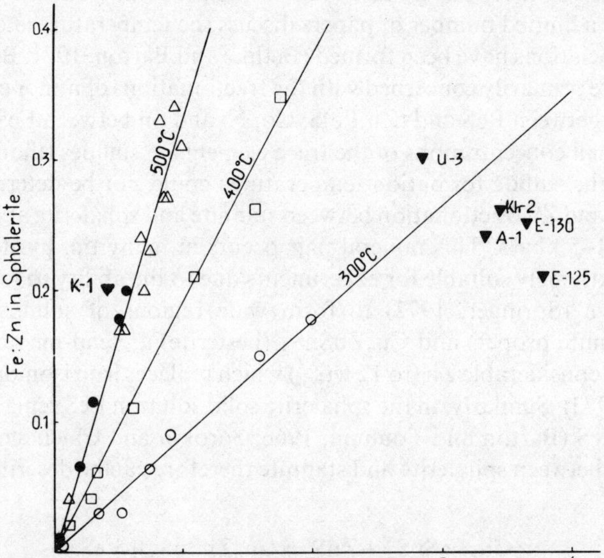

FIG. 1. Fe and Zn partitioning between stannite and sphalerite. Open circles, squares and triangles are for 300°, 400° and 500°C, P = 1 kbar. Solid circles are for 500°C and 5 kbars. Solid triangles represent natural samples from the deposits: U—Ulakhan-Egelakh; Kh—Khetin; A—Arseniev; E—Ege-Khajak (the Dalnaja Zone); D—Deputatsky; K—Kester.

The distribution constant K_D = (Fe/Zn)sph: (Fe/Zn)st of reaction (1) is equal to the reaction constant, *K*, and is independent of the compositions of sphalerite and stannite solid solutions. Allowing for the nearly ideal behavior of iron in sphalerite (Fleet, 1975), the constant of reaction (1) can be considered thermodynamic to a first approximation. Temperature dependence of the distribution and reaction (1) constants is given by $\lg K = -1274/T + 1.174$. Entropy effect of reaction is 5.4 kcal/mod. grad.

From the free energy of reaction (1), $G_p = -RT \ln K$, we calculated the free energy of kesterite (Cu_2ZnSnS_4) formation for 300°, 400° and 500°C: −131.9; −124.3; −115.9 kcal/mole, respectively (standard state of sulfur was S_2 gas). The presence of tin suggests the stannite–sphalerite solid solution in sphalerite, along with the described exchange reaction. However, this solid solution does not affect strongly the foregoing relationships owing to low miscibility.

The obtained pressure-independent relationships of the constant of Fe and Zn fractionation between sphalerite and stannite can be used as a geothermometer, when two members of an association in ore deposits have been formed simultaneously.

We have tried to use the obtained experimental evidence for estimating the temperatures at which ore and polymetallic deposits, bearing stannite and sphalerite, have been formed. The studies were done with ore samples from the cassiterite–quartz and cassiterite–sulfide formations in the north-east U.S.S.R. and Primorie. Analyses of coexisting stannite and sphalerite (Table 1) reveal the marked scatter of Fe and Zn fractionation coefficient values for the minerals from different deposits. These data suggest that the temperature at which stannite and sphalerite have been formed in cassiterite–quartz deposits was within 580–380°C, and within 300–220°C in cassiterite–sulfide deposits. These findings agree reasonably with the temperature estimations by decrepitation and homogenizing (Jakovlev, 1971).

TABLE 1. THE COMPOSITION OF COEXISTING STANNITE AND SPHALERITE FROM TIN ORE DEPOSITS IN THE U.S.S.R.

Deposits	Numbers of samples	Minerals	wt. % Zn	Fe	Sn	Cu	S	Total	K_D	t°C
Ege-Khay	111	sphalerite	51.97	14.87	0.22	0.78	31.18	99.02	0.05	242
		stannite	1.97	12.19	28.03	29.13	27.12	98.44		
	125	sphalerite	52.94	11.29	0.17	0.62	33.49	98.51	0.06	259
		stannite	2.78	10.52	26.90	28.50	28.30	97.00		
	130	sphalerite	53.30	13.53	0.26	0.42	29.88	97.39	0.07	274
		stannite	3.38	12.32	26.80	28.00	27.40	97.90		
	273	sphalerite	51.35	15.45	0.12	0.29	31.27	98.48	0.06	259
		stannite	2.75	12.92	24.55	28.02	32.00	100.24		
	282	sphalerite	53.42	11.61	0.31	0.37	34.37	100.08	0.03	199
		stannite	1.62	12.32	27.00	31.17	27.60	99.71		
Kester	1	sphalerite	55.80	11.24	0.18	0.53	32.02	99.77	0.49	589
		stannite	10.16	4.13	26.38	27.32	29.19	97.18		
Ulakhan-Egelakh	3	sphalerite	51.14	15.62	0.34	0.87	30.03	98.00	0.11	324
		stannite	4.15	11.76	26.38	30.02	28.70	101.01		
Arseniev-skoe	1	sphalerite	51.00	12.20	0.20	0.20	35.90	99.55	0.07	274
		stannite	3.10	10.30	28.50	27.10	30.20	99.20		
Khetinskoe	2	sphalerite	52.16	13.77	0.34	0.79	31.12	98.18	0.08	288
		stannite	3.12	10.70	26.90	28.33	29.12	98.17		

References

BARTH, T. F. W. (1951) The feldspar geologic thermometer. *Neues Jahrb. Miner. Abh.*, vol. 82.

BARTON, P. B. and TOULMIN, P. (1966) Phase relations involving sphalerite in the Fe–Zn–S system. *Econ. Geol.* **61**, 815–849.

BESMEN, N. I., TICHOMIROVA, B. I. and KOSOGOVA, B. P. (1975) Pyrite–pyrrhotite geothermometer: partition of nickel and cobalt. *Geokhimiya*, No. 5, 700–714.

BETHKE PH. B. and BARTON, P. B. (1971) Distribution of some minor elements between coexisting sulphide minerals. *Econ. Geol.* **66**, 140–163.

CLARK, S. P. (ed.) (1966) Handbook of Physical Constants. Geol Soc. Amer., memoir no. 97, 587 pp.

FLEET, M. E. (1975) Thermodynamic properties of (Zn,Fe)S solid solutions at 850°C. *Amer. Miner.* **60**, 466–470.

JAKOVLEV, JA. V. (1971) Temperature of hydrothermal solutions. *Geology and Genesis of Tin-bearing Deposits of Yakutya,* Moscow.

PERCHUK, L. L. (1969) Stavrolite–garnet thermometer. *Dokl. Akad. Sci. U.S.S.R.* **186**, no. 6.

PETRUK, W. (1973) Tin sulphides from the deposit of Brunswick tin mines limited. *Can. Miner.* **12**, 46–54.

SAXENA, S. K. (1969) Silicate solid solutions and geothermometer. I. Use of the regular solution model. *Contr. Mineral. Petrol.* **21**, no. 4, 338–345.

SOROKIN, V. I. and CHICHAGOV, A. V. (1974) The Zn–Fe–S system sulphides in equilibrium with acqueous solution of NH_4Cl at 400°C and 1000 kg/cm². *Ocherky physiko-chim. Petrologii* no. 4, 176–185.

SPRINGER, G. (1972) The pseudobinary system Cu_2FeSnS_4 Cu_2ZnSnS_4 and its mineralogical significance. *Can. Miner.* **11**, 535–541.

La Distribution Globale des Principaux Eléments Métalliques dans l'Ecorce Terrestre des Continents

IVAN G. MAGAKIAN

Institut des Sciences Géologiques de l'Académie des Sciences de l'Arménie, Erevan, U.S.S.R.

Abstract

1. The concentrations of the metals in the crust are very irregularly distributed and form regions with immense reserves (Cu–Mo in the Cordillera and the Andes, Au–U in the Witwatersrand and others).

2. The conditions of preconcentrations are of different types: endogene, exogene or metamorphogene. The initial material of the endogene deposits is often of exogene or metamorphogene origin and, to the contrary, the source of exogene or metamorphogene concentrations can be the destruction of endogene concentrations. The circulation of the elements in the natural processes occur in that way.

3. The size and the associations of the metal-concentrations depend on the type of the structure, the intensity of the stages of its development and the age of the structure.

The geotectonic position of the region in a shield or a folded belt, or an area of activization is also a decisive parameter.

4. The shield structures contain more frequently metamorphogene concentrations of Fe, Mn and Au–U.

Two different types of concentrations are recognized in the folded belts: (a) the internal zones with moderately acid granitoids and basic–ultrabasic intrusions with Cr, Fe, Ti, Pt, Cu–Mo, Pb–Zn, Au, Ag and Hg mineralizations; (b) the external zones with acid granitoids and Sn, W, Mo, U, Pb–Zn, Au and Sb mineralizations.

The areas of activization are accompanied by alkaline–ultrabasic intrusions, and in connection with them, by great concentrations of carbonatites (Nb, Ta, Zr, REE), but also of Pb–Zn, Au, U, Hg–Sb and fluorine.

5. The different metals have a definite distribution in times the Precambrian structures in Fe, Au, U, Ni, Pt-oides, Ti; the Paleozoic structures in Cr, Pt, Cu, Sn; the Cimmerian in Sn, W, Au, Sb and the Alpine are remarkable rich in Cu and Mo, Ag, Hg, Mn (the last of sedimentary type).

Résumé

Les concentrations des principaux éléments métalliques à la surface de l'écorce terrestre sont distribuées très inégalement. L'échelle des concentrations et le mode d'association des métaux dans des parties distinctes de l'écorce terrestre sont mis en relation avec le type de structure (bouclier, ceinture plissée, région activée), avec l'intensité des stades de développement de la structure et avec son âge.

Les structures précambriennes sont particulièrement riches en Fe, Au, U, Pt-oïdes, Ni, Ti; les structures d'âge paléozoïque sont riches en Cr, Pt, Cu, Sn; les structures cimmériennes sont surtout riches en Sn, W, Sb, Au et les structures alpines sont spécialement riches en Cu et Mo, Ag, Hg, Mn (ce dernier dans le type sédimentaire).

Les concentrations des principaux éléments métalliques dans la croûte continentale et sa couverture sont distribuées de manière très inégale. Comme exemples, on peut citer les gisements de fer de Kursk et Krivoï Rog, ceux de la région d'Itabira au Brésil, du Lac Supérieur aux U.S.A., de l'Australie Occidentale et de 'Inde Centrale, avec des réserves colossales. Il en va de même pour les régions très riches en étain et en wolfram de l'Asie du Nord-Est et du Sud-Est ou de la Bolivie ainsi que pour les régions à cuivre et molybdène des Cordillères des Andes et de la Méditerranée Orientale, ou encore pour les conglomérats aurifères à uranium du Witwatersrand en Afrique du Sud et pour bien d'autres gisements.

Les conditions de formation des grandes concentrations d'éléments métalliques dans l'écorce terrestre sont variées, faisant intervenir des processus endogènes (magmatogènes), exogènes (surtout sédimentaires) et métamorphogènes. Il est assez fréquent que le matériel initial des concentrations exogènes et métamorphogènes de métaux soit le produit de la destruction, de la sédimentation et du remaniement métamorphique de concentrations endogènes antérieures. Inversement, les magmas et les solutions minéralisées empruntent souvent une partie de leur matériel aux minéraux de concentrations exogènes et métamorphogènes antérieures. C'est ainsi que se développe un cycle fermé des éléments dans les processus naturels.

Qu'il s'agisse de processus de minéralisations endogènes ou exogènes, une signification importante s'attache aux exhalaisons volcaniques qui dégagent, à partir de foyers magmatiques, des quantités énormes de Fe, Mn, Al, Cu et d'autres éléments et forment de grands gisements du type exhalatif-sédimentaire (volcano-sédimentaire).

Le niveau de concentration et l'expression paragénétique des métaux dans les différents domaines de l'écorce terrestre sont régis par certains facteurs tels que le type de structure, l'evolution géologique et l'intensité des étapes de cette évolution, l'âge de la structure, et d'autres encore.

La profondeur de formation des unités magmatiques et des gisements associés ayant le même faciès de profondeur, se traduit par des caractères métallogéniques visibles (types de gisements, nature de la zonalité, etc.).

Dans l'Asie du Nord-Est et du Sud-Est, par exemple, les gîtes à quartz-cassitérite–wolframite sont liés aux granites acides de relativement grande profondeur et, à distance de ces intrusions, les gisements polymétalliques, puis les gisements de stibine et de mercure. Aux granites hypabissaux d'acidité modérée des mêmes régions sont liés les gisements du type à sulfures–cassitérite (dont les polymétalliques à cassitérite). Enfin aux extrusions de quartz-porphyres sont liées de petites concentrations de cassitérite de faciès "étain de bois". Au Nord de la Bolivie, les granites abyssaux sont accompagnés d'une série de gisements disposés en succession zonale autour des massifs: pegmatites à W–Sn–Mo, gisements hypothermaux de minerais de W–Bi–Sn, gisements épithermaux de minerais polymétalliques à Au–Ag–Sb–Hg. En Bolivie centrale et, en partie, du Sud, la brusque courbure de la structure des Andes est accompagnée de grandes fractures qui ont permis l'ascension de magmas granitiques en intrusions proches de la surface ou en extrusions.

Ici, bien que le stock métallique reste le même, le caractère de la minéralisation est tout différent. La zonalité n'est que vaguement exprimée; les minerais ont une composition complexe (Sn–W–Bi accompagnés de Sb–Au–Ag–As) et les métaux ne s'isolent pas, non seulement au niveau des zones et des gisements, mais même dans les minerais. Enfin, les sulfosels sont communs (stannite, teallite, cylindrite, franckeite et autres).

L'influence de la profondeur de l'érosion sur le caractère de la minéralisation, base de la théorie batholitique de W. Emmons, est rejetée dans l'interprétation de l'auteur à partir des travaux des géologues soviétiques. Mais on ne peut nier que lorsqu'une série de gisements composant une structure métallogénique zonée correspondent à des étapes définies du développement de la structure, la profondeur de la surface d'érosion joue souvent un rôle important à l'égard de la distribution zonale des métaux.

Nous pouvons citer un exemple bien connu de l'auteur: celui des régions métallifères de Zangezur et Vajk en Arménie. En fonction du stade d'érosion de cette ceinture minéralisée en Cu–Mo, le Vajk fait apparaître des minéraux polymétalliques avec des traces de Cu et Mo et c'est dans le Zangezur, où la coupe est plus profonde, que se trouvent les minerais de Cu et Mo. Dans le Petit Caucase, riche en amas de pyrite cuprifère (Alaverdi, Bolnissi, Kaphan et autres), nous voyons une zonalité verticale avec la succession, de haut en bas, des types de minéralisation suivants: barytine et gypse hypogène, minerais polymétalliques, minerais de cuivre et pyrite, exposés selon la profondeur de la coupe d'érosion.

En Kazakstan Central, les parties hautes des filons de quartz sont riches en wolframite et les parties moyennes en molybdénite tandis que les parties profondes n'ont qu'un intérêt médiocre. C'est selon la profondeur de la coupe d'érosion que se découvre l'une de ces trois zones. Si la profondeur de l'érosion est correctement évaluée, la zonalité de la minéralisation peut servir dans beaucoup de cas comme guide des travaux de recherche et de prospection.

Ainsi, les régions de Vajk et de Zangezur avec une minéralisation polymétallique en surface offrent des perspectives pour les minéralisations Cu–Mo en profondeur. De même, les terrains à gypse hypogène et barytine dans le périmètre des gisements du type amas de pyrite cuprifère du Petit Caucase, offrent des perspectives pour des minerais polymétalliques, du cuivre et de la pyrite en profondeur. De même, les formations minéralisées en quartz-wolframite du Kazakstan Central offrent des perspectives de minéralisation à molybdénite en profondeur. On pourrait ajouter beaucoup d'exemples de ce genre et souligner que, dans plusieurs cas, des travaux de recherche ont apporté une vérification expérimentale de la régularité du changement de la minéralisation avec la profondeur. Néanmoins, il faut constater que les règles de la zonalité et de la profondeur de l'érosion ne s'appliquent qu'à un type de minéralisation bien défini. Par conséquent, il y a peu de chances de trouver, dans les niveaux profonds d'un gisement du type amas de pyrite cuivreuse, des minéralisations à Cu–Mo ou, encore moins, à Sn–W qui sont caractéristiques d'autres conditions géologiques.

Ce sont les traits lithologiques et stratigraphiques des roches encaissantes qui fixent le type morphologique et l'intensité de la minéralisation, tout au moins à l'intérieur d'un même type de formations métallifères.

Ainsi, la minéralisation à Cu–Mo du Petit Caucase est représentée dans les massifs granitoïdes par le type à *veinules disséminées,* dans les contacts avec les roches carbonatées par le type *skarn* et dans les zones de cisaillement des porphyrites par des minerais à Cu–Mo dans *les roches broyées.*

Au Nord-Est de l'Asie, dans les régions formées de roches gréso-schisteuses, les gisements d'étain sont constitués par des veines et des stockworks, mais au Sud-Est de l'Asie, surtout en Chine, dans les régions riches en roches carbonatées, les gisements sont à sulfures–cassitérite du type skarn. On notera que dans les deux cas, le bore est un compagnon très caractéristique de la minéralisation, sous forme de tourmaline dans les veines et stockworks et sous forme de ludwigite, axinite et datolite dans les skarns.

C'est aussi très régulièrement que la minéralisation de type pyrite cuprifère (avec baryte et minerais polymétalliques) du Petit Caucase se localise dans les roches volcaniques d'âge essentiellement jurassique, tandis que les minéralisations épigénétiques de manganèse se localisent dans les roches volcano–sédimentaires d'âge crétacé supérieur. Dans les deux cas, le contrôle lithostratigraphique de la minéralisation est due à ses liens génétiques avec les foyers magmatiques dont proviennent à la fois les minerais et leurs roches encaissantes.

La morphologie des gisements est étroitement liée à la lithologie des roches encaissantes.

Les minerais en stocks de pyrite cuprifère massive ou broyée (cette dernière structure est métasomatique) se développent dans les kératophyres broyés; les stockworks et les veines dans les porphyrites et porphyres massifs. Les amas de minerai de manganèse se développent de manière métasomatique dans les roches carbonatées, les minerais broyés dans les tuffites broyées, les veines dans les porphyrites. Ce rôle du facteur lithostratigraphique ne s'exerce pas seulement pour les minéralisations endogènes mais aussi pour les minéralisations exogènes (surtout sédimentaires). Ainsi les minerais sédimentaires de fer oolithique coïncident avec les couches littorales terrigènes d'âge ordinairement jurassique (Lorraine) ou tertiaire (Kertch); les minerais de manganèse avec les couches à silex d'âge tertiaire (Nikopol, Tchiatura, Polunotchnoe et autres); les bauxites dans l'intervalle entre deux couches de roches carbonatées d'âges différents (Krasnaya, Schapotchka et autres).

Une signification décisive s'attache à l'appartenance de la partie concernée de l'écorce à un bouclier, à une ceinture plissée ou à une région activée, ainsi qu'à l'intensité des étapes évolutives du magmatisme et de la minéralisation, ou encore à l'âge géologique de la structure.

Il est clair qu'au cours des étapes du développement de la structure et en fonction de la mutation corrélative des types de foyer métallifères magmatiques, les types de gisements changent dans le temps et s'isolent dans l'espace. Il s'ensuit la profondeur de formation des complexes magmatiques et des minéralisations, le stade atteint par l'erosion, la lithologie et la stratigraphie des roches encaissantes des gisements et, finalement, l'âge de la structure, marqueront de leur empreinte le caractère de la minéralisation. Il faut souligner qu'aux types définis de structures et à chacune des étapes de leur développement, correspondent des types de gisements et des séries de formations métallifères, déterminées à la fois dans leurs conditions géologiques spécifiques, leurs particularités de compositions et leur importance économique.

Pour les boucliers constitués de roches métamorphiques et magmatiques d'âge précambrien, les gisements caractéristiques sont d'une part les énormes concentrations de minerais de fer et de manganèse de type métamorphogène, des conglomérats et quartzites à or et uranium, des schistes à rutile, et d'autre part les gisements endogènes (hydrothermaux) d'uranium et d'or, ainsi que les gisements de type ségrégation de cuivre et de nickel, auxquels sont alliés le cobalt et les platinoïdes.

Aux roches volcanogènes du Précambrien, fréquemment schistifiées (leptites et autres), sont étroitement liés de grands gisements de pyrite cuprifère (Outokumpo, Boliden, Broken-Hill, Mount-Isa, Flin-Flon, Sherrit-Gordon, Noranda, Sullivan et bien d'autres) dont l'origine est probablement volcano–sédimentaire et dont le contenu est important en Cu, Zn, Pb, et, accessoirement, en As, Cd, In, Au, Ag et autres métaux.

Il faut noter que le matériel initial des gisements métamorphogènes de fer et de manganèse provient en partie de la destruction des roches des continents et en partie aussi, probablement, d'une intense activité exhalative-sédimentaire accompagnant le volcanisme d'âge principalement protérozoïque.

Les accumulations énormes de conglomérats et quartzites à or et uranium des gisements de Witwatersrand, Blind-River, Jakobina et autres sont liés à l'érosion de socles archéens riches en filons de quartz aurifères et en pegmatites uranifères. Enfin, les concentrations de minerais de cuivre et de nickel du type ségrégation sont étroitement liés aux massifs basiques–ultrabasiques mis en place dans les roches précambriennes et riches en sulfures de Fe, Ni, Cu (avec aussi Co et les platinoïdes).

Pour les ceintures plissées, on doit distinguer deux types de minéralisations:

(a) Celles qui sont particulières aux zones intérieures des ceintures, représentant les parties axiales des anciens géosynclinaux avec croûte de type océanique et prédominance du magmatisme basique-ultrabasique aux stades précoces puis intermédiaire-acide aux stades moyens du développement des structures.

Au magmatisme basique–ultrabasique sont liées les grandes concentrations de type proprement magmatique (chromite, titanomagnétite, groupe du platine), ainsi que des gisements hydrothermaux et exhalatif-sédimentaires de pyrite cuprifère, minerais polymétalliques et barytine. Aux stades moyens, en liaison avec des granitoïdes intermédiaires-acides se sont formés de grands gisements de fer et de wolfram (scheelite) de type skarn, ainsi que des gisements méso et épithermaux de minerais à cuivre-molybdène, polymétalliques, ou à sulfures d'argent et de mercure.

Ordinairement, les étapes précoces et moyennes se développent dans les mêmes régions, formant des ceintures métallifères caractéristiques telles que l'Oural, le Caucase-Transcaucasie, les Balkans, la Cordillère des Andes, le Japon, etc.

(b) Celles qui sont particulières aux zones externes des ceintures, touchant aux boucliers, avec une croûte de type continental et une prédominance des granitoïdes du stade tardif. Ici, en liaison avec les granitoïdes acides se sont développées des concentrations de minerai à étain–wolfram, molybdène (avec wolfram mais sans cuivre), or, antimoine, uranium, plomb et zinc.

Ordinairement les régions de développement de l'étape tardive sont séparées des régions de développement des étapes moyenne et précoce; elles forment des ceintures métallifères celles du Nord-Est et du Sud-Est de l'Asie, du Kazakstan Central, de la Bolivie, etc. Compte tenu du développement limité dans l'espace de l'étape tardive, on peut présumer qu'elle correspond à des conditions tectoniques spéciales, marquant les périodes de tension maximale dans des zones de flexion et d'effondrement de l'écorce sialique, suivies de la formation de granites acides et d'une minéralisation spécifique.

Pour *les régions réactivées des boucliers* (le long des zones de rift et des linéaments), *les massifs intermédiaires* et *les régions à plissement déjà achevé,* le magmatisme bien représenté est de type alcalin et alcalin-ultrabasique, en partie de type granitoïde. Ici sont caractéristiques, les gisements du type carbonatites (avec des concentrations énormes de Nb, Ta, Zr, Terres Rares et autres), ainsi que des concentrations assez intéressantes de minerais de type hydrothermal polymétallique ou à or, uranium, antimoine-mercure et fluorine.

Dans les limites des boucliers et de leurs plates-formes activées se développent des intrusions alcalines–ultrabasiques et des kimberlites. En liaison avec ces dernières sont disposées les grandes régions diamantifères (Afrique, Brésil, Indes, Sibérie et autres) tandis qu'au long des linéaments à complexes ultrabasiques alcalins se forment les carbonatites.

En ce qui concerne les massifs intermédiaires et les zones plissées réactivées, ce sont de grandes concentrations de minerais à étain et wolfram, polymétalliques, à antimoine–mercure, et aussi à or, uranium et fluorine qui sont caractéristiques. Comme exemples de ce type de région, on peut citer la Meseta Ibérique, les massifs Central Français, Armoricain, Bohémien, les Rhodopes, etc.).

Conclusion

Les séries métallifères, loin d'être isolées dans la nature, forment des séries à peu près synchrones, étroitement liées à certains types de régions et caractéristiques de ces types.

Le terme "formation métallifère" désigne habituellement un groupe de gisements métalliques dont l'unité s'exprime par une communauté de composition minéralogique, de traits génétiques, de conditions géologiques et par la ressemblance de leur importance économique.

Comme les éléments chimiques et les minéraux se groupent en associations d'éléments et en paragenèses minérales, les formations métallifères se groupent en séries tirant leur origine de certaines positions géologiques au cours des divers stades de développement d'un complexe tectono–magmatique donné.

Récemment, l'auteur a tenté de répartir en dix séries distinctes les formations métallifères de types magmatogène, exogène et métamorphogène, proposant la classification reproduite dans le tableau 1.

TABLEAU 1

No de la série	Composition des formations métallifères de la série	Position géotectonique de la série	Régions typiques
I	Chromite, platine, titanomagnétite (diamant)	stade précoce des zones plissées du type Oural, basites-ultrabasites	Oural, Petit Caucase, plateau Koriak
II	Cuivre-nickel sulfuré (avec Co, Ptoïdes), magnésioferrite, carbonatite (avec Nb, Zr, HTR), diamant	plate-formes et leurs parties réactivées; trappes complexes alcalins ultrabasiques, kimberlites	Plate-formes (Afrique, Canada, Bresil, Sibérie, etc.)
III	Apatite-magnétite, apatite, loparite	parties réactivées des boucliers (massifs)	massif Baltique
IV	Pegmatites à métaux rares, greisen (Sn–W–Mo–Be), or, polymétallique, antimoine–mercure	stade tardif des zones plissées de type Transbaïkal; granite acide	le NE et le SE de l'Asie, Kazakstan
V	Skarns métallifères, cuivre–molybdène, or et sulfures, polymétallique, antimoine-mercure	stades moyens des zones plissées de type Oural et Asie centrale; granitoïdes d'acidité moyenne	Oural, Asie Centrale, Petit Caucase, Caucase septentrional, Cordillère et Andes
VI	Pyrite cuprifère, cherts à Fe, cherts à Mn	stades précoces des zones plissées de type Oural; spilites kératophyres et porphyrites	Oural, Balkan, Caucase, Transcaucasie
VII	Or-argent, mercure-antimoine–ferberite (scheelite) réalgarorpiment étain, rhyolitique	stades tardifs des zones plissées de différents types, extrusions de felsite, rhyolites, dacites	Caucase, Transcaucasie, Transbaïkal, Cordillère et Andes
VIII	Latérites ferrifères et alumineuses, minerais silicatés de nickel (souvent avec asbolane)	manteau d'érosion des roches basiques–ultrabasiques, acides et alcalines	Oural, Cuba, Guinée, Brésil, etc.
IX	Minerais sédimentaires de fer (oolitique) et de manganèse, sédimentaires de bauxite	parties littorales bassins géosynclinaux, bassins lacustres sur les plate-formes	Kertch Lorraine (Fe); Nikopol, Tchiatou (Mn); Krasnaia Schapotchka, Tichvinskoie (Al)
X	Minerais métamorphogènes de fer et de manganèse, conglomérat et quarzite urani–aurifères, schistes métamorphiques avec disthène, sillimanite, andalousite, parfois rutile	socle Précambrien des boucliers de l'Afrique, Canada, Brésil, Inde, Chine, Australie, U.R.S.S., etc.	Krivoï Rog, Itabira, Lac Supérieur (Fe); Morro de Minas, Bihar (Mn); Witwatersrand, Blind-River, Jakobine (Au, U); Mexique, U.S.S.R. (rutile, disthène)

FIG. 1. Distribution globale des principaux éléments métalliques dans l'écorce terrestre des continents. (1) Boucliers et leurs plateformes; (2) Zones plissées d'âge calédonien; (3) Zones plissées d'âge hercynien; (4) Zones plissées d'âge cimmérien; (5) Zones plissées d'âge alpin.

La signification essentielle de cette distinction de "séries de formations métallifères" correspond, pour l'auteur, à la nécessité d'établir des unités naturelles, dans un cadre géologique défini, de manière à déceler, le cas échéant, les chaînons absents de la série.

Les provinces de même type ont, malgré des différences d'âge, des minéralisations très semblables et quasi-identiques lorsque l'âge est le même, indépendamment de la distance qui peut les séparer, c'est pourquoi il y a une grande ressemblance entre les métallogénies alpines du plateau Koriak et du Kamtchatka d'une part et du Petit Caucase d'autre part. En revanche, deux provinces du même âge mais de type différent présentent, même si elles sont voisines (par exemple Tchukotka et Kamtchatka, ou Bolivie et Chili), des différences essentielles dans le caractère de leur minéralisation.

L'échelle et l'intensité de la minéralisation varient en fonction de l'âge géologique des structures. Ainsi les structures du Précambrien sont particulièrement riches en Fe, Au, U, Ni, Pt-oïdes, Ti. Les structures d'âge paléozoïque sont riches, selon le type de minéralisation, en Cr, Pt-oïdes, Cu ou en Sn, W. Les structures cimmeriennes sont surtout riches en Sn, W, Sb, Au et les structures d'âge alpin sont spécialement riches en Cu et Mo, Ag, Hg, Mn (ce dernier dans le type sédimentaire).

Les zones de concentration de certains métaux se disposent autour de centres ou s'allongent en ceintures métallifères le long de fractures régionales ou de structures plissées. Leur disposition, en même temps qu'elle souligne les traits spécifiques de la métallogénie, donne une idée de la géographie économique des ressources mondiales en matières minérales. La distribution des grandes concentrations de métaux de l'écorce terrestre en structures d'âges et de types différents est illustrée par la figure 1.

Zur Geochemie der Sedimentite des Permosiles im Südteil der D.D.R.

JOHANNES RENTZSCH und ARIBERT KAMPE

Zentrales Geologisches Institut, Invalidenstrasse 44, 104 Berlin, D.D.R.

Abstract

Since the beginning of this century the late Variscan red bed formations were said to be the source of the metals of the Upper Permian "Kupferschiefer"-mineralization. Up to the last years there was a lack of regional investigations of the geochemistry of these rocks. To prove the above mentioned theory, we determined the Fe-, Corg-, S-, Cu-, Pb-, Zn-, Ni-, Co-, Mn-, B-, Ba-, Sr-, Rb-, V-, Cr-, Ti-, Zr- and Sn-contents of more than 2000 samples of red, grey and bleached psephitic, psammitic and pelitic sedimentary rocks and of some coal samples.

In all investigated rock types the average Fe-, Ni- and Cr-contents are higher in the red-coloured rocks than in the grey and bleached ones; the average B-, Rb- and Ba-contents are equal in the red, grey and bleached rocks; the average Corg-, S-, Cu-, Pb- and Zn-contents are higher in the grey rocks than in the red-coloured ones. Only in the last group of elements the average contents are higher than the planetary Clarke of each rock type. Special investigations were made about geochemical facies, distribution of base metals, diagenetic and epigenetic influences on the distribution of base metals, base metal mineralizations and the source of elements. The late Variscan volcanism was proved to be the main source of elements in the red bed formations. The sedimentary and magmatic rocks of the molasse stage are Pb-specialized, therefore the Central European Variscan belt forms a lead province.

1. Einleitung

Obwohl seit Jahrhunderten kleine Cu- und Pb–Zn-Vorkommen vom Red Bed-Typ in den permosilesischen Molassesedimenten bekannt sind und seit Jahrzehnten der Metallgehalt der Erzlagerstätten des Kupferschiefers von den meisten Bearbeitern aus diesen Sedimentiten abgeleitet wird, setzte ihre stoffliche Bearbeitung erst nach dem Bekanntwerden von Urananreicherungen in den Kohlen und klastischen Sedimentiten ein. Die erste und wichtigste Arbeit ist die von Rösler durchgeführte geochemische Untersuchung der permosilesischen Kohlen (Leutwein und Rösler, 1956).

Im Rahmem der Zunahme der Intensität metallogenetisch-geochemischer Forschungen wurden die Sedimente der verschiedenen spätvaristischen intraorogenen Molassesenken in den letzten fünfzehn Jahren intensiv untersucht, so daß jetzt für den Bereich der Innersudetischen Senke (Cadkova, 1971), für das Riesengebirgsvorland (Malkovski *et al.*, 1974), für das Saar-Nahe-Gebiet (Herzberg, 1962; Oppermann, 1967) und für den Saale-Trog und einige der übrigen intraorogenen Senken im Südteil der D.D.R. ein vergleichbares Bild vorliegt.

Die Sedimentgesteine des Permosiles setzen sich vorwiegend aus fein- bis grobklastischen Sedimenten von grauer und roter Farbe und untergeordnet aus Kohlen und Brandschiefern zusammen. Daneben treten in ganz geringem Umfang Karbonatgesteine und dichte Kieselgesteine (Hornsteine) auf. In der vorliegenden Arbeit wurden die Sandsteine, Grauwacken und Konglomerate zur Gruppe der "Psammite" zusammengefaßt und die Tonsteine und Siltsteine zur Gruppe de "Pelite", Brandschiefer und Kohlen bilden die Gruppe der "kohligen Gesteine". Grauschwarze, graue, grüngraue und gebleichte klastische Sedimente werden als "grau"; rote, rotbraune und rotgefleckte klastische Sedimente als "rot" bezeichnet.

Neben den von Rösler untersuchten Kohlen kamen Proben aus dem Thüringer Wald (Saaletrog), dem nordöstlichen Saaletrog, dem Ilfelder Becken, der Flechtingen-Roßlauer-Scholle, Nordwestsachsen, dem Siles des Osterzgebirges und aus der Lausitz zur Bearbeitung und Auswertung. Die Untersuchungen wurden im Zentralen Geologischen Institut und an der Bergakademie Freiberg durchgeführt. Über die Ergebnisse der Untersuchungen an den einzelnen regionalen und stratigraphischen Einheiten wird an anderer Stelle berichtet (Hoth *et al.*, 1973; Lützner und Rentzsch, 1975; Grunewald, 1976)

Insgesamt wurden die geochemischen Daten von 2161 Sedimentproben ausgewertet. Es lagen 886 Psammite, 760 Pelite, 490 kohlige Gesteine, 10 Karbonatgesteine und 15 Hornsteine vor.

Naßchemisch wurden Co_2, C_{org}, $Fe_{ges.}$, Fe^{++}, $S_{ges.}$ und Cu, Pb und Zn bei Gehalten $>0{,}1\%$ bestimmt. Mittels der optischen Spektralanalyse sind die Spurenelemente B, Co, Cr, Cu, Mo, Ni, Pb, Sn, Ti und V nachgewiesen worden. Die Elemente Ba, Rb, Mn, Sr, Zr, und Zn wurden röntgenfluoreszenzspektrometrisch bestimmt. Die rechnerische Auswertung der Daten wurde im Rechenzentrum des ZGI durchgeführt. Es sind die Mittelwerte und die Streuung berechnet worden und der Verteilungstyp geschätzt. Zur Klärung der Bindungsart der Elemente sind Faktoren- und Komponentenanalysen berechnet worden.

2. Zur Geochemie einiger Neben- und Spurenelemente der permosilesischen Sedimente

Die zusammengefaßten Ergebnisse der geochemischen Untersuchungen werden in der Tabelle 1 und den Abb. 1a und 1b dargestellt. Um Konzentrationskoeffizienten (Verhältnis Gehaltswert/Clarke-Wert des Lithotyps) berechnen zu können, werden in Tabelle 2 die Clarke-Werte für die Lithotypen Grauwacke, Tonschiefer und Karbonatgesteine angegeben. Die Gruppe der "Psammite" wird mit dem Grauwacken-Clarke, die der "Pelite" mit dem Tonschiefer-Clarke und die Karbonatgesteine mit dem Karbonatgesteins-Clarke verglichen.

2.1. *Eisen*

Die Fe-Gehalte der Psammite und Pelite liegen knapp unterhalb der jeweiligen Clarke–Werte (Konz.-Koeff. 0,69–1,02). Die seltenen geringmächtigen Karbonatgesteine heben sich mit ihren Fe-Gehalten von 0,8 bis 3,7% (Konz.-Koeff. 2,1–9,7) von denen ihrer klastischen Nebengesteine ab. In den Rotsedimenten sind die Gehalte meist einige Zehntel Prozent höher als in den Grausedimenten. Wenn in Grausedimenten einzelner Teilgebiete höhere Fe-Gehalts auftreten als in den Rotsedimenten, so ist das auf eine erhöhte Sideritführung der Grausedimente in der Nachbarschaft von Kohlenflözen zurückzuführen. Eisen ist in Grausedimenten an Karbonate (Fe^{2+}: Siderit, Ankerit, Fe-Kalzit,), Chlorit (Fe^{2+} und Fe^{3+}) und detritische femische Silikate (Fe^{2+}), Sulfide (Fe^{2+}:Pyrit, Markasit,

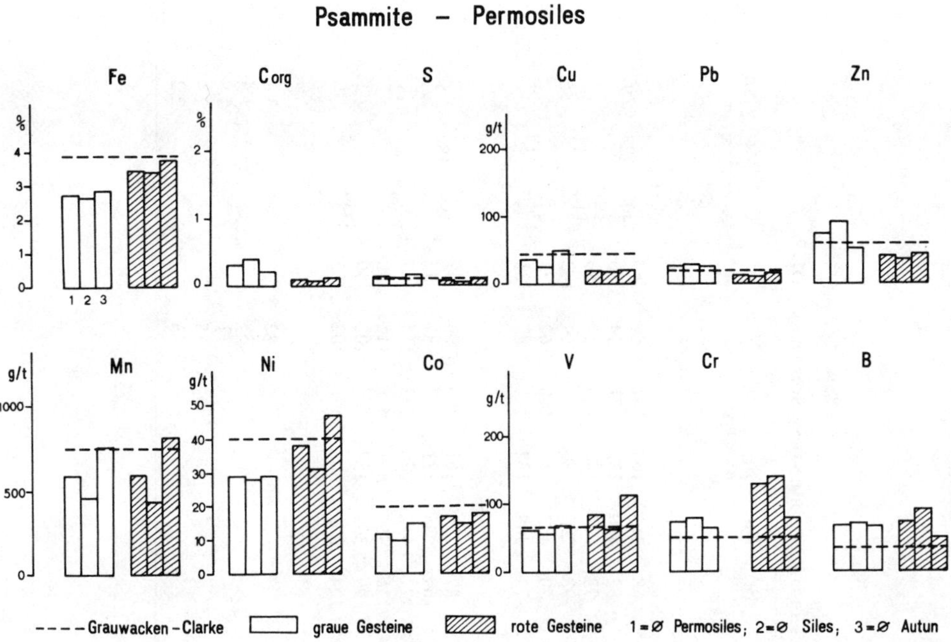

ABB. 1a. Durchschnittsgehalte einiger Neben- und Spurenelemente in den permosilesischen Psammiten des Südteils der D.D.R.

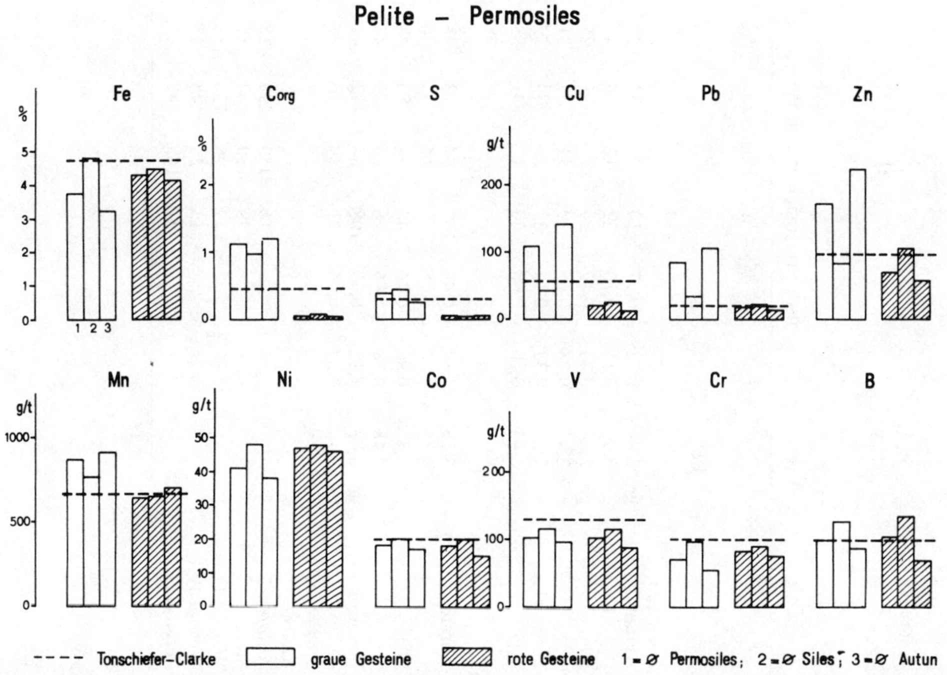

ABB. 1b. Durchschnittsgehalte einiger Neben- und Spurenelemente in den permosilesischen Peliten des Südteils der D.D.R.

Tabelle 1. Durchschnittsgehalte der Permosilesischen Grau- und Rotsedimente im Sudteil der DDR

		Fe%	C_{org}%	S%	Cu	Pb	Zn	Ni	Co	Mn	B	Ba	Sr	Rb	V	Cr	Ti	Zr	Sn
					g/t														
Psammite																			
Autun	gr	2,84	0,22	0,17	50	27	53	29	15	759	68	413	108	131	69	67	3404	227	9
	r	3,75	0,11	0,10	22	16	44	47	18	812	52	480	132	171	114	80			
Siles	gr	2,66	0,40	0,12	26	29	91	28	10	459	72	393	140	165	57	80	3114	211	8
	r	3,42	0,08	0,06	19	11	37	31	15	431	93	297	180	160	64	142			
Permosiles	gr	2,74	0,32	0,14	37	28	74	29	12	590	70	402	116	140	62	74	3242	219	9
	r	3,43	0,10	0,08	20	13	41	38	17	588	74	368	132	171	87	131			
Pelite																			
Autun	gr	3,25	1,19	0,46	142	107	223	38	17	917	87	326	161	166	96	59	3388	183	5
	r	4,15	0,06	0,05	13	14	59	46	15	701	69	228	121	228	87	75			
Siles	gr	4,81	0,96	0,26	43	34	83	48	20	773	127	412	70	133	115	97	3796	189	12
	r	4,48	0,08	0,06	26	23	105	48	20	648	135	460	n.b.	n.b.	115	90			
Permosiles	gr	3,73	1,12	0,39	108	84	172	41	18	873	99	352	157	164	102	70	3535	186	7
	r	4,29	0,07	0,06	21	19	70	47	18	643	104	339	n.b.	n.b.	102	83			
Kohlen																			
Autun					16	330	299	15	11						32				3,2
Stefan					78	180	67	19	8						39				1,4
Westfal D					19	96	136	31	18						11				1,9
Westfal B					13	30	48	15	8						35				2,4
Permosiles					27	145	149	25	15						23				2,2

TABELLE 2. CLARKE-GEHALTE DER TONSCHIEFER, GRAUWACKEN UND KARBONATGESTEINE IN g/t

	Tonschiefer	Grauwacke	Karbonatgestein
B	100	35	20
Ba	800	380	10
Co	20	20	0,1
Cr	100	50	11
Cu	57	45	4
Fe	47 200†	38 400	3800
Mn	670	750	1100
Ni	95	40	20
Pb	20	n.b.	9
Rb	200	80	3
S	3000	1200	12000
Sr	450	200	610
Sn	6†	n.b.	0,X (0,2)
Ti	4500	3800	400
V	130	67	20
Zn	95†	60	20
Zr	200	450	19
C_{org}	4500	n.b.	2000

Karbonatgesteins-Clarkes nach Turekian-Wedepohl (1961) Tonschiefer-Clarkes nach Vinogradov (1962) mit Ausnahme der mit † gezeichneten Werte. Diese stammen ebenfalls von T + W. Grauwacken-Clarkes nach Wedepohl (1968).

Pyrrhotin, Chalkopyrit u.a.) und an Hämatit (Fe^{3+}) gebunden. Die Verteilung der Bindungsformen des Eisens in den Grausedimenten ist wie in Tabelle 3.

TABELLE 3

	Psammite	Pelite	Karbonatgest.
Karbonatisch Fe^{2+}	28–70%	14–58%	48–63%
Leichtlösliche Silikate Fe^{3+}	–13%	1-31%	—
Oxidisch Fe^{3+}	–6%	–16%	—
Sulfidisch Fe^{2+}	–30%	–33%	14–18%
Schwerlösliche Silikate Fe^{2+}	14–30%	6–77%	9–38%

In den Rotsedimenten ist Eisen vorwiegend an Hämatit und Fe-Hydroxide (Fe^{3+}), adsorptiv an Tonminerale (Fe^{3+}), an detritische femische Silikate (Fe^{2+}) und sehr selten an Karbonate (Fe^{2+}) gebunden. Die Bindungsformen des Eisens in den Rotsedimenten sind wie in Tabelle 4.

TABELLE 4

	Psammite	Pelite
Karbonatisch Fe^{2+}	—	–3%
Leichtlösliche Silikate Fe^{3+}	15–20%	5–10%
Oxidisch Fe^{3+}	57–68%	63–70%
Sulfidisch Fe^{2+}	—	—
Schwerlösliche Silikate Fe^{2+}	14–26%	22–30%

In den sulfidführenden Konkretionen des Acanthodes-Horizontes der Goldlauterer Schichten (Thüringer Wald) sind nach Lützner und Rentzsch (1975) 52–75% des Eisens karbonatisch, bis 11% an leichtlösliche Silikate, 9–34% sulfidisch und bis 12% an schwerlösliche femische Silikate gebunden.

2.2. *Organischer Kohlenstoff*

Organischer Kohlenstoff ist in den permosilesischen Sedimenten an allochthone und autochthone organische Substanz gebunden. Die Grausedimente führen durchschnittlich die 2–5-fache (Psammite) bzw. die 12–20-fache Menge (Pelite) an organischem Kohlenstoff wie die Rotsedimente. In den Grausedimenten ist die organische Substanz als Milieubildner von größter Bedeutung für die Anreicherung von Buntmetallen und anderen Spurenelementen. Bei den Psammiten gilt das vor allem für die Stadien der Diagenese und Epigenese. Für die grauen Pelite beträgt der durchschnittliche Konz.-Koeff. 2,49, für die roten nur 0,16.

2.3. *Schwefel*

Die S-Gehalte der permosilesischen Grausedimente schwanken um den jeweiligen Clarke-Wert. Bei den Rotsedimenten ist im Falle der Pelite der durchschnittliche Konz.-Koeff. 0,20 und im Fall der Psammite 0,67.

Während der Schwefel in den Grausedimenten vorwiegend sulfidisch (Pyrit, Markasit, Pyrrhotin, Buntmetallsulfide) auftritt, ist es in den roten Gesteinen im wesentlichen in Form von Erdalkalisulfaten vorhanden. Die zur Sulfidfällung in den Grausedimenten notwendigen Sulfidionen werden lokal durch Fäulnisbakterien bei der Eiweißzersetzung und im Rahmen der Sulfatreduktion durch die desulfurizierenden Bakterien aus der geringen Menge der Erdalkalisulfate produziert.

2.4. *Kupfer, Blei und Zink*

Die Buntmetalle gelangen in ionarer Lösung und adsorptiv an Tonminerale gebunden in die limnischen Becken. Sie werden in den Grausedimenten gegenüber den Rotsedimenten angereichert. In den grauen Psammiten schwanken die Gehalte um den jeweiligen Clarke–Wert. In den grauen Peliten sind die Buntmetallgehalte parallel zu den erhöhten C_{org}-Gehalten durchschnittlich deutlich angereichert (Konz.-Koeff. Cu = 2,5; Pb = 5,35; Zn = 2,35). Die Kohlen des Autun sind die durchschnittlich Pb–Zn-reichsten Gesteine des Permosiles.

Kupfer liegt in den Grausedimenten z.T. in sulfidischer Bindung (Chalkopyrit, Tennantit u.a.) vor. Außerdem ist es adsorptiv an die Tonminerale gebunden. Aus der Faktorenanalyse läßt sich eine schwache Bindung des Kupfers an die Feldspat-Glimmerfraktion ableiten. Blei ist in den Grausedimenten sulfidisch als Galenit, an die Feldspatfraktion und adsorptiv an Tonminerale gebunden. Zink tritt in Form von Sphalerit, adsorptiv an Tonminerale und an detritische femische Silikate gebunden auf. In den Kohlen sind die Buntmetalle an die Kohlensubstanz und gegenwärtig vor allem sulfidisch gebunden. Die in den grauen und roten Psammiten nur geringfügig unterschiedlichen Pb-Gehalte beweisen die vorherrschende silikatische Bindung des Bleis in diesen Gesteinen.

In den Rotsedimenten sind die Buntmetalle im wesentlichen adsorptiv an die Fe- und Mn-Hydroxide, die Tonmineral- und Feldspatfraktion gebunden.

2.5. *Nickel und Kobalt*

Die Ni- und Co-Gehalte der permosilesischen Sedimente liegen in den meisten Fällen unterhalb der Clarke-Werte, d.h. die Konz.-Koeff. sind fast immer <1. Die grauen Psammite sind Co- und Ni-ärmer als die roten. Das gilt bei den Peliten auch für das Ni. Die Co-Gehalte in den roten und grauen Peliten sind gleich.

Das zeigt, daß sowohl in den Rot- als auch in den Grausedimenten adsorptive Bindungen überwiegen. In den Rotsedimenten dominiert die adsorptive Bindung an Mn- und Fe-Hydroxide, die an Tonminerale tritt demgegenüber etwas zurück. Sowohl in Grau- als auch in Rotsedimenten kommt dazu noch eine Bindung an detritische femische Silikate. In den Grausedimenten treten Co und Ni vorwiegend adsorptiv an Tonminerale gebunden auf. Beim Vorhandensein von Sulfiden kommt eine sulfidische Bindung hinzu. Wenn Fe-haltige Kupfersulfide, wie z.B. Chalkopyrit vorhanden sind, werden Co und Ni praktisch quantitativ in diese Sulfide eingebaut. Sie substituieren das Fe^{2+} dieser Minerale isomorph. Die in Grausedimenten sonst häufige Bindung an die organische Substanz tritt hier weitgehend zurück. In den Kohlen kommen Ni und Co meist in sulfidischer Form an Pyrit, Millerit (NiS), Linneit $(Co, Ni)_3S_4$ und Bravoit $(Fe, Ni)S_2$ gebunden vor.

2.6. *Mangan*

Die Mn-Gehalte der Psammite liegen durchschnittlich unterhalb des Clarke-Wertes (Konz.-Koeff. 0,78) Bei den Peliten sind die Grausedimente etwas Mn-reicher (Konz.-Koeff. 1,3) als die Rotsedimente (Konz.-Koeff. 0,96). Im Süßwasser wird Mn vom Fe etwas getrennt, da Mn-Ionen länger in Lösung beständig sind als die Fe-Ionen. Letztere gehen in Solform über. Fe-Hydroxidgele flocken bereits randnahe aus. In den Grausedimenten ist Mangan zusammen mit Fe^{2+} und einem Teil des Strontiums an die Karbonate gebunden. In den Rotsedimenten tritt es im wesentlichen in Form von Mn-Hydroxid auf.

2.7. *Vanadim und Chrom*

Die Vanadium- und Chromgehalte der roten Psammite (V: Konz.-Koeff. 1,3; Cr: Konz.-Koeff. 2,62) sind gegenüber den grauen Psammiten (V: Konz.-Koeff. 0,93; Cr: Konz.-Koeff. 1,48) etwas erhöht.

In den grauen und roten Peliten sind die V- und Cr-Gehalte fast gleich und erreichen den jeweiligen Clarke-Wert durchschnittlich nicht. Die wichtigste Bindungsform beider Elemente ist die adsorptive an die Tonminerale. Die Bindung des V an die organische Substanz tritt demgegeñuber stark zurück. In den Rotsedimenten sind V und Cr neben den Tonmineralen auch an die Fe-Hydroxide adsorptiv gebunden.

2.8. *Bor*

Die durchschnittlichen B-Gehalte der Psammite sind gegenüber dem Clarke-Wert erhöht (Konz.-Koeff. ~2,0). Die Gehalte der Grau-und Rotsedimente sind etwa gleich. Lediglich die autunischen Grausedimente sind gegenüber den Rotsedimenten etwas angereichert, was auf eine beginnende Eindampfung hindeutet (Karbonat- und schwache Sulfatführung). Die beim Clarke-Wert liegenden B-Gehalte der Relikte sind in Grau- und Rotsedimenten ebenfalls gleich. Im Bereich des Osterzgebirges und der Elbtalzone liegt eine negative B—Anomalie mit Konz.-Koeff. <0,5 sowohl in Peliten als auch in Psammiten vor. Bor ist an die Tonminerale und dabei vor allem an den Illit geknüpft.

2.9. *Barium, Strontium, Rubidium*

Die Bariumgehalte der Psammite entsprechen dem Clarke-Wert, die der Pelite sind deutlich niedriger (Konz.-Koeff. ~0,4). Das spricht dafür, das in diesen Sedimenten die Bindung des Ba an die Tonminerale gegenüber der Bindung an die K-haltigen Feldspäte zurücktritt. Die Strontiumgehalte der Psammite und Pelite liegen unterhalb der Clarke-Werte. Sr ist sowohl adsorptiv an Tone, als auch an Feldspäte und Ca-Karbonat gebunden. Während Rubidium in den Peliten etwa in der Größenordnung des Clarke-Wertes vorhanden ist, treten in den Psammiten erhöhte Gehalte auf (Konz.-Koeff. ~2,0). Rb ist im wesentlichen an die Feldspäte gebunden.

2.10. *Titan*

Die Titangehalte der Psammite und Pelite liegen knapp unter den jeweiligen Clarke-Werten. Sie sind an Tonminerale und vor allem an detritischen Rutil und Anatas gebunden.

2.11. *Zirkon*

Die Zirkongehalte der Psammite liegen weit unter dem Clarke-Wert (Konz.-Koeff. ~0,5). In den Peliten wird dieser knapp erreicht. Zirkon tritt in Form detritischen Zirkons im Schwermineralbestand auf.

2.12. *Zinn*

Die Sn-Gehalte der permosilesischen Sedimente liegen im allgemeinen im Bereich des Tonschiefer-Clarkes. Erhöhte Gehalte in einigen Teilgebieten, die sowohl in Psammiten als auch in den Peliten und Kohlen nachzuweisen sind, können auf anomale Sn-Gehalte in den Liefergebieten zurückgeführt werden. Sn ist vorwiegend an detritischen Kassiterit gebunden.

2.13. *Zusammenfassende Darstellung der Geochemie der Einzelelemente*

Bei der gemeinsamen Betrachtung aller Elemente lassen sich in den Gesteinstypen jeweils drei Elementgruppen aushalten. In der ersten Gruppe sind die Gehalte der roten Gesteine höher als die der grauen, in der zweiten die der grauen und der roten Gesteine gleich und in der dritten die der grauen höher als die der roten.

Bei den Psammiten gehören die Elemente Fe, Ni, Co (siderophil), V und Cr (lithophil) zur ersten Gruppe und die Elemente Mn, B, Sr, Rb und Ba (lithophil) zur zweiten. In der dritten Gruppe treten neben C_{org} die chalkophilen Elemente S, Cu, Pb und Zn auf.

Fe ist in den roten Gesteinen vorwiegend an Hämatit und Fe-Hydroxide, an Tonminerale, femische Silikate und sehr selten an Karbonate gebunden. Bei den Spurenelementen Ni, Co, V und Cr überwiegt die adsorptive Bindung an die Fe-Hydroxide und die Tonminerale. Nur die Cr-Gehalte übersteigen allgemein den Clarke-Wert.

Mn ist in den roten Psammiten an Oxide und Hydroxide und in grauen an Karbonate gebunden. Die übrigen Elemente der zweiten Gruppe (B, Sr, Rb und Ba) treten vorwiegend in Feldspäten und Tonmineralen auf. In den grauen Gesteinen ist Sr auch an das Kalziumkarbonat gebunden.

In der dritten Gruppe kommt bei reduzierenden Bedingungen und etwas erhöhten Buntmetall- und S-Gehalten neben der adsorptiven Bindung der Buntmetalle an die Tonminerale noch eine sulfidische Bindung hinzu.

In den Siltsteinen sind in den roten Gesteinent neben Fe nur noch Ni und Cr stärker angereichert als in den grauen. Die Bindungsformen dieser Elemente entsprechen der in den Psammiten. In der zweiten Gruppe treten zu B, Ba und Rb noch V und Co hinzu, da hier neben der adsorptiven Bindung an Fe–Mn-Hydroxide und Tonminerale (Rotsedimente) bei Co die sulfidische Bindung und beim V die Bindung an die organische Substanz (Grausedimente) hinzukommt. In der dritten Gruppe sind neben den Buntmetallen, die sich wiederum zusammen mit C_{org} und S anreichern, noch Mn und Sr vorhanden, die vorwiegend an die Karbonate gebunden sind. Nur die Elemente der dritten Gruppe sind gegnüber dem jeweiligen Clarke-Wert angereichert.

Die Gehalte der an detritische Schwerminerale gebundenen Elemente Ti und Zr liegen im allgemeinen knapp unterhalb der jeweiligen Clarke-Werte. Wo erhöhte Sn-Gehalte auftreten, lassen sie sich von erhöhten Gehalten der Liefergebiete ableiten.

3. Komplexe Auswertung der Ergebnisse

3.1. *Zur geochemischen Fazies der permosilesischen Sedimente*

Über geochemisch-fazielle Untersuchungen an permosilesischen Sedimenten wurde von Lützner und Rentzsch (1975) berichtet. Die Autoren wählten als Modellfall die Ablagerungen der limnisch-fluviatilen Goldlauterer Schichten des Autun im Thüringer Wald, da deren Sedimentationsraum sowohl lithologisch-paläogeographisch als auch geochemisch am intensivsten untersucht wurde. Sie stellten fest, daß von den Spurenelementverhältnissen noch am ehesten das Verhältnis V/Cr zur Faziesanalyse geeignet erscheint (Tabelle 5). Das kann bei der Betrachtung aller permosilesischen

TABELLE 5. SPURENELEMENTVERHÄLTNISSE DER PERMOSILESISCHEN GRAU- UND ROTSEDIMENTE IM SÜDTEIL DER D.D.R.

		V/Cr	Mn/Fe	Ni/Co
Psammite				
Autun	gr	1,0	0,027	1,9
	r	1,4	0,022	2,6
Siles	gr	0,7	0,017	2,8
	r	0,4	0,013	2,1
Permosiles	gr	0,8	0,021	2,4
	r	0,7	0,017	2,2
Pelite				
Autun	gr	1,6	0,028	2,2
	r	1,2	0,017	3,1
Siles	gr	1,2	0,016	2,4
	r	1,3	0,014	2,4
Permosiles	gr	1,5	0,024	2,3
	r	1,2	0,015	2,6

Sedimente im Südteil der D.D.R. bestätigt werden. Die durchschnittlichen V/Cr-Verhältnisse der grauen Pelite und Psammite sind allgemein größer als die der roten. In den grauen Peliten schwanken sie von 0,9 .... 3,3; in den roten von 0,6 ... 2,0. Die entsprechenden Werte der grauen Psammite schwanken von 0,4 ... 2,3, die der roten von 0,2 ... 2,0. Die im allgemeinen zwischen 1,0 und 2,0 liegenden V/Cr-Verhältnisse der grauen Pelite sprechen nach Krejci-Graf (1966) für Ablagerungen der Gyttjafazies.

Wie bereits Lützner und Rentzsch nachwiesen, ist das Ni/Co-Verhältnis zur Faziesanalyse ungeeignet. Ni/Co-Verhältnisse, die meist $>2{,}0$ sind, treten sowohl in grauen als auch in roten Sedimenten auf. Nur in chalkopyritführenden grauen Sedimenten sind die Ni/Co-Verhältnisse $>1{,}5$ ... $>1{,}0$. Die Ursache dafür liegt in der im kontinentalen und kontinental-marinen Übergangsbereich in exogenen Kupferanreicherungen dominierenden Elementparagenese Cu–Co.

Die günstigste Möglichkeit zur Bestimmung des geochemischen Milieus, das während der Sedimentation und Diagenese herrschte, bietet die Untersuchung der Bindungsformen des Eisens. Slavin (1961) fand an Hand der Häufigkeitsverhältnisse der Bindungsformen des Eisens folgende Milieubereiche:

Typ 1
oxidierendes Milieu $= Fe^{3+} > Fe^{2+}_{HCl} \gg Fe^{2+}_{sulf}$

Typ 2
schwach reduzierendes Milieu $= Fe^{2+}_{HCl} > Fe^{3+} > Fe^{2+}_{sulf}$

Typ 3
reduzierendes Milieu $= Fe^{2+}_{HCl} > Fe^{2+}_{sulf} > Fe^{3+}$

Typ 4
stark reduzierendes Milieu $= Fe^{2+}_{sulf} > Fe^{2+}_{HCl} \gg Fe^{3+}$

Lützner und Rentzsch wiesen nach, daß im Becken der Goldlauterer Schichten (Autun, Thüringer Wald) in der Schwemmfächerfront im Verzahnungsbereich psephitisch-psammitischer und pelitischer Sedimente sich spätestens im Gefolge der Diagenese ein oxidierendes Milieu einstellte. Die dunklen C_{org}- und sulfidführenden Pelite der Goldlauterer (Acanthodes-Horizont) und Manebacher Schichten (Autun, Thüringer Wald) wurden unter reduzierenden Bedingungen sedimentiert und verfestigt. Das Gleiche gilt für die sulfidreichen Konkretionen bei Goldlauter. Auch das von Borchert und Krejci-Graf (1959) verwendete Mn/Fe-Verhältnis eignet sich zur geochemischen Faziesanalyse, wenn einheitliche Lithotypen verglichen werden. In Richtung vom Rand zum Becken reichert sich Mn gegenüber dem Fe an. Allgemein sind in den grauen "Becken"-Sedimenten größere Mn/Fe-Verhältnisse als in den roten "randnahen" Sedimenten vorhanden (Tabelle 5).

3.2. *Die Buntmetallverteilung*

3.2.1. *Syngenetisch gesteuerte Buntmetallverteilung*. Die Verteilung der Buntmetalle in den Sedimenten gibt Auskunft über geochemische Migrationsprozesse, die während der Sedimentation, Diagenese und Epigenese der Sedimente abliefen.

Man kann davon ausgehen, daß limnischen Becken die Metallgehalte im wesentlichen durch Oberflächenwässer zugeführt werden. Die Zonalität der Buntmetallverteilung ist auf die Küstenlinie und dabei ganz speziell auf die fluviatilen Schwemmfächer ausgerichtet. Man kann beim Studium der regionalen Buntmetallverteilung eine weitgehende Parallelität der Grenzen der Lithotypen und der Linien gleicher Gehalte beobachten. Die laterale Zonalität der Buntmetallverteilung in limnischen Becken ist folgendermaßen:

Küstenlinie mit fluviatiler Lösungszufuhr → Fe^{3+} → Cu → Pb → Zn (Becken). Im Vertikalprofil ist von unten nach oben parallel zur Entwicklung der Lithotypen in einem Sedimentationszyklus zuerst die "transgressive" Zonalität Cu→Pb→Zn und weiter die "regressive" Zonalität Zn → Pb → Cu zu beobachten. Das bezieht sich zunächst auf die pelitischen C_{org}-reichen Gesteinstypen. Bei Sulfidionenunterschuß wird diese Zonalität durch die unterschiedlichen Löslichkeitsprodukte der Buntmetallsulfide gesteuert (Tischendorf und Ungethüm, 1965). Da die Menge der adsorptiv an Tonminerale und Eisenoxide gebundenen Buntmetalle relativ gering ist, sind bei reduzierenden Verhältnissen die Buntmetalle in Abhängigkeit vom Schwefelangebot sulfidisch gebunden. Die hier beschriebenen Verhältnisse wurden in den autunischen Manebacher und Goldlauterer Schichten des Thüringer Waldes und in den Unteren Ilfelder Schichten des Harzes z.T. sehr detailliert nachgewiesen.

3.2.2. *Diagenetische und epigenetische Einflüsse auf die Buntmetallverteilung.* Die syngenetisch angelegte regionale Metallverteilung wird durch diagenetische und epigenetische Lösungsmigration mehr oder minder intensiv modifiziert. Das ist darin begründet, daß sich in den permosilesischen Kontinentalsedimenten vom Red Bed-Typ die Redoxgrenzen in den Porenwässern im Laufe der Lithifikation und der gesamten weiteren geologischen Entwicklung verschiebt. In randnahen Bereichen der limnischen Becken, wo Rotsedimente weitaus häufiger sind als Grausedimente, werden unter reduzierenden Verhältnissen abgelagerte Pelite bereits während der Diagenese sekundär oxidiert. Die Gesteine werden dabei rotfleckig oder vollkommen rotfarbig. Während dieser diagenetischen Schichtoxidation in Sinne von Perelman (1972) gehen mit den Gehalten an organischem Kohlenstoff auch die Buntmetallgehalte stark zurück.

Entsprechende Beobachtungen wurden an Übertage- und Bohraufschlüssen der autunischen Goldlauterer Schichten im Raum Friedrichroda-Tabarz gemacht. In den klastischen Sedimenten im Hangenden und Liegenden der limnischen Kohlenflöze und der weiter beckenwärts gelegenen kohlig-bituminösen Pelithorizonte bildete sich bereits frühdiagenetisch bei Vorhandensein von Schwefelwasserstoff eine Reduktionsbarriere aus, an der der Metallgehalt der intraformationellen Wässer sulfidisch gefällt wurde. Als Beispiel können die galenitführenden Sandsteine zwischen unbedeutenden Kohlenflözen der autunischen Manebacher Schichten im Thüringer Wald (z.B. Aufschluß Sachsendelle) genannt werden. In den bituminösen Peliten wird der Buntmetallinhalt der während der Diagenese migrierenden Porenwässer an koprolithogenen Konkretionen durch bei der Eiweißversetzung produziertem Schwefelwasserstoff sulfidisch gefällt (Rösler, 1964). An der Hangend- und Liegendgrenze der Kohlenflöze treten bisweilen Maxima der Buntmetallführung auf, wie das von Kaemmel (1972) in Plötz (Stefan) und im Rahmen dieser Untersuchungen in Ilfeld (Autun) nachgewiesen worden ist Es kann jedoch nicht entschieden werden, ob diese Maxima bereits diagenetisch aus dem Metallgehalt intraformationeller Wässer oder epigenetisch aus späteren exogenen Lösungszuflüssen entstanden sind. Wenn sich die laterale und vertikale Zonalität der Buntmetallverteilung in den durch postsedimentäre Lösungmigration beeinflußten Profilen der syngenetisch angelegten regionalen Metallverteilung unterordnet, dürften diagenetische Beeinflussungen vorliegen. Erst wenn die Zonalität der Metall- und Parageneseverteilung auf den heutigen Erosionsanschnitt orientiert ist, bzw. Metallabscheidungen in porösen wasserführenden Horizonten, die primär praktisch buntmetallfrei sind, auftreten, kann man sicher von exogen-epigenetischen Buntmetallanreicherungen sprechen. Der erste Fall tritt in den Oberen Gehrener Schichten (Autun) von Brattendorf im Thüringer Wald auf, wo sich in einer stratiformen diagenetisch gebildeten Chalkopyrit–Pyrit–Paragenese parallel zum Erosionsanschnitt zementativ eine Chalkosin–Hämatit–Paragenese ausbildete. Der zweite Fall wurde in Zwickau beobachtet, wo im Grenzbereich klastischer Autunsedimente mit rhyolithoiden Vulkaniten Abscheidungen von gediegenem Kupfer auftreten (v. Hoyningen, 1959). Selbstverständlich ist bei dem in den kontinentalen Ablagerungen herrschenden Sulfidionenunterschuß bei sulfidischer Bindung der Buntmetalle auch in epigenetischen Bildungen die Zonalität Cu→Pb→Zn zu beobachten, die dann aber in den meisten Fällen unabhängig von der primären lithofaziellen Entwicklung ist.

3.3. *Zur Stoffherkunft*

Umfangreiche Ermittlungen zur Klärung der Herkunft der Buntmetalle in den permosilesischen Sedimenten wurden von Lützner und Rentzsch (1975) und von Rentzsch *et al.* (1976) angestellt.

In den prävulkanitischen silesischen Sedimentiten dominieren im Geröllanteil Abtragungsprodukte altpaläozoischer Schiefer mit Cu:Pb:Zn-Verhältnissen, die denen des Tonschiefer-Clarkes (1:0,35:1,7) ähneln. In den postvulkanitischen silesischen und vor allem unterpermischen Sedimenten herrschen im Geröllanteil deutlich die permosilesischen

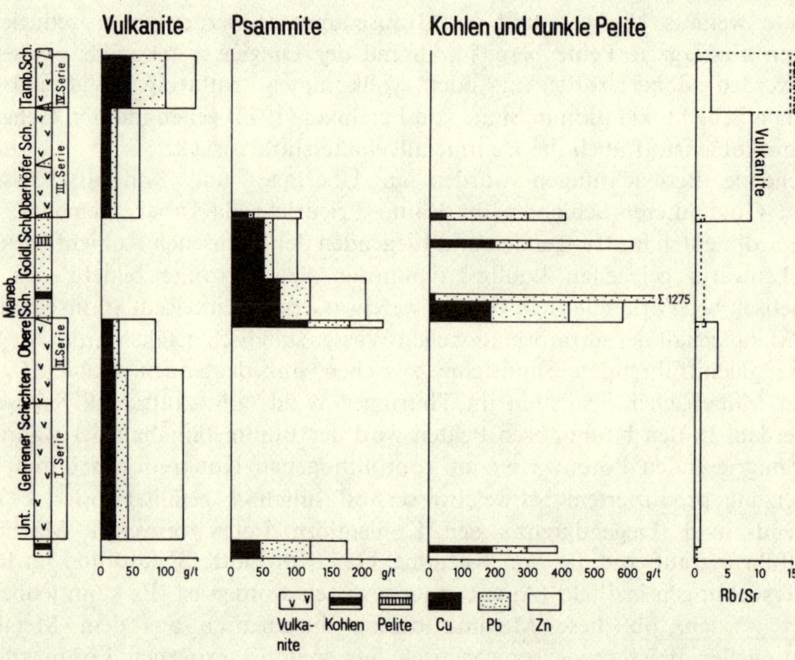

ABB. 2. Schema der Buntmetallverteilung und der Rb–Sr-Verhältnisse der permosilesischen Vulkanite und Grausedimente des Thüringer Waldes.

Vulkanite mit ihrem durchschnittlichen Cu:Pb:Zn-Verhältnis von 1:1,6:4,0 vor. Am klarsten läßt sich die Metallherkunft am Schema der Buntmetallverteilung in den permosilesischen Vulkaniten und Grausedimenten des Thüringer Waldes ableiten (Abb. 2). Die durchschnittlichen Buntmetallgehalte der Vulkanite gehen von der I. zur III. autunischen Eruptivserie deutlich zurück und steigen erst in der mengenmäßig unbedeutenden saxonischen IV. Eruptivserie wieder an. Das gilt besonders für die Pb-Gehalte (Schirmer, 1972). In den Psammiten der postvulkanischen autunischen Sedimente gehen die Buntmetallgehalte von den Oberen Gehrener über die Manebacher zu den Goldlauterer Schichten deutlich zurück. Die Abnahme der absoluten Cu- und Pb-Gehalte ist in den Psammiten von den Oberen Gehrener zu den Oberhöfer Schichten kontinuierlich. Entsprechendes ist an den grauen Peliten zu beobachten.

Eindeutige Angaben über den Einfluß vulkanogenen Materials auf die sedimentären Schichtenfolgen des Molassestockwerkes ermöglicht der Vergleich der Rb/Sr-Verhältnisse der Vulkanite (Schirmer, 1972) und Sedimentite. Das zeigt eine Gegenüberstellung dieser Verhältnisse am Beispiel der permosilesischen Schichtenfolge des Thüringer Waldes (siche Tabelle 6).

TABELLE 6

Stratigraph. Einheit	Eruptive-serie	Rb/Sr Vulkanite	Rb/Sr Sedimentite
Tambacher Schichten	IV	2,3	14,3
Oberhöfer Schichten	III	8,6	
Oberhöfer Schichten			
Unt. Protriton-Horizont		—	1,3
Goldlauterer Schichten		—	0,9
Manebacher Schichten		—	1,4
Ob. Gehrener Schichten		—	1,7
	I+II	1,4	
	II	3,5	
Unt. Gehrener Schichten	I	0,8	
Gehrener Basissedimente		—	0,3

In den psammitischen Sedimentiten findet man etwa die Rb/Sr-Verhältnisse der jeweils älteren Vulkanitserien wieder. Während in den Manebacher Schichten Abtragungsprodukte der I. und II. Eruptivserie auftreten, dominieren in den Goldlauterer Schichten die der I. Serie. Besonders deutlich tritt der Vulkanismus der III. Serie als Stofflieferant für die Sedimente der Tambacher Schichten auf.

Die bereits mehrfach beschriebene Abhängigkeit der Buntmetallgehalte der permosilesischen Kohlen von ihrem stratigraphischen Alter ist ebenfalls vom vulkanischen Geschehen abhängig. Parallel zur Intensität des Vulkanismus steigen vom Siles zum Autun die Buntmetallgehalte der kohligen Gesteine genauso wie die der grauen Pelite deutlich an und erreichen im Autun ihr Maximum. Das gilt vor allem für das Element Pb. Von den buntmetallreichen autunischen Kohlen ist die einzige prävulkanitsche (Ilfeld) die buntmetallärmste.

	Cu (g/t)	Pb (g/t)	Zn (g/t)
ϕ Autunkohlen	16	330	299
Ilfeld	—	150	63

Der permosilesische Vulkanismus ist damit der wichtigste Buntmetallieferant für die permosilesischen Sedimentite. Wo in Grausedimentiten Buntmetallanreicherungen auftreten, können sie im allgemeinen von anomalen Buntmetallgehalten in den Vulkaniten der Liefergebiete abgeleitet werden. Das gilt besonders für die Pb-Anreicherungen.

3.4. *Die geochemische Spezialisierung der permosilesischen Sedimente*

Die geochemische Spezialisierung der permosilesischen Sedimente läßt sich am besten im Metalltypendreieck Cu–Pb–Zn (Knitzschke, 1965) darstellen. Als Vergleichsbasis dienen der Tonschiefer- und der Grauwacken-Clarke. Außerdem wurden noch Daten von permosilesischen Sedimenten aus dem Innersudetischen Becken, dem Riesengebirgsvorland und dem Saar-Nahe-Gebiet ausgewertet. In den Tabellen 4 a, b werden die Cu-, Pb- und Zn-Gehalte, die Summe Cu + Pb + Zn und der prozentuale Anteil jedes Metalls an der Summe Cu + Pb + Zn für die permosilesischen Sedimente des Südteils der D.D.R und der obengenannten Vergleichsgebiete aufgeführt.

In der Abb. 3 kommen die permosilesischen Sedimente des Südteils der D.D.R. im Metalltypendreieck Cu–Pb–Zn zur Darstellung. Die Daten der Sedimente aus dem Innersudetischen Becken stammen von Cadkova (1971 a, b), die aus dem Riesengebirgsvorland von Malkovsky u.a. (1974) und aus dem Saar-Nahe- Gebiet von Herzberg (1962) und Oppermann (1967).

Es zeigt sich im Vergleich mit den Metalltypen des Tonschiefer-und des Grauwacken-Clarkes, daß die permosilesischen Sedimente des Südteils der D.D.R. und des Riesengebirgsvorlandes stärker Pb-betont sind als die des Saar-Nahe-Gebietes und des Innersudetischen Beckens. Das gilt besonders für die Grausedimente und für die stefanischen und autunischen Kohlen. Auf die Pb–Zn-Spezialisierung sowohl eines großen Teils der spätvaristischen Vulkanite, Intrusiva und Sedimentite als auch der Vererzungen an der Zechsteinbasis (Typ Kupferschiefer) war bereits von Rentzsch *et al.* (1976) hingewiesen

TABELLE 7a. METALLTYPEN DER SILESISCHEN SEDIMENTE

		g/t				% Anteil		
Südteil der D.D.R.		Cu	Pb	Zn	Cu,Pb,Zn	Cu	Pb	Zn
Psammite	gr	26	29	91	146	18	20	62
	r	19	11	37	67	28	16	56
Pelite	gr	43	34	83	160	27	21	52
	r	26	23	105	154	17	15	68
kohlige Gesteine								
Stefan		78	180	67	325	24	55	21
Westfal D		19	96	136	251	8	38	54
Westfal B		13	30	48	91	14	33	53
Innersudetisches Becken (Stefan)								
Pelite		58	26	100	184	32	14	54
kohlige Gesteine		150	520	720	1390	11	37	52
Saar-Nahe-Gebiet (Stefan)								
Psammite	gr	80	9	70	159	50	6	44
	r	3	8	40	51	6	16	78
Pelite	gr	165	45	50	260	64	17	19
	r	25	15	70	110	23	14	63

TABELLE 7b. METALLTYPEN DER AUTUNISCHEN SEDIMENTE

Südteil der D.D.R.		g/t Cu	Pb	Zn	Cu,Pb,Zn	% Anteil Cu	Pb	Zn
Psammite	gr	50	27	53	130	38	21	41
	r	22	16	44	82	27	19	54
Pelite	gr	142	107	223	472	30	23	47
	r	13	14	59	86	15	16	69
kohlige Gesteine		16	330	299	645	2	51	47
Riesengebirgsvorland								
Psammite		23	24	28	75	31	32	37
Pelite		36	26	54	116	31	22	47
Innersudetisches Becken								
Pelite		70	30	146	246	29	12	59
Saar-Nahe-Gebiet								
Psammite		47	—	40	87	54	—	46
Pelite		90	—	70	160	56	—	44
kohlige Gesteine		325	593	443	1361	24	44	32

worden. Die vorliegenden detaillierteren Untersuchungen beweisen, daß die permosilesischen Sedimentite des Südteils der D.D.R. Pb-spezialisiert sind. Da das auch für die Magmatite gilt und außerdem zahlreiche Pb–Zn–Lagerstätten und Pb-führende Lagerstätten vorhanden sind, kann man sagen, daß das zentrale Mitteleuropa eine spätvariszische Pb-Zn-Provinz darstellt.

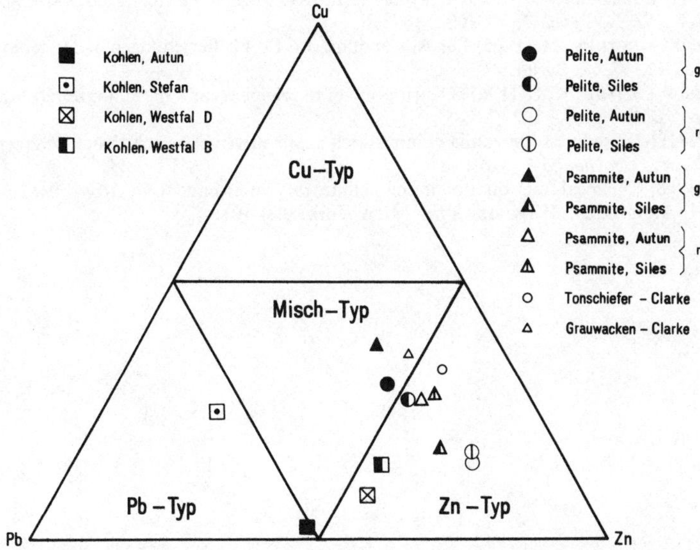

ABB. 3. Metalltypen der permosilesischen Sedimente des Südteils der D.D.R.

Verweisungen

BORCHERT, H. und KREJCI-GRAF, K. (1959) *Spurenmetalle in Sedimenten und ihren Derivaten,* Bergbauwiss., Goslar.

CADKOVA, Z. (1971a) Distribuce kovovijch pruku v permskych sedimentech vnitrosudetske panve. *Sbornik Geol. VED, loziskova geol.* **14**, 91–121, Praha.

CADKOVA, Z. (1971b) Zonal distribution of copper, lead and zinc in Permian sediments of the Intra-Sudetic Basin. *Cas. Min. Geol.* **16**, 1–7, Praha.

GRUNEWALD, V. (1976) Metallogenetische Untersuchung des sedimentären Permosiles im Bereich der Halle-Wittenberger Scholle. *Freib. Forschh.* C 311, Leipzig.

HERZBERG, W. (1962) Die Uranführung der Unterrotliegend-Sedimente, insbesondere der Mittleren Lebacher Schichten. Ein Beitrag zur Genesis der Uranerze in den Porphyren und Sedimenten der Saar-Nahe-Senke. Unveröff. Diss. TU, Berlin.

HOTH, K., PAECH, H. J. und KAMPE, A. (1973) Hinweise auf Intrusivkörper im nördlichen Teil der Flechtingen-Roßlauer Scholle. *Z. geol. Wiss.* **1**, 861–866, Berlin.

HOYNINGEN, E. v. (1969) Zur Geologie der Kupfervorkommen im Rotliegenden von Zwickau. *Freib. Forschh.* C 58, 93–106, Berlin.

KAEMMEL, TH. (1972) Zur präoberstefanischen Verwitterungsrinde im Thüringer Wald im Rahmen der permosilesischen Verwitterungserscheinungen. *Jb. Geol.* **4**, 81–96, Berlin.

KNITZSCHKE, G. (1965) Zur Typisierung des Kupferschiefers im südöstlichen Harzvorland. *Freib. Forschh.* C. 193, 123–141, Leipzig.

KREJCI-GRAF, K. (1966) Geochemische Faziesdiagnostik. *Freib. Forschh.* C 224, Leipzig.

LEUTWEIN, F. und RÖSLER, H. J. (1956) Geochemische Untersuchungen an paläozoischen und mesozoischen Kohlen Mittel- und Ostdeutschlands. *Freib. Forschh.* C 19, Berlin.

LÜTZNER, H. und RENTZSCH, J. (1975) Sedimentation und Metallogenie in einem intermontanen Becken der variszischen Molasse. *Z. geol. Wiss.* **3**, 1473–1490, Berlin.

OPPERMAN, H. (1967) Uran- und Spurenelementführung von Kohlen und anderen Sedimenten des Stefan C und des Unterrotliegenden der Saar-Nahe-Senke. Unveröff. Diss. TU, Berlin.

PERELMAN, A. J. (1972) *Geochemie epigenetischer Prozesse,* Berlin.

RENTZSCH, J., SCHIRMER, B., RÖLLIG, G. und TISCHENDORF, G. (1976) On the metal source of non-ferrous mineralisations at the base of Zechstein (Kupferschiefer type). *Sbor. Geol. Inst.* 171–188, Warsaw, (The current metallogenic problems of Central Europe.)

RÖSLER, H. J. (1964) Eine Vererzung vom Kupferschiefertyp im Rotliegenden Thüringens. *Ber. geol. Ges. DDR* **9**, 217–226, Berlin.

SCHIRMER, B. (1972) Beitrag zur Gliederung und Geochemie der permosilesischen Vulkanite des Thüringer Waldes. *Ber. geol. Ges. DDR* **17**, A, 881–896, Berlin.

SLAVIN, P. S. (1961) Tipy okislitelno-vosstanovitelnych obstanovok v melovych i teticnych otlozenijach Turkmenii. *Izv. Akad. Nauk SSR, ser. geol.* **7**, 88–100. Moskau.

TISCHENDORF, G. und UNGETHÜM, H. (1965) Zur Anwendung von Eh-Ph-Beziehungen in der geologischen Praxis. *Z. angew. Geol.* **11**, 57–67, Berlin.

TUREKIAN, K. K. und WEDEPOHL, K. H. (1961) Distribution of the elements in some major units of the earth's crust. *Geol. Soc. Amer. Bull.* **72**, 175–192.

VINOGRADOV, A. P. (1962) Srednie soderzanija chimiceskich elementov v glavnych tipach izverzennych gornych porod zemnoj kory. *Geochimija* **7**, Moskva.

WEDEPOHL, K. H. (1968) Chemical fractionation in the sedimentary environment. In *Origin and Distribution of the Elements,* L. H. AHRENS, ed., Pergamon Press, New York, 999–1016.

Pyrite–Pyrrhotite Geothermometer. Distribution of Cobalt, Nickel and Tin

I. J. NEKRASOV and N. I. BESMEN

Institute of Experimental Mineralogy, Academy of Sciences of the U.S.S.R., Moscow

Abstract

Co, Ni and Sn partitioning between FeS and $Fe_{1-x}S$ has been studied and equations for temperature dependence of their partitioning coefficient have been obtained for 300–800°C and P = 1 kb. The temperatures at which the pyrite-pyrrhotite associations have been formed in tin and pyrite deposits of the U.S.S.R., C.S.S.R. and Sweden are estimated.

Experimental study on Co, Ni and Sn partitioning between the coexisting pyrite and pyrrhotite resulted in creating mineralogical thermometers from which sulphide formation temperatures in pyrite, polymetal and cassiterite–sulphide deposits can be found.

The partitioning of these elements was studied in the NH_4Cl (5–10%) solution. Runs were in titan autoclaves at 300–500°C and $P_{H_2O} = 1$ kb. Whether equilibrium had been attained was indicated by the constant value of the distribution coefficient throughout longer runs and by homogeneity of FeS_2 and $Fe_{1-x}S$ grains analysed with JXA-3 and M-46 "Cameca" X-ray microanalysers. Equilibrium in the systems was reached within 15 days at 500°C, with 35 days at 400°C and within 50 days at 300°C. Cobalt and nickel contents in one-mineral samples FeS_2 and $Fe_{1-x}S$ were given by spectrophotometric and Sn content by spectral analyses. Using the exchange equilibrium:

$$FeS + MeS_2 = MeS + FeS_2$$

which describes the minor element concentrations in a pair of coexisting minerals, we can find a distribution coefficient (K_D^{Me}) for any element present as an impurity:

$$K_D^{Me} = \left(\frac{\text{Me at.\%}}{\text{Fe at.\%}}\right)_{\text{pyrrhotite}} : \left(\frac{\text{Me at.\%}}{\text{Fe at.\%}}\right)_{\text{pyrite}}.$$

According to the estimates, both pressure (P_{H_2O}) and isomorphous impurities, which amount to 1 wt.% in FeS_2 and $Fe_{1-x}S$, have little effect on K_D^{Me} and they can be neglected.

The distribution isotherms show linear relationship over the examined temperature range at 1.5 wt.% Ni and Co in FeS_2 and $Fe_{1-x}S$ (Figs. 1 and 2).

The Sn distribution isotherms depart from ideality at Sn content in FeS_2 and $Fe_{1-x}S$ as low as the tenths and hundredths of % (Fig. 3). Table 1 gives the experimental distribution coefficients for Ni, Co and Sn in FeS_2 and $Fe_{1-x}S$.

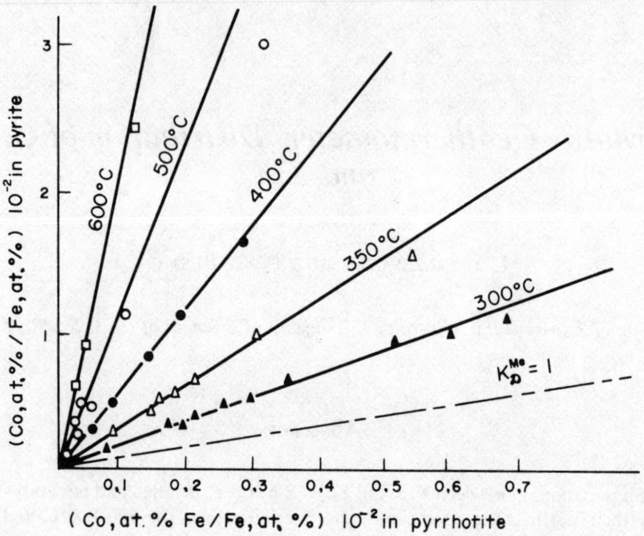

FIG. 1. Experimental isotherms for cobalt fractionation between pyrite and pyrrhotite.

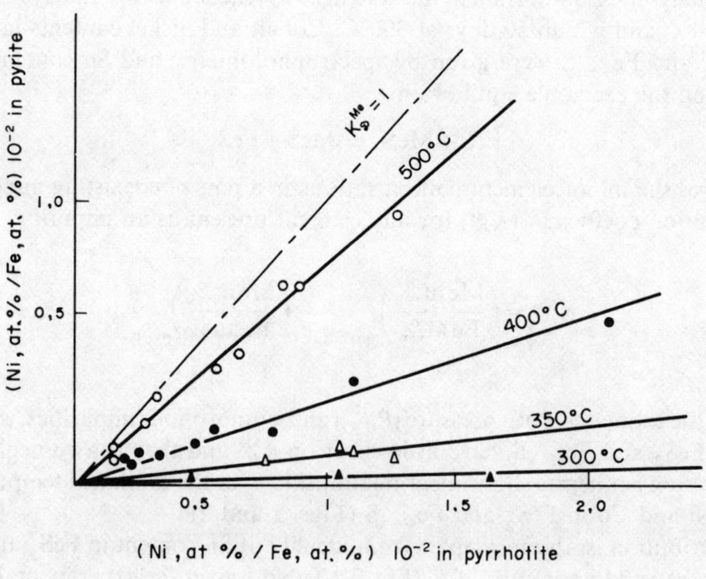

FIG. 2. Experimental isotherms for nickel fractionation between pyrite and pyrrhotite.

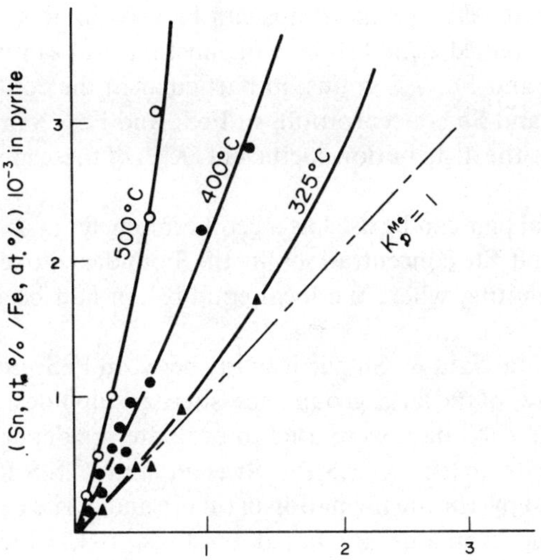

FIG. 3. Experimental isotherms for tin fractionation between pyrite and pyrrhotite.

TABLE 1. THE COEFFICIENTS OF Ni, Co AND Sn DISTRIBUTION BETWEEN COEXISTING PYRITE AND PYRRHOTITE. EXPERIMENTAL VALUES

Temperature, t°C	Distribution coefficient K_D^{Co}	K_D^{Ni}	K_D^{Sn}
300	0.489	17.734	—
325	—	—	0.863
350	0.290	6.836	—
400	0.155	3.098	0.506
500	0.074	1.149	0.254
600	0.035	—	—

K_D^{Me} temperature dependence is given by the following equations:

$$\text{the } K_D^{Ni}\ t°\text{C} = \frac{1000}{1.315 + 0.356 \lg K_D^{Ni}} - 273,$$

$$\text{the } K_D^{Co}\ t°\text{C} = \frac{1000}{1.907 - 0.538 \lg K_D^{Co}} - 273,$$

$$\text{the } K_D^{Sn}\ t°\text{C} = \frac{1000}{(0.083x^2 + 0.217x + 1.671) + (0.521x^2 - 0.106x + 0.727) \lg K_D^{Sn}} - 273$$

were x is the Sn content in FeS_2 (wt.%). Experimental data show that at 300–500°C the distribution of the three metals between the coexisting pyrrhotite and pyrite is temperature dependent. With increasing temperature, Co, Ni and Sn undergo redistribution from monosulphides into disulphides.

The experimentally established relationships can be used in geothermometry, provided FeS_2 and $Fe_{1-x}S$ have attained equilibrium. For most natural samples we observed local equilibria between FeS_2 and $Fe_{1-x}S$ grains, in particular in the contact zones. Therefore, despite differing Co, Ni and Sn concentrations in FeS_2 and $Fe_{1-x}S$ in the samples from the same deposit or ore body, the distribution coefficient (K_D^{Me}) of these metals changes regularly along one isotherm.

That is why this mineral pair can be used as a geothermometer in spite of the zoning in the minerals. The Co, Ni and Sn concentrations in the boundary zones, i.e. on the contact between pyrite and pyrrhotite, where the local equilibrium had been attained become of primary importance.

We used our experimental data on Sn partitioning between FeS_2 and $Fe_{1-x}S$ to evaluate the formation temperatures of the large group of cassiterite–sulphide and polymetal deposits in the U.S.S.R., the Co and Ni data were used to evaluate ore deposition temperatures in pyrite and skarn deposits in the U.S.S.R., Sweden and C.S.S.R. Table 2 gives the temperatures of pyrite and pyrrhotite formation in tin ore and pyrite deposits in the U.S.S.R. and other countries. The estimates are based on Figs. 1–3, which bear experimental isotherms for Co, Ni and Sn partitioning between FeS_2 and $Fe_{1-x}S$. Table 2 shows that the pyrite–pyrrhotite association had been formed in endogenic deposits at temperatures from 510 to 230°C, depending on the genetic type of a deposit.

TABLE 2. DISTRIBUTION OF Co, Ni AND Sn BETWEEN FeS_2 AND $Fe_{1-x}S$ IN PYRITE AND TIN DEPOSITS

Locality	Content Me wt.%				
	$Fe_{1-x}S$	FeS_2	K_D^{Me}	t°C	Literature
	1. Sn in cassiterite–quartz deposits				
Polyrnoe (U.S.S.R)	0.058	0.180	0.396	510	Nekrasov (1977)
Odinokoe (U.S.S.R.)	0.007	0.018	0.289	505	Nekrasov (1977)
Choxchur (U.S.S.R.)	0.066	0.108	0.307	470	Nekrasov (1977)
Kere-Uryh (U.S.S.R.)	0.047	0.019	0.324	465	Nekrasov (1977)
	2. Sn in cassiterite–silicate–sulphide deposits				
Ilin-Tass (U.S.S.R.)	0.012	0.019	0.470	400	Nekrasov (1977)
Deputatsky (U.S.S.R.)	0.110	0.138	0.594	360	Nekrasov (1977)
Kirgillah (U.S.S.R.)	0.087	0.105	0.617	355	Nekrasov (1977)
	3. Sn in cassiterite–sulphide deposits				
Ulachan-Egelax (U.S.S.R.)	0.051	0.036	1.056	300	Nekrasov (1977)
Hoton-Hay (U.S.S.R.)	0.048	0.029	1.235	290	Nekrasov (1977)
Kenzuha (U.S.S.R.)	0.047	0.032	1.096	275	Nekrasov (1977)
	4. Co in pyrite deposits				
Edak-Mein (Sweden)	0.031	0.100	0.220	370	Gavelin *et al.* (1947)
Burleden (Sweden)	0.020	0.070	0.223	370	Gavelin *et al.* (1947)
Abkhasia (U.S.S.R.)	0.014	0.006	1.69	220	Tvalchrelidze (1974)
Abkhasia (U.S.S.R)	0.016	0.011	1.860	230	Tvalchrelidze (1974)
Slovovize (C.S.S.R.)	0.034	0.054	0.470	310	Hegeman (1943)
	5. Ni in pyrite deposits				
Rouliden (Sweden)	0.020	0.010	1.500	440	Galevin (1947)
Rouliden (Sweden)	0.010	0.003	2.490	410	Galevin (1947)
Linskold-Mein (Sweden)	0.010	0.007	1.060	485	Galevin (1947)
Gelpa (C.S.S.R.)	0.103	0.0118	6.56	350	Cambel (1967)
Gelpa (C.S.S.R.)	0.090	0.0066	10.14	330	Cambel (1967)

The temperature of pyrite and pyrrhotite formation is estimated to have been 500 to 350°C in cassiterite–quartz and cassiterite–sulphide–silicate deposits. The pyrite ores in contact skarn deposits in Sweden have been formed at 440 to 370°C and at 350 to 220°C in stratiformic volcanic–sedimentary deposits in the C.S.S.R. and in Upland Abkhasia (U.S.S.R.). Other methods including homogenizing and decrepitation gave similar formation temperatures of pyrite–pyrrhotite association in these deposits. We believe that the experimental relationships of Co, Ni and Sn fractionation between FeS_2 and $Fe_{1-x}S$ could be useful in temperature estimations in a large group of deposits. This geothermometer has certain advantage over other geothermometers, because the composition of pyrrhotite, coexisting with pyrite, does not interfere with temperature determination.

References

CAMBEL, B. and JARKOSKY, J. (1967) *Geochemic der Pyrite einiger Lagerstatten der Tschechoslowakei*, Monographie Vydav Slov. akad. vied. Bratislavia.

GAVELIN, S. and GABRIELSON, O. (1947) Spectrochemical investigation of sulfide minerals from ores of the Skellett district. *Sver. Geol. Undersokn.* Ser. C, 491, Asbrok 41, 42, Stockholm.

HEGEMAN, F. (1943) Die geochemische Bedeutung Von Kobalt und Nikel in Pyrit. *Zeit. angew. Mineral.* **4**, 2–3.

NEKRASOV, I. J. (1977) *The Phase Relations in Tin-bearing Systems*, "Nauka", M.

TVALCHRELIDZE, A. G., EREMIN, N. I. and BESMEN, N. I. (1974) On the conditions of pyrrhotite mineralization in Abkhasia, as determined by laser microanalysis. *Geokhimiya*, No. 4.

SECTION IX

GEOCHEMISTRY AND HEALTH

Geochemistry and Physics of Plutonium Migration

A. T. JAKUBICK

Gesellschaft für Kernforschung, Abteilung Behandlung Radioaktiver Abfälle, Projekt Nukleare Sicherheit, Postfach 3640, D–75 Karlsruhe, Federal Republic of Germany

Abstract

Similarities in the distribution pattern of the worldwide Pu-239 fallout and in the filtered industrial release of the same nuclide were demonstrated. This analogy led to the assumption that predictions concerning the subsequent fate of the minute quantities of the HEPA-filtered plutonium released by industrial operations may be drawn by studying the environmental transfer of the worldwide fallout plutonium.

The history of Pu-239, 240 concentration in the ground-level air and its deposition known from previous investigations for the vicinity of Heidelberg was brought up to date by new values. After subtracting the wet deposition from the total amount of deposition found in the soil a dry deposition velocity for PuO_2-particles of 0.05 cm/s was obtained under the meteorological conditions of the area studied. Differences were found in the residence time of the worldwide fallout in the troposphera and in the filtered airborne PuO_2 discharge in the ground-level air. Nevertheless, the analogous distribution patterns of the PuO_2 particles and the natural aerosols suggest that in the case of dry deposition velocity, the results obtained can be applied for PuO_2 stack discharge too.

The decrease of PuO_2 contamination in the biologically active top soil was estimated by computer simulation. The Pu-239, 240 fallout deposition was used as the input function and the measured Pu-239, 240 depth distribution as the result of the combination of irregular input and continous dissipation. The downward migration found out of a 2-cm-thick layer of Rhine valley loess loam was proceeding at a rate of approximately 50% contamination per 3.5 years. By comparing this migration with the movement of an HTO-tracer a water-transport response as small as 10^{-3} was estimated.

Unlike the stable PuO_2, hydrolytic changes can be expected for $Pu(NO_3)_4$ upon contact with an aqueous environment after release. The equilibrium concentration of all valence states of plutonium was calculated for the whole range of natural waters. The sum of all potentially dissolved plutonium oxidation states may reach up to 1 mol/l under extreme but naturally occurring outdoor conditions.

1. Analogous Patterns of Worldwide Fallout and Filtered Airborne PuO_2 Discharge

In industrial handling of plutonium the formation of small amounts of aerosols cannot be entirely avoided.

Most of the aerosols produced are composed of PuO_2, but some of the plutonium released may come from evaporation of $Pu(NO_3)_4$. In the latter case $Pu(NO_3)_4 \times 5H_2O$ is the probable chemical form encountered up to 40°C. Above this temperature a change to PuO_2 starts; this chemical alteration is completed at 250°C.

By various operations aerosols of different AMAD† particle sizes are produced. In a reprocessing plant the sizes mainly range from 0.1 to 1.0μm. In fuel element fabrication the particle-size distribution peaks between 1 to 5 μm. In research and development facilities the usual size seems to be 1 to 2 μm (Elder *et al.*, 1974). In routine work plutonium is exclusively

† AMAD = activity median aerodynamic diameter.

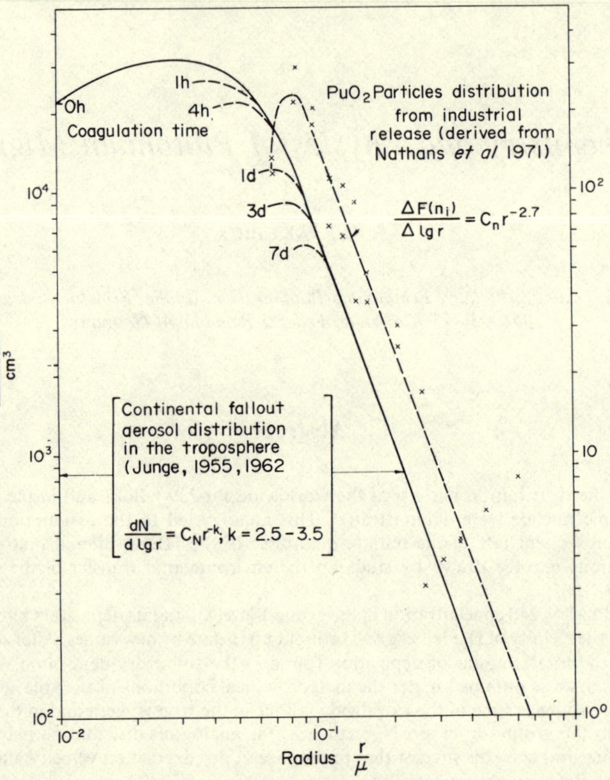

FIG. 1. Comparison of continental fallout aerosol and PuO_2 release from a nuclear facility.

released through HEPA-filters† with removal gradient of 10^{-4} to 10^{-5}. A combination of several filters will result in a removal gradient of 10^{-7}. Experiments indicate that the highest penetration is due to the 0.3-μm-diameter particles.

For an environmental impact control of plutonium-handling facilities, considerations have to be made concerning the subsequent fate of the minute quantities of plutonium released.

One approach to the task of predicting how the filtered plutonium release from nuclear facilities will behave is to evaluate what happened to the worldwide plutonium fallout from atmospheric nuclear tests. In the course of these tests from 1950 to 1962 about 500 kCi Pu-239 were injected into the atmosphere (Hardy *et al.*, 1973). The size of the particles immediately after an explosion may have ranged from 0.01 to 1000 μm (Junge, 1962). On their way through the stratosphere and upper troposphere the fallout particles underwent a filtration which resulted in a size range of 0.02 to 0.3 μm in the lower troposphere (Junge, 1962).

The HEPA-filtered PuO_2 released by industrial plutonium handling (Nathans and Holland, 1971) and the worldwide fallout PuO_2 filtered by transport seem to have comparable properties after having reached the ground-level air (Fig. 1). The particle size

†HEPA-filters = high efficiency particulate air filter.

distribution between 0.1 to 1 μm of the released PuO_2 can be described in both cases by

$$\frac{dF(n_i)}{d\lg r} = Cr^{-k}, \tag{1}$$

$F(n_i)$ = integral distribution of n_i,
n_i = frequency class i,
r = aerosol radius,
C = constant of the distribution function,

with $k = 2.7$ for the industrial airborne discharge which is well in the range $k = 2.5$ to 3.5 observed for fallout aerosols by Junge (1962). The difference mainly consists in a slight shift to larger particles of the maximum of PuO_2 released by industrial operations.

2. Plutonium in the Atmosphere

Due to the worldwide bomb fallout of this nuclide, the theoretical Pu-239 concentrations in the ground-level air in Heidelberg given in Table 1 and Fig. 2 prevailed from 1954 to 1975. The values have already been published and discussed (Jakubick, 1975) for the period from 1954 to 1971. The supplementary data for 1972 to 1975 were derived from measurements performed in Munich (Hötzl *et al.*, 1970–1975).

It was also shown previously (Jakubick, 1975) that the following empirical relationship

TABLE 1. EVALUATED FALL-OUT DATA OF Pu-239, 240 (HEIDELBERG, SW GERMANY)

Year	Concentration in the surface air (fCi/m³)	Wet deposition rate (mCi/km²a)	Dry deposition rate (mCi/km²a)	Cumulative total deposition (mCi/km²)
1954	0.086	0.052	0.0014	0.054
1955	0.11	0.059	0.0017	0.12
1956	0.14	0.090	0.0022	0.21
1957	0.14	0.097	0.0022	0.31
1958	0.24	0.16	0.0038	0.47
1959	0.35	0.17	0.0055	0.64
1960	0.037	0.024	0.00059	0.67
1961	0.036	0.025	0.00057	0.69
1962	0.083	0.040	0.0013	0.73
1963	1.13	0.53	0.018	1.28
1964	0.75	0.36	0.012	1.65
1965	0.23	0.20	0.0036	1.85
1966	0.072	0.051	0.0011	1.90
1967	0.036	0.022	0.00057	1.93
1968	0.065	0.052	0.0010	1.98
1969	0.039	0.026	0.00062	2.01
1970	0.039	0.027	0.00062	2.03
1971	0.042	0.020	0.00066	2.06
1972	0.015	0.0088	0.00023	2.064
1973	0.013	0.0079	0.00021	2.07
1974	0.023	0.018	0.00040	2.09
1975	0.015	0.0080	0.00024	2.10

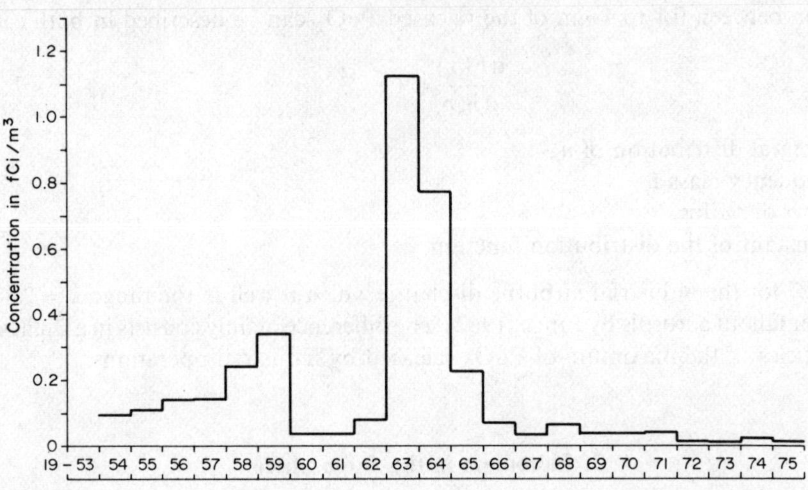

FIG. 2. Theoretical Pu-239 in ground layer air concentration in Heidelberg (SW Germany).

holds between the plutonium concentration in the air on a yearly basis and the tropospheric removal by wet fallout under average atmospheric conditions in Heidelberg:

$$\frac{R_w}{c(t)} = 8.0 \times 10^2 N(t) - 7 \times 10^3, \tag{2}$$

R_w = wet fallout deposition rate,
$c(t)$ = Pu-239, 240-concentration in the ground level air,
$N(t)$ = precipitation.

The resulting wet fallout deposition values (Table 1) are equivalent to a mean deposition velocity of about 0.18 cm/s.

From the sum of the wet plutonium fallout deposition values and plutonium concentration in the air, the dry fallout deposition velocity can be computed if the total cumulative deposition is known. Measurement of plutonium concentration in the soil from a site containing almost all of plutonium fallout (section 3) permitted evaluation of the dry deposition velocity of PuO_2 by

$$A_{\text{total}} = v(t)_d \int_{t_0}^{t} c(t)\mathrm{d}t + \int_{t_0}^{t} (8.0 \times 10^2 N(t) - 7 \times 10^3)\, c(t)\mathrm{d}t, \tag{3}$$

A_{total} = total amount of Pu-239 in the soil,
$v(t)_d$ = average dry deposition velocity in cm/s.

The value of $v(t)$ was calculated by:

$$\overline{v(t)_d} = (A_{\text{total}} - D_w) / \sum_i c(a) \tag{3a}$$

where D_w stands for the second term of equation (3).

The resulting value of $v_d = 0.043$ cm/s is valid for the ground level of the troposphere under average meteorological conditions in Heidelberg (mean wind velocity 59.12% 1 m/s; 35.03% 2 m/s; 5.21% 3 m/s; 0.5% 4 m/s).

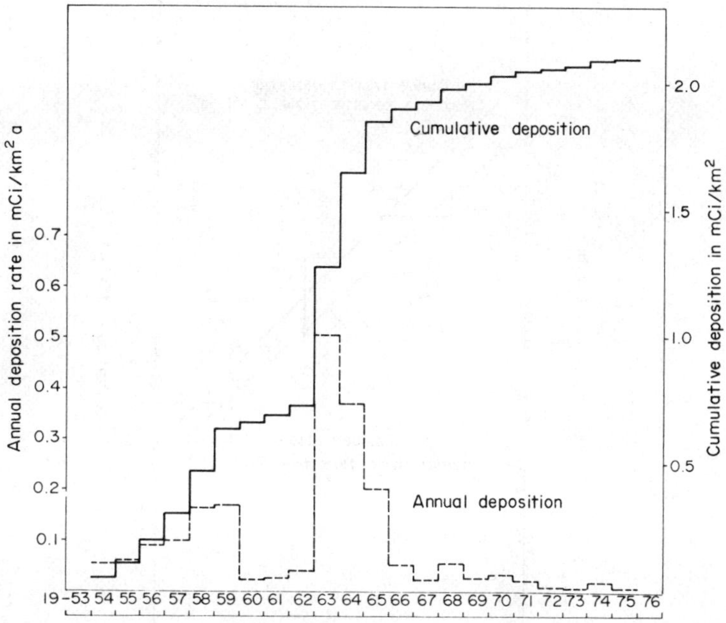

FIG. 3. Theoretical Pu-239 deposition in Heidelberg (SW Germany).

Taking into consideration that a small amount of fallout remains at the grass level, a slightly higher value of $\overline{v(t)_d} = 0.05\,cm/s$ was chosen for calculations of the total cumulative deposition, D_{total}:

$$D_{\text{total}} = 0.05 \sum_i c(a) + \sum_i 8.0 \times 10^2 N(t) - 7 \times 10^3 c(a), \tag{4}$$

$c(a)$ = annual average of Pu-239 concentration in the ground-level air.

The results are shown in Fig. 3.

During the period after the atmospheric nuclear test ban the ground level air plutonium concentration decreased with a time constant $t_c = 1.1$ and the plutonium deposition was decreased with $t_D = 1.2$ (Fig. 4). From these values a tropospheric residence time of about 36 days can be deduced for plutonium particles.

For routine industrial airborne discharge of PuO_2 into the ground-level air a residence time of no longer than 2 days can be expected. Apparently, even in spite of this difference in residence times, the relevant physical characteristics (particle size distribution, density) do not differ substantially. This could be attributed to the greater extent of aerosol coagulation in the ground-level air and to the tendency of aerosols to preserve their size distribution (Friedländer, 1965), which in the case of Pu-239 is certainly also supported by α-recoil.

Thus, considering the dry deposition velocity, the bomb fallout PuO_2 can be regarded as a simulator of the filtered airborne PuO_2 discharge.

3. Plutonium in the Soil

For soil sampling, a meadow some 9 km south-west of Heidelberg (Gauangelloch,

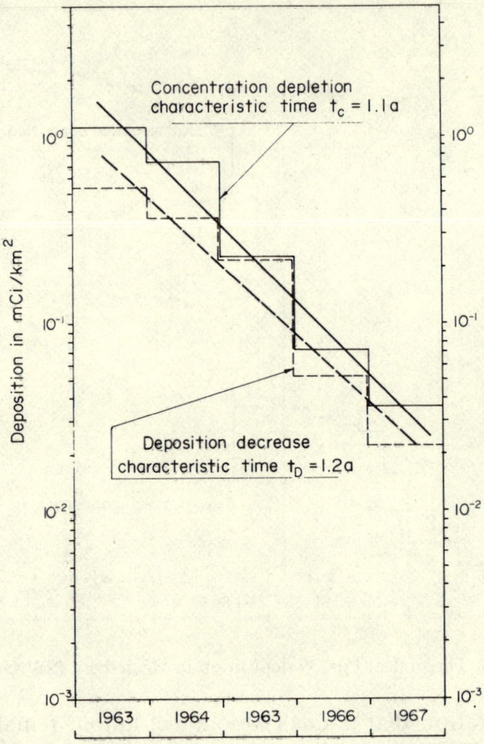

FIG. 4. Decrease of Pu-239 concentration in ground-layer air and Pu deposition, respectively, in Heidelberg after the atmospheric nuclear-test-ban treaty (1962).

property of Mr Treiber) appeared to be suitable. The site had no trees or bushes on it. Since the beginning of the history of bomb fallout, the site has not been disturbed by man.

From a surface area of 1500 cm^2, samples were taken from 2-cm-thick soil layers each consisting of not less than 3000 g. The major problem in sample collection and processing was to avoid a cross-contamination (the uppermost layers having a much higher Pu-239 concentration than the zone below) and to obtain samples which were representative of the depth under investigation.

The analysis of Pu-239, 240 was performed by the HASL†-method (Hasl, 1972) applied by LFE‡) (Major *et al.*, 1971). The Pu-239 distribution found is presented in Table 2.

As the present Pu-239 depth distribution pattern (Fig. 5) is a result of the complex effects of irregular input by fallout and continuous dissipation in depth, a computer simulation of these simultaneous processes was carried out according to a model for accumulation of radionuclides (Bunzl, 1972) which are fixed on the soil grain surfaces in a reversible manner:

$$M(t) = M_o \exp\left(-\frac{t}{\tau}\right) + \sum_i R_i \tau \left[1 - \exp\left(-\frac{t-t_i}{\tau}\right)\right] H(t-t_i)$$

$$-\left[1 - \exp\left(-\frac{t-t_{i+1}}{\tau}\right)\right] H(t-t_{i+1}), \quad (5)$$

† Health and Safety Laboratory, New York.

‡ LFE Environmental Analysis Laboratories, Richmond CA.

TABLE 2. AMOUNT OF Pu-239, 240 CONTAMINATION IN SOIL (GAUANGELLOCH, SW GERMANY)

Depth (cm)	Soil density (g/cm³)	Pu-239, 240 contamination (mCi/km²)	Standard mean error (%)
0– 2.75	0.81	0.32	±2
– 4.75	1.05	0.34	±2
– 6.87	1.03	0.45	±2
– 8.62	1.19	0.39	±2
–10.50	1.20	0.29	±2
–12.33	1.16	0.13	±3
–14.26	1.32	0.065	±4
–16.66	1.37	0.035	±3
–18.66	1.21	0.012	±9
–20.76	1.50	0.074	±11
–22.96	1.39	0.0051	±12
–25.96	1.37	0.0058	±16
–27.16	1.28	0.0016	±17
–28.42	1.13	0.0016	±11
–30.52	1.37	0.0027	±12

$M(t), M_o$ = amount of contamination in mCi/km² at time t and $t = 0$, *resp.*,
R_i = annual deposition rate in mCi/km² a or contamination transferred for the next layers,
i = calculation step of the time period considered,

$$H(t-t_i) = \begin{cases} 0; \text{ for } t \leqq t_i \\ 1; \text{ for } t > t_i \end{cases}$$

τ = residence time of PuO_2 particles in the soil layer.

The mean residence time τ was used as a transport parameter.

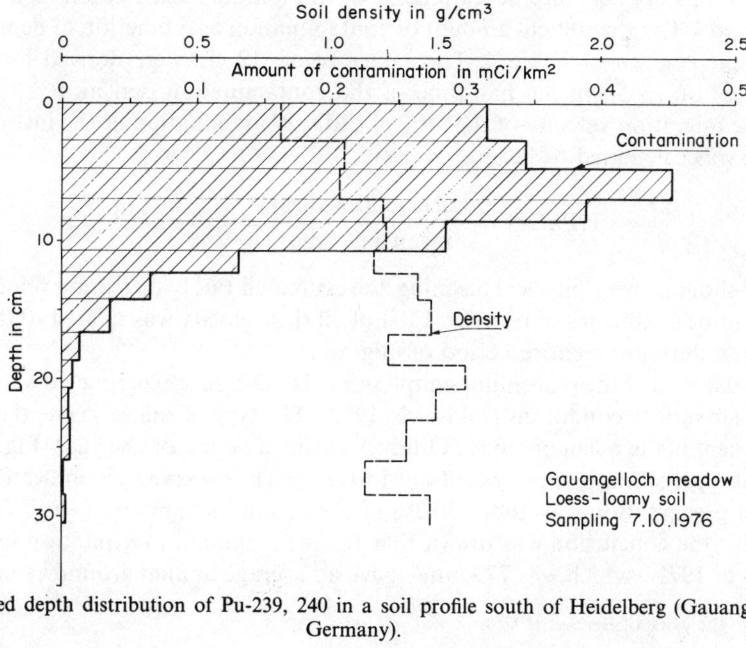

FIG. 5. Measured depth distribution of Pu-239, 240 in a soil profile south of Heidelberg (Gauangelloch, SW Germany).

The first choice of τ was a random one followed by checking on a trial-and-error basis and continued by a step-by-step approximation.

The results for the first soil layer were obtained by carrying out the repetitive procedure of equation (5) with the whole sequence of the total annual deposition input data R_i. For each subsequent deeper layer the plutonium discharge of the previous layer was taken as the input.

TABLE 3. MEAN RESIDENCE TIME OF Pu-239, 240 IN A LOESS–LOAMY SOIL

Depth† (cm)	Soil layer thickness (cm)	Porosity (%)	Mean residence time (a)
2.75	2.75	68	6.2
4.75	2.0	59	4.2
6.87	2.12	57	5.0
8.62	1.75	53	5.2
10.50	1.88	53	5.9
12.33	1.83	55	4.2
Average	2.055	57.5	5.12
S.M.D.	0.33	5.15	0.76

† Estimations below $\simeq$ 12 cm were not sufficiently certain.

The validity of τ (Table 3) was proved by comparing the calculated contamination values with the measured amount of contamination in the field (Table 2). Estimations of τ below about 12 cm were somewhat uncertain as the contamination discharge-to-background ratio was no longer clearly distinguishable from the contamination-to-background fluctuation. This rather high activity fluctuation is caused by the particular character of PuO_2 contamination.

Then the history of the development of plutonium distribution since 1954 was reconstructed.‡ The theoretical amount of contamination as a function of depth calculated for each year is given in Table 4. The values below 12 cm were derived for the sake of completeness only. Using the half-time of the contamination depletion $T_{1/2} = 0.693 \times \tau = 3.5$ a, the migration velocity of the 50% of PuO_2 contamination at the investigated loess loamy site was calculated to be

$$v(PuO_2) = \frac{\text{soil layer thickness}}{\text{mean half-time}} = 0.59 \text{ cm/a}.$$

If $v(PuO_2)$ should have a physical meaning, the estimated PuO_2-migration velocity has to be related to some conditions of transfer. First of all this velocity was related to the infiltrated precipitation during the entire period of migration.

Some 500 m west of the plutonium sampling site HTO-tracer experiments were carried out under similar surface conditions (Jakubick, 1972). The type of soil was essentially the same. The movement of the mean of the HTO-tracer during a period of one year (Fig. 6) showed a vertical water percolation velocity of about 130 cm/a. The observed groundwater recharge of the annual precipitation was about 30.2% of the annual precipitation (Fig. 7).

From this the conclusion was drawn that the mean annual precipitation for the period from 1954 to 1975—which was 779 mm—gave an average annual groundwater recharge of

‡ Similar to the work of Bennett (1975).

Table 4. Calculated Time History of Pu-239, 240 Distribution in a Loess–Loamy Soil (Gauangelloch, SW Germany)

Time:	1954	–55	–56	–57	–58	–59	–60	–61	–62	–63	–64	–65	–66	–67	–68	–69	–70	–71	–72	–73	–74	–75	–76
Depth in cm	Contamination in mCi/km²																						
0–2	0.038	0.087	0.152	0.220	0.325	0.432	0.421	0.382	0.359	0.701	0.976	1.053	0.975	0.856	0.770	0.685	0.608	0.537	0.467	0.405	0.359	0.315	0.274
2–4	0.011	0.020	0.037	0.056	0.092	0.128	0.137	0.160	0.184	0.299	0.342	0.375	0.409	0.445	0.474	0.478	0.471	0.445	0.431	0.403	0.376	0.345	0.314
4–6	0.003	0.006	0.012	0.021	0.037	0.054	0.067	0.090	0.111	0.163	0.188	0.229	0.273	0.316	0.357	0.390	0.419	0.441	0.454	0.461	0.461	0.454	0.442
6–8	0.001	0.002	0.005	0.008	0.013	0.020	0.027	0.039	0.050	0.073	0.087	0.114	0.140	0.170	0.201	0.233	0.266	0.297	0.325	0.350	0.372	0.389	0.402
8–10			0.001	0.002	0.004	0.007	0.010	0.015	0.021	0.031	0.038	0.053	0.067	0.085	0.105	0.127	0.152	0.178	0.205	0.232	0.259	0.285	0.310
10–12						0.001	0.001	0.003	0.005	0.008	0.011	0.017	0.021	0.029	0.038	0.047	0.059	0.071	0.085	0.100	0.116	0.132	0.148
12–14†									(0.001)	(0.002)	(0.003)	(0.005)	(0.007)	(0.011)	(0.014)	(0.019)	(0.025)	(0.031)	(0.038)	(0.047)	(0.056)	(0.065)	(0.076)
14–16†												(0.001)	(0.002)	(0.004)	(0.006)	(0.008)	(0.011)	(0.014)	(0.018)	(0.023)	(0.029)	(0.035)	(0.042)
16–18†														(0.001)	(0.001)	(0.002)	(0.003)	(0.005)	(0.006)	(0.008)	(0.010)	(0.012)	(0.015)
18–20†																(0.001)	(0.001)	(0.002)	(0.003)	(0.005)	(0.006)	(0.007)	(0.009)
20–22†																		(0.001)	(0.002)	(0.003)	(0.004)	(0.005)	(0.006)
22–24†																			(0.001)	(0.002)	(0.004)	(0.006)	(0.008)
24–26†																					(0.001)	(0.002)	(0.002)
26–28†																						(0.002)	(0.003)
28–30†																						(0.002)	(0.005)
Σ Contamination	0.053	0.115	0.207	0.307	0.471	0.642	0.663	0.689	0.731	1.277	1.645	1.847	1.894	1.917	1.966	1.990	2.015	2.032	2.035	2.039	2.035	2.056	2.056

† The values of contamination below 12 cm are not sufficiently certain.

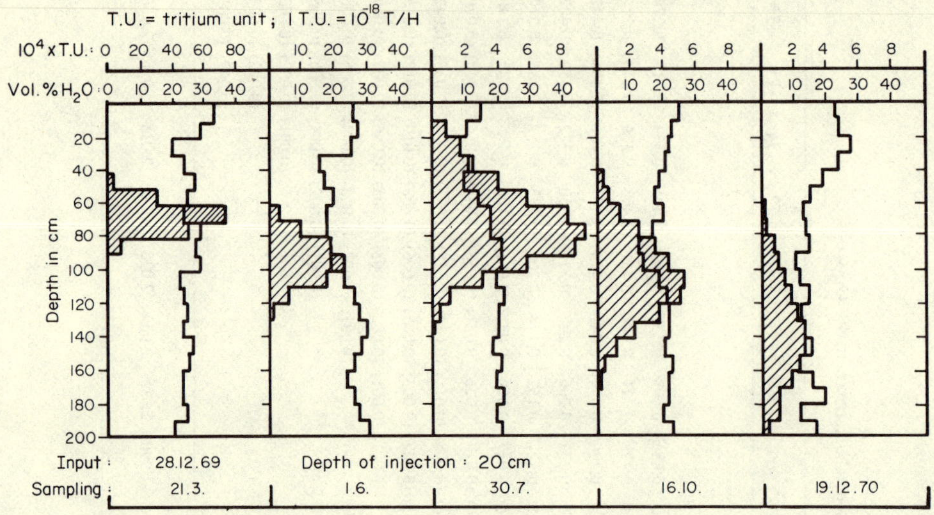

FIG. 6. Movement of a HTO-tracer in a loess-loamy soil (Gauangelloch, near Heidelberg, SW Germany, precipitation during the period: 1000 mm).

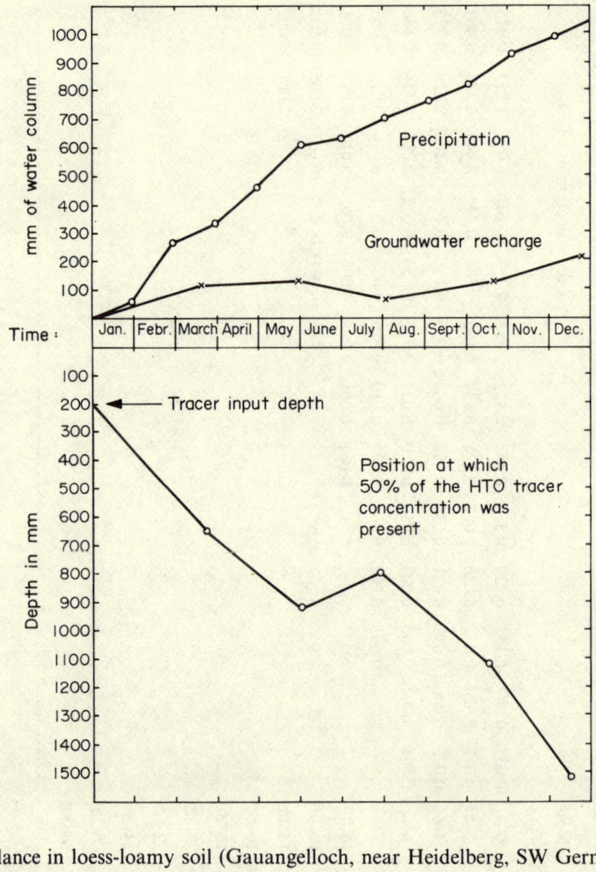

FIG. 7. Soil water balance in loess-loamy soil (Gauangelloch, near Heidelberg, SW Germany, during 1970).

235 mm. With a mean moisture content of 0.235 an average water velocity of about 100 cm/a was calculated.

The resulting dimensions less relative PuO_2 migration velocity $v_r(PuO_2)$ was:

$$v_r(PuO_2) = \frac{v(PuO_2)}{v(HTO)} = 5.9 \times 10^{-3}.$$

The small ratio indicates that the response of PuO_2 migration to water percolation is not high; this means that the effect of water transport must be considered in conjunction with other factors controlling the transfer.

Particle transfer in the soil can be effected by (1) bio-redistribution, (2) oscillation of the soil temperature (diurnal and annual) water content (wetting, drying), soil volume (microscale shrinking/swelling) and, of course, by (3) water movement.

The first type of process has a random mixing effect. The rate of this slow turnover of soil depends on the number, size and species of soil organisms, but on the average a turnover time of 10 years can be taken for the 2-cm-thick soil layers considered.

The oscillation of physical properties is probably responsible for the breaking of the weak bonds between the soil and PuO_2 particles.

Soil water percolation presumably gives the momentum and the water stream lines the direction of their vectors to the free PuO_2-particles. As the percolation in a partially saturated soil is, as a whole, a one-dimensional vertical flow, the resulting net PuO_2 movement is a downward displacement.

An evaluation of the mobility of $Pu(NO_3)_4$ microconcentrations in the field can be achieved according to the migration theory of Inoue and Kaufman (1963) when applied to the partially saturated zone (Inoue and Akaki, 1971):

$$v_r(Pu(NO_3)_4) = \left(1 + \rho K_d \frac{1-p}{ps}\right)^{-1}, \tag{6}$$

v_r = relative migration velocity,
p = porosity,
s = saturation,
ρ = soil density in g/cm^3,
K_d = distribution factor.

Accurate estimations of the distribution factor K_d for $Pu(NO_3)_4$ exist as a result of the experiments made by Prout (1959). For the pH value of 6.5† measured a nearly 100% absorption and $K_d = 6.4 \times 10^3$ for the Pu(IV) form, which is the most probable one (Section 4). Using the average physical properties of the sampling site soil, $p = 0.57$ (Table 3), $s = 0.35$ (Fig. 6), $\rho = 1.23$ (Table 2; Fig. 5), we obtain for the relative migration velocity

$$v_r(Pu(NO_3)_4) = 5.9 \times 10^{-5}.$$

A superficial examination already shows that the migration of solid PuO_2 particles by a filtrational type of transport is about 100 times faster than transportation by a reversible linear adsorption with permanent equilibrium type of migrational mechanisms.

† Average value for soils in Central Europe.

TABLE 5. VALUES OF DISSOLVED Pu UNDER Eh AND pH CONDITIONS OF NATURAL WATERS

$\frac{Eh}{V}$ \ ^{c}Pu $M/1$	pH: 1.0	2.0	3.0	4.0	5.0	6.0	7.0	8.0	9.0	10.0	11.0	12.0
1.1												
1	2.16×10^{-2}	2.15×10^{-2}	2.15×10^{-2}									
0.9	1.56×10^{-3}	2.51×10^{-5}	2.49×10^{-5}	2.49×10^{-5}	2.49×10^{-5}							
0.8	7.16×10^{-2}	7.52×10^{-6}	3.37×10^{-7}	3.36×10^{-7}	3.36×10^{-7}	3.36×10^{-7}						
0.7		3.51×10^{-4}	4.19×10^{-8}	6.79×10^{-9}	6.79×10^{-9}	6.79×10^{-9}	6.79×10^{-9}	6.79×10^{-9}				
0.6		1.72×10^{-2}	1.72×10^{-6}	3.10×10^{-10}	1.38×10^{-10}	1.38×10^{-10}	1.38×10^{-10}	1.38×10^{-10}	1.38×10^{-10}	1.38×10^{-10}		
0.5		8.43×10^{-1}	8.43×10^{-5}	8.43×10^{-9}	3.66×10^{-12}	2.82×10^{-12}	2.82×10^{-12}	2.82×10^{-12}	2.82×10^{-12}	2.82×10^{-12}	2.82×10^{-12}	
0.4			4.13×10^{-3}	4.13×10^{-7}	4.13×10^{-11}	6.18×10^{-14}	5.76×10^{-14}	5.76×10^{-14}	5.76×10^{-14}	5.76×10^{-14}	5.76×10^{-14}	5.76×10^{-14}
0.3			2.02×10^{-1}	2.02×10^{-5}	2.02×10^{-9}	2.04×10^{-13}	1.20×10^{-15}	1.18×10^{-15}	1.17×10^{-15}	1.18×10^{-15}	1.17×10^{-15}	1.17×10^{-15}
0.2				9.92×10^{-4}	9.92×10^{-8}	9.92×10^{-12}	1.02×10^{-15}	2.41×10^{-17}	2.40×10^{-17}	2.40×10^{-17}	2.40×10^{-17}	2.40×10^{-17}
0.1				4.86×10^{-2}	4.86×10^{-6}	4.86×10^{-10}	4.86×10^{-14}	5.35×10^{-18}	4.90×10^{-19}	4.89×10^{-19}	4.89×10^{-19}	4.89×10^{-19}
0					2.38×10^{-4}	2.38×10^{-8}	2.38×10^{-16}	2.38×10^{-16}	3.38×10^{-20}	9.99×10^{-21}	9.98×10^{-21}	9.98×10^{-21}
−0.1					1.17×10^{-2}	1.17×10^{-6}	1.17×10^{-14}	1.17×10^{-14}	1.17×10^{-18}	3.20×10^{-22}	2.04×10^{-22}	2.04×10^{-22}
−0.2					5.73×10^{-1}	5.73×10^{-5}	5.73×10^{-9}	5.73×10^{-13}	5.73×10^{-17}	5.73×10^{-21}	4.73×10^{-24}	4.15×10^{-24}
−0.3						2.81×10^{-3}	2.81×10^{-7}	2.81×10^{-11}	2.81×10^{-15}	2.81×10^{-19}	2.82×10^{-23}	8.76×10^{-26}
−0.4							1.38×10^{-5}	1.38×10^{-9}	1.38×10^{-13}	1.38×10^{-17}	1.38×10^{-21}	1.39×10^{-25}
−0.5									6.75×10^{-12}	6.75×10^{-16}	6.75×10^{-20}	6.74×10^{-24}
−0.6											3.31×10^{-18}	3.31×10^{-22}
−0.7												
−0.8												
−0.9												
−1.0												
−1.1												

4. Plutonium in Natural Waters

Unlike the high-fired PuO_2, $Pu(NO_3)_4$ belongs to the soluble compounds of plutonium. This is true to some extent even for the products of the $Pu(NO_3)_4$ reaction with natural waters if one bears in mind that sufficient time is available to establish the solubility equilibrium in most of the hydrological systems: the turnover time of water is generally from days to weeks in rivers, small lakes, reservoirs and estuaries; from several decades up to a hundred years in large lakes, seas and typical groundwaters; and thousands of years in oceans.

The value of $\sqrt{\Phi(Pu^{IV})} = 2.1$ ($\Phi(Pu^{IV})$) being the ion potential of the most common tetravalent state of plutonium), which is less than the limit usually given for the existence of dissolved cations in natural water $\sqrt{\Phi_{lim}} = 3.2$, indicates that diluted $Pu(NO_3)_4$ solution (concentration $\leq 10^{-9}$mol/1) will hydrate and hydrolysate in natural water wherever a pH-gradient occurs (Keller, 1971):

$$Pu^{4+} \underset{H^+}{\overset{OH^-}{\rightleftharpoons}} Pu(OH)^{3+} \rightleftharpoons Pu(OH)_2^{2+} \rightleftharpoons Pu(OH)_3^{+} \rightleftharpoons Pu(OH)_4$$

The same behaviour is observed for the tri- and hexavalent forms Pu^{3+} and $PuO_2{}^{2+}$ which hydrate and precipitate as $Pu(OH)_3$ and $PuO_2(OH)_2$.

Finally, $Pu(OH)_4$ ages to become more stable PuO_2 according to their free energies (Keller, 1971):

$$\Delta G_f{}^\circ\,(PuO_2) = -240.2; \qquad \Delta G_f{}^\circ\,(Pu(OH)_4 = -340.$$

However, evaluations of the hydroxide solubility are necessary for the long period until this chemical alteration takes place.

The question of mobilization by dissolution of different plutonium hydroxides can be answered by solving the following set of equations given by Polzer (1971):

$$\frac{[Pu^{3+}][PuO_2{}^{+}][H^{+}]^4}{[Pu^{4+}]} = 6.56 \times 10^{-4}, \tag{7}$$

$$\frac{[Pu^{3+}][PuO_2{}^{2+}]}{[Pu^{4+}][PuO_2{}^{+}]} = 1.28 \times 10, \tag{8}$$

$$\frac{[PuOH^{3+}][H^{+}]}{[Pu^{4+}]} = 3.0 \times 10^{-2}, \tag{9}$$

$$\frac{[H^{+}]}{[Pu^{4+}]} = 1.66. \tag{10}$$

TABLE 6. STANDARD POTENTIALS E° BETWEEN DIFFERENT OXIDATION STATES OF PLUTONIUM (after data from Polzer, 1971)

$\frac{E^\circ}{V}$	Pu (III)	Pu (IV)	Pu (V)	Pu (VI)
Pu (III)	—			
Pu (IV)	0.9827	—		
Pu (V)	1.0761	1.1694	—	
Pu (VI)	1.0228	1.0429	0.9164	—

The relation between equilibrium concentrations of two valence states being:

$$\frac{[\text{Pu}\,(n+1)]}{[\text{Pu}\,(n)]} = \exp\left(2.3026 \times \frac{\text{Eh} - E^\circ_{n+1,n}}{0.05916}\right). \tag{11}$$

The standard potentials $E^\circ_{n+1,n}$ between two oxidations states $n, n+1$ are well known (Table 6). Thus the equilibrium concentration of each valence state (including the pentavalent form) follow from equations (7) to (11).

The sum of the computed concentrations

$$m_{\text{Pu}} = \sum_{n=3}^{6} c_{\text{Pu}(n)} \tag{12}$$

m_{Pu} = amount of dissolved Pu,
c_{Pu} = concentration of the valence state n in the solution, are given for the whole redox and pH range of natural waters (Table 5).

Figure 8 shows more comprehensively that plutonium concentrations up to 1 mol/l may be dissolved under extreme but existing conditions (Baas Becking *et al.*, 1960).

Of course, natural waters are seldom free from complexing agents (CO_3^{-2}, PO_4^{-3}, humic-, fulvic acid and porphyrin being the most important natural ligands for plutonium). They react according to:

$$\text{Pu}\,(\text{OH}_2)_a \rightarrow \text{Pu}\,(\text{anion})\,(\text{OH}_2)_{a-1},$$

a = coordination number, which is 8–12,

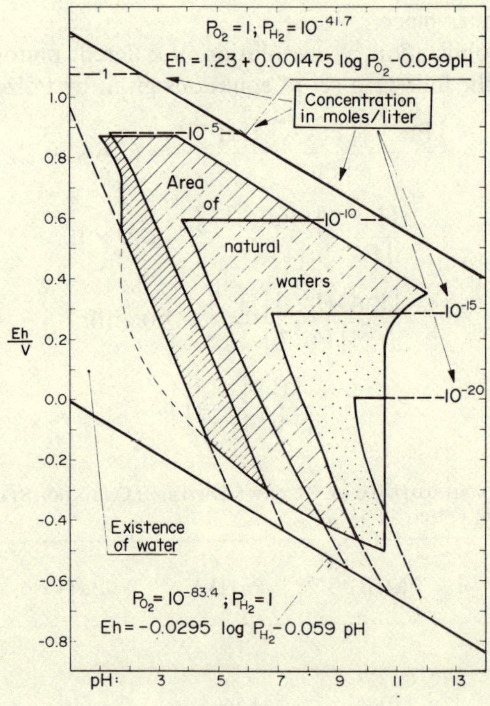

FIG. 8. Potentially soluble amounts of $Pu(OH)_4$ in natural waters.

and can distort considerably the potential solubility field given in Fig. 8. The solubility values of Table 5 should be considered as a field of reference concentration values which allow the dissolved plutonium amounts to be matched. By relating the measured values to them the plutonium undersaturation or chelation capacity of the system investigated can be judged.

5. Conclusions for Plutonium Health Physics

From some similarities between efficiently filtered PuO_2 stack discharge from nuclear facilities and worldwide bomb fallout shown here, we conclude that specific questions regarding the environmental impact of nuclear industry can be answered by investigating the behaviour of the bomb fallout plutonium.

Because the deposition of the plutonium activity released and the flux of natural aerosols seems to be proportional, the value of $v_d = 0.05$ cm/s, derived here for the dry fallout deposition, applies under comparable meteorological conditions for the routine PuO_2 release of industry as well.

Although the concordance is sufficiently good for routine release aspects, the analogy is incomplete for accidentally released PuO_2; the differences in particle size and shape cannot be ignored in this case.

After deposition the availability of PuO_2 in the biologically active top soil decreases due to migration into the depth. This takes place at a rate of 50% of the contamination per 3.5 years from a 2-cm-thick layer of Rhine valley loess-loam. This is about 10^{-3} of the water percolation velocity but there is no doubt that other mechanisms than percolation are involved too.

The potential solubility mobilization of the hydrolysation products of $Pu(NO_3)_4$ by natural water may reach under pH and corresponding redox conditions of rain water and of water in rivers and lakes values from 1.2×10^{-15} up to 4.1×10^{-3} mol/l. If the conditions are more reducing like for deep ground water which is not in contact with the atmosphere, the solubility will be even smaller: 1.2×10^{-15} to 1.2×10^{-6} mol/l.

Nevertheless, these values must not be interpreted as limits to the plutonium amounts dissolved in a given real water system, but much more as reference levels to which monitoring estimations should be related.

Acknowledgements

I wish to thank Mr H. Krause, Mr T. Dippel and Mrs H. Gottschalk for reading and correcting the manuscript. The work was sponsored by Project Nuclear Safety (PNS) and Project Reprocessing and Waste Treatment (PWA), respectively. But most of all I am grateful to Mrs I. Kahl, who gave me technical assistance throughout the entire time.

References

BAAS BECKING, L. G. M., KAPLAN, I. R. and MOORE, D. (1960) Limits of the natural environment in terms of pH and oxidation-reduction potentials, *J. Geol.* **68**, 243–284.

BENNETT, B. G. (1975) Transuranic element pathway to man, *Transuranium Nuclides in the Environment*, IAEA, Vienna.

BUNZL, K. (1972) Kinetic theory for the accumulation of a radioactive substance in a solid, applied at a variable rate to the surface. *Z. Phys. Chem. Neue Folge* **81**, 177–188.

ELDER, J. C., GONZALES, M. and ETTINGER, H. J. (1974) Plutonium aerosol size characteristics. *Health Phys.* **27**, 45–53.

FRIEDLÄNDER, S. K. (1965) Theory of self-preserving size distributions in a coagulating dispersion. *Radioactive Fallout from Nuclear Weapons Tests*, AEC, 253–259.

HARDY, KREY and VOLCHOK (1973) Global inventory and distribution of fallout plutonium. *Nature* 241.

HASL (1972) *Procedure Manual: Chemical Procedures E-Pu-02-01*, HASL-300.

HÖTZL, H., ROSNER, G. and WINKLER, R. (1970–1975) *Künstliche Radioaktivität der bodennahen Luft und des Niederschlags München-Neuherberg 1970–1975*, GSF-Bericht S. 413, Gesellschaft für Strahlen- und Umweltforschung mbH, München.

INOUE, Y. and AKAKI, F. (1971) On the performance of radionuclides in the aerated zone *Proceedings of the 26th Annual Meeting of Japan Society of Civil Engineers* II-228, 633–34.

INOUE, Y. and KAUFMANN, W. J. (1963) *Health Phys.* **9**, 705.

JAKUBICK, A. T. (1972) Water movement in partially saturated soils—tritium tracer method for estimation of the ground water recharge, Thesis, Heidelberg.

JAKUBICK, A. T. (1975) *Migration of Pu in Natural Soils*, IAEA, Vienna, 1975. Transuranium nuclides in the environment.

JUNGE, C. E. (1962) *Radioaktive Aerosole, Nuclear Radiation in Geophysics*, Springer Verlag, Berlin—Göttingen–Heidelberg.

KELLER, C. (1971) *The Chemistry of the Transuranium Elements*, Verlag Chemie, Weinheim.

MAJOR, W. J., KIM, D., LEE, R. A., WESSMANN, L. and LEVENTHAL (1971) Rapid dissolution of large soil samples for Pu-239 and Am-241 analysis, TLW-6102.

NATHANS, M. W. and HOLLAND, W. D. (1971) Analysis of Pu-239 particles collected near the Rocky Flats facility, TLW-6105.

POLZER, W. L. (1971) Solubility of plutonium in soil/water environments, CONF-710401, 411–429.

PROUT, W. (1959) Adsorption of fission products by Savannah River plant soil, Rep. TID-4500.

THEIS, W. (1969) Untersuchungen über das Vorkommen von Plutonium in atmosphärischen Proben, Thesis, Heidelberg.

Discussion

J. GONI:

Avez-vous travaillé sur plusieurs types de sols? Vos expériences ont-elles été faites seulement sur la zone non saturée?

A. T. JAKUBICK:

Yes, we analysed other soils for Pu too (a clayed and a sandy soil). But the distribution doesn't seem to be very different. The results presented do apply to partially saturated soil. The sampling was done above the ground water level.

Aspects of Geochemistry and Health in the United Kingdom

I. THORNTON and J. S. WEBB

Applied Geochemistry Research Group, Department of Geology, Imperial College of Science and Technology, London, England

Abstract

Health and production of crops and livestock in the U.K. are frequently influenced by the distribution of trace elements in soils and in the bedrock and transported overburden from which they are derived. Relationships between human disease and geochemical parameters are less well defined though several empirical studies have been made. The paper reviews recent progress and current research in the U.K. and in particular describes the compilation of the Geochemical Atlas of England and Wales at Imperial College together with research into the applications of regional geochemical maps to problems in agriculture, public health and medicine.

Problems affecting the health and production of agricultural crops and livestock due to trace-element deficiency or excess are frequently reported in particular areas or regions associated with specific geological formations or with transported material derived from them. Regional variations in the prevalence of human disease are also recognized though relationships with the distribution of trace elements in the environment are in many cases controversial and in general are less well founded than those with agricultural disorders. This is not surprising as, with the exception of developing and emergent countries, human populations no longer live "close to the ground". In the United Kingdom, with the exception of home-grown vegetables, sources of food-stuffs are widespread and frequently imported; water and milk supplies are rarely local.

This paper reports on the present state of knowledge concerning the relationships between geochemical parameters and the occcurrence of agricultural disorders and human disease in the U.K. The applications of geochemical reconnaissance surveys to problems of health and disease are also described.

Agriculture

Fifteen or more elements present in rock and soils in very small amounts are essential for plant and animal nutrition. It has been recognized for some time that copper, cobalt, fluorine, iodine, iron, manganese, molybdenum, selenium and zinc are required by animals; more recently nickel, silicon, tin and vanadium have been added to the list (Mertz, 1974). Boron, copper, iron, manganese, molybdenum, silicon, vanadium and zinc are required by plants. Excessive amounts of some of the trace elements may be toxic to plants and/or animals or may affect the quality of foodstuffs for human consumption; these include arsenic, boron, cadmium, copper, fluorine, lead, mercury, molybdenum, nickel, selenium and zinc.

The main source of trace elements in soils is the parent material from which they are derived, including weathered bedrock or overburden transported by wind, water or ice. Soils developed from these materials tend to reflect their chemical composition, though this relationship may be modified to varying degrees by the processes of soil formation leading to the mobilization and redistribution of trace elements within the soil profile and between neighbouring soil types. Secondary distribution is primarily related to soil drainage, redox potential, pH and organic matter and varies with the mobility of specific trace elements.

The optimum range of trace elements in soils is narrow, and serious deficiencies or excesses can result in crop failure or death of livestock. Less severe imbalance may lead to a reduction in crop and livestock production without obvious clinical symptoms in the plant or animal. These subclinical problems are thought to be more extensive than the clinical and may frequently remain undetected.

Deficiencies of one or more trace elements giving rise to crop and animal disorders are widespread in the U.K. and well documented. Copper deficiency in cereal crops leading to a marked reduction in grain yield is recognized on a variety of soils derived from peaty, sandy and calcareous parent materials (Pizer *et al.*, 1966; Caldwell, 1971; Mitchell, 1974; Reith,

TABLE 1. RANGE AND MEAN CONCENTRATIONS FOR TOTAL COPPER IN SOME AGRICULTURAL SOILS RELATED TO DIFFERENT GEOLOGICAL FORMATIONS IN ENGLAND AND WALES

Geological formation	Cu ppm		Geological formation	Cu ppm	
Recent			Lower Greensand		
Marine silt	9–50	(e)	Hythe Beds	3–13	(e)
(East Anglia)	22 (29)		(Sussex & Hants)	8 (28)	
			Sandgate Beds	4–14	(e)
Pleistocene			(Sussex & Hants)	9 (26)	
Breckland	2–20	†	Folkstone Beds	2–17	(e)
(chalk-sand drift)	8 (35)		(Sussex & Hants)	8 (28)	
(East Anglia)			Weald Clay		
			(Sussex & Hants)	10–25	(e)
Tertiary				16 (4)	
Bagshot Beds	4–37	(a)			
(Dorset)	13 (22)		*Jurassic*		
			Kimmeridge Clay	9–27	(e)
Cretaceous			(Berks & Oxon)	16 (21)	
Chalk	7–28	(e)	(South-west England)	11–40	(c)
(Wiltshire)	15 (128)			18 (30)	
(Sussex & Hants)	12–17	(e)	Corrallian Sandstone	5–19	(e)
	14 (3)		(Berks & Oxon)	11 (30)	
(Berks & Oxon)	10–17	(e)	Corrallian Limestone		
	13 (7)		(Berks & Oxon)	10–24	(e)
Upper Greensand				16 (15)	
(Sussex & Hants)	10–20	(e)	(South-west England)	11–26	(e)
	14 (8)			17 (11)	
(Berks & Oxon)	7–14	(e)	Oxford Clay		
	11 (3)		(Berks & Oxon)	10–26	(e)
(Wilts)	4–18	(e)		20 (12)	
	10 (28)		(S.W. England)	10–25	
Gault				18 (10)	
(Sussex & Hants)	10–17	(e)	Cornbrash	15–20	(c)
	13 (9)		(S.W. England)	17 (4)	
(Berks & Oxon)	14–25	(e)	Lower Lias	30–40	(d)
	19 (6)		(Gloucestershire)	35 (6)	

TABLE 1. *Continued*

Geological formation	Cu ppm		Geological formation	Cu ppm	
(Somerset)	6–60	(a)	(Monmouth)	21 (4)	
	32 (30)		Brownstonone	5–30	(d)
			(Monmouth)	18 (37)	
Permo-Triassic			Tintern Sandstone	5–20	(d)
Permian Sandstone	5–40	(d)	(Monmouth)	13 (24)	
(Cumberland)	15 (18)		Ditton Sandstone	20–50	(d)
Triassic Sandstone	4–16	(d)	(Hay-on-Wye)	32 (56)	
(Cumberland)	10 (17)				
Keuper & Bunter Sandstone	10–40	(d)	*Silurian*		
(L. Severn Valley)	20 (13)		(Denbigh Upland)	9–90	(b)
Keuper Marl	10–300	(d)		22 (99)	
(L. Severn Valley)	48 (80)		(Denbigh Moorland)	6–32	(b)
				15 (45)	
Carboniferous Shales			(Herefordshire)	20–50	(d)
(Staffordshire)	40 (68)	(f)		31 (15)	
Devonian			*Soils contaminated by mining and smelting in the Tamar Valley area of south-west England*		
Devonian Slates	14–42	(b)			
(S. Devon)	26 (44)				
Devonian Sandstones	2–31	(b)	Upland Soils	29–2000	(g)
(N. Devon)	14 (37)			314 (28)	
Old Red Sandstone	13–40	(d)	Alluvial Soils	35–2000	(g)
Downton Marls	27 (60)			620 (12)	
(Herefordshire)	16–60	(d)			
	30 (56)				
Raglan Marl	20–40	(d)			
(Monmouth)	28 (19)				
St. Maughans Gp.	10–40	(d)			

Source of data: (a) Thornton (1968)
(b) Keeley (1972)
(c) Thomson (1971)
(d) Wood (1975)
(e) Jordan (1975)
(f) Fletcher (1965)
(g) Colbourn *et al.* (1975)

† No. of samples in parenthesis

1975). Deficiencies have also been reported in fruit trees (Bould *et al.*, 1953) and seedling Sitka spruce (Benzian and Warren, 1956). Absolute copper deficiency in cattle is occassionally reported on sandy soils. Total copper levels in British soils range from 2 ppm Cu in Breckland sands in East Anglia to 2000 ppm Cu in soils contaminated by mining in south-west England (Table 1). Within individual sandstone formations, the copper content of the soils derived from the coarser-grained rock is lowest with concentrations increasing with decreasing grain-size (Wood, 1975). A total soil copper content of 2 ppm or less has proved a useful indication of soils giving rise to deficiency in cereals (Purves and Ragg, 1962), though frequently problems are found on soils with larger amounts of copper where the metal is unavailable to the plant. EDTA-extractable copper is widely used as a diagnostic aid for the identification of deficient soils in Britain (Pizer *et al.*, 1966; Mitchell, 1971).

Cobalt deficiency in sheep, known as "pine", results from a deficiency of vitamin B_{12} of which cobalt is a part. First reported on soils derived from the Dartmoor and Bodmin Moor granites in south-west England (Patterson, 1938), it is most commonly found on soils derived from sandstones, limestones, Silurian and Ordovician shales and acid igneous rocks in the west and north of England and in Wales (Archer, 1971) and on a variety of soils derived from acid igneous and arenaceous rocks in Scotland (Mitchell, 1974). The total cobalt content of soils ranges from 1 to 100 ppm Co and deficient herbage is usually found on soils containing 10 ppm Co or less; a more reliable estimate of availability has been obtained by extraction with acetic acid (Mitchell, 1964).

Agricultural problems have also been related to deficiencies of boron, iodine, manganese, molybdenum and zinc in part associated with geochemical parameters, and there is current interest in nutritional deficiency of selenium in sheep and cattle. It is thought possible that the sandy soils in England and Wales deficient in copper and cobalt will also be deficient in selenium as selenium deficiency in stock has already been associated with arenaceous soils in Scotland (Mitchell, 1974).

Potential toxic amounts of trace metals in soils may be derived either from metal-rich source rocks or from contamination from mining, smelting and other industrial or urban sources. Soils derived from ultra-basic rocks containing nickel-rich ferromagnesium minerals in parts of Scotland provide an example of a natural source of metal excess, leading under poor drainage conditions to nickel toxicity in cereal and other crops (Mitchell, 1974). Of particular importance to agriculture in Britain, excess molybdenum in soils and pastures gives rise to molybdenosis or molybdenum-induced copper deficiency in cattle. Molybdenum toxicity was originally described on calcareous soils containing 20 ppm Mo or more derived from inter-bedded shales and limestones of the Lower Lias formation in Somerset (Ferguson *et al.*, 1943; Lewis, 1943; Le Riche, 1959). However, it is now recognized that soils with 5 ppm Mo or more may support herbage containing in excess of 2 ppm Mo, which may in turn lead to scouring, loss of production and growth retardation in cattle due to reduced copper absorption and utilization. The mean molybdenum content of Scottish soils developed from different rock types ranges from less than 1 ppm (andesite, granite, granitic gneiss, shale, sandstone) to 5 ppm (quartz–mica schist) (Mitchell, 1971), while in England and Wales soils derived from marine black shales of Cambrian, Ordovician, Silurian, Jurassic, Carboniferous, Cretaceous and Recent age contain from 1 to 100 ppm Mo compared to the majority of soils developed from other parent materials which contain less than 2 ppm Mo (Thomson *et al.*, 1972; Thornton and Webb, 1976). Clinical bovine hypocuprosis has been recognized in a number of localities underlain by marine black shale (Fig. 1) and sub-clinical disorders are being increasingly reported each year.

Soils developed in mineralized areas frequently contain high concentrations of one or more of the heavy metals, copper, lead, zinc and cadmium, and arsenic. Metalliferous mining in Britain commenced in Roman times or earlier and reached a peak in the latter part of the nineteenth century. Workings were extensive and smelters frequently located within or near the mines. Present-day mining is far more restricted with five operational tin mines in Cornwall and lead mined as a by-product of the fluorspar industry in Derbyshire and the North Pennines. The present-day output is about 3000 tonnes of lead per year compared to a peak of 60,000 tonnes around 1875.

In the Tamar area of south-west England there is considerable contamination of both upland and alluvial agricultural soils with heavy metals and arsenic (Table 2), with peak values up to 2500 ppm As and 2000 ppm Cu reflecting mining and smelting activities in an area of predominantly copper–tin–arsenic mineralization (Colbourn *et al.*, 1975). The variable influence of mine-waste, dust and smelter fume together with the effects of

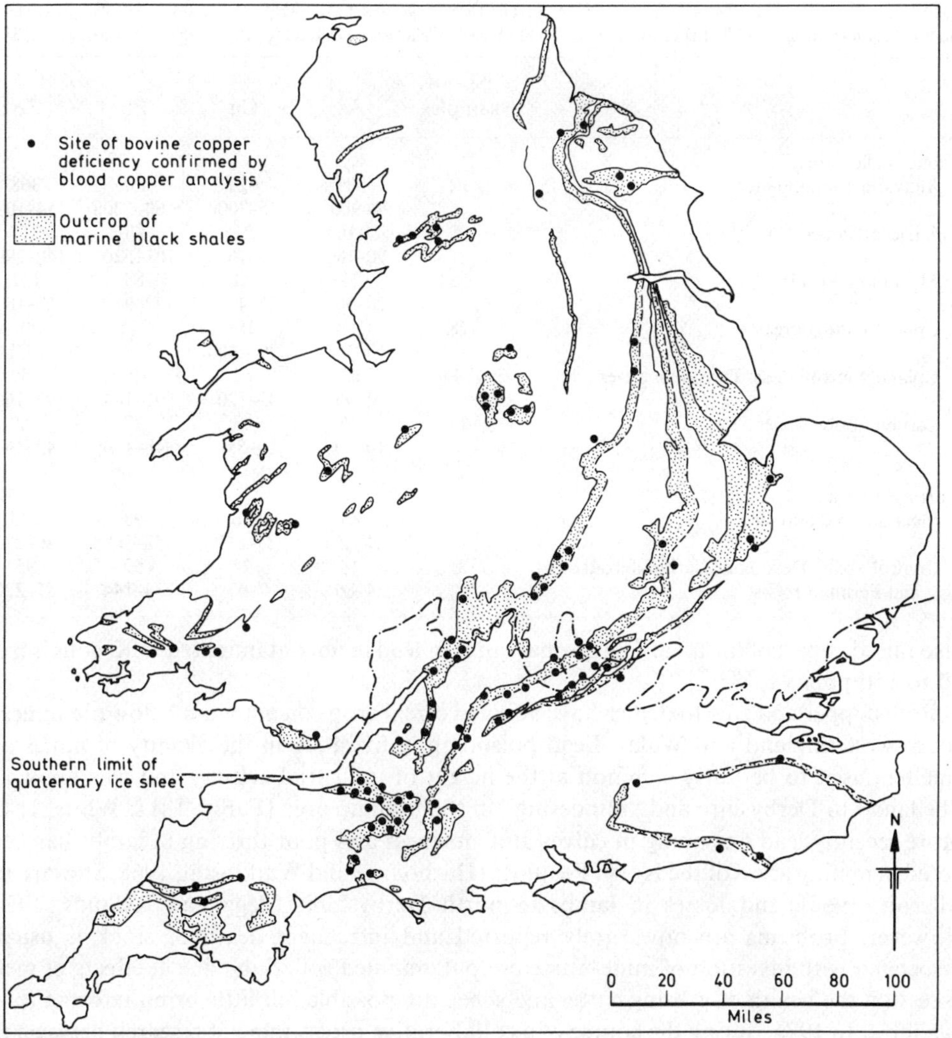

FIG. 1. Known distribution of bovine hypocuprosis in localities underlain by black shale in England and Wales. (Information from Agricultural Development and Advisory Service and veterinary surgeons; geology after Institute of Geological Sciences.)

underlying metal-rich material results frequently in large variations in the metal content of soils on a local scale. Similar extensive contamination has been shown over some 250 km^2 in the west of the Cornish peninsula (Thornton, 1977). In a neighbouring part of south-west England, raised levels of arsenic and copper in soils developed from metamorphosed rocks around the Dartmoor granite (Table 2) are thought to reflect hydrothermal activity at the time of the granitic intrusion (Aguilar-Ravello, 1974).

Soils in the mineralized area of Derbyshire of some 250 km^2 comprising the eastern part of the Carboniferous limestone and neighbouring beds contain from several hundred to several thousand ppm of lead with maximum values exceeding 3% Pb at sites close to old workings and smelters (Thornton and Webb, 1975; Colbourn 1976). Levels of zinc and cadmium are

TABLE 2. RANGES AND MEAN TOTAL CONCENTRATIONS OF ARSENIC AND SELECTED HEAVY METALS IN AGRICULTURAL SOILS (0–15 cm) FROM TAMAR VALLEY AND DARTMOOR AREAS (ppm) (from Colbourn *et al.*, 1975)

	No. of samples	As	Cu	Pb	Zn
Tamar Valley area					
Alluvium contaminated	12	309	620	435	368
		90–900	35–2000	60–1008	114–1020
Adjacent slopes	5	110	71	370	217
		90–180	35–76	101–520	140–290
Alluvium control	5	25	44	85	121
		20–30	36–40	72–96	67–180
Upland-mining areas	28	385	314	215	207
		60–2500	29–2000	62–850	58–380
Upland control areas: Devonian slates	8	27	62	121	139
		4–95	42–120	91–144	92–166
Carboniferous shales	13	26	43	77	92
		10–40	16–55	48–120	42–149
Dartmoor area					
Metamorphic aureole	17	88	76	98	133
		24–250	20–213	58–168	40–220
Control areas: Devonian, Carboniferous and Permian rocks	21	18	24	59	95
		4–80	7–47	34–144	51–220

also raised in part of the area. The normal range of lead in uncontaminated U.K. soils is from 10 to 150 ppm Pb.

Both copper and zinc toxicities have affected cereal crops on alluvium below old mines in south-west England and Wales. Lead poisoning in livestock in the vicinity of mines and smelters used to be fairly common at the height of industrial activity and was known as "belland" in Derbyshire and "mindering" in the Mendip area (Farey, 1811; White, 1842). More recently lead poisoning in calves and lameness and poor thriving in lambs has been noted in mining areas of the North Pennines (Harbourne and Watkinson, 1968; Stewart and Allcroft, 1956), and losses in lambs in north Derbyshire (Clegg and Rylands, 1966). However, problems are now rarely reported and infrequent death of stock is usually associated with ingestion of mine-waste or contaminated soil. Sub-clinical effects of metal excess on the health of grazing cattle and sheep are possible but little or no information is available. In 1976, during the course of a collaborative programme of research between the Veterinary Investigation Service and the Applied Geochemistry Research Group (AGRG), at Imperial College, samples taken from 180 cattle in Derbyshire showed blood lead to reflect clearly lead levels in the soil, with herd means around 10 μg Pb/100 ml blood on background farms with 100–200 ppm Pb in the soil, compared with 30 μg Pb/100 ml or more on high-lead farms with 1000–2000 ppm Pb in the soil. It is of interest that within the grazing season (May to October) there were no significant differences in pasture lead contents between farms with background or contaminated soils (Table 3) and it would seem that the major pathway of lead into the animal is in the form of ingested soil. Previous work on land heavily contaminated with mine-waste in south-west England has demonstrated that soil may range from 1 to 11% of the dry matter intake of cattle; in this instance it was concluded that cattle could ingest up to ten times the amount of arsenic, copper and lead directly from soil compared to that incorporated in the plant fraction of the dietary intake (Thornton, 1974). However, the availability of soil metal to the animal is not known, and will depend on its form and extractability and absorption in the alimentary trace.

TABLE 3. MEAN LEAD CONTENT OF PASTURE HERBAGE (ppm dry matter) ON THREE* GROUPS OF FARMS IN DERBYSHIRE

		Low soil Pb (100–200 ppm)	Medium soil Pb (600–800 ppm)	High soil Pb (1000–2000 ppm)
June	1976	9	7	16
August	1976	18	18	22
October	1976	26	36	24

* Each group consisted of four forms.

The metal content of pasture herbage usually reflects only to a small degree the extremely high levels in the contaminated soils of U.K. mining areas, and it would seem that metals, particularly lead, copper and arsenic, are either present in forms relatively unavailable for plant uptake or that regulatory processes within the plant limit uptake and/or translocation (Colbourn, 1976; Thoresby, 1977; Thornton, 1977).

Food-crops

The poor availability of metal contaminants to grass is mirrored to a degree by uptake in food-crops. In contaminated mining areas of south-west England and north Wales, the edible tissues of barley, strawberries and lettuce showed only small increases in copper, lead and arsenic compared to those sampled at control sites. In the Tamar area of south-west England the arsenic content of barley grain increased with soil content over the range 20 to 300 ppm As, though in no case did the grain contain more than 0.4 ppm As in the dry material (Fig. 2). Barley from twenty control sites in other parts of England and Wales ranged from less than 0.001 to 0.015 ppm As (D.M.) on soils containing 4–13 ppm As.

Winter vegetables grown in Derbyshire also reflected the metal content of the garden soils to only a small extent, even when the soil contained 1% or more lead (Table 4). Lead was found in larger amounts in outer compared to inner leaves of cabbages, perhaps due to soil contamination. The pH of these Derbyshire soils ranged from 7.0 to 7.9, and the high pH may account for the low uptake of cadmium even from soils containing 20 ppm or more. Moderate increases in the uptake of lead by salad and root crops from soils contaminated by mining in Wales and south-west England have been reported by other workers (Warren *et al.*, 1967; Alloway and Davies, 1971; Davies, 1973; Davies and Roberts, 1975).

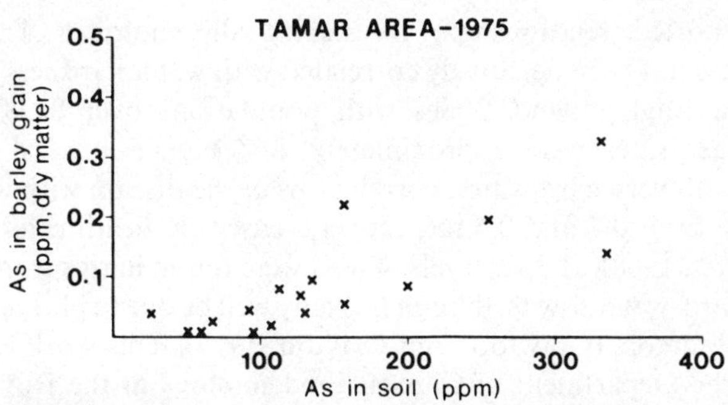

FIG. 2. Arsenic content of barley grain (ppm, dry matter) in relation to arsenic in the soil (ppm, 0–10 cm) in the Tamar area of south-west England.

TABLE 4. LEAD, ZINC AND CADMIUM CONTENT OF GARDEN SOILS* (0–15 cm) AND VEGETABLES (dry matter) IN DERBYSHIRE VILLAGES (from Thornton and Webb, 1975) (Range and arithmetic mean (ppm), no. of samples in parentheses)

Soil	Brussels sprouts	Cabbage, inner leaves	Cabbage, outer leaves
Lead			
209–806	5	5–8	7–14
360 (6)	5.0 (3)	6.4 (5)	10.6 (5)
1580–4075	5–6	4–7	7–17
2723 (8)	5.7 (4)	5.3 (7)	13.1 (7)
5038–15240	5.7	7–9	14–19
9012 (5)	6.0 (2)	8.0 (3)	15.7 (3)
Zinc			
185–733	51–78	31–60	21–45
365 (8)	61 (4)	45 (6)	32 (6)
1035–1970	53–62	41–66	33–73
1466 (7)	57 (3)	53 (7)	50 (7)
2110–4500	100,108	104,123	133–196
3568 (4)	104 (2)	113 (2)	164 (2)
Cadmium			
3–7	0.8–1.6	0.7–1.6	1.0–2.8
5 (8)	1.1 (4)	1.1 (6)	1.6 (6)
11–18	0.8–1.0	0.6–1.0	1.1–1.8
15 (6)	0.9 (3)	0.8 (5)	1.6 (5)
22–41	1.2, 1.7	0.8–1.2	1.1–3.8
30 (5)	1.5 (2)	1.0 (4)	2.1 (4)

* Soil pH (in water) ranged from 7.0 to 7.9

Human Health

Although there is strong evidence linking the distribution of trace elements in the environment and the prevalence of certain human diseases, actual causal relationships are few and probably limited to those between fluorine in water supplies and dental caries, iodine distribution and endemic goitre and, in an area of molybdenum mining in Russia, where molybdenum excess occurs in soils, foods and waters accompanied by 30% incidence of endemic gout.

In the U.K., reported relationships are nearly all empirical. The prevalence of cardiovascular disease has been negatively correlated with water hardness in sixty-one large county boroughs of England and Wales with populations over 80,000. In this study cardiovascular disease rates were approximately 40% higher in very soft-water towns compared to those with very hard water; correlations of the disease with water calcium and carbonate were high ($r = 0.7$ and 0.6 respectively, based on death rates at ages of 45–64 years) (Crawford *et al.*, 1968). Higher levels of lead were found in ribs from three soft-water compared to three hard-water towns, though this may well be due to plumbo-solvency rather than to geochemical factors (Crawford and Clayton, 1973). This work is being continued and broadened by the Department of Clinical Epidemiology at the Royal Free Hospital, London, which is currently examining causes of death from some seventy diseases including cardiovascular problems and cancers in over 200 towns with populations of 50,000 or more in England, Wales and Scotland. Information on the quality of water supplies is being obtained by the Water Research Centre, Medmenham, for comparison. Following this investigation, it is intended to select 10,000 people in a number of the towns for detailed studies on morbidity in relation to environmental factors.

An abnormal distribution of cancer in several villages in south-west England (Allen-Price, 1960) has led to a number of studies on trace metals in mineralized and neighbouring areas (Pinsent, 1968; Davies, 1973). However, large amounts of arsenic and heavy metals in stream sediments and soils in parts of the area examined did not provide clearcut geochemical differences between high cancer and control villages. The incidence of stomach cancer in North Wales has been positively correlated with the zinc:copper ratio in garden soils (Stocks and Davies, 1964), though the relationship is entirely empirical. The Royal College of General Practitioners is currently funding a reconnaissance survey of medical data from the Second National Morbidity Survey, together with statistics from sixty practices recording morbidity data, for comparison with environmental data from published sources, including regional geochemical data from the Geochemical Atlas of England and Wales, compiled by AGRG (B. E. Davies, personal communication).

Variations in the prevalence of dental caries have been recorded in comparisons between the high molybdenum "teart" area on the Lower Lias formation in south-west England (on which molybdenosis is recognized in grazing cattle) and nearby control areas, though differences in the composition of drinking water, 3 μg/l Mo in the former compared to 1.8 μg/l in the latter, were not considered sufficient to account for the caries incidence (Anderson, 1969). Examination of 12–year-old children in the mineralized Tamar Valley in south-west England showed a similar dental caries prevalence to other children of similar age in other parts of south-west England and Wales, with the exception of those living on the Bere Peninsula where children had a raised caries level (Anderson *et al.*, 1976). It has been suggested that this increase in caries prevalence is related to extensive soil contamination with lead. Comparative studies by the same workers in Cardiganshire, Wales, have confirmed high caries prevalence in areas of historical lead mining (B. E. Davies, personal communication). Investigations will shortly be undertaken, funded by the Medical Research Council, to study the possible effects of metal pollution on dental caries in 5- and 12-year-old children in the past mining areas of the Mendips and Halkyn (North Wales) compared to a control area in the Cotswold Hills (B. E. Davies, personal communication).

Regional geochemical maps, to be discussed in the following paragraphs, have been used to select areas in which to compare soil lead content with the lead burdens of 2–4-year-old children in villages contaminated by past mining and in control villages in Derbyshire. A

TABLE 5. LEAD CONTENT OF GARDEN SOIL, HOUSE DUST, BLOOD AND HAIR
(from Barltrop *et. al.*, 1975)
(Range and geometric mean, no. of samples in parenthesis)

		Blood (ug/100 ml)		Hair (ppm)
Soil (ppm)	House dust (ppm)	Child	Mother	Child
130–900	190–2450	15–33	9–40	2–36
420 (24)	531 (20)	21 (29)	14 (25)	8 (29)
1050–9100	390–25000	13–45	11–44	2–62
3390 (40)	1564 (36)	24 (43)	19 (41)	11 (43)
10000–28000	1000–5100	21–43	13–19	10–39
13969 (8)	2582 (8)	29 (10)	15 (8)	20 (10)

collaborative study between the Paediatric Unit, St. Mary's Hospital Medical School, and AGRG showed that when households were grouped according to the lead content of their garden soils, lead in children's blood and hair increased with that in soil and housedust (Table 5). However, none of the lead values in children were high enough to be considered hazardous even though the amounts in soil and dust were extremely high, peaking at 2.8 and 2.5% Pb respectively (Barltrop *et al.*, 1975). The childrens' lead status does, however, reflect that of the environment and, in the absence of evidence linking the human burden with lead in foodstuffs or water supplies, it has been suggested that the major pathway of lead into the children is through inhalation and involuntary ingestion of dust particles (Barltrop *et al.*, 1975).

Trace Element Maps

The present interest in relationships between human disease together with problems of public health and the distribution of trace elements in the environment, on the one hand, and the growing awareness of agricultural problems due to trace-element deficiency and excess, on the other hand, highlights the need for systematic information on the regional distribution of trace elements. Ideally trace-element maps for agriculture would be based on the systematic sampling and analysis of soil or vegetation. In Britain and Ireland this has so far proved impracticable because of the cost and time necessary to take the numerous close-spaced samples required to account for differences due to the complex geology and soil types. In parts of Scotland, information on trace elements in soils has been acquired by the Macaulay Institute during the course of soil survey and advisory work. In England, Wales and Northern Ireland, however, there are few systematic data on either total or available levels of trace elements in soils. However, the need for trace element maps has been met in part by geochemical reconnaissance surveys undertaken by AGRG. These primary reconnaissance surveys use stream sediment sampling, a method based on the premise that active stream sediment represents a composite sample of the erosion products of rock and soil upstream from the sample-site. Such surveys are now standard practice in mineral exploration, where together with other procedures, they are used to detect concealed mineralization and, on a broad scale, to delineate mineralized districts and geochemical provinces. Exploratory work in Africa, followed by trial surveys over 2800 square miles in England and Wales led to the recognition of broad-scale patterns of metal distribution reflecting both the composition of bedrock and soil and indicating specific geochemical facies within individual geological formations (Webb *et al.*, 1978; Nichol *et al.*, 1970 a, b, 1971).

The Geochemical Atlas of England and Wales

Following a trial survey of the 5000 mi.^{2} of Northern Ireland (Webb *et al.*, 1973), nearly 50,000 stream sediment samples were taken from tributary drainage at stream/road intersections over a 10-week period in the summer of 1969. Samples were not taken from urban areas or obviously contaminated drainage. The average sampling density was one sample per 2.5 km^2 from drainage with catchment areas of less than 25 km^2. At each site duplicate grab samples were taken of approximately 100–200 g of active mid-stream sediment, each sample comprising two subsamples taken approximately 20 m apart. Where possible unsorted sediment was taken, avoiding large pebbles and collapsed bank material. These samples were oven-dried at 80°C in the field, one of which was sieved to 200 μm and the other retained for reference purposes.

Analysis of the minus 200 μm fraction for Al, Ba, Ca, Co, Cr, Cu, Fe, Ga, K, Li, Mn, Ni, Pb, Sr, Sn and V was carried out using a 40-channel direct-reading emission spectrograph linked to an automatic card punch; Cd and Zn were determined by atomic absorption spectrometry and As and Mo by colorimetric procedures. Data have been presented by computer as grey-scale line-printer maps at scales ranging from 1 inch:4 miles to 1:2 million and again as lasergraphic plots printed on to diazo film. These maps have been on open-file at Imperial College since 1974 and will be published in 1978 (Webb *et al.*, in press).

It is stressed that from an agricultural viewpoint, these maps do no more than focus attention on patterns of low and high trace-element distribution wherein to concentrate more detailed, time-consuming and costly surveys based on soil, plant and animal tissue. Only these subsequent detailed surveys can provide information on a farm or field scale. Research at Imperial College over the past 12 years, aimed at the interpretation and use of these maps for agriculture, has confirmed relationships between regional geochemical data and the known distribution of agricultural disorders, and also has indicated further suspect areas requiring further investigation (Webb, 1964; Webb *et al.*, 1971; Thornton and Webb, 1970; Thomson *et al.*, 1971; Jordan *et al.*, 1975). The maps have delineated areas of low cobalt, copper, manganese and zinc and high arsenic, cadmium, copper, lead, zinc, nickel, chromium and molybdenum. From a pollution viewpoint, these geochemical surveys have been shown to provide a useful catalogue of base-line information related to the geochemistry of bedrock and soil parent materials; at the same time, the maps clearly show patterns reflecting metal contamination on a regional scale from past and present mining and smelting activities and from industrial and urban sources (Thornton, 1975; Thornton and Webb, 1975).

The map for cadmium (Fig. 3) serves as a useful illustration, showing anomalies due to three main sources of the metal: (a) areas associated with sulphide mineralization contaminated by base-metal mining activity, including parts of south-west and central England and Wales; (b) areas contaminated by present-day and recent industrial processes including those near Bristol and the Swansea Valley affected by recent smelting of lead and zinc; (c) areas underlain by cadmium-rich marine shales of Carboniferous age (containing up to 100 ppm Cd) in which stream sediments and soils derived from the rock are enriched in the metal. Soils in each of these types of anomaly have been shown to contain up to 20 ppm or more cadmium and present a potential hazard to agriculture.

Research into the application of regional geochemical maps to water-quality assessment has also proved encouraging and has demonstrated relationships between metals in drainage sediments and waters, particularly in areas affected by contamination from past mining and smelting (Aston *et al.*, 1974; Thornton and Webb, 1977). A knowledge of the chemical composition of the sediment can prove useful in the prediction of sites where, under certain conditions of rainfall and runoff, the metal load in associated waters may exceed medically recommended limits. Thus, the sediment may prove a more stable indication of metal excess than one-off sampling of the waters themselves. At the same time multi-element sediment analysis may help in identifying those metals that require monitoring in the course of more costly and time-consuming on-going programmes of water analysis. It is considered that the Geochemical Atlas of England and Wales can play a useful contributory role in assisting the

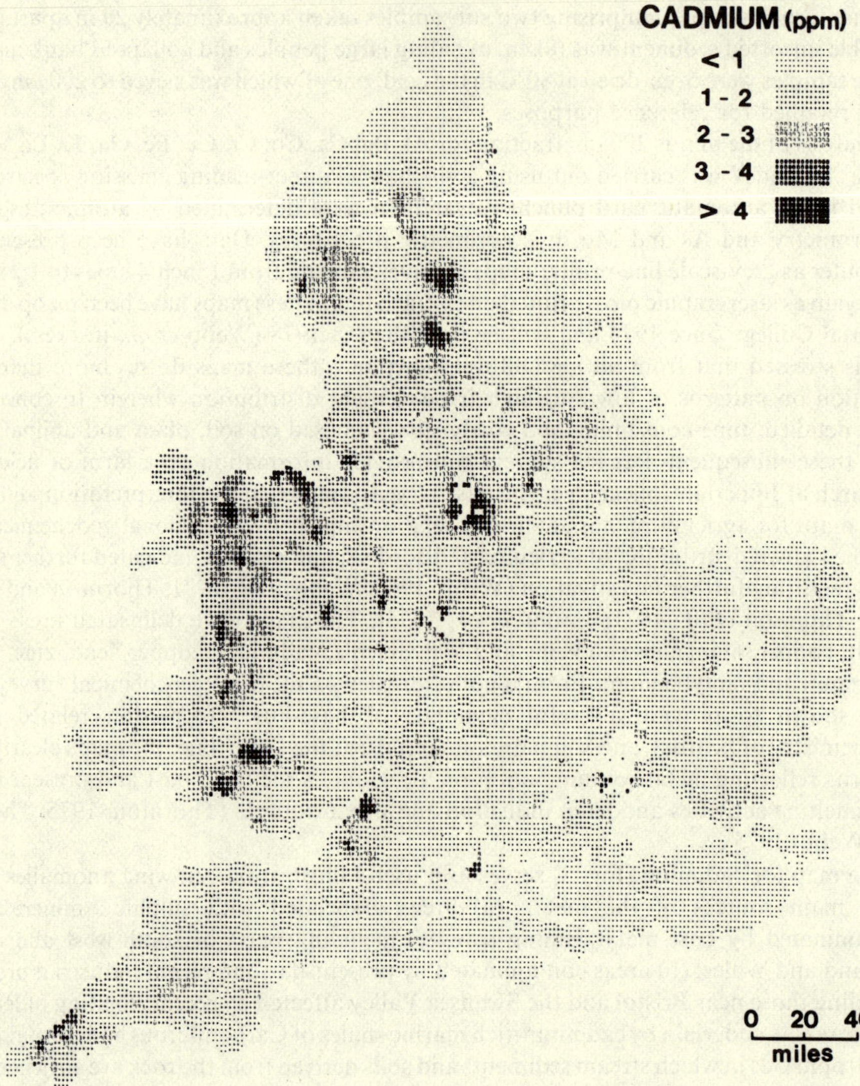

FIG. 3. Map showing the distribution of cadmium in stream sediment in England and Wales (compiled by AGRG as part of *The Geochemical Atlas of England and Wales*, Webb *et al.*, 1978, in press).

water industry in the selection of key sites for quality control. In addition, the reflection of water composition in the sediment has obvious value in the consideration of environmental factors possibly affecting human and animal health.

In addition to water supplies, geochemical reconnaissance maps have also been shown to have obvious applications in the selection of key areas for detailed surveys of soils and foodstuffs. Though investigations into possible relationships between geochemistry and human health are as yet in their infancy in the U.K., it is considered that the geochemical atlas provides a unique source of multi-element data for the siting of medical surveys and also as a data bank for epidemiological studies.

Conclusions

Relationships between geology, the distribution of trace elements in soils and their parent materials and deficiency and excess in crops and livestock are well documented in the U.K. However, sub-clinical problems affecting agricultural production due to marginal trace-element imbalance is less well defined and may be widespread and of considerable economic significance. Reported relationships between human disease prevalence and trace elements in the environment are mainly empirical and need more detailed study. The national requirement for trace-element maps as a background to health investigations is partly met by geochemical reconnaissance mapping in England, Wales and Northern Ireland. Similar surveys are being undertaken by the Institute of Geological Sciences in Scotland and the Irish Agricultural Institute in parts of Eire. It must be emphasized that one of the prime contributions of multi-element regional geochemical atlases, such as that for England and Wales, is that they show the relative distributions of a number of elements and, in this respect, provide a unique source of information bearing on relevant interactions that could well be of considerable importance in agricultural and epidemiological surveys and to problems of public health.

Acknowledgements

The Geochemical Atlas of England and Wales was financed by a generous grant from the Wolfson Foundation. A number of examples are cited from the continuing programme of the Applied Geochemistry Research Group in environmental geochemistry, financed by the Agricultural Research Council and the Natural Environment Research Council. The authors are grateful to their colleagues and postgraduate students who contributed to the research.

References

AGUILAR-RAVELLO, A. N. (1974) The regional distribution of arsenic in south-west England. Unpub. Ph.D. Thesis, Univ. of London.

ALLEN-PRICE, E. D. (1960) Uneven distribution of cancer in west Devon. *Lancet*, 4 June 1960, 1235–1238.

ALLOWAY, B. J. and DAVIES, B. E. (1971) Heavy metal content of plants growing on soils contaminated by lead mining. *J. Agric. Sci.* **76**, 321–323.

ANDERSON, R. J. (1969) The relationship between dental conditions and the trace element molybdenum. *Caries Res.* **3**, 75–87.

ANDERSON, R. J., DAVIES, B. E. and JAMES, P. M. C. (1976) Dental caries prevalence in a heavy metal contaminated area of the west of England. *Brit. Dent. J.* **141**, 311–314.

ARCHER, F. C. (1971) Factors affecting the trace element content of pastures. In *Trace Elements in Soils and Crops. Min. Agric. Fish. Fd. Tech. Bull.* **21**, 150–157, H.M.S.O., London.

ASTON, S. R., THORNTON, I., WEBB, J. S., PURVES, J. B. and MILFORD, B. L. (1974). Stream sediment composition: an aid to water quality assessment. *Water, Air and Soil Pollution* **3**, 321–325.

BARLTROP, D., STREHLOW, C. D., THORNTON, I and WEBB, J. S. (1975) Absorption of lead from dust and soil. *Postgraduate Med. J.* **51**, 801–804.

BENZIAN, B. and WARREN, R. G. (1956) Copper deficiency in Sitka spruce seedlings. *Nature, Lond.* **178**, 864–865.

BOULD, C., NICHOLAS, D. J. D., TOLHURST, J. A. H. and POTTER, J. M. S. (1953). Copper deficiency of fruit trees in Great Britain. *J. Hort. Sci.* **28**, 268–277.

CALDWELL, T. H. (1971) Copper deficiency in crops. In *Trace Elements in Soils and Crops, Min. Ag. Fish. Fd. Tech. Bull.* **21**, 62–87, H.M.S.O., London.

CLEGG, F. G. and RYLANDS, J. M. (1966) Osteoporosis and hydronephrosis of young lambs following the ingestion of lead. *J. Comp. Path.* **76**, 15–22.

COLBOURN, P., ALLOWAY, B. J. and THORNTON I. (1975). Arsenic and heavy metals in soils associated with regional geochemical anomalies in south-west England. *The Science of the Total Environment* **4**, 359–363.

COLBOURN, P. (1976) The application of geochemical reconnaissance data to trace metal pollution in agriculture. Unpub. Ph.D. Thesis, Univ. of London.

CRAWFORD, M. D. and CLAYTON, D. G. (1973) Lead in bones and drinking water in towns with hard and soft water. *Brit. Med. J.* **2**, 21.

CRAWFORD, M. D., GARDNER, M. J. and MORRIS, I. N. (1968) Mortality and hardness of water supplies. *Lancet* **1**, 827–831.

DAVIES, B. E. (1973). Occurrence and distribution of lead and other metals in two areas of unusual disease incidence in Britain. In *Proceedings of the International Symposium on Environmental Health Aspects of Lead*, 125–134. Luxemburg: Commission of the European Communities.

DAVIES, B. E. and ROBERTS, L. J. (1975) Heavy metals in soils and radish in a mineralised limestone area of Wales, Great Britain. *Sci. Total Env.* **4**, 249–261.

FAREY, J. (1811) *General view of the Agriculture and Minerals of Derbyshire*, Vol. III. London: MacMillan.

FERGUSON, W. S., LEWIS, A. H. and WATSON, S. J. (1943). The teart pastures of Somerset. I. The cause of teartness. *J. Agric. Sci.* **33**, 44–51.

FLETCHER, W. K. (1968) Geochemical reconnaissance in relation to copper deficiency in livestock in the Southern Pennines. Unpub. Ph.D. Thesis, Univ. of London.

HARBOURNE, J. F. and WATKINSON, J. (1968) An unusual outbreak of lead poisoning in calves. *Vet. Rec.* **83**, 515–517.

JORDAN, W. J. (1975) The application of regional geochemical reconnaissance to arable cropping in England and Wales. Unpub. Ph.D. Thesis, Univ. of London.

JORDAN, W. J., ALLOWAY, B. J. and THORNTON, I. (1975) The application of regional geochemical reconnaissance data in areas of arable cropping. *J. Sci. Fd. Agric.* **26**, 1413–1424.

KEELEY, H. C. M. (1972) Cobalt, copper and manganese in relation to geochemical reconnaissance and agriculture. Unpub. Ph.D. Thesis, Univ. of London.

LE RICHE, H. H. (1959) Molybdenum in the Lower Lias of England in relation to the incidence of teart. *J. Soil. Sci.* **10**, 133.

LEWIS, A. H. (1943). The teart pastures of Somerset II. The relation between soil and teartness. *J. Agric. Sci.* **33**, 52–57.

MERTZ, W. (1974) The newer trace elements, chromium, tin, vanadium, nickel, and silicon. *Proc. Nutr. Soc.* **33**, 307–313.

MITCHELL, R. L. (1964) Trace elements in soils. In *Chemistry of the Soil*, 2nd ed., F. E. BEAR, ed., Reinhold Pub. Co., New York. 320–368.

MITCHELL, R. L. (1971) Trace elements in soils. In *Trace Elements in Soils and Crops. Min. Agric. Fish. Fd. Tech. Bull.* **21**, 8–20, H.M.S.O., London.

MITCHELL, R. L. (1974) Trace element problems in Scottish soils. *Neth. J. Agric. Sci.* **22**, 295–304.

NICHOL, I., THORNTON, I., WEBB, J. S., FLETCHER, W. K., HORSNAIL, R. F. and KHALEELEE, J. (1971) Regional geochemical reconnaissance of part of Devon and North Cornwall. Rep. No. 71/2, Inst. Geol. Sci.

NICHOL, I., THORNTON, I., WEBB, J. S., FLETCHER, W. K., HORSNAIL, R. F., KHALEELEE, J. and TAYLOR, D. (1970a) Regional geochemical reconnaissance of the Derbyshire area. Rep. No. 70/2, Inst. Geol. Sci.

NICHOL, I., THORNTON, I., WEBB, J. S., FLETCHER, W. K., HORSNAIL, R. F., KHALEELEE, J. and TAYLOR, D. (1970b) Regional geochemical reconnaissance of the Denbighshire area. Rep. No. 70/8, Inst. Geol. Sci.

PATTERSON, J. B. E. (1938) Some observations on a disease of sheep on Dartmoor. *Emp. J. Expt. Agric.* **6**, 262.

PINSENT, R. J. F. H. (1968) Some problems of environment and disease. *Report of the Welsh soils Discussion Group* No. 9, 120–129.

PIZER, N. H., CALDWELL, T. H., BURGESS, G. R. and JONES, J. L. O. (1966). Investigations into copper deficiency in crops in East Anglia. *J. Agric. Sci. Camb.* **66**, 303–314.

PURVES, D. and RAGG, J. M. (1962) Copper deficient soils in south-east Scotland. *J. Soil Sci.* **13**, 241–246.

REITH, J. W. S. (1975). Symptons of copper deficiency in plants with effects on crop yield. *Proc. Copper in Farming Symposium*, Copper Development Ass. London.

STEWART, W. L. and ALLCROFT, R. (1956) Lameness and poor thriving in lambs on farms in old lead mining areas in the Pennines. 1. Field investigations. *Vet. Rec.* **68**, 723–728.

STOCKS, P. and DAVIES, R. I. (1964) Zinc and copper contents of soils associated with the incidence of cancer of the stomach and other organs. *Brit. J. Cancer* **18**, 14–24.

THOMSON, I. (1971) Regional geochemical studies of black shale facies with particular reference to trace element disorders in animals. Unpub. Ph.D. Thesis, Univ. of London.

THOMSON, I., THORNTON, I. and WEBB, J. S. (1972) Molybdenum in black shales and the incidence of bovine hypocuprosis. *J. Sci. Fd. Agric.* **23**, 871–891.

THORESBY, P. (1977) The distribution of Cu, Pb, Zn, Cd and As in soils and crops in mineralised and non-mineralised areas. Ph.D. Thesis, Univ. of London. In preparation.

THORNTON, I. (1968) The application of regional geochemical reconnaissance to agricultural problems. Unpub. Ph.D. Thesis, Univ. of London.

THORNTON, I. (1974) Biogeochemical and soil ingestion studies in relation to the trace element nutrition of livestock. In *Trace Element Matabolism in Animals*, Vol. 2, W. G. HOEKSTRA, J. W. SUTTIE, H. H. GANTHER and W. MERTZ, eds., Baltimore: University Park Press, 451–454.

THORNTON, I. (1975). Applied geochemistry in relation to mining and the environment. In *Minerals and the Environment*, M. J. JONES, ed., London: Institution of Mining and Metallurgy, 87–102.

THORNTON, I. (1977) Geochemical aspects of heavy metal pollution and agriculture in England and Wales. *Proceedings of Conference on Inorganic Pollution and Agriculture*, H.M.S.O., London, in press.

THORNTON, I. and WEBB, J. S. (1970) Geochemical reconnaissance and the detection of trace element disorders in animals. In *Trace Element Metabolism in Animals, Proc. WAAP/IBP International Symposium*, C. F. MILLS, ed., London: Livingstone.

THORNTON, I. and WEBB, J. S. (1975) Trace elements in soils and surface waters contaminated by past metalliferous mining in parts of England. In *Trace Substances in Environmental Health*, Vol. IX, D. D. HEMPHILL, ed., Columbia, Missouri: Univ. of Missouri, 77–88.

THORNTON, I. and WEBB, J. S. (1976) Distribution and origin of copper deficient and molybdeniferous soils in the United Kingdom. *Proc. Copper in Farming Symposium*, Copper Development Ass., London.

THORNTON, I. and WEBB, J. S. (1977) Application in the water industry of regional geochemical maps of England and Wales. *J. Inst. Wat. Engrs. Scient.* **31**, 11–25.

WARREN, H., DELEVAULT, R. E. and CROSS, C. H. (1967) Possible correlations between geology and some disease patterns. *Ann. N.Y. Acad. Sci.* **136**, 657–710.

WEBB, J. S. (1964) Geochemistry and life. *New Sci.* **23**, 504–507.

WEBB, J. S., LOWENSTEIN, P. L., HOWARTH, R. J., NICHOL, I. and FOSTER, R. (1973) *Provisional Geochemical Atlas of Northern Ireland*, Applied Geochemistry Research Group, Technical Communication No. 60.

WEBB, J. S., THORNTON, I., HOWARTH, R. J., THOMPSON, M. and LOWENSTEIN, P. L. (1978) *The Geochemical Atlas of England and Wales*, Oxford Univ. Press, in press.

WEBB, J. S., THORNTON, I. and NICHOL, I. (1971) The agricultural significance of regional geochemical reconnaissance in the United Kingdom. In *Trace Elements in Soils and Crops, Min. Agric. Fish. Fd. Tech. Bull.* **21**, H.M.S.O., London.

WHITE, J. (1842) *A Compendium of Cattle Medicine*, 6th ed., London: Longman, Brown, Green & Longman.

WOOD, P. (1975) Regional geochemical studies in relation to agriculture in areas underlain by sandstones. Unpub. Ph.D. Thesis, Univ. of London.

Photoelectron Spectroscopy Analysis of Asbestos Dissolution in Acidic Media of Biological Interest

J. GONI,† J. H. THOMASSIN,‡ M. C. JAURAND,§ and J. C. TOURAY,‡

† Service geologique national (B.R.G.M.), B.P. 6009, 45018 Orléans Cédex, France.

‡ Institut de recherches "Ressources et matériaux minéraux", Université d'Orléans, 45045 Orléans Cédex, France.

§ Départment de recherches sur les affections respiratoires et l'environnement (Dr. J. Bignon), Laboratoire de biopathologie pulmonaire, C.H.U. Henri Mondor, avenue du Maréchal de Lattre de Tassigny, 94010 Créteil, France.

Abstract

Many authors have studied asbestos dissolution, in relation with physical and chemical properties and/or biological effects. Most often, the experiments have been carried out under conditions poorly related to health problems. A possible conclusion is a correlation between the reactivity of leached mineral surfaces and the cancerigenic power.

In earlier works (Thomassin *et al.* 1977) we have studied the dissolution kinetics of chrysotile using XPS analysis of the solids, after leaching in non-stirred media by strong reagents (0.1 N oxalic acid and 6 N HCl). It appears that the two possible rate-limiting steps are:

— diffusion of Mg^{++} through an external silica gel,
— chemical reaction.

Complementary experiments carried out with 0.01 N HCl favor a rate mechanism involving a chemical reaction, followed by diffusion through a silica layer and a chemical reaction after some collapse or corrosion of the ordered silica gel.

Using XPS, we have studied three aspects of asbestos dissolution related to biological problems:

1. comparative leaching study of various asbestos minerals (chrysotile, crocidolite, amosite and anthophyllite) by 0.1 N chemical reagents from the Krebs cycle (pyruvic, citric and oxalo-acetic acids);
2. study of the dissolution mechanism of chrysotile at very low concentrations of oxalic acid (10^{-4} and 10^{-6} N);
3. study of the effects of grinding:

The most soluble asbestos is chrysotile, the best of the investigated reagents being pyruvic acid, which is not so active as oxalic acid at the same concentration. At low concentrations, the dissolution of chrysotile in oxalic acid is limited to the first brucite layer; grinding of crocidolite favors significantly its dissolution.

These few experiments enable us to suggest that in the human body, except in the stomach, asbestos dissolution is very low, favored by grinding and leaves a surface layer having ordered silanols groups, the high reactivity of which has been demonstrated in the case of chrysotile (Pacco *et al.*, 1976).

Introduction

Minerals of the asbestos group are presently well known for their fibrogenic carcinogenic and cocarcinogenic effect in humans, as well as in animals (for review see Bignon, 1974). Toxicity has been demonstrated in isolated cells by the release of specific factors (Allison, 1971; Davies *et al.*, 1974) and/or by cellular lysis related to both the experimental conditions and type of fiber used; their effects seem to be dependent on the characteristics of the fibers, i.e. nature, length, diameter, substances chemically or physically adsorbed, trace elements, etc.

The toxicity of fibers has been assessed for various cell types and cell lines: red blood cells (Schnitzer *et al.*, 1970; HARINGTON *et al.*, 1971), macrophages (Parazzi *et al.*, 1968; BECK *et al.*, 1971; Miller *et al.*, 1972), fibroblasts (Hext *et al.*, 1976), transformed cells (Wade *et al.*, 1976) and embryonic cells (Lavappa *et al.*, 1975). Recent data have shown the chemical instability of chrysotile *in vitro* (Morgan *et al.*, 1971; Langer *et al.*, 1972; Jaurand *et al.*, 1976) and *in vivo* (Jaurand *et al.*, 1977). Toxicity mechanisms related to mineral–cell interactions must be now considered, taking into account physico-chemical modifications in the fibers due to the possible action of the biological medium. Physico-chemical studies of chrysotile leaching have already been performed (Clarke *et al.*, 1960; Fripiat *et al.*, 1968; Chowdhury, 1975; Harris *et al.*, 1975; Thomassin *et al.*, 1976). The experimental conditions used for these studies (nature of dissolving agent, pH, temperature) were quite different from the characteristics of the biological media.

Six natural fibrous silicates are conventionally called "asbestos": fibrous serpentine (chrysotile or "white asbestos"); fibrous amphiboles: actinolite, amosite ("brown asbestos") anthophyllite, crocidolite ("blue asbestos") and tremolite. Many fibrous (or acicular) minerals are familiar to the mineralogists but they are not called asbestos because they do not have the notable physical properties (thermal stability, tensile strength, elasticity, etc.) and chemical properties (unreactivity under acid or alkali attack except for chrysotile) of the six silicates mentioned above.

Some years ago, asbestos was considered only as a highly toxic component of air pollution often associated with other pollutants such as SO^x, NO^x, CO^x, polycyclic hydrocarbons and metals such as Ni, Cr, Cd, Pb, etc. Since 1971 asbestos fibers have also been found in water and even in various generally consumed beverages; thus, fiber analysis may be an important factor for water potability.

Many studies have been performed on asbestos dissolution; among dissolution mechanisms in acid media the following have been suggested:

— congruent dissolution (amosite, crocidolite) in short time experiments (Choi *et al.*, 1974);

— incongruent dissolution as following:
diffusions of protons till reaction sites,
protonation reaction,
diffusion of cations through surface layer and interface,
transport in the bulk of solution.

According to some works, in a weakly chelating acid medium (EDTA 10^{-4} M) dissolution appears to be limited to the outside of the fibers, for example, to the first brucitic layer in the case of chrysotile (Papirer *et al.*, 1976). Other works have shown that it was possible in more aggressive solutions to achieve a more important extraction (e.g. 0.1 N oxalic medium (Johan *et al.*, 1974), 6 N hydrochloric medium (Bleiman and Mercier, 1975)); extraction amounts then exceed 80% of the total magnesium content in chrysotile. It is even possible to prepare fibrous gels in which magnesium is completely absent. Under other conditions (e.g. 6 N HCl medium + isopropanol, Fripiat and Mendelovici, 1968), intermediate percentages of magnesium were extracted. All these experiments have been carried out under specific conditions which cannot be found in biological media.

The aim of the present work was to study chrysotile solubility under conditions consistent with the properties of the intra-cellular medium, i.e. by using, on the one hand, acids from some metabolic pathways and, on the other hand, by choosing concentrations with low molarity. Thus acids were used from the glycolysis and Krebs cycle and oxalic acid was used in concentrations down to 10^{-6} M.

Most of the studies published up to the present time were based upon general chemical analysis of solutions or solids after reaction and only in special cases were fibers morphology and composition studied with finer methods. It seemed important to study the asbestos dissolution phenomenon by photoelectron spectrometry (XPS) Thomassin *et al.*, 1976) which allows chemical surface analysis to a depth of approximately 100Å. The instrument used (AEI ES 200 is fitted with a magnesium anticathode (radiations MgKα = 1253.6 eV) and the experimental data are obtained by measuring peaks heights or areas for Mg_{2s} and Si_{2p}).

A. Formulation of a Dissolution Mechanism for Chrysotile from Thetford (Canada) in Slightly Acid Medium

In an earlier work (Thomassin *et al.*, 1977) we studied by XPS spectrometry the kinetics of chrysotile dissolution in 0.1 N oxalic medium. The results can be interpreted either on the hypothesis of a magnesium diffusion from an outer leached layer governing kinetics (apparent $D = 7.5 \times 10^{-18}$ cm² s^{-1} at 25°C; activation energy 19 k cal mole^{-1}) or on the hypothesis that the exchange reaction itself is the slowest step. The rate of progression of the reaction barrier is then calculated as (v = 3Å h^{-1} at 25°C). On the hypothesis of diffusion-controlled kinetics, the results obtained agree with those of Bleiman and Mercier, (1975); however, on the basis of our results it is not possible to chose between the two proposed mechanisms so we have therefore made other experiments with lower acid concentrations.

1. *Kinetics of Chrysotile Alteration in* HCl 10^{-2} M *at 40° C Medium*

The experiments were carried out according to a previously described procedure (Thomassin *et al.*, 1977). It should, however be kept in mind they were made in an unshaken and not renewed medium. The results are shown in Fig. 1. Three parts can be seen on the experimental curve: a quite steep slope, then a plateau followed by a parabolic ending.

It seems impossible to discuss this curve theoretically; in contrast to experiments in 0.1 N oxalic medium where the $C^{2+}_{Mg} \simeq 0$ limiting condition in the solution is constantly maintained, resulting from an excess of reagent and from the sequestration of most of the magnesium in chelated form, this series of experiments was very probably carried out with a magnesium concentration which was variable at the solid–solution interface, thus it is not possible to solve the diffusion differential equation. A qualitative interpretation can, however, be made on the basis of the nature of the curve and especially on the position of the plateau. Referring to the theoretical fiber structure (Fig. 2), it is possible to estimate that the magnesium atoms in the second layer are approximately 10Å under the surface.

The relation $R\ (Mg_{2s}/Si_{2p}) = 0.78\ e^{-d/\lambda}$ makes it possible to calculate the theoretical signal corresponding to an entirely protonated fiber on the first layer (d = 10Å; λ = 25 to 30 Å; Colombin *et al.*, 1976)†. Calculation gives the following results for: λ = 25Å, R = 0.52; for λ = 30Å, R = 0.56. The hatched area of Fig. 1 indicates the two extreme values of R.

† d represents the average distant of an Mg^{++} of the second surface layer and λ is the mean free path of the Mg_{2s} and Si_{2p} electrons.

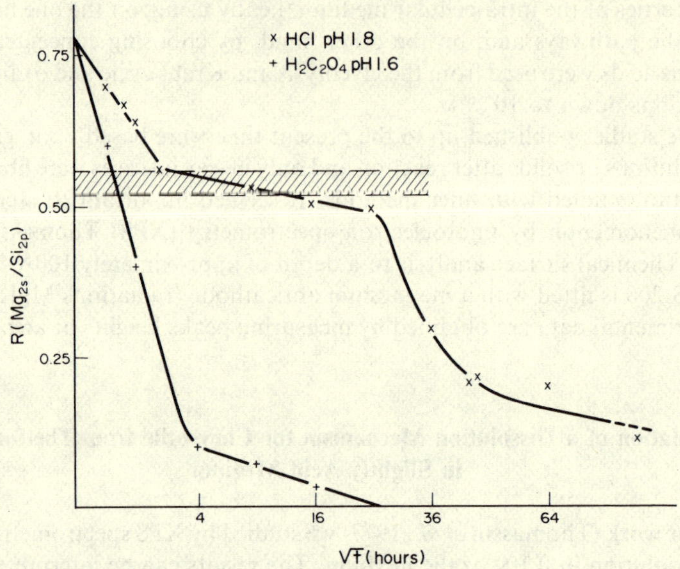

FIG. 1. Deterioration kinetics of chrysotile in HCl 10^{-2} M medium. For comparison, dissolution curve with oxalic acid (same pH).

The position of the plateau in Fig. 1 therefore corresponds, taking into account possible errors, to the end of the leaching of the first layer. This plateau which is between approximately 4 h and 30 h indicates an extremely slow stage and it is reasonable to suppose that the kinetics is then controlled by diffusion through the residual silicic gel corresponding to the first leached layer.

If our interpretation is correct, then the first part of the curve reflects the exchange reaction between the H^+ from the solution and the Mg^{++} of the first brucitic layer. At this stage of the exchange reaction is rate controlling. After approximately 30 hours, the observed slope change probably indicates a change in the dissolution mechanisms which may be related to the structure of the fibrous silica gel.

This series of experiments suggests the existence of a "screen effect" which considerably slows down chrysotile leaching when the first layer exchange is achieved. This conclusion should be further strengthened by the results of an experiment at pH 1.8 in the presence of HCl and glycine (1 g per 100 ml).

The curve (Fig. 3) shows another characteristic slope change near the value Mg/Si = 0.55 but there is no equally clear plateau and the first part of the curve is no longer linear to $\sqrt{t}$ after a certain latency period the leaching of the first layer seems very rapid. Curves for the dissolution of magnesium from chrysotile fibers in organic and hydrocloric media (Fripiat and Mendelovici, 1968; Zapata *et al.*, 1972) showing an organic radicalgraft reaction do not look like curves obtained with hydrochloric media alone. Our experiment suggests that the protonation of chrysotile is complicated by a glycine adsorption phenomenon.

2. Kinetics of Chrysotile Alteration in Diluted Oxalic Media

In order to reproduce concentrations of biological interest, the kinetics of chrysotile leaching in 10^{-6} M and 10^{-4} M at 40°C media were examined. The results are shown in Fig. 4.

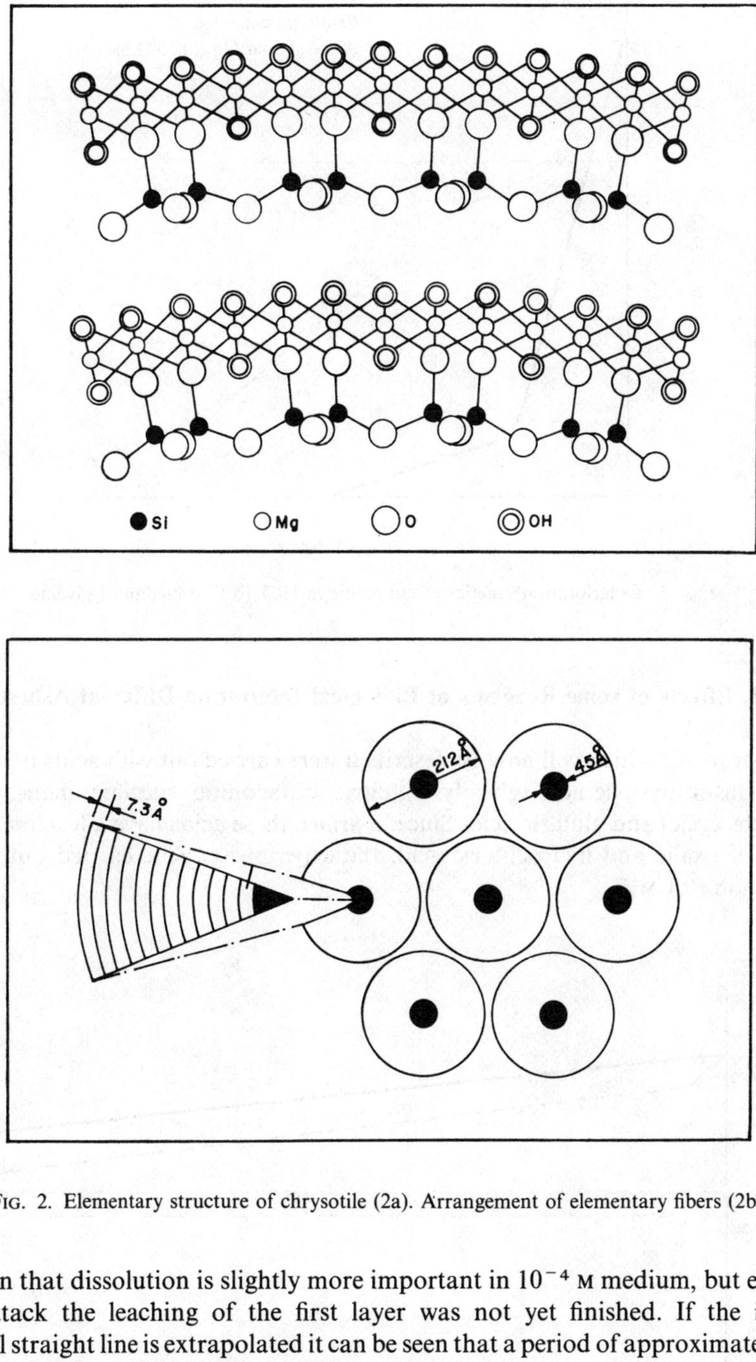

(a)

(b)

FIG. 2. Elementary structure of chrysotile (2a). Arrangement of elementary fibers (2b).

It can be seen that dissolution is slightly more important in 10^{-4} M medium, but even after 100 h of attack the leaching of the first layer was not yet finished. If the indicated experimental straight line is extrapolated it can be seen that a period of approximately 8 days would be necessary to leach the first layer.

This group of experiments shows that in weak solutions the protonation of chrysotile leads more or less rapidly to the alteration of fibrils to a state in which only the outer layer has entirely lost its magnesium.

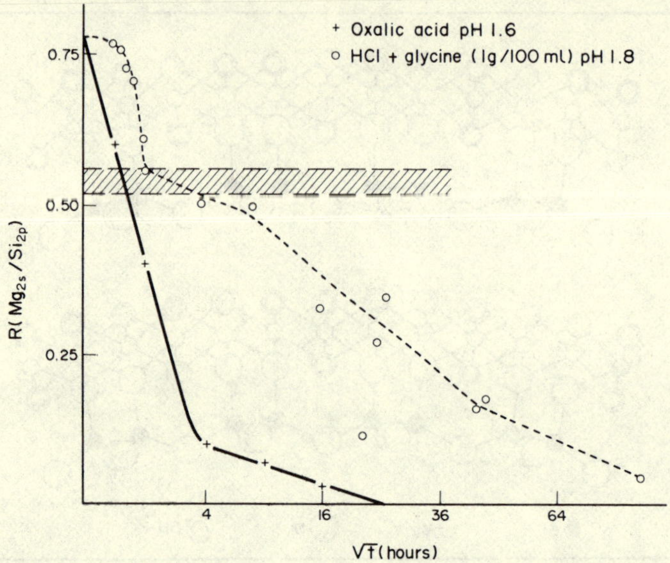

FIG. 3. Deterioration kinetics of chrysotile in HCl 10^{-2} M medium + glycine.

B. Effects of some Reagents of Biological Interest on Different Asbestos

The experiments which will now be described were carried out with acids involved in the cell metabolism: pyruvic acid (glycolysis), citric, cis-aconitic, succinic, malic, oxaloacetic acids (Krebs cycle) and glutaric acid. Since, *a priori*, these acids have a leaching power less than that of oxalic and hydrochloric acid, the experiments were carried out at a higher concentration (0.1 M).

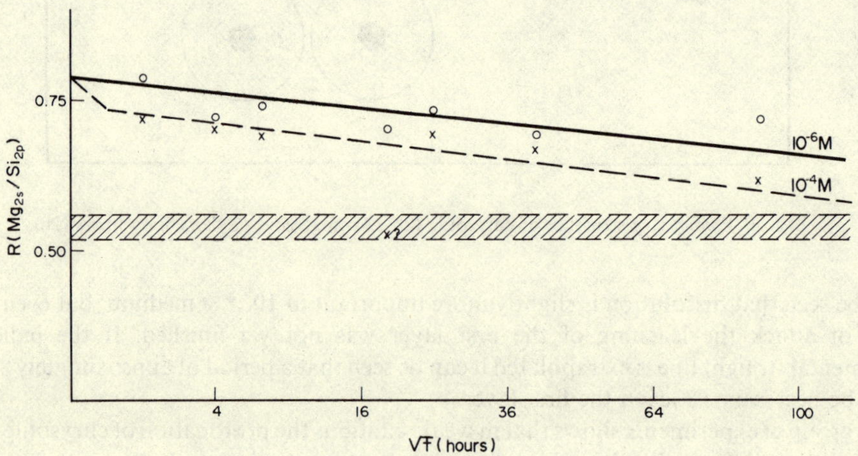

FIG. 4. Deterioration kinetics of chrysotile in dilute oxalic medium.

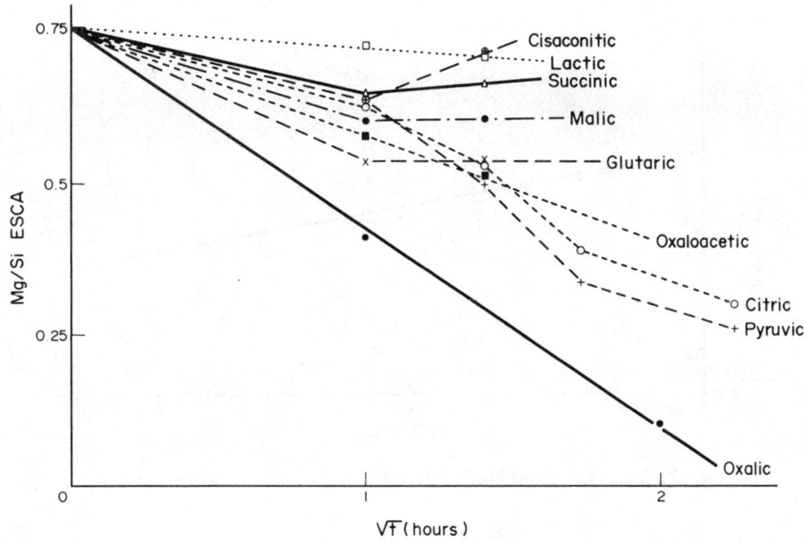

FIG. 5. Solubility of chrysotile with various organic acids (0.1 N).

The results for chrysotile are shown in Fig 5. It can be observed that the activity of the various acids used is very different, but always less than oxalic acid at the same concentration. These results show that pyruvic and citric acids may play a part in chrysotile dissolution in biological media. Some results obtained with other asbestos fibers are presented in Figs. 6 and 7.

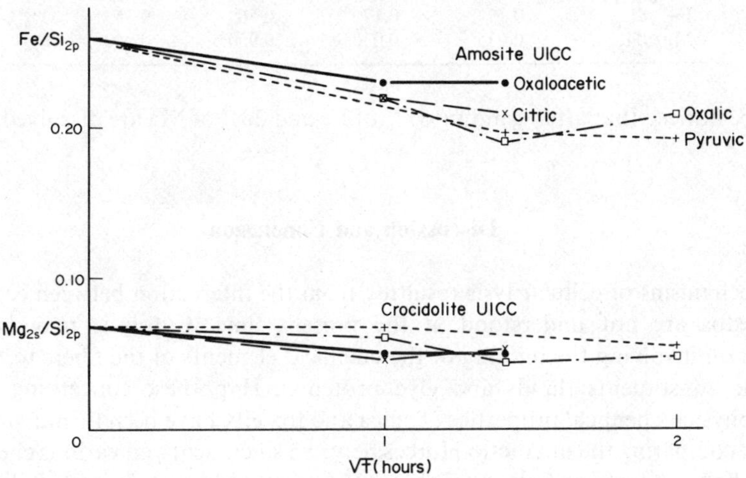

FIG. 6. Solubility of amosite and crocidolite UICC with various organic acids (0.1 N).

Lastly, the study of each type of asbestos should be considered separately. For example, the effects of grinding† play a more important part in crocidolite than in chrysotile

† Manual grinding in an agate mortar.

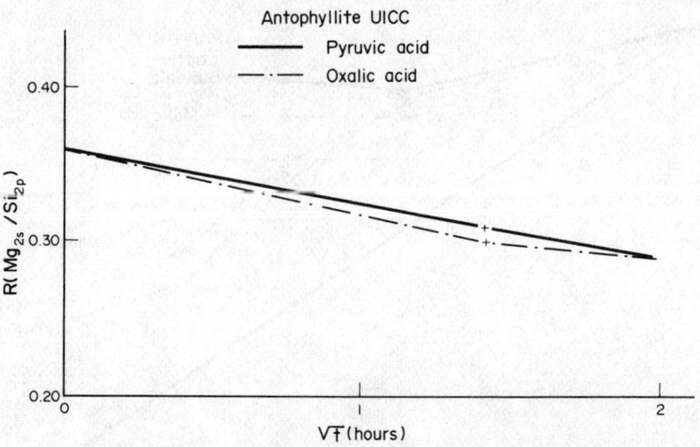

FIG. 7. Solubility of anthophyllite with various organic acids (0.1 N).

dissolution. Experimental data for the dissolution of a crocidolite (from South Africa) in 0.1 N at 40°C oxalic medium are as shown in Table 1.

TABLE 1.

	Before grinding		After grinding		
	unleached	leached	unleached	leached	
Na_{2s}/Si_{2p}	0.15	0.16	0.06	0.06	—
Fe_{2p}/Si_{2p}	0.17	0.17	0.56	0.35	37%
Mg_{2s}/Si_{2p}	0.035	0.030	0.050	0.040	20%

It can be noticed that after grinding 37% of Fe and 20% of Mg are dissolved after 4 hours of attack.

Discussion and Conclusion

The mechanisms of cellular lysis resulting from the interaction between cell membranes and asbestos are not understood at the present time. Cytolysis may be a physical phenomenon involving the binding of the chemical elements of the fibers to the groups of membrane constituents (lipids and glycoproteins). Hypothese concerning the relations between physico-chemical properties of silica and toxicity have been formulated by Nash *et al.* (1966); comparing the interaction forces between silicic acid and various chemical groups (peptidic link, phosphate link, phosphate ester), they suggest the possibility of linkage between phospholipids and silica; then formation of hydrogen bonds could initiate alteration of the cell membrane preceding the release of intracellular constituents.

In accord with this hypothesis, some studies have indicated a protective action by polyvinyl pyridine N-oxide (Schnitzer and Pundsack, 1970; Voisin *et al.*, 1960; Schnitzer and Bunsescu, 1970). In the case of asbestos fibers, and especially chrysotile, the reactivity may be related to linkage of magnesium with the cell membrane (Harington *et al.*, 1971). To support this hypothesis studies have shown an increase in hemolytic activity as a direct function of the MG/Si ratio; moreover, EDTA acts as a protective agent relative to hemolysis (Schnitzer and Pundsack, 1970; MacNab and Harington, 1967; Schnitzer and Bunsescu, 1970). The erythrocyte membrane groups involved in these interactions could be sialic acid; Harington *et al.* (1971) observed on the one hand, inhibition of hemolysis by enzymatic cleavage of this residue and on the other hand, by treatment of chrysotile with sialic acid. However, it must be observed that particles with the chemical composition, $Mg(OH)_2$, are not hemolytic and that we observed (unpublished results) an increasing hemolytic activity on rabbit red blood cells, by action of chrysotile completely leached by oxalic acid.

However, any hypothesis on the nature of interactions between membrane groups and the atoms of minerals should take into account the accessibility of these groups. The stability of the structure obtained also depends on the number of bonds formed, thus low-energy interactions may be important factors of stability.

Apart from these physico-chemical hypotheses, toxicity due to asbestos could be of a chemical nature. Peroxidation of unsaturated bonds constituting fatty acids of membrane lipids has been demonstrated (Gabor and Anca, 1975) This hypothesis has recently been discussed (Kilroe-Smith, 1974) and, to our knowledge, no precise mechanism has been proposed. This phenomenon could be a stage of the cell lysis.

Lastly, asbestos fiber reaction could be due to specific functional alteration resulting from a modification in the cellular micro-environment. The "*in-vivo* leaching" of certain asbestos fibers could then play an important part because of local variations in medium composition. The use of ionophores now leads to a better understanding of the role of ions in controlling cellular functions (Reed and Lardy, 1972; Pfeiffer and Lardy, 1976). In a recent paper, Hand *et al.* (1977) showed the importance of magnesium influx in alveolar macrophage activation.

The use of asbestos fibers with a succession of controlled physical and chemical states resulting from treatment by biological acids in low concentration should therefore contribute to the understanding of the mechanisms of toxicity.

On chrysotile, we made experiments with very efficient selective agents (oxalic and hydrochloric acids) at low concentrations. In addition, experiments carried out at the concentration 0.1 N allowed us to compare various acids present in the cell and oxalic acid. It seems possible to us to extrapolate the results of these two types of experiments to the acids of biological interest at low concentrations. This extrapolation suggests that the crystallo-chemical stability of fibers leached only on their outer layer (i.e. for approximately 13% of the total magnesium) is high.

It is also known that the fibrous silica gel which is then formed presents ordered silanol radicals (Pacco *et al.*, 1976). Moreover, the chemical reactivity of this surface was demonstrated by organic molecule graft experiments (Fripiat and Mindelovici, 1968) and we propose as a working hypothesis that the adsorption properties of this ordered gel are able to disturb structure and function of the cell membranes.

References

ALLISON, A. C. (1971) Lyosomes and the toxicity of particulate pollutants. *Arch. Intern. Med.* **128**, 131–139.

BECK, E. G., HOLT, P. F. and NASRALLAH (1971) Effects of chrysotile and acid-treated chrysotile on macrophages cultures. *Brit. J. Ind. Med.* **28**, 179–185.

BIGNON, J. (1974) Effets biologiques des polluants particulaires inorganiques. *XVIIème Congrès national de la tuberculose et des maladies respiratoires,* Clermont-Ferrand.

BLEIMAN, C. and MERCIER, J. P. (1975) Attaque acide et chloration de l'asbeste chrysotile. *Bull. Soc. Chim. Fr.* **3–4**, 529–534.

CHOI, I., MALGHAN, S. G. and SMITH, R. W. (1974) The dissolution kinetics of fibrous amphibole minerals in water. *Proceedings International Symposium on Water–Rock Interaction*, Tchécoslovaquie, 395–403.

CHOWDHURY, S (1975) Kinetics of leaching of asbestos mineral at body temperature. *J. Appl. Chem. Biotechnol.* **25**, 347–353.

CLARK, S. G. and HOLT, P. F. (1960) Dissolution of chrysotile asbestos in water, acid and alkali. *Nature (Lond.)* **185**, 237.

COLOMBIN, L., JELLY, A., VERBIST, J., CHARLIER, H., CAUDANO, R., PIREAUX, J. J., RIGA, J. and TENRET-NOEL, C. (1976) Methods for surface studies and chemical profiling in glass. *Physics in Industry*, Pergamon Press, Oxford, 113–116.

DAVIES, P., ALLISON, A. C., ACKERMAN, J., BUTTERFIELD, A. and WILLIAMS, S. (1974) Asbestos induces selective release of lysosomal enzymes from mononuclear phagocytes. *Nature*, **251**, 423–425.

FRIPIAT, J. J. and MENDELOVICI, E. (1968) Dérivés organiques des silicates. I. Le dérivé méthylé du chrysotile. *Bull. Soc. Chim.* **2**, 483–491.

GABOR, S. and ANCA, Z. (1975) Effect of asbestos on lipid peroxidation in the red cells. *Brit. J. Ind. Med.* **32**, 39–41.

HAND, W. L., KING, N. L., JOHNSON, J. D. and LOWE, D. A. (1977) Requirement of magnesium influx in activation of alveolar macrophages mediated by ionophore A 23 187. *Nature*, **265**, 543–544.

HARINGTON, J. S., MILLER, K. and MACNAB, G. (1971) Hemolysis by asbestos. *Environ. Res.* **4**, 95–117.

HARRIS, A. M. and GRIMSHAW, R. W. (1975) The leaching of ground chrysotile. *3rd Int. Conf. on the Chemistry and Physics of Asbestos*, Laval University Québec **4**, 17, 1–12.

HEXT, P. M. and RICHARDS, R. J. (1976) Biochemical effects of asbestiform minerals on lung fibroblast cultures. *Brit. J. Exp. Path.* **57**, 281–285.

JAURAND, M. C., GONI, J., JEANROT, P., SEBASTIEN, P. and BIGNON, J. (1976) Solubilité du chrysotile *in vitro* et dans le poumon humain. *Rev. Fr. Mal. Resp.* **4**, suppl. 2. 111–120.

JAURAND, M. C., SEBASTIEN, P., BIGNON, J. and GONI, J. (1977) Leaching of chrysotile asbestos in human lungs. Correlation with *in vitro* studies using rabbit alveolar macrophages. *Environ. Res.* (sous presse).

JOHAN, Z., GONI, J., SARCIA, C., BONNAUD, G. and BIGNON, J. (1974) Influence de certains acides organiques sur la stabilité du chrysotile. *Adv. Organic Geochemistry*. 1973, Paris, Technip, 893–903.

KILROE-SMITH, T. A. (1974) Peroxidative action of quartz in relation to membrane lysis *Environ. Res.* **7**, 110–115.

LANGER, A. M., RUBIN, I. B., SELIKOFF, I. J. and POOLEY, F. D. (1972) Chemical characterization of uncoated asbestos fibers from the lungs of asbestos workers by electron microprobe analysis. *J. Histochem. Cytochem.* **20**, 735–740.

LAVAPPA, K. S., FU, M. M. and EPSTEIN, S. S. (1975) Cytogenetic studies on chrysotile asbestos. *Environ. Res.* **10**, 165–173.

MACNAB, G. and HARINGTON, J. S. (1967) Haemolytic activity of asbestos and other mineral dusts. *Nature* **214**, 522–523.

MILLER, K. and HARINGTON, J. S. (1972) Some biological effects of asbestos on macrophages. *Brit. J. Exp. Path.* **53**, 397–405.

MORGAN, A., HOLMES, A. and GOLD, G. (1971) Studies of the solubility of constituents of chrysotile asbestos *in vivo* using radioactive tracer techniques. *Environ. Res.* **4**, 558–570.

NASH, T., ALLISON, A. C. and HARINGTON, J. R. (1966) Physico-chemical properties of silica in relation to its toxicity. *Nature* **210**, 259–261.

PACCO, F., VAN GANGH, L. and FRIPIAT, J. J. (1976) Etude par spectroscopie infrarouge et résonance magnétique nucléaire de la distribution homogène des groupes silanols d'un gel de silice fibreux. *Bull. Soc. Chim. Fr.* **7–8**, 1022–1026.

PAPIRER, E., DOVERGNE, G. and LEROY, P. (1976) Modifications physico-chimiques du chrysotile par attaque chimique ménagée. *Bull. Soc. Chim. Fr.* **5–6**, 651–657.

PARAZZI, E., PERNIS, B., SECCHI, G. C. and VIGLIANI, E. C. (1968) Studies on "*in vitro*" cytotoxicity of asbestos dusts. *Med. Lav.* **59**, 561–576.

PFEIFFER, D. R. and LARDY, H. A. (1976) Ionophore A 23187: the effect of H^+ concentration on complex formation with divalent and monovalent cations and the demonstration of K^+ transport in mitochondria mediated by A 23187. *Biochemistry* **15**, 935–943.

REED, P. and LARDY, H. A. (1972) A 23187: a divalent cation ionophore. *J. Biol. Chem.* **247**, 6970–6977.

SCHNITZER, R. J. and BUNSESCU, G. (1970) Polymers as selective antagonists of hemolytic. *Asbestos Fibers* **20**, 481–482.

SCHNITZER, R. J. and PUNDSACK, F. L. (1970) Asbestos hemolysis. *Environ. Res.* **3**, 1–13.

THOMASSIN, J. H., GONI, J., BAILLIF, P. and TOURAY, J. C. (1976) Etude par spectrométrie ESCA des premiers stades de la lixiviation du chrysotile en milieu acide organique. *C.R. Acad. Sci. Paris*, **283**, série D, 131–134.

THOMASSIN, J. H., GONI, J., BAILLIF, P., TOURAY, J. C. and JAURAND, M. C. (1977) An XPS study of the dissolution kinetics of chrysotile in 0,1 N oxalic acid at different temperatures. *Physics and Chemistry of Minerals* **1**, 385–398.

VOISIN, C., AERTS, C., WATTEL, F. and TONNEL, A. B. (1960) Cytotoxicité des poussières de silice et des poussières mixtes pour les macrophages alvéolaires de cobaye en survie in vitro. Essai d'appréciation quantitative. *Ann. Inst. Pasteur Lille* **20**, 183–192.

WADE, M. J., LIPKIN, L. E., TUCKER, R. W. and FRANK, A. L. (1976) Asbestos cytotoxicity in a long-term macrophage-like cell culture. *Nature* **264**, 444–446.

ZAPATA, *et al.* (1972) Dérivés organiques des silicates. II. Les dérivés vinyliques et allyliques du chrysotile et de la vermiculite. *Bull. Soc. Chim. Fr.* **1**, 54–63.

Discussion

U. ASWATHANARAYANA:

Cement–asbestos sheets are being used extensively in India to provide cheap, fire-proof housing. Dr. Goni's work draws attention to the health hazard involved.

R. A. BERNER:

Have you studied other Mg-silicate minerals (e.g. talc) and their possible toxic effects so as to ascertain whether chrysotile toxicity is due to its unique structure rather than due to a general property of leached Mg-silicate surfaces?

J. GONI:

L'effet toxique de l'amiante-chrysotile est en grande partie dû à "l'effet fibre". Le talc ou autres silicates de magnésium non fibreux ne sont pas toxiques.

B. KUBLER:

Si le Mg n'est pas la source de toxicité, il reste donc la couche superficielle organisée des tétraèdres de silice. Comme on a remplacé les filtres à amiante par des filtres à perlites, Kieselgur, etc. qui sont formés de silice, on va peut-être changer de toxicité.

J. GONI:

En effet, il semblerait que la cyto-toxicité du chrysotile soit due à l'action d'une couche organisée pauvre en silice, et riche en magnésium sur la surface de la fibre.

Biogeochemical Distribution of Rare Earths and Other Trace Elements in Plants and Soils†

J. C. LAUL, W. C. WEIMER and L. A. RANCITELLI

Physical Sciences Department, Battelle-Northwest, Richland, Washington 99352, U.S.A.

Abstract

The rare-earth concentrations in vegetables (corn, potatoes, peas and butternut squash) have been found to be extremely low: 10^{-8} g/g–10^{-10}g/g. The chondritic normalized vegetable REE patterns are fractionated, including a negative Eu anomaly, and behave as a smooth function of the REE ionic radii. These patterns may be governed by the geochemistry of accessory minerals in the host soils. The depletion factors for various elements by vegetables relative to bulk soils are $\sim 10^{-4}$ for REE, Hf, Ta, Th and U; $\sim 10^{-3}$ for Al, As, Ba, Cr, Fe, Mn, Sc, Se and Sr; $\sim 10^{-2}$ for Co, Cs, Na, Ni and Sb; $\sim 10^{-1}$ for Rb; and ~ 1 for K, Zn and Br.

Introduction

We have investigated the uptake of thirty–five trace elements, including the rare-earth elements (REE) by vegetables from their host soils. The uptake study of the REE is particularly important as it may provide an indicator of the longterm behaviour of their transuranic analogs in the natural environment. In the past there has been little known about the concentration levels of REE, their behaviour, and their uptake by plants from soils. Robinson (1943) and Robinson *et al.* (1958) reported extremely high concentrations of the total REE (2300 ppm) in hickory leaves. The hickory leaves are exceptional in the plant kingdom in the sense that they are excellent accumulators of the REE. However, in general the REE levels in plants are very low, with the result that their precise measurements by classical methods have been rather limited (Cossa, 1870; Scribner, 1939).

In this study, we used the highly sensitive technique of neutron activation analysis (NAA) for elemental determinations. While our work focuses mainly on the REE, the biogeochemical distribution of other trace elements in vegetables and soils is also addressed. Plausible mechanisms in soils which control the availability of the REE and other trace elements are discussed in this paper.

Experimental

Samples

The samples we analyzed were received from the following locations:

1. North Eastham, Massachusetts. We received from E. P. Hardy (Health and Safety

† This paper is based on work performed under the U.S. Energy Research and Development Contract EY-76-C-06-1830.

Laboratory, HASL, New York) six soil layers from different depths and the interior portions of four vegetables grown on the same soil. These were: Soils: S–1958 (0–5 cm), S–1959 (5–10 cm), S–1960 (10–15 cm), S–1961 (15–20 cm), S–1962 (20–25 cm), S–1963 (25–30 cm). Vegetables: F–4383 (corn), F–4385 (potatoes), F–4387 (peas), F–4410 (butternut squash).A detailed description of these vegetable samples and the cultivation and harvesting techniques used has been given by Hardy and Bennett (1977).

2. Richland, Washington. We received form the PNL Arid Lands Ecology Rcscrvc one surface soil (S–0004) and a sample of cheatgrass (F–0004).

Sample preparation was performed in a clean room. About 0.2–0.5g of a sample aliquot was weighed into a clean 2/5–dram polyethylene vial and the 2/5–dram vial was heat sealed. The 2/5–dram vial was then placed in a 2–dram polyethylene vial which was also heat sealed.

Procedure

Our instrumental and radiochemical neutron activation analysis (INAA and RNAA) scheme used here has been previously reported by Laul and Rancitelli (1976). In brief, our sequential INAA consisted of two irradiation periods and four counting intervals to optimize the detection sensitivities of individual elements. Samples and standards were irradiated in a TRIGA reactor at Washington State University. The irradiated samples were counted on a 130–cm^3 Ge (Li) detector (FWHM 1.8 keV for 1332 keV γ of ^{60}Co). The γ-ray spectra from samples and standards were stored on a magnetic tape and the data reduced by a PDP–15 computer. A USGS BCR–1 sample was used as an internal control in our INAA and RNAA experiments.

Our radiochemical scheme consists of Ba(U), Sr and REE group separations. The irradiated samples were fused with Na_2O_2 + NaOH mixture, with the Ba, Sr and REE mixed carriers already dried in Ni crucibles. The fusion cake was decomposed with H_2O and neutralized with HC1. The Ba and Sr were precipitated as sulfates with H_2SO_4, and the supernate containing the REE was precipitated as a group hydroxide. The REE $(OH)_3$ was dissolved in HC1 and precipitated with NH_4HF_2 and HF. The REE fluoride was dissolved in H_3BO_3 and HNO_3 and precipitated as $REE(OH)_3$ with NH_4OH. The fluoride and hydroxide cycle was repeated twice prior to their counting. The Ba and Sr precipitate was converted to carbonate by boiling with K_2Co_3. The carbonate was then converted to nitrate with fuming HNO_3. Using $Fe(OH)_3$ scavenging, Ba was precipitated as $BaCrO_4$ at pH 5 and Sr as oxalate from the supernate.

The REE group aliquots were counted twice on the Ge(Li) system to optimize the detection sensitivities for relatively long-lived isotopes of Ce, Nd, Eu, Gd, Tb and Tm. Chemical yields of the REE were determined by reactivation. The Ba and Sr were also counted on the Ge(Li) system. In Ba counting, U was determined via fission products of 12.8d ^{140}Ba–^{140}La (816, 1597 keV) radionuclide. Chemical yield of Sr was measured by reactivation.

Results and Discussion

Our INAA and RNAA data for thirty-five elements are shown in Table 1. In INAA, the estimated errors due to counting statistics based on BCR–1 are: $\pm$1–3% for Al, Fe, Mn, Na, Sc, Co, Rb, Ba, Sr, La, Sm, Eu, Yb, and Lu and Hf; $\pm$5% for Br, Ca, K, Cs, Ta, Th and Zn; and $\pm$5–10% for As, Cr, Ni, Sb, Se, Ce, Nd and Tb. In RNAA, the estimated errors due to

TABLE 1. ELEMENTAL ABUNDANCES (ppm) BY INAA AND RNAA IN SOILS AND PLANTS*

	Eastham, MA. Soils, Vegetables: HASL no.											Richland, Wa		
Elements	S-1958 0–5 cm	S-1959 5–10 cm	S-1960 10–15 cm	S-1961 15–20 cm	S-1962 20–25 cm	S-1963 25–30 cm	Mean Soil	F-4383 Corn	F-4385 Potatoes	F-4387 Peas	F-4410 Squash	S-0004 Surface soil	F-004 Cheat grass	BCR-1 USGS STD.
Al	3.4E+4	3.0E+4	3.2E+4	3.6E+4	—	—	3.3E+4	70	50	60	110	9.3E+4	720	7.2E+4
Fe	1.3E+4	1.4E+4	1.1E+4	1.1E+4	1.3E+4	1.3E+4	1.3E+4	40	40	90	50	5.1E+4	260	9.5E+4
Mn	350	350	340	340	—	—	345	6.0	5.0	10	10	1130	50	1400
Na	4200	4500	3900	3900	4500	4800	4300	40	50	50	70	1.5E+4	100	2.4E+4
K	1.1E+4	1.1E+4	—	—	—	—	1.1E+4	1.2E+4	2.4E+4	2.3E+4	2.5E+4	—	4170	1.4E+4
Sc	2.9	3.0	2.7	2.4	2.8	2.8	2.8	5.0E−3	5.0E−3	5.0E−3	7.0E−3	17.0	0.074	32.0
Cr	130	140	130	90	120	140	120	0.60	0.32	0.31	0.73	60	0.43	18
Co	2.4	2.5	2.2	1.7	2.2	2.4	2.2	0.015	0.13	0.10	0.17	18.5	0.18	36.0
Ni	40	40	40	30	40	50	40	<0.5	0.80	1.5	1.1	<15	<1	<10
Zn	40	35	30	30	35	30	33	34	21	38	37	160	12	130
Br	5.1	6.7	5.4	6.1	6.0	7.7	6.1	0.90	4.2	3.3	9.5	1.3	2.8	—
As	3.8	4.7	4.3	4.1	3.5	2.2	3.8	<0.03	<0.03	<0.03	<0.06	3.3	0.07	—
Sb	0.25	0.28	0.22	0.23	0.25	0.44	0.28	<2E−3	<2E−3	0.014	0.010	0.60	0.029	0.70
Se	4.7	3.2	3.3	2.4	2.6	5.1	3.5	<0.03	<0.03	<0.03	<0.05	1.7	<0.03	—
Rb	32	34	24	24	33	34	30	3.0	5.4	3.4	12.7	55	0.9	46
Cs	0.76	0.82	0.65	0.64	0.80	0.74	0.73	<2E−3	8.0E−3	3.5E−3	0.013	2.7	0.016	0.92
Ba	140	170	180	150	200	250	180	0.050	0.12	0.12	0.28	640	20	670
Sr	40	45	40	45	60	60	50	<0.2	<0.2	0.62	0.76	290	5.4	330
La	23.0	28.9	28.2	33.2	32.0	35.0	30	5.0E−3	2.0E−3	5.1E−3	0.018	36.1	0.17	26.0
Ce	47.5	65.7	50.4	72.4	70	74	63	0.010	2.5E−3	6.4E−3	0.016	74.7	0.33	54.0
Pr	4.9	8.2	6.8	7.9	—	—	7.0	<1E−3	<2E−3	<2E−3	<1E−3	8.4	0.040	6.7
Nd	21.3	34.8	26.9	35.0	27	30	29	<0.010	<0.010	<0.010	<0.010	34.6	0.15	30.0
Sm	3.68	7.00	4.26	5.34	5.30	5.70	5.2	7.4E−4	2.1E−4	5.5E−4	1.0E−3	6.60	0.035	6.60
Eu	0.39	0.57	0.43	0.49	0.47	0.44	0.46	1.0E−4	3.0E−5	4.0E−5	2.0E−4	1.55	8.4E−3	1.90
Gd	3.2	5.2	3.4	4.3	—	—	4.0	<2E−3	<2E−3	<2E−3	<3E−3	6.2	0.037	6.7
Tb	0.46	0.87	0.49	0.64	0.51	0.55	0.58	<1E−4	1.0E−4	<1E−4	2E−4	0.95	8.6E−3	1.0
Ho	0.68	1.2	0.84	0.99	—	—	0.93	1.2E−4	8.7E−5	6.2E−5	1.4E−4	1.5	<0.02	1.2
Er	1.8	2.6	2.0	2.4	—	—	2.2	<5E−4	<2E−3	<5E−4	<5E−4	3.9	<0.5	3.5
Tm	0.28	0.40	0.29	0.37	—	—	0.33	<3E−4	<1E−4	<2E−4	<2E−4	0.59	0.05	0.52
Yb	1.8	2.5	2.0	2.4	2.6	2.3	2.3	2.0E−4	0.80E−4	9E−5	4.2E−4	3.5	0.020	3.5
Lu	0.27	0.38	0.30	0.37	0.39	0.32	0.34	2.7E−5	<2E−5	1.3E−5	4E−5	0.51	3.0E−3	0.52
Hf	8.9	9.5	8.0	8.0	11.1	9.7	9.2	<3E−3	<2E−3	<3E−3	<4E−3	8.1	0.026	4.7
Ta	2.7	2.2	2.1	2.0	1.9	3.0	2.3	<1E−3	<1E−3	<1E−3	<6E−3	1.1	5E−3	0.80
Th	14	14	10	14	14	15	14	<5E−3	8E−3	0.015	0.020	8.2	0.040	6.0
U	3.2	3.3	2.4	3.5	—	—	3.1	8E−4	8E−4	<1E−3	<1E−3	2.5	0.012	1.7

* E is base to 10.Ba, Sr, REE (La–Lu) and U elements except in S-1962 and S-1963 were obtained by RNAA.

counting statistics based on BCR–1 are relatively small: $\pm$1–5%. In cases where the concentrations are very low, especially in plants, the error due to counting statistics is somewhat larger: $\pm$5–20%. The detection limits in Table 1 are 3σ of the peak areas. The concentrations for many elements, in particular Cs, Ba, Sr, Th and REE, are extremely low in plants. These concentrations represent the lowest values ever reported for vegetables.

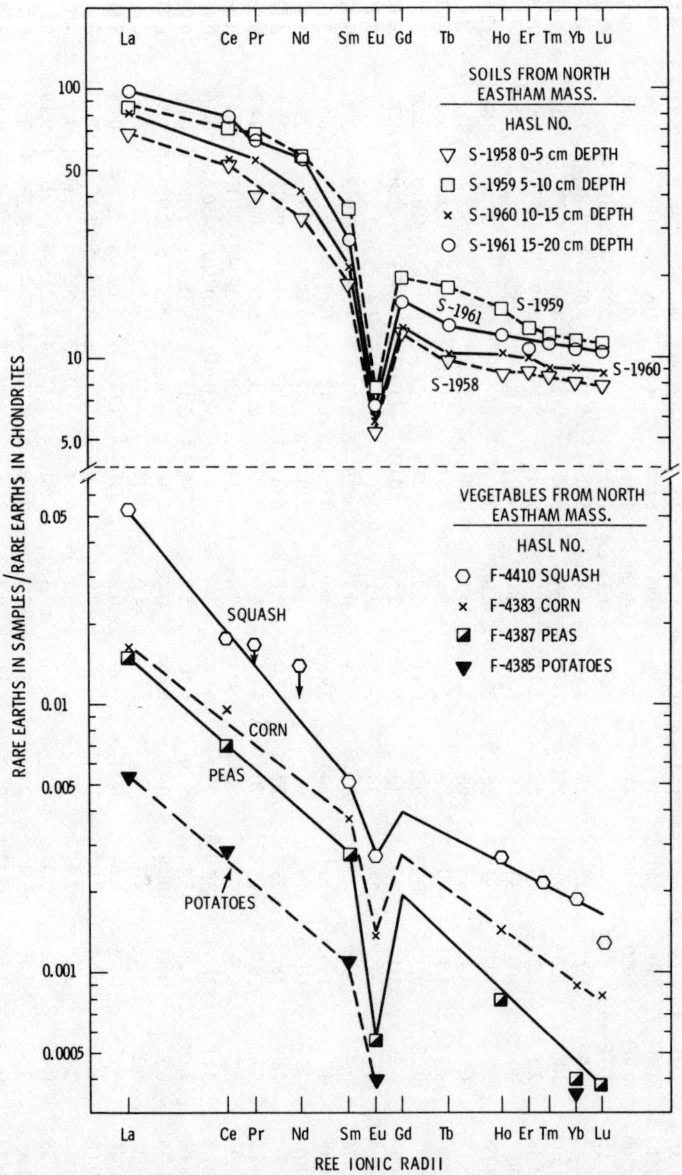

FIG. 1. The REE abundances in HASL vegetables and soils are normalized to chondrite meteorites. The chondritic abundances (ppm) used for normalization are La 0.34, Ce 0.91, Pr 0.12, Nd 0.64, Sm 0.195, Eu 0.073, Gd 0.26, Tb 0.047, Ho 0.078, Er 0.20, Tm 0.032, Yb 0.22 and Lu 0.034 (Laul and Schmitt, 1975).

The REE data for vegetables and their host soils are normalized to chondrites, and these REE patterns are shown in Figs. 1 and 2. The North Eastham, Massachusetts, HASL soils are fairly uniform in their bulk chemical compositions within the depth profile. This is expected as these are cultivated soils. The HASL soils fall in the range of granitic-type rocks. This classification is supported by the fact that the chondritic normalized REE patterns show steep negative Eu anomalies which are typical only of some terrestrial granites. The HASL soils are considerably enriched in As, Br and Se. The Richland soils S–0004 also shows a slight negative Eu anomaly. However, based on major, minor and trace elements, the soil seems to consist of about equal proportions of granitic and basaltic components.

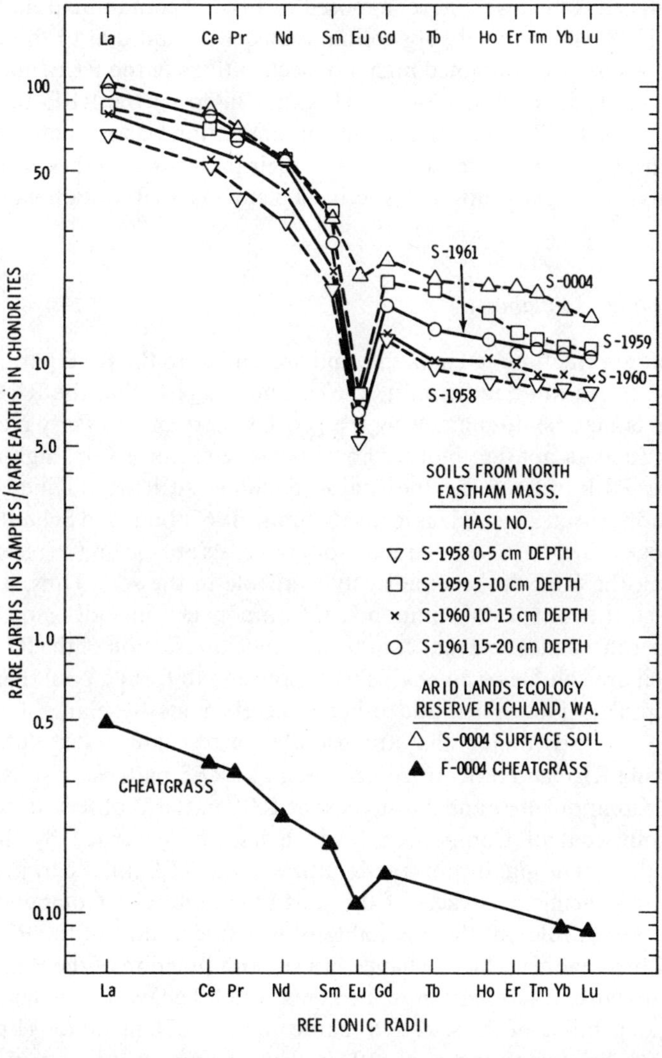

FIG. 2. The REE abundances normalized to chondrite meteorites. See Fig. 1 for chondritic values.

Rare-earth Patterns

The most striking feature observed in our plant study is that the chondritic normalized REE patterns behave as a smooth function of the REE ionic radii. The chondritic normalized REE patterns (Fig. 1) in HASL vegetables relative to the host soil are somewhat fractionated. All the vegetables show similar REE patterns, including the negative Eu anomaly, and a steeper slope from the light REE to the heavy REE with chondritic normalized La/Lu ratio of ~ 25 as compared to a La/Lu ratio of ~ 10 in the host soil.

The possibility of soil contamination in the HASL plants is remote. The REE concentrations in these vegetables relative to the host soils are extremely low ($\sim 10^4$ times); 2–18 ppb for La to 0.02–0.04 ppb for Lu, and these concentrations most likely represent indigenous contents in plants. Furthermore, these samples represent the interior portions of the vegetables produced, which were never exposed to any soil contamination. On the other hand, cheatgrass (F–0004) was probably contaminated with soil due to the nature of its growth characteristics. It also contained higher concentrations of the REE, and displayed a REE pattern identical to its host soil S–0004 (Fig. 2). Based on the REE and other trace elements (Sc, Cr, Sb, Cs, Hf, Th and U), the soil contamination was estimated to be 0.5% in the cheatgrass. Thus, the REE concentrations and their patterns, together with other trace elements, provide us a strong indication of whether a plant has been contaminated by its host soil.

Mechanisms Controlling Availability

The REE patterns are fractionated in plants and are similar to the REE pattern observed in the mineral apatite (Haskin and Schmitt, 1967). This suggests that the REE uptake by plants from host soils may be dominated by the geochemistry of accessory materials. The accessory minerals such as apatite, biotite, hornblende, etc., have very high distribution coefficients for the REE relative to the major minerals such as olivine, plagioclase, orthopyroxene, clinopyroxene, etc. (Haskin and Schmitt, 1967; Laul and Schmitt, 1975). On weathering of the rock minerals in soil, the accessory minerals are the first to become loose or decompose releasing the REE to become freely available in the soil. Thus, a soil can be considered to consist of two types of components. Component 1 includes mostly accessory minerals and demineralized material that contain a significant fraction of the REE and other trace elements which are soluble or amenable to dissolution to form a pool that serves as a reservoir for the uptake of the REE and other trace elements by plants. Component 2 includes the minerals (mostly major) that are insoluble or resistant to dissolution and also contains the remaining REE and other trace elements. The REE patterns in soils (Figs. 1 and 2) are comprised of Components 1 and 2, whereas the REE patterns observed in plants (Fig. 2) are perhaps reminiscent of Component 1 which may be governed by the accessory minerals, provided there is no significant fractionation of the REE uptake by plants. We are pursuing aqueous and chemical extracts of the soils to characterize Component 1.

At this stage we cannot rule out the possibility of selective uptake of the REE by plants from Component 1 in a manner which behaves as a smooth function of the REE ionic radii. Plants are known to cause fractionation in elements which behave geochemically alike in terrestrial systems, depending of the selective charge and size of a metal ion (Epstein, 1972; Mortvedt *et al.*, 1973; Schneiderman *et al.*, 1975). This is supported by our data on other trace elements (Table 1), where the elemental ratios of geochemically alike pairs such as

Rb/Cs, Fe/Mn, Sb/As, Ba/Sr, etc., are different in soils than in plants. However, the fractionation of REE by plants is less likely due to the fact that the physical and chemical properties of the REE, unlike other geochemical pairs, are very much alike. This is supported by our data on a 9% hydrogen peroxide extract from the soil. The chondritic normalized REE pattern in the peroxide fraction is similar to those observed in plants. This suggests strongly that the plants are probably not fractionating the REE during their uptake. We estimated the % REE extracted by H_2O and 9% H_2O_2 fractions from the soil. Based on La, for example, the H_2O fraction contains 3.0×10^{-3}% La and the H_2O_2 fraction contains 0.56% La relative to the total La concentration in the bulk soil. These estimates yield 0.90 ppb La in the H_2O fraction/gram of soil, and 170 ppb La in the H_2O_2 fraction/gram of soil, which are potentially available for the uptake by plants from the Component 1 pool. The mean La concentration in HASL plants is 7 ppb, which corresponds to 800% (essentially complete) uptake from the H_2O extract fraction and 4% uptake from the H_2O_2 extract fraction, respectively. Our future experiments should enable us to investigate more on this aspect.

Hogdahl (1966) has reported REE data for sea water, which suggests that Ce in aqueous systems can exist as the +4 state. Further, Robinson *et al.* (1958) in their work on REE in hickory leaves reported large fluctuations in the proportion of Ce relative to La and Nd, and Ce seemed to correlate with oxidation-reduction conditions in the host soil profile. They attributed this anomalous behaviour of Ce to the presence of Ce^{+4} which is strongly retained in oxygenated subsoil, and thus is less available to plants relative to the other REE that do not undergo valence change under soil conditions. However, based on the smooth REE patterns observed in soils and plants, and H_2O and 9% H_2O_2 soil extracts, we note that Ce most likely exists in the +3 state in the Component 1 pool of the soil and during its uptake by plants.

The solubility of the REE in the pool is another factor which may govern the observed REE patterns during their uptake by plants. The REE in aqueous systems at pH 5–8 form $REE(OH)_3$, and their solubility products vary decreasingly from La $= 1 \times 10^{-19}$ to Lu $= 2.5 \times 10^{-24}$ (Stevenson and Nervik, 1961). If we plot the REE solubilities in a fashion similar to that shown in Figs. 1 and 2, the pattern is not at all a smooth function of the REE ionic radii. This sugggests that either the REE uptake is via different chelating systems, in which case the hydroxide solubility products are not applicable, or the REE concentrations in the pool are well below the hydroxide solubility products.

Depletion Factors

We have plotted in Fig. 3 the concentration ratios of the REE and other trace elements (Table 1) in HASL plants relative to the bulk host soil. These ratios may change somewhat, depending on the elemental concentrations in Component 1 of the soil. Nevertheless, the gross pattern shows the relative degree of biogeochemical distribution of various elements by plants, assuming the proportion of Components 1 and 2 remains constant during the uptake. The elements in Fig. 3 are arranged in decreasing order. The depletion factors in major groups are $\sim 10^{-4}$ for REE, Hf, Ta, Th and U; $\sim 10^{-3}$ for Al, As, Ba, Cr, Fe, Mn, Sc, Se and Sr; $\sim 10^{-2}$ for Co, Cs, Na, Ni and Sb; 10^{-1} for Rb; and ~ 1 for K, Zn and Br. The elements K and Zn are essential nutrients in plants. Bromine is monovalent and very mobile, and thus is easily taken up by plants. The high uptake factor for Rb is due to substitution for K (Epstein, 1972). The Cs, being a larger ion (1.78Å) than Rb (1.56Å), has less tendency for

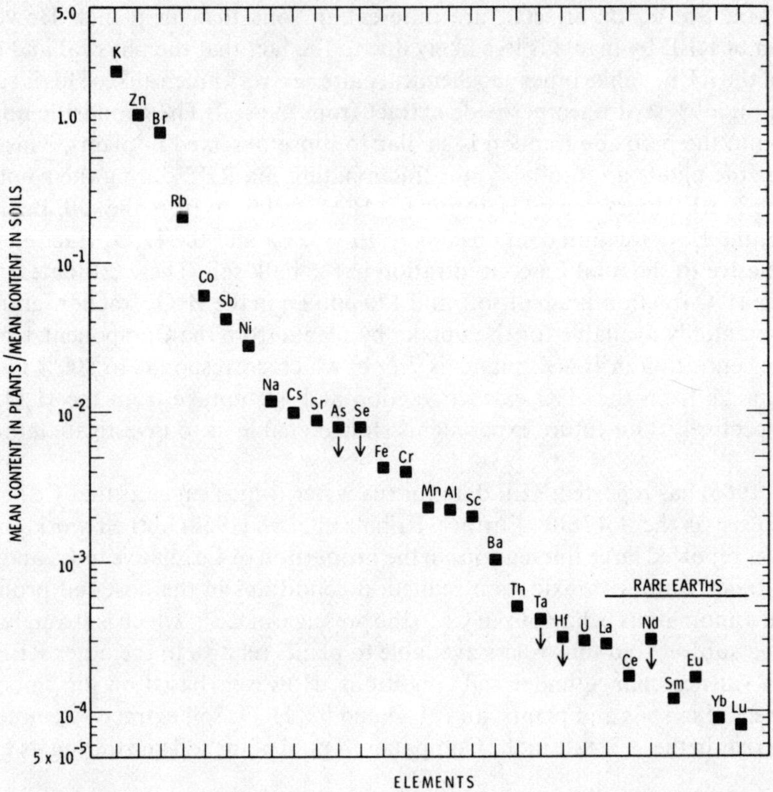

FIG. 3. Average elemental concentration ratios in HASL vegetables to HASL bulk soil.

substitution and thus less uptake. The higher oxidation states +3 and +4 metal ions have in general extremely low (10^{-3} to 10^{-4}) uptake factors. The uptake mechanism by plants is quite complex and selective for different elements. Our gross pattern (Fig. 3) supports the notion that the selective charge and size of an ion play a dominant role in plant uptake from Component 1 in soils. However, the REE are probably taken up without significant fractionation by plants from Component 1. Under steady state conditions, the depletion factors essentially yield the maximum uptake limits by HASL plants from their host soil. We should point out, however, that our results for thirty-five trace elements in peeled vegetables are a limited data base at this stage and thus need not represent a mean of all plants. We plan to pursue our work on other plants grown on other soils and test our general proposed plausible mechanism of the uptake of REE and other trace elements by plants.

Acknowledgements

We are highly indebted to R. W. Perkins, J. S. Webb and H. L. Cannon for their valuable suggestions. We thank M. R. Smith for his assistance in the experiment. This paper is based on work performed under U. S. Energy Research and Development Administration Contract EY–76–C–06–1830.

References

COSSA, A. (1970) Sulla diffusione del cerio, del lantanoe del didimio. *Gaz, Chim. Ital.* **9**, 118–140.

EPSTEIN, E. (1972) *Mineral Nutrition of Plants: Principles and Perspectives,* John Wiley & Sons, Inc.

HASKIN, L. A. and SCHMITT, R. A. (1967) Rare earth distributions. *Researchers in Geochemistry* **2**, John Wiley & Sons, Inc., 234–258.

HARDY, E. P. and BENNETT, B. G. (1977) Radionuclide uptake by cultivated crops. USERDA Report HASL–321, 19–37.

HOGDAHL, O. (1968) *Distribution of the REE in Sea Water*, Scientific Affairs Division, Paris, NATO Report No. 6.

LAUL, J. C. and RANCITELLI, L. A. (1976) Multielement analysis by sequential instrumental and radiochemical neutron activation. *Proceedings 1976 International Conf. on Modern Trends in Activation Analysis,* **2**, 1074–1083.

LAUL, J. C. and SCHMITT, R. A. (1975) Dunite 72417: A chemical study and interpretation. *Proc. 6th Lunar Sci. Conf.* 1231–1254.

MORTVEDT, J. J., GIORDANO, P. M. and LINDSAY, W. L. (1973) *Micronutrients in Agriculture,* Soil Science Society of American, Inc.

ROBINSON, W. O. (1943) The occurrence of rare earths in plants and soils. *Soil Sci.* **56**, 1–6.

ROBINSON, W. O., BASTRON, H. and MURATA, K. J. (1958) Biogeochemistry of the rare earth elements with particular references to hickory trees. *Geochim. Cosmochim. Acta* **14**, 55–67.

SCHNEIDERMAN, G. S., GARLAND, T. R., CATALDO, D. A., WILDUNG, R. E. PELROY, R. A. and DRUCKER, H. (1975) Fate of heavy metals and heavy metal complexes in soils and plants. *Battelle Progress Report.*

SCRIBNER, B. F. (1939) Spectroscopic detection of rare earths in plants. *Proceedings 6th Summer Conference on Spectroscopy and its Applications,* The Technology Press, New York.

STEVENSON, P. C. and NERVIK, W. E. (1961) The radiochemistry of the REE, Sc, Y and Ac. *Nuclear Sciences Series* NAS–NS–3020.

Discussion

H. L. CANNON:

Were the plant/soil ratios in the last slide based on data from a single plant species or were they a value obtained from all of your kinds of vegetables?

J. C. LAUL:

The vegetables used were corn, squash, peas and potatoes. The mean values of a trace element in these four vegetables were taken and divided by the mean value of six soils from different depths to establish the depletion factor.

U. ASWATHANARAYANA:

Is there any particular reason for your choosing REE for the plant uptake studies?

J. C. LAUL:

REE provide an ideal choice for the transuranic elements. The long-term behaviour of the transuranic elements in the natural environment can be predicted from the uptake study of rare earths by plants.

J. GONI:

Are the rare earths biologically toxic on the long-term basis in the environment?

J. C. LAUL:

It is hard to say at this stage as we have begun our rare-earth studies recently. However, rare earths are being used as stable tracers in various studies concerning air, aqueous and sedimentary transport systems. Based on such evidence, I would think that the rare earths are probably not biologically toxic.

J. GONI:

What is the biological effect of the uptake of REE upon plants?

J. C. LAUL:

Our initial experiments show that the physicochemical effects of the REE is negative. But we do not know its effect in the long run. It is too early to tell.

Research in the United States Relative to Geochemistry and Health

WILLIAM L. PETRIE† and HELEN L. CANNON‡

†*National Academy of Sciences, 2101 Constitution Avenue, Washington, U.S.A.*

‡*U.S. Geological Survey, Federal Center MS 915, Box 25046, Denver, U.S.A.*

Abstract

Increasing concern regarding the effects of the geochemical environment on health in the United States has fostered research studies in a number of universities and government agencies. The necessity to evaluate the effects of natural and man-made elemental excesses in the environment on health requires the establishment of requirements and tolerance limits for the various elements in water and crops. Maps of the geographic distribution of these elements in rocks, surficial materials and ground and surface waters are also essential for comparison with the occurrence of disease. Funding support for research projects that relate to various parameters of these problems emanates largely from a few federal agencies, and much of the work is conducted at government, university and private facilities. An example of the latter is the National Academy of Sciences–National Research Council, which has several components that are addressing a variety of comparative studies of the geochemical environment related to health; studies involve specific trace elements (like selenium and magnesium), diseases (like cancer, urolithiasis and cardiovascular disease), other health factors (like aging and nutrition) and links with timely major problems (like the health effects of greatly increasing the use of coal).

Introduction

Nutritionists, toxicologists and physicians have long been aware of certain benefits and hazards of particular elements to the health of plants, animals and humans. Within the past 10–15 years, however, a rapidly expanding group of scientists with other specialties, including geochemists, have brought their expertise to bear on the formidable problems and opportunities for meaningful research in areas vital to the health and well-being of man. Man-made pollutants, as well as the naturally occurring elements of the geochemical environment, have been the subject of a greatly accelerated scientific and medical interest as the effluents and emissions of numerous industries have come under the scrutiny of environmentalists and regulators throughout the U.S.A. Not so well recognized are the degree and extent of the natural hazards and benefits of the geochemical environment, which need much more attention. This is particularly true of baseline studies, which help distinguish those pollutants that should be controlled more stringently, from those that might be harmlessly, or even beneficially, relaxed.

This survey highlights some of the early beginnings of environmental geochemistry and health in the United States and an overview of the research beginning with pertinent activities of the National Academy of Sciences–National Research Council, the government and then universities.

Background

Early Conferences and Symposia

One of the earlier group efforts of geochemists with other specialists was a symposium on *The Relation of Geology and Trace Elements to Nutrition* held on November 17, 1963, in New York at the Annual Meeting of the Geological Society of America (GSA) and sponsored by the Geochemical Society (Table 1). The proceedings volume for the 1963 symposium was published as a special paper by GSA (Cannon and Davidson, 1967). Researchers in geochemistry, botany, biology, soil science and public health all emphasized the need for continued contact between disciplines to explore interrelationships of trace elements in the environment. A second symposium was held on October 7, 1964, in New York in connection with meetings of the Public Health Association, and a third in Montreal on December 28, 1964, at the American Association for the Advancement of Science (AAAS) meetings.

A particularly important series of Annual Conferences on *Trace Substances in Environmental Health* (TSEH) was begun on July 10–11, 1967, by D. D. Hemphill at the University of Missouri Environmental Health Center in Columbia, Missouri (Hemphill, 1967, 1969–1976). Now in its tenth year, this series provides the outstanding annual event on this subject in the United States, which regularly attracts top environmental and biochemical scientists from many countries.

On December 30, 1968, a very significant *Symposium on Environmental Geochemistry in Health and Disease* was held during the AAAS meetings in Dallas. Papers from the symposium were published by the GSA as a memoir (Cannon and Hopps, 1971). Afterward, thirty-five scientists from specialities like geochemistry, chemistry, soil science, geography, epidemiology, pathology, biochemistry, nutrition and dentistry considered ways to further interdisciplinary research on possible casual relations between chemical elements in the environment and animal as well as human health. Their enthusiasm led to the organization of an *ad hoc* Committee for continued multidisciplinary investigation of relationships between health and geochemical environment, plans for acquiring recognition and support for operational funds, and also the early beginnings of a new scientific society.

The Society for Environmental Geochemistry and Health

At the December 30, 1968 meeting, the need also became evident for a society to which this heterogeneous group of scientists could adhere for purposes of communication and for the development of a common identity. The Society for Environmental Geochemistry and Health (SEGH), whose members range from the earth and environmental sciences on the one hand to the biomedical and nutritional sciences on the other, traces its origin to that meeting. On June 13, 1972, at its charter meeting in Columbia, Missouri, Delbert D. Hemphill was elected the first President of SEGH. The Society's official publication, *Interface*, a quarterly which began to appear in February 1972, carries both substantive journal articles, critical evaluations, book reviews, and newsletter items of interest to a widely divergent membership of nearly 500 (SEGH, 1972).

The Formation of GERHD

In 1969 the *ad hoc* Committee found a home in the National Academy of Sciences—National Research Council's Division of Earth Sciences, which led to the establishment of

TABLE 1. ENVIRONMENTAL GEOCHEMISTRY AND HEALTH ACTIVITIES IN THE UNITED STATES (1962–1968)

Dates	Activity	Sponsor/Affiliate	Reference
November 1962	Informal Discussion in Houston	Geochemical Society	Cannon and Davidson, 1967
November 17, 1963	Symposium on The Relation of Geology and Trace Elements to Nutrition, in New York	Geological Society of America	Cannon and Davidson, 1967
October 7, 1964	Symposium in New York	Public Health Association	Cannon and Davidson, 1967
December 28, 1964	Symposium in Montreal	American Association for the Advancement of Science	Cannon and Davidson, 1967
July 10–11, 1967	First Annual Conference on Trace Substances in Environmental Health, in Columbia, Missouri	University of Missouri and the National Science Foundation	Hemphill, 1967
July 16–18, 1968	Second Annual Conference on Trace Substances in Environmental Health, in Columbia, Missouri	University of Missouri and the National Science Foundation	Hemphill, 1969
December 30, 1968	Symposium on Environmental Geochemistry in Health and Disease, in Dallas	American Association for the Advancement of Science	Cannon and Hopps, 1971

the *Subcommittee on the Geochemical Environment in Relation to Health and Disease* (GERHD) under the U.S. National Committee for Geochemistry (USNC/G). Thus, GERHD, the NRC frontrunner in environmental geochemistry and health, held its first meeting in March 1970, under the chairmanship of Helen Cannon, and conducted its first symposium on *Minor Metals of the Geochemical Environment, Health and Disease* in December 1970, cosponsored by the GSA and the American Geophysical Union (AGU) at the AAAS meetings in Chicago (Cannon and Hopps, 1972).

In October 1971, some thirty papers were presented at a 3-day GERHD *Conference on the Geochemical Environment in Relation to Health and Disease*, held by the New York Academy of Sciences (NYAS) in New York. The 352-page proceedings was unfortunately published by NYAS without appropriate reference to GERHD (Hopps and Cannon, 1972).

Research Programs and Projects

National Research Council

Assembly of Mathematical and Physical Sciences. From 1972 to 1974 GERHD conducted a trilogy of week-long workshops, each with approximately forty to fifty participants. These well-received, forward-thinking, stimulating forums brought many new ideas and possibilities to the attention of diverse groups of scientists not generally knowledgeable in each others' fields:

The *Asilomar Workshop,* chaired by Helen Cannon, was held on February 7–12, 1972, at Pacific Grove, California. It brought together some thirty-eight invited participants from geochemistry and such widely divergent fields as soil science; biochemistry; plant, animal, and human nutrition; epidemiology; human and veterinary medicine; analytical chemistry; physics; statistics; and water hygiene. The Workshop which included two well-known overseas participants—E. J. Underwood of CSIRO in Perth, Australia, and John S. Webb of Imperial College, London—primarily surveyed the information available on the amounts of trace elements naturally occurring in rocks, soil and water for eleven elements—flourine, iodine, chromium, lithium, cadmium, zinc, lead, selenium, tellurium, copper and molybdenum—the forms to which they are changed during weathering; additions to the environment caused by man's activity; the uptake and concentration of trace substances in plants; absorption and retention of trace elements by animals and man; and the effects of trace elements on health and disease. *Geochemistry and the Environment,* Volume I, was published by NAS in 1974 (Cannon, 1974).

The *Capon Springs Workshop,* chaired by Walter Mertz, was held on May 6–11, 1973, at Capon Springs, West Virginia. Approximately fifty invited participants from many different disciplines met to extend the work of the Asilomar Workshop by examining further the various ways the geochemical environment influences the processes of health and disease in man.

Their report considers seven more elements–beryllium, magnesium, nickel, silicon, strontium, tin and vanadium—under Trace Elements Related to Health and Disease and several selected studies under Geochemical Environment and Man, which reports on important case studies, published by NAS in 1977 as *Geochemistry and the Environment,* Volume II (Mertz, 1977).

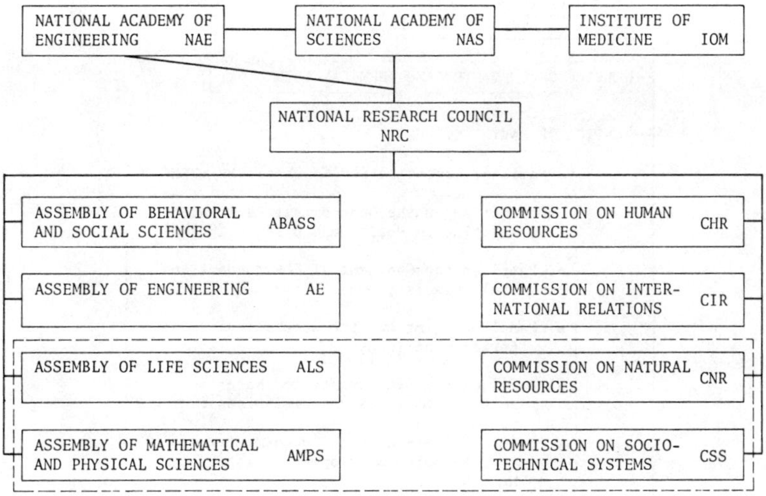

FIG. 1 NRC components with research projects concerning environmental geochemistry and health (within dashed lines).

In the NRC reorganization of 1974, several NRC divisions, including the Division of Earth Sciences, were disbanded and some of their responsibilities were transferred to the newly formed Assembly of Mathematical and Physical Sciences (AMPS) (Fig. 1).

The *Captiva Island Workshop,* chaired by Howard C. Hopps, was held on October 6–11, 1974, at Captiva Island, Florida. GERHD once again brought together approximately fifty invited participants from diverse disciplines, but instead of emphasizing additional elements, it began to apply previously compiled information toward the identification of causal relationships and associations of specific elements and geochemical conditions in the environment to certain selected diseases. The diseases selected for study were cancer of the esophagus, stomach and colon-rectum; cardiovascular diseases including myocardial infarction, hypertension and stroke; and urolithiasis, more specifically the geochemical circumstances surrounding the formation of kidney stones.

Their report, *Geochemistry and the Environment,* Volume III, is currently undergoing review, and publication is expected in late 1977 (Hopps, 1977, in preparation).

However, effective as they were in numerous ways, the workshops were also subject to cumbersome delays which led GERHD to try projects with smaller *ad hoc* panels. Thus, in 1975, GERHD embarked on a series of panel studies on timely topics with six to ten member specialists. Four such panels are now in operation, each contemplating a report within a year or two (Fig. 2). Described briefly in their organizational order, they are:

- The *Panel on the Geochemical Environmental and Urolithiasis* (POGU) is a five-member panel chaired by G. K. Davis (a nutritionist) and is made up of a urologist, a water geochemist, an epidemiologist and a nephrologist. POGU first met on April 20, 1976 to evaluate evidence relating the existence of a "stone belt" in the U.S. to the conditions conductive to the formation of renal stones in humans. POGU has met twice and its report is in preparation.
- The *Panel on Aging and the Geochemical Environment* (PAGE) is a six-member panel chaired by H. C. Hopps (a pathologist) and is made up of a retired soil scientist, an emeritus

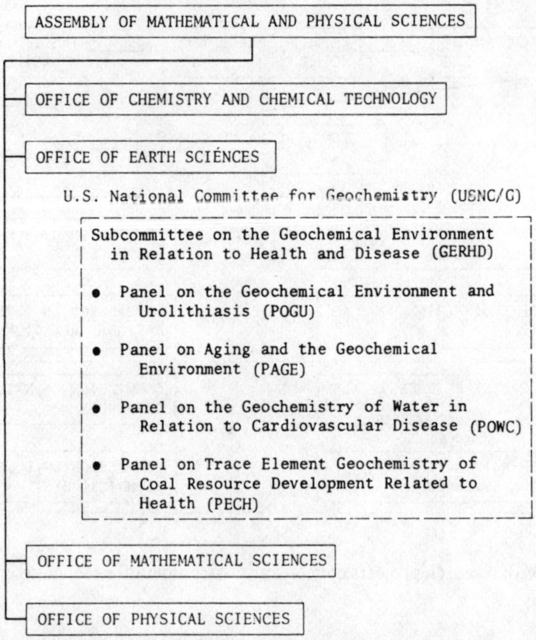

FIG. 2 Selected activities of the assembly of mathematical and physical sciences.

geologist, a groundwater geochemist, a medical statistician and a gerontologist. PAGE first met on June 7, 1976, to review and evaluate existing evidence of environmental effects on aging and to discern and report any new causative relationships that may be developed. PAGE has also met twice and its report is in preparation.

● The *Panel on the Geochemistry of Water in Relation to Cardiovascular Disease* (POWC) is an eight-member panel chaired by E. E. Angino (a water geochemist) and is made up of a medical nutritionist, three epidemiologists, a trace-element geochemist, and two pathologists. POWC first met on September 13, 1976, to consider the chemical components of water, especially those that may relate to cardiovascular diseases in man. Although POWC has met but once, some portions of their report have already been drafted.

● The *Panel on the Trace-element Geochemistry of Coal Resources Development Related to Health* (PECH) is a nine-member panel chaired by B. G. Wixson (a biogeochemist) and is made up of a soil geochemist, a biochemical engineer, an environmental geologist, a trace-element toxicologist, a coal petrographer and geochemist, an industrial hygienist, an analytical geochemist, and a trace-element coal geologist. PECH first met on December 16, 1976, to study possible health-related effects on man from greatly increased use of coal, particularly with regard to those caused by mining, processing and usage.

Assembly of Life Sciences. The ALS program is divided into three main categories: Special Programs under the Executive Office, Division of Biological Sciences and Division of Medical Sciences (Fig. 3) (ALS: Assembly of Life Sciences, 1976).

Under the ALS Executive Office, the twelve person *Committee on Toxicology* advises government agencies on broad matters of policy involving toxicology, recommends interim environmental exposure standards, suggests specific toxicology testing or research, and

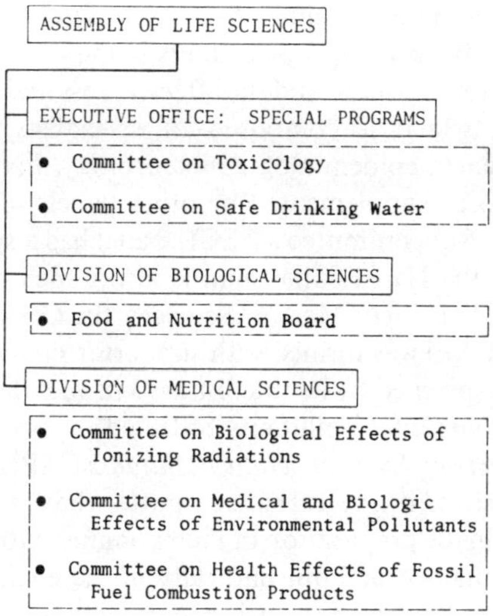

FIG. 3 Selected activities of the assembly of Life Sciences.

advises on and participates in field studies on toxicity problems. It has Subcommittees on: Fuels, Fuel Additives and Their Combustion Products; Fire Toxicology; Lead in Paint; and Principles and Procedures for Evaluating the Toxicity of Household Substances (ALS, 1976, p. 15).

Also under the ALS Executive Office, the fourteen-person *Committee on Safe Drinking Water* studies the adverse effects on health attributable to contaminants in drinking water, in conjunction with the Environmental Protection Agency, to serve as the scientific basis for revision of the national interim drinking water regulations, according to the timetable prescribed in the Safe Drinking Warer Act, effective December 16, 1974. There are ten Subcommittees on: Margin of Safety and Extrapolation, Metalloids, Microbiology, Morbidity and Mortality, Organic Contaminants, Particulate Contaminants, Pesticides, Radioactivity, Special Ions and Trace Metals (ALS, 1976, p. 21).

Under the ALS Division of Biological Sciences, the usually fifteen-person *Food and Nutrition Board* has several committees with environmental geochemistry and health-related interests including the FNB Committees on: Clinical Nutrition, Dietary Allowances, Food Protection, GRAS (generally recognized as safe) List Survey—Phase II, International Nutrition Programs, Nutrition and Dental Health, Nutritional Misinformation, Nutrition of the Mother and Preschool Child, and Sodium-restricted Diets (ALS, 1976, p. 64).

Under the ALS Division of Medical Sciences, the twenty-person *Committee on the Biological Effects of Ironizing Radiations* has developed information on radiation health risks and is concerned with problems of regulation and regulatory responsibility. It has four Subcommittees on: Concepts, Medical Usage or Ironizing Radiations, Nuclear Energy (Power) and Other Uses of Ironizing Radiations. The latter has studied radiation exposure as occupational, consumer products and services, and both natural and man-made environmental exposures (ALS, 1976, p. 81).

Also under the ALS Division of Medical Sciences, the twelve-person *Committee on Research Needs on the Health Effects of Fossil Fuel Combustion Products* (HEFF) is assessing the validity of existing data to determine what critical information is lacking, and to recommend a research program to obtain it. Among the variables being considered are: the health and environmental effects of the by-products of the generation of energy, the genetic and somatic risk posed by these effects, and the effectiveness and cost of measures that could reduce the risk. HEFF, which is composed of specialists in biochemistry, biology, biostatistics, clinical medicine, epidemiology, meteorology, toxicology and air and water-pollution engineering is to be supplemented by other panels of experts as required (ALS, 1976, p. 99). The GERHD Subcommittee's PECH panel has a special working relationship with HEFF, inasmuch as the HEFF Chairman is also a member of PECH.

In 1970, also under the Division of Medical Sciences, the *Committee on Biologic Effects of Atmospheric Pollutants* (BEAP) was formed with support from the Environmental Protection Agency to consider topics specified by its title. In 1973 its mandate was expanded to include medical effects and environmental pollutants reflected in its new title, *Committee on the Medical and Biologic Effects of Environmental Pollutants* (MBEEP). Although dominantly biomedical, the Committee and its several subcommittees have made use of geochemists and related environmentalists in the preparation of their commendable set of reports on chemical elements and substances that occur both naturally in the environment and as man-made pollutants. In order of their appearance, their reports are as follows: Asbestos, Fluorides, Lead, Particulate Polycyclic Organic Matter, Manganese, Chromium, Vanadium, Nickel, Vapor-phase Organic Pollutants, Selenium, Chlorine and Hydrogen Chloride, Copper and Ozone and Other Photochemical Oxidants. Other studies to examine the problems of discerning potential harm from long-term and low-level exposure to natural and man-made substances occurring in the human environment are underway and reports are expected shortly on: Airborne Particles, Ammonia, Arsenic, Carbon Monoxide, Hydrogen Sulfide, Iron, Nitrogen Oxides, Platinum-group Metals and Zinc. An appointed subcommittee conducts the study of each pollutant and its Chairman is a member of the parent Committee. Upon completion of their reports, the subcommittees are discharged (ALS, 1976, p. 111).

Commission on Natural Resources. The CNR encompasses four Boards (Fig. 4) on: Agriculture and Renewable Resources (BARR), Energy Studies (BES), Environmental Studies (ESB) and Mineral Resources (BMR), which covers most of its responsibilities. Several of them are of interest in our present context (NAS: National Academy of Sciences, 1976).

Under BARR (Fig. 4), the eight-person *Committee on Animal Health* is concerned with public policy issues with respect to animal health, veterinary public health, and the adequacy and saftey of foods of animal origin. There are two Subcommittees on: Animal Disease Eradication and Brucellosis Research (NAS, 1976, p. 126).

Also under BARR, the nine-person *Committee on Animal Nutrition* is concerned with all aspects of animal nutrition and the interrelationship of animal nutrition with human health and the agricultural enterprise. The main emphasis is the preparation of the Nutrient Requirement Series on economically important domestic animals (NAS, 1976, p. 127).

Under BES, the thirteen-person *Committee on Radioactive Waste Management* conducts studies at the request of federal agencies on long-range waste-management plans and programs for an expanding nuclear energy industry; its advisory services include a general assessment of the adequacy of present and projected technology to meet long-range health, safety and other environmental requirements, and the identification of new research needs. The Committee has two panels (NAS, 1976, p. 130).

Also under BES, the twelve-person *Committee on Energy and the Environment* reviews and assesses the relationships, impacts and trade-offs of environmental laws and regulations on various aspects of energy production and consumption. The Committee has two panels (NAS, 1976, p. 130).

Another BES component is the fifteen-person *Committee on Chemistry of Coal Utilization,* which is updating and extending the reference work entitled *Chemistry of Coal Utilization,* originally prepared in 1945 and supplemented in 1967 by NRC Committees.

Under the ESB, the eight-person *International Environmental Programs Committee* serves as U.S. National Committee for the Scientific Committee on Problems of the Environment (SCOPE) of the International Council of Scientific Unions (ICSU). The Committee provides consultative service to the U.S. Department of State and other federal agencies, maintains contacts with international organizations and national institutions in other countries concerned with environmental problems, and undertakes special studies of critical environmental problems having an international impact (NAS, 1976, p. 132).

Also under ESB, the nine-person *Committee on a Multimedia Approach to Municipal Sludge Management* is examining sludge disposal or reuse in water, on land and in the air. The need for examining scientific and technical information about these options is indicated by the variety of management alternatives available and the necessity for EPA to develop a sludge management program. Economic, legal and institutional data is included to broaden the perspective on possible solution (NAS, 1976, p. 133).

Another ESB component is the twelve-person *Committee on Scientific and Technical Assessments of Environmental Pollutants* (STAEP), which was established to provide EPA with assessments and evaluations of currently scientific and technical information on present and potential environmental pollutants (NAS, 1976 p. 134).

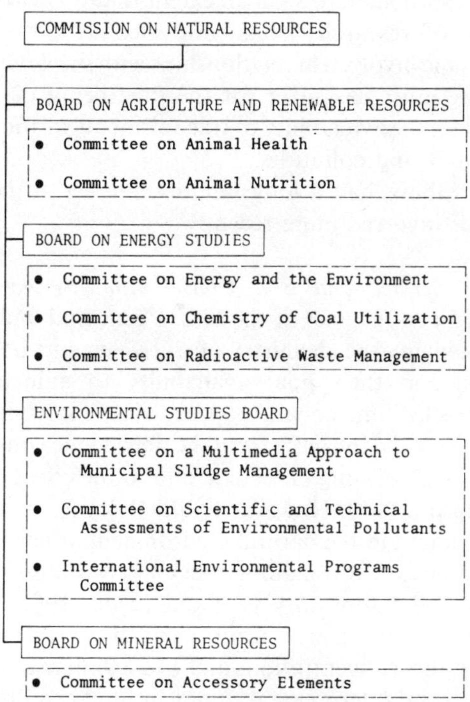

FIG. 4 Selected activities of the Commission on Natural Resources.

Under the BMR (Fig. 4), the eleven-man *Committee on Accessory Elements* is studying the redistribution of accessory elements and compounds which would result from future mining and utilization on a large scale of coal deposits, uranium ores, oil shales, iron ores and nonbauxite aluminium ores, such as anorthosite and aluminium-bearing clays (NAS, 1976, p. 134).

Commission on Sociotechnical Systems The CSS includes the activities of five major Boards dealing with research in Transportation Building, National Materials, Maritime Transportation and Military Personnel Supplies. Of the several extra-board committees in CSS, two are of interest: the ten-man *Committee on Processing and Utilization of Fossil Fuels* evaluates the status of technology, research and development for utilizing and processing coal in liquefaction, low-Btu gasification, direct combustion, advanced power cycles and coal mining. There are five *ad hoc* panels that specialize in the foregoing topics (NAS, 1976, p. 160).

Also under CSS is the seven-man *Committee on Evaluation of Sulfur Oxides Control Technology*, which evaluates current and prospective technology for the control of sulfur oxide emissions from industrial processes and fossil fuel combustion sources. There are three *ad hoc* panels on: Identification of Sulfur Oxide Sources; Application of Sulfur Oxides Control; and Status of Sulfur Oxides Control Technology (NAS, 1976, p. 161).

Federal Government Research

During the past 10 years, interest in possible relationships between the environment and health and disease has mushroomed to such an extent that it is virtually impossible to give more than an overview of research being conducted at major federal agencies and universities that have become involved in relationships with the *natural* environment of rock, soils and plants. Thus we must pass over for the time being the enormous amounts of research being conducted on the availability of trace elements to animals (including man), as well as studies directly involving pollution.

Several federal agencies have had a longstanding interest in the natural environment, while others have become involved more recently.

Agricultural Research Service and Soil Conservation Service, U.S. Department of Agriculture. Members of both agencies at the U.S. Plant, Soil and Nutrition Laboratory, Ithaca, New York, are studying the chemistry of trace elements in soils, as related to their concentration in plants, and their final availability to animals. Research has been concentrated mainly on selenium, cobalt, copper, zinc, molybdenum, iron, nickel and chromium. These studies proceed logically from the dependence on the composition of rock substrate, to the effect of soil-forming processes, and to the effect of plant response on the plant composition (Hodgson, 1969; Hodgson *et al.*, 1971).

The distribution of selenium in the natural environment at levels toxic and deficient for animal health has been studied throughout the United States by scientists of the Soil Conservation Service at the Laboratory (Kubota *et al.*, 1967). Their data have been compared with heart disease and cancer occurrences by Shamberger *et al.*, (1971, 1975). Deliberate control of selenium concentrations at 0.1 to 2 ppm in animal and human diets by additions to soil, feed, salt and interregional shipment has been suggested (Allaway, 1968).

Current studies at the Laboratory include environmental effects on the form, concentration, nutritional value and distribution of essential elements in plants, and the soil–plant–animal cycling of selenium, chromium and other essential trace elements.

The interactions between metals as they affect uptake by plants have been a particular concern at the Agricultural Research Service in Beltsville, Maryland. Dr Lagerwerff especially was concerned with the uptake of toxic metals (Lagerwerff and Biersdorf, 1971).

U.S. Geological Survey, U.S. Department of the Interior. Major efforts by the USGS in studying possible relationships between the geochemical environment and health and disease have had five principal thrusts: (1) involvement with committees and panels of the National Research Council, as well as with government agencies, (2) areal studies of trace element distribution in water, soil, and plants, (3) reviews of environmental geochemistry of particular elements, (4) trace-element studies as related to a particular disease and (5) studies of trace element distribution with reference to point sources of contamination.

The Survey has been well represented on the NRC Subcommittee on the Geochemical Environment in Relation to Health and Disease, the NRC Committee on Medical and Biologic Effects of Environmental Pollutants, and on several workshops of the Institute of Environmental Health in the U.S. Public Health Service. The Survey has also cooperated on Environmental Protection Agency projects, and worked closely with state surveys on environmental problems.

Areal studies of trace element distribution in the United States have included analyses of soils collected at 50-mile intervals throughout the conterminous United States (Shacklette and others, 1971a, 1973, 1974), strontium in water (Skougstad and Hem, 1962), trace elements in municipal water supplies (Durfor and Becker, 1964), factors that affect the availability of iron and manganese in water (Hem, 1970), flourine in groundwaters of the United States (Fleischer, 1967), selenium in groundwater near Golden, Colorado (Moran, 1976), selenium in ground and surface waters of Nebraska (Engberg, 1973), a study of the minimal variation in the trace metal content of soils and sagebrush of the Powder River Basin of Wyoming (Connor *et al.*, 1976), and lithium in sediments and plants of the Basin and Range Province of the western United States (Cannon *et al.*, 1975). Baselines or geochemical background values were established for 147 landscape units (Connor and Shacklette, 1975). Studies of the trace-element exchange between water and sediment by E. A. Jenne and a geochemical survey of western energy regions are in progress. A survey of metals in plant foodstuffs throughout the United States has not yet been released. Another major source of background data is the Exploration Research Branch where large sampling programs of soils and plants are carried out in mineral districts and also in connection with the assessment of mineral potential in wilderness areas.

Review papers on the geochemistry of mercury (USGS, 1970) and of lead (Lovering, 1976) in rocks, soils, stream sediments, water, plants and the atmosphere have been published by the Survey, and one on the environmental impact of cadmium compiled under the auspices of the National Institute of Environmental Health Sciences includes Survey data (Fleischer *et al.*, 1974).

Several geochemical studies directly related to the evidence of a particular disease have been made or are in progress in cooperation with the Public Health Service. These include a geochemical study of plants and soils in areas of high and low heart disease rates in Georgia (Shacklette *et al.*, 1970) and in areas of high and low cancer rates in Maryland (Cannon, 1977). A comparison was made of disease incidence on the Gila River Indian Reservation in relation to the geochemical environment in cooperation with the Phoenix Medical Center (Sievers and Cannon, 1974). A major study of the environmental geochemistry of the entire state of Missouri as related to cardiovascular-renal disease in humans and congenital defects in swine has been made in cooperation with the University of Missouri (Tidball, 1971; Tidball and Sauer, 1975; Selby and Tidball, 1976). Unpublished studies include projects on a correlation of water hardness with cardiovascular-renal disease in North Carolina; gallstone chemistry and water supply in Denver; and the relation of water quality to dental health.

National Cancer Institute. The National Cancer Institute has become interested in possible environmental effects on the incidence of cancer and has recently published atlases or maps showing cancer mortality by county and economic area for many types of cancer from 1950–1969 for both whites and nonwhites (Mason *et al*., 1975, 1976).

National Science Foundation The National Science Foundation has supported a study of metals in the environment at the Oak Ridge National Laboratory for several years. This research has resulted in studies of cadmium in plants (Francis and Rush, 1973). NSF support has provided a workshop on cadmium, which resulted in a comprehensive report on cadmium in the environment (Fulkerson and Goeller, 1973). Extensive annotated bibliographies on arsenic (Copenhaver *et al*., 1973a), mercury (Caton *et al*., 1972) and cadmium (Copenhaver *et al*., 1973b) have been compiled.

Environmental Protection Agency. The U.S. Environmental Protection Agency has established Health Effects Research Laboratories which are issuing a series of publications on metallic elements. An example is a study published in 1975, "Molybdenum—a toxicological appraisal" (Friberg *et al*., 1975).

Cooperative Water Studies. The occurrence of trace elements in water has been studied by several government agencies. In particular, the areal distribution of certain elements in water supplies and the effect of treatment on the trace-element content has been studied by the U.S. Geological Survey. Samples of raw and finished waters have been analyzed for the past 19 years through a surveillance and monitoring program conducted by the Federal Water Pollution Control Administration (Kopp, 1970). The Environmental Protection Agency, which has also been conducting surveillance programs, has recently updated the U.S. Public Health Service drinking water standards (Federal Register, 1975).

University Programs

Universities in the United States are effectively organized to implement interdisciplinary environmental programs and, as a result, many have set up special environmental research centers with programs largely funded by the government.

The Department of Geology at the *University of Kansas* has shown the infeasibility of using Kansas salt mines as a repository for radioactive wastes, and has suggested a permanent repository under the Antarctic ice cap (Zeller *et al*., 1973). Arsenic as a water-pollution hazard and its existence in detergents has been studied (Angino *et al*., 1970).

Geochemists in the Department of Geosciences at *Pennsylvania State University* have been conducting research on the amounts and behavior of heavy metals dissolved or suspended in soil moisture and in ground and surface waters (Suarez and Langmuir, 1975; Back and

Langmuir, 1974). Of particular interest is heavy metal sorption and solution precipitation reactions.

The geochemical behavior of selenium in natural waters has been studied at the *University of Georgia* (Howard, 1972).

An ambitious 7-year interdisciplinary study of lead was sponsored by NSF-RANN at three universities: *Colorado State University, University of Missouri-Rolla* and *University of Illinois*. These studies both individually and jointly have resulted in several comprehensive publications including a study of environmental lead contamination, edited by H. W. Edwards (1974) at Colorado State University, and *Lead in the environment*, Boggess and Wixson (1977, in press). The lead studies have been followed by similar studies of zinc, cadmium and copper as related to lead mining, milling and smelting in southeastern Missouri. Research findings have been used by industry for pollution control and by state and federal agencies for the establishment of new guidelines and standards. Similarly, studies of lead at the University of Illinois have been followed by studies of cadmium and zinc in plants (Miller and Koeppe, 1970; Miller *et al.*, 1976; Root *et al.*, 1975).

D. J. Horvath at *West Virginia University* has investigated the availability of manganese and iron to plants and animals (1972).

The Department of Environmental Health, *University of Cincinnati,* on grants from the Public Health Service, has related trace elements in soils and plants to the use of snuff and the development of antral cancer (Baumslag and Keen, 1972); and also related the concentrations of metals in bituminous coals to the incidence of pneumoconiosis (Sorenson and others, 1974).

A comprehensive and interdisciplinary project on molybdenum in the environment was sponsored by NSF RANN for several years at the *University of Colorado* and *Colorado State University*. The geochemical aspects of the project were under the direction of D. D. Runnells (Runnells and others, 1975; La Gendre and Runnells, 1975; Runnells, 1973; Jorden and Meglen, 1973; Katz and Runnells, 1974). The studies were followed by an interdisciplinary symposium on molybdenum held in Denver. Volume 1 (*The Biology of Molybdenum in the Environment*) was released in 1976 (Chappell, 1976). Volume 2 (*The Geochemistry, Cycling and Industrial Uses of Molybdenum*) is scheduled for publication in 1977 (Chappell and Peterson, 1977).

A study of the origins, fate and control of mercury in water and in the biogeochemical environment was also funded by NSF at *Stanford University* (Stanford, 1973). The project has contributed much valuable information on the occurrence of mercury in a mercury district, its binding and eventual release by iron sulfide, the biological formation and decomposition of methyl-mercury, and solubility of inorganic and organic mercury in water (Burkstallar *et al.*, 1975; MacNaughton and James, 1974).

Departments on two separate campuses of the *University of California* have been actively involved in studying the effects of toxic levels of metals on plants. The Department of Soil Science and Agricultural Engineering at the University of California at *Riverside* has reported significant findings related to the uptake of cadmium by a wide number of plant species from soils of different chemical and physical properties. Their research provides basic data for the recognition of environmental situations where cadmium levels in forage or foods might be dangerous (Page and Bingham, 1973; Bingham *et al.*, 1976).

The Laboratory of Nuclear Medicine and Radiation Biology at the *UCLA* Campus, has reported the results of a transuranic-element studies program, studies of the effects at toxic levels copper (Berry, 1977), chromium (Wallace *et al.*, 1976), tin (Romney *et al.*, 1975) on plants; comparative toxicities and interactions between metals in plants (Berry, 1977); causes of iron deficiency in soils of different chemistry (Jones, 1976); and finally the potential toxicity of trace elements as related to fossil fuels (Berry and Wallace, 1974). Present studies involve environmental assessments in the environs of a coal-fired powerplant and a geothermal test site.

T. F. Yen of the *University of Southern California* has studied vanadium chelates in ancient and recent sediments, and recently reported that the shale oil and also the spent shale from the Green River oil shales contain considerable amounts of polycyclic aromatic hydrocarbons (Yen, 1972; T. F. Yen, written communication, January 21, 1977).

Sources of Research Support

To the Research Applied to National Needs (RANN) portion of the National Science Foundation (NSF) goes the major credit in the United States for encouraging and supporting the beginnings of environmental geochemistry and health (Table 2). The NSF Trace Contaminants Program of the Research Applications Division of Environmental Systems and Resources provided early support to the Subcommittee on the Geochemical Environment in Relation to Health and Disease at the National Academy of Sciences—National Research Council, the annual Trace Substances in Environmental Health Conferences at the University of Missouri-Columbia, the Trace Contaminants Program at the Oak Ridge National Laboratory; the Lead Project at Colorado State University, the University of Missouri-Rolla and the University of Illinois; the Molybdenum Project at the University of Colorado and Colorado State University; and numerous other grants to individual researchers.

Other government agencies soon expanded pre-existing environmental geochemistry and health programs of in-house research as well as newly extended grant or contract work through other government agencies, state, universities and private organizations. Major sponsoring agencies in these areas include: the U.S. Public Health Service, the Environmental Protection Agency, the National Cancer Institute, the U.S. Atomic Energy Commission [now the Energy Research and Development Administration], the U.S. Bureau of Mines, and the U.S. Geological Survey. Agencies with strong statutory interest in the subject include: the Food and Drug Administration, the National Institute of Environmental Health Sciences, the National Heart, Lung and Blood Institute, the National

TABLE 2. PRIMARY SOURCES OF FINANCIAL SUPPORT FOR ENVIRONMENTAL GEOCHEMISTRY AND HEALTH

Department/Agency	Bureau/Division
National Science Foundation	Research Applications (RANN)
Environmental Protection Agency	National Environmental Research Center
Department of Health, Education and Welfare	U.S. Public Health Service National Cancer Institute
U.S. Atomic Energy Commission (Now Energy Research and Development Administration)	Division of Biomedical and Environmental Research, Washington, D.C.
Department of the Interior	U.S. Bureau of Mines
Department of the Interior	U.S. Geological Survey

Institute of Arthritis, Metabolism and Digestive Diseases, and the National Institute of Aging. ERDA is reportedly now planning a large, million-sample, sampling program at universities.

Summary

This overly brief coverage of a broad subject has been further summarized in Table 3. The summaries, we have given, however, inevitably omit mention of work that should have been included and undoubtedly include some that might have been properly omitted. We accept the responsibility for both situations and regret that space does not permit us to carry on this review to the ultimate conclusion—the effects of trace substances present in the environment on man.

TABLE 3. SUMMARY OF ORGANIZATIONS WITH RESEARCH PROGRAMS AND PROJECTS IN ENVIRONMENTAL GEOCHEMISTRY AND HEALTH

National Academy of Sciences–National Research Council	University
Assembly of Mathematical and Physical Sciences	Kansas
Assembly of Life Sciences	Pennsylvania State
Commission on Natural Resources	Georgia
Commission on Sociotechnical Systems	North Carolina
	New York at Binghamton
	Colorado State
	Missouri
	Illinois
Federal Government	West Virginia
Department of the Interior	Cincinnati
Department of Health, Education, and Welfare	Colorado
Department of Agriculture	Stanford
National Science Foundation	California at Riverside
Environmental Protection Agency	California at Los Angeles
	Southern California

References

ALLAWAY, W. H. (1968) Control of the environmental levels of selenium. In *Trace Substances in Environmental Health*, Vol. II, D. D. HEMPHILL, ed., University of Missouri, Columbia, Mo. 181–206.

ANGINO, E. E., MAGNUSON, L. M., WAUGH, T. C., GALLE, O. K. and BREDFELDT, J. (1970) Arsenic and water pollution hazard. *Science* **170**, 871–872.

Assembly of Life Sciences (1976) *1975–1976 Report*, National Academy of Sciences, Washington, D.C. 146 pp.

Assembly of Mathematical and Physical Sciences (1976) *Annual Report of the Assembly of Mathematical and Physical Sciences for fiscal year 1976 (July 1, 1975–June 30, 1976)*, National Academy of Sciences, Washington, D.C. 161 pp.

BACK W. and LANGMUIR, D. (1974) Influence of near-surface reactions and processes on chemical character of ground water: A review of selected experiences in the United States. *Mem. Ass. Int. Hydrogéologiques* **10**, no 1, 31–37.

BAUMSLAG, N. and KEEN, P. (1972) Trace elements in soils and plants and antral cancer. *Arch. Envir. Health* **25**, 23–25.

BERRY, W. L. (1977) Comparative toxicity of V, Cr, Mn, Co, Ni, Cu, Zn, and Cd to lettuce seedlings. In *Environmental Chemistry and Cycling Processes Proc.*, ERDA Symposium series, D. C. ADRIANO and I. L. BRASKIN, eds. (in press).

BERRY, W. L. and WALLACE, A. (1974) Trace elements in the environment—their role and potential toxicity as related to fossil fuels: A preliminary study. University of California Lab of Nuclear Medicine and Radiation Biology, UCLA no. 12–946. AEC Contract PT(04–1) GEN–12 (Nat. Tech. Inform. Service).

BINGHAM, F. T., PAGE, A. L., MAKLER, R. J. and GANJE T. J. (1976) Cadmium availability to rice in sludge-amended soil under "flood" and "non-flood" culture. *Soil Sci. Soc. Am. J.* **40**(5), 715–719.

BOGGESS, W. R. and WIXSON, B. G., eds. (1977, in press) *Lead in the Environment*, National Science Foundation, Washington, D. C.

BURKSTALLER, J. E., MCCARTHY, P. L. and PARKS, G. A. (1975) Oxidation of cinnabar by acid mine waters. *Envir. Sci. Techn.* **9**, 676–678.

CANNON, H. L., ed. (1974) *Geochemistry and the Environment*. Vol. I: *The Relation of Selected Trace Elements to Health and Disease, Proc. of Workshop of the Subcommittee on GERHD at Asilomar, Pacific Grove, Calif., February 7–12, 1972.* National Academy of Sciences, Washington, D.C. 113 pp.

CANNON, H. L. (1977) The geochemical environment as related to the distribution of cancer in Washington County, Maryland. Inst. for Environmental Studies. *Int. Heavy Metals Conferences*, Vol. II.

CANNON, H. L. (in press) *The Geochemical Environment as Related to the Distribution of Cancer in Washington County, Maryland*, University of Toronto, Canada.

CANNON, H. L. and DAVIDSON, D. F. eds., (1967) Relation of geology and trace elements to nutrition. Symposium at Geol. Soc. Am. Meeting, New York, 1963. *Geological Society of America Special Paper* No. 90, 67 pp. ($4.25).

CANNON, H. L. and HOPPS, H. C., eds. (1971) Environmental geochemistry in health and disease. Symposium at AAAS Meeting, Dallas, 1968. *Geol. Soc. Amer. Memoir* No. 123, 230 pp. ($14.00).

CANNON, H. L. and HOPPS, H. C., eds. (1972) Geochemical environment in relation to health and disease. Symposium at AAAS Meeting, Chicago, 1970. *Geol. Soc. Am. Special Paper* No. 140, 77 pp. ($4.00).

CANNON, H. L., HARMS, T. F. and HAMILTON, J. C. (1975) Lithium in unconsolidated sediments and plants of the Basin and Range Province, Southern California and Nevada. *U.S. Geol. Surv. Prof. Paper* 918. U.S. Government Printing Office, Washington, D.C., 23 pp.

CATON, G. M., OLIVEIRA, D. P., OEN, C. J. and ULRIKSON, G. U. (1972) *Mercury in the Environment—an Annotated Bibliography*, ORNL–EIS–71–8. Oak Ridge National Laboratory, Oak Ridge, Tenn., 271 pp.

CHAPPELL, W. R. and PETERSON, K. K. eds., (1976) *The Biology of Molybdenum*, Vol. 1, *Molybdenum in the Environment*, University of Colorado.

CHAPPELL, W. R. and PETERSON, K. K., eds. (1977) *The Geochemistry, Cycling, and Industrial Uses of Molybdenum*, Vol. 2 of *Molybdenum in the Environment*, Int. Symp. Proc., held in Denver, Co. Marcel Dekker, New York.

CONNOR, J. J. and SHACKLETTE, H. T., with sections on field studies by EBENS, R. J., ERDMAN, J. A., MIESCH, A. T., TIDBALL, R. R. and TOURTELOT, H. A. (1975) Background geochemistry of some rocks, soils, plants, and vegetables in the conterminous United States. *U.S. Geol. Surv. Prof. Paper* 574–F. U.S. Government Printing Office, Washington, D. C., 168 pp.

CONNOR, J. J., KEITH, J. R. and ANDERSON, B. M. (1976) Trace-metal variation in soils and sagebrush in the Powder River Basin, Wyoming and Montana. *J. Res. U.S. Geol. Surv.* **4**(1), 49–59.

COPENHAVER, E. D., ULRIKSON, G. U., NEWMAN, L. T. and VAN HOOK, R. J. JR. (1973a) *Arsenic in the Environment, an Annotated Bibliography,* ORNL–EIS–73–16. Oak Ridge National Laboratory, Oak Ridge, Tenn., 310 pp.

COPENHAVER, E. D., ULRIKSON, G. U., NEWMAN, L. T. and FULKERSON, W. (1973b) *Cadmium in the Environment, an Annotated Bibliography,* ORNL–EIS–73–17. Oak Ridge National Laboratory, Oak Ridge, Tenn., 451 pp.

DURFOR, C. N. and BECKER, E. (1964) Public water supplies of the 100 largest cities in the United States, 1962. *U.S. Geol. Surv. Water Supply Paper* No. 1812. U.S. Government Printing Office, Washington, D.C., 264 pp.

EDWARDS, H. W., ed. (1974) *Environmental Contamination Caused by Lead*, Colorado State University, Fort Collins, Colo.

ENGBERG, R. A., (1973) Selenium in Nebraska's groundwater and streams. *Nebraska Water Survey Paper* No. 35, 42 pp.

Federal Register, Wednesday, December 24 (1975) National interim primary drinking water regulations. U.S. Government Printing Office, Washington D.C., pp. 59566–59588.

FLEISCHER, M., (1967) Fluoride content of ground water in the conterminous United States. In *Relation of Geology and Trace Elements to Nutrition,* H. L. CANNON and D. F. DAVIDSON, eds., *Geol. Soc. Am. Special Paper* No. 90. Geological Society of America, New York, NY, 65.

FLEISCHER, M., SAROFIM, A. F., FASSETT, D. W., HAMMOND, P., SHACKLETTE, H. T., NISBET, I. C. T. and EPSTEIN, S. (1974) Environmental impact of cadmium: A review by the Panel on Hazardous Trace Substances. *Envir. Health Perspectives* No. 7, 253–323.

FRANCIS, C. W. and RUSH, S. G. (1973) Factors affecting uptake and distribution of cadmium in plants. In *Trace Substances in Environmental Health*, Vol. VII, D. D. HEMPHILL, ed., University of Missouri, Columbia, Mo., 75–81.

FRIBERG, L., BOSTON, P., NORDBERG, G., PISCATOR, M. and ROBERT, K-H. (1975) *Molybdenum—A Toxicological Appraisal*, EPA–600/1–75–004. Office of Research and Development, Environmental Protection Agency, Research Triangle Park, NC.

FULKERSON, W. and GOELLER, H. E., eds. (1973) *Cadmium, the Dissipated Element*, ORNL–NSF–EP–21. Oak Ridge National Laboratory, Oak Ridge, Tenn. 473 pp.

HEM, J. D., (1970) Chemical factors that influence the availability of iron and manganese in aqueous systems. In *Geochemical Environment in Relation to Health and Disease*, H. L. CANNON and H. C. HOPPS, eds., *Geol. Soc. Am. Special Paper* No. 140, 17–24.

HEMPHILL, D. D., ed., (1967, 1969–1976) Trace substances in environmental health. *Proc. of Annual Conferences at the University of Missouri, Columbia.*

HODGSON, J. F., (1969) Chemistry of trace elements in soils with reference to trace element concentration in plants. In *Trace Substances in Environmental Health*, Vol. III, D. D. HEMPHILL, ed., University of Missouri, Columbia, 45–57.

HODGSON, J. F., ALLAWAY, W. H. and LOCKMAN, R. B. (1971) Regional plant chemistry as a reflection of environment. *Geol. Soc. Amer. Mem.* **123**, 57–72.

HOPPS, H. C., ed., (1977, in prep.) *Geochemistry and the Environment*, Vol. III: *Distribution of Trace Elements Related to the Occurrence of Certain Cancers, Cardiovascular Diseases, and Urolithiasis*, Subcommittee on the Geochemical Environment in Relation to Health and Disease, National Academy of Sciences, Washington, D.C.

HOPPS, H. C. and CANNON, H. L., eds. (1972) Geochemical environment in relation to health and disease. Symposium by the New York Academy of Sciences, New York, October 4–6, 1971. *Ann. N.Y. Acad. Sci.* **199,** 1–352.

HORVATH, D. J. (1972) Availability of manganese and iron to plants and animals. In *Geochemical Environment in Relation to Health and Disease*, H. L. CANNON and H. C. HOPPS, eds., *Geol. Soc. America Special Paper* 140, 33–47.

HOWARD, J. H., III (1972) Control of geochemical behavior of selenium in natural waters by absorption on hydrous ferric oxides. In *Trace Substances in Environmental Health*, Vol. V, D. D. HEMPHILL, ed., University of Missouri, Columbia, 485–495.

JONES, J. B., JR., ed. (1976) A series of papers from one laboratory on iron deficiency in plants and its correction. *Comm. Soil Sci. Plant Anal.* **7**(1), 1–127.

JORDEN, R. M. and MEGLEN, R. R. (1973) Aqueous release of molybdenum from non-point and point sources. In *Trace Substances in Environmental Health*, Vol. VII, D. D. HEMPHILL, ed., University of Missouri, Columbia, 105–116.

KATZ, B. G. and RUNNELLS, D. D. (1974) Experimental sorption of molybdenum by desert, agricultural and alpine soils. In *Trace Substances in Environmental Health*, Vol. VIII, D. D. HEMPHILL, ed., University of Missouri, Columbia, 107–114.

KOPP, J. F., (1970) The occurrence of trace elements in water. In *Trace Substances in Environmental Health*, Vol. III, D. D. HEMPHILL, ed., University of Missouri, Columbia, 59–73.

KUBOTA, J., ALLAWAY, W. H., CARTER, D. L., CARY, E. E. and LAZAR, V. A. (1967) Selenium in crops in the United States in relation to the selenium–responsive diseases of livestock. *J. Agric. Food Chem.* **15**, 448–453.

LA GENDRE, G. R. and RUNNELLS, D. D. (1975) Removal of dissolved molybdenum from waste waters by precipitates of ferric iron. *Envir. Sci. Techn.* **9,** 744–749.

LAGERWERFF, J. V. and BIERSDORF, G. T. (1972) Interaction of zinc with uptake and translocation of cadmium in radish. In *Trace Substances in Environmental Health*, Vol. V, D. D. HEMPHILL, ed., University of Missouri, Columbia, 515–522.

LOVERING, T. G., ed. (1977) Lead in the environment. *U.S. Geol. Surv. Prof. Paper* No. 957. U.S. Government Printing Office, Washington, D.C., 90 pp.

MASON, T. J., MCKAY, F. W., HOOVER, R., BLOT, W. J. and FRAUMENI, J. F. JR. (1975) *Atlas of Cancer Mortality for U.S. Counties: 1950–1969,* DHEW Pub. No. (NIH) 75–780. U.S. Department of Health, Education, and Welfare, Washington, D.C., 103 pp.

MASON, T. J., MCKAY, F. W., HOOVER, R., BLOT, W. J. and FRAUMENI, J. F. JR. (1976) *Atlas of Cancer Mortality among U.S. Nonwhites: 1950–1969*, DHEW Pub. No. (NIH) 76–1204. U.S. Department of Health, Education, and Welfare, Washington, D.C., 142 pp.

MACNAUGHTON, M. G. and JAMES, R. O. (1974) Adsorption of aqueous Hg(II) complexes at the oxide/water interface. *J. Coll. Interf. Sci.* **47,** 431.

MCDUFFIE, B., KEMPNER, S. M. and GOLTZ, R. D. (1974) Cardiovascular death rates in three neighbouring upstate New York communities with differing water hardness. In *Trace Substances in Environmental Health*, Vol. VIII, D. D. HEMPHILL, ed., University of Missouri, Columbia, 75–80.

MERTZ, W., ed. (1977) *Geochemistry and the Environment*, Vol. II: *The Relation of Other Selected Trace Elements to Health and Disease*, Subcommittee on the Geochemical Environment in Relation to Health and Disease, National Academy of Sciences, Washington, D.C., 163 pp.

MILLER, J. E., HASSETT, J. J. and KOEPPE, D. E. (1976) Uptake of cadmium by soybeans as influenced by soil cation exchange capacity, pH, and available phosphorus. *J. Envir. Qual.* **5** (2), 157–160.

MILLER, R. J. and KOEPPE, D. E. (1971) Accumulation and physiological effects of lead in corn. In *Trace Substances in Environmental Health*, Vol. IV, D. D. HEMPHILL, ed., University of Missouri, Columbia, 186–193.

MORAN, R. E. (1976) Geochemistry of selenium in ground water near Golden, Jefferson County, Colorado. *Geol. Soc. Am. Abstracts with Programs,* 1018.

National Academy of Sciences—National Academy of Engineering—Institute of Medicine (1976) *Organization and Members, 1976–1977,* National Academy of Sciences, Washington, D.C., 239 pp.

PAGE, P. L., and BINGHAM, F. T. (1973) Cadmium residues in the environment. *Residue Rev.* **48**, 1–43.

ROMNEY, E. M., WALLACE, A. and ALEXANDER, G. V. (1975) Responses of husk bean and barley to tin applied to soil and to solution culture. *Plant and Soil* **42**, 585–589.

ROOT, R. A., MILLER, R. J. and KOEPPE, D. E. (1975) Uptake of cadmium—its toxicity and effects on the iron ratio in hydroponically grown corn. *J. Envir. Qual.* **4** (4), 473–476.

RUNNELLS, D. D., (1973) Detection of molybdenum enrichment in the environment through comparative study of stream drainages, Central Colorado. In *Trace Substances in Environmental Health*, Vol. VII, D. D. HEMPHILL, ed., University of Missouri, Columbia, 99–104.

RUNNELLS, D. D., CHAPPELL, W. R. and MEGLEN, R. (1975) The molybdenum project: Geochemical aspects. *Geol. Soc. America Special Paper* **155**, 61–72.

SELBY, L. A., and TIDBALL, R. R. (1977, in press) The environmental geochemistry of agricultural soils and its association with malformations in domestic swine. In *Trace Substances in Environmental Health*, Vol. X, D. D. HEMPHILL, ed., University of Missouri, Columbia.

SHAMBERGER, R. J. and WILLIS, C. E. (1971) Selenium distribution and human cancer mortality. *CRC Crit. Rev. Clin. Lab. Sci.* **2,** 211–221.

SHAMBERGER, R. J., TYTKO, S. A. and WILLIS, C. E. (1975) Selenium and heart disease. In *Trace Substances in Environmental Health*, Vol. IX, D. D. HEMPHILL, ed., University of Missouri, Columbia, 15–22.

SHACKLETTE, H. T., SAUER, H. I. and MIESCH, A. T. (1970) Geochemical environments and cardiovascular mortality rates in Georgia. *U.S. Geol. Surv. Prof. Paper* No. 574-C. U.S. Government Printing Office, Washington, D.C., C1–C39.

SHACKLETTE, H. T., BOERNGEN, J. G. and TURNER, R. L. (1971a) Mercury in the environment—surficial materials of the conterminous United States. *U.S. Geol. Surv. Circular* 644. U.S. Government Printing Office, Washington, D. C., 5 pp.

SHACKLETTE, H. T., HAMILTON, J. C., BOERNGEN, J. G. and BOWLES, J. M. (1971b) Elemental composition of surficial materials in the conterminous United States. *U.S. Geol. Surv. Prof. Paper* No. 574–D. U.S. Government Printing Office, Washington, D.C., 71 pp.

SHACKLETTE, H. T., BOERNGEN, J. G., CAHILL, J. P. and RAHILL, R. L. (1973) Lithium in surficial materials of the conterminous United States and partial data on cadmium. *U.S. Geol. Survey Circular* 673. U.S. Government Printing Office, Washington, D.C., 8pp.

SHACKLETTE, H. T. BOERNGEN, J. G. and KEITH, J. R. (1974) Selenium, fluorine, and arsenic in surficial materials of the conterminous United States. *U.S. Geol. Survey Circular* 692. U.S. Government Printing Office, Washington, D.C., 14 pp.

SIEVERS, M. L. and CANNON, H. L. (1974) Disease patterns of Pima Indians of the Gila River Indian Reservation of Arizona in relation to the geochemical environment. In *Trace Substances in Environmental Health*, Vol. VII, D. D. HEMPHILL, ed., University of Missouri, Columbia, 43–55.

SKOUGSTAD, M. W. and HORR, C. A. (1963) Chemistry of strontium in natural water. *U.S. Geol. Survey Water Supply Paper* 1496. U.S. Geological Survey, Reston, Virginia, 97 pp.

Society for Environmental Geochemistry and Health (1972) ff. *Interace, the Quarterly Newsletter of the Society*, D. J. HORVATH, ed., Univ. West Virginia, Morgantown.

SORENSON, J. R., KOBER, T. E. and PETERING, H. G. (1974) The concentration of Cd, Cu, Fe, Ni, Pb, and Zn in bituminous coals from mines with differing incidences of coal workers' pneumoconiosis. *Amer. Indust. Hygiene Ass. J.* 93–98.

Stanford University (1973) Trace elements in water: origin, fate, and control. 1. *Mercury; Report of progress, March 1, 1972–February 1, 1973.* G1–32943. Stanford University, Stanford, Calif.

SUAREZ, D. L., and LANGMUIR, D. (1976) Heavy metal relationships in a Pennsylvania soil. *Geochim. Cosmochim. Acta* **40**, 589–598.

TIDBALL, R. R., (1971) Geochemical variation in Missouri soils. In *Trace Substances in Environmental Health*, Vol. IV, D. D. HEMPHILL, ed., University of Missouri, Columbia, 15–25.

TIDBALL, R. R. and SAUER, H. I. (1975) Multivariate relationships between soil composition and human mortality rates in Missouri. *Geol. Soc. America Special Paper* 155, 41–59.

U.S. Geological Survey, (1970) Mercury in the environment. *U.S. Geol. Survey Prof. Paper* No. 713. U.S. Government Printing Office, Washington D.C., 67 pp.

VOORS, A. W., (1971) Atherosclerotic heart disease and drinking water trace metals in North Carolina. In *Trace Substances in Environmental Health*, Vol. V, D. D. HEMPHILL, ed., University of Missouri, Columbia. 523–534.

WALLACE, A., SOUFI, S. M., CHA, J. W. and ROMNEY, E. M. (1976) Some effects of chromium toxicity on bush bean plants grown in soil. *Plant and Soil,* **44**, 471–473.

YEN, T. F., (1972) Vanadium chelates in recent and ancient sediments. In *Trace Substances in Environmental Health*, Vol. VI, D. D. HEMPHILL, ed., University of Missouri, Columbia, 347–353.

ZELLER, E. J., SAUNDERS, D. F. and ANGINO, E. E. (1973) Putting radioactive waste on ice. *Bull. Atomic Scient.*, Jan., 4–52.

Discussion

I. THORNTON:

Has the National Academy of Sciences set up a body to make recommendations on analytical methods, etc., to ensure compatibility and avilability of analytical data between the many institutes and laboratories involved in multidisciplinary studies on geochemistry and health?

W. L. PETRIE:

To an extent, yes. Although the Subcommittee on the Geochemical Environment in Relation to Health and Disease frequently addresses such problems, as do other Committees from time to time, there is no one body with this as an exclusive function.

Sources and Sediment Associations of Heavy Metals in Polluted Coastal Regions

ULRICH FÖRSTNER

Institut für Sedimentforschung der Universität, Im Neuenheimer Feld 236, D-69 Heidelberg, West Germany

Abstract

Vertical profiles and lateral distribution patterns of heavy metals in Recent sediments are increasingly employed in the assessment of the major sources of pollution in coastal regions, including estuaries and marginal seas. After a general review of studies from typical locations, examples from New Bedford Harbor, Mass./U.S.A., and from the southern German Bight in the North Sea are described in more detail.

Chemical extraction procedures are used to evaluate different forms of metal associations in aquatic sediments: (i) heavy metals are transported and deposited as major, minor and trace constituents in the natural rock detritus, mostly in inert positions; (ii) slight alkalinity in sea water results in the formation and precipitation of hydroxides, carbonates, sulphides and phosphates of heavy metals; (iii) sorption and cation exchanges takes place in particular on fine-grained substances with large surface areas, e.g. on manganese dioxide, humic acids, hydrous iron oxides and clay minerals; (iv) organic substances play an important role in the transfer of metals between dissoved and solid phases, particularly by the process of flocculation; (v) under oxidizing conditions, the redox-sensitive Fe- and Mn-hydroxides and -oxides are significant "sinks" of heavy metals in aquatic systems. The characteristic rise in the residual content of lead, copper and cadmium in the polluted samples points toward the strong influence of atmospheric dust particles.

Heavy metals immobilized in the bottom sediment constitute a potential hazard to water quality and aquatic life since they may be released as a result of chemical changes. In relation to the marine environment, the following mechanisms should be taken into account: (i) increased salinities at the inflow of rivers into the sea water will affect the partial exchange of metal cations sorbed onto the surface of suspended particles; (ii) oxygen deficiency in the bottom sediments may lead to the dissolution of hydrous iron and manganese oxides and to a release of co-precipitated heavy metals; (iii) complexing substances of both natural and synthetic origin increases the solubilization of metals; (iv) microbial activities can convert inorganic metal compounds to organic molecules and may thereby increase toxicity, as exemplified in organo-mercurial transformations.

Introduction

Pollutant concentrations in sediments serve as a convenient means of obtaining an indication of the state of associated waters. Vertical sediment profiles preserve the historical sequence of pollution intensities; lateral distributions are used to determine and evaluate local sources of pollution. Sediment analyses, however, are not only useful when selecting critical sites for routine water sampling, but they can reveal the fate of contaminants under varying environmental conditions. This is particularly true of the heavy metals: their availability for metabolic processes is closely related to their chemical species in particulate substances. In the last 10 years the sediment method has successfully been used in the detection of pollution sources in lakes, rivers and marine coastal areas, especially in North

America, Western Europe and Japan. The latest studies deal increasingly with the differentiation of the chemical bonding types in investigations of polluted sediments. The present review covers the sources, distribution and possible changes of characteristic forms of trace metal compounds associated with sediments in polluted coastal regions.

Sources and Distribution of Heavy Metals in Coastal Sediments

The metal contamination of coast areas, including estuaries and marginal seas, can generally be attributed to one of the following sources (see Weichart, 1973): (i) the direct input of effluents from industries and communities near the coast; (ii) the dumping of waste from ships and effluents from communities (and, more seldom from industries) via sewer outfalls; (iii) contamination as a result of atmospheric fallout; (iv) input of soluble and suspended loads from rivers.

In the following section, typical examples of individual pollution sources will be given, in particular those which are responsible for the distribution of the dangerous metal poisons—mercury and cadmium.

1. Direct Input of Heavy Metals from Effluent

The most tragic example of industrial metal pollution is that of Minamata, Japan, where forty-six people died and many others still suffer from permanent damage to their health due to mercury discharges into the sea. According to Kitamura's (1968) investigations (Fig. 1a) more than 2000 mg/kg of mercury were in the sediment at the channel inflow to Minamata Bay. The sediments of the central part of the bay contained mercury concentrations between 40 and 60 mg/kg of Hg. In the estuary leading to the open sea, sediment samples taken approximately 1 to 1.5 km away from the discharge point contained 22 mg of Hg per kg and 12 mg of Hg per kg respectively. In the coastal zone adjacent to Minamata Bay, 0.4 to 3.4 mg of Hg per kg sediment were measured (Takeuchi, 1972).

In a study of sediments from the New Haven Harbor (Conn.), Applequist *et al.* (1972) attempted to asses the sources of pollution in a typical industrial harbor from the distribution of mercury in fine-grained sediments. As was the case in a number of other investigations, a correlation could be established between the mercury and the silt/clay fraction rich in organic matter. A more important feature, however, was the direct correlation between the highest mercury concentrations and the locations of major municipal sewage inflows (Fig. 1b).

Such unregulated effluents from communities and industries from harbor areas, estuaries and marine coastal regions are quite widespread. Tests with trace metals from bottom sediments were performed† in a number of bays and harbors in Japan,[(a)] Australia,[(b)] North America[(c)] and Europe[(d)].

† Examples of sediment studies in coastal areas:

[(a)] *Japan*: Saroma, 1972; Goto, 1973; Nisimura, 1974; Ito, 1975; Murakami *et al.*, 1975; Momoayama *et al.*, 1975; Matsumoto and Yukota, 1976.

[(b)] *Australia*: Talbot *et al.*, 1976; Knauer, 1976.

[(c)] *North America*: (West Coast): Young *et al.*, 1973; Bothner and Carpenter, 1974; Crecelius and Carpenter, 1974; Crecelius *et al.*, 1975; Grieve and Fletcher, 1975; Bradford, 1976; Bertine and Goldberg, 1977; Moyer and Budinger, 1974; Schell and Nevissi, 1977. (East Coast): Bopp and Biggs, 1973; Capuzzo and Anderson, 1973; Segar and Pellenbarg, 1973; Holmes *et al.*, 1974; Villa and Johnson, 1974; Brinkman and Iverson, 1975; Carpenter *et al.*, 1975; Lindberg *et al.*, 1975; Armstrong *et al.*, 1976; Helz, 1976; Goldberg *et al.*, 1977; Greig and McGrath, 1977; Greig *et al.*, 1977.

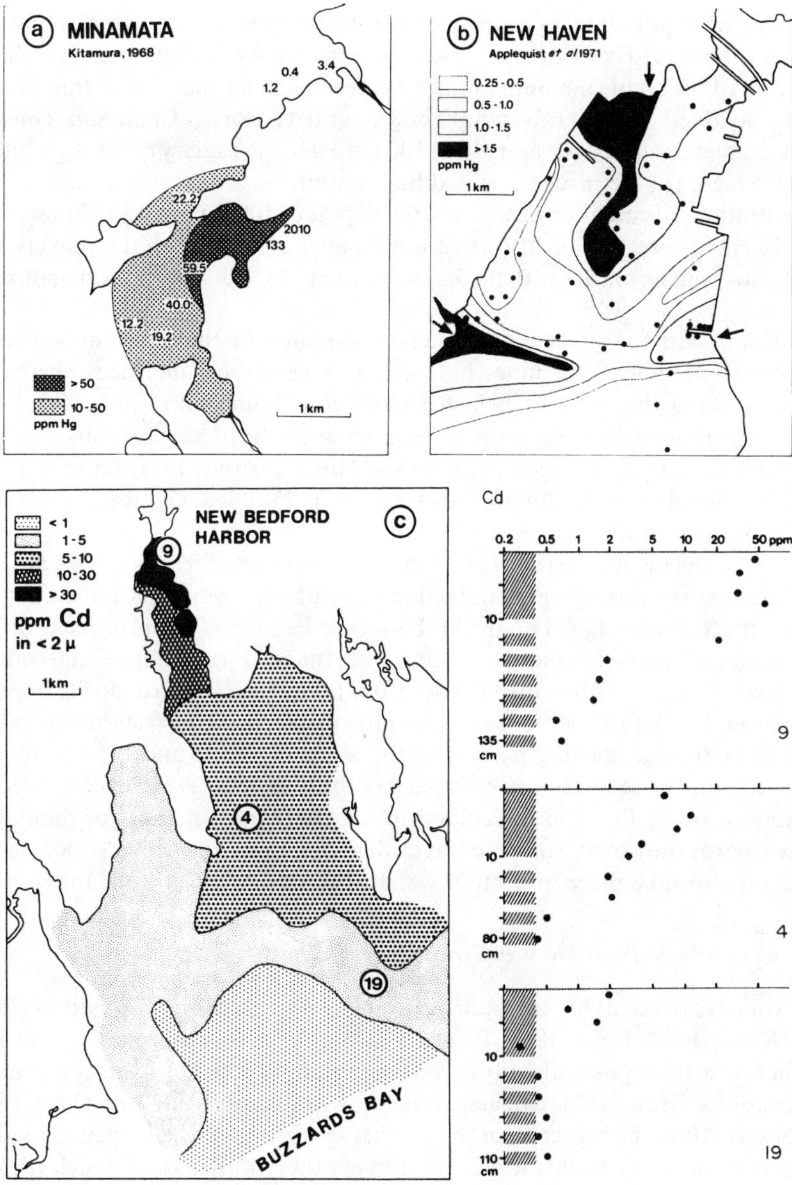

FIG. 1. Sediment concentrations of mercury and cadmium from unregulated effluents ([c]: unpubl. data, Inst. f. Sedimentforschung, Heidelberg).

(d) *Europe* (Great Britain) Elderfield *et al.*, 1971; Covill, 1971/2; Bloxam *et al.*, 1972; Perkins *et al.*, 1973; Smith *et al.*, 1973; Thornton and Webb, 1973; Abdullah and Royle, 1974; Boyden, 1975; Thornton, 1975; Bevan *et al.*, 1975. (Southern North Sea): De Groot *et al.*, 1971; Martin *et al.*, 1973; Little-Gadow and Schäfer, 1974; Nauke in Caspers, 1975; Dörjes *et al.*, 1976; Wollast, 1976, Salomons and De Groot, 1977. (Baltic Sea): Hallberg, 1973; Hasselrot and Björklund, 1974; Olausson, 1975; Suess and Erlenkeuser, 1975; Zink-Nielsen and Grogh, 1976; Cato, 1977; Olausson *et al.*, 1977. (Mediterranean): Cumont *et al.*, 1972; Renzoni *et al.*, 1973; Aubert, 1975; Grancini *et al.*, 1976; Roth and Hornung, 1977.

The severe metal pollution, presumed to be from metallurgical industries along the coastal area off New Bedford Harbor (Mass.), was investigated by Stoffers *et al.* (1977). Figure 1c shows the distribution of cadmium in the clay-sized (smaller than 2 μm) fraction of surface sediments collected from grab samples and gravity cores. Cadmium concentrations gradually decrease seaward for more than 100 mg/kg in the estuary of the Acushnet River to less than 0.5 mg/kg at the edge of Buzzard Bay. Within the core profiles, marked increases in the concentrations of cadmium occur at burial depths of 100 cm in core 9, 50 cm in core 4 and at about 20 cm in core 19. The history of waste metal accumulation thus covers a period of about 80 years which concurs with the known history of Cu discharge from industrial plants in this area.

When dealing with the direct input of metal pollutants into the sea, mention must be made of the disposal of tailings from mines and smelters. One of the main places where this is done at present is along the coast of British Colombia (Thompson *et al.*, 1974; Ellis, 1975), although strong metal accumulation from mine waste has also been observed in coastal sediments off southwestern England (Bryan and Hummerstone, 1971; Thornton *et al.*, 1975; Yim, 1976), as well as in the Philippines (Lesaca, 1975), New Guinea (Brown, 1974) and Tasmania (Thrower and Eustace, 1973).

Particular problems may arise when mine and other metalliferous wastes are released in fjords where the exchange of water and sediments with the open sea is limited. Examples are given from the Sagueney Fjord on the St. Lawrence Estuary which is injected with mercury from industrial effluents (Lorning, 1975) and from the Ide Fjord on the Skagerrak where the mercury level of surface sediments increased 100-fold compared to deposits from 100 years ago (Olausson, 1975). Forty-five tons of lead enter the Sörfjord in Western Norway daily and 60 tons of zinc from a smelting plant whereby levels of 7% Pb and 12% Zn by weight are reached in the bottom sediments in some parts of the fjord (Skei *et al.*, 1972, 1973). In the case of the Agfardlikavsa Fjord in Greenland into which large amounts of mine waste were introduced, it was shown that, under unfavorable conditions, the strong enrichment of heavy metals could also take place in bottom water (Thompson, 1975; Bondam *et al.*, 1976).

2. *Dumping of Waste from Ships and via Sewer Outfalls*

Large volumes of solids are transported to the sea by barges or disposed of from sewage outfalls (Koch, 1960; Føyn, 1965; Brown and Smith, 1969; Andrulewica and Portmann. 1975). These coastal deposits already exceed the amount caused by natural sedimentation in several locations. Gross (1971) calculated that waste discharges into Long Island Sound are, at 10^6 tons/yr, 10 times higher than the discharge of river-borne suspended sediment.

The amount of heavy metals which accompany these wastes is very high (Pearce, 1969; gross, 1972; Ali *et al.*, 1975). Studies on metal distribution in the sediments of the New York Bight established that in an area of approximately 60 km^2, the concentration of copper, Similar investigations on sewage sludge disposals and the effects on the metal composition of marine sediments were carried out in the Black River Estuary (Md.) and the Delaware Bay of marine sediments were carried out in the Back River Estuary (Md.) and the Delware Bay of the eastern coast of the U.S.A. (Helz *et al.*, 1975; Szucs and Oostdam, 1975), in the Liverpool Bay, Swansea Bay, Tor Bay and Thames Estuary of Great Britain (Wood *et al.*, 1973/74; Clifton and Vivian, 1975; Taylor, 1974; Shelton, 1971), and in the vicinity of the Athens sewage outfall (Papakostidis *et al.*, 1975). Metal analyses from bottom sediments of an area in the Firth of Clyde in Scotland (Mackay *et al.*, 1972; Halcrow *et al.*, 1973) where sludge

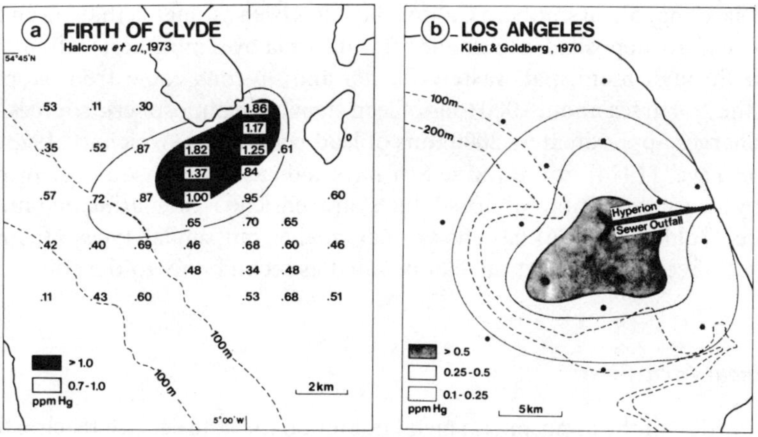

FIG. 2. Sediment concentrations of mercury from sewage disposal.

from the city of Glasgow is dumped into the sea confirmed the findings of Applequist *et al.* (1972) that the mercury content may serve as an indicator of sewage discharges (Fig. 2a): based on a background value of approximately 0.1 mg of Hg per kg in the less polluted parts of the Firth of Clyde, an increase in mercury concentrations by a factor of 20 and more could be detected in the centre of the disposal area.

Further evidence for the "discrete association of mercury with sewage sludge" (MacKay *et al.*, 1973; Halcrow *et al.*, 1973) can be found in sediments near sewer outfalls. Figure 2b gives an example from the southern Californian coast which was studied by Klein and Goldberg (1970). Near the outfall, the mercury content of the sediment is as much as 50 times higher than the presumably uncontaminated sediments further away. From sediment studies, Young and his team (1975) demonstrated that the annual inputs of mercury, DDT and PCB from municipal waste waters are of an order of magnitude higher than those from direct industrial discharges and storm runoff.

Industrial metal waste is a particular problem. In the southern North Sea, on the east coast of the United States and in the Mediterranean, several million tons of acid iron waste is annually dumped from special barges (Weichart, 1973; Gross *et al.*, 1971; Naeve, 1974); the question of dumping alkaline "red mud" from bauxite extraction is still unresolved due to the harmful effects it would have on various filter and suspension feeders (Rosenthal *et al.*, 1973).

3. *Metal Enrichment from Atmospheric Fallout*

Murozumi *et al.* (1969) presented classic proof of the world-wide lead contamination by atmospheric fallout from studies of core profiles in ice in northern Greenland. This pollution can be attributed to coal burning in the past and to the increased use of lead additives in gasoline in the last 40 years. Further studies by Chester and Stoner (1973[a,b]) clearly showed that low-level marine atmospheric particulates are strongly enriched in lead; particularly heavy lead concentrations were measured in the northern latitudes between 30° and 50° over the eastern Atlantic.

Chow *et al.* (1973) demonstrated that by their isotope ratios the lead accumulations in sediment cores from Baja California can be derived from lead additives such as are present in the type of gasoline sold in southern California; it has been estimated than on an annual basis 180 tons of lead are deposited into a zone off California by atmospheric fallout, 200 tons are discharged through municipal waste effluents and 90 tons come from storm and river runoff. In the North sea about 1000 tons of lead stem from atmospheric sources, whereas the Rhine discharges approximately 2000 tons of lead per annum (Weichart, 1973). Studies by Erlenkeuser *et al.* (1974) in the Baltic Sea have indicated that a number of other metals released by the combustion of fossil fuels are enriched in marine sediments via the atmosphere. Goldberg (1976) has shown that smaller but similar types of emissions, as a consequence of cement production, can be rated as second only to the effects of fossil fuel burning.

4. Metal Input from Rivers

A large portion of the atmospheric metal fallout on the land surface reaches the rivers via soil erosion and is transported to the sea together with waste materials from municipal and industrial effluents. In many rivers the anthropogenic metal discharges strongly exceed the metal concentrations from natural sources; in the lower Rhine River, for example, only 2% of the sediment-associated cadmium originates from natural weathering products, 98% of the Cd concentrations in the particulate matter is introduced either from direct waste effluent, or indirectly from surface erosion and from sewage treatment plants (Förstner and Müller, 1974).

In the past two decades there has been a general trend towards an increased level of heavy metals, particularly of the most toxic ones—mercury and cadmium—in river sediments and consequently higher amounts will also be found in estuarine and coastal sediments from the mouth of the Rhine River during the decade from 1960 to 1970; for cadmium the increase was a high as 94% (De Groot *et al.*, 1973). Müller and Förstner (1976) measured a tripling of the mercury content of clay size sediments in the Elbe estuary over a 3-year period (Jan. 1973–Oct. 1975).

Metal contamination can extend great distances from the mainland into the sea with the sediment load of rivers. Figure 3a gives the examples of metal enrichments in a sediment core taken off Helgoland in the southeastern part of the North Sea (Förstner and Reineck, 1974); the actual levels of zinc, cadmium, lead and mercury were 7 to 10 times those of the pre-cultural levels of that area, and were probably caused by the metal load of the Elbe and Weser Rivers (Gadow and Schäfer, 1973). Similar effects have been observed in certain estuaries in Great Britain, e.g. in the Mersey Estuary (Craig and Morton, 1976), in the Bristol Channel and Severn Estuary (Butterworth *et al.*, 1972; Abdullah and Royle, 1974; Chester and Stoner, 1975), in the Thames Estuary (Smith *et al.*, 1971) and in the Humber Estuary (Jaffé and Walters, 1977); in the U.S.A. considerable metal pollution is reported from various tributaries of the Chesapeake Bay (Cross *et al.*, 1971; Pheiffer, 1972; Huggett *et al.*, 1972; Wolfe *et al.*, 1973).

In many estuaries, a characteristic lowering of metal levels in the suspended particulates and in sediments can be observed seaward, particularly among those metal species which are strongly enriched by anthropogenic influences. From observations in the Elbe Estuary (Fig. 3b shows the distribution of mercury in pelitic sediments of rivers in northern Germany), we concluded that the mixing of heavily polluted river sediment with relatively "clean" marine sediments is the main cause of the decrease of heavy metal concentrations within the solid phases (Müller and Förstner, 1975).

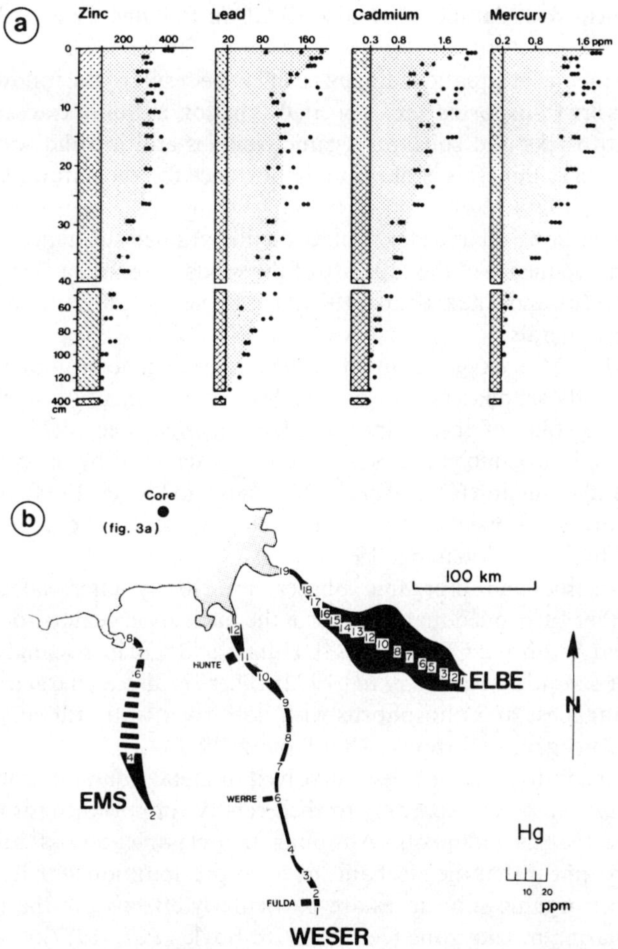

FIG. 3. Enrichment of mercury and other heavy metals in pelitic sediments (fraction $<2\ \mu m$) from rivers in northern Germany and the adjacent North Sea ([a]: Förstner and Reineck, 1974; [b]: Förstner and Müller, 1974).

Examples of true remobilization effects will be given below (section 2). A series of studies performed in the last 10 years have shown that in the estuary zone, a number of elements tend to leave the soluble phase and go into the particulate phase (Burton and Liss, 1976). This behavior is certainly uncontested for iron as it has been seen in estuaries in Canada (Williams and Chan, 1966; Bewers *et al.*, 1974), the U.S.A. (Coonley *et al.*, 1971; Windom *et al.*, 1971; Boyle *et al.*, 1974), England (Holliday and Liss, 1976) and the Netherlands (Duinker and Nolting, 1976); other elements have also been known to waver from the metal/chlorinity line—such as cadmium, copper and zinc—as seen in the mouth of the Rhine River (Duinker and Nolting, 1977) and zinc in the Firth of Clyde (MacKay and Leatherland, 1976). It is often the case, however, as Brewer and Spencer (1975) stated, that "non-conservative" effects in trace element chemistry may well be hard to distinguish through direct observations. One way of solving this difficulty would be the study of the trace metal bonding forms in particulate matter.

Sediment Associations of Heavy Metals in Polluted Coastal Areas

The bonding of metals in aquatic sediments can be the result of the following mechanisms:

(i) Heavy metals are transported and deposited as major, minor or trace constituents in the detrital minerals of rocks and soils, in organic residues and in solid waste material. The composition of the clay minerals is of prime importance for the natural background levels (Hirst and Nicholls, 1958).

(ii) Sorption and cation exchange takes place on fine-grained substances with large surface areas. A generalized sequence of the capacity of the solids to sorb heavy metals was found to occur in the order (Guy and Chakrabarti, 1975): manganese oxides > humic acids > hydrous iron oxides > clay minerals.

(iii) A rise in pH and the oxygen content in the estuarine zone promotes the formation of metal hydroxides and carbonates. Hydrous Fe/Mn oxides constitute significant "sinks" of heavy metals by the effect of co-precipitation (Jenne, 1968; Lee, 1975).

(iv) In waters rich in organic matter solubilizaton is achieved by the combined processes of complexation and reduction (Ong *et al.*, 1970; Theis and Singer, 1974); incorporation into the sediment involves the mechanisms of adsorption, flocculation, polymerization and precipitation (Saxby, 1973; Jonasson, 1977).

The various organic and inorganic phases are closely interrelated: clay minerals, carbonates and other suspended materials form the nucleation centers for the deposition of Fe/Mn hydroxides (Aston and Chester, 1973). Humic acids are increasingly adsorbed to clay minerals at higher salinities (Rashid *et al.*, 1972). There is also a characteristic relationship between iron, manganese and phosphorus with both river water (dissolved) and seawater (partly flocculated) organic substances (Sholkovitz, 1976).

Table 1 shows the individual processes involved in metal bonding as well as the various substrates which are classified according to their relative importance for the enrichment of heavy metals in estuarine sediments. Although this classification is still not considered conclusive, it does appear that the mechanisms of co-precipitation with hydrous iron oxides and flocculation of organic substances are particularly effective in the bonding of heavy metals in the estuarine mixing zone (Jenne, 1976; Boyle *et al.*, 1977).

TABLE 1. PHASES AND PROCESSES INFLUENCING THE CONCENTRATION OF HEAVY METALS IN ESTUARINE SEDIMENTS

	Detrital particles	Interactive organics	Hydrous Fe-oxides	Calcium carbonate	Heavy metal precipitates
Incorporation in ± inert positions	× ×				
Adsorption = physico-sorption	(×)	×	×	(×)	(×)
Cation exchange = chemi-sorption	×	×	×	(×)	(×)
Precipitation					× ×
Co-precipitation			× × ×	×	
Complexation + flocculation		× × ×			

1. Selective Extraction of Metal Compounds in Sediments

The question as to the specific bonding of chemical constituents in particulate substances is one which, regardless of its current application in environmental investigations, has been raised for some time. Problems related to soil fertility led to the development of procedures for determining the leachability of the mineral or organic components. In the field of sedimentology, interest has focused on differentiating the metal content in the detrital and non-detrital fractions of carbonate rocks (Hirst and Nicholls, 1958) as well as of pelagic sediments (Goldberg and Arrhenius, 1958). The next step was the chemical separation of those trace elements incorporated in ferromanganese concretions; the "acid-reducing" method of Chester and Hughes (1967) found particularly wide application in marine geochemistry (e.g. Chester and Hughes, 1969; Horowitz and Cronan, 1976). Combinations of different extraction procedures were first used by Nissenbaum (1972) for sediments from the Okhotsk Sea and by Gibbs (1973) for analyses of trace metals in suspended solids of the Amazon and Yukon Rivers. A summary of the most important extraction methods for the determination of the various forms of metal associations in both natural and polluted sediments is shown in Table 2; examples of combinations of methods can be seen in Table 3.

TABLE 2. SELECTIVE EXTRACTION PROCEDURES FOR TRACE METALS IN RECENT SEDIMENTS

Form of chemical association	Extraction procedures	References
Detrital/authigenic phases	Treatment with complexing agents (EDTA) or dilute acids (0.1 M HCl or HNO_3)	Goldberg and Arrhenius (1958), Piper (1971), Jones (1973), Malo (1977)
Reducible phases (sorption, carbonate, Fe/Mn-hydrate)	1 M $NH_2OH.HCl$ + acetic acid "acid-reducing agent"	Chester and Hughes (1967)
Exchangeable cations	$BaCl_2$; $MgCl_2$; NH_4OAc	Jackson (1958)
Carbonate phases	CO_2-treatment	Patchineelam (1975)
Easily reducible phases (carbonate, Mn-oxides)	0.1 M hydroxylaminhydrochloride + 0.01 M HNO_3	Chao (1972)
Moderately reducible phases (hydrous Fe-oxides)	Na-dithionite/citrate 0.04 M $NH_2OH.HCl$ + 25% Ac	Holmgren (1967), Gupta and Chen (1975)
Oxidizable phases (sulphides, organic substances)	30% H_2O_2 (95°), extraction with 1 N ammonium acetate	Engler *et al.* (1974)
Humates	0.5 N NaOH; 0.1 N NaOH	Rashid (1971)
Lipids, bitumen	Alkohol–benzene extraction	Volkov and Fomina (1974)
Organic residues	Na-hypochlorite, citrate–dithionite extraction	Gibbs (1973)

(i) Differentiation of "oxidizable," "reducible" and the "residual phase" was performed by Presley *et al.* (1972) on the sediments from a fairly unpolluted reducing fjord—Saanich Inlet in British Columbia; these sediments displayed high concentrations of copper associated with sulfides or organics. The comparison of the data from the Soledad Basin off California with those from Whites Point located near a sewer outfall (both examples were studied by Bruland *et al.* 1974) shows the definite increase of the oxidizable and, in particular, the reducible fraction in man-made contamination of copper and zinc.

(ii) This development concurs with the results of studies by Brannon *et al.* (1976) and by Gupta and Chen (1975) on the contaminated harbor sediments from the east and west coasts of the U.S.A., both groups using further differentiation of the reducible fraction: When anthropogenic metal enrichment increase, the residual components, i.e. the predominantly inertly fixed metal content, decreases. Both studies show very low amounts of metal in exchangeable positions.

TABLE 3. TRACE METAL ASSOCIATIONS IN MARINE COASTAL SEDIMENTS
(Examples from (a) Presley *et al.*, 1972; (b) Bruland *et al.*, 1974; (c) Brannon *et al.*, 1976; (d) Gupta and Chen, 1975; (e) Patchineelam and Förstner, 1977)

A	*Saanich Inlet,*[(a)] British Columbia			*Soledad Basin,*[(b)] California			*Whites Point*[(b)] California		
Fraction %	Co	Cu	Zn	Co	Cu	Zn	Co	Cu	Zn
Oxidizable	9	52	19	10	17	18	20	49	8
Reducible	6	5	20	4	11	18	17	31	85
Residual	85	43	61	86	72	64	63	20	7
Total (ppm)	46	42	88	4	34	72	14	610	1800

B	*Mobile Bay,*[(c)] Alabama			*Los Angeles*[(d)] *Harbor*, Calif.			*Bridgeport,*[(c)] Connecticut		
Fraction %	Ni	Zn	Cd	Ni	Zn	Cd	Ni	Zn	Cd
Exchangeable cations	<1	—	—	<1	<1	—	<1	<1	<1
Organics, sulfides	7	58	17	10	28	38	26	53	96
Easily reducible	2	4	3	9	26	14	2	13	4
Moderately reducible	—	—	—	18	23	34	3	17	—
Residual fraction	90	38	80	63	23	14	69	17	—
Total (ppm)	127	243	3.6	47	612	2.2	183	1067	15.2

C	*German Bight* near Helgoland (southeastern North Sea)[(e)]									
Fraction	Mn		Zn		Cu		Pb		Cd	
%	*u*	*l*	*u*	*l*	*u*	*l*	*u*	*l*	*u*	*l*
Cation exchange	2	1	2	2	9	7	<1	<1	36	29
Humates	<1	<1	4	3	13	13	2	4	10	18
Carbonates	75	65	45	49	31	52	<1	<1	36	17
Fe/Mn-oxides	15	11	17	17	12	17	65	93	9	14
Residual phases	8	23	31	29	33	11	32	2	10	22
Total (ppm)	1070	370	180	120	22	10	59	17	0.82	0.35

(iii) The historical changes in bonding forms under the influence of increased pollution can best be seen in the sediment cores.

Table 2c compares the bonding types of selected heavy metals from the German Bight (Patchineelam and Förstner, 1977) (see Fig. 3a for metal concentrations in $< 2\mu m$ fraction) in metal-rich samples from the upper part of the profile and uncontaminated samples from 1 m depth. The extraction of carbonate-associated metals showed extremely high concentrations of Mn, Zn, Cu and Cd; the high amount of cadmium in exchangeable form is also of note. As the metal concentrations rise, manganese and cadmium shift clearly from residual bonding to associations with carbonates. The unexpected rise in the residual content of the lead in the polluted sample can most likely be attributed to the influence of lead-rich dust particles.

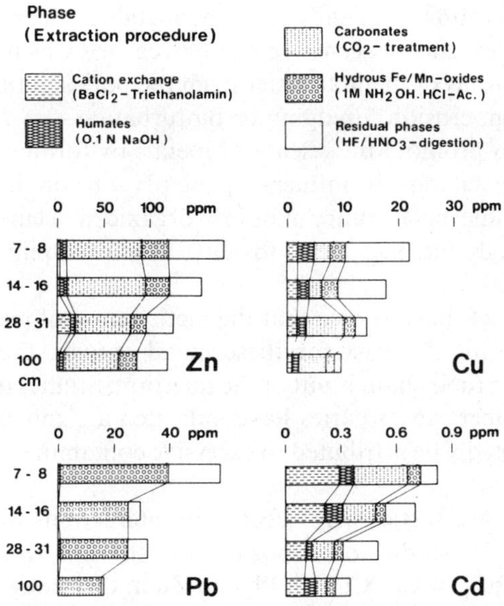

FIG. 4. Heavy metal associations in bulk sediment samples of a core profile from the German Bight, North Sea (data from Patchineelam and Förstner, 1977; see Fig. 3a for metal concentrations in fractions $<2\ \mu m$)

2. Remobilization of Metals from Polluted Coastal Sediments

Heavy metals which are "immobilized" in the bottom sediments of rivers, lakes and seas constitute a potential hazard to water quality and aquatic life as they may be released as a result of physico-chemical changes. Of the possible variables (Förstner and Patchineelam, 1976), the *lowering of pH* in the inflowing waters would have little influence since the interstitial solutions in the marine sediments are well buffered (Ben-Yaakov, 1973). Similarly, an increase of *salinity* in the river/sea interface, which leads to competition between dissolved cations and adsorbed heavy metal ions can, at first, cause a release of the metals (which has been proved in laboratory experiments and radioisotope studies, Kharkar *et al.*, 1968; Bachman, 1961). Nonetheless, the immobilization effects predominate (Turekian, 1971, 1977). Two mobilization mechanisms, however, seem to play an important role in estuaries:

(i) *A change in the redox conditions* usually is caused by the increased input of nutrients. Oxygen deficiency in the sediments leads to an initial dissolution of hydrated manganese oxide, followed by that of iron compounds. Since these metals are readily soluble in their divalent states, any co-precipitates with metallic coatings are partially remobilized.

The experiments of Lu and Chen (1977) indicated that under reducing environments, Fe and Mn were released to the interfacial seawater in significant quantities, while other metals (Cd, Cu, Ni, Pb and Zn) decreased in comparison with the original seawater. This might be related to the formation of sulfide solids. Because the measured concentrations of some heavy metals, e.g. Cu and Zn, in the interstitial waters exceeds the calculated stabilities of the respective metal sulfides, it has been suggested that organic metal complexing is competing with the inorganic precipitation (Hallberg, 1973; Cline and Upchurch, 1973). In the pore waters of sediments from the Conway Estuary, Elderfield and Hepworth (1975) found an 10^2- to 10^4-fold increase of Cu, Ni and Co compared to the respective seawater values;

presumably, this concentration gradient allows the metals in the sediments' pore water to diffuse upwards until a new chemical environment is reached which sets new limits on their solubility. Other mechanisms of metal transfer from the pore solutions into the open water are sediment compaction, erosion, dredging or bioturbation.

(ii) *Microbial activities* promote the release of metals by forming inorganic compounds capable of complexing metal ions, by influencing the pH–Eh condition of the environment, and by converting inorganic metal compounds into organic molecules. This last mechanism, in particular, may strongly increase metal toxicity, as is exemplified in organo-mercurial transformations.

Olson and Cooper (1974) have shown that the methylation of mercury can take place *in situ* in estuarine sediments. At least for these metal species, the possibility of further enrichment in the aquatic food chain is out of the question. Studies of mussels and fish from European and North American estuaries have indicated a significant rise in the mercury levels which can undoubtedly be attributed to excessive contamination (Bourget and Cossa, 1976; Krüger *et al.*, 1975).

Biochemical reactions are particularly effective in interactions with sewage particulates. Rohatgi and Chen's (1976) studies on deposits from the Los Angeles Hyperion outfalls showed that more than 30% of Cd, Cu, Ni, Pb and Zn in the sewage sludge particulates are mobilized when they reach the sea. For this reason, important results can be expected from future studies on bonding types and the remobilization of metals from particulates.

Acknowledgements

My thanks to J. Broadbridge for her assistance in preparing the English version of this paper. The German Research Society supported our investigations on marine pollution with material assistance.

References

Abdullah, M. I. and Royle, L. G. (1974) Cadmium in some British coastal and fresh water environments. Proc. Int. Symp. *Problems of the Contamination of Man and his Environment by Mercury and Cadmium*, Luxembourg, July 3–5, 1973, 69–81.

Abdullah, M. I. and Royle, L. G. (1974) A study of the dissolved and particulate trace elements in the Bristol Channel. *J. Mar. Biol. Ass. U.K.* **54**, 581–597.

Ali, S. A., Gross, M. G. and Kishpaugh, J. R. L. (1976) Cluster analysis of marine sediments and waste deposits in New York Bight. *Environ. Geol.* **1**, 143–148.

Andrulewica, E. and Portmann, J. E. (1975) Determination of metal content of sediments in relation to dumping and discharge control. *ICES Papers and Reports* No. E: **22**, 9 pp.

Applequist, M. D., Katz, A. and Turekian, K. K. (1972) Distribution of mercury in the sediments of New Haven (Conn.) Harbor. *Environ. Sci. Technol.* **6**, 1123–1124.

Armstrong, P. B., Hanson, G. M. and Gaudette, H. E. (1976) Minor elements in sediments of Great Bay Estuary, New Hampshire. *Environ. Geol.* **1**, 207–214.

Aston, S. R. and Chester, R. (1973) The influence of suspended particles on the precipitation of iron in natural waters. *Estuar. Coast. Mar. Sci.* **1**, 225–231.

Aubert, M. (1975) Le problème du mercure en Méditerranée. *Rev. Int. Océanogr. Méd.* **37/38**, 215–231.

Bachman, R. W. (1961) Zn-65 in studies of the water–zinc cycle. In *Radioecology*, V. Schultz and A. W, Klement eds., Proc. First Nat. Symp. Radioecol., Colo. State Univ., 485–496.

Ben-Yaakov, S. (1973) pH buffering of pore water of recent anoxic marine sediments. *Limnol. Oceanogr.* **18**, 86–94.

Bertine, K. K. and Goldberg, E. D. (1977) History of heavy metal pollution in southern California coastal zone—reprise. *Environ. Sci. Technol.* **11**, 297–299.

Bevan, C. D., Harbison, S. A., Nelson, L. A. and Lakey, J. R. A. (1975) A trace element study in the Thames Estuary. *Proc. Int. Symp. Impacts Nucl. Releases Aquat. Environ.*, IAEA, Vienna, 93–106.

BEWERS, J. M., MACAULAY, I. D. and SUNDBY, B. (1974) Trace metals in the waters of the Gulf of St. Lawrence. *Can. J. Earth Sci.* **2**, 939–950.

BLOXAM, T. W., AURORA, S. N., LEACH, L. and REES, T. R. (1972) Heavy metals in some river and bay sediments near Swansea. *Nature (Phys. Sci.)* **239**, 158–159.

BONDAM, J., ASMUND, G. and SCHRØDER, S. (1976) Miljøkontrol ved Marmorilik in Nordvest Grønland. In *Cadmium Forskning i Danmark*, Danmarks Tekniske Højskole, 21–32.

BOPP, F. and BIGGS, R. B. (1973) Trace metal environments near Shell Banks in Delaware Bay. *Delware Bay Rept. Ser.* **3**, 23–69.

BOTHNER, M. H. and CARPENTER, R. (1974) Rate of mercury loss from contaminated estuarine sediments in Bellingham Bay, Washington. *Proc. First Annu. NSF Trace Contam. Conf. 1973*, 198–210.

BOURGET, E. and COSSA, D. (1976) Mercury content of mussels from the St. Lawrence Estuary and northwestern Gulf of St. Lawrence, Canada. *Mar. Poll. Bull.* **7**, 237–239.

BOYDEN, C. R. (1975) Distribution of some trace metals in Poole Harbour, Dorset. *Mar. Poll. Bull.* **6**, 180–187.

BOYLE, E., COLLIER, R., DENGLER, A. T., EDMOND, J. M., NG, A. C. and STALLARD, R. F. (1974) On the chemical mass-balance in estuaries. *Geochim. Cosmochim. Acta* **38**, 1719–1738.

BOYLE, E. A., EDMOND, J. M. and SHOLKOVITZ, E. R. (1977) The mechanism of iron removal in estuaries. *Geochim. Cosmochim. Acta* **41**, 1313–1324.

BRADFORD, W. L. (1976) Distribution and movement of zinc and other heavy metals in South San Francisco Bay, California. *U.S. NTIS PB Rept. PB-251111*, 64 pp.

BRANNON, J. M., ENGLER, R. M., ROSE, J. R., HUNT, P. G. and SMITH, I. (1976) Distribution of manganese, nickel, zinc, cadmium and arsenic in sediments and the standard elutriate. *U.S. Army Engs. Waterways Experim. Station Misc. Paper D-76-18*, 39 pp.

BREWER, P. G. and SPENCER, D. W. (1975) Minor element models in coastal waters. In *Marine Chemistry in the Coastal Environment*, T. M. CHURCH ed., *Am. Chem. Soc. Symp. Ser.* **18**, 80–96.

BRINCKMAN, F. E. and IVERSON, W. P. (1975) Chemical and bacterial cycling of heavy metals in the estuarine system. In *Marine Chemistry in the Coastal Environment*, T. M. CHURCH ed., *Amer. Chem. Soc. Symp. Ser.* **18**, 319–337.

BROWN, M. J. F. (1974) A development consequence—disposal of mining waste on Bougainville, Papua, New Guinea. *Geoforum* **18**, 19–27.

BROWN, R. P. and SMITH, D. D. (1969) Marine disposal of solid wastes. An interim summary. *Mar. Poll. Bull.* **1**, 12–16.

BRULAND, K. W., BERTINE, K., KOIDE, M. and GOLDBERG, E. D. (1974) History of metal pollution in Southern California coastal zone. *Environ. Sci. Technol.* **8**, 425–432.

BRYAN, G. W. and HUMMERSTONE, L. G. (1973) Brown seaweed as an indicator of heavy metals in estuaries in south-west England. *J. Mar. Biol. Ass. U.K.* **53**, 705–720.

BURTON, J. D. and LISS, P. S. (eds.) (1976) *Estuarine Chemistry*, Academic Press, 229 pp.

BUTTERWORTH, J., LESTER, P. and NICKLESS, G. (1972) Distribution of heavy metals in the Severn Estuary. *Mar. Poll. Bull.* **3**, 72–74.

CAPUZZO, J. M. and ANDERSON, F. E. (1973) The use of modern chromium accumulation to determine estuarine sedimentation rates. *Mar. Geol.* **14**, 225–235.

CARMODY, D. J., PEARCE, J. B. and YASSO, W. E. (1973) Trace metals in sediments of New York Bight. *Mar. Poll. Bull.* **4**, 132–135.

CARPENTER, J. H., BRADFORD, W. L. and GRANT, V. (1975) Processes affecting the composition of estuarine waters (HCO_3, Fe, Mn, Zn, Cu, Ni, Cr, Co and Cd.) In *Estuarine Research*, L. E. CRONIN ed., Vol. 1, Pergamon Press, 137–152.

CASPERS, H. (ed.) (1975) *Pollution in Coastal Waters: an interim report on the results of the priority programme from 1966 to 1974. A report of the German Research Society*, Boldt Verlag, 142 pp.

CATO, I. (1977) Recent sedimentological and geochemical conditions and pollution problems in two marine areas in south-western Sweden. *Striae, Societas Upsaliensis pro Geologia Quaternaria*, Vol. 6, 158 pp.

CHAO, L. L. (1972) Selective dissolution of manganese oxides from soils and sediments with acidified hydroxylamine hydrochloride. *Soil Sci. Soc. Amer. Proc.* **36**, 764–768.

CHESTER, R. and HUGHES, M. J. (1967) A chemical technique for the separation of ferro-manganese minerals, carbonate minerals and adsorbed trace elements from pelagic sediments. *Chem. Geol.* **2**, 249–262.

CHESTER, R. and HUGHES, M. J. (1969) The trace element geochemistry of a North Pacific pelagic clay core. *Deep-Sea Res.* **16**, 639–654.

CHESTER, R. and STONER, J. H. (1973) Average trace element composition of low level marine atmospheric particulates. *Nature* **246**, 138–139.

CHESTER, R. and STONER, J. H. (1975) Trace elements in sediments from the lower Severn and Bristol Channel. *Mar. Poll. Bull.* **6**, 92–96.

CHOW, T., BRULAND, K. W. BERTINE, K. K., SOUTAR, A., KOIDE, M. and GOLDBERG, E. D. (1973) Lead pollution: records in southern California coastal sediments. *Science* **181**, 551–552.

CLIFTON, A. P. and VIVIAN, C. M. G. (1975) Retention of mercury from an industrial source in Swansea Bay sediments. *Nature* **253**, 621–622.

Cline, J. T. and Upchurch, S. B. (1973) Mode of heavy metal migration in the upper strata of lake sediment. *Proc. 16th Conf. Great Lakes Res.*, 349–356.

Coonley, L. S., Baker, E. B. and Holland, H. D. (1971) Iron in the Mullica River and in Great Bay, New Jersey. *Chem. Geol.* **7**, 51–63.

Covill, R. W. (1971/72) The quality of the Forth Estuary. *Proc. Roy. Soc. Edinburgh* (B) **71**, 143–170.

Craig, P. J. and Morton, S. F. (1976) Mercury in Mersey estuary sediments, and the analytical procedure for total mercury. *Nature* **261**, 125–126.

Crecelius, E. A., Bothner, M. H. and Carpenter, R. (1975) Geochemistry of arsenic, antimony, mercury and related elements in sediments of Puget Sound. *Environ. Sci. Technol.* **9**, 325–333.

Crecelius, E. A. and Carpenter, R. (1974) Arsenic distribution in waters and sediments of the Puget Sound region. *Proc. 1st Ann. NSF Trace Contam. Conf. 1973*, 615–625.

Cross, F. A., Duke, T. W. and Willis, J. N. (1970) Biogeochemistry of trace elements in a coastal plain estuary: distribution of manganese, iron, and zinc in the sediments, water and polychaetous worms. *Chesapeake Sci.* **11**, 221–234.

Cumont, G., Viallex, G., Lelievre, H. and Bobenrieth, P. (1972) Contamination des poissons de mer par le mercure. *Rev. Int. Océanogr. Méd.* **28**, 95–129.

Dörjes, J., Little-Gadow, S. and Schäfer, A. (1976) Zur Schwermetallverteilung in litoralen Sedimenten der ostfriesischen Küste. *Senckenbergiana Marit.* **8**, 91–102.

Duinker, J. C. and Nolting, R. F. (1976) Distribution model for particulate trace metals in the Rhine Estuary, the Southern Bight and Dutch Wadden Sea. *Neth. J. Sea Res.* **10**, 78–102.

Duinker, J. C. and Nolting, R. F. (1977) Dissolved and particulate trace metals in the Rhine estuary and the Southern Bight. *Mar. Poll. Bull.* **8**, 65–71.

Elderfield, H. and Hepworth, A. (1975) Diagenesis, metals and pollution in estuaries. *Mar Poll. Bull.* **6**, 85–87.

Elderfield, H., Thornton, I. and Webb, J. S. (1971) Heavy metals and oyster culture in Wales. *Mar. Poll. Bull.* **2**, 44–47.

Ellis, D. V. (1975) Pollution controls on mine discharges to the sea. Abstr. *Conf. Heavy Metals Environment, Toronto*, A-21/23.

Engler, R. M., Brannon, J. M. and Rose, J. (1974) A practical selective extraction procedure for sediment characterization. *168th Meeting Amr. Chem. Soc. Atlantic City, N.J.*, 14 pp.

Erlenkeuser, H., Suess, E. and Willkomm, H. (1974) Industrialization affects heavy metal and carbon isotope concentrations in recent Baltic Sea sediments. *Geochim. Cosmochim. Acta* **38**, 823–842.

Förstner, U. and Müller, G. (1974) *Schwermetalle in Flüssen und Seen.* Springer, 225 pp.

Förstner, U. and Patchineelam, S. R. (1976) Bindung und Mobilisation von Schwermetallen in fluviatilen Sedimenten. *Chemikerztg.* **100**, 49–57.

Förstner, U. and Reineck, H. -E. (1974) Die Anreicherung von Spurenelementen in den rezenten Sedimenten eines Profilkerns aus der Deutschen Bucht. *Senckenbergiana Marit.* **6**, 175–184.

Føyn, E. (1965) Disposal of waste in the marine environment and the pollution of the sea. *Oceanogr. Mar. Biol. Ann. Rev.* **3**, 95–114.

Gadow, S. and Schäfer, A. (1973) Die Sedimente der Deutschen Bucht: Korngrössen, Tonmineralien und Schwermetalle. *Senckenbergiana Marit.* **5**, 165–178.

Gibbs, R. (1973) Mechanisms of trace metal transport in rivers. *Science* **180**, 71–73.

Goldberg, E. D. (1976) *The Health of the Oceans,* UNESCO Press, 172 pp.

Goldberg, E. D. and Arrhenius, G. O. S. (1958) Chemistry of Pacific pelagic sediments. *Geochim. Cosmochim. Acta* **13**, 153–212.

Goldberg, E D., Gamble, E., Griffin, J. J. and Koide, M. (1977) Pollution history of Narragansett Bay as recorded in its sediments. *Estuar. Coast. Mar. Sci.* **5**, 549–561.

Goldberg, E. D., Hodge, V., Koide, M. and Griffin, J. J. (1976) Metal pollution in Tokyo as recorded in sediments of the Palace Moat. *Geochem. J.* **10**, 165–174.

Goto, M. (1973) Inorganic chemicals in the environment—with special reference to the pollution problems in Japan. *Environ. Qual. Safety* **2**, 72–77.

Grancini, G., Stievano, M. B., Girardi, F., Guzzi, G. and Pietra, R. (1976) The capability of neutron activation for trace element analysis in sea water and sediment samples of the northern Adriatic Sea. *J. Radioanal. Chem.* **34**, 65–78.

Greig, R. A. and McGrath, R. A. (1977) Trace metals in sediments of Raritan Bay. *Mar. Poll. Bull.* **8**, 188–192.

Greig, R. A., Reid, R. N. and Wenzloff, D. R. (1977) Trace metal concentrations in sediments from Long Island Sound. *Mar. Poll. Bull.* **8**, 183–188.

Grieve, D. and Fletcher, K. (1975) Trace metals in Fraser Delta sediments. *Geol. Surv. Can. Paper* **75–1**, 161–163.

Groot, A. J. De, Allersma, E. and Driel, W. Van (1973) Zware metalen in fluviatiele en mariene ecosystemen. *Symp. Waterlopkunde in Dienst von Industrie en Milieu*, Publ. 110N, 27 pp.

Groot, A. J. De, Goeij, J. J. M. De and Zengers, C. (1971) Contents and behaviour of mercury as compared with other heavy metals in sediments from the rivers Rhine and Ems. *Geol. Mijnbouw* **50**, 393–398.

Gross, M. G. (1972) Waste-solid disposal in coastal waters of North America. In *Man's Impact on Terrestrial and Oceanic Ecosystems*, W. H. Matthews *et al.* eds., M.I.T. Press, 252–260.

GROSS, M. G. (1972) Geologic aspects of waste solids and marine waste deposits, New York Metropolitan region. *Geol. Soc. Amer. Bull.* **83**, 3163–3176.

GROSS, M. G., BLACK, J. A., KALIN, R. J., SCHRAMEL, J. R. and SMITH, R. N. (1971) Survey of marine waste deposits, New York Metropolitan region. *Marine Sciences Research Center, State Univ. New York, Stony Brook, Techn. Rept.* No. 8, 72 pp.

GUPTA, S. K. and CHEN, K. Y. (1975) Partitioning of trace metals in selective chemical fractions of nearshore sediments. *Environ. Lett.* **10**, 129–158.

GUY, R. D. and CHAKRABARTI, C. L. (1975) Distribution of metal ions between soluble and particulate forms. Abstr. *Conf. Heavy Metals Environment, Toronto* D-29/30.

HALCROW, W. MACKAY, D. W. and THORNTON, I. (1973) The distribution of trace metals and fauna in the Firth of Clyde in relation to the disposal of sewage sludge. *J. Mar. Biol. Ass. U.K.* **53**, 721–739.

HALLBERG, R. O. (1973a) Paleoredox conditions in the eastern Gotland Basin during the last 400 years. *Contr. Askö Lab. Univ. Stockholm* **2**, 89–117.

HALLBERG, R. O. (1973b) The microbiological C–N–S cycles in sediments and their effect on the ecology of the sediment–water interface. *Oikos Suppl.* **15**, 51–62.

HASSELROT, T. B. and BJÖRKLUND, I (1974) Undersökning i samband med Uddevallavarvets muddringsarbeten i Byfjorden. *Statens Naturvårdsverk SNV PM 512*, 7 pp.

HELZ, G. R., HUGGETT, R. J. and HILL, J. M. (1975) Behavior of manganese, iron, copper, zinc, cadmium and lead discharged from a waste water treatment plant into an estuarine environment. *Water Res.* **9**, 631–636.

HELZ, G. R. (1976) Trace element inventory for the northern Chesapeake Bay with emphasis on the influence of man. *Geochim Cosmochin. Acta* **40**, 573–580.

HIRST, D. M. (1962) The geochemistry of modern sediment from the Gulf of Paria. II. The location and distribution of trace elements. *Geochim. Cosmochim. Acta* **26,** 1147–1182.

HIRST, D. M. and NICHOLLS, G. D. (1958) Techniques in sedimentary geochemistry: 1. Separation of the detrital and non-detrital fractions of limestones. *J. Sediment. Petrol.* **28**, 461–468.

HOLLIDAY, L. M. and LISS, P. S. (1976) The behaviour of dissolved iron, manganese and zinc in the Beaulieu Estuary, S. England. *Estuar. Coast. Mar. Sci.* **4**, 349–353.

HOLMES, C. W., SLADE, E. A. and MCLERRAN, C. J. (1974) Migration and redistribution of zinc and cadmium in marine estuarine system. *Environ. Sci. Technol.* **8**, 255–259.

HOLMGREN, G. S. (1967) A rapid citrate–dithionite extractable iron procedure. *Soil Sci. Soc. Amer. Proc.* **31**, 210–211.

HOROWITZ, A. and CRONAN, D. S. (1976) The geochemistry of basal sediments from the North Atlantic Ocean. *Mar. Geol.* **20**, 205–228.

HUGGETT, R. J., BENDER, M. and SLONE, D. H. (1972) Mercury in sediments from three Virginia estuaries. *Chesapeake Sci.* **12**, 280–282.

ITO, K. (1975) Heavy metal pollution in the sediments of Nagoya harbor. *Kogai To Taisaku* **11**, 650–659 (in Japanese).

JACKSON, M. L. *Soil Chemical Analysis*, Prentice Hall., 498 pp.

JAFFÉ, D. and WALTERS, J. K. (1977) Intertidal trace metal concentrations in some sediments from the Humber Estuary. *Sci. Tot. Environ.* **7**, 1–15.

JENNE, E. A. (1968) Controls on Mn, Fe, Co, Ni, Cu and Zn concentrations in soils and water: the significant role of hydrous Mn and Fe oxides. *Amer. Chem. Soc. Adv. Ser.* **73**, 337–387.

JENNE, E. A. (1976) Trace elements sorption by sediments and soils—sites and processes. In *Symposium on Molybdenum in the Environment* W. CHAPPEL and K. PETERSEN eds., Marcel Dekker, 425–553.

JONASSON, I. R. (1977) An overview of the problems of metal-sediment-water interactions in fluvial systems. Proc. Workshop *Fluvial Transport of Sediment-Associated Nutrients and Contaminants,* H. SHEAR and A. E. P. WATSON eds., IJC/PLUARG, Windsor, Canada, 255–271.

JONES A. S. G. (1973) The concentration of copper, lead, zinc and cadmium in shallow marine sediments, Cardigan Bay, Wales. *Mar. Geol.* **14**, M1-M9.

KHARKAR, D. P., TUREKIAN, K. K. and BERTINE, K. K. (1968) Stream supply of dissolved silver, molybdenum, antimony, selenium, chromium, cobalt, rubidium and cesium to the oceans. *Geochim. Cosmochim. Acta* **32**, 285–298.

KITAMURA, S. (1968) Determination of the mercury content in bodies of inhabitants, cats, fishes and shells in Minamata District and in the mud of Minamata Bay. In *Minamata Disease*, Shuhan Co., 257–266.

KLEIN, D. H. and GOLDBERG, E. D. (1970) Mercury in the marine environment. *Environ. Sci. Technol.* **4**, 765–768.

KOCH, P. (1960) Discharge of wastes into the sea in European coastal areas. In Proc. 1st Int. Conf. *Waste Disposal in the Marine Environment,* E. A. PEARSON, ed., Pergamon Press, 122–162.

KNAUER, G. A. (1976) Immediate industrial effects on sediment mercury concentrations in a clean coastal environment. *Mar. Poll. Bull.* **7**, 112–115.

KRÜGER, K. E., NIEPER, L. and AUSLITZ, H.- J. (1975) Bestimmung des Quecksilbergehaltes der Seefische auf den Fangplätzen der deutschen Hochsee- und Küstenfischerei. *Arch. Lebensmittelhyg.* **26**, 201–240.

LEE, G. F. (1975) Role of hydrous metal oxides in the transport of heavy metals in the environment. In *Heavy Metals in the Aquatic Environment,* P. A. KRENKEL ed., Pergamon Press, 137–147.

LESACA, R. M. (1975) Heavy metal monitoring in Philippine rivers, bay waters and lakes. Abstr. *Int. Conf. Heavy Metals Environ, Toronto,* C-105/107.

LINDBERG, S. E., ANDREN, A. W. and HARRISS, R. C. (1975) Geochemistry of mercury in the estuarine environment. In *Estuarine Research,* Vol. 1, L. E. CRONIN, ed., Pergamon Press, 64–107.

LITTLE-GADOW, S. and SCHÄFER, A, (1974) Schwermetalle in den Sedimenten der Jade: Bestandsaufnahme und Vergleich mit der inneren Deutschen Bucht. *Senckenbergiana Marit.* **6**, 161–174.

LORING, D. H. (1975) Mercury in the sediments of the Gulf of St. Lawrence. *Can. J. Earth Sci.* **12**, 1219–1237.

LU, C. S. J. and CHEN, K. Y. (1977) Migration of trace metals in interfaces of seawater and polluted surficial sediments. *Environ. Sci. Technol.* **11**, 174–182.

MACKAY, D. W., HALCROW, W. and THORNTON, I. (1972) Sludge dumping in the Firth of Clyde. *Mar. Poll. Bull.* **3**, 7–11.

MACKAY, D. W. and LEATHERLAND, T. M. (1976) Chemical processes in an estuary receiving major inputs of industrial and domestic wastes. In *Estuarine Chemistry,* J. D. BURTON and P. S. LISS eds., Academic Press, 185–218.

MARTIN, J. M., KULBICKI, G. and GROOT, A. J. DE (1973) Terrigenous supply of radioactive and trace elements to the ocean. In *Proceedings of the International Symposium on the Hydrogeochemistry and Biogeochemistry, Tokyo, Sept. 7–9, 1970,* Vol. 1, The Clarke Co., 463–478.

MATSUMOTO, E. and YOKOTA, S. (1976) History of heavy metal pollution in Tokyo Bay and Osaka Bay. *Kaguku* **46,** 182–184.

MOMOYAMA, K., KOBAYASHI, T., YOSHITSUGU, K. and TAKAYAMA, S. (1975) Sediment of the fishing grounds of Yamaguchi Prefecture in the Inland Sea of Japan (in Japanese). *Yamaguchi-ken Naikai Suisan Shinkenjo Hokoku* **5**, 79 pp.

MOYER, B. R. and BUDINGER, T. F. (1974) Cadmium levels in the shore-line sediments of San Francisco Bay. In *Trace Substances in Environmental Health,* D. D. HEMPHILL, ed., Vol. VIII, Univ. of Missouri, 127–135.

MÜLLER, G. and FÖRSTNER, U. (1975) Heavy metals in sediments of the Rhein and Elbe Estuaries: mobilisation or mixing effect? *Environ. Geol.* **1**, 33–39.

MÜLLER, G. and FÖRSTNER, U. (1976) Schwermetalle in den Sedimenten der Elbe bei Stade: Veränderungen seit 1973. *Naturwiss.* **63**, 242–243.

MURAKAMI, T., KIDA, A., NAKAI, M. and MATSUNAGA, S. (1975) Heavy metal elements in sea sediments. I. Selection of control values (in Japanese). *Eisei Kaguku* **21**, 275–281.

MUROZUMI, J., CHOW, T. J. and PATTERSON, C. (1969) Chemical concentration of pollutant lead aerosols, terrestrial dusts and sea salts in Greenland and Antarctic snow strata. *Geochim. Cosmochim. Acta* **33**, 1247–1294.

NAEVE, E. (1974) Sources of pollution in the Mediterranean and its effects on living resources and fishing. *Rev. Int. Océanogr. Méd.* **35/36**, 5–20.

NISSENBAUM, A. (1972) Distribution of several metals in chemical fractions of sediment core from the Sea of Okhotsk. *Israel J. Earth Sci.* **21**, 143–154.

NISIMURA, H. (1974) Hg in fish and sediments in Tokuyama Bay. Cit. p. 13 in *Review of Environmental Control of Mercury in Japan, Rept. 3-WP-76-7,* Environ. Protection Service, Canada, 81 pp.

OLAUSSON, E. (1975) Byfjorden: sediment, sedimentation och geokemi. *Statens Naturvårdsverk SNV PM 564,* 30 pp.

OLAUSSON, E. (1975) Man-made effect on sediments from Kattegat and Skagerrak. *Geol. Fören. Stockholm Förhand.* **97,** 3–12.

OLAUSSON, E., GUSTAFSSON, O., MELLIN, T. and SVENSSON, R. (1977) Current level of heavy-metal pollution and eutrophication in the Baltic proper. *Medd. Maringeol. Lab. Göteborg* **9**, 28 p.

OLSON, B. H. and COOPER, R. C. (1974) *In situ* methylation of mercury in estuarine sediment. *Nature* **252**, 682–683.

ONG, H. L., SWANSON, V. E. and BISQUE, R. E, (1970) Natural organic acids as agents of chemical weathering. *U.S. Geol. Surv. Prof. Paper 700-C,* 130–137.

PAPAKOSTIDES, G., GRIMANIS, A. P., ZAFIROPOULOS, D., GRIGGS, G. B. and HOPKINS, T. S. (1975) Heavy metals in sediments from the Athens sewage outfall. *Mar. Poll. Bull.* **6**, 136–138.

PATCHINEELAM, S. R. (1975) *Untersuchungen über die Hauptbindungsarten und die Mobilisierbarkeit von Schwermetallen in fluviatilen Sedimenten.* Unpubl. Diss. Univ. Heidelberg, 137 pp.

PATCHINEELAM, S. R. and FÖRSTNER, U. (1977) Bindungsformen von Schwermetallen in marinen Sedimenten. Untersuchungen an einem Sedimentkern aus der Deutschen Bucht. *Senckenbergiana Marit.* **9**, 75–104.

PEARCE, J. B. (1969) The effect of waste disposal on the New York Bight—interim report for January 1, 1970. *U.S. Bureau of Sport Fisheries and Wildlife,* Sandy Hook Marine Laboratory, 103 pp.

PERKINS, E. J., GILCHRIST, J. R. S., ABBOTT, O. J. and HALCROW, W. (1973) Trace metals in Solway Firth sediments. *Mar. Poll. Bull.* **4**, 59–61.

PHEIFFER, T. H. (1972) Heavy metal analyses of bottom sediments in the Potomac River estuary. *EPA, Annapolis Field Office Techn. Rept.* **49**, 22 pp.

PIPER, D. Z. (1971) The distribution of Co, Cr, Cu, Fe, Mn, Ni and Zn in Framvaren, a Norwegian anoxic fjord. *Geochim. Cosmochim. Acta* **35**, 531–550.

PRESLEY, B. J., KOLODNY, Y., NISSENBAUM, A. and KAPLAN, I. R. (1972) Early diagenesis in a reducing fjord,

Saanich Inlet, British Columbia. II. Trace element distribution in interstitial water and sediment. *Geochim. Cosmochim. Acta* **36**, 1073–1090.

RASHID, M. A. (1971) Role of humic acids of marine origin and their different molecular weight fractions in complexing di- and tri-valent metals. *Soil Sci.* **111**, 298–305.

RASHID, M. A., BUCKLEY, D. E. and ROBERTSON, K. R. (1972) Interactions of a marine humic acid with clay minerals and a natural sediment. *Geoderma* **8**, 11–27.

RENZONI, A., BACCI, E. and FALCIAI, L. (1973) Mercury concentrations in the water, sediments and fauna of an area of the Tyrrhenian coast. *Rev. Int. Océanogr. Méd.* **31/32**, 17–47.

ROHATGI, N. K. and CHEN, K. Y. (1976) Fate of metals in waste water discharge to ocean. *J. Environ. Engng. Div. ASCE* **102** (EE3), 675–685.

ROSENTHAL. H., DETHLEFSEN, V. and TIEWS, K. (1973) Rotschlamm in die Nordsee? *Umschau Wiss. Technik* **73**, 118–121.

ROTH, I. and HORNUNG, H. (1977) Heavy metal concentration in water, sediments, and fish from Mediterranean coastal area, Israel. *Environ. Sci. Technol.* **11**, 265–269.

SALOMONS, W. and GROOT, A. J. DE (1978) Pollution history of trace metals in sediments, as affected by the Rhine River. Proc. 3rd Int. Symp. *Environmental Biochemistry*, Wolfenbüttel/Ger., 149–162.

SAROMA, A. M. (1972) Heavy metal distribution in bottom sediments of lagoons and bays in Hokkaido. *Chikyu Kaguku* **25**, 149–158.

SAXBY, J. D. (1973) Diagenesis of metal–organic complexes in sediments: formations of metal sulphides from cystine complexes. *Chem. Geol.* **12**, 241–248.

SCHELL, W. R. and NEVISSI, A. (1977) Heavy metals from waste disposal in Central Puget Sound. *Environ. Sci. Technol.* **11**, 887–893.

SEGAR, D. A. and PELLENBARG, R. E. (1973) Trace metals in carbonate and organic rich sediments. *Mar. Poll. Bull* **4**, 138–142.

SHELTON, R. G. J. (1971) Sludge dumping in the Thames Estuary. *Mar. Poll. Bull.* **2**, 24–27.

SHOLKOVITZ, E. R. (1976) Flocculation of dissolved organic and inorganic matter during the mixing of river water and seawater. *Geochim. Cosmochim. Acta* **40**, 831–845.

SKEI, J. M., PRICE, N. B., CALVERT, S. E. and HOLTEDAHL, H. (1972) The distribution of heavy metals in sediments of Sörfjord, West Norway. *Water Air Soil Poll.* **1**, 452–461.

SKEI, J. M., PRICE, N. B. and CALVERT, S. E. (1973) Particulate metals in waters of Sörfjord, West Norway. *Ambio* **2**, 122–124.

SMITH, J. D., NICHOLSON, R. A. and MOORE, F. J. (1971) Mercury in water of the tidal Thames. *Nature* **232**, 393–394.

SMITH, J. D., NICHOLSON, R. A. and MOORE, P. J. (1973) Mercury in sediments from the Thames Estuary. *Environ. Poll.* **4**, 153–157.

STOFFERS, P., SUMMERHAYES, C. P., FÖRSTNER, U. and PATCHINEELAM, S. R. (1977) Copper and other heavy metal contamination in sediments from New Bedford Harbor, Massachusetts: a preliminary note. *Environ. Sci. Technol.* **11**, 819–821.

SUESS, E. and ERLENKEUSER, H. (1975) History of metal pollution and carbon input in Baltic Sea sediments. *Meyniana* **27**, 63–75.

SUMMERHAYES, C. P., ELLIS, J. P., STOFFERS, P., BRIGGS, S. R. and FITZGERALD, M. G. (1977) Fine-grained sediment and industrial waste distribution and dispersal in New Bedford Harbor and Western Buzzard Bay, Massachusetts. *Rept. 76-115*, Woods Hole Oceanographic Institution, 110 pp.

SZUCS, F. K. and OOSTDAM, B. L. (1975) A comparative study of heavy metals in sediments of ocean dump sites. Abstr. Int. Conf. *Heavy Metals Environ., Toronto*, C-52/54.

TAKEUCHI, T. (1972) Distribution of mercury in the environment of Minamata Bay and the Inland Ariake Sea. In *Environmental Mercury Contamination*, R. HARTUNG and B. D. DINMAN eds., Ann Arbor Science Publishers, 79–81.

TALBOT, V., MAGEE, R. J. and HUSSAIN, M. (1976) Distribution of heavy metals in Port Phillip Bay. *Mar. Poll. Bull.* **7**, 53–55.

TAYLOR, D. (1974) The natural distribution of trace metals in sediments from a coastal environment, Tor Bay, England. *Estuar. Coast. Mar. Sci.* **2**, 417–424.

THEIS, T. L. and SINGER, P. C. (1974) Complexation of iron (II) by organic matter and its effect on iron (II) oxygenation. *Environ. Sci. Technol.* **8**, 569.

THOMPSON, J. A. J. (1975) Copper in marine waters—effects of mining wastes. Abstr. Int. Conf. *Heavy Metals Environ., Toronto*, C-165/6.

THOMPSON, J. A. J. and MCCOMAS, F. T. (1974) Copper and zinc levels in submerged mine tailings at Britannia Beach, B.C. *Fish. Res. Bd. Canada Techn. Rept. No. 473*, 33 pp.

THOMPSON, J. A. J. and PATON, D. W. (1975) Chemical delineation of a submerged mine tailings plume in Rupert and Holberg inlets, B.C. *Techn. Rept. Fish. Mar. Serv. Canada No. 506*, 24 pp.

THORNTON, I., WATLING, H. and DARRACOTT, A. (1975) Geochemical studies in several rivers and estuaries for oyster rearing. *Sci. Tot. Environ.* **1**, 325–345.

THORNTON, I. and WEBB, J. S. (1973) Environmental geochemistry: some recent studies in the United Kingdom. In

Trace Substances in Environmental Health, Vol. III, D. D. HEMPHILL, ed., Univ. of Missouri, 89–98.

THROWER, S. J. and EUSTACE, I. J. (1973) Heavy metal accumulation in oysters grown in Tasmanian waters. *Fd. Technol. Aust.* **25**, 446–553.

TUREKIAN, K. K. (1971) Rivers, tributaries and estuaries. In *Impingement of Man on the Oceans*, D. W. HOOD, ed., John Wiley, 9–73.

TUREKIAN, K. K. (1977) The fate of metals in the oceans. *Geochim. Cosmochim. Acta* **41**, 1139–1144.

VILLA, O. JR. and JOHNSON, P. G. (1974) Distribution of metals in Baltimore Harbour sediments. *EPA, Annapolis Field Office Tech Rept.* **59**, EPA- 903/9-74-012, 71 p.

VOLKOV, I. I. and FOMINA, L. S. (1974) Influence of organic material and processes of sulfide formation on the distribution of some trace elements in deep-water sediments of the Black Sea. *Amer. Ass. Petrol. Geol. Mem.* **20**, 256–476.

WEICHART, G. (1973) Pollution of the North Sea. *Ambio* **2**, 99–106.

WILLIAMS, P. M. and CHAN, K. S. (1966) Distribution and speciation of iron in natural waters: transition from river water to a marine environment, British Columbia, Canada. *J. Fish. Res. Bd. Canada* **23**, 575–593.

WINDOM, H. L., BECK, K. C. and SMITH, R. (1971) Transport of trace elements to the Atlantic Ocean by three Southern streams. *Seast. Geol.* **12**, 169–181.

WOLFE, D. A., CROSS, F. A. and JENNINGS, C. D. (1973) The flux of Mn, Fe and Zn in an estuarine system. In *Radioactive Contamination of the Marine Environment*, IAEA, Vienna, 159–175.

WOLLAST, R. (1976) Transport et accumulation de pollutants dans l'estuaire de l'Éscaut. In *Projet Mer,* Vol. 10, *L' Estuaire de L'Escaut,* J. C. J. NIHOUL and R. WOLLAST, eds., Progr. Polit. Scient. Bruxelles/Belgium, 191–218.

YIM, W. W. S. (1976) Heavy metal accumulation in estuarine sediments in historical mining of Cornwall. *Mar. Poll. Bull.* **7**, 147–150.

YOUNG, D. R., JOHNSON, J. N., SOUTAR, A. and ISAACS, J. D. (1973) Mercury concentrations in dated varved marine sediments collected off Southern California. *Nature* **244**, 273–274.

YOUNG, D. R., MCDERMOTT, D. J., HEESEN, T. C. and JAN, T. K. (1975) Pollutant inputs and distributions off Southern California. In *Marine Chemistry in the Coastal Environment*, T. M. CHURCH, ed., *Amer. Chem. Soc. Symp. Ser.* **18**, 424–439.

ZINK-NIELSEN, I. and KROGH, O. (1976) Orienterende undersøgelser af sedimentet fra Lille Baelt for indhold af tungmetaller. In *Sediment Undersøgelser, Miljøstyrelsen Baeltprojektet*, 45–50.

Discussion

I. THORNTON:

Have your studies extended to examining the bioavailability of the trace elements and their movement in the food chain?

U. FÖRSTNER:

We have not considered these aspects in the present work on coastal pollution, but there is presently at this institute research being done on the factors influencing the adsorption of metals by organisms in polluted freshwater sediments. We are carrying out laboratory experiments in which the bioavailability of trace metals of clearly different forms of chemical association is investigated from various substrates. In addition to those organically bound, metal contents could play an important role in carbonate associations since the latter are strongly influenced by the pH of micro-environments in the organisms. We therefore attribute great importance to the determination of carbonate-associated trace metals in sediments. We are presently developing a process in which, with the help of acidic ion exchangers, a selective extraction of these metal contents can be accomplished. Another important factor for the bioavailability is the grain-size of the sediment particle, pointed out already by several authors. However, this influence factor can be thought of as being a part of a more or less direct relationship to the forms of chemical associations.

Author Index

Subject Index